ENCYCLOPAEDIA

of

FOOD SCIENCE
FOOD TECHNOLOGY

and

NUTRITION

ENCYCLOPAEDIA *of* FOOD SCIENCE FOOD TECHNOLOGY *and* NUTRITION

Volume Six

pH–Soya Milk

Edited by

R. Macrae
University of Hull, Hull, UK

R. K. Robinson
University of Reading, Reading, UK

M. J. Sadler
Consultant Nutritionist, Ashford, UK

ACADEMIC PRESS

Harcourt Brace Jovanovich, Publishers
London San Diego New York
Boston Sydney Tokyo Toronto

ACADEMIC PRESS LIMITED
24/28 Oval Road
LONDON NW1 7DX

United States Edition published by
ACADEMIC PRESS INC.
San Diego, CA 92101

This book is printed on acid-free paper

A catalogue record for this book is available from the British Library

Library of Congress Cataloging-in-Publication Data

Encyclopaedia of food science, food technology, and nutrition/edited
by R. Macrae, R. K. Robinson, M. J. Sadler.
p. cm.
1. Food—Encyclopedias. 2. Food industry and trade—
—Encyclopedias. 3. Nutrition—Encyclopedias. I. Macrae, R.
II. Robinson, R. K. (Richard Kenneth). III. Sadler, M. J. (Michèle
J.)
TX349.E47 1993
664′.003—dc20 92-15557
CIP

ISBN Volume 1 0-12-226851-2
Volume 2 0-12-226852-0
Volume 3 0-12-226853-9
Volume 4 0-12-226854-7
Volume 5 0-12-226855-5
Volume 6 0-12-226856-3
Volume 7 0-12-226857-1
Volume 8 0-12-226858-X

ISBN Set 0-12-226850-4

EDITORIAL STAFF

Managing Editor
Gina Fullerlove

Assistant Editor
Sarah Robertson

Production Editors
David Atkins
Sara Hackwood

Production Controller
Anne Doris

Editorial Assistants
Jill Farrow
Lara King
Martina Tuohy

Copy Editors
Len Cegielka
Rich Cutler
Alison Walsh

Cross-referencer
Jane Wells

Proof Reader
Alison Woodhouse

Indexer
Merrall-Ross International

Editorial and Technical Support Manager
Christopher Gibson

Head of Book Production
Helen Whitehorn

Typeset by The Alden Press, Oxford
Printed in Great Britain by Butler and Tanner Ltd, Frome, Somerset, UK

D. B. Emmons
Food Research Centre, Ottawa, Canada
O. R. Fennema
University of Wisconsin-Madison, Madison, USA
D. L. Georgala
AFRC Institute of Food Research, Reading, UK
M. J. Gibney
Trinity College Medical School, Dublin, Eire
C. Gopalan
The Nutrition Foundation of India, New Delhi, India
P. Gray
Commission of the European Communities, Bruxelles, Belgium
M. S. Y. Hadaddin
University of Jordan, Amman, Jordan
B. Hallgren
Swedish Nutrition Foundation, Goteborg, Sweden
J. M. Hutchinson
FAO, Rome, Italy
B. Jarvis
HP Bulmer Ltd, Hereford, UK
M. E. Knowles
MAFF, London, UK
J. A. Kurmann
Institut Agricole de Fribourg, Posieux, Switzerland
R. A. Lawrie
University of Nottingham, Loughborough, UK
F. A. Lee
Cornell University, New York, USA
D. X. Lin
Wuxi Institute of Light Industry, Wuxi, People's Republic of China
V. Loureiro
Technical University of Lisbon, Lisbon, Portugal
P. Lunven
FAO, Rome, Italy
E. J. Mann
International Dairy Federation, Reading, UK
V. Marks
University of Surrey, Guildford, UK
H. E. Nursten
University of Reading, Reading, UK
R. Paoletti
Universita di Milano, Milan, Italy
K. L. Parkin
University of Wisconsin-Madison, Madison, USA
G. T. Prance
Royal Botanic Gardens, Kew, UK
A. Reps
Institute of Food Biotechnology, Kortowo, Poland
D. R. Richardson
Nestle, Croydon, UK
P. Richmond
AFRC Institute of Food Research, Norwich, UK
A. Rougereau
Robert Debre Hospital, Amboise, France
J. D. Schofield
Kings College, University of London, UK
N. E. Schwartz
National Centre for Nutrition and Dietetics, Chicago, USA
R. L. Sellars
Chr. Hansen's Laboratory Inc., Wisconsin, USA
J. Solms
Swiss Federal Institute of Technology, Zurich, Switzerland
D. A. T. Southgate
AFRC Institute of Food Research, Norwich, UK
K. R. Spurgeon
South Dakota State University, South Dakota, USA
M. Stasse-Wolthuis
Netherlands Nutrition Foundation, Wageningen, The Netherlands
P. S. Steyn
National Food Research Institute, Pretoria, South Africa
P. Trayhurn
Rowett Research Institute, Aberdeen, UK
A. S. Truswell
University of Sydney, Sydney, Australia
G. Varela
Universidad Complutense de Madrid, Madrid, Spain
J. V. Wheelock
University of Bradford, Bradford, UK
E. M. Widdowson
Addenbrooke's Hospital, Cambridge, UK
R. B. Wills
Food Industry Development Centre, Kensington, Australia

GUIDE TO USE OF THE ENCYCLOPAEDIA

Structure of the Encyclopaedia

The material in the Encyclopaedia is arranged as a series of entries in alphabetical order. Some entries comprise a single article, whilst entries on more diverse subjects consist of several articles that deal with various aspects of the topic. In the latter case the articles are arranged in a logical sequence within an entry.

To help you realize the full potential of the material in the Encyclopaedia we have provided three features to help you find the topic of your choice.

Contents Lists

Your first point of reference will probably be the contents list. The complete contents list appearing in Volume 8 will provide you with both the volume number and the page number of the entry. Additionally each volume has a list of entries and articles to be found within that particular volume, which allows you to locate material within the volume without having to return to the complete contents list each time a new topic is required. On the opening page of an entry a contents list is provided so that the full details of the articles within the entry are immediately available.

Alternatively you may choose to browse through a volume using the alphabetical order of the entries as your guide. To assist you in identifying your location within the Encylopaedia a running headline indicates the current entry and a running footer indicates the current article within that entry.

You will find 'Dummy Entries' where obvious synonyms exist for entries or where we have grouped together related analytical techniques or commodities. Dummy entries appear in both the contents list and the body of the text. For example, a Dummy Entry appears for Vitamin C which directs you to Ascorbic Acid, where the material is located.

Example

If you were attempting to locate material on Oranges via the Contents List.

Volume 5

Oesophageal Cancer	9721
Offal	
Types of Offal	9726
Dietary Importance	9731
Olives	9736
Onions and Related Crops	9741
Oral Contraceptives and Nutritional Status	9746
Oranges *see Citrus Fruits*	
Organic Foods	9751

Then once you have been directed to the correct location in the contents list this would then provide the page number.

Volume 2

If you were trying to locate the material by browsing through the text and you looked up Oranges then the following is the information you would be provided.

ORANGES

See Citrus Fruits

Alternatively, if you were looking up Citrus Fruits the following information would be provided.

CITRUS FRUITS

Contents

Cross References

All of the articles in the Encyclopaedia have been extensively cross referenced.

The cross references, which appear at the end of a paragraph, have been provided at three levels:

1. To indicate if a topic is discussed in greater detail elsewhere.

Groups at Risk of Vitamin C Deficiency (Scurvy)

The major determinant of vitamin C intake is the consumption of fruit and vegetables, and the range of intakes in healthy adults in Britain reflects this. The 2·5 percentile intake is 19 mg per day (men) and 14 mg per day (women), while the 97·5 percentile intake is 170 mg per day (men) and 160 mg per day (women). Deficiency is likely in people whose habitual intake of fruit and vegetables is very low. Smokers may be more at risk of deficiency; there is some evidence that the rate of ascorbate catabolism is two-fold higher in smokers than in non-smokers. Clinical signs of deficiency are rarely seen in developed countries. ***See*** Scurvy

2. To draw the reader's attention to parallel discussions in other articles.

Fruit Morphology
Grapefruit is composed of three distinctly different morphological parts. The epicarp consists of the coloured portion of the peel and is known as the flavedo. In the flavedo are cells containing the carotenoids which give the characteristic colour to the fruit. The oil glands, also found in the flavedo, are the raised structures in the skin of the fruit that contain the essential oil, naringin, characteristics of the grapefruit. ***See*** Citrus Fruits, Oranges; Citrus Fruits, Lemons

3. To indicate material that broadens the discussion.

Plasma Concentration of Ascorbate
At intakes below 30 mg per day the plasma concentration of ascorbate is extremely low, and does not reflect increasing intake to any significant extent. As the intake rises, so the plasma concentration begins to increase sharply, reaching a plateau of 55–85 μmol l^{-1} at intakes between 70 and 100 mg per day, when the renal threshold is reached and the vitamin is excreted quantitatively with increasing intake. ***See*** Dietary Reference Values

Index

The index will provide you with the volume number and page number of where the material is to be located, and the index entries differentiate between material that is a whole article, is part of an article or is data presented in a table. On the opening page of the index detailed notes are provided.

Colour Plates

The colour figures for each volume have been grouped together in a plate section. The location of this section is cited both in the contents list and on the opening page of the pertinent articles.

Contributors and Referees

A full list of contributors and a full list of referees appears in Volume 8.

CONTENTS

pH – PRINCIPLES AND MEASUREMENT

pH measurements have been, and continue to be widely used as a rapid, accurate measure of the acidity of fluids of all sorts. There are two methods for measuring pH: colorimetric methods using indicator solutions or papers, and the more accurate electrochemical methods using electrodes and a millivoltmeter (pH meter). It was the development of the glass electrode, which is convenient to use in a variety of environments, and the development of the pH meter, which has enabled the widespread application of pH measurement and control to take place. The determination, and hence the control of pH is of great importance in the food industry.

Basic Theory

In water, molecules (H_2O) are in equilibrium with hydrogen ions (H^+) and hydroxide ions (OH^-) (eqn [1]).

$$H_2O \rightleftharpoons H^+ + OH^- \quad (1)$$

This ionization is of great importance in the chemistry of aqueous systems as it enables water to give or take H^+ ions as required by other dissolved substances.

Applying the law of mass action to eqn [1]:

$$[H^+][OH^-]/[H_2O] = \text{constant}, \quad (2)$$

where [ ] indicates concentration in units of moles per cubic decimetre (mol dm^{-3}).

In pure water and dilute solutions the concentration of the undissociated water may be considered constant, hence $[H^+][OH^-] = K_w$, where K_w is a constant called the ionic product of water.

The ionic product varies with temperature, but at about 25°C its value is 10^{-14} mol^2 dm^{-6} ($K_w(0°C) = 10^{-14.9}$, $K_w(25°C) = 10^{-14.0}$, $K_w(60°C) = 10^{-13.0}$). This means that in pure water the ionization is very small. As the concentration of H^+ ions and OH^- ions are equal in pure water, and as $[H^+][OH^-] = 10^{-14}$ mol^2 dm^{-6} at 25°C, $[H^+]$ (or $[OH^-]$) $= 10^{-7}$ mol dm^{-3}.

To be strictly correct eqn [1] should be written as eqn [3],

$$2H_2O \rightleftharpoons H_3O^+ + OH^- \quad (3)$$

where the H^+ ion is attached to a water molecule to form the oxonium ion (H_3O^+). However, the symbol H^+ will be used throughout this article.

When, in a solution, there are an equal number of H^+ and OH^- ions the solution is said to be neutral, when there is an excess of H^+ ions ($> 10^{-7}$ mol dm^{-3} at 25°C) it is acidic and when there is an excess of OH^- ions ($[H^+] < 10^{-7}$ mol dm^{-3}) it is basic or alkaline.

The pH Scale

Although concentrations of H^+ and OH^- ions in acidic and alkaline solutions can be expressed in these molar concentrations, a much more convenient method was introduced by S. P. L. Sörensen in 1909. He proposed the use of the H^+ ion exponent pH, defined by the relationship $pH = -\log_{10}[H^+]$. (This may also be expressed as $[H^+] = 10^{-pH}$). To be strictly correct this equation is $pH = -\log_{10}([H^+]/[1])$, as a logarithm must be dimensionless. From this it can be seen that pH does not have units – the often used expression 'pH unit' is wrong.

Using the pH scale, a change of 1 corresponds to a 10-fold change in the H^+ ion concentration, a change of 2 corresponds to a 100-fold change, etc. The pH scale has the advantage that all solutions from 1 mol dm^{-3} acid to 1 mol dm^{-3} alkali can be expressed by positive numbers from 0 to 14. Thus, at 25°C a neutral solution, which has $[H^+] = 10^{-7}$ mol dm^{-3}, has a pH of 7, a 1 mol dm^{-3} solution of a strong (completely ionized) acid such as hydrochloric acid, which has $[H^+] = 1 = 10^0$ mol dm^{-3}, has a pH of 0, and a 1 mol dm^{-3} solution of a strong (completely ionized) alkali such as sodium hydroxide, which has $[OH^-] = 1$ mol dm^{-3}, hence $[H^+] = K_w/[OH^-] = 10^{-14}$ mol dm^{-3}, has a pH of 14. pH values below 0 and above 14 occur in solutions of strong acids and alkalis of strengths greater than 1 mol dm^{-3}.

As K_w varies with temperature, pH measurements should be made at about 25°C. The pH of a neutral solution at 0°C is 7·45 ($K_w = 10^{-14.9}$, $[H^+] = \sqrt{10^{-14.9}} = 10^{-7.45}$ mol dm^{-3}) and at 60°C the pH is 6·5 ($K_w = 10^{-13.0}$, $[H^+] = \sqrt{10^{-13.0}} = 10^{-6.5}$ mol dm^{-3}).

To measure pH values Sörensen used the electrochemical cell:

$$\text{Pt, } H_2(g) \mid \text{Solution X} \mid \text{Salt bridge} \mid Hg_2Cl_2 \mid Hg$$

This representation indicates a hydrogen electrode (hydrogen gas passed over a platinum metal electrode) in a solution X and a calomel electrode (metallic mercury in contact with calomel (Hg_2Cl_2), in contact with a chloride ion solution) separated by a salt bridge (which allows ionic conduction to occur, but prevents solution X and the chloride ion solution from mixing). For such a cell the difference in electromotive force (emf)

$(E_1 - E_2)$ between two cells in terms of the H^+ ion concentrations $[H^+]_1$ and $[H^+]_2$ of the two solutions is given by the equation

$$E_1 - E_2 = (RT/F)\log_e([H^+]_2/[H^+]_1), \quad (4)$$

where R is the gas constant, T the temperature and F the Faraday constant. If $[H^+]_2$ is 1 mol dm^{-3}, E_2 will have a definite value, E^0. Therefore, $E_1 - E^0 = (RT/F)\log_e(1/[H^+]_1)$, or $E_1 = E^0 + (RT\log_e 10/F)p_sH_1 = E^0 + 0{\cdot}059\,16\, p_sH_1$ V (at 25°C) (p_sH is used here for Sörensen's pH). Sörensen assumed that $[H^+]_1 = \alpha_1 M_1$ where α_1 is the degree of dissociation of the acid in solution 1, and M_1 is its molality. $E^0 = 0{\cdot}337\,6$ V at 18°C.

After this initial work of Sörensen it was realized that the emfs of cells depend on activity rather than concentration, and other assumptions were incorrect. Sörensen and Linderström-Lang (1924) proposed $p_aH = -\log_{10} a_{H^+} = -\log_{10}[H^+]y_{H^+}$ where a_{H^+} is the activity, $[H^+]$ is the concentration of the H^+ ion in mol dm^{-3} and y_{H^+} is the activity coefficient of the H^+ ion on the molarity scale. (To be strictly correct a ratio of concentrations, as earlier, is needed here to give a dimensionless quantity.) These p_aH values can be shown to be related to the original p_sH values by the relationship:

$$p_aH = p_sH + 0{\cdot}04 \quad (5)$$

Operational Definition of pH

In the previous section some of the problems of the theory of pH were discussed. Careful study has shown that the numbers obtained depend in a complex manner on H^+ ion activity of electrolytes in solution and it became clear than an operational definition of pH and a standard scale were needed to unify the great variety of pH measurements being made by research scientists and by industry and commerce. Hence, we now have an operational definition which has been endorsed by the International Union of Pure and Applied Chemistry (IUPAC). This definition is

$$pH(X) = pH(S) + (E_S - E_X)F/RT\log_e 10 \quad (6)$$

where E_X is the emf of the galvanic cell

Reference electrode | KCl solution | Solution X | H_2(g),Pt

and E_S is the emf of the same cell with solution X of unknown pH(X) replaced by solution S of standard pH(S).

In practice a glass electrode is almost always used in place of the platinum/hydrogen gas electrode. The reference electrode is usually mercury/mercury(I) chloride (calomel), silver/silver chloride, or thallium amalgam/thallium(I) chloride. The standard reference pH is that of an aqueous solution of potassium hydrogenphthalate of molality 0·05 mol kg^{-1} at 25°C. This has a pH of 4·005.

Food and pH

There are many instances when the food scientist needs to measure or control pH. For example, pH control is vital in the clarification and stabilization of fruit and vegetable juices, in the use and control of enzymes and microorganisms, in the preparation of foods by the fermentation of fruit or cereals, in the control of the texture of jams and jellies, and in the stability of the colour and flavour of fruits.

Properties (e.g. emulsification and foaming ability) of many colloidal systems are affected by pH – these include proteins, pectins and gums, all of which occur in foods. Chemical reactions that can occur in foods, particularly hydrolysis, are catalysed by hydrogen ions, and can therefore be controlled by controlling the pH of the system. *See* Colloids and Emulsions

Hydrogen ions affect the rate of growth of moulds, yeasts and, particularly, bacteria. Usually there is a pH value at which optimum growth occurs, above and below which growth is inhibited. The relationship of pH to sterilization by heat is of particular significance during the canning of foodstuffs. *See* Canning, Principles; Spoilage, Bacterial Spoilage; Spoilage, Moulds in Food Spoilage; Spoilage, Yeasts in Food Spoilage

pH is a rough measure of the maturity of fruit. Young fruits initially have a pH close to that of the plant, but this rapidly decreases as the acidity increases. The pH of acidic fruits such as lemons and limes is about 2·0, that of mildly acidic fruits about 4·0. These values can be compared with vegetables which have pH values of 5–6. The pH range of a number of fruits and other foods is listed in Table 1. *See* Ripening of Fruit

Table 1. pH values of some foods

Food	pH	Food	pH
Limes	1·8–2·0	Bananas	4·5–4·7
Lemons	2·2–2·4	Cheese	4·8–6·4
Gooseberries	2·8–3·0	Carrots	4·9–5·3
Plums	2·8–3·0	Spinach	5·1–5·7
Pickles	3·0–3·4	Potatoes	5·6–6·0
Grapefruit	3·0–3·4	Peas	5·8–6·4
Oranges	3·0–4·0	Tuna	5·9–6·1
Rhubarb	3·1–3·2	Corn	6·0–6·5
Cherries	3·2–4·0	Salmon	6·1–6·3
Pineapples	3·4–3·7	Butter	6·1–6·4
Pears	3·6–4·0	Chicken	6·2–6·4
Apricots	3·6–4·0	Drinking water	6·5–8·0
Tomatoes	4·0–4·4		

Buffer Solutions

Buffer solutions are of prime importance in pH determinations as they are the standards for both colorimetric and for electrochemical methods of determining pH.

Buffer solutions are solutions that resist a change in H^+ ion concentration (pH) when an acid or alkali is added to them. They are made up of a weak acid or weak base and its salt.

The equilibria in a solution of, say, a weak acid and its sodium salt are shown in eqns [7] and [8].

$$HA \rightleftharpoons H^+ + A^- \quad (7)$$

$$NaA \rightleftharpoons Na^+ + A^- \quad (8)$$

As HA is a weak acid, equilibrium (7) at moderate concentration will lie over to the left, i.e. as undissociated acid HA. The equilibrium of the sodium salt (8) is a source of A^- ions. If acid (i.e. H^+ ions) is added to this weak acid and salt solution, the H^+ ions will be removed as they will combine with the A^- ions to form undissociated HA. If base (i.e. OH^- ions) is added, the OH^- ions will be removed as they will combine with the H^+ ions to form water and the equilibrium (7) will move to the right to supply the H^+ ions required. Hence the concentration of H^+ ions in the solution (i.e. the pH) will not significantly change. This is the buffer effect.

Common buffer solutions are made from potassium hydrogenphthalate with hydrochloric acid or sodium hydroxide (for pH 3–5), potassium dihydrogenphosphate with sodium hydroxide (for pH 6–8) and boric acid with sodium hydroxide (for pH 8–10).

Biologically, the buffer effect is often more important than the absolute value of pH. For example, buffer chemicals affect the way in which enzymes and microorganisms respond when the pH is changed. Juices from plant tissue have the ability to buffer pH changes to a greater or lesser degree, due to the presence of organic acids, acid–salt systems, proteins and acid phosphates.

Measurement of pH

Colorimetric Methods

Determination of pH using the colour of acid–base indicators is a very simple technique that can be carried out rapidly and reproducibly. Approximate pH values can be obtained very quickly using pH papers. Under optimum conditions accurate values of pH can be obtained by colorimetric methods. These methods will no doubt be used for many years to come, but must become less favoured with the advent of portable, cheap, easy-to-use electrochemical pH meters. ***See*** Spectroscopy, Visible Spectroscopy and Colorimetry

The Colour Change of Acid–Base Indicators

Indicators are natural or artificial dyestuffs, which are weak acids or bases (acids and bases that are only partially dissociated), and which have different colours in their acidic and basic forms.

If the indicator is written as HIn and it dissociates (eqn [9])

$$HIn \rightleftharpoons H^+ + In^- \quad (9)$$

then the equilibrium (or ionization) constant is

$$K_{In} = [H^+][In^-]/[HIn], \quad (10)$$

and

$$[H^+] = K_{In}([HIn]/[In^-]). \quad (11)$$

Using $pH = -\log_{10}[H^+]$,

$$pH = pK_{In} + \log_{10}([In^-]/[HIn]); \quad (12)$$

pK_{In} is $-\log_{10} K_{In}$ and is called the indicator constant. The ratio $[In^-]/[HIn]$ determines the colour of the indicator in the solution, and eqn [12], therefore, directly relates the colour to the pH of the solution.

Equation [12] shows that when $pH = pK_{In}$ the concentration of the indicator in each form is equal (i.e. $[HIn] = [In^-]$), and that at all pH values both the acid (HIn) and basic (In^-) forms of the indicator are present in the solution. Our eyes are unable to detect the colour of less than about 10% of one form of the indicator in the presence of the other and the solution will appear to be the 'acid' colour when $[In^-]/[HIn] < 1/10$, and the 'alkaline' colour when $[In^-]/[HIn] > 10$. Therefore, the solution will be the colour of the acid form of the indicator until $pH = pK_{In} - 1$, when the colour will appear to change until it is that of the alkaline form of the indicator from $pH = pK_{In} + 1$. That is the colour changes over approximately 2 in terms of pH. In practice the range for the colour change is about 1·6–2 in terms of pH; this corresponds to a 40–100-fold change in the H^+ ion concentration. By choosing indicators with appropriate pK_{In} values the whole of the pH range may be covered – a selection of some common pH indicators is given in Table 2.

Colorimetric Measurements of pH

Indicator Papers. For approximate determination of pH values, indicator papers may be used. Of particular value are 'non-bleeding' indicator papers. These contain dyestuffs strongly bound to the cellulose, so that the dye does not bleed into the test solution. These indicator papers are available covering very narrow pH ranges, enabling moderately accurate pH values to be obtained very cheaply, conveniently and quickly.

Comparison Method. This is carried out by adding similar quantities of an indicator to the test solution and a reference solution, and if the two solutions have the same colour they are assumed to have the same pH. The accuracy of the method, therefore, depends on the accuracy of the pH of the reference solution.

Table 2. Some pH indicators and their colour changes and ranges

Indicator	Colour in acid solution	Colour in alkaline solution	pH range	pK_{In}
Thymol blue (acid)	Red	Yellow	1·2–2·8	1·7
Bromophenol blue	Yellow	Blue	2·8–4·6	4·0
Methyl orange	Red	Yellow	3·1–4·4	3·7
Methyl red	Red	Yellow	4·2–6·3	5·1
Litmus	Red	Blue	5·0–8·0	
Bromothymol blue	Yellow	Blue	6·0–7·6	7·0
Phenol red	Yellow	Red	6·8–8·4	7·9
Thymol blue (base)	Yellow	Blue	8·0–9·6	8·9
Phenolphthalein	Colourless	Red	8·3–10·0	9·6
Universal[a]	Red, orange, yellow	Green, blue, violet	3·0–11·0	
Full-range[a]	Red, orange, yellow	Green, blue, violet	1·0–14·0	

[a] These are mixtures of indicators. They give a continuous colour change over a wide pH range.

The indicator used must be one which has a colour at the pH of the test solution intermediate between the acidic and basic forms.

The approximate pH of the solution is first detemined using a 'universal, or full-range, indicator, when an appropriate indicator for the measurement can be chosen. A quantity of the test solution is measured into a tube and a measured amount of indicator added. Tubes of buffer solutions covering the pH range of the indicator are treated in the same way. The colours of the tubes are viewed through the length of the tube against a white background. The accuracy of the method will depend on the differences of pH of the buffer solutions. These are typically 0·2 and the pH of the unknown solution can then be measured to within $\pm 0{\cdot}1$.

The Lovibond Comparator. The need to use buffer solutions is eliminated by using a Lovibond 2000 comparator, in which the colour of the test solution is compared with coloured glasses. Standard coloured glass rings are fitted into a disc that can be rotated and the coloured glasses compared with the test solution. For accurate results the sample solution and discs must be viewed against north daylight or in white light. Indicator discs for 14 different indicators are available. Special discs are available for the determination of the pH of blood.

With the Nesslerizer attachment the solution is viewed through the depth of the liquid instead of through its thickness. The proportion of indicator can therefore be considerably reduced, resulting in a significant gain in accuracy, particularly when unbuffered or slightly buffered solutions are being tested.

Sources of Error. There are a number of possible sources of error when making colorimetric measurements.

(1) *Slightly buffered solutions.* As indicators are weak acids or bases they will change the pH of the solution if it is unbuffered or only slightly buffered. Samples of relatively pure water, and solutions of a salt of a strong acid and strong base are the most frequently encountered solutions of this type. It is recommended that an appropriate pH meter is used for such solutions.

(2) *The salt effect.* As discussed earlier, equilibrium constants depend on the activities of the components rather than concentrations and it is only in very dilute solutions that activity equals concentration. Buffer solutions normally used have ionic strengths of up to 0·2 M. If the solution under test has a lower ionic strength than the buffer the determined pH will be too small. The salt correction that must be applied is also affected by the types of ion present and the indicator used.

(3) *The effect of colloids.* Colloid particles in the solution may preferentially absorb either the acid or the alkaline form of the indicator, giving totally erroneous results.

(4) *The effect of proteins.* pH values obtained in the presence of proteins are unreliable. The effect of a protein on the determination depends on both the type of protein and the indicator used. Usually the effect is greater when the protein is on the acid side of its isoelectric point.

(5) *The effect of finely divided particles.* Chemical reactions may occur between the indicator and particles in the solution, giving erroneous results.

(6) *Coloured solutions.* Colour comparison is not possible if the test solution is coloured.

(7) *Nonaqueous solvents.* Literature data on indicators applies to aqueous solutions; the indicator equilibria, hence the colour, will be different in other solvents. The colour of an indicator in, say, a water–alcohol mixture should be taken as only a rough guide to the true pH.

The Electrochemical Method

In order to measure pH by the electrochemical method the emf of a reference cell is determined (see earlier). In this cell the potential of one electrode (the hydrogen electrode) changes with pH, and the potential of the second, reference, electrode does not. Historically, the hydrogen gas electrode was used as the pH electrode, but the discovery by M. Cremer in 1906 that the potential difference across a glass membrane depends on the H^+ ion concentration on either side of it led to the development of the glass electrode. This is the electrode that is now always used for measuring pH.

The Glass Electrode

In its simplest form the glass electrode consists of a thin glass bulb containing a solution of constant pH, usually hydrochloric acid or a phosphate buffer with potassium chloride. In this solution is an electrode, usually silver coated with silver chloride.

The exchange of ions in the glass membrane for H^+ ions in the solutions on either side of it to form silanol (≡Si—OH) layers, is the major factor in determining the pH response of the electrode, and special sodium and lithium glasses are used. As this surface layer is crucial to the correct functioning of a glass electrode the pH-sensitive tip is stored in a pH 7 buffer solution, preferably containing 0·1 mol dm^{-3} potassium chloride. As it is most important that the tip does not dry out, commercial electrodes are supplied with a storage teat containing the buffer plus potassium chloride solution.

Glass electrodes give good linear response in the pH range 2–9, and up to pH 14 if the salt concentration in the test solution is not too high.

Combination Electrodes

Historically, the separate glass and reference electrodes were placed side by side in the test solution. Such electrodes are still available but are now used for special applications. Almost all pH measurements are made with combination electrodes in which the glass and reference electrodes are combined together. Although there are many details that vary from one electrode to another the basic design is as shown in Fig. 1. The glass membrane is at the bottom and is surrounded by the reference electrode. This is usually a silver/silver chloride electrode (silver wire coated with silver chloride in saturated potassium chloride solution) and the reference–test solution junction is at a porous glass frit near the bottom of the electrode. The potassium chloride in the reference cell slowly bleeds through the frit into solutions being tested, and must be topped up when necessary through the filling hole near the top of the electrode.

For many purposes this electrode is perfectly satisfactory, but under adverse conditions the positive flow of reference electrolyte is reversed and the sample solution contaminates the reference cell. This can be avoided by

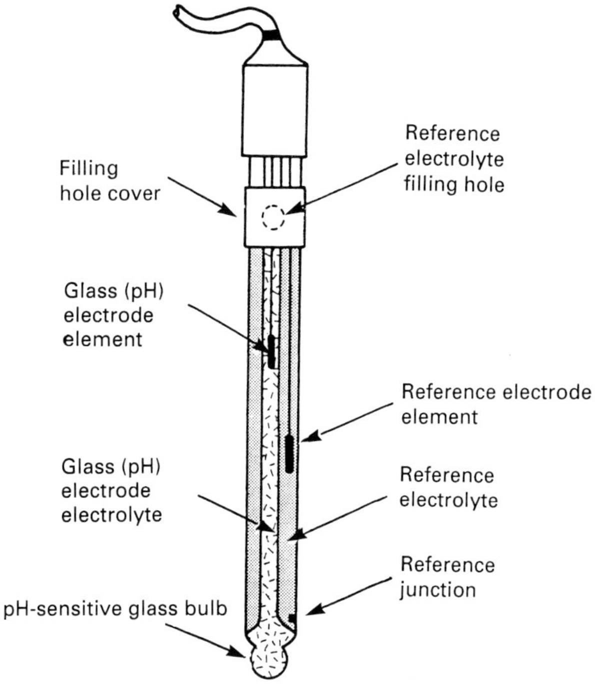

Fig. 1 A combination pH electrode.

using a double-junction reference system in which the reference cell is connected through a glass frit with an intermediate electrolyte which is in turn in contact, through a glass frit, with the test solution. Such an electrode also eliminates clogging due to precipitation of silver chloride at the glass frit. Also, the intermediate electrolyte can be changed if that being used reacts with the test solution at the glass frit junction.

Special Glass Electrodes

The Amplified Electrode. The major experimental difficulty in measuring the emf of a pH cell is the high electrical resistance of the glass membrane, commonly 10–500 MΩ. This requires a high-impedance millivoltmeter and short, screened connections to the electrode. One solution to this problem is a glass electrode with a built-in amplifier (Hanna Instruments), where the high-impedance circuitry is in an integrated circuit encapsulated at the top of the electrode together with a mercury battery. This electrode is ideal for industrial pH measuring and monitoring as long, unshielded cables can be used from the electrode to the pH meter.

Other Electrodes. pH glass and combination electrodes of many types are available, these include electrodes in plastic bodies (for durability), microelectrodes for small volumes of test solution, electrodes resistant to proteins, and flat electrodes for pH measurement at surfaces.

pH Meters

A pH meter is a high-impedance millivoltmeter that is designed to convert millivolts to pH. There are many types available at a range of prices. Temperature compensation is important and many meters have a temperature probe facility for this purpose. Many meters contain microprocessors and are very easy to use.

To make a pH measurement it is necessary to calibrate the meter first. This is done by using two buffer solutions that, ideally, span the pH of the test solution. After rinsing the electrode this is placed in the test solution and the pH read from the meter.

Accurate pH measurements can be made without difficulty on test solutions containing dissolved acids, bases or salts, If, however, the sample is virtually pure water (e.g. tap, rain or boiler feed water) pH measurements are unreliable. This is because (1) the glass electrode requires a long time to stabilize in a low ionic-strength solution, (2) the solution has a low buffering capacity and is therefore susceptible to drift as atmospheric carbon dioxide is absorbed, (3) the low conductivity of the solution allows the pick up of electromagnetic noise and (4) there is electrical noise created at the reference junction. Special low-conductivity electrode buffer kits have been developed (Russell pH Ltd) for such measurements. In these kits the glass electrode is of extremely low resistance and the calibration errors are reduced by using a buffer of low ionic strength.

Very small, portable pH meters are now available. A good example is the Piccolo (Hanna Instruments). This meter weighs only 100 g and uses an interchangeable amplified electrode in a plastic container. It has a built-in temperature sensor that compensates for temperatures from 0 to 70°C. It measures pH in the range 1–13 with an accuracy and resolution of ±0·01.

Bibliography

Anonymous (1987) *pH Values*, 8th edn. Poole: BDH, Ltd.

Bates RG (1973) *Determination of pH. Theory and Practice*, 2nd edn. New York: Wiley.

Covington A (1989) *pH and its Measurement*. London: Royal Society of Chemistry (a chemistry cassette and workbook).

David Webster
Hull University, Hull, UK

PHENOLIC COMPOUNDS

The name 'phenol' specifically refers to the monohydroxy derivative of benzene, but it is applied generally to all derivatives of benzene and its relatives having nuclear hydroxy groups. According to the number of hydroxy groups present, the compounds are termed monohydric, dihydric (catechol, resorcinol, quinol) and trihydric (pyrogallol, phloroglucinol, 1,3,4-trihydroxybenzene) phenols. Natural phenols in foods encompass a diverse group of compounds, including simple phenols, derivatives of benzoic and cinnamic acids, and polyphenols. Their levels vary dramatically, and are influenced especially by factors such as germination, ripening, storage and type and extent of processing. Polyphenols are widespread but the simple compounds are relatively uncommon. In addition to potential toxicological concern, these compounds have been implicated as influencing the functional, nutritional and sensory properties of foods with which they are associated.

Simple Phenols

Besides phenol itself, alkyl-substituted phenols (*o*-, *m*- and *p*-cresols, *o*-, *m*- and *p*-xylenols, 2- and 4-ethylphenols, 3- and 4-*n*-propylphenols, *o*- and *p*-*s*-butylphenols, 2,3,5-trimethylphenol, 2,6-di-*t*-butylmethylphenol, *p*-isopropyl-*m*-cresol), alkenyl-substituted phenols (4-vinylphenol and 4-allylphenol), alkoxy-substituted phenols (*o*-methoxyphenol (guaiacol) and its 4-methyl-, ethyl-, propyl-, vinyl-, allyl (eugenol)- and propenyl (isoeugenol)-substituted derivatives, *m*- and *p*-methoxyphenols, 2,6-dimethoxyphenol (syringol) and its 4-ethyl-, iso-propyl-, propyl-, vinyl-, methyl- and allyl-substituted derivatives) and sulphur-containing phenols (thiophenol and its 2-methyl- or t-butyl-substituted derivatives) have been reported to occur in varying amounts (parts per million level) in foodstuffs such as cocoa, coffee, tea, alcoholic beverages, meat, poultry products, nut products, vegetables and other diverse commodities. The acidity and volatility of the above phenols vary with the nature and location of the substituents they carry. In general, the effect of substituents in the *m* position of phenol is limited, while the effect in the *o* and *p* positions depends on the resonance interactions between the substituent and the phenolic hydroxyl group and its inductive effect. Electron-withdrawing substituents enhance while electron-donating groups suppress the acidic and hydrogen bond formation characteristics of phenols. Both neutral and ionized phenols are ambient nucleophiles and react at *o* or *p* centres with neutral or charged electrophiles.

Decarboxylation of phenolic carboxylic acids and thermal degradation of lignin appear to be the primary pathways for the formation of simple phenols in food systems. Secondary pathways include bacterial, fungal, enzymatic and glycosidic reactions. Alicyclic rings with oxygen functions may be dehydrogenated to phenols.

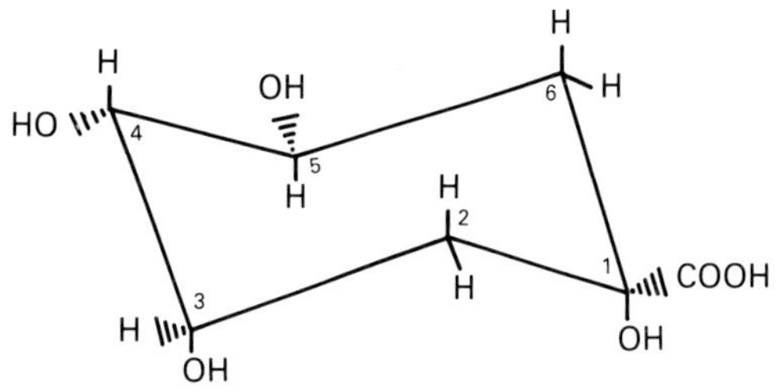

(**1**)

Chlorogenic Acids

The main phenolic acids isolated from cocoa, coffee, tea, alcoholic beverages, cereal grains, oil seeds, fruits, vegetables and other foods are the derivatives of hydroxybenzoic and cinnamic acids. These acids may either be present in the free state or found in a bound state. Thus, benzoic acid derivatives (*o*- and *p*-hydroxy, 3,4-dihydroxy (protocatechuic acid), *p*-methoxy (*p*-anisic acid), 4-hydroxy-3-methoxy (vanillic acid), 4-hydroxy-3,5-dimethoxy (syringic acid), 2,5-dihydroxy (gentisic acid), 3,4,5-trihydroxy (gallic acid) derivatives, ellagic acid) are mainly present in the form of glucosides while cinnamic acid derivatives (*p*-hydroxy (*p*-coumaric acid), *o*-hydroxy (*o*-coumaric acid), 3,4-dihydroxy (caffeic acid), 4-hydroxy-3-methoxy (ferulic acid), 3-hydroxy-4-methoxy, 4-hydroxy-3,5-dimethoxy (sinapic acid) derivatives) are frequently present as simple esters with a variety of carboxylic acids such as malic, tartaric, citric, tartronic, shikimic, galactaric, glucaric, gluconic and quinic acids or as glycosides. In general, *o*-acetylated flavonoid glycosides (especially anthocyanins and flavones) are widespread in nature, wherein the acetylating component on the sugar is mainly *p*-coumaric or caffeic acid. ***See*** Colours, Properties and Determination of Natural Pigments

Hydroxycinnamic acids and their derivatives are capable of existing in *Z* (*cis*) and *E* (*trans*) forms. While there is evidence that the natural forms are all *E*, isomerization inevitably occurs during extraction, and mixtures of isomers are frequently obtained. Oxidation in the *o* position can also occur during extraction and isolation.

Simple esters of hydroxycinnamic acids with quinic acid are termed chlorogenic acids. Prior to the new International Union of Pure and Applied Chemistry (IUPAC) recommendations, the nomenclature of quinic acid isomers was very confusing in the literature. In the IUPAC nomenclature, quinic acid is now treated as cyclitol (**1**). In the preferred configuration the carboxyl group and the C4 and C5 hydroxy groups are equatorial, with the C1 and C3 hydroxy groups being axial.

Green arabica coffee beans contain 6–7% 'chlorogenic acids' and robusta 8–9%. The chlorogenic acids isolated or detected in coffee beans to date include 3-(neochlorogenic acid), 4-(cryptochlorogenic acid) and 5-(chlorogenic acid) caffeoylquinic acids, 3,4-, 3,5- and 4,5-dicaffeoylquinic acids, 4-caffeoyl-5-feruloylquinic acid and 5-caffeoyl-4-feruloylquinic acid. 5-Caffeoylquinic acid (chlorogenic acid) predominates. During roasting, the content of chlorogenic acids declines. Also, 3-, 4- and 5-coumaroyl or feruloyl or galloyl quininates have been detected in other foodstuffs. ***See*** Coffee, Green Coffee

The general mesomeric enhancement by the hydroxy group to electrophilic substitution in an aromatic ring (*o*-/*p*-directing) leads to ready reaction of phenols with weak electrophiles, and the reactions are more pronounced in di- and trihydric phenols. Phloroglucinol can be substituted by electrophilic carbons in reactions with α,β-unsaturated aldehydes. Dihydric phenols show a greater tendency to ketonize than phenol since the energy barriers between the two forms are lower. In *m*-dihydric phenols, oxo forms mediate in many reactions, especially in basic media.

Aromatic rings may be hydroxylated *in vivo* by monooxygenases. Such reactions are often encountered in aromatic compounds derived from the shikimate–prephenate pathway. Thus, phenylalanine is *p*-hydroxylated to tyrosine by phenylalanine monooxygenase using molecular oxygen, and cinnamic acid is hydroxylated to *p*-hydroxycinnamic acid and to di- and trihydroxy acids, e.g. caffeic and gallic acids, with adjacent hydroxy functions.

Polyphenols (Including Tannins)

Polyphenols confer on fruits, vegetables and other plant foods qualities both desirable and undesirable. Flavonoids form a broad major group and are characterized by the presence of a C_6-C_3-C_6 carbon skeleton consisting of two aromatic rings linked by an aliphatic three-carbon chain. This skeleton is made up of two biogenetically distinct fragments: the C_6-C_3 fragment that contains the B ring and the C_6 fragment forming the A ring. The classification is made on the basis of the state of oxidation of the aliphatic fragment of the basic skeleton. The hydroxylation pattern of the A and B nuclei and the degree of polymerization of the C_6-C_3-C_6 unit results in a large number of different polyphenolic compounds. The major flavonoids of plants that are of wide occurrence are the anthocyanins, catechins, procyanidins, flavones and flavonols. Accompanying these major flavonoids in certain plant groups are the minor flavonoids, chalcones, flavanones, aurones, isoflavones and biflavonoids. Among the anthocyanins which are derivatives of the basic flavylium cation structure, pelargonidin, cyanidin, delphinidin, penonidin, petunidin and malvidin are important in foods. An anthocyanin pigment is composed of an aglycone (an anthocyanidin) esterified to one or more sugars. The most important flavonols are kaempferol, quercetin and myricetin. Apigenin, luteolin and tricetin are the predominant flavones. The above aglycones usually exist as glycosides. The basic building block of procyanidins is a flavonol (usually catechin or epicatechin) forming a dimer through a C4–C8 or a C4–C6 linkage, but trimers and higher polymers are common. All produce an anthocyanidin, when heated in the presence of a mineral acid. The term 'tannin' as used in foods includes two types of compounds. The first type is the 'condensed tannins' which may be C4–C8 or C4–C6 oligomers and polymers as described above. Compounds having other linkages, such as C2–O7 or C4–O7, have also been reported. The second type is the 'hydrolysable tannins', including the gallotannins and the ellagitannins. These give gallic or ellagic acid on hydrolysis and consist of phenolic moieties substituted into a central sugar. Thus, commercial 'tannic acid' consists primarily of pentagalloyl glucose. The theaflavin 'tannins' of black tea are produced by enzymatic oxidation of colourless catechins, gallocatechins and their galloyl esters to give a coloured benzotropolone nucleus which may also be esterified with gallic acid residues. *See* Tannins and Polyphenols

Sensory Properties

The most important attributes of a food are its sensory characteristics (texture, flavour, aroma, shape and colour). *See* Sensory Evaluation, Sensory Characteristics of Human Foods

Colour

Some colourless phenols can affect anthocyanin colour by copigmentation and copolymerization. They are even more important in many practical instances as precursors of colour formation or discoloration. Colour formation may result from combination with metallic cations such as those of iron or copper. Browning and blackening, the most significant types of phenolic discoloration, involve the production of many overlapping chromophores to produce the required absorbance through much or all of the visible region. Phenols contribute most obviously to browning when enzymatically oxidized to quinones, which polymerize to relatively stable polymers. The polymerization appears to be a random free radical type of process, even when initiated by enzymatic oxidation. Reactions of phenols to give coloured products are not necessarily oxidative, particularly those that involve reaction with the phloroglucinol portion of flavonoids (phenol–aldehyde condensations). Benzotropolone formation and xanthylium salt formation are two oxidative reactions that give brightly coloured products. It appears that these products and flavilium salts can be units within large phenolic polymers and contribute to the colour. Reactions between furfural and phenols would give extra colour during caramelization-type browning reactions. The polymeric brown pigments from phenol oxidation include phlobaphenes, humic acids and other complex products. The characteristic phenolic pigments of beetroot (red) and the spice turmeric (yellow) are betalaines and curcumin, respectively.

Flavour

Some simple phenols (4-vinylphenol, 2-methoxy-4-vinyl-phenol, *m*-cresol, *p*-cresol, *p*-ethylphenol, isoeugenyl esters, 4-ethylguaiacol, 2,6-dimethoxyphenol) in foods have unique sensory properties. Most are found in foods in the low parts per million range. The guaiacol derivatives (capsaicin and dihydrocapsaicin in capsicum species, gingerols and shogaols in ginger) and eugenol derivatives (vanillyl side-chain in cloves) in spices are pungent or aromatic.

Astringency

This sensation is registered within the oral cavity generally, as well as on the tongue. Phenolic acids have historically been classified as possessing sensory properties described as being astringent and bitter, sensations that in some products are not considered very desirable. Maga and coworkers have reported taste thresholds of various phenolic compounds (10–90 ppm). Based upon their respective threshold values, they concluded that vanillic, *p*- and *o*-coumaric acids and ferulic acids were present in high enough concentrations to be organoleptically detectable in wheat. From a structural point of view, it can be said that functional groups in the *m*-position increase threshold sensitivity whereas the same functional group in the *o* position decreases sensitivity. For benzoic acid compounds, a methoxy group resulted in a lower threshold than a hydroxy group. Mixtures of phenolic compounds have a synergistic effect. Phenolic compounds are thought to be important in the development of the typical beer flavour, contributing both volatile aroma and nonvolatile astringency. Astringency is usually regarded as a desirable characteristic of fruit and cider, but it is in red wine and tea that it is considered the most important. In both these beverages it is associated with the high content of those polyphenolic substances which are also involved in colour. These are procyanidins, anthocyanins and their oxidation products, which are collectively described as 'tannins'. In black tea, it is the galloyl groups of the theaflavin gallates and other polyphenols that are most important. ***See*** Tea, Chemistry

Phenolic substances in the form of flavonoids are important sources of bitterness in citrus juices. The best known is naringin, a glycoside of the flavanone naringenin with the disaccharide neohesperidose, which occurs in grapefruit and Seville oranges. ***See*** Citrus Fruits, Composition and Characterization

Taints

Basically, an off-flavour in a food can either be derived from reactions within the foodstuff caused by poor handling of the product or it can be caused by the contact of the food with an alien material and migration of substances into the food. Complications occur when the foodstuff is exposed to a perfectly safe and almost odourless compound but which reacts with components in the product to form obnoxious compounds (taints). Chlorophenols and their simple derivatives have become a major cause of taints. They may be derived from intermediates in the synthesis of herbicides (e.g. 2,4-dichlorophenol), disinfectants (e.g. chlorine-substituted phenols, cresols or xylenols), soft wood (which has been treated with pentachlorophenols as preservatives) used as ship container flooring or packing material for foods or chlorination of simple phenols in foods from water used in processing or cooking. Chlorophenols themselves are not especially unpleasant, but they are methylated and partially dechlorinated by various soil fungi (or bacteria) to form chlorine-substituted anisoles, which produce obnoxious odours. 2,4,6-Trichloroanisole and 2,3,4,6-tetrachloroanisole are well known for their ability to cause musty/mouldy taints in foods (chicken, milk and other dairy products, rice stored in contaminated jute bags) and beverages (cork taints in wines, and brandy in contact with cork oaks that had been treated with pentachlorophenol-containing preservatives) (2·4–27 $\mu g\ kg^{-1}$). Pentachloroanisole, although not itself a recognized tainting agent, is frequently present in foods contaminated with lower homologues.

Unfortunately, it has been found that no packing material is particularly effective as a barrier to these chemicals. Permeability studies have shown that 6-chloro-*o*-cresol readily passes through all packing materials, and only PVDC-coated polypropylene offers any resistance to 2,3,4,6-tetrachloroanisole.

Analysis of Phenolic Compounds

Phenols and related compounds form a class which, from the chemical standpoint, has become relatively difficult to investigate. The high reactivity of these compounds and the numerous forms in which they can exist in a variety of foods have made reliable data difficult to obtain. The traditional extraction techniques usually result in the coextraction of numerous other classes of compounds, making further isolation or purification difficult. As a result, many methods indicate only the presence of a 'phenol' or 'phenolic fraction' or 'total phenols', which makes further characterization almost impossible. The application of a solvent extraction technique followed by a preliminary clean-up procedure using adsorption or ion exchange chromatography achieves reasonable separation of phenolics. Subsequently, the characterization or quantification of individual phenolic compounds can be performed using colorimetry directly or thin-layer chromatography (TLC) and paper chromatography (PC). Recent advances in gas chromatography (GC) and high-performance liquid chromatography (HPLC) have resulted in the successful analysis of many phenolic compounds. Recent reports include the application of GC coupled with mass spectrometry (MS) and nuclear magnetic resonance (NMR). ***See*** Chromatography, Principles; Chromatography, Thin-layer Chromatography; Chromatography, High-performance Liquid Chromatography; Chromatography, Gas Chromatography; Mass Spectrometry, Principles and Instrumentation; Spectroscopy, Nuclear Magnetic Resonance

Colorimetry

Various reactions of phenols have been utilized to yield products absorbing visible light. These can be used for qualitative detection (spray reagent in PC or TLC) and quantitative analysis. Coloration with neutral ferric chloride, nitrous acid–mercuric nitrate, Gibbs reagent, diazonium coupling, vannillin with *p*-toluene sulphonic acid, ferric ferricyanide, ammonium vanadate and the antipyrine reaction have been used for quantitative analysis. In our laboratory, several sensitive and selective visible spectrophotometric (colorimetric) methods have been developed for the determination of some simple phenols, chlorogenic acid and polyphenols based on colour development under specified experimental conditions with different aminophenol, phenylenediamine or *N*-1-naphthylethylenediamine plus oxidant combinations. The resulting coloured species appeared to be either oxidative coupling products or charge transfer complexes involving *in situ* intermediate oxidation products (from aminophenol or phenol) and original phenol or aminophenol or *m*-phenylenediamine or *N*-1-naphthylethylenediamine. The pairs of reagents used include *m*-aminophenol–periodate (catechol, eugenol, guaiacol, quercetin, chlorogenic acid), *p*-*N*-methylaminophenol sulphate (metol)–periodate (pyrogallol, gallic acid, propyl gallate, phloroglucinol), *p*-aminophenol–dioxygen (guaiacol, catechol, morin), *p*-*N*,*N*-dimethylphenylenediamine–periodate (thymol, catechol), *m*-phenylene diamine–periodate (catechol, guaiacol), metol–Cr^{6+} (resorcinol and phloroglucinol, tannins, morin), *p*-aminoacetophenone–, thiosemicarbazide– or isonicotinic acid hydrazide–periodate (catechol, guaiacol, eugenol), *N*-1-naphthylethylenediamine–periodate or Cr^{6+} (catechol, guaiacol). ***See*** Spectroscopy, Visible Spectroscopy and Colorimetry

Extraction and Isolation

The method of extraction has to be designed keeping in mind the nature of the food material and the type of phenolic compounds present (or expected to be present) in the sample. In general, the simple phenols and their derivatives, phenolic glycosides, catechins and procyanidins are soluble in water while flavones and flavonols are only sparingly soluble. Solvents generally employed for extraction of total phenolic compounds are alcohol and acetone. Anthocyanins are stable as salts and are generally extracted with acidic alcohol. Some sort of preliminary fractionation (column or ion exchange chromatography) is generally necessary before the application of PC, TLC, GC or HPLC techniques.

A general procedure for the extraction of phenolics from plant materials is as follows.

The powdered plant material is extracted with separate portions of boiling alcohol. The combined extract is concentrated in a rotary evaporator under reduced pressure. The concentrate is extracted first with light petroleum to remove chlorophyll and waxy matter if present, then with diethyl ether (to separate catechins and chlorogenic acids) and finally with ethyl acetate (to obtain flavones, flavonols and procyanidin oligomers). The different extracts are then subjected to further separation using adsorption or ion exchange chromatography, if necessary, followed by characterization of individual phenols with PC, TLC, GC or HPLC.

Paper and Column Chromatography

Ascending, descending and two-dimensional modes of development in paper chromatography are routinely used. The main solvents used for paper chromatography include Forestal (acetic acid:water:hydrochloric acid, 30:10:3) and toluene–acetic acid for aglycones and butanol–acetic acid–water for glycosides. The use of an acetonitrile and aqueous ammonium acetate system has been found to be relatively successful. By varying the ratios of appropriate solvents in the mobile phase together with the pH, the system is adjusted for maximum separation of phenolic acids and polyphenols.

Column chromatography has been used for purification of foodstuff extracts. Column packings have included silica gel, alumina, Sephadex, MN polyamide SC-6, and polyethylene glycol dimethacrylate. Caffeic, *p*-coumaric and quinic acid conjugates have been separated on silica gel using a gradient from cyclohexane–acetonitrile to *t*-butanol–acetonitrile. Flavonoid compounds have been separated on columns using gradients of solvents including sequences with petroleum ether, benzene, chloroform, butanol and ethanol.

A few reports have included ion exchange columns in the separation of phenolic acids. The acidic and neutral (*o*-dihydroxy and non-*o*-dihydroxy) phenolic compounds can be separated by use of weak anion exchange columns.

Gas Chromatography

Both capillary and packed columns can be used. The packing material on which the stationary liquid phase (methyl silicone, OV-1, phenyl silicone with a medium content of phenyl groups, OV-17, phenyl silicone with a high content of phenyl groups, OV-25, methyl cyanopropyl phenyl silicate, OV-25 or silicone gum rubber with 5% phenyl, SE-52) is coated is usually a porous particle such as Chromosorb W or Gas Chrome P. Analysis of nonvolatile compounds by GC requires the derivatization of the compounds prior to analysis. The two most common derivatives are methylsilyl and trimethylsilyl derivatives. The use of di-(chloromethyl)-tetramethyldisilazane and chloromethyldimethylchlorosilane to form silylated phenolic acid has been suggested for greater sensitivity. C_{13} and C_{14} n-alkanes have been selected as internal standards because they are unaffected by silylation and by column derivatization. Several examples of excellent GC separations of phenolic acids may be found in the literature. The procedures are difficult to compare, not only because different conditions are used but because different compounds are included in the analyses, depending upon the particular interest of the researcher.

The mass spectra of silylated phenolic acids show molecular ion peaks of high relative intensity. The compounds would therefore be readily amenable to selected ion monitoring.

High-performance Liquid Chromatography

The use of HPLC for the analysis of phenolics has become very popular, mainly because of the substantial shortening of the time required for analysis compared to other chromatographic procedures. Phenolic compounds are polar and, because some may irreversibly bind to columns used in adsorption chromatography, most of the procedures described in the literature employ reversed-phase columns. The most popular are the μ Bondapak C_{18} and LiChrosorb RP-8 columns. Other columns, of various types, that have been used include Zorbax ODS, μ Bondapak alkyl phenyl, Mercosorb Si-60, LiChrosorb-10 diol and Radial Pak-A. Excellent separations of hydroxybenzoic acids and cinnamic acids, including chlorogenic acids and their derivatives such as glucose esters and tartrates, have been reported. The solvents used are combinations of methanol, water and acetic acid. Techniques using both isocratic and gradient conditions have been developed. Ultraviolet absorbance detectors (both fixed- and variable-wavelength models) are widely used in the analysis of phenolic compounds.

For the separation of flavonoids and their glycosides, the columns used include μ Bondapak C_{18}, μ Bondapak alkyl phenyl, μ Porasil, Mercosorb Si-60, Zorbax ODS, Bondapak C_{18} in series with μ Bondapak C_{18}/Corasil, Vydac ODS and LiChrosorb RP-8.

Bibliography

Coultate TP (1989) *Food, The Chemistry of its Components*, 2nd edn, pp 126–184. London: Royal Society of Chemistry.

Hardin MJ and Stutte CA (1984) Chromatographic analysis of phenolic and flavonoid compounds. In: Lawrence JF (ed.) *Food Constituents and Food Residues—Their Chromatographic Determination*, pp 295–322. New York: Marcel Dekker.

Haslam E (1989) *Plant Polyphenols*. Cambridge: Cambridge University Press.

Haslam E and Lilley TH (1988) Natural astringency in food stuffs. *CRC Critical Reviews in Food Science and Nutrition* 27: 1–40.

Herrmann K (1989) Occurrence and content of hydroxycinnamic acid and hydroxybenzoic acid compounds in foods. *CRC Critical Reviews in Food Science and Nutrition* 28: 315–347.

Macheix J (1990) *Fruit Phenolics*. Boca Raton: CRC Press.

Maga JA (1978) Simple phenol and phenolic compounds in food flavor. *CRC Critical Reviews in Food Science and Nutrition* 10: 323–371.

Mathew AG and Parpia HAB (1971) Food browning as a polyphenol reaction. *Advances in Food Research*, vol. 19, pp 75–145. New York: Academic Press.

Ranganna S (1986) Polyphenols. *Handbook of Analysis and Quality Control for Fruit and Vegetable Products*, 2nd edn, pp 66–104. New Delhi: Tata-McGraw Hill.

Rao KE and Sastry CSP (1984) Spectrophotometric determination of some phenols with *m*-phenylenediamine and sodium metaperiodate. *Mikrochimica Acta* 1(5–6): 313–319.

Rao KE Gurucharandas V and Sastry CSP (1985) New spectrophotometric method for the determination of catechol and guaiacol using *N*-(1-naphthyl) ethylenediamine. *Analusis* 13(2): 90–91.

Sastry CSP, Rao KE and Prasad UV (1982) Spectrophotometric determination of some phenols with sodium metaperiodate and amino phenols. *Talanta* 29: 917–920.

Sastry CSP, Gurucharandas V and Rao KE (1985) Spectrophotometric methods for the determination of *o*-hydroxy benzene derivatives. *Analyst* 110: 395–398.

Sastry CSP, Rao KE, Vijaya D, Rao AR and Rao MV (1988) Spectrophotometric methods for the determination of morin, quercetin, tannin and chlorogenic acid. *Journal of Food Science and Technology (India)* 25(3): 156–157.

Saxby MJ (1982) Taints and off flavours in foods. In: Morton D and MacLeod AJ *Developments in Food Science*, pp 439–457. New York: Elsevier.

Whiting DA (1979) Phenols. In: Steddart JF (ed.) *Comprehensive Organic Chemistry—The Synthesis and Reactions of Organic Compounds*, vol. I, pp 707–797. New York: Pergamon Press.

C.S.P. Sastry and B.S. Sastry
Andhra University, Visakhapatnam, India

PHOSPHOLIPIDS

Contents

Properties and Occurrence

Phospholipids have been scientifically studied since the 1700s and became commercially available as 'lecithin' in the 1930s. Their primary commercial source today is the soya bean, but phospholipids can be found in all living cells as part of the cellular membranes. This article covers the properties and occurrence of phospholipids as commercial lecithins, their chemistry, manufacture, composition, specifications, and their potential use in food systems.

Occurrence

The International Lecithin and Phospholipid Society defines lecithin as 'a complex mixture of glycerophospholipids obtained from animal, vegetable or microbial sources, containing varying amounts of substances such as triglycerides, fatty acids, glycolipids, sterols, and sphingophospholipids'. Lecithins are natural surfactants primarily derived from soya beans and eggs. They are found most abundantly in seeds and nuts, eggs, brains, and cell walls, in a concentration range of 0·5–2%. *See* Eggs, Structure and Composition; Fatty Acids, Properties; Soya Beans, Properties and Analysis; Triglycerides, Structures and Properties

Properties of Phospholipids (Lecithins)

There are three types of properties necessary to define phospholipids and lecithins: (1) chemical, (2) physical and (3) functional.

Chemical Properties

The chemical composition of de-oiled and liquid soya bean lecithin is shown in Table 1. There are approximately 17 different compounds in commercial lecithin, including carbohydrates, phytosterols and minor phytoglycolipids. The three major phospholipids are phosphatidylcholine, phosphatidylethanolamine and phosphatidylinositol.

Structure of The Major Phospholipids

The chemical backbone of the major phospholipids is a diacylglycerol molecule with the third carbon attached to a phosphate molecule. Choline, ethanolamine, serine and inositol can be attached to the phosphate group to change the physical and functional properties, leading to the formation of phosphatidylcholine, phosphatidylethanolamine, phosphatidylserine and phosphatidylinositol, respectively. The groups attached to positions 1 and 2 (α or β) are C_{14}–C_{18} fatty acids with double bonds associated with the lecithin source. The second carbon of the glycerol molecule, the β-position, usually contains linoleic acid.

Physical Properties

There are two major physical classes of lecithins: (1) fluid, and (2) waxy solids.

Properties and Occurrence

The fluid lecithins can have viscosities from 5000 to 100 000 cP, depending on processing conditions and diluents. The low-viscosity products are made through the addition of fatty acids and vegetable oil, depending on the function and stability required. Divalent metal ions like calcium can be added during drying to decrease viscosity. The moisture content can also make a difference. Water levels above 1% will increase the viscosity, eventually to a plastic state.

De-oiled lecithins are waxy solids that can be ground to various particle sizes. They are stable, free-flowing granules or powders.

Functional Properties

Lecithins are multifunctional agents. They can be used for many purposes in a food system, as shown in Table 2. The most popular functionalities are discussed below.

Anti-dusting Agents. Lecithins reduce static electricity by wetting dusty particles. They can be used alone or in conjunction with vegetable oils. Oils can be selected for the degree of shelf life required. *See* Vegetable Oils, Applications

Crystal Formation Modifier. Lecithins retard nucleation in fats and even monoglycerides, reducing graininess in texture.

Emulsifiers. Lecithins are most often used as amphoteric emulsifiers. They promote stable formation of oil-in-water and water-in-oil emulsions by reducing the interfacial surface tension between immiscible liquids. *See* Emulsifiers, Organic Emulsifiers

Mixing and Blending Aids. Lecithins decrease time and increase efficiency of mixing of unlike ingredients such as sugar and shortening by providing lubricity as well as viscosity reduction at the contact surfaces of the incompatible solids.

Release Agents. Lecithins provide easy release from metallic surfaces by attaching to the metal surface during hot or cold cooking. They assist in the cleaning of hot surfaces where proteins or batters are applied. They also reduce sticking between frozen food products.

Separating Agents. Lecithins prevent the adhesion of products that normally stick when in contact, like cheese slices and caramel confectionery.

Viscosity Modifiers. Lecithins reduce viscosity by coating particles to reduce particle–matrix friction such as in chocolates.

Wetting Agents. Lecithins provide complete wetting of fatty or hydrophilic powders in aqueous systems. The fatty acids are attracted to the fatty portion and the hydrophilic portions of the molecules actively imbibe water and control the hydration of the powder.

Table 1. Chemical composition of soya bean lecithin

	Granular lecithin	Typical liquid lecithin
Phosphatides (acetone insolubles) (%)	95 (minimum)	60 (minimum)
Soya bean oil (%)	2–3	39
Moisture (%)	1	0.7
Fat (grams per 100 grams of product)	90	93
Monounsaturated (oleic acid) (%)	9·2	17·9
Polyunsaturated (linoleic, linolenic acids) (%)	65·9	60·7
Saturated (palmitic, stearic acids) (%)	24·9	20·3
Carbohydrates (grams per 100 grams of product)	8	5
Approximate composition (100 gram sample)		
Fatty acid content (g)	50	66
Fatty acid content (relative composition) (%)		
Linoleic	58·9	54·0
Linolenic	7·0	6·7
Oleic	9·2	17·9
Palmitic	20·3	15·6
Stearic	4·6	4·7
Other Fatty Acids	0·0	1·1
Total	100·0	100·0
Primary acetone insolubles (g)		
Phosphatidylcholine	23	15
Phosphatidylethanolamine	20	12
Phosphatidylinositol	14	9
Elemental analysis (mg)		
Calcium	65	40
Iron	2	1
Magnesium	90	60
Phosphorus	3000	2000
Potassium	800	440
Sodium	30	10

From Central Soya, Chemurgy Division (1989) *The Lecithin Book*. Fort Wayne: Central Soya.

Manufacture

The majority of phospholipids are solvent extracted from their source. Usually, nonpolar hydrocarbons, like hexane, are used. Soya beans, for example, are cleaned, cracked and dehulled before flaking and extraction. The hexane is removed from the solvent micelle and the crude soya bean oil is cooled for further refining. Figure 1 shows a flow diagram for the degumming of crude soya bean oil and the production of lecithin. Approximately 2–3% water is added to the crude oil and agitated for at least 30 min. The phosphatides hydrate, swell, and are separated by centrifugation. The wet gums with 50% moisture are dried through a thin-film drier. The important points here are careful drying temperatures and cooling of the product below 20°C. Lightening the colour of the product can be achieved with hydrogen peroxide in the gum stage. The hydrogen peroxide is removed through drying.

Modification

The chemistry and functionality can be altered by simple chemical additions with acids and bases as well as hydrogen peroxide and acetic anhydride. These modifications increase the dispersion and hydration properties of the lecithin. Enzyme modification is also possible with lipases and phospholipases. These changes markedly affect the functionality in emulsification.

Properties and Occurrence

Table 2. Functionality of lecithins

Adhesion aid
Antibleed agent (as in fat bloom)
Anticorrosive
Antidusting agent
Antioxidant
Antispatter agent
Biodegradable additive
Biologically active agent
Catalyst
Colour intensifier
Conditioning agent
Coupling agent
Dispersing agent, mixing aid
Emollient, softening agent
Emulsifier or surfactant
Flocculant
Grinding aid
Lubricant
Liposomal encapsulating agent
Machining aid
Modifier
Moisturizer
Nutritional supplement, vitamin source
Penetrating agent
Plasticizer
Promoter
Release agent, antisticking agent
Spreading agent
Stabilizer
Strengthening agent
Suspending agent
Synergist
Viscosity modifier
Water repellent
Wetting agent

From Schmidt JC and Orthoefer FT (1985) In Szuhaj BF and List GR (eds) *Lecithins*, p 187. Champaign: American Oil Chemists' Society.

Composition of Lecithins

The composition of lecithins and their phospholipids will vary depending on their source: vegetable, animal or bacterial. There are minor differences within a class but major differences between the sources. Vegetable lecithins are high in phosphatidylcholine, phosphatidylethanolamine, phosphatidylinositol and phosphatidic acid, but very low in phosphatidylserine and contain no sphingomyelin. Animal lecithins are high in phosphatidylcholine, phosphatidylethanolamine, phosphatidylserine and sphingomyelin but contain no phosphatidylinositol. Microbes have phospholipids similar to the plant kingdom with high levels of phosphatidylethanolamine and phosphatidylcholine and phosphatidylserine or sphingomyelin.

Specifications of Lecithins

Lecithins may be qualified in several ways, chemically, physically and functionally, but there are also specifications used to assess quality and purity (see Table 3). These include acetone insolubles (AI), acid value (AV), hexane-insoluble matter (HI), peroxide value (PV), moisture, colour, free fatty acids (FFA), divalent metals (DVM), iodine value (IV) and phosphorus. Most of these analytical methods are found in the American Oil Chemists' Society (AOCS) *Official Methods and Recommended Practices*, Section J.

Acetone Insolubles. Phospholipids are nearly insoluble in cold acetone. This quantitative method should measure the active ingredients in lecithin. Depending on the type of lecithin the range is 35–98%

Acid Value. The phosphorus group in lecithins have a titratable acidity that is measured with this volumetric method. Free fatty acids are also measured in this test and should not be confused with phospholipid acidity. The AV range is 20–36 mg of potassium hydroxide per gram.

Hexane Insolubles. In the processing of soya bean lecithin, particulate matter finds its way through some processes and gives the product a hazy appearance. The HI can be determined by dissolving the product in hexane, centrifuging and gravimetrically measuring the insolubles. The HI range for soya bean lecithins is 0·05–0·3%.

Peroxide Value. The PV is a measure of oxidation in fats and oils. In lecithin, however, the PV usually measures the residual hydrogen peroxide from processing. The range in unbleached products is 0–10 meq kg^{-1} and in bleached products is 10–75 meq kg^{-1}.

Moisture. The water content of lecithins is quite low at 0·1–1·0%. It can be measured by oven drying but is more accurately determined by the Karl–Fischer method. This water content is so low that there is no measurable water activity for microbial growth. Moisture contents above 1% will change the viscosity from a fluid to a plastic state. ***See*** Water Activity, Principles and Measurement

Viscosity. The Brookfield viscosity at 25°C will have a range of 150–20 000 cP. Diluents and divalent metals can alter the viscosity to usable levels.

Free Fatty Acids. This method measures the true fatty acid levels in lecithins which range from 1 to 5 mg of potassium hydroxide per gram. Fatty acids are added as additives to adjust the viscosity. ***See*** Fatty Acids, Analysis

Iodine Value. The IV is a traditional method for qualifying lecithin sources. The more unsaturated fatty

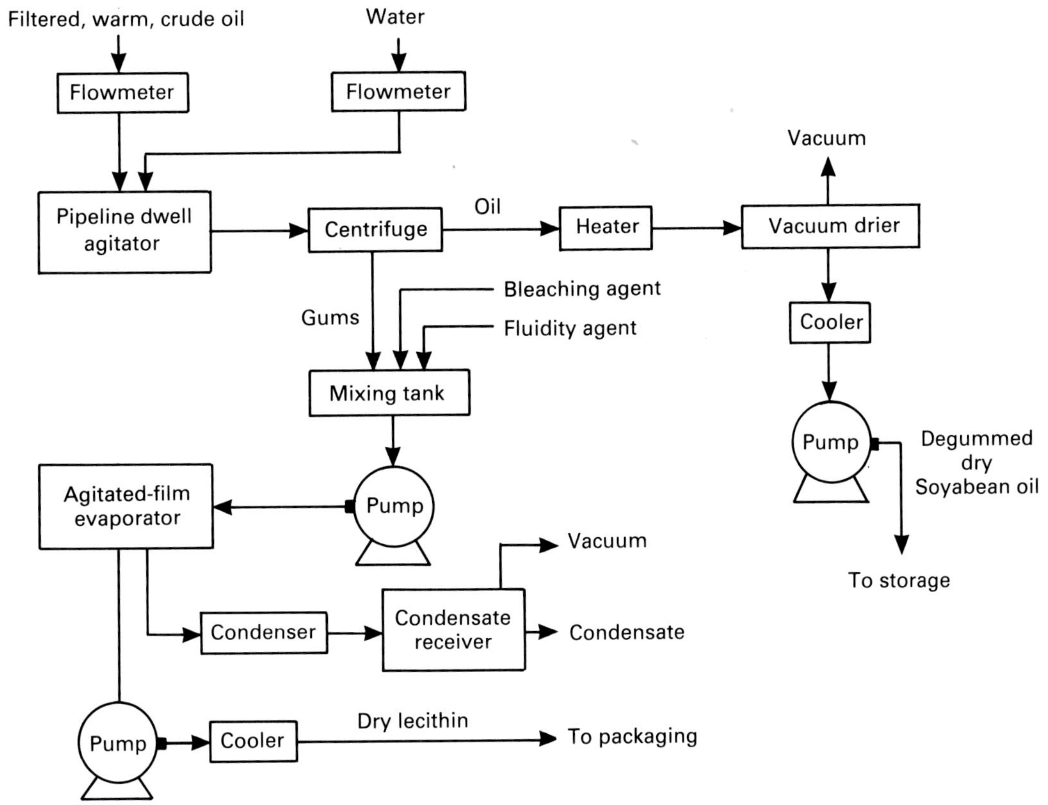

Fig. 1 Flowsheet for degumming and crude lecithin production. From List GR (1989) In (ed.) Szuhaj BF *Lecithins: Sources, Manufacture & Uses*, p 149. Champaign: American Oil Chemists' Society.

acids are found in soya bean lecithins, which have a range of 95–110 in natural fluid lecithins, to 80–90 in de-oiled lecithin.

Phosphorus. This wet chemical method is an indirect way of measuring the phospholipid content. The typical level of phosphorus is 2·0% in fluid lecithin and 3·0% in de-oiled products. The AOCS method Ca 12–55 has an approximation for converting per cent phosphorus to the phosphatides in soya bean oil. The equivalent phosphatides content is equal to per cent phosphorus ×30.

Uses of Lecithin

There are many uses of phospholipids in the food industry. As seen from the functional properties, there are multiple functions for lecithins. The following is a listing of the major areas of use:

- margarines – emulsifier, stabilizer and antispatter;
- confectionery and snack foods – crystallization control, viscosity control, antisticking;
- instant foods – wetting and dispersing agent, emulsifier;
- commercial bakery products – crystallization control, emulsifier, wetting agent, release agent;
- cheese products – emulsifier, release agent;
- meat and poultry processing – browning agent, phosphate dispersant;
- dairy and imitation products – emulsifier, wetting and dispersing agent, antispattering and release agent;
- packaging aid – release agent, sealant;
- processing equipment – internal or external release agent, lubricant.

See individual foods

Applications of lecithins in foods are clearly supported by the Food Chemicals Codex, the European E322 regulations, and they are considered as 'generally regarded as safe' substances by the US Food and Drugs Administration.

Storage and Handling

Lecithins are very stable products. They are shipped in drums or in bulk containers. They can be stored at ambient temperatures for up to 2 years without loss of activity or becoming rancid or spoiling. The water

Table 3. Specifications range for commercial lecithins

Analysis	Typical range		
AI (acetone insolubles) (%)	35–98		
AV (acid value) (mg KOH g^{-1})	20–36		
HI (hexane insolubles) (%)	0·05–0·3		
PV (peroxide value) (meq kg^{-1})			
Unbleached	0–10		
Bleached	10–75		
Moisture (%)	0·1–1·0		
Viscosity (Brookfield, 25°C) (cP)	150–20 000		
FFA (free fatty acids) (mg KOH (g^{-1})	1–5		
IV (iodine value) (cg I g^{-1})			
Natural	95–110		
Oil-free	80–90		
	Fatty acid composition (%)		
	Soya bean oil	Natural	Oil-free
$C_{16:0}$	10·3	15·6	20·3
$C_{18:0}$	4·4	4·7	4·6
Total saturates	14·7	20·3	24·9
$C_{18:1}$	24·5	17·9	9·2
$C_{18:2}$	53·8	54·0	58·9
$C_{18:3}$	7·0	6·7	7·0
Total unsaturates	85·3	78·6	75·1
Unsaturated: saturated ratio	5·8:1	3·9:1	3·0:1

From Central Soya, Chemurgy Division (1990).

activity is so low that no microbial growth can occur. The products may be heated to 25°C for easier application. ***See*** Water Activity, Effect on Food Stability

Bibliography

Burner D (ed.) (1991) *Official Methods and Recommended Practices*. Champaign: American Oil Chemists Society.

Central Soya, Chemurgy Division (1990) *The Lecithin Book*. Fort Wayne: Central Soya.

Charalambous G and Doxastakis G (1989) *Food Emulsifiers: Chemistry, Technology, Functional Properties and Applications*. New York: Elsevier.

Hanin I and Pepeu G (1990) *Phospholipids: Biochemical, Pharmaceutical, and Analytical Considerations*. New York: Plenum Press.

Szuhaj BF (ed.) (1989) *Lecithins: Sources, Manufacture & Uses*. Champaign: American Oil Chemists' Society.

Szuhaj BF and List GL (eds) (1985) *Lecithins*. Champaign: American Oil Chemists' Society

Bernard F. Szuhaj
Central Soya Company Inc., Fort Wayne, USA

Determination

Determination

Phospholipids are a well-known class of lipids that have been thoroughly analysed over the past three centuries. Their complete analysis was facilitated by the great advances in separation science and qualitative procedures that have occurred in the last 50 years. This paper will cover the analysis of phospholipids from their structure and composition, extraction techniques, qualitative and quantitative assays, and industrial methodology. ***See*** Lipids, Classification

Structure of Phospholipids

There are at least a dozen compounds that fall into the class of phospholipids. They have a basic structure of a diacylglycerol backbone with a phosphate ester on the α or third carbon of the glycerol molecule. Usually another compound is attached that characterizes the phospholipid. Figure 1 shows examples of the structures of the major phospholipids. These phospholipids (and their common abbreviations) are:

phosphatidylcholine (PC, PtdCho);
phosphatidylethanolamine (PE, PtdEth);
phosphatidylserine (PS, PtdSer);
N-acylphosphatidylethanolamine (NAPE, *N*-acylPtdEth);
phosphatidylinositol (PI, PtdIns); phosphatidic acid (PA, PtdA);
phosphatidylglycerol (PG, PtdGro);
plasmologen (PM);
diphosphatidylglycerol (DPG, diPtdGro);
lysophosphatidylcholine (LPC, lysoPtdCho);
lysophosphatidylethanolamine (LPE, lysoPtdEth).

The proper nomenclature for phospholipids has been defined by the 1976 revised recommendations of the International Union of Pure and Applied Chemistry (IUPAC) and the International Union of Biochemistry (IUB) Committee on Biochemical Nomenclature. For example, the term 'lecithin' is permitted for phosphatidylcholine but the systematic name is 1,2-diacyl-*sn*-glycero-3-phosphorylcholine. The generic name of 3-*sn*-phosphatidylcholine could be used. The abbreviation PtdCho is also allowed. This article will use the common names listed above since the literature has thousands of references with this terminology.

Composition of Phospholipids

The composition of phospholipids depends on the source of the phospholipids. Those from animal, plant and microbial sources will have different compositions, depending on the nature of the tissue from which the lipids are extracted – for example brain, liver or blood. In plants, it will vary on whether they are from soya beans, corn, cotton, rapeseed, sunflower, etc. In microbial sources it depends upon the organism.

The phospholipid classes are similar within a species, but differ primarily in fatty acid acyl composition around the 1 and 2 positions on the glycerol backbone.

Fatty Acids

The fatty acid chain length is commonly from C_4 to C_{26}, with different degrees of unsaturation from one to six double bonds, which may be at different locations on the acyl group. However, there is a pattern that is relevant to the present discussion. ***See*** Fatty Acids, Properties

In animals the primary fatty acids range from $C_{12:0}$ (lauric acid) to $C_{24:0}$ (tetracosanoic acid). Again, depending on the species and tissue extracted, the fatty acids can be variable.

In the plant kingdom, the primary diacyl groups on the phospholipids will range from $C_{12:0}$ to $C_{18:3}$, i.e. lauric to linolenic acid. There are usually no C_{20} fatty acids and higher as in the animal kingdom. The degree of unsaturation depends on the origin of the crop, i.e. from temperate or tropical regions. Also, the climate within the zone can make a seasonal difference.

Microbial phospholipid acyl groups are more similar to those found in the plant kingdom than they are to those from the animal kingdom. The predominant acyl fatty acids range from $C_{16:0}$ to $C_{18:3}$. There are more odd carbon numbered fatty acids in microorganisms than elsewhere.

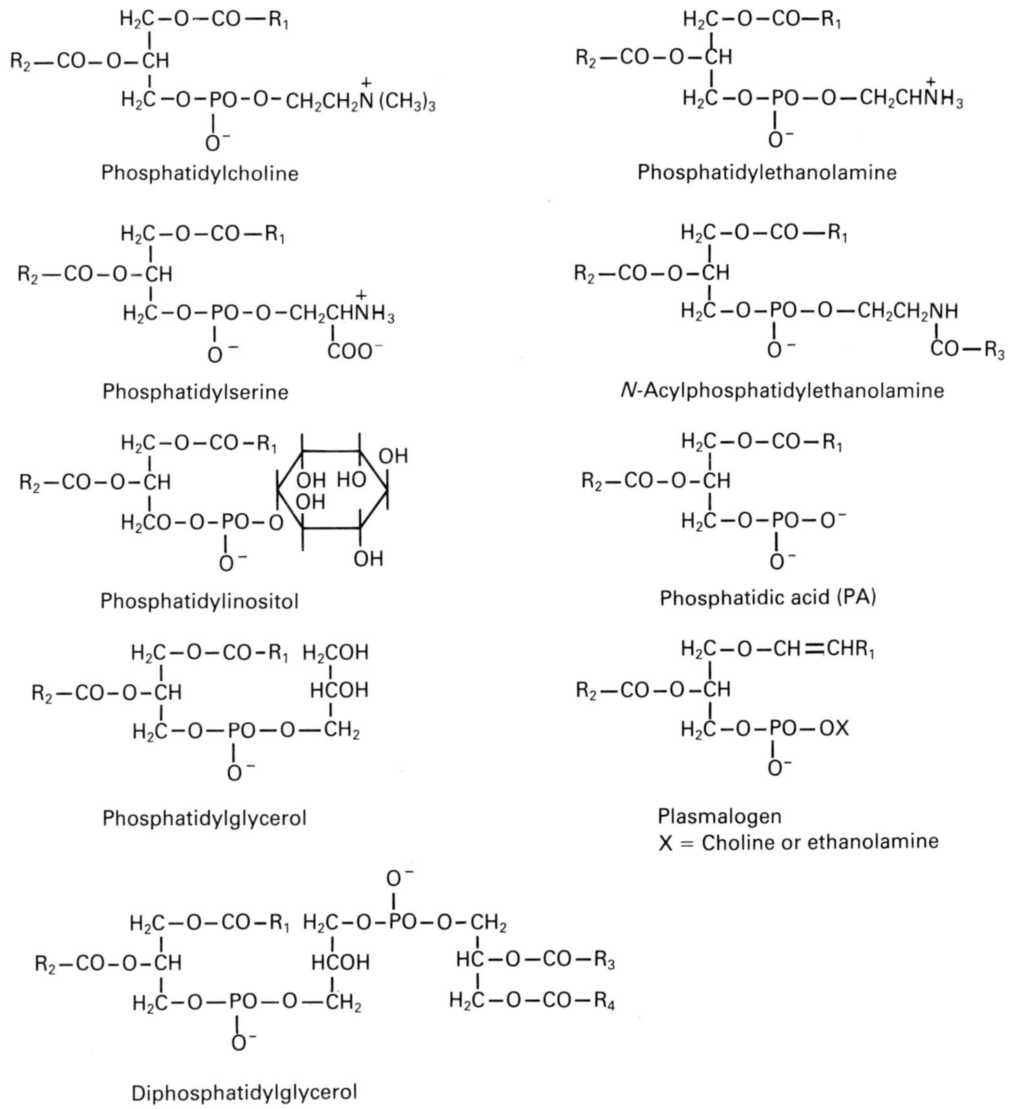

Fig. 1 Structures of the major phospholipids. From Scholfield CR (1985) In: Szuhaj BF and List GR (eds) *Lecithins* p3. Champaign: American Oil Chemists' Society.

A further factor that makes phospholipid composition more complex is the ability to have different fatty acids on the 1 and 2 positions of the molecule. Most research has shown that polyunsaturated fatty acids are usually in the 2 position.

Extraction Techniques

One of the most important factors in phospholipid analysis is the initial extraction procedure. If the analysis is on a finished commercial lecithin there is no problem, but if the analysis is from tissue samples or food samples the extraction technique will be critical in obtaining meaningful results.

Tissue Samples

For many years the Folch extraction of tissue homogenates with chloroform/methanol 2:1 (v/v) has been the method of choice by most researchers. Some found that the use of chloroform/methanol 1:1 (v/v) was preferable and some have used a biphasic system of butanol/methanol with dilute hydrochloric acid. Some have used hexane/2-propanol 3:2 (v/v). Which solvent system used depends largely on the required accuracy, but in most cases chloroform/methanol 2:1 (v/v) is the best solvent to try initially.

Food Samples

Since food samples may have lecithin or phospholipid added rather than being incorporated into the tissues of the food matrix, the dried sample can be ground and extracted with petroleum ether. If the lipids are bound in the product through processing, chloroform/methanol 2:1 (v/v) can be used. Because of environmental considerations di- or trichloroethane should be used in place of chloroform.

Drying of high-moisture products is required, but not by oven drying or air drying, as oxidation of the fatty acids can occur. Freeze drying of the product is preferred.

Qualitative Analysis

There are several ways to detect the presence of phospholipids. Traditional instrumental methods using ultraviolet and infrared are not used as often since thin-layer chromatography (TLC) can give a qualitative and semiquantitative result in one assay. Also, the use of conformational chemical sprays on the TLC plates can further identify the products.

Determination

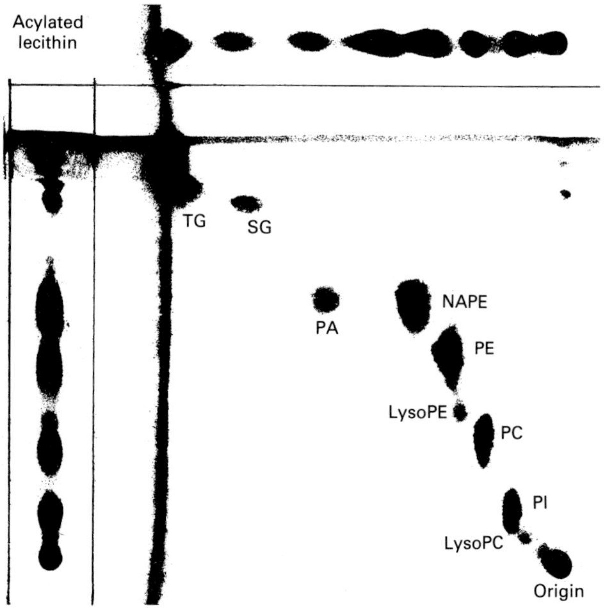

Fig. 2 TLC of soya bean lecithin in two dimensions: triglycerides (TG), sterol glucosides (SG), phosphatidic acid (PA), *N*-acylphosphatidylethanolamine (NAPE), phosphatidylethanolamine (PE), lysophosphatidylethanolamine (LYSO PE), phosphatidylcholine (PC), phosphatidylinositol (PI), lysophosphatidylcholine (LYSO PC). Silica gel plate; first dimension chloroform/methanol/acetic acid/water, 85:15:15:3 (v/v/v/v); second dimension chloroform/acetone/methanol/acetic acid/water (10:4:2:2:1) (v/v/v/v/v). Courtesy of J. Yaste, Central Soya, Food Research, Fort Wayne, Indiana, USA.

Thin-layer Chromatography

There are several types of silica gel plates available for TLC. Silica gel G and H are the most useful. Phospholipids may be separated on a 20 × 20 cm plate in one or two directions. A polar solvent and a nonpolar solvent system are used. The polar system is chloroform/methanol/water (65:25:4, v/v/v) and the nonpolar solvent is petroleum ether/diethyl ether/acetic acid (90:10:1, v/v/v). See the American Oil Chemists' Society (AOCS) recommended practice Ja 7-86 for alternative methodology. ***See*** Chromatography, Thin-layer Chromatography

These TLC plates are air or oven dried after separation of 20–50 μg of sample and are sprayed with 10% sulphuric acid and heated to char the phospholipids. Alternatively, they may be sprayed with a phosphorus spray containing molybdenum blue. Phospholipids stain a deep blue on heating the TLC plate. An example is shown in Fig. 2.

Nondestructive visualization techniques can be used if the phospholipids are to be determined or the fatty acid composition is to be run. Ultraviolet light and

2′,7′-dichlorofluorescein easily detects lipids on TLC. On prep plates, the bands are scraped off and extracted with chloroform/methanol (1 : 1, v/v) and the fatty acids converted to methyl esters using boron trifluoride and then determined using gas–liquid chromatography (GLC).

Quantitative Analysis of Phospholipids

Column Chromatography

Column chromatography precedes TLC in the separation of phospholipids. The techniques are slow and require good skill with column preparation, flow rates, and solvent removal. Commercial lecithins can be separated by dissolving the crude mixture in petroleum ether and passing it through a deactivated silica gel column with petroleum ether. The phospholipids are adsorbed and do not pass through the column, whilst triglycerides, sterol esters, etc., are eluted. The phospholipids are subsequently quantified by TLC and wet phosphorus analysis.

High-performance Liquid Chromatography (HPLC)

Newer technologies have found that HPLC can separate and quantify phospholipids more quickly and accurately. Separation is carried out on several types of columns, including silica gel and an amino group bonded to the silica surface (μBondapak-NH_2). The columns are eluted with chloroform/methanol gradients, acetonitrile/methanol/85% phosphoric acid, or acetonitrile/methanol/water. The eluent is measured at 205 nm or detected with flame ionization. Figure 3 shows an HPLC separation of commercial lecithin, using ultraviolet detection. The mass detector, an evaporative analyser, has also been successfully used for the HPLC determination of phospholipids. ***See*** Chromatography, High-performance Liquid Chromatography

Densitometry

Densitometric scanning has been used as an indirect method for determining phospholipid content on TLC plates. While the method has some promise, a problem is the quanitative charring of the phospholipid spots. Each phospholipid has a different charring density and this depends on fatty acid composition. Only with proper standards can this method be useful.

Thincography

Thin-rod TLC combines TLC with quantification by flame ionization detection. Rods are used rather than plates but controversy still exists over the suitability of the technique for routine lipid analyses.

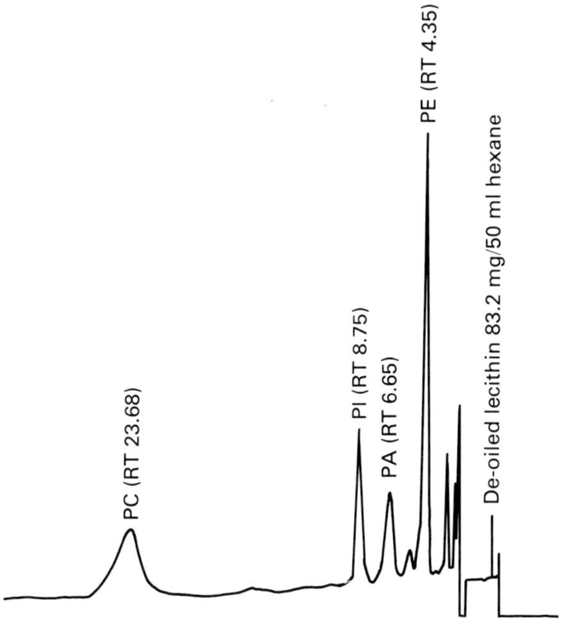

Fig. 3 HPLC of de-oiled soya bean lecithin. Column, μ Porasil 10μ 3·98×300 mm; mobile phase, hexane/2-propanol/acetate (8:8:1, v/v/v), buffer pH 4·2; detection, ultraviolet (206 nm); injection, 10 μl; flow rate, 1 ml min^{-1}. Retenton time (RT) in minutes. Courtesy of P. Balazs, Central Soya, Food Research, Fort Wayne, Indiana, USA.

Phosphorus Analysis on Phospholipids

Phosphorus analysis is an indirect method for the quantification of phospholipids because the qualitative composition of the sample must be known, if accurate values are to be obtained. With pure phosphatides this will work well, but most separation techniques give mixed phospholipids. The preferred method for phosphorus in lecithins is the AOCS method Ja 5-55. This determines the total phosphorus content of the sample. For commercial lecithins a multiple factor of 30 is used to convert total phosphorus values to acetone insoluble value.

There are various methods to determine phosphorus through molybdenum blue and molybdovanadophosphate yellow. To improve reproducibility many factors need to be evaluated. This includes the digestion method, chromogenesis and sensitivity. The AOCS method is the most straightforward and should be used especially in the food area.

Industrial Methods of Analysis

Phospholipids are characterized by a different set of assays than the determination of compounds used for academic or biochemical use. Most commercially available phospholipids come from soya beans and from eggs, and the methods outlined below can be used to qualify or categorize the products.

Acetone Insolubles (AI). This method determines the content of phospholipids in commercial lecithins. The method employs AOCS method Ja 4-46. The AI is an approximation of the active ingredients in formulations. Cold lecithin-saturated acetone must be used in this test.

Acid Value (AV). This method determines the phosphatide and fatty acid content of commercial phospholipids. The method utilized is AOCS method Ja 6-55. Phosphatide acidity is often confused with fatty acid addition to commercial lecithin. It is a combination of organic acids and phosphoric acid.

Peroxide Value (PV). A measurement of oxidative state of commercial phospholipids. It measures the milliequivalents of peroxide per kilogram of sample which oxidize potassium iodide. It also measures the residual peroxide used in process stabilization and bleaching. AOCS method Ja 8-87 is used.

Free Fatty Acids (FFA). This method utilizes AOCS method Ca 5a-40. When run on the acetone-soluble portion of the AI method, it gives the added fatty acids.

Phosphorus Content. The determination of total phosphorus is an indirect method for quantifying phospholipids. This method (AOCS method Ja 5-55) quantifies phospholipids through a molybdate reaction to a chromophore quantitated by phenolphthalein titration.

Gas Chromatography (GC). This method is commonly used to measure the fatty acid composition and does not quantify the phospholipids themselves (AOCS method Ce 1-62).

High-performance Liquid Chromatography (HPLC). This is gradually replacing the older techniques for qualifying and quantifying particular phospholipids in commercial mixtures. A uniform technique is being addressed by the AOCS.

Phospholipid analyses have come a long way since their study by Theuticum, circa 1800. Each analyst must choose the best method depending on constraints of accuracy and time.

Bibliography

Burner D (ed.) (1991) *Official Methods and Recommended Practices.* Champaign: American Oil Chemists' Society.

Central Soya, Chemurgy Division (1990) *The Lecithin Book.* Fort Wayne: Central Soya.

Charalambous G and Doxastakis G (1989) *Food Emulsifiers: Chemistry, Technology, Functional Properties and Applications.* New York: Elsevier.

Hanin I and Pepeu G (1990) *Phospholipids: Biochemical, Pharmaceutical, and Analytical Considerations.* New York: Plenum Press.

Szuhaj BF (1989) *Lecithins: Sources, Manufacture & Uses.* Champaign: American Oil Chemists' Society.

Szuhaj BF and List GL (1985) *Lecithins.* Champaign: American Oil Chemists' Society.

Bernard F. Szuhaj
Central Soya Company Inc., Fort Wayne, USA

Physiology

Physiology

Synthesis

Phospholipids are complex lipids consisting of a glycerol or sphingosine backbone, one or two hydrocarbon chains, a phosphate and, usually, an esterified alcohol, such as choline (Fig. 1). They are the major structural constituents of cell membranes and are constantly synthesized as a result of the metabolic turnover of membranes. Phospholipids are also essential components of plasma lipoproteins and lung surfactant; they are important in signal transduction pathways; they serve as a source of arachidonic acid, and are used as emulsifiers in total parenteral nutrition and as vehicles

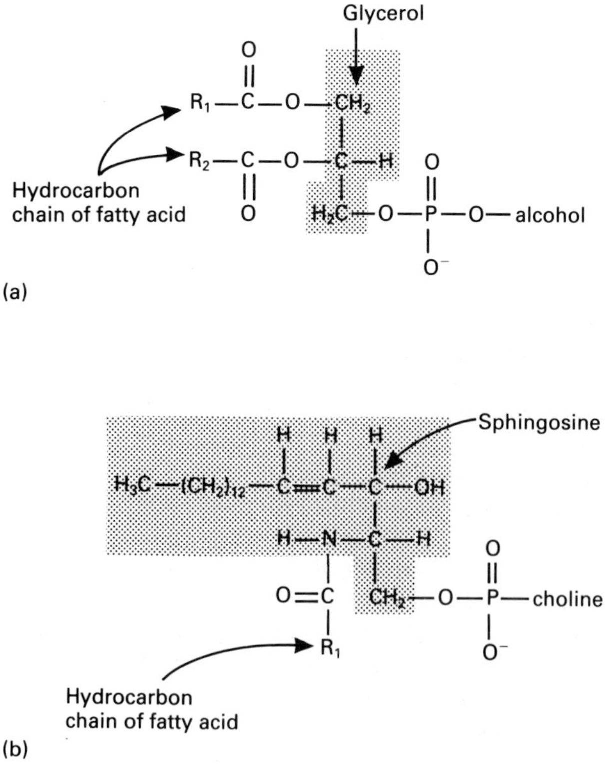

Fig. 1 Basic structure of (a) glycerophospholipids and (b) sphingolipids.

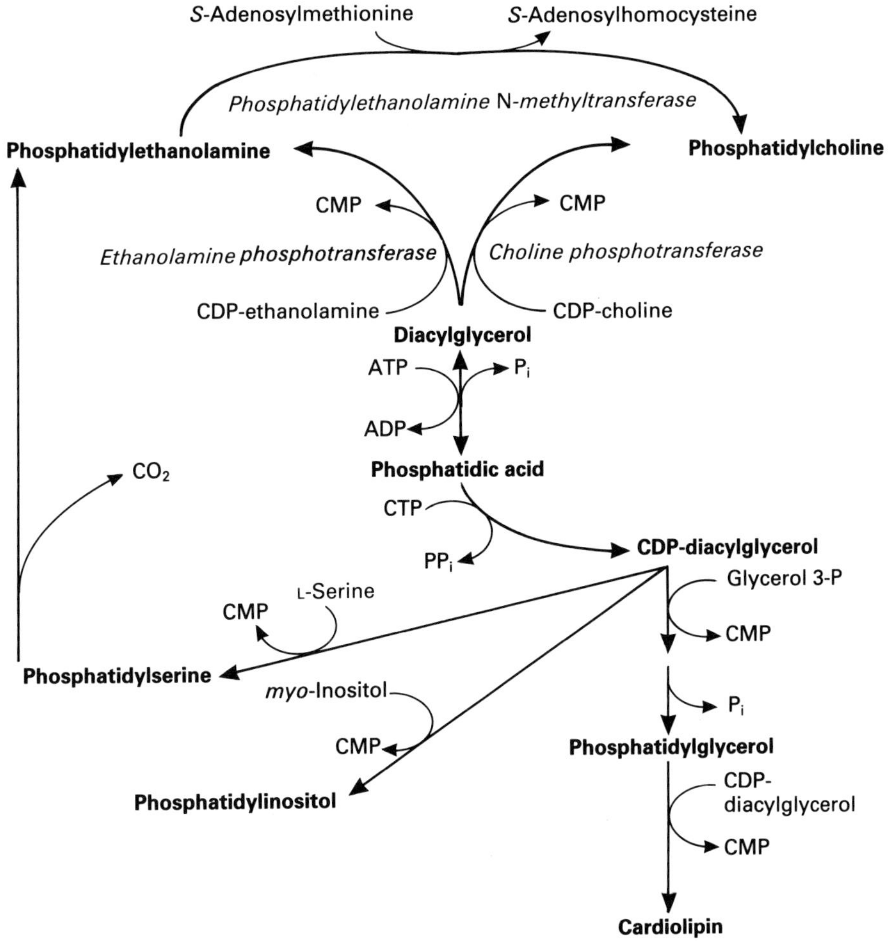

Fig. 2 Phospholipid biosynthesis – CMP: cytidine monophosphate; CDP: cytidine diphosphate; CTP: cytidine triphosphate; ATP: adenosine triphosphate; ADP: adenosine diphosphate; P_i inorganic phosphate; PP_i pyrophosphate; glycerol-3-P: glycerol-3-phosphate; CDP-ethanolamine: cytidine diphosphoethanolamine; CDP-choline: cytidine diphosphocholine; CDP-diacyglycerol: cytidine diphosphate-diacyglycerol.

for the delivery of lipid-soluble drugs. ***See*** Lipids, Classification

Glycerophospholipids are the most abundant subclass of phospholipids and include phosphatidic acid, phosphatidylcholine, phosphatidylethanolamine, phosphatidylserine, phosphatidylinositol, phosphatidylglycerol, plasmalogens, platelet activating factor, cardiolipin, and lysophospholipids. Phosphatidic acid is the least complex glycerophospholipid and may be synthesized from glycerol 3-phosphate or dihydroxyacetone phosphate, or by phosphorylating diacylglycerol.

Phosphatidylethanolamine and phosphatidylcholine (also called lecithin) are synthesized from diacylglycerol and an activated derivative of ethanolamine or choline (cytidine diphosphoethanolamine (CDP)-ethanolamine) or cytidine diphosphocholine (CDP-choline) (Fig. 2). Phosphatidylethanolamine can also be synthesized by the decarboxylation of phosphatidylserine. The only pathway for *de novo* synthesis of choline molecules in higher organisms is by successive methylation of phosphatidylethanolamine to form phosphatidylcholine. The enzyme phosphatidylethanolamine *N*-methyltransferase, found primarily in liver, performs three successive methylations using *S*-adenosylmethionine as the methyl donor. Phosphatidylserine, phosphatidylinositol and phosphatidylglycerol are synthesized from activated diacylglycerol (CDP-diacylglycerol) and serine, inositol, and glycerol 3-phosphate, respectively.

Plasmalogens, which are abundant in membranes of muscle and nerve cells, differ from other glycerophospholipids in having an alkenyl ether bond between a hydrocarbon chain and glycerol, usually at the first position. Cardiolipin is formed by the condensation of CDP-diacylglycerol and phosphatidylglycerol and is a common component of mitochondrial and bacterial membranes. The action of phospholipase A_2 results in the formation of lysophospholipids which have only one acyl group, usually at the first position.

Sphingomyelin is a sphingolipid found in plasma lipoproteins, cell membranes and is an important com-

Table 1. Phospholipid content of membranes

	Rat liver (percentage of total)			
Phospholipid	Plasma membrane	Mitochondrial membrane	Nuclear membrane	Microsomal membrane
Phosphatidylcholine	46·1	40·5	57·3	59·1
Phosphatidylethanolamine	24·7	34·7[a]	26·1	24·1[a]
Phosphatidylserine	4·2	—	5·5	—
Phosphatidylinositol	6·7	6·6	3·9	9·2
Sphingomyelin	16·8	2·4	6·3	4·2
Lysophosphatidylcholine	0·5	1·4	—	2·0
Cardiolipin	1·0	14·8[b]	—	1·0[b]

[a] Includes phosphatidylserine.
[b] Includes phosphatidic acid.
Taken from the data of Ansell GB *et al.* (1973).

ponent of the myelin sheath surrounding nerve fibres. Sphingosine, a long-chain aliphatic amine found in animal cells is acylated on its amine group with a fatty acid to form ceramide, which combines with phosphocholine from phosphatidylcholine to synthesize sphingomyelin.

Healthy humans probably ingest 6–12 g of phosphatidylcholine per day. It is often added as an emulsifier to processed foods in the form of a partially purified phospholipid mixture called lecithin (this provides an average per capita exposure to phosphatidylcholine of 1·5 mg per kg of bodyweight for adults). Lecithin is added to foods such as nondairy creamer, margarine, ice cream and chocolate bars where it keeps lipids in an emulsion. Eggs, fish, grains, lean meats and organ meats, especially brain and liver, are rich in phospholipids with a high concentration of choline phosphatides (phosphatidylcholine, sphingomyelin, lysophosphatidylcholine). Oil seeds (soya bean and sunflower) also contain a high concentration of glycerophospholipids, especially phosphatidylcholine. ***See*** Emulsifiers, Organic Emulsifiers

Biological Functions

Phospholipids in Membranes

The major structural component of all biological membranes is the phospholipid bilayer. In an aqueous environment, the polar head groups (phosphate and esterified alcohol) and the nonpolar tails of the fatty acyl chains cause the phospholipids to spontaneously arrange in a bilayer or micelle formation which serves to limit the contact of the hydrocarbon chains with water. The polar phosphate groups arrange themselves on the outer surface of the vesicle with the acyl chains on the inside. The hydrophobic barrier formed by the acyl chains does not allow hydrophilic molecules to pass through membranes without a transport system such as a carrier or channel. Aside from a structural role, the phospholipids also serve as a source of energy for the cell and in intracellular signalling. Phosphatidylcholine, phosphatidylethanolamine, sphingomyelin, phosphatidylserine and phosphatidylinositol are the major membrane phospholipids and are found in varying concentrations in cell and organelle membranes (Table 1). In addition, the phospholipid composition of each side of the membrane bilayer is different. In erythrocytes the outside layer contains more phosphatidylcholine and sphingomyelin and the inside more phosphatidylethanolamine and phosphatidylserine. The asymmetric arrangement may be attributable to the unidirectional nature of their synthesis, preferential association with specific membrane proteins, and/or the differences between the intra- and extracellular environment. The phospholipid membrane is a dynamic structure in which the lipids and proteins are able to undergo rapid lateral motion; however, transverse motion across the bilayers occurs very slowly. ***See*** Cells

Phospholipids in Bile and Lipoproteins

Liver exports phospholipids in two secretions: bile and plasma lipoproteins. The amount of phosphatidylcholine exported per day in humans is approximately 10–20% of the liver phosphatidylcholine pool, divided equally between the two secretions. Phospholipids (mainly phosphatidylcholine) secreted in bile (12 g per day) play a role in the micellar solubilization of cholesterol, free fatty acids, 2-monoglycerides, steroids and fat-soluble vitamins. The liver packages triglycerides for export as very-low-density lipoproteins (VLDLs) which are surrounded by a lipoprotein envelope rich in phosphatidylcholine and phosphatidyletha-

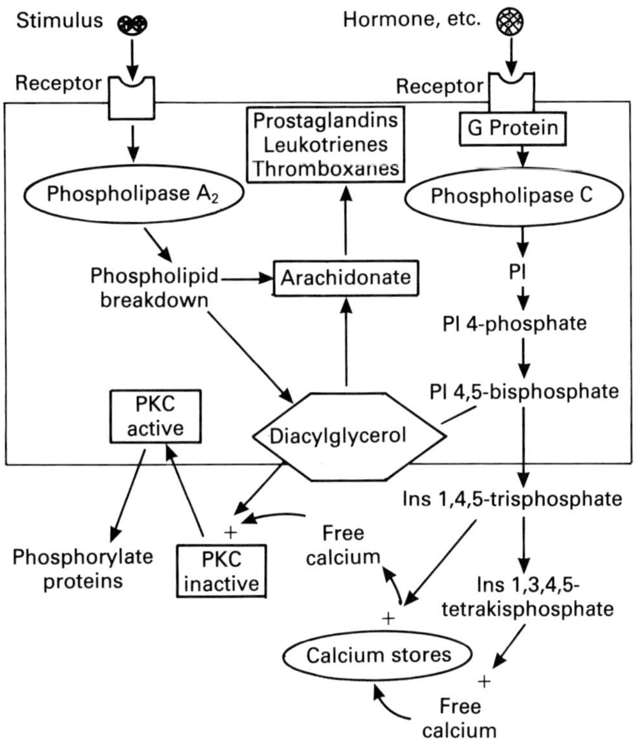

Fig. 3 Inositol-phospholipid breakdown and signal transduction – PKC: protein kinase C; PI: phosphatidylinositol; PI 4-phosphate: phosphatidylinositol 4-phosphate; PI 4,5-bisphosphate: phosphatidylinositol 4,5-bisphosphate; Ins 1,4,5-phosphate: Inositol 1,4,5-trisphosphate; Ins 1,3,4,5-tetrakisphosphate: Inositol 1,3,4,5-tetrakisphosphate.

nolamine. Circulating high-density lipoproteins (HDLs) transfer apoprotein C to VLDL; this apoprotein activates lipoprotein lipases in capillary endothelia, resulting in the hydrolysis of triglyceride. As triglycerides are removed from VLDL, the ratio of envelope (and thus phosphatidylcholine) to triglyceride increases and a transport lipoprotein of intermediate size is formed. The excess surface components (phospholipids, proteins and cholesterol) are released and enter HDL. The intermediate size lipoproteins are converted to low-density lipoproteins (LDLs) by the liver. Both LDL and HDL are rich in phosphatidylcholine content. *See* Bile; Lipoproteins

Phospholipids and Respiratory Function

Phospholipids, proteins and mucopolysaccharides constitute the surfactant system of the lung which reduces the cohesive force between water molecules at the alveolar surface to prevent the air spaces from collapsing at low lung volumes. Between 50% and 60% of lung surfactant is a specialized form of phosphatidylcholine, dipalmitoylphosphatidylcholine, in which both fatty acids are palmitic acid. Other phospholipids include phosphatidylglycerol (10% of total lipid) and small amounts of phosphatidylinositol, phosphatidylserine, phosphatidylethanolamine and sphingomyelin. Specialized type II alveolar cells of the lung synthesize the phospholipids. Once secreted, the surfactant lipids coat the air–water interface as a phospholipid monolayer to lower the surface tension within the alveoli. The full complement of enzymes needed to synthesize surfactant phosphatidylcholine is not expressed until near term in humans. For this reason, premature infants often do not synthesize sufficient surfactant and develop collapsed lung alveoli (respiratory distress syndrome). This syndrome can be treated by infusion of phosphatidylcholine into the lung.

Phospholipids and Signal Transduction

Phospholipids have an important role in cellular signal transduction following hormone–receptor interaction (Fig. 3). Phosphatidylinositol located in the plasma membrane is sequentially phosphorylated by phosphatidylinositol kinase to yield phosphatidylinositol 4-phosphate and phosphatidylinositol 4,5-bisphosphate. Upon receptor activation, phospholipase C is activated through a linking guanosine triphosphate (GTP)-binding protein, and phosphatidylinositol 4,5-bisphosphate is hydrolysed to yield inositol 1,4,5-phosphate and diacylglycerol. Inositol 1,4,5-phosphate acts as a second messenger to stimulate release of calcium ions (Ca^{2+}) from intracellular Ca^{2+} stores, which, in combination with diacylglyerol, activates protein kinase C. Protein kinase C is able to phosphorylate a wide variety of cellular proteins, resulting in increased or decreased enzymatic activity of specific proteins. Inositol 1,4,5-phosphate is further phosphorylated by a 3-kinase to yield inositol 1,3,4,5-tetrakisphosphate. This compound may also act as a second messenger to make free Ca^{2+} available to the inositol 1,4,5-phosphate-sensitive endoplasmic reticulum store. Upon receptor activation, a phospholipase C is also activated; this breaks down phosphatidylcholine, generating additional diacylglycerol and enhancing the phosphatidylinositide-based signalling system. The signalling is ended when sphingomyelin breakdown generates sphingosine, an inhibitor of protein kinase C.

In addition to activating protein kinase C, diacylglycerol is further metabolized by phospholipase A_2 to release arachidonic acid (20:4 n−6) which may also be produced by the direct action of phospholipase A_2 on membrane phospholipids. Arachidonic acid is the substrate for lipoxygenase and cyclooxygenase, leading to the synthesis of leukotrienes, thromboxane A_2 or prostaglandins (Fig. 3). These eicosanoids modulate protective immune responses and cell functions in addition to physiological functions of the cardiovascular, pulmonary and reproductive systems. ***See*** Fatty Acids, Metabolism

Platelet Activating Factor

Platelet activating factor (PAF or 1-*O*-alkyl-2-acetyl-sn-glyceryl-3-phosphorylcholine) is a potent, pharmacologically active phospholipid released when macrophages, neutrophils, eosinophils, vascular endothelium and platelets are stimulated. Its physiological effects include the following: platelet aggregation and release of histamine, serotonin and factor 4; microvascular leakage; protein extravasation; bronchoconstriction and bronchial hyperresponsiveness; decreased mucociliary clearance; PAF is important in anaphylaxis, inflammation and asthma.

Phospholipids in Total Parenteral Nutrition and Drug Administration

The ability of phospholipids to form bilayers and micelles is exploited in their use as fat emulsifiers in total parenteral nutrition (TPN) and in the formation of liposomes employed as drug carriers. Fat emulsions in TPN usually contain egg or soya lecithins, as the emulsifying agent, and a source of triacylglycerol. The phospholipid composition is critical for the stability of fat emulsions. When phosphatidylcholine and phosphatidylethanolamine are the major components, small amounts of phosphatidylserine, phosphatidic acid, lysophosphatidylcholine, lysophosphatidylethanolamine and sphingomyelin must also be present. These minor components ionize at the oil–water interface and stabilize the emulsion. Once in the circulation, the fat particles mimic chylomicrons, as they pick up apoproteins C-II and C-III from HDL and are metabolized by lipoprotein lipase. The phospholipid composition determines their ability to acquire the necessary apoproteins and therefore their ability to be efficiently metabolized. These same fat emulsions can also be used to administer lipid-soluble drugs. The advantages of this system include the protection of the drug from hydrolysis, lower incidence of side-effects, and good tolerance owing to their resemblance to chylomicrons.

Bibliography

Ansell GB, Hawthorne JN and Dawson RMC (eds) (1973) *Form and Function of Phospholipids*. New York: Elsevier Science Publishers.

Berridge MJ (1987) Inositol lipids and cell proliferation. *Biochimica Biophysica Acta* 907: 33–45.

Hanin I and Ansell GB (ed.) (1986) *Lecithin: Technological, Biological and Therapeutic Aspects*. New York: Plenum Press.

Hanin I and Pepeu G (ed.) (1989) *Phospholipids: Biochemical, Pharmaceutical and Analytical Considerations*. New York: Plenum Press.

Weihrauch JL and Son Y (1983) The phospholipid content of food. *Journal of the American Oil Chemists Society* 60: 971–978.

Acknowledgements

This work was supported by a grant from the National Institutes of Health, USA (HD26553).

Elizabeth M. Rohlfs, Kaaren M. Haldeman, Melinda Fine and Steven H. Zeisel
The University of North Carolina, Chapel Hill, USA

PHOSPHORUS

Contents

Properties and Determination

The extensive and varied chemistry of phosphorus transcends the traditional boundaries of inorganic chemistry because of its vital role in the biochemistry of all living organisms. It was first isolated from urine by Hennig Brandt in 1669 as a white waxy substance. The spontaneous chemiluminescent reaction of white phosphorus with moist air was the first observed property and was also the origin of its name (Greek: *phos*, 'light'; pherein, 'bringing').

Chemical Properties

Phosphorus is a typical nonmetal placed in group VA of the periodic table. The phosphorus atom consists of 15 electrons distributed as $1s^2 2s^2 2p^6 3s^2 3p_x^1 3p_y^1 3p_z^1$, with atomic energy levels as shown in Fig. 1. Thus, three unpaired electrons, together with the availability of low-lying vacant 3d orbitals, account for the predominant oxidation states III and V in phosphorus chemistry. Phosphorus exists in many allotropic forms which reflect various modes of catenation. At least five crystalline polymorphs are known and there are several 'amorphous' or vitreous forms. The most common forms of phosphorus are white, red and black. Of these, α-white is the most volatile and reactive solid which is thermodynamically the least stable. All forms, however, melt to give the same liquid, which consists of symmetric P_4 tetrahedral molecules, $P–P = 2{\cdot}21 \pm 0{\cdot}02$ Å. The same molecular form exists in the gas phase at $> 800°C$ and low pressure. The bonding orbitals in P_4 have only 2% of 3s and 3d character. Therefore, the bonds in P_4 are bent with $\angle P–P–P = 60°$ and a strain energy of 22·8 kcal-mol^{-1}, which accounts for its high reactivity. Electronic spectral studies suggest a resonance structure with strong π bonds. The regions of high electron density in the P_4 molecule are shown in Fig. 2. The moment of inertia about any of its three major axes is $2{\cdot}5 \times 10^{-40}$ g cm^2.

The only naturally occurring isotope, ^{31}P, has a nuclear spin $I = \frac{1}{2}h/2\pi$ and a large magnetic moment (1·13 NM) but no quadrupole moment. It is suitable for nuclear magnetic resonance (NMR) spectroscopy. In a 10 kG field ^{31}P resonates at 17·24 MHz. Phosphorus forms compounds with all elements except tin, bismuth and the inert gases. It also reacts readily with heated aqueous solutions to give a variety of products. The bonding and stereochemistry of the phosphorus atom is varied, essentially due to empty d orbitals. It is known in at least 14 coordination geometries with coordination numbers up to 9, though the most frequently met coordination numbers are 3, 4, 5 and 6. Some typical geometries along with orbitals used in the formation of bonds are given in Table 1. ***See*** Spectroscopy, Nuclear Magnetic Resonance

Forms in Foods

Plants need phosphate for healthy growth, especially for the development of roots, flowers, fruits and seeds,

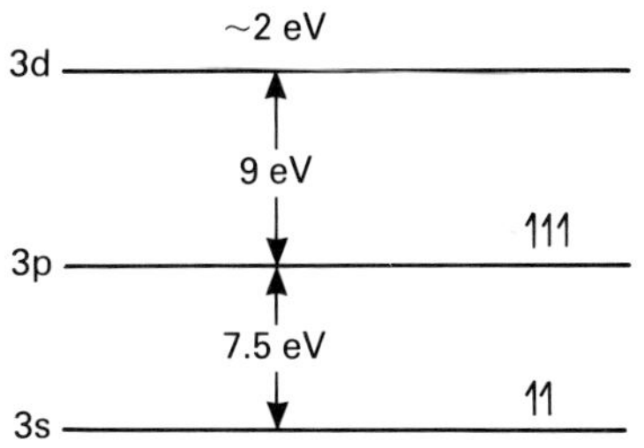

Fig. 1 The atomic energy levels in phosphorus.

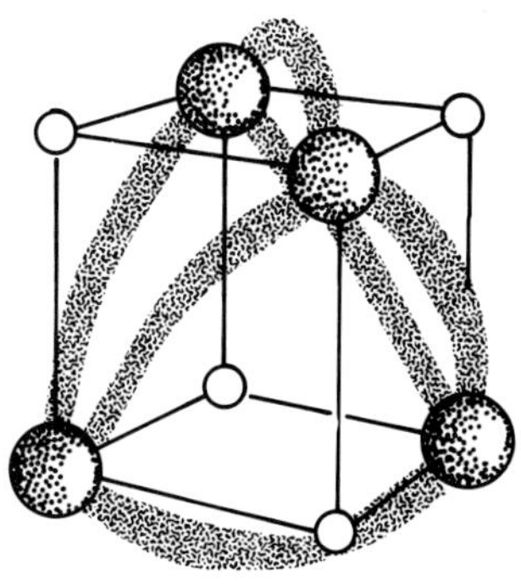

Fig. 2 A sketch demonstrating the three-dimensional distribution of valence shell electron density in the P_4 molecule. Courtesy of Hart, Robin and Kuebler (1965) *Journal of Chemical Physics* 42: 3631–3638

Table 1. Types of phosphorus compounds and their molecular structure

Number of bonds	Hybridization	Geometry		Examples
3	sp^3 (with one lone pair		Pyramidal	PH_3
4	sp^3		Tetrahedral	PH_4^+, $POCl_3$ ($d\pi$–$p\pi$ bonding in compounds with PO)
5	sp^3d	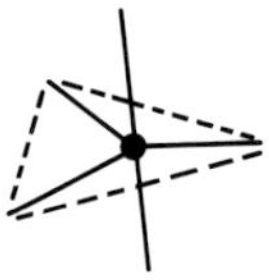	Trigonal bipyramid	PCl_5,PF_5
6	sp^3d^2		Octahedral	PF_6^-

though the requirement is less compared to that for nitrogen and potassium. All foods contain phosphorus in the form of the phosphate anion (PO_4^{3-}), and is consumed by living organisms as such. Very few natural compounds contain phosphorus in any other form. It is picked up from the soil, where it is present as organic and inorganic phosphates (soluble as well as insoluble) in nucleoproteins, nucleic acids and the coenzymes NAD (nicotinamide adenine dinucleotide), NADP (nicotinamide adenine dinucleotide 2-phosphate), ATP (adenosine triphosphate) and other high-energy phosphates. Organic phosphates include sugar phosphates such as glucose 6-phosphate (**1**), phospholipids and pigments.

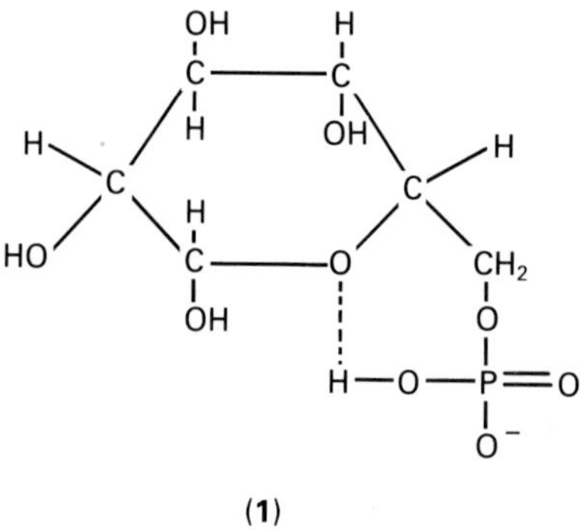

(**1**)

Most of the phosphorus in foods is in the form of organic phosphates which are digested in the intestines to form inorganic phosphates of sodium, calcium and potassium. In egg yolk and fish it occurs in the form of phospholipids, phosphatidylcholine and phosphatidylethanolamine. In certain foods such as cereal grains and proteins from vegetable sources, 50–80% of the phosphorus occurs in the form of phytin, which is usually the calcium/magnesium salt of phytic acid (the hexaphosphate ester of inositol). Natural starch, particularly of potato, contains phosphoric acid as an ester (**2**). ***See*** Phospholipids, Properties and Occurrence; Phytic Acid, Properties and Determination; Starch, Structure, Properties and Determination

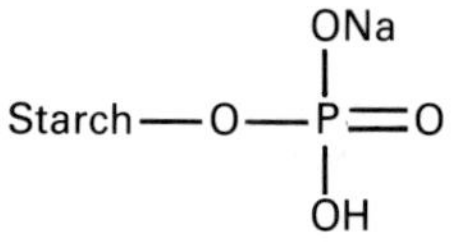

(**2**)

The phosphate cycle in water is controlled by the biocycle. The inorganic part of the cycle consists of HPO_4^{2-} in solution in equilibrium with $H_2PO_4^-$, PO_4^{3-} and H_3PO_4. Planktonic algae can absorb inorganic phosphates. Some algae also have a method of breaking down polyphosphates and even utilizing phospholipids.

Properties and Determination

Occurrence

Phosphorus is the 11th element in order of abundance in the earth's crust and occurs to the extent of 1120 ppm. It is an essential constituent of every known tissue and cell in the body and accounts for about 1% of the body-weight. Ordinarily, about 70% of phosphorus ingested in foods is absorbed by the body. As phosphate is a major constituent of all plant and animal cells, it is present in all natural foods, mostly in the organic form.

Heavy concentrations of phosphorus are found in the meristematic regions of actively growing plants, where it is involved in the synthesis of nucleoproteins. Phospholipids, along with proteins, are significant constituents of cell membranes. Generally, phosphorus is more concentrated in seeds. In plant tissues and juices it exists as phosphoric acid and the phosphate anion. In some citrus fruits it occurs, up to 2·7%, as dihydrogen phosphate anion ($H_2PO_4^-$).

The phosphorus contents in various foods are: cereal grains, 0·1–0·5%; pulses 0·3–0·6%; leafy vegetables 0·02–0·5%; nuts and oilseeds, 0·01–1·0%; condiments and spices, 0·01–0·5%; fruits, 0·01–0·5%; fish and seafood, 0·2–2·0%; meat and poultry, 0·15–0·8%; milk and dairy products, 0·1–1·0%; fats and edible oils, 0·01–0·2%; sugars, 0·01–0·4%; beverages, 0·01–0·1%. Semi-refined brown sugar contains a high level of phosphorus but refined sugar is completely stripped of the element. In the refinement of cereals 50–60% of phosphorus is lost. Total food supplies in the UK and USA provide 1·5 g of phosphorus per person per day, of which about 10% is added artificially. During pregnancy and the lactation period, however, the requirement for phosphorus is increased. In muscle tissues phosphorus is present as phosphocreatine, which synthesizes ATP. Its deficiency in adults may occur with excessive use of alcohol, prolonged vomiting, liver disease or hyperparathyroidism. ***See*** individual foods

Plants lacking phosphorus may develop necrotic areas on the leaves, petioles or fruits; they may have a general overall stunted appearance and the leaves may have a characteristic dark to blue-green coloration.

Properties of Phosphates

All phosphates are salts of oxyacids which contain a P=O group and at least one P–OH group that ionizes. Some species also have a P–H group where the hydrogen atom is not ionizable. Phosphates of metal ions and other cations, mixed metal phosphates and condensed phosphates are well known because of their commercial and technical importance. Many phosphates, especially long-chain polyphosphates, are known for their toxicity as they adversely affect the osmotic pressure of body fluids and prevent absorption of mineral nutrients. Phosphates are capable of interacting with many of the constituents of food systems, and inactivate metal ions, and are thus important in food processing.

Monosodium phosphate (NaH_2PO_4) is water soluble and used as a phosphatizing agent on steel surfaces. Its acidic property is used in effervescent laxative tablets and as a leavening agent in baking powder. Monopotassium phosphate (KH_2PO_4) crystals show piezoelectric effects and are used in submarine sonar systems. Disodium and dipotassium phosphates are used as buffering agents to maintain pH. This property is used for stabilizing meat. These salts are also used as sequestering agents in the food industry. Sodium orthophosphate (Na_3PO_4) is highly alkaline and finds a use in industrial hard surface cleaners. Its aqueous solution is a valuable constituent of scouring products, paint strippers and grease saponifiers. Its complex with sodium hypochlorite ($(Na_3PO_4 \cdot 11H_2O)_4 \cdot NaOCl$) releases active chlorine when wetted; this combination of scouring, bleaching and bacterial action makes the adduct valuable in automatic dish washing powder formulations. Potassium orthophosphate (K_3PO_4) is used to regulate the rate of polymerization of styrene–butadiene rubber. Mono- and diammonium phosphates are used as fertilizers and nutrients. An important property of ammonium phosphate is as a flame-retardant agent for cellulose materials. The action depends on its dissociation to ammonia and orthophosphoric acid when heated. Acid so generated catalyses the decomposition of cellulose to char and smother the flame. Urea phosphate is generally used to flameproof cotton fabrics. A dilute solution of diammonium phosphate, with an initial pH of 7·85, upon boiling evolves ammonia and the pH drops to 5·78 in 2 h. This property is used for the precipitation of colloidal dyes on wool fabrics. Dicalcium phosphates are used in pharmaceutical tablets as supplements. Natural phosphate minerals are all orthophosphates, the major one being fluoroapatite; hydroxoapatites, partly carbonated, make up the mineral part of teeth. These are important constituents of bones. Calcium orthophosphates are particularly important in fertilizer technology. ***See*** Leavening Agents; Stabilizers, Types and Function

Many phosphate complexes of transition metal ions are known. Ce^{4+}, Th^{4+}, Zr^{4+}. U and Pu form insoluble phosphates from fairly strong acid solution (3–6 M nitric acid). Condensed phosphates contain more than one phosphorus atom and P–O–P bonds with three main building units – the end unit (**3**), middle unit (**4**) and branching unit (**5**). These units can be readily distinguished by reactivity with water and ^{31}P NMR. These units can also be incorporated into linear or cyclic polyphosphates. Linear polyphosphates are the salts with the general formula $[P_nO_{3n+1}]^{(n+2)-}$, such as $M^I_4P_2O_7$ and $M^I_5P_3O_{10}$. Cyclic polyphosphates are the salts with the general formula $[P_nO_{3n}]^{n-}$, such as $M_3P_3O_9$

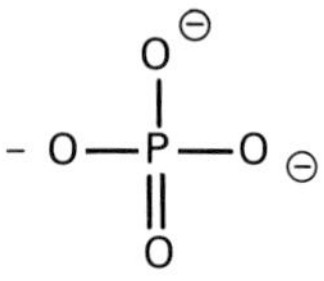

(3)

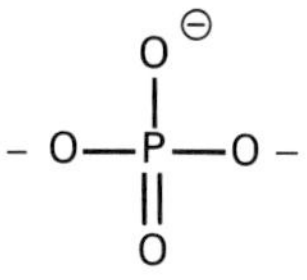

(4)

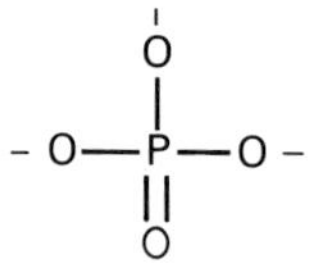

(5)

and $M_4P_4O_{12}$. Condensed phosphates form soluble complexes with many metals. Chain phosphates are used as water softeners in industry. Polyphosphates aid in controlling the microbiological population on the surface of poultry meat. Some metaphosphates having an infinite chain length are also known, e.g. KPO_3.

Organic phosphates contain phosphate groups linked through OH groups of organic compounds (–C–O–P linkage) such as sugars. Large numbers of phosphate esters are known. Some of these are of industrial importance, particularly for solvent extraction of metal ions from aqueous solutions, e.g. tri-*n*-butylphosphate. Organic derivatives of fluorophosphoric acid ($FP{=}O(OH)_2$) have promising properties as insecticides. The rate of hydrolysis of phosphate esters and of triphosphate esters such as ATP is important in biological systems.

Analysis

Phosphorus is generally detected on the basis of the reaction between orthophosphoric acid and the molybdate ion (MoO_4^{2-}), which gives a yellow-coloured precipitate in a strongly acid solution.

Total Phosphorus

Any solution containing phosphorus is fumed with aqua regia almost to dryness, followed by heating with 1 M nitric acid, whence lower oxidation states are oxidized to the orthophosphate (PO_4^{3-}) form. The resultant solution can be used for the estimation of total phosphorus by gravimetric, titrimetric or spectrophotometric methods. Some of these have been recommended by the Association of Official Analytical Chemists for the analysis of total phosphorus in vegetables, fruits, cereals and other foods. These are dried in a silica/platinum crucible and heated over a low bunsen flame to volatilize organic matter. A 10% sodium bicarbonate solution is added and the contents evaporated to dryness. Oil may be burnt off at a low temperature without smoking and, finally, ashing is carried out in a muffle furnace at 500°C. The contents are dissolved in concentrated nitric acid, heated to dryness and then dilute hydrochloric acid added.

Gravimetric methods include the formation and weighing of phosphorus as ammonium phosphomolybdate, ammonium magnesium phosphate or pyrophosphate and quinoline molybdophosphate. On addition of ammonium molybdate solution (12·5 g dissolved in 75 ml of water is slowly added to another solution containing 125 g ammonium nitrate in 125 ml of water and 175 ml of nitric acid diluted to 500 ml) a yellow-coloured precipitate is obtained which is filtered, dried at 105°C and weighed as $(NH_4)_3PO_4 \cdot 12MoO_3 \cdot H_2O$. On further heating at 450°C a dark greenish blue complex, $P_2O_5 \cdot 24MoO_3$, is obtained which may be weighed. In another method, the ammonium molybdate is replaced by a reagent containing sodium molybdate and quinoline so that quinoline molybdophosphate, $(C_9H_7N)_3[PO_4 \cdot 12MoO_3]$, is precipitated. Alternatively, magnesia reagent (magnesium chloride and ammonium chloride in ammoniacal solution) may be used as the precipitating reagent. The precipitate may be weighed as $Mg(NH_4)PO_4 \cdot 6H_2O$, or after heating in a muffle furnace at 1000°C as $Mg_2P_2O_7$.

The most common titrimetric method consists of precipitation as $Mg(NH_4)PO_4 \cdot 6H_2O$, filtration, washing and dissolution of the precipitate in excess dilute hydrochloric acid. An excess of standard EDTA solution is added and its pH is adjusted to 10. Excess EDTA is titrated with a standard solution of magnesium chloride or magnesium sulphate using eriochrome black T as the indicator. Alternatively, the precipitate of ammonium molybdophosphate may be titrated with standard sodium hydroxide solution using phenolphthalein indicator.

Several spectrophotometric methods such as the molybdenum blue method and those using ammonium vanadate and 1-amino-2-naphthol-4-sulphonic acid have been used. A solution containing orthophosphate and molybdate ions condense in acid solution to give phosphomolybdic acid, which upon reduction with hydrazinium sulphate produces a blue colour due to molybdenum blue of uncertain composition. It exhibits a maximum at 620–630 nm. When phosphate, ammonium vanadate and ammonium molybdate react, a bright yellow-coloured complex of phosphovanadomolybdate is formed with a λ_{max} of 460–480 nm. In yet another method, molybdate reagent and 1-amino-2-naphthol-4-sulphonic acid solution (0·5 g with 30 g of sodium bisulphite and 6 g of sodium sulphite made up to 250 ml) are added. After standing for 10 min the absorbance is measured at 650 nm. In all these methods, a blank is always taken and a standard calibration curve is constructed from which the concentration of phosphorus is calculated. ***See*** Spectroscopy, Overview

Phosphate is also determined nephelometrically. On adding molybdate strychnine reagent (in two parts: solution A of molybdenum trioxide in 5 M sulphuric acid, solution B of 1·6 g of strychnine sulphate in 500 ml of water; both solutions are mixed before use) a white-coloured turbidity is obtained. The phosphorus content is determined from the calibration graph.

In recent years, several instrumental methods, viz. inductively coupled plasma source mass spectrometry (ICPSMS), high-performance liquid chromatography (HPLC), flow injection analysis with continuous microwave oven digestion, an enzyme sensor system and X ray fluorescence (XRF) have been used. With the availability of nuclear reactors, more sensitive methods employing radioactivity measurements have been employed. On thermal neutron irradiation, ^{32}P (β emitter, 1·71 MeV) is formed. A gas flow proportional counter is used with an aluminium absorber (thickness 27 mg cm^{-2}). Only a small sample size ($\sim$50 mg) is required. A derivative activation analysis and those employing Cerenkov and Bremsstrahlung counting have also been employed. ***See*** Chromatography, High-performance Liquid Chromatography; Mass Spectrometry, Principles and Instrumentation

Organic Phosphorus

Phosphorus-containing organic compounds are first extracted into benzene or any other suitable solvent. An aliquot is transferred to a beaker and the solvent is evaporated slowly on a water bath. The residue is heated with concentrated nitric acid and then with a small amount of potassium chlorate to dryness. After cooling, concentrated hydrochloric acid is added repeatedly until a clear solution is obtained. A 2·5% ammonium molybdate and potassium iodide/sodium carbonate solution is added. The flask is immersed in a steam bath for 15–20 min. After cooling, 0·5% sodium sulphite solution is added dropwise until the iodine colour disappears. Finally, the absorbance is measured at 650 nm.

Organophosphorus compounds can also be combusted in a Schöninger oxygen flask to give orthophosphate, which can be absorbed by either sulphuric or nitric acid. Phosphorus can then be determined spectrophotometrically. Recently, malonyl dihydrazide has been proposed for spectrophotometric extraction of organophosphorus compounds in vegetables. Chromatographic separations employing aluminium oxide and thin-layer chromatography (TLC) have also been employed. Gas chromatography–mass spectrometry (GCMS) has been used for the confirmation of organophosphorus pesticide residues in foods. A TLC enzymatic method has been suggested for the determination of organophosphorus compounds down to a detection limit of 10^{-10} g. ***See*** Chromatography, Thin-layer Chromatography; Chromatography, Gas Chromatography

Other Forms

Phosphorus may also be found as phosphite (PO_3^{3-}), which may be determined as mercurous chloride or ammonium magnesium phosphate. In the first case, an acid solution of phosphite reduces mercury(II) chloride solution (eqn [1]) and mercury(I) chloride is weighed, the latter being directly proportional to the phosphite concentration.

$$2HgCl_2 + H_3PO_3 + H_2O \rightarrow Hg_2Cl_2 + H_3PO_4 + 2HCl \quad (1)$$

In the other case, phosphite is oxidized by nitric acid to phosphate, which is determined as $Mg(NH_4)PO_4 \cdot 6H_2O$ or as $Mg_2P_2O_7$.

Yet another form of phosphorus is hypophosphite (PO_2^{3-}). It is oxidized quantitatively by excess ceric(IV) sulphate in sulphuric acid solution in 30 min at 60°C; the excess Ce^{4+} is titrated with ferrous ammonium sulphate to a permanent red end-point using ferroin indicator (eqn [2]).

$$4Ce^{4+} + H_3PO_2 + 2H_2O \rightarrow 4Ce^{3+} + H_3PO_4 + 4H^+ \quad (2)$$

It is possible to determine both hypophosphite and phosphite by affecting complete oxidation to phosphate. These different forms may also be differentiated by ion chromatography.

Bibliography

Corbridge DEC (1980) *Phosphorus, An outline of its Chemistry, Biochemistry and Technology*, 2nd edn. Amsterdam: Elsevier.

Ellinger RH (1975) Phosphates in food processing. In: Furia TA (ed.) *Handbook of Food Additives*, pp 617–780. Cleveland: CRC Press.

Emseley J and Hall D (1976) *The Chemistry of Phosphorus*, pp 1–76. London: Harper and Row.

Toy ADF (1973) Phosphorus. In: Bailar JC, Nyholm RS and Trotman-Dickenson AF (eds) *Comprehensive Inorganic Chemistry*, pp 389–545. Oxford: Pergamon Press.

Weginwar RG, Samudralwar DL and Garg AN (1989) Determination of phosphorus in biological samples by thermal neutron activation followed by β^- counting. *Journal of Radioanalytical and Nuclear Chemistry Articles* 133: 317–324.

A. N. Garg and R. G. Weginwar
Nagpur University, Nagpur, India

Physiology

Phosphorus, an essential nutrient, is required for many different functions in body tissues, both intracellularly and extracellularly. This element, which exists in biological systems as phosphates, is used by cells to make structural molecules and outside of cells to make the crystals of bones and teeth; it serves as a component of intracellular regulatory molecules; it serves as a buffering component in both intra- and extracellular fluids;

Physiology

and it is an important factor in cellular energetics, especially in the mitochondria where it is used to make most of the high-energy bonds needed for cellular activities. Lastly, the metabolism of inorganic phosphate is closely linked to that of calcium, and this review will, therefore, also deal to some extent with calcium and its homeostatic control. Adequate phosphorus and calcium intakes are critical not only for skeletal growth, but also for growth and development of soft tissues, especially in neonates. ***See*** Bone; Calcium, Physiology; Cells; Energy, Measurement of Food Energy

Phosphorus, both in the inorganic phosphate (Pi) and organic phosphate forms (Po), is abundant in nearly all foods traditionally consumed. This element is especially rich in animal products, such as meats, fish, poultry, eggs, milk and other dairy products, but it also exists in cereal grains and most vegetables in good quantities. Because Pi ions are so readily absorbed across the small intestine, with an efficiency of roughly 65–75% in adults and with a somewhat higher efficiency in children, the prompt rise in blood Pi after a meal or snack influences calcium homeostasis, when Pi ions enter the blood from the gut, essentially unaccompanied by the more slowly absorbed calcium ions. An elevation in the plasma concentration of Pi tends to depress the serum calcium concentration, mainly Ca^{2+}, through the formation of a complex of the two ions, and this decline in Ca^{2+} stimulates parathyroid hormone (PTH) release. Although the specific mechanism through which Pi exerts this effect on plasma Ca^{2+} has not been fully established, many experimental data support the existence of this phenomenon.

Dietary phosphorus deficiency, although highly unlikely because of the abundance of this element in foods, contributes to low serum Pi concentration and thereby limits bone mineralization via osteoblasts and the total amount of bone mineral mass deposited in the skeleton. Furthermore, Pi deficiency increases bone turnover, which, during infancy, can lead to rickets.

Dietary Sources of Phosphorus

Although both organic and inorganic forms of phosphorus are widely distributed in foods, 75% or so of the phosphorus in foods is converted to Pi following the various digestive steps in the stomach and upper small intestine. The organic phosphorus usually remains associated with the fat-soluble dietary molecules which are absorbed without digestion of the phosphate groups from these molecules, such as is the case with phosphatidylcholine (lecithin).

Phosphate additives in processed foods and in beverages can also make an important contribution to total phosphorus intakes. Concern has been raised about the quantity of phosphate added to cola-type soft drinks, which usually contain approximately 60 mg phosphate per 12-oz. container in the United States, but in reality much greater amounts of phosphates are added to other processed foods commonly consumed in developed nations. Phosphate additives are used most by the food industry in baked goods, meats, cheeses, and milk products in the United States.

Mean phosphorus intakes of American women, according to recent surveys, range between 900 and 1200 mg per day, depending on age and caloric intake; older women consume less phosphorus than younger women and active women consume more than sedentary individuals. Men usually consume closer to an average of 1500 mg per day, and their intakes also decline with age and with declining activity and caloric intake. The US Recommended Dietary Allowance (RDA) for each sex over 24 years of age is 800 mg phosphorus per day, which is readily achieved by nearly all adults. The ratios of Ca:P in typical diets of American adult females is approximately 0·5:1, which raises concern among nutritionists because of the potential adverse effects of nutritional secondary hyperparathyroidism (see below). ***See*** Dietary Requirements of Adults

Table 1 gives representative values of the total phosphorus composition of selected foods from the major food groups compared to the calcium content of these same foods. ***See*** individual foods

Intestinal Absorption of Phosphates

The net absorption of Pi is highly efficient; and the Pi efficiency is more than twice that of calcium absorption, which is usually stated as being between 25% and 30% in adults. The efficiency of Pi absorption by infants has been reported to be as high as 80–90%. Typical meals which contain representative items from all food groups, including dairy, have ratios of Ca:Pi approaching 0·7:1·0. If the actual amount of calcium in the meal is 350 mg and the amount of phosphorus is 500 mg, then 75% of the phosphorus, or 375 mg, will be absorbed, most of it within the first postprandial hour, but only 30% or 105 mg of calcium will be maximally absorbed and this process usually takes several hours to be completed. The net effect is that the rise in blood Pi concentration tends to depress the serum Ca^{2+}, perhaps through a reciprocal adjustment in the serum ion concentration product, i.e. $[Ca^{2+}] \times [Pi] = \text{constant}$, a mechanism proposed in the 1940s. Parathyroid hormone (PTH) secretion from the parathyroid glands responds to the depressed serum Ca^{2+} concentration as long as absorbed Pi continues to lower the Ca^{2+} concentration. ***See*** Hormones, Thyroid Hormones

Pi absorption occurs by two major routes, transcellular and paracellular, throughout the small intestine and probably also the large intestine. The transcellular route

Table 1. Phosphorus and calcium composition of representative foods of the basic food groups

Food item	Weight (g)	Calcium (mg)	Phosphorus (mg)
Dairy			
Milk, whole (3·3% fat)	244	291	228
Milk, low-fat (2%)	244	297	232
Milk, low-fat (1%)	244	300	235
Milk, nonfat (skim)	245	302	247
Cheddar cheese	28	204	145
Egg, whole, poached	50	28	90
Meat, fish and poultry			
Salmon, pink canned	85	167	243
Shrimp, canned	85	98	224
Beef, ground, lean with 10% fat	85	10	196
Liver, beef, fried	85	9	405
Pork roast, cooked	85	9	218
Chicken breast, cooked, boneless	85	18	210
Fruits			
Avocado, raw, skinless	251	23	30
Orange, whole, peeled	131	54	26
Pineapple, raw, dried	155	26	12
Raisins	14	9	14
Watermelon	926	30	43
Breads and Cereals			
White bread, soft	25	21	24
Whole-wheat bread, soft	28	14	24
Bran flakes cereal	35	19	125
Corn flakes cereal	25	1	9
Shredded wheat cereal	25	11	97
Macaroni, cooked	130	14	85
Noodles (egg), cooked	160	16	94
Rice, white, instant	165	5	31
Spaghetti, cooked	130	14	85
Legumes, nut, seeds			
Almonds, shelled	130	304	655
Beans, Lima, cooked	190	55	293
Peas, black-eye, cooked	250	43	238
Peas, split, cooked	200	22	178
Vegetables			
Asparagus, cooked	145	30	73
Beets, cooked	100	14	23
Beet greens, cooked	145	144	36
Broccoli, cooked	180	158	112
Cabbage, cooked	145	64	29
Carrots, cooked	155	51	48
Corn, sweet, cooked	140	2	69
Kale, cooked	110	206	64
Pepper, sweet, cooked	73	7	12
Potatoes, baked	156	14	101
Squash, summer, cooked	210	53	53
Sweet potatoes, baked	114	46	66
Tomatoes, raw	135	16	3
Miscellaneous			
Tomato soup with milk	250	168	155
Tomato soup with water	245	15	34
Split pea soup with water	245	29	149

Source: *Home and Garden Bulletin* No. 72, USDA.

Table 2. Serum fractions of phosphates and calcium

Phosphate form	Percentage of total	Calcium	Percentage of total
Free HPO_4^{2-}	44	Free Ca^{2+}	48
Free $H_2PO_4^{-}$	10	Protein-bound	46
Protein-bound	12	Complexed	3
Cation-bound	34	Other	3

Adapted from Bringhurst FR (1989).

is considered the more significant one, but not much is known about the paracellular pathway in terms of location in the small intestine or the quantitative contribution of this component. The transcellular route involves at least two distinct mechanisms of entry at the brush border membrane (mucosa) and probably as many as part of the exit step at the basolateral membrane (serosa). Much of the Pi absorption is considered passive because of co-transport with Na^+ or other cations. Because of the co-transport mechanism with Na^+ following a meal, Pi ions may be much more rapidly absorbed than if a separate and independent mechanism for Pi absorption *per se* were required. Understandings of the mechanisms of Pi absorption remain limited. The hormonal form of vitamin D, 1,25-dihydroxyvitamin D, increases Pi absorption via the intracellular route, but much less is known of this pathway than the vitamin D-mediated Ca^{2+} absorption. ***See*** Cholecalciferol, Physiology

Some secretions of Pi ions into the gastrointestinal tract occur at every level, i.e. salivary glands, stomach, intestine, pancreas, liver and large intestine. Thus, net Pi absorption represents the difference between total Pi absorption and endogenous Pi secretion. Pi ions not absorbed by the gut pass into the stools.

Blood Concentrations of Phosphates

The distribution in human blood serum is compared to the distribution of calcium in Table 2. Po in the lipid fraction of blood plasma generally represents a fairly stable value in individuals on diets with considerable amounts of phosphorus, i.e. typical Western diets, but it will decline if the diet becomes deficient or much lower in phosphorus. The reason for this change is not clear, but it may result from a homeostatic adaptation to the low-phosphorus intake through which membrane-bound Po groups are cleaved and released to the blood in order to keep the plasma Pi concentration at or near its 'set' level. The set level, which is determined genetically in a species, is regulated by various homeostatic mechanisms; in the case of serum Pi, several hormones, especially PTH, are involved in maintaining serum Pi at or near its set level.

The hormonal form of vitamin D – 1,25-dihydroxyvitamin D – is thought to enhance the intestinal absorption of Pi, as it does for Ca, but relatively little is known about the cellular mechanism of this action. For example, Pi-carrier proteins have been identified at the brush border membrane, but no intracellular Pi-binding proteins have been found after vitamin D treatment in animal models.

Pi ions in the blood or extracellular fluids are distributed to all tissues in the body to meet cellular needs and to be taken up by the bone fluid compartment for incorporation in hydroxyapatite crystals. During tooth development, Pi ions are taken up by cells (odontoblasts, ameloblasts) and transferred into the extracellular compartment of the developing hydroxyapatite crystals.

Phosphate Homeostatic Mechanisms

The serum Pi concentration is not as rigorously regulated at any time during the lifecycle as that of calcium. Several hormones are involved in Pi homeostasis. PTH, calcitonin, and the hormonal form of vitamin D are considered the major regulators, but many other hormones also affect Pi homeostasis, including insulin, glucagon, growth hormone, oestrogens, adrenaline, and adrenal corticosteroids. ***See*** Hormones, Steroid Hormones

Calcitonin may play a critical role in the postprandial conservation of Ca^{2+} through a direct action on bone cells which overrides any effect of an elevated PTH on these same cells during this time frame. Such an action favours the retention of absorbed Ca^{2+} in the bone fluid compartment and, thus, keeps the plasma $[Ca^{2+}]$ low, which favours PTH-mediated renal Ca^{2+} reabsorption and, hence, Ca^{2+} conservation.

An elevation of PTH acts to stimulate renal Ca^{2+} reabsorption while acting to block renal Pi reabsorption during this postprandial period, and at the same time PTH is thought to be relatively inactive on bone cells because of the dominance of calcitonin. The net effect of PTH actions on different tissues is to try to conserve Ca^{2+} in the face of an elevated plasma Pi. However, if PTH remains elevated after the blocking effect of calcitonin decays, then an increased transfer of Ca^{2+} from the bone fluid compartment to blood and PTH-stimulated osteoclastic resorption can together slowly deplete bone when the dietary phosphorus intake greatly exceeds calcium on a long-term basis. This potential mechanism, then, is considered a highly likely way in which low-calcium diets contribute to osteopenia and subsequent fractures which characterize osteoporosis. The name given to this condition is nutritional secondary hyperparathyroidism (see below). ***See*** Osteoporosis

Physiology

Regulation of the plasma Pi concentration is not well understood, but PTH and other hormones are involved through their renal, intestinal and skeletal activities. High Pi intakes are handled readily and promptly by healthy individuals, primarily through increased renal excretion. The problem of a high-Pi intake coupled with low-calcium consumption in a typical dietary pattern, however, is handled at the expense of bone mineral through a chronic elevation of PTH and 1,25-dihydroxyvitamin D. Whether 1,25-dihydroxyvitamin D actually exerts a major enhancing effect on intestinal Pi absorption is not clear, but it does increase cellular transfer of Ca^{2+} from the gut lumen to the blood.

Other hormones that also have an influence on Pi homeostasis are important metabolic regulators, such as insulin following carbohydrate ingestion. Insulin stimulates glucose uptake in many extrahepatic tissues and Pi moves into the cells along with glucose, probably in a 1:1 molar ratio. On the other hand, during situations when glucagon is elevated, glucagon enhances renal excretion of Pi, probably through a mechanism that blocks renal tubular Pi reabsorption. Calcitonin is also phosphaturic. Growth hormone and other anabolic hormones and local tissue factors generally stimulate Pi incorporation into organic structures, much like insulin, to meet the needs of growth, cell division, or structural requirements for tissue maintenance. ***See*** Carbohydrates, Requirements and Dietary Importance

Another aspect of Pi homeostasis involves membrane-bound organic (Po) molecules, such as the phospholipids. Animal models fed low-phosphorus intakes have increased degradation of these Po membrane components, which release Pi to the blood and helps to maintain the blood Pi concentration at its set level. These findings strongly suggest that membrane phospholipids have a role in Pi homeostasis.

Oestrogen may also enhance Pi transfer into cells because plasma Pi becomes somewhat elevated in postmenopausal women and in women who have had an oophorectomy. This rise may also reflect bone turnover and Pi release to blood from the skeleton. Oestrogen replacement therapy in postmenopausal women causes a slight decline in serum Pi.

Excess Pi in cells can be stored in various organelles, such as mitochondria and endoplasmic reticulum, along with Ca^{2+} as calcium phosphates, which can be solubilized in times of need to be retrieved for cellular requirements.

Functional Roles of Phosphates

Phosphorus as Pi and/or Po exists in all tissues, both intracellularly and extracellularly. In extracellular sites phosphorus exists in the hydroxyapatite crystals of bones and teeth and in phosphorylated proteins in diverse extracellular matrices. The phosphorylation of Type I collagen in bone may trigger the mineralization process. Phosphates also serve as a buffer system in blood and other extracellular fluids.

Intracellularly, phosphates also serve as important buffers, and Po is a component of numerous classes of molecules, including membranes, high-energy molecules, regulatory proteins, regulatory phospholipids, and nucleoproteins. In the phospholipids of nervous tissue, Po is a critical component of many diverse molecules. Phosphorylation of inositol to phosphatidylinositol and cleavage of inositol triphosphate represent an important regulatory mechanism in cells. In addition, phosphorylation of certain enzymes by a variety of protein kinases and the dephosphorylation of these same enzymes (proteins) by phosphatases is central to activation and inactivation of key regulatory enzymes controlling specific metabolic pathways in cells. Pi uptake by cells for the synthesis of regulatory peptides and regulatory phospholipids is also important in metabolically active cells. Thus, phosphorus is much more widely distributed within cells than is calcium and it serves many diverse roles.

Pi uptake by cells is enhanced by insulin, but other hormones also increase the uptake of this anion, including oestrogens, adrenalin, calcitonin, and many growth factors, such as IGF-1. Once in the cytosol, Pi ions are used for phosphorylating glucose and related intermediate molecules derived from glucose in a meal. In addition, Pi ions are shuttled across intracellular organelle membranes for use or storage. For example, in mitochondria Pi ions are essential if oxidative phosphorylation is to be adequately coupled. Mitochondria also store roughly 20% of the cells' Pi as calcium salts. Similarly, the endoplasmic reticulum (ER) uses and stores Pi for phosphorylation of various proteins. Also, the ER contains approximately 30% of total cellular Pi for storage and use in phosphorylation of proteins and other molecules. The nucleus, Golgi complex, and lysosomes contain the remainder of the total Pi.

Phosphate in Health and Disease

The balance of phosphorus is determined by the difference between input and output, and it is kept remarkably constant in healthy individuals until late in the life cycle when lean tissue loss accelerates. Balance is clearly negative prior to death in ill patients because of the death of numerous cells without renewal. Pi absorption declines slightly late in life, i.e. sometime after the fifties, because of a reduced efficiency of absorption and, because food energy intake also declines in the elderly. Thus, the overall input of Pi is lowered. Pi excretion also declines slightly in healthy older individuals, but it declines more so if renal function is seriously compromised. Normally, urinary excretion of Pi approximates 67% of Pi consumed in the diet, and this percentage holds throughout life in healthy individuals. Unabsorbed phosphorus makes up nearly all of the remainder of faecal Pi elimination, although sweat and skin losses do contribute a small percentage to the total excretion of Pi.

Several issues of Pi homeostasis need further explanation because of their potential impact on health and disease in populations of developed nations.

Ageing and Renal Function. Among a cohort of women from age 40 to almost 90 years who were defined as healthy, i.e. with normal renal function, the serum Pi concentration did not decrease with age. The range of Pi across this broad age span was 2·71–4·20, with the 50th centile at 3·49 mg dl^{-1}. In comparison with a sample of women from the general population, whose renal function was not normal, women with decrements in renal function with ageing had almost identical values of serum Pi. Thus, no major age-related changes in serum Pi could be shown in either group of women; the same finding was observed for serum calcium. Modest declines in renal function do not affect serum Pi and calcium concentrations in women, although it is well known that individuals with severe renal dysfunction do not adequately adapt to elevations in serum Pi concentrations (see below). ***See*** Kidney, Structure and Function

Nutritional Secondary Hyperparathyroidism. This disorder has only been clearly demonstrated in animal models, but several human studies suggest that it may exist in a subset of women who typically consume low-calcium, high-phosphorus diets ($<$0·5:1·0 ratios). A 4-week study of young adult females placed on a diet with a 0·25:1·0 Ca:P ratio had persistently elevated serum PTH concentrations and mildly elevated 1,25-dihydroxyvitamin D levels over the course of the study. No changes in bone mineral density could be detected over such a short period, but, if women, even premenopausal women, remain on such diets for many years, decrements in bone density should be detectable with the sensitive measurement devices available today. Further research is needed to establish the significance of this disorder in human populations.

Figure 1 illustrates the potential pathophysiological changes that occur in response to a chronic low-calcium, high-phosphorus diet characteristic of nutritional secondary hyperparathyroidism.

Renal Secondary Hyperparathyroidism. When renal function becomes compromised to such an extent that creatinine, other nitrogenous metabolic products, and Pi are retained abnormally and excessively by the body, then several pathophysiological adaptations occur which have serious effects on health. One of the

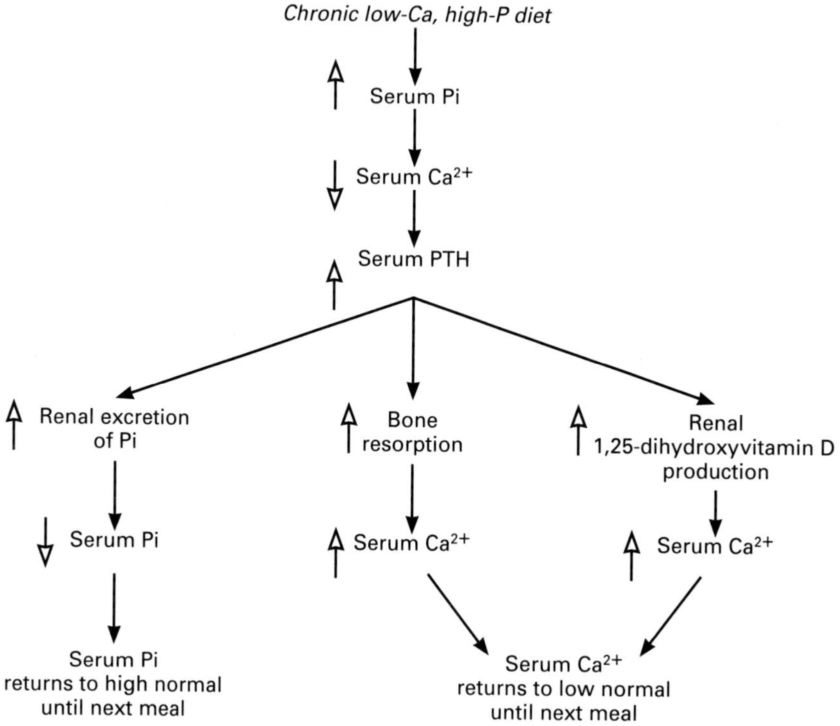

Fig. 1 Illustration of steps in development of long-term nutritional secondary hyperparathyroidism: chronic low-calcium, high-phosphorus diet.

important adverse effects of the retention of Pi is the rapid and progressive loss of mineral mass. The chronic elevation of serum Pi causes a decline in serum Ca^{2+}, which triggers PTH secretion. The net effect is a constantly elevated PTH concentration which continues to act on bone tissue, i.e. resorption, to try to raise $[Ca^{2+}]$ to its homeostatic set level. Since Pi is also released from bone along with Ca^{2+} during resorption, the serum Pi concentration also increases. Because the kidneys cannot eliminate Pi adequately, $[Ca^{2+}]$ can never be raised to its set level and bone tissue continues to be degraded as part of an unending vicious cycle.

Various dietary manipulations have been tried to control the loss of bone mass, but no one regimen has been very successful. Reductions of both dietary protein, especially animal protein, and phosphorus have been moderately successful in slowing the progress of chronic renal disease, but the diets are not very palatable or satisfying.

Conclusions

Pi metabolism is much more complex than that of calcium because of the many intracellular pathways which utilize Pi ions at one stage or another. The cytosolic utilization of Pi is closely linked with that of glucose for the formation of glucose 6-phosphate and for triglyceride synthesis through glycerol 3-phosphate formation, as well as with other molecules, during the postprandial period. Pi is utilized by cells for many diverse molecules, including regulatory peptides and phospholipids. Extracellular regulation of Pi is closely associated with that of calcium through PTH and other calcium-regulating hormones. Under typical dietary conditions of excessive phosphorus intake compared to calcium, i.e., low Ca:P ratio, nutritional secondary hyperparathyroidism and the long-term development of osteopenia are likely to result. Renal secondary hyperparathyroidism, a serious consequence of renal functional impairment, produces severe bone loss because of altered homeostatic regulation of Pi.

Bibliography

Anderson JJB (1991) Nutritional biochemistry of calcium and phosphorus. *Journal of Nutritional Biochemistry* 2: 300–307.

Arnaud CD and Sanchez SD (1990) Calcium and phosphorus. In: Brown, ML (ed.) *Present Knowledge in Nutrition*, 6th edn. Washington, DC: International Life Sciences Institute, Nutrition Foundation.

Avioli LV (1988) Calcium and phosphorus. In: Shils ME and Young VE (eds) *Modern Nutrition in Health and Disease*, 7th edn. Chap. 5. Philadelphia: Lea & Febiger.

Bringhurst FR (1989) Calcium and phosphate distribution, turnover, and metabolic actions. In: DeGroot LJ (ed.) *Endocrinology*, 2nd edn, vol. 2. Philadelphia: WB Saunders.
Calvo MS, Kumar R and Heath H III (1990) Persistently elevated parathyroid hormone secretion and action in young women after four weeks of ingesting high phosphorus, low calcium diets. *Journal of Clinical Endocrinology and Metabolism*, 70: 1340–1344.
Cruz MLA and Tsang RC (1992) Introduction to infant mineral metabolism. In: Tsang RC and Mimouni F (eds) *Calcium Nutrition for Mothers and Children. Carnation Nutrition Education Series*, vol. 3. Raven Press, New York: Carnation Co.
Greger JL and Krystofiak M (1982) Phosphorus intakes of Americans. *Food Technology* 36 (1): 78–84.
Kotowicz MA, Melton LJ, Cedel SL, O'Fallon WM and Riggs BL (1990) Effect of age on variables relating to calcium and phosphorus metabolism in women. *Journal of Bone and Mineral Research* 5: 345–352.
Massey LK and Strang MM (1982) Soft drink consumption, phosphorus intake, and osteoporosis. *Journal of the American Dietetic Association* 80: 581–583.
Muhlbauer RC and Fleisch H (1990) Inverse relation between plasma inorganic phosphate and phospholipids in mice: Effect of dietary inorganic phosphate, fasting and glucagon. *Mineral and Electrolyte Metabolism* 16: 341–347.
National Research Council (1989) *Recommended Dietary Allowances*, 10th edn, pp 184–187. Washington, DC: National Research Council, National Academy Press.
Neer RM (1989) Calcium and inorganic phosphate homeostasis. In: DeGroot LJ (ed.) *Endocrinology*, 2nd edn, vol. 2. Philadelphia: WB Saunders.
Talmage RV, Cooper CW and Toverud SU (1983) The physiologic significance of calcitonin. In: Peck WA (ed.) *Bone and Mineral Research Annual I*, pp 74–143. Amsterdam: Excerpta Medica.
USDA (1978) *Nutritive Value of Foods. Home and Garden Bulletin* No. 72. Washington, DC: US Department of Agriculture.

John B. Anderson
University of North Carolina, Chapel Hill, USA

PHYLLOQUINONE

See Vitamin K

PHYSICAL PROPERTIES OF FOODS

Rheology: Definition and Scope

Rheology was defined in 1928 as the study of the deformation and flow of matter. It emerged in order to meet the demands of developments in technology, as a science to analyse and rationalize the complex mechanical behaviours displayed by most real materials, for which the classical Hookean elastic solid and the Newtonian viscous fluid represent only limiting behaviours, seldom encountered in practice.

Rheology combines different aspects:

(1) Measurement. The design and application of patterns of mechanical stimulus, adapted to the material and to the problem studied, to which the samples are submitted under completely defined geometrical and physical conditions, allowing the expression of mechanical excitation and responses in terms of variables having a physical meaning intrinsic to the material: stress, strain and time.
(2) Phenomenology. The description of the rheological behaviour of the material under defined conditions. This amounts to the study of the relations between stress, strain and time; in the simplest cases these relations can be expressed in the form of a mathematical equation called the constitutive equation of the material.
(3) Structural interpretation. The analysis of the rheological behaviour in terms of the structural characteristics of the material. This implies a relationship between the macroscopic scale of the rheological properties as measured and the molecular or colloidal scale of the underlying structural and physical chemical mechanisms.

This article contains a survey of the different types of rheological behaviour and of the main classical rheological tests which can be used to study them in the case of isotropic materials. The rupture properties, which are often included in rheology, will not be considered, although they are involved in the evaluation of food texture.

Relevance of Rheology to Food Science and Technology

Rheological properties of ingredients and products are involved in all stages from formulation to consumption of food, and their knowledge is therefore important to many aspects of food science and technology:

(1) The texture or consistency of a food product, which is essentially the expression of its rheological behaviour through the mediation of our senses, affects its acceptability by the consumer and is part of its overall quality. An extreme example is that of some gelled desserts, the rheological properties of which contribute greatly to their value. Instrumental evaluation of texture and its correlations with sensory evaluation is a specific field not dealt with in this article. ***See*** Sensory Evaluation, Texture
(2) The study of the rheological behaviour of such complex materials as concentrated macromolecular and colloidal systems of common occurrence in the field of food science allows an insight into their structure and their modifications under the influence of different factors; for example the storage modulus of a gel is related to the density of junction zones, and the appearance of a yield stress indicates flocculation in a dispersion. ***See*** Colloids and Emulsions
(3) During storage, processing or packaging, raw materials, ingredients, and intermediary and final products are submitted to mechanical stresses and deformations. Knowledge of the rheological behaviour of materials is necessary to model operations such as pumping, mixing and heat exchange, and to design the required machines.
(4) Rheological measurements are necessary to control processes, to control quality of input and output products, and to study functional relationships between composition, processing, etc., and the final texture.

Basic Concepts

The Response to Mechanical Stimulus

A material subjected to a mechanical stimulus will undergo processes through which the mechanical energy introduced is dissipated or stored. These processes will depend on the modes of the mechanical excitation on the one hand, and on the structure of the material on the other. They result in different physical responses of the material: mechanical or thermal, in some cases, optical or electrical; the primary reponse is mechanical, and the present discussion is restricted to this.

Energy dissipation occurs upon deformation through friction between the structural elements constituting the material. It is instantaneous and complete for purely viscous liquids, and results in nonrecoverable deformation (flow). In contrast, purely elastic solids are characterized by their capacity to store the mechanical energy input, which is recoverable by completely reversible deformation. Many materials in real conditions show a combination of viscous and elastic properties.

Stress and Strain

Mechanical stimuli and responses are expressed in terms of stress and strain, which may be defined in the case of a very simple type of deformation: planar shear.

Consider a homogenous, isotropic material, maintained at a constant temperature between two parallel horizontal plates of area A, and view it as a stack of infinitely thin layers parallel to the plates, able to glide over each other without mass transfer between them (laminar deformation). When a constant force F is applied horizontally to one of the plates, two layers separated by a vertical distance dz will slide relative to each other by a distance of dy in the direction of the force during the time dt. By definition, the (shear) stress applied to the material is $\sigma = F/A$ and the resulting deformation is measured by the strain $\gamma(z,t) = dy(z,t)/dz$ (nondimensional), which is, in the general case, a function of both the vertical position z in the gap and of the time t of observation.

These simplified definitions can be extended as approximations in the case of other shear geometries. Handling and processing of materials imply complex deformations, but classical rheological measurements are performed in shear or in uniaxial extension or compression; in this latter type of deformation, the case of, say, a rod pulled in the direction of its axis, stress is normal to the section and strain $\varepsilon = dL/L$ is the relative change in length.

Solids and Liquids

Two different deformation responses can be observed when a stress σ is applied:

(1) The strain eventually reaches a finite limit γ_e with respect to time: the material is a solid; the equilibrium strain γ_e depends on the value of σ.

(2) The strain increases infinitely with time: the material is a liquid, it flows. Flow is characterized by the rate of strain $\dot{\gamma}(z,t) = d\gamma(z,t)/dt = dv(z,t)/dz$ (v is the relative velocity of the two layers), which has the dimension of reciprocal time. Viscosity is the physical property which implies a relation between σ and $\dot{\gamma}$.

The behaviour of a solid is described by the relation between the variables σ, γ and t; that of a liquid by the relation between σ, $\dot{\gamma}$ and t. When time does not appear in the constitutive equation, the behaviour is said to be time-independent.

Limiting Behaviours

Time-independent response is displayed by the ideal elastic solid and the ideal viscous liquid, which are abstractions, but this property can be approached under restricted conditions in real materials.

The ideal elastic solid instantaneously reaches its equilibrium strain on loading, which vanishes instantaneously upon removal of the load. If, in addition, the response is linear, i.e. if the strain is proportional to the stress, the behaviour is said to be Hookean, and is completely defined by the proportionality constant $G = \sigma/\gamma$ or $E = \sigma/\varepsilon$, the modulus of elasticity in shear and in uniaxial deformation, respectively; E is also called Young's modulus. The ratio E/G varies between 2 and 3, depending on the compressibility under hydrostatic pressure; $E/G = 3$ for incompressible materials.

Real solids can show approximately Hookean behaviour when strain and observation time are sufficiently small. Some typical values (in newtons per square metre, at 20°C) are $E = 3 \times 10^9$ for dry spaghetti; for bread crumb E varies from $0{\cdot}25 \times 10^3$ to 25×10^3 as the result of staling; starch retrogradation, which is the main process underlying staling, increases E from $0{\cdot}5 \times 10^6$ to 3×10^6 for 50% wheat starch gels; $G = 2 \times 10^5$ for a 20% gelatine gel.

The ideal viscous liquid instantaneously reaches a steady state flow whatever the applied stress, and flow stops instantaneously on removal of the load; the strain is totally irrecoverable. Linearity corresponds to proportionality between stress and shear rate; the liquid is then said to be Newtonian and is characterized by its viscosity coefficient ('viscosity' in short) $\eta = \sigma/\dot{\gamma}$ (elongational viscosity is beyond the scope of this article). Simple liquids such as water, solvents, oils, solutions of minerals and of small organic molecules show Newtonian behaviour over very wide stress ranges. At 20°C, viscosity (in millipascal seconds) is 1 for water, 84 for olive oil and 1490 for glycerol.

Dilute solutions of polymers such as polysaccharides, and dilute dispersions like milk, can also be considered practically as Newtonian.

Complex Behaviours

Under usual conditions, most real materials show behaviours which deviate from the Hookean and the Newtonian ones. Two types of deviation occur either separately or together.

Time-dependent Behaviours

The response depends on the duration of the excitation; for example, to reach equilibrium, strain or steady state shear rate will require a certain length of time after the stress has suddenly been imposed.

Since the time-dependent response is necessarily studied under transient or dynamic mechanical regimes, inertial effects in the experiments can be taken for or superimposed on it. Rheological time-dependent effects should also be distinguished from chemical or physical ageing phenomena corresponding to changes in the structure of the material which do not result from the mechanical history, even if they could be affected by it, but cause progressive modifications in the rheology of the system. For systems undergoing such changes, the study of the rheological time-dependent behaviour would require the rheological characteristic time to be much smaller than the characteristic time of the kinetics, or the system to be quenched; it is therefore not always feasible.

Nonlinearity

Real solid or liquid materials behave linearly only over a limited range of strains or shear rates, and the limit is usually low. For time-dependent behaviours, linearity means that at any given time σ and γ (or $\dot{\gamma}$) are proportional.

The Viscoelastic Behaviour in Classical Rheological Tests

Viscoelasticity is the time-dependent behaviour of materials simultaneously displaying elastic and viscous properties. Elasticity dominates at short observation times; the behaviour appears increasingly viscous as the time-scale broadens. Viscoelastic materials have a fading memory: their state of stress or strain depends on their mechanical history, but the dependence decreases as the mechanical events become more and more remote in time. Typical viscoelastic materials are macromolecular networks, such as concentrated solutions of high-molecular-weight polysaccharides, polysaccharide or protein gels, or gluten. Their response spans large time-scales because spatial correlations between the structural elements exist with different ranges, and because junction zones show at the time-scale of the experiment with a permanent or transient character according to their nature. Simple patterns of mechanical excitation are used to characterize viscoelastic behaviour.

Creep and Recovery

A constant stress is 'instantaneously' applied to the relaxed sample at time zero till time t_1 (creep experiment); at time t_1 the stress is 'instantaneously' removed (creep recovery). The resulting strain is monitored as a function of time. The sequence of creep and recovery constitutes the retardation test. The results are expressed in terms of creep compliance $J(t,\sigma)=\gamma(t,\sigma)/\sigma$, the strain/stress ratio, which is a unique function of the time when the behaviour is linear.

Figure 1(a) and (b) shows the typical responses of viscoelastic bodies as compared to the ideal solid and liquid. The compliance of a viscoelastic solid grows during creep towards an asymptotic limit J_e; during creep recovery it decreases asymptotically to zero. The deformation is completely recoverable. For a viscoelastic liquid, $J(t)$ is the sum of a recoverable term $J_r(t)$, and of the contribution of flow which will not be recovered after removal of stress, $J(t)=J_r(t)+(t/\eta)$. As time increases, $J_r(t)\rightarrow J_e$ and the behaviour becomes dominated by flow: the creep curve shows a terminal linear region, the slope of which is the reciprocal of the viscosity.

The distinction between solid and liquid viscoelastic behaviour is not always easy in practice; it can depend on the performance of the instrument and on the patience of the operator.

Stress Relaxation

A fixed strain γ is 'instantaneously' imposed on the relaxed sample at time zero, and the progressive decay of the resulting stress is followed in time. The results are expressed in term of $G(t,\gamma)=\sigma(t,\gamma)/\gamma$, the relaxation modulus, which is a unique function of the time when the behaviour is linear. For a Hookean solid, $G=1/J$. For a viscoelastic material $G(t)\neq 1/J(t)$, but the two linear viscoelastic functions are mathematically related; however, creep and relaxation are not always equivalent tests in practice. Figure 1(c) shows typical relaxation curves.

Stress relaxation of viscoelastic liquids can alternatively be studied consecutively to sudden removal of a shear rate previously applied till steady state flow is reached.

Dynamic Measurements

A steady state sinusoidal strain or stress (frequency $f=\omega/2\pi$) is imposed on the sample. If the behaviour is linear, the response (stress or strain) is itself sinusoidal with the same frequency; but stress is in advance of strain by a phase ('loss') angle ϕ. For a Hookean solid $\phi=0$ and for a Newtonian liquid $\phi=\pi/2$; for a viscoelastic material ϕ $(0<\phi<\pi/2)$ and the amplitude ratio σ_0/γ_0 depend on ω. The storage and loss moduli are defined as $G'(\omega)=(\sigma_0/\gamma_0)\cos\phi$ and $G''(\omega)=(\sigma_0/\gamma_0)\sin\phi$; they are the real and imaginary parts of the complex modulus $G^*(\omega)=G'+iG''$ and related to the amount of energy stored and dissipated per cycle, respectively. The results can be expressed alternatively

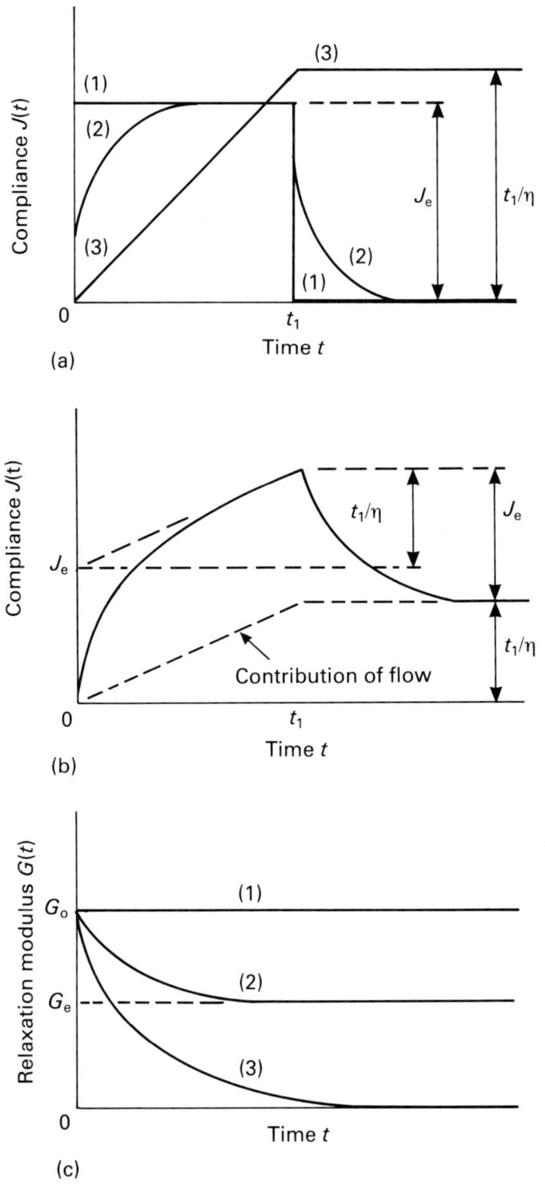

Fig. 1 Schematic responses in transient regimes: (a) creep and recovery of an ideal elastic solid (1), a viscoelastic solid (2) and an ideal viscous liquid (3); (b) creep and recovery of a viscoelastic liquid; (c) stress relaxation of an ideal elastic solid (1), a viscoelastic solid (2) and a viscoelastic liquid (3).

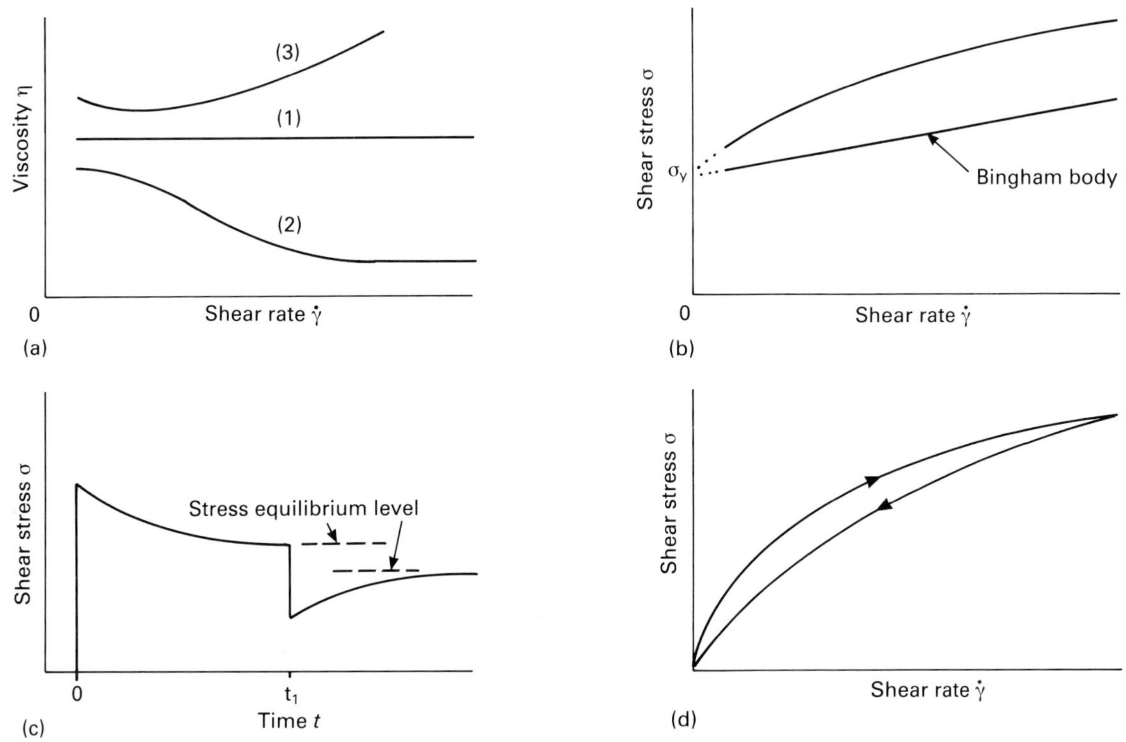

Fig. 2 Typical flow behaviours: (a) time-independent flow curves (Newtonian (1), shear thinning (2), shear thickening (3)); (b) plastic flow curves; (c) schematic thixotropic response of an inelastic material to a rectangular step of shear rate imposed at $t=0$ and stopped at $t=t_1$; (d) hyteresis loop of a thixotropic material submitted to a triangular variation of shear rate.

in terms of the complex compliance $J^*(\omega)=J'-iJ''$: $J'=G'/(G'^2+G''^2)$, $J''=G''/(G'^2+G''^2)$. The behaviour of a viscoelastic material will be described by the variations of both G' (or J') and G'' (or J'') with ω over the entire frequency scale ('mechanical spectrum' of the material). In principle, $G^*(\omega)$ and $J^*(\omega)$ are equivalent to $G(t)$ and $J(t)$, with ω equivalent to t^{-1}. However, because of the practical frequency range usable, dynamic characterization is limited to behaviour in the short-time domain, $\sim 10^{-3}$ to ~ 200 s. It is thus complementary to creep or relaxation tests, which allow characterization from $\sim 0{\cdot}1$ s to very long times. It must be remembered that dynamic measurements are valid only for linear behaviour, whereas creep and relaxation can be used outside the linearity range of conditions, compliance and relaxation moduli then being functions of both time and stress or strain.

As dynamic measurements at a fixed frequency only takes a short time, it is a convenient method to follow the kinetics of structural changes, such as sol–gel transitions.

Typology of Flow Behaviours

Flow behaviour is usually studied using viscometric experiments: a shear rate (or stress) is imposed on the liquid and the resulting response (shear stress or rate) is measured. Viscometry is primarily concerned with the steady state response.

Time-independent Flows

The steady state response is reached very quickly, possible transient viscoelastic effects usually being short lived except at low $\dot{\gamma}$ values. The flow behaviour is then characterized by the flow curve, i.e. the biunivocal relation between stress, or apparent viscosity $\eta(\dot{\gamma})=\sigma/\dot{\gamma}$, and shear rate. When the condition that $\sigma=0$ for $\dot{\gamma}=0$ is added, three different types of behaviours are possible (Fig. 2(a)). Shear thinning is the most frequent one; shear thickening is displayed by some very concentrated suspensions.

Non-Newtonian flow generally results from entanglement or interactions between macromolecules or particles. It can also be due to alignment of highly asymmetric rigid molecules or particles with the direction of flow. At a given value of stress or shear rate, an equilibrium is reached very quickly between the break up and the restoration of the 'structure' or orientation. A change in $\dot{\gamma}$ immediately shifts the equilibrium.

Time-dependent Flows

When this equilibrium is not quickly established, it will take a certain time to reach a constant response after a change in the imposed value of $\dot{\gamma}$ or σ. A gradual decrease of viscosity under shear followed by its recovery when shear is stopped is called thixotropy (Fig. 2(c)), the opposite phenomenon being antithixotropy. When not completely reversible, these behaviours are especially difficult to study. Often testing is restricted to a comparative test in which $\dot{\gamma}$ is increased linearly from zero to a given maximum value in a given time and then decreased to zero in the same way (Fig. 2(d)). Viscoelasticity can be superimposed on these viscous memory effects.

Thixotropy, generally associated with a shear thinning equilibrium flow curve, is a common behaviour, typical of aggregated dispersions (sauces, soups, etc.); it corresponds to the progressive break up with time of aggregates under shear.

Plastic Behaviours

Some materials behave either as a solid or a liquid as the stress value is below or above a threshold σ_y, the yield stress. Their flow curves $\sigma(\dot{\gamma})$ do not pass through the origin, $\eta \rightarrow \infty$ when $\dot{\gamma} \rightarrow 0$; a special case is the Bingham body, which behaves linearly for $\sigma > \sigma_y$ (Fig. 2(b)).

The yield stress is classically determined by extrapolation of the $\sigma(\dot{\gamma})$ curve to $\dot{\gamma} = 0$; the result depends strongly on the range of shear rates used, and not infrequently its very existence could become questionable if measurement were carried on at lower shear rates. An alternative safer way to determine σ_y is through creep experiments at different stresses.

Concentrated multiphase systems, such as butter, chocolate, tomato purée or mayonnaise, are plastic materials. When $\sigma < \sigma_y$, interparticulate interactions stabilize a weak gel structure; when $\sigma > \sigma_y$ the cohesive interactions are overcome and flow occurs. Yield stress is often associated with thixotropy; plastic materials show viscoelasticity, at least when $\sigma < \sigma_y$.

Bibliography

Barnes HA, Hutton JF and Walters K (1989) *An Introduction to Rheology*. Amsterdam: Elsevier.

Bourne MC (1982) *Food Texture and Viscosity: Concept and Measurement*. New York: Academic Press.

Ferry JD (1980) *Viscoelastic Properties of Polymers*, 3rd edn. New York: Wiley.

Prentice JH (1984) *Measurements in the Rheology of Foodstuffs*. London: Elsevier.

Singh RP and Medina AG (eds) (1989) *Food Properties and Computer-Aided Engineering of Food Processing Systems*. Dordrecht: Kluwer.

Whorlow RW (1980) *Rheological Techniques*. Chichester: Ellis Horwood.

Jacques Lefebvre
INRA, Nantes, France

PHYTIC ACID

Contents

Properties and Determination

Phytic acid is commonly called *myo*-inositol hexaphosphoric acid or, technically, *myo*-inositol 1,2,3,5/4,6-hexakis(dihydrogen phosphate) (Fig. 1(a)). It usually occurs in seeds, tubers and certain roots as mixed potassium, magnesium and calcium salts (phytins/phytates), where its primary function is believed to be as a store of phosphate and trace metals. Since the phytate molecule is highly charged with six phosphate groups, it is an excellent chelator, forming complexes with mineral cations and proteins. The great concern over the presence of phytate in certain plant foods arises mainly from the fact that it complexes with certain dietary minerals and decreases their bioavailability. ***See*** Bioavailability of Nutrients

This article summarizes the structure and physicochemical properties of phytic acid, its occurrence and form in foods, and its chemical interactions with other components in fresh and processed foods; the methods for analysis of phytic acid are also discussed.

Structure and Physicochemical Properties

Phytic acid is *myo*-inositol 1,2,3,5/4,6-hexakis (dihydrogen phosphate), abbreviated to $InsP_6$. *myo*-Inositol is a *meso* form having a plane of symmetry through C2 and C5. Evidence for the structure and conformation of phytic acid is provided by the following sequence of physical methods: (1) X ray crystallography provides an absolute determination of configuration, thus in solid *myo*-inositol the hydroxyls are disposed equatorially, except for that on C2 which is axial; (2) Raman spectroscopy, applicable to both solids and solutions, shows that the solution conformation of *myo*-inositol is the same as the solid; (3) a comparison of the ^{13}C nuclear magnetic resonance (NMR) spectra of solutions of *myo*-inositol and un-ionized phytic acid shows that their conformations are equivalent, i.e. one axial and five equatorial hydroxyls or phosphates, respectively (Fig. 1(b)).

Phytic acid, $Ins(PO_4H_2)_6$, ionizes to a dodecyl anion, $Ins(PO_4)_6^{12-}$, each phosphate having two exchangeable protons. When phytic acid is titrated with sodium hydroxide two major points of inflexion are noted at about pH 6 and pH 10, corresponding to 8 and 12 mol of base per mole of phytic acid. The approximate extent of hydrogenation is 6 ($Ins(PO_4H)_6^{6-}$) from pH 2–6·3, 4 (octavalent) from pH 6·3–10, and almost full dissociation at pH > 10. The pK_a values of each proton of $InsP_6$ can be obtained more accurately from the points of inflexion of the ^{31}P-NMR chemical shift titration plot. The chemical shift at a particular pH being the weighted average of shifts for protonated and deprotonated forms, deprotonation causes a downfield shift. Using dilute solutions (0·1 M $InsP_6$) and sodium counter-ions, each phosphate can be shown to have one strongly acidic proton (pK_a 1–2). Each of the remaining protons show multiple inflexions with a weakly acidic component (pK_a 5–6) and a very weakly acidic component (pK_a 8·6–9·6). The corresponding ^{13}C spectra show one pattern from pH 2 to 8 and a different pattern at pH > 10, indicating a switch in conformation at the point where the three least acidic protons dissociate. The Raman spectrum of this alkaline form corresponds to that of the solid Na_{12} hydrated salt, which, from the X ray diffraction pattern, is five-axial/one-equatorial (Fig. 1(c)). Thus, both chair conformations of phytic acid are

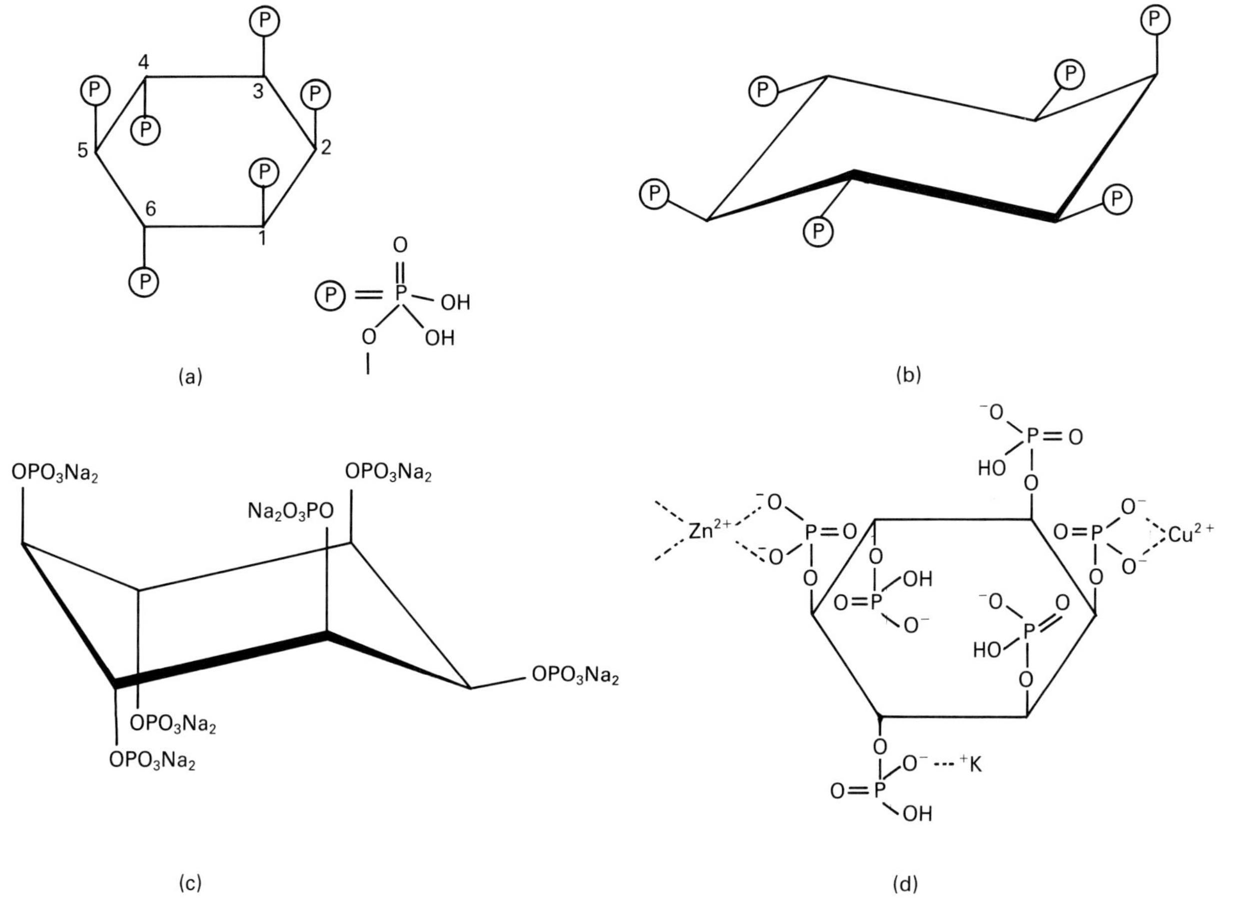

Fig. 1 (a) Structure of phytic acid – note that the phosphate groups on C4 and C6 are below the plane of the ring; (b) the structure and configuration of phytic acid, *myo*-inositol 1,2,3,5/4,6-hexakis(dihydrogen phosphate) (c) the five-axial/one-equatorial configuration of the sodium salt; (d) phytic acid chelate(s) at neutral pH.

possible. The precise pK_a and pH values at which the conformation switches will depend on which particular counter-ion is involved.

Chemical Properties Relevant to Food

At pH values normally encountered in foods or in the gut, phytic acid exists predominantly in the five-equatorial/one-axial conformation as the hexa- or octa-valent anion. Phytic acid is therefore a strong chelator of cations and, at the appropriate pH and concentration, causes precipitation of polyvalent cations and proteins. Precipitation causes reduced bioavailability of essential trace minerals and reduced digestibility of proteins. This is a major cause of its significance as an antinutrient. With multivalent cations a variety of structures is possible, forming a chelate within one phosphate group, or between two phosphates of the same phytate ion or intermolecularly (see Fig. 1(d)), though few have been identified by X ray crystallography. The *in vitro* studies will be considered here, and the effects of cooking and processing will be discussed below.

Chelation of Minerals

One mole of phytate can complex with up to 6 mol of either Cu^{2+} or Co^{2+}, up to 5 mol of Ca^{2+}, but only 3·5–4 mol of Zn^{2+} and 4 mol of Fe^{3+}. Binding of Ca^{2+}, Zn^{2+} or Cu^{2+} requires the presence of at least one fully deprotonated phosphate per divalent cation bound. The ${M^{2+}}_{4-6}$ phytate complexes are appreciably less soluble than ${M^{2+}}_{1-3}$ complexes and will precipitate at a pH above about 3·5.

The major site of mineral absorption is the duodenum, which has a pH of 6–7·4. Within this pH range the ${M^{2+}}_{4-6}$ phytate complex is at its minimum solubility and therefore would be unavailable. The individual binding affinities of nutritionally important ions are ranked as follows: $Cu^{2+} > Zn^{2+} > Co^{2+} > Mn^{2+} > Fe^{3+} > Ca^{2+}$. In the diet there are interactions between different mineral ions and phytate. There is evidence that the synergism between phytate, calcium and zinc reduces zinc bioavailability more than phytate – zinc chelation on its own because the dimetal complex appears to be more stable than the monometal complex. The low bioavailability of zinc and iron from plant foods rich in phytates, such as wheat bran and whole grain products, is of particular concern. ***See*** Minerals, Dietary Importance ***See also*** individual minerals

Chelation of the Cations of Metalloenzymes

A number of metalloenzymes containing Zn^{2+} or Ca^{2+} at their active site are present in the small intestine and are necessary for the digestion of food, viz. the carboxypeptidases, aminopeptidases, amylase and alkaline phosphatase. An inhibitory effect of phytate has been found *in vitro* on the Zn^{2+}-containing enzyme carboxypeptidase A. In the presence of Cu^{2+}, phytate catalyses the exchange of zinc for copper at the active site, causing a loss of 95% of enzyme activity. A similar effect has been found with intestinal alkaline phosphatase – also a zinc metalloenzyme. The potential for impairing digestion by this mechanism could be significant, but it has yet to be studied *in vivo*.

Chelation of Iron – Antioxidant Properties

Phytic acid has unique antioxidant properties as an iron chelator. It catalyses the oxidation of Fe^{2+} by molecular oxygen but prevents Fe^{2+} from catalysing the production of hydroxyl radicals from hydrogen peroxide (Fenton reaction). The reaction requires one iron coordination site to be accessible to water but all are coordinated by phytate at pH 7·4 over a wide range of iron:phytate ratios. The effect is obtained at low concentrations of phytate and iron without requiring precipitation. Consequently, phytate has several potential applications as a food preservative. It prevents iron-catalysed lipid peroxidation and so protects food against rancidity. It also prevents discoloration and prolongs the shelf life of fruit and vegetables. ***See*** Antioxidants, Natural Antioxidants

Chelation of Protein

$InsP_6$ is known to interact and complex with proteins at both acidic and alkaline pH. Below pH 3 phytate is only slightly ionized and therefore it has little binding capacity. ***See*** Protein, Functional Properties

In the pH range 3–5 phytate is mostly in the hexavalent anionic form and proteins are below their isoelectric point so the basic residues such as lysine, histidine, arginine and the *N*-terminus bear a positive charge. Insoluble complexes form which will only dissolve below pH 3. The complex may be regarded as a salt-like linkage: Ins–O–$P(O_2H)$–$O^-\cdots{}^+NH_3$–CH_2–protein. For this reason soya protein isolates prepared by isoelectric precipitation (pH 4·2) may contain as much as 60–70% of the original phytate of the defatted soya meal.

The inhibitory effect of phytate on proteases can likewise be explained. *In vitro* studies show that phytate causes no inhibition of pepsin activity above pH 4·5, irrespective of the isoelectric point of the substrate protein, which might vary between p*I* 4–7. As the pH is lowered, protease activity decreases rapidly to a minimum at pH 2–3, and increases below pH 2. Inhibition follows the extent of protonation of carboxyl groups, suggesting that phytate binding requires, in addition to an overall positive charge, the protonation of carboxyl groups.

In the pH range 7–10 both phytate and most proteins carry a net negative charge so complete dissociation is expected. But in the presence of multivalent cations soluble ternary complexes (protein–cation–phytate) occur with the phytate binding through a bridging cation. The proposed protein-binding sites being the imidazole group of histidine and possibly ionized carboxyl groups. The feasibility of such ternary complexes has been demonstrated using *N*-acetyl amino acids as models for the α-carboximide of protein backbones. Changes in ^{13}C and ^{31}P chemical shifts (different from those of the three binary complexes: phytate–cation, phytate–amino acid, cation–amino acid) indicate that ternary complexes formed at pH 7 in solutions of Ca^{2+}, Cu^{2+} and Zn^{2+} with phytate and *N*-acetyl histidine but not with *N*-acetyl glutamate. Using again the example of soya protein concentrates, glycinin has numerous histidine-binding site for zinc. In the presence of phytate and zinc ternary complexes form, which precipitate at higher phytate concentrations. The processing of defatted soya flour to obtain a protein concentrate by precipitating and dehydrating neutralized extracts may result in reduced availability of zinc.

Occurrence in Raw and Processed Foods

Only certain plant foods are rich in phytates, the levels in animal products being insignificant. The levels and form of $InsP_6$ in raw plant material will be considered first, and then the effects which food preparation have on the form of phytic acid consumed.

Occurrence in Raw Plant Material

Phytates are present in seeds of all kinds. In the following descriptions $InsP_6$ contents are expressed on a dry weight basis. Cereal grains have variable $InsP_6$ content (e.g. oats 0·8–1·0%, barley 1·0–1·2%, rice 0·9%, rice (polished) 0·2–0·3%, maize 0·9–1·0%, wheat 0·8–1·2%). Phytate is concentrated in the aleurone and germ at levels of 3–6%; the endosperms contain very small amounts. In wheat bran the $InsP_6$ content ranges from 3 to 6·5% (3·8% in AACC wheat bran). The $InsP_6$ content of whole wheat bread is about 0·6%. The $InsP_6$ content of legume seeds is also variable (e.g. peas 1·2%, navy beans 0·7–1·6%, black gram 1·5%, green gram 0·7%, brown bengal gram 0·2% (immature), 1·2% (mature), soya beans 1·0–1·5%), and the phytates are mainly in the cotyledons, where they are localized in protein-coated bodies, called globoids. The $InsP_6$ content of soya flour is about 2·0%, and the levels in soya protein concentrates range from 1·2 to 2·2%. Oil seeds generally have higher levels compared with grains and legumes (e.g. peanuts 0·7%, rapeseed 2–4%, cottonseed flour 2·9%, sesame seed 4·7%). Roots and tubers usually contain lower levels (e.g. carrot 0·5%, parsnip 0·8%); but levels are significant in the tubers (e.g. cassava 6%, cocoyam 9%). The $InsP_6$ content of leaves, stems and fruit pulps is negligible. Thus, the InsP content depends largely on the plant organ, and also on its stage of maturity at harvest. ***See*** individual foods

Recent work has confirmed that phytates have a marked inhibitory effect on nonhaem iron absorption, which accounts for 85–90% of the total dietary iron intake. It is of interest that only small amounts of phytates (5–10 mg of phytate phosphorus) are needed to reduce iron absorption by half and that the effect of phytates is much less marked in a meal supplemented with ascorbic acid and/or meat. High intake of phytate-rich foods can also reduce the availability of zinc. In a balanced Western diet, where fibre-rich foods account for 20–30 g of daily dietary fibre intake, the effect of phytate is generally counteracted by the animal protein in the diet. However, the effect of a high intake of fibre, from bran-enriched products, on iron and zinc absorption may be undesirable for infants, children and young adolescents, and recommendations for dietary fibre intakes in these groups should be different from those in adults. ***See*** Dietary Fibre, Effects of Fibre on Absorption

Phytic acid occurs as the salt ‘phytin’, which is a noncrystalline solid, nonstoichiometric with respect to the cation. The associated cations are predominantly K^+ and Mg^{2+} but Ca^{2+}, Fe^{3+}, Mn^{2+} and Zn^{2+} are also found. Phytin is also nonstoichiometric with respect to phosphate groups, being mainly $InsP_6$, but also containing $InsP_5$ and $InsP_4$. In general, in immature seeds there are higher proportions of $InsP_1$ to $InsP_5$.

Synthesis and Metabolic Lability

$InsP_6$ is synthesized by successive phosphorylations of *myo*-inositol. Phytate degradation is a natural and rapid consequence of germination. Catabolism is by nonspecific acid phosphatases and leads to a loss of cation-binding capacity. Partly dephosphorylated forms have many possible isomers and each would differ in its binding characteristics and solubility. For example, $InsP_5$, $InsP_4$ and $InsP_3$ have 2,3,4, *meso* diastereoisomers and 2,6,8 enantiomeric pairs, respectively, but only a certain subset would be formed metabolically. Wheat phytase digests $InsP_6$ to L-Ins(1,2,3,4,5)P_5. Subsequent digestions are by consecutive removal of phosphates adjacent to free hydroxyls.

Form in Prepared Foods

Phytate-rich foods are rarely consumed raw (except for wheat bran, some nuts and seeds). An important function of food preparation is to encourage degradation of phytate either by activation of endogenous phytases or by the phytases of fermentative microorganisms.

Mineral deficiencies are more prevalent in societies that are heavily dependent on cereals and legumes as sources of protein and minerals. Traditional food preparation practices for reducing phytate levels include: soaking seeds and grains in water overnight or for 3 days to allow sprouting and some catabolism of phytic acid; soaking flours in water to form a dough then leaving for up to 3 days to ferment naturally or more quickly with added yeast. Tubers, such as cassava, which have a high water content can also be fermented after squeezing excess water from the pulp. These procedures can result in the degradation of most of the phytic acid. A combination of germination (3 days), milling and fermentation (8 days) of sorghum results in complete loss of phytate ($InsP_6 + InsP_5$).

The completeness of phytic acid catabolism during fermentation depends on the extent of concomitant acidification, pH 5 being the optimum for phytase activity. Dough pH can drop to 3 in certain fermentations, thereby inactivating the phytase. Much less of the phytic acid is degraded during the preparation of soda breads and unleavened whole-grain breads such as chapatti. ***See*** Chapatis and Related Products

Phytic acid is stable to heat, but phytases are labile. Extrusion cooking results in most of the phytate being undegraded since the phytases are inactivated. Cooking dry seeds or grains in boiling water results in the loss of only 16–32% of phytic acid. Cooking of legume seeds may result in a reduction of phytate levels. However, some legume seeds require long boiling times to soften and this 'hard-to-cook' phenomenon has been correlated with a lack of phytate. The phytic acid in tuber tissues has been found to be more labile, with 35% being lost on oven drying and 67% on cooking in boiling water. As an example of the extent of $InsP_6$ degradation in processed food, wheat bran is flaked and toasted for 2–3 min at 288°C to prepare the breakfast cereal product in which the inositol phosphate content (mmol kg^{-1}) is 0·6 $InsP_2$, 1·4 $InsP_3$, 2·6 $InsP_4$, 6·4 $InsP_5$ and 8·8 $InsP_6$. ***See*** Extrusion Cooking, Chemical and Nutritional Changes

Isolation and Analysis

Extraction

Phytic acid is present in many different forms in food samples. The first steps in isolation are drying (105°C), fine milling samples to 0·45 mm and ether extraction for samples with >5% fat. Efficient extraction requires solubilization of all phytate salts by dilute acids. The following have proved useful: 0·5 M nitric acid, 0·18 M trichloroacetic acid, 0·2–0·65 M hydrochloric acid for 1–3 h with continuous shaking. Recently, an ultrasonic liquid processor (Heat Systems-Ultrasonics Inc, New York) has been reported to give a much more rapid (1–3 min) extraction into 0·5 M hydrochloric acid. The extract is then centrifuged and/or filtered to give a clear solution.

Analysis – Ferric Phytate Method

Until 1982 methods usually relied on the presumed exclusive precipitation of phytic acid from dilute (0·15 M) acid by ferric chloride. Quantitative precipitation is best achieved in the presence of a slight excess of ferric chloride. The mixture is heated at 100°C for 30 min then cooled and centrifuged. The precipitate is washed with 0·2 M hydrochloric acid to remove coprecipitated inorganic phosphate and most of the lower inositol phosphates. Assuming a stoichiometry of $4Fe:InsP_6$, the amount of phytic acid precipitated can be quantified colorimetrically on the basis of:

(1) Ferric ions remaining in solution; by, for example, reaction with ammonium thiocyanate, extraction into amyl alcohol and measuring the absorbance of the pink complex $Fe(SCN)_3$ at 465 nm. This is probably the least reliable of the methods.
(2) Ferric ions in the precipitate; this can be achieved by adding 1·5 M sodium hydroxide to solubilize the phytate and precipitating Fe^{3+} as ferric hydroxide, dissolving this precipitate in hot 3·2 M nitric acid and reacting with thiocyanate then proceeding as above.
(3) Phosphate in the precipitate; phosphate reacts with 2·5% ammonium molybdate in 0·5 M sulphuric acid to form phosphomolybdic acid, which on reduction gives molybdenum blue. There are many variants of this method. In the most widely used, the phytate is hydrolysed by boiling in a mixture of sulphuric acid and nitric acids to release the phosphate and then boiling in water to hydrolyse any pyrophosphate which may have formed. The phosphate is converted to phosphomolybdic acid, as above, and reduced with 1-aminonaphthol sulphonic acid. The absorbance (640 nm) is read after 15 min. ***See*** Spectroscopy, Visible Spectroscopy and Colorimetry

Generally, procedures (2) and (3) give reasonably good estimates of phytic acid, but the methods have to be improved since the stoichiometry of the precipitate

$4Fe:InsP_6$ cannot be relied upon. Inositol phosphate esters lower than $InsP_6$ also precipitate, leading to overestimates. It is not sufficient to measure the inositol and phosphate content of the precipitate, since this provides only an average stoichiometry.

Chromatography

The first significant improvement over ferric precipitation methods was the development of anion exchange chromatography. This is more sensitive to low levels of phytic acid (< 0·2%). The Association of Official Analytical Chemists recommended method uses a 2 g sample, and an extraction with 40 ml of 0·65 M hydrochloric acid for 3 h with shaking. The acid extract is diluted 10-fold in water before applying to an AG1-X4 100–200 mesh Cl^- form resin, washing off inorganic phosphate and lower inositol phosphates with water and 0·1 M sodium chloride then eluting $InsP_6$ with 0·7 M sodium chloride. This eluate is acid digested and estimated for phosphate as method (3) above. *See* Chromatography, Principles

For studies of processed foods and digesta especially, it is important to know the remaining content of $InsP_6$ and $InsP_5$, which have the greatest ion-chelating capacity, and the extent of phytate breakdown from the amount of $InsP_4$, $InsP_3$, etc. Methods using a combination of anion exchange chromatography and high-performance liquid chromatography have been developed to separate inositol polyphosphates into congeneric groups but they have yet to be generally accepted because of problems of reproducibility. There is a need in nutritional research for reliable methods based on the identification of isomers. *See* Chromatography, High-performance Liquid Chromatography

NMR Spectroscopy

This allows the determination of phytic acid and its degradation products without fractionation. Phytic acid at pH 4–5·1 gives narrow ^{31}P signals with ^{1}H decoupling. At higher pH values (up to 12) signals broaden due to proton exchange. By adding excess EDTA to chelate paramagnetic ions, sodium hydroxide to adjust the pH to 4·5, and 0·25% phosphoric acid as an internal standard, to a sample of the acid extract and acquiring the spectrum with gated ^{1}H decoupling, phytate can be quantified by comparing the total area of the phytate signals with that of the internal standard. *See* Spectroscopy, Nuclear Magnetic Resonance

^{31}P-NMR spectroscopy has been used to follow the hydrolysis of phytate during dough fermentation and should be applicable to other cooked and processed foods.

Bibliography

Ersöz JW, Akgün H and Aras NK (1990) Determination of phytate in Turkish diet by phosphorus-31 Fourier transform magnetic resonance spectroscopy. *Journal of Agricultural and Food Chemistry* 38: 733–735.

Graf E (1986) *Phytic Acid: Chemistry and Applications.* Minneapolis: Pilatus Press.

Lehrfeld J (1989) High performance liquid chromatography analysis of phytic acid on a pH-stable macroporous polymer column. *Cereal Chemistry* 66: 510–515.

Loewus FA (1990) Cyclitols. In: Dey PM and Harborne JB (eds) *Methods in Plant Biochemistry 2. Carbohydrates*, pp 219–233. London: Academic Press.

Martin CJ and Evans WJ (1989) Phytic acid-enhanced metal ion exchange reactions: the effect on carboxypeptidase A. *Journal of Inorganic Biochemistry* 35: 267–288.

Reddy NR, Sathe SK and Salunkhe DK (1982) Phytates in legumes and cereals. *Advances in Food Research* 28: 1–92.

P. Ryden and R. R. Selvendran
Institute of Food Research, Norwich, UK

Nutritional Impact

Phytic acid, myo-inositol hexakisphosphoric acid, is the major phosphorylated compound present in seeds. In some literature the term is also applied to phytin, the mixed calcium-magnesium salt. The following is a discussion of its occurrence and food composition, nutritional impact and effects on food preparation.

The primary role of phytate in the mature seed is considered to be that of a storage form of phosphate for the developing embryo. The total phosphorus in maturing wheat seeds remains constant, but the fraction present as phytate phosphorus increases as maturation proceeds. Upon germination or sprouting, phosphate is released through mediation of the enzyme phytase. The hydrolysis catalysed by phytase proceeds in a stepwise sequence, yielding inositol phosphates with lower degrees of phosphorylation.

Inositol polyphosphates besides the hexaphosphate have been identified in both plant and animal tissues. Specific position isomers of inositol tri, tetra-, and penta-phosphates are involved as biological messenger molecules, and even the hexaphosphate has been found in rat brain. Avian red blood cells contain inositol pentaphosphate. The role of these compounds in cellular function is an area of active research and is not the subject of this article.

Occurrence and Food Composition

Phytate is found in all seeds. The concentration reported in a few selected cereal grains, legumes and nuts is shown

Table 1. Phytate in foods

Food	Moisture (%)	Phytate (g per kg dry weight)
Cereal grains and products		
Bread, whole wheat	36·4	6·13
Bread, white	35·8	1·07
Corn meal	12·0	9·43
Macaroni, dry	10·4	2·90
Oatmeal, dry	8·9	10·35
Popcorn, popped	4·0	6·40
Rice, polished	10·3	2·84
Rye flour	15·0	10·81
Sesame seed	5·5	17·1
Wheat flour, whole grain	12·0	9·60
Wheat bran	10·4	33·6
Legumes		
Bean, navy, raw	10·9	10·2
Chickpea, raw	10·7	8·17
Lentil, raw	12·2	4·94
Pea, raw	13·3	9·82
Peanut, toasted	6·6	9·99
Soya bean, raw	10·0	25·8
Soya flour	8·0	15·2
Soya isolate	8·0	13·4
Nuts		
Almond	5·0	13·5
Cashew	5·2	19·7
Hickory nut	3·3	16·7
Walnut, English	23·5	9·93

Selected data from Harland BF and Oberleas D (1987).

in Table 1. Distribution of phytate among different morphological structures of a seed is often reflected in foods derived from that seed. The aleurone layer which is separated into the bran stream during milling contains about 85% of the phytate in wheat. Consequently, bread made with white wheat flour contains 20% or less of the phytate found in wholemeal bread, and wheat bran may contain as much as 5% by weight of phytate (Table 1). Corn germ contains a higher concentration of phytate than does the corn endosperm. Thus foods derived from numerous cereal grains vary in phytate content depending on the milling fraction(s) from which they are derived. Many different soya bean products are now available for use in the manufacture of human foods. The phytate content of soya protein isolates or concentrates varies depending on the processing procedure used for their preparation. ***See*** Cereals, Contribution to the Diet; Wheat

Although it occurs in low concentrations, phytate is reported to be found in some fruits and vegetables and tuberous food sources. For example, potatoes reportedly contain about 2 g of phytate per kg dry matter. Vegetable oils expressed from the oil seeds contain no phytate.

Autoclaving in acid medium will bring about hydrolysis of phytate to inositol and phosphate, but hydrolysis of phytate under usual household cooking conditions has not been documented. During the fermentation steps in breadmaking, phytase from the wheat and yeast will produce hydrolysis of phytate. One study, using the step-gradient ion exchange analysis procedure, demonstrated a 15% decrease in phytate in wholemeal bread over a 2-h fermentation period. However, another study, in which the inositol phosphates were separated, showed over 50% reduction in the amount of inositol hexaphosphate in the same time frame. The amount of phytate hydrolysed can be greatly increased by extending the fermentation time beyond that generally used. Almost twice as much phytate is recovered in ileostomy contents of patients consuming an extruded wheat bran product compared to the amount recovered from the same patients when consuming a raw wheat bran product. The phytase endogenous to the bran in the extruded product is presumably inactivated by the moist heat generated during the extrusion process, and less phytate is hydrolysed. In addition to plant and yeast sources of phytase, numerous moulds and bacteria produce phytase(s). Figure 1 illustrates the hydrolysis of phytate in peanut press cake treated with phytase from *Rhizopus*. The fermentation leading to tempeh from soya milk results in some hydrolysis of phytate by the fungal phytase. Germination of seeds brings about induction of phytase activity. Harvested bean sprouts contain less phytate than the unsprouted mature seeds. ***See*** Fermented Foods, Fermentations of the Far East

Nutritional Impact of Phytate

Binding of Minerals

Phytate forms complexes with di- and polyvalent cations of most metals; of particular nutritional importance are calcium, magnesium, zinc and iron. Metabolic balance studies performed 50 years ago in the UK showed that the faecal excretion of calcium, magnesium and iron was greater when human subjects consumed brown bread made from wholemeal or high-extraction wheat flour than when they consumed bread made from white flour lower in phytate concentration. The nutritional significance of zinc for humans was not fully appreciated at that time and the effect on zinc was not measured. The potential effect of phytate on dietary zinc utilization by humans came to light as the aetiology of human zinc deficiency in the Middle East was studied. An important source of energy was unleavened bread baked from high-extraction wheat flour. Balance trials showed that zinc balance was improved by bread made

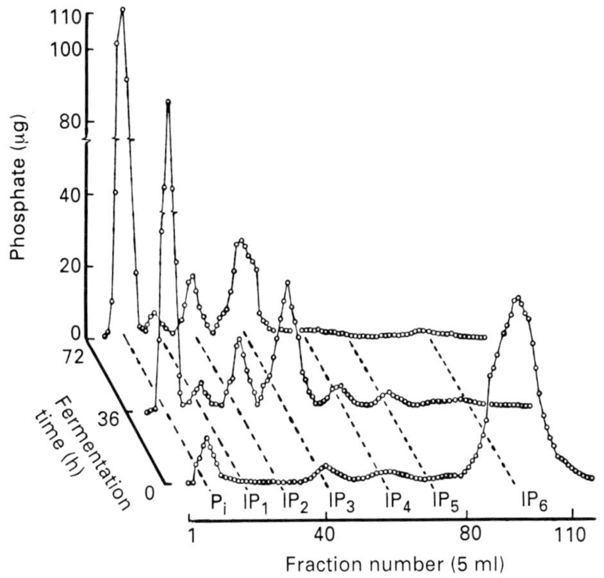

Fig. 1 Action of phytase from *Rhizopus* on phytate in peanut press cake. IP_6, inositol hexaphosphate; IP_5, inositol pentaphosphate, etc; P_i, inorganic phosphate. (Illustration by courtesy of *Journal of Food Science*.)

from low phytate flour or by yeast leavening, a process which reduced the phytate intake. ***See*** Bioavailability of Nutrients

Calcium and Magnesium

The impact of dietary phytate on calcium and magnesium nutriture of humans is uncertain. Human balance studies using purified phytate, wheat bran and oatmeal, and absorption studies using isotopic labels have demonstrated that phytate will reduce dietary calcium bioavailability. However, all these studies have been of relatively short duration, ranging from days to a few months. The ability of the body to regain calcium balance over extended time has led some investigators to conclude that phytate will have little effect on calcium nutriture and that the level of intake is of greater importance in maintaining calcium homeostasis. Although balance studies have shown that high phytate increases faecal excretion of magnesium, decreased urinary excretion provides homeostatic adjustment and overall balance is generally not affected. ***See*** Calcium, Physiology; Magnesium

Iron

The relationship of phytate to iron nutriture has been studied most extensively in wheat-based and soya-bean-based foods. Whole wheat and foods containing wheat bran impair absorption of nonhaem iron in humans. The chemical form of iron in wheat bran is thought to be monoferric phytate, but this accounts for only a fraction of the total phytate in wheat bran. Although isolated monoferric phytate is water-soluble, a high ionic strength solution is required to remove it from wheat bran. The iron of monoferric phytate is readily bioavailable to rats, and equilibrates with the miscible nonhaem iron pool of meals eaten by humans. Iron in the form of insoluble di- and tetraferric phytate, or mixed calcium-ferric phytate, is poorly utilized by rats. With increasing use of soya bean protein in human foods and knowledge of its phytate content, concern was expressed when nonhaem iron absorption in humans was demonstrated to be reduced by soya protein. Some recent research indicates that removal of the phytate from soya protein alleviates the impairment of iron absorption by soya products. ***See*** Iron, Physiology

Zinc

The deleterious effect of dietary phytate on zinc nutriture may be partially mediated by the binding of endogenous zinc secreted into the intestine. Regardless of whether the effect is on the binding of ingested dietary zinc, endogenous zinc or both, the molar ratio of phytate to zinc in the diet is a useful index of zinc bioavailability in rats. A dietary phytate:zinc molar ratio of 15 or more results in indicators of deficient zinc status in growing rats if the dietary zinc concentration is at or near the minimal requirement. The ratio of 15 will not, however, result in zinc deficiency signs if the dietary zinc concentration is sufficiently high. Balance studies and isotope absorption studies have likewise indicated that a phytate:zinc ratio greater than 15 may impair zinc utilization by humans. As in animals, the impact of the phytate:zinc ratio on zinc nutriture of humans may reside wholly with the absolute level of zinc intake. The dietary calcium concentration also influences the response of animals to high phytate:zinc ratios. High calcium exacerbates the effect of phytate in reducing zinc bioavailability in animals. Analysis of foods consumed by women in Nepal for phytate and zinc content indicated a phytate:zinc ratio high enough that human zinc deficiency signs might be expected. However, no indications of zinc deficiency were found in the population studied. Further dietary analysis showed that the calcium intakes were low. Under these conditions, a high phytate:zinc ratio may have little effect on zinc utilization by humans. ***See*** Zinc, Physiology

Effect of Inositol Tetra- and Pentaphosphate

Studies in the past decade have demonstrated that, for both animals and humans, inositol pentaphosphate inhibits zinc utilization, but to a lesser degree than hexaphosphate. Tetraphosphate also inhibits, but to a lesser degree than the pentaphosphate. A similar pattern also occurs for iron absorption in humans. As noted above, almost all food composition and dietary analysis data available in the literature were obtained by methodologies which did not differentiate between the inositol phosphates. For some cooked foods, particularly when either endogenous phytase or microbial or fungal phytases may be present, hydrolysis products of phytate may be present and the available phytate values may be in error. Therefore future analytical studies should be conducted using methodology which will quantify at least the inositol tetra-, penta- and hexaphosphates individually.

Bioavailability of Phytate Phosphorus

Faecal excretion of phytate by humans varies from 10% to 80% of that ingested. Whether or not the hydrolysis of phytate in the human gastrointestinal tract is the result of intestinal phytase activity is uncertain. Studies with ileostomy patients suggest that the stomach and small intestine exhibit little of either phytase or phosphatase activity. Greater than 90% of the phytate consumed by ileostomy patients was recovered in the intestinal contents when consuming phytase-inactivated wheat bran. When unprocessed raw bran possessing endogenous phytase activity was eaten, slightly less than 50% of consumed phytate was recovered. In the absence of endogenous food phytase, colonic microbial action may be solely responsible for any phytate hydrolysis in the intact gastrointestinal tract of humans. Intestinal phytase activity in several monogastric species has been demonstrated, but clear differentiation of phytase activity from phosphatase activity is absent. The ability of monogastric species to utilize phytate as a dietary source of phosphate is of practical importance for poultry and swine feeding. Both species utilize a larger fraction of the phytate phosphorus from a wheat source than they would from corn, which contains no endogenous phytase. Pretreatment with microbial phytase improves utilization of phytate phosphorus by poultry and swine, but commercial application of the process is still limited.

Inhibition of Digestive Enzymes

Under certain conditions, phytate will inhibit activity of the enzymes amylase, carboxypeptidase a, and alkaline phosphatase. The action on amylase is thought to be through binding of Ca^{2+}, a cofactor for full activity. *In vitro*, phytate promotes exchange of zinc in the native protein of carboxypeptidase and alkaline phosphatase, with another metal inactivating the enzyme. Whether or not this occurs in the milieu of the intestinal lumen is unknown. One report found no effect of phytate on biological value of a soya protein, but phytate was negatively correlated with biological value, using both rat bioassay and *in vitro* indices, for a wheat bran protein concentrate.

Effects of Phytate in Foods

About 90% of the phytate in corn germ is extracted by water, whereas the extractability of phytate in soya beans is 70% and in sesame meal 13%. Less than 10% of wheat bran phytate is soluble in cold water, but 70% is extracted by 1·2 M ammonium acetate. The differential solubilities indicate that phytate exists bound to proteins and/or other cellular components in many different ways. Protein phytate fractions have been isolated and partially characterized from beans and soyabeans. The technology of preparing soya protein isolates is covered in a separate article.

Studies on the relationship of cooking quality of beans and phytate content have yielded ambiguous findings. For those studies reporting an increase in cooking time with decrease in phytate content, the indications are that phytate binds calcium ion, preventing the interaction of calcium and pectin in the cell walls. The usual cooking methods produce little change in phytate content of foods unless some opportunity is provided for endogenous phytase to act.

Another suggested role of phytate in seeds is to serve as a natural antioxidant by complexing ferric ion. Reportedly, the phytate–ferric ion complex has no free coordination site for the hydroxyl ion, thus preventing free radical formation. Potential antioxidant action of phytate has been extended to the retardation of carcinogenesis. Formation of induced colon tumours in rats was suppressed by adding sodium phytate to the drinking water. Extrapolation of these studies to humans must be tempered by consideration of the possible adverse effect on mineral nutriture. The aetiology of low incidence of certain types of cancers in human populations consuming diets high in dietary fibre may in part be mediated through the antioxidant action of phytate. This aspect deserves further consideration. ***See*** Cancer, Diet in Cancer Prevention ***See also*** Dietary Fibre

Bibliography

Cosgrove DJ and Irving GCJ (1980) *Inositol Phosphates, their Chemistry, Biochemistry and Physiology*. Amsterdam: Elsevier Science Publishers BV.

Fardiaz D and Markakis P (1981) Degradation of phytic acid in oncom (peanut fermented press cake). *Journal of Food Science* 46: 523–525.

Graf E (ed.) *Phytic Acid – Chemistry and Applications*. Minneapolis: Pilatus Press.
Harland BF and Oberleas D (1987) Phytate in foods. *World Review of Nutrition and Dietetics* 52: 235–259.
Isbrandt LR and Oertel RP (1980) Conformational states of myo-inositol hexakis(phosphate) in aqueous solution. A ^{13}C NMR, ^{31}P NMR, and Raman spectroscopic investigation. *Journal of the American Chemical Society* 102: 3144–3148.
Lasztity R and Lasztity L (1990) Phytic acid in cereal technology. *Advances in Cereal Science and Technology* 10: 309–371.
Lehrfeld J (1989) High-performance liquid chromatography analysis of phytic acid on a pH-stable, macroporous polymer column. *Cereal Chemistry* 66: 510–515.
Mazzola EP, Phillippy BQ, Harland BF *et al.* (1986) Phosphorus-31 nuclear magnetic resonance spectroscopic determination of phytate in foods. *Journal of Agricultural and Food Chemistry* 34: 60–62.
Morris ER and Ellis R (1989) Usefulness of the dietary phytic acid/zinc molar ratio as an index of zinc bioavailability to rats and humans. *Biological Trace Element Research* 19: 107–117.
Phillippy BQ and Bland JM (1988) Gradient ion chromatography of inositol phosphates. *Analytical Biochemistry* 175: 162–166.
Reddy NR, Pierson MD, Sathe SK and Salunkhe DK (1989) Phytates in cereals and legumes. CRC Press, Boca Raton, Florida, USA.

Eugene R. Morris
US Department of Agriculture, Beltsville, MD, USA

PICKLING

Pickling is one of the oldest, and most successful, methods of food preservation known to humans. This article reviews the origins of pickling, the various methods of pickling employed commercially, and the nature of preservative action. The optimization of pickle quality depends on maintenance of proper acidity, salt concentration, temperature and sanitary conditions.

History and Tradition

It is difficult to suggest a date for production of the first pickled foods, but it is known that both vinegar and spices were being used during Biblical times. Fermentation of plant and animal foods was known to the early Egyptians, and fish were preserved by brining in prehistoric times. By the 3rd century BC, Chinese labourers were recorded to be consuming acid-fermented mixed vegetables while working on the Great Wall. In about 2030 BC, northern Indians brought the seed of the cucumber to the Tigris Valley. The Koreans created 'kimchi' from acid-fermented Chinese cabbage, radish and other ingredients many centuries ago. Corn, cassava and sorghum were fermented and became staples of the African diet. In the West, acid fermentation of cabbage and cucumbers produced sauerkraut and pickles, products that are still popular today. Early explorers carried kegs of sauerkraut and pickles that prevented scurvy on their voyages.

Peterson in 1977 defined pickling, in a broad sense, as 'the use of brine, vinegar or a spicy solution to preserve and give a unique flavour to a food'. In a 1936 document, the US Department of Agriculture described cucumber pickles as:

> immature cucumbers properly prepared, without taking up any metallic compounds other than salt, and preserved in any kind of vinegar, with or without spices. Pickled onions, pickled beets, pickled beans and other pickled vegetables are vegetables, prepared as described above, and conform in name to the vegetables used.

Literature references concerning the technology of acid fermentation began to appear in the Western press in the early 1900s. In 1919, Orla-Jensen isolated strains of *Betacoccus arabinosaceus* from sour potatoes, sour cabbage and sour dough. Pederson, in a number of classic studies in the 1930s, enumerated and identified the sequence of microorganisms involved in sauerkraut fermentation. *Leuconostoc mesenteroides* was identified as being one of the most important microorganisms for initiation of vegetable fermentation. Numerous investigators have carried out studies on acid-fermented vegetables over the past century. ***See*** Fermented Foods, Beverages from Sorghum and Millet

Advent of New Pickling Methods

Brining vegetables in salt, and the resultant lactic acid fermentation, is an ancient form of preservation. Two new methods of pickling cucumbers, which represent the largest volume of a single vegetable preserved by pickling, have been developed during the 1900s. Both methods use lower salt concentrations and result in a milder product. The first new method, pasteurization and direct acidification, was developed and began commercial production in the 1940s. The second, refrigeration and direct addition of acid and preservative, was introduced in the 1960s. Relative quantities of cucumbers preserved by the three pickling methods in 1984 were: brine fermentation, 43%; pasteurization, 43%; and refrigeration, 14%.

Pickling remains a major form of preservation in many countries because it: (1) yields desirable organoleptic qualities, (2) provides a means for extending the processing season of fruits and vegetables and (3) requires relatively little mechanical energy input.

Outline of the Process

Pickled products may be produced by one of three methods, i.e. fermentation, acidification followed by pasteurization or refrigeration. Because they are relatively straightforward, the methods for producing pasteurized and refrigerated pickled products will be described first.

Pasteurized Pickles

Vegetables which are fresh or only partially fermented may be preserved by the addition of vinegar or acetic acid and subsequent pasteurization. Vinegar alone is not sufficient to ensure product safety, and so requires an additional form of preservation such as heat or refrigeration. The steps involved in producing pasteurized or 'fresh-pack' pickles are the following:

- slice, cube or dice product;
- place in clean container;
- mix water, salt, vinegar, sugar, spices and bring to boil;
- add hot brine to container;
- seal and pasteurize.

Two possible pasteurization methods include: (1) heating such that the centre of the jar or can reaches 75°C, and holding for 15 min, followed by a prompt cooling to 35°C or below, or (2) heating at 70°C for 10 min, followed by prompt cooling. ***See*** Vinegar

The pasteurization process essentially destroys spoilage microorganisms and prohibits fermentation from occurring. Both acid-forming bacteria, which are active in brine fermentation, and yeasts, which cause gas production, are destroyed by pasteurization. The pasteurization process inhibits polygalacturonase, the enzyme responsible for pickle softening. Enzyme activity may also be controlled through use of appropriate salt concentrations. Calcium chloride is often added to brine to aid in firming cucumber pickles. Pasteurized pickled products have steadily gained in popularity, and have a very different flavour and texture than that of fermented pickled products. ***See*** Pasteurization, Principles

Refrigerated Pickles

Refrigerated pickle products, which represent the latest development in pickling technology, are produced by direct acidification and addition of sodium benzoate or another preservative. In these nonfermented pickles, the preservative takes the place of pasteurization in preventing spoilage of the product. This process is essentially the same as that used for pasteurized pickles, but instead of pasteurizing, the sealed containers are refrigerated. It is essential that this type of pickle is kept refrigerated throughout the production process and during subsequent consumption.

Fermented Pickles

Cucumbers are by far the most common vegetable fermented; therefore, most of the following processes will be described using cucumbers as an example. There are three general methods which may be used for cucumber fermentation; (1) salt stock, (2) genuine dill and (3) overnight dill.

Salt Stock

The first method, salt stock, involves fermentation in 5–8% sodium chloride until all fermentable sugars have been converted to acids and other end products. The cucumbers are then stored in open tanks containing 10–16% salt to maintain stability for up to 1 year. Figure 1 outlines the general process for salt stock pickles. Desalting to an acceptable organoleptic level (2–2·5% salt) is carried out by leaching in water. Salt stock pickles make up the largest percentage of fermented pickle products.

The controlled fermentation of pickles has been a goal of the industry for years. In 1973, such a procedure was introduced in the USA. Although the industry has yet to adopt the entire process, addition of some of the modifications outlined has improved pickle quality in general. In particular, most processors now acidify and purge tanks after brine is added. Acidification inhibits the growth of acid-sensitive Gram-positive and Gram-negative bacteria and, therefore, favours the growth of lactic acid bacteria. Purging decreases the incidence of bloating, which results from carbon dioxide production by both the fermenting microorganisms and the cucumber itself (Fleming, 1984).

Genuine Dill

These pickles are fermented in 4–5% sodium chloride, to which dill weed, garlic and other spices have been added.

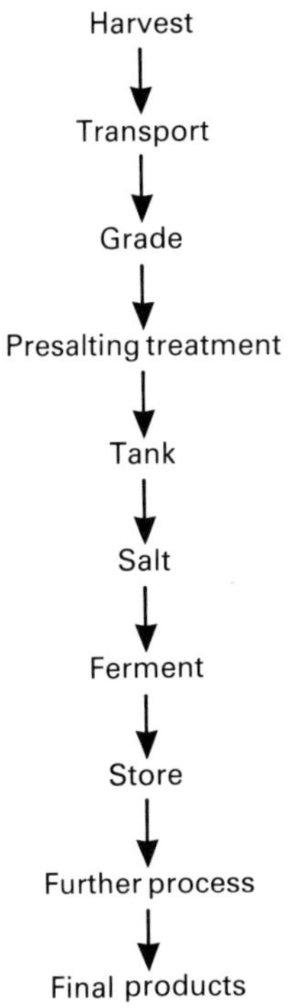

Fig. 1 General flow chart for preservation of vegetables by brining. Taken from Fleming HP and McFeeters RF (1981).

It takes approximately 3–6 weeks for fermentation to reach completion, where the lactic acid content is 1·0–1·5% and the salt content is 3–3·5%. This type of pickle does not require desalting, but may be sold as such with the filtered fermentation liquor. Genuine dill pickles are somewhat susceptible to scum yeast development and should be consumed within 12 months of manufacture.

Overnight Dill

Overnight dill pickles are fermented in 2–4% sodium chloride, containing dill weed and garlic, until the desired level of acidity (0·75–1% titratable acidity as lactic acid) is reached, which should take approximately 1 week. The product must then be refrigerated, and should not be kept longer than 6 months due to its extreme perishability.

Nature of Preservative Action

Two components of the pickling process, acid and salt, are key participants in the preservation of perishable products. Acid, which may be added directly or produced through microbial conversion of indigenous sugars to acids, will lower the pH of the product and inhibit spoilage microorganisms. Only undissociated acid molecules are active in inhibiting microorganisms; therefore, it is important that the acidity of pickles be pH 3·5 or below, when most of the acid present will be in the undissociated form. Salt also acts to inhibit the growth of undesirable bacteria and to delay enzymatic softening. ***See*** Acids, Natural Acids and Acidulants; pH – Principles and Measurement

In fermented pickles, microorganisms ferment sugars to lactic acid and also produce enzymes which modify pickle texture. The absence of fermentable carbohydrates is a deterrent to undesirable secondary fermentations which can be initiated by yeasts at pH values below 3·8. Residual sugars can also cause gas production and brine turbidity in finished products if yeast and bacteria growth continues (Fleming and McFeeters, 1981).

Lactic acid bacteria are the primary microorganisms involved in preservation of fermented pickled products. Although these microorganisms represent only a smaller proportion of the total microbial flora present on the surface of plant materials, they predominate under acidic conditions. In cucumbers, *Leuconostoc mesenteroides* typically grows until the pH begins to drop, then *Pediococcus pentosaceus* predominates, followed by *Lactobacillus brevis* and finally *Lac. plantarum*. ***See*** Lactic Acid Bacteria

Acidity, salt concentration, temperature and sanitary conditions are the primary environmental factors which control fermentation. Most vegetables range in pH from 6·5 to 4·6, while fruits range from 4·5 to 3·0. In fruits and fruit juices, yeasts and moulds predominate in the more acidic environments. Salt inhibits the growth of undesirable microorganisms and, in addition, it withdraws water and nutrients from plant tissues and allows these to become substrates for lactic acid bacteria.

Low temperatures inhibit the growth of lactic acid bacteria and thus slow fermentation. At 7·5°C, *L. mesenteroides* will grow, but the growth of *Lactobacillus* and *Pediococcus* species is very low. At temperatures in the range 18–23°C, *Lac. brevis* and *Lac. plantarum* exhibit active growth while, at 32°C, *Lac. plantarum* and *Pediococcus pentosaceus* predominate. Pasteurization is often the final step in pickle processing. The US Department of Agriculture recommends that all pickled products should be pasteurized for safety.

Sensory and Nutritional Attributes

Pickling imparts unique characteristics to fruits and vegetables. Desirable changes in flavour, texture and colour take place in fermented, pasteurized and refrigerated pickles, and are carefully monitored. However, some of the same bacteria involved in normal fermentation, such as the lactic acid formers, may cause spoilage if not destroyed. Selected pickling problems which may affect sensory quality are listed in Table 1.

Table 1. Selected pickling problems

Problems	Cause
Soft, slippery slimy pickles (discard pickles, spoilage is occurring)	Hard water
	Acid level too low
	Cooked too long or at too high a temperature
	Water bath too short, bacteria not destroyed
	Jars not airtight
	Jars in too warm a resting place
Shrivelled, tough pickles	Pickles overcooked
	Syrup too heavy
	Too strong a brine or vinegar solution
	Pickles not fresh enough at outset
	Fruit cooked too harshly in vinegar/sugar mixture
Dark, discoloured pickles	Iron utensils used
	Copper, brass, iron, or zinc cookware used
	Hard water
	Metal lid corrosion
	Too great a quantity of powdered and dried spices used
	Iodized salt used

Taken from McNair JK (1975).

Pickled cucumbers are composed primarily of water (Table 2). They are not a good source of protein or fat, but are a fairly rich source of vitamin A, vitamin C and phosphorus. Salt and sugar additions to the brine will affect the sodium and carbohydrate contents. ***See*** Sensory Evaluation, Sensory Characteristics of Human Foods ***See*** individual nutrients

Specific Examples

Pickled cucumbers are by far the most abundant pickled product available in the Western world today. Other common pickled products include sauerkraut, pickled pears, peaches and plums, pickled nuts, relishes, cured meats, fish and poultry, and such speciality items as pickled mushrooms and cherries.

Salt stock is used to prepare sour cucumber pickles, which typically have a final acidity not lower than 2·5%. Sweet pickles are prepared in a similar fashion, except that a sweet, spiced vinegar solution is added to the salt stock.

Sauerkraut is produced through salt-controlled bacterial fermentation. Cabbage selected for sauerkraut should have at least 3·5% sugar to ensure an adequate carbohydrate source for bacteria. Shredded cabbage is mixed with salt (2·25% by weight) and the final product contains an average of 1·5–2·0% lactic acid.

Table 2. Nutritive analysis of pickles (The composition of 100 g of edible portion (approx. 1 large dill pickle or 1/2 cup of fresh cucumber pickle slices)

	Fermented dill pickles	Sweet pickles	Sour pickles	Fresh pack cucumber pickles
Water (%)	93	60·7	94·8	78·7
Food Energy (J)	46·2	613·2	42	306·6
Protein (g)	0·7	0·7	0·5	0·9
Fat (g)	0·2	0·4	0·2	0·2
Carbohydrate (g)	2·2	36·5	2·0	17·9
Ash (g)	3·6	1·7	2·5	2·3
Calcium (mg)	26·0	12·0	17·0	32·0
Iron (mg)	1·0	1·2	3·2	1·8
Vitamin A (iu)[a]	100	90	100	140
Thiamin	Trace	Trace	Trace	Trace
Riboflavin (mg)	0·02	0·02	0·02	0·03
Vitamin C (mg)	6·0	6·0	7·0	9·0
Phosphorus (mg)	21·0	16·0	15·0	27·0
Potassium (mg)	200·0	—	—	—
Sodium (mg)	1428·0	—	1353·0	673·0

Taken from McNair JK (1975).
1 iu=0·6 μg β-carotene.

Bibliography

Binsted R, Dewey JD and Dakin JC (1939) *Pickle and Sauce Making*. London: Food Trade Press.

Fleming HP (1984) Developments in cucumber fermentations. *Journal of Chemical Technology and Biotechnology* 34B: 241–252.

Fleming HP and McFeeters RF (1981) Use of microbial cultures: vegetable products. *Food Technology* 1: 84–88.

McNair JK (ed.) (1975) *All About Pickling*. San Francisco: Ortho Books.

Pederson CS (1930) Floral changes in the fermentation of sauerkraut. *NYSAES Technical Bulletin* 168.

Pederson CS (1979) *Microbiology of Food Fermentations*, 2nd edn. Westport: AVI.

Peterson MS (1977) Pickles. In: Desrosier NW (eds) *Elements of Food Technology*, pp 690–691. Westport: AVI.

Steinkraus KH (1983) *Handbook of Indigenous Fermented Foods*. New York: Marcel Dekker.

US Department of Agriculture (1936) *Service and Regulatory Announcements*, No. 2, rev. 5. Washington, DC: Food and Drug Association.

Diane M. Barrett
Oregon State University, Corvallis, USA

PIGMENTS

See Colours

PILCHARDS

See Fish

PINE KERNELS

Pine nuts, also known as Indian nuts, piñons or pignolias, have been an important food crop in some areas of the world since prehistoric times. The 'nut' is, in fact, the seed of different species of pine (*Pinus*), nearly all of which belong to a group known as 'soft' or 'white' pines or their relatives. These species are evergreen, coniferous trees whose cones are softly woody, have few scales, and two large seeds per scale which lack a wing (see Fig. 1). The kernel consists of the endosperm tissue of the seed containing the stored food material and the developing embryo (germ). The 'shell' surrounding it is the testa. This must be removed before the kernel can be eaten.

Sources

Pine nut kernels are almost all obtained from wild forest trees. Piñons from the drier areas of southwest USA and Mexico form the largest wild source, but the Italian pignolia tree has been cultivated and protected in the Mediterranean region for centuries. Almost all the other species grow in upland mountainous areas but information about them is scanty (Table 1). The harvest of pine nuts is only a fraction of the production of cultivated nuts such as pecan, macadamia, walnut or filbert, and the crop is generally irregular, with bumper harvests occurring approximately every 5 years. Harvesting is chiefly by an itinerant labour force which moves into the pine forests during the autumn. Green cones are cut from the trees, dried in the sun and open to release the

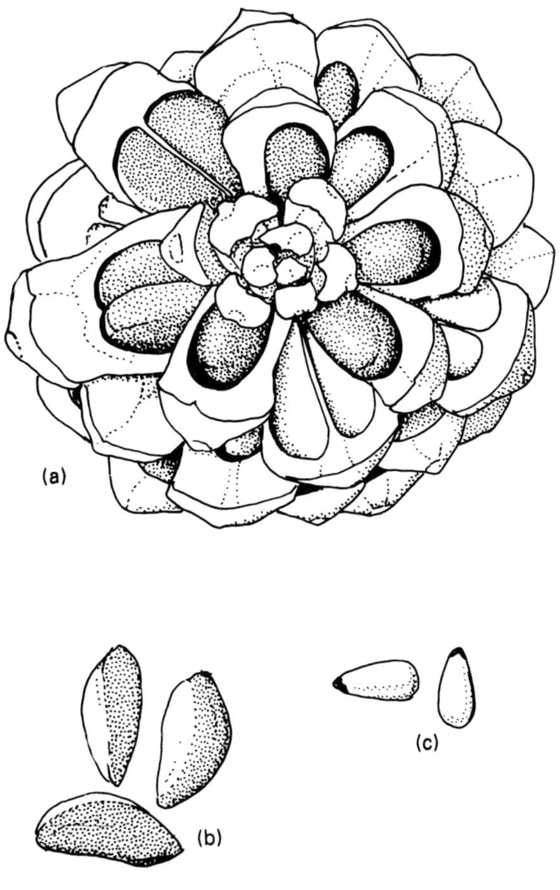

Fig. 1 Pignolia (*Pinus pinea*): (a) cone; (b) seeds; (c) polished kernels; (all natural size).

Table 1. Sources of pine kernels (*Pinus* species and their distribution)

Group	Latin name	Distribution
Piñon pines	*Pinus edulis*	Southwestern states of USA and Mexico
	P. monophylla	
	P. quadrifolia	
	P. maximartinezii	Mexico
	P. cembroides	
Stone pines		
Italian (pignolia)	*P. pinea*	Mediterranean basin and Turkey
Japanese	*P. pumila*	Northeast Asia, Siberia to Korea and Japan
Korean	*P. koraiensis*	Korea, southeast Siberia and Japan
Siberian	*P. siberica*	Western Russia to Siberia and Mongolia
Swiss	*P. cembra*	Alps and Carpathian Mountains
Chilgoza pine	*P. gerardiana*	Eastern Afghanistan to northern India and Pakistan

Table 2. Comparison of the nutritional value of pine kernels and some commercial nuts

	Food content		
Type of nut	Protein (%)	Fat (%)	Carbohydrate (%)
Colorado piñon, *Pinus edulis*	14	62–71	18
Single leaf piñon, *P. monophylla*	10	23	54
Mexican piñon, *P. cembroides*	19	60	14
Parry piñon, *P. × quadrifolia*	11	37	44
Digger pine, *P. sabiniana*	30	60	9
Pignolia pine, *P. pinea*	34	48	7
Siberian stone pine, *P. sibirica*	19	51–75	12
Chilgoza pine, *P. gerardiana*	14	51	23
Pecan, *Carya illinoensis*	10	73	11
Peanut, *Arachis hypogaea*	26	39	24
English walnut, *Juglans regia*	15	68	12
Almond, *Prunus dulcis*	21	54	7
Brazil nut, *Bertholletia excelsa*	16	69	8

Data from Lanner RM (1981); Botkin CW and Shires LB (1948). Typical energy value per 100 g = 556 kcal

seeds, or the seeds are collected from beneath the trees. The collection and later preparation can at best be considered as a cottage industry. The kernels of only two or three of the piñon pines and the pignolia pine are important in commercial trading on a large scale. They may be brought in health food shops or delicatessens, but are considered expensive, £12·50 per kg in the UK.

Kernels of most of the other pines may be found on sale in local markets near the site of collection. There are no reliable statistics on yields of these wild crops, but Little (1938) estimated that about a million kilograms of Colorado pine nuts were collected annually in southwest USA. This quantity has probably not changed a great deal since then.

Storage and Preparation

After harvesting nuts should be kept unshelled, in a dry, cool, well-ventilated place in cloth or paper bags. Tannins in the shell and in the seed coat of the kernel may function as antioxidants in preserving the fat and oil in the nuts. Fresh kernels stored in closed containers soon go mouldy or rancid, but after drying they can be safely preserved and have excellent keeping qualities for up to 3 years. They freeze well when fresh. Shelling of the nuts can be done domestically by lightly crushing the nuts on a cloth with a rolling pin. Those of the thin-shelled, single-leaf pine can be easily removed with the fingers. The thicker Colorado piñon, stone pine and pignolia nuts must be cracked mechanically to release the kernels. ***See*** Antioxidants, Natural Antioxidants; Oxidation of Food Components

Although the kernels can be eaten raw, roasting is necessary to bring out their full flavour. They may be roasted with or without the shell, the time depending on shell thickness and moisture content.

Composition and Nutritional Value

The richness and flavour of pine kernels is well known and they rank highly among all other nuts in food value. There are, however, considerable differences in the properties of the kernels of different species. A comparison of the nutritional value of pine kernels and some commercial nuts is provided in Table 2.

Refuse and Wastage

When compared with other nuts, piñons have low percentages of shell waste or refuse. Shell thickness varies considerably: 30–35% of seed weight of the very-thin-shelled single-leaf piñon is made up of shell, whilst the thicker-shelled Colorado pine has a waste factor of 42%. Mediterranean pignolia has a particularly thick shell which must always be removed before it is sold. Thus pine kernels of piñons constitute some 58–70% of the edible portion. The average is lower than that found in all other commercial nut crops except peanuts.

Protein

The average protein content is about 15% for piñon nut kernels and about 34% for pignolias. It is higher than for pecan nuts and about the same as that for English walnut and Brazil nut. It is only much exceeded by peanuts. A study in Yugoslavia has shown that kernels of pignolias are richer in proteins than pork and goose meat. Kernel protein has a digestibility value almost as good as that of beef and is distinctly better than almost all other nuts. It is near the proportion required for a balanced diet. Colorado and single-leaf kernels are especially rich in tryptophan and cystine. ***See*** Protein, Requirements

Fat (Lipid) and Oil

Piñon pine kernels average about 60% of fatty materials. This is lower than pecan, English walnut and Brazil nut, which approximate to 70%. Pignolia kernels in Yugoslavia are reported to contain 48% of lipid, which is again higher than for fat pork (37%) and goose meat (44%) (Kaic, 1985). As Table 2 indicates, the Siberian stone pine is also very rich in fats and oil. This is processed commercially for the production of cooking oil, but no published details of production methods or quantities are available. The fats of piñons, particularly Colorado and single-leaved pine, contain monounsaturated oleic, and polyunsaturated linolenic and linoleic acids. Budzynska (1964) reports that pignolia kernels from the Mediterranean region contain up to 50% of linoleic acid. Commercially produced pine kernels studied contained 46·40 g of fat by weight, of which 6·12 g is saturated (fatty acid not named). ***See*** Fats, Requirements

Carbohydrate

The average oily piñon kernel of Colorado pine contains 19% carbohydrate, but this can rise to as high as 54% in the more starchy kernels of the single-leaf pine. In the Parry pine it is 44%, but the former is much preferred. ***See*** Carbohydrates, Requirements and Dietary Importance

Other Substances

Pine kernels are extremely rich in phosphorus (6040 mg kg^{-1}) which is about the same as for soya bean, and in iron (53 mg kg^{-1}). They also contain significant amounts of vitamin A, thiamin, riboflavin and niacin. ***See*** individual nutrients

Uses

In the past, pine kernels were a staple in the diet as well as a subsistence food but, because of their enormous versatility, they now hold a high profile in modern cuisine among people of the developed nations, particularly in the USA, Europe, Asia and North Africa. As roasted nuts they are very widely used in the preparation of soups, sauces and dressings, and may be served as a garnish for fish entrées, in mixtures with cooked meat and as part of side-dishes with rice. They form suitable ingredients in cakes, puddings and biscuits, and as a garnish for ice cream. They are also highly appreciated in fruit desserts and vegetable salads. The kernels are so nutritious that a number of recipes have recently appeared under the label of 'Backpackers Friends'. These products bear such names as Piñon Pemmican and Granola, and may be carried as bars in the rucksack as emergency rations or snacks. Pine kernels will probably become even more popular as the rise in vegetarianism continues.

Bibliography

Botkin CW and Shires LB (1948) *The Composition and Value of Piñon Nuts.* Bulletin 344, New Mexico Experiment Station.

Budzynska J (1964) Note sur la composition en acides gras de l'huile des noix de *Pinus pinea*. *Revue Francaise des Corps Gras* 11(3): 143–145.

Kaic M (1985) Prehrambenotenoloske osobine sjemenki oraha (*Juglans regia*), crnog oraha (*Juglans nigra*) i pinije (*Pinus pinea*). *Sumarski List* 109 (7–8): 325–328.

Lanner RM (1981) *The Piñon Pine.* Nevada: University of Nevada Press.

Little EL (1938) *Food Analyses of Piñon Nuts.* Research notes, Southwestern Forest and Range Experiment Station, Arizona.

B.T. Styles
University of Oxford, Oxford, UK

PINEAPPLES

The pineapple (*Ananas comosus* (L.) Merr.), botanically a member of the ornamental Bromeliaceae family, originated in tropical South America but is now widely grown in all tropical and subtropical areas of the world. In Spanish-speaking countries the fruit is known as *pina*; in Portuguese-speaking countries as *abacaxi*; as *ananas* in Dutch- and French-speaking former colonies, and as *nanas* in Southern Asia. More than $4 \cdot 5 \times 10^6$ t, both fresh and canned, are marketed worldwide each year from at least nine major countries (see Table 1), with the major cultivar by far being the large juicy-fruited smooth-leafed cultivar Smooth Cayenne. Wild pineapple varieties still grow in the tropical savanna of South America but most have small, seedy, fibrous fruits. ***See*** Fruits of Tropical Climates, Commercial and Dietary Importance

The Plant, its Appearance and Physiology

The pineapple is a perennial, monocotyledonous, xerophytic plant, up to 1·5 m high, of herbaceous, lily-like habit, but with tough, spiny tipped leaves that are waxy on the upper surface and possess a fragile dusty bloom on the underside. The leaves in all but a few cultivars, such as the important Smooth Cayenne, also have numerous formidable barbs along the edges, which make cultivation hazardous. In all varieties the concave leaves channel any precipitation into the plant centre for absorption by spongy leaf tissue and roots. Other features which enhance the shallow-rooted plant's adaption to low rainfall include leaves that do not wilt, and its crassulacean acid metabolism (CAM), in which the stomata open at night to take up carbon dioxide rather than in the day, greatly reducing water loss. Malic acid is accumulated during the night and is decarboxylated during the day.

The fruit of the pineapple, botanically a sorosis or a syncarp, comprises spirals of fused fleshy fruitlets radiating from a fibrous but succulent core and topped by a leafy 'crown' or 'top', an extension of the peduncle (central plant stem) of the plant which is also used for commercial reproduction if available (Fig. 1). The commercial fruit is normally seedless, because of its requirement for cross-pollination (i.e. pollination with a variety of the same species which is substantially genetically different) to set seeds. In the absence of humming birds, reportedly the natural pollinator, small hard seeds, 1–2 mm across, will occasionally form if the plant is growing near other varieties. The plant produces one fruit 14–24 months after planting, but two or more vegetative shoots (suckers, slips) subsequently produce additional 'ratoon' crops. Fruits become smaller with successive ratoons as a result of overcrowding or disease, and usually a maximum of two ratoon crops are commercially viable.

Because pineapple plants cannot tolerate frost or prolonged cold, production is restricted to coastal or near-coastal areas of low or moderate elevation. They tend to thrive on tropical and subtropical islands, where the surrounding water mass maintains a more ideal constant temperature and moderate humidity.

Table 1. Major commercial pineapple production and consumption ($\times 10^3$ t) for 1985

	Production	Export	Processed	Import
USA	540[a]	7	399	
Mexico	350	13	87	
Formosa	121	1	30	
Thailand	1800	6	536	
Philippines	903	149	759	
Malaysia	177	13	152	
Ivory Coast	275	175	81	
South Africa	223	3	195	
Australia	123	—	105	
Total	4512	369	2499	
USA				54
Canada				11
Japan				129
EEC				190

EEC, European Economic Community.
[a] Presumably all Hawaiian; mainland USA produces negligible quantities.
Data from Naville R (1987) Commerce mondial de l'ananas. *Fruits* 42 (1): 25–41.

Varieties

The Cayenne variety (of which Smooth Cayenne is a cultivar) is one of five recognized varieties of *A. comosus*. (A cultivar is a plant selected from, and genetically very close to, a particular variety; a variety is a species originating from a particular geographical location.) The other four – Spanish, Queen, Pernambuco and Perolera – are more disease-resistant but are usually only grown where the more productive Smooth Cayenne variety cannot be economically produced. Under humid tropical conditions the Smooth Cayenne

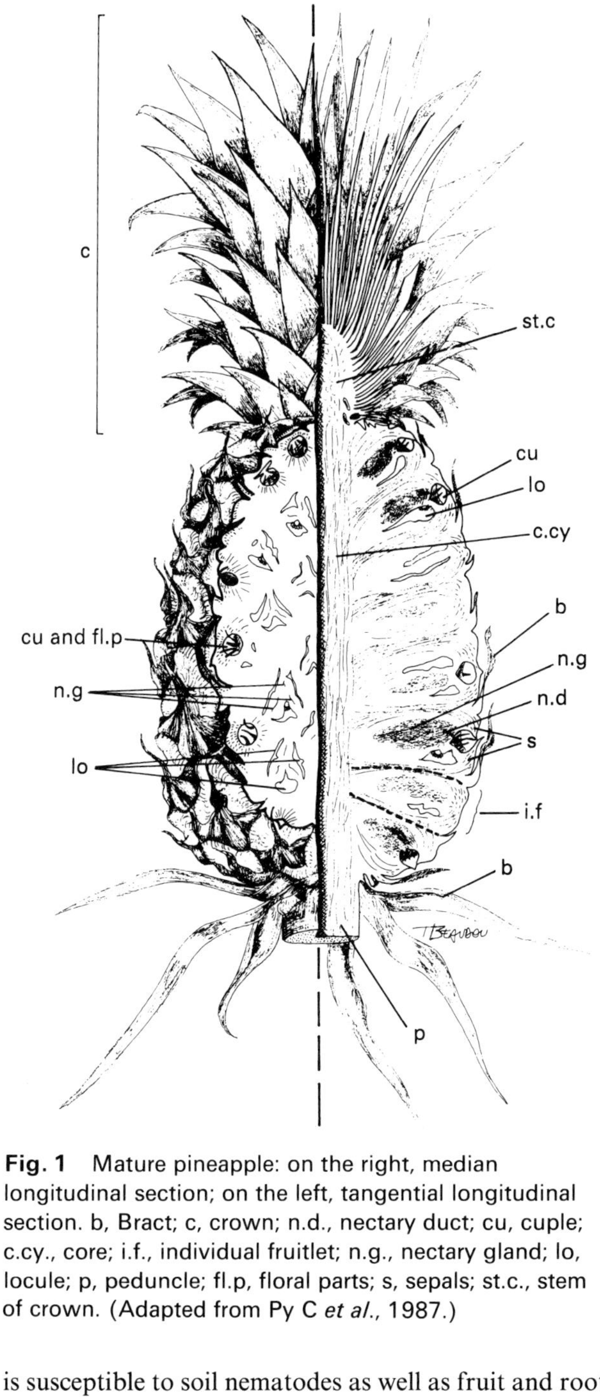

Fig. 1 Mature pineapple: on the right, median longitudinal section; on the left, tangential longitudinal section. b, Bract; c, crown; n.d., nectary duct; cu, cuple; c.cy., core; i.f., individual fruitlet; n.g., nectary gland; lo, locule; p, peduncle; fl.p, floral parts; s, sepals; st.c., stem of crown. (Adapted from Py C *et al.*, 1987.)

is susceptible to soil nematodes as well as fruit and root disease, and the large fruit has a tendency to lodge because of its weight. The five varieties are compared in Table 2. Smooth Cayenne fruits are the largest and most juicy, and their fruitlets usually possess a flat surface, allowing a thinner peel and substantially better flesh recovery than varieties having 'deep set' fruitlets. Other varieties tend to have a more aromatic flavour, and a crisper or drier flesh. Numerous cultivars of all groups have been selected in different production areas. Cross-breeding and selection have been undertaken to produce an improved cultivar, combining the fruit size and quality of Smooth Cayenne with the disease resistance of the other varieties, but it has been found that it is impossible to fix and refine desirable attributes by cross-breeding because of the self-incompatibility of similar genotypes. After nearly 50 years of attempts at improvement, little success has been achieved. The Smooth Cayenne variety of modern commerce is virtually the same as that originally selected by native Americans.

Composition and Nutritional Value

The pineapple fruit is typically a supplementary food rather than a staple. The pineapple's very sweet and sour taste, its mild aromatic flavour and firm succulent flesh in a large and attractive form makes it somewhat unique among foods. While very palatable by itself, it is equally used as a flavouring component, both cold and hot, making many other less attractive but nutritionally sound foods more edible. The dietary composition of Smooth Cayenne, both fresh and canned, is shown in Table 3. The nutrient profile of the pineapple is, in general, similar to many other fruits, containing high levels of carbohydrate and low levels of fat and protein. Dietary fibre constitutes about 14% of the dry matter and, as with most fruits, can be incorporated into a cholesterol-lowering diet. Its vitamin C content is about half that in citrus, and the level of retinol equivalents (3 per 100 g) is low compared to those of pawpaw (papaya) (153) and mango (356). ***See*** individual nutrients

The total sugar content is the major quality determinant of the fruit for both fresh and canned markets (see *Factors Affecting Pineapple Eating Quality*, below) and in commercial practice sugar content is often regularly monitored, using a refractometer calibrated in degrees Brix which, in pineapples, is virtually identical with percentage total sugar. Fresh Smooth Cayenne fruit above 15° Brix would be considered of excellent eating quality by most tasters and fruit of 9–12° Brix of marginal quality; however, taste preferences towards Smooth Cayenne of a particular Brix level vary considerably between individuals and also between ethnic groups. In canned pineapple the sugar content of the fresh fruit makes less difference to consumer quality.

Variability of Organoleptic Parameters

Wide variations occur in the total sugar content and titratable acidity in fresh pineapples. Not only is the bottom of the fruit always about a 3° Brix higher than the top but, more importantly, there is wide variation between samples. For example, the 8% total sugars from Queensland fruit in Table 3 is very low compared to fruit from Hawaii or Ivory Coast and is either from immature summer fruit (average mature fruit is 15–18° Brix) or is from winter-grown fruit (6–11° Brix). Fruit grown

Table 2. Characteristics of the major pineapple varieties

	Cayenne[a]	Spanish	Queen	Pernambuco[b]	Perolera
Main production	Worldwide	Caribbean Malaysia	South Africa Australia	Brazil Venezuela	Colombia Ecuador, Peru
Weight (kg)	1·8–3·5	1–1·8	0·5–1·2	1–1·8	1·8–3·5
Shape	Cylindrical	Spherical	Conical	Conical	Cylindrical
Ripe skin colour	Orange-yellow	Reddish yellow	Bright yellow	Green-yellow	Bright red-yellow
Ripe flesh	Near translucent	Near translucent	Opaque	Opaque	Opaque
Flesh fibrosity	Nonfibrous	Fibrous	Crisp	Nonfibrous	Crisp
Flesh colour	Pale yellow	Near white	Bright yellow	White-yellow	Bright and pale yellow
Flavour	Sweet and acid	Spicy	Aromatic	Low acidity	Low sugar and acidity
Wilt resistance[c]	*****	*	**	**	**
Nematodes[c]	*****	*	**	**	**
Uses	Fresh, canned Major export	Fresh Minor export	Fresh Minor export	Fresh Minor export	Fresh Minor export

[a] Includes Smooth Cayenne, which is but one variety of the Cayenne group.
[b] Other name, 'Abacaxi', but this should be avoided as it means 'pineapple' in Portuguese.
[c] *=resistant, *****=very susceptible.
Based on data from Py C *et al.* (1987).

under cooler conditions is lower in sugar and higher in acid, although the sour taste of winter-grown pineapples is caused by the low sugar content rather than the high acidity.

Post-harvest Changes

The post-harvest changes in various parameters in fruit held near 20°C during 15 days after harvest are illustrated in Fig. 2. Total sugar concentration and eating quality remain relatively constant after harvest; acidity and carotene content increase moderately, while the shell colour (degree of yellowness of the skin) and ester concentration increase substantially.

Skin Colour and Ripeness

Although the fruit normally changes from green to yellow as the fruit ripens on the plant, skin colour is a poor indication of palatability in Smooth Cayenne pineapples grown under commercial practices. Green-skinned fruits can taste quite acceptable. Such 'green-ripe' pineapples can either be physiologically 'unripe' (meaning physiologically not fully developed) but have sufficient sugar to taste acceptable, or their flesh may be prematurely ripened before the skin as a result of specific climactic (i.e. climate, nothing to do with the nonclimacteric fruit physiology) conditions during growth. This latter type of 'green-ripe' fruit tend to taste insipid and the condition is now regarded as a physiological disorder. On the other hand, attractive yellow-skinned fruit can taste sour (see below). ***See*** Ripening of Fruit

One way of selecting intact Smooth Cayenne fruit of uniform ripeness is on the basis of fruit specific gravity. Fruits that float in water are less ripe than those that sink. Fruits that sink in a 3% common salt solution are very ripe. This method works well for Queensland and Hawaiian fruit but reportedly not South African. The difference has not been explained.

Fresh Fruit Selection

Probably as a result of the variability of eating quality, there is a degree of folklore applied in the determination of when a harvested pineapple is 'ripe', with potential purchasers plucking crown leaves, or squeezing, smelling and tapping the fruit. Leaf plucking is completely useless, but fruits that feel heavy for their size will be riper than lighter fruit of the same size. However, the post-harvest nonripening character of pineapple means that once the fruit is picked the eating quality is fixed. There is no advantage for consumers in waiting for the harvested fruit to yellow, as delay only results in the fruit losing volatile flavours and going 'stale'.

Factors Affecting Pineapple Eating Quality

At its best, Smooth Cayenne is a most delightful fresh fruit to eat, but the eating quality of fruits produced from different seasons and growing areas are extremely *variable*, much more so than other pineapple varieties. The eating quality of Smooth Cayenne is probably the most variable of any commercial fruit. It is difficult, if not impossible, even for an expert, to predict the eating quality, even when the fruit is cut. The eating quality or

Table 3. Composition of 100 g of pineapple flesh

	Canned[a] heavy syrup, drained	Canned[a] pineapple juice, drained	Fresh[a], peeled	Fresh[b] peeled
Proximate				
Water (g)	74·8	84·6	86·0	80–86
Energy (kJ)	350	188	158	—[c]
Protein (g)	0·8	0·7	1·0	0·2
Fat (g)	0·0	0·0	0·1	0
Carbohydrate, total (g)	20·4	10·2	8·0	—
Sugars, total (g)	20·4	10·2	8·0	10–18
Starch (g)	0·0	0·0	0·0	—
Ash (g)	0·2	0·3	0·5	0·3–0·6
Cholesterol (mg)	0	0	0	—
Acids, total (g)	—	—	—	0·5–1·6
Total nitrogen (mg)	—	—	—	45–120
Pigments, xanthophylls (mainly carotenoids) (g)	—	—	—	0·2–0·3
Dietary fibre (g)	1	2	2	—
Minerals (mg)				
Sodium	1	4	2	14
Potassium	76	140	180	11–330
Calcium	5	6	27	3–16
Magnesium	10	14	11	10–19
Iron	0·3	0·5	0·3	0·05–0·3
Zinc	0·2	0·3	0·2	0
Vitamins				
Retinol equivalents (μg)	3	3	4	—
Retinol (μg)	0	0	0	—
β-Carotene equivalents (μg)	18	18	25	—
Thiamin (μg)	40	30	40	69–125
Riboflavin (μg)	30	30	30	20–88
Nicotinic acid equivalents (mg)	0·3	0·3	0·3	—
Nicotinic acid (mg)	0·2	0·2	0·1	0·2–0·3
Vitamin C (mg)	15	14	21	3–25

[a] Queensland fruit; data from *Composition of Foods Australia*. Australian Food Publishing Service, Canberra. 1989.
[b] From Py C *et al.* (1987); data presumably based on fruit from Ivory Coast.
[c] Data not given.

flavour of pineapple is almost totally dependent on the percentage of sugar in the flesh rather than the degree of visual ripeness. The sugar content is strongly affected by the temperature and light intensity during the last 3–4 weeks of growth, and, as a consequence, the sugar content in the fresh fruit is greatly affected by the season, climate, the degree of maturity at harvest and, to a lesser extent, the farm production methods.

Smooth Cayenne grown in equatorial regions generally have consistent and excellent eating quality. In the subtropical regions, where significant quantities of the Smooth Cayenne of fresh commerce are grown, the eating quality is most often very variable as a result of the daily and seasonal variations of temperature. Many subtropical production areas have developed because they are closer to areas of major urban populations and consumption, but in these areas the seasonal inconsistency of flavour is an important problem, particularly in fresh-market pineapples.

Fruit Physiology

A unique feature of the pineapple is that flowering (with subsequent fruit set) is controllable by growth hormones over an extremely wide range of plant maturity. Modern commercial production practices can regulate date of harvest to fall virtually any day of the year; thus fruit production can be precisely programmed.

Pineapples are nonclimacteric fruits and do not get better to eat (i.e. do not ripen) after harvest in spite of any subsequent colour change (see Fig. 2). Other

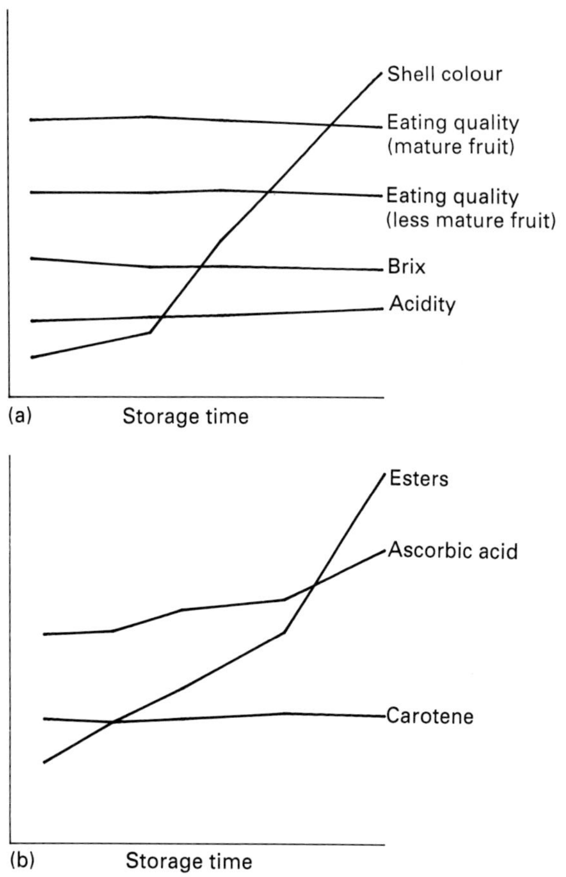

Fig. 2 Post-harvest changes in chemical and sensory parameters over 15 days. Eating quality data from the author (unpublished). Other data from Singleton (1959) as quoted by Dull GG (1970).

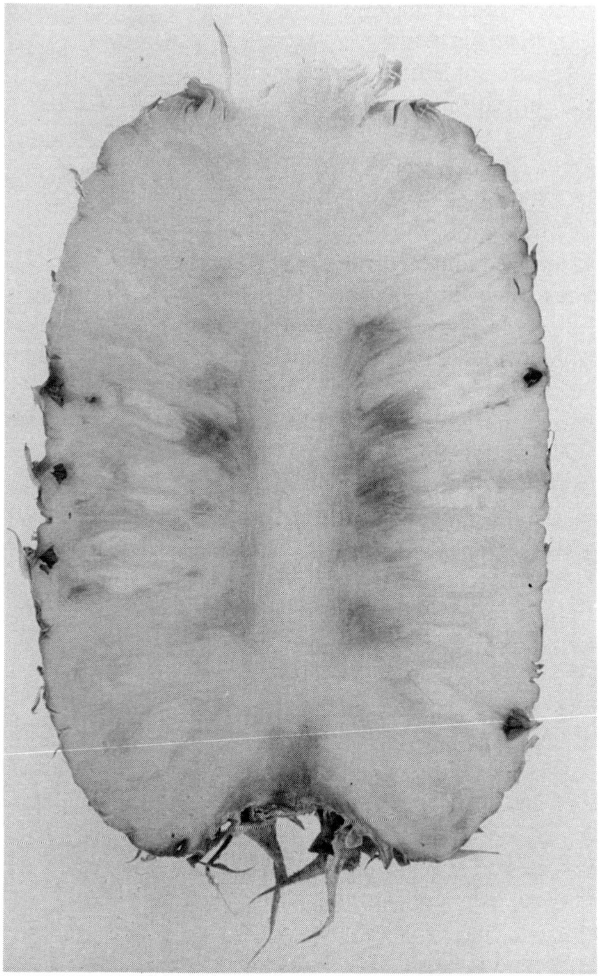

Fig. 3 Blackheart or internal browning in pineapple, a serious disorder in fresh-market pineapples which can develop following refrigeration, but which also naturally occurs in subtropical production areas. It is a physiological disorder (i.e. not a pathogenic disease), known as chilling injury and caused by prolonged exposure to temperatures below 21°C.

members of this same group of fruits include oranges, strawberries, grapes and cherries. The absence of starch in the fruit is often cited as the cause of the nonripening post-harvest behaviour but many climacteric fruits also contain no starch. The physiology involved needs to be better clarified.

Handling, Storage, Refrigeration, and Chilling Injury

Handling

In spite of its tough appearance, Smooth Cayenne is very susceptible to injury and needs to be handled with care. The fruits will readily bruise if they fall more than 20 cm and, once bruised or injured, a ubiquitous fungal disease caused by *Ceratocystis paradoxa* (*Thielaviopsis paradoxa*), which generally spreads quickly throughout the fruit, can readily develop. Commercially, fruit are commonly treated with fungicides such as thiabendazole, benomyl, or triadimefon. In contrast to Smooth Cayenne, Queen fruits are very resistant to fruit rots and in many ways are ideal fresh-market fruits. They keep well without refrigeration and, if free from developing blackheart, have a much longer shelf life than Smooth Cayenne. ***See*** Fungicides

Storage and Refrigeration

Although pineapples can keep reasonably well without refrigeration, particularly if the humidity is kept high to prevent desiccation, refrigeration is necessary to maintain a glossy, fresh-looking skin and to reduce the incidence of rots. However, pineapples do not out-turn well if kept refrigerated for longer than 2 weeks. They can suffer from 'chilling injury', a physiological injury caused by cool temperatures, resulting in disruption of the cellular enzymic processes. It is a disorder characteristic of many tropical fruits which will often develop within 2–3 days after removal from refrigeration. In pineapples, chilling injury is manifest most commonly by the fruit developing brown spots adjacent to the core, which are sometimes called endogenous brown spots. With increasing severity, the spots darken further, and in severe cases the complete centre of the fruit can become black, giving rise to the other name for the disorder, blackheart (Fig. 3).

Chilling injury is a constant threat when pineapples are shipped under refrigeration, and intermittent heavy losses occur. Storage for 3–12 days at temperatures as high as 21°C have been shown to induce blackheart, but injury incidence is unpredictable, possibly dependent on the intensity of available sunlight during the 3–4 weeks prior to harvest. Temperatures of 15–18°C induce the greatest injury but, as the fruits are usually shipped at 8°C for 7–10 days from tropical production areas, they are usually consumed before any injury develops.

Fruits from the Queen group are very susceptible to chilling injury but a Hawaiian variety, known as 53/116, appears to be entirely resistant. (This hybrid variety lacks characteristic pineapple flavour, however, and is not widely planted.) Dipping the fruit in a carnauba-based wax, or applying a lipid-carboxymethyl cellulose coating, can control blackheart. These skin coatings function by reducing the oxygen concentration in the flesh, necessary for the melanin pigment to develop, but the anaerobic flesh can develop an alcoholic flavour. Heating the fruit to 40°C for 24 h either immediately before or after cool storage also reduces blackheart and improves pineapple eating quality, but is not commercially used because of logistical limitations.

Pineapples that are cool-stored for prolonged periods can also develop other symptoms of chilling injury such as flaccid watery flesh, skin necrosis, crown wilting and crown necrosis, but fruit with such injuries are often already damaged by internal chilling. Overcoming chilling injury in fresh-market pineapples is currently under investigation.

Controlled-atmosphere Storage

Controlled-atmosphere (CA) storage does not prolong the life of the cool stored pineapples and is not used for international shipping. ***See*** Controlled Atmosphere Storage, Applications for Bulk Storage of Foodstuffs

Field Blackheart

Pineapples from cooler production areas are particularly susceptible to blackheart, where it can develop in the field following periods of cool overcast weather. Because the disorder shows no external symptoms, affected fruit cannot be identified. In Queensland and Hawaii pineapple canneries close or restrict intake during such periods and thus avoid excessive losses.

Pineapple Products

Canned pineapple is a commodity of major international trade; figures showing exports and imports are shown in Table 1. Pineapples are also processed into a number of other products other than the canned fruit. Pineapples are often used as a component of commercially marketed fresh fruit salad, either fresh or frozen. Less ripe fruits freeze best as they have firmer flesh. Soft ripe fruits can develop off-flavours when frozen as well as turning out very flaccid. Fried pineapple pieces are commonly used in Asian cuisine. Dried pineapple pieces are easily prepared and are being sold in increasing quantities as they retain a more attractive appearance than other fruits that strongly darken when dried. Sliced thin and dried, pineapple 'chips' are also sold as a snack food. Glacé pieces are readily prepared from either the fresh product or, more easily, from canned rings. Pineapple juice is a major item of commerce, as fresh or frozen, full strength or as a concentrate. Reasonably high in carbohydrates, fermented juice is used to make vinegar. It does not make a good-quality wine but, fermented and distilled, it makes a potent, if somewhat disreputable, alcoholic spirit, available in several Pacific island countries. Cannery waste (crushed skins, etc.) makes a good cattle feed supplement, while old plantation fields are often let out to cattle. Pineapple leaves are used to make cloth or rope in both Taiwan and the Philippines, while use of the whole plant as a source of energy is under current investigation. ***See*** Canning, Quality Changes During Canning; Freezing, Structural and Flavour Changes

Bromelain

Pineapple fruits contain a proteolytic enzyme, bromelain, which can constitute nearly half the measured protein in the fruit. Much higher concentrations of the basic iso-enzyme (the fruit form is acidic) exist in the stem. Stem bromelin is commercially extracted and sold, as a buff-coloured powder, slightly soluble in water and practically insoluble in alcohol, chloroform and ether. Bromelin is used in the pharmaceutical industry for digestive and anti-inflammatory products, in the manufacture of cattle feed and to 'chill-proof' beer (to prevent formation of a haze of proteinaceous material which can form when brewed products are refrigerated). Some recent pharmacological reports indicate that bromelin may have value in modulating tumour growth and blood coagulation, as well as debridement of third-degree burns.

Commercial Farm Production

Modern commercial culture of pineapples is generally capital intensive, requiring substantial land preparation, machinery for planting, boom (fertilizer) spray and harvesting, plus bulk equipment for cannery fruit and/ or harvester-packers for fresh market fruit.

By using a combination of plant hormones and appropriate planting practices, the harvest date, and the fruit size and quality can be preset and regulated as required. This makes modern pineapple production function with almost clockwork precision. On modern fresh-fruit farms a year-round cycle is maintained of weekly, if not daily, planting and harvesting. Combined with modern planting machinery, modern herbicides, and large mobile harvesting equipment, production of 1000–2000 t of fruit per man year is common.

Being parthenocarpic (seedless, but otherwise normal), pineapple progagation is vegetative and can be problematical as only limited planting material is available from any one plant. The best planting material, that producing the most uniform plants and fruits, is the fruit crown, but if the fruit is sold as fresh fruit (as opposed to canning) tops are not available, so that suckers and other vegetative material must be collected and used. Up to 80 000 pieces of planting material (either tops, suckers or slips) are required per hectare, planted in double rows. The time of planting and the size and type of planting material affects the date of harvest and fruit size, and is predetermined. Weeds can be readily controlled as the plant is resistant to several effective herbicides (e.g. diuron, fluazifopbutyl, ametryne) which are applied over the rows of growing plants with a boom spray. Ten to fourteen months after planting, a flower-inducing hormone (commonly the ethylene releasing compound Ethrel®, 2-chloroethyl-phosphonic acid) is applied, and the treated plants uniformly produce a single, blue-petalled inflorescence about 8 weeks later. Fruit development occurs without further interruption, but as sunburn can be a serious injury, fruit may be individually covered (bagged) or sprayed with a reflective bentonite clay suspension.

Harvesting pineapples at the correct stage of internal ripeness is important as it is the major factor that determines the fruit quality at consumption. As skin colour is an inconsistent guide to the eating quality, pickers have to use past experience and market demands.

Sodium naphthaleneacetic acid (NAA) is sometimes applied as a spray on the fruit to increase fruit size by delaying degreening (becoming yellow). Adverse effects on fruit quality from using NAA are reportedly minor but are under current debate. Ethrel® is also sometimes sprayed on both fresh and cannery crops to increase flesh yellowness and to 'accelerate ripening'. While very effective and cost-saving on cannery fruit, its use on fresh-market fruit has recently been shown to be of questionable value as the variability of the eating quality is greatly increased.

Bibliography

Bartholomew DP and Paull RE (1986) Pineapple. In: Monselise SP (ed.) *Handbook of Fruit Set and Development*, pp 371–388. Boca Raton, Florida: CRC Press.

Dull GG (1970) The pineapple: general. In: Hulme AC (ed.) *The Biochemistry of Fruits and their Products*, vol. 2, pp 303–324. London: Academic Press.

Py C, Lacoeuilhe JJ and Teisson C (1987) *The Pineapple: Cultivation and Uses*. Paris: GP Maisonneuve et Larose.

Smith LG (1988) Indices of physiological maturity and eating quality in Smooth Cayenne pineapples. *Queensland Journal of Agricultural and Animal Sciences* 45 (2): 213–228.

Lyndall G. Smith
Queensland Department Primary Industries, Hamilton, Australia

PITUITARY HORMONES

See Hormones

PLANTAINS

See Bananas and Plantains

PLANT DESIGN

Contents

Basic Principles

The primary concerns in food plant design that distinguish it from the design of other manufacturing facilities are protection of the food from contamination so as to protect the consumer and protection of the facility from the harsh conditions encountered in cleaning and sanitizing. There is great diversity within the food industry regarding the nature of raw materials, processes, products and packages, and so individual food plants can differ greatly in the ways that they protect foods and themselves. This article discusses some of the general principles that apply widely.

Typical Hazards to Food

The most common hazards against which a food plant must protect include pathogenic microorganisms that could cause disease or even death in consumers; foreign matter that could cause injury or, at least, reduced acceptance; spoilage, which causes economic loss and consumer dissatisfaction; and other forms of quality deterioration, which have less obvious, but still significant, economic impacts such as loss of market position and profit margin. ***See*** Contamination, Types and Causes; Spoilage, Chemical and Enzymatic Spoilage; Spoilage, Bacterial Spoilage; Spoilage, Moulds in Food Spoilage; Spoilage, Yeasts in Food Spoilage

In addition to the defensive function of protecting against these hazards, a properly designed food plant must accomplish its stated purpose. That is, a biscuit bakery must produce biscuits at the intended rate, cost and quality; a meat plant must kill and process the appropriate animals; an ice cream plant must make the desired volume and quality of product, and so forth. This is a significant aspect of food plant design, because there is great variety in the processes used, even for similar products; the technologies used in processes change frequently; and the purpose of a given plant is quite likely to change several times over its useful life. The general hazards will be considered first, and the process implications will be discussed separately.

Pathogenic Microorganisms

Many different bacteria occur naturally in food raw materials and can live easily in food products. Most food-preservation processes are designed to reduce greatly the population of such bacteria and other microorganisms that can cause disease or spoilage. However, foods are vulnerable to contamination after they have been processed. Such contamination can occur by human contact; through vermin, such as insects, birds or rodents; and from the environment. Proper food plant design defends against all such threats. ***See*** Insect Pests, Insects and Related Pests; Insect Pests, Problems Caused by Insects and Mites; Preservation of Food

One fundamental principle is to avoid contact between raw and processed foods – directly or indirectly – by avoiding the use of common equipment, storage space or containers. It also is good practice to isolate possibly contaminated packaging material from ingredients in processing areas. This may require special provisions for transferring materials to clean containers or for cleaning packages such as bags.

People coming into direct contact with foods must be in good health and practise good personal hygiene. Good design practices include there being no direct entry from toilets to processing areas, convenient hand-washing stations, and equipment designed to minimize direct human contact. Food plant workers often wear special uniforms, should always have hair covering (hair is a particularly unsanitary contaminant), and, when required, should use disposable gloves.

Food plants are designed to exclude birds and rodents by use of screens, tightly shutting doors, proper sealing of joints, and maintenance of clear areas at the perimeter both inside and out. Rats and mice travel along walls at night because they have poor sight and find their way by feel. If they encounter objects in their path, they are deflected into the main part of a plant, where they have good chances to contaminate food and equipment. Traps should be placed along paths near walls and checked regularly.

Space along outside walls is kept clear of grass and shrubs to permit easy inspection for small openings and to allow easy trapping of rodents. Also, grass and shrubs can attract insects. Insects are attracted to ultraviolet lights, which can be used to lure them to electric traps, which are often placed near entrances to prevent invasion of the main part of the plant. Also, proper use of outdoor lighting can attract insects away from a plant by using lamps which give off ultraviolet radiation (mercury vapour) near the perimeter of the property. Complete removal of infestation may require chemical fumigation, but the effective chemicals are highly toxic and pose hazards of their own to foods and workers. Increasingly, food plants are designed to be heat treated to kill insects. This requires careful design of plant components to withstand the heat. ***See*** Fumigants

Finally, contamination can be transferred through the air. To minimize this threat, all incoming air should be filtered, and air from areas where raw materials are stored should not be used where processed food is unprotected. Microorganisms require moisture to grow, so it is good practice to keep food plants as dry as possible. Trench drains provide places that are hard to clean where bacteria can thrive and be transferred to the air; it is better to use hub drains which reduce the opportunity for such problems. (Trench drains may be unavoidable where large quantities of water are used in food processing, but even in such cases, it is better to reduce the water use!) ***See*** Water Activity, Effect on Food Stability

All spaces within a food plant should be ventilated to reduce the build-up of humidity which encourages growth of moulds and yeast. This includes the head-space of large storage tanks, especially for syrups; the space above false ceilings; and refrigerated spaces, where humidity is especially high.

Foreign Matter

Many strange objects find their way into food products, usually by human error or equipment failure, but good plant design can prevent many incidents. For example, painted surfaces should be minimized in a food plant because paint can flake and contaminate food. Cold pipes should be insulated to minimize condensation, which can carry dissolved material or microorganisms into food. Corrodible materials should not be found in food plants, because corrosion products, such as rust, could contaminate the products. (In practice, this means that food plants use a lot of stainless steel and plastic.) ***See*** Corrosion Chemistry

There should not be glass in a food-processing area. This may mean using plastic in windows or partitions, or minimizing such openings. Use of wood should be minimized in food plants. Paints, floor finishes, cleaning materials, lubricants, pesticides and other chemicals used in food plant construction or maintenance must be approved for food use. In the United States, the regulatory authorities include the US Department of Agriculture (which regulates meat, poultry and seafood processing), the Food and Drug Administration (which regulates all other food processing), and the Environmental Protection Agency (which regulates pesticides). In Europe, the EC has similar agencies and regulations, but it is not safe to assume that materials and practices approved in one country or by one agency will be acceptable to another. ***See*** European Economic Community, International Developments in Food Law

Spoilage

Foods are often preserved by cooling or freezing, so parts of a food plant are often designed to maintain low temperatures. Failure to do so can cause loss by spoilage. This requires proper insulation, reliable power supply, and efficient mechanical design of refrigeration equipment. ***See*** Freezing, Principles

Moisture can cause spoilage by causing clumping in hygroscopic materials and by encouraging microbial growth. Thus control of humidity is usually important in food plants. This is achieved by cooling air to low temperatures to condense moisture or by absorption in concentrated salt solutions. One advantage of the absorption process is that the air is also cleaned. Totally dry air may not be desirable because it can cause static discharge, which can ignite dusts and can affect the behaviour of packaging materials, so humidification may be required in some packaging areas to achieve optimum performance.

Other forms of spoilage may be specific to the foods involved, including damage by light, temperature fluctuations, inadequate packaging, and extraneous odours or flavours. Proper plant design requires awareness of such issues and protection against them.

Quality

A significant impact on quality and consistency of food products is made by worker attention and care. Design of facilities to encourage worker enhancement of quality involves thoughtful use of colour and space, provision for attractive working areas, consideration of worker comfort and safety, and overall signalling of the importance of sanitation and quality in performance.

Hazards to the Plant

Most food plants are designed to be washed frequently with water and strong chemicals, such as caustic soda, acids and chlorine. Water alone can be destructive to materials and structures not intended to be wet. Thus most food plants are built from impervious materials, such as glazed tile, brick, concrete, fibreglass, stainless steel, and epoxy-painted masonry.

Good design requires the choice of economic materials of construction in light of their intended service. For example, brick or tile floors are common in food plants because they wear well and can be repaired when damaged. However, they are expensive because the materials and labour are costly. Monolithic polymeric coatings, usually filled with minerals, are cost-effective substitutes in many cases. Care must be taken in specifying such coatings to ensure that the temperature expansion characteristics are suited to the conditions, because they can be vulnerable to hot water and steam if not properly formulated.

Whenever possible, food plants should be maintained dry, as this minimizes microbial growth and corrosion. However, dry conditions can create dust, which is difficult to clean and poses an infestation and contamination hazard to food. To minimize dust, equipment must be tight, recirculating air should be filtered, and flat surfaces should be minimized. Also, to make dry cleaning easier, floor/wall intersections should be coved, column/floor intersections should be coved, structural members should be tubular or enclosed and capped, and flat beams and ledges should have slanted surfaces.

Concrete block should be thoroughly filled and sealed, and masonry joints should be continuous rather than staggered. This allows moisture to drain smoothly.

Concrete is a good material of construction for food plants because it is durable, easy to maintain and provides a smooth and impervious surface. Precast or poured-in-place roof decks are more expensive than steel bar joist supports, but are preferable because they have fewer surfaces to gather dust. Steel supports are acceptable in storage areas. However, it is well to remember that storage areas often become desirable expansion space in food plants. Thus, an overall master plan is essential before final decisions are made about materials and forms of construction.

Other General Considerations

It is rarely economically sound to construct more space than is needed in the immediate future, but it is critical to plan for future construction from the beginning. In this way, utilities such as boilers, electrical gear, and machine rooms can be located so they do not block logical expansion. Likewise, truck docks, offices and employee support facilities, such as lockers and lunch rooms, are best located where they need not be moved.

One of the dilemmas faced in plant design is where and how to receive raw materials and ship finished goods. Most plants rely heavily on trucks, but some raw materials and finished goods are shipped by rail, especially flour and grains. The location of a rail spur on a property will dictate much about plant siting. Likewise, it is good practice to separate truck traffic from employee and visitor automobile traffic, for safety and efficiency. Usually, entrance to a plant site is controlled, for security, and often trucks arriving and leaving are weighed. A staging area may be necessary to control trucks waiting for access to docks. Wherever trucks have access to the plant there needs to be an office for a clerk and a waiting room complete with toilet facilities for drivers, who generally are restricted from access to the main part of the plant. If shipping and receiving are on opposite sides of a plant, there need to be two such arrangements and two clerks, at least. Receiving of packaging material and containers is often arranged close to the point of use, so as to minimize storage and handling.

Usually, storage areas are adjacent to shipping and receiving docks. Some flexibility is achieved if these are close to each other, recognizing that raw materials and finished goods should not be stored together, but that space used for one purpose can be converted to another when conditions change.

Whether to build a single or multistorey facility is often an issue, especially when material handling might be assisted by gravity. Multistorey food plants were once common, including meat packing plants in which cattle and hogs walked up ramps to a killing floor at the top and left as roasts or hams from the bottom floor. Such a design is much less common today. Single-storey buildings are more flexible, permit better communication among workers, and can more easily be converted to other purposes if necessary. They are also less expensive to construct; however, they require more land, which can represent a significant cost in urban areas. For certain industries, especially breakfast cereals and pet foods, a processing tower may still be appropriate.

Additional considerations for food plant design include drainage, provisions for fire protection, water supply, liquid and solid waste handling, and impact on the neighbourhood. Obviously, provisions to remove storm water must be made, as standing water can be a nuisance and attractive to pests. In addition, storm water can become contaminated by improper plant operation and discharge could actually be illegal in some areas.

In remote areas, storage of fire protection water may be necessary, along with a pump and stand-by generator. If municipal water has insufficient pressure, a pump is necessary. Exact requirements are dictated by insurance companies and codes.

An adequate supply of potable water must be available to a food plant for use as an ingredient, for clean-up and for cooling and heating. Regulatory authorities require certification of the water composition and regular checks of its continued suitability. If a plant must supply its own water, for example from wells, it must perform regular analyses and maintain files to document the water's safety. *See* Water Supplies, Water Treatment

Liquid and solid wastes can represent significant costs in many areas as legal requirements become more stringent. Disposal of solids in landfills is increasingly difficult, especially for food wastes, because they decompose quickly to generate odours and potential leachate solubles. Many solid food wastes find use as animal feeds or as raw materials for other processes. Before a site is selected, it is necessary to investigate the availability of adequate options for waste treatment and disposal. Most food plants need some liquid waste pretreatment, such as screening, grease removal and, perhaps, primary settling, even if the discharge is sent to a municipal treatment plant. If adequate sewage treatment is not available, the plant may need to provide its own, which represents not only a significant investment but also an ongoing operating cost. *See* Effluents from Food Processing, On-site Processing of Waste; Effluents from Food Processing, Microbiology of Treatment Processes; Effluents from Food Processing, Disposal of Waste Water; Effluents from Food Processing, Composition and Analysis

Many food plants have successfully applied their liquid wastes to land, which may even be planted with pasture or other cover crops. This, obviously, requires substantial property, may not work well in very cold climates, and must be monitored so that it does not contaminate water supplies.

Finally, food plants interact with their neighbours in several ways, for example by emitting odours; by receiving odours and dust from other plants; by providing employment; by creating truck and automobile traffic; by purchasing materials locally; and by creating waste streams. Some interactions may be governed by local laws and codes; others may impact performance of the plant, negatively or positively, and even impact performance of the parent company. Proper design before construction should anticipate such interactions and attempt to reduce the negative and enhance the positive so far as possible.

In conclusion, food plant designs have many elements in common, regardless of the specific product and process, which arise simply because the plant produces food. While this article describes a few such elements, the application to any given case is a challenging exercise for specially trained and experienced architects and engineers, who, while rare, should be sought for consultation. *See* Cleaning Procedures in the Factory, Overall Approach

Bibliography

Gould WA (1990) *CGMP'S/Food Plant Sanitation.* Baltimore, Maryland: CTI Publications.

Imholte TJ (1984) *Engineering for Food Safety and Sanitation.* Crystal, Minnesota: Technical Institute of Food Safety.

Jowitt R (ed.) (1980) *Hygienic Design and Operation of Food Plant.* Westport, Connecticut: AVI.

Valentas KJ, Levine L and Clark JP (1991) *Food Processing Operations and Scale-Up.* New York: Marcel Dekker.

J. Peter Clark
A. Epstein and Sons International, Inc., Chicago, USA

Designing for Hygienic Operation

As with facility design, discussed in the previous article, process and equipment design has many elements specific to the food being processed and others which represent common principles. This article discusses a number of those common principles, types of equipment and common processes.

Overall Considerations

Equipment used in food-processing plants must be appropriate, effective, noncontaminating, easy to clean and easy to inspect. It must also be safe for the workers using it. In the United States, equipment used in meat, poultry and fish processing must be specifically approved in advance by the US Department of Agriculture (USDA). Equipment used in dairy processing should be listed in 3A standards (a voluntary industry standard-setting group), and equipment used in baking should satisfy Baking Industry Sanitation Standards Committee (BISSC) requirements. There are other standard-setting groups in other countries and for other industries.

In general, such standards address the earlier-mentioned general requirements in light of the specific conditions. For example, USDA emphasizes cleaning and inspection, because most meat processing equipment is cleaned daily. Dairy equipment is often cleaned in place, and so surface finish and absence of crevices is important. Baking equipment is less often cleaned with water, but is vulnerable to insect infestation. ***See*** Cleaning Procedures in the Factory, Overall Approach; Insect Pests, Insects and Related Pests

As a result of nearly universal sanitation requirements, much food equipment is constructed of stainless steel, usually type 304, but often 316 and occasionally some other alloy. The alloys differ in their weldability and resistance to stress corrosion cracking, which is accelerated by chloride ions found in many foods and cleaning materials. Stainless steels are nonmagnetic and so fragments which may occur due to wear or misuse are not removed by magnets. Metal detectors, which function by measuring changes in inductance, can detect stainless steels and are often used on final packages of foods. ***See*** Corrosion Chemistry

Many polymers are acceptable for food use, especially Teflon, Neoprene, polyethylene and polypropylene. Carbon steel is acceptable where conditions will not promote corrosion and contamination. Copper is used occasionally, for special purposes such as candy cooking.

Equipment designed for food processing can usually be disassembled with few tools and uses wide screw-threads or flanged connections to avoid creation of places where residues can be caught. Wherever possible, pipelines are welded, taking care to make inside surfaces of welds smooth and flush. Long runs of pipe or tubing must have inspection access and must be installed carefully so that drainage is complete. Dead spots are avoided in pipes so that systems can be cleaned thoroughly.

Clean-in-place

The concept of cleaning in place by flowing cleaning solutions and rinse water through a system in place of food liquids has significantly increased the productivity of food plants. Previous to its wide use, dairies and other food plants were disassembled and washed completely by hand. This consumed so much time and labour that it limited the practical size of plants.

The elements of a typical clean-in-place (CIP) system are chemical storage, solution supply and recovery tanks, supply and return pumps (usually centrifugal), spray devices in tanks, air-controlled automatic valves, manual connection stations (flowverters), additional piping to complete circuits, and automatic control devices such as programmable logic controllers (PLC). Empirical design rules have evolved from experience for flow velocities, spray intensities and solution strengths for effective cleaning of typical soils from food plants.

Most experience is with dairy and other liquid-product plants, such as soft drink, fruit juice and syrup plants. Minimum velocities of $1{\cdot}5\ \mathrm{m\ s^{-1}}$ are generally required in pipelines to achieve sufficient turbulence. Tees and dead legs should not exceed three pipe diameters length from the flowing stream so as to ensure cleaning. Tanks are sprayed at about 37 litres per minute per metre of circumference. Often such flow rates surpass the normal process flow rates, and so special pumps may be needed just for CIP. Sometimes it may be possible to use dual-speed process pumps, with a low speed for process flow and a higher speed for CIP. Such a technique saves investment cost and space.

It is common to design CIP systems so that cleaning solutions and final rinse water are recovered for re-use, to save costs and reduce discharges. Caustic soda solutions (1·5%), nitric, phosphoric and citric acids are common cleaning agents. Sometimes detergents are used, alone or with acids or bases. In addition, sanitizing agents, such as iodophores, quaternary ammonium compounds and chlorine-releasing agents, may be used, usually as a final rinse which is allowed to sit overnight. In the morning, the sanitizing agent is displaced with fresh water and then with product. ***See*** Cleaning Procedures in the Factory, Types of Detergent; Cleaning Procedures in the Factory, Types of Disinfectant; Cleaning Procedures in the Factory, Modern Systems

With a properly designed CIP system, most of the plant waste water should be discharged during cleaning from the CIP unit. This permits monitoring and pretreatment, if necessary.

Some manual labour is involved in cleaning even highly automated systems. Positive-displacement pumps must be disassembled, bypassed and cleaned by hand because their construction prevents complete cleaning by flushing. It is common practice to require manual connections at flowverters to prevent accidental contamination of food with cleaning solutions. Connections are commonly verified by electronic signals to the controller.

It is difficult to separate CIP design from process system design in many cases; they are best developed together. Physical arrangements, details of nozzles on tanks, valve placement, pump utilization, pipe runs and other details are significantly affected by CIP.

Food Plant Layout

Overall process flow is dictated in part by location of shipping and receiving and also by constraints on equipment. For example, cookie and cracker baking ovens are as long as 100 metres and must run in a straight line, because they use metal bands as baking surfaces and partially baked pieces cannot easily be transferred. Hydrostatic sterilizers for cans and jars are usually located on outside walls for ease of access and maintenance. Cooling tunnels for chocolate enrobing are usually long and straight.

Dry equipment and processes should not be physically adjacent to wet ones; cold process areas should be separated from hot ones; 'dirty' areas should be separated from clean ones, and so forth. Such separations are not merely physical, in the sense of walls and doors, but also include separate air-handling units, and perhaps limited access by people from other areas.

Vehicle traffic should be separate from people traffic. Often it is desirable to keep visitors, who may be customers, off the production floor, in part for their own safety, but also to reduce exposure of food to humans. At the same time, if visitors are likely, it is desirable to offer a decent and controlled view of the process. One solution is an elevated, enclosed walkway with viewing windows. If the process side of the windows is refrigerated, as many food plants are, the windows need to be heated to prevent condensation.

Plant layout can encourage or inhibit communication among workers; with fewer people running modern food processes, good communication is essential. A serpentine layout may be helpful compared to a long, straight arrangement. Clever use of closed-circuit television can overcome distance and obstructions.

Typical Unit Operations

Some unit operations are so common and important to food processing that they deserve individual discussion.

Mixing

Processes of mixing solids with solids, solids with liquids, and liquids with liquids are all quite common in food processing. For hygienic operation, mixers must be designed to empty completely, to be easily cleaned, and to be easily inspected. Access for addition of minor ingredients and for sampling may also be necessary. Doughs and pastes, such as bread dough, cookie batter and confection fondants are prepared in specially designed machines with relatively high-powered motors.

Incorporation of gas, especially of air, may be desirable or undesirable for a given case; the mixer design changes accordingly. (High speed, giving high shear, leads to gas incorporation.)

Dispersion of powders into liquids, such as starch or gums into water or syrup, requires high-speed agitation to avoid formation of partially wet agglomerates, called 'fish eyes'. One technique is to pour solids into a deliberately inefficient centrifugal pump which is circulating liquid from and to a tank. The inefficiency means that energy is applied to dispersing the solids rather than pumping the liquid.

Dissimilar liquids, such as oil and water, do not form solutions but can be made to form stable suspensions or emulsions by formation of very small droplets of oil and by using surfactants. The small droplets are formed by high-shear agitation and by pumping under high pressure through small orifices or special valves in a homogenizer. Many foods contain natural surfactants, such as proteins and polysaccharides, which reduce surface tension at the interfaces between liquids and so help maintain an emulsion. Milk, ice cream and salad dressings are examples of foods containing oils and aqueous solutions in stable suspension. ***See*** Colloids and Emulsions

Heat Transfer

Heating and cooling of foods is critical to many preservation and processing techniques, including pasteurization, freezing, baking, and other types of cooking. One of the more common heat exchangers used in food processing is the plate-type. This has dimpled or otherwise embossed plates separated by gaskets and held together by mechanical pressure on a sturdy frame. Hot and cool fluids pass on opposite sides of the plates, providing a large heat transfer surface in a rather compact space. The major advantage of the plate exchanger for food service is the ease with which it can be disassembled for inspection and maintenance. It can also be reconfigured easily by adding or removing plates on the same frame (up to the limit of the frame). ***See*** Heat Transfer Methods

Shell and tube, concentric tube and spiral tube heat exchangers are also found in food plants. Direct steam injection is also common.

Cooling and freezing are performed in several ways. Wiped-surface heat exchangers are common for freezing flowable fluids, such as ice creams, and for cooling fluids that have been heated for sterilization. Refrigerated plates are used for contact freezing of packages of food, such as vegetables or preplated dinners. Direct contact with very cold vapours or snow from liquid nitrogen or carbon dioxide is used in cryogenic freezing, usually on continuous belts passing through tunnels. Air cooled by refrigeration can also be used in similar tunnels. Cryogenic equipment is less expensive than mechanical refrigeration equipment, but is more expensive to operate. The exact balance depends on local costs, but cryogenics are a good choice when minimizing capital is key; mechanical freezing is more common for established, ongoing processes, where operating costs are more critical. ***See*** Freezing, Principles; Freezing, Freezing Operations; Freezing, Blast and Plate Freezing; Freezing, Cryogenic Freezing

Steam heating of food in sealed containers, such as cans, jars and pouches, is commonly performed in pressure vessels known as retorts. Batch retorts may be horizontal or vertical, and may have water in addition to steam or steam alone for heating. For containers such as jars and pouches, which cannot tolerate internal pressure, air is added during the cooling phase, which uses water, to counter the internal pressure built up during cooking.

There are several approaches to continuous cooking of foods in containers. One of the more efficient is the hydrostatic retort, in which containers are transported in carriers on a chain through a 'U'-shaped tower in which the vertical leg of water serves as a seal to contain a high-pressure chamber. (In both batch and continuous retorts, pressures to reach 120°C are used so as to shorten the time required to sterilize food without giving it overcooked taste and texture.) ***See*** Sterilization of Foods

Another approach uses a helical track inside a horizontal cylindrical shell. The track rotates, transporting metal cans. Cans enter and leave through a star wheel valve. A similar shell and track is used to cool the cans under pressure.

Conveying

Many foods are solids or are made from solids, and so solids material handling is critical to food processing. Pneumatic conveying, by pressurized air or vacuum, is commonly used to move flour, sugar and grains in food plants. Pressure conveying permits movement from one source to several delivery points, such as flour to several dough mixers. Vacuum conveying works well for several sources to one delivery point, as in dust collecting. It is also often used to unload bulk material delivery vehicles, such as rail cars or trucks. Dense-phase pneumatic conveying is a variation used for fragile materials, such as sugar, which might be damaged by the high velocities encountered in pressure or vacuum conveying. Typically, for dense-phase conveying, powders are blown from a holding vessel in slugs rather than being entrained in conveying gas.

Screw conveyors are used to move pasty or granular materials, often up inclines. They can be difficult to clean and are very dangerous, but for some purposes they are uniquely appropriate.

Belt conveyors can be used to move loose food materials, but are best for packages and containers. They can also pose safety hazards, with numerous pinch points and motors which start and stop under remote control.

Vibratory conveyors are well suited to many food applications, especially for fragile materials such as fried snacks, because they can be quite gentle and are easily cleaned.

No matter what solids conveying system is used, it rarely can be properly designed without specific measurement of the physical properties of the actual materials involved. Solids used in food plants vary widely in the properties that influence conveying and handling, and these properties are difficult to predict from theory or correlations. Examples of key properties include bulk density, particle size distribution, angle of repose, cohesiveness, abrasiveness and moisture content.

Pumping

Many foods are quite viscous and non-Newtonian, so proper design of pumps is challenging. Positive-displacement pumps, especially those using close-fitting lobes and screw-shaped rotors, are common. Such pumps are usually made of stainless steel and designed for easy disassembly. Centrifugal pumps are more conventional, but also likely to be stainless steel and built to be taken apart by hand with few tools.

A special case of pumping, which combines the functions of mixing and heating, is screw extrusion. Extruders may be single or double screw, with heating or cooling in jackets, and with a wide variety of die designs. When starchy materials, such as corn, are extruded under pressure through a die, an expanded foam is formed, which has a desirable texture and density. Pet foods, cereals and snacks are made in large quantities in this way. ***See*** Extrusion Cooking, Principles and Practice

Typical Food Processes

Many food processes have a number of steps in common, including mixing of ingredients, forming of pieces or shapes, filling of containers or packages, cooking and cooling, packaging for shipment, and unitizing. The specific details obviously depend on the product and circumstances. However, some issues are almost universal.

Designing for Hygienic Operation

Formulation

In order to be mixed, a formula must be weighed or measured; this may involve feeding and scaling of particulate solids, metering of liquids and addition of small quantities of key ingredients. Most formulas for foods contain three categories of ingredients: major (percentages in the tens), minor (single-digit percentages), and micro (less than 1%). (Even these distinctions may vary from case to case.) It is common to prepare formulas in batches of several thousand kilograms, even if subsequent processing is continuous. Issues that arise include whether to use one scale, which must then be capable of weighing the entire batch, or multiple scales, sized for accuracy of each category. Cycle time of batching is longer the fewer scales are used, but costs are lower. Liquid ingredients can be metered using any of several devices. Sometimes ingredients are simply added by units, such as entire bags or drums, but this is less precise than scaling, because usually such units are slightly heavier than their label weight.

Critical issues in any automatic formulating system are reliable feeding and flow of highly variable solids; even such common commodities as flour and sugar can clump, cake and bridge in hoppers and chutes, disrupting flow and upsetting operations.

Forming

Forming operations are highly specific, but can include such examples as flaking, extrusion, sheeting (rolling into thin layers), cutting, laminating, decorating, dicing, and many others. Usually these operations are continuous and so need careful metering of a mix, which may have difficult flow properties (doughs or pastes, for instance). Forming operations usually work best when operated at a smooth and constant rate, so surge capacity ahead of the operation and reliable removal of output are important.

Filling

Primary containers may be filled before or after such processing as cooking, baking or sterilization. If the material is liquid or a suspension, it can probably be filled volumetrically; if it is free flowing solid, it may be filled volumetrically or it may be scaled. Weighing of valuable particulate solids is often done with multicompartment digital scales, in which a computer scans compartments to select several whose summed weight most closely approaches the target. Such scales are very accurate and justify their considerable cost by reducing overweight packages.

It is common for food packages, such as cans of soup, to be filled in several successive steps, with the final 'mixing' occurring in the package. Chicken noodle soup, for instance, has noodles, chicken and broth each added separately using specialized fillers.

Cooking

The meaning of cooking varies from food to food but almost always involves some heating, directly or indirectly, to cause desirable chemical reactions, remove excess moisture, develop colour (usually browning), and reduce microbial populations. Heat may be applied by flames, steam, hot air, oil, microwaves or hot water. After heating, cooling often occurs by exposure to ambient air, but may be accelerated by refrigeration. Cooking almost always involves more than one reaction, some of which may be undesirable, so control is complex. In frying, for example, water is removed and largely replaced by fat, volume expands, and browning occurs. ***See*** Browning, Nonenzymatic; Cooking, Domestic Techniques

Packaging

In addition to a primary package, which may be filled before or after cooking, there are usually additional layers of package to provide additional protection during storage and shipping. These may include cartons and cases. Cartons are usually made of heavy paper and are part of the consumer package, so usually have multicolour printing of brand names, ingredients, use instructions, weight and nutritional information. Shipping cases are usually removed in a store, rarely are seen by consumers and so can be made of sturdy corrugated cardboard with minimum printing and graphics.

Plastics are increasingly important as components of food packaging because they are lighter in weight than metal or glass, can be used in microwave ovens to reheat contents, and do not break so easily as glass. Plastic containers usually cost more than the metal or glass containers they replace, but are perceived to add value to the consumer. Recycling of food containers is an increasingly important concern to consumers and complicates the selection of packaging material. Some of the most effective plastic containers are composed of several types of plastic, each providing particular properties, such as barrier to oxygen or strength. Such multicomponent materials are less easily recycled than single-component materials.

Unitizing

Shipping cases of food are usually assembled into larger units, often of about 450 kg on wooden pallets, for

Designing for Hygienic Operation

storage and shipment. These pallet loads may be wrapped in stretch plastic, wrapped with heat-shrinkable plastic, strapped or stabilized by glueing boxes or bundles together. Alternatively, large cardboard sheets known as slip sheets may be used to hold a unit load. Special equipment is needed to pick up and move loads on slip sheets, in contrast to the common fork-lift truck for which wooden pallets are designed, but the slip sheets are less expensive than pallets and take less space in storage and shipping.

In automatic storage systems and racks, used to achieve higher volumetric density in warehouses, special captive or slave pallets may be used, which are not shipped with the product load. This may require an additional transfer operation.

Some short-shelf-life foods are unitized on special carts, baskets and trays suited for storage on small delivery trucks for direct shipment to stores. Fluid milk, cultured dairy products and baked goods are often handled this way.

Analysis of the best packaging and unitizing procedure and equipment is a specialized area of engineering and requires balancing investment cost, operating cost and delivery system costs to achieve the best overall arrangement.

Conclusion

Design of food plants for hygienic and effective operations requires consideration of the interaction of the building with the process, selection of the proper equipment, development of an efficient layout, and consideration of the roles of people, materials, and processes. The goals are safety, quality and cost efficiency. Many special pieces of equipment have been developed to aid in reaching these goals.

Bibliography

Matz SA (1976) *Snack Food Technology*. Westport, Connecticut: AVI.

Matz SA (1988) *Equipment for Bakers*. McAllen, Texas: Pantech International.

Mercier C, Linko P and Harper JM (eds) (1989) *Extrusion Cooking*. St. Paul, Minnesota: American Association of Cereal Chemists, Inc.

Sacharow S and Griffin RC (1980) *Principles of Food Packaging*. Westport, Connecticut: AVI.

Woodroof JG and Phillips GF (1981) *Beverages: Carbonated and Noncarbonated*. Westport, Connecticut: AVI.

J. Peter Clark
A. Epstein and Sons International, Inc., Chicago, USA

Process Control and Automation

Food processes, like other industrial operations, are measured and controlled in order to obtain consistent, safe and efficient results. Wherever possible, response to measurements is automated, to reduce human labour and the risk of error. Modern processes rely heavily on computers to monitor operations, record data, and generate reports. Increasingly, computing power is in the form of microprocessors, which are dedicated components of instruments, relieving control computers of functions, simplifying installation, and increasing versatility. This article discusses techniques for measuring typical parameters, especially those found most often in food processes, and the ways in which measurements are used to control processes. The food industry has adopted measurement and control techniques from other industrial processes, including computer-integrated manufacturing (CIM), which is discussed briefly at the end of the article.

Typical Parameters

Weight

Weight, usually of solids, but also of mixes, packages and storage vessels, is one of the most important and common measurements made in a food plant. Mechanical lever arrangements with counterweights and indicating dials are still widely used, but more common is the use of electronic strain gauges or load cells. These have quick response and give a signal which is converted to a digital signal which can be read directly by a computer. Load cells need to be calibrated and their limitations understood; they can be sensitive to temperature, moisture and vibration and they must be selected for the total weight range they will experience.

The combination of an electronic scale with a variable-speed screw or vibrating cone creates a feeder which can be programmed to deliver a given rate or a given total amount of material. Such a feeder may operate by tracking the loss in weight (LIW) from a hopper or by weighing the receiving bin (gain in weight). Loss in weight is most common. Periodically, the feed bin must be refilled; during this time, the feeder continues to run at the last speed it had. To minimize the time in which this 'volumetric' feeding occurs, refilling is kept to 5–10% of the cycle time. This means the refill flow rate is 10–20 times the controlled feed rate; for high flow rates, this can be quite substantial and represents a significant shock to the receiving bin. Refilling can also aerate solids and make them hard to control or force them out of the feeder. Thus, bin sizing, delivery system and flow control are all integral to successful weighing and delivery of solids.

Checkweighers are special scales placed in line with a package conveyor to weigh individual packages. Usually, a checkweigher controls a kick-out device to remove over- or under-weight packages. Data is also accumulated to calculate averages, standard deviations and other quality statistics, which are used for statistical process control (SPC). SPC involves identifying process capabilities by measuring performance, then using such data to detect abnormalities and gradually to improve performance. Occasionally, a checkweigher is used to control a filler or other process affecting package weight.

When storage or process tanks and bins are to be weighed, they must be carefully isolated from other supports so that the load cells bear the true weight. This requires flexible connections from delivery pipes and other gear which may be attached to a typical tank. Sometimes load cells are installed to permit calibration of other instruments, such as meters, which are regularly used to control formulations.

Flow of Fluids

The traditional method of measuring flow rates of liquids and gases is by measuring pressure drop across an orifice (a small hole in a plate inserted in a pipe). For food fluids, an orifice plate may pose an unacceptable obstruction to suspended solids and may also be hard to clean in place. For homogeneous, clean fluids, this approach is reliable and inexpensive, but it provides a limited turn-down ratio of about 3:1.

Better for measuring homogeneous fluids (liquids or gases) is the vortex meter, which detects small changes in capacitance created as vortices are shed from a bluff body inserted in the fluid path. The vortices, or swirls of fluid, are created and shed in direct proportion to the flow rate. Turn-down ratio can be 10:1 or greater and the measurement is independent of temperature, density and viscosity. Installation is simple and obstruction of the flow path is slight.

An increasingly common technique, which has no flow obstruction and does not require long runs of straight pipe before measurement, is the mass flow meter. A tube is inserted into a pipe as part of the flow path. The tube is vibrated at a resonant frequency. As fluid flows through the vibrating tube, it changes the phase difference of vibration at two points on the tube (or it imparts a twist to a bent tube) which can be detected and converted to a signal proportional to the flow rate. Magnetic or optical detectors measure the location of the tube and electronic devices convert the signals to flow rates.

Other flow rate measuring devices count turns of rotors, vanes or some other signal of flow rate. For electrically conducting fluids, changes in a magnetic field due to flow can be measured without any obstruction to the flow path. Positive-displacement meters use the flow directly to generate a signal by driving a lobe or piston connected to a shaft; rotations are proportional to flow. If a measurement is used to control flow, it usually signals an air or electrically operated valve, but it could signal a variable-speed drive on a pump.

Temperature

Temperature is a critical measurement in many food processes because it controls sterilization and cooking operations; too high or too low a temperature may be harmful. Thermocouples are inexpensive and reliable. They generate a small but consistent voltage difference at the junction of dissimilar metals, which can be correlated with temperature.

Most common for accurate measurement over larger ranges are resistance temperature detectors (RTD), which use the change in resistance of platinum with temperature to measure temperature. A common standard is to set the resistance at 100 ohm for 0°C.

Bimetallic thermometers are commonly used as temperature indicators – the familiar dial thermometer – but they do not provide an electronic signal. Various solid-state devices, such as thermistors, are used to measure temperature because they give a signal directly compatible with electronic equipment, are compact and do not require special wiring, as do thermocouples; however they are not very accurate.

When temperature must be controlled, the signal usually drives a steam, gas or water valve. Control of temperature in multizone equipment, such as baking ovens, can become very complex, as there are many temperatures to measure and many valves to control, most of which interact with each other. Safety interlocks are also required, to detect temperature excursions, to trigger alarms, to activate fire-suppression equipment, to control exhaust blowers (preventing build-up of gas), and to indicate restoration of operating conditions.

Mercury-in-glass and alcohol-in-glass thermometers are still found in canning operations, as calibration standards for retorts, but are rarely used in other food plants, because of the risk of contamination. ***See*** Canning, Principles; Mercury, Toxicology

Temperature can be measured at a distance by infrared instruments. These are useful for checking insulation on overhead lines, roofs, and tall tanks.

A difficult temperature measurement challenge is to follow the temperature history of a container in a continuous sterilizer, such as a hydrostatic retort. The solution is a solid-state measuring and recording device which fits within or on a container and is plugged to a computer when it is recovered. The data are recorded in semiconductor memory which is read by computer or special instruments. Similar devices are used to track the temperature history of refrigerated and frozen foods in the distribution chain.

Traditionally, sterilization and pasteurization processes were required to keep ink tracings of temperature history continuously, usually on familiar circular instrument charts. Such records are still seen, but electronic data collection is increasingly accepted. It is critical to ensure the integrity of such data if it is to displace the ink recorder. ***See*** Pasteurization, Principles; Sterilization of Foods

Pressure

Solid-state strain gauges, similar to those at the heart of electronic scales, can also be used to measure pressure, since both parameters involve force. Sanitary pressure gauges involve flush-mounted diaphragms whose deflection generates a signal. The same device can serve as a level indicator in a tank.

Traditional Bourdon tube pressure indicators are found in food plants, especially on utilities such as steam. If used in direct contact with food, they need to be designed to prevent contamination by the food or of the food by any filling material.

Level

Level of fluids is measured by weight or pressure at the bottom, but also can be detected by floats, position detectors or conductivity switches. Level of solids is more difficult to measure because hydrostatic pressure is not directly proportional to solids height in a bin. A variety of devices, using plum bobs, ultrasonics, and light are used to measure distance to the top surface of solids, but these suffer from the tendency of solids to remain in a pile with an uneven surface. Bin weight is the best way to control inventory of solids.

Probes can be used to measure capacitance change as liquids or solids cover more or less of the length, thus giving level indication around the location of the probe. Vibrating reeds or 'tuning forks' generate a change in signal when they are covered, thus indicating the presence of solids or liquid at their location in a bin. Capacitance detectors can also be used as point indicators of level, as can devices which detect change in resistance to rotation of a small paddle.

Composition

Refractive index correlates with solids content of many liquid food materials, such as syrups, milk, and juices. Thus instruments to measure refractive index can report results in terms of composition and can be used to control blending operations. Other properties found to correlate with solids content include viscosity and density. On-line instruments exist to measure these properties also.

Direct measurement of composition is quite difficult except by versions of analytical instruments, such as gas chromatographs, mass spectrometers and pH meters. Ion content can be detected and quantified by ion-specific electrodes, and salt content can be measured by conductivity. Relatively few of these techniques are found directly connected to processes; more often they are used as off-line quality control devices. ***See*** Chromatography, Gas Chromatography; Mass Spectrometry, Principles and Instrumentation; pH – Principles and Measurement

Moisture

Moisture is very important in foods and is usually measured by drying a sample. Instruments exist to give moisture determination in minutes, but this is still too slow for direct process control. Water activity can be detected by a humidity sensor over a sample in a closed container, but again it requires minutes. Changes in radiofrequency signals or of microwaves can be used for noncontact, on-line sensing of moisture in some foods.

Moisture, fat and protein content can be measured quickly for meats and other foods using infrared spectrometery. This has been an off-line technique, but it can be done on the process floor, giving results in time to modify a batch, if necessary. Reflected near infrared (NIR) is used on-line to measure some composition changes, including moisture and oil content. ***See*** Protein, Determination and Characterization; Spectroscopy, Near-infrared; Water Activity, Principles and Measurement

Colour can be detected and quantified on-line and has been used to control baking ovens, and also to help sort out burnt potato chips (crisps). ***See*** Colours, Properties and Determination of Natural Pigments; Colours, Properties and Determination of Synthetic Pigments

Other Parameters

Counting, most often of packages and cases, can easily be done simply by breaking a light beam to a photo detector or by using proximity switches, which detect changes in capacitance or magnetic fields due to the presence of an object. Such switches are also used to detect position and control starting and stopping of machinery, such as accumulation conveyors.

Seal integrity, most often of heat-sealed polymer packages, such as snack foods, is difficult to detect quickly. Visual inspection is the most frequently used technique. Often such packages have a small amount of inert gas, such as nitrogen, injected to displace oxygen; the 'pillowing' effect that this creates can be used to detect presence of leaks in seals by lightly compressing the package and measuring its resistance to deformation.

Integrity of cans and jars is measured by detecting the deflection of lids due to internal vacuum. The vacuum is created when hot filled contents cool or by steam flushing of the head space. The deflection can be measured physically or by tapping to create a distinctive tone; 'duds' are rejected as likely to have leaked.

Vision systems are complex combinations of optical devices and computers to observe, quantify, and report on shapes. Such systems are becoming less expensive and more useful as the cost of computing power continues to decline. A vision system can check pizzas for the proper count and placement of garnish, such as pepperoni slices; it can look for and discard 'doubles' in a candy bar process; it can detect the presence and proper placement of labels on containers. Special devices can read bar-codes on cases and report data to an inventory management system.

Vision systems will become more complex and useful with time and will assume many duties now performed by humans, such as sorting, grading, trimming and assembling.

Integration of Processes

Integration of processes means the linking of data from various sources to maintain smooth operation and the use of such data to improve operations. At a primitive level, such techniques as signalling downstream or upstream operations about stoppages can avoid waste and prevent damage to equipment. On a more sophisticated level would be recording data about stoppages for later analysis. A computer can be used to coordinate the signalling and to do the analysis. Even more, it can write the report, recommending process improvements! ***See*** Computers – Use in Food Processing

Computer-integrated manufacturing (CIM) encompasses both levels cited plus more. The basic idea is to avoid ever having to re-enter data or information once it exists in a computer. For example, direct process data, such as case counts or batch weights are available in a process control computer. It should not be necessary to obtain a report and re-enter that data in another computer in order to calculate daily, weekly, or monthly production results. Nor should it be necessary to correlate data from two or more different sources to identify production problems and detect improvements. In a CIM environment, all data is available on a common data base shared among various computers linked electronically.

Computers included in such a network range from mainframes and minicomputers through desktop or microcomputers to factory-floor devices such as programmable logic controllers (PLC). The power found in each of these devices is increasing as the cost decreases, so the capabilities available to those running food plants are constantly improving. Current practice in modern food plants is to control nearly all equipment (pumps, conveyors, ovens, sterilizers, feeders, etc.) through operator-interface terminals (OIT) which have colour graphic displays and keyboards, keypads or touch screens.

The displays, of which there may be several dozen, usually show flow diagrams of the process and status of the equipment (on or off, temperature, etc.). Time histories of any measured parameter can be plotted on command, deviations from desired ranges flagged, and correlations tested among measured parameters.

Standard reports are generated and data can be summarized for use at the next level of management. For example, on the operational level, the instantaneous rate of production is a concern (cans per minute, kilograms per hour, etc.); at the next level, the interest might be cases or pallets per day (calculated by summing all the actual cans or kilograms or cases or pallets counted); at the plant and corporate levels, totals for weeks, months, quarters and years are the only concern. If all the results, at whatever level, are calculated from the same basic data, with little human intervention, they will be consistent and available quickly, and people will be relieved of tedious and unnecessary time-consuming labour.

Barriers to this ideal situation have included incompatibilities among computers, languages, programs and communications media. Additional barriers have been low computer literacy among workers involved and institutional resistance to sharing of information. Evolution and cooperation among vendors have led to the emergence of common protocols for exchanging information electronically, so that many of the previous technical barriers have fallen. Training and the use of computers in early school grades is reducing the literacy problem. The institutional resistance may be the most difficult, but as the significant benefits of computer integration are appreciated, this too will disappear.

Conclusion

The overwhelming importance of computers in process control and business systems, which are increasingly linked together, has significant impact on food plant design. The need to have data compatible with computers influences the selection of instruments and measurement techniques. Instruments are being made 'smarter' by incorporating computing power in them with microprocessors, leaving only simple connections to control computers. Such smart instruments calibrate themselves, control drift, correct for changing conditions and are inexpensive to install. The ability of fewer people to manage and control large processes reduces the labour requirements while raising the education and training requirements of the employees.

Bibliography

The Instrument Society of America (ISA) and the Institute of Electrical and Electronics Engineers (IEEE) have technical groups devoted to the food industry, which provide additional sources of information on instrumentation and control.

Bernard JW (1989) *CIM in the Process Industries*. Research Triangle Park, North Carolina: Instrument Society of America.

Food and Process Engineering Institute (1990) *Food Processing Automation Conference Proceedings*. St. Joseph, Michigan: American Society of Agricultural Engineers.

Schwartzberg H and Rao MA (1990) *Biotechnology and Food Process Engineering*. New York: Marcel Dekker.

J. Peter Clark
A. Epstein and Sons International, Inc, Chicago, USA

PLANT TOXINS

Contents

Trypsin Inhibitors

Trypsin inhibitors occur in a wide range of foods, the most important of which are the grain legumes such as chickpeas, mung beans and soya beans. Grain legumes are a major source of protein in the diets of many people throughout the world and are thought to provide about 10% of the world's dietary protein. They are used extensively as a protein source for animal feeds. The existence of a number of antinutritive factors in legume plants, especially a group of protease inhibitors, has limited the use of legume seed protein. Several different types of protease inhibitors may be present in the same tissue. ***See*** Legumes, Legumes in the Diet; Legumes, Dietary Importance

Structure and Mechanism of Action

The most common inhibitors in legumes act on serine proteases, a group of proteolytic enzymes which includes trypsin and chymotrypsin. Serine protease inhibitors are proteins that form very stable complexes with these digestive enzymes, reducing their activity to very low levels. Some nonprotein inhibitors have also been identified.

Protease Inhibitors

Protease inhibitors have been classified into families based on homologous sequences of amino acids at the reactive inhibitory sites. The molecular structure of the inhibitor affects both strength and specificity of the inhibitor. The two main groups of protease inhibitors found in legumes are soya bean trypsin inhibitor (Kunitz) and soya bean protease inhibitor (Bowman–Birk). ***See*** Soya Beans, Properties and Analysis

Kunitz Soya Bean Trypsin Inhibitor (STI)

The first protease inhibitor to be isolated and characterized was STI. Molecular weights (M_r) ranging from 18 to 24 kDa and isoelectric points of 3·5–4·4 have been reported. This wide variation may be attributable in part to isoforms of the inhibitor. Taking into account all the variations, it appears that the M_r of most STIs is approximately 21 kDa. Soya bean trypsin inhibitor consists of 181 amino acid residues linked by two disulphide bonds; the reactive site is located at residues 63 and 64 (Fig. 1).

A competitive inhibitor, STI binds to the reactive site of trypsin, in a similar manner to the substrate protein, causing hydrolysis of the peptide bonds between reac-

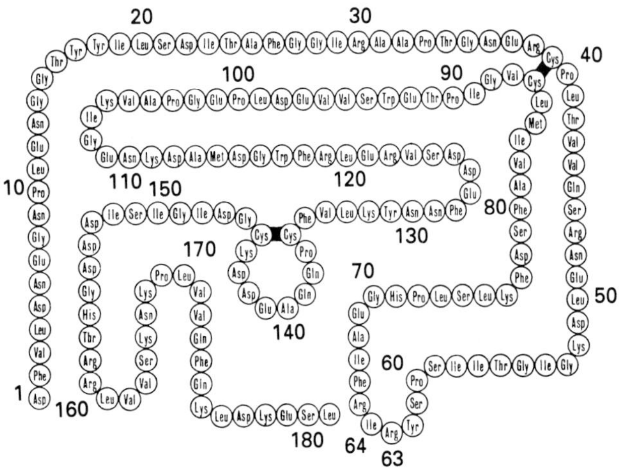

Fig. 1 Structure of the Kunitz soya bean trypsin inhibitor (Koide T and Ikenaka T, 1973).

tive site residues of the inhibitor or substrate. Inhibitors differ from substrate proteins in that the reactive site residues are held between disulphide bonds. After hydrolysis, the modified inhibitor is held together with the same conformation, owing to the disulphide bond. This forms a stable enzyme–inhibitor complex.

Bowman–Birk Soya Bean Proteinase Inhibitor (BBI)

The second proteinase inhibitor to be isolated and characterized, the BBI differs from STI in four characteristics. It is a relatively small molecule, M_r 8 kDa, and is a 'double-headed' inhibitor with independent binding sites for chymotrypsin and trypsin. As it contains seven disulphide bonds, the BBI is rich in cystine (20%). It is very stable in the presence of heat, acid, alkali and proteolytic enzymes such as pepsin. This is probably attributable to the stabilizing effect of the disulphide bonds on the whole structure (Fig. 2).

Limited proteolysis results in separation of the inhibitor into two active fragments, one with trypsin inhibitor activity and the other with chymotrypsin activity. Further study of protease inhibitors has revealed that the BBI is a prototype for a whole family of homologous inhibitors found in grain legumes and in some other plant families. The sequence of amino acids surrounding the two reactive sites of the BBI are remarkably similar, not only to each other but also to homologous active sites of other legume inhibitors isolated from other common grain legumes. Comparison of the BBI with inhibitors isolated from groundnut and chickpea show that these are double-headed inhibitors, although the mechanism by which they inhibit trypsin may differ slightly owing to the spatial arrangement of the reactive sites. As many as five protease inhibitors that have properties similar to the BBI may be present in soya beans. ***See*** Peanuts; Pulses

Determination of Activity

The most commonly used procedures to quantify trypsin inhibitors are based on the method described by Kakade *et al.* (1974). The basic method has been improved to increase its accuracy and sensitivity and has been further simplified and miniaturized. Many other modifications to the basic method have been made, the most common are the use of different substrates, e.g. casein instead of the synthetic substrate benzoyl-DL-arginine-*p*-nitroanilide (BAPA), or the use of other types of trypsin, e.g. porcine rather than bovine. These modifications make it difficult to compare the results published by different authors.

The trypsin inhibitor affinity assay, described by Roozen and de Groot (1988), only determines protein type inhibitors forming complexes with trypsin, but is highly sensitive at low concentrations.

Immunological methods developed for the detection and characterization of specific types of protease inhibitors involve the use of specific protease antibodies.

Interfering Substances

With samples low in trypsin inhibitor activity (TIA), the accuracy of the values obtained can be distorted by nonspecific interference from other proteins and compounds in the reaction mixture. In particular, tannins in coloured legume seed coats, free fatty acids released in fermented products such as miso and tempeh, and indigestible polysaccharides can complex with trypsin and contribute to the total TIA under assay conditions.

Distribution within Plants

Trypsin inhibitors are found most often in the seed but their location is not necessarily restricted to this part of the plant. In some legumes, such as the mung bean and the field bean, high levels of TIA are found in the leaves as well.

A higher level of TIA has been found in the outer part of the cotyledon of soya beans, kidney beans and chickpeas, and in five different cultivars of peas four times as much TIA was found in the cotyledon than in the hull or seed coat. Similar findings have been reported for soya beans. In contrast, *Vicia faba* (faba bean) seeds were found to have twice as much TIA in the hull than the cotyledon. ***See*** Peas and Lentils

Levels in Grain Legumes

Trypsin inhibitors are widely distributed across many genera and species in the Leguminoseae family and

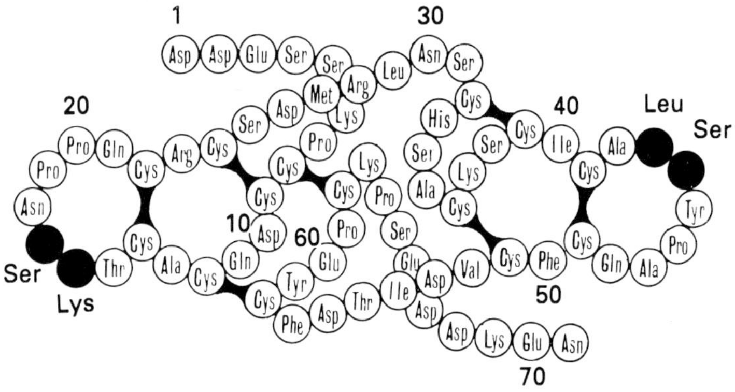

Fig. 2 Structure of the Bowman–Birk inhibitor (Odani S and Ikenaka T, 1973).

Table 1. Reported distribution of TIA in some common legumes

	Valdebouze P *et al.* (1980) (TIU per mg)	Bhatty RS (1974) (TIU per mg)	Hove EI and King S (1979) (mg of trypsin inhibitor per g)
Lupins	<1		0·38–0·47
Peas	2·9–5·5		0·72–3·5
Soya beans			26·2
Faba beans	5·6–11·8	2·2–32·1	
Red kidney beans			11·6
Pinto beans			3·4
Navy beans			10·8
Lima beans			20·2

TIU, trypsin inhibitor unit.

many other plant families; TIA has also been found in a range of legumes, including red gram, kidney beans, navy beans, black-eyed peas, peanuts, field beans, French beans and sweet peas, and in all varieties tested of cowpeas, mung beans, lima beans, winged beans, chickpeas and rice beans. In addition, TIA was found in lentils, but there was almost no TIA in lupins.

Although trypsin inhibitors are found in most legumes, the levels present tend to vary considerably. Most legume species contain less than 50% of the TIA of soya beans. Particularly low levels are present in broad beans, peas, mung beans, lupins and a few varieties of kidney beans. Those species with at least 75% of the TIA in soya beans include cowpeas, pinto beans, pigeon peas, kidney beans, moth beans and navy beans. Common legumes that contain levels higher than those of soya bean include lima beans, winged beans and black beans.

Assessing the variations in inhibitor activity between legumes is made very difficult by the fact that TIA is not determined using a standard procedure. This can be seen from the range of results obtained from different laboratories (Table 1). It is evident, however, that there are significant differences in TIA between genera of the Leguminoseae family.

Differences in TIA between cultivars have been observed, suggesting the possibility of breeding for low TIA. For example, Valdebouze *et al.* (1980) found that winter cultivars of peas were twice as active as summer cultivars. Considerable variability has also been reported amongst different strains of cowpeas.

Stage of maturity has also been shown to influence TIA. As soya beans mature the amount of TIA increases, although the magnitude of increase differs between varieties.

Effect of Processing

Processing can generally be divided into methods that are commonly used to prepare legumes for human consumption, and those that are used in the production of animal meals. The popularity of grain legumes in human diets is no doubt attributable to the ease with which TIA can be destroyed by many of these methods. However, destruction of TIA inevitably results in loss of some of the nutritive value of the legume as well.

Table 2. Effect (percentages of original activity) of various types of heat treatment on the TIA of legumes

Legume	TIA (%; soya TIA is 100%)	TIA (units per g)	Wet heat: heating extract in boiling water-bath 30 min	Wet heat: heating extract in boiling water-bath 60 min	Auto-claved extract (103 kPa for 15 min)	Dry heat: Roasting for 15 min at: 75°C	Dry heat: Roasting for 15 min at: 100°C	Dry heat: Roasting for 15 min at: 125°C	Dry heat: Roasting for 2 min at 200°C
Moth bean (*Phaseolus aconitifolus*)	27	1443	57·8	11·6	2·4	80·8	42·5	Nil	7·6
Cow pea									
Vigna catjung	79	4172	59·9	49·3	4·8	72·2	72·2	65·0	Nil
Vigna sinensis	64	3391	96·7	36·0	3·7	94·2	91·2	71·1	8·1
Red gram (*Cajanus cajan*)	60	3187	71·5	64·5	12·3	80·5	54·5	17·9	11·7
French bean (*Phaseolus vulgaris*)	80	4239	61·6	43·2	6·2	76·1	73·5	72·0	10·6
Pea (*Pisum sativum*)	25	1334	20·8	12·5	Nil	42·1	27·6	4·3	Nil
Lentil (*Lens culinaris*)	25	1313	8·5	Nil	Nil	73·1	69·4	65·2	20·4
Green gram (*Phaseolus aureus*)	37	1977	11·2	5·6	Nil	97·4	69·5	24·7	Nil
Black gram (*Phaseolus mungo*)	52	2735	28·4	16·2	Nil	69·4	57·6	39·4	11·9
Bengal gram (*Cicer arietinum*)	66	3472	16·0	3·2	Nil	62·0	40·9	14·2	7·6
Soya bean (*Glycine max*)		5296	61·9	22·1		58·9			

From Rackis JJ *et al.* (1986).

Boiling

Cooking presoaked (24 h) winged beans in boiling water for 30 min was found to destroy most TIA. It was also reported that mature soya beans required a presoak of 24 h and cooking for at least 20 min to completely eliminate TIA. Immature seeds do not require presoaking to eliminate TIA. This, and the fact that oven heating has very little effect on TIA, suggests that the moisture content of the seed plays an important role in the destruction of trypsin inhibitors. *See* Cooking, Domestic Techniques

It has been suggested that the extent to which TIA is destroyed by heating is a function of the temperature, duration of heating, particle size and moisture content, variables that are closely controlled in the industrial processing of soya bean meals in order to obtain a product maximum nutritional value. The nonprotein TIA content would also determine the reduction in TIA achieved by the heating process. A summary of the effects of various heat treatments on TIA is given in Table 2.

Oven or Dry Heating

Heating at temperatures below 75°C appears to have no effect on winged beans. Increasing the temperature to 100°C resulted in a 5% decrease in TIA of winged beans. The addition of 0·03 M sodium sulphite to soya flour and heating for 1 h at 75°C is reported to have completely destroyed TIA, leaving no sulphite residues in the proteins. By increasing the temperature to 80°C, Kadam and Smithard (1987) found that TIA in white winged beans decreased by 25% in 5 min and 45% after 30 min. Dry heating (177°C for 20 min) peanuts and soya beans decreased TIA by 7% and 20%, respectively.

Autoclaving

Autoclaving has been shown to decrease TIA significantly. In peanuts and soyabeans autoclaved at 121°C for 20 min TIA decreased by 80% and 86% respectively. Autoclaving whole winged beans and winged bean meal decreased TIA by 80%.

Infrared

Infrared treatment for 30 s on winged beans was reported by Kadam and Smithard (1987) to have destroyed most TIA in the seeds.

Microwave Radiation

Microwave radiation of broad beans for 30 min was found to destroy 90% of their TIA. Kadam and Smithard (1987), found that microwave treatment (time not known) had no effect on TIA in winged beans. Microwave treatment for 1·5–3·0 min of presoaked soya beans destroyed 85–90% of the TIA.

Germination

The germination of a number of grain legumes appears to increase their nutritive value, although the effect on

TIA seems to be quite variable. There have been no reports of a significant decrease in TIA on germination of lentils, chickpeas, and navy beans but there was a decrease in TIA when field peas were germinated. Other studies have found that after germination of *Vicia faba* for 5 days TIA decreased by 63%; TIA decreased by 50% in red kidney beans after 10 days' germination, and TIA in three soya bean varieties decreased by an average of 13% over 3 days' germination.

The loss of TIA in some cases may be accounted for by leaching during soaking and washing procedures. Soaking chickpeas, broad beans and mung beans reduced TIA levels by 58–92%.

Physiological Effects

The majority of experiments which have considered the physiological effects of grain legume TIA have been carried out on animals, and inferences made for humans. The effects of TIA found in soya beans have been predominantly studied as the availability of pure forms of STI and BBI have made the task of elucidating the inhibitor actions much easier. The efficient digestion of proteins in the digestive system requires the action of a number of proteolytic enzymes, each hydrolysing peptide chains at specific points. These enzymes are produced and stored in the pancreas in a precursor form known as zymogens. The efficient digestion of proteins requires simultaneous activation of all the zymogens. Trypsin is thought to act as the common activator of pancreatic zymogens. ***See*** Protein, Digestion and Absorption of Protein and Nitrogen Balance

Proteolytic Activity and Growth

Owing to the fact that trypsin activates proteolytic enzymes, an overall decrease in proteolytic activity would be expected from the action of trypsin inhibitors. *In vitro* studies have shown that STI, BBI and lima bean inhibitor completely inhibited trypsin and chymotrypsin activity of human and rat pancreatic juice. The TIA caused up to 50–60% reduction in total proteolytic activity, and the residual activity observed was attributed to carboxypeptidase activity.

Reduction in growth owing to loss of proteolytic activity appears to vary between species fed raw soya bean diets. Growth is reduced in the rat, mouse, chick and young guinea pig, whereas the adult guinea pig, dog, pig, calf, monkey and, probably, the human appear to grow normally when fed these diets.

There are a number of factors that might contribute to the differences in proteolytic activity and growth between different species. One is that gastric juices can inactivate protease inhibitors. Krogdahl and Holm (1981) showed that human gastric juice almost eliminated the protease inhibitor activity of STI when incubated for 24 h. They found that lima bean inhibitor activity was only slightly affected by incubation in gastric juice, suggesting that inhibitors homologous to lima bean inhibitor are nutritionally more important than inhibitors from the STI family.

Another reason for the varying activity of protease inhibitors between species may be differences in the specific activity of species-specific trypsin. In humans, proteolytic enzymes hydrolyse casein at a lower rate than in many animals, and the effect of inhibitors may therefore be weaker in humans. Soya bean trypsin inhibitor was shown to have a much greater effect on trout trypsin than bovine trypsin, probably owing to a higher specific activity of trout trypsin.

Human trypsin is known to exist in two forms. The cationic form constitutes two thirds of the total trypsin secreted and is only very weakly inhibited, whereas the anionic form is inhibited stoichiometrically. Infusion of raw soya bean extract into the duodenum of rats and humans caused the secretion of a modified inhibitor-resistant trypsin, which was resistant to typical serine protease inhibitors. The BBI was more potent than STI in producing TIA-resistant enzymes. Whether or not this was just the cationic form of trypsin is unclear. The ability of the pancreas to adapt to trypsin inhibitors may determine the effects that they have on different species.

Toxicological Effects

The main toxicological effect observed from the ingestion of trypsin inhibitors is a marked hypertrophy of the pancreas. Pancreatic hypertrophy is thought to be caused by TIA stimulating excess enzyme secretion, which draws essential amino acids away from other body functions; in some cases this results in death of the animal.

Cholecystokinin (CCK) stimulates pancreatic secretion and the existence of free trypsin and chymotrypsin in the small intestine inhibits CCK. Therefore, when trypsin inhibitors form enzyme–inhibitor complexes, CCK is released and pancreatic secretion occurs. This negative-feedback control is known to occur in the rat, pig and calf, as well as in humans, suggesting that other factors are also involved in pancreatic hypertrophy.

Bibliography

Bhatty RS (1974) Chemical composition of some faba bean cultivars. *Canadian Journal of Plant Science* 54: 413–421.

Bowman DE (1946) Fractions derived from soy beans and navy beans which retard tryptic digestion of casein. *Proceedings of the Society of Experimental Biology and Medicine* 63: 547–550.

Brandon DL, Bates AH and Friedman M (1988) Enzyme-linked immunoassay for soybean kunitz trypsin inhibitor using monoclonal antibodies. *Journal of Food Science* 53: 102–106.

Hafez YS and Mohamed AI (1983) Presence of non-protein trypsin inhibitors in soy and winged beans. *Journal of Food Science* 48: 75–76.

Hove EI and King S (1979) Trypsin inhibitor content of lupin seeds and other legumes. *New Zealand Journal of Agricultural Research* 22: 41–42.

Kadam SS and Smithard RR (1987) Effects of heat treatments on trypsin inhibitors and haemagglutinating activities in winged beans. *Plant Food for Human Nutrition* 37: 151–159.

Kakade ML, Rackis JJ, McGhee JE and Puski G (1974) Determination of trypsin inhibitor activity of soy products: a collaborative analysis of an improved procedure. *Cereal Chemistry* 51: 376–382.

Koide T and Ikenaka T (1973) Studies on soybean trypsin inhibitors. 3. Amino acid sequence of the carboxyl-terminal region and the complete amino-acid sequence of soybean trypsin inhibitor (Kunitz). *European Journal of Biochemistry* 32: 417–431.

Krogdahl A and Holm H (1979) Inhibition of human and rat pancreatic proteinases by crude and purified soybean proteinase inhibitors. *Journal of Nutrition* 109: 551–558.

Krogdahl A and Holm H (1981) Soybean proteinase inhibitors and human proteolytic enzymes: selective inactivation of inhibitors by treatment with human gastric juice. *Journal of Nutrition* 111: 2045–2051.

Kunitz M (1947) Crystalline soybean trypsin inhibitor. II General properties. *Journal of General Physiology* 30: 291–310.

Liener IE (1980) Protease Inhibitors. In: Liener IE (ed.) *Toxic Constituents of Plant Foodstuffs*, pp 7–57. New York: Academic Press.

Odani S and Ikenaka T (1973) Studies on soybean trypsin inhibitors. VIII. Disulfide bridges in soyabean. Bowman–Birk proteinase inhibitors. *Journal of Biochemistry* 74: 697–715.

Rackis JJ, Wolf WJ and Baker EC (1986) Protease inhibitors in plant foods: content and inactivation. In: Friedman M (ed.) *Advances in Experimental Medicine and Biology: Nutritional and Toxicological Significance of Enzyme Inhibitors in Foods*, pp 299–337. New York: Plenum Press.

Roozen JP and de Groot J (1988) Analysis of residual inhibitor in feed flour. In: Huisman J, van der Poel TFB and Liener IE (eds) *Recent Advances of Research in Antinutritional Factors in Legume Seeds*, pp 114–117. Wageningen: Pudoc.

Savage GP (1988) The composition and nutritive value of lentils (*Lens culinaris*). *Nutrition Abstracts and Reviews (Series A)* 58: 319–343.

Valdebouze P, Bergeron E, Gaborit T and Delort-Iaval J (1980) Content and distribution of trypsin inhibitors in some legume seeds. *Canadian Journal of Plant Science* 60: 695–701.

Van Oort MG, Hamer RJ and Slager EA (1988) The trypsin inhibitor assay: improvement of an existing method. In: Huisman J, van der Poel TFB and Liener IE (eds) *Recent Advances of Research in Antinutritional Factors in Legume Seeds*, pp 110–113. Wageningen: Pudoc.

G. P. Savage and P. R. Elliott
Lincoln University, Canterbury, New Zealand

Haemagglutinins

Haemagglutinins

The presence of heat-labile toxic factors in plant products, mainly in leguminous seeds, makes them unsuitable as food for humans unless they are properly cooked. There is evidence, however, that leguminous seeds were an important part of the diet of prehistoric man, and the use of fire for cooking must have had an important bearing on the safe use of leguminous seeds in the diet. In the late eighteenth century it was noticed that plant lectins, or phytohaemagglutinins (haemagglutinins) occurred in castor bean seeds and that these compounds could agglutinate red blood cells. ***See*** Legumes, Dietary Importance

The ingestion of raw or incompletely cooked beans regularly leads to incidents which at first appear to be cases of food poisoning. When no causative organism appears to be involved, high concentrations of haemagglutinins are usually implicated. Vomiting and diarrhoea usually occur within 2–3 h after consumption. As few as four or five beans are often the cause. In many instances the beans have been soaked for a few hours and then eaten without further cooking.

General Properties

Haemagglutinins are protcins that possess a specific affinity for certain sugar molecules. Since carbohydrate units exist in most animal cell membranes, the haemagglutinins may attach to these receptor groups. This attachment will occur only if the lectin molecule has at least two active groups. Haemagglutinins are characterized and detected by their action on red blood cell membranes, causing the blood cells to clump together. Other cells can also be affected. The receptor site on the cell surface must be exposed in order to react with a specific lectin. Pretreatment of red blood cells with a protein-digesting enzyme, such as papain, trypsin or pronase, activates the cells and presumably exposes the receptor sites.

Structure

All plant lectins are proteins. Some contain a range of covalently bound sugars and can therefore be called glycoproteins. Lectins from different legumes show a remarkable range of specificity towards various animal red blood cells, which suggests that there is a considerable range of structures in this group. ***See*** Protein, Food Sources

Location

Most haemagglutinins of higher plants are found in seeds, but tubers and plant sap are also sources of lectins. Lectins are most commonly found in leguminous seeds. They have been observed in potatoes and cereals

Table 1. Haemagglutinin content of a number of legumes

Legume	Haemagglutinin (units per g dry wt)	Haemagglutinin removed (%) on soaking in water for 18 h
Phaseolus vulgaris		
Red kidney beans	37 000–53 000	22–66
White kidney beans	17 000–43 000	18–36
Rose coco beans	39 000	18
Lens culinaris		
Red lentils	77 000	22
Green lentils	18 000	19
Split peas	1000–16 000	11
Pisum sativum		
Garden peas	5100	65
Vigna sinensis		
Black-eye beans	1000	100

Source: Bender AE (1983).

but the amounts in these foods do not appear to have any adverse effects.

Lectin Content of Edible Leguminous Seeds (see Table 1)

The seeds of a wide range of edible legumes are known to contain proteins which agglutinate red blood cells. Some of these haemagglutinins have been suggested to contribute to the poor nutritive value of raw beans. The level of toxicity of kidney beans has been shown to be directly related to the lectin content and hence haemagglutinating activity. Using rat growth experiments and haemagglutinin tests on a range of leguminous seed lines available in the UK, leguminous seeds have been classified into four broad groups. In this study, group A seeds from most varieties of kidney (*Phaseolus vulgaris*), runner (*P. coccineus*) and tepary (*P. actifolius*) beans showed high reactivity with red blood cells from a range of different species and were also highly toxic (none of the rats survived the feeding experiment). Group B contained seeds from lima or butter beans (*Phaseolus lunatas*) and winged bean (*Phosphocarpus tetagonolobus*) and they agglutinated only human and pronase-treated rat erythrocytes. The seeds did not support proper growth, but the animals did survive the feeding experiment. Seeds included in group C were lentils (*Lens culinaris*), peas (*Pisum sativum*), chickpeas (*Cicer arietinum*), black-eye peas (*Vigna sinensis*), pigeon peas (*Cajanus cajan*), mung beans (*Phaseolus aureus*), field or broad beans (*Vicia faba*) and adzuki beans (*Phaseolus angularis*). These generally had low reactivity with all cells and were nontoxic. Group D contained soya (*Glycine max*) and pinto (*Phaseolus vulgaris*) beans which generally had low reactivity with all cells but caused growth depression at certain levels of inclusion in the diet. It was thought that the growth depression in this group was due to antinutritive factors other than lectins. The experiment showed that no single erythrocyte type could be used as a sole indicator of toxicity. The potential toxicity could, however, be predicted from the response to various erythrocytes. The usefulness of animal feeding trials was clearly shown by these experiments. ***See*** Beans; Peas and Lentils

Assay Methods

The detection of lectins in plant extracts is generally performed by a serial dilution technique followed by visual estimation of the end-point. Washed red blood cells are activated by pretreatment with a suitable proteinase (pronase, trypsin, papain) and the test is always carried out in saline solution (0·9%). A more quantitative method is based on the photometric measurement of the density of red blood cells that have not been agglutinated by the extracted lectin. Haemagglutinin activity is usually expressed as haemagglutinin units per milligram of sample. One unit is usually defined as the lowest amount of sample required for agglutination under the test conditions, but Grant *et al.* (1983) defined one unit as the amount of material per millilitre in the last dilution giving 50% agglutination.

The rather subjective nature of this test makes it quite difficult to compare results between different research centres, but it gives good results in the hands of experienced workers. Many workers use a standard reference material to calibrate their own experiments. Wide use of a reference sample between research centres would make comparison of published data more feasible.

Biological Effects of Haemagglutinins

The *in vitro* precipitation of red blood cells does not explain the *in vivo* effects of lectins. Lectins cause agglutination or 'clumping together' of cells by binding to the saccharides branching from lipid and protein molecules of the cells' outer surface. A definite relationship between the lectin content of legume-containing diets and the nutritive value has been made. The feeding of lectin preparations from some legumes has been shown to cause definite growth retardation and in extreme cases death of the animals. In contrast, it has been shown that white pea haemagglutinin has no growth-depressing or toxic effects on rats when fed at 1% level in the diet.

Many ingested lectins cause toxicity by binding to the surface of intestinal epithelial cells, which results in the reduced absorption of nutrients across the intestinal wall. Suppression of intestinal disaccharidase activity and proteolytic activity has also been observed. Since the disaccharidases are involved in the breakdown of carbohydrates to monosaccharides prior to absorption in the small intestine, their reduced activity would be expected to lead to reduced digestibility of dietary carbohydrates. The decreased proteolytic activity would also lead to reduced availability of amino acids and peptides for absorption. The growth inhibition caused by lectins can therefore be attributed to the reduction in absorption of nutrients due to their disruption of the absorptive area of the intestine. ***See*** Carbohydrates, Digestion, Absorption and Metabolism

Detoxification

The detoxification of plant lectins is usually achieved by traditional methods of household cooking. There are conditions where this type of processing is not completely effective. When leguminous seeds are used in the manufacture of animal feeds, cooking is usually omitted, although steam pelleting of the final mixed feed may have some effect in the degradation of antinutritive factors. The time and the temperature of the feed during the pelleting process is clearly insufficient to degrade a significant proportion of the haemagglutinins. As manufacturers do not want reduced food utilization occurring with their products they usually limit the amount of leguminous seed protein incorporated in their mix.

Lima bean lectin has been observed to be stable between pH 5 and 7 at elevated temperatures. Addition of *N*-acetyl-L-cystine (NAC) inactivated lectin at a lower temperature than in its absence. This suggests that NAC reacted with the disulphide linkage of the lectin to bring about inactivation. NAC had no effect on lectins extracted from soya bean flour; this supports the proposed role of NAC, as it is known that soya bean lectins do not contain any disulphide bridges. Addition of NAC prior to heat-processing of some legume flours may reduce the amount of heat-processing required and improve the overall quality of the final product.

Haemagglutinins are almost completely destroyed by boiling for 60 min at atmospheric pressure. Autoclaving at 105°C for 30 min is also effective. If leguminous seeds are previously soaked for 18 h, the haemagglutinins can be completely destroyed by boiling for 2–5 min. Other traditional techniques used to improve the nutritional value of legume seeds involve soaking and germinating. It has been shown that germination of lima beans for 6 days results in a 50% reduction in activity. Soaking for the same time results in an overall 34% reduction in activity. It should be noted that soaking, though not as effective as germination, would only be the first step in preparation; normal cooking would destroy further lectin in the beans.

Dry heat treatment (roasting) to destroy haemagglutinins is remarkably ineffective. For example, winged beans (*Psophocarpus tetragonolobus*) heated at 100°C for 2 h show less than 5% reduction in lectin activity. The generally low but quite variable levels of haemagglutinins in this seed could easily be destroyed by autoclaving at 120°C for 5 min.

The treatment of whole cow peas (*Vigna unguiculata*) with microwaves (2400 MHz) for 6 min was almost ineffective in destroying haemagglutinins, while dry roasting (160°C for 50 min) led to between 28% and 47% reduction in lectin content. Treatment of cow peas with mild alkali (0·5% sodium bicarbonate for 12 h) led to an 80% reduction in haemagglutinin content. ***See*** Cooking, Domestic Use of Microwave Ovens

Infrared heating of winged bean (*Psophocarpus tetragonolobus*) for 60 s was remarkably effective in reducing haemagglutinating activity of this seed, whereas oven and microwave heating had little effect.

Conclusions

Extensive research has been carried out on the chemical, physical and biochemical properties of haemagglutinins, but the understanding of their toxicological and nutritive properties is limited. These properties are an important aspect of our knowledge as legumes provide a vital source of protein in developing countries. The preparation of protein isolates from leguminous seeds and the use of these in quick cooking food products needs careful consideration of the possible antinutritional effects of residual haemagglutinins. It is possible that these toxic effects may be only partially destroyed and the resulting, slightly reduced food utilization may be difficult to detect. It is not clear what the long-term effect of consuming low levels of these compounds will be.

Haemagglutinins occur in a wide range of leguminous seeds. Lupins are the only common legume that do not appear to contain any lectins. Legumes contain a wide variety of toxins, including haemagglutinins, that are heat-labile and should therefore present no health hazard when properly cooked. Thus, it is surprising that outbreaks of poisoning following the consumption of incompletely cooked leguminous seeds occur in an otherwise well-informed society.

Bibliography

Bender AE (1983) Haemagglutinins (lectins) in beans. *Food Chemistry* 11: 309–320.

Boyd WC (1963) The lectins: Their present status. *Vox Sanguinis* 8: 1–32.

Golstein IJ, Monsigny M, Osawa T and Sharon N (1980) What should be called a lectin? *Nature* 285: 66.

Grant G, More LJ, McKenzie NH, Stewart JC and Pusztai A (1983) A survey of the nutritional and haemagglutination

properties of legume seeds generally available in the U.K. *British Journal of Nutrition* 50: 207–214.
Hill GH (1977) The composition and nutritive value of lupin seed. *Nutrition Abstracts and Reviews (Series B)* 42: 511–529.
Huprikar SV and Sohonie K (1965) Haemagglutinin in Indian pulses. 2. Purification and properties of haemagglutinin from white pea *Pisum sativum*. *Enzymologia* 28: 300–345.
Jaffé WG (1960) Effect of feeding lectins and peas on the intestinal and hepatic enzymes of albino rats. *Arzneimittel Forschung* 10: 1012–1016. (Cited by Jindal *et al.*, 1982.)
Jaffé WG (1969) Haemagglutinins. In: Liener IF (ed.) *Toxic Constituents of Plant Food Stuffs* 1st edn, pp 69–74. New York: Academic Press.
Jaffé WG (1980) Hemagglutinins (lectins). In: Liener IF (ed.) *Toxic Constituents of Plant Food Stuffs* 2nd edn, pp 73–102. New York: Academic Press.
Jaffé WG and Vega Lette CL (1968) Heat-labile growth-inhibiting factors in beans (*Phaseolus vulgaris*). *Journal of Nutrition* 94: 203–210.
Jaffé WG, Moreno R and Wallis V (1973) Amylase inhibitors in legume seeds. *Nutrition Reports International* 7: 169–173.
Jindal S, Soni GL and Singh R (1982) Effect of feeding of lectins from lentils and peas on the intestinal and hepatic enzymes of albino rats. *Journal of Plant Foods* 4: 95–103.
Kadam SS, Smithard RR, Eyre MD and Armstrong DG (1987) Effects of heat treatments of antinutritional factors and quality of proteins in winged bean. *Journal of the Science of Food and Agriculture* 39: 267–275.
Liener IE (1955) The photometric determination of the hemagglutinating activity of soyin and crude soybean extracts. *Archives of Biochemistry and Biophysics* 54: 223–231.
Liener IE (1962) Toxic factors in edible legumes and their elimination. *American Journal of Clinical Nutrition* 11: 281–298.
Liener IE (1979) The nutritional significance of plant protease inhibitors. *Proceedings of the Nutrition Society* 38: 109–113.
Muzquiz M, Burbano C, Gorospe MJ and Rodenas I (1989) A chemical study of *Lupinus hispanicus* seed – toxic and anti-nutritional components. *Journal of the Science of Food and Agriculture* 47: 205–214.
Ologhobo AD and Fetuga BL (1984) Haemagglutinin extracts from raw, sprouting and differentially processed lima beans. *Journal of Agricultural Science (Cambridge)* 102: 241–244.
Pusztai A and Palmer R (1977) Nutritional evaluation of kidney beans (*Phaseolus vulgaris*): the toxic principle. *Journal of the Science of Food and Agriculture* 28: 620–623.
Reaidi GB, McPherson L and Bender AE (1981) Toxicity of red kidney beans (*Phaseolus vulgaris*). *Journal of the Science of Food and Agriculture* 32: 846–847.
Sales MG, Maia GA, Teles FFF *et al* (1984) Cowpea solids prepared by mild alkali treatment, dry roasting, and microwave heating. *Nutrition Reports International* 30: 973–981.
Tan NH, Rahim ZHA, Khor HT and Wong KC (1983) Winged bean (*Psophocarpus tetragonolobus*) tannin level, phytate content, and hemagglutinating activity. *Journal of Agricultural Food Chemistry* 31: 916–917.
Wallace JM and Friedman M (1985) Inactivation of haemagglutinins in lima bean (*Phaseolus lunatus*) flour by *N*-acetyl-L-cysteine, pH and heat. *Nutrition Reports International* 32: 743–748.

G. P. Savage
Lincoln University, Canterbury, New Zealand

Detoxification of Naturally Occurring Toxicants of Plant Origin

Proteins of plant origin, particularly oil seeds and legumes, provide a valuable source of protein for humans and animals despite the fact that they contain substances that adversely affect the nutritional quality of the protein (Table 1). Fortunately, in many cases, these substances are heat-labile and can be readily eliminated or inactivated by the heat treatment involved in domestic cooking, commercial processing or by treatment with chemicals. In other cases, potential toxicants remain relatively innocuous until they are acted upon by enzymes of endogenous origin. Paradoxically, advantage can sometimes be taken of the action of these enzymes to detoxify certain plants employing such traditional modes of food preparation as germination or fermentation. ***See*** Legumes, Legumes in the Diet; Legumes, Dietary Importance;

Protease Inhibitors

Proteins capable of inhibiting mammalian digestive enzymes such as trypsin and chymotrypsin are widely distributed in nature and have been shown to retard the growth of animals by virtue of their ability to interfere with the digestion of dietary protein. The animal tends to adapt to this situation by undergoing an enlargement of the pancreas which, in the extreme case of the long-term ingestion of these inhibitors by rats, leads to pancreatic cancer.

On account of its economic importance, the soya bean has received the most attention with respect to the means whereby these inhibitors can be inactivated. In general, destruction by heat treatment is a function of the temperature, duration of heating, particle size and moisture content, conditions which are carefully controlled during the commercial processing of soya beans in order to obtain a product possessing maximum nutritional value. An example of the relationship between the destruction of the trypsin inhibitor and the concomitant improvement in the nutritional quality of the protein is shown in Fig. 1. It is reassuring to note that most commercially available soya bean products intended for human consumption such as tofu, soya milk, soya protein isolates and concentrates, and meat analogues have received sufficient heat treatment to reduce the trypsin inhibitor activity to nontoxic levels. ***See*** Soya Beans, The Crop; Soya Beans, Processing for the Food Industry; Plant Toxins, Trypsin Inhibitors

Although the treatment of soya beans and other legumes with live steam (toasting) is the most commonly

Table 1. Examples of naturally occurring toxicants of plant origin, their distribution and physiological effects

Toxicant	Distribution	Physiological effect
Protease inhibitors	Most legumes	Impaired growth Enlarged pancreas Pancreatic cancer
Lectins	Most legumes	Impaired growth/death
Goitrogens	Cabbage family	Hyperthyroidism
Cyanogens	Lima beans Cassava	Respiratory failure
Phytate	Most plants	Interference with mineral availability
Tannins	Most legumes	Interference with protein digestibility
Oligosaccharides	Most legumes	Flatulence
Vicine/convicine	*Vicia faba*	Favism
β-*N*-oxalyl-α,β-diaminopropionic acid	*Lathyrus sativus*	Lathyrism

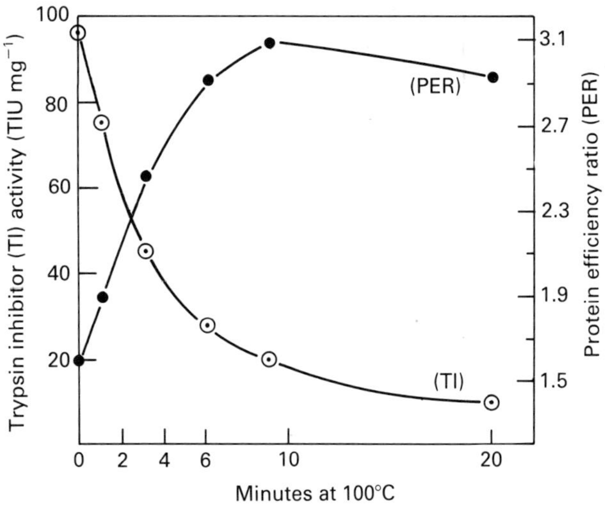

Fig. 1 Effect of heat treatment on the trypsin inhibitor (TI) activity and protein efficiency ratio (PER) of soya beans. PER=gain in weight (g)/protein intake (g). Courtesy of Academic Press, New York.

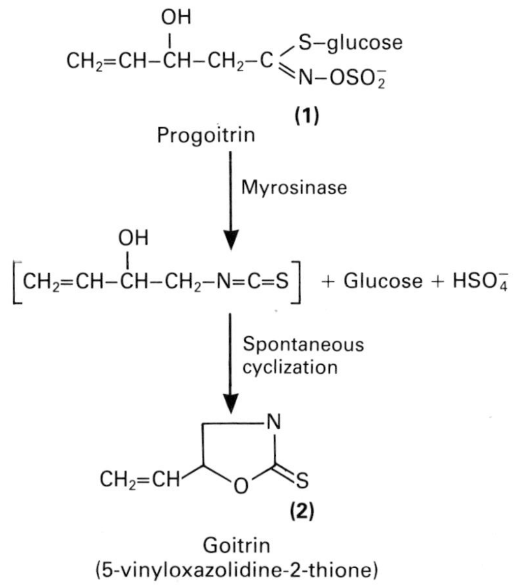

Fig. 2 Structure of one of the goitrogenic compounds present in the cabbage family. Progoitrin (**1**) is biologically inactive but, upon hydrolysis by the enzyme myrosinase, the active principle goitrin (**2**), is released.

employed method for inactivating the trypsin inhibitor, other modes of heat treatment or processing have proved to be equally effective. These include boiling in water, dry roasting, microwave cooking, γ irradiation and infrared radiation.

Traditional methods of preparing various foods derived from beans generally result in products which are quite low in protease inhibitor activity. Tofu is comprised mainly of protein which has been precipitated from a hot water extract of soya beans with calcium–magnesium salts, and soya milk is simply a hot water extract of soya beans which may have been clarified by filtration. In both cases the boiling or steaming of the soya beans prior to extraction with water serves to inactivate the inhibitors. The same holds true for fermented preparations of soya beans and other legumes such as tempeh, miso and natto in which the beans have been subjected to boiling prior to the fermentation step. ***See*** Fermented Foods, Fermentations of the Far East; Soya Cheeses; Soya Milk

The trypsin inhibitors are protein molecules whose activity is retained only if the disulphide bonds of the cystine residues remain intact. Cleavage of these disulphide bridges by treatment of soya beans with reducing agents (cysteine, *N*-acetylcysteine, mercaptoethanol or reduced glutathione) or with sodium metabisulphite causes inactivation of the inhibitors.

Detoxification of Naturally Occurring Toxicants of Plant Origin

Lectins

Paralleling the distribution of protease inhibitors among the oil seeds and legumes is a class of proteins referred to as 'lectins'. These are proteins which exhibit the unique property of being able to bind to specific sugars which comprise the structure of glycoproteins. Their nutritional significance lies in the fact that, by binding to the glycoproteins located on the surface of the cells lining the small intestine, they interfere with the absorption of nutrients; the result is a failure in growth and eventual death. ***See*** Protein, Chemistry

The toxic effects of the lectins can be effectively eliminated by heat treatment, essentially under the same conditions as those which inactivate the protease inhibitors. However, there are documented cases where beans which have been eaten raw or only partially cooked have led to serious outbreaks of gastrointestinal illness. For example, bean-containing stews and casseroles, when cooked in a slow cooker where a relatively low temperature is maintained for a long period of time, retain sufficiently high levels of lectin to cause illness. ***See*** Beans; Pulses

Goitrogens

Goitre-causing agents in the form of glycosides (also referred to as 'glucosinolates') are found in members of the cabbage family, which includes not only cabbage but also broccoli, brussels sprouts, cauliflower, turnips, rapeseed and mustardseed. These compounds are biologically inactive as long as they remain bound to glucose ((**1**), Fig. 2), but an enzyme, myrosinase, present in the same plant serves to release the active goitrogenic principle ((**2**), Fig. 2). The glucosinolates found in the cabbage family appear to pose little risk to human health since the enzyme responsible for the release of goitrin is inactivated by household cooking. The glucosinolates may also be removed to a great extent by leaching out into the cooking water. The common usage of rapeseed in the feeding of livestock may prove to be toxic unless it has been treated with moist heat. Alternative methods of detoxification of rapeseed may involve prior extraction of the glucosinolates with water or acetone or by decomposition with iron salts or soda ash. These procedures, however, do not preclude the possibility that some goitrin produced enzymatically prior to processing may still remain in the meal. A more effective means of detoxification is one in which an aqueous slurry of the meal is deliberately allowed to undergo autolysis, which serves to liberate virtually all of the goitrin; the latter is then removed by extraction with water or acetone. Lactic acid fermentation or treatment with a specific fungus (*Geotrichum candidum*) has also been reported to be an effective means for the biological destruction of the glucosinolates in rapeseed meal. The immobilization of myrosinase on a solid matrix offers a promising biotechnological approach for the hydrolysis of the glucosinolates, provided the goitrogenic end products are subsequently removed by extraction as described above. ***See*** Glucosinolates; Goitrogens and Antithyroid Compounds

Cyanogens

Many plants are potentially toxic because they contain glycosides, principally linamarin ((**3**), Fig. 3) from which hydrogen cyanide may be released by the action of an endogenous enzyme, linamarinase, when the plant tissue is macerated. Among the cyanogenic plants most likely to be eaten by humans are the lima bean and cassava. Although certain tropical varieties of the lima bean may contain toxic levels of cyanide, those varieties consumed in the USA and in Europe generally have levels of cyanide far below the dosage known to be toxic in humans. Cassava is a stable food item in the tropics which may contain toxic levels of cyanide unless properly processed. The traditional method of preparing cassava involves the removal of the peel, which is

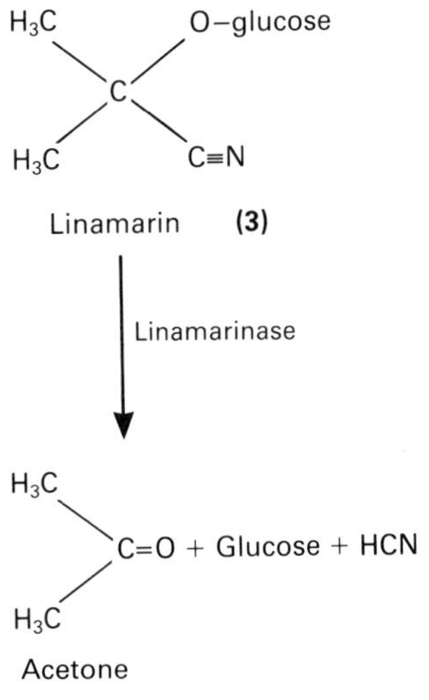

Fig. 3 Structure of the major cyanogen present in lima beans and cassava. Enzymatic hydrolysis of linamarin (**3**) releases hydrogen cyanide.

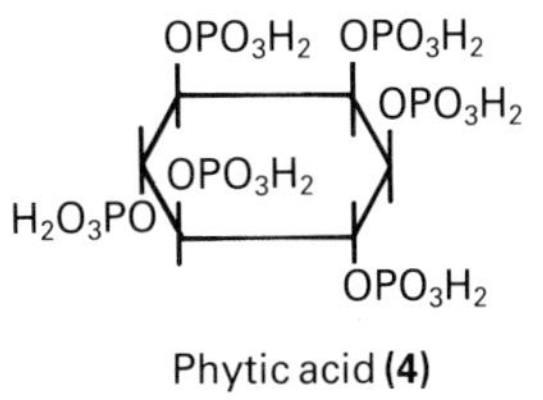

Fig. 4 Structure of phytic acid (**4**). Enzymatic removal of the phosphate groups by phytase serves to eliminate its metal-binding property.

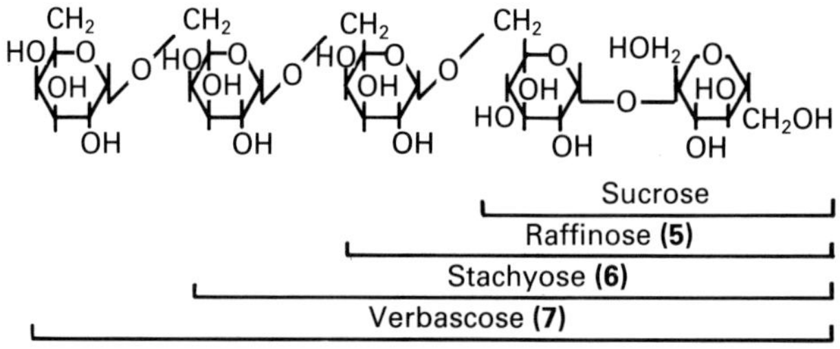

Fig. 5 Structure of the oligosaccharides (**5**, **6** and **7**) responsible for flatulence. In the absence of enzymes in the small intestine of humans which are capable of hydrolysing α-1,6-galactosidic linkages, these compounds are metabolized by the microflora of the gut into gaseous end products.

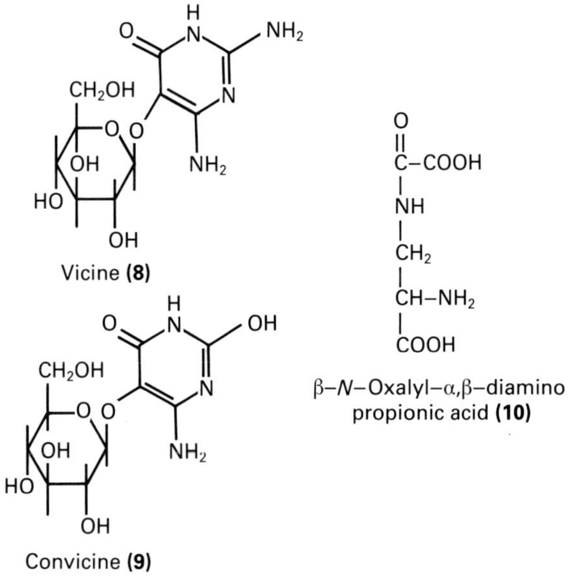

Fig. 6 Causative factors of favism (vicine (**8**) and convicine (**9**)) and lathyrism (β-*N*-oxalyl-α,β-diaminopropionic acid (**10**)).

particularly rich in linamarin, followed by thorough washing of the pulp with running water. A further reduction in toxicity is achieved by the application of heat (boiling, roasting or sun drying) which serves to inactivate linamarinase and to volatilize any hydrogen cyanide which may have been produced. The cyanide content of cassava can also be reduced by microbial fermentation to produce a native Nigerian dish called gari. The hydrogen cyanide released by the enzymes elaborated by the fermenting microorganisms is then eliminated by frying. The use of an exogenous source of linamarinase derived from one of several species of fungi has also been proposed for enhancing the detoxification of gari. *See* Cassava, The Nature of the Tuber

Phytate

Phytic acid (**4**, Fig. 4) or its salt phytate is inositol combined with six phosphate groups and is a common constituent of most plants. Its antinutritional effect lies in the fact that it forms chelates with metal ions such as calcium, magnesium, zinc and iron to form poorly soluble compounds which are not readily absorbed from the intestine, thus interfering with the bioavailability of these essential minerals. The ability of phytate to bind metal ions is lost when the phosphate groups are removed by hydrolysis through the action of phytase. Heat alone is relatively ineffective in reducing the phytate content of plant materials, but the phytate level can be reduced by taking advantage of the endogenous enzyme phytase which accompanies the phytate in separate compartments of the plant tissue. For example, by the simple expedient of allowing ground beans to undergo autolysis under conditions which are optimal for phytase activity, one can obtain a significant reduction in phytate content. An exogenous source of phytase is commercially available as a feed additive to livestock and poultry diets. This not only serves to eliminate the mineral binding ability of the phytate that may be present from other plant ingredients of the diet, but also makes more phosphate available to the animal.

Traditional fermented dishes derived from such plants as soya beans, cassava, rice, maize, cowpeas or yams have reduced levels of phytate, presumably due to the action of phytase produced by various moulds, bacteria or yeasts involved in the fermentation. Even the use of yeast in breadmaking serves to reduce the phytate content of wheat flour. The germination of the mature seeds of various legumes results in a great increase in phytase activity with a concomitant reduction in phytate. Other techniques for removing phytate such as ultrafiltration and ion exchange chromatography have also been proposed, but these are unlikely to replace the simpler methods already described. *See* Phytic Acid, Nutritional Impact

Tannins

Oil seeds and legumes contain appreciable levels of polyphenolic compounds broadly referred to as tannins. The negative nutritional effects of tannins are diverse and incompletely understood, but the major effect is to interfere with the digestibility of dietary protein. This may be duc to the binding of the tannins to the protein to form enzyme-resistant substrates or to a direct binding to the digestive enzymes themselves.

Since tannins are located primarily in the seed coat of dry seeds, the physical removal of the seed coat by dehulling or milling markedly reduces the tannin content with a resultant improvement in the nutritional quality of the protein. Soaking in water or salt solution prior to household cooking also causes a significant reduction in tannin, provided the cooking broth is discarded. Treatment with a variety of chemicals such as ammonia, hydrogen peroxide, formaldehyde, polyethylene glycol or polyvinylpyrrolidone has also proved to be effective means for reducing the phytate content of plant sources of protein. Germination causes more than a 50% loss in tannins in such legumes as the chick pea, green gram, mung bean and black gram, presumably due to the action of polyphenol oxidase of endogenous origin.

Flatulence-producing Factors

One of the important factors limiting the use of legumes in the human diet is the production of intestinal gas (flatulence). Flatulence is generally attributed to the inability of the gut enzymes to hydrolyse the α-1,6-galactosidic linkages of such oligosaccharides as raffinose, stachyose and verbascose ((**5**), (**6**) and (**7**), respectively (Fig. 5)). In their unhydrolysed forms these oligosaccharides become metabolized by the microflora of the large intestine into carbon dioxide, hydrogen and methane, gases which are responsible for the characteristic features of flatulence, namely nausea, cramps, diarrhoea and the social discomfiture associated with the release of rectal gases. Advantage may be taken of the endogenous galactosidases present in the plant tissue by allowing a slurry of the raw ground beans to undergo autolysis under suitable conditions or by treatment with an exogenous source of these enzymes derived from moulds or bacteria. Such traditional dishes as tofu and tempeh which have undergone fermentation by microorganisms likewise have little flatus activity. Protein isolates from which these oligosaccharides have been eliminated during the course of their isolation are essentially devoid of these offending oligosaccharides. Preparations of galactosidases from microbial sources which have been immobilized as continuous flow reactors or in the form of a hollow-fibre dialyser have also been employed for removing flatulence-producing factors. ***See*** Microflora of the Intestine, Role and Effects

Lathyrism/Favism

A few plants are known to cause specific diseases in humans. For example, the ingestion of the chickling vetch (*Lathyrus sativus*) causes a neurological condition known as lathyrism, a disease which is common in India and Bangladesh. The consumption of the broad bean (*Vicia faba*) causes a disease characterized by haemolytic anaemia (favism) in certain genetically disposed individuals. Although the causative factors of favism ((**8**) and (**9**), Fig. 6) and lathyrism ((**10**), Fig. 6) have been tentatively identified, only in the case of lathyrism has an effective means of detoxification been found. In the latter case, over 90% of the toxin can be eliminated by the simple expedient of soaking the seeds overnight in an excess of water followed by steaming, roasting or sun drying.

Bibliography

Cheeke PR (ed.) (1989) *Toxicants of Plant Origin*, vol I–IV. Boca Raton: CRC Press.

Friedman M (ed.) (1986) *Nutritional and Toxicological Significance of Enzyme Inhibitors in Foods*. New York: Plenum Press.

Liener IE (ed.) (1980) *Toxic Constituents of Plant Foodstuffs*, 2nd edn. New York: Academic Press.

Liener IE (1987) Detoxifying enzymes. In: King RD and Cheetham PSJ (eds) *Food Biotechnology*, vol. 1, pp 249–271. London: Elsevier.

Liener IE (1989) Antinutritional factors. In: Mathews RH (ed.) *Legumes: Chemistry, Technology, and Human Nutrition*, pp 339–382. New York: Marcel Dekker.

Liener IE (1989) Control of antinutritional and toxic factors in oilseeds and legumes. In: Lusas EW, Erikson DR and Nip W-K (eds) *Food Uses of Whole Oil and Protein Seeds*, pp 249–271. Champaign, IL: American Oil Chemists Society.

Irvin E. Liener
University of Minnesota, St Paul, USA

PLUMS AND RELATED FRUITS

The European plum (*Prunus domestica*) probably originated in eastern Europe or western Asia around the Caucasus and the Caspian Sea. It has been known in Europe for more than 2000 years. Prunes are the most important subgroup. Others are the greengage, yellow egg, imperatrice and lombard groups. The Japanese plum (*P. salicina*) originated in China and was domesticated in Japan. An American fruit breeder, Luther Burbank, developed many commercial cultivars from it after its introduction into the USA around 1870. Native plum species occur in some countries, such as North America (e.g. *P. americana*, *P. hortulana*, *P. subcordata*) and the UK (*P. spinosa*, blackthorn sloe; *P. institia*, damson or bullace), but these are not extensively grown commercially. Some are used in breeding programmes with Japanese plums. Other plum types, used as rootstocks or in plum breeding programmes, or occurring in home orchards, are *P. cerasifera* (myrobalan, cherry plum), *P. simoni* (apricot plum) and *P. institia*.

Plums are cultivated over a wide range of climatic conditions. In 1989, world production of plums was estimated at 6 523 000 t (Table 1), representing about 10% of the world production of deciduous fruit. Twelve countries produced more than 100 000 t each, or more than 83% of the total production. The ratio of Japanese plum to European plum is approximately 99:1 for Japan, 60:40 for South Africa and Israel, 40:60 for Canada (Ontario) and Italy, and 0:100 for Yugoslavia.

Anatomy and Morphology of the Fruit

The plum is a succulent fruit, called a drupe, and is formed by thickening of the ovary wall following fertilization. The floral remnants become senescent and drop off. The skin (epicarp) consists of a layer of elongated living cells (epidermis) covered by a thin film of cutin (cuticle). It protects the underlying tissue and permits exchange of metabolites or gases with the external environment through openings (lenticels). Wax deposited on the cuticle forms a light-grey bloom which gives the skin a dull appearance and makes it impermeable to water. The thick, fleshy, edible portion (mesocarp) consists of parenchyma cells which have an active protoplast where all metabolic reactions occur. The protoplast is enclosed by a pectinaceous cell wall. Middle lamellae, which occur between adjacent cell walls, have a high calcium content and play a vital role in maintaining cohesion between cells, thus providing structural rigidity. The single seed is enclosed by a hard stone (endocarp) consisting of isodiametric cells with thick, lignified cell walls (sclerenchyma). The mesocarp usually clings to the endocarp, although some cultivars are freestone (i.e. the mesocarp is free from the endocarp).

Japanese plums are large and attractive, round or heart-shaped, with or without a prominent apex. The surface of some cultivars is covered by a heavy bloom. The flesh and skin colour of ripe fruit may vary from yellow to blood red. The skin of the red-coloured cultivars changes during ripening from green to red, usually starting at the apex. Fruit texture varies from firm and nonmelting to soft and melting. European plums are usually oval, with bulging of the central side, and compressed bilaterally. Their skin colour is blue or purple and they have a thick, meaty, freestone flesh. Fruit texture is usually firm and semimelting.

Physiology of Plum Fruit

Cultivar, climatic, environmental and cultural factors significantly affect physiology. Cell division, cell multiplication and differentiation of tissue occur during the first few weeks after fertilization. Cell enlargement and maturation follow. Numerous biochemical changes occur in plum fruits during the last few weeks prior to harvest, whilst the fruits are still attached to the tree. Many of these physiological changes are consistent and predictable and can be used as indices for determining harvesting maturity. Substances synthesized in the leaves through photosynthesis are translocated to the fruit and transformed into products which eventually determine the quality and nutritional value of the fruit.

The sugar and total soluble solids content increase throughout the period of growth. Organic acids accumulate in the fruit during the early stages of growth and then gradually decrease. The phenolic content of the fruit is high during the early stages, decreases and then remains contant until harvest. Volatiles, which determine the flavour or aroma, are produced. Wax develops on the skin of the fruit. The fruit attains its full size and optimum maturity although it is still unripe. ***See*** Phenolic Compounds; Sensory Evaluation, Aroma; Sensory Evaluation, Taste

During ripening, pectic substances in the cell walls change from an insoluble to a soluble form, resulting in softening of the fruit. The chlorophyll (green pigments) content of the skin decreases, and the carotenoid (yellow and red pigments) content increases. Based on its respiration pattern, the plum is a climacteric fruit. The respiration rate is high during and immediately after cell division. As maturity approaches, it decreases to the pre-climacteric minimum and then increases irreversibly to a maximum (climacterium) during ripening. At the climacterium, the fruit are soft and sweet, with a characteristic flavour, and are ideal for eating. Subsequently, senescence sets in, whereupon the respiration rate decreases and the fruits become overripe and decayed. ***See*** Ripening of Fruit

Table 1. World production of plums during 1989

Country	Production ($\times 10^3$ t)
Countries producing more than 10 000 t	
Africa	
Algeria	31[a]
Egypt	40[a]
Morocco	40[a]
South Africa	18
North and Central America	
Mexico	86[a]
USA	786
South America	
Argentina	50[a]
Chile	86
Asia	
Afghanistan	37[a]
China	770[a]
India	38[a]
Iraq	33[a]
Israel	15[a]
Japan	70[a]
Korea	40[a]
Lebanon	12[a]
Pakistan	50[a]
Syria	31
Turkey	166
Europe	
Albania	13[a]
Austria	82
Bulgaria	139
Czechoslovakia	49
France	146
FRG	382[b]
Hungary	180
Italy	133
Norway	13[a]
Poland	75
Romania	765[a]
Spain	145
Switzerland	31[a]
UK	15[b]
Yugoslavia	819
Oceania	
Australia	20
Commonwealth of Independent States	1050[a]
Countries producing less than 10 000 t	67
Total	**6523**

[a] FAO estimate.
[b] Unofficial figure.
Data from FAO (1990) *FAO Yearbook for 1989: Production.* FAO Statistics Series No. 43. Rome: Food and Agriculture Organization.

Chemical Composition and Nutritional Value

Food composition tables generally do not specify the cultivar analysed nor whether a sample was fresh or cold-stored before analysis. Nevertheless, they are useful guides to the composition and nutritional value of fruit. A typical composition table for plums and related fruits is given in Table 2. ***See*** individual nutrients

Taste is largely determined by a balance between the sugar and acid contents. Low acid and high sugar contents result in a bland taste, and high acid and low sugar contents in a sour taste. The contents of individual sugars and acids vary between cultivars. For example, in Australia, the cultivar Santa Rosa contains approximately twice as much malic acid as Mariposa, but has a lower soluble solids content. ***See*** Acids, Natural Acids and Acidulants

Harvesting, Handling and Storage

Plums are harvested in summer, from about December to February in the southern hemisphere and from about July to October in the northern hemisphere. It is important to harvest at the correct stage of maturity. If plums are harvested while still immature, the characteristic flavour will not develop and fruit quality will be inferior. Plums harvested too ripe are very susceptible to injuries and fungal infection during harvest and post-harvest handling, and their storage life will be short.

Dessert Plums

Dessert plums are harvested when they are physiologically mature, but unripe. Ripening to a stage at which the fruit has a pleasant taste and aroma occurs during and, mostly, after storage. Plums which are to be sold soon after harvest must be picked at a riper stage than usual, or cold-stored for a few days to promote proper ripening. They require extra care during handling.

Total soluble solids content, measured by means of a refractometer, and mesocarp firmness, measured by

Table 2. Nutrient and mineral content of fresh plums, damsons, greengages and dried prunes

	Content (per 100-g edible portion)			
Nutrient	Plum[a]	Damson[b]	Greengage[b]	Prune[a]
Moisture (%)	85·2	77·5	78·2	32·4
Energy value (kJ)	230·0	162·0	202·0	1000·0
Carbohydrate (g)	11·0	9·6	11·8	55·5
Dietary fibre (g)	2·0	4·1	2·6	7·2
Protein (g)	0·8	0·5	0·8	2·6
Total fat (g)	0·6	Trace	Trace	0·5
Nicotinic acid (mg)	0·5	0·3	0·4	2·0
Pantothenic acid (mg)	0·18	0·27	0·20	0·46
Riboflavin (mg)	0·10	0·03	0·03	0·16
Thiamin (mg)	0·04	0·10	0·05	0·08
Folic acid (μg)	2·0	3·0	3·0	4·0
Vitamin A (μg)	192·0	220·0	—[c]	1194·0
Vitamin B_6 (mg)	0·08	0·05	0·05	0·26
Vitamin C (mg)	10·0	3·0	3·0	3·0
Vitamin E (mg)	0·65	0·70	0·70	—[c]
Sodium (mg)	0	2·0	1·0	4·0
Potassium (mg)	172·0	290·0	310·0	745·0
Calcium (mg)	4·0	24·0	17·0	51·0
Magnesium (mg)	7·0	11·0	8·0	45·0
Phosphorus (mg)	10·0	16·0	23·0	79·0
Iron (mg)	0·1	0·4	0·4	2·5
Copper (mg)	0·04	0·08	0·08	0·43
Zinc (mg)	0·10	0·10	0·10	0·53

[a] Data from Langenhoven ML *et al.* (1991).
[b] Data from Paul AA and Southgate DAT (1978).
[c] No data available.

means of a probe (penetrometer) forced into the flesh, are good indicators of maturity. However, these tests are destructive and pickers use skin colour, in combination with these tests, as a maturity index. Fruit size is an additional criterion used to determine optimum maturity stage.

Fruits are picked by hand into picking bags and carefully transferred to containers with a capacity of about 500 kg. Plums bruise easily and must be handled carefully during and after harvest. Filled containers are transported to a packhouse. Bruised, cut or decayed fruit and culls are removed. The plums are sorted according to size and either hand- or mechanically packed into cardboard or wooden containers. These usually have a capacity ranging from about 5 to 20 kg. Trays or padding material immobilize the fruit in the container to prevent transit injuries. Packed containers are unitized (palletized) to facilitate handling and protect the fruit.

To extend the storage life of the fruit, its temperature must be lowered as soon after harvest as possible. Precooling, which removes field heat from the fruit, commences immediately after packing and palletizing. Cold air is distributed through the pallets in such a way that it comes into contact with all fruit. The ideal storage conditions for plums are a temperature of −0·6°C to 0°C (31–32°F) and a relative humidity of 85–90%. The storage period depends on the cultivar. Most cultivars cannot be stored for longer than 3–4 weeks, but others can be stored for up to 3 months. The fruit of some cultivars are susceptible to chilling injury. To prevent chilling injury the temperature is raised from −0·6°C to about 7°C after 7–10 days and this temperature is maintained for the rest of the storage period. This is called dual-temperature storage. Controlled- or modified-atmosphere storage increases the storage life of some cultivars. ***See*** Controlled Atmosphere Storage, Effect on Fruit and Vegetables; Storage Stability, Mechanisms of Degradation

Prunes

Prunes are harvested when fully ripe and in an ideal condition for fresh consumption. Overmature fruit discolour and break down during drying. The fruit of some cultivars drop from the trees naturally and must be gathered promptly to prevent losses. Soluble solids content and flesh firmness are important maturity indices. As in the case of fresh plums, these indices are related to skin colour.

Most prunes are dried rather than sold fresh. Fresh prunes can be cold-stored, but their storage life is shorter than that of dessert plums. Sorted prunes are dipped into a hot caustic solution (generally 0·5% sodium hydroxide at boiling point) for sufficient time to remove the wax layer and cause minute cracks in the skin. This facilitates moisture loss and speeds up drying. The prunes are then packed onto trays and dried in the sun. Prunes can also be dried in mechanical dehydrators. The fruits are washed and placed on trays in the dehydrator. Hot air circulates through the trays and drying takes place within 24 h. Prunes are gathered when dry but still pliable, with a moisture content of about 20%. ***See*** Drying, Drying Using Natural Radiation; Drying, Theory of Air Drying

Market Disorders and Diseases of Plum Fruit

Disorders

Disorders are transit- or storage-related. Transit-related disorders include bruising as a result of rough handling, freezing injury owing to temperatures below the freezing point of plums, and shrivelling owing to moisture loss caused by low relative humidities in the storage atmosphere. These disorders can be controlled by effective post-harvest management.

Storage-related disorders are similar to transit-related disorders, but include internal breakdown or internal browning. When a fruit is cut open, internal browning can be seen as a reddish-brown discoloration which is more intense around or near the endocarp. It is responsible for extensive losses during cold storage. It usually occurs at low temperatures and can be prevented by storing fruit at slightly elevated temperatures. Some cultivars are more susceptible than others. Dual-temperature storage is used where plums have to be shipped over long distances.

Diseases

Fungi cause market diseases. The most important post-harvest diseases of plums and prunes are brown rot (*Monilinia laxa*, *M. fructicola*), blue mould rot (*Penicillium expansum*), grey rot (*Botrytis cinerea*), mucor rot (*Mucor piriformis*) and rhizopus rot (*Rhizopus* spp.). These fungi cause decay in a wide range of commodities and their occurrence is not limited to plums. Fungal spores present in the orchard on plant debris or in the air are dispersed to fruit by wind or insects. The spores of *B. cinerea* and *Monilinia* spp. can infect uninjured fruit in the orchard, especially during the 1–2 weeks prior to harvest, when the susceptibility of fruit to infection increases. However, disease symptoms are only expressed during cold storage. The other fungi are wound pathogens which penetrate fruit through skin breaks caused by rough handling or insects. Decay symptoms are related to the fungus involved. Mucor and rhizopus rots develop rapidly, and affected plums become soft and watery, and are covered by black spore masses. Brown rot starts as a small, water-soaked spot which enlarges rapidly and becomes brown or black. The skin becomes leathery but remains intact. Fruit infected by *B. cinerea* are firm, spongy, and only slightly moist. The skin covering lesions readily slips away when slight pressure is applied. Decay development is slower and, in advanced stages of decay, grey spores cover the affected areas. Blue mould rot lesions are wet and soft, and covered with blue-green mould growth. ***See*** Spoilage, Moulds in Food Spoilage

Control of post-harvest decay involves the integration of pre-harvest factors (soil preparation, spray programmes, orchard hygiene, etc.) with sound post-harvest crop management. Fruit produced under optimal conditions possess a high natural resistance to infection. Sanitation in the orchard and packshed keeps the number of spores to a minimum and is an important precautionary measure. All fungi causing post-harvest decay, except *Rhizopus* spp., can grow at 0°C, and cold storage does not prevent decay development. Chemicals can be used to control decay, but only a few are registered for post-harvest use. ***See*** Fungicides

Industrial Uses

Plums have a wide variety of industrial uses. Japanese plums are mostly eaten fresh, although a small percentage is dried. European plums are mostly dried (prunes), cooked (greengages), canned (greengages, yellow egg, damsons, other European plums), or used in jams (plums, greengages, damsons) or jellies (American plums). Prune juice is used as a laxative and plum purée as a baby food. Some plums are used as rootstocks (e.g. marianna, myrobalan) or in breeding programmes (*P. simoni*) and are planted as ornamentals in home gardens. In some parts of Europe, plums are fermented and distilled into a 'brandy'. In Romania, Hungary and Yugoslavia it is called slivovitz, and in other parts of Europe the name depends on the cultivar used (e.g. Quetsch or Mirabelle).

Bibliography

Chandler WH (1965) *Deciduous Orchards* 3rd edn. Philadelphia: Lea and Febiger.

Eksteen GJ and Combrink JC (1987) *Manual for the Identification of Post-harvest Disorders of Pome and Stone Fruit.* Stellenbosch, South Africa: Van der Stel Printers.

Harvey JM, Smith Jr WL and Kaufman J (1972) *Market Diseases of Stone Fruits.* US Department of Agriculture, Agriculture Handbook 414. Washington, DC: US Government Printing Office.

Langenhoven ML, Kruger M, Gouws E and Faber M (1991) *MRC Food Composition Tables* 3rd edn. Parow, South Africa: South African Medical Research Council.

Paul AA and Southgate DAT (1978) *McCance and Widdowson: The Composition of Foods* 4th edn. Oxford: Elsevier Biomedical Press.

Ramming DW and Cociu V (1990) Plums (*Prunus*). *In*: Moore JN and Ballington Jr JR (eds) *Genetic Resources of Temperate Fruit and Nut Crops*, vol. 1, pp 235–281. Wageningen, The Netherlands: International Society for Horticultural Science.

Salunkhe DK, Bolin HR and Reddy Nr (1991) *Storage, Processing, and Nutritional Quality of Fruits and Vegetables* 2nd edn, vol. 1. Boston, Massachusetts: CRC Press.

Taylor HV (1949) *The Plums of England.* London: Crosby Lockwood and Son.

Teskey BJE (1978) *Tree Fruit Production* 3rd edn. Westport, Connecticut: AVI Publishing.

Wills RBH, Scriven F and Greenfield H (1983) Nutrient composition of stone fruit (*Prunus* spp.) cultivars: apricot, cherry, nectarine, peach and plum. *Journal of the Science of Food and Agriculture* 34: 1383–1389.

J. C. Combrink
Stellenbosch Institute for Fruit Technology, Stellenbosch, South Africa

POLITICS AND NUTRITION

Politics can be defined as both the art and science of government, as well as the play of competing economic and ideological interests. This article reviews some aspects of government nutrition policies, and the effects of disparate interests on nutrition in developing and developed countries.

Nutrition Policies

Governments in both developed and developing countries have over the last two decades increasingly focused attention on nutrition through explicit national nutritional policies. *See* National Nutrition Policies

Hunger and malnutrition were put on the international agenda by the League of Nations in the 1930s and remained an important focus of the United Nations (UN) technical agencies, the Food and Agriculture Organization (FAO), the World Health Organization (WHO), and the United Nations Children's Emergency Fund (UNICEF), which were created immediately after World War II. The type of nutrition interventions and development policies supported were shaped by the World Food Surveys conducted every decade by the FAO – these estimated the extent and causes of malnutrition – and by post-colonial theories of economic development. During the 1950s and 1960s the prevailing policy encouraged industrialization and large-scale agriculture in order to increase economic wealth, from which improved nutrition would 'trickle down' rather than be achieved by improved distribution. Nutrition continued to be approached mainly through piecemeal interventions, including child-feeding programmes and nutrition education, but also via some integrated, village-level 'Applied Nutrition Programme' which addressed the economic, educational and food resource constraints on good nutrition. *See* World Health Organization

By the early 1970s it had become clear that rapid economic growth had led to increased inequalities, as well as increased absolute poverty in some countries, such as Pakistan, Nigeria and Brazil, resulting in greater levels of undernutrition in the impoverished.

Dissatisfaction with the effectiveness of past approaches to reducing malnutrition led to the concept of National Nutrition Planning, based on the coordination of sectoral policies to focus on nutrition so as to remove the barrier to economic growth of malnutrition that results in increased morbidity and mortality, poor educational achievement, low work capacity, increased absenteeism and low productivity. This integrated, systematic, intersectoral approach was endorsed and promoted by the FAO and the United States Agency for International Development (USAID), which have assisted several countries to prepare explicit integrated nutrition policies. *See* Malnutrition, The Problem of Malnutrition; Malnutrition, Malnutrition in Developing Countries

In this international context the need for integrated national food and nutrition policies was also expressed in the developed countries, where the malnutrition of poverty had been largely eliminated but where malnutrition of affluence, reflected in increased rates of coronary heart disease, diabetes, obesity and cancer, is the current concern. Several countries developed national dietary goals in the 1960s and 1970s to encourage their populations to consume a more healthy diet, but Norway was the first to pass an intersectoral Food and Nutrition Policy in 1975. ***See*** Malnutrition, Malnutrition in Developed Countries

Types of Nutrition Interventions

The types of interventions that form part of national nutrition policies tend to be limited to palliative measures, such as vitamin supplementation, nutrition education and child-feeding programmes, because the underlying political issues that lead to malnutrition, whether undernutrition or overnutrition, involve fundamental economic and political interests. These are much more difficult and contentious issues to address and rarely overtly feature in nutrition policies. ***See*** Community Nutrition

Famine

One extreme example of the political aspects of nutrition is famine, with mass starvation and death, often attributed to climatic causes such as drought or flooding. Such factors may precipitate famine but it is only the most vulnerable members of society that starve in this type of famine. To quote Amartya Sen, 'starvation is the characteristic of some people not *having* enough food to eat. It is not the characteristic of there *being* not enough food to eat'. He argues that poverty and famines are related to lack of 'entitlement' or the ability of individuals to obtain food through the legal means available in the society, including production possibilities, trade opportunities, state provisions and other methods of acquiring food. When individuals can no longer obtain enough food in exchange for their products or services, they suffer from famine, as happens when food prices rise owing to local shortages and hoarding. Several examples exist where aggregate food supplies were adequate but certain groups could not afford to buy them, such as the Irish famine of 1846, the Bengal famine of 1943, and more recent famines in the Sahelian countries.

War

War is an extreme example of political crisis and is an increasingly common contributor to the causes of famine through the blockade or diversion of food supplies, the destruction of cropland and the migration of farmers away from their lands in the war zones. War contributed to the famines in Bengal, The Netherlands and Warsaw (World War II), and more recently in Ethiopia, Timor, Kampuchea, Laos, Vietnam, Afghanistan, Chad, Uganda and the Sudan. War and its aftermath also have a strong impact on nutrition, short of famine. A recent example is the 1990–1991 Gulf War, which resulted in extensive destruction of Iraq's infrastructure and economy so that widespread malnutrition, particularly of children, has occurred.

Refugees

War and political crises also create refugees. At least 35 million people in the world have either fled their country as refugees, or been displaced internally, mainly as a result of civil war. Their numbers have doubled in the 1980s. Refugees suffer from the same type of diseases as other vulnerable groups in developing countries, the main killers being measles and diarrhoea, to which they are more susceptible because of malnutrition. They often receive food that is inadequate in quantity and quality. If prolonged, this deprivation leads to starvation and debilitating outbreaks of scurvy, pellagra, beriberi and other deficiency diseases that are now common. This is attributable to the dependency of refugees on the food provided by governments and aid donors, which has little diversity and is inadequately planned. ***See*** Refugees – Nutritional Management; Scurvy

Poverty

Famine is only the tip of the iceberg of the much larger problems of poverty and access to resources which affects nutrition on a longer-term basis. Various economic models or political views exist on the causes of poverty and wealth creation, ranging from simplistic models of idleness and ignorance to systematic models such as Marxist, which views poverty as an inevitable and even necessary part of the capitalist economic system. These models form the framework of what policymakers believe can be done to break out of poverty and malnutrition, what the implications are for the rest of the economy, and therefore the type of interventions they support.

Equity

Poverty is a major, but not the only, determinant of undernutrition. At the same level of family or national income some individuals or countries do better than others in terms of nutrition. Internationally, there is a clear relationship between gross national product

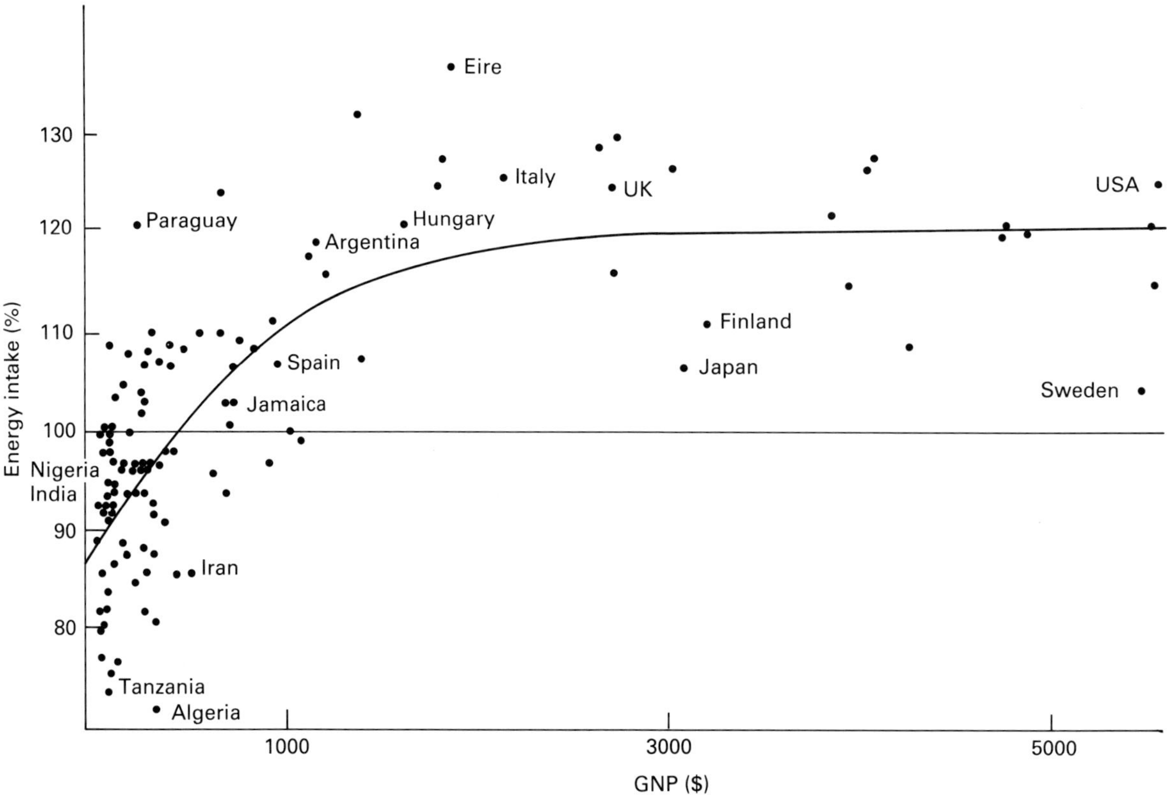

Fig. 1 Energy intakes and gross national product (GNP). Energy intakes are strictly food moving into consumption, and are expressed as a percentage of requirements; GNP is expressed as $ per head per year. (Source: Geissler and Miller, *Food Policy* 191–206.)

(GNP) and several indicators of health and nutrition, including food available for consumption, infant mortality, etc. (see Fig. 1). Countries that have done best in improving nutrition in recent years tend to be those where there is greater equity or where policies have concentrated on ensuring the satisfaction of basic needs, including adequate food. These include a wide spectrum of political ideologies, from communist China to capitalist South Korea. China is the classic example of a country that is still poor but has conquered millenial problems of malnutrition and famine – except during the short but disastrous Great Leap Forward policy of rural industrialization at the end of the 1950s – through effective organization of food production and distribution. Other examples are Kerala state in India, Sri Lanka, and Thailand, which have better nutrition conditions than other countries with similar GNPs. In contrast, some countries have extensive chronic malnutrition, e.g. Bangladesh despite massive aid, and Brazil despite rapid economic growth.

International Relations

Poverty and malnutrition are controlled not only by national policies and resource distribution but also by the power play of international economic relations. Again, opposing political views shed very different lights on their effects.

On one side are those who believe that the rich industrialized countries and the international agencies do much to assist poor developing countries through technical advice and monetary or food aid, while international corporations provide them with employment opportunities.

On the other side are those who believe that the odds are stacked against poor countries breaking out of poverty as the countries that industrialized first now hold economic power internationally, dominating trading and banking systems. In this view, national elites of many countries assist the world powers to dominate their economies through concessions to multinational companies, the acceptance of aid, and requests for finance from the international banking system.

During the 1980s the latter view has received wider acceptance and has led to calls for a 'new economic order' by such groups as the Brandt Commission, which warned that the inequalities of opportunity between the North (the rich industrialized countries) and the South (the poor developing countries) could lead to political and environmental instability and it is therefore in the interests of both to allow the South greater control of its destiny.

The questions of international food trade, food aid and the role of international agencies in relation to nutrition are therefore highly contentious political issues that can be touched on only briefly here.

Food Trade

The economies of the developing countries are affected by the policies of the major producers of food and other essential commodities. For example, the 'world food crisis' in 1973, which was marked by soaring prices of grain and other food commodities on the world market, was the result of several factors, including climate and the food policies of the USA, which was deliberately reducing its grain stocks by taking land out of production to maintain producer prices, and of the USSR, which had suddenly increased its grain purchases from the world market to support its policy of livestock production. This and the oil crisis caused great economic hardship for many developing countries that depended on imports for their food supply, and led to a spate of policies to increase food self-sufficiency.

Similarly, the Common Agricultural Policy (CAP) of the EEC has a controversial impact on the Third World through price fluctuations caused by periodic dumping of surplus food on the world market and through tariff barriers that restrict imports. Some trade concessions have been made by the EEC to the developing countries through the Lome Convention, but this provides little protection in practice. ***See*** European Economic Community, International Developments in Food Law; European Economic Community, The Common Agricultural Policy

These protectionist policies are the subject of international bargaining for free trade that is carried out in periodic rounds of talks in the council of the General Agreement on Trades and Tariffs (GATT). However, these mainly reflect the interests of the seven most economically powerful nations, the USA, Canada, FRG, France, Italy, the UK and Japan (the G7), and have only in the most recent (1986–1992) round included food commodities, prompted by the food price policies which are detrimental to the food trade of the USA and other countries.

Food Aid

Food and other international aid has often been used as a political weapon, given to politically friendly countries, such as Southeast Asia in the late 1960s and to Egypt and Syria in 1974, and withdrawn from those that do not conform to the policies of the donor countries, such as Chile in 1971 and Mozambique in 1981. ***See*** Food Aid for Emergencies

Food aid has also received much criticism on the grounds that it has been used by the donor countries to dispose of surpluses and to penetrate food markets of developing countries in order to create a long-term trade demand, that it is unreliable, involves excessive opportunity costs, reinforces expensive subsidy programmes, provides opportunities for corruption, and is used by governments to keep urban prices down, thereby acting as a disincentive to domestic agriculture and maintaining rural poverty and malnutrition.

Food aid and capital-intensive agricultural policies are also seen to contribute to the rapid rural urban migration which exists in so many developing countries, where cities are unable to cope with the rate of influx to provide adequate housing, sanitation, water supply and employment. As a result, the migrants can often only scrape a living through petty trading, begging and low-wage employment. This influx undermines minimum wage legislation or enforcement and keeps the cost of labour and, therefore, incomes low. The consequent inability of the urban poor to purchase adequate diets, and the insanitary living conditions, result in extensive urban malnutrition in many developing countries. Such effects are seen to negate the short-term benefits of food aid used in famine situations, and of child-feeding and food-for-work projects that act as welfare benefits or resource transfers to the poor.

The structure of food aid has altered since it became a permanent international transfer mechanism in 1954 when the USA enacted the public law PL480, referring to sales to 'friendly nations', to be paid initially with local money deposited as counterpart funds for use by the USA or with their approval (Title I), and donations for famine relief and projects (Title II). The latter represents a small proportion, contrary to the popular belief that all aid is a gift. Initially, the USA supplied nearly all aid, but now Canada, Australia, the EEC and other countries contribute about 50%. About 25% of food aid is channelled through the multilateral World Food Programme for feeding projects; this reduces political bias. However, much is still bilateral and several countries, such as Bangladesh and Egypt, remain dependent on food aid and, therefore, politically dependent on the donor.

Structural Adjustment

Over the last decade or so, many developing countries have experienced severe economic crises owing to rapid changes in oil prices, falling and unstable prices of export commodities, rapidly increasing rates of interest, and increasing dependence on foreign borrowing, resulting in reduced foreign exchange reserves, and inability to service debts.

To deal with the crisis, countries have had to implement a variety of 'macroeconomic adjustment policies', including reductions in government spending. These conditionalities are rigidly imposed by the International Monetary Fund (IMF) and the World Bank (International Bank for Reconstruction and Development, or IBRD) to obtain new financial loans. These institutions are funded by quotas from members who have voting rights in proportion to their contribution, assessed according to economic status. The USA has 20% of the votes, the European countries 5–7% each, and others very small percentages, so that decisions are effectively in the hands of the major industrialized countries, especially the USA. This banking structure means that the policies of borrowing countries are dictated by the industrialized nations.

The primary aim of macroeconomic adjustment policies is to improve the balance of payments. Therefore short-term effects on the poor are usually ignored unless they threaten political stability, e.g. through urban riots. Adjustment frequently includes changes that are of particular concern to the poor, such as increased food prices and decreased expenditure on social programmes. The effects of these policies on health care, food consumption, incomes and prices have led to a serious deterioration in indicators of nutrition, health status and school achievement in several countries. Efforts are now being made by UNICEF and other bodies to buffer vulnerable groups from these effects.

Affluence

In most developed countries the problems of poverty-related malnutrition have been largely overcome by higher incomes and a safety net of welfare benefits, and replaced by problems associated with affluent diets. The politics of nutrition in this case are very different from those associated with poverty. Only a small proportion of the population is involved in primary food production (2–3% in the USA and UK) and the majority of the population depends for its food supply on the food industry, which is a major industry in economic terms in Europe and the USA: in the UK agriculture, food processing and distribution contribute more than 7% to the gross domestic product (GDP). Therefore the agricultural and food industry lobbies are potentially important in food and nutrition policy formulation. In food policy, nutrition has generally a very low priority. For example, the EEC CAP has no explicit nutritional component, its main aims being to support farm incomes and food security for Europe in terms of adequate supplies and stable prices. The emphasis is therefore on quantity, not on nutritional quality. The concept of food quality lags behind current values and generally refers to the purity, hygiene and richness of nutrients. These are obviously important but now that nutritional deficiencies have been largely overcome, the concept of a healthy diet can no longer be related only to purity and abundance. Government policies have also lagged behind current needs such that food legislation and producer incentives have continued to favour a high-fat diet, including premium producer prices for high-fat milk and inappropriate grading systems for livestock.

Efforts to translate general nutritional advice on healthy eating into explicit policies have been met with considerable concern and opposition from some sections of the food industry that have been threatened with change. The debate has included the following arguments: the evidence for the relationship between diet and disease is not adequately conclusive; the level of risk to the population as a whole for certain dietary components, such as salt, does not warrant blanket nutritional policies, only targeted interventions to those at risk; and, in a democratic society, the individual must be free to make his or her own choice of food consumption without coercive pressures from a 'nanny state'. ***See*** Diseases, Diseases of Affluence

This resistance of the food industry and of government has been viewed in some quarters as a conspiracy of vested interests between food producers, government policy makers and advisory committees. However, others view it as a natural cautious attitude which is necessary because of the widespread industrial and economic changes entailed. Despite this resistance the food industry has responded to pressures for change in various ways, including research into leaner livestock, the formulation of a variety of low-fat and low-energy products, and voluntary nutritional labelling. ***See*** Legislation, Labelling Legislation

Much stronger political support exists for interventions based only on informed choice, and little for coercive measures to improve the diet, such as price manipulation through consumer subsidies and taxation, even though these instruments are used for other purposes. There is therefore much effort put into legislation on food labelling.

The food industry has become progressively more vertically integrated, from production through processing to retailing, by the development of multinational agribusiness and large chain stores which increasingly order food directly from the producers. This means that the buyers for these food chains become exceedingly important arbiters of the national diet in deciding the choice available to the consumer.

Conclusion

Food and nutrition involve highly political issues that determine both access to sufficient nutrients and the commercial and consumer pressures on a balanced diet.

Bibliography

Berg A (1973) *The nutrition factor. Its role in national development.* Washington DC: The Brookings Institution.

Biswas H and Pinstrup-Anderson P (eds) (1985) *Nutrition and Development.* Oxford: Oxford University Press.

BNF (1982) *Implementation of Dietary Guidelines: Obstacles and Opportunities.* London: British Nutrition Foundation.

Brandt W (1980) *North–South: a Programme for Survival.* The report of the independent commission of international leaders and statesmen on international development issues under the chairmanship of Willy Brandt. London: Pan World Affairs.

George S (1976) *How the Other Half Dies.* London: Pelican.

George S (1984) *Ill Fares the Land. Essays on Food, Hunger and Power.* London: Writers and Readers Publishing Cooperative Society.

Jolly R and Cornia GA (eds) (1984) *The Impact of World Recession on Children.* UNICEF report. New York: Pergamon Press.

NACNE (1983) *The Health Education Council.* A discussion paper on proposals for nutritional guidelines for health education in Britain. London: National Advisory Committee on Nutrition Education.

ODI (1986) *The CAP and its Impact on the Third World.* ODI briefing paper. London: Overseas Development Institute.

Sen A (1982) *Poverty and Famines: an essay on entitlement and deprivation.* Oxford: Oxford University Press.

Wheelock JV (1986) *The Food Revolution.* Maidenhead: Chalcombe Publications.

Catherine Geissler
Kings College, University of London, UK

POLYCYCLIC AROMATIC HYDROCARBONS

Polycyclic aromatic hydrocarbons (PAHs) are ubiquitous in the environment and occur in a variety of foodstuffs. They are characterized by the presence of three or more, fused five- or six-membered aromatic rings, relatively high molecular weights, and hydrophobicity. Nomenclature is complex and the terminology used here is that of the International Union of Pure and Applied Chemistry (1979). Structures of some common PAHs are shown in Fig. 1.

Mutagenic and carcinogenic effects of PAHs are major concerns. Oxygen, sulphur or nitrogen can substitute for carbon in one or more rings to yield heterocyclic compounds. These will not be considered here other than to note that they arise from the same sources as

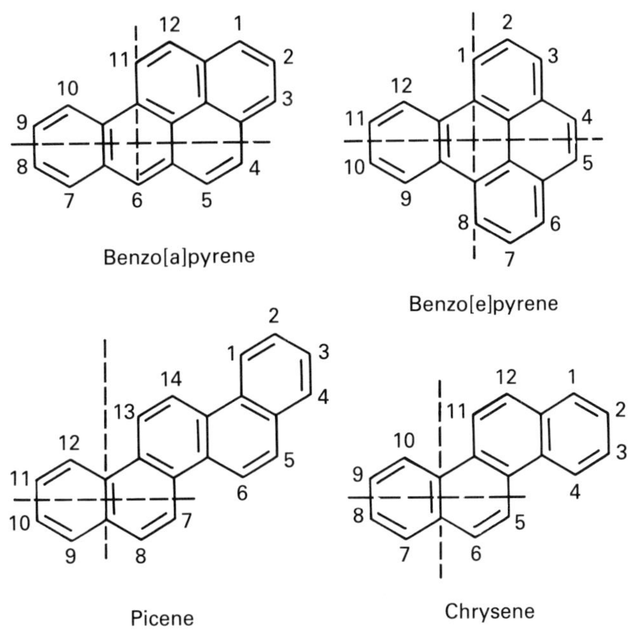

Fig. 1 Commonly encountered PAHs.

Sample (10–30g) ↓	Saponify with 1.5 M KOH –EtOH and water; extract with *n*-hexane
n-Hexane layer ↓	Wash with water; extract with DMSO
DMSO layer ↓	Add water; extract with *n*-hexane
n-Hexane layer ↓	Wash with water; dehydrate with anhydrous Na_2SO_4; concentrate to 5 ml
Column chromatography ↓	Upper, silica gel; lower, alumina. 1st fraction elutes with *n*-hexane, 2nd fraction elutes with 15% Et_2O in *n*-hexane
HPLC-FD	2nd fraction

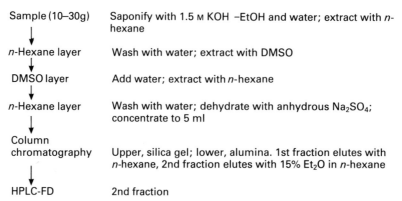

Fig. 2 A common procedure for sample preparation of PAHs following the procedure described by Obana H *et al.* (1981) *Bulletin of Environmental Contamination and Toxicology*, 26: 613–620. DMSO=dimethyl sulphoxide; HPLC-FD=high-performance liquid chromatography with fluorescence detection.

PAHs, occur with PAHs in the environment, and include carcinogenic/mutagenic members.

Sources of PAHs

Various natural and synthetic combustion or pyrolytic processes form complex mixtures of PAHs. Major natural PAH sources are volcanoes, forest fires, biosynthesis (of certain congeners), and geothermal processes. Synthetic sources include the use of fossil fuels and other carbonaceous materials as energy and feedstock materials. PAHs result from incomplete combustion. Generally, PAH formation increases as the oxygen/fuel ratio decreases. Pyrolytic processing, e.g. the destructive distillation of bituminous coal or catalytic cracking of petroleum, yields products with high concentrations of PAHs. Higher temperatures lead to simpler, nonalkylated PAHs, while cooler processes, such as petroleum formation, lead to more complex, alkylated PAHs. Special sources of human exposure are tobacco smoking, improper smoking of foods, certain cooking methods, and some foodstuffs.

Analysis of PAHs

PAH measurement in foodstuffs is complicated because specimens contain complex PAH mixtures, so that efficient separation methods coupled with specific detections systems are required; concentrations are in the low $\mu g\ kg^{-1}$ range; and some portion of tissue PAHs may be bound to cellular constituents.

Several comparative studies have demonstrated large interlaboratory differences in measurements of common nonalkylated PAHs. Fortunately, experience in food regulatory laboratories has shown that with careful quality management procedures an interlaboratory relative standard deviation of approximately 10% can be achieved for measurement of benzo[*a*]pyrene at a concentration of $10\ \mu g\ kg^{-1}$ wet weight. Generally, several PAH compounds must be measured. Popular analytical methods are capillary gas chromatography – mass spectrometry and reversed-phase liquid chromatography with UV absorption or fluorescence detection. Sample treatment varies somewhat, with saponification/extraction/adsorption or gel filtration chromatography being popular for foodstuffs. A typical scheme is shown in Fig. 2. Photodegradation can be a problem with certain PAHs. ***See*** Chromatography, Gas Chromatography; Mass Spectrometry, Principles and Instrumentation

Environmental Inputs of PAHs

Owing to both the complex nature of PAH-containing materials and the fact that PAH formation is dependent upon process efficiency, estimates of PAH inputs to the environment are imprecise. Some estimates are available. Mishandling and inefficient combustion of fossil and other fuels is the major input. The annual global release of PAHs to the atmosphere in the mid-1970s was approximately 100 000 tonnes (5000 tonnes of benzo[*a*]-pyrene). In terms of residential space heating, the relative annual hydrocarbon emissions in the USA from gas, kerosene/fuel oil, coal, and wood were approximately 5%, 5%, 2% and 88%, respectively. Approximately 6×10^9 tonnes of asphalt (containing 2–3 μg PAHs per gram) are present on road surfaces in the United States. The environmental significance of this massive amount of PAHs is unknown, since their release may be too slow to be significant. Approximately 950 000 tonnes of creosote/coal tar, containing 5–10% PAHs, are used annually in the United States as wood preservative.

Environmental Occurrence of PAHs

Most environmental samples show about the same qualitative distribution of nonalkylated PAHs. The

Table 1. PAHs in smoked foods ($\mu g\ kg^{-1}$)

Compound	Ham	Kippered cod	Smoked whiting	Barbecued beef	Hot sausage
Benz[*a*]anthracene	1·3,9·6	–	–	13·2	0·5
Benzo[*a*]pyrene	0·7,0·7	4·0	6·6	3·3	0·4
Benzo[*e*]pyrene	–	–	–	1·7	–
Benzo[*ghi*]perylene	–	2·2	2·4	4·3	–
Fluoranthene	0·6,2·9	–	–	2·0	–
Pyrene	0·2,0·9	0·6	<0·5	3·2	1·5
4-Methylpyrene	–	–	–	–	1·9
Chrysene	0·5,2·6	1·4	–	9·6	1·0
Perylene	—	0·4	0·7	–	–
6-Methylchrysene	–	–	–	–	0·5

Data abstracted from Malanoski AJ *et al.* (1968) *Journal of the Association of Official Analytical Chemists* 51: 114.

fluoranthene/pyrene ratio is near to unity with relative abundances of phenanthrene, fluoranthene, and pyrene being approximately 12%, 16%, and 15%, respectively. Benzo[*a*]pyrene, the PAH of greatest toxicological interest, generally represents 5% or less of total PAHs. The evidence shows that most environmental PAHs are of terrestrial and anthropogenic origin. PAH levels in terrestrial soils correlate with types and intensities of nearby industries. Soil concentrations of benzo[*a*]-pyrene fall in the range 40–1300 $\mu g\ kg^{-1}$. Studies of dated sedimentary cores show that the qualitative PAH distribution has been constant with respect to distribution of alkylated PAHs and their percentage composition over the past 125 years. The total PAH concentration has increased since the start of this period, suggesting an anthropogenic input consistent with fossil fuel use. PAH concentrations in older (deeper) cores are much lower and the composition is more characteristic of forest-fire input. PAH distributions in foodstuffs vary, depending on the source of PAH contamination.

As expected from their hydrophobicity, PAH concentrations in drinking water are rarely very high. One estimate is that individuals obtain only 0·1% of their total PAH consumption from drinking water.

Many plants and their products, including leafy vegetables and root crops, contain PAHs from three sources: endogenous synthesis, uptake from contaminated air, and uptake from contaminated soil.

Animals exposed to environmental PAHs may or may not show accumulation depending upon exposure, uptake, and degradation rates. Many higher animals metabolize PAHs to water-soluble excreta and do not accumulate them. Lower animals, most notably shellfish such as crustaceans and bivalves, do not metabolize PAHs appreciably and accumulate startlingly high PAH concentrations. Crustaceans, such as the American lobster (*Homarus americanus*) with a large, lipid-rich digestive gland (mid-gut gland, 'hepatopancreas'), do not readily depurate PAHs when placed in a clean environment. Considering that shellfish live in industrial harbours and that such harbours also contain large amounts of creosoted structures, e.g. pilings, it is not surprising to find shellfish containing high PAH concentrations. Mussels (*Mytilus edulis*) in contact with creosoted piles within an industrial harbour contained up to 215 μg benzo[*a*]pyrene per kg wet weight. Concentrations decreased with increasing distance from the source. ***See*** Shellfish, Contamination and Spoilage of Molluscs and Crustacea

Food Occurrence of PAHs

PAHs, including carcinogenic ones, are present in many foodstuffs. Levels of individual compounds generally range from below 1 $\mu g\ kg^{-1}$ wet weight to a few $\mu g\ kg^{-1}$ wet weight. The tolerance for benzo[*a*]pyrene in smoked meats in the Federal Republic of Germany is 1·0 $\mu g\ kg^{-1}$ wet weight. Some idea of PAHs, and their concentrations found in foods, may be obtained from the data presented in Tables 1 and 2. Food intake of PAHs can surpass that from tobacco smoking. Dietary PAH input depends on many factors, e.g. intakes of various foodstuffs, PAH concentrations in foodstuffs, and the efficiency of PAH uptake from different foodstuffs. Fats, carbohydrates and proteins can generate PAHs when improperly processed, requiring regulations and recommendations regarding certain processing and cooking methods.

Plants

PAHs have been detected in a variety of unprocessed vegetables, e.g. lettuce, tomatoes, leeks and cabbages in Germany and kale in the Netherlands. Uptake from the

Table 2. PAHs found in smoked and charcoal-broiled food products (μg kg^{-1})

Foods	BbFL	BaA	BeP	BaP	DBahP	IP	PR	CR
Smoked								
Bologna	–	–	5·0	2·0	0·5	–	–	–
Frankfurters	–	1·5	–	2·0	–	–	–	2·0
Salami	–	1·5	–	2·0	–	–	–	2·0
Pepperoni	–	0·2	2·0	–	–	1·0	–	–
Various sausages	–	0·5	2·5	1·0	8·0	1·0	–	–
Ham	–	0·2	0·2	2·0	1·0	0·5	–	–
Westphalian ham	–	–	5·0	2·0	–	–	–	–
Bacon	–	0·5	–	0·5	1·0	–	–	4·0
Smoked beef	–	8·0	–	–	–	0·5	–	–
Smoked pork	–	–	–	0·2–0·3	–	–	–	–
Smoked herrings	–	20·0	2·0	15·0	–	0·2–9·0	–	8·0
Various smoked fish	–	1·0	1·5	0·5	5·0	–	–	10·0
Canned smoked fish	–	–	1·0	–	–	–	1·0	–
Canned (smoked) oysters	25·0	30·0	16·0	2·0	–	5·0	20·0	–
Canned (unsmoked) oysters	–	15·0–30·0	1·0	–	–	2·0	–	–
Charcoal-broiled								
Porterhouse steak	–	1·0	4·0	3·0	–	–	2·0	
Barbecued chicken	–	–	2·0	–	–	10·0	2·0	–
Hamburger	30·0	50·0	20·0	20·0	–	–	–	–
Frankfurter	2·0	2·0	5·0	5·0	–	–	–	–

BbFL=benzo[*b*]fluoranthene; BaA=benz[*a*]anthracene; BeP=benzo[*e*]pyrene; BaP=benzo[*a*]pyrene; DBahP=dibenzo[*a,h*]pyrene; IP=indeno[1,2,3-*cd*]pyrene; PR=perylene; CR=coronene.
Data abstracted from Panalaks T (1976) *Journal Environmental Science Health Bulletin* 4: 299.

environment is probably the major source of plant PAH, although there is some evidence for endogenous synthesis of certain PAHs. Plants grown in areas of aerial PAH pollution have elevated PAH levels compared to plants grown in rural areas. Uptake from both air and soil has been shown. Surface contamination by PAHs is evident, and rinsing with water removes only a small portion.

Processing of food products can result in either an increase or a decrease in PAH concentration in the products. Milling partitions PAHs – for example, PAH levels ranged 5–10 times higher in wheat bran than other milled fractions. Processed wheat breakfast cereals contained higher PAH concentrations than products from other grains. PAHs concentrate in oily tissues. Peeling can remove PAHs. Certain processing materials, if contaminated, e.g. petroleum-based solvents and waxes, can add PAHs to products. PAH levels in these materials are regulated and PAH levels were insignificant in recent studies of commercial food-grade solvents and waxes. Certain processing procedures such as decolorization and deodorization of oils can remove PAHs. Direct (flue) drying, particularly with certain fuels and operating conditions, increases the PAH level in the final product far higher than can be accounted for by dehydration alone. Indirect drying reduces PAH formation, at least in wheat and rye. ***See*** Drying, Theory of Air Drying; Vegetable Oils, Refining

Meat and Poultry

Owing to the ability of higher animals to metabolize PAHs, unprocessed meat and poultry do not contain appreciable PAH levels. Processing can result in inadvertent contamination; however, the major source of PAHs in edible flesh products comes from improper smoking and certain types of cooking. ***See*** Meat, Preservation; Smoked Foods, Principles

Investigation of smoked foods was prompted initially by the hypothesis that the higher incidence of stomach cancer in certain groups in Iceland and England was related to the consumption of smoked foodstuffs. Ultimately, the concentrations of PAHs in a smoked product depend on the smoke generation system (fuel and system smouldering temperature), smoking time, and treatment of the smoke before contact with the goods. PAH concentrations in smokes increase with increasing smouldering temperatures. Smouldering hardwoods produce fewer PAHs than smouldering softwoods. PAH smoke concentrations can be lowered by cooling the smoke and discarding the condensate, washing the smoke with water, or filtration, i.e. by using processes that remove larger soot particles. ***See*** Stomach Cancer

PAHs from smoke accumulate on product surfaces, although migration into the interior occurs. The extent of migration depends upon the nature of the product and its storage time. Certain casing materials retain PAHs but allow passage of flavour components of smoke. Artificial casings retain both groups.

Cooking methods which expose foods to higher temperatures or smoke generated from both the heat source and drippings falling onto the fire can drastically increase PAH content in the cooked food. Broiling thick beefsteak immediately above an intense heat source onto which drippings fall resulted in benzo[*a*]pyrene concentrations as high as 50 $\mu g\ kg^{-1}$ wet weight. Lower concentrations resulted when the steak was cooked to the same internal temperature, but at a greater distance from the heat source. The production of carcinogenic PAHs can be minimized by avoiding contact of the food with cooking flames, cooking at lower temperatures for longer periods, and using meats with a low fat content.

Fish

Higher fish species (teleosts) metabolize PAHs, resulting in freshly caught fish having low PAH tissue levels. Processing and cooking, as described above for meats, can increase PAH levels in the edible product. Canned sardines, for example, can contain appreciable PAH levels from the oils and smoking used in processing. ***See*** Fish, Processing

Molluscan products (mussels, oysters) frequently contain benzo[*a*]pyrene at concentrations greater than 5 $\mu g\ kg^{-1}$ wet weight. Products from smaller crustaceans contain lower, if detectable, levels, probably because, generally, only the tail muscle is used in contrast to molluscs where the entire soft tissues are consumed. Large crustaceans, e.g. American lobster (*Homarus americanus*), in which various body parts, such as the digestive gland, are eaten, can contain surprisingly high PAH concentrations in fatty tissues.

Lobsters captured offshore of Nova Scotia had only trace concentrations of benzo[*a*]pyrene in tail meat, while lobsters taken nearshore contained 0·2–0·9 μg benzo[*a*]pyrene per kg wet weight. Lobsters that had been held in a tidal storage pond of creosote-treated timber construction contained 7·4–281 μg benzo[*a*]-pyrene per kg wet weight in tail meat. Digestive gland concentrations were about ten times higher. The highest value recorded was 2300 μg benzo[*a*]pyrene per kg wet weight. Similarly, digestive gland tissue from lobsters taken from the south arm of Sydney Harbour, Nova Scotia, Canada, contained 433–2240 μg benzo[*a*]pyrene per kg wet weight. The south arm of Sydney Harbour received drainage from a coal-coking plant. It has been closed to lobster fishing since 1982. As expected, relative concentrations of other PAHs were present in all samples. Although there are presently no tolerances for PAHs in lobster tissue in any country, the magnitude of this contamination can be contrasted to other foodstuffs, where levels of benzo[*a*]pyrene were detectable in 32 of 60 food products with a maximum of 7 μg benzo[*a*]pyrene per kg wet weight. In Canada, pond owners have now removed creosote-treated timbers.

Degradation of PAHs

PAHs are destroyed in the environment through a variety of chemical and biochemical oxidative mechanisms. Photooxidation is the most important pathway quantitatively. PAHs are also susceptible to oxidation by ozone, peroxides and nitrogen and sulphur oxides. Microorganisms can metabolize PAHs, in particular the lower-molecular-weight ones. PAHs on particulate matter, e.g. from stack emissions, are less readily oxidized than nonparticulate PAHs.

Metabolism of PAHs in higher organisms is a complex process depending upon the animal's genetic nature (species and strain), tissue, age, sex and nutritional status. Most higher animals have some basal level of the enzyme system responsible for the initial steps in the metabolism of PAHs, i.e. aryl hydrocarbon hydroxylase. This enzyme system catalyses the addition of oxygen to the aromatic structure to yield epoxides. These rapidly hydrate to the dihydrol, are oxidized further to a diol epoxide, and are then hydrated to yield a tetrahydroxide derivative. A variety of aryl hydrocarbon hydroxylases are known, so that oxidation of a single PAH compound such as benzo[*a*]pyrene leads to an array of stereospecific and positional isomeric oxidized products. Conjugating enzymes promote further reaction with hydrophilic compounds to form highly polar derivatives, i.e. glucuronides, glycosides and mercapturates, which are excreted in the bile and urine. Pretreatment of the animal with PAHs (or other inducers) leads to induction of higher enzyme activity, which results in rapid metabolism of compounds like benzo[*a*]pyrene. Some conversion products of aromatic hydrocarbons tend to accumulate in tissues and their concentrations might remain high after exposure.

Toxicology of PAHs

PAHs, particularly those with larger numbers of rings, are not very acutely toxic. The Ames test has shown the mutagenicity of PAHs. Mutagenicity is greatest in fractions containing four and five-membered rings. ***See*** Mutagens

The ability of certain PAH compounds and PAH mixtures to induce cancers has been a research topic since the original observation of Pott in 1775 of increased scrotal cancer incidence among chimney sweeps and later the induction of skin tumours in rabbits by dermal application of coal tar for 150 days. Induction of tumours in animals is not a straightforward procedure in defining human health hazard since, besides species differences, questions can be raised regarding dosing levels, dosing procedures, exacerbating and ameliorating factors, routes of administration, and target organs.

The initial reaction leading to cancer is believed to be the reaction of a metabolic intermediate with a cellular constituent such as DNA. The concentration of this 'active' intermediate depends on its rate of formation and degradation, reaction with other non-cancer-inducing cellular constituents, and competition by other enzymatic routes for its precursor. Given the complexity of the aryl hydrocarbon hydroxylase and conjugation systems, it is probably not surprising that such cancer-inducing chemicals have been divided into several activities: (1) compounds that induce cancer by themselves, e.g. benzo[*a*]pyrene, 3-methylcholanthrene; (2) cocarcinogens that are synergistic with carcinogens; (3) tumour initiators that when used as a pretreatment followed by a low dose of another compound lead to cancer; (4) promoters that are not carcinogens, but that act with initiators to cause cancer; and (5) anticancer agents that are antagonists to carcinogens. Various PAH compounds have been shown to have one or more of these five activities. Some common PAH compounds that have been identified as carcinogens are benzo[*a*]-pyrene, 3-methylcholanthrene, benz[*a*]anthracene, and dibenz[*a,h*]anthracene. Many individual PAHs have been also identified as cocarcinogens, initiators, promoters, or, infrequently, even as anticancer agents. ***See*** Carcinogens, Carcinogenic Substances in Food

Probably as a result of the complexity of both PAH exposure and metabolism, very few countries have imposed fixed tolerances for total PAH or benzo[*a*]-pyrene concentrations in foodstuffs. However, all national health agencies recommend minimal exposure to PAH-containing materials because of their cancer-inducing roles.

Bibliography

Andelman JB and Snodgrass JE (1974) Incidence and significance of polynuclear aromatic hydrocarbons in the water environment. *CRC Critical Reviews in Environmental Control* 4: 69–83.

Bjorseth A (1983) *Handbook of Polycyclic Aromatic Hydrocarbons*. New York: Marcel Dekker.

Blummer M (1976) Polycyclic aromatic hydrocarbons in nature. *Scientific American* 234: 35–45.

Dunn BP and Fee J (1979) Polycyclic aromatic hydrocarbons in commercial seafoods. *Journal of the Fisheries Research Board of Canada* 36: 1469–1476.

Haenni EO (1978) Recommended methods for the determination of polycyclic aromatic hydrocarbons in foods. *Pure and Applied Chemistry* 50: 1763–1773.

Howard JW and Fazio T (1969) A review of polycyclic aromatic hydrocarbons in foods. *Journal of Agriculture and Food Chemistry* 17: 527–531.

International Union of Pure and Applied Chemistry (1979) *Nomenclature of Organic Chemistry. Section A. Hydrocarbons*. Oxford: Pergamon Press.

Lo MT and Sandi E (1978) Polycyclic aromatic hydrocarbons (polynuclears) in foods. *Residue Reviews* 69: 35–86.

Malins DC and Hodgins HO (1981) Petroleum and marine fishes: A review of uptake, disposition, and effects. *Environmental Science and Technology* 15: 1272–1280.

National Academy of Sciences of the USA (1972) *Particulate Polycyclic Organic Matter*. Washington DC: National Academy of Sciences.

National Research Council of Canada (1983) *Polycyclic aromatic hydrocarbons in the Aquatic Environment: Formation, Sources, Fate, and Effects on Aquatic Biota*. Ottawa: NRCC Associate Committee on Scientific Criteria for Environmental Quality.

Neff JM (1979) *Polycyclic Aromatic Hydrocarbons in the Aquatic Environment*. Barking: Applied Science Publishers.

Stahl W and Eisenbrand G (1988) In: Macrae R (ed.) *HPLC in Food Analysis*, 2nd edn. London: Academic Press.

Uthe JF (1979) The environmental occurrence and health aspects of polycyclic aromatic hydrocarbons. *Fisheries and Marine Service Technical Report* No. 914. Halifax, Nova Scotia: Department of Fisheries and Oceans.

C. J. Musial
C. Musial Consulting Chemist Ltd., Halifax, Canada
J. F. Uthe
Department of Fisheries and Oceans, Halifax, Canada

POLYPHENOLS

See Tannins and Polyphenols

POLYSACCHARIDES

See Carbohydrates

POPULATION, DEVELOPMENT AND NUTRITION

Population growth affects food security and nutrition through a number of pathways. The major linkages are shown in Fig. 1. The effects of development policy on population and food security and nutrition are likewise both direct and indirect. The purpose of this article is to discuss briefly each of the linkages illustrated in Fig. 1.

Population, Food Production and Food Supply

In the late 1960s and 1970s worldwide attention was focused on the problem of food shortages in developing countries. Rapid population growth exacerbated shor-

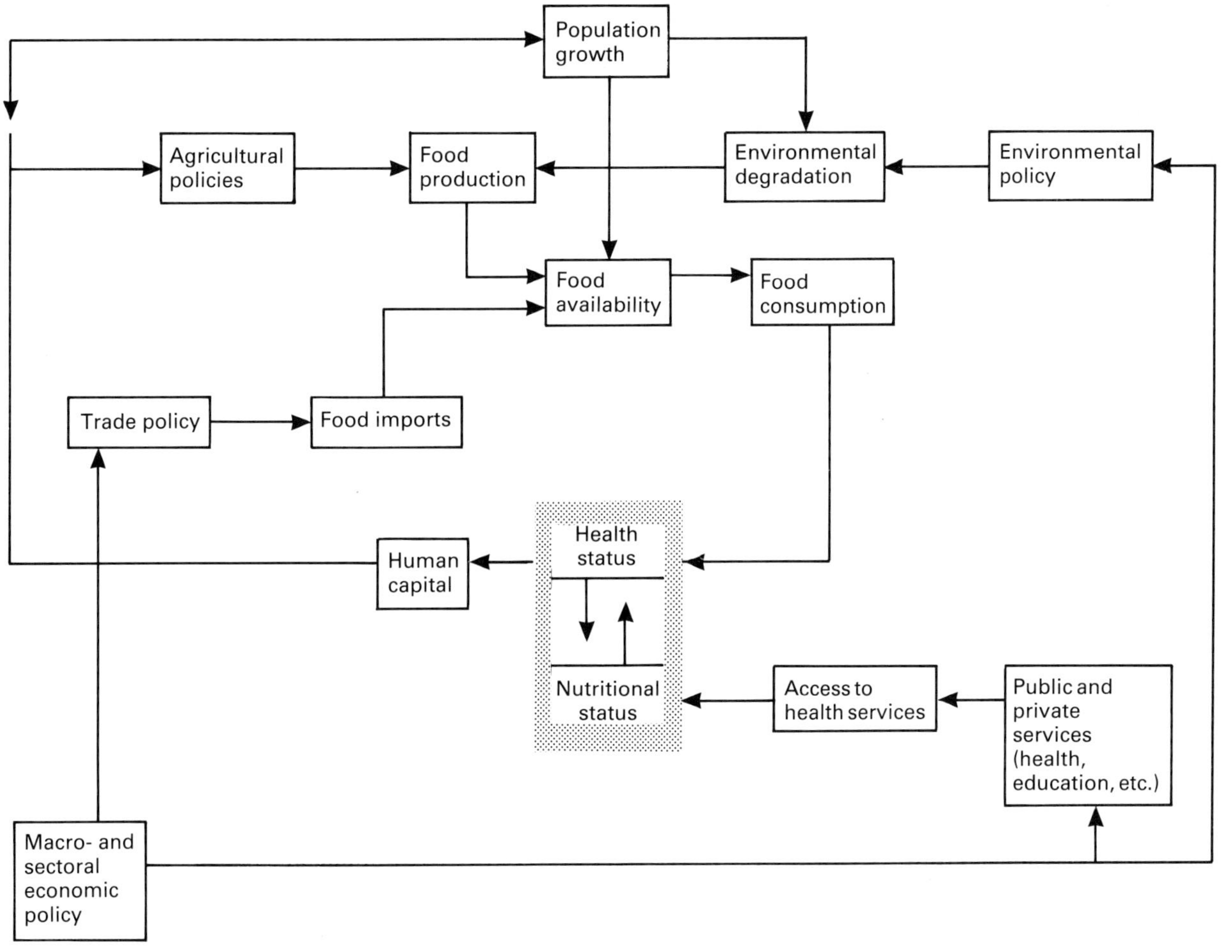

Fig. 1 Population, development and nutrition: the main linkages.

tages caused by the underlying constraints to food production and distribution in many parts of the developing world. These food shortages were related to lack of natural resources, incentives to produce, purchasing power of the poor, and appropriate institutions and infrastructure.

During the period 1961–1981, the world population grew by 1·9% annually, compared to an annual growth rate in agricultural output of 2·4%. Supply increases were more rapid than population increases. However, these statistics are misleading since they mask large differences between developed and developing countries. In developed countries as a whole, populations grew 0·9% annually, with food production increasing by 2·2% per year. In sub-Saharan Africa the picture is reversed; the population grew by 2·9% compared to food production, which grew by only 1·7% annually. For developing countries as a whole, food production increased slightly faster than the overall rate of population growth but, here again, this masks the deficits in individual countries and regions.

The future food production and food supply situation for developing countries does not appear encouraging. Over the coming decades the global population is expected to increase by 90×10^6 people per year, with 95% of this growth occurring in the developing countries of Africa, Asia and Latin America. The dynamics of rapid population growth and urban expansion in regions which are primarily poor and rural suggest that per capita food production must increase dramatically if food consumption levels are to be maintained. However, many countries, both developed and developing, are currently witnessing a rapidly shrinking agricultural resource base on which to feed their growing populations.

Ultimately, food production has to be raised through a combination of increased yields and increased land devoted to cultivation. However, at the global level, increases in crop yields seem to be levelling off. In addition, there is less scope to increase the area of land cropped in order to compensate. Moreover, efforts to overcome these constraints often degrade the environment and conflict with the longer-term sustainability of agriculture. The relentless pressure to increase food production to keep pace with population growth has forced many countries to accept short-term solutions which ultimately threaten the sustainability of initial food production gains. These solutions, such as more intensive use of increasingly fragmented land-holdings, or expansion of cultivated land into ecologically fragile areas, are compounded by the effects of encroaching urbanization and deforestation. It is estimated that 6–7×10^6 ha of Third World agricultural land are made unproductive each year because of soil erosion; waterlogging, salinization and alkalinization from irrigation damage another $1{\cdot}5 \times 10^6$ ha.

Of all the developing regions, Africa faces the most alarming prospects. With an annual population increase of 3% – unprecedented in history for an entire continent – per capita food production has declined to the point that Africa has ceased to be a net exporter of food staples. It is anomalous that in a continent where more than 75% of the population is engaged in smallholder food production, over 20% of food staples consumed are imported, and that this percentage is rising. If current production growth trends continue, food output in sub-Saharan Africa alone could fall short of projected demand by almost 50×10^6 t by the year 2000.

The situation in Asia is not much better. Unlike land-abundant Africa, certain subregions within Asia are densely populated. In addition, in parts of Asia the number of landless households is increasing. Asia, home to over half the world's population and producer of over 90% of its rice, currently has an estimated 82% of its potential cropland already under production. Thus the possibility of significantly increasing food production by bringing more land into cultivation – historically the most common response to increased food demands – is very remote, especially in the densely populated areas of south Asia. A 1987 UN study ranked the food and nutrition problems of south Asia as 'one of the world's most serious issues of human welfare'. Even in the rapidly industrializing countries of east Asia, where rising incomes have resulted in significant improvements in food consumption since the1970s, the risks of food insecurity are great: nonfarm uses claim roughly 500 000 ha of cropland each year.

The food availability statistics for Latin America as a whole are positive. However, gains in food supplies were not uniform across the region. While the gains in energy consumption were positive for the region as a whole for the period 1960–1985, the Andean countries (Bolivia, Ecuador and Peru) and the Southern Cone countries had a negative rate of growth in energy consumption.

Changing Patterns of Food Consumption

Not only has food production per capita been declining in a number of areas of the developing world, but the patterns of food consumption have been changing. The diets of rural populations throughout the world are based heavily on cereals and grains. However, there is a movement away from the traditional, locally produced coarse grains (millet, sorghum, maize) to nontraditional grains such as wheat and rice. This shifting consumption pattern is particularly prominent in West Africa, where rice consumption has gone from 12% to 21% of total cereal consumption in the period 1961–1983, and wheat consumption has increased from 3% to 10% during the same time period. ***See*** Cereals, Contribution to the Diet

The shift from traditional to nontraditional grains is of concern for several reasons. First, wheat and rice cannot be produced efficiently in many of the countries where consumption of these cereals is increasing the most rapidly because the agroecological settings are not conducive to their cultivation. Thus countries are forced to rely more heavily on commercial food imports which must be financed through limited foreign exchange reserves. The increasing dependence on imported commodities suggests increasing vulnerability to sudden shifts in world grain prices.

In addition, wheat and rice tend to cost more per unit energy purchased than millet, sorghum and maize. As households switch to more expensive sources of energy, the deficits in energy intake may be further exacerbated.

Urbanization is an additional factor that has influenced food consumption patterns in developing countries. Migration from rural to urban areas is associated with changing food patterns because households move from a diet based largely on home production to a diet based predominantly on purchased foods. One obvious change is the trend towards more highly processed foods. In urban areas households typically have a wider variety of products available and this influences consumption patterns.

Urban food supplies can be supplied by domestic production. However, data suggest that in areas of the developing world experiencing rapid urbanization, commercial imports have also increased, suggesting that the additional food supplies are not being provided by domestically produced food.

Urbanization is typically associated with higher household cash incomes, which result in greater demand for purchased food. Even in countries which have experienced very positive rates of growth in food production, increasing household incomes have necessitated increased food imports.

Implications of Demographic Shifts for Food and Nutrition Security

The increasing population throughout parts of the developing world has meant that the agricultural sector, historically the dominant employer for labour, has been unable to absorb the growing rural workforce. As a result, the pace of rural-to-urban migration and seasonal rural-to-rural migration has increased. One consequence of this is a growing number of households which are, in effect, headed by women. These households are termed *de facto* (as opposed to *de jure*) female-headed households. The growing proportion of households headed by women is most common in Africa and Latin America, and is a lesser phenomenon in Asia. Estimates of the number of female-headed households range from 50% in Botswana, to 33% in Jamaica, and at least 10 percent in most Arab Middle Eastern countries.

The phenomenon of increasing numbers of female-headed households is of concern because these households tend to be poorer, own less land, have less access to services, and have fewer assets. Despite the lower levels of income in many female-headed households, these households do not necessarily experience greater food insecurity or higher levels of malnutrition. Some recent evidence from Kenya and Malawi suggests that at very low levels of income, *de facto* female-headed households have lower rates of preschool malnutrition than other types of households at similar levels of income. Several reasons could account for this, including the tendency for many female-headed households to allocate a higher proportion of their income to food, and the fact that women in these households allocate a higher proportion of household energy to the children.

However, female-headed households in almost all countries tend to be more susceptible to climatic and economic stress than their male counterparts. This has negative implications for absolute levels of food security. The continued trend toward increasing number of female-headed households suggests that for many countries the food insecurity problem will be increasing.

Links between Food Production/ Consumption and Nutrition

Food availability at the national, regional or local level is but one factor (Fig. 1) that affects household level food consumption, and it is not necessarily the most important. Until recently, an inordinate emphasis was placed on national or regional food self-sufficiency as a means of achieving food security for the population. Policy makers believed this strategy would alleviate malnutrition. There are two major flaws with this reasoning. First, food insecurity and malnutrition are primarily problems of distribution, not production. It is common to have 20–30% of a country's population consuming less than 80% of energy requirements, even though national level food availability is at or greater than 100%. It is the household's ability and desire to obtain food that is critical in ensuring household food security. The ability and desire of the household to obtain food is related to the household's access to acceptable (in a cultural sense) food. As the purchasing power of households increase, access to food increases.

The second flaw in the argument relates to the links between hunger and malnutrition. Hunger and malnutrition are not synonymous. Food is only one of a series of inputs into the production of health and nutritional status. Factors such as the health and sanitation environment, including hygienic practices within the household, the availability of health services, food distribution within the household, cultural practices related to weaning, and child feeding can have a greater influence on nutrition than simply the amount of food available at the household level.

This is not to argue that macroeconomic policies are not important in determining household food security; indeed, many would argue that they are more influential than agriculture sector policies. Clearly, national level development policies have a direct effect on market prices, in particular market prices for food, household income, and availability of publicly provided health and social services. These community and household level factors, then, have an impact on the health and nutritional status of individual household members. However, in order to understand the ultimate impact of public policy on an individual's nutritional status, one needs to understand how households react to the range of policies and programmes implemented in country and community.

Nutritional status should be viewed not simply as an outcome measure of overall welfare, but also as a key input into agricultural productivity. The argument, put simply, is that better nourished individuals have a higher physical and cognitive labour productivity, which results in higher earnings and better food security. For the nutrition–physical-productivity link, the empirical evidence is quite strong, but evidence for the nutrition–cognitive-performance link is less strong. ***See*** Food Aid for Emergencies

Macroeconomic Policy, Food Security and Nutrition

The deteriorating economic situation in the 1980s forced many developing countries to adopt a series of macroeconomic policy reforms that are typically categorized under the heading of 'economic adjustment'. Economic adjustment policies typically involve a combination of economic stabilization and structural adjustment measures. Simply put, stabilization measures are intended to curb short-run expenditures, while adjustment measures are intended to promote the production of commodities that can generate the maximum foreign exchange on world markets. Specific measures include the following: a devaluation of the local currency; an emphasis on export crop production; the reduction or elimination of input subsidies, including agricultural inputs such as fertilizer; the reduction or elimination of consumer food price subsidies; the reduction or freezing of civil servant wages and/or hiring; reduced government expenditures in general; and the liberalization or privatization of markets.

Food Price Subsidies

Food price subsidies, aimed at increasing the food consumption of poor households, were, until recently, a popular and common type of intervention in developing countries. Subsidized food items were provided at below-market prices and thus low-income households were able to purchase more food at current income levels. A multicountry study in 1989 found that food subsidy programmes could have a significant impact on household energy consumption of low-income households. However, given that subsidies generally covered a large portion of the population, a disproportionate share of the benefits were accruing to nonvulnerable households. This therefore impaired the financial sustainability of the subsidies while simultaneously rendering governments vulnerable to political unrest lest the subsidies be removed.

Economic adjustment programmes have generally involved the elimination of poorly targeted food subsidy programmes as one way of decreasing government expenditures. Food subsidies had involved from 3% to 20% of government expenditures. Evidence to date suggests that the total elimination of food subsidies, without a simultaneous increase in the incomes of the low-income population or an improved targeting effort, has had a negative effect on household consumption.

Export or Cash Crops

Increased commercialization of semisubsistence agriculture – more commonly called cash cropping – is an important element of most structural adjustment policies. Many developing countries have actively encouraged the production of export crops as a way to generate foreign exchange earnings and fiscal revenues, to increase the incomes of the small landholder, and to provide employment for the rural landless poor. A recent synopsis of the food security and nutrition impacts of cash-cropping schemes in six countries reports that incomes of farmers participating in cash-cropping schemes increased significantly, which led to an improvement in household food consumption. However, the links between household income and the nutritional status of children were much weaker. The increases in household incomes did not decrease morbidity and as a result there was no dramatic improvement in children's nutritional status. One conclusion from the six-country study was that income-generating schemes such as cash crop production need to be linked with health and sanitation activities in order to result in significant improvements in nutritional status.

Economic Adjustment and Government Social Spending

There are few, if any, hard facts on the impact of economic adjustment (sometimes called economic reform) upon the nutritional status of households and of women in particular. Arguments that cuts in government health and education spending will increase burdens on poor women and children have failed to separate out the potentially negative nutritional impacts of adjustment from the economic crisis that preceded it. Furthermore, adjustment has often been a prelude to foreign financing from international financial institutions and bilateral donors which have sustained government expenditures in social services. Moreover, social security expenditure reductions need not hurt the poor. Given sufficient political will, the potential exists for significant reductions in expenditure and the introduction of user fees in the areas of advanced curative hospital procedures and university education.

Development Policy for the Year 2000

If governments are concerned about the food security, health and nutritional status of their populations, certain issues must be addressed. First, even the most robust rates of growth in agricultural production will be eroded by rapid population growth. Therefore, in countries with rapid population growth rates, population programmes must be part of the solution to sustainable food security.

Optimal economic growth in developing countries will continue to be associated with growth in employment; this growth in employment will result in increased demand for food. Thus the agricultural sector must be equipped to provide this increased food either from a more productive agriculture or through continued reliance on commercial food imports and/or food aid. The latter two options are seen by many governments as less attractive because they are prone to external shocks. A more promising strategy for achieving adequate levels of food security is one that intensively utilizes the abundant input, i.e. labour, to improve the level and pattern of productivity gains in agriculture. In addition, institutions and infrastructure that are intended to serve agriculture must be developed and improved.

In summary, increased food consumption is a necessary but not sufficient condition for alleviating much of the malnutrition that exists in developing countries. Policies designed to promote (or, at the very least, to avoid punishing) agriculture need to be combined with initiatives that attack the underlying morbidity-linked causes of malnutrition. ***See*** Malnutrition, The Problem of Malnutrition

Bibliography

Alba M (1991) *Early childhood factors as determinants of education and earnings: the case of four rural villages in Guatemala.* PhD thesis, Stanford University.

Deolalikar A (1988) Nutrition and labor productivity in agriculture: estimates for rural South India. *Review of Economics and Statistics* 70(3): 406–413.

FAO (1987) *The Fifth World Food Survey.* Rome: Food and Agricultural Organization.

Haddad L and Bouis H (1991) The impact of nutritional status on agricultural productivity: wage evidence from the Philippines. *Oxford Bulletin of Economics and Statistics* 53(1): 45–68.

Hoddinott J and Haddad L (1991) *Household expenditures, child anthropometric status and the intrahousehold division of income: evidence from the Côte d'Ivoire.* Washington, DC: International Food Policy Research Institute.

Kennedy E and Alderman H (1987) *Comparative analyses of nutritional effectiveness of food subsidies and other food related interventions.* Washington, DC: International Food Policy Research Institute.

Kennedy E and Peters P (1992) Influence of gender of head of household on food security, health and nutrition. *World Development* (in press)

Kumar S and Alderman H (1989) Food consumption and nutrition effects of consumer-oriented food subsidies. In: Pinstrup-Andersen P (ed.) *Food Subsidies in Developing Countries.* Baltimore: Johns Hopkins University Press.

Leslie J (1988) Women's work and child nutrition in the Third World. *World Development* 16(11): 1341–1362.

Mellor J (1988) Global food balances and food security. *World Development* 16(9): 997–1011.

Mellor J, Delgado C, and Blackie M (eds) (1987) *Accelerating Food Production in Sub-Saharan Africa.* Baltimore: Johns Hopkins University Press.

Pan American Health Organization (1990) *Food and nutrition issues in Latin America and the Caribbean: A Joint PAHO/IDB Position Paper on Policy and Strategy in Support of Member Governments, HPN 90.1.* Washington DC: Pan American Health Organization.

Pinstrup-Andersen P (ed.) (1989) *Food Subsidies in Developing Countries: Costs, Benefits and Policy Options.* Baltimore: Johns Hopkins University Press.

Rosenhouse S (1989) *Identifying the poor: Is 'headship' a useful concept?* Living Standards Measurement Study Working Paper No. 58. Washington, DC: The World Bank.

Sadik N (1990) *The State of the World Population, 1990.* New York: United Nations Fund for Population Activities.

Sahn D and Haddad L (1992) Comment: gender and economic reform in Africa. *American Journal of Agricultural Economics* (in press).

Stark O (1991) *The Migration of Labour.* Oxford: Basil Blackwell.

Von Braun J and Kennedy E (1992) *Commercialization of Agriculture and Nutrition.* Washington, DC: International Food Policy Research Institute. (In press)

World Bank (1986) *Poverty and Hunger: Issues and Options for Food Security in Developing Countries.* A World Bank Policy Study. Washington, DC: World Bank.

Worldwatch Institute (1990) *State of the World 1990: A Worldwatch Institute Report on Progress Toward a Sustainable Society.* New York: WW Norton.

World Resources Institute (1990) *World Resources 1990–91: Guide to the Global Environment.* New York: Oxford University Press.

World Resources Institute and International Institute for Environment and Development (1988) *World Resources 1988–89: An Assessment of the Resource Base that Supports the Local Economy.* New York: Basic Books.

Eileen Kennedy and Lawrence Haddad
International Food Policy Research Institute, Washington, DC, USA

PORK

Development of the Swine Industry

The earliest known records of swine domestication are from China and date to 4900 BC. Christopher Columbus brought the first pigs to the USA via the Canary Islands in 1493. The early colonists of the USA brought livestock with them from England throughout the 1600s. Gradually, as more grain was grown, larger herds of pigs developed and swine production became a true industry.

By the early 1800s, swine slaughtering plants had emerged at several large population centres in the USA. The largest of these was at Cincinnati (also known as 'Porkopolis' at that time) because of its prime location on the Ohio river. Early slaughtering plants were located near rivers for four reasons: rivers provided ice to keep the meat cold, transportation for products to be shipped, a source of water and a place to deposit plant waste materials such as blood.

The advent of mechanical refrigeration led to industry expansion as pork could be processed year round and kept fresh longer. Development of rail systems and the use of refrigerated rail carriages boosted industry growth as both livestock and meat could be more widely distributed. In the 1800s, pigs were often allowed to roam free in pastures and were fed garbage or what little grain was available. Today, most pigs are raised in large numbers in environmentally controlled buildings with a very specific diet designed to maximize growth.

The first slaughter operations were separate from packing operations. By the 1850s, the two had started to become integrated. Today, many plants have reverted back to specializing in either slaughter and fabrication or pork processing. This specialization has allowed modern-day slaughtering facilities to process around 1000 head per hour.

The pigs in the USA today are thought to have descended from two wild stocks, the European wild boar (*Sus scrofa*) and the East Indian pig (*Sus vittatus*). From these two genetic bases, modern pigs have been refined into several breeds which each have distinguishing characteristics. In addition to hair colour and marking differences, breeds have been developed for specific purposes such as prolificacy, bacon production, ham production and others.

In general, pigs have become leaner as the demand for lard has decreased and the demand for leaner meats has increased. The modern market pig is approximately 71·5% carcass while the remaining 28·5% is mostly by-products used in the medical and animal feed industries.

Slaughter, Fabrication and Processing

Procurement

Although pigs of all sizes can be slaughtered, most swine-slaughtering plants have equipment designed to handle pigs marketed at about 105 kg. Swine producers can sell their pigs to slaughter plants in various ways. Pigs can be marketed at a central location such as an auction or a packer buying station or they can be purchased from the farm by a packer buyer. The latter method is the most popular today. A new way to market pigs that has some popularity is contract producing for a specific packer.

After pigs arrive at the slaughtering plant, they are slaughtered within 2–3 h of arrival. If a substantial delay is expected, then they will usually be held in pens for 12–24 h. It is important to allow pigs time to become acquainted with their environment as high-stress conditions can cause undesirable changes in muscle colour and juiciness. In pigs with a certain genetic make-up, severe stress can cause the meat to be pale and watery (known as pale, soft and exudative, PSE). Long-term stress, on the other hand, can lead to a condition where the muscle is dark and very dry-appearing (known as dark, firm and dry, DFD). The physiological basis for these conditions is an abnormal muscle pH change between the time of death and the completion of rigor mortis. Therefore, it is necessary to minimize stress by proper handling of pigs prior to slaughter. Feed is usually kept from the animals during this period to decrease stomach fill. This improves percentage of carcass dressing, aids evisceration during slaughter, helps minimize contamination and improves colour. ***See*** Meat, Slaughter

Slaughter

All meat sold interstate in the USA must be inspected by the US Department of Agriculture, Food Safety and Inspection Service (USDA-FSIS). Inspection begins *ante-mortem* and continues throughout the entire slaughter, fabrication and processing system. This helps ensure the wholesomeness of meat products offered to the consumer. If the meat will be sold intrastate only, some states allow inspection by state-employed inspectors. This inspection, by law, must be equal to Federal Inspection. Similar inspection regimes apply in Europe and other developed regions of the world.

At the time of slaughter, pigs are herded into a chute where they are stunned to render them unconscious. Stunning can be accomplished by one of several methods, but is mandatory to comply with the US Humane Slaughter Act of 1958. If stunning is not performed correctly, the incidence of poor meat quality problems could increase. An example of this is a condition referred to as blood splashing. Blood splashing occurs when small capillaries in muscle tissue burst during stunning, leaving spots of blood in the muscle. The most common method of stunning is accomplished using an electrical current. Other methods include a blow from a captive bolt to the head, moving pigs through a carbon dioxide gas chamber to cause suffocation, or cardiac arrest stunning. The goal of stunning is to render the animals unconscious while allowing the heart to continue to beat and pump blood out of the body.

After stunning, the pig is immediately shackled, hoisted by the hind leg and the jugular vein is severed with a knife to allow blood to exit the body. Bleeding will remove around 50% of the blood in the body. Much of the remaining blood is removed with the organs during evisceration.

Pigs can either be skinned or they can be scalded and dehaired. To dehair carcasses, the pig is put in a scalding vat of hot water (60°C) with lye to loosen the hair follicles. The pig is removed from the scalding tank and dehaired mechanically and/or by hand using blades to scrape away the hair. Whole skins from skinned pigs can be saved and made into pigskin for use in garments and for gelatin production.

After dehairing or skinning, the abdominal and thoracic cavities are opened and the internal organs removed. Certain organs are washed, inspected and saved as edible by-products. These include the liver, heart, tongue and brain. Other organs may also be saved as inedible by-products, and will be discussed later in this article.

The carcass is washed, weighed and placed in a cooler for chilling. The cooler temperature is maintained at about −2 to 5°C so that internal areas of the carcass will be adequately chilled in 24 h. While in the cooler, most of the remaining stored energy in the muscle will be depleted and the carcass will undergo rigor mortis. Some carcasses are processed before they go through rigor mortis. This is called 'hot boning'. Although fabrication is more difficult, the cuts can be chilled faster than whole carcasses, so product that has been hot boned can be moved through the system faster.

Fabrication

Pig carcasses are fabricated into wholesale cuts, using knives, saws and other mechanical equipment. The wholesale cuts include the ham, loin, Boston shoulder, picnic shoulder, belly and spare ribs. The wholesale cuts may be shipped fresh to further processors, food service operations or to the retail market. Some plants will continue to fabricate the wholesale cuts into retail cuts and trimmings and sell these items. Many plants that slaughter will carry out further processing such as curing and smoking bellies and hams in the same plant. *See* Curing

Processing

Further processing of pork can involve many different processes. Processing encompasses many steps, including particle size reduction, the addition of nonmeat ingredients, cooking, smoking and a variety of packaging procedures. Patties can be made by grinding pork trimmings, with a specified fat level, with salt and other seasonings. Restructured products often have salt and phosphates added to flaked or chunked meat. These ingredients are then mixed and formed into steaks, chops or roasts. Sausage is another type of further processed meat. Sausages can be cured, smoked, dried or cooked. Some sausages are coarsely ground while others are finely chopped, forming what is sometimes called an emulsion or batter. Bratwurst, frankfurters and bologna are sausages that typically contain pork. *See* Meat, Sausages and Comminuted Products

The hams and bellies from pork carcasses are usually cured and smoked. The curing may be done by packing in a dry cure or, more commonly, by pumping or injecting a solution containing the cured ingredients into the meat. The curing solution may contain any or all of the following ingredients: water, salt, sugar, phosphate, nitrite, erythorbate or spices and seasonings. Water is used to add moisture to the product as well as acting as the solvent in a curing solution. Salt solubilizes proteins in meat so the meat particles bind together better. Salt also adds flavour and, at high levels, retards microbial spoilage. Sugar can be used for flavour, to promote browning and to act as a food source for desirable bacteria in some fermented meat products. Phosphates are added to help retain water in the product and to prevent oxidative rancidity from occurring. Nitrite is the curing agent. Nitrite binds to meat pigments and causes the meat to develop a stable pink colour. It also lends flavour, acts as an antimicrobial agent and helps to prevent oxidative rancidity. Erythorbate is used to speed up the curing process, which is economically important to pork processors. Spices and seasonings are added to develop flavours unique to a particular product. *See* Smoked Foods, Principles

Some sausages are cured and may also contain acidulants, starter cultures or extenders. Acidulants and starter cultures of bacteria can be used to lower the pH of products in order to develop certain flavours and protect the product from spoilage. Nonmeat extenders,

Table 1. Composition of various raw pork cuts (g kg^{-1} raw meat)

	Moisture	Fat	Protein	Ash
Ham slice	730	55	205	10
Bacon	370	530	95	5
Loin chop	710	70	210	10
Spare ribs	575	245	170	10

such as soya proteins or nonfat dry milk, may be used as inexpensive protein sources.

The US Department of Agriculture (USDA) has regulations regarding the composition of processed meat products and the proper use of some ingredients used in processing. Hams must be labelled according to the amount of protein they contain on a fat-free basis. The amount of phosphates, nitrites, erythorbate and nonmeat ingredients added are also limited. These regulations, and similar ones in other countries, are designed to provide consumers with high-quality, safe meat products.

Pork is also regulated by the USDA in an attempt to control the disease trichinosis. Trichinosis is caused by the organism *Trichinella spiralis* and can be transmitted from pig to man by ingestion of infected muscle tissue. Simply put, the regulation states that pork for use in products that are not fully cooked in the home must be certified as 'trichina-free'. Pork may become certified trichina-free by being subjected to specific heat or freezing to kill the organism. ***See*** Zoonoses

Chemical Composition

Fresh pork muscle is 70–75% water, and the protein content ranges from 18 to 22%. There are three main types of protein in the pig, myofibrillar (skeletal), stromal (connective tissue) and sarcoplasmic (pigments). Lipid or fat is another major constituent of fresh pork. It makes up between 5 and 7% of the muscle tissue. Lipids include phospholipids and triacylglycerols (also known as triglycerides). The carbohydrate content of meat is negligible, generally less than 1%. Vitamin and mineral content of fresh pork is usually about 1–2%. ***See*** Lipids, Classification; Meat, Structure; Meat, Dietary Importance; Protein, Chemistry

Different cuts in the carcass will have different compositions. This is largely due to the varying fat level in different areas of the carcass. Table 1 contains some example cuts and their typical raw composition.

Nutrient Value and Dietary Significance of Pork

Pork supplies many nutrients essential for maintenance and growth.

As with other meat, pork is an excellent source of protein. A single 85 g serving of pork contributes 41% of the daily protein requirement for a normal adult male. Not only does pork contain a large amount of protein, this protein is of good quality. ***See*** Protein, Requirements

Pork also contains lipids and fats. About 34% of pork fatty acids are saturated and 66% are unsaturated. Cholesterol is another lipid found in pork. Cholesterol is found in cell membranes in the animal body and is synthesized in the liver of humans and animals. Consumption of animal products, therefore, provides a dietary source of cholesterol which can be used in the body. Cholesterol, like saturated fats, has been associated with increased risk of developing heart disease. The relationship is not well understood, but the American Heart Association recommends keeping dietary intake of cholesterol to less than 300 mg per day. One 85 g serving of pork provides about 79 mg of cholesterol or about 26% of the recommended 300 mg. ***See*** Cholesterol, Role of Cholesterol in Heart Disease; Fatty Acids, Properties

Pork is an excellent food source for several vitamins and minerals. It supplies large amounts of thiamin, vitamin B_{12}, niacin, riboflavin and zinc. Pork is also a good source of vitamin B_6, phosphorus and iron. Dietary iron can be classified into two types, haeme and nonhaeme. Haeme iron, which is the major type found in pork, is absorbed more easily and better utilized by the body. Iron is a component of the molecule haemoglobin which is the major carrier of oxygen in our bloodstream. Intake of haeme iron is especially important in warding off anaemia, which may result from a low level of haemoglobin in the blood. ***See*** individual vitamins and minerals

Pork, when consumed in moderation, is an excellent source of many important dietary nutrients.

Microbiological and Other Hazards

Muscle is essentially sterile prior to death. However, meat destined for human consumption is cross-contaminated with microorganisms by equipment and handling at the time of slaughter and processing. Just as pork is an excellent source of nutrients for our bodies, muscle or meat is also an excellent growth medium for microorganisms. Controlling the growth of microorganisms on pork by acidifying, curing, salting, modified-atmosphere packaging, drying, cooking or refrigerating is essential. ***See*** Meat, Preservation

Food poisoning can result from consuming pork that has been mishandled, allowing certain microorganisms to grow. Causative organisms of food poisoning may include *Staphylococcus aureus*, *Bacillus cereus*, *Salmonella* spp., *Listeria monocytogenes*, *Yersinia enterocoli-*

tica, *Clostridium botulinum*, *C. perfringens* and *Campylobacter fetus* ssp. *jejuni*. Food poisoning is quite common and, in its mildest forms, is often mistaken for influenza because the symptoms are very similar. Generally, foodborne illnesses are relatively short lived and more uncomfortable than harmful. However, food poisoning can be a very serious matter. It can be debilitating or even fatal for those with poor immunological defences such as infants or the elderly. Like other microorganisms, pathogens which cause food poisoning are well controlled by heat, refrigeration, chemicals or other means mentioned earlier. However, undercooking, improper cooling or recontamination of cooked food by raw food are common ways that pathogens appear in the food supply. ***See*** *Bacillus*, Food Poisoning; Campylobacter, Properties and Occurrence *Clostridium*, Food Poisoning by *Clostridium perfringens*; *Clostridium*, Botulism; *Listeria*, Properties and Occurrence; *Staphylococcus*, Food Poisoning

Of particular concern in pork is the parasitic nematode *T. spiralis*. This organism forms a cyst in porcine muscle. The organism can be transmitted to humans who consume the contaminated pork, but is readily destroyed by heating the muscle to 62°C. Processing plants that sell pork which is not likely to be cooked again are required to heat or freeze the meat to certify that it is trichina-free.

Pork-slaughtering and -processing plants have rigid sanitation programmes that allow production of safe food. Good sanitation at the plant and proper handling throughout the food chain help keep microbial growth under control. Plants producing pork must keep processing temperatures below 10°C or stop production and sanitize the equipment every 8 h. Most plants keep their working temperature low enough to require cleaning and sanitizing only once every 24 h. Cleaning and sanitizing are not the same thing, but they are most effective when carried out together. Generally speaking, a plant is cleaned with soap and water to remove residual meat particles. Then the soap is rinsed off and sanitizer is applied to kill microorganisms that survived the cleaning. Common sanitizers used can be chlorine, ammonia or iodine based. ***See*** Sanitization

A great deal of effort is spent to clean and sanitize the plant thoroughly, as well as educate employees about the importance of good hygiene in reducing food contamination. Microbiological status of the processing area is monitored daily. In addition, the plant must pass a sanitation inspection by a USDA inspector before the day's production can begin.

Despite all the in-plant efforts to control microorganisms, pork can still be contaminated or growth of microorganisms already present can occur as a result of product abuse in the warehouse, on the delivery truck, in the retail outlet or in the home. Perishable foods should always be frozen or refrigerated at temperatures below 4°C. Once cooked, pork should be kept above 60°C or quickly cooled to under 4°C. Many microorganisms grow rapidly in the temperature range 4–60°C. Two very common mishandling problems that occur in the home are failing to refrigerate leftovers promptly and recontaminating cooked product by using the same utensils used with the raw product.

Table 2. Pork storage recommendations[a]

	Refrigerator (2–4°C)	Freezer (−18°C)
Fresh pork	4 days[b]	3–6 months[c]
Cured pork	7 days	2 months

[a] Packaging and handling prior to the consumer will greatly impact shelf life of pork.
[b] Ground meat, 2 days.
[c] Ground meat, 1–2 months.

To minimize microbiological hazards and maximize eating quality, the recommendations in Table 2 have been devised as maximum limits for storage of pork. Moulds and yeasts are of little concern in fresh pork because the high water activity allows bacteria to dominate. In dried pork items such as pepperoni, moulds may grow on the surface. However, mould growth is retarded by a potassium sorbate dip applied by the manufacturer or by vacuum packaging. ***See*** Spoilage, Bacterial Spoilage; Spoilage, Moulds in Food Spoilage; Spoilage, Yeasts in Food Spoilage; Water Activity, Effect on Food Stability

Very few other hazards exist with the consumption of pork. Muscle from pigs is regularly monitored for drug and pesticide residues by the USDA. Incidences of contaminated meat have been isolated and total far less than 1% of the pork supply.

Food Uses and Products

Pork is a very versatile food that can be prepared in many ways. Popular entrée items include pork roast, ham steaks, barbecued spare ribs and grilled pork chops. Pork is also a common component of many foods such as bacon on pizza or ham dices in a chef's salad.

Pork trimmings can be ground and made into patties or sausages of all types, including frankfurters, pepperoni and Italian sausage.

Waste or By-product Utilization

People in the swine industry say that no part of the pig is wasted. This statement is not far from the truth. The major waste product from the live pig is manure, which is a good nitrogenous fertilizer.

Waste products from swine slaughter and processing are perhaps better referred to as by-products because they are seldom wasted. Blood, bones and inedible viscera from pig slaughter are usually dried and ground for use in animal feeds. Gelatin is made from collagenous proteins found in pig skin. Industrial lubricants, plastics and rubber are made from fat trimmed off pig carcasses. Pig skin fabric can be produced from cured pork skins. Even the pig's hair is sometimes used for brushes or insulation.

Perhaps the most important by-products of swine slaughter and processing are those used in medicine. Hormones such as pig insulin or heparin can be prepared for human use. Heart valves from pigs have been used to replace damaged or diseased human heart valves and are often used in heart disease research. Skin grafts from pig skin have been successfully used on human burn victims. These are only a few examples of many medicinal uses of by-products from the swine industry.

Acknowledgements

Appreciation is expressed to Doreen Blackmer for research and editorial assistance.

Bibliography

Mandigo R (1989) *Pork Operations in the Meat Industry.* Washington, DC: American Meat Institute.

National Academy of Sciences, National Research Council, Food and Nutrition Board (1986) *Recommended Daily Dietary Allowances*, 10th edn. Washington, DC: National Academy of Sciences.

National Live Stock and Meat Board (1973) *Hog is Man's Best Friend.* Chicago: National Live Stock and Meat Board.

National Live Stock and Meat Board (1977) *Lessons on Meat.* Chicago: National Live Stock and Meat Board.

National Live Stock and Meat Board (1987) *Exploring Diet and Health.* Chicago: National Live Stock and Meat Board.

Troller J (1983) *Sanitation in Food Processing.* Orlando: Academic Press.

US Department of Agriculture (1983) Composition of Foods; pork products, raw, processed, prepared. *USDA Handbook*, Nos. 8–10. Washington, DC: Human Nutrition Information Service.

R. W. Mandigo
University of Nebraska, Lincoln, USA

PORT

Contents

The Product and its Manufacture

Definition

Port is the name given to a number of related types of dessert wines originally developed in the area around the Douro valley in the north of Portugal. The characteristic flavours of the product are to a large part dependent on the varieties of grape used and the conditions which prevail in the area in which they are grown. In spite of this a limited number of imitations are produced in other countries and described by names, including the word 'port', but this does not seem to be a serious problem for the Portuguese producers. In this article the term 'port' refers only to the Portuguese product.

Under Portuguese law, grapes for port are grown in a delimited area covering the Eastern part of the River Douro and its tributaries. A body elected by the Farmers' Association with powers delegated by the government is based at Regua. It classifies vineyards in which it is proposed to grow port grapes according to quality and, each year, taking into account anticipated world demand, it licences each grower to assign a certain percentage of his grapes for port wine. The percentage authorized for each grower depends on the quality of his vineyard. This complicated system seems to be quite effective in keeping total port stocks in a sensible relation to the market.

The export of port wine can take place only from lodges situated in a defined area in Vila Nova de Gaia or in the demarcated area. It is regulated by the Port Wine Institute situated just across the river in Oporto. The Institute regulates the authenticity of wines, certifying shipments after tasting and analysis. This body also oversees maturation, much of which takes place in the Gaia lodges, though some occurs in the lodges inland in the upper Douro valley.

The Vineyards

Many of the vineyards are situated on the steep valley sides of the River Douro and its tributaries, though some, especially in the very hot area close to the Spanish border, are on shallower slopes. The better area is considered to be that east of the River Corgo.

The River Douro and its tributaries have cut down through a granite plateau into a layer of schist. It is this schistous rock which forms the vineyards. The slopes are unstable and must be terraced. The older terraces are very narrow, supported by retaining walls with two or three closely spaced vines, and it is difficult to mechanize the cultivation. More recent terraces are somewhat wider, allowing the use of special tractors and other plant. The rock weathers quickly, releasing many essential nutrients, but little humus forms.

In most years rainfall is largely confined to the period between November and April; heavy falls are needed if the water table is to remain high enough during the summer – irrigation is not usual. All else being equal, heavy winter rainfalls give good grapes. In some years, summer rains are heavy, and fungus is a problem, *Oidium*, *Peronospora* and *Botrytis* spp. all take their toll and a heavy and expensive spraying programme is essential. Weeds also become a problem, and herbicides are needed on those parts of the terraces, such as the walls, inaccessible to cultivation.

Temperatures in the upper Douro valley are quite extreme. In winter, heavy frosts occur. In summer, temperatures sometimes exceed 40°C, and the mean maximum shade temperature is about 36°C.

Planting, Training and Pruning

Plant lice of the genus *Phylloxera* are endemic in the area. Vines are field grafted onto virus-free hybrid rootstocks, which are not affected: R99, R110 and Rupestris de Lot are frequently used. Vines are trained onto two wires about 50 cm and 1 m above the ground, respectively. Vines are often short spur pruned: two spurs of about four buds are retained, but this is adjusted in line with the area and the vigour of the variety. Alternatively, many modern plantations use a double-cordon training system.

Grape Varieties

Much of the flavour and character of port wine derives from substances present in the individual grape varieties. Fifty varieties of red grapes and 37 of white are known to be grown in the Douro, but extensive studies by some of the leading port shippers indicate that a more limited selection is desirable. Red varieties regarded as desirable include Touriga Nacional, Tinta Roriz, Touriga Francesa, Tinta de Barca, Tinta Barocca, Tinta Cao and Mourisco. Of these, Touriga Nacional probably has a flavour more characteristic of traditional port wine than any other. The best clones of Tinta Cao produce wines of excellent flavour, but inferior clones are present in many locations. Mourisco is a higher-yielding variety which imparts less colour and flavouring but is invaluable when cheaper, quick maturing wines are required. The most useful white varieties are Codega, Rabigato and Malvasia Fina. Many older vineyards are planted unsystematically with a mixture of varieties which are harvested and vinified together. Recent vineyards are planted with the varieties in blocks so that each can be harvested at optimum ripeness.

The Vintage

The vintage may begin as early as 31st August in a hot year in a hot area well inland and may begin as late as 25th October in a cool year in a cooler area, especially in the westerly lower Corgo. It lasts for about 3 weeks in any one area. The topography makes mechanical harvesting impractical.

Most port is red wine made from red grapes though a percentage of white grapes may be incorporated in lesser-quality wines intended for early maturation. White port exclusively from white grapes is also made.

Colour and Tannin Extraction

The aim is to stop the fermentation at a suitable sweetness with brandy. The skill comes in extracting into the liquid sufficient tannin, colour and flavour during the relatively brief fermentation. ***See*** Tannins and Polyphenols

If the work is to be done by the most traditional methods, enough uncrushed grapes are placed in stone troughs, known as lagars, to fill them to a depth of about 50 cm. The lagars are made of granite, are rectangular in shape, the sides being 3–6 m long and about 60 cm deep. They contain enough grapes to make from about 2500 to 11 000 litres of must. The grapes are trodden with bare feet and legs, preferably one person for each 1000 litres of must – 3–4 h periods with 2 h rests each day until fermentation is complete. Often, the temperature of the grapes is 15°C when first placed in the lagar, so that if the first fermentation of the season is being carried out it might be 3 or 4 days before fermentation commences.

Only the natural yeast flora is used, the population of which is initially quite low.

Although completely traditional methods are still used by some farmers, it is more usual to pass the grapes through a roller crusher before filling the lagars to facilitate the treading process, which at best is cold, demanding work. Treading with or without preliminary crushing is a very effective way of ensuring that the alcohol-containing fermenting liquid comes into intimate contact with the tannin- and colour-containing skins, thus maximizing extraction. The method is still used in the upper Douro for a proportion of port processing, particularly for the best quality grapes. Quality is unsurpassed by the best mechanical methods. While treading is not taking place the skins are carried to the surface by the carbon dioxide generated from the fermentation, where they are effectively insulated by bubbles of gas from the main bulk of the liquid.

An alternative to treading is to crush the grapes and then to place them in a lagar. Some of the liquid is then withdrawn and sprayed forcibly over the crushed grapes. Suitable machines consisting of a portable filter and pump are available.

However, the larger shippers make most of their wine in tanks by methods similar to those used in other wine-making areas. Fermentations are carried out in epoxy-lined concrete or stainless steel tanks, colour extraction being by pumping over the surface skins as described above, by manually pushing the skins below the liquid surface or by using a device, 'a heading down board', to hold the skins below the surface. Essentially, the difficulty is to prevent local overheating within the bulk of the skins, which permits the multiplication of undesirable microorganisms. ***See*** Wines, Production of Table Wines

Also widely used are autofermenters of the type developed in Algeria in the 1950s. These devices work well enough but are difficult to control. Local overheating occurs and adequate cleaning is difficult. Roto tanks, developed in Australia, are being tried and may offer some advantages.

Limited use is made of thermovinification, a process whereby, immediately after crushing, the liquid is separated, flash heated to some 80°C, poured over the solids to extract colour and tannin, cooled to about 32°C, yeasted and allowed to ferment without the skins. The process works well if sulphur dioxide levels are well regulated. It is particularly successful in poor years when there are many mouldy grapes but is viewed with suspicion by traditionalists and little used for the better quality wines.

At one time white port was made by fermenting the grapes in contact with the skins as for red wine, but nowadays most is made by pressing the grapes before fermentation and fermenting the blended free-run and first press musts.

Fermentation

The specific gravity of grape must for red port before fermentation is between 1·090 and 1·100; for white port, 1·085–1·095. Fermentation is generally allowed to proceed until the specific gravity is about 1·045, although some drier wine is produced and for blending purposes some must is fortified after little if any fermentation. If the natural yeast flora is used there is a succession as in other areas; apiculate yeasts dominate at first before the *Saccharomyces* species take over. The *Saccharoymces cerevisiae* strain, at one time called *S. steineri*, is said to be unusually plentiful in the area. However, some of the larger manufacturers use cultured yeast, if only to speed the process and free fermentation vessels, which tend to be in short supply. The strain of yeast used does not seem to be critical. The Montrachet strain of *S. cerevisiae* of Californian origin is often used and 25°C seems to be about the ideal temperature; mechanical refrigeration is required in conjunction with the larger vessels. Initial temperatures are often too low; some form of preheating might be useful but is rarely available.

Fortification

This must be carried out with the special brandy supplied for the purpose by government authority. It is distilled at a strength of approximately 75% alcohol by volume and consequently is by no means neutral in flavour. Little is produced in the Douro and often it is not Portuguese in origin. Only enough brandy is made available to enable each grower to fortify the amount of must he is authorized to make into port wine.

Since the brandy is sold at a rather inflated price this amounts to a form of taxation. The style and quality varies, and users have some choice of the brandy they accept, but often it is made available so late that there is little time for comprehensive analysis.

The still fermenting wines are racked off from the skins, etc., before fortification to obviate loss of alcohol into the solids. Usually the fortification is carried out in vats or tonels (large wooden containers), and approximately 210 litres of brandy is added to each 1000 litres of must at a specific gravity of 1·045; less for drier musts. This takes the strength to approximately 19% by volume, normally just sufficient to prevent the development of lactobacilli. These bacteria can prove troublesome in young port wines, causing acetification, especially in years when there is mould on the grapes. ***See*** Lactic Acid Bacteria

Jeropigas, which are musts which are fortified almost as soon as fermentation begins to keep the wine as sweet as possible, are also fortified to approximately 19% alcohol by volume.

Maturation

The wine is stored in the valley of the River Douro throughout the winter after the vintage. It is stored in epoxy-lined concrete tanks, stainless steel tanks, large vessels of oak or 600 litre casks. These containers are kept as full as possible: aeration is avoided. Racking is carried out between November and March, as the weather permits. The wine is fortified to a minimum of 18%, more usually 19·5–20·5%. It is not filtered at this stage and a high proportion of red wine is shipped to Vila Nova de Gaia. White port often remains longer in the Douro.

The frequency of racking and the ambient temperature (higher in the Douro) influence development of the wine.

Racking regimes vary from shipper to shipper; typically every 3–6 months for 3 years, after which less frequent racking is needed.

Commercial Styles of Port

Vintage Port

Red wines selected as being of superior quality, and all from one vintage are blended after about 2 years and filled into bottles after little or no fining and filtering and no refrigeration. Vintage ports require at least 10 years in the bottle, some wines needing much more time; they will throw a heavy deposit in the bottle. The bottles must be carefully cellared at lowish temperatures and decanted before consumption. Wines from different shippers, and different years develop distinctive flavours. It is wine for the informed consumer, in short supply and expensive.

Crusted Port

Crusted port is fairly similar to vintage port except that it is not all from the same year and it spends approximately 4 rather than 2 years in wood. Blended from wines not quite as high in quality as vintage port, it is nevertheless a very acceptable substitute and much cheaper.

Late Bottled Vintage Port

Red wines of superior quality and all from one vintage are matured in wood for 5 or 6 years in cool conditions with infrequent racking. It is bottled after fining, filtration and possibly refrigeration, and generally offered for immediate consumption. It probably will not need decanting. It is a good heavy red port, smooth and with a little development towards tawny.

Ruby Port

Ruby port is a standard to good quality red port blended from several wines with ages between 3 and 5 years in oak pipes under cool conditions, fined, filtered and refrigerated, bottled and offered for immediate consumption. It should not need decanting. The younger wines usually described as 'fine ruby' are fresh and fruity, and the tannin is evident. The older wines, described as, for example, 'special reserves' are still fruity but smoother and are beginning to acquire a brown tint. The best are quite similar to and rather cheaper than late bottled vintage wines.

Tawny Port

This is a wine matured in the cask long enough to have developed a brown rather than a red colour. The cheaper wines are not necessarily older than rubies, but are put together from wines from predominantly lighter grapes such as the Mourisco, and grapes from the lower Corgo. They are matured under warmer conditions. Better quality tawnies are simply rubies which have been aged much longer in the cask, they are sold as '10-year-old', etc. Many shippers sell tawnies of mixed and unspecified age under proprietary labels. They are often of considerable age, superior quality and better value than those bearing an age claim. Tawny wines are fined, filtered and, in the case of younger wines, refrigerated and should not require decanting.

White Ports

In effect, white ports are, or were, ruby ports but made from white grapes. The traditional ones, made by fermenting in contact with the skins, have relatively high tannin levels and are as sweet as the red wines. More recently, there has been a tendency towards less sweet versions with tannin levels no higher than table wines. These are made by pressing the must before fermentation.

Bibliography

Bradford S (1969) *The Englishman's Wine: The Story of Port*. London: Macmillan.

Goswell RW and Kunkee RE (1977) Fortified wines. In: Rose AH (ed.) *Alcoholic Beverages. Economic Microbiology*, vol. 1, pp 477–535. London: Academic Press.

Moreira de Fonseca, Galhano A, Serpa Pimental E, Rosas J R-P (1981) *Port Wine: Notes on its History, Production and Technology*. Porto: Instuto do Vinho do Porto.

Robin W. Goswell
John Harvey and Sons Ltd, Bristol, UK

Composition and Analysis

Port is made from grapes grown in the Duoro region in Northern Portugal. Both red and white ports are made, but the red port has by far the larger market share. There are two main styles of port both made from red grapes: ruby ports are red, full bodied and often still quite fruity in character, reminiscent of the grapes; tawny ports generally have an amber colour and a flavour developed during prolonged ageing in mature wood, which is best described as pale, with an oaky note and giving a dry impression. Within these two main styles many different ports are produced, each having its own sensory characteristics. The manufacturing procedures for the ports are all very similar. Besides the climate and the geological influences, both the choice of fruit and the maturation parameters determine to a great extent the chemical composition and the final quality of the product, and will be discussed in this article.

Although white port is essentially made in the same way as red and ruby ports, there is very little information published on this style. This article concentrates specificically on red ports.

Organoleptic Assessment

Ports are matured in old oak vats or casks and are usually blended during their maturation period. Sensory analysis during this period plays an important role, and is used to help with the classification of the ports in quality categories and for the blending procedure. Although ports contain up to 20% alcohol, young ports are susceptible to bacterial spoilage and tasting serves as one of the methods to detect early symptoms of spoilage, usually expressing itself in the form of increased volatile acidity. The ports are generally tasted just before the racking procedure, which takes them off any lees into a new vat or cask. The lees consists of yeast cells from the fermentation which have since precipitated, tartrate deposit and some colour components. Racking normally takes place three times in the 1st year, twice in the 2nd year and only once in subsequent years. Usually, a small panel of well-trained expert tasters is used. The quality characteristics regarding colour, aroma and flavour are scored on a scale from 1 (very poor) to 10 (excellent), using full-strength samples. Tulip-shaped glasses which are only part filled are used to allow the maximum development of volatile components above the sample for sniffing. To assess the colour, natural northern daylight is preferred. Another way of tasting the ports is by sensory profiling. In contrast to the quality tasting of the product, this method aims to describe the product as fully as possible, without including a quality judgement. During training sessions the tasters develop a set of terms, often referred to as 'profiling language', with which a wide range of ports can be accurately described. A typical language developed by a panel of expert port tasters is shown in Table 1. ***See*** Barrels; Sensory Evaluation, Sensory Characteristics of Human Foods; Sensory Evaluation, Practical Considerations; Sensory Evaluation, Sensory Rating and Scoring Methods; Sensory Evaluation, Descriptive Analysis; Sensory Evaluation, Appearance; Sensory Evaluation, Texture; Sensory Evaluation, Aroma; Sensory Evaluation, Taste

Variations with Processing and Origin

The different styles of port quality are due to the choice of grape varieties, their composition as a result of climatological and geological influences, and the fermentation followed by small variations in maturation, causing different modifications to the primary grape flavour. These factors will be discussed separately.

Grapes

Many different grape varieties are used for the production of port. The quality and variety of the grape will influence the quality and composition of the port. In most vineyards a mixture of cultivars is planted. These are all picked and processed at the same time, with no allowance for variations in requirements necessary to achieve optimum grape maturity. Some of the more

Table 1. Profiling terms developed to describe ports

Colour	Aroma	Flavour
Intensity	Soft	Intensity
Clarity	Clean	Sweetness
Depth of colour	Bland/dull	Acidity
Redness	Complex	Bitterness
Brownness	Violet	Astringency
Yellowness	Blackcurrants	Tannin
Blueness	Raspberries	Woody
Purpleness	Toffee/caramel	Fruity
	Burnt/liquorice	Raisin/liquorice
	Chocolate	Body/viscosity
	Nutty	Balance
	Biscuity	Persistency
	Tannin/stalky	Crispness
	Oaky/vanilla	Musty
	Volatile acid	Spirity
	Pungent/peppery	Metallic
	Estery/banana	
	Spirit	
	Wet earth	

important cultivars are Touriga Nacional, Tinta Barroca, Tinto Cao, Touriga Francesa, Tinta Roriz, Tinta Francisca and Mourisco. Recently planted vineyards are likely to include a selection of these, usually planted with each variety in an identifiable block or row.

Fermentation and Fortification

All ports are made by crushing the grapes in the presence of sulphur dioxide, normally no more than 150 mg l^{-1}. The stalks are removed to prevent the extraction of large quantities of phenols, which would give a harsh, astringent taste, requiring a long maturation period to allow the ports to develop a mellower character.

The fermentation of the grape mash, including the skins, takes place using natural yeasts or by inoculation with a selected yeast strain. A typical red grape mash has a total acidity of 0·45–0·75 g per 100 ml expressed as tartaric acid, a specific gravity of 1·090 to 1·100 and a pH value of 3·5–3·9. The fermentation on the skins is short, generally less than 2 days.

Halfway through the fermentation to dryness at a specific gravity of about 1·045 the fermenting mash is pressed and the juice is fortified, with brandy containing 77% (v/v) alcohol, to 19% (v/v) alcohol to arrest any further fermentation. The brandy is a grape spirit, normally distilled from wines made in the south of Portugal or the Douro region. Its character is certainly not neutral and has a distinct influence on the quality of the port. The most prominent compound present is acetaldehyde, which greatly influences the maturation processes in ports. Higher alcohols and esters also contribute to the quality of the brandy. ***See*** Brandy and Cognac, Brandy and its Manufacture

The young fortified ports may still contain a residual concentration of sulphur dioxide (20–100 mg l^{-1}) used during crushing, which will exert an influence over the ageing mechanisms occurring in the young ports.

Anthocyanin Content of Grapes and Young Ports

The colour of a young port is to a great extent due to anthocyanins, which are water-soluble pigments responsible for the attractive colours of many flowers, leaves and fruits. In the *Vitis vinifera* grapes used for the production of ports of all the anthocyanins are 3-glucosides. In all but three cultivars the anthocyanins are located in the skins. In both grapes and young ports, anthocyanins based on malvidin generally predominate. Malvidin 3-glucoside is the major pigment (33–94%), followed by malvidin 3-*p*-coumarylglucoside (1–51%) and malvidin 3-acetylglucoside (1–18%). Peonidin 3-glucoside (1–39%) is prominent in four cultivars, but delphinidin 3-glucoside (1–13%), petunidin 3-glucoside (2–12%) and cyanidin 3-glucoside (trace to 6%) are present at low concentrations. ***See*** Colours, Properties and Determination of Natural Pigments

The ratio malvidin 3-acetylglucoside: total malvidin 3-glucosides (the latter being defined as the sum of percentages of malvidin 3-glucoside, malvidin 3-acetylglucoside and malvidin 3-*p*-coumarylglucoside) is characteristic of cultivar, independent of site and a useful aid to identification of grape cultivars.

The distribution of anthocyanins in the grape differs from the young ports. Percentages of malvidin 3-glucoside are higher in the ports than in the skins, the percentages of malvidin 3-acetylglucoside are slightly higher or the same whereas, excepting two cultivars, the percentages of malvidin 3-*p*-coumarylglucoside are lower in ports. Hydrolysis of malvidin 3-*p*-coumarylglucoside during fermentation would account for such a loss, which would be accompanied by an increase in its hydrolysis product malvidin 3-glucoside, the latter being indeed present in higher concentrations in ports.

The fresh juice from two of the three cultivars containing coloured pulp (Tinta Santarem and Viera da Natividade) contain peonidin 3-glucoside (41–76%) as a major component, whereas the corresponding grape skins contain much less (23–36%). The third cultivar (Souzao) shows only a small difference, 10% peonidin 3-glucoside in the juice and 6% in the grape skins. These differences in composition between the skins and the pulp presumably reflect differences in enzymatic activities related to pigment biosynthesis.

The total concentration of anthocyanins in young single variety ports ranges from 143 to 1080 mg l^{-1}, with an average of 330 mg l^{-1}. Souzao gives by far the most strongly coloured port, containing at least twice as much anthocyanins as any other port. The concentrations of total anthocyanins in young ports made from single cultivars from the same site over 3 years may vary twofold. This variation is as great as the variation in the same cultivar from different sites in any given year. Cultivars contain consistently more anthocyanins when grown in the upper Corgo than when grown in the lower part of the region.

Colour of Grapes and Young Ports

The colour of young ports depends on the anthocyanin concentration but also on other parameters such as pH value, presence of other phenols and the degree of polymerization which has occurred during processing. Variation in total pigments (a measure of total colour potential independent of the pH value of the port) is affected much more by cultivar than by season. Cultivar variations can be up to 12-fold in total pigments, but only 3·6-fold in total phenols. Seasonal variations can be up to two-fold in total pigments and 1·6-fold in total phenols. *See* Phenolic Compounds

Souzao, Tinta da Barca and Touriga Nacional ports are the most coloured. Mourisco and Tinto Cao are the least coloured. Only small differences are observed between upper Corgo and lower Corgo ports.

Changes During Maturation

Ports are generally matured in wood for at least 3 years to produce young rubies, but a considerably longer maturation period (up to 20 years) in wood is needed to produce some of the tawny styles.

The colour, aroma and flavour of a young port are due to compounds extracted from the grape during fermentation. However, to develop a product with attractive and complex sensory attributes the port is allowed to mature. The chemical processes underlying the changes occurring during maturation are complex and not fully understood. The colour becomes browner, changing from a deep red with a purple edge at the rim of the glass, to a brick red or even an amber tawny colour. Accompanying these colour changes are alterations in perceived astringency due to modifications of the phenols. At the same time the bouquet of the port changes from a grapey aroma to that of a typical port, depending on the length of maturation and whether the port has matured in wood or in bottles.

Many factors affect the chemical processes occurring during maturation; however, research has mainly been directed towards determining those influencing the colour characteristics. The changes in colour depend on concentrations of anthocyanins and other phenols in the grapes and the ports, processing parameters including storage conditions such as temperature, the degree of aeration the product receives during normal racking procedures, concentrations of acetaldehyde and sulphur dioxide, and the pH value of the port.

Climate and Bottle Versus Vat

Differences in perceived quality are due to a number of factors. Variations in growing conditions, such as an ideal growing season, allow a small part of the production to be declared as 'vintage'. In such years the quality of the grapes grown at certain top-ranking vineyards is considered outstanding, and likely to result in very high-quality ports. These ports will be bottled after a maturation period of no longer than 2 years, followed by maturation in the bottle for at least 10 years or more. The resulting sensory properties of such a product are quite different from ports of comparable age which have been aged in oak vats: this difference is to an extent attributable to the nonoxidative ageing in the bottle. These ports generally appear darker red, with a mature fruit aroma, full and tannic and with a distinctive flavour typical of bottle age.

To counterbalance annual differences in climate on the quality of the grapes, most other ports on the market are usually blends of ports made in different years and regions to maintain a consistent quality over the years. The indication of age on those products is an average age.

Colour Changes During Maturation

The colour intensity of a port increases markedly for several months during its early maturation, reaches a maximum, generally referred to as 'closing up', and then declines. In contrast the colour of unfortified red table wine remains relatively stable or falls slowly. The initial increase in colour during port ageing is due to the formation of aldehyde-containing oligomeric or polymeric pigments which are more coloured at port pH values and less coloured under acid conditions than the anthocyanins from which they are derived. These polymeric pigments are also less susceptible to bleaching by sulphur dioxide than the monomeric anthocyanins. When the colour maximum is reached, the polymeric pigments may have attained their maximum colour intensity or the colour loss by precipitation of the largest polymers may have been compensated by formation of freshly formed oligomers. Subsequent colour loss can be explained by a combination of factors, e.g. precipitation of the largest polymers now exceeding the formation of new oligomers and the polymers which remain soluble become less coloured by incorporation of other colourless phenols.

Role of Aldehyde During Maturation

An important difference between port and red table wine is the presence of free aldehyde, largely acetaldehyde, in the former. The aldehyde concentration exerts an important influence on the ageing processes in ports and is involved in the polymerization reactions occurring during maturation. Port aldehyde originates from the brandy spirit used for fortification and from aldehydes formed during fermentation. Acetaldehyde can also be produced during maturation by coupled oxidation of ethanol, which is known to occur even at low concentrations of sulphur dioxide. The predominant aldehyde and also the most reactive one is acetaldehyde. Since methods used for determining acetaldehyde concentrations will usually also determine small concentrations of other aldehydes present, the term 'aldehyde' is used here.

Aldehyde binds reversibly, but strongly, in equimolar concentrations to sulphur dioxide. Only free aldehyde, defined as aldehyde not bound to sulphur dioxide, participates in the polymerization reactions. However, as the sulphur dioxide concentration decreases due to oxidation reactions, bound aldehyde is gradually released as free aldehyde. Freshly fortified ports contain free aldehyde because total aldehyde is always in molar excess over sulphur dioxide.

The mechanism of formation of aldehyde-containing oligomeric pigments is probably similar to that described in model systems, with the formation of complexes containing an acetaldehyde link between anthocyanin and flavan-3-ol.

Maturation Rate

Direct condensation between anthocyanins and other phenols (flavan-3-ols) also occurs; this ageing mechanism is more prominent in red table wine due to the much lower acetaldehyde content in this product. The colour changes in ports are interpreted largely in terms of these two reactions: aldehyde ageing which is superimposed upon direct condensation. The greater the free aldehyde content of the port, the more prominent is the aldehyde-induced reaction and vice versa. The course of anthocyanin condensation and polymerization can be represented by a gradual transition from monomeric anthocyanins through oligomers to polymeric pigments. The onset of the polymerization occurs already during fermentation. Immediately after fermentation between 17 and 69% of the colour is due to polymeric pigments.

Losses of anthocyanins are logarithmic with time. The rate of loss of total anthocyanins reflects the rate of anthocyanin participation in polymerization reactions. During the initial ageing in the presence of free aldehyde the molar loss of anthocyanins is higher than the molar loss of aldehyde, indicating that both aldehyde condensation and direct condensation, not involving aldehyde, are taking place. Increasing acetaldehyde content increases the rate of loss of anthocyanins. Aldehydes are still reacting when only very low concentrations of anthocyanins remain, indicating the formation of complex branched polymers.

Volatile Flavour Compounds

During the long maturation period the volatile components in ports undergo considerable changes. Little information is available on the chemistry of these changes, although some data available on red table wines may be relevant to ports.

In ports, volatile components have been identified whose formation can be traced back to products of yeast fermentation, or to chemical changes occurring during maturation, such as acetylation, esterification, oxidation and carbohydrate degradation as well as wood extraction. Older ports tend to have higher concentrations of volatile components resulting from carbohydrate degradation and wood extraction, such as furfural and oak lactone.

Of the more than 200 volatile components thus far detected in ports, 141 have been wholly or partially identified. These consisted of 14 alcohols, one diol, two phenols, two alkoxyalcohols, two alkoxyphenols, five acids, five carbonyls, three hydroxycarbonyls, 81 esters, two lactones, nine dioxolanes, two oxygen heterocyclics, one sulphur-containing component, four nitrogen-containing components, six hydrocarbons and two halogen compounds. A number of the identified volatile component are present in high enough concentrations to contribute to the overall port aroma. The relatively large concentrations of succinates (approximately 100 $\mu g\ g^{-1}$) should contribute a winey and fruity aroma to the overall port bouquet. Esters from 2-phenethanol have fruity sweet aromas, contributing to the bouquet. β-Methyl-γ-octalactone, described as a coconut flavour, probably contributes to the nutty, oaky note of the aroma. Furan derivatives, which are readily derived from carbohydrate degradation, have generally sugary, oxidized aromas and probably contribute to the overall port aroma.

Oxidation of alcohols during storage may account for the presence of aldehydes, which are also partly present as acetals. Although acetaldehyde will influence the maturation of the nonvolatile components in the ports, the role of acetaldehyde as part of the overall port aroma is not clear.

Bibliography

Bakker J (1986) HPLC of anthocyanins in port wines; determination of ageing rates. *Vitis* 25: 203–214.

Bakker J and Timberlake CF (1985) The distribution of anthocyanins in grape skin extracts of port wine cultivars as determined by high performance liquid chromatography. *Journal of the Science of Food and Agriculture* 36: 1325–1324.

Bakker J and Timberlake CF (1985) The distribution and content of anthocyanins in young port wines as determined by high performance liquid chromatography. *Journal of the Science of Food and Agriculture* 36: 1315–1333.

Bakker J and Timberlake CF (1986) The mechanism of colour changes in ageing port wine. *American Journal of Enology and Viticulture* 37: 288–292.

Bakker J, Bridle P, Timberlake CF and Arnold GM (1986) The colour, pigment and phenol content of young port wines; effects of cultivar, season and site. *Vitis* 25: 40–52.

Bakker J, Preston NW and Timberlake CF (1986) Ageing of anthocyanins in red wines; comparison of HPLC and spectral methods. *American Journal of Enology and Viticulture* 37: 121–126.

Goswell RW (1986) Microbiology of fortified wines. In: Robinson RK (ed.) *Developments in Food Microbiology*, vol. 2, pp 1–19. London: Elsevier.

Goswell RW and Kunkee RE (1977) Fortified wines. In: Rose AH (ed.) *Alcoholic Beverages Economic Microbiology*, vol. 1, pp 477–535. London: Academic Press.

Maarse H and Visscher CA (eds) (1989) *Volatile Compounds in Food, Qualitative and Quantitative Data*, 6 edn, vol. II. Zeist, Netherlands: TNO CIVO Food Analysis Institute.

Williams AA, Lewis MJ and May HV (1983) The volatile components of commercial port wines. *Journal of the Science of Food and Agriculture* 34: 311–319.

Jokie Bakker
AFRC Institute of Food Research, Reading, UK

POSTHARVEST DETERIORATION

See Spoilage

POTASSIUM

Contents

Properties and Determination

Potassium is an alkali metal with an atomic weight of 39·098 and an electronic orbital structure of $1s^2 2s^2 2p^6 3s^2 3p^6 4s^1$. Having a single electron in the 4s orbital, the metal is easily ionized to the cation (K^+) through the loss of the outer electron. Consequently, potassium is never found as a free metal in nature. Potassium is extremely abundant and makes up about 2·4% of the earth's crust. Like its sister alkali metal sodium, potassium is essential for life and as such is usually found as the chloride salt. It is abundant and essential in both plant and animal tissues, whereas sodium is primarily found in animal tissues, being rather toxic to many plants. The normal dietary requirement for potassium is easily satisfied by most food intake because of its ubiquitous occurrence in both plant and animal foodstuffs. Normal ingestion of potassium in food is about 2–3 g per day. Although sodium and potassium are very similar in their chemical characteristics, they play very different roles in animal tissues. Potassium is actively maintained as the predominant intracellular cation while sodium is actively transported from the interior of cells into the extracellular space, where it is the predominant extracellular cation. As an intracellular ion, potassium functions as a regulator of numerous important enzymes. It attaches to negatively charged sites on enzyme proteins, changing the conformation of the molecule and its catalytic activity. Potassium is also important in the regulation of cell division. Finally, potassium transport and permeability play an important role in the generation of transmembrane potentials that are necessary in the functioning of cells such as those of nerve and muscle. Because the gradients of potassium are linked to heart muscle and nerve function, the regulation of extracellular potassium concentrations is also very important in human physiology. Extracellular potassium is regulated primarily through excretion by the kidneys. ***See*** Sodium, Physiology

Potassium chloride is the principal ingredient in salt substitutes used to satisfy the appetite for salt, while restricting the intake of the sodium ion, which has been associated with some forms of hypertension. However, serious cardiac arrhythmias have been linked to the excessive ingestion of potassium chloride-containing salt replacements. It is clear that hypo- or hyperkalaemia (high or low blood levels of potassium) can bring about serious cardiovascular or neuromuscular problems. Correction of blood levels generally results in resolution of these conditions. ***See*** Hypertension, Hypertension and Diet

Preparation of Samples for Analysis

Potassium compounds were first prepared by the evaporation in pots of aqueous extracts of plant ashes. The name 'potash', from which 'potassium' is derived, relates to this process and indicates that potassium salts are rather water-soluble. Indeed, like the salts of most alkali metals, potassium salts are generally highly soluble in either acid or alkali aqueous solutions. However, simple extraction procedures are not always adequate to provide complete extraction of the ion from food or biological samples. Potassium can be compartmentalized or tightly bound to tissue components. For these reasons, more drastic procedures than simple homogenization followed by aqueous extraction are often required. Freeze–thawing, boiling, ultrasonic treatment, or extraction in acid or alkaline solutions can all be used to extract nonsequestered potassium. When sequestering of potassium exists, it is often necessary to destroy the sample by heat or by strong acid digestion. When samples are ashed using heat, the ash can then be extracted as described above. ***See*** Sodium, Properties and Determination

Gravimetric Analysis

One of the earliest methods for determining potassium content of extracts or other samples was by use of a gravimetric procedure: the potassium would be precipitated with chloroplatinate or tetraphenylboron. The precipitate was separated by centrifugation, washed, dried and weighed.

When optical methods, based on flame emission and flame absorption were introduced, gravimetric analysis was not completely abandoned but became less frequently used. Even when spectral analysis is used, samples sometimes contain substances that interfere with emission properties of potassium. In such cases, the interfering substance can be removed by quantitative potassium precipitation using chloroplatinate.

Other methods of potassium analysis are mechanistically identical to methods of analysis for sodium. These include flame emission spectrophotometry, flame absorption spectroscopy, helium glow photometry, ion-selective electrodes, electron probe microanalysis, and nuclear magnetic resonance spectroscopy. Radioactive dilution is one additional procedure which is often used to determine potassium distribution.

Since the physical principles underlying these methods have been discussed in some detail in the article on sodium, the reader is referred to that article for a review. Specific properties relating to potassium itself will be presented in the remainder of this article. ***See*** Sodium, Properties and Determination

Flame Emission and Flame Absorbance Spectroscopy

Potassium, in biological materials, is usually determined by flame emission spectroscopy. Commercially available instruments, using the dominant emission line of 766·5 nm, are capable of measuring potassium concentration down to 5×10^{-12} M. While the spectral line for potassium is different from that of sodium, the presence of the latter is reported to enhance the potassium spectral intensity at low sodium concentration. Since the effect of sodium plateaus at concentrations above 1·25 mM sodium, some investigators add at least 1·25 mM of sodium to all samples to ensure that the effect of sodium is constant. Flame atomic absorption spectroscopy is equally appropriate for measuring potassium in biological samples but is not as commonly used.

When very low concentrations of potassium are being measured, care must be taken to minimize contamination. Two common sources of contamination are room dust or smoke. Ideally, one should use the flame photometer in a dust-free room when measuring low levels of potassium. Since this precaution cannot be implemented in many laboratories, stirring of dust should be minimized by restricting movements of individuals in the room where analyses are being done. Smoking of tobacco is an even more significant problem because smoke from tobacco contains large amounts of potassium carbonate. It is not unusual for the intensity of the potassium emission to reach very high levels when smoke is exhaled by a person smoking in the room.

Helium Glow Photometry

Helium glow photometry of potassium is in principle the same as for sodium. The only difference is that a different photodetection system must be used. Since smoke and dust also affect the helium glow technique, as well as flame emission and flame absorption techniques, efforts should be made to prevent contamination from these sources.

Electron Microprobe X Ray Analysis

This modern technique potentially measures any element and has been particularly useful for measuring sodium and potassium in biological tissues. The method of detection depends on the characteristics of X rays emitted when a high-energy electron beam impinges on a very small area of a sample. This analysis is usually done in an electron microscope and allows the quantification and localization of the element in the sample, even in subcellular compartments of plant or animal tissue. The method is not practical for determinations of overall potassium concentrations of relatively large samples of foodstuffs or biological tissues. The preparation of the sample is rather complex, involving nearly all the steps necessary to prepare a sample for electron microscopy.

Potassium-selective Electrodes

Potassium ion-selective electrodes offer a convenient and nondestructive means of measuring potassium activity in solution. These electrodes are based on the development of ion-selective materials and have allowed potentiometric measurements of numerous ions, including potassium and sodium. While ion-selective glass electrodes, sensitive to potassium, were the first to be developed, their use has never been very extensive, due to rather poor ion selectivity. A second type of potassium-sensitive electrode uses ion exchange compounds dissolved in organic solvents. These electrodes make up a large proportion of potassium electrodes available commercially, although they show significant inteferences from sodium and other cations, which are likely to be present in many solutions.

Valinomycin, a neutral carrier compound, is very selective for potassium and is the basis for the third class of potassium-sensitive electrodes. While organic solvent solution membranes of valinomycin show low interference from other ions, the response time for these electrodes is much slower than ion-exchanger electrodes. Neither the valinomycin nor the ion exchanger may be used in organic solvents. It should be remembered that electrodes measure ion activities and thus do not sense either ions that are bound to other molecules or potassium that is not in solution.

Nuclear Magnetic Resonance (NMR) of Potassium

NMR of potassium holds great promise as a means of nondestructively measuring the ion in biological tissues. Although the NMR receptivity of the ^{39}K nuclide is about 2000 times less than that of a hydrogen atom, the 93% natural abundance of the nuclide, the relatively high concentrations in biological tissues, and the relatively short longitudinal relaxation time make it easily detectable in biological samples. Furthermore, the use of cell membrane impermeable chemical shift reagents have made it possible to measure simultaneously intracellular and extracellular tissue concentrations of potassium. A recently developed method allows intracellular potassium concentration determinations without the use of shift reagents. The principal concern about NMR spectroscopic measurement of potassium in tissues is the apparent NMR invisibility of 60% of the potassium known to be present by other means of analysis. The nature of this NMR invisibility now seems to be understood, but still presents a problem to the quantification of potassium in biological tissues. ***See*** Spectroscopy, Nuclear Magnetic Resonance

Dilution of Radioactive Isotopes of Potassium

On occasion it is important to determine not only the overall content of potassium in a whole organism or tissue but to establish the concentration and distribution of the amounts in the various compartments of the sample. This is particularly important in whole animals where there are great differences in the concentration of potassium between the extra- and intracellular environments. Dilution of radioactive isotopes is used to establish the compartment volumes of the tissue or organism. From these determinations, and the measurements of potassium concentrations or specific activities in each of the spaces, the desired potassium concentrations and distributions can be calculated. Both animals and plants have potassium compartments that exhibit different abilities to accumulate the ion. In animals, there are usually three compartments: (1) plasma and extracellular fluid, (2) cells and (3) space in which potassium is tightly bound or complexed. In plants, there are only two major compartments to consider since the extracellular space is relatively small. The two compartments are cytoplasm and vacuoles in those cells that contain these compartments. Cellulous walls can also contain or bind the ion.

The measurement of spaces in animals is complex because the isotope is not only distributed among compartments but is being lost simultaneously from the intact animal through excretion in urine, faeces and other bodily fluids. The ideal measurement consists of injecting a known amount of the tracer isotope of potassium into the extracellular compartment (plasma) and then following changes in the plasma isotope concentration until equilibrium is reached. For these measurements to be accurate the marker must come to equilibrium with all the compartments. In the first phase after injection, the radioactive material is distributed throughout the extracellular compartment; during the second phase there is equilibrium between the cellular and extracellular compartments; and, in the third phase, equilibration of the ion with the sites binding the ion. At equilibrium the distribution of the potassium can be expressed by the relationship $Q_{inj} = Q_{ECF} + Q_{cell} + Q_{bound}$, where the subscripts represent, respectively, injected, extracellular fluid, cell and bound quantities of the potassium. The ECF volume (V_{ECF}) can be estimated as the volume of dilution of some solute other than potassium that is restricted completely to plasma and interstitial fluids. The best tracer solutes are usually polymers, such as inulin or polyethylene glycols, that cannot penetrate cell membranes. The total distribution space, which is the sum of the extracellular and intracellular space, can be estimated by using a material, such as labelled water (tritiated or deuterated) that distributes in all extracellular and cellular compartments. Cellular volume can be estimated by subtracting the previously measured volume of ECF from the total distribution space. The quantity of potassium in the extracellular compartment can then be calculated from the total V_{ECF} times the specific activity of the isotope in extracellular fluid. The cellular potassium is easily calculated from the cell volume and the amount of radioactive potassium in the cellular compartment. The bound quantity of potassium is then defined by subtracting the sum $Q_{ECF} + Q_{cell}$ from Q_{inj}.

Potassium content and distribution in plant tissue is also very complex. The amount of cellulose surrounding cells varies with the location of the plant cell, as does the relative size of the cell vacuole. In most (but not all) plants, vacuoles accumulate potassium. Vacuolar accumulation occurs by a hydrogen ion – potassium exchange transport mediated by an H^+/K^+-dependent ATPase that transports hydrogen and potassium in opposite directions.

Separating vacuolar volume and cytoplasmic volume is difficult by isotope dilution kinetics. Although it is theoretically possible to calculate the volume of distribution and concentration of potassium in plants, this is not the method of choice for obtaining this information. Cytoplasmic and vacuolar potassium concentrations are probably more easily obtained using ion-selective microelectrodes or X ray microprobe analysis. X Ray microprobe analysis can also be used for determining potassium content of the cellulose wall. In short, there are better methods for analysing plant cell compartments for potassium than radioactive dilution methods.

Other Methods

As with the analysis of sodium, there are several other methods by which potassium can be analysed. These include most of the methods used for sodium, including inductively coupled plasma optical emission or inductively coupled plasma atomic fluorescence, ion-scattering spectrometry, and other methods.

Although these methods may be useful for particular tissues or when measurement of numerous other atomic species are also being made, they are not commonly used for routine analysis of potassium.

Bibliography

Ammon D (1986) *Ion Selective Microelectrodes*. New York: Springer-Verlag. N.Y.

Davis RE, Gailey KE and Whitten KW (1985) *Principles of Chemistry*. Philadelphia: Saunders.

Hamilton EI (1979) *The Chemical Elements and Man*. Springfield: Charles C. Thomas.

Kare MR, Fregly MJ and Bernard RA (1980) *Biological and Behavioral Aspects of Salt Intake*. New York: Academic Press.

Kernan RP (1965) *Cell K*. Washington, DC: Butterworths.

Phipps DA (1976) *Metals and Metabolism*. London: Oxford University Press.

Riordan JF and Valles BL (1988) Metallobiochemistry. *Methods in Enzymology*, vol. 158. San Diego: Academic Press.

Springer C (1987) Measurement of metal cation compartmentalization in tissue by high-resolution metal cation NMR. *Annual Review of Biophysics and Biochemistry*, pp 375–399.

William R. Galey and Sidney Solomon†
University of New Mexico School of Medicine, Albuquerque, USA

Physiology

Potassium, found in all living tissues, is the major intracellular cation in the human body. It is an element of great importance in a number of cellular metabolic functions.

The study of potassium metabolism comprises aspects related to intake, absorption, transport, distribution and excretion, as well as homeostatic mechanisms that prevent modifications in plasma potassium concentration, hypokalaemia and hyperkalaemia.

Role in the Body

Potassium is characterized by its multifunctionality, affecting functions of the cardiovascular, digestive, endocrine, respiratory, renal and neurological systems. It is a cofactor for enzymes involved in carbohydrate storage, energy transduction, cellular growth, and others. Potassium deficiency induces growth retardation, with pronounced decrease in circulating somatomedin C and concomitant inhibition of protein synthesis. ***See*** Coenzymes

† Deceased

Many of the functions of potassium are related to its ionic (K^+) character, whereby it generates gradients of concentration, potential and pressure; K^+ participates in regulation of the acid–base equilibrium. By virtue of being the predominant osmotically active species within cells, K^+ plays a major role in the distribution of fluid inside and outside the cell, and in the maintenance of cellular volume. ***See*** Electrolytes, Analysis

The transmembrane concentrations of sodium (Na^+) and K^+ are primary determinants of the transmembrane electrochemical gradient (sum of the concentration gradient and the electrical potential difference). Sudden changes of permeability for Na^+ and K^+ in nerve and muscle cells contribute to transmission of nervous impulses and contraction of muscles, including that of the heart. ***See*** Sodium, Physiology

Experimental and clinical evidence suggests that potassium plays an important role in determining the morbidity and mortality of patients with systemic hypertension or chronic heart failure. Whereas a high-potassium diet has little or no effect in normotensive subjects, it decreases blood pressure under hypertensive conditions. The following mechanisms for this action have been suggested: the natriuretic properties of potassium increasing urinary sodium; the suppression of the sympathetic nervous and renin–angiotensin systems; the prevention of the development of vascular injury, and a direct vasodilatory effect on resistance vessels. A high-potassium diet probably preserves the integrity of endothelial cells even when they are under great tension in hypertensive subjects, thereby preventing artery wall lesions with subsequent cerebral haemorrhage and infarcts. In normal individuals a high potassium intake is also associated with low risk of stroke and sudden death. ***See*** Hypertension, Physiology

Finally, in chronic heart failure potassium can modify both the mechanical and electrical properties of the heart, reducing the frequency and complexity of potentially lethal ventricular tachyarrhythmias.

Requirements and Daily Intakes

Before 1983 there were recommended dietary potassium intakes only in FRG, USA, Australia and the USSR, for some groups of the population. The UK report (COMA, 1991) on Dietary Reference Values (DRVs) set a Reference Nutrient Intake of 3500 mg per day for adults. The Nutrition Working Group of International Life Sciences Institute Europe suggested for a healthy adult male aged 20–59 years a Recommended Daily Intake (RDI) of potassium equivalent to the actual molar sodium intake (100 mmol per day), i.e. 3900 mg per day. This recommendation was made because a reduction in sodium intake and an increase in potassium intake is desirable in most European countries in order to provide a molar Na:K ratio of 1, which has been defined as beneficial in the prevention of several pathologies.

In infancy, potassium supply is particularly important, since the kidneys are immature. Some authors argue that pregnancy and lactation do not increase the need for potassium, while others consider that an extra amount is needed for the body stores or milk production. In the elderly, depletion or excess of potassium is frequent and it appears that the minimal satisfactory potassium intake is about 2350 mg per day. During exercise, potassium requirements may be elevated owing to losses in sweat.

The normal daily dietary potassium intake in Western countries is estimated to range between 0·75 and 1·25 mmol kg^{-1}. Fruits – in particular citrus fruits – vegetables and meats contain high amounts of potassium. Milk is the largest dietary potassium source for infants (Table 1).

Food processing has little impact on dietary potassium, because it is an intracellular cation. Nevertheless, processes such as boiling or canning seem to decrease the potassium content of the food.

Absorption

About 90% of the ingested potassium is absorbed in the gastrointestinal tract, somewhat more in individuals who are short of potassium.

Transport in the small intestine is predominantly a process of passive diffusion, reaching equilibrium between plasma and lumen in the jejunum and ileum. The movements of water promote absorption of Na^+ and K^+ in the duodenum and upper jejunum (solvent drag effect). Both of them are absorbed through pores of the epithelium in the form of hydrated ions. However, the solvent drag effect does not occur in the ileum, which may be explained by the fact that the pores in the ileum are smaller than in the duodenum. The transmembrane potential difference in the small intestine is near zero; thus it represents a minor role in K^+ transport.

Like the distal nephron, the colon is the only segment of the intestine that modifies K^+ transport in response to variations in potassium status. The proximal colon secretes K^+, predominantly via a transcellular pathway: active K^+ uptake across the basolateral membrane mediated by (Na^+-K^+)-ATPase (Na^+-K^+-adenosine triphosphatase), and passive exit across the apical membrane via a K^+ conductance pathway. Potassium can also be secreted via a paracellular pathway driven by the negative transmembrane potential difference across the lumen. The distal colon can both secrete and absorb K^+. Secretion via the transcellular pathway resembles that of the proximal colon, but in the distal colon basolateral K^+ uptake may also use a Na–K–chloride

Table 1. Sodium and potassium content of various potassium-rich foods

	Content (mg per 100 g edible portion)	
	Sodium	Potassium
Meat		
Beef, veal, lamb	90–100	330–355
Liver	85–95	280–320
Chicken	80	257
Turkey	66	315
Fish		
Halibut	67	446
Herring, mackerel	95	380
White tuna	35	426
Trout	40	465
Vegetables		
Potato	3	443
Spinach	65	633
Parsley	33	1000
Haricot bean, lentils	2–4	800–1300
Soya bean	4	1740
Peanut	5	706
Fruit		
Orange, mandarin	2	200
Banana	1	393
Avocado pear	3	503
Raisins	21	782
Fig (dried)	40	856
Plum (dried)	8	824
Various		
Cows' milk	48	157
Chocolate	19	397

(Cl) transporter. Absorption of K^+ is an active process and appears to be partly Cl-dependent. *See* Colon, Structure and Function

Bioavailability

In contrast with other elements, the bioavailability of potassium (the proportion of the ingested element that is utilized in the normal metabolic pathways) is high, 90–95%. Few dietary factors have been described that modify potassium bioavailability. Olive oil is one of the fat sources favouring potassium uptake, while certain ion exchange resins bind appreciable amounts of K^+ and, not being absorbed, are useful where the amount of potassium in the body must be decreased. *See* Bioavailability of Nutrients

Physiological status also affects the bioavailability of potassium. During the second half of pregnancy, faecal potassium and sodium decrease, whilst urinary potassium output increases in parallel to retention of Na^+ by the kidney. Nevertheless, whole-body potassium is not affected and the amount of potassium retained is enough to accomplish gestation.

Pathological alterations of the electrolyte balance, e.g. causing diarrhoea or vomiting, can modify potassium bioavailability.

Transport

The concentration of K^+ in the cell is 30 times higher than that in extracellular fluid. Such a concentration gradient is maintained by a pump-leak system in which loss of K^+ from cells by diffusion is balanced by active uptake (Fig. 1). This pump is dependent on the (Na^+-K^+)-ATPase bound in the basolateral membrane which extrudes $3Na^+$ outwards in exchange for $2K^+$ (in some conditions this ratio can be slightly different). The process is electrogenic, because one net positive charge is removed from the cell per pump cycle. Therefore, the pump opposes the electrochemical gradient for both Na^+ and K^+.

Another mechanism for active K^+ transport is the K^+ pump, which extrudes hydrogen ions (H^+) in exchange for K^+. This transport has an important role in the reabsorption of K^+ in the distal tubules of the nephron.

Passive transport of potassium occurs via intracellular or paracellular pathways. The intracellular pathway involves a specific channel which is located in both apical and basolateral sides of the membrane. The number of channels in 'open' or 'closed' form determines the intensity of the transport. An increase in cell calcium (Ca^{2+}), intracellular pH, or cyclic adenosine monophosphate (cAMP) activates the K^+ channels.

Other processes that contribute to the transport of potassium are Na–Cl–K cotransport and K–Cl cotransport. A transport protein plays the role of carrier across the membrane. These systems maintain electroneutrality and are not directly energy-dependent. Nevertheless, in some cells there appears to be a requirement for ATP, presumably to phosphorylate the transport protein.

Transport of potassium is dependent on factors such as the transmembrane electrochemical gradient, the activity of (Na^+-K^+)-ATPase, pH, extracellular bicarbonate concentration, osmotic pressure and hormonal factors.

Distribution and Storage

Total body potassium, determined by chemical methods or by measuring the naturally occurring stable isotope

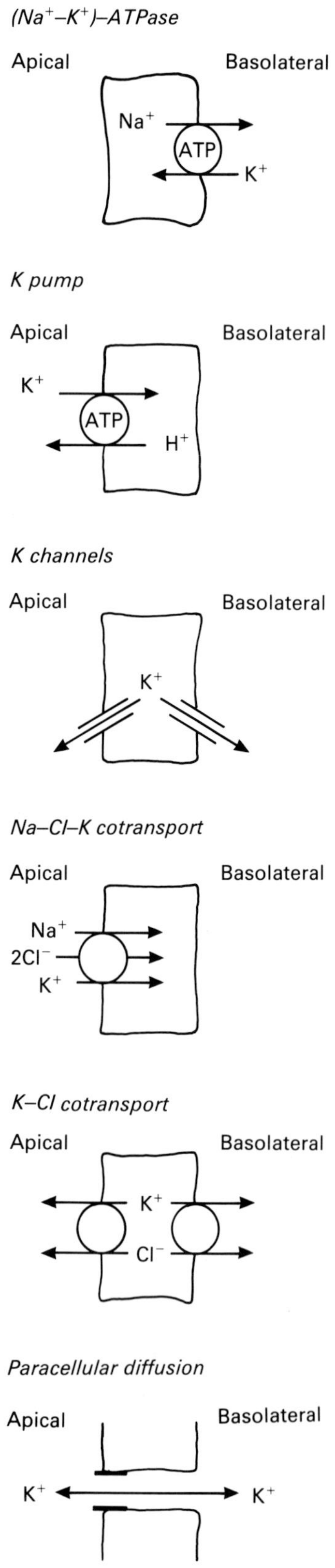

Fig. 1 Potassium transport via intracellular and paracellular pathways. (From Stanton B and Giebisch G (1990) Renal regulation of potassium transport. In: *Renal Physiology, Handbook of Physiology*. 2nd edn. New York: Oxford University Press).

^{40}K, is 45–55 mmol per kg of bodyweight. A 70-kg adult man contains approximately 135 g of potassium. Total body potassium is a function of age, weight, height and sex. Females contain 35–40 mmol per kg of bodyweight, whereas the values during childhood and in the elderly are lower than those for young women.

The majority of potassium is located in the cells (110–150 mmol per l of cells) and only 2% is in the extracellular compartment. Plasma concentrations vary between 4 and 5 mmol l^{-1} (156–195 mg l^{-1}), being only 10–20% of the potassium in plasma bound to proteins. More than 95% of the potassium pool is exchangeable. In erythrocytes, potassium is slowly changeable; in the skeleton, it is not readily exchanged, owing to the small amount present.

An increase of potassium in the extracellular fluid equivalent to 1% of total body potassium may double the concentration of potassium in plasma. Nevertheless, the storage of such a quantity in the cells would result in only minimal intracellular concentration changes, with little change to the potassium concentration difference across the cell membranes. To accomplish this 'buffer' function, the major reservoir of potassium is the muscle, followed by the liver and erythrocytes, although each cell, including those of kidney tubules, possesses the capacity for accumulating potassium.

Excretion

Potassium excretion is carried out primarily via the kidneys. The faeces contains approximately 10% of ingested potassium. Sweat losses are small, provided that the climate and exercise conditions are not extreme.

Renal Potassium Handling

The glomerular membranes freely filter the cation. The amount filtered is equivalent to the product of plasma potassium concentration and the filtration rate. Usually, only 5–15% of the amount of potassium filtered is excreted. ***See*** Kidney, Structure and Function

Along the proximal tubule there is net K^+ reabsorption in both the earliest part of the proximal convoluted tubule and in the later parts, approximately 60–70% of the total filtered. Under normal conditions, similar fractions in Na^+, K^+ and water are reabsorbed in this tubule. However, it seems that K^+ reabsorption is at least partially independent of the movement of these substances. Two mechanisms have been suggested – active transport or a movement of K^+ across the epithelium against an electrochemical potential gradient. The second model assumes that the K^+ concentration in the interstitial fluid is low, which may be attributable to the (Na^+-K^+)-ATPase pump of the basolateral membrane and to the limited diffusion of K^+ from plasma back to the interstitial fluid. This model allows for absorption and secretion of K^+ against electrochemical potential gradients, associated with the interstitial K^+ concentration.

The concentration of K^+ increases as the filtrate passes through the loop of Henle, in which permeability for K^+ is very high. At the end of the descending limb the amount of K^+ present usually exceeds that of the glomerular filtrate. There is a net secretion of K^+, which arises from reabsorption in the collecting tubule and, partly, in the ascending limb. These pathways of K^+ transport constitute a potassium recycling in the renal medulla.

The K^+ movement in the ascending limb appears to be linked to cotransport with Na^+ and Cl^-, and to electrically conductive diffusion across the luminal membrane. Both Na–Cl–K cotransport and the presence of (Na^+-K^+)-ATPase in the basolateral membrane have been demonstrated. Some K^+ can also be transported by a paracellular pathway owing to the positive electrical gradient across the lumen.

The distal tubule, as a target of multiple regulatory factors affecting K^+ secretion, is a major determinant of urinary K^+. It usually receives fluid with a low K^+ concentration from the thick ascending limbs. The concentration of K^+ rises along the distal tubule, as a result of the absorption of fluid and K^+ secretion, although complete suppression of K^+ secretion, or even net reabsorption, can be the response to particular situations.

The distal convoluted tubule, like the thick ascending limb, secretes K^+ but not as rapidly as the next downstream subsegments. Some experiments suggest that K^+ secretion is accomplished by a K–Cl cotransporter located in the luminal membrane. Active Na^+ and K^+ reabsorption can result under conditions of maximal K^+ conservation. The connecting tubule has the major capacity for K^+ secretion, probably because its transepithelial voltage is strongly lumen-negative.

The cortical collecting tubule is also characterized by the secretion of K^+ against significant concentration differences. The electrical potential difference of this tubule is quite similar to that of the late distal tubule, becoming more negative along the tubule; this is associated with Na^+ reabsorption. Potassium is secreted in the luminal membrane through K^+ channels. The Na^+ movements through the apical membrane generate a potential difference that tends to make the cell interior electrically positive; thus K^+ diffuses in the opposite direction, towards the lumen. Simultaneously, in the basolateral membrane, the electronegativity owing to the loss of Na^+ to the interstitial fluid favours the movement of K^+ into the cell and, finally, its secretion. Potassium secretion in this segment is also mediated by the K–Cl cotransporter in the apical membrane.

The medullary collecting duct does not greatly modify K^+ secretion in normal conditions, but in states of K^+ deprivation it may play a significant role in regulating K^+ excretion. In such situations active K^+ reabsorption has been observed, resulting from H–K exchange driven by a (H^+-K^+)-ATPase.

Regulation of Potassium Homeostasis

The extracellular potassium concentration must be maintained within narrow limits. This is achieved by the 'internal balance', i.e. distribution of K^+ between the intracellular and extracellular spaces, carried out by the (Na^+-K^+)-ATPase, and 'external balance', i.e. intake–output equilibrium, which takes place principally in the kidney but also in the colon. Both internal and external balances are under hormonal control and modified by factors such as dietary K^+ intake, plasma K^+ concentration, fluid flow rate, Na^+ excretion and acid–base status.

In response to a high-potassium diet there is an enhanced cellular K^+ uptake, predominantly in muscle. The external balance is also modified by increasing the colon K^+ secretory capacity and, mainly, the renal K^+ secretion. Both are independent of the concomitant increase in aldosterone, although this hormone is needed for the whole regulatory mechanism. A low-potassium diet acts in the opposite fashion.

Within the physiological range, Na^+ concentration does not alter K^+ secretion. However, concentrations lower than approximately 15–30 mmol l^{-1} reduce K^+ secretion by the cortical collecting duct. This is because most of the urinary K^+ is secreted in exchange for Na^+, which is reabsorbed in the distal and collecting tubules in response to mineralocorticoids.

An elevation in flow rate increases K^+ excretion, an effect that is dependent on potassium intake. When potassium intake is high the effect is intense but there is little effect when potassium intake is low. This influence on flow rate is often related to Na^+ excretion, since increases of flow are normally accompanied by an increase in the delivery of Na^+.

Plasma K^+ levels are elevated in acute metabolic acidosis and respiratory acidosis, and depressed in metabolic and respiratory alkalosis. Acidosis induces a shift of K^+ from the intracellular to the extracellular fluid and a decrease of the distal and tubular K^+ secretion by the kidney, whilst alkalosis exerts opposite extrarenal and renal actions.

Hormonal Regulation

Several hormones contribute to potassium homeostasis, acting at different levels (Fig. 2). Insulin, epinephrine

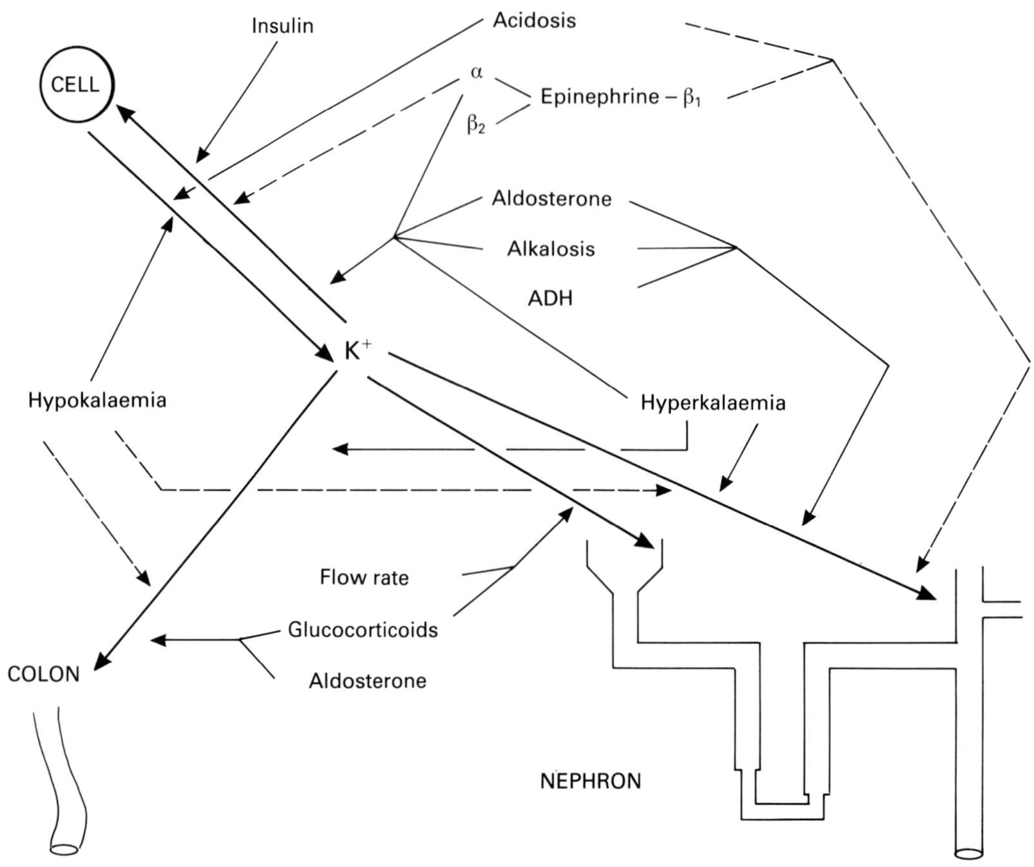

Fig. 2 Potassium homeostasis. —— Activation; - - - - inhibition; ADH, antidiuretic hormone.

(by β_2-adrenergic activity) and aldosterone enhance cellular potassium uptake. α_1-Adrenergic stimulation increases serum potassium. ***See*** Hormones, Adrenal Hormones

Serum K^+ is also controlled by regulating excretion of potassium. Glucocorticoids and mineralocorticoids stimulate K^+ secretion in the colon and therefore elevate K^+ content in faeces, but these hormones mostly affect kidney function. Glucocorticoids produce kaliuresis, even on a low sodium and normal potassium intake. They increase glomerular filtration rate, urinary flow rate and Na^+ excretion, and indirectly affect the distal tubule.

The distal and collecting tubules, where primary control of K^+ excretion takes place, are governed by antidiuretic hormone (ADH), catecholamines and principally, aldosterone: ADH, by direct or indirect action, favours K^+ secretion; epinephrine, by β_1-adrenergic activity, decreases renal K^+ excretion at the level of the late distal tubule and the cortical collecting tubule independently of perfusate K^+ concentration; and, in the absence of adequate levels of mineralocorticoids, the secretory capacity of the kidney for K^+ declines. Aldosterone increases K^+ excretion and Na^+ reabsorption and, although both effects can be independent, the increase in urinary K^+ depends on an adequate supply of Na^+ to the distal nephron. The mechanisms of action of this mineralocorticoid consist of an increase of basolateral uptake of K^+ by both active and electroconductive pathways, followed by an increase of the electrochemical gradient favouring cell-to-lumen movement, and, finally, an increase in the luminal membrane permeability for K^+.

The processes of regulation indicated above permit the organism to maintain a normal level of K^+ because K^+ exerts a feedback regulation over the various hormones. For example, hyperkalaemia induces enhancement of the aldosterone secretion and hypokalaemia induces inhibition. The immediate regulatory response occurs at the cellular level, by modifying intracellular K^+, and the definitive response corresponds to the excretion system.

Hypokalaemia

Plasma potassium concentration below 3·5 $mmol\,l^{-1}$ (136 $mg\,l^{-1}$) is considered abnormally low, and levels of 2·5 $mmol\,l^{-1}$ or less may be associated with severe potassium depletion (total body potassium loss). Depletional or nondepletional hypokalaemia is characterized by an increased extracellular–intracellular K^+ gradient.

A decrease of the plasma potassium concentration to 50% of the normal values produces hyperpolarization of the membrane, affecting the activity of nerves and muscles. Hypokalaemia with potassium depletion also presents a decrease in cell pH; thus both K-sensitive and pH-sensitive enzyme systems can be affected. Consequently, the clinical features of potassium depletion include the following: cardiac arrhythmias, skeletal and smooth muscle paralysis, impaired glucose tolerance with impaired insulin secretion, impaired protein anabolism, metabolic alkalosis, impaired urine concentration, and aciduria. The aetiology of hypokalaemia without potassium depletion is as follows:

- Alkalosis
- Insulin excess
- β-Adrenergic catecholamine excess
- Intoxications (theophylline, barium, toluene)
- Familial hypokalaemic periodic paralysis
- Thyrotoxic periodic paralysis.

The aetiology of hypokalaemia with potassium depletion involves the following *renal* causes:

- Diuretic conditions
- Syndrome of chloride depletion
- Renal tubular acidosis
- Mineralocorticoid excess
- Glucocorticoid excess
- Batter's syndrome
- Magnesium depletion
- Antibiotic therapy.

Extrarenal causes are as follows:

- Insufficient potassium intake (e.g. fast, anorexia nervosa)
- Gastrointestinal losses (e.g. diarrhoea, laxative abuse)
- Sweat losses.

Hypokalaemia can be produced by a K^+ shift from the extracellular to the intracellular compartment with maintenance of total body potassium. In general, factors that increase activity of the (Na^+-K^+)-ATPase pump increase cellular K^+ uptake. Potassium therapy does not alleviate this type of hypokalaemia.

When total body potassium is decreased, there is a greater loss of K^+ from cells than from extracellular fluid. The number of (Na^+-K^+)-ATPase pumps lowers in various tissues, mainly skeletal muscle, thereby donating K^+ to the extracellular space by passive diffusion. Treatment of depletional hypokalaemia involves administration of K^+ (KCl) slowly, at a uniform rate, in order to prevent a hyperkalaemic state, which can be easily induced because the number of pump sites remains low for several days after K^+ replacement.

Hyperkalaemia

Plasma potassium levels above 6 mmol l^{-1} (234 mg l^{-1}) indicate a hyperkalaemic state, and 7 mmol l^{-1} or more must be considered an emergency. The most important manifestations of hyperkalemia are on the heart, ultimately resulting in ventricular fibrillation and cardiac arrest. Other consequences are flaccid paralysis, natriuresis, and some minor endocrine effects.

Acidosis and digitalis intoxication promote the exit of K^+ from the cells by decreasing the activity of the Na–K pump. In states of insulin deficiency, a combination of mechanisms have been suggested, including acidosis, hyperglycaemia, hyperglucagonaemia and hyperosmolarity. During acute maximal physical performance, plasma potassium can rise considerably. This may be explained by the electrical phenomena that initiate muscle contraction. However, physical training attenuates exercise-induced hyperkalaemia, presumably by increasing the quantity of Na–K pumps in the skeletal muscle.

The aetiology of hyperkalaemia without potassium excess is as follows:

- Acidosis
- Insulin deficiency
- β-Adrenergic blockers
- Digitalis intoxication
- Hyperkalaemic periodic paralysis
- Arginine hydrochloride
- Succinylcholine
- Vigorous exercise.

The renal causes of hyperkalaemia with potassium excess are as follows:

- Renal failure
- Aldosterone deficiency
- Primary tubule dysfunction (acquired disorders, congenital pseudohypoaldosteronism, potassium-sparing diuretics).

In hyperkalaemia with potassium excess a large quantity of potassium must have first filled the relatively high intracellular pool of potassium, and the renal function must be unable to excrete excess K^+. The majority of cases of potassium excess therefore result from abnormalities in urinary K^+ elimination owing to renal failure or mineralocorticoid insufficiency. This is the case in infant or elderly patients, in whom there is often also a lack of response to aldosterone.

Bibliography

Brenner BM and Rector FC (1991) *The Kidney* 4th edn, pp 283–317, 805–840. Philadelphia: WB Saunders.

Brown RS (1986) Extrarenal potassium homeostasis. *Sidney International* 30: 116–127.

COMA (1991) *Dietary Reference Values for Food Energy and*

Nutrients for the United Kingdom. Department of Health, Committee on Medical Aspects of Food Policy. Report on Health and Social Subjects 41. London: Her Majesty's Stationery Office.
Munson RD (1985) *Potassium in Agriculture*. Proceedings of an International Symposium, pp 577–594, 619–633. Madison: American Society of Agronomy.
Nutrition Working Group (ILSI) (1990) *Recommended Daily Amounts of Vitamins and Minerals in Europe*. Nutrition Abstracts and Reviews (Series A) 60: 827–842.
Sterns RH and Spital A (1987) Disorders of internal potassium balance. *Seminars in Nephrology* 7: 206–222.
Wright FS (1987) Renal potassium handling. *Seminars in Nephrology* 7: 174–184.

M. Pilar Navarro and M. Pilar Vaquero
Instituto de Nutrición y Bromatología (CSIC), Madrid, Spain

POTATOES AND RELATED CROPS

Contents

The Root Crop and its Uses

The potato (Irish potato, white potato) is of ancient origin. It was first domesticated in South America, but has spread around the world during the past 400 years. It gained recognition as an inexpensive and nutritive food in the eighteenth century. It is now among the 10 major food crops of the world and grown in 140 countries.

Geographical Distribution

The potato is largely grown in cool regions where the mean temperature during the growing season does not exceed 18°C. It is grown in 71% of the countries in the world, and represents about 47% of the total tuber crops produced. It is the fourth most important food crop in the world, next to wheat, maize and rice in global tonnage (Table 1). The principal producers are the European countries, the USSR, North America, Australia and the Andean countries of Latin America (Table 2). Northern Europe and the USSR account for nearly 70% of the world acreage and 75% of the total world output. The USA and Canada together account for 3% of world acreage and 6% of production.

Table 1. World production of potatoes compared to major tuber crops and cereals

Crop	Area harvested ($\times 10^3$ ha)	Yield (kg ha^{-1})	Production ($\times 10^3$ t)
Potato	17 854	15 098	269 561
Cassava	15 635	10 084	157 656
Yam	2928	10 057	29 447
Sweet potato	11 910	11 058	131 707
Wheat	231 548	2570	595 149
Rice	145 776	3557	518 508
Maize	129 116	3682	475 429

Source: FAO Production Year Book (1990) Vol 44. Rome: Food and Agriculture Organization of the United Nations.

Varieties

There are more than 150 wild species of potato found in Central America, Mexico and the USA. *Solanum tuberosum* L., (tetraploid) represents the cultivated species, and there are seven other cultivated species, including *S. ajanhuiri*, *S. goniocalyx*, *S. phureja*, *S. stenotomum* (diploids), *S.* × *juzepczukii*, *S.* × *chaucha* (triploids), *S.* × *curtilobum* (pentaploid), which are grown in different parts of Peru, Bolivia, Ecuador and Venezuela. Many improved local varieties have been developed in recent years by breeding, and these may show increased yield, disease resistance, better tuber shape, texture and quality for processing as well as flavour. In India, Kufri varieties have been introduced by the Central Potato Research Institute, Simla, to suit different agroclimatic conditions.

Table 2. Potato production in parts of the world.

	Area harvested ($\times 10^3$ ha)	Yield (kg ha^{-1})	Production ($\times 10^3$ t)
World	18 070	14 923	276 740
Africa	779	8634	6722
North Central America	735	28 352	20 837
South America	912	12 560	11 453
Asia	4607	13 330	61 416
Europe	4790	21 502	102 984
Oceania	48	27 685	1328
USSR	6200	11 613	72 000
Developed countries	11 875	16 916	200 881
Developing countries	6195	12 245	75 860

Source: FAO Production Year Book (1989) Vol 43. Rome: Food and Agriculture Organization of the United Nations.

Table 3. Pattern of potato utilization in some European countries

	Percentage used for:		
Country	Livestock feeding	Industrial purposes	Human consumption
Poland	20–60	3·0	12·0
FRG	44	4·0	39·0
The Netherlands	20	48	24·0
UK	—	1·0	65·0

Source: Adapted from Harris PM (1978).

Commercial Importance

The potato crop produces more dry matter and protein per hectare than the major cereal crops of the world. It is a major food crop in the temperate zone and is either a staple food or merely a vegetable in tropical and subtropical countries. Tubers are consumed in various forms after cooking or processing, and are also used for propagation. The pattern of potato utilization in a few countries is shown in Table 3.

In the UK, India and the USA, 65–90% of the production is consumed as food, and a small quantity is used for industrial purposes. Potatoes are generally boiled, cooked in oil, or baked. Dehydrated, frozen and canned products are also popular. In the USA, 50% of the potato crop is consumed in a processed form, the most relished being potato chips. The frozen products include 'hash browns' (patties), French fries, croquettes (puffs) and mashed potato, while the dehydrated products include granules, flakes, diced chunks and 'julienne' strips. The common canned products are 'new' potatoes, 'hash', stews and soups. 'Papa seca' and 'chuno' are the traditional dried potatoes which form a vital part of the diet in the highland areas of Peru and Bolivia.

Starch, alcohol, glucose and dextrin are the industrial products of potato. World production of potato starch is about 2×10^6 t. The waste water collected from peeling, cutting and blanching in the processing industries has been used to recover highly nutritive protein of 80–85% purity. The residual water can also be used in preparing a growth medium for single-cell protein production. ***See*** Starch, Sources and Processing

Morphology of the Tuber

The potato is an underground, modified, fleshy stem, with a shortened axis, developing from the subapical region of a diageotrophic stolon or lateral shoot. The tuber end that is attached to the stolon is called the 'stem end' or 'heel end', and the other is the 'bud end' or 'rose end'. There are as many as 20 'eyes' in the axils of leaf scars (eye brows) arranged spirally around a tuber. Each eye has one main bud and several small lateral buds. Physiologically, the youngest bud is the last-formed apical bud (Fig. 1a). Tubers sprout after completing the inherent dormancy period which lasts from 6 to 16 weeks after harvest, depending upon the variety. Studies have shown that the balance between the two hormones, abscisic acid (ABA) and gibberellic acid (GA_3), in the tubers determines the extent of the dormancy period. Tubers exhibit apical dominance. They can have round, oval, or elongated shapes, and the size, shape, skin colour and depth of 'eye' of the tuber determine its acceptability for various end-uses.

Anatomy

The potato tuber is protected by the outermost skin or periderm, consisting of six to ten layers of suberized cells. Skin colour varies from light yellow to black or dark purple depending upon the anthocyanin concentration in the periderm and peripheral cortex. The active periderm of a young tuber can be easily removed, and tuber enlargement is associated with sloughing-off of the periderm, which is then replaced by a new 'cork' layer, formed from beneath. Similarly, 'wound healing' takes place in damaged tuber tissue by the formation of a 'wound periderm', which is more impervious than normal skin. A number of 'lenticels' are found in the periderm, and these pores facilitate the exchange of gases and the entry of pathogens.

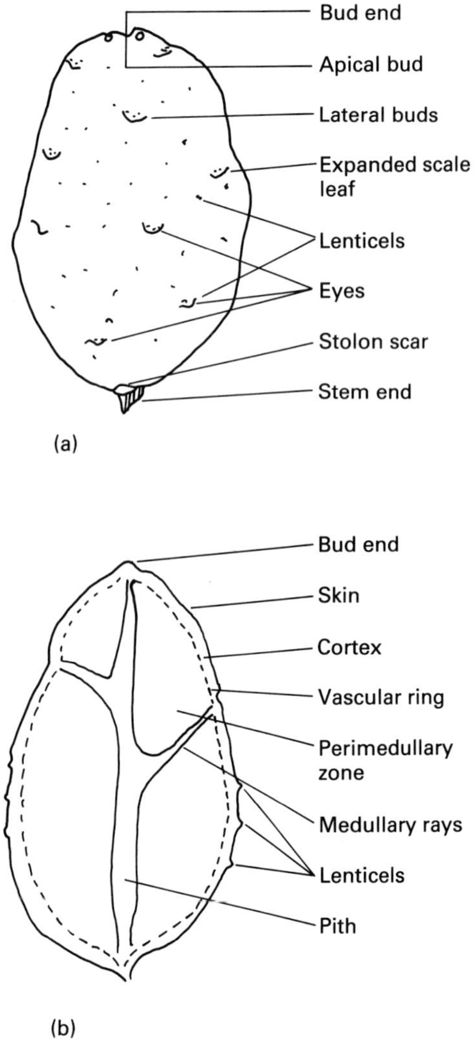

Fig. 1 A potato tuber: (a) morphological details; (b) longitudinal section.

Inner to the periderm is the parenchymatous cortex (0·3–1·0 cm thick) in which food material is stored in the form of starch granules. The parenchymatous perimedullary zone is seen between the vascular ring and the medullary ray (Fig. 1b). Xylem is visible as a ring, while phloem forms many bundles in the cortex and perimedullary zones. Medullary rays run from the stem end to the 'eyes'. The pith present at the centre of the tuber has less starch and is translucent.

Chemical Composition and Nutritive Value

The chemical composition of potato varies with variety, soil type, cultural practice, maturity stage and storage conditions. The data in Table 4 show the potato to be a good source of carbohydrates, proteins, vitamins and minerals.

Table 4. Chemical composition of the dry matter (22·5%) of the potato tuber

	Content[a]
Crude fibre	2·1
Starch	74·2
Total sugar	1·3
Reducing sugar	0.6
Fat	1·0
Total nitrogen (N)	1·2
Protein N	1·0
Protein fractions	
(% of total protein N)	
Albumin	48·9
Globulin	25·9
Prolamin	4·3
Glutelin	8·3
Minerals	
Calcium	0·02
Magnesium	0·08
Potassium	1·47
Sodium	0·02
Iron (ppm)	15·70
Vitamins (mg per 100 g)	
Thiamin	0·73
Ascorbic acid	92·08
Nicotinic acid	10·08
Riboflavin	0·12

[a]Values given as percentages, unless otherwise stated.
Data adapted from Rastovski A *et al.* (1987).

Carbohydrates

Starch is the major carbohydrate of potato (16–20% on a fresh-weight basis; fwb), constituting 60–80% of the dry matter, and it is composed of amylose and amylopectin in a 3:1 ratio. Potato starch gelatinizes above 70°C. The total sugar content ranges from 0·1% to 0·7% (fwb), and this is chiefly associated with maturity, senescence and sprouting. The major sugars of potato are glucose, fructose and sucrose. Trace amounts of melibiose, raffinose, stachyose, glycerol, galactinol and glucosyl have also been identified. Tubers containing more than 2% (fwb) reducing sugars give dark-coloured chips owing to Maillard reactions and are not considered good for processing. The non-starchy polysaccharides, such as cellulose, hemicellulose and pectic substances, constitute about 1·2% (fwb) and are present in the cell walls and middle lamellae. They contribute to the final texture of cooked potato, and act as a source of dietary fibre. ***See*** Browning, Nonenzymatic; Carbohydrates, Requirements and Dietary Importance;

Carbohydrates, Classification and Properties; Starch, Structure, Properties and Determination

Protein

Potato is considered to be low in protein (2% fwb), but is rich in lysine compared to cereal proteins, while the concentration of sulphur amino acids is less than in cereals. The protein is concentrated more in the cortex and pith. The levels of proteins such as albumin, globulin, prolamine and glutelin are 48·9%, 25·9%, 4·3% and 8·3% of total protein respectively. Two thirds of the nonprotein-nitrogen fraction is composed of free amino acids, and 21 of them have been identified. Sprouting, storage, diseases and fertilizer treatments influence the concentration of free amino acids in the tuber. ***See*** Protein, Chemistry; Protein, Requirements

Vitamins and Minerals

Potato contains substantial quantities of vitamins B and C (Table 4). Vitamin C is present in both oxidized and reduced forms. Freshly harvested tubers may contain 20 mg of ascorbic acid per 100 g; losses of vitamin C occur during long-term storage and during cooking. The vitamin B group comprises thiamin, riboflavin, nicotinic acid and pyridoxine; folic acid, along with pantothenic acid, is also present in potato. The ash content is about 1% (fwb), which is equivalent to 4–6% of the dry matter content. The major elements present are phosphorus, potassium and calcium. Others, such as boron, copper, zinc, iodine, aluminium, arsenic, nickel and molybdenum, are found in trace amounts. ***See*** individual minerals and vitamins

Lipids and Organic Acids

Approximately 0·1% (fwb) lipid is found in potato; it is concentrated in the periderm. Linoleic, linolenic and palmitic acids are the major fatty acids. A number of organic acids are present in potato in varying quantities; they contribute to the flavour and also buffer the potato sap. The major organic acids are citric, oxalic, fumaric and malic acids; phytic, nicotinic and chlorogenic acids have also been reported. Chlorogenic acid reacts with ferric iron, forming a complex, which causes darkening after cooking. Enzymatic browning in cut and homogenized potato tissue is caused by the oxidation of tyrosine. Other phenolic compounds found in potato are the flavones, anthocyanins and tannins. ***See*** Acids, Natural Acids and Acidulants; Fatty Acids, Properties; Lipids, Classification; Phenolic Compounds; Tannins and Polyphenols

Enzymes and Pigments

The enzyme systems reported in potato include amylase, glyoxalase, phosphorylase, tyrosinase, catalase, aldehydrase, peroxidase, polyphenoloxidase, dehydrogenase, phosphatase, sistoamylase and zymohexase. The phosphorylase and amylase systems form sugars at low temperatures. The yellow colour of potato flesh is attributable to carotenoids such as violaxanthin, lutein and lutein-5,6-epoxide. Neoxanthin occurs in low concentrations. Chlorophyll is present in tubers exposed to light, and these green potatoes lose their market value. ***See*** Carotenoids, Properties and Determination; Chlorophyll; Colours, Properties and Determination of Natural Pigments; Enzymes, Functions and Characteristics

Antinutritional Factors

The synthesis of toxic glycoalkaloids such as α-solanine and α-chaconine, which are believed to be a part of a disease resistance mechanism, takes place in damaged and light-exposed tubers; a normal tuber contains an insignificant amount (5–10 mg per 100 g) of these alkaloids. They are concentrated more in the peel, sprouts and around the 'eyes'. Tubers develop bitterness and off-flavours when the alkaloid content is 25–80 mg per 100 g, and consumption of tubers with more than 20 mg of glycoalkaloid per 100 g causes fatal illness. The nutritional significance of two other toxic substances, proteinase inhibitors and lectins, has not been much studied, but they are reported to be heat-labile. ***See*** Alkaloids, Toxicology

Nutritive Value

Potato is one of the richest sources of energy and has a greater capacity to supply energy and protein than any other food crop. In some developing countries, potato is considered to be merely a vegetable rather than a staple food because excessive consumption causes flatulence in humans. Experiments have shown that cooked potato starch is easily digestible, and hence forms a valuable food even for infants. Potato has a slightly lower energy content (335 kJ per 100 g) than other roots, tubers, cereals and legumes (Table 5), but this is advantageous in overcoming the problem of obesity in the developed world. On the other hand, large quantities of potato have to be consumed in developing countries to meet their populations' daily energy needs. It has been calculated that 100 g of potato can supply 5–7% of the daily energy and 10–12% of the daily protein needs of children aged 1–5 years. The production of weaning foods containing potato has been recommended. Potato

Table 5. Composition of potato (raw and dried) and other major cereals and root crops (per 100 g edible portion)

Crop	Energy (kJ)	Moisture (%)	Crude protein (g)	Fat (g)	Total carbohydrate (g)	Crude fibre (g)	Ash (g)
Potato (raw)	335	78·0	2·1	0·1	18·5	2·1	1·0
Potato (dried)	1343	11·7	8·4	0·4	74·3	8·4	4·0
Rice	1523	12·0	6·8	0·5	80·2	2·4	0·6
Wheat	1389	12·3	13·3	2·0	70·9	12·1	1·7
Sweet potato	485	70·2	1·4	0·4	27·4	2·5	1·1
Yam	444	72·0	2·2	0·2	24·2	4·1	1·0
Cassava	607	62·6	1·1	0·3	35·2	5·2	0·9

Source: Woolfe JA (1987).

protein has a biological value equal to that of soya bean protein, and the ratio of total essential amino acids to total amino acids is so balanced that it can meet the needs of infants and small children. It has been calculated that 100 g of potato can supply 3–6% of the daily, human (adult) protein requirement (depending upon sex and bodyweight), and one medium-sized potato provides 15 mg of vitamin C, which is about 20% of the recommended allowance (75 mg) per person per day. Consumption of tuber with the skin increases the dietary fibre intake. It is also a most valuable food for those who suffer from excess acidity of the stomach as it has an alkaline reaction. Potato fat is too low to have any nutritional significance, but it does contribute towards palatability. The tuber also provides most of the trace elements needed to maintain good health, and although potato is not a primary source of iron, 100 g of cooked potato can supply 6% and 12% of the daily iron requirement for children and adults respectively. ***See*** Amino Acids, Metabolism; Ascorbic Acid, Physiology; Dietary Fibre, Physiological Effects; Energy, Measurement of Food Energy; Iron, Physiology; Protein, Quality

Handling and Storage

In the majority of countries, potatoes are harvested only at certain times of the year, and the harvested tubers must be stored for at least a few months. Harvesting, and subsequent processes such as assembling, grading, bagging, transport and marketing, may cause damage to the tubers; such tubers lose water quickly and are highly susceptible to microbial spoilage. Thus curing, a process which promotes wound healing, is necessary before the storage of the potatoes. Complete curing takes 3–6 days at 20°C and 85–90% relative humidity (RH), but is faster at higher temperatures.

Good storage conditions should prevent excessive sprouting, root development, moisture loss, sugar accumulation and temperature damage. For short periods, the tubers can be stored in clamps, in specially designed simple sheds, pits, cellars or buildings with controlled temperature and humidity. Long-term storage with minimum loss of moisture is possible at 4–5°C, but temperatures below 6°C lead to the formation of reducing sugars, and this is not a desirable feature for industrial processing. Tubers can be reconditioned by transfer to a higher temperature, but the 'sweetness' developed by long-term, low-temperature storage ('senescent sweetening') cannot be eliminated completely. The freezing of tubers takes place at −1°C to −2°C, and frozen tubers soon become soft and unusable. Potato stores are artificially heated using oil or gas heaters when the outside air temperature is lower than the required temperature. On the other hand, when potatoes are stored above 30°C, an accumulation of carbon dioxide leads to the death of cells, causing 'black heart'. The recommended temperatures for tubers to be used for different purposes are as follows:

- Seed potato, 2–4°C
- Fresh consumption, 4–5°C
- Chipping, 7–10°C
- French frying, 5–8°C
- Granulation, 5–7°C.

See Storage Stability, Mechanisms of Degradation

In temperate regions, where the ambient temperature falls below 4°C, potatoes are cooled with outside air. In the tropics and subtropics, refrigeration is used. Sprouting increases weight loss and softens the tuber, thus decreasing marketability and nutritive value. Apart from low-temperature storage, other methods of achieving sprout inhibition include the use of maleic hydrazide (MH), isopropyl-*N*-phenylcarbamate (IPC) and isopropyl-*N*-(3-chlorophenyl)-carbamate (CIPC), and γ irradiation. Time, method of application and concentration of the chemical are important factors, as treatment may have adverse effects such as inhibition of wound healing, increasing the reducing sugar content, and the problem

Table 6. Sprout yield in naphthaleneacetic acid-treated tubers at the end of 3 months of storage

Storage temperature (°C)	Sprout wt (fresh) (g per 100 tubers)	
	Control	Treated
2±1	17·9	6·2
10±1	13·9	4·8
25±5	26·0	2·5

Source: Rama MV (1985).

of toxic residues. The sodium salt of naphthaleneacetic acid (1000 ppm) is an effective and economical post-harvest sprout retardant for storing potatoes under tropical ambient conditions (Table 6). Ware potatoes (those potatoes intended for human consumption in contrast to seed potatoes) can be kept sprout-free for a long time by irradiation, but its economic feasibility remains to be ascertained. There is only one industrial, potato irradiation plant operating at present, in Japan. Irradiated tubers need to be stored below 20°C, and irradiation may also lead to increased *Fusarium* attack and discoloration after cooking. Sprout inhibition beyond 3 months at tropical temperatures (22–35°C) is possible by 'vapour' heat treatment (at 60°C, RH 95% for 60 min). This method is superior to manual desprouting, and avoids the use of toxic chemicals. ***See*** Irradiation of Foods, Applications

Evaporative cooling systems can be used effectively in tropical conditions to store potatoes. The containers needed for this storage can be fabricated locally using brick, sand, bamboo and metal sheets; it is a cheaper method of storing potatoes, with reduced water loss, at farm level. These containers maintain a high RH (85–95%) in the atmosphere and the tuber temperature, which is close to the wet bulb temperature, is the lowest possible for evaporative cooling. Seed potatoes are often exposed to natural or artificial light during storage, as light retards physiological ageing, inhibits sprout growth and increases resistance to fungal infections. On the other hand, exposure of ware potatoes to light should be avoided, as it leads to the formation of the toxic α-solanine and the green pigment, chlorophyll; the market value of green potatoes is also low.

Avoiding damage to potatoes and dry, cool storage are necessary to prevent bacterial and fungal attack. Benzimidazole compounds are used to combat fungal diseases. Efficient ventilation in any store is a must for eliminating excess moisture, which otherwise may enhance 'rotting'. ***See*** Fungicides; Spoilage, Bacterial Spoilage

Bibliography

Anonymous (1984) *A Potato Store Run on Passive Evaporative Cooling*. Technical Bulletin No. 11, pp 1–14. Simla, India: Central Potato Research Institute.

Harris PM (ed.) (1978) *The Potato Crop*. London: Chapman and Hall.

Pushkarnath (1976) *Potato in Sub-Tropics*. New Delhi: Orient Longman.

Rama MV (1985) *Studies on the effect of various post-harvest treatments for controlling shrivelling, sprouting and spoilage of potatoes during storage*. PhD thesis, University of Mysore, India.

Rama MV and Narasimham P (1986) Heat treatments for the control of sprouting of potatoes during storage. *Annals of Applied Biology* 108: 597–603.

Rama MV and Narasimham P (1989) Control of potato (*Solanum tuberosum* L. cv. kufri jyoti) sprouting by sodium naphthyl acetate during ambient storage. *Journal of Food Science and Technology* 26(No. 2): 83–96.

Rama MV and Narasimham P (1991) Evaporative cooling of potatoes in small naturally ventilated chambers. *Journal of Food Science and Technology* 28(No. 3): 145–148.

Rastovski A, Van Es A *et al.* (1987) *Storage of Potatoes, Post Harvest Behaviour, Store Design, Storage Practice, Handling*. Wageningen, The Netherlands: Centre for Agricultural Publishing and Documentation.

Smith O (1977) *Potatoes, Production Storing, Processing*. Westport, Connecticut: AVI Publishing.

Van der Zaag (1990) Recent trends in development, production and utilization of potato crop in the world. *Asian Potato Journal* 1: 1–11.

Woolfe JA (1987) *The Potato in the Human Diet*. Cambridge: Cambridge University Press.

Yamaguchi M (1983) *World Vegetables: Principles, Production and Nutritive Value*, p 415. Westport, Connecticut: AVI Publishing.

M. V. Rama and P. Narasimham
Central Food Technological Research Institute, Mysore, India

Fruits of the Solanaceae

The family Solanaceae includes about 75 genera and 2000 species of herbs, shrubs and small trees distributed in the tropical and temperate regions of the world. Important vegetable crops, such as tomato, brinjal, pepper, tree tomato and husk tomato, are included in this family. Potato and tobacco are also of this family. This article reviews the morphology, anatomy, chemical composition, nutritive value and uses of some solanaceous fruits.

Morphology and Anatomy

The important fruits of the solanaceous family are shown in Table 1. Tomato (*Lycopersicon esculentum*) is the most popular and widely cultivated vegetable and has been consumed by the inhabitants of Central and

South America since prehistoric times. It is considered to be a native of the Peruvian and Mexican regions. The fruit is round, lobed or pear-shaped; the diameter ranges from 1 to 12 cm. It is a two- to many-loculed berry with fleshy placenta and many small, kidney-shaped seeds, covered with short, stiff hairs. The pericarp thickness varies with the cultivar. ***See*** Tomatoes

The genus *Lycopersicon* includes red-fruited edible species with carotenoid pigmentation, and green-fruited species with anthocyanin pigmentation. *Lycopersicon esculentum* and *L. pimpinellifolium* are red-fruited cultivars, while *L. pissisi*, *L. peruvianum*, *L. hirsutum*, *L. glandulosum* and *L. cheesmanii* are green-fruited. Varieties evolved by pure-line selection and hybridization are also cultivated in many countries. ***See*** Colours, Properties and Determination of Natural Pigments

The edible *Physalis* species are grown widely in the warmer parts of the world. The fruit of *Physalis pruinosa* (husk tomato) is round, yellowish and acidic-sweet. *Physalis peruviana* (Cape gooseberry) is a native of the Andes, and grows from Venezuela to Chile; the fruit is a spherical or ellipsoidal, smooth berry, measuring about 4 cm long and 3 cm wide. Skin colour is greenish-yellow. *Physalis ixocarpa* (tomatillo or ground cherry) is of Mexican origin; the fruit is a round, green or purplish berry, with a high ascorbic acid content. ***See*** Ascorbic Acid, Properties and Determination

Tree tomato (*Cyphomandra betacea*) is a native of Peru, but is grown in India, Sri Lanka and other countries. The fruit is greenish or purple in the early stages and turns reddish at maturity. It has a musky, acid taste and tomato-like flavour. The fruit rind is rough with a disagreeable flavour.

Several species included in the genus *Solanum* are prominent vegetable crops. *Solanum melongena* (eggplant or aubergine) is a popular vegetable crop in Asia; the leading country for its cultivation is China, followed by Japan. In the warm areas of the Far East, including India and the Philippines, it is grown extensively. In temperate zones it is an annual crop and perennial in the tropics. The fruit can be round, globose, long or pear-shaped. Most cultivars have purple to blackish skin, while some have white-green or mottled green skin, with white flesh; in the white-green types, the mature fruit has yellow skin. There are regional preferences for colour. *Solanum melongena* has three main varieties; the details are given in Table 2.

Table 1. Principal fruits of the family Solanaceae

Scientific name	Common name
Lycopersicon esculentum	Tomato
L. pimpinellifolium	Redcurrant or grape tomato
Solanum melongena	Eggplant, brinjal, berenjana, aubergine, guinea squash
S. macrocarpon	African eggplant
S. nigrum	Garden huckleberry, wonderberry
S. muricatum	Pepino, melon pear
S. quitoense	Naranjillo, lulo
S. gilo	Jilo
Capsicum annuum	Bell pepper
C. frutescens	Pimiento, chilli, aji
Physalis peruviana	Cape gooseberry, uchuba
P. ixocarpa	Tomatillo, ground cherry
P. pruinosa	Husk tomato
Cyphomandra betacea	Tree tomato, tamarillo

Table 2. Common cultivated varieties of *Solanum melongena* and their fruit characters

Variety	Fruit characters
Var. *melongena* (Wees)	Large, pendent, ovoid, oblong berries, 5–30 cm long, shining purple, white, yellowish or striped
Var. *depressum* (Baily)	Long, pyriform to ovoid, 10–12 cm long
Var. *serpentium* (Deeft)	Greatly elongated, 30 cm long, 2–5 cm in diameter, end curled

Solanum melongena var. *incarrum* (Linn) is a non-edible

African eggplant (*S. macrocarpon*) is a perennial crop grown in the Ivory Coast region of Africa. The fruit resembles the eggplant. Garden huckleberry (*S. nigrum*) is a native of North America with wide distribution in temperate and tropical regions; the colour may be red, yellow or black. Jilo (*S. gilo*) is a major crop of Nigeria and a minor crop in central and southern Brazil. The fruit has an orangish-red colour and a spherical to oval shape, 4 cm in diameter and 6 cm long.

Naranjillo or lulo (*S. quitoense*) is native to Ecuador, and has a spherical fruit, 3–5 cm in diameter; the ripe fruit has a yellow skin. The pulp is green and acidic in taste. Pepino or melon pear (*S. muricatum*) is an ancient cultivated crop of the Andes. The fruit is long, ovoid or ellipsoidal; the colour varies from light green to pale yellow. It is cultivated to a limited extent in northern Argentina, Chile, New Zealand and Australia.

Pepper is the second most important fruit of the Solanaceae. The important species are *Capsicum annuum* (annual with flowers borne singly in leaf axils) and *C. frutescens* (perennial with flower clusters in leaf axils); the former is widely cultivated. Varieties of this species will have differing fruit size, shape and pungency. The unripe fruit is commonly green, but fruits with cream, greenish-yellow, orange, purple and purplish-black skin also exist. Ripe fruits are usually red, but sometimes yellow or orange. The fruit is a pod-like berry with a cavity between the placenta and fruit wall. The bell pepper or sweet pepper (*C. annuum*), which is either mildly pungent or nonpungent, has a thick pericarp; its use is found in flavouring vegetable preparations. The

Table 3. Macronutrient content (per 100 g edible portion) of some solanaceous fruits

Crop	Energy (J)	Water (g)	Protein (g)	Fat (g)	Carbohydrates (g)
Tomato (green)	97	93·0	1·9	0·1	3·6
Tomato (ripe)	84	93·8	1·2	0·3	4·2
Eggplant	109	92·0	1·6	0·3	5·6
Chilli pepper	487	65·4	6·3	1·4	24·8
Bell pepper	109	92·0	1·3	0·2	6·0
Tree tomato	202	85·9	1·5	0·3	11·3
Naranjillo	118	92·0	0·7	0·1	6·8
Tomatillo	134	92·0	0·4	1·0	6·3
Cape gooseberry	223	83·0	1·8	0·2	11·1
Husk tomato	105	92·0	0·7	0·6	5·8
Pepino	134	92·0	0·4	1·0	6·3

Sources: FAO (1972) *Food Composition Table for Use in East Asia*. Rome: Food and Agriculture Organization.
Yamaguchi M (1983), p 299.
Aykroyd WR (1966).

Table 4. Vitamin and mineral content (per 100 g edible portion) of some solanaceous fruits

	Vitamin (mg)					Mineral (mg)			
Crop	A (iu)	B_1	B_2	Nicotinic acid	C	Calcium	Iron	Magnesium	Phosphorus
Tomato (green)	320	0·07	0·01	0·4	31	20	1·8	15	36
Tomato (ripe)	385	0·06	0·04	0·6	23	7	0·6	12	30
Eggplant	124	0·08	0·07	0·7	6	22	0·9	16	37
Chilli pepper	576	0·37	0·51	2·5	96	86	3·6	24	120
Bell pepper	530	0·07	0·08	0·8	103	12	0·9	13	34
Tree tomato	—	0·04	0·04	1·4	17	13	0·8	34	24
Naranjillo	170	0·06	0·04	1·5	65	8	0·4	—	14
Tomatillo	80	0·05	0·05	2·1	2	7	0·4	23	40
Cape gooseberry	2380	0·05	0·05	0·1	180	10	2·0	31	67
Husk tomato	380	0·05	0·02	2·1	0·2	7	0·4	23	40
Pepino	200	0·08	0·04	0·5	32	18	0·8	—	14

Source: FAO (1972) *Food Composition Table for Use in East Asia*. Rome: Food and Agriculture Organization.
Yamaguchi M (1983), p 299.
Aykroyd WR (1966).

highly pungent fruit of *C. frutescens* has a thin, smooth pericarp and is used as a condiment.

Chemical Composition and Nutritive Value

Much variation is found in the chemical composition of the fruits of this family (Tables 3 and 4). The stage of maturity also influences the chemical composition. Glucose and fructose are the principal sugars in tomato with small amounts of sucrose. Ripe fruits also contain raffinose; the glucose concentration increases with ripening, while the starch content decreases. The texture and firmness of the fruit is mainly controlled by the pectic constituents. In the ripening process, the predominant proto-pectin in the green fruit decreases and the pectin increases (Table 5). The protein content is about 1·0% in ripe fruit, and all the essential amino acids except tryptophan are present; other amino acids identified are tyrosine, aspartic acid, glutamic acid, serine, glycine, α-aminobutyric acid and pipecolic acid. Citric acid is the principal organic acid of tomato, along with a small amount of malic acid; traces of acetic, formic, lactic and succinic acids have also been reported. The chief colouring materials of tomato are carotene and lycopene; the changes in the content of carotenoids during ripening are listed in Table 6. ***See*** Acids, Natural Acids and Acidulants; Amino Acids, Properties and

Table 5. Pectic constituents of tomato at different stages of maturity

Stage of maturity	Pectin (%)	Protopectin (%)
Green	1·02	2·32
Yellow	2·34	2·21
Pinking red	3·10	1·62
Overripe	2·29	1·17

Source: *The Wealth of India* (1962), p 193.

Table 6. Colouring matter content (mg per 100 g) in ripe, half-ripe and green tomato

Pigment	Green	Half-ripe	Fully ripe
Lycopene	0·11	0·84	7·85
Lactoene	0·16	0·43	0·73
Xanthophyll	0·02	0·03	0·06
Xanthophyll ester	0·00	0·02	0·10

Source: *The Wealth of India* (1962), p 194.

Table 7. Maturity stages of tomato

Stage	Character
Immature	Seeds and the jelly-like substance surrounding the seeds not yet formed
Mature green	Fully grown fruit shows a brownish ring at stem scar, and blossom end turns yellowish green. Seeds are surrounded by jelly-like substance
Turning (breaker stage)	One quarter of the surface at blossom end shows pink
Pink	Three quarters of the surface shows pink
Hard ripe	Nearly all red or pink, but flesh is firm
Over ripe	Fully red and soft

Occurrence; Carotenoids, Properties and Determination; Ripening of Fruit

Tomato is regarded as an essential protective food. Research has shown that the inclusion of tomato in the diet can avoid deficiencies of vitamins and minerals. One medium-size, raw tomato provides 47% of the B vitamins, 33% of vitamin A and 1% of the energy of the daily dietary requirement of an average person, apart from calcium, potassium, magnesium and iron, the tomato contains zinc, boron, iodine, cobalt and aluminium in trace amounts. ***See*** Retinol, Physiology ***See*** individual minerals

The tomato contains a glycoalkaloid, tomatine and traces of solanine. Narcotine is present in unripe fruit. The tomatine content is lowest in the pink stage of ripeness, and increases slightly in the fully ripe tomato.

Brinjal or eggplant contains about 4% carbohydrate, composed of sucrose, glucose and fructose. Brinjal protein has a high biological value. It contains amino acids such as arginine, histidine, lysine, tryptophan, leucine, isoleucine and valine. The protein is relatively poor in lysine, tryptophan, isoleucine and methionine. Brinjal contains a higher percentage of vitamins than many other vegetables. However, it is not a rich source of vitamin B_2. Dark-purple-skinned varieties contain more vitamins than those with white skins. The main pigment of the fruit is an anthocyanin – nasunin; lycopene and lycoxanthin are also present. The seed oil is reported to be rich in linoleic acid. ***See*** Carbohydrates, Classification and Properties; Riboflavin, Physiology

The fruit peel contains a bitter principle, solasonine. The phenolic compounds present in the fruit are chlorogenic acid, neochlorogenic acid, scopoletin and caffeic acid; a trace of hydrocyanic acid is also present. ***See*** Phenolic Compounds

Husk tomato, tomatillo and Cape gooseberry are also important sources of vitamins and minerals (Table 4). Tree tomato contains substantial amounts of carbohydrates and minerals. It is rich in vitamin C and also a good source of provitamin A. ***See*** Ascorbic Acid, Physiology

Pepper is a good source of vitamins A and C, and is superior to tomato and eggplant in this respect. Vitamin C content varies with the variety and maturity stage. The pungent principle of chilli is capsaicin, the concentration of which is more in the inner walls; a highly irritating vapour is liberated on heating capsaicin. The colouring matter of ripe fruit contains capsanthin, capsorubin, zeascanthin, cryptoxanthin, α- and β-carotene.

Handling and Storage

The tomato is harvested at different stages of maturity (Table 7) depending upon its use, and it is often harvested at the mature green stage for the market and ripened either in transit or during storage. For canning or for juice extraction, the fruit is harvested at the ripe stage. The post-harvest losses in tomato may range from 5% to 50%; the highest loss is attributable to mechanical damage, but other causes may be physiological disorder, diseases and heat and chilling injury.

Fruits are transported to the market in bamboo baskets, corrugated board cartons or wooden boxes. Mature green tomatoes can be stored at 13–18°C and relative humidity (RH) 85–90% for 3 to 4 weeks. Fruits ripen at this temperature without chilling injury; above 18°C, fruits ripen faster. The ripe tomatoes can be preserved for 7–10 days at 3–5°C and RH 85–90%. Storage at a higher temperature (above 18°C) results in early ripening which leads to a rapid deterioration in the quality of the fruit, while at about 30°C the red pigment formation is hindered and the ripe fruit develops an orange to yellow colour.

The shelf life of the tomato could be prolonged by controlled-atmosphere and hypobaric storage. It is reported that post-harvest treatment with gibberellic acid markedly retards ripening, but 2,4-dichlorophenoxyacetic acid, naphthaleneacetic acid and ethylene hasten ripening, and this facilitates marketing of mature green tomatoes. Chemicals such as capton, dithane, and thiram have the greatest potential as fungicidal treatments on tomatoes. ***See*** Controlled Atmosphere Storage, Effect on Fruit and Vegetables; Fungicides

Brinjal, being a heavy vegetable, requires care in handling, even though it is not as perishable as the tomato. The fruit becomes edible from the time it attains one third of the full growth size. Overmature fruits are dull, seedy and fibrous, sometimes tasting bitter. After harvest, the graded fruits are packed in special crates or bushel baskets. Information on the extent of post-harvest losses of eggplants and the cause of the losses is meagre. However, in tropical countries, chilling injury, certain pests and diseases cause high losses of eggplants. They can be stored successfully for 2 to 3 weeks at 10·0–12·8°C and 92% RH. Controlled-atmosphere storage was not found to be beneficial in extending the storage life of the eggplant. Small fruits are not well suited for long-term storage. Chilling injury is more common in tender fruits. Prepacking of brinjal in adequately ventilated polyethylene bags (100-gauge) increases the shelf life. Treating brinjal with a fungicidal wax emulsion also extends the shelf life by 30–40%.

Bell peppers are harvested when they attain full size and are still green. Chilli meant for use as a raw vegetable is harvested green but, for processing, only ripe (red) chilli is harvested. The important post-harvest losses in chillies are weight loss, chilling injury and microbial spoilage. At the recommended storage condition of 7·2°C and 85–90% RH, green peppers can be stored for 2–3 weeks, and ripe peppers at 5·6–7·2°C and 90–95% RH can be stored for about 2 weeks. Storage life can be increased by wrapping the fruits in plastic film. Peppers are commonly wax-coated to reduce moisture loss before shipment. Ripe fruit can be preserved at room temperature (20°C) for only a week, but for 12 to 14 weeks at 3–4°C after dipping in hot water at 50°C for 10 min. ***See*** Spoilage, Bacterial Spoilage

Uses of Solanaceous Fruits

Fresh, ripe tomato fruits are refreshing and appetizing, and are consumed raw in salads or after cooking. Unripe fruits are usually cooked and eaten. In southern European countries and in California, large quantities are processed. Tomatoes can be processed into juice, purée, paste, ketchup, sauce, soup and powder. Special varieties have been evolved to meet these processing requirements. The tomato cultivars for juice should possess a bright colour, rich flavour and high total acidity. They should be juicy and not meaty. Tomatine, a glucoalkaloid present in the fruit, yields – on hydrolysis – tomatidine, which can be transformed into hormones such as progesterone and testosterone. The oil of tomato seeds is used in the soap and paint industry. The pressed cake can form a feed for livestock or a fertilizer.

Brinjal is valued more as a vegetable during autumn when other vegetables are scarce. It is a fairly good source of calcium, iron, phosphorus and vitamin B. Treatment with brinjal is recommended in liver complaints. It is reported to stimulate the intrahepatic metabolism of cholesterol, and produce a marked drop in blood cholesterol level.

The garden huckleberry (*S. nigrum*) leaf has great medicinal value. The tender leaves are boiled like spinach and eaten in many parts of India. Ripe fruits are used in pies and preserves; they are also used as a substitute for raisins in plum puddings. The berry is considered to possess tonic, diuretic and carthartic properties and is useful in the treatment of heart disease. The berries also find use as a domestic remedy for fever, diarrhoea, ulcers and eye trouble.

Peppers are used green, or after ripening, in a great variety of ways. The red pepper of commerce consists of the fruit of small, pungent varieties, which is dried and ground to a fine powder. Pepper sauce, prepared in a variety of ways, consists of the fruit of pungent varieties preserved in brine or strong vinegar. Tabasco sauce is prepared from the juice of a pungent variety, expressed by applying pressure. Paprika, a Hungarian condiment, is made from fruit ground after removing the seeds. Pepper is used in pickles of various kinds. The sweet varieties are stuffed and baked. *Capsicum* preparations are used as a counter-irritant in neuralgia and rheumatic disorders.

The melon pear fruit is cooked and eaten when immature, while the mature fruit has fleshy pulp with an aroma and flavour similar to the cucumber. Lulo is cultivated mainly for the juice, especially in Ecuador, Peru, Columbia and Central American countries. The young shoots of jilo are finely chopped and used in soups in Nigeria. Tomatillo gives a good chilli sauce and dressing for meats in Mexico. Ground cherry, which is acidic-sweet, is used for preserves and sometimes for sauces. The Cape gooseberry is eaten raw, or preserved as a pickle.

Bibliography

Aykroyd WR (1966) *Nutritive Value of Indian Food and Planning of Satisfactory Diets*, pp 51–137. New Delhi: Indian Council of Medical Research.

Richards AA (1967) *Tomatoes and Cucumber*, pp 11–12. Revised by HG Schoffer. London: WH and L Collingridge.

Salunkhe DK and Desai BB (1984) *Post Harvest Biotechnology of Vegetables*, vol. II, pp 39–58. Boca Raton, Florida: CRC Press.

Salunkhe DK and Desai BB (1984) *Post Harvest Biotechnology of Vegetables*, vol. I, pp 55–82. Boca Raton, Florida: CRC Press.

The Wealth of India (1950) A Dictionary of Indian Raw Materials and Industrial Products, vol. II, pp 69–73. New Delhi: Council of Scientific and Industrial Research.

The Wealth of India (1962) A Dictionary of Indian Raw Materials, and Industrial Products, vol. VI, pp 187–197. New Delhi: CSIR.

The Wealth of India (1972) A Dictionary of Indian Raw Materials and Industrial Products, vol. IX, pp 383–393. New Delhi: CSIR.

Thompson CH (1939) *Vegetable Crops*, pp 453–486. New York: McGraw-Hill.

Yamaguchi M (1983) *World Vegetables, Principles, Production and Nutritive Value*, pp 291–310. Westport, Connecticut: AVI Publishing.

M. V. Rama and P. Narasimham
Central Food Technological Research Institute, Mysore, India

Processing Potato Tubers

The Potato as Food

The species of potato most widely utilized for consumption as human food is *Solanum tuberosum* L. The tuber, an enlarged stolon (underground stem), is the edible portion of the potato plant. Potato tubers are composed of approximately 78% water, 20% carbohydrates and 2% protein. Starch makes up approximately 80% (by weight) of potato carbohydrates. Potato starch is composed of amylopectin (75–79%) and amylose (21–25%). Potato protein is comprised of all the essential amino acids, with methionine and cystine being somewhat limited for human nutritional requirements. Potatoes are an excellent source of vitamin C and they offer substantial amounts of riboflavin, nicotinic acid and thiamin. The potato is also a good source of iron and magnesium. ***See*** individual nutrients

This article reviews certain key aspects involved when raw potatoes are to be further utilized as a food by processing tubers into various food products.

Storing Potatoes for Processing

Nutritionally, potatoes are quite stable and have become available year-round through modern storage and processing technology. Long-term storage for processing has become an essential part of the potato industry. Typically, a specific type of potato is produced and stored long-term under specific storage conditions to supply a particular type of processing market. From initial suberization and preconditioning procedures through long-term holding and final reconditioning, the storage requirement is to maintain the highest internal and external quality possible while minimizing disease risk and moisture loss in large quantities of bulk potatoes. This is achieved by providing the temperature, humidity, gaseous composition and airflow required to maintain a nonstressful environment.

Properly designed and operated storage facilities are successfully storing quality raw product for up to 7 months for chips (crisps) and up to 11 months for French fries (chips). Such a storage facility must be structurally adequate, properly insulated, and provided with a precisely controllable ventilation and humidification system. ***See*** Controlled Atmosphere Storage, Applications for Bulk Storage of Foodstuffs

Raw Product Quality

Heightened consumer awareness of quality aspects in finished products has required that additional attention be given to the quality of the raw materials for processing, primarily the raw potato tubers.

Although physical characteristics of raw potato tubers such as size, shape, smoothness, defects (greening, bruising), diseases (rots) and sprouts directly influence the quality of the finished potato product, biochemical composition of the raw tuber has become more critical to finished product quality in all forms for processing.

Dry-matter content (usually measured as specific gravity) directly affects the economics of processing because of its influence on final product yield. Dry-matter content also affects oil absorption in frying processes. In both instances, higher dry matter (specific gravity) is desirable for greater yield and reduced oil absorption. Examples of desirable ranges of specific gravity in raw potatoes for making selected processed products are given in Table 1.

The amounts of reducing sugars (glucose and fructose) in the raw tubers also become critically important

Table 1. Examples of desirable ranges of specific gravity in raw potatoes for making selected processed products

Processed product	Specific gravity (sp. gr.)	Dry-matter content (%)[a]
Chips (crisps)	1·075–1·100	19·2–24·4
Flakes	1·075–1·100	19·2–24·4
French fries (chips)	1·075–1·095	19·2–23·4

[a] Calculated by percentage dry matter=24·182+211·04 (sp. gr. −1·0988).
From von Scheele C *et al.* (1935).

whenever the process also includes amino acids in the presence of high-temperature frying because finished product colour can be adversely affected (Maillard reaction). Typically, a glucose content less than 0·035% (by weight) in the raw tubers is necessary to assure production of light-coloured potato chips (crisps). Slightly higher concentrations of sugar are acceptable when producing French fries. ***See*** Browning, Nonenzymatic

Processing potatoes into finished food products involves a sequence of operations depending on the products being produced. Although a number of new processed potato products have been developed in recent years, the majority of potatoes continue to be processed into frozen French fries (chips), chips (crisps) and dehydrated products (flakes, granules).

Figure 1 shows a typical sequence of operations for processing the three major types of finished potato products.

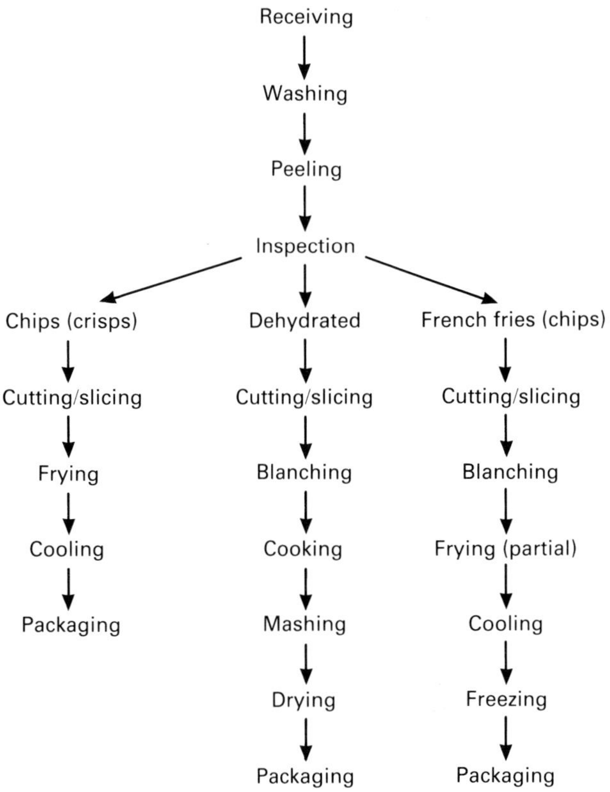

Fig. 1 Sequence of primary processing operations for three types of potato products.

Receiving

Receiving is the initial operation and is common to all three product processes. Incoming lots of raw potatoes for processing should be evaluated for quality attributes, such as size and shape, specific gravity, defects, flesh colour and sugar content. Samples for this purpose may be drawn from the incoming controlled-temperature highway trucks and railroad cars used to deliver raw potatoes to processing plants. Typically, receiving operations at a plant consist of either wet fluming or dry conveying systems. Fluming systems, in particular, must be carefully designed for effective transport of potato tubers.

Washing

Washing systems remove dirt and extraneous materials from the tubers prior to processing. Pressurized water sprays, usually combined with rotating brush rollers in either transverse or longitudinal arrangements, in specially designed washers are used to dislodge dirt from the tubers. Equipment for dirt, stone and trash removal are generally part of the receiving and washing operations.

Peeling

After the raw tubers are washed, the next step in the sequence of preparation operations is peeling (Fig. 1). Abrasive, caustic and steam peelers are common, and there is a recent trend towards steam peeling. Abrasive peelers physically erode the peel from the tubers with abrasive rollers, discs or drums in the presence of water sprays to flush away the debris. Caustic peelers provide a 1–3-min immersion of the tubers in a 10–15% sodium hydroxide solution heated to 95–100°C to soften the intact peel. A separate item of equipment then provides water sprays and brushes to remove the softened peel from the tubers. Steam peelers soften the peel on the tubers through the action of controlled heat and pressure (steam at approximately 1585 kPa) in an enclosed vessel. Again, a separate washing and brushing unit performs the actual removal of the softened peel. Depth of peel softening is critically controlled by correct time and temperature settings in the caustic and steam peeling processes. Peel removal normally ranges from 0% to 10% (by weight) of raw product in chipping (crisp) plants and from 10% to 25% in French fry (chip) plants. Steam and abrasive peeling result in more easily biodegradable waste materials than caustic peeling.

Inspection

The final preparation procedure, common to the three major types of processes shown in Fig. 1, is the inspection operation. Potatoes are presented for inspection on equipment that may both rotate and translate the tubers for viewing by inspectors under heightened lighting conditions. Blemishes such as bruised and discoloured portions of tubers may be trimmed away or entire tubers removed. In some plants, electromechanical equipment senses and diverts those tubers exhibiting off-grade surface characteristics. The number of human inspectors and/or mechanical devices depends on a combination of product volume and defects. In some processing plants, the peeled tubers may be sized via spaced rollers in order to feed more effectively any electromechanical sorting equipment and subsequent cutting and slicing operations. In some plants, a second inspection may be provided after the cutting and slicing operation.

Cutting and Slicing

When comparing the sequence of operations in the three different types of processes shown in Fig. 1, a major divergence occurs at the cutting and slicing operation. Depending on the product being produced, a variety of desired sizes and shapes of slabs and strips may be obtained from raw tubers through the use of an industrial-type rotary vegetable slicer. In some plants, unique water knives are used; these cut peeled tubers into strips for French fries while being carried by flowing water in pipes. Specially designed equipment provides for proper orientation of tubers relative to cutter blades and results in the longest possible cut strips.

French fry strips may vary from 0·65 to 1·25 cm in thickness. Crisp slices usually range from 0·125 to 0·175 cm. Slabs for production of dehydrated products will also vary from 0·65 to 1·25 cm in thickness, while diced pieces may reflect a combination of sizes within this same general range. Plant capacity requirements are normally met through the use of multiple slicing units.

Blanching

Blanching (soaking the cut slices or strips in water at 70–80°C) during French fry and flake or granule production gelatinizes starch granules to improve product texture. Blanching also limits enzyme activity, reduces subsequent cooking oil absorption and leaches reducing sugars from raw potato strips. Leaching of reducing sugars by blanching is used in French fry plants to minimize the production of dark-coloured French fries, but the practice is not normally used in chip processing. Cooling in water (at approximately 10°C) immediately following blanching completes the starch gelatinization process in potato flake or granule producing plants, but is not needed in French fry production.

Frying

Potato chips (crisps) are fried in cooking oil to reduce their moisture content rapidly to less than 2%. Typical times and temperatures for chip frying may range from 1·5 to 4 min and from 160°C to 190°C; the shorter times and higher temperatures are associated with continuous fryers, the longer times and lower temperature with batch (kettle) frying. When crisps exit the fryer, they are cooled at ambient conditions on conveyor belts in the plant.

French fry (chip) frying times depend upon slice thickness and oil temperature, but commonly vary from 1 to 2·5 min in oil ranging from 160°C to 190°C to complete a 'par fry' (partially fried by removing moisture to less than 30–40%). Partially fried strips may undergo removal of excess surface oil by passing through a stream of air at 70°C immediately upon exiting the fryer. The fried strips are then quick-frozen. This operation rapidly reduces the product temperatures to between −15°C and −7°C with refrigerated air or by application of liquid nitrogen (−50°C) directly to the product surface. Final cooking of partially fried French fries is accomplished by the end-product user.

Dehydrating

Potato tuber pieces being prepared for dehydrated products require cooking following the blanching and cooling operations. Cooking times and temperatures are selected to accomplish cell aggregation with minimal cell rupture. Typically, live steam (100°C at atmospheric pressure) is provided in the cooker for 20–30 min depending on the potato slab thickness and potato solids content. Cell rupture is also minimized at this point by preparing the necessary mash or slurry immediately after cooking.

Emulsifiers are added to the mash to prevent stickiness by complexing free starch. Chelating agents may be added to prevent after-cooking darkening by complexing metal ions. Antioxidants to prevent oxidation of potato components may also be incorporated. Mixing the mash or slurry blends the cooked potato tissue with the additives. Flakes are made by drying (to <6% moisture) a thin layer of the mashed potato material on a steam-heated rotating drum dryer. A sheet of dried potato tissue ranging from 0·015 to 0·025 cm thick results and is reduced to flakes by size reduction devices. *See* Antioxidants, Natural Antioxidants; Antioxidants, Synthetic Antioxidants; Emulsifiers, Uses in Processed Foods

Production of granules is similar, but the cooked potato slurry is cooled and conditioned (held at approximately 15–25°C and with some previously dried granules included for about 1 h). After conditioning, the mash or slurry is mixed and dried in a spray drier to less than 6% moisture. The spray-drying process creates aggregates (granules) which are screened for sizing. *See* Drying, Spray Drying

Finished Product Quality

Monitoring of finished product quality in relation to some set of standards is practised in most potato processing plants. For example, chips (crisps) may be monitored for colour, French fries (chips) for oil content and dehydrated flakes for moisture content. Product colour can be evaluated by visual comparisons to colour reproductions of finished chips or fries or by spectrophotometric determinations of light reflected from a sample of the product. Sensory analysis is used to monitor flavour, texture and appearance of products. *See* Sensory Evaluation, Sensory Characteristics of Human Foods

Oil content of fried products may be determined by refractometry or Soxhlet extraction. The oil content of chips and French fries will be influenced by raw product dry-matter content, cut sizes, fry times and temperatures, but will normally range from 30% to 45% by weight in chips and from 4% to 7% in partially fried French fries. The moisture content of potato tissue during various stages of processing can be determined rapidly by infrared or microwave drying of a sample or by slower but more complete drying in a vacuum oven.

Packaging

Packaging of processed potato products is generally performed by fully automated systems. Form-fill or carton-fill equipment automatically fills retail or institutional packages with frozen French fries. Polyethylene and corrugated cardboard materials are commonly used for packaging frozen potato products. Automated fillers place potato flakes into institutional or retail packages made of moisture- and vapour-proof materials.

Potato chips are packaged with form-fill equipment that currently utilizes foil laminates to eliminate light and oxygen. Immediately before sealing, nitrogen is injected into the package to serve both as an oxygen displacer and as a cushion for the contents. Marketing requires numerous types, sizes and shapes of packages in the potato processing industry. These individual packages are normally combined into pallet loads for handling and transporting.

Other Products

Some potato products (such as starch, flour and canned potatoes) may be either processed as primary products in a plant or accumulated and prepared as by-products in larger processing plants. In addition, variously configured products made by forming, drying and freezing pieces of potato tissue from the main processing line may be important by-products in frozen French fry processing operations. Other types of processing plants may specialize in reconstituting dehydrated potatoes into manufactured (extruded or expanded) snacks such as chips and puffs. However, all of these products together constitute only a numerically minor portion of the potatoes that are processed annually. Frozen French fries, potato crisps and dehydrated flakes or granules make up the greatest portion (85–90%) of products obtained through the processing of potato tubers. *See* Extrusion Cooking, Principles and Practice; Snack Foods, Range on the Market

Bibliography

Gould WA and Plimpton S (1985) Quality evaluation of potato cultivars for processing. *Ohio Agricultural Experiment Station Bulletin* 1172.

Lisinska G and Leszczynski W (1989) *Potato Science and Technology*. New York: Elsevier Science Publishers.

Lulai EC (1983) Interrelationships of quality testing methods and relative effectiveness of surfactants on potato flake quality. *American Potato Journal* 60: 441–448.

Lulai EC and Orr PH (1979) Influence of potato specific gravity yield and oil content of chips. *American Potato Journal* 56: 379–390.

Orr PH and Graham CK (1983) A generalized model for determining least-cost sources of processing potatoes. *Transactions of the American Society of Agricultural Engineers* 26(6): 1875–1878.

Schaper L, Orr PH, Yaeger EC, Smith N and Hunter JH (1987) *Hydraulic Transport of Potatoes*. USDA Technical Bulletin No. 1727. Washington, DC: US Department of Agriculture.

Shallenberger RS, Smith O and Treadway RH (1959) Role of the sugars in the browning reaction in potato chips. *Journal of Agriculture and Food Chemistry* 7: 274–277.

Processing Potato Tubers

Sowokinos JR and Preston DA (1988) Maintenance of potato processing quality by chemical maturity monitoring (CMM). *Minn. Agr. Expt. Sta. Bull.* AD-SB-3441.
Talbert WF and Smith O (1987) *Potato Processing* 4th edn. New York: Van Nostrand Reinhold.
van Es A and Hartmaus KJ (1987) Starch and sugars during tuberization, storage and sprouting. In: Rastovski A *et al.* (eds) *Storage of Potatoes*. Wageningen: Pudoc.
von Scheele C, Svensen G and Rasmussen J (1935) Du bestamming an patatisens. Starelse och Torrsubstanshalt wed tilhjap av dess specifka vkt. *Novd. Jordbr. Forsk:* 12–37.

Paul H. Orr
US Department of Agriculture Potato Research Laboratory, East Grand Forks, USA
Jerry N. Cash
Michigan State University, East Lansing, USA

POULTRY

Contents

Chicken

Chicken is one of the most widely accepted muscle foods in the world. Its high-quality protein, relatively low fat content, new products, and generally low selling price because of favourable feed conversion make chicken a high-demand food in the marketplace. Furthermore, the absence of cultural or religious taboos allows increased chicken production and consumption worldwide.

This article reviews specific characteristics of chicken, its chemical composition and nutritive importance, production advances, slaughter and further processing, food uses and products, and microbiological problems.

The USA leads the world in poultry meat production, followed by the USSR, China, and Brazil. Consumption of chicken in the USA has increased dramatically over the last 20 years. Table 1 shows the changes in chicken consumption and compares those changes with other meats. Broiler consumption in the USA continues to increase, 1990 consumption surpassing that of beef and pork on a retail-weight basis.

Chemical Composition, Nutrition, and Dietary Significance

Chicken is a very digestible source of high-quality proteins, its muscles differentiated into light meat (primarily the breast) and dark meat (the legs). These muscles vary in myoglobin content as well as in fat content. Table 2 lists the chemical composition of selected chicken parts, and shows some of the proximate changes that occur on cooking. ***See*** Protein, Food Sources

Chicken meat is low in total fat and saturated fat, and light meat is lower in fat than dark meat (Table 2). Most of the fat in poultry is deposited under the skin and is easily removed by pulling the skin from carcasses or parts. Because chicken fat has a high ratio of unsaturated to saturated fatty acid (approximately 30% saturated), the fat melting point is relatively low and liquefies easily. Chicken fat is also prone to oxidative rancidity. Although most consumers believe otherwise, sodium and cholesterol contents in chicken are similar to those in most meats. Light meat contains more protein, and dark meat contains higher levels of fat and cholesterol. Older chickens generally contain more fat and less moisture. Cooking tends to concentrate the level of cholesterol and protein because water is lost in the heating process. ***See*** Cholesterol, Properties and Determination; Fatty Acids, Properties; Sodium, Properties and Determination

Chicken contains all the essential amino acids, and is a good source of B vitamins and minerals such as iron

Table 1. Consumption (kg per capita) of red meat (carcass weight) and poultry (ready-to-cook weight) in the USA, 1967–1989

	1967	1972	1977	1982	1987	1989
Broiler	14·7	17·2	18·5	22·5	27·4	30·0
Turkey	4·0	4·1	4·1	4·9	6·9	7·7
Beef	49·0	52·5	56·1	47·1	47·0	44·4
Pork	32·8	32·1	27·5	28·4	28·4	30·0

Source: Putnam JJ (1990).

Table 2. Proximate analysis of chicken broilers (per 100 g edible portion, wet-weight basis)

Nutrient	Breast meat with skin (raw)	Breast meat without skin (raw)	Breast meat with skin (roasted)	Leg meat with skin (raw)	Leg meat with skin (roasted)
Water (g)	69·46	74·76	62·44	69·91	60·92
Protein (g)	20·85	23·09	29·80	18·15	25·96
Energy (kJ)	772	462	827	785	974
Lipid					
Total (g)	9·25	1·24	7·78	12·12	13·46
Saturated (g)	2·66	0·33	2·19	3·41	3·72
Monounsaturated (g)	3·82	0·30	3·03	4·89	5·24
Polyunsaturated (g)	1·96	0·28	1·66	2·65	3·00
Cholesterol (mg)	64	58	84	83	92
Ash (g)	1·01	1·02	0·99	0·85	0·92
Sodium (mg)	63	65	71	79	87

Source: Posati LP (1979).

and phosphorus. Poultry meat also provides potassium, calcium, magnesium, and copper, but carbohydrates and fibre are negligible. Light meat contains more nicotinic acid than dark meat, although both are good sources. Conversely, dark meat contains more riboflavin, iron, and zinc than light meat. ***See*** Amino Acids, Properties and Occurrence ***See also*** individual vitamins and minerals

Economics of Fast-growing Broiler Chickens

Chicken broilers have been bred to a uniform size specifically for meat yield. A common broiler strain combines Cornish genes for conformation and fleshing with the White Plymouth Rock for white feathers and faster growth. White-feathered birds are the norm for commercial broiler production because of their cleaner-looking defeathered carcasses.

Many factors have contributed to the efficiency and economics of broiler production. Advances in breeding, nutrition, disease control, and management practices have enabled the broiler industry to produce a chicken weighing 1·8 kg in 6–7 weeks. Vaccines, antibiotics, confinement rearing, and computer-balanced rations assist in producing broilers with a feed conversion ratio of less than 2 kg of feed per kg gain. Vertical integration in the broiler industry has increased efficiency even more by combining most of the activities in producing, processing, and marketing broilers under the same ownership and management.

Chicken Processing

Chicken slaughter and processing have evolved into a sophisticated and automated procedure whereby a processing plant can handle as many as 20 000 broilers per h. Transforming live chickens into a ready-to-cook form involves live bird catching, crating and hauling, unloading, hanging on a conveyor line, stunning, slaughtering and bleeding, scalding, defeathering, eviscerating, inspecting, chilling, grading, packing, and shipping. ***See*** Meat, Slaughter

Assembly of Live Birds

Broilers should be taken off feed 8–12 h before the time of slaughter to reduce faecal contamination during processing. Bruising in chickens can be minimized by removing feeders and waterers prior to the arrival of the catching crew, and by careful handling of the birds as they are caught and placed into coops. Novel mechanical methods (e.g. herding, sweeping, and vacuum systems) for catching chickens are also available and commercially used by a few companies in Europe.

Hauling and Unloading

After the birds are loaded into wooden or plastic crates or specially built compartments, they are transported via open-sided trucks to the slaughter plant. Large-volume, slow-speed fans or evaporative cooling provide good ventilation and comfort for the birds in the holding shed.

Trailer loads move into the plant unloading area as needed. Various methods (e.g. manual bird removal, or dumping the birds through side-doors) unload the crates. Birds are hung by their shanks on shackles attached to an overhead monorail conveyor for transfer through the slaughtering operations. The unloading area should be dimly lit, and a breast rub bar can be used to calm the birds. According to UK regulations, birds should be slaughtered within 3 min after they are suspended from shackles.

Stunning, Killing and Bleeding

Stunning is used to enhance bleeding and feather release. The heads of the birds are usually dragged through an electrically charged water bath. For consistent stunning, the water should contain from 0·1% to 1·0% sodium chloride. A fine mist of water spray should be directed at the bird's feet to provide a positive electrical contact prior to entering the stunner. The shackles complete the circuit and cause an electric current to run through the body of the bird as a result of the applied voltage (approximately 50 V). Overstunning can rupture blood vessels and break carcass bones due to excess muscle contraction. It is essential that the electric shock does not kill the bird; bleeding must be the cause of death.

Killing can be either manual or mechanical. In the manual operation, a skilled worker with a sharp knife severs the jugular vein and carotid artery by cutting across the side of the neck at the base of the bird's head. In mechanical neck cutting, a guide bar with grooved rollers holds the head rigid and extends the neck in preparation for the cuts. The bird's head is guided across a single, revolving, circular blade or between a pair of revolving blades.

Bleed time should be 1·5–2 min, and the blood lost accounts for approximately 4% of the live weight. The shackles convey the birds through a blood tunnel, so that the blood can be collected and disposed of properly.

Scalding

After bleeding, the birds are scalded to loosen the feathers by immersion in agitated hot water. Wing flapping and struggling should have ceased by the time birds enter the scalder. In general, a soft- or semiscald temperature of 50–54°C is used, with an immersion time of 1·5–2·5 min. For the fresh chilled broiler market, low-temperature scalding permits retention of the yellow cuticle on the chicken skin and retards skin drying.

Chickens are subscalded at 59–60°C for cuticle removal if there is a preference for lighter-skinned birds, if the carcasses are to be frozen, or if the parts are to be battered and breaded. Higher temperatures greatly facilitate feather removal, and necks and wings of semiscalded birds are often scalded at the subscald conditions to achieve cleaner carcasses. In the USA, the US Department of Agriculture (USDA) recommends a minimum overflow from the scalder of 1 l per bird to reduce build-up of contamination. Various chemicals added to scald tanks assist feather removal by reducing surface tension and enhancing wetting of the feathers. An alternative to immersion scalding with separate defeathering machines is a combined spray-scalding and defeathering system. This type of scalding may reduce contamination spread from bird to bird.

Defeathering

After the carcasses leave the scalder, they enter a series of on-line defeathering machines. The picking machines consist of banks of counter-rotating stainless steel discs or drums, with rubber 'fingers' mounted on them. As the birds are conveyed through the rotating picker fingers, the feathers are rubbed or plucked from the carcass. Continuous water sprays flush away the feathers.

Remaining pinfeathers are removed by hand. In the USA, the birds pass through a gas flame to singe the fine hairs. A final step in the defeathering area is an outside spray wash.

The head removal operation may take place in the defeathering area between the picker and the outside bird washer, or in the eviscerating room to facilitate crop removal. The bird's neck passes through a device which restrains the head as the overhead conveyor pulls the body forward, separating the neck vertebrae at the base of the skull. Some head removal devices can include a set of rollers to separate the neck bones and a rotating knife to sever the neck skin instead of tearing it; yield is improved with this system.

The birds pass through an automatic hock cutter which severs the shanks at the hock joint and causes the carcass to drop onto a conveyor for transfer to the evisceration area. In the UK, the feet of the broilers are cut off just above the spur by means of a rotating knife. The severed feet remain on the shackles and are removed mechanically on the return line.

The scalding, defeathering, and other operations occur in a separate portion of the plant, apart from the evisceration and final processed bird area. The transfer conveyor takes the defeathered carcasses into the evisceration area, where the carcasses are rehung on the clean evisceration line.

Evisceration

The first cutting operation in the evisceration area involves removing the preen or oil gland, located on the top side of the tail next to its base. A sharp, short-bladed knife is used to remove the gland manually, or a machine with a cutting blade can be used on-line.

An opening cut into the body cavity is then made using a knife, a special drill-like vent cutter with a rotary blade, or an on-line automated machine. Making an incision in the abdomen and removing the vent can be

Fig. 1 Automatic chicken eviscerator.

performed manually or by machine to complete the opening process for evisceration.

Removing or drawing the viscera from inside the bird can be performed by hand or by machine. To retain the identity of the viscera within the carcass and be in a position for easy inspection, the viscera are draped uniformly to one side of the tail but remain attached to the carcass until the carcasses have passed the inspection for wholesomeness.

Mechanized removal of the viscera is very common in large processing plants (Fig. 1). An eviscerating spoon is mechanically inserted into the cavity and withdraws the viscera. Back-up workers may be necessary to remove viscera missed by the machine.

Mandatory USDA inspection of every bird for wholesomeness is performed under the supervision of an inspector employed and trained by the government. If evidence of unwholesomeness is detected, the carcass or affected parts are trimmed or condemned.

Workers remove the giblets (heart, liver, and gizzard) from the viscera of inspected birds and clean and chill them. Then the lungs are removed with a vacuum lung gun, or they can be withdrawn with the viscera on a mechanical eviscerator. The neck is generally cracked with shears, the crop and windpipe are removed, and the neck is then pulled off by stationary guide bars.

Finally, the shackles convey the carcasses through a bird washer where spray nozzles rinse the inside and outside of the carcasses.

Chilling

The most common chilling operations in the USA immerse carcasses in long flow-through tanks containing agitated chlorinated water or slush ice. The USDA regulations require that a chiller overflow rate be maintained at 2 l of water per chilled bird, in order to minimize microbial build-up in the chill water.

The chickens are first placed in a prechiller, containing water at 10–18°C, and then into a slush ice chiller at 0–1°C. Chickens must be chilled to 4°C or lower within 4 h to meet USDA requirements; the chilling time generally needed to obtain 4°C carcasses is about 40–60 min.

After chilling, carcasses are hung by one leg on a drip line for 2·5–4 min for draining, and are then conveyed to the packing area. Although the meat absorbs some moisture during washing and chilling, the drained moisture level is strictly regulated. The drip line shackles are usually weighing devices which drop the sized birds into appropriate bins.

Air chilling of eviscerated carcasses is used extensively in Europe. Spraying the carcasses with water at intervals avoids weight loss during evaporative chilling. ***See*** Meat, Preservation

Grading

Carcass grading in the USA is a voluntary practice, but is often required by purchasers. Government graders sort the birds into A, B, and C categories according to body conformation, fleshing, fat cover, deformities, bruises, and defects such as pinfeathers, disjointed or broken bones, missing parts, etc. Europe uses similar grading standards.

Packing

Giblet packs containing a heart, liver, gizzard, and neck are stuffed into the body cavity of chilled, sized carcasses. The carcasses then go into overwrapped trays which are heat-sealed, or into polyethylene bags which are clipped or taped shut ready for market. Another common method in the USA for distributing broiler carcasses from the processing plant is to put them into corrugated, wax-impregnated boxes. After boxing, ice covers the birds to keep the skin moist during shipment. Nearly half of the broilers in the USA leave the processing plant as ice-packed birds (see Table 3). An

Table 3. Proportion of processors' chicken broiler volume sold by marketing form in the USA

Marketing form	1962	1970	1978	1985	1989
Ice-packed	90·3	78·6	66·4	54·7	44·1
Whole, uncut	83·0	65·3	43·7	21·2	12·4
Cut, parts	7·3	13·3	22·7	33·5	31·7
Prepacked, chilled	—	13·1	20·5	26·0	22·6
Whole, uncut	—	5·8	7·5	9·3	6·7
Whole, cut	—	3·2	3·3	1·5	1·5
Part packs	—	4·1	9·7	15·2	14·4
Frozen[a]	9·7	8·3	8·2	7·0	7·1
Whole, uncut	4·1	1·8	3·5	0·9	1·3
Cut, parts	5·6	6·5	4·7	6·1	5·8
CAP[b], bulk	—	—	3·2	4·7	4·1
Further-processed[c]	—	—	—	6·2	9·3
Other[d]	—	—	1·7	1·4	12·8

Source: Anonymous (1990).
[a] Mostly for export.
[b] Controlled-atmosphere packaging.
[c] Hot dogs, patties, nuggets, strips, frozen–fried parts, etc.
[d] Includes mechanically deboned meat.

increasingly popular packing method is chill-packing, whereby a −6°C air blast lowers carcass temperatures to between −2°C and −1°C. Refrigeration then maintains that temperature during marketing of the birds.

Waste Products

Poultry processing results in large amounts of waste waters, semisolids, and solids, which require separation and treatment before being discharged into the environment. Waste material that can be reclaimed and, for example, used for animal feeds or fertilizer helps to reduce the overall load for disposal.

There are several ways to handle poultry offal (heads, feet, viscera, inedible parts, condemned whole birds, feathers, and blood). Usually, all offal except the blood and condemned birds is floated in water from the processing areas to an accumulation area for removal by trucks. Screens and rotating drums remove water and separate offal components. Because of its high oxygen demand for decomposition, blood is usually handled separately.

Most of the offal solids go to a rendering plant, while liquid generally goes through a primary and secondary lagoon treatment system. Some plants can discharge the liquid effluent into a municipal water system after removing the fat and as many solids as possible.

Another method of waste treatment is spray irrigation or spreading the processing effluent on the land. This type of waste disposal requires large fields to prevent overloading. Runoff into streams is a major concern. *See* Effluents from Food Processing, Composition and Analysis; Effluents from Food Processing, Disposal of Waste Water

Food Uses and Products

Most broilers in the USA are sold as fresh (unfrozen) carcasses, but frozen deboned meat is used in further processing. A decline in the sale and consumption of whole chickens in the USA (from 11·4 kg per person in 1975 to 5·7 kg per person in 1990) has been counterbalanced by a steady increase in the sale of cut-up carcasses and parts. More than half (53%) of the processed carcasses in 1990 were marketed as cut-up broilers and parts.

Cut Up

The cut-up operation is usually mechanized with motor-driven equipment and shielded circular blades. Individual cut-up stations or on-line automated machines can be used. A popular method of packaging is to place the cut-up parts in a tray containing an absorbent pad to collect seepage and then overwrap the tray. The bagged whole carcasses and tray packs may be prepriced before delivery to the retail market.

Deboning

The spectacular growth in the sale of value-added, further-processed poultry products in recent years has placed a heavy demand on the production of deboned meat. Poultry meat has traditionally been removed from the carcass by hand with a sharp knife, while the carcasses hang from special shackles on a slow-moving line or are positioned on static or moving cones.

Several automatic deboning machines are also available for removing breast fillets and thigh and drumstick meat. One system operates by holding a particular chicken part in position above a contoured recess in a base plate. The machine forces the meat from the bone into the depression. Another meat deboning system pushes the bone lengthwise out of the carcass part (e.g. drumstick) and strips off the meat in the process.

Currently, deboning is primarily done on aged chilled carcasses. Hot stripping the muscles and skin from defeathered but uneviscerated carcasses is being studied. Some researchers are also testing hot deboning, the removal of meat afer evisceration but prior to chilling. Generally, hot deboned meat is tough, which greatly limits its potential uses. Some poultry meat, especially that of mature hens, is cooked prior to hand deboning.

After the major muscles are deboned from the carcass, the remaining frames (also necks, backs, and low-value parts) can be more completely deboned by special machines. In the first stage of mechanical deboning, a grinder reduces bone and meat particle size. In the second step, a pressure system squeezes the ground meat and bone against a perforated screen or microgrooved cylinder. Advances in mechanical deboning have eliminated pregrinding and produced a more texturally attractive product. An alternative system is a batch operation. Pressure from a ram forces meat to flow from the bones through a sieve screen, the bone cake is ejected, and a new batch is introduced.

Further-processed Products

Much of the hand- and mechanically-deboned chicken meat is directed into value-added products. Deboned meat can be marketed 'as is' fresh, as whole breast or split breast fillets, strips, chunks, etc. Alternatively, deboned meat can be tumbled, marinated, chopped, formed, ground, emulsified, or prepared in a number of ways for sale.

Many nonmeat ingredients are used in further processing. Salt helps to extract proteins for improved binding and texture, acts as a preservative, and enhances flavour. Up to 0·5% phosphate increases water-holding capacity and final product yield. Other ingredients, such as sweeteners, spices, binders, and curing salts, are used in the wide array of poultry products on the market. ***See*** Curing

The driving force behind poultry product development is the consumer. Fat content in meats is a current concern; many people want healthier diets, and the low fat in poultry is a major reason for its popularity. Fat pads are removed from many whole carcasses in response to consumer desires. As new poultry products are designed, diet and health, microwave ovens, and an ageing population cannot be ignored.

The fast food restaurants have had a tremendous influence on new poultry product popularity for items such as nuggets, tenders, marinated breast fillets, and frozen–fried parts. Ethnic foods are popular in the USA, especially Italian, Mexican, and Chinese. Different flavours in poultry products are in demand, including cajun, barbecue, and honey. Spicy chicken wings are an example of a popular seasoned finger-food appetizer. In addition to less fat, smaller portions, and different flavours, consumers want less salt and cholesterol in their chicken products. The food must be of high quality, convenient, safe, and nutritious.

In the retail supermarket, low-fat chicken products such as patties, rolls, sausages, and sliced luncheon meats are selling well. Poultry frankfurters and bologna are generally lower in fat than similar red meat products and are a healthy alternative. ***See*** Meat, Sausages and Comminuted Products

Most recent entries among the new poultry products are boneless chicken breasts stuffed with ingredients such as cheese and broccoli or wild rice and mushrooms. These products are sold in a variety of packages and combinations, but are almost always ready for cooking in the microwave. Other gourmet items include Chicken Kiev (breaded chicken breasts filled with garlic butter and parsley), Chicken Cordon Bleu (filled with ham and cheese), and Herb Roasted Chicken Dinner (low-fat, low-cholesterol, low-sodium meal). Even the luncheon market has new chicken 'lunchable' products, which are individual packages containing items such as cooked sliced poultry, cheese slices, and crackers.

Microbiological Concerns

Modern poultry husbandry and processing techniques have greatly improved the quality of poultry meat. However, chicken can carry many kinds of organisms. The two major concerns are control of spoilage organisms which cause consumers to reject the product due to odour or flavour, and the minimization of pathogenic organisms which may (under faulty handling) lead to a health hazard.

Spoilage Microorganisms

Although poultry is refrigerated, and even frozen, for shelf life extension, spoilage will invariably occur owing to the growth and metabolic activities of specific types of bacteria. The psychrophiles, such as *Pseudomonas*, can generally grow at refrigerator temperatures and are responsible for most of the spoilage. Spoilage organisms are present in high numbers when an off-odour becomes apparent (10^7 cells per cm^2) and even greater when slime formation occurs from coalescence of colonies (*c.* 10^8 cells per cm^2). ***See*** Spoilage, Bacterial Spoilage

Pathogenic Microorganisms

Poultry is considered to be a major source of salmonella, which can spread during processing, and even after cooking if raw meat preparation surfaces were not cleaned before placing a cooked product on the same surface. However, normal cooking procedures destroy salmonella and it will not grow well under refrigeration. The most common problems are undercooked poultry or poultry contaminated after cooking. According to data on the vehicle of foodborne salmonellosis outbreaks in the USA between 1973 and 1987, only 3·8% (30 of 790) of the cases were caused by chicken.

In addition to salmonella, other pathogens sometimes found in poultry include staphylococci, campylobacter, *Listeria* spp., clostridia, and coliforms. Careful product handling, proper refrigeration, and adequate cooking will almost always assure product safety. ***See*** Campylobacter, Campylobacteriosis; *Clostridium*, Food Poisoning by *Clostridium perfringens*; *Clostridium*, Botulism; *Listeria*, Listeriosis; *Staphylococcus*, Food Poisoning

Sources of Bacteria

Microbes in poultry processing plants are ultimately found on the product and come from three main sources: the birds (feet, feathers, intestinal contents), the environment (water, air, supplies), and the workers. Careful management at the production site and adherence to good manufacturing practices at the processing plant (e.g. filtered air, cool temperatures, thorough cleaning and sanitation, and good worker hygiene) will minimize final product contamination with bacteria.

Bibliography

Anonymous (1990) *Broiler Industry Marketing Practices, Calendar Year 1989*. Washington, DC: National Broiler Council.

Bean NH and Griffin PM (1990) Foodborne disease outbreaks in the United States, 1973–1987: pathogens, vehicles, and trends. *Journal of Food Protection* 53(9): 804–817.

Brant AW, Goble JW, Hamann JA, Waback CA and Walters RE (1982) *Guidelines for Establishing and Operating Broiler Processing Plants*. Agriculture Handbook 581. Washington, DC: Agricultural Research Service, US Department of Agriculture.

Cunningham FE and Cox NA (1987) *The Microbiology of Poultry Meat Products*. Orlando: Academic Press.

Mead GC (1989) *Processing of Poultry*. Essex: Elsevier Science Publishers.

Mountney GW (1976) *Poultry Products Technology* 2nd edn. Westport, Connecticut: AVI Publishing.

North MO (1984) *Commercial Chicken Production Manual* 3rd edn. Westport, Connecticut: AVI Publishing.

Posati LP (1979) *Composition of Foods. Poultry Products, Raw, Processed, Prepared.* Agriculture Handbook 8-5. Washington, DC: Science and Education Administration, US Department of Agriculture.

Putnam JJ (1990) Food consumption, prices, and expenditures, 1967–88. Statistical Bulletin No. 804. Washington, DC: Economic Research Service, US Department of Agriculture.

Stadelman WJ, Olson VM, Shemwell GA and Pasch S (1988) *Egg and Poultry-Meat Processing*. Cambridge: VCH Publishers.

Arthur J. Maurer
University of Wisconsin, Madison, USA

Ducks and Geese

Ducks and geese (waterfowl) are a delicacy to many people, while others object to the higher amounts of fat in the carcasses compared to broilers and turkeys. However, breeding and mass selection programmes are improving the meat: bone and meat: fat ratios in waterfowl. Annual duck consumption on a ready-to-cook basis in the USA is only about 0·2 kg per person. Goose consumption is somewhat less at approximately 0·01 kg per person per year. Although chickens and turkeys dominate the world poultry industry, in parts of Asia ducks are commercially more important than broilers (chickens), and there are more geese than turkeys in areas of Europe. This article reviews duck and goose processing and preparation, meat composition and nutrition, food uses and products, waste products, and microbiological problems.

Processing and Preparation

With a few exceptions, duck and goose processing is similar to that of broilers and turkeys. For a more complete slaughter and evisceration procedure, please refer to the preceding article.

In the USA, White-Pekin-type ducks are generally slaughtered at 7 weeks of age and weigh about 3·2 kg live. White Muscovy ducks are also raised for meat, but they are a slower-growing breed and commercially less popular than the White Pekin. Geese are 3–5 months old at slaughter, and their average live weight is 4·5–6·8 kg. The average ready-to-cook weight of geese varies, but is generally about 4·5–5 kg. Feather maturity (lack of pinfeathers) is an important factor in determining the best time to slaughter waterfowl. ***See*** Meat, Slaughter

Processors transport ducks and geese to slaugher plants on open trucks or trailers. The loading and unloading process is unique as the birds are not usually caught by their legs; they are herded onto the conveyance, and off into holding pens at the processing plant. Processors use turkey-sized crates when there are relatively few geese to transport.

Ducks and geese are driven from holding pens through chutes onto scales for weighing and then into the shackle hanging area. Care in herding the waterfowl is essential to avoid pile-ups, skin scratches, damaged legs, and smothered birds.

It is important to minimize struggling by the birds when hanging them on motorized conveyor line shackles as excessive flapping can result in bruises. In smaller processing plants, killing funnels can be used to restrain and position the birds for slaughter.

The birds are usually stunned with an electric current

and then bled. Sometimes a cut inside the mouth is used, but a common method of killing the birds is to cut the outside of the throat on the left side at the base of the jaw, severing the left jugular vein and carotid artery.

Following bleeding, both ducks and geese can be scalded or dry-picked. However, the latter method is slow and laborious. Generally, ducklings proceed through immersion scalders containing agitated water at 58–63°C. Geese can be scalded in a similar commercial scalder, or they can be hand-scalded for a small-scale operation. Scald water temperature for geese should be 63–66°C. The duration of the scald varies from 1·5 to 3 min, usually longer for geese than for ducks. The time and temperature will also change depending on the age of the bird, time of year (season), and density of feathering. The lower the temperature, the longer the scald. A little detergent or an alkaline defeathering agent can be added to facilitate thorough wetting of the feathers. Processors can hand-scald waterfowl by pulling the bird repeatedly through the water against the lay of the feathers.

After scalding, the birds may be rough-picked by hand, picked by a conventional rubber-fingered picking machine (on-line), or placed in a spinner-type picker. Because pins and down remaining on the carcass are difficult to remove, it is common practice to finish rough-picked birds by dipping each bird into melted wax specially formulated for this purpose.

The defeathered birds should be surface-dried (a jet of compressed air can be used) just long enough to allow the wax to adhere. The waxing operation is usually mechanized and often includes conveying the ducks in and out of two wax tanks. A resin-based microcrystalline wax is used for greater resilience and improved stripping.

Good wax penetration and adhesion are achieved in the first tank with immersion in 90°C wax for 15 s. The second wax dip at 71°C puts a heavier coating on the birds to thicken the wax for good pulling power and cleaner stripping. After waxing (and sometimes between wax dips), a cold-water spray over the birds or a dip into a tank of cold water will cool and harden the wax to a tacky state. The wax is then removed by hand or by a rubber-fingered wax stripper. Some processing plants dip the wax-picked birds into water at 82°C to tighten the skin for easier manual removal of any remaining pinfeathers. Grasping pinfeathers between the thumb and a dull knife will assist in this operation.

The wax is reclaimed by remelting and straining out the pins, down, and feathers. Occasionally, the used wax should be 'cooked down' to remove all water that may have been mixed with it by emulsion.

After completely removing feathers, pins, and down, processors eviscerate the carcasses by making an opening cut in the abdomen and pulling out the viscera. Waterfowl viscera are somewhat difficult to withdraw, so that evisceration is most often done manually. However, processing plants in Europe are using automatic duck evisceration and automatic head and trachea pulling with processing rates of 4000 ducklings per h.

Each carcass on the eviscerating line is inspected for wholesomeness before the withdrawn viscera are detached. After the inedible viscera and lungs are discarded, the heads, tracheas and feet are removed; the giblets are cleaned and wrapped for later placement in the body cavity. In many processing plants, only the two lobes are cut from gizzards because the inner lining is difficult to peel.

Following evisceration, birds receive an inside and outside spray wash before proceeding through a chilling treatment of cold tap water or ice and water. Air-agitated cooling vats, a continuous immersion chiller, or air-spray chilling can be used.

After carcass cooling and draining in commercial waterfowl plants, each carcass is graded for conformation, fleshing, fat covering, and defects such as pinfeathers and exposed flesh. Processors then vacuum-package carcasses in barrier bags, heat-shrink the bags, and quick-freeze the packaged carcasses. Young ducks and geese also may be sold fresh, unfrozen. ***See*** Meat, Preservation

Composition and Nutrition

Dressing percentages and meat yield of ducks and geese will vary with breed, age, sex, weight, and grade. The processing loss from live to ready-to-cook (carcass, neck, and giblets) for ducks is approximately 30%, 25–32% for geese, and 24% for broilers.

As the data in Table 1 show, raw ducks and geese contain more fat, less water, and less protein than are found in broiler chickens. Roasting causes fat and some water loss, thereby concentrating (increasing) the protein content. Because of the relatively high fat content of waterfowl, the carcasses become rancid more easily, and frozen shelf life is shorter than that of broilers or turkeys. ***See*** Fats, Digestion, Absorption and Transport; Protein, Food Sources

Skinning the carcass can remove much of the fat in poultry. For example, the fat content of raw duckling meat with skin is 39·3%; without skin it is 6·0%. Values for goose are 33·6% with skin and 7·1% without skin. The fat in ducks and geese is highly unsaturated, as is the case in all poultry. Duck and goose skin (including separable fat) makes up 34–38% of the ready-to-cook carcass, but broiler skin is about 15% of the carcass. Conversely, the meat yield from ready-to-cook carcasses for ducks and geese is 34–47%, compared to 52% for broilers.

Duck is a good source of thiamin, and goose is an excellent source of phosphorus. The other vitamins and

Table 1. Proximate analysis of ducks, geese, and broilers (100 g edible portion, wet-weight basis)

Nutrient	Duck meat with skin (raw)	Duck meat with skin (roasted)	Goose meat with skin (raw)	Goose meat with skin (roasted)	Broiler meat with skin (raw)
Water (g)	48·50	51·84	49·66	51·95	65·99
Protein (g)	11·49	18·99	15·86	25·16	18·60
Energy (kJ)	1697	1415	1558	1281	903
Lipid					
Total (g)	39·34	28·35	33·62	21·92	15·06
Saturated (g)	13·22	9·67	9·78	6·87	4·31
Unsaturated (g)	23·77	16·55	21·53	12·77	9·47
Cholesterol (mg)	76	84	80	91	75
Ash (g)	0·68	0·82	0·87	0·97	0·79
Sodium (mg)	63	59	73	70	70

Source: Posati LP (1979).

minerals are also present in ample amounts. Iron content is especially high in waterfowl, contributing to the darker colour of the breast meat. ***See*** Phosphorus, Properties and Determination; Thiamin, Properties and Determination

Food Uses and Products

Most duck and goose meat is consumed in whole carcass form. The carcasses are roasted in the same way as other poultry, but they yield less cooked edible meat. Because waterfowl contain more fat than broilers or turkeys, they do not require basting during roasting. If the carcasses are excessively fat, it helps to puncture or scratch the skin to permit some fat to cook out during roasting. Thorough cooking of waterfowl (to an internal temperature of at least 85°C) is important to attain a crispy skin and for complete customer satisfaction.

Food manufacturers sell some further-processed waterfowl products. Duckling breast portions, semi-boneless halves, marinated breasts, fully cooked roast half duckling, and other gourmet items are available to the institutional trade. A few shops market speciality duck and goose products, such as smoked duck, boneless breast of duckling, smoked goose leg, or goose liver sausage.

Goose livers are popular in delicacies such as pâté de foie gras; the product must contain at least 30% goose liver. For a period of time in Europe and in parts of the USA, many people force-fed, or noodled, their geese. The starchy noodling ingredients (corn, wheat, barley, and rye) used in this frequent forced feeding produced an extra large liver (weighing nearly 1 kg each), which was sold at premium prices.

Goose fat is sometimes used in place of butter. It can also be used in frying, baking, cooking, and preparation of gravies, broths, and soups.

Waste Products

Duck and goose slaughtering produces the usual waste products of blood, feathers, feet, heads, inedible viscera, grease, debris, and cleaning water. The control and handling of these waste materials depends on the standards set by individual countries and local authorities. The methods of handling can include disposal to a public sewer after the fat and as many solids as possible have been removed, or treatment on-site and disposal of the filtered effluent to a lagoon or spray irrigation system.

In general, the first stage of any effluent system is the removal of coarse solids by screening. A fat trap or dissolved air flotation can then remove fine solids, fat, and grease. The effluent is further cleaned by either anaerobic digestion or aeration. ***See*** Effluents from Food Processing, Composition and Analysis; Effluents from Food Processing, Disposal of Waste Water

In contrast to other poultry processing waste treatment problems, ducks and geese provide several valuable by-products. The Far East is a good market for frozen duck feet, where they are stuffed with pork and considered a delicacy. Duck tongues are also valued by orientals as hors d'oeuvres. Some parts (heads and offal) go into mink or other animal food.

Other important by-products are duck and goose down and feathers, used chiefly by the bedding and clothing industries. The poultry slaughter plant rinses and centrifuges the wet feathers to decrease the water content prior to shipping them to a feather processing plant. A machine separates the down (15–25% of the feather mixture) and feathers, and washes and dries them. About five ducklings or three goslings are needed to produce 0·45 kg of dry feathers.

On a smaller scale, home processors can wash feathers in soft, lukewarm water which includes either a mild detergent or a little borax and washing soda. After rinsing, feathers are spread out to dry.

Microbiological Problems

In general, microbiological problems in ducks and geese are similar to those in other poultry. Pseudomonads are the main spoilage organisms. However, because of the very low duck and goose meat consumption, few illnesses have been attributed to waterfowl. The combination of scalding at 60°C, followed by immersion in molten wax at *c*. 90°C to aid final removal of feathers, appears to have a beneficial effect on the microbial quality of the finished product. ***See*** Spoilage, Bacterial Spoilage

Bibliography

Cunningham FE and Cox NA (1987) *The Microbiology of Poultry Meat Products*. Orlando: Academic Press.

Mead GC (1989) *Processing of Poultry*. Essex: Elsevier Science Publishers.

Mountney GW (1976) *Poultry Products Technology* 2nd edn. Westport, Connecticut: AVI Publishing.

Orr HL (1978) *Duck and Goose Raising*. Publication 532. Ontario: Ministry of Agriculture and Food.

Posati LP (1979) *Composition of Foods. Poultry Products, Raw, Processed, Prepared*. Agriculture Handbook 8-5. Washington, DC: Science and Education Administration, US Department of Agriculture.

Stadelman WJ, Olson VM, Shemwell GA and Pasch S (1988) *Egg and Poultry-Meat Processing*. Cambridge: VCH Publishers.

Arthur J. Maurer
University of Wisconsin, Madison, USA

Turkey*

Chickens and turkeys dominate the world poultry industry. Increasing popularity of poultry meat, including turkey, comes from low cost (value for money), a healthy nutritious image, and availability in a variety of convenient forms. Efficiencies in integrated turkey production, processing, and marketing have helped maintain favourable retail prices. This article reviews specific characteristics of turkey, processing of turkeys, food uses and products, microbiological concerns, waste products, and the nutritional profile of turkey meats and products.

* The colour plate section for this article appears between p. 3896 and p. 3897.

Specific Characteristics

Turkey consumption has increased in many countries throughout the world with Israel the leader (9·5 kg per person in 1986). The 1990 per capita turkey consumption in the USA was 8·2 kg, and the projected amount for 1991 is even higher, at 8·8 kg. Less than one third of this per capita consumption (2·7 kg) are whole-carcass birds. Clearly, the growth in turkey popularity has come in cut-up parts and more fully prepared products.

Turkey hens in the USA, sold at about 16 weeks of age, provide almost 6 kg of ready-to-cook carcass. Toms (male turkeys) are slightly older, at 20–24 weeks, and yield 10–12 kg of marketable carcasses. Desirable characteristics of turkeys include heavy bird weights without excess fat, a high dressing yield and an ample proportion of valuable parts. The breast (white meat) is 35–40% of the ready-to-cook carcass. Leg meat, 25–30% of the carcass, contains more myoglobin and is darker than breast meat.

Other desirable turkey meat attributes include tenderness, bland flavour (allows further processing and seasoning), and good functional properties, such as protein extraction, protein gelation, water holding, meat binding, and emulsification. Turkey meat is similar to broiler meat in its composition and qualities, easily digestible, high in protein, and low in fat.

Processing of Turkeys

Turkey processing refers to slaughtering, feather removal, evisceration, and chilling. Other operations, such as inspection and packaging, are also important steps in processing turkeys for market. Most of the processing procedures are similar to those used for broilers.

Procurement

Special trailers or trucks with built-in cages haul turkeys to the slaughter plant. If crates are used to transport turkeys, they are considerably higher than those used for chickens. Handlers must catch, load, and unload turkeys very carefully to minimize bruises and broken bones. Escalator-type loaders are sometimes used to elevate turkeys to the built-in truck cages. The turkeys should be off feed 8–12 h prior to slaughter to minimize faecal contamination in the processing plant. Large-volume, slow-moving fans or evaporative cooling provide good ventilation and comfort for the birds in holding sheds at the processing plant.

Because turkeys are large and heavy, shackle line height in the unloading area and truck cages must be adjusted to the same level to minimize lifting required by the hangers. Workers hang the turkeys by both feet to the overhead moving shackles.

A dimly lit hanging area discourages the birds from struggling and flapping. Some processors install a smooth plastic bar parallel to and slightly below the overhead conveyor line so that the turkey's breast rubs against it to provide a soothing effect. The turkeys should move from the hanging area to the stunner within 6 min.

Stunning

The shackle line drags the heads of the birds through 0·1–1·0% saline water which contains a submerged electrode. Careful control of the voltage and current is vital. Too little current will not immobilize the birds, and too much current can cause a violent muscle contraction, often resulting in broken clavicle fragments in the muscle tissue.

Slaughter

Killing is usually performed manually by cutting across the side of the neck at the base of the bird's head, severing a jugular vein and carotid artery. Bleed time should be at least 2 min in a blood tunnel to collect the blood for proper disposal. Workers in some turkey slaughter plants cut both sides of the neck for more complete bleeding. ***See*** Meat, Slaughter

Scalding

After bleeding, processors scald turkeys by immersion in agitated hot water to facilitate feather release. For some fresh chilled markets, they semiscald turkeys at 50–52°C to retain the epidermal cuticle which reduces skin dehydration. Most processors, however, scald turkeys at approximately 60°C for 2–2·5 min. The USA has a requirement for an overflow from the scald tank of approximately 1 l of water per bird to help float debris from the water.

Defeathering

After scalding, the shackle line carries the birds through a series of picking machines containing rotating rubber fingers on discs or drums which rub or pluck the feathers from the carcasses. Workers remove the remaining pinfeathers by hand.

The birds pass through a gas flame to singe off the filoplumes or hairs protruding from the skin surface. The final process on the kill line is a thorough washing of the external surface of the carcass with pressurized water jets.

The carcasses then pass through a shank cutting station. In some plants, an automatic tendon puller removes the shanks and pulls up to nine of the main sinews from the drumstick. At this point, a conveyor transfers the carcasses from the killing, scalding and defeathering area into another room for evisceration.

Evisceration

For evisceration, turkeys hang from both legs with their heads also placed in a centre slot of the shackle, creating a three-point suspension. This presents the birds horizontally, breast up, for easier cutting and eviscerating. Plants equipped with mechanical eviscerating equipment use the two-point leg suspension.

Evisceration begins with an incision made through the abdominal wall. The cut continues around the vent and enlarges in the abdomen for easier removal of the viscera. In some plants, a bar (transverse) cut in the abdomen allows evisceration and later trussing of the legs. A mechanical vent cutter with a revolving cylindrical blade is another method to make an opening cut and remove the vent.

Processors withdraw the viscera through the abdominal opening, taking special care to avoid damaging the intestines or spilling the contents. The viscera are left attached to the carcass and draped over the outside of the bird for inspection. An inspector then examines each bird (inside, outside, and viscera) for wholesomeness. Carcasses can be trimmed to remove damaged parts or dressing defects, or condemned if unfit for food.

Workers harvest, clean, and save the giblets (heart, liver, and gizzard). They clean the gizzard by splitting it, washing away the contents, peeling off the hard lining, and washing. They discard the inedible viscera, and remove the lungs with a special tool or vacuum lung gun.

As one of the last operations, the head is cut or pulled off, and the oesophagus and crop are removed from the front of the carcass. Finally, the skinless neck is cut off, washed, and retained for packing. Following evisceration, the birds must be washed inside and out before chilling.

It is possible to perform some of the slaughter and evisceration operations mechanically, but the variability in turkey carcass sizes presents unique problems. Fully automated equipment for turkey processing is still being developed.

Chilling

Most processors chill poultry meat in cold water or water and ice, but they use air chilling extensively in Europe, where carcasses are more often soft-scalded and sold fresh. The most-used chillers in the USA drop the birds from the evisceration line into a prechiller which also serves as a very effective washer. The agitated water has a temperature of less than 18°C. The birds travel to a second chiller with a water temperature of less than 2°C. Inspectors carefully monitor chiller water overflow, and carcass exit temperature and water uptake.

After chilling, graders evaluate the carcasses according to conformation, fleshing, fat covering, and defects such as pinfeathers and exposed flesh. *See* Meat, Preservation

Packaging

In some plants, turkey carcasses move directly to a cut-up or deboning line. If turkeys are sold as whole carcasses, workers place a giblet packet in the crop cavity and a neck in the body cavity. Processors truss the legs into the bar cut, or use a metal or plastic hock lock. These turkeys will probably be frozen, in which case they are placed in an oxygen-impermeable, shrink-film bag. After the air is evacuated, the bags are clipped shut and passed through a hot-water shrink tunnel prior to freezing.

The freezing process is very rapid, with the first step being a brine or blast freezer to set the carcass surface colour. Final freezing and storage occur in a holding freezer, where the birds can be stored for relatively long periods of time until they are moved into market channels.

Food Uses and Products

About 75% of the turkeys produced in the USA are cut up or further processed. In some countries, such as Israel, France and Italy, the proportion of cut-up or further-processed turkey is 90% or more. However, whole birds are still popular at holiday times. Some of the whole carcasses are injected with various flavoured solutions to create a self-basting, juicier carcass when roasted.

Deboning

Deboning of chilled turkey carcasses usually occurs as they hang from special shackles on a slow-moving line, or sit on deboning cones which may be static or moving. Precise cuts are made to remove parts, breast meat, thigh meat, and trimmings from the skeleton. The drumsticks (and also the thighs) are often deboned on automated machines, whereby the drumstick bone may be pushed lengthwise out of the meat; thigh pieces may be compressed in a specially designed mould, squeezing the meat away from the bone. More deboning automation will occur in the near future.

After removal of intact muscles using knives or semiautomated procedures, the remaining frames, necks, and backs are usually mechanically deboned. One system uses an auger principle, pressing the meat and bone against a perforated cylinder screen or microgrooved cylinder. The alternative system is a hydraulic-press-type batch design which squeezes the meat through a series of stationary filter rings. Depending on the incoming meat materials (meat:bone ratio), screen perforation size, and machine adjustments, the mechanically deboned meat can be a very fine purée or have a particle size of 5 mm, ideal for some sausages.

Further Processing

Hand- and mechanically-deboned meats are useful in many further-processed products. The term 'further processing' encompasses procedures such as deboning, size reduction, injection, tumbling, massaging, reforming, and emulsifying. The principles of water-holding, protein extraction, protein gelation, and meat binding are very important in making further-processed products. Other processes, such as battering, breading, cooking, and freezing, may take place. Further processing also refers to whole carcasses that are basted, marinated, or smoked.

The types of products now available are increasing, ranging from cut-up portions to reformed roasts, breasts, rolls, steaks, hams, burgers, frankfurters, bolognas, coarse-ground sausages, salamis and bacons. More recently, ready-prepared meals have utilized an increasing amount of turkey meat. *See* Meat, Sausages and Comminuted Products

Consumers with changing careers and households are the driving force behind trends in the consumption of poultry products. Smaller families often lack time but can afford the convenience of further-processed poultry. Diet and health, microwave ovens, and an ageing population are important factors in new product development. Ethnic foods, especially Italian, Mexican, and Chinese, are popular in the USA. Different flavours and marinades, such as honey and barbecue, are well liked in turkey products. Smaller portions, less fat, less salt, and low cholesterol are in vogue. Poultry appetizers, finger foods, sliced luncheon meats, and centre-of-the-plate, cooked, vacuum-packaged, chilled boneless breasts are very popular. New turkey products could soon include *sous vide* (vacuum-sealed, cooked, refrigerated, ready-to-heat) entrées, or possibly surimi (minced washed gelled protein, usually from fish) seafood-style favourites.

Microbiological Concerns

Turkey meat quality is highest immediately after processing. The maintenance of acceptable quality depends on initial microbial levels and measures taken to minimize organism growth and prevent further contamination. The two major concerns are spoilage organisms which cause odours or off-flavours, and pathogenic organisms which may, under faulty handling such as undercooking or temperature abuse, lead to a health hazard. The cutting, deboning, handling, mixing, and packaging of turkey meat also increases possible microbial contamination and growth. ***See*** Spoilage, Bacterial Spoilage

Although turkey and turkey products are refrigerated or frozen for shelf life extension, spoilage can occasionally occur as a result of the growth and metabolic activities of specific types of bacteria. Psychrophiles, such as *Pseudomonas*, can grow at refrigerator temperatures and cause problems. Turkey has reached spoilage conditions when an off-odour becomes apparent (10^7 cells per cm^2) or when slime formation occurs (10^8 organisms per cm^2). It is important to maintain refrigeration temperatures at 0–4°C to minimize microbial growth in turkey and turkey products.

Food pathogens are more serious than spoilage organisms because the food product may not look or smell spoiled. Some of the pathogens of concern in turkey are *Salmonella*, *Staphylococcus*, *Campylobacter*, *Listeria*, and coliforms. Recently, salmonella has been a major worry for consumers, and turkey carcasses do harbour the organism. However, a 1973–1987 survey in the USA has shown that of 790 foodborne salmonellosis outbreaks, only 36 were caused by turkey. To prevent foodborne illness from poultry, it must be kept refrigerated, and cooked properly, and cross-contamination or post-cooking contamination from unclean utensils or equipment must be avoided. ***See*** Campylobacter, Campylobacteriosis; *Listeria*, Listeriosis; *Staphylococcus*, Food Poisoning

Bacteria can come from many sources. At the processing plant, bacteria arrive on the feet and feathers of the birds; they are present in the intestinal contents, and can also come from the workers and the environment (air and water supplies). Bacterial problems can be minimized by following good production and manufacturing practices such as feeding clean feed and keeping the litter dry at the production site, using clean hauling equipment, filtering incoming air at the processing plant, monitoring the water supply, eviscerating carefully, chlorinating chiller water, insisting on good worker hygiene, and using an approved plant clean-up and sanitation programme.

Utilization of Waste Products

Poultry processing results in large amounts of highly polluting waste waters, semisolids, and solids, which must be separated and treated before being discharged into the environment. Where practical, the use of waste products for livestock food or fertilizer reduces the overall load for disposal. Rendered poultry by-products as an animal feed ingredient can provide 50–60% protein. Feather meal and dried blood also have value as a feedstuff. Such products must be carefully processed before being recycled as animal feed, to avoid microbial contamination.

The types of waste and by-products differ at varying stages of processing. Manure, feathers, blood, viscera, flesh debris, grease, and cleaning water are examples of the pollutants to be treated and either used or discarded.

The methods of disposal are to a public sewer, or treatment on-site followed by disposal to a water course or to fields. Preliminary treatment of turkey processing wastes using a coarse-solids screen separator as well as a fatty-matter trap, or chemical flocculation combined with a dissolved air flotation system, will significantly reduce the pollution potential of the effluent before discharge or further biological treatment. Based on the strength of the effluent, a secondary biological treatment uses a mixed culture of microorganisms for anaerobic digestion or aerobic treatment. The cheapest and most cost-effective options for the disposal of stabilized processing sludges arising from the biological treatment of poultry processing effluents are to spread them on the land or to discharge them at land-fill sites. ***See*** Effluents from Food Processing, Composition and Analysis; Effluents from Food Processing, Disposal of Waste Water

Nutritional Significance

Table 1 shows the nutrient composition of selected raw and cooked turkey meats, and Table 2 lists the composition of several further-processed turkey products.

Turkey and its products have a favourable reputation as nourishing and healthy foods. The composition of raw turkey meat depends on factors such as diet, age, sex, and growth environment. Processed product nutrition (Table 2) is a result of incoming meat and nonmeat ingredients as well as cooking processes and formula variations by different manufacturers. Salt and fat can vary markedly, depending on whether salt was needed for protein extraction and binding, or whether skin (with adhered fat) was added for juiciness and flavour.

Table 1. Proximate analysis of turkey (per 100 g edible portion, wet-weight basis)

Nutrient	Breast with skin (raw)	Breast with skin (roasted)	Leg with skin (raw)	Leg with skin (roasted)	Light meat only (raw)
Water (g)	70·05	63·22	72·69	61·19	73·82
Protein (g)	21·89	28·71	19·54	27·87	23·56
Energy (kJ)	659	794	605	874	483
Lipids					
Total (g)	7·02	7·41	6·72	9·82	1·56
Saturated (g)	1·91	2·10	2·06	3·06	0·50
Unsaturated (g)	4·32	4·25	3·89	5·59	0·69
Cholesterol (mg)	65	74	71	85	60
Ash (g)	0·91	1·03	0·89	0·99	1·00
Sodium (mg)	59	63	74	77	63

Source: Posati LP (1979).

Table 2. Proximate analysis of selected turkey products (per 100 g edible portion, wet-weight basis)

Nutrient	Ham	Roll (light)	Salami	Frankfurter	Loaf (breast)
Water (g)	71·38	71·55	65·86	62·99	71·85
Protein (g)	18·93	18·70	16·37	14·28	22·50
Energy (kJ)	538	617	823	949	462
Lipids					
Total (g)	5·08	7·22	13·80	17·70	1·58
Saturated (g)	1·70	2·02	—	—	0·48
Unsaturated (g)	2·67	4·24	—	—	0·73
Cholesterol (mg)	—	43	82	107	41
Ash (g)	4·23	2·00	3·42	3·53	4·18
Sodium (mg)	996	489	1004	1426	1431

Source: Posati LP (1979).

Consumers recognize turkey as a good protein source. Turkey meat is easily digestible, contains all the essential amino acids, and is a good source of the B vitamins and iron. ***See*** Amino Acids, Metabolism; Vitamins, Overview

Turkey has a relatively low fat content, and the fat is only about 30% saturated. Because the fat is highly unsaturated (more than beef and pork), it is a softer type of fat and prone to oxidation. As turkeys grow, they deposit more fat under the skin. The fat content is higher in dark meat, and protein content is greater in light meat. Moisture and some fat are lost during heating, and protein is concentrated in cooked turkey.

Turkey products are somewhat similar to other processed meat products in protein content but are higher in moisture and lower in fat and energy. This favourable nutritional profile for turkey is one of the major reasons for its increasing popularity. ***See*** Protein, Food Sources

Bibliography

Bean NH and Griffin PM (1990) Foodborne disease outbreaks in the United States, 1973–1987: pathogens, vehicles, and trends. *Journal of Food Protection* 53(9): 804–817.

Cunningham FE and Cox NA (1987) *The Microbiology of Poultry Meat Products*. Orlando: Academic Press.

Mead GC (1989) *Processing of Poultry*. Essex: Elsevier Science Publishers.

Mountney GW (1976) *Poultry Products Technology* 2nd edn. Westport, Connecticut: AVI Publishing.

Nixey C and Grey TC (1989) *Recent Advances in Turkey Science*. Oxford: Butterworth–Heinemann.

Posati LP (1979) *Composition of Foods. Poultry Products, Raw, Processed, Prepared.* Agriculture Handbook 8-5. Washington, DC: Science and Education Administration, US Department of Agriculture.

Stadelman WJ, Olson VM, Shemwell GA and Pasch S (1988) *Egg and Poultry-Meat Processing*. Cambridge: VCH Publishers.

Arthur J. Maurer
University of Wisconsin, Madison, USA

POWDERED MILK

Contents

Milk Powders in the Market Place

Today, a wide variety of milk-based powders with well-defined physical, functional and compositional characteristics are available for use by the food and associated industries. This article will deal mainly with the development, manufacture and use of whole-milk powder (WMP; also known as dry whole milk or full cream milk powder) and skim milk powder (SMP or nonfat dry milk), the most widely used dried milk products. The following article deals with the characteristics of milk powders; other dried milk-based products are covered under specific product entries. *See* Casein and Caseinates, Methods of Manufacture; Infant Foods, Milk Formulas; Lactose; Whey and Whey Powders, Production and Uses

Types of Powder

Originally produced in response to the desire to preserve the nutrients in milk and the economic need to utilize further the by-products of processing, e.g. skim milk and buttermilk (BM) from buttermaking, milk powders are now prized for their functional, nutritional and organoleptic characteristics in food applications and form a major component of the food proteins market. Our growing understanding of the function and importance of individual milk components coupled with advances in processing technology and equipment has resulted in a growing range of 'tailor-made' milk powders being available to the food industry. WMP and SMP, however, remain the most important products in terms of production and usage, although there is a continual search for cheaper protein alternatives with comparable or better functional characteristics.

WMP and SMP are available in both roller-dried and spray-dried forms, the latter being the most widely preferred. Spray-dried milk powders are available in two types, viz. ordinary or nonagglomerated (non-instant) powders and instant or agglomerated powders. Depending on their composition, ordinary milk powders generally consist of small, single particles of high bulk density. Ordinary SMP also tends to be rather dusty. Reconstitution of ordinary powder is difficult because the powder particles tend to clump together on the surface of the reconstituting liquid and have poor wettability. In the case of ordinary WMP, the wettability problem is compounded because of 'free fat' forming a hydrophobic film on the surface of the powder particles.

The purpose of instantizing milk powders is to enhance their reconstitution properties in cold liquids by improving one or more of the following powder properties: wettability, sinkability, dispersibility, rate of hydration, and solubility. The basis of instantizing is the formation of porous aggregates of dried milk particles by a process of agglomeration. These porous agglomerates, when placed in contact with the reconstituting liquid, facilitate the wetting of individual powder particles, which subsequently sink into the body of the liquid, disperse and finally dissolve. The extent of the 'instant' character of the powder depends on the degree of agglomeration, which in turn depends on the nature of the product and the equipment and process used.

Depending on the level of agglomeration, the instantizing process generally causes a significant reduction in the bulk density of the powder, resulting in increased packaging, storage and transport costs. The heat stability of the powder is also affected. For powders which are destined for reconstitution in hot beverages, such as coffee or tea, instant character is of less concern than the heat stability of the powder. The combination of high temperature and the acidity of the tea or coffee will, if the powder is not heat stable, cause coagulation of the milk protein, giving rise to the condition known as 'feathering' in which particles of coagulated protein are visible in the beverage.

Commercial SMP and WMP are required to meet a number of national or international standards similar to those adopted by The Food and Agriculture Organization (FAO) of the United Nations, the International Dairy Federation (IDF) or the American Dairy Products Institute (ADPI). These standards are used to grade powders, having due regard for the drying method used, and to define the powder in terms of selected components, usually moisture and fat, and chemical, microbiological and sensory requirements. These specifications seek to ensure freedom from adulteration,

Table 1. Typical composition (g per 100 g) of milk powders

Component	WMP	Instant SMP	Ordinary SMP	Sweet cream BMP
Moisture	2·5	4·0	3·2	3·0
Milk fat	26·7	0·7	0·8	5·8
Protein	26·3	35·1	36·2	34·3
Lactose	38·4	52·2	52·0	49·0
Ash	6·1	8·0	7·9	7·9

From Wong NP, Jenness R, Keeney M and Marth EH (eds.) (1988) *Fundamentals of Dairy Chemistry*, 3rd edn. New York: Van Nostrand Reinhold.

undue microbial contamination and defects, and that the powder has the correct composition. Internationally recognized standards, such as those defined by ADPI, which are based on specific and detailed analytical procedures, provide an objective and uniform measurement of powder quality and facilitate international trade in these products. Data on the typical composition of milk powders are given in Table 1, and the ADPI specification for extra grade powders is shown in Table 2. ***See*** Adulteration of Foods, History and Occurrence; Contamination, Types and Causes

Milk powders are also frequently classified on the basis of the heat treatment which the milk has received prior to evaporating and drying. This heat classification provides an indication of the suitability of the powder for specific end uses, such as recombining, and is traditionally based on measurement of the undenatured whey protein in the powder (expressed as mg per g powder) and indicated by the Whey Protein Nitrogen Index (WPNI). Seasonal variations in the level of whey protein limit the usefulness of the WPNI classification, particularly in the classification of low-heat powders, and alternative classification schemes such as the Casein or Heat Number or the Cysteine Number have been proposed. These heat classification schemes, together with typical temperature–time regimes used to achieve the desired degree of heat treatment, are given in Table 3. ***See*** Whey and Whey Powders, Protein Concentrates and Fractions

Membrane technology, and ultrafiltration (UF) in particular, has been widely accepted by the dairy industry for the processing of milk. The use of UF, either alone or in conjunction with diafiltration, to concentrate and modify the composition of milk prior to drying has made it possible to produce an almost infinite array of skim and whole milk powders 'tailor-made' for specific applications. Such powders include lactose-reduced or lactose-free powders for use with lactose-intolerant populations, and high-protein or retentate powders for use in recombined cheesemaking or as a food ingredient or nutritional supplement. The composition and properties of these powders vary markedly from those of normal milk powders and will depend on the concentration factor and/or the number of diafiltration steps employed during manufacture. UF/diafiltration technology can also be used to produce high-protein powders containing both casein and whey proteins, with protein levels approaching those of caseinate or coprecipitate but without the requirement of isoelectric precipitation or pH adjustment in combination with heat treatment and their concomitant adverse effect on protein composition and properties. ***See*** Membrane Techniques, Principles of Ultra-filtration

Selective or gross manipulation of the mineral levels in milk can be effected by the application of techniques such as electrodialysis, ion exchange, ultrafiltration or diafiltration prior to drying. This has allowed the

Table 2. ADPI specifications for extra grade milk powders. All figures are maximum permitted except where indicated

	Spray-dried			Roller-dried	
	WMP	Instant SMP	Ordinary SMP	WMP	SMP
Milkfat (g per 100 g)	min 26·00	1·25	1·25	min 26·00	1·25
Moisture (g per 100 g)	4·50	4·50	4·00	4·50	4·00
Titratable acidity (%)	0·15*	0·15	0·15	0·15*	0·15
Solubility index (ml)	1·00	1·00	1·25	15·00	15·00
Bacterial estimate (per g)	50 000	30 000	50 000	50 000	50 000
Coliform (per g)	10	10	90	10	90
Scorched particles (mg)	15·0	15·0	15·0	22·5	22·5
Dispersibility (%)	–	min 85·0	–	–	–
Copper (mg per kg)	1·5*	–	–	1·5*	–
Iron (mg per kg)	10·0*	–	–	10·0*	–

From *Standards for Grades of Dry Milks Including Methods of Analysis* (American Dairy Products Institute, Bulletin 916 (Revised), 1990).
* Optional tests.

Milk Powders in the Market Place

Table 3. Heat classification of milk powders

Heat classification	WPNI[a] (mgN per g powder)	Casein number[b] (%)	Cysteine number[c] (%)	Heat treatment (°C, time in min)
Extra low heat	–	–	24–31	72, 0·25
Low heat	⩾6·0	⩽80	32–38	72, 0·25–1
Medium heat	5·99–1·51	80·1–83·0	39–48	82, 2–5
Medium high heat	–	83·1–88·0	49–62	82, 12–15
High heat	⩽1·5	⩾88·1	>62	82, 30

[a] Data from ADPI Bulletin 916.
[b] Data from provisional IDF Standard 114: 1982.
[c] Data from IDF Experts Group E17.

development of powders with defined mineral profiles that meet specific nutritional requirements, e.g. low-sodium powders. Similarly, a range of lactose hydrolysed milk powders can also be prepared by enzymatic hydrolysis of the lactose in milk with β-galactosidase prior to drying. The milk fat level can also be manipulated to produce a range of powders of varying fat content.

Production of Milk Powders

A schematic diagram for the manufacture of milk powders is given in Fig. 1. The removal of water from milk to produce a powder involves two discrete operations. The first involves the production of a concentrate of 40–50% total solids (TS) content, the final TS level being determined by the nature of the starting material and the type of atomizer and drying process to be used, e.g. whole milk, nozzle atomizers and one-stage drying systems generally use concentrates of lower TS than do skim milk, centrifugal atomizers or multistage drying systems. Concentration is generally achieved by vacuum evaporation, although concentrates can be prepared by membrane processing techniques, either alone or in conjunction with evaporation, as occurs in the preparation of retentate powders. The evaporation step removes around 90% of the water in the milk.

The remaining moisture is removed in the second or drying operation to produce a powder of 2–5% moisture content, the final moisture depending on the nature of the product and/or buyer specifications. Commercially drying is effected by spray drying or roller drying, the former being both the most widely used and the method of choice, particularly where high powder solubility and high throughput are required. The spray drying process may be either a single-stage or a multistage (spray plus fluidized-bed drying) process using nozzle or centrifugal atomization. The drying conditions used will depend on the nature of the product, the method of atomization and the type of drying process. Typically one- and two-stage spray driers are operated at outlet temperatures of 75–85°C and at inlet temperatures of 180–200°C and 180–230°C and concentrate feed solids of 42–48% and 45–50% for WMP and SMP or BMP, respectively. Higher inlet temperatures (250–300°C) can be used for SMP in three-stage drying systems without impairing powder quality. *See* Drying, Theory of Air Drying; Drying, Spray Drying

One of the most critical operations in the production of milk powders is preheating or forewarming, during which many of the physical, chemical and functional properties, and hence potential end uses, of the final powder are established. This is achieved primarily through the controlled thermal denaturation of the whey proteins prior to concentration. Typical time–temperature regimes used and the corresponding heat classifications and levels of whey protein denaturation achieved are given in Table 3. In the production of high- and medium-high heat powders, HTST preheating regimes (e.g. 120°C, 1–2 min) and direct steam injection are frequently used to reduce holding capacity requirements. Properties affected by preheating include colour and flavour; mineral and nitrogen distribution; pH; bacteriological quality; use in cheesemaking; viscosity on reconstitution; heat stability and baking properties and resistance to oxidation. *See* Pasteurization, Principles; Protein, Interactions and Reactions Involved in Food Processing

In the manufacture of WMP, the milk must be standardized to a fat:solids-not-fat (SNF) ratio of 1:2·76 prior to evaporation to ensure that the final powder meets the legal requirement of a minimum content of 26% milk fat. Whole milk is normally given a high preheat treatment to inactivate lipases and to develop natural antioxidant activity through the formation of sulphydryl groups, thereby improving the keeping quality of the powder. Two-stage homogenization of the concentrate immediately prior to drying reduces the free fat content of the powder and further improves powder stability. Ordinary WMP should also be cooled as soon as possible after drying to minimize free fat. *See* Antioxidants, Natural Antioxidants

Milk Powders in the Market Place

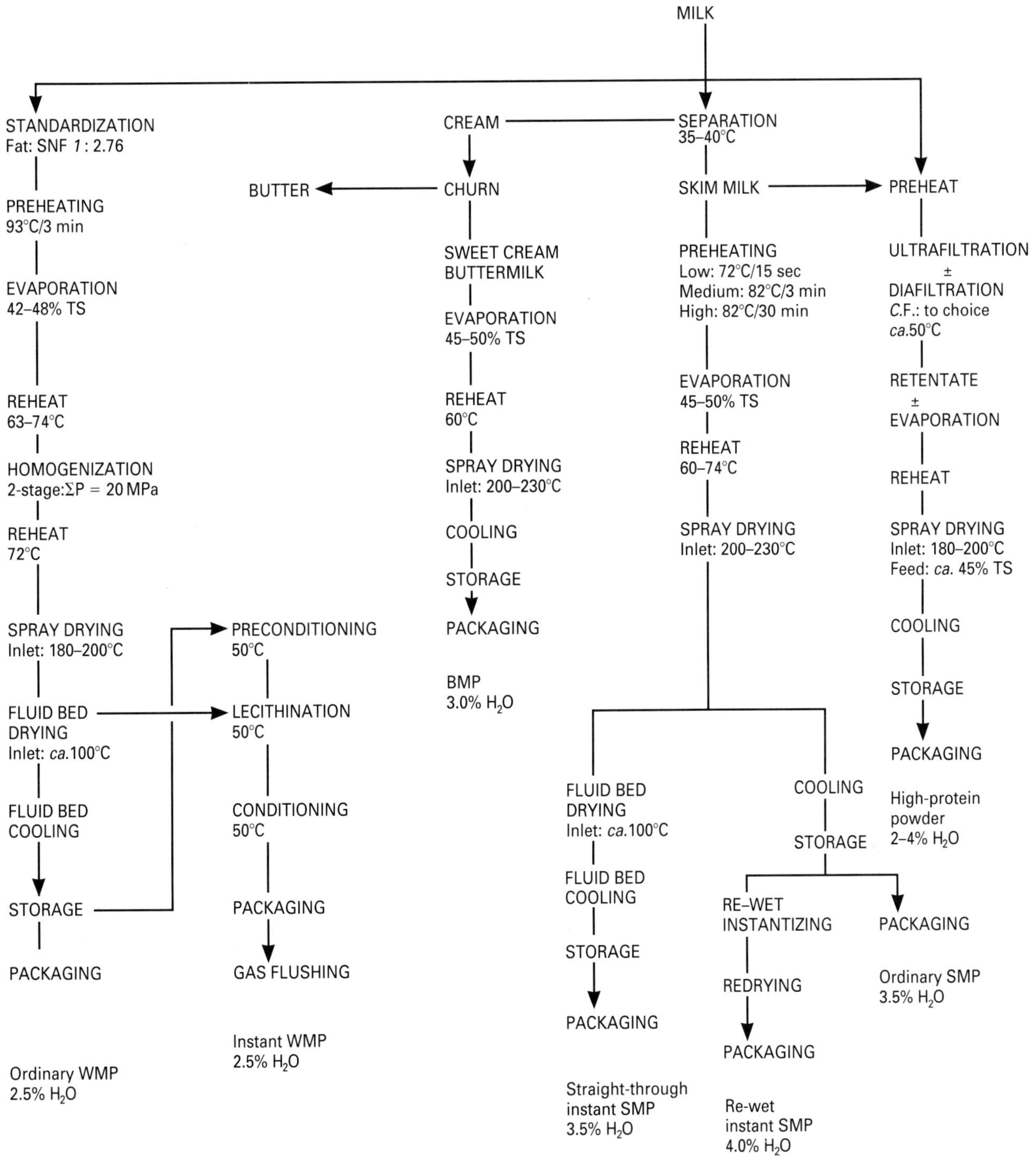

Fig. 1 Schematic diagram of the production of spray-dried milk powders.

Instant SMP is generally produced by either a re-wet agglomeration process which forms large agglomerates and thus imparts a high degree of instant character, or more commonly by a 'straight-through' instantizing process, with or without fines return, providing low-level agglomeration. As the re-wet process involves a further heat treatment of the milk solids, a low-heat powder should be used to avoid cooked flavour production in the redried powder and to minimize agglomerate shattering, as is typically found with agglomerates formed from high-heat powders.

Although agglomerating the powder particles helps to make WMP instant in warm water (>45°C), it is also necessary to improve the wettability of the powder to provide cold-water instant properties. This is most commonly achieved by spraying a warm (70°C) 1:1 mixture of the natural surfactant lecithin and butter oil onto the powder to provide a final lecithin content of

0·2% in the powder. Lecithination may be done in-line or as a separate operation, but in either case the base powder should have high particle density and minimum free fat content (< 4% of total fat).

The storage stability of milk powders is directly related to the moisture content of the powder. Deteriorative changes such as Maillard browning, oxidation and free fatty acid formation are at a minimum at moisture contents of 4% or less for SMP and 2·5–3% for WMP. For WMP, decreasing the moisture content below 2·5% reduces the stability to oxidation owing to removal of the protective monomolecular moisture layer. Increases in moisture content and/or storage temperature accelerate these deteriorative changes with a resultant loss of shelf-life and change in functional behaviour of the powder. *See* Browning, Nonenzymatic; Oxidation of Food Components; Storage Stability, Parameters Affecting Storage Stability

As milk powders are hygroscopic, moisture uptake must be controlled by the use of moisture-proof packaging. Commercial SMP is normally packed in 25-kg multiwall paper bags with an inner polyethylene liner bag or in plastic-lined corrugated paper board or metal tote bins. Similar packaging can be used for bulk WMP, while inert-gas-flushed metal containers are used for consumer packs.

SMP with moisture levels less than 4% is relatively stable for 12–18 months at temperatures up to 20°C. At storage temperatures around 30°C, however, there is a significant reduction in powder solubility, pH, flavour, available lysine, heat stability and cheesemaking characteristics and an increase in viscosity characteristics, thereby reducing the usefulness of the powder. Increases in powder moisture can have a similar adverse effect during extended storage.

WMP, like SMP, must have a low moisture content to minimize deterioration on storage. However, even at low moistures, fat oxidation limits the effective shelf life of WMP to approximately 6 months at normal temperatures. Elevated storage temperatures and exposure of the powder to light will further reduce this. Preheating at elevated temperatures to liberate natural antioxidants in the milk and inert gas packing are used to help control oxidative changes in the powder. Inert gas packing is used particularly in instant WMP production, where hot lecithinated powder is packed directly into metal cans and the oxygen in the headspace is reduced to a maximum of 2% by evacuation and gas flushing. Inert gas packing can also be used for bulk WMP storage in tote bins.

Applications in the Food Industry

Milk powders have long been used as ingredients by the food industry because of their nutritional, functional and organoleptic properties. The functional attributes of milk powders are essentially those of the milk proteins and include water absorption and binding properties, solubility, gelation, emulsification and foaming characteristics and viscoelasticity, while the organoleptic properties are primarily a function of the milk fat component. Because of these innate properties, milk powders have applications in the following areas. *See* Protein, Functional Properties

Confectionery Products

The functional properties of interest in confectionery are viscosity, gelation, emulsification and fat binding. The milk protein is essential for the development of flavour and colour of items such as toffees, caramels, nougats, fudges and milk chocolate. Milk proteins aid the emulsification and mixing of ingredients and influence mix viscosity. During cooking and subsequent processing of the mix, the milk proteins unfold and cross-link through S–S bonds to form a viscoelastic network which imparts structure and texture to the product. The firm, chewy texture of several confections, such as candy, is related to the binding of water by casein. Casein also prevents ‘cold flow’ in toffees.

WMP is used in milk chocolate at 12–25% by weight of chocolate and contributes to the colour, flavour and nutritional value of the chocolate. Flavour compounds contributed by the WMP include aldehydes, ketones and other carbonyls formed by autoxidation and lipolysis of milk fat. Milk fat is perhaps the most functional component of WMP in chocolate as it contributes to the smooth flavour and texture of the chocolate, aids in preventing ‘bloom’ and readily mixes with the cocoa butter without altering its properties. Roller-dried WMP is more functional than spray-dried WMP in chocolate because it has a much higher (*c.* 90%) level of free fat than spray-dried WMP (3–6%). Consequently, when milk chocolate is made from spray-dried WMP, additional expensive cocoa butter has to be used to achieve the same final product quality. Browning reactions involving the lactose component of the powder also contribute to the flavour and colour of the products. *See* Cocoa, Chemistry of Processing

Bakery Products

The functional properties of interest in baking are water binding, foaming and emulsification. The addition of SMP increases the water absorption capacity of the dough roughly in direct proportion to the amount of SMP added. High-heat SMP is used, particularly for breadmaking, as low-heat SMP reduces the dough extensibility and gives a poor loaf volume due to the presence of a volume-depressant factor in low-heat powders. In general, the addition of SMP to bread

Milk Powders in the Market Place

improves crumb texture, flavour and product shelf life. This is attributable to the water-binding properties of the milk proteins. Continuous breadmaking processes generally use SMP at levels below 2% as higher levels produce weak doughs and bread with poor loaf volume. The inclusion of SMP in cake mixes improves foam structure and texture. The addition of milk powders to bakery products also improves crust colour owing to browning. BMP is also added to bakery products to improve their flavour and shelf-life. ***See*** Bread, Breadmaking Processes

Meat Products

High-heat SMP has been used for many years in the manufacture of comminuted meat products to enhance emulsification, water binding, texture, colour and flavour characteristics. While the SMP has a positive effect on flavour, the performance of the milk protein is not optimum for emulsifying purposes as it is in a micellar form. Furthermore, the solubilization of the myofibrillar meat proteins, which is essential for the stabilization of finely comminuted products, is inhibited by the ionic calcium from the added SMP. Consequently, there has been a trend towards using soluble milk protein isolates to help overcome these problems. Calcium-reduced SMP can be prepared by ion exchange or membrane techniques and has been used successfully in meat products. ***See*** Meat, Sausages and Comminuted Products

Dairy Foods

The dairy industry itself is one of the larger users of milk powders, with SMP being used in the manufacture of products such as ice cream, cultured milk products, modified milks, cottage cheese, filled products and low-calorie spreads. A major and rapidly expanding use of milk powders is in the recombination of milk and milk products. Considerable research has been undertaken in identifying the requirements and properties of milk powders for specific recombining applications. For example, powders for recombined evaporated milk production must be heat stable to withstand in-can sterilization regimes. This is achieved by giving the milk a high preheat treatment prior to drying. Similarly, powders for recombined sweetened condensed milk should be medium-heat classified in order to meet final product viscosity requirements. It should be noted that while the powder heat classification gives an indication of its suitability for a particular end use, this should be verified by appropriate functionality testing. A wide array of recombined products can now be satisfactorily prepared, including a variety of cheeses. Of particular interest in relation to recombined cheesemaking is the use of retentate or high-protein powders. The use of the UF process opens up the possibility of producing a range of powders specifically designed for cheesemaking. Although reconstitution of WMP to prepare a range of fat-containing products is feasible, recombination of SMP and anhydrous milk fat is still favoured for these products at present because of oxidative flavour problems and costs associated with the use of WMP. ***See*** Recombined and Filled Milks

Future Directions

Although SMP and WMP will continue to be used as functional ingredients by the food industry, there is a clear trend towards finding cheaper alternatives with comparable performance. Considerable research effort is currently being directed towards whey and plant proteins to this end. Increasing knowledge and understanding of milk and its components and the potential of membrane and associated processing technologies is likely to see a continuing dissection of milk to produce more specialized powdered ingredients for the food industry.

Bibliography

Hansen R (ed.) (1985) *Evaporation, Membrane Filtration and Spray Drying in Milk Powder and Cheese Production.* Copenhagen: North European Dairy Journal.

International Dairy Federation (1990) *Recombination of Milk and Milk Products*, Special issue No. 9001. Brussels: International Dairy Federation.

Kyle WSA and Rich BR (1986) *Principles of Milk Powder Production.* Melbourne: Victorian College of Agriculture and Horticulture.

Robinson RK (ed.) (1986) *Modern Dairy Technology*, vol. 1, *Advances in Milk Processing.* Barking: Elsevier Applied Science Publishers.

W.S.A. Kyle
Victoria University of Technology, Werribee, Australia

Characteristics of Milk Powders

Advances in evaporation and drying technology and equipment design during the past few decades have meant that today's manufacturer is able to produce an array of milk powders with well-defined characteristics. These can be divided into general characteristics which define the powder in terms of compositional and quality parameters and are traditionally identified by the powder specifications, and specific characteristics which define the powder in terms of its physical and functional properties and defects and are primarily influenced by processing technology. This article will review the more important powder characteristics, together with some chemical and microbiological aspects of milk powders. *See* Drying, Theory of Air Drying

Physical and Functional Characteristics

The manufacture of milk powders to defined specification requires an understanding both of the powder characteristics themselves and of the factors affecting them. Many of the powder characteristics are a function of a number of interacting factors (Table 1), while some characteristics are influenced by other characteristics. Powder-specific characteristics can be divided notionally into physical and functional properties. The physical characteristics define the structure and appearance of the powder and include bulk density, particle density and size distribution and faults such as scorched particles and flecks. The functional characteristics define the powder in terms of its suitability for particular applications and include instant properties, heat classification and thermal stability, hygroscopicity, flowability, solubility and free fat. The distinction between physical and functional characteristics is not clear-cut; for example, bulk density, while primarily a physical characteristic is also an important functional one in relation to instant properties. Powder-specific characteristics, such as solubility index and scorched particles, are often specified in powder standards, e.g. ADPI (American Dairy Products Institute) specifications, while others are included as buyer specifications. *See* Physical Properties of Foods

Table 1. Factors affecting powder characteristics

Factor	Main characteristics affected
Raw milk quality	Solubility index, heat stability and viscosity
Milk composition	Particle density, bulk density, heat classification and hygroscopicity
Preheating	Heat classification, heat stability, viscosity and bulk density
Degree of concentration	Solubility index, bulk density and particle size distribution
Drying conditions	Solubility index, scorched particles, free fat, free moisture, bulk density and particle size distribution
Type of processing equipment used	Solubility index, scorched particles, free fat, instant properties, hygroscopicity and bulk density

Characteristics of Milk Powders

Particle Size Distribution

This is a measure of the average powder particle diameter and of the spread of sizes on either side of the average. It is generally not a grading requirement but is of importance in relation to reconstitution properties, flow characteristics and appearance. It is primarily influenced by the atomization process, with the average particle size being decreased by increasing the speed of rotation of the wheel of centrifugal atomizers, reducing the concentrate feed rate to the wheel, or by increasing the feed pressure in nozzle atomization. It is also decreased by lowering the concentrate viscosity and by lowering the temperature differential between the droplet and drying air during spray drying. Roller-dried particles tend to be irregular in shape and give a broader size distribution. *See* Drying, Spray Drying

Bulk Density, Particle Density and Microstructure

Bulk density is a measure of the mass of powder which will fit into a specific volume under defined packing conditions. It is generally expressed in g ml^{-1}. Economically it is an important property as it influences transportation and packaging costs and high-bulk-density powders are preferred in this regard. Functionally it is important in relation to the instant characteristics of powder, with agglomeration causing a significant reduction in bulk density.

The bulk density is a very complex property and is influenced primarily by the particle density and by the content of interstitial air (i.e. the way in which the particles pack together). The particle density depends on the product material density and the amount of air occluded within the particles. The product material density depends on the composition of the product and the densities of the individual components making up the product; for example, the density of milk fat is significantly lower than that of SNF (solids not fat) and so the bulk density of WMP (whole milk powder) is normally less than that of SMP (skim milk powder). Powder material density cannot be changed without changing the product composition, and so for any given product is constant. For a given product, therefore, occluded air becomes the prime factor influencing particle density.

Occluded air within the powder particles lowers both particle and bulk densities. It arises from air which is incorporated into the concentrate feed during passage from the evaporator through to atomization. Consequently, to achieve high particle density it is important to minimize air incorporation. Air is incorporated into the atomizer droplets through the action of the atomizer itself. In general, more air is incorporated with centrifugal atomizers than with nozzles, although with the use of

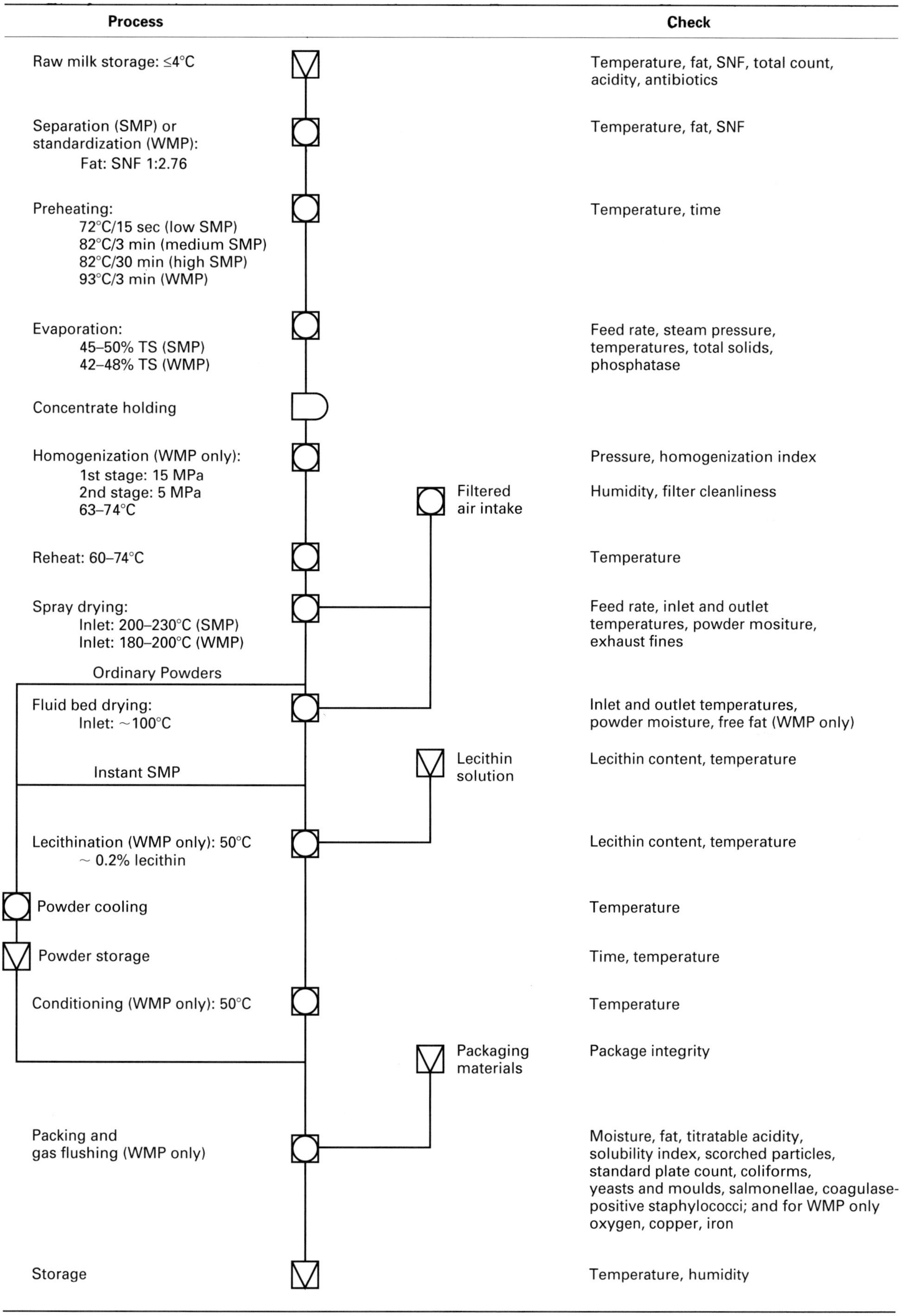

Fig. 1 Process flow chart – ordinary and instant milk powders.

Table 2. Hazard audit sheet – powder manufacture

Hazard analysis			
Critical operation	Potential risk	Critical control point	Prevention, control and monitoring measures
Raw milk storage	Growth of spoilage and pathogenic organisms leading to increased solubility index; toxin and heat stable enzyme production; flavour defects	Hygiene and sanitation of silos and ancillary equipment	Clean and sanitize to set procedures (QA manual); inspect silos regularly (at least once per week) with UV light
		Chilling operation	Ensure milk chilled to $\leqslant 4°C$ immediately on receipt and held at $\leqslant 4°C$ for minimum period ($\leqslant 12$ hours)
Separation/standardization	Failure to meet fat specification	Separator efficiency	Ensure temperature of separation, milk and cream flow rates are as prescribed; ensure fat:SNF ratio is 1:2·76 (WMP); on-line analysis of standardized milk
Preheating	Ineffective pasteurization; failure to meet microbiological specifications, oxidative deterioration (WMP); failure to meet prescribed heat classification (SMP)	Hygiene and sanitation of preheaters	Ensure prescribed CIP procedures are followed (QA manual)
		Time/temperature	Ensure milk heated for prescribed time/temperature for at least pasteurization or to meet heat classification required; regular checks on thermometers (including calibration) and flowmeters; monitor and record preheat process
Evaporation/concentrate holding	Post-pasteurization contamination and growth	Plant hygiene and sanitation	Ensure prescribed CIP procedures are followed (QA manual); frequent changing and cleaning of concentrate holding tanks
		Holding time/temperature	Minimize volume and time concentrate held. If held, ensure holding temperature is as prescribed (QA manual)
Homogenization (WMP only)	Post-pasteurization contamination and growth	Plant hygiene and sanitation	Ensure prescribed CIP procedures are followed (QA manual)
	Product stability	Homogenization index	Operate at prescribed pressures (1st stage: 15 MPa, 2nd stage: 5 MPa)
Drying/cooling	Post-pasteurization contamination and growth	Plant hygiene and sanitation; air supply; structural integrity of plant	Ensure cleaning and sanitation of plant and premises (internal and external) is as prescribed (QA manual); ensure correct location of air intakes; regular inspection, cleaning and sanitizing of air filters as per QA manual; regular inspection for structural defects e.g. dryer cracks
	Out of specification moisture, solubility index and scorched particles	Inlet and outlet temperatures	Regular checks on temperature sensors and controllers; automatic control of dryer; in-line monitoring of powder moisture

continued

Table 2. *continued*

Hazard analysis			
Critical operation	Potential risk	Critical control point	Prevention, control and monitoring measures
Lecithination (WMP)	Lecithin level in powder out of specification	Lecithin concentration in spray solution	Check lecithin concentration based on free-fat:total fat in base powder
		Incorrect dosing	Check dosing equipment and conditions (QA manual)
Packaging materials	Contamination by micro-organisms and foreign matter	Hygiene	Ensure appropriate storage and handling (QA manual)
		Package integrity	Check materials for damage, defects, etc.
Packing	Contamination by micro-organisms and foreign matter	Hygiene	Ensure prescribed standards of cleaning, operator awareness, personal hygiene and product sampling are met (QA manual); segregation of packing operation; provision of clean environment
		Package integrity	Ensure integrity of seams
	Oxidative deterioration (WMP)	Removal of air	Ensure oxygen level $\leqslant 2\%$ by gas flushing
	Incorrect fill weight	Scales	Check calibration of scales; check weigh packages as per prescribed statistical testing plan (QA manual)
Storage	Production deterioration through chemical and microbiological changes	Hygiene	Application and monitoring of GMP
		Temperature	Store at 15–20°C
		Humidity	Control and monitor humidity; ensure packaging material prevents moisture uptake

large-capacity nozzles the difference is not significant. Because of this problem, special centrifugal atomizers have been designed. The occluded air content is influenced by a number of factors as follows.

(1) *Feed system and atomizer design.* Concentrate, as it leaves the evaporator, is effectively de-aerated. The action of pumping the feed to the atomizer and the atomization process itself can re-incorporate significant amounts of air. Centrifugal atomizers, because of the large air–feed interface area within the wheel and the action of the wheel itself, cause significant air incorporation into the feed during atomization. To counter this, special wheels, such as the milk wheel (curved vane or high-bulk-density wheel) and the steam-swept wheel, have been introduced. The latter seeks to replace the air–feed interface by a steam–feed interface, and while effective in preventing the creation of vacuoles in spray-dried particles, and hence increasing bulk density, does cause some deterioration in powder quality, solubility index and process economy.

(2) *Properties of the feed.* The amount of air incorporated into the feed depends on the ability of the feed to form a stable foam. This is influenced by the protein content, by the nature of the proteins and by the presence of anti-foaming agents. High protein content increases the tendency of the feed to foam. Undenatured whey proteins possess high foaming ability and so the occluded air content will be related to the preheat treatment applied prior to evaporation (WPNI). High heat treatment reduces occluded air content and increases bulk density. Fat-containing concentrates exhibit much less tendency to foam than do skim concentrates, owing to their lower protein content and free fat acting as a foam destabilizer. Highly concentrated or warm feeds have less foaming ability than those of low

TS (total solids) or at low temperatures. ***See*** Protein, Functional Properties

(3) *Drying conditions.* During spray drying the removal of water causes a considerable reduction in the weight, volume and diameter of the atomized droplet. The extent of this shrinkage will depend on the drying conditions, the TS in the feed and the presence of air in the droplet. During the drying process there is a gradual solidification of the droplets with the evaporated water creating vacuoles within the particle. This gradual solidification causes the particle to warm up and any occluded air to expand and fill the vacuoles, preventing or limiting further shrinkage. The occluded air content in the particle increases with the drying temperature, with a high temperature differential between the droplet and drying air causing rapid air expansion and particle solidification and preventing shrinkage. Excessive overheating can create balloon-like particles, further reducing particle density. Sometimes the solidified crust cannot withstand the pressure created by the expanding air and the particles fracture, giving rise to fines which are hard to recover in the powder/air separation systems.

If the temperature of the surrounding air during the critical stages of drying is kept low, such as occurs in multistage drying, air expansion will be minimized, shrinkage will be maximized and particle density will be improved. Thus, if all other conditions are the same, the two-stage drying process will produce a powder with higher particle density than single-stage drying. Roller-dried powders also tend to have less occluded air and higher bulk density than spray-dried powders.

The content of interstitial air, or the way in which the powder particles pack together, depends on the particle size distribution, degree of agglomeration and flowability. The proper particle size distribution ensures that space between the large particles is filled with smaller ones, thereby increasing bulk density. Agglomeration, on the other hand, reduces bulk density, the extent depending on the degree of agglomeration. The ability of the powder to flow during bulk density testing will also influence the value obtained. Good flow characteristics are favoured by a large particle size.

As with bulk density, the microstructure of the powder is influenced by the nature of the feed material, the method of atomization and the type of drying process. Ordinary spray-dried SMP particles are spherical in shape with a wrinkled surface due to the casein in the milk contracting on the particle surface. These particles contain large vacuoles which determine the bulk density. The particle shape will be influenced by the extent of shrinkage occurring during drying and hence by the amount of occluded air. Thus, particles produced from a multistage drying process tend to be more dense and more deflated and more irregular than those produced by a single-stage process.

Solubility Index

The solubility index is considered one of the most important quality criteria and is a measure of the extent of denaturation of the proteins (principally casein) in milk powders. It is determined as the amount (expressed in millilitres) of insoluble material recovered from a sample of the powder reconstituted and centrifuged under defined conditions. Poor solubility (high solubility index) in milk powders is caused by subjecting the milk to high temperatures, particularly at high TS levels, during processing. This effect is further enhanced by increasing acidity in the milk. Thus high solubility indices may be caused by:

1. Poor milk quality: a high acid content due to bacterial activity increases the sensitivity of the caseins to thermal denaturation.
2. High drying temperatures: increases in concentrate viscosity, e.g. through age-gelation following high-temperature evaporation and holding, leading to poor atomization and larger droplet formation, will require the use of higher temperatures to achieve drying. One-stage drying processes with their higher drying temperatures are more likely to cause higher solubility indices than multistage processes.
3. Failure to cool the powder before bulk storage.

Roller-dried powders, because of the higher product temperatures attained, have significantly greater solubility indices than spray-dried powders. ***See*** Protein, Interactions and Reactions Involved in Food Processing

The term 'insolubility index' has been proposed by the IDF as an alternative to the reverse scale of solubility index. The occurrence of white flecks in powders is of similar origin to the insoluble sediment and can be visually detected.

Moisture Content

The moisture content is traditionally defined in the powder specifications and is important from both a powder quality and process economy point of view. Economically, it is important to control the moisture content as close as possible to the maximum, consistent with maintenance of powder quality. In multistage drying, control of the moisture in the powder leaving the main chamber is important as it influences the bulk density, particle density, solubility index, agglomeration, etc., of the powder from the fluid-bed drier/cooler. Moisture is influenced by a large number of factors, including drying air conditions, atomization and feed conditions. Because of its importance, in-line measurement of powder moisture is becoming increasingly common. ***See*** Water Activity, Principles and Measurement

Scorched Particles

Scorched particles are overheated or burnt particles generally arising from deposits within the drying chamber. They can arise through defective atomization, the use of inadequately maintained inlet air filters, incorrect drier shut-down procedures, dirt or sediment in the raw milk, or burnt-on deposits in the evaporator. Scorched particles therefore reflect the quality of the processing conditions. They are measured by filtration of a sample of milk reconstituted under defined conditions and comparison of the filter pad with standards.

Hygroscopicity

This depends to a large extent on the quantity and physical state of the lactose in the powder. The amorphous form of lactose is responsible for the hygroscopic nature of milk powders and their tendency to 'cake' during storage, and the higher the level of this form of lactose the greater the problem, e.g. in whey powders. Crystallization of the lactose to the α-monohydrate form reduces its hygroscopicity. To prevent 'caking', powders should be adequately protected from moisture and stored at normal temperatures. ***See*** Lactose

Chemical and Microbiological Aspects

Maillard browning and oxidation are perhaps the two major chemically based problems associated with milk powders, and both contribute to flavour and other deteriorative changes during storage.

Browning Reaction

The Maillard browning reaction, a reaction between lactose and the free α- and ε-amino groups of milk protein, is promoted during the preheating, evaporation, drying and storage phases of powder production. This reaction produces a number of chemical compounds including carbon dioxide, hydroxymethylfurfural (HMF), maltol and formic acid that cause discoloration and off-flavours ('cardboard or gluey') in the product. It also causes a decrease in the powder solubility and loss of nutritive value by a reduction in the available lysine through complex formation with lactose. Moisture is a controlling factor in browning, which has a maximum at intermediate values of water activity (5–10% moisture), values close to commercial powder moisture contents. To minimize off-flavour production and discoloration, powders should be stored under conditions of low temperature (15–20°C) and humidity. ***See*** Browning, Nonenzymatic

Oxidation

Oxidation of fat is the main shelf-life-determining factor in WMP. The oxidation reaction is a free-radical reaction involving hydroperoxides and leading to the production of a variety of aldehydes, ketones, hydroxy-acids and hydrocarbons, many of which are volatile and have strong odours and off-flavours variously described as 'stale', 'tallowy' or 'cardboard'. ***See*** Oxidation of Food Components

The rate of lipid oxidation is influenced by the water activity of the powder and is rapid at the water activities found in dehydrated foods. In most powders the best protection against fat oxidation is achieved when the moisture level is sufficient to form a monomolecular layer of water over the dried particles, hence providing a barrier to oxygen contact with the fat. The storage temperature is also vital, with every 10°C rise in temperature doubling the oxidation rate. Processing conditions which encourage the formation of free fat will also accelerate oxidative deterioration. ***See*** Water Activity, Effect on Food Stability

Approaches used to control the oxidative changes in WMP include high preheat treatment of the milk to generate antioxidants (SH groups) *in situ* and inert gas flushing of retail cans to reduce the residual oxygen to less than 2%. The effective shelf life of WMP at normal temperatures and humidity is around six months. ***See*** Antioxidants, Natural Antioxidants; Storage Stability, Parameters Affecting Storage Stability

Microbiology

General

Basic microbiological requirements for milk powders are generally stipulated in the powder specifications and tested according to nationally established standard methods. The presence of microorganisms in milk powder arises from two sources – those surviving the manufacturing process and those arising from recontamination of the product.

The processes involved in the manufacture of milk powders have varying influences on the survival of the microorganisms present in raw milk. The spray-drying process itself is not lethal to all microorganisms and many may survive the relatively short-time, low-outlet-temperature conditions that powder particles are subjected to in the drying process. There is considerable evidence that heat-sensitive organisms, including *Salmonella* spp. can survive the spray-drying process. The preheating step has the most significant effect on reduction of microbial numbers, with the level of reduction increasing with the severity of preheating. Evaporation, as might be expected, results in a general increase in count, with the increase in TS affording increased protection to the organisms against heat treatment.

The main types of microorganisms surviving the powder manufacturing processes are the thermoduric, thermophilic and spore-forming organisms. Furthermore, a number of heat-stable enzymes elaborated by psychrotrophs and toxins produced by other organisms, e.g. *Staphylococcus aureus* during storage and processing, may also survive the process and influence powder quality and safety. Consequently, storage times and temperatures must be adequately controlled to minimize the production of these. ***See*** *Staphylococcus*, Properties and Occurrence

Theoretically, powder produced under good conditions should contain only those organisms which have survived processing. Many of the microorganisms isolated from milk powder, however, are due to contamination of the powder during or after drying. These contaminating organisms may include both spoilage and pathogenic organisms, and so the microbiological tests carried out on in-line samples would routinely include standard plate count, thermoduric count, coliform count, and *S. aureus* together with frequent testing for *Salmonella*. ***See*** *Staphylococcus*, Detection; Spoilage, Bacterial Spoilage

Salmonella

Salmonella is a pathogenic organism which has been implicated in a number of documented food poisoning outbreaks associated with dried milk products. Because of this, the prevention of *Salmonella* in milk powders remains one of the biggest concerns of the milk powder industry. Research has shown that *Salmonella* spp. can survive spray drying, although the survival rate decreases with increasing drier-outlet temperature. Thus, the drying process cannot be relied upon to destroy these organisms. Since all salmonellae are readily destroyed at pasteurization temperatures, their presence in powder is indicative of post-evaporation contamination. The ubiquitous nature of this pathogen presents a major challenge to the production of *Salmonella*-free powders.

The requirements for effective *Salmonella* control, and thus good manufacturing practice (GMP), comprise the following.

1. Prevention of entry of *Salmonella* by control of all potential carriers, e.g. people, pests, materials, vehicles and environment.
2. If entry does occur, containment and prevention of growth by effective cleaning and sanitization of plant and facilities, segregation of wet and dry processing areas, controlled movement of personnel and materials, etc.
3. Heat treatment, equivalent to at least HTST pasteurization, of all raw materials used and application of appropriate drying temperatures.
4. Prevention of post-pasteurization contamination by segregation of wet and and dry production areas and packaging areas, controlled personnel and material movement, and maintenance of plant and premises in a clean and hygienic condition – prevention of product deposits, routine crack testing of driers, maintenance of processing and environmental air quality, air filters etc.
5. Continuous knowledge and appraisal of the pathogen status of staff, facilities, environment (internal and external) and product.

Effective product and environmental testing is essential in reducing the risk of *Salmonella* contamination. For milk powders, various sampling levels are recommended by different authorities, with perhaps that proposed by the International Commission on Microbiological Specifications for Foods (ICMSF) (60×25-g samples per batch; aggregates allowed, 15×100 g or 3×500 g) being the most widely accepted. All sampling schemes demand absence of *Salmonella* from any sample tested.

In the event of *Salmonella* being detected in a batch of milk powder, the following may occur: (1) the product may be destroyed; (2) the product may be reprocessed and retested; (3) the product may be resampled at an increased specified testing rate. If positive, then proceed with (1) or (2). If the samples are negative, the product may be released.

The continuous production of *Salmonella*-free powder requires the application and monitoring of GMP and continual vigilance by the manufacturer.

Trends in Quality Control

Traditionally, quality control (QC) of milk powders relied heavily on end-product testing and conformance of the powder with defined specifications before release; being an 'end-of-line' operation, it did little, if anything, to prevent production of defective products. Pressure on companies by more discerning customers saw a change from a QC approach to one of quality assurance (QA) where the process itself was controlled to give a high degree of confidence that quality products would be produced at the 'end-of-line'. End-product testing, of course, still remains an integral part of QA. In order to assure quality in the final product, the potential hazards to achieving quality within the process must be known, and so the Hazard Analysis and Critical Control Point (HACCP) concept evolved. HACCP is now widely used as a recognized QA technique in milk powder production. An example of an HACCP process flow chart and hazard audit sheet for milk powder manufacture is given in Figure 1 and Table 2 respectively. These are indicative only. In establishing a QA programme each manufacturer needs to establish a process flow chart and hazard audit sheet for their own particular operation based on their defined product specifications. An integral part of this QA programme is also the establishment of a QA manual to which the audit sheet will make reference. The QA manual will also identify the frequency of checking the items identified in the flow chart, the person responsible for each process step and where the check is recorded.

Further details on the principles and practice of HACCP are given in the relevant entry. *See* Hazard Analysis Critical Control Point; Quality Assurance and Quality Control

Bibliography

Hansen R (ed.) (1985) *Evaporation, Membrane Filtration and Spray Drying in Milk Powder and Cheese Production.* Copenhagen: North European Dairy Journal.

International Commission on Microbiological Specifications for Foods (ICMSF) (1986) *Micro-organisms in Foods 2. Sampling for Analysis: Principles and Specific Applications.* Toronto: University of Toronto Press.

MacCarthy D (ed.) (1986) *Concentration and Drying of Foods.* London: Elsevier Applied Science.

Mether AE (1989) Pathogens in milk powders – Have we learned the lessons? *Journal of the Society of Dairy Technology* 42: 48–55.

Robinson RK (ed.) (1990) *Dairy Microbiology*, vol. 1, *The Microbiology of Milk*, 2nd edn. London: Elsevier Applied Science.

W. S. A. Kyle
Victoria University of Technology, Werribee, Australia

POWER SUPPLIES

Contents

Use of Electricity in Food Technology

In general, the uses of electricity in food technology are broadly similar to those found in other process industries, being used mainly for motive power and heating. Many subdivisions of these major headings occur. For example, the motive power may be used to drive the compressor of a refrigerator or heat pump, in which case the final effect is one of heating and cooling; or the purpose of heating may be to dry a product, in which case the final effect is one of mass transfer. *See* Heat Transfer Methods

This article reviews typical areas where electricity is used in food processing, covering principally motive power and heating applications.

Electricity Supply

Although the end user of electricity need not be concerned with the technical details of the generation and distribution of electricity, it may be helpful to review briefly that process.

Electricity is almost invariably generated as three-phase alternating current at a frequency of 50 or 60 Hz (cycles/second). The reason for generating and distributing electricity as three-phase is basically that it makes the most economic use of generating and distribution plant. After generating, the supply is transformed to a very high voltage (again, to make the most economic use of plant) and distributed in a 'trunk network' or 'grid'. Many generators contribute to the supply of this grid and, from this very high voltage network, transformers reduce this voltage to a somewhat lower one for area distribution and, from this area distribution network, the supply is transformed down still further, ultimately to the familiar 120- or 240-volt supply fed to smaller commercial and domestic premises.

Large industrial or commercial premises may be supplied at a voltage higher than 'end use' voltage and the final transformation is done at the customer's premises. The economics of this choice will depend on local circumstances, but in the United Kingdom, higher voltages would be supplied for installed loads above about 10^6 volt-amps (1 MVA).

Comparison Between Single-phase and Three-phase Supplies

All industrial and larger commercial premises are furnished with a three-phase supply. This simply means that there are three Line (or live) terminals and a Neutral which is nominally at earth potential. (The neutral conductor is bonded to earth at some point in the distribution system). Each line terminal is supplied with a sinusoidal voltage but each sine wave is displaced by 120°.

The supply voltage is invariably expressed as the root mean square (RMS) value, which is the value of the dc (direct current) voltage which would give the same heating effect in a given load as the sine wave AC source which is actually being used. For example, if we supply an electric fire from a 240-V RMS supply, it gives out exactly the same amount of heat as if it were connected to a 240-V DC supply (e.g. a battery).

Most single-phase loads (e.g. lighting, small appliances) are connected between line and neutral and, in the United Kingdom, this Phase voltage is nominally 240 V.

Appliances of more than a few kilowatts, and most motors except fractional-horsepower types, operate from a three-phase supply. In this case, the supply voltage is often referred to by the Line voltage, which is the voltage between each of the three Line terminals. Its value is $\sqrt{3}$ times the Phase voltage so that, in the United Kingdom, the line voltage is 415 V.

There are many reasons why, for all but the smallest of loads, a three-phase supply is preferable. Because the total power is shared by all three phases, the current in each is only one-third of that which would be required by a single phase supply, which means lighter-duty switchgear and control systems. If the load is completely balanced, there is no neutral current, resulting in a reduction in supply system losses. While single-phase induction motors of up to a few kilowatts are available, they are less efficient than their three-phase counterparts, they are not inherently self-starting (and therefore need special starting gear), and they are not easily reversible.

Quality of the Load

Ideally, all loads would be resistive (i.e. pure heating effect) and draw exactly equal currents from all three supply phases. The instantaneous value of the load current would also follow the supply voltage exactly – i.e. the current would be sinusoidal.

Practical loads fall short of these ideals and electricity companies apply financial pressure to their customers in an effort to persuade them to improve their loads as far as possible. In certain cases, there are statutory limits on the 'load quality' to prevent excessive disturbance to other consumers connected to the same supply.

The main 'load defects' are poor power factor, phase unbalance, nonsinusoidal loads (harmonic generation), flicker, and peak demands substantially above mean demand. Poor power factor can be caused by excessive inductive reactance in the load, or by the use of thyristor phase-angle control, or both. If the nature of the load is constant and the problem is caused solely by inductive reactance, then the power factor can be well corrected by connecting capacitors across the load. However, if the problem is caused principally by the use of phase-control, then connection of parallel capacitors is of only limited benefit and can only partially correct the power factor over a narrow load range.

Unbalanced loads are caused by a preponderance of single-phase loads unequally distributed across the three phases and can usually be corrected satisfactorily by judicious redistribution. Voltage unbalance can cause overheating of motors and transformers.

Nonsinusoidal waveforms have become far more significant with the advent of thyristor control, in which the load current is discontinuous. The effect is to produce high-order harmonics on the supply voltage that may interfere with other equipment in the same premises and with other consumers connected to the same point of common connection. Furthermore, overheating of transformers can result.

Flicker is caused by the rapid on-off switching of a load, or by sudden and rapid changes in load impedance; the latter is unlikely in the food industry but the former may occur when loads are burst-fired.

Excessive peak demand may be controlled by process management.

Situations where the above problems may arise are highlighted in the following sections.

Motive Power

For local motive power, electric motors are unsurpassed and have virtually no competition from any other form of power. They are used for a great variety of purposes, from small mixers and blenders to extruders, rotary toasters, pumps and compressors.

When operated correctly, electric motors can have efficiencies exceeding 90% regardless of type. However, incorrect selection of the motor, or highly intermittent loads, can substantially reduce the overall operating efficiency.

The great majority of industrial electric motors are induction motors operating from three-phase supplies; these are quiet, efficient and need minimal maintenance. For practical purposes they must be regarded as constant-speed motors, whose speed is dependent on the supply frequency. Thus, the maximum speed of an induction motor is slightly less than 3000 or 3600 rpm when operated from 50-Hz or 60-Hz supplies, respectively. Proper fractions of these speeds can be achieved by increasing the number of pole-pairs in the motor; for example a 4-pole motor runs at just below 1500 rpm and a 6-pole one at just below 1000 rpm from a 50-Hz supply.

Induction motors also, in general, have a relatively poor starting performance in which the starting torque is much less than the maximum obtained at just below normal operating speed. Although the rotor construction can be modified to achieve high starting torque, this is usually at the expense of some worsening of full-load efficiency. The degree of difficulty in starting depends upon the type of load. For example, in a refrigeration plant the fan motors would rarely cause any difficulty (since the load torque of the fan varies as the square of the speed), but the motor driving a reciprocating compressor may require to be specially chosen, or some external 'starting aid' connected, in order to start the compressor satisfactorily.

Two problems of starting may be encountered, singly or in combination: the motor may not provide sufficient starting torque; or the motor may draw excessive current from the supply. The latter problem is especially serious in rural areas (on farms, for example) where the electricity supply rating is adequate for normal loads but may be overloaded when a motor attempts to start a difficult load.

If this problem is encountered, the supply can be reinforced (which may be prohibitively expensive), or some form of starting device can be connected. In general, starting devices take two forms:

- A 'soft starter', which limits the motor current but does not increase starting torque; this is only suitable if the load can be started at this reduced current.
- An inverter, which changes both frequency and voltage supplied to the motor and enables an infinite speed range to be obtained together with a starting torque equalling the maximum torque. The inverter can be controlled from a small signal source so the system can be incorporated in a control loop, for example.

The advent of satisfactory inverter-fed induction motor drives has almost entirely displaced other types of motor in this industry sector. Nevertheless, inverters are expensive and, if the problem is only one of starting rather than of variable speed control, then solutions other than the use of an inverter may be sought; for example, a compressor may have a bypass system to unload the system when starting.

It has long been recognized that many electric motors operate in a manner that leads to a poor overall efficiency, for example when the mechanical power is much lower than the rated output of the motor for much of the time with occasional peak loads requiring a rated power much greater than the mean. This has led to the development of 'high efficiency' induction motors in which operation at low load factors does not degrade the overall operating efficiency to such an extent. This is achieved by the use of a larger-section magnetic circuit with high-quality laminations and operating with a lower magnetic flux density (to reduce the iron losses) and larger-section copper windings to reduce the copper losses. These motors are therefore larger, heavier and more expensive than their conventional counterparts, but the energy savings can more than offset this when they are used in appropriate applications.

Control of Motors

In many cases induction motors can be used 'as is', running at a constant speed (having regard to the possible starting difficulties referred to above) but there are other cases where precise, variable speed control is required.

Induction motors are essentially constant-speed devices whose speed (in rpm) is the supply frequency divided by the number of pole-pairs; for example a 2-pole motor supplied with 50 Hz runs at a nominal 50 rev s^{-1} or 3000 rpm. Speed control can be achieved by pole switching (coarse steps), voltage control (limited application), or variable frequency (inverter fed).

For precise, stepless control, the inverter system is unsurpassed and has the added advantage that maximum rated torque can be achieved at any speed. The inverter supplies a motor voltage which is proportional to the frequency; if this were not done the motor would draw excessive current at low speed and overheat. In addition, the maximum motor speed can usually exceed, to some extent, that which can be achieved when operating directly from the mains supply. The only disadvantages of inverter control are relatively high cost and a very slight reduction of motor efficiency.

Voltage control is only applicable in a very few cases where the load torque rises continuously with speed, such as fan loads. However, the control of speed is very imprecise and the technique has little industrial application.

Motors can sometimes be reconnected so that, for example, a two-pole motor can operate as a four-pole one, halving its speed. Again, this is of limited applicability and many of the techniques for speed control that were common some years ago have been displaced by inverter control.

Other Motors

Because they are virtually maintenance free, induction motors are almost universally employed for large drives, but there are some low-power applications where DC motors are used. One example is in product feeders, where a DC motor controlled by an electronic speed controller is frequently used: excellent speed control is obtained at moderate cost.

Electricity for Heating

Although it is commonly supposed that electric heating is more expensive to operate than fuel-fired systems, this is not necessarily always true. Electric heating has an efficiency approaching 100% at the point of use and, combined with the inherent controllability and the absence of combustion air and exhaust gases, this can make electric heating more economic in certain circumstances.

Water and Aqueous Fluid Heating

Two basic techniques for electric liquid heating are available: resistance heater elements immersed in the liquid, or passing current directly through the liquid. Both of these techniques are applicable to the heating of static liquid in a tank or as it flows through the heater continuously. The first technique is the well-known 'immersion heater' and the second is typified by the 'electrode boiler'.

While both these techniques are widely used for heating water and dilute solutions they may not be universally suitable for heating of foodstuffs even in aqueous state. Immersion elements generally have a relatively small surface area for a given power output and hence a high surface heat flux, leading to the possibility of large temperature gradients close to the surface of the element and the possibility of burn-on or other degradation.

Electrode boilers operate by passing current directly through the liquid and, while in conventional form they are not suitable for foodstuff heating (owing to surface interactions at the electrodes), the principle has been adapted very successfully in 'ohmic heating' (see below). ***See*** Heat Treatment, Electric Process Heating

Specialized Heating Methods

While many conventional forms of heating can use electricity as their heat source (for example, infrared, conduction and convection heating), there are certain specialized techniques which use electricity in unique ways and have particular applications in food heating. These include dielectric heating and ohmic heating.

Dielectric heating, which includes radiofrequency and microwave techniques, makes use of the dielectric properties of the material being heated rather than electrical conduction and so is employed mainly in the processing of dry material or materials having low electrical conductivity such as fats and oils.

Ohmic heating is, quite simply, the bulk heating of a foodstuff by the passage of an electric current through it; the material must therefore have a reasonable electrical conductivity, and this implies a pumpable aqueous-based fluid which may contain solid particles, such as fruit lumps in syrup or meat in gravy. While the technology is related to the electrode boiler referred to above, the construction and electrode materials are much more sophisticated so as to ensure complete absence of fouling and consequent degrading of the product. Because the heating is uniform throughout the bulk of the product and does not rely on heat transfer from a surface, the necessary sterilization temperature can be guaranteed without the need for overcooking or excessive holding, which leads to improvements in product quality and production rates.

Control of Heating Loads

Electric heaters are frequently controlled by conventional thermostats, switching the heater elements through contactors. This technique is satisfactory if the time constant (or thermal inertia) is large, the variation of temperature about the set point is acceptable, and possible electrical interference from the contactor can be tolerated. This form of control is therefore entirely acceptable for applications such as noncritical temperature control of large vats or tanks.

For more critical control, such as when temperatures have to be precise, or the time constant is short, some form of proportional control is necessary and this requires that the heater voltage is continuously controlled or that it is switched on and off in rapid succession (at time intervals much shorter than the time constant of the process). Invariably these operations are carried out by a solid-state controller which may either control the voltage by means of phase control, or switch the heater on and off by burst firing.

Phase control operates by only supplying a part of each cycle of the supply to the load: exactly as is done in the familiar household light dimmer. It is very efficient, but degrades the power factor and generates harmonics in the supply, which may reach a level exceeding statutory limits. Burst firing is a technique which uses zero-voltage switching, where a solid-state device is used to switch the load on and off only at the moment when the instantaneous voltage is zero. Therefore, harmonic production is eliminated entirely but, in turn, the problem of flicker is raised. Flicker is said to occur when the supply voltage changes suddenly at regular intervals, usually caused by a significant load being switched on and off. As its name suggests, its effect is particularly noticeable in causing flickering effects in lighting and is most severe when the interval between switching is less than 1 second.

The severity of these problems (and many of the other 'load defects' referred to earlier) depends not only on the load itself but the characteristics of the supply to that load and its relationship with other loads in the premises. For example, a 'poor' load may have an insignificant effect when operating in a large factory, but may be quite unacceptable if it is the dominant load in the premises and may even cause interference to the supplies to other consumers. In all cases the advice of the electricity company should be sought.

Conclusion

Electricity is a highly processed fuel demanding wise and efficient use. In countries where electricity is generated principally from fossil fuels, the overall efficiency of the process of generating and distributing electricity to the final user is of the order of 30–35%; in other words, the user only receives 30–35% of the energy content of the fuel which was fed to the power station boiler. This is usually reflected in the price of electricity relative to 'raw' fuels.

However, the efficiency of the electrical 'appliance' at the point of use is often very high (for example, for water heating) and this often overcomes the apparent inefficiency of the generating and distribution process.

Electricity is clean, highly controllable and, because of its controllability and some unique techniques only available with electricity, can often be the most effective and economic solution.

Bibliography

Say MG (ed.) (1985) *Electrical Engineer's Reference Book*, 14th edn. Sevenoaks: Butterworths.

Stirling R (1987) Ohmic heating – a new process for the food industry. *Power Engineering Journal* November: 365–371.

J.T. Griffith
EA Technology, Capenhurst, UK

Gas and Other Fossil Fuels

Ever since the discovery of fire by early man, heat for the preparation of food has been an important factor in human survival. Today the preparation, manufacture and preservation of foodstuffs for the consumer is carried out on a large scale in factories where energy is used for cooking, cleaning, chilling, drying, evaporation and dehumidification. These operations consume about 100 PJ (1 PJ $=1\times10^{15}$ J) per year in the UK (Energy Efficiency Office, 1989). Fossil fuels (gas 41% of the total energy, oil 26% and coal 10%) are the predominant means of heat energy, and electricity (22%) provides mainly power.

Natural gas utilization in the food industry has increased significantly since its discovery in the North Sea in the 1960s. Natural gas has significant advantages: environmental friendliness, with low emissions compared with other fossil fuel combustion; ease of handling; and availability.

Steam

Historically the widespread use of steam for heating both processes and buildings became established when coal was the major fuel available. A centralized boiler-house was used to accommodate the associated coal handling and ash removal facilities. Since then, boilers using oil or gas, which are easier fuels to handle, have become predominant. Steam is now responsible for 75% of the heat energy required for the food industry.

In this industry, steam is used for numerous processes, which include preparation (e.g. blanching, peeling, mixing and blending), cooking, evaporation, concentration and drying, pasteurization and sterilization. It is considered to be cheap, easy to handle, clean and versatile. High heat transfer rates at constant temperature (an inherent characteristic of steam) allow the use of compact heat exchangers with little risk of product damage due to local overheating. Unfortunately, it has been found that gross thermal efficiencies of, typically, only 50% are achieved (Jones, 1988). ***See*** Cooking, Domestic Techniques; Heat Transfer Methods; Pasteurization, Principles; Sterilization of Foods

Point of Use Equipment

Greater thermal efficiency and control are possible if the heat from combustion products is used in place of steam, either in direct contact with the food or indirectly by means of heat exchangers. Such equipment is used in washing, drying, baking, chilling and evaporating processes.

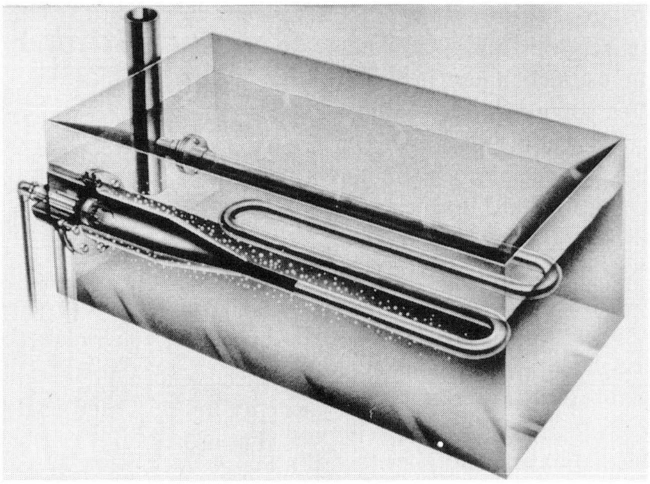

Fig. 1 Direct gas-fired high-intensity small-bore immersion tube system.

Washing

There is a need for large quantities of hot water to clean food prior to, or during, its preparation. Traditionally, steam passed through tubes immersed in the water has provided a simple method of heating liquids. A more efficient method is to use fossil fuel combustion products instead of steam in the tubes. This has been achieved with large-diameter, gas-fired, immersion tubes, which can produce efficiencies up to 70%, although they can occupy a large portion of the tank.

Higher-firing-intensity small-bore immersion tube burners (Fig. 1) are available which produce 80% gross thermal efficiency. They have been used in many installations for food washing and meat scalding. Other gas-fired heat exchangers have been developed from this basic concept, and are used in the pasteurization of milk and in the mashing and wort boiling operations in small breweries (Jones, 1989). ***See*** Beers, Ales and Stouts, Preparation of Wort

Even higher thermal efficiencies can be realized if the exhaust temperature of the combustion products is reduced below their dewpoint. This causes water vapour from the combustion process to condense, which releases more heat to the process. Owing to the clean nature of the products of combustion of natural gas, it is possible to obtain this effect by intimately mixing these products with the fluid being heated. Submerged combustors (Page, 1987) use this technique to attain a very high efficiency when operating temperatures of less than 60°C are required. At higher temperatures, the heat energy mainly evaporates water from the liquid surface and does not raise its temperature.

To achieve gross thermal efficiencies of over 95%, with water temperatures of almost 100°C, a direct-contact water heater can be used. One example (Fig. 2) uses a burner firing into a large-diameter immersion tube. The exhaust combustion products then pass through a tower filled with a packing material. Cold water enters the top of the tower and removes sensible and latent heat from the flue gases.

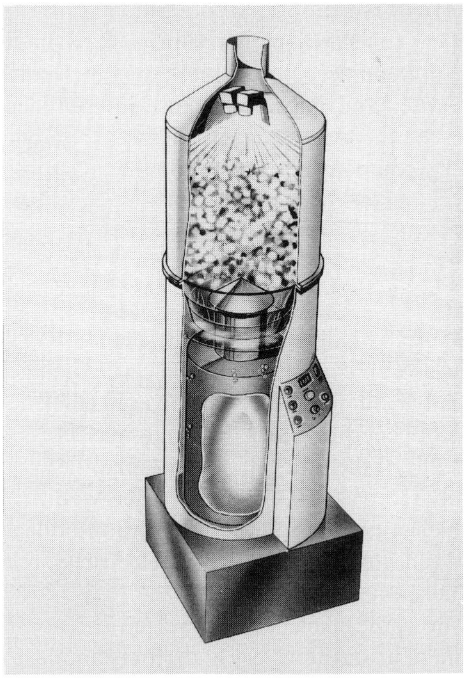

Fig. 2 The improved direct contact water heater.

This heater can be used for washing food and cleaning work surfaces. It is not recommended in circumstances where the water is used directly in the preparation of foodstuffs (Heap, 1991).

Drying

Many food products are produced in a dehydrated form or require drying after washing. This requires heat to remove moisture from the product, which is normally carried out convectively by a hot air stream, or by direct contact with combustion products. Heat can be transferred to the surface of the wet solids to evaporate the water. A wide range of gas-fired driers are used in the food industry, for example, conveyor, tray, fluidized-bed, rotary, spray, cascade and others (Pritchard *et al.*, 1977). ***See*** Drying, Spray Drying

Hygiene in food preparation is very important; hence some sensitive foods may need to be heated indirectly, without contact with the combustion products. Where it is possible for combustion products to be in contact with food, careful control of the combustion process must be maintained. Natural gas has advantages over other fossil fuels as it has a consistent quality and, with good burner design, it produces clean combustion products as demonstrated in the direct-fired drying of milk and malt (McDonald, 1974; Leaver, 1985).

Heat pumps (ETSU, 1983) can be used very efficiently to carry out convective drying duties without the direct contact of combustion products with the food product. This is achieved by recovering latent as well as sensible heat from the wet exhaust air, and upgrading this to the drier working temperature. Types of unit which could be used with gas as a fuel include engine-driven vapour compression systems and absorption-based units. The former are already available and have the advantage of being able to provide hot water and steam from the engine waste heat. This can be used in the drier or in other parts of the process. Absorption heat pumps are currently being developed in various countries. Their principle of operation is such that they are powered by heat rather than a compressor. They therefore potentially offer similar benefits to vapour compression units but with the advantages of fewer moving parts, and they contain no environmentally harmful CFC refrigerants.

Dehumidification is often required in the food industry to improve the quality of the manufacturing environment in terms of hygiene. Desiccants extract moisture without generating water condensate in which mould can grow. Moreover, the desiccant can have a bactericidal action, so bacteria carried on droplets in the atmosphere are killed. For example, for the room temperature storage of popcorn, which becomes sticky when damp, a gas-fired desiccant wheel (Pearson and Thompson, 1990) has been used successfully. In this device, a gas burner is used to regenerate the wet desiccant.

Baking

The baking oven used for bread, cakes, biscuits and crispbreads has the prime function of causing heat and mass exchange between the uncooked products and the oven's atmosphere. This can be achieved by conduction, natural and forced convection, and radiation. It is estimated (Energy Efficiency Office, 1989) that over 20 PJ per year is used for baking in the UK, with more than 50% being supplied by gas.

Traditionally, baking was carried out in batch ovens, which transfer heat by radiation. The first continuous ovens relied on the same principles, but as increased throughputs were required, convection or turbulence has been introduced to improve oven efficiency. Direct gas-fired ovens are used extensively in the biscuit and breadmaking industries. ***See*** Biscuits, Cookies and Crackers, Methods of Manufacture; Bread, Breadmaking Processes

Baking ovens tend to be simple. Many have high thermal inertia and heat from the exhaust gases, which can be at 300°C, is not often recovered for use in the baking process. Burners are usually controlled manually (Fovargue, 1977), although the techniques of control are becoming more sophisticated (Mowbray and Stamper, 1977), with growing emphasis on automation. Simultaneously the monitoring of temperature and humidity are becoming increasingly complex and refined.

With the increasing trend towards computer-controlled plant, it is vital that process control systems can be easily applied. This has been achieved efficiently with gas-fired forced-convection ovens, which have a fast response time and are suitable for automation. Published data (Gales and Petty, 1960) for the manufacture of biscuits suggest that forced-convection ovens use 25–30% less energy than radiant types. Other advantages appear to be rapid start-up times and shorter baking times.

Chilling and Freezing

The long-term storage of food, by chilling or freezing, has become essential in modern society. This can be carried out by either vapour compression or absorption techniques. ***See*** Freezing, Freezing Operations

The energy required for both these techniques could be supplied by a combined heat and power (CHP) system. These generate electricity locally, by reciprocating engines or gas turbines, and heat is produced from the combustion process. These prime movers can be used in space-heating buildings, or to provide hot water and steam for industrial processes.

For the chilling and freezing application, vapour compression requires electrical power to drive a conventional refrigeration cycle; the absorption chiller uses waste heat or a gas flame to boil off a refrigerant in a similar manner to an absorption heat pump.

CHP technology, together with the necessary control equipment, heat-recovery devices and ancillaries, has advanced to such a stage that a wide range of CHP

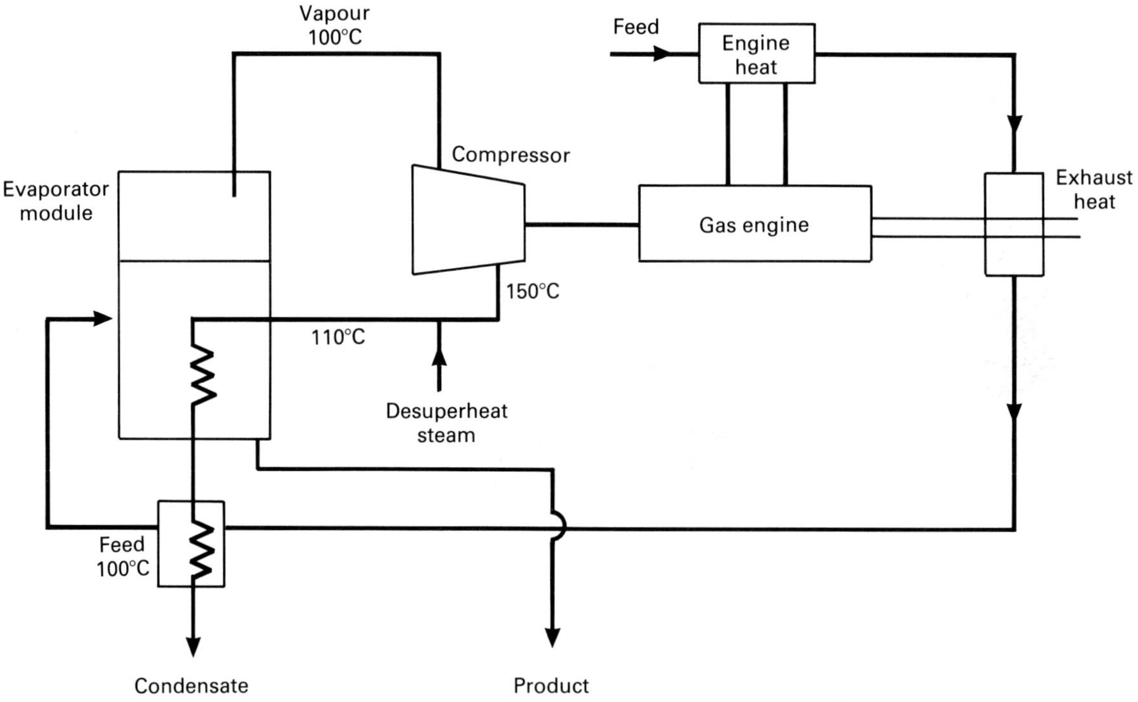

Fig. 3 Conceptual arrangement for gas-driven MUR applied to a single-effect evaporator.

equipment is now readily available (Masters *et al.*, 1990). There are, however, some fundamental guidelines that must be followed before CHP can be successfully installed. The process must operate for a large proportion of the year and there must be a simultaneous need for the heat and power generated by the prime mover. The system must use all of the heat generated and preferably all the power, although electricity can be imported or exported to the national grid.

These conditions are common in the food industry, and consequently CHP is ideally suited to it. Sugar has been produced for many years using steam turbines for power generation and the exhausted steam for the refining process. Recently, gas turbines have been used in the manufacture of biscuits and general foods, where they have replaced steam boilers and imported power. A further example, in maize processing, has seen the use of combined cycle CHP. In this case, the exhaust from the gas turbine produces steam in a waste heat boiler, which is used to drive a steam turbine, thus producing more power.

Evaporation and Distillation

The removal of water or solvent to concentrate foodstuffs is carried out in evaporators. These usually consist of heat exchangers in which the feedstock is heated. This causes vapour to be boiled off from the product, thus concentrating the feed. The energy requirement is high if one heat exchanger is used, and multiple units which reuse heat from the previous unit are common. Normally the energy is provided by steam. Direct gas can be used for this duty providing that the solvent is not flammable.

The energy requirements can be reduced by reclaiming heat from the expelled vapour. Compression will increase the temperature of the vapour, which can be returned via a heat exchanger into the same evaporator or into another unit. This technique, known as mechanical vapour recompression (MVR), uses the least energy for any given evaporation, and there are a number of installations operating worldwide. Most of these use an electric motor to drive the compressor. A more efficient option, however, is to use a gas engine or gas turbine. In this case (Fig. 3) waste heat from the prime mover can be recovered to preheat the feed or to generate additional steam or hot water for use in the process. This technique could be applied in a similar manner to the reboiler of a distillation column.

Bibliography

Energy Efficiency Office (1989) *Industrial Energy Markets*. Energy Efficiency Series No. 9. Energy Efficiency Office:

A waste recovery system incorporating a gas engine driven heat pump installed on two malt kilns. *Energy Conservation Demonstration Projects Scheme*, Energy Technology Support Unit, Harwell F/21/83/36.

Fovargue P (1977) Energy management in the baking industry. *Cake and Biscuit Alliance Technologists Conference*, pp 1–37. London.

Gales DR and Petty MH (1960) Oven developments – the forced convection oven. *Food Trade Review*, March: 56–64.

Heap CR (1991) Direct improvements in water heating. IGasE Midlands Section Young Engineer Award.
Jones DA (1988) *Efficient Gas Firing – An Alternative to Steam*. London: Mechanical Engineering Publications.
Jones DA (1989) Energy saving techniques for liquid heating. *Institute of Energy, Industrial Energy Management Conference*, Birmingham.
Leaver GA (1985) Gas fired low NOx combustion systems and air heaters for direct food processing. *Journal of the Society of Dairy Technology* 38(2): 42–45.
McDonald DP (1974) Fuels and heating systems. *Brewing and Distilling International* 4(3): 20–24.
Master J, Pugh RR and Weller GB (1990) Combined heat and power – generating new markets. IGasE, MRS E 582.
Mowbray WR and Stamper JR (1977) The trends and pressures in oven design. *Baking Industries Journal* February: 11–15.
Page MW (1987) Advanced applications of natural gas in low temperature heating processes. *Second World Basque Congress*, Bilbao, MRS E 512.
Pearson J and Thompson K (1990) Gas fired dehumidification for building energy services. Institute of Energy, MRS E 598.
Pritchard R, Guy JJ and Conner N (1977) *Industrial Gas Utilisation*, pp 693–709. Epping: Bowker.

D. A. Jones
British Gas plc, Solihull, UK

PRADER–WILLI SYNDROME – NUTRITIONAL MANAGEMENT

The Syndrome and its Management

Prader–Willi Syndrome (PWS) is a congenital disorder difficult both to diagnose and manage. It is a complex multi-system condition in which manifestations begin prenatally with decreased fetal movement; infants present at birth with lethargy, weak cry and poor reflexes including poor sucking. Special feeding techniques, e.g. tube feeding, are usually necessary, initially, but artificial ventilation is not. Failure to thrive is common in the first year of life. Hyperphagia presents at any time between 1 and 6 years of age, resulting in central obesity, unless energy intake is restricted. ***See*** Inborn Errors of Metabolism

The placid infant and smiling friendly child can, as he or she enters teenage years, or even earlier, begin to exhibit behavioural problems – primarily temper tantrums, stubbornness and manipulative tendencies.

Motor and language development are delayed; learning difficulties may be mild or severe; adults are not able to live completely independent lives owing to the obsession with food.

Characteristics of the Syndrome

Many (approximately 70%) typical patients have a characteristic chromosome 15 (q11–q13) deletion on high-resolution cytogenic analysis from the parentally derived 15. Some of the remainder have been found on molecular analysis to have submicroscopic deletions or maternal disomy; the aetiology of the rest who present as cytogenically normal is unknown.

The range and severity of clinical characteristics are similar in all sufferers. They can be considered under three main headings: structural, functional and cognitive.

The dysmorphological features include the following:

- Narrow face
- Almond-shaped eyes
- Small, downturned mouth
- Poor dentition
- Abnormally 'sticky' saliva
- Eye problems such as strabismus and myopia
- Short stature
- Hypogonadism (undersized or nonexistent genital organs)
- Small hands and feet
- Straight ulnar borders
- Obesity, especially around trunk, hips and thighs
- Scoliosis and/or lordosis
- Fair colouring.

Functional abnormalities present include the following:

- Hypotonia (poor muscle tone)
- Hyperphagia (overeating)
- Sleep disturbances such as apnoea and daytime sleeping
- Decreased motor skills and speech difficulties
- A high pain threshold
- Skin picking and scratching
- Absence of or decreased vomiting
- Decreased levels of (physical) activity

Prader–Willi Syndrome – Nutritional Management

- Decreased expenditure of energy leading to a decreased energy requirement
- Abnormal hormonal secretions; limited sex hormone production and higher levels of cholecystokinin during eating
- Poor body temperature control.

Cognitive features are:

- Low intelligence quotient (IQ); the majority of individuals have levels between 70 and 100
- Failure to function to actual IQ levels
- Difficulties with arithmetic
- Stubbornness
- Rapid alteration in mood related to behaviour problems
- Exhibiting unusual skills with jigsaw puzzles.

Gross obesity has resulted in the past, leading to premature death; however, as awareness of the syndrome has increased in medical and other caring professionals, appropriate individual treatment plans can be followed which not only keep weight within reasonable limits, but also raise self-esteem and result in longer lifespan. Dietary restriction for life, in conjunction with individual behaviour management programmes carried out in a loving, caring and supportive environment, can assist syndrome sufferers to live full and worthwhile lives. ***See*** Obesity, Treatment; Bulimia Nervosa

Food stealing is an unusual and poorly understood phenomenon of the syndrome; surreptitious eating is common.

Thus PWS is clearly an eating rather than a weight disorder.

Nutritional Management

Early appropriate intervention is essential. Distressingly many individuals still do not receive correct nutritional and dietary management at an early stage. Whenever a diagnosis is confirmed (or even suspected) dietary treatment should commence.

Poor feeding leading to failure to thrive and hyperphagia leading to rapid onset of weight gain are the complete antithesis. Yet in individuals with PWS both are present at different stages of development; nutritional management must therefore encompass these two extremes.

Newborn infants, with little suck reflex, normally require special methods of feeding, e.g. nasogastric tube feeding. There are no reports of successful breast-feeding; difficulties are still encountered with bottle feeding. Some mothers choose to express breast milk; others find this option impossible to maintain. These parents must not be made to feel guilty when choosing artificial milk but given sympathetic understanding. All infants with PWS sleep well and rarely wake for feeds; when awake they may cry feebly but are normally unresponsive to both their own needs and their surroundings. In addition to diminished suck, poor head and neck control, as a result of poor muscle tone, occurs and this exacerbates the difficulties encountered during feeding.

Nevertheless transition from tube to bottle feeding should not be delayed longer than necessary; slow bottle feeding – as little as 30 ml h^{-1} has been reported – can be speeded up either by use of larger holes in the teat or by using those normally reserved for premature babies; continuous tapping of the base of the bottle has been reported to help as it not only ensures a continuous flow of milk but also helps to keep the infant awake; sometimes it may be preferable to give liquids from a spoon. Small frequent feeds are advisable. It may be necessary to increase the energy density of the feed to ensure not only an adequate energy level is supplied but that all necessary nutrients are present. Advice relating to vitamin and mineral supplements needs to be given on an individual basis.

Introduction of solids may commence around 4 months; parents/carers may need reminding of risks of commencing solids too early. Infants with PWS take solids from a spoon better than liquid from a bottle. Sugars and other concentrated carbohydrate, e.g. honey and sweetened 'baby drinks', should be avoided to both lessen the risk of a 'sweet tooth' developing and to limit energy intake.

Carers (normally parents) are usually delighted when the infant develops an 'appetite' and takes food easily. At this stage of development, energy intake needs to be adequate but not excessive; comprehensive nutritional assessment by a qualified professional who understands the syndrome is necessary.

The high palate present in some syndrome sufferers may affect chewing/eating; if so, it is helpful for both speech therapists and dietitians to work closely together with carers.

Weight gain can be extremely rapid; there are reports of excess fat deposits preceding the increase in appetite; regardless of which occurs first weight measurements should be accurately recorded regularly – at a minimum monthly, but in infants under 12 months weekly weighing is recommended. Length or height measurements are necessary. Results should be plotted on standard height–weight charts; thus developmental records quickly show any deviation from normal patterns of growth.

Some research has indicated that children with PWS require less energy intake than a normal child for growth; some may require as little as 50% of normal, while others develop on 75%. This factor appears to be related to reduced energy expenditure level; research has not yet established if this is solely related to general reluctance to participate in physical exercise, or whether there are other metabolic reasons. ***See*** Energy, Energy Expenditure and Energy Balance

To follow any dietary regime is difficult; adhering to a strict regime for life is restrictive. For people with PWS not only is an energy-controlled diet necessary, but it is made more difficult because they have an apparent lack of satiety: as long as food is available, many syndrome sufferers will continue to eat; reasons for this are unknown but may be related to a hypothalamic disorder. Recent research into the nature of the eating disorder suggests that during periods when food is consumed, levels of cholecystokinin are raised. ***See*** Diet

All people with the syndrome must adhere to weight-reducing diets/weight control regimes. Additionally, individuals usually have a total obsession with food, e.g. requiring to know what is for lunch and supper at breakfast time. There is evidence to suggest that younger children can be taught the following guidelines:

1. What is or is not acceptable, e.g. quantities of food consumed at mealtimes.
2. What is a 'good' food and what (because of energy density and/or low nutrient value) should be avoided.
3. Normal social behaviour, e.g. it is acceptable to want something to eat late afternoon but not immediately after lunch.
4. To eat slowly, replacing cutlery on the plate between each mouthful and not gorge.

Strict calorie counting is advisable, until carers and individuals are familiar with appropriate quantities. Some carers and individuals prefer to weigh food, at least initially.

There is no single successful diet; the one that should be followed is that which is acceptable to sufferer and carer, fits in with the family eating pattern, provides all macro- and micronutrients and an energy level which maintains growth without obesity developing.

Very-Low-Calorie Diets should never be used without medical supervision. The use of such diets is controversial; there may be a place for short-term use when there are indications that this is the best option. Nevertheless, those treating individuals with PWS must remember that this is not a simple weight problem.

There have been incidences where parents have been over-cautious about the energy level and linear growth has ceased. Weekly weighing of children during the growing years, with accurate but less frequent height measurements, allows for proper dietary advice to be given promptly. Excessive weight gain in short periods of time is common and requires immediate action. Teenagers with the syndrome do not exhibit the normal pubertal growth spurt and, as has already been mentioned, short stature is common. Dietary control is critical at this time.

It is possible for older people with PWS to lose weight and return to an acceptable agreed level, provided that they are in a suitable environment with understanding and sympathetic support. As weight reduces, mobility increases; increased self-esteem must not be undervalued.

Other Factors

Successful management of PWS is not restricted to nutritional considerations but includes other equally important factors.

Accessibility of Food

Food stealing is a phenomenon which is hard for the uninitiated to accept. It is emotive and considered the most difficult of all characteristics. It may be necessary to take measures which conflict with normally acceptable social behaviour. These may include locking food cupboards, refrigerators and freezers, ensuring that no food is eaten other than at the table at mealtimes, removing the fruit bowl from the lounge, ensuring that other family members/residents do not keep food, sweets or drinks in their bedrooms. Some carers, both in the home and other community environments, have found that light-sensitive burglar alarms are extremely useful on staircases and at kitchen entrances, when for safety reasons locking of internal doors would be potentially dangerous.

Some individuals will steal and consume any food they find, including some normally not eaten by humans – e.g. cat and dog food, berries and other garden produce, frozen products and discarded food scraps in dustbins.

Carers frequently find that syndrome sufferers deny consuming 'missing' food; excuses given often appear amusing and/or devious; research now indicates that this denial is related to cognitive learning problems and is not deliberate deception. Thus carers must learn to handle such situations not only appropriately but sympathetically.

Pica may occasionally be exhibited. In such instances appropriate interventions must be introduced.

Each individual and his or her carers should know how the problem of consumption of extra energy will be handled; the strategy must be implemented, e.g. miss all or part of the next meal. Inconsistency in such circumstances is best avoided. Nevertheless a challenge to the unacceptable behaviour must be considered.

Special Occasions, Holidays and Treats

People with PWS should be included in social activities; they are a normal part of life. Carers do have the additional responsibility of providing supervision; realistic planning and forethought may well avoid behaviour problems completely – these are not only distressing to the affected individual but also to observers and can spoil the event entirely. It is preferable to reduce overall energy intake prior to such events rather than afterwards. This then permits inclusion of additional calories which will not have a detrimental effect.

People with PWS can have as rough a day as anyone else and may benefit from an occasional extra treat.

Reinforcers/Tokens

Reinforcers are frequently used when teaching or training people with learning difficulties. Traditionally, these revolve around food, usually high-energy and detrimental in the case of people with PWS. Suitable alternatives are preferable – books, games, make-up, etc. Food items should be restricted but not necessarily excluded, e.g. sugarless gum, low-energy drinks or exotic fruit can usually be worked into the diet with other treats, e.g. inclusion of a small quantity of sweets or 'trail mix' or a visit to a salad bar or suitable 'grill' used when particular targets have been reached.

Rules and Routine

Consistency in both rules and routine is essential – individuals do not respond well to alterations to routine or inconsistency in overall care. Occasionally, rules have to be made which affect others, e.g. eating only at mealtimes. Any management programme must ensure all carers, both direct and indirect – family, friends, teachers, social workers, neighbours, welfare assistants, club organizers – involved in care realize the importance of decisions and adhere to them.

Exercise

Parents can assist with improving muscle development from a very early age. The individual programme may be devised by a physiotherapist or be adapted from a particular scheme, e.g. Portage. All syndrome sufferers, not only children, must be encouraged to participate in physical exercise daily to assist with improving both muscle tone and metabolism. The latter also helps weight control. As well as activities that include families and/or peers, e.g. games, dancing, outings, swimming, etc., all efforts should be made to increase less obvious ways of exercising, e.g. walking the dog, delivering newspapers or leaflets, tenpin bowling or walking to the cinema where the prime motive is not exercise.

As many children have difficulties with balance, cycling – modified for example, with stabilizers, tandems or exercise bikes – trampolines and other floor equipment that the individuals can roll or climb over, may be helpful.

Education

Many children commence at a mainstream infant school and progress to a similar junior or primary school. It is acknowledged that the pace of learning falls off from about the age of 8 years after which time most PWS children, though not all, have difficulties in keeping up with their peers. Few individuals complete education without additional support (welfare assistants or extra individual teaching) in a mainstream setting. A move to a different school, e.g. a school for those with mild or moderate learning difficulties, may be the better option. Others may require specialist help for other developmental difficulties, e.g. physical handicap or speech delay best found in specialist schools.

Regrettably, after compulsory schooling ceases, training places meeting the individual and specialist needs of these adults are rare; individuals are often directed to situations unable to meet all aspects of care. Rapid weight gain often accompanies the frustration and anger of these young adults.

Vocational training is difficult due to the necessity of ensuring access to food is minimal. Many young adults have false expectations – some of which educators have encouraged – which are rarely met and therefore lead to further failure.

In the right environment some people with PWS have achieved success with qualifications in academic and/or craft or practical subjects.

Behavioural Management

Frequent references have been made to behaviour management programmes. Each individual must have needs assessed by the appropriate professionals and an individual programme must be established; regular reviews are essential including dietary factors.

Prognosis

Medical Complications

Diabetes mellitus has been reported; this is almost entirely type II (non-insulin-dependent); nevertheless, where weight control is poor insulin therapy may be indicated.

The recorded incidence of osteoporosis appears to be on the increase; this may be due entirely to longer lifespan, influenced by lack of hormonal secretions in females or simply be more readily recognized as a complication of PWS. ***See*** Osteoporosis

Skin picking can be severe. Infected sores, which were originally minor spots, bites or even operation scars, remaining as open wounds for months or even years, may be related to reduced pain threshold.

Oxygen deprivation leads to sleep disturbances and somnolence; control of weight helps prevent the severity.

There are reports that sudden onset of severe chest infection and/or pneumonia can result in early death.

Some individuals are reported to suffer mental illness which requires specialist treatment.

As well as affecting life expectancy, many of these medical complications not only affect health but can precipitate an alteration in living conditions; crisis management rarely provides the best alternative living environment.

Independence

Adults with PWS are no different from others; they have a right to independent living. What is different is the nature of the eating disorder which limits normalization. Complete independence has not yet proved successful. Unsupervised access to food leads to a rapid onset of weight; excessive weight increase decreases mobility, in addition to affecting the severity of other social factors – obtaining clothing and shoes, the ability to climb on and/or off public transport, or to sit comfortably at the cinema or training centre, etc.

Adults with the syndrome may appear eminently capable of looking after themselves in all areas. Carers therefore have a duty to plan appropriate community care in a loving and safe environment. This may be a group home (the US Prader–Willi Syndrome Association has supported the development of homogeneous PWS group homes where individuals live harmoniously together), care village, hostel or sheltered housing. Each environment must be considered on its own merits as to whether it can handle the complex problems of PWS alongside other residents and whether it is suitable for a particular individual. Placements which do break down are generally those where insight into and belief of the difficulties are lacking.

Realistic expectations are essential and no carer should have guilt feelings about restricting access to food.

There are, at the time of writing, support groups for parents and professionals in 20 countries; in May 1991 an International Association was established with a representative from each of these.

Bibliography

Bray GA, Dahms WT, Swedloff RS *et al.* (1983) The Prader–Willi syndrome: a study of 40 patients and a review of the literature. *Medicine* 62(2): 59–79.

Cassidy SB (1984) *Prader–Willi Syndrome. Current Problems in Pediatrics*, vol. XIV, no. 1. Chicago: Year Book Medical Publishers.

Cassidy SB (ed.) (1991) *Prader–Willi Syndrome and Other Chromosome 15q Deletion Disorders – Proceedings of Workshop May 1991.* NATO Advanced Studies Institute Series.

Caldwell ML and Taylor RL (eds) (1988) *Prader–Willi Syndrome, Selected Research and Management Issues.* New York: Springer-Verlag.

Greenswag LR and Alexander RC (eds) (1988) *Management of Prader–Willi Syndrome.* New York: Springer-Verlag.

Holm VA, Sulzbacher SJ and Pipes PL (eds) (1981) *Prader–Willi Syndrome.* Baltimore: University Park Press.

PWSA (1987) *Development of Proper Placement for Individuals with Prader–Willi Syndrome.* Minneapolis: Prader–Willi Syndrome Association.

Waters J (1990) *Prader–Willi Syndrome and the Older Person. A Handbook for Parents and Professionals.* Prader–Willi Syndrome Association (UK).

Margaret S. N. Gellatly
Honorary Dietary Adviser PWSA (UK), Chelmsford, UK

PRAWNS

See Shellfish

PREGNANCY

Contents

Metabolic Adaptations and Nutritional Requirements

Predicted Energy and Nutrient Costs

Women during pregnancy are generally believed to be nutritionally 'at risk'. There are obvious energy and nutrient costs arising not only from the growth of the fetus and placenta, but also from enlargement of the maternal reproductive tissues and the deposition of a substantial energy reserve in the form of fat. These have been determined by direct chemical analyses and by indirect measurements (Table 1), and have been used as the basis for predicting additional nutritional needs for a successful outcome in pregnancy.

The Importance of Body Fat

The relationship between nutrition and reproduction, however, begins at an earlier stage. In late childhood body fat accounts for about 12% of bodyweight in both boys and girls. With the onset of puberty, the proportion of body fat rises in girls to reach around 17% at menarche, and 24% in the physiologically mature woman some 12 months later. The attainment of this additional fat deposit, amounting on average to 226 MJ (54 000 kcal) of stored energy, is believed to be the major determinant of fertility in normal healthy women. Young women who for professional reasons must maintain a low proportion of body fat, such as track athletes and ballet dancers, are frequently infertile, and the cessation of menstruation is an early symptom of the wasting disease anorexia nervosa. Should conception occur, total body fat is calculated to be sufficient to meet the energy costs of pregnancy (excluding further deposition of fat during gestation) and to sustain a 3-month period of lactation. ***See*** Anorexia Nervosa; Fats, Requirements; Fertility and Nutrition; Lactation

Table 1. Calculated energy and nutrient costs of pregnancy

Energy/nutrient	Cost
Energy (fat reserve, new tissue synthesis increase in basal metabolic rate)	335 MJ (80 000 kcal)
Protein (maternal reproductive tissues, conceptus)	910 g
Calcium (fetal skeleton)	28 g
Iron	300–400 mg

Effects of Undernutrition

The validity of this argument was revealed in a study of the effects of acute severe malnutrition that occurred in the winter of 1944 and in the following spring in cities of the western regions of The Netherlands. Immediately before the famine, the nutritional status of the population had been generally satisfactory. Thereafter, however, the situation deteriorated rapidly, and food from all sources provided no more than 2·1–2·5 MJ (500–600 kcal) per day. The incidence of amenorrhoea rose dramatically in the young female population. At the height of the famine, the number of infants conceived fell to 50% of the previous conception rate. Nevertheless, in the most vulnerable women, exposed to famine in the third trimester, average birthweight declined to a maximum of about 300 g below the prefamine level. Since the body of a healthy newborn infant contains more than 500 g of fat, most if not all of the deficit in birthweight may have been attributable to an inability to accumulate fat, priority being given to lean tissue growth.

Acute severe malnutrition, however, occurs rarely among normally well-nourished populations, and might give a misleading picture of the interaction of undernutrition and reproductive performance. In most developing countries chronic moderate malnutrition is commonplace, and nutritional surveys invariably report mean birthweights significantly below the average for Europe and North America. Although this has commonly been thought to reflect the poor nutritional status of the mothers, it has been impossible to isolate the putative influence of maternal diet on fetal growth from that of other environmental factors associated with endemic malnutrition, perhaps the most important being the smaller stature of the mothers. Even within an affluent society, an association between birthweight and maternal height and weight-for-height is found. ***See*** Malnutrition, Malnutrition in Developed Countries; Malnutrition, Malnutrition in Developing Countries

Nutrition Intervention and Pregnancy Outcome

A number of well-controlled nutrition intervention studies were carried out in developing countries in the early 1970s, no doubt with the aim of demonstrating the need for, and benefit to be derived from, diet supplementation during pregnancy. The subjects lived in countries in which protein–energy malnutrition was endemic. Energy intakes ranged from 5·86 to 7·53 MJ (1400–1800 kcal) per day. Given the small stature of the mothers, such a marginal plane of nutrition would just satisfy the needs of a nonpregnant woman. The dietary supplements provided energy alone or energy with protein, and were administered on a scale that often exceeded the estimated additional costs for energy and protein during pregnancy. The findings were as surprising as they were informative.

Only in one study, conducted in rural Guatemala, was a significant increase in mean birthweight achieved, amounting to 117 g. Although most women showed no improvement in reproductive performance, the response was very variable, those with the lowest customary energy intakes experiencing the greatest benefit, and the proportion of low-birthweight infants (<2500 g) was consequently reduced.

It was evident from these studies that there exists an 'energy threshold', around 6·7–7·1 MJ (1600–1700 kcal) per day, above which diet supplementation has no effect, but below which fetal growth may be stimulated, the improvement being directly related to the degree of energy deprivation. It was also revealed that the major determinant of fetal growth was the maternal energy supply; the inclusion of protein in the supplements had no effect independent of the energy it provided.

Dietary Habits in Well-nourished Populations

The results of the various intervention studies clearly indicate that if a woman is adequately nourished then pregnancy entails no appreciable additional energy (food) cost. It could be argued that the fetus is remarkably protected from the influence of maternal malnutrition, and that women exposed to generations of deprivation have simply adapted to a limited food supply. Small stature is often cited (incorrectly) as an example of adaptation. Studies of the dietary habits of women living in affluent circumstances, however, have shown that this is not so.

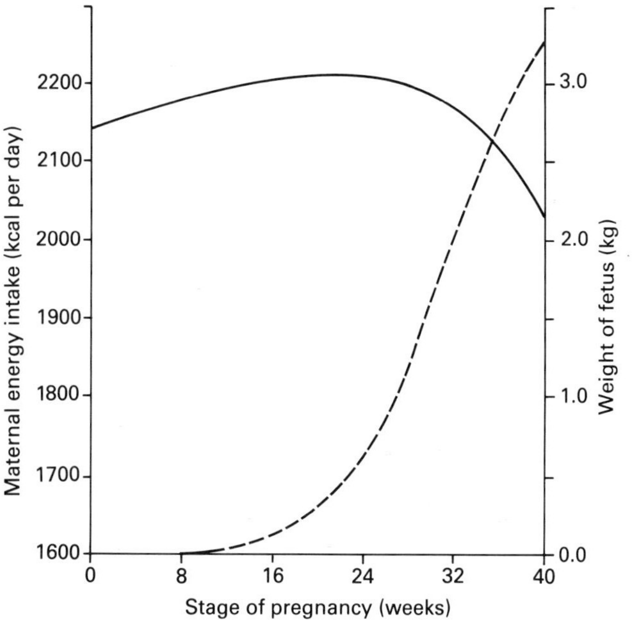

Fig. 1 Diagram showing changes in maternal energy intake (—) and weight of the fetus (- - -) at different stages of pregnancy.

As long ago as 1953, analyses of the diets of 2300 well-nourished women in Nashville, USA failed to show an association between maternal food intake and birthweight. Furthermore, on average, no increase in food consumption during pregnancy was apparent (Fig. 1). During the second trimester an increase of around 251 kJ (60 kcal) per day was noted, but during the last trimester, when a marked acceleration in fetal growth occurs, energy intake fell by 753 kJ (180 kcal) per day. This surprising discovery, ignored for 30 years, has been amply confirmed by studies in the 1980s in Paris, Cambridge and Glasgow.

How then is the obvious requirement for increased nutrition during pregnancy (Table 1) to be reconciled with the mother's failure, even under the most favourable environmental circumstances, to increase food intake? Only a small fraction of the fat stored in the first two trimesters of pregnancy is mobilized before term, and there exists no *ad hoc* store of minerals and water-soluble vitamins. The answer to this question is to be found in the complex mechanisms whereby nutrients are transported from the maternal circulation to the fetus, and in the adjustments that occur in the maternal physiology in response to the gravid state.

Metabolic Adaptations and Nutritional Requirements

The Role of the Placenta

The role of the placenta is discussed in the subsequent article. It is primarily an organ for nutrition of the fetus, but functions also as an endocrine gland throughout the greater part of gestation, regulating nutrient utilization by the mother.

Glucose provides the energy for fetal growth, the fetus being unable to oxidize fatty acids. Transfer of glucose across the placenta is by 'facilitated diffusion', and is therefore influenced by the maternal blood glucose concentration. A prolonged low blood glucose concentration, as may occur in mothers in the developing world engaged in hard physical work, or an abnormally high concentration associated with gestational diabetes can therefore lead to growth retardation, or to excessive growth and fat deposition (macrosomia). From early pregnancy the fasting blood glucose concentration is reduced, a change that may favour transfer to the fetus since the placenta is particularly effective at extracting nutrients from low concentrations in the blood. Glucose tolerance following a glucose load is also progressively relaxed so that the glucose concentration remains elevated for a longer period, again facilitating placental uptake. Whether or not the metabolic response to carbohydrate consumed as part of a meal is similarly affected is unclear. ***See*** Glucose, Maintenance of Blood Glucose Level; Glucose, Glucose Tolerance and the Glycaemic Index

Fat-soluble vitamins also cross the placenta by diffusion, maternal blood levels being maintained by dietary input and mobilization of tissue stores. In contrast, amino acids, minerals and the water-soluble vitamins are actively transported, the concentrations in the fetal circulation being much higher than in the maternal circulation. Furthermore, there is evidence that some vitamins are chemically altered by the placenta to a form which is unable to diffuse back into the maternal bloodstream, so reinforcing the one-way transport mechanism. Thus the fetus has prior claim on the nutrient supply, much of which is maintained, by homeostatic regulation, within narrow concentration limits in the maternal plasma. In adverse nutritional circumstances it is the mother, not the fetus, who is 'at risk'. ***See*** Amino Acids, Metabolism; Minerals, Dietary Importance; Vitamins, Overview

Metabolic Adjustments in Pregnancy

Under normal circumstances the integrity of the tissues is assured by homeostatic regulation of the absorption, excretion and catabolism of the nutrients. During reproduction these control mechanisms are adjusted, under the influence of hormones secreted by the placenta, in order to make available an additional nutrient supply for growth of the conceptus.

Absorption of the Nutrients

The macronutrients, fat, carbohydrate and protein, are almost completely digested and absorbed in the small intestine, so that no appreciable change occurs in pregnancy. The absorption of minerals, however, is carefully regulated. Most Western diets provide a considerable excess of calcium, and no more than around 30% is normally absorbed. From early in pregnancy the blood concentration of the physiologically active form of vitamin D (1,25-dihydroxycholecalciferol) is elevated, leading to an increased uptake of calcium from the gut, thus allowing the mother to satisfy the needs of her developing fetus without recourse to her own skeletal calcium or the need to increase her dietary intake. Only when her customary calcium intake or the biosynthesis of vitamin D is severely restricted is calcification of the fetal skeleton compromised. ***See*** Calcium, Physiology; Cholecalciferol, Physiology

As with calcium, the proportional absorption of iron from a diet providing 12–14 mg per day is very small (about 10%), but, as gestation advances, iron uptake increases progressively, and may exceed 40%. This adjustment in absorption, in association with a reduction in iron loss resulting from the cessation of menstruation, is sufficient to satisfy the iron cost of pregnancy. Judged by normal standards the characteristic fall in haemoglobin concentration that is noted in pregnancy would indicate a state of anaemia. Blood volume, however, is considerably increased so that the total amount of haemoglobin in circulation, and consequently the oxygen-carrying capacity of the blood, is actually raised. Supplementation of the diet of women who begin pregnancy in an iron-sufficient state may therefore be injurious as it has been shown to increase red cell size and could adversely affect blood flow in the small capillaries. ***See*** Iron, Physiology

Other minerals, such as copper and zinc, may also show improved absorption. ***See*** Copper, Physiology; Zinc, Physiology

Urinary Excretion of Nutrients

The glomerular filtration rate is increased during pregnancy. It is suggested that the presence of traces of glucose in the urine at some stages of pregnancy in most women merely indicates that more glucose has been filtered from the plasma than can be reabsorbed by the kidney tubules at that time. Some women, however, may lose large amounts of glucose in their urine, particularly those who develop gestational diabetes. ***See*** Kidney, Structure and Function

The urinary excretion of most nutrients is altered, but whether or not this reflects specific modifications in kidney function is unclear since the picture is obscured by withdrawal of nutrients from circulation by the placenta, which changes with time. The pattern of excretion of amino acids is most consistent. In nonpregnant women, about 1% of total nitrogen excretion is accounted for by free amino acids, but urinary loss rises in early pregnancy, reaching a plateau in midpregnancy at a level approximately double that found in the nongravid state. Most of the increase is accounted for by the nonessential amino acids, possibly reflecting selective uptake by the placenta. Since the plasma amino acid concentrations show little change, the large increase in urinary excretion cannot be attributed to kidney overload and is most likely the result of altered hormonal status affecting kidney function. The return to normal levels of excretion during lactation supports this conclusion.

One example of altered kidney function that favours conservation has clearly been identified. The amino acid taurine, a derivative of the nonessential amino acid cystine, shows a dramatic and sustained reduction in excretion from early pregnancy. It has been suggested that taurine may act as a membrane stabilizer, as an inhibitory neurotransmitter or neuromodulator, and as a growth modulator in fetal tissues. Tissue concentrations are particularly high in the fetus, and the suppression of urinary excretion of taurine during pregnancy is seen as a means of satisfying the needs of the developing fetus which lacks the ability to synthesize taurine *de novo*.

The Metabolism of Protein

The rate of accretion of nitrogen by the fetus and maternal reproductive tissues is known to rise 10-fold between the first and last quarters of gestation. Since, on average, no appreciable change in food intake, and thus protein intake, occurs in pregnancy, then one would expect to find a positive nitrogen balance, rising continuously throughout pregnancy. In the few nitrogen-balance studies carried out on pregnant women, however, no differences have been found between the values for nitrogen retention measured at different stages of gestation. This apparent paradox was resolved by the study of reproduction in the rat. ***See*** Protein, Digestion and Absorption of Protein and Nitrogen Balance

During the first 2 weeks of pregnancy, when competition between the dam and her fetuses for nutrients is minimal, a substantial reserve of protein is built up in the muscles (the 'anabolic phase'). During the final week, the period of rapid fetal growth, the protein reserve is broken down, the amino acids released being taken up by the placenta for growth of the conceptus. This 'catabolic phase' is not influenced by the protein content of the maternal diet, and is controlled by the hormone progesterone. ***See*** Hormones, Steroid Hormones

Evidence for a similar biphasic system of protein metabolism in women was provided by measurement of the excretion of the amino acid 3-methylhistidine. This amino acid is found predominantly in muscle. Histidine undergoes methylation after its incorporation into the peptide chains of the contractile proteins actin and myosin, and, in the course of muscle protein turnover, the 3-methylhistidine released is not reutilized, but is quantitatively excreted in the urine. In pregnancy the excretion of 3-methylhistidine rises sharply during the last trimester, indicating the hydrolysis of an amount of protein approximating the estimated protein cost of pregnancy.

Such a redistribution of amino acids from maternal to fetal tissues would not be detected in measurements of nitrogen balance.

This cyclic course of protein metabolism has important nutritional implications. The protein cost of pregnancy is distributed over the whole of the gestational period, and the influence of acute or chronic maternal undernutrition on fetal growth is minimized.

Nitrogen retention occurs when the intake of protein-nitrogen exceeds nitrogen losses from the body, largely the products of amino acid catabolism in the urine (mostly urea) and undigested dietary protein in the faeces along with cells shed from the gut epithelium. If, during pregnancy, the catabolism of amino acids were in part suppressed, then the fraction spared could be used for fetal protein synthesis. In the case of a woman existing on 40 g of protein per day, as in many developing world populations, a reduction of less than 9% in amino acid oxidation would spare enough protein to cover the entire estimated requirement for pregnancy.

The hypothesis was tested in the rat. The activities of two rate-limiting enzymes (alanine aminotransferase and argininosuccinatesynthetase), which regulate the oxidation of amino acids and the conversion of amino-nitrogen to urea, were measured in the livers at different stages of gestation. The activities of both enzymes were markedly depressed by the end of the first week, and declined even further (by around 50%) as pregnancy advanced. A parallel change in plasma urea was also noted.

Evidence for a similar adjustment in amino acid catabolism in pregnant women was later obtained in a metabolic study using urea labelled with a stable isotope of nitrogen. Measurements made in the last trimester and in the postpartum postlactational period showed a reduction of 30% in the rate of urea synthesis in late pregnancy and a similar fall in the plasma urea concentration.

The mechanism for protein sparing, which operates progressively throughout pregnancy, combined with the biphasic system of early storage and later breakdown of stored protein ensures a supply of amino acids commensurate with the demands of the growing fetus. The suppression of hepatic amino acid oxidation was also shown to be induced by the placental hormone progesterone. The fetoplacental unit thus indirectly controls its supply of amino acids as well as its energy needs.

The Metabolism of Energy

As stated earlier, the dietary energy supply is the major determinant of fetal growth. In healthy pregnant women energy balance becomes positive during the first trimester, probably in response to the rising secretion of progesterone. The purpose of the augmented fat reserve, amounting, on average, to some 4 kg of fat, is primarily to subsidize the high energy cost of lactation, but a small amount is mobilized in late pregnancy to provide an alternative fuel for use by the maternal tissues and enhance the availability of glucose for use by the fetoplacental unit. The human fetus derives its energy almost exclusively from the oxidation of glucose. ***See*** Energy, Energy Expenditure and Energy Balance

The discrepancy between the measured energy intakes of healthy pregnant women and the values predicted for energy expenditure has yet to be explained. There is no doubt that some saving is made in energy expenditure from a reduced level of physical activity, but this is difficult to quantify.

The alterations in carbohydrate metabolism that are characteristic of pregnancy could also lead to the sparing of energy. In pregnancy there is an increased output of insulin in response to a glucose stimulus, and reduced glucose uptake by the peripheral tissues (muscle and adipose tissue). Insulin antagonism has been attributed to the action of placental lactogen. Consequently, more of the ingested carbohydrate is directed to the liver, the organ that maintains the blood glucose concentration. The conversion of carbohydrate to fat is a very costly process. Approximately 20% of the energy that is available from the direct oxidation of glucose is lost as heat if the glucose is first converted to fat and the fat is later oxidized to produce energy. This adjustment in carbohydrate utilization, therefore, not only conserves energy for anabolic purposes, but also safeguards the fetal energy supply. ***See*** Carbohydrates, Digestion, Absorption and Metabolism

Dietary Recommendations for Pregnancy

Growth of the fetus is little affected by transient or prolonged moderate undernutrition of the mother. Relatively small deficits in birthweight may be accounted for by a lower proportion of body fat resulting from maternal dietary energy restriction in late pregnancy, from the diversion of glucose (the fetal fuel) to the muscles should the mother be engaged in hard physical work throughout pregnancy, or from malfunction of the placenta.

The security of the fetus is provided by the placenta, an organ designed not only to ensure a priority claim on all nutrients present in the maternal circulation but, by the secretion of hormones, to modulate the homeostatic regulation of nutrient utilization at all levels – absorption, excretion and catabolism, in order to augment the nutrient supplies. No government committee responsible for devising dietary guidelines would be so incautious as to suggest that women during pregnancy require no more food than when in the non-gravid state, although all evidence points very clearly to that conclusion. One obvious *a priori* condition would be that nutritional status should be satisfactory before conception and throughout pregnancy.

Attention has been focused on the latter half of pregnancy, when fetal growth is most rapid, the need for nutritional input is at its greatest, and the effects of maternal food deprivation most apparent. These considerations led a Joint FAO/WHO (Food and Agriculture Organization, World Health Organization) Ad Hoc Expert Committee in 1973 to propose an additional 628 kJ (150 kcal) per day for the first trimester, rising to 1464 kJ (350 kcal) per day for the second and third trimesters, the stage of pregnancy when women, unencumbered by professional advice, would spontaneously reduce their food consumption. In the light of continuing research, however, estimates of energy and nutrient requirements are being revised in a downward direction. In 1974 the US report on Recommended Daily Allowances proposed an additional daily supplement of 30 g of protein, 400 mg of calcium and 30–60 mg of iron for the pregnant woman, acknowledging that such a high intake of iron could not be met by the iron content of habitual US diets. The implication was that pregnancy was a clinical condition that required therapeutic intervention. One decade later, the proposed increments in energy and calcium were little changed, but the supplement of iron was reduced to 15 mg, and the protein allowance was changed to anticipate the pattern of accretion by the fetus, rising from 1·2 g per day in the first trimester to 10·7 g in the final trimester of pregnancy. ***See*** Dietary Requirements of Adults

The UK dietary recommendations for pregnancy have consistently been on a less generous scale. Over a similar period the daily allowance for energy has fallen from 1004 kJ (240 kcal) in the second and third trimesters to 837 kJ (200 kcal) during the third trimester only. Protein is unchanged at 6 g per day throughout pregnancy, but the recommendation for calcium has fallen from 700 mg per day in the third trimester to zero. Likewise, no recommendation was made for an increase in iron intake in the 1991 report on Dietary Reference Values, compared with the small increase of 1 mg per day in the earlier 1979 edition.

There is no doubt that as scientific opinion changes, other values will also be reduced, and nutritional guidelines for pregnancy will ultimately correspond to the dietary practices of healthy women satisfying their natural appetites on a well-balanced diet.

Bibliography

Campbell DM and Gilmer DG (eds) (1983) *Nutrition in Pregnancy*. London: Royal College of Obstetricians and Gynaecologists.

Department of Health (1991) *Dietary Reference Values for Food Energy and Nutrients for the United Kingdom*. London: Her Majesty's Stationery Office.

Hytten F and Chamberlain GVP (eds) (1980) *Clinical Physiology in Obstetrics*. Oxford: Blackwell Scientific Publications.

Naismith DJ (1983) Maternal nutrition and fetal health. In: Chiswick ML (ed.) *Recent Advances in Perinatal Medicine*, vol. I, pp 21–39. Edinburgh: Churchill Livingstone.

National Research Council (1989) *Recommended Daily Allowances* 10th edn. Washington, DC: National Academy Press.

Donald J. Naismith
King's College, London, UK

Role of Placenta in Nutrient Transfer

During intrauterine life the developing fetus effectively receives all of its nutrition across the placenta, which in the human has a chorionic villus structure in which maternal blood directly superfuses the external surface, the trophoblast of the placenta. This trophoblast forms an unusual epithelium separating the maternal plasma from the fetal extracellular space of the villus core through which the fetal capillary circulation flows. For the fetus the trophoblast is thus equivalent to the epithelium lining the small intestine of the newborn, since from a nutritional perspective it must absorb those molecules required for both growth and maintenance of the organism. The trophoblast also has an important role in acting as the lung for the fetus (since all gas exchange between mother and baby must occur across this surface), as the fetus' kidney (since excretion from the conceptus to the mother occurs across this structure), as well as being an important endocrine tissue secreting peptide and steroid hormones into the mother. The placenta itself also plays a very substantial role in intermediary metabolism, so that it cannot be assumed that there is no metabolism of absorbed nutrients during solute movement from mother to fetus.

As an epithelium the trophoblast has two surfaces, the one facing the mother and the other facing the fetus. These surfaces differ structurally; for example, microvilli are found projecting into the maternal bloodstream at the apical surface, where they form a brush border, whereas the basal surface facing the fetus does not have this surface specialization. From the functional point of view it is the differences in the distribution of transport membrane proteins (channels, carriers and pumps) that determines the overall transport of nutrients from mother to baby, at least as far as water-soluble molecules are concerned. In the placenta all transport appears to be across the trophoblast since, uniquely, the cells that compose this structure are fused to form a syncytium; this is in contrast to other epithelia where transport between adjacent epithelial cells allows a functional paracellular route to lie in parallel to the route through the epithelial cells themselves (the transcellular route). The question in the human as to whether there is a special route available for the transport of large molecules (MW 5000) is not yet resolved; certainly at term the human placenta does appear able to permit the transport of larger molecules at a slow rate from baby to mother. This route is unlikely to be of nutritional significance, since the rate of transport is low, but it may be important with regard, for example, to immunological sensitization.

The placenta, together with the growing fetus, has a very substantial metabolic energy requirement and, in the human, ATP synthesis appears to be met by the very substantial rate of glucose delivery from the mother. The glucose transporter systems that are found in both the apical and basal surfaces of the human trophoblast appear to be the type named GLUT1; in other words, they are sodium-independent facilitated transporters which are not regulated by insulin. The K_m for glucose transport at both surfaces is relatively high (approximately 30 mM) and the V_{max} is very substantial. The result of this is that glucose delivery across the brush border membrane of the placenta will be in a direction and at a rate that is dependent solely on the chemical driving force from mother to baby (maternal–fetal plasma glucose concentration). It seems likely that this fundamental property is the basis for the macrosomia ('large-for-dates') found in the babies of mothers with elevated plasma glucose concentration as typically found in diabetes mellitus. Transport of glucose, which is stereospecific, may be inhibited by glucose analogues which share the chemical structure of *d*-glucose at carbon 1; for example both 3-*O*-methylglucose and 2-deoxyglucose are transported, whereas α-methylglucoside (with a methyl group on carbon 1) is not. The transport of other monosaccharides has been studied rather little; in the human, fructose is transported much more slowly than glucose itself (in contrast to other nonprimate species). The question of regulation of carbohydrate delivery across the placenta is not fully resolved since some glucose transporters may be regulated by phosphorylation and the gradient for transplacental glucose delivery will also depend upon factors regulating glucose utilization and production by the placenta itself; little is known of the physiological regulation of either of these processes. ***See*** Carbohydrates, Digestion, Absorption and Metabolism; Glucose, Maintenance of Blood Glucose Level

In contrast to glucose transport, amino acid transport can be powered. The overall gradient of amino acid between maternal plasma and fetal plasma varies for individual amino acids. Typically, individual amino acid concentrations are twice as high in fetal as in maternal plasma. Current understanding of the mechanisms responsible for this relates to the distribution of the membrane transport proteins between the two faces of the trophoblast and in particular to the distribution of sodium-coupled transporters which are found predominantly (although not exclusively) in the brush border. Recent work using isolated membranes which reseal to form artificial structures (vesicles) has been useful in establishing the numbers and properties of such transporters. In addition to the direct effect of sodium ions in moving amino acids into the trophoblast across the brush border membrane against a concentration gradient, these transporters are often electrogenic and are thus also driven physiologically by the membrane potential. One example of such a process is the transport system called A, which uses alanine, serine and proline as transported substrates and accumulates these amino acids in the trophoblast against a concentration gradient. These amino acids then leave the trophoblast across the basal membrane by a different transport system. Other amino acids may be transported via tertiary active transport; for example, leucine is found in higher concentration in fetal than maternal plasma but it is not itself a substrate for sodium-coupled transport; rather it appears to exchange with amino acids such as alanine which have been accumulated in the trophoblast as just described.

The cationic and anionic amino acids are unusual in that, having their own charge, they will be accelerated or retarded by the membrane potential in crossing each of the plasma membrane surfaces of the trophoblast. For cationic amino acids (lysine, arginine, histidine), entry into the placenta appears to be largely by system y^+ (Na^+-independent), whereas exit into the fetus involves an electroneutral system (y^+L) which exchanges the cationic amino acid for a neutral amino acid (e.g. leucine) and a sodium ion, thus effectively solving the problem of permitting positively-charged amino acids to exit against an inside-negative membrane potential. For anionic amino acids (glutamate and aspartate) very high intratrophoblast concentrations are achieved by a transport system which is coupled to K^+ efflux as well as Na^+ entry. Essential amino acid requirements for the fetus are different from those of adults. However, it is not clear whether transport of specific amino acids across the placenta ever becomes rate-limiting for fetal growth. In the human, intrauterine growth retardation not associated with other disease has been shown to be associated with reduced placental delivery of amino acids through specific systems, e.g. associated with decreased function of system A. No compelling evidence is yet available to show whether these effects are due to primary or secondary events. In certain unusual metabolic disorders (e.g. maternal phenylketonuria, PKU) maternal levels of one particular amino acid may be elevated; this results in competition between this amino acid and others which share the same transporters. Some of the abnormalities found in the developing babies of such mothers may be a consequence of nutritional deprivation of, for example, tyrosine owing to competition by raised maternal phenylalanine levels for the delivery of this amino acid across the placenta. The fact that amino acid transport across the placenta involves a family of transport proteins with overlapping substrate (amino acid) specificities means that the nutritional consequences for the fetus of changing the level of one amino acid in the mother will be complex. This follows because, in contrast to placental glucose transporters, the K_m and V_{max} of the amino acid transporters are relatively low.

Lipid transport across the placenta in relation to human nutrition has been studied less rigorously, in part because it is likely to be perfusion- rather than membrane-limited since the lipid-soluble nature of such substrate allows ready transmembrane transport. Nutritonally the nervous system of the developing fetus requires substrate delivery of precursors for myelin synthesis. Studies suggest that placental binding proteins may provide a pool of essential fatty acids for fetal utilization. ***See*** Amino Acids, Metabolism; Fatty Acids, Metabolism

Transport of the inorganic cations of sodium and potassium involves both channels and transporters. Sodium transport into the trophoblast is coupled to the entry of those solutes (which include both organic and inorganic molecules) powered by secondary active transport. The extrusion of sodium across the basal surface of the trophoblast is likely to be a result of sodium pumping by Na^+/K^+ ATPase activity. In contrast, potassium, accumulated in the trophoblast as in other epithelia by the sodium pump requires channel-mediated release to account for its movements between mother and baby. Potassium channels have recently been shown to be sensitive to modulation in this tissue (e.g. by G proteins, by arachidonic acid and by pH). These regulatory factors may themselves be controlled by circulating factors in both mother and fetus. It is clear that fetal plasma potassium is carefully regulated by control of placental transport of this cation.

The divalent cation of calcium is found in higher concentration in fetal than in maternal plasma. Active transport processes are therefore involved in placental calcium metabolism and these are unusual in that regulation of transport is clearly precisely controlled. It seems likely that calcium extrusion across the basal surface of the trophoblast is ATP-driven and regulated by calmodulin. Entry of calcium across the brush border may be via calcium channels which are themselves regulated, for example, by membrane potential. Fetal parathyroid hormone is likely to regulate all of these events.

Iron is also transported from mother to fetus in a regulated fashion and again there is a greater concentration of iron in the fetal circulation. In some species transport of iron has been shown to be active in that it is inhibited by anoxia. The role of the transferrin receptor found in the human placental microvillus membrane is probably related to the mechanism of ion entry into the cytosol. From this compartment iron has to leave, but the mechanisms responsible for this are not fully understood. ***See*** Calcium, Physiology; Hormones, Thyroid Hormones; Iron, Physiology; Potassium, Physiology; Sodium, Physiology

Anion entry into the placenta has been studied using isolated membrane preparations. These studies show that for monovalent ions (chloride and other halides) two routes are available, an exchange and a conductive route, the latter likely to be via channels. The anion exchange system appears to be functionally linked to the transport of organic anions (bicarbonate, lactate) from placenta into the maternal circulation. Phosphate transport is also regulated and appears to involve a sodium-dependent cotransporter at the maternal-facing surface and an efflux mechanism (possibly driven by the membrane potential) at the basal surface of the tissue. As for calcium transport, phosphate delivery is regulated by parathyroid hormone concentration in the fetus.

Trace element delivery across the placenta also involves specific placental binding proteins analogous to those found in adult liver; however, membrane transport is also required to allow such ions to gain access to and from the trophoblast. The nature of such transporters varies greatly. For zinc there is evidence that the histidine amino acid transporter is responsible for delivery of a histidine–zinc complex, whereas for transition-metal oxides it appears that anion exchange is important. The transport of iodide across the human placenta is also likely to be by anion exchange since SCN can inhibit it, but the mechanisms responsible for the concentration of this element in the fetal compartment are not clear. Selenate appears to share a pathway with sulphate for entry across the brush border and the transport of both of these ions is inhibited by blockers of anion exchange. This pathway appears to be shared with those available for transport of the trace elements chromium and molybdenum as chromate and molybdate. ***See*** Trace Elements

Vitamin transport also is highly specific for individual substrates; thus for ascorbic acid a sodium-independent transporter for the reduced form of this nutrient has been described in the brush border membrane; this transporter appears to be functionally coupled to a placental system which maintains ascorbate in this chemical form.

Bibliography

Bain K *et al.* (1990) Permeability of human placenta *in vivo* to four non-electrolytes. *Journal of Physiology* 431: 505–513.

Boyd CAR and Kudo Y (1990) Inhibition of placental tyrosine transport in maternal PKU. *Journal of Inherited Metabolic Disease* 13: 17–28.

Boyd CAR and Kudo Y (1990) Placental amino acid transport. *Biochimica et Biophysica Acta* 121: 169–174.

Jones CT and Rolph TP (1985) Substrates for foetal nutrition. *Physiological Reviews* 65: 357–430.

Shennan DB and Boyd CAR (1986) Placental ion transport *Biochimica et Biophysica Acta* 117: 1–30.

Shennan DB and Boyd CAR (1988) Trace element transport in placenta. *Placenta* 9: 333–343.

Sibley CP and Boyd RDH (1988) Control of transfer across the placenta. *Oxford Reviews of Reproductive Biology* 10: 382–435.

Smith CH *et al.* (1992) Nutrient pathways across the epithelium of the placenta. *Annual Reviews of Nutrition* 12: 183–206.

C.A.R. Boyd
University of Oxford, Oxford, UK

Safe Diet

Would-be pregnant and pregnant women have heightened concerns about their diet which can be summarized by the common question, ‘Will it harm my baby?’. Because the question is so common, many sources of advice are available in the lay press, not all as scrupulously researched and scientifically based as they might be. Another source of potentially misleading advice is the food industry, the agents of which possess a lack of objectivity that may not be obvious to the casual reader. Articles proliferate with titles such as ‘How to have a beautiful baby’ and ‘How to have a perfect baby’ (*sic*). These tend to contain a mixture of dietary and life-style advice with the implication that, if followed, the undesirable outcomes of pregnancy such as miscarriage, congenital malformation and fetal death are avoidable. Professionals in the field of nutrition must be aware of this for the following reason. If a woman learns of such advice only after her pregnancy has failed in some way there is a potential for a lifelong burden of guilt. For this reason purveyors of advice should restrict themselves to that which has been demonstrated in scientifically valid experiments. In particular, they should be guarded in extrapolation from observational studies and animal experiments.

This article highlights the principal areas in which women have dietary concerns. Some of these are dealt with in more detail elsewhere and are cross-referenced.

Preconceptional Nutrition

The human appears to be one of those species in which an adequate energy reserve is a requirement before conception can take place. Frisch has observed that at the completion of growth the body of a well-nourished woman contains about 26–28% fat and about 52% water. In contrast, a man's body contains about 14% fat and 61% water. The woman's fat store is thought to be a reserve against famine in pregnancy and lactation. The minimum fat content for the establishment and maintenance of ovulation was found to be 22%. Thus a loss of body fat of about 10–15%, through dieting, exercise or illness, leads to a reversible state of anovular infertility. ***See*** Body Composition

Whilst many claims are made about the importance of preconceptional micronutrients both in deficient and sufficient women, particularly in possible prevention of congenital malformations, only the link between multivitamins and neural tube defects bears scrutiny.

Nutritional Management of Common Symptoms in Pregnancy

Heartburn

Heartburn is thought to be caused by gastrooesophageal reflux. Although occasionally experienced in the first trimester, it is generally more common in the last trimester and occurs in 30–50% of women.

Small, frequent meals or snacks are usually tolerated better than large, well-spaced meals. Common foods cited as causing heartburn are spicy and fatty foods, fizzy drinks, citrus fruits, fruit juices, cucumber and bananas. Milk and milk products can help to relieve symptoms but antacids are frequently used.

Nausea and Vomiting

Psychological factors, changing hormone levels, hunger, altered carbohydrate metabolism and vitamin deficiencies have all been proposed as possible causes, but none has been confirmed.

Symptoms may start before the woman knows she is pregnant or in later pregnancy but commonly they are worst between weeks 6 and 10, and subside by about 16 weeks. Nausea is experienced at any time of day or night and can be either slight or severe. It often becomes worse when the stomach is empty, and it is relieved by eating small, frequent meals based on starchy carbohydrates. Morning sickness is common and this can be relieved by consuming dry biscuits or toast before getting up.

Nausea can also be triggered by travelling, fried and spicy foods and smells such as coffee, perfume and cigarette smoke.

Some women just feel nauseated while others actually vomit as well; this may cause minor weight loss but rarely causes nutrient deficiency. Women need reassurance that not eating proper cooked meals or losing some weight and their taste alterations will cause no problems for their developing fetus. The more severe cases of pregnancy vomiting (hyperemesis gravidarum) require hospital admission, intravenous fluids and sometimes parenteral nutrition.

Constipation

Constipation is common at all stages of pregnancy. It may be related to a general reduction in motility in the gastrointestinal tract, with prolonged transit times and increased water resorption from the stool. General advice about constipation is also suitable for pregnancy, i.e. increased intake of fibre, particularly cereal fibre, and increased fluid intake.

Constipation may be aggravated by consumption of iron tablets; if it is not appropriate to reduce or stop these, bulking agents may be prescribed.

Qualitative Aspects of Diet

The following section refers to common questions of dietary safety; some arise because of suspicion of harm when items are included in the diet, others for the paradoxical reason that their omission from the diet might be dangerous.

Sodium Chloride

Because of the relationship between chronic salt intake and hypertension the idea arose that a high salt intake might be a risk factor in pre-eclampsia. This question is reviewed elsewhere. ***See*** Hypertension, Hypertension and Diet

Alcohol

A well-defined group of anomalies referred to as the 'fetal alcohol syndrome' is now recognized. The major

Table 1. Congenital malformation rates and maternal alcohol consumption

Drinks per day	Malformation rates (per 1000)[a]
None	78
Less than one	77
One to two	83

[a] No statistically significant difference.
From Mills JL and Graubard BI (1987) Is moderate drinking during pregnancy associated with an increased risk for malformations? *Pediatrics* 80: 309–314.

defects of affected infants are weight and length below the 10th centile for gestational age, and microcephaly. The latter is a condition of small head size associated with an underdeveloped brain. Such children are likely to be mentally retarded and their physical growth is stunted. The syndrome has been reported in up to 40% of the infants of women drinking more than 6 units of alcohol per day. (One unit is 15 g of absolute alcohol, e.g. 0·28 l of beer, one glass of wine or one measure of spirits.) ***See*** Alcohol, Metabolism, Toxicology and Beneficial Effects

Women drinking more than 2 but less than 6 units per day may produce infants with a milder form of the syndrome sometimes referred to as 'fetal alcohol effects'. There is no evidence of harm from occasional drinking or the consumption of less than 2 units per day (see Table 1). Despite this, many women do give up alcohol when trying to conceive or whilst pregnant, and this seems a sensible but not mandatory practice.

Caffeine

Caffeine is present in tea, coffee, cola drinks and many over-the-counter remedies for colds and allergic symptoms. Animal experiments with high doses have shown that some congenital malformations may be induced but there is no evidence of harm in humans. A relationship between the consumption of doses in excess of 400 mg of caffeine per day and retarded growth of the fetus has been suggested. (This is the equivalent of three cups of filter coffee or six cups of tea per day.) Further analysis of the data and correction for smoking and demographic variables made it unlikely that caffeine had any independent effect on fetal growth. Other infant outcomes are compared with coffee consumption in Table 2. ***See*** Caffeine

Over-the-counter Remedies

As a general rule, self-prescribing in pregnancy should be kept to a minimum. Those drugs which have been used most widely and have been shown to be safe should be preferred to anything new or untested. Small doses of simple analgesic drugs such as aspirin and paracetamol appear to be safe, as do common cold remedies which often have these drugs as their active contents.

No problems have been reported with the use of homeopathic remedies in pregnancy. There has been much interest in possible therapeutic benefit from the consumption of oil of evening primrose in pregnancy. It has been hypothesized that the linoleic acid and γ-linolenic acid in the compound may stimulate production of vasodilatory prostaglandins and be of benefit in lowering blood pressure in abnormal states of pregnancy such as pre-eclampsia. The limited studies so far reported do not support this hypothesis.

Recent reports have compared women on diets rich in natural fish oil, the Faroese, with women on the Danish mainland whose diets contain much less fish oil. The Faroese had fewer premature babies, fewer problems with blood pressure, and heavier babies than the mainlanders. Unfortunately, these studies cannot be interpreted as showing uniform benefit from fish oil supplements as the Faroese also lost more babies as stillbirths. A further report, comparing women in Aberdeen with women in the North Sea Orkney Islands, showed the latter to have average birthweights 250 g higher. After correction, a small but significant proportion of the difference was found to be related to genetic or environmental factors. The latter included a diet in Orkney containing 30% more fish than in Aberdeen.

Ginger root may act as an antiemetic by a local effect in the stomach. No evidence of benefit or harm in pregnancy exists.

Garlic is widespread in the human diet. There has always been a recognition of possible benefits of garlic supplements on a wide range of disorders, including hypertension, hyperlipidaemia and thrombosis. No studies of supplementation in pregnancy have been reported. ***See*** Fatty Acids, Gamma Linolenic Acid; Fish Oils, Dietary Importance

Vitamin and Mineral Supplementation

Some pregnant women benefit from iron and/or folic acid supplements when dietary deficiency provokes anaemia. Although self-medication with vitamin and mineral supplements is widespread, there is a good case during pregnancy to restrict such supplements to those prescribed by doctors. Vitamin A and its analogues can cause congenital malformations in megadosage, and vitamin D supplementation is thought to have been the cause of idiopathic hypercalcaemia. This was a congenital syndrome consisting of abnormal facial features, mental retardation and abnormal calcium metabolism. It was common in the UK in the late 1950s and early

Safe Diet

Table 2. Infant outcomes (%) and maternal coffee consumption

	No coffee or tea	Coffee (cups per day)				
		0	1	2	3	⩾4
Birthweight < 1500 g	7·4	7·3	8·0	8·8	8·0	10·4
Preterm birth < 37 weeks completed gestation	7·2	7·3	6·7	8·0	8·5	9·7
Major congenital malformations	2·5	2·7	3·1	2·1	2·8	2·0
Minor congenital malformations	6·7	6·5	6·0	5·5	8·2	5·4
Low Apgar score (birth asphyxia)	7·6	7·7	7·9	7·6	8·5	6·4
Stillbirth	0·5	0·5	0·7	0·7	0·3	0·7
Respiratory problems	5·2	5·0	5·2	6·2	4·6	4·9
Newborn jaundice	19·1	19·1	20·5	19·9	18·6	18·2

From Linn S, Schoenbaum SC, Monson RR *et al.* (1982) No association between coffee consumption and adverse outcomes of pregnancy. *New England Journal of Medicine* 306: 141–145.

1960s when vitamin D supplementation of a variety of foods was practised. (The relationship of periconceptional multivitamins to neural tube defects is discussed elsewhere.) ***See*** Cholecalciferol, Physiology; Folic Acid, Physiology; Iron, Physiology; Retinol, Physiology

It seems clear that some members of dark-skinned races are at risk of vitamin D deficiency when resident in temperate climates. Their newborns may be at risk of hypocalcaemia, a cause of convulsions in the newborn period. These children may also have delayed growth in the first year of life. Both problems may be corrected by vitamin D supplementation in pregnancy.

At various times, maternal zinc deficiency has been proposed as a cause of both congenital malformation and retarded fetal growth. No valid evidence has yet emerged of any benefit derived from maternal zinc supplementation with regard to either outcome in humans. ***See*** Zinc, Physiology

Food Safety in Pregnancy

Pregnant women share the concerns of the general public in matters such as salmonellosis and other types of food poisoning. The risk of bovine spongiform encephalopathy (BSE), however remote, is a particular source of concern unless vertical transmission is shown not to occur. Listeriosis is also a significant problem in pregnancy because infection of the mother can cause fetal death following transmission of the organism across the placenta. It is not certain at the present time whether the apparent increase in fetal loss from this infection represents improved ascertainment or a genuine increase in disease frequency. There is a suggestion in the UK that altered dietary habits and poor food storage technique – including increased consumption of soft cheeses, pâtés and 'cook–chill' meals, and inadequate temperature control in display cabinets and domestic refrigerators – might be contributory factors. ***See*** Bovine Spongiform Encephalopathy (BSE); *Listeria*, Listeriosis

Toxoplasma is a protozoan organism which can cross the placenta when primary infection occurs in pregnancy. It can lead to fetal death, mental handicap and/or blindness in an estimated 30% of offspring of infected mothers. Sources of the infective form of the organism are raw or undercooked meat, particularly pork and mutton, and domestic cat faeces. Pregnant women should therefore avoid undercooked meat and should always wear protective gloves when gardening or handling cat litter trays. ***See*** Parasites, Illness and Treatment

Other concerns of pregnant women in the area of food safety include the hazards of fungicide and pesticide residues, and possible harmful effects of natural food toxicants. At present it seems unlikely that there are toxic residues on foods which can be associated with harm in pregnancy, but this observation should not be taken as grounds for complacency. One of the main difficulties for scientists wishing to explore this matter is the limited availability of reproducible assays for the minute quantities to be studied. Similarly, the study of natural food toxicants for possible harmful effects on the fetus is in its infancy. The hypothesis that glycoalkaloids from green potatoes might be causative of neural tube defects has not been substantiated. ***See*** Alkaloids, Toxicology; Fungicides; Pesticides and Herbicides, Toxicology

Conclusion

It is important to monitor continuously any effects our dietary habits may be having on child-bearing and child development, but we should be able to achieve this without raising unnecessary scares. Most women who consume a good general diet, and are not addicted to tobacco, alcohol or other drugs, can contemplate pregnancy without concern that their dietary habits can 'harm the baby'. ***See*** Smoking, Diet and Health

Safe Diet

Bibliography

Drife J and McNab G (1986) Mineral and vitamin supplements. *Clinics in Obstetrics and Gynaecology* 13: 253–267.

Frisch RE (ed.) (1990) *Adipose Tissue and Reproduction. Progress in Reproductive Biology and Medicine*, vol. 14. Basel: Karger.

Karen G (ed.) (1990) *Maternal-Fetal Toxicology: A Clinicians Guide*. New York: Marcel Dekker.

R. B. Fraser and F. A. Ford
Northern General Hospital, Sheffield, UK

Maternal Diet, Vitamins and Neural Tube Defects

Clinical Features, Aetiology and Epidemiology of Neural Tube Defects

Neural tube defects (NTDs) are generally considered to include anencephaly, encephalocoele, myelocoele (myelomeningocoele), meningocoele and complicated spina bifida occulta. Isolated hydrocephalus or simple spina bifida occulta, conditions in which an NTD is part of a chromosome abnormality such as trisomy 18 or triploidy, part of a recognized malformation syndrome such as Meckel–Gruber syndrome, or part of fetal disruption (amniotic band) syndrome are excluded. Anencephalics are nearly all stillborn and account for about 50% of NTDs; there is usually absence of the vault of the skull and disorganized brain tissue overlying the base of the skull. Encephalocoeles are relatively uncommon; they present with a posterior skull or upper cervical defect, allowing membranes, often with abnormal brain, to protude as a skin-covered sac. Such infants may survive but are often mentally impaired. In myelocoeles the defect is lower on the neuraxis, mostly in the lumbar or sacral region. It is a very variable condition, often with the unclosed neural tube exposed, and is generally associated with hydrocephalus, paralysis and incontinence, as well as leg and foot deformity. Untreated it has a poor prognosis for survival and even with treatment the majority who survive grow up wheelchair-bound and to some extent mentally impaired. In meningocoele, an uncommon condition, only membranes are involved, leaving the spinal cord largely intact. If operated upon, such infants grow up to be near normal. Cases of spina bifida occulta are not included in most epidemiological studies, partly because of difficulty in diagnosis.

With modern antenatal surveillance, in the form of maternal α-fetoprotein screening and ultrasound during midtrimester, the majority of cases of anencephaly and myelocoeles are now detectable, and 'secondary prevention' is possible by selective abortion of affected pregnancies. However, 'primary prevention', i.e. preventing the malformation from occurring in the embryo in the first place, is dependent on knowing the causes.

The NTDs are different end products of the same general process, which is now widely accepted to have a multifactoral aetiology. A major genetic component, thought to be polygenic, renders the embryo susceptible to interference by intrauterine environmental factors during the fourth week after conception when the neural tube develops. There are likely to be a number of such factors, perhaps with different ones being of importance in populations with a high incidence. As there is little likelihood of effectively modifying the genetic component, 'primary prevention' of NTD is dependent on being able to identify the environmental triggers, so that they can be either removed from the environment or avoided. Although NTDs can be produced in experimental animals by numerous and very differing environmental insults, epidemiological and clinical studies in humans have produced evidence of one specific environmental factor of importance, i.e. poor maternal nutrition and, more specifically, insufficient folic acid. In humans, a few cases seem to be precipitated by maternal anticonvulsant therapy with sodium valproate, poorly controlled maternal diabetes and hyperthermia caused by virus and other infections, and possibly steroids.

Poor Maternal Nutrition as a Cause of NTD

Up to the later 1960s in the UK the incidence of NTD was between 8 per 1000 and 3·5 per 1000 births depending on the region. In most of continental Europe, North America, most of eastern Asia and in many negroid populations the incidence is less than 1 per 1000. Poor maternal nutrition has for some time been suspected as an aetiological factor in the UK at least, as surveys have shown a higher incidence of NTD in social classes IV and V than in classes I and II. In addition, the more deprived regions, such as the Irish Republic, Northern Ireland, South Wales, the northwest of England and the southwest of Scotland, had the highest incidences, and there were seasonal variations, with a higher incidence amongst conceptions occurring in winter and spring when fresh foods tend to be less plentiful and more expensive.

In the last two decades, not only has the general incidence of NTD fallen dramatically, especially in the higher-incidence regions, but also social class differences and seasonal variations have become less pronounced. These changes may well be attributable to improving social conditions, although secular trends have been noted in the past and have usually been difficult to interpret. However, the sharp rise in the incidence of NTD in Boston, USA, during the depression years has been interpreted as the result of poor maternal nutrition prevailing at the time. In this connection in China, where the incidence of NTD is in general relatively low, a study of the births in seven townships in the north (Shuniyi area) found an incidence of almost 7 per 1000, as high as that in South Wales two decades ago. It is of special interest that in China conceptions most at risk are those occurring in winter, when the mothers are apparently on a particularly inadequate diet.

South Wales Dietary Studies

Studies were carried out in South Wales into the nutritional status and dietary intake of women who had had pregnancies with an NTD and who normally run a recurrence risk of about 1 in 20. Suitable women living in the counties of Glamorgan and Gwent were identified. They were questioned about their usual diet at the time of the survey (interpregnancy diet) and about diet in the first trimester of previous pregnancies.

The diet sheet used was designed to reveal any deficiency in the consumption of foods containing significant amounts of folic acid during an average week. The diets were classified as good, fair or poor. Good diets were those expected to provide generous intakes of all the nutrients (including folic acid) and dietary fibre. Poor diets were expected to provide marginal intakes of vitamins and minerals. The diets judged to be fair were intermediate between these extremes, but in addition were thought to provide low intakes of one or more nutrients. This classification was carried out after the end of the study by an independent research worker who was unaware of the outcome of the pregnancies. *See* Dietary Fibre, Properties and Sources; Folic Acid, Properties and Determination

At the start of and during the study, serum and red cell folic acid concentrations and serum vitamin B_{12} were measured; haematocrit was also estimated. *See* Cobalamins, Physiology

The women in the western area of Glamorgan were counselled to improve their diets and to stop smoking not later than the time when contraceptive precautions were stopped. Those in the eastern area were not specifically counselled.

Of 415 subjects in the study, only 65 reported a good interpregnancy diet, 197 a fair diet and 149 a poor diet; 4 diet histories could not be analysed. The mean serum folate concentration and red cell folate for these groups (Table 1), as well as the gradient of the measurements, suggest that the dietary assessment had some validity. Serum B_{12} concentration did not correlate with this classification of diet. In 174 women, there were 186 subsequently confirmed pregnancies (12 reporting twice) – 103 from west Glamorgan and 71 from east

Table 1. Women with a previous fetal NTD: quality of interpregnancy diet and mean concentration of serum folate, red blood cell (RBC) folate and serum B_{12}

Quality of diet	Subjects n	Subjects %	Mean serum folate ($\mu g\,l^{-1}$)	RBC folate (μg per l of red cells)	Serum B_{12} ($\mu g\,l^{-1}$)
Good	65	16	8·9	295	283
Fair	197	47	5·4	236	272
Poor	149	36	4·9	197	285
Not known	4	1			
All women	415	100	5·9	238	277

Table 2. First dietary study: outcome of project pregnancies by quality of diet in first trimester

Outcome of pregnancy	Quality of diet: Good	Fair	Poor	All pregnancies
Normal child	53	85	22	160
Recurrent NTD	—	—	8	8
Miscarriage	—	3	15	18
All outcomes	53	88	45	186

Glamorgan and Gwent. In previous epidemiological studies, no obvious differences in the social and demographic characteristics of the women in the two areas had been found and their diets in the interpregnancy period and during the first trimester of previous pregnancies were similar. In the western area, almost three quarters (71%) of women (counselled) improved their diet by their next pregnancy. Over 80% of the uncounselled women did not change the pattern of their diet. The outcome of 109 pregnancies in the western-area women was 10 miscarriages and 3 recurrences of NTD; amongst 77 in the eastern-area women there were 8 miscarriages and 5 recurrences. The ratio of the risk of recurrence in East Glamorgan compared to West Glamorgan was 2·4, although this did not differ statistically significantly from 1·0. All eight recurrences occurred in women whose diets were considered to be poor during the first trimester of that pregnancy (Table 2; $p < 0{\cdot}001$). Of the three women with recurrences who had received counselling, one had ignored the advice but two suffered nausea and vomiting during the first month of conception which had continued into hyperemesis gravidarum.

A second dietary study with women who had at least one previous pregnancy with NTD included two control groups – the sisters of the index cases who had had only normal babies, and a group of professional mothers. The dietary investigation was based on a modified form of Burke's dietary history.

Of the index women, 41% had a poor diet, compared with 26% amongst their sister controls and 6% of the upper-class controls. All the women were given dietary counselling. A total of 176 of the index women had further pregnancies, with only 32 (18%) remaining on a poor diet and with 5 NTDs (2·8%), about half the number of recurrences expected. All five recurrences occurred in women on poor diets, a rate of 16% ($p<0.01$).

Folate Deficiency as a Possible Cause of NTD

The effect of poor maternal nutrition on the developing embryo may be due to a number of vitamin deficiencies acting together to interfere with the closure of the neural tube, or to allow some as yet unidentified teratogen to have an influence. However, based on experimental, embryological and clinical evidence, it seems more likely that the main cause, in the UK at least, is a lack of folate.

Neural tube defects can be induced in the rat by exposing developing embryos to a variety of teratogens, after the 8th day following fertilization. In the rat the neural tube begins to form about the 8th day, closure being complete towards the middle of the 12th day, the stage when the rat embryo's chorioallantoic placenta begins to develop; the embryonic heart starts to perfuse it and the pace of development and energy requirement increase significantly. During this time there are also considerable changes in the metabolic processes. Before the neural tube is formed, the rat embryo derives its energy largely from glycolysis and the pentose phosphate shunt, but from that time there is a progressive switch to the tricarboxylic acid cycle. In the chick embryo it has been shown that neural tube closure is critically dependent on an oxygen gradient in the neural crests producing a narrow band of hypoxic surface cells along the opposing neural crests. Lack of sufficient folate at that stage, or a general metabolic disturbance, could then interfere with the normal closure process and either the crests do not meet correctly or if the hypoxic necrosis progresses too far, they meet and then dehisce again. Translated to the human embryo, all these processes take place between the 21st and 25th day after fertilization, and there is every reason to suppose that they differ only in detail and in timing from those in the chick and the rat. ***See*** Folic Acid, Physiology

Both the serum and the red cell folate levels are low in women who are on an unsatisfactory diet (Table 1), and Smithells *et al.* (1977) found similar results in women from social classes III, IV and V, both when not pregnant and during the first trimester, but especially at the beginning of a pregnancy that produced an infant with a NTD. Women who had previously given birth to one infant with a NTD possessed significantly lower serum and red cell folate levels, and more frequently showed evidence of disturbed folic acid metabolism, than matched controls. Those who had previously given birth to two or more infants with NTDs possessed even lower folic acid levels. This suggests that there might be an inherited disorder of folic acid metabolism as a predisposing factor. Finally, there is the small human experience with aminopterin, a folic acid antagonist in precipitating NTD.

Table 3. Folic acid supplementation trial: outcome of pregnancy by treatment group

Outcome of pregnancy	Folate supplement: Compliers	Folate supplement: Noncompliers[a]	Placebo supplement	All cases
Normal fetus	44	14	59	117
Neural tube defect	0	2	4	6
All outcomes	44	16	63	123

[a] Serum folate <10 $\mu g\ l^{-1}$.

South Wales Folic Acid Supplementation Trial

A double-blind randomized controlled trial of folate treatment compared with an inert placebo given before conception and continuing until at least the 12th week of pregnancy was started in South Wales in 1969 and continued until 1974. Those who agreed to take part were randomized to two treatment groups, taking either 4 mg of folic acid per day or the identical tablet of inert filler without folic acid. These women were instructed to take the tablets for at least 28 days before intending to become pregnant and to taken them until they missed a period, and for treatment to be continued up to 12 weeks of pregnancy.

There were 123 pregnancies which met the study criteria: the women had to have been taking tablets for 28 days before missing the first period, had to be seen within 11 weeks of the last menstrual period and had to provide a pregnancy dietary history; the serum folate concentration had to be reported by the laboratory. There were 60 pregnancies in the folate group and 63 in the placebo group. There were 4 recurrences of NTD out of 63 (6·4%) in the placebo group and 2 recurrences out of the 60 (3·3%) in the folate group (Tables 3 and 4). Although the relative risk was 2·0, or a possible 50% reduction in recurrences, this could not be considered statistically significant as the numbers were so small. However, the 2 recurrent cases in the folate group were both in noncompliant women (serum folate $<10\ \mu g\ l^{-1}$). Counting 16 noncompliers with placebo group, on the basis of biochemical outcome rather than treatment, 6 recurrences occurred in 79 unsupplemented

Table 4. Folic acid supplementation: outcome of pregnancy by quality of diet in first trimester, and by folate supplementation received[a]

Outcome of pregnancy	Quality of diet							
	Good		Fair		Poor		All diets	
	N	NF	N	NF	N	NF	N	NF
Normal fetus	18	25	16	32	10	16	44	73
Neural tube defect	0	0	0	0	0	6	0	6
All outcomes	18	25	16	32	10	22	44	79

[a] F, folate supplemented; NF, folate not supplemented; 16 non-compliers in folated-treated group included in NF.

pregnancies (7·6%) and none in the 44 truly supplemented pregnancies (Fisher's exact test $p=0{\cdot}04$). Thus the data suggest that there was a biological effect of folate which could be demonstrated, although in practice it was difficult to ensure that women took the tablets and so achieved satisfactory results.

The specific effect of taking folate had to be separated from the more general effect of diet. There were no recurrences among 91 pregnancies who received good or fair diet ($p<0{\cdot}001$, Fisher's exact test; Table 4). However, within the high-risk group of women on poor diets, there were no recurrences in the 10 who had taken folate supplementation but 6 in the 22 who had not taken supplementation ($p=0{\cdot}004$ Fisher's exact test). Thus, although there may have been some bias owing to women who were receiving an inadequate diet, the preventive effect of folic acid supplementation could still be detected despite an inadequate diet.

Conclusions from the South Wales Trials

It was demonstrated that in the circumstances applying in South Wales at the time of the trials (1969–1974), it was difficult to establish a trial of preventive treatment in pregnancy as there was no guarantee that the subjects would comply with a protocol. However, the following conclusions were reached:

1. Women who had previously had a fetal NTD were more likely to be eating a poor diet.
2. It is possible to persuade women to change their diet.
3. All the recurrences occurred in women reported to have a poor diet.
4. No recurrences occurred in women who took folate tablets, whether or not they had an adequate diet; similarly, no recurrences occurred in women who had an adequate diet, whether or not they were supplemented; all the recurrences occurred in the small group who had poor diets and were not supplemented. This does not exclude the possibility that there are other nutrient deficiencies or some abnormality of folic acid metabolism which can be swamped by an excessive intake of folate.
5. The supplementation study seems to demonstrate that an inadequate supply of folic acid to the embryo during the first 4 weeks of development is an important aetiological factor in NTD in South Wales at least, and that dietary supplementation with folic acid seems to prevent recurrences. It is not necessary for diets in the first trimester of women who have a fetus with an NTD to be frankly deficient in foods containing folic acid. Much of the folic acid content of food is often destroyed – in storage, in processing (e.g. canning) and in food preparation. Some women cannot eat their usual diet as they suffer from nausea and vomiting almost from the time they conceive. Analogous to the absorption of dietary zinc, that of folates may be interfered with by excessive amounts of refined carbohydrates and fats which are so common in the diet of women in South Wales. In other women a disordered metabolism of folic acid may also play a part, for there seems to be a suggestion that a greater proportion of women who have just given birth to a child with a NTD show evidence of impaired folic acid metabolism, in comparison with women who have a normal outcome. Because of potential problems with folic acid absorption and metabolism, it was thought that a pharmacological dose of folic acid, i.e. about 10 times the daily requirement for a pregnant woman, might swamp any block. Such a dose is thought to be entirely safe and without side-effects. *See* Zinc, Physiology

Other Supplementation Studies

Periconceptional Supplementation with Multivitamins (Pregnavite Forte F)

Smithells and his colleagues (1977) in a careful dietary study of almost 200 first-trimester pregnant women found that the dietary intakes of vitamins in those from social classes III, IV and V and those under the age of 20 were poorer than in women from social classes I and II. They also found lower levels of red cell folic acid, leucocyte vitamin C and serum vitamin A, and, to a lesser extent, red cell riboflavin in nonpregnant women from classes III, IV and V and more so in those who were pregnant. Because these findings paralleled the population incidence of NTD in the UK, Smithells *et al.* concluded that subclinical deficiencies of these micronutrients probably contributed to the cause. This view was reinforced by discovery of particularly low levels of these vitamins in the first trimester of pregnancies complicated by a fetus with an NTD. *See* Ascorbic Acid, Physiology; Retinol, Physiology; Riboflavin, Physiology

Table 5. Outcome of pregnancies supplemented with Pregnavite Forte F

	Supplemented	Unsupplemented
Number of outcomes	459	529
Fetuses with NTD	3	24
Miscarriages not examined	30	19
Normal fetuses or infants	426	486
Recurrences (%)	0·7	4·7

Smithells *et al.* (1981) tried to launch a placebo-controlled supplementation trial of the effectiveness of a vitamin cocktail, containing folic acid, in the prevention of NTD. Smithells was not able to obtain ethical permission for the use of placebo, so he had to make do with less satisfactory controls. His index patients, recruited mostly from local genetic counselling centres, came from Yorkshire, Lancashire, Cheshire, Northern Ireland and the southeastern part of England, were volunteer women who had had one or more pregnancies with an NTD and were planning further pregnancies but were not pregnant at the time. They were asked to take three Pregnavite Forte F tablets* per day, not less than 28 days before conception, and to continue taking these until the date of the second missed period. ***See*** individual nutrients

Control women were from the same areas but mostly from Northern Ireland, who were already pregnant at the time of recruitment, or who declined to take the tablets. The results of 454 supplemented mothers and 519 controls from two cohorts showed 3 and 24 recurrences respectively (Table 5), suggesting that the risk of recurrence was reduced from 4·7% to 0·7% by the supplements.

There is possibly some doubt as to whether or not the folic acid content of Pregnavite Forte F (0·12 mg per tablet or 3·6 mg per day) is sufficient to overcome any metabolic block or absorption disorder that might be present. In addition, there is currently no firm evidence that all the substances present in Pregnavite Forte F – which include two vitamins (A and D) – that are at least potentially teratogenic are in any way effective in preventing NTD. Finally, supplementation with Pregnavite Forte F is almost 30 times more expensive than with folic acid alone.

Other trials of taking folic acid to prevent NTD were carried out in Boston, Cuba and South Wales, all suggesting that folic acid alone had a protective effect.

* Pregnavite Forte F (Bencard). Daily recommended intake (in three tablets per day): vitamin A, 4000 iu; vitamin D_2, 400 iu; vitamin B_1, 1·5 mg; vitamin B_2, 1·5 mg; vitamin B_6, 1 mg; nicotinamide, 15 mg; vitamin C, 40 mg; dried ferrous sulphate, 252 mg; calcium phosphate, 480 mg; folic acid, 0·36 mg.

Medical Research Council's (MRC) Multicentre Vitamins Trial

The results of both the South Wales folic acid supplementation trial and the five-centre periconceptual multivitamin trial appeared to give encouraging results. However, statistical opinion was that the results of neither study were sufficiently clear to justify offering supplementation on a wide scale. Both the supplementation studies had statistical flaws. The South Wales study, although a double-blind, randomized, placebo-controlled trial, was of only a relatively small number of pregnancies and was complicated by a high rate (27%) of noncompliance with folate-taking. The periconceptional multivitamin study, although of considerable numbers, lacked proper controls in so far as the noncompliant or the women already pregnant were used. There are strong reasons to believe that the index women differ substantially from noncompliant women or those who do not enter trials.

Because it was felt that statistically valid data had to be available on the part that multivitamins on the one hand and folic acid on the other might play in the prevention of NTD, the MRC in the UK first considered mounting a randomized, placebo-controlled trial at the end of 1981. However, this aroused considerable professional and public opposition because of the use of placebo when an apparently effective preventive treatment was already available. Eventually, the trial started with volunteer women who had had at least one previous NTD being randomized into one of four groups, those in each being asked to take a capsule of different content. The four arms of the trial were as follows: minerals† plus vitamins; minerals plus 4 mg of folic acid; minerals plus vitamins plus 4 mg folic acid; minerals only (placebo). The first patient was not recruited until July 1983.

Collaborating centres in the UK in Europe, Israel, Canada and Australia numbered 33. By April 1991, when a statistically significant result became evident, over 1800 women had been recruited and they had almost 1200 informative completed pregnancies with 23 recurrences (Table 6). Preconceptional supplementation with 4 mg of folic acid was shown to have a 72% protective effect. The vitamins alone did not show a statistically significant protective effect.

Conclusions

Effective prevention of NTD will depend on a variety of approaches. Undoubtedly, the most important way to achieve 'primary prevention' is through health educa-

† Dried ferrous sulphate, 252 mg; calcium phosphate, 480 mg.

Table 6. MRC vitamin study: informative completed pregnancies

Treatment group				
Folic acid	Other vitamins	Completed pregnancies	Recurrences	
+	−	298	2	5 in 514
+	+	295	3	
−	−	300	11	18 in 517
−	+	307	7	

tion. As folic acid seems to be the factor which is lacking in the 'inadequate' diet in the UK, population supplementation with folic acid (without other vitamins) should be considered for the higher-risk populations perhaps by adding folic acid to one or two staple items of the diet (as has been done for the prevention of endemic goitre by adding iodine to salt). The main considerations will be the acceptability of fortification, the amount of fortification required for effectiveness and the half-life of folic acid in foodstuffs rather than the cost, for folic acid itself is a relatively inexpensive substance. ***See*** Food Fortification

Serum α-fetoprotein and high-resolution ultrasound screening should be available for all pregnancies in order to detect NTDs early enough for selective termination to be offered. All women recognized to be at increased risk for NTD should be offered genetic, preconceptional and dietary counselling, as well as prenatal diagnosis in the form of high-resolution ultrasound scan and, possibly, an amniocentesis followed by the usual amniotic biochemical investigations.

It must not be forgotten that there are certainly a number of environmental factors that trigger NTD; inadequate folic acid may be an important factor in the UK; other factors may be relatively more important elsewhere. The search for further triggers must continue.

Bibliography

Hibbard ED and Smithells RW (1965) Folic acid metabolism and human embryopathy. *Lancet* i: 1254.

Laurence KM (1990) Genetics and the prevention of neural tube defects and of 'uncomplicated' hydrocephalus. In: Emery AEH and Rimoin DL (eds) *Principle and Practice of Medical Genetics* 2nd edn, vol. 1, pp 323–346. Edinburgh: Churchill Livingstone.

Laurence KM (1990) Diet and pregnancy, folic acid and vitamin supplementation in the prevention of neural tube defects. In: Chamberlain G (ed.) *Modern Antenatal Care of the Fetus*, pp 301–326. Oxford: Blackwell Scientific Publications.

Laurence KM (1991) Folic acid to prevent neural tube defect. *Lancet* 338: 379.

Laurence KM, James N, Miller M and Campbell H (1980) Increased risk of recurrence of neural tube defects to mothers on poor diets and the possible benefits of dietary counselling. *British Medical Journal* 283: 1542–1544.

Laurence KM, James N, Miller M, Tennant GP and Campbell H (1981) Double-blind randomised controlled trial of folate treatment before conception to prevent recurrences of neural tube defects. *British Medical Journal* 282: 1509–1511.

Laurence KM, Campbell H and James NE (1983) The role of improvement in the maternal diet and preconceptual folic acid supplementation in the prevention of neural tube defects. In: Dobbing J (ed.) *Prevention of Spina Bifida and other Neural Tube Defects*, pp 85–106. London: Academic Press.

Milusky A, Jick H, Jick SS *et al.* (1989) Multivitamin/folic acid supplementation in early pregnancy reduces the prevalence of neural tube defects. *Mtd Ass* 262: 2847–2852.

MRC Vitamin Study Steering Group (1991) Prevention of neural tube defects: results of the Medical Research Council's Vitamin Study. *Lancet* 338: 131–137.

Smithells RW (1983) Prevention of neural tube defects by vitamin supplements. In: Dobbing J (ed.) *Prevention of Spina Bifida and other Neural Tube Defects*, pp 53–84. London: Academic Press.

Smithells RW, Sheppard S and Schorah CJ (1976) Nutritional deficiencies and neural tube defects. *Archives of Disease in Childhood* 51: 944–950.

Smithells RW, Ankers C, Carver ME *et al.* (1977) Maternal nutrition in early pregnancy. *British Journal of Nutrition* 38: 497–506.

Smithells RW, Sheppard S, Schorah CJ, Nevin NC and Seller MJ (1981) Trial of folate therapy to prevent recurrences of neural tube defects. *British Medical Journal* 282: 1793.

Smithells RW, Sheppard S, Schorah CJ *et al.* (1984) Neural tube defects and vitamins: the need for a randomised clinical trial. *British Journal of Obstetrics and Gynaecology* 91: 516–523.

Thiersch JB (1952) Therapeutic abortions with folic acid antagonist, 4-amino pteroylglutamic acid (4 amino PGA) administered by oral route. *American Journal of Obstetrics and Gynecology* 63: 1298–1304.

Vergel RG, Saucher LR, Heredoro BL, Rodriguez PL and Martinez AY (1990) Primary prevention of neural tube defects with folic acid supplementation: Cuban experience. *Prenatal Diagnosis* 10: 149–152.

Yates JRW, Ferguson-Smith MA, Shenkin R, Gugman-Rodriques M, White M and Clarke BT (1987) Is disordered folate metabolism the basis for the genetic predisposition to neural tube defect? *Clinical Genetics* 31: 279–287.

K. M. Laurence
University of Wales College of Medicine, Cardiff, UK

Pre-eclampsia and Diet

Definition and Description

Hypertensive disorders of pregnancy are among the serious medical complications of pregnancy, complicating up to 8% of all pregnancies. Pre-eclampsia (formerly

Table 1. Clinical classification of hypertensive disorders of pregnancy

Disorder	Definition
Gestational hypertension and/or proteinuria	Hypertension or proteinuria developing in pregnancy, labour or the puerperium in a previously normotensive nonproteinuric woman Gestational hypertension (without proteinuria) Gestational proteinuria (without hypertension) Gestational proteinuria hypertension (pre-eclampsia)
Chronic hypertension and chronic renal disease	Hypertension or proteinuria in pregnancy in a woman with chronic hypertension or chronic renal disease diagnosed before, during or after pregnancy Chronic hypertension (without proteinuria) Chronic renal disease (proteinuria with or without hypertension) Chronic hypertension with superimposed pre-eclampsia (proteinuria developing for the first time in pregnancy in a woman with known chronic hypertension)
Unclassified	Hypertension and/or proteinuria found either (1) at first examination after 20 weeks' gestation in a woman without known chronic renal disease or chronic hypertension, or (2) during pregnancy, labour or the puerperium where information is insufficient to permit classification
Eclampsia	Generalized convulsions during pregnancy, labour or within 7 days of delivery not caused by epilepsy or other convulsive disorders

Davey D and MacGillivray I, 1988.

called pre-eclamptic toxaemia) is the combination of hypertension and proteinuria with or without oedema. Hypertension is defined as a diastolic blood pressure of $\geq$110 mmHg on any one occasion or $\geq$90 mmHg on two or more consecutive occasions 4 h or more apart, and proteinuria indicating $\geq$300 mg protein excretion in 24 h assuming pre-existing causes of hypertension and proteinuria have been excluded (Table 1). It is primarily a nonrecurrent disease of women in their first term pregnancy, occurring in the latter half of pregnancy, usually in the last 3 months, which resolves usually within days of delivery. Pre-eclampsia occurs more frequently in women with a placenta that is larger than normal, which may occur in diabetics or with twins, and also in those women with pre-existing hypertension from any cause. Family studies have illustrated a genetic predisposition, with daughters of women with eclampsia having a higher incidence of pre-eclampsia than daughters-in-law. ***See*** Hypertension, Physiology

Clinical Features

Most frequently, women are asymptomatic, with clinical examination revealing the combination of hypertension, proteinuria and oedema. The condition is classified as mild or severe.

In severe cases the woman may complain of headaches, visual disturbances or, uncommonly, upper abdominal pain (owing to acute distension of the liver capsule). Central nervous system irritability may be detected with increased reflexes. Abnormal renal function and abnormal coagulation may develop. The condition is progressive with symptoms and signs deteriorating over hours, days or weeks. Central nervous system irritability may extend to convulsions, in which case the term 'eclampsia' is used.

Pathophysiology and Dangers to Mother and Fetus

Hypertensive disorders of pregnancy have consistently been one of the two most important causes of maternal death over the last 15 years, the other being pulmonary embolism. In the triennium 1985–1987, hypertension was associated with 19·4% (27 of 139) maternal deaths in the UK. Death occurs as a result of kidney, liver or adrenal failure, pulmonary oedema or widespread coagulation disorders.

The essential pathological feature of pre-eclampsia is widespread spasm of small blood vessels throughout the body. There is a reduction in circulating plasma volume with correspondingly increased blood viscosity and increased sensitivity of the blood vessels to substances controlling blood pressure. The rise in blood pressure initially compensates for vascular spasm, maintaining blood flow, but subsequently fails to compensate.

Renal biopsies of women with pre-eclampsia have shown a characteristic abnormality with swelling of the cells lining the capillaries. In severe cases degeneration of the kidney tissue occurs. The reduction in the ability of the kidney to filter blood caused by the blockage of the capillaries causes reduced sodium filtration and, thus, sodium and water retention. In turn, this causes widespread oedema, causing swollen hands, feet and face and, in severe cases, swelling of liver and brain. ***See*** Sodium, Physiology

Abnormalities of coagulation occur initially, leading to occlusion of small blood vessels, but uncontrolled bleeding may subsequently occur, resulting in haemorrhage into the kidneys, adrenal gland, pituitary gland, liver or brain, any of which may be fatal.

Jaundice and liver damage can occur with haemorrhage and destruction occurring throughout the liver substance, possibly culminating in total liver failure.

Placental blood flow is reduced, again owing to vascular spasm of the uterine arteries, with deposition of fatty tissue (likened to 'acute arteriosclerosis') in the arteries.

In the absence of widespread tissue destruction, the pathological changes are completely reversible after delivery of the baby, and there does not seem to be a relationship between pre-eclampsia and the subsequent development of hypertension or kidney disease.

Dangers to Fetus

The increase in perinatal mortality rate associated with pre-eclampsia is attributable partly to an increase in prematurity (provoked by the need for early delivery in the maternal interest) and partly to intrauterine asphyxia caused by reduced placental blood flow. The latter results in an increase in both stillbirths and deaths after birth. However, there is no rise in the death rate for those babies of women with mild pre-eclampsia. Indeed, in the majority of cases, the reduced placental blood flow is insufficient to compromise the oxygenation or nutrition of the fetus, especially if the disease is short lived, as is usually the case.

In the neonatal period, there is a higher rate of neurological and respiratory problems than normal, but no additional risk over and above that of premature or asphyxiated babies from other causes. Long-term follow-up reveals no long-term complications specific to pre-eclampsia itself.

Nutritional Factors in Aetiology and Treatment

Interest in nutritional factors in the aetiology of pre-eclampsia stemmed from three sources. First, reported geographical differences (since reported to be false owing to difficulties in accurate data recording, especially in developing countries) in the 1930s suggested that diet may be important. For example, meat-eating Muslims were reported to have a higher incidence of pre-eclampsia than Hindus (vegetarian). Second, studies of the incidence of pre-eclampsia during wartime (a reduction during World War I and during the Dutch Hunger Winter 1944–1945, followed by a postwar increase) were assumed to be caused by dietary deficiencies during wartime: other demographic changes, such as the number of births from a first pregnancy postwar, which in itself could explain the changes, were not considered. The fall in pre-eclampsia in 1915 preceded the food shortage, as the rise in 1919 preceded its alleviation; thus dietary restrictions cannot be implicated. The third source of interest came from an apparent association of pre-eclampsia with social class, with the assumption that the pre-eclampsia was associated with a poorer diet among lower social classes. This association is now doubtful, and again other factors, such as availability of antenatal care, were not considered, but gave support to proponents of a dietary aetiology. ***See*** Hypertension, Hypertension and Diet; Hypertension, Nutrition in the Diabetic Hypertensive

The current scientific knowledge is that dietary manipulations have no effect on prevention of pre-eclampsia. The evidence for each nutritional intervention will be discussed.

Maternal Weight

Maternal obesity is associated with an increased risk of essential hypertension, and, possibly secondary to this, pre-eclampsia. (Obese women must have blood pressure measurements performed with a large cuff; otherwise, artificially high readings will be obtained.)

Underweight women (at least 20% below ideal bodyweight-for-height) may also be at increased risk, although this has still to be proved scientifically.

Maternal weight gain in pregnancy has attracted a lot of attention, with widespread advice to restrict weight gain in the hope of preventing pre-eclampsia. A higher weight gain in the second half of pregnancy (and possibly also in the first half of pregnancy) is associated with an increased risk of pre-eclampsia, but the overlap between those remaining normotensive and those with pre-eclampsia is so large that weight gain is not a very useful predictive factor.

This association has led to attempts at prevention of weight gain, with dramatic improvements in pre-eclampsia rates claimed with intensive antenatal care and dietary advice (high-protein, low-carbohydrate, low-weight-gain diet). There was no evidence that it was the dietary advice, and not increased antenatal surveillance which caused the reduction.

Controlled trials have failed to show any effect of dietary restriction. A comparison of two groups of women, one given dietary advice to gain an 'ideal' weight and the other with no dietary advice, resulted in the advice group achieving the ideal weight but with no reduction in pre-eclampsia. Similarly, no benefit in pre-eclampsia rates can be shown with weight reduction in high weight gainers, or in obese women, even though a reduction in weight gain can be achieved.

Energy Intake

There is no evidence that the overall nutritional intake or energy intake differs between women with and those without pre-eclampsia. Interventions aimed at ensuring an optimal diet improved the quality of the maternal diet, but failed to affect the incidence of pre-eclampsia. *See* Energy, Measurement of Food Energy

Protein

Pre-eclampsia is associated with low protein levels in the maternal blood, which has an effect on fluid balance in the body. This has led to interest in the effect of protein supplementation in prevention and treatment of the disease. *See* Protein, Requirements

Dietary surveys and biochemical markers of protein intake and development of pre-eclampsia have failed to show an association between protein intake and the development of pre-eclampsia.

Studies on dietary supplementation also provide little evidence. An influential early report concluded that a high-protein diet did not increase the incidence of pre-eclampsia, and this led to support for dietary supplementation to prevent pre-eclampsia. However, despite attempts since then, there have been no scientifically adequate studies establishing or refuting a relationship between protein supplementation and pre-eclampsia.

Salt

Dietary salt is considered to be an important factor in the development of essential hypertension in nonpregnant subjects. It therefore seems likely that interest will be generated again in the role of salt in the aetiology of pre-eclampsia. At present there is no evidence to suggest a causative role. On the contrary, in theory, salt restriction may exacerbate pre-eclampsia, by further reducing plasma volume.

Salt restriction was practised widely early this century, for the treatment of oedema in pregnant and nonpregnant subjects, and, subsequently, for the prevention of pre-eclampsia, without any scientific basis. Two studies comparing high and low salt intakes failed to show any difference in outcome; one study showed fewer cases of pre-eclampsia in the *high*-salt group.

Vitamins and Minerals

Dietary surveys have shown no convincing relationship between any single vitamin or mineral and pre-eclampsia, but supplementation studies suggest possibilities for further research.

The role of zinc may be important: low plasma zinc concentrations have been found in women with severe pre-eclampsia, and a lower incidence of pre-eclampsia among women who had received zinc supplementation, despite the fact that this supplementation had not raised plasma zinc concentrations. However, this does not imply dietary deficiency as a causative factor and further study is necessary to evaluate this. *See* Zinc, Deficiency

Calcium is considered to be important in the regulation of blood pressure, with the flow of calcium ions into cells being essential for contraction of the muscle cells in blood vessel walls. Blockage of this flow with calcium-blocking drugs may be used for the control of blood pressure both in pregnant and nonpregnant subjects. It would therefore seem theoretically possible that calcium restriction may prevent pre-eclampsia. *See* Calcium, Physiology

Despite this theory, there is recent evidence that women with pre-eclampsia may have low calcium intakes, and that calcium supplementation may reduce the incidence of pre-eclampsia. At present, there is insufficient data to conclude that supplementation is of benefit, and further controlled trials need to be performed.

Supplementation with folic acid or vitamin B has been shown to be ineffective, but multivitamin and mineral supplementation may be useful. One study failed to show any beneficial effect of supplementation (cereal plus vitamin plus vitamin D) but two studies have suggested a reduction in pre-eclampsia in the group supplemented with multivitamins and minerals. This area requires further scientific investigation. *See* individual nutrients

Polyunsaturated Fatty Acids

Interest has been raised in the relationship between prostaglandins and the development of pre-eclampsia. Alteration in the ratio of these naturally occurring compounds in the body could cause the spasm of blood vessels seen in pre-eclampsia.

Experimentally, linoleic acid supplements result in a reduction in sensitivity of blood vessels to substances controlling blood pressure. Deprivation of essential fatty acids in pregnant rabbits has been shown to increase this sensitivity; hence there is a theoretical possibility that linoleic acid supplementation may prevent pre-eclampsia. However, when this study was extended to pregnant humans, linoleic acid supplementation failed to alter the incidence of pre-eclampsia, despite a slight lowering of blood pressure. ***See*** Fatty Acids, Gamma Linolenic Acid

Circumstantial support for a role of prostaglandins comes from the Faroe Islands where a diet rich in fish oils, and hence high in prostacyclin, one of the natural compounds which relaxes small blood vessels, is consumed. The incidence of pre-eclampsia in the Faroes is lower than on the mainland, but again a causative role has yet to be proved. Dietary surveys among pregnant women with and without pre-eclampsia have failed to show any difference in fatty acid intake. ***See*** Fish Oils, Dietary Importance

Conclusions

Advocates of a nutritional basis for pre-eclampsia have had considerable influence in the past, with weight restriction in particular becoming an integral part of antenatal prevention of pre-eclampsia. This view is now not widely held, and has minimal influence on current antenatal practice. Greater influence has been exerted by Thomas Brewer, of California; he maintains that pre-eclampsia is a disease of maternal malnutrition, whose symptoms can largely be eliminated by eating a well-balanced diet with adequate protein, energy, vitamins and minerals. Weight restriction should be avoided. This theory has not been substantiated scientifically, but Brewer's work has been a major influence and source of information, not least through women's self-help groups. It seems certain that in general terms Brewer's advice has been of some benefit in improving the diet of pregnant women, and in eliminating widespread weight restriction or salt or water restriction, but there is no evidence at present that it has led to a reduction in the incidence of pre-eclampsia.

Despite numerous attempts to prove linkage, at present there is no scientific data on which to base a nutritional theory for the prevention or treatment of pre-eclampsia.

Bibliography

Davey D and MacGillivray I (1988) The classification and definition of the hypertensive disorders of pregnancy. *American Journal of Obstetrics and Gynecology* 158: 892–898.

Green J (1989) Diet and prevention of preeclampsia. In: Chalmers I, Enkin M and Keirse MJNC (eds) *Effective Care in Pregnancy and Childbirth*, vol. 1, pp 281–300. Oxford: Oxford University Press.

MacGillivray I (1983) *PreEclampsia. The Hypertensive Disease of Pregnancy*. London: WB Saunders.

Winick M (1989) *Nutrition Pregnancy and Early Infancy*, pp 114–119. Baltimore: Williams and Wilkins.

Katharine Stanley and Robert Fraser
University of Sheffield, Sheffield, UK

Nutrition in Diabetic Pregnancy

Nutrition in Diabetic Pregnancy

Diabetes in Pregnancy

Diabetes complicates 1–2% of all pregnancies within the UK. The majority of diabetic pregnancies are in women who become diabetic for the first time during pregnancy. This type of diabetes is called gestational diabetes mellitus (GDM) and results from the changes in maternal carbohydrate metabolism that occur with pregnancy. Women with established diabetes, either insulin-dependent diabetes mellitus (IDDM) or noninsulin-dependent diabetes mellitus (NIDDM), account for less than 25% of all diabetic pregnancies.

Maternal and fetal morbidity and mortality rates are increased in pregnancies complicated by diabetes. Pregnancy outcome in diabetic women is closely correlated with maternal blood glucose control. To ensure optimal maternal blood glucose control, all pregnant diabetic women require dietary treatment and some of these women will, in addition, require insulin. The diet prescribed needs to be tailored to the individual and will depend on her type of diabetes, her pre-pregnancy weight and the given trimester.

The Effect of Pregnancy on Carbohydrate Metabolism

Maternal carbohydrate metabolism undergoes physiological changes throughout pregnancy. In nondiabetic pregnancies fasting blood glucose concentrations fall and postprandial glucose concentrations rise. The mechanism for the fall in fasting maternal blood glucose concentration is unknown as it occurs in the first 10 weeks of pregnancy before any rise in plasma insulin concentration is seen. By the second trimester, fasting plasma insulin concentrations and glucose-stimulated insulin levels increase, and continue to rise throughout pregnancy. The increase in blood glucose after a meal facilitates glucose transfer across the placenta to the fetus. The increased postprandial insulin level helps to promote maternal fat synthesis and hence the laying down of maternal fat stores. ***See*** Glucose, Maintenance of Blood Glucose Level

Decreased insulin sensitivity is a feature of normal pregnancy and is due to circulating placental hormones which are antagonistic to the action of insulin. During pregnancy, glucose homeostasis can only be maintained if insulin secretion increases sufficiently to overcome the physiological diminution in insulin sensitivity. By the third trimester, following a meal, insulin release needs to increase two- to threefold to prevent maternal hyperglycaemia.

IDDM is caused by a destruction of pancreatic β cells which leads to a deficiency in insulin production. This absolute insulin deficiency can only be corrected with insulin therapy. In pregnancy, when insulin demand is increased, insulin treatment has to be increased accordingly.

NIDDM is characterized by both a reduction in insulin secretion to oral glucose and a reduced sensitivity to insulin action. Obesity, which decreases insulin sensitivity, is a further contributing factor to the development of NIDDM. Many nonpregnant women with NIDDM can be treated with dietary therapy alone. During pregnancy, when insulin sensitivity decreases further, insulin treatment, in addition to diet, is usually needed to prevent hyperglycaemia. The use of diabetic tablet therapy in pregnancy is usually avoided as most commonly prescribed agents cross the placenta and can cause neonatal hypoglycaemia.

Compared to pregnant women with normal glucose tolerance, women with GDM have decreased insulin production and decreased insulin sensitivity. This deficiency in insulin production is less than that of women with other forms of diabetes; thus treatment with insulin is not always required, provided that dietary management and weight control have been introduced early in pregnancy.

Adverse Effect of Diabetes on Fetal Development

Hyperglycaemia is teratogenic to the developing fetus. There is a fourfold increase in congenital abnormalities in infants born to IDDM mothers, and a twofold increase in infants born to NIDDM mothers. A number of studies have shown that poor maternal glycaemic control at the time of conception and organogenesis is associated with an increased incidence of congenital abnormalities. In clinics where diabetic women are offered preconceptual counselling, and blood glucose values are kept within the normal range during the periconceptual period and the first trimester, the incidence of fetal abnormalities is reduced.

Apart from congenital abnormalities, the fetus of the diabetic mother is at risk of abnormal intrauterine growth. Insulin is a potent stimulus for fetal growth, and fetal hyperinsulinaemia occurs as a result of maternal hyperglycaemia. Infant birthweight in excess of 4000 g (macrosomia) is more common in poorly controlled diabetic pregnancy and is associated with obstetric complications, most notably birth trauma. By controlling maternal hyperglycaemia the incidence of macrosomic infants born to diabetic mothers is reduced. Other complications of fetal hyperinsulinaemia include neonatal polycythaemia, respiratory distress and hypoglycaemia. In addition to the adverse effects of maternal hyperglycaemia on fetal outcome, it may also predispose to the earlier presentation of NIDDM in the infant of a diabetic mother in later life.

Whilst the detrimental effects of maternal hyperglycaemia to the fetus are well recognized, the effects of maternal hypoglycaemia are less clear. Insulin does not cross the placenta and hence the fetus will be protected from brief periods of maternal hypoglycaemia due to excessive insulin treatment. However, prolonged maternal hypoglycaemia (from whatever cause) removes glucose from the fetal circulation, which can have a detrimental effect on fetal neurological development.

In addition to the harmful effects of hyperglycaemia on fetal development, ketosis has also been shown to be teratogenic in animal studies. Maternal ketosis has been implicated in the poor neurological and psychological development of infants born to diabetic mothers. Although a direct effect of maternal ketosis on neurological development is questioned by some workers, it is universally accepted that ketosis should be avoided in all pregnant diabetic women.

Principles of Dietary Management in Diabetic Pregnancy

Dietary compliance in pregnancy is usually good. The principles for the diet in both the nondiabetic and pregnant diabetic woman are similar. Both the quantity and quality of carbohydrate, protein and fat which make up the diet are important. It is generally accepted that the diet be composed of 50–60% carbohydrate, 12–20% protein and 20–35% fat. Dietary carbohydrates should be mostly unrefined with a high fibre content, the protein should be from fish and white meat, whilst fat should be predominately polyunsaturated. High-fibre diets have a beneficial influence on carbohydrate metabolism in both diabetic and nondiabetic pregnancies. Dietary fibre improves glucose metabolism in nondiabetic subjects by reducing small-bowel glucose absorption, thereby lowering peak postprandial glucose and insulin levels. Insulin sensitivity is increased when fibre is substituted for refined carbohydrate in the diet. An increase of dietary fibre from 10 to 50 g has been shown to improve glucose tolerance in the latter half of pregnancy. The Royal College of Obstetrics and Gynaecology currently recommend a diet which includes 30 g of fibre per day for all pregnancies. The benefits of a high-fibre diet in diabetic pregnancies are well recognized, and refined carbohydrates should be reduced to a minimum, being used principally to treat symptomatic hypoglycaemic symptoms.

Individual calorific requirements vary during pregnancy according to the pre-pregnancy bodyweight. In the nonobese diabetic woman a diet of approximately 30 kcal per kg of bodyweight per day will be required to give an expected pregnancy weight gain of about 12·5 kg. In obese diabetic women ($\geq 120\%$ of ideal bodyweight) optimal pregnancy weight gain is less, and a diet of approximately 24 kcal per kg of bodyweight per day, calculated on pre-pregnancy weight, should be sufficient to meet nutritional needs whilst preventing excessive weight gain. ***See*** Dietary Fibre, Effects of Fibre on Absorption

The exact meal plans for the insulin-dependent and noninsulin-dependent diabetic mother treated with insulin need to be individually assessed, to prevent hypoglycaemia occurring at a time when insulin therapy is being intensified. The dietary management of the gestational diabetic woman needs to be based around the basic principles of a diabetic diet, i.e. optimizing glycaemic control and imposing calorie restraint where necessary to prevent excessive weight gain. The gestational diabetic woman should be instructed in a healthy eating pattern. This should continue after pregnancy as the majority of these women will develop NIDDM in later years.

Management of the Pregnant Insulin-dependent Diabetic Woman

Before the availability of insulin therapy 50 years ago, IDDM was associated with low fertility. When pregnancy did occur, both neonatal and maternal morbidity were high. With the introduction of insulin treatment, the prognosis for pregnancy improved dramatically. However, the incidence of congenital abnormalities among the infants of diabetic mothers remains unacceptably high. It is hoped that with improved glycaemic control in the early weeks of pregnancy this fourfold increase in congenital abnormalities will be reduced. Ideally, all insulin-dependent mothers should be encouraged to seek pre-conceptual counselling, so that diet and insulin treatment can be intensified to achieve exemplary glycaemic control during this critical period.

All mothers with IDDM need to be referred to a hospital antenatal clinic as soon as pregnancy is diagnosed; then specialized medical, nursing and dietetic care can be provided throughout pregnancy. Meticulous care is needed to ensure that maternal blood glucose concentrations are kept between 4 and 6 mmol per litre throughout the day. Patient education and cooperation is essential if this degree of glycaemic control is to be achieved. The insulin requirements of the mother with IDDM will rise two- to threefold during pregnancy; therefore the prescribed insulin regimen needs to be flexible, allowing for the anticipated increases in insulin dosage. Suitable insulin regimens in pregnancy include twice daily injections of a short- and intermediate-acting insulin, or a long-acting insulin given at bedtime with short-acting insulin given prior to each meal. The latter regimen has the advantage that the short-acting insulin dosage can be adjusted according to the size and content of each meal; in addition, the exact timing of the meals is less critical than for a twice daily insulin regimen. If blood glucose levels are to remain in the normal range the mother will have to test her own blood glucose before and two hours after each meal and learn to anticipate the insulin dose required to cover each meal.

During the first trimester, insulin-dependent diabetic women are at particular risk of both hypoglycaemia and ketosis, both of which are exacerbated by the nausea and vomiting which are common in early pregnancy. Each patient requires an individual meal plan which allows for small snacks between meals to provide adequate carbohydrate to prevent hypoglycaemia and ketosis. During the second trimester, insulin requirements will increase and larger doses of insulin need to be given at night to prevent fasting hyperglycaemia. However, it is important to ensure that adequate carbohydrate is given at night to prevent nocturnal hypoglycaemia, a common problem during the second and third trimester. During the third trimester, fetal growth is at its maximum and in the presence of hyperglycaemia accelerated fetal growth can occur, resulting in the birth of a macrosomic infant. Maternal energy requirements are at their greatest during this time, being 200–300 kcal per day above non-pregnancy levels. A delicate balance exists during the third trimester between maintaining normal blood glucose levels and sustaining adequate maternal nutrition, whilst at the same time preventing hypoglycaemia and excessive maternal weight gain. This can only be done using individualized meal plans that take account of the patient's changing insulin regimen, level of activity and individual susceptibility to hypoglycaemia. ***See*** Glucose, Glucose Tolerance and the Glycaemic Index

Management of the Pregnant Noninsulin-dependent Diabetic Woman

The pregnant woman with NIDDM presents a different management problem to the insulin-dependent diabetic woman. The mother with NIDDM is likely to be obese and prior to pregnancy will have been treated with either dietary therapy alone, or diet and oral agents. Ideally, those mothers treated with oral agents will have been changed to insulin therapy prior to their pregnancy and instructed in the methods of home blood glucose monitoring. The increased incidence in congenital abnormalities with NIDDM is a reflection of hyperglycaemia at the time of conception and during the first trimester. Improved glycaemic control at this time should reduce this incidence. The majority of women with NIDDM treated with diet alone prior to pregnancy will require insulin treatment during pregnancy. Insulin treatment should begin early in these women to allow for instruction about the necessary injection techniques and principles of home blood glucose monitoring. The dietary advice will be similar to that for the insulin-dependent diabetic mother. However, the mother with NIDDM is less vulnerable to hypoglycaemia and ketosis; therefore less emphasis needs to be given to encouraging carbohydrate snacks between meals. Similarly, in the management of the gestational diabetic woman, excessive maternal weight gain is to be discouraged. With the use of high doses of insulin in the latter half of pregnancy the appetite of these already obese women is likely to be stimulated and great care needs to be taken to ensure that calorie intake does not exceed that prescribed.

Management of Gestational Diabetic Women

The treatment of the gestational diabetic woman is directed to achieving normal maternal blood glucose concentration throughout pregnancy. Many women can be successfully treated by diet alone. When blood glucose concentrations cannot be kept within the normal range (4–6 mmol l^{-1}) with diet alone, additional insulin therapy will be needed.

The gestational diabetic woman differs from the insulin-dependent and noninsulin-dependent diabetic woman in many important ways. The diagnosis of GDM will have been made during pregnancy, usually in the second or third trimester, depending on the screening policy of the antenatal clinic. The mother with GDM will not previously have been on any specific diet or have received any education concerning diabetes. The majority of these women are obese and many will have already had excessive pregnancy weight gain by the time the diagnosis is made. When the diagnosis of GDM is made, the mother will have to be educated in both the principles of a diabetic diet and the methods of home blood glucose monitoring.

As obesity has a deleterious effect on diabetes by reducing insulin sensitivity, the prescribed diet for the gestational diabetic must control blood glucose and limit excessive maternal weight gain. It is now recognized that the optimal maternal weight gain for the obese mother is half that for the nonobese mother. The obese diabetic mother is at particular risk of having a macrosomic infant as both poor glycaemic control and maternal obesity independently contribute to this risk. The nutritional recommendation for pregnancy as outlined by the National Academy of Science in 1990 considered a maternal weight gain of 6·5 kg to be appropriate for obese women. The prescribed diet for the mother with GDM should be similar to that for the noninsulin-dependent diabetic with a degree of calorie restriction if necessary. A diet of 1500–1800 cal per day (or 20 cal per kg of bodyweight per day) is frequently required for the treatment of the gestational diabetic woman to achieve glycaemic control and prevent both excessive maternal weight gain and accelerated fetal growth patterns. The woman with GDM is less vulnerable to becoming hypoglycaemic or ketotic than a mother with IDDM, and these obese mothers appear to tolerate mild calorie restriction well. Since these diets are being prescribed in the third trimester, adequate maternal energy stores will have been laid down already in early pregnancy, thereby ensuring sufficient fetal fuels for the remainder of pregnancy.

With delivery, glucose intolerance corrects spontaneously but the majority of women with GDM will experience a deterioration in glucose homeostasis at some time in the future, many women becoming frankly diabetic. Pregnancy is therefore an ideal opportunity to instruct and advise women of the need to continue a healthy eating pattern after pregnancy; it is hoped that this will delay the onset of future diabetes, principally by curtailing obesity.

Bibliography

Dornhorst A, Nicholls JSD and Johnston DG (1990) Diet and diabetes in pregnancy. In: Franks S (ed.) *Ballière's Clinical Endocrinology and Metabolism*, vol. 4, no. 2, pp 291–311. London: Ballière Tindall.

Gillmer MDG (1983) Obesity in pregnancy, clinical and metabolic effects. Nutrition in pregnancy. In: Campbell DM and Gillmer MDG (eds) *Proceedings of the Tenth Study Group of the Royal College of Obstetricians and Gynaecologists*, pp 313–230. London: RCOG.

Hollingsworth DR and Ney D (1988) Dietary management of diabetes during pregnancy. In: Reece EA and Coustan DR (eds) *Diabetes Mellitus in Pregnancy, Principles and Practice*, pp 285–311. New York: Churchill Livingstone.

Holman SR (1989) Nutritional management. In: Hare JW (ed.) *Diabetes Complicating Pregnancy*, pp 69–80. New York: Alan R. Liss.

Spellacy WN (1983) Carbohydrate metabolism in pregnancy. In: Fuchs F and Klopper A (eds) *Endocrinology of Pregnancy*, pp 161–175. Philadelphia: Harper and Row.

Anne Dornhorst and Jonathan S.D. Nicholls
St Mary's Hospital Medical School, London, UK

PREMENSTRUAL SYNDROME

See Menstrual Cycle and Premenstrual Syndrome – Nutritional Aspects

PRESERVATION OF FOOD

Definition

'Food preservation' is an umbrella term covering any measures that make food keep well over a reasonable period of time. The objective is to minimize damage or, ideally, to avoid damage altogether. Food preservation ensures that the consumer is provided with enjoyable, good-tasting food which as far as possible retains its full nutritional value. Above all, food preservation prevents health hazards caused by contamination of foods with pathogenic microorganisms.

Food preservation does not extend to combating losses of harvested food resulting from attack by rodents, insects and other pests, although such losses are at least as high as those caused by food spoilage in the true sense of the term. ***See*** Insect Pests, Insects and Related Pests

Necessity for Food Preservation

In many parts of the world, cereals, fruit and other foods of plant origin are available fresh only during a brief harvest period. As a result, inhabitants of those regions have always needed to ensure that harvested food has a long shelf life. In other regions, food preservation is made necessary by periods of rain or drought.

With the passage of time, food came to be preserved not only for reasons of natural necessity but also because the pattern of civilization so required. People are increasingly living in towns and cities, where self-sufficiency in food on any large scale is out of the question. Fewer and fewer people have the responsibility of obtaining and producing food for more and more other people. A development of this kind is possible only if foodstuffs can be preserved for a sufficiently long time.

Without food preservation it is impossible, in either the developed or the underdeveloped countries, to supply the population with food in any methodical manner. Only if food has a certain minimum storage life can it be transported over long, or indeed relatively short, distances. The desire to enjoy foods from foreign countries also makes food preservation essential.

Notwithstanding these facts, the trend in public opinion is increasingly towards 'fresh' products, the term 'fresh' in this context also being taken to include products which are not, strictly speaking, fresh but are given short-term stability by the latest methods of preservation. Although their nutritive value and other quality aspects are on a par with those of fresh products, they themselves are not fresh in the true sense.

Definition of the Term 'Food Spoilage'

In its broadest sense, food spoilage is any change in a foodstuff that appreciably reduces its value, especially its nutritive value, its sensory quality, palatability or usability. Reduction, however, is a relative concept governed by convention and linguistic usage. Whether or not a change in a foodstuff is termed 'spoilage' sometimes depends on the extent of the change and the course of the reaction. The biochemical, enzymatic and microbiological processes in the ripening of cheese are one example of this. Depending on how far it progresses, the microbiological process is termed ripening (desired) or spoilage (undesired).

Types of Food Spoilage

Essentially, food spoilage includes chemical, enzymatic, physical and microbiological changes.

Important *chemical changes* are undesired oxidation, nonenzymatic browning reactions and chemically governed degradation reactions of food ingredients. Some begin during the processing of the food and continue during storage. The main ingredients affected by such reactions are proteins and lipids, vitamins, aroma compounds, and colorants. They are either gradually destroyed, e.g. by oxidation, or changed by chemical reactions to such an extent that they cease to be of any nutritional value, as is the case, for example, with some vitamins. One reaction of particular importance is the Maillard reaction, which converts certain amino acids into a form that cannot be utilized, as well as producing brown pigmentation. Furthermore, the chemical reactions mentioned may produce compounds with undesirable sensory properties. ***See*** Browning, Nonenzymatic; Oxidation of Food Components; Spoilage, Chemical and Enzymatic Spoilage

Among the *enzymatic reactions* the most important ones are reactions caused by hydrolases like lipases, proteases, and oxidoreductases. Enzymatic spoilage occurs primarily after the comminution of foods because the comminution process enables the enzyme and substrate to come into direct contact. One of many examples is the brown coloration of vegetable raw materials caused by the activity of polyphenol oxidases.

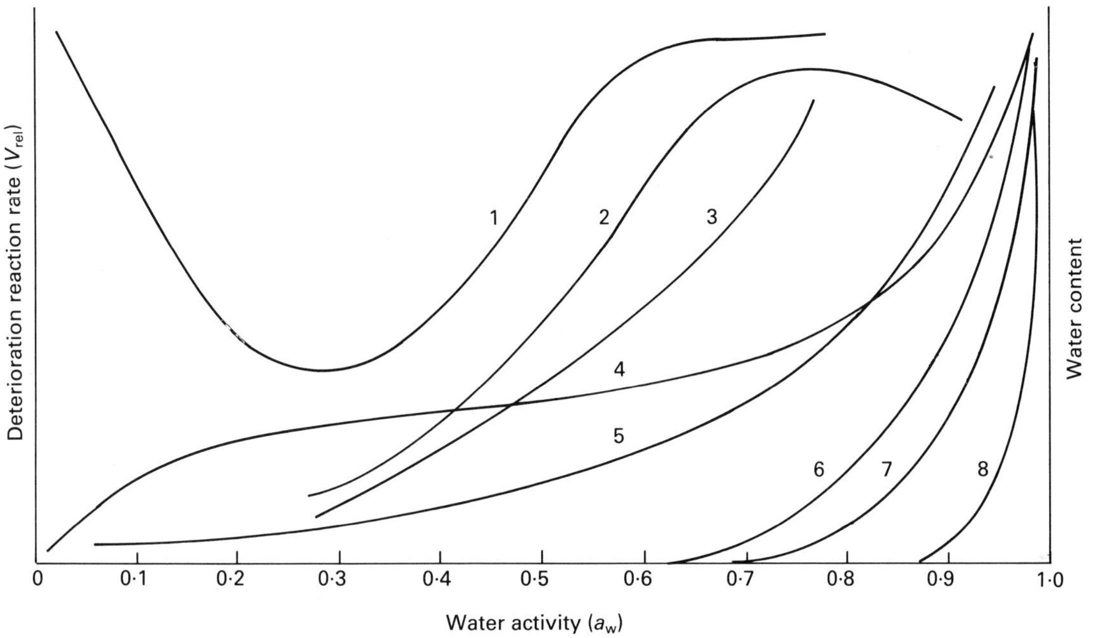

Fig. 1 Generalized deterioration reaction rates (V_{rel}) in food systems as a function of water activity (a_w) at room temperature. 1, lipid oxidation; 2, nonenzymatic browning; 3, hydrolytic reactions; 4, sorption isotherm (a_w related to water content); 5, enzyme activity; 6, growth of moulds; 7, growth of yeasts; 8, growth of bacteria.

The most important changes in *physical properties* are signs of deposits, destabilization of emulsions and foams, undesired water absorption and undesired loss of water. Staling of bakery products is caused by retrogradation of the starch and is a form of physicochemical change.

Microbiological food spoilage is by far the most important type. It is caused by bacteria, yeasts and moulds. Microorganisms need to be present on the food for spoilage to occur; even if they have been inactivated by thermal treatment, their enzymes, which have migrated into the food, may still cause spoilage. Concerning foodstuffs in general, the microorganisms need to have nutrients available in a suitable form and adequate quantity, as well as favourable living conditions as far as temperature, water activity, presence or absence of oxygen, redox potential and pH value are concerned. If a foodstuff is consumed before undesired microbial growth takes place, there will be no food spoilage. ***See*** Spoilage, Bacterial Spoilage; Spoilage, Moulds in Food Spoilage; Spoilage, Yeasts in Food Spoilage

Food spoilage is manifested by a reduction in the aroma and flavour, texture and nutritive value of foodstuffs. In extreme cases the foodstuffs become totally unpalatable. In addition to the economic damage caused, it is now well established that many microorganisms release toxins that may cause damage to health. Botulism poisoning, which is caused by *Clostridium botulinum*, has long been familiar. Even today it still causes some fatalities. It has also been known for a few decades that some moulds form carcinogenic mycotoxins and can release these on foods. ***See*** *Clostridium*, Botulism; Mycotoxins, Occurrence and Determination

All forms of spoilage are heavily dependent on the water activity (a_w) of a foodstuff. Low water contents encourage some forms of spoilage and inhibit others, and the same is true of high water contents (Fig. 1).

Methods of Preventing Chemical, Enzymatic and Physical Food Spoilage

Oxidation phenomena can be restricted by withholding or inactivating the oxygen that is in contact with the food. This is achieved by using oxygen-impermeable packaging materials. The air contained in the packs can be replaced by inert gases or removed by vacuum packaging. Another method is to use antioxidants, which scavenge free radicals produced during oxidative processes. Finally, it is possible to eliminate the oxygen chemically or enzymatically by using palladium with hydrogen or glucose oxidase. Chemical changes in foods can be delayed by cold storage, because *chemical reactions* are slowed down by lowering the temperature. Some changes, however, cannot be avoided in practice and have to be accepted. The only possibility in such cases is to add the relevant substances, e.g. vitamins or the sweetener aspartame, in comparatively large quantities at the outset as a way of compensating in advance for the losses. ***See*** Antioxidants, Natural Antioxidants; Antioxidants, Synthetic Antioxidants; Chilled Storage, Packaging Under Vacuum

In the case of *enzymatic reactions*, blanching by heat is the preferred method for keeping spoilage under control. Once inactivated, the enzyme can no longer cause spoilage. A low water content limits enzyme activity. Certain additives, such as citric acid or sulphur dioxide, inactivate enzymes chemically.

Physical changes, typified by precipitation reactions and destabilization, are prevented by additives such as emulsifiers and thickeners. The water content can be controlled by adding humectants. To some extent these also prevent drying out, which can be kept within limits by the use of suitable coating agents. The most effective agent for preventing losses of water, however, is water-impermeable packaging. ***See*** Emulsifiers, Uses in Processed Foods

Methods of Preventing Microbiological Food Spoilage

The microorganisms that cause food spoilage can be combated in two essentially different ways: physical and chemical. The physical methods involve the use of heat, cold, drying and irradiation. The chemical methods involve adding some kind of preservative to the food.

The chemical methods of food preservation give long-term protection, albeit not unlimited. With a few exceptions the chemicals remain in the food and are therefore also effective against any microorganisms that may subsequently find their way into the food. In comparison, the physical methods rely on there being no subsequent infection, as in the case of heat treatment or irradiation; otherwise, they need to be continued throughout the entire storage period, as in the case of refrigeration. If these preconditions are not met, spoilage will follow.

Heat Treatment

For more than 10 000 years, humans have known how to make food more palatable by the use of fire. Interestingly, however, the value of the heat treatment for improving the keeping properties of foods was not known in early prehistory except for the use of smoking, which combines heat with a chemical method of preservation. The preservation of meat by heating in a pressure vessel in the absence of air, as pioneered by Papin, dates back only to the eighteenth century. The founder of modern preservation technology, however, was the French cook, Appert, who, for the first time in 1807, preserved meat in sealed glass jars by the action of heat for the French Navy, long before Pasteur worked out the theoretical principles underlying this method. ***See*** Curing; Heat Treatment, Chemical and Microbiological Changes; Smoked Foods, Principles

Preservation by heat involves killing all or certain microorganisms. Sterility in this context implies the absence of microorganisms in any viable form. So-called commercial sterility – the absence of microorganisms that may impair food under normal conditions, and also the absence of pathogenic and toxin-forming microorganisms – is usually considered adequate.

The extent to which microorganisms are killed depends primarily on the temperature and the duration of its action, but also on the pH value, water content, presence or absence of certain food ingredients or additives, and other factors. The killing of microorganisms is a first-order reaction. It is expressed as the decimal reduction time, known for short as the D value. The D value is the time in minutes taken to kill 90% of particular microorganisms or spores at any given temperature. The higher the D value, the greater is the heat tolerance of the relevant microorganism.

The heat treatment may have not only an antimicrobial action on foodstuffs but also other effects, desired or otherwise. The treatment actually cooks the foods to a greater or lesser extent. Furthermore, sensitive ingredients, such as vitamins and aromas, may be damaged. In principle, the duration of the heat treatment plays a more important role than the treatment temperature in reducing the extent of these reactions.

In practice, a distinction is drawn between various processes, according to the extent of the heat treatment. Sterilization is the killing of practically all microorganisms and spores. The sterilization effect is expressed by the F value, which is derived from the word 'Fahrenheit'. The F value is the time in minutes at 250°F (121·1°C) necessary to achieve commercial sterility. This F value is defined as the time at 250°F required to reduce the amount of *Clostridium botulinum* spores by 10^{12}. In pasteurization, which is employed in preference for acidic liquid and pasty goods, only vegetative pathogenic microorganisms are killed. Depending on the temperature–time conditions, a distinction is drawn between the now little-used holder pasteurization (30 min at 62–65°C), flash pasteurization (15 s at 71–74°C) and high-temperature, short-time pasteurization (2–4 s at 85–90°C), whereas the ultra-high-temperature process (2–10 s at 130–150°C) represents a quality-preserving variant of the sterilization process and produces a product which is commercially sterile. ***See*** Heat Treatment, Ultra-high Temperature (UHT) Treatments; Pasteurization, Principles; Sterilization of Foods

Refrigeration and Freezing

It has been known for thousands of years that certain foods can be kept longer if stored at low temperatures. This is why food used to be stored in cool cellars and caves. Only since the invention of the compression cooler by Linde in the 1870s, however, have refrigeration and freezing methods been used on a relatively widespread scale.

Unlike heat treatment, refrigeration does not kill microorganisms but merely inhibits their growth. The lower the ambient temperature, the greater is the inhibitory effect. The ingredients of food are nearly always damaged less by refrigeration than by heat treatment. However, refrigeration and freezing may influence the physical structure of food according to its type. In the case of foods such as meat, one controlling factor is the freezing rate. Slow freezing leads to the formation of bigger ice crystals between the cells, which are damaged as a result, so that the thawed meat loses its juiciness. ***See*** Freezing, Structural and Flavour Changes

A basic distinction is drawn between chilling or cooling, i.e. storage at temperatures above freezing point, and freezing, which implies storage at temperatures below freezing point. Deep freezing is storage at temperatures below $-18°C$. The temperature can be reduced directly or indirectly, using ice, cooling brines, cooling plates or other installations. ***See*** Freezing, Principles

Dehydration

Dehydration or drying is probably the oldest food preservation method of all. It is used, for example, to preserve cereals, vegetables, fruit, fish, and liquids such as milk or coffee. ***See*** Drying, Theory of Air Drying

In the drying process the microorganisms are not killed but merely inhibited in their growth because they lack the free water necessary. The factor that determines the extent of microorganism inhibition is not the absolute water content of the food but the a_w value (Fig. 1). Water activity is a measure of the quantity of available water contained in food and corresponds to the relative humidity of the ambient air. It lies between 0 and 1, the latter value being the a_w of pure water. The various classes of microorganism differ in their water requirement. Only very few bacteria grow at a_w values below 0·9; many yeasts cease growing below 0·83, and most moulds do so at below 0·80. Only highly xerophilic yeasts and moulds tolerate a_w values of 0·60.

The drying process usually causes a far-reaching change of the structure of the foods. The physical characteristics of the foods are influenced to such an extent that only in exceptional cases, e.g. after lyophilization, can they be restored to products resembling their original structure by the addition of water. Drying also causes damage to many of the food ingredients.

Foods can be dried by merely being left in the air (air-drying or sun-drying), by the direct or indirect action of heat in equipment of widely differing kinds. The best results in respect of quality retention can be achieved by spray-drying or lyophilization. The latter process involves removing the water from a previously deep-frozen product and has a particularly low damage factor. The duration of the drying process depends on the nature of the food to be dried, the ingredients of the food and the changes to be expected in it during storage (Fig. 1). Recently, *intermediate-moisture foods* have been gaining increasing importance; these are foods with an a_w of 0·75–0·90. In structure they are more similar to undried foods than intensively dried products, but their resistance to microbiological spoilage is greater than that of undried foods. ***See*** Drying, Drying Using Natural Radiation; Drying, Spray Drying; Intermediate Moisture Foods

Intermediate-moisture foods are susceptible to mould attack, however, and as a result are often treated with sorbic acid as an additional preservative measure.

Irradiation

Irradiation, a method of food preservation that has been developing over the last 50 years, has not yet gained legal approval for use in all countries. Irradiation is based on the fact that certain electromagnetic radiations kill microorganisms. The radiation used for the purpose consists of ultraviolet light, X rays or γ rays. The effect of the irradiation depends on the amount of energy absorbed. A measure for the dosage of radiation is the gray (Gy); $1\ Gy = 1\ J\ kg^{-1}$). *Radappertization* is high-dosage irradiation (10–50 Gy) and comparable to sterilization. *Radicidation* is the elimination of pathogenic bacteria, especially *Salmonella*, e.g. in poultry, by radiation dosages between 0·1 and 10 Gy. *Radurization* is irradiation with similarly low dosages and is used to extend the shelf life of food such as meat or fish at low temperatures. The effects of irradiation on the structure and chemical composition of the food are small if the radiation dosage is correct. *Ultraviolet irradiation* is used less for direct food preservation than for eliminating microorganisms from the ambient air. ***See*** Irradiation of Foods, Basic Principles

Use of Food Preservatives

The use of additives for food preservation is a preservation method with a long tradition.

Substances that Lower Water Activity

Substances that lower a_w, especially common salt and sugar, are comparable to drying in their action. In a similar manner to the way that water is removed by drying, the free water is eliminated from the microorganisms by being bound to the substance concerned.

The a_w of foods is thereby reduced to values that cease to afford the microorganisms the conditions for life, or which at least make these conditions far less favourable. Salt was known as a preservative as long ago as the days of the Ancient Romans and has been used to preserve meat and fish for centuries. ***See*** Fish, Processing; Meat, Preservation

Substances that act by altering the a_w must be added in relatively high concentrations (of 1% at least) if they are to have any appreciable effect. Consequently, they influence the flavour of the products that they are intended to preserve, which may or may not be desirable. In addition, they significantly influence the products' physical structure, which is one explanation for the fact that such preservatives cannot be used universally. Alcohol, too, belongs to this class of substances, although it also has a direct microbicidal action. In concentrations of more than about 18% it completely inhibits the growth of microorganisms.

Substances that Lower the pH

Among the substances that act as preservatives by influencing the pH the most important are acetic acid and lactic acid. These alter the pH in foods to values at which certain microorganisms, especially pathogenic bacteria, are unable to live or reproduce. These values have no inhibitory effect on many yeasts and moulds, so that lowering the pH value is often insufficient by itself as a preservative measure. In common with substances that act by altering the a_w, the acidifying agents are added in concentrations of 0·5–5%, which means that their taste is perceptible, but in many cases this is highly desirable. ***See*** pH – Principles and Measurement

Other Food Preservatives

One ancient method of preservation is the use of smoke, which contains many individual components with an antimicrobial action in its gas phase. Chief among these are phenols. Sulphur dioxide was already known to the Ancient Romans and since the Middle Ages its use has expanded greatly, especially in the preservation of wine. In the nineteenth century chemical food preservation was boosted by developments in chemistry and medicine. People discovered increasing numbers of substances with an antimicrobial action that could be used in medicine and also food preservation. One of these, benzoic acid, is still used today. The latest development is sorbic acid, manufactured industrially since the 1950s and now the most important preservative in the food sector. Although chemical food preservatives as a whole have always been regarded with great suspicion, their action against toxin-forming organisms means that any danger involved in their use is less than that which exists if they are not used. This is particularly true of sorbic acid.

The preservatives work primarily by inhibiting enzyme reactions that occur in the microorganism and destroy the cell wall or cell membrane, or disrupt the transport mechanism for nutrients within the cell. These reactions do not take place in the human organism, since the concentrations that occur in the body's cells are smaller by many powers of ten than those in the microorganisms. The preservatives are used in concentration ranges varying between a few parts per million and 0·2%. Consequently, they usually have little or no effect on the sensory properties of the foods concerned.

The main preservatives employed are sorbic acid, benzoic acid and sulphur dioxide. Other preservatives, too, are permitted in many countries for special purposes. Mention should be made here of nitrite, for example. This is not only a pickling agent but also an important preservative, as it suppresses the growth of *Clostridia* in meat products. ***See*** Vinegar

Modified-atmosphere Packaging

The antimicrobial action of protective gases derives from the fact that in many cases these gases keep out the oxygen vital to microbial growth. In addition, they give protection against oxidation. The most important protective gases are nitrogen and carbon dioxide, the latter of which also demonstrates a marked antimicrobial action of its own by intervening in the respiratory metabolism of various microorganisms. Its action against moulds is particularly marked. Modified-atmosphere packaging is expensive, however, since oxygen must not diffuse into the pack and neither must the protective gas diffuse out of it. ***See*** Chilled Storage, Use of Modified Atmosphere Packaging

Combined Methods

As the preceding descriptions have explained, there are many methods of food preservation, which are distinguished by fundamental differences. All have advantages and drawbacks. Consequently, there has always been a tendency to combine certain methods with one another with a view to maximizing the advantages and minimizing the drawbacks. Not only are preservatives combined with one another; so, too, are physical preservation processes; and physical processes are also used alongside chemical preservation processes. In the literature, Leistner has used the terms 'hurdle concept' or 'hurdle technology' to describe this approach: it is better to keep microorganisms under control by various hurdles than by a single measure. Moderate drying, for example, can be combined with preservatives; refrigeration can be used alongside or after heating processes; or irradiation can be used together with refrigeration.

Aseptic Processing

Whereas the emphasis in chemical food preservation used to be on combating microorganisms, aseptic processing has recently become an increasingly important factor. Efforts are being made to use aseptic conditions in which to pack food that has previously been rendered largely or completely free of microorganisms by pasteurization or other forms of heat treatment. The process is expensive and time-consuming. Moreover, asepsis in conjunction with food is harder to implement in practice than is the case in medicine. In the latter the main aim is to avoid pathogens that are present only in limited numbers, whereas with foods the spoilage-causing organisms that need to be kept at bay are present everywhere. This necessitates, amongst other things, sterile filtration of the air, as well as sterility of equipment, clothing and other utensils. Nevertheless, aseptic processing is likely to be used on a wider scale in the future.

Bibliography

Branen AL and Davidson PM (1983) *Antimicrobials in Foods.* New York: Marcel Dekker.

Diehl JF (1989) *Safety of Irradiated Foods.* New York: Marcel Dekker.

Fellows P (1988) *Food Processing Technology.* Chichester: Ellis Horwood. Weinheim: VCH.

Labuza TP (1(971) Kinetics of lipid oxidation in foods. *Critical Reviews in Food Technology* 2: 355.

Leistner L (1987) Shelf-stable products and intermediate moisture foods based on meat. In: Rockland LB and Beuchat LR (eds) *Water Activity: Theory and Applications to Food*, pp 295–327. New York: Marcel Dekker.

Lueck E (1980) *Antimicrobial Food Additives. Characteristics, Uses, Effects.* Berlin: Springer.

Richardson T and Finley JW (eds) (1985) *Chemical Changes in Food during Processing.* Westport, Connecticut: AVI Publishing.

Seow CC, Teng TT and Quah CH (eds) (1988) *Food Preservation by Moisture Control.* Essex: Elsevier Science Publishers.

Thorne S (1986) *The History of Food Preservation.* Carnforth: Parthenon.

Erich Lück
Hoechst Aktiengesellschaft, Frankfurt am Main, FRG

PRESERVATIVES

Contents

Classification and Properties

Preservatives are substances added to food to inhibit microbial spoilage. Chemical food spoilage by enzymatic and nonenzymatic mechanisms may be controlled by specific additives, e.g. antioxidants and antibrowning agents, which are described in the relevant articles. Substances such as common salt, sugar, vinegar or spices, which are effective antimicrobial agents under appropriate conditions, are not regarded as food preservatives. Reactions of food preservatives, which are not part of the chemistry of food preservation, but which nevertheless affect food quality, will also be considered. The chemical reactions which lead to cured meat colour and flavour are reviewed elsewhere. ***See*** Antioxidants, Natural Antioxidants; Antioxidants, Synthetic Antioxidants; Browning, Nonenzymatic; Curing; Spoilage, Chemical and Enzymatic Spoilage; Spoilage, Bacterial Spoilage; Spoilage, Moulds in Food Spoilage; Spoilage, Yeasts in Food Spoilage

Need for Preservatives

The use of preservatives to extend the life of foods and improve their safety has been recognized for thousands of years. Traditional methods of preserving foods have relied on fumigation by sulphur dioxide, curing with brines containing nitrites and nitrates, and smoking, in which phenolic antimicrobial agents are added to the food. These processes not only ensure that stored food is of acceptable microbiological quality but frequently also provide it with unique organoleptic characteristics. ***See*** Fumigants

Table 1. Substances recognized as suitable for use as food preservatives. Those which are approved for use within the EEC have their E-number shown alongside. The presence of any substance in this list does not imply that its use is permitted throughout the world. Biphenyl, orthophenylphenol and thiabendazole are used for surface treatment

Sorbic acid and Na^+, K^+, Ca^{2+} salts	E200–E203
Benzoic acid and Na^+, K^+, Ca^{2+} salts	E210–E213
Ethyl, methyl and propyl *p*-hydroxybenzoate and Na^+ salts	E214–E219
Sulphur dioxide, Na^+ and Ca^{2+} sulphites and hydrogen sulphites (bisulphites), Na^+ and K^+ disulphites (metabisulphites or pyrosulphites)	E220–E227
Biphenyl	E230
Orthophenylphenol	E231
Thiabendazole	E233
Formic acid and Na^+, Ca^{2+} salts	E236–E238
Na^+ and K^+ nitrite and nitrate	E249–E252
Acetic acid and Na^+, K^+, Ca^{2+} salts	E260–E263
Lactic acid	E270
Propionic acid and Na^+, K^+, Ca^{2+} salts	E280–E283
Hydrogen peroxide	
Nisin	

Food preservatives ensure the safety and organoleptic quality of food when sold in the type of marketing environment now found in the developed countries. A particular problem associated with multiuse packs of food, e.g. beverages and spreads, is that an initially sterile, pasteurized or hygienically-assembled, food becomes infected with bacteria, fungi and yeasts, and begins to spoil. One cause of instability might be that the food has been formulated in such a way that it has little or no natural protection against microorganisms. For example, a low-fat spread contains significantly larger water droplets than butter and can, therefore, more readily support microbial growth. A preservative is particularly useful in such a product. A relatively short increase in storage life can be of significant economic and practical advantage. This is well illustrated with advantages gained by extending the shelf life of bread by a few days. ***See*** Water Activity, Effect on Food Stability

Food preservatives are often used in conjunction with other methods of food preservation, e.g. some preservatives allow milder thermal treatment to be applied for the preservation of food with consequent improvement in textural and nutritional properties.

Structures of Food Preservatives

Substances commonly used to preserve foods are listed in Table 1. These represent a wide range of molecular structures and tendency to form ions. The carboxylic acids sorbic, benzoic, formic, acetic, lactic and propionic are monobasic. Aqueous solutions of sulphur dioxide are commonly referred to as sulphurous acid, despite the fact that H_2SO_3 does not exist in measurable amounts. It is established that SO_2 does not interact significantly with water molecules and the term 'molecular SO_2' is more appropriate. Aqueous solutions of SO_2 behave as solutions of a dibasic acid (eqn [1]).

$$\begin{aligned} SO_2 + H_2O &\rightleftharpoons HSO_3^- + H^+ \\ HSO_3^- &\rightleftharpoons SO_3^{2-} + H^+ \end{aligned} \qquad (1)$$

Hydrogen sulphite ion is also in equilibrium with disulphite ion (eqn [2]).

$$2\ HSO_3^- \rightleftharpoons S_2O_5^{2-} + H_2O \qquad (2)$$

In dilute aqueous solution ($<0{\cdot}1$ M) the position of equilibrium lies well over to the left. Significant amounts of disulphite ion are formed only in concentrated systems. When sodium or potassium disulphite is used as a food preservative the salt is hydrolysed to hydrogen sulphite ion.

The nitrites are salts of nitrous acid, which is also the source of the nitrosating species N_2O_3 and oxides of nitrogen, particularly NO.

Regardless of the chemical form of a preservative when it is added to food, its ionic state is determined largely by the pH of the food and the pK_a of the acid. The pK_a values of those preservatives which have an acid group capable of ionizing within, or close to, the pH range of foods are given in Table 2. The fraction, α, of an acid which remains undissociated at a given pH may be found using,

$$\alpha = (10^{pH - pK_a} + 1)^{-1}$$

Since significant changes in α occur in the range $(pK_a - 2) < pH < (pK_a + 2)$, the proportions of ionized and unionized forms of the acids listed in Table 2 change substantially over the pH range of foods. On the other hand, the hydroxybenzoate esters do not ionize signifi-

Table 2. pK_a values of commonly used food preservatives

	pK_a
Sorbic acid	4·76
Benzoic acid	4·18
Sulphur dioxide	1·86
Formic acid	3·75
Nitrous acid	3·40
Acetic acid	4·76
Lactic acid	3·08
Propionic acid	4·88

cantly in the pH range of foods and are, therefore, uncharged. ***See*** pH – Principles and Measurement

An essential feature of food preservatives is that their molecules should have an appropriate balance of lipophilic and hydrophilic behaviour, so as to be capable of traversing relatively nonpolar membranes and yet be sufficiently soluble in the aqueous environment of the microorganisms. This applies to the undissociated carboxylic acids, which will consequently tend to *partition* between the aqueous and oil phase of food. The ionized acids are relatively insoluble in nonaqueous environments. A partition coefficient, P, is defined by

$$P = c_{\text{oil}}/c_{\text{aq}}$$

where c refers to concentration of undissociated acid in each phase. Values of P vary from 0·17 and 2·5 for propionic and sorbic acids, respectively, to 26 and 88 for the ethyl and propyl esters of *p*-hydroxybenzoic acid. The latter two preservatives therefore show a marked bias for the oil phase. Numerous surfactants exist in solution as micellar structures which are also found in the lipid bilayers of cell membranes. The interior of such structures is markedly hydrophobic and carboxylic acids are able to partition into micelles. When expressed in the same way as the oil–water partition coefficient, but using mole fractions instead of concentrations, values of the micellar partition coefficients for benzoic and sorbic acids are in the range 1000–2000. The presence of an oil phase and/or surfactants in solution can significantly reduce the availability of a food preservative.

Mechanism of Antimicrobial Action

Carboxylic Acids and Esters

The antimicrobial action of a food preservative depends on the concentrations of both the ionized and unionized forms, and the specific efficacy of each. In general, the undissociated acid is the better antimicrobial agent but the presence of the ionized form cannot be neglected particularly when, at high pH, its concentration may be much larger than that of the undissociated acid. The minimum concentration of an undissociated acid required to inhibit microorganisms is generally one to two orders of magnitude smaller than that of the corresponding anion.

There is no general theory to explain the mechanisms whereby these carboxylic acid and ester preservatives exert their effect. Transport of these preservatives across cell membranes is passive, but accumulation in the cell causes a reduction in the protonmotive force through the membrane and to a reduction in pH within the cell. These contribute to cessation of cell growth or cell death. Certain preservatives could also exert specific effects on metabolic enzymes. Sorbic acid is alleged to react with the sulphydryl groups of fumarase in catalase-positive bacteria, moulds and yeasts, and aspartate, succinic and yeast alcohol dehydrogenase. It is suggested that this might apply to sulphydryl-containing enzymes in general. Benzoate inhibits enzymes in oxidative phosphorylation and at various stages in the tricarboxylic acid cycle. ***See*** Enzymes, Functions and Characteristics

Microorganisms are able to metabolize some food preservatives when present at sublethal concentrations. A significant example is the conversion of sorbic acid to hexadienol by certain strains of lactic acid bacteria. This product can react with ethanol to form 1-ethoxy-2,4-hexadiene and 2-ethoxy-3,5-hexadiene, which give rise to a geranium-type odour, occasionally detected in wines treated with the preservative.

Sulphur Dioxide and Sulphites

Molecular sulphur dioxide passes into yeast cells by passive transport, and possibly into some other microbial cells by active transport. As with carboxylic acids, it interferes with membrane transport processes. However, sulphites are much more chemically reactive than any of the carboxylic acids referred to above. Thus, sulphite ion acts as a powerful nucleophile causing the cleavage of disulphide bonds of proteins; it reacts with coenzymes (NAD^+), cofactors and prosthetic groups (flavin, thiamin, haem, folic acid and pyridoxal). Consequently, a broad range of metabolic enzymes is inactivated and structural proteins which contain disulphide bonds may be denatured. Prior to the death of yeast cells treated with sulphite there is a rapid decrease in ATP content. This has been attributed to the inactivation of glyceraldehyde-3-phosphate dehydrogenase by sulphite species. The additive also reacts with carbonyl constituents of the metabolic pool, to form hydroxysulphonates. When treated with sublethal concentrations of sulphite, yeasts tend to excrete increased amounts of acetaldehyde. This is due to the trapping of this metabolic intermediate as the particularly stable hydroxysulphonate, thereby preventing its conversion to ethanol. Glycerol is formed instead of ethanol by reduction of glyceraldehyde-3-phosphate to glycerol-3-phosphate which is subsequently dephosphorylated.

Nitrite Ion

Nitrite ion is believed to exert its action on the phosphoroclastic system of enzymes which causes the conversion of pyruvate to acetate and is a source of ATP. Evidence for this involvement comes from the fact that in the presence of nitrite, ATP concentration in the cell rapidly decreases and pyruvate is excreted. Nitrite ion is in equilibrium with nonionic forms including NO. This is a good ligand for iron and the mechanism of inhibition is likely to be the reaction of NO with the non-haem iron-containing pyruvate:ferredoxin. In aerobic bacteria the haem iron of cytochrome oxidase is also a likely target for NO. Inhibition of enzymes with sulphydryl groups as a result of the formation of *S*-nitroso products is also possible at relatively high nitrite concentrations. ***See*** Nitrosamines

When nitrite-containing bacteriological media are heated they become more inhibitory towards *Clostridium botulinum* than media which have been heated prior to the addition of the preservative. This so-called Perigo effect can be modelled using mixtures of cysteine, nitrite and iron(II) salts. When heated, such mixtures give rise to iron–sulphur bridged complexes, of which Roussin's black salt is a well known example. These compounds have been shown to be effective inhibitors of clostridial spores. ***See*** *Clostridium*, Occurrence of *Clostridium botulinum*

Nisin

Nisin is a polypeptide (molecular weight 3500) which usually exists as a dimer. It is produced by *Lactococcus lactis* and may be formed naturally in cheese. The polypeptide chain contains L-amino acids and the unusual sulphur-amino acids lanthionine and β-methyllanthionine. A specific enzyme, nisinase, has been widely reported and its formation is possibly one reason why many lactic acid bacteria inactivate nisin. The peptide is also susceptible to α-chymotrypsin but not to pepsin, trypsin and carboxypeptidase A, among other proteolytic enzymes. ***See*** Nisin

Chemical Reactivity of Preservatives Towards Food Components

Chemical reactions between food preservatives and components of microbial cells, or with food components where there are implications with regard to antimicrobial action, have been described above. However, some food preservatives, and particularly sorbic acid, sulphur dioxide, sulphites and nitrite ion are capable of more extensive reactivity with food components. This may lead to the formation of reaction products of toxicological importance and a reduction in the concentration of available preservative.

Sorbic Acid

The conjugated dienoic acid structure of sorbic acid makes it susceptible to nucleophilic attack. In foods the most significant nucleophiles are thiols and sorbic acid reacts with low-molecular-weight thiols such as cysteine and glutathione to form the 5-substituted 3-hexenoic acid. Such adducts are labile under acid conditions and may not be detected during analysis of sorbic acid in foods when the sample is acidified before extraction. Low-molecular-weight alkyl thiols form diadducts which are stable towards hydrolysis. Sorbic acid can also undergo autoxidation to malonaldehyde, acetaldehyde and β-carboxyacrolein. It is suggested that the double bond in position-4 is most susceptible to oxidative attack. The preservative is also capable of browning, particularly in the presence of amino compounds, probably as a result of the formation of reactive carbonyl compounds following autoxidation.

Sulphur Dioxide and Sulphites

It is well known that the concentration of sulphite species in a food decreases with time and the shelf life may be limited by this reactivity. The conversion of sulphite to sulphate through autoxidation is catalysed by transition metal ions and involves the formation of oxidizing free radicals. There is evidence to suggest that the facile autoxidation of ascorbic acid, which is less easily controlled by antioxidants, may promote the oxidation of sulphite in foods containing the vitamin. The free-radical intermediates ($^{\bullet}OH$, $^{\bullet}O_2^-$) in sulphite autoxidation are known for their ability to cause the oxidation of unsaturated organic compounds in the absence of suitable antioxidants. Free-radical sulphonation of unsaturated organic compounds by $^{\bullet}SO_3^-$ is also possible. An important nucleophilic reaction of sulphite ion is its addition to the α,β-unsaturated carbonyl moiety of 3,4-dideoxyosulos-3-enes formed as reactive intermediates in Maillard and ascorbic acid browning. This reaction can cause a considerable depletion of the additive in foods susceptible to these forms of nonenzymatic browning.

Nitrite

The reaction of secondary amines with nitrosating species derived from nitrite, to give *N*-nitroso compounds, is very well known and has been given much publicity on account of the possible carcinogenicity of the products. The reaction proceeds at a maximum rate in the pH ranges 2·25–3·4 and is catalysed by weakly acid anions, notably thiocyanate. In meat products the reaction is most probably associated with the adipose tissue and it is suggested that nonionic amine and the nitrosating species N_2O_3 partition into the nonaqueous phase where a facile reaction takes place. A similar explanation has been suggested for the catalytic effect of surfactants and cell membranes on nitrosation, but in this instance the reaction takes place within the micellar environment. *C*-Nitrosation of phenolic components of food, particularly in smoked products, leads to the formation of nitrosophenols which can readily oxidize to the corresponding nitro compounds. The *C*-nitrosation of activated methylene groups is also established; suitable reactants in food include the 3-deoxyosuloses formed as intermediates in Maillard and ascorbic acid browning reactions, when the product is an oxime. *S*-Nitroso compounds are readily formed by the nitrosation of thiols and represent a reversibly bound form of the preservative.

Chemical Interactions Between Preservatives

Sulphite species, nitrite and sorbic acid are chemically incompatible. Sulphite ion reacts readily with nitrite ion to form imidodisulphonate ion. The mechanism of the reaction between sorbic acid and sulphite species depends on whether or not oxygen is present. Under aerobic conditions, a sulphite-mediated oxidation of sorbic acid takes place. Under anaerobic conditions a much slower nucleophilic addition of sulphite ion gives 5-sulpho-3-hexenoic acid. Mixtures of sorbic acid and nitrite ion give rise to ethylnitrolic acid and 1,4-dinitro-2-methylpyrrole, both of which give a positive result in the Ames mutagenicity test.

Bibliography

Branen AL and Davidson PM (eds) (1983) *Antimicrobials in Food.* New York: Marcel Dekker.

Branen AL, Davidson PM and Salminen S (eds) (1990) *Food Additives.* New York: Marcel Dekker.

Gould GW (ed.) (1989) *Mechanisms of Action of Food Preservation Procedures.* London: Elsevier Applied Science.

Lueck E (1980) *Antimicrobial Food Additives. Characteristics, Uses, Effects.* Berlin: Springer-Verlag.

Wedzicha BL (1984) *Chemistry of Sulphur Dioxide in Foods.* London: Elsevier Applied Science.

Wedzicha BL (1988) Distribution of low-molecular weight food additives in dispersed systems. In: Dickinson E and Stainsby G (eds) *Advances in Food Emulsions and Foams*, pp 329–371. London: Elsevier Applied Science.

B.L. Wedzicha
University of Leeds, Leeds, UK

Food Uses

Factors Affecting Choice

General Considerations

The type of preservative applicable to a particular need is determined by the composition of the food, the type of microbial spoilage which takes place and the desired shelf life. Important compositional variables include pH, since the efficacy of a given preservative generally decreases with increase in pH. The maximum pH at which a preservative is useful is often quoted. Some components of food can also potentiate the action of preservatives, e.g. chelating substances such as EDTA or citric acid can render preservatives more effective and at the same time protect the food from other forms of spoilage. Salt is a common food component which acts synergistically with several food preservatives. Ideally, the use of preservatives should be regarded within a 'preservative systems' approach to include interactions, rather than in terms of individual substances. This concept is extended to include other methods of food preservation, e.g. the use of thermal processes in conjunction with chemical preservatives, to optimize the use of each. Such effects of food components and processing on antimicrobial activity also make theoretical predictions of the efficacy of a given preservative unreliable and the levels of application are usually decided upon empirically. ***See*** pH – Principles and Measurement

Preservatives are also chosen for their specific physical properties such as solubility in particular foods and ease of handling. Calcium sorbate, for instance, is sparingly soluble in water (1·2 g per 100 g water) and is suitable for surface treatment of foods such as cheeses as it does not quickly dissolve in surface moisture or migrate into the cheese. Chemical stability is important with regard to the handling of sulphites; the disulphites (metabisulphites) tend to be the most stable sulphite species towards autoxidation and dissolve in water to give solutions with pH values in the middle of the pH range of foods. ***See*** Oxidation of Food Components

Cost may be an important consideration; sulphites are the cheapest to use whilst sorbic acid, sorbates and the *p*-hydroxybenzoate esters are the most expensive.

Spectrum of Antimicrobial Action

In general, benzoic acid, its salts and esters of *p*-hydroxybenzoic acid have the broadest spectrum of antimicrobial activity and are useful against many spoilage bacteria, fungi and yeasts. Benzoic acid cannot, however, be relied upon to preserve foods against bacteria effectively because, whilst food poisoning and spore-forming bacteria are inhibited at normal levels of use in low pH foods, many spoilage bacteria are more resistant. Benzoic acid is most suitable for foods at pH 2·5–4 whilst the esters of *p*-hydroxybenzoic acid may be effective to pH 7 or higher. However, the latter tend to be less effective against bacteria and particularly Gram-negative organisms. Thus, in relatively acidic products where yeasts and moulds are the greater causes of spoilage, benzoic acid and its salts offer an effective means of preservation. ***See*** Spoilage, Bacterial Spoilage; Spoilage, Moulds in Food Spoilage; Spoilage, Yeasts in Food Spoilage

Sorbic acid and its salts are useful additives against yeasts and moulds but are less effective against bacteria. They provide antimicrobial activity up to pH 6·5 and are good preservatives for foods with a high fat content and pH, e.g. low-fat spreads, processed cheeses; the *p*-hydroxybenzoate esters tend to be too soluble in the nonaqueous component of such foods whilst benzoate is relatively ineffective at the higher pH values.

Sulphur dioxide and sulphites also represent broad-spectrum antimicrobial agents in acidic foods, particularly beverages. In general, this preservative is more effective against bacteria than moulds and yeasts, but Gram-positive bacteria are less susceptible than Gram-negative bacteria. The effect against yeasts is greatest around pH 3·5 and falls off markedly as pH is raised above this value. There is a tendency for desirable fermentation yeasts to show a greater resistance to this preservative than undesirable 'wild yeasts' and lactic acid bacteria. This is exploited in the use of sulphites during wine-making. Despite the relatively high pH associated with meat products such as the British sausage (often pH 6–6·5), sulphites tend to inhibit enterobacteriaceae, and particularly salmonellae, in this environment. In some situations sulphites are added also to inhibit chemical spoilage, notably enzymatic browning caused by oxidation of *o*-diphenols, other forms of oxidative spoilage and nonenzymatic browning of reducing sugars and ascorbic acid. ***See*** Browning, Nonenzymatic; Spoilage, Chemical and Enzymatic Spoilage

Propionic acid and its salts are effective against moulds despite the fact that some species of *Penicillium* can grow on media containing as much as 5% propionic acid. Yeasts and most bacteria are generally less affected. However, a specific target for propionic acid is the rope-forming *Bacillus subtilis* in bread, which is effectively controlled at pH 6. The preservation of a wide range of foods with vinegar (acetic acid) has been known since ancient times; acetic acid is more effective against yeast and bacteria than moulds. Its effect on bacteria is, however, predominantly due to the associated reduction in pH. *Acetobacter* and some lactic acid bacteria are unaffected by this preservative. Lactic acid is similarly a preservative associated with traditionally prepared and particularly fermented foods. As with acetic acid, its antimicrobial action is mainly due to a reduction of pH to below that at which bacteria can grow. It does, however, inhibit spore-forming bacteria at pH 5·0 but is ineffective against yeasts and moulds. ***See*** Vinegar

The most important effect of nitrite ion in food is its inhibition of *Clostridium botulinum* in cured meat products. Whilst it also has an inhibitory effect towards a wide range of different species of bacteria some, and particularly *Salmonella*, lactobacilli and *Clostridium perfringens*, tend to be more resistant. Nitrite and nitrate are essential for the development of cured meat colour and flavour. ***See*** *Clostridium*, Occurrence of *Clostridium perfringens*; *Clostridium*, Occurrence of *Clostridium botulinum*; Curing

Thiabendazole, orthophenylphenol and biphenyl inhibit fungi and are suitable only for surface application.

Specific Applications

Beverages

Nonalcoholic beverages, including fruit juices, are usually preserved with benzoic acid and its salts, esters of *p*-hydroxybenzoic acid, sorbic acid and its salts, or the sulphites. Typical concentrations of benzoic acid required are in the range 500–1000 mg kg^{-1}. In the event that these levels lead to a noticeable taste, benzoic acid is replaced wholly or partly by mixtures of methyl and propyl esters of *p*-hydroxybenzoic acid (usually 2:1, methyl:propyl) at a combined concentration of 300–500 mg kg^{-1}, or sorbic acid and its salts at 200–1000 mg kg^{-1}. Whilst sulphites are effective antimicrobial agents in beverages at concentrations of 50–500 mg SO_2 per kg, they tend to undergo significant oxidation when beverage containers are repeatedly opened and air admitted. A combination of sulphite with another preservative, e.g. sorbic or benzoic acid is frequently used for fruit juices such that the sulphite acts to control forms of chemical spoilage, and lactic and acetic acid fermentation, whilst the second preservative acts as a longer-lasting agent against yeasts and moulds. A significant disadvantage of adding sulphites to beverages coloured with anthocyanins is that these are decolorized, even at low concentrations of preservative, and additional sulphite-stable food-colour agents are required. Fruit juice concentrates used for the manufacture of normal-strength juice may contain any of the preservatives used in the final beverage. In practice it is not feasible to add sufficiently high concentrations of the carboxylic acid or ester preservatives to the concentrate to provide an acceptable level of additive in the finished product. In general, such concentrates need to be protected against browning, which demands high levels (e.g. 2·5–10 g SO_2

per kg) of sulphite; this also acts as a preservative. In any case the concentration of preservative in the finished product is a result of further addition after dilution and blending.

Sulphites are the major preservatives for alcoholic beverages. Their most important function is the control of the so-called wild yeasts in grape juice and after fermentation the preservative is required at levels of 50–150 mg SO_2 per kg to inhibit further microbial action. A specific problem associated with the use of sulphites to preserve wine, cider and perry is that fruit which has been affected by moulds may contain significant amounts of acetaldehyde, pyruvic acid, α-ketoglutaric acid and other aldehydes and ketones which are able to bind the preservative reversibly as hydroxysulphonate adducts. When present in this form, sulphite is unable to act as a preservative; however, the concentration of hydroxysulphonates is included in the analytical concentration of the preservative in the beverage. When grapes affected by *Botrytis* are used for wine-making, as much as 80% of the total preservative can be found as hydroxysulphonate adducts. During fermentation in the presence of sulphites, yeasts form sulphite-binding compounds and particularly acetaldehyde. Sorbic acid is also a good preservative for wine and, despite its chemical reactivity towards sulphite, mixtures of the two are advocated; in such a situation sorbic acid serves to prevent further fermentation, whilst the sulphite protects against chemical and bacterial spoilage. However, depletion of sulphite through chemical reactions could result in malo-lactic bacteria metabolizing sorbic acid to intermediates which ultimately lead to off-flavours. In general, potassium sorbate is the preferred salt on account of its higher solubility than that of the sodium salt. In some wines, however, potassium tartrate may precipitate, in which case the sodium salt is used. ***See*** Perry; Wines, Production of Table Wines; Wines, Production of Sparkling Wines

Fruit Products

Sulphur dioxide is an effective fumigant in the control of post-harvest decay of grapes caused by the fungus *Botrytis cinerea*. Raspberries may be similarly treated for *Botrytis* and *Cladosporium*. Another unique application of sulphite is for the storage, as pulp, of soft fruit to be used for jam manufacture. A typical ‘sulphite liquor’ would contain up to 3000 mg SO_2 per kg solution and is capable of preserving fruit for up to 2 years with a high retention of ascorbic acid. Whilst the anthocyanins are bleached in this process, the colour of the fruit reappears as the preservative is ‘lost’ during jam making. An endopolygalacturonase enzyme from *Rhizopus sexualis* is not inactivated by sulphite and can cause the breakdown of strawberries in pulp. Jam and related products, e.g. fruit purées and pie fillings, may be successfully preserved with benzoic acid (1000 mg kg^{-1}), *p*-hydroxybenzoate ester (methyl:propyl ester, 3:1; 700 mg per kg), sorbic acid (800–1500 mg kg^{-1}) and sulphite (100 mg SO_2 per kg in jam). ***See*** Fumigants

Dried fruits are frequently prepared with the aid of sulphite as an antibrowning agent. In such products the concentration of additive may be as high as 2500 mg SO_2 per kg, and this will also effectively preserve the product. When a relatively moist unsulphited product is desired, a preservative is necessary, and a potassium sorbate dip or spray to give 200–500 mg sorbic acid per kg, or similar application of sodium benzoate at 1000 mg benzoic acid per kg, are effective.

Vegetable Products

One of the most common traditional methods of preserving vegetables is by pickling with vinegar, acetic acid contributing to the reduction of pH, a specific antimicrobial effect and a characteristic taste. Typically, raw vegetables are immersed in 0·5–3% acetic acid solution, but some yeasts and moulds are still capable of causing long-term spoilage. The combined use of acetic acid with benzoic acid, *p*-hydroxybenzoate ester mixtures (methyl:propyl, 2:1) or sorbic acid (all up to 1000 mg kg^{-1}) allows lower acetic acid concentrations and safeguards against spoilage. In sweet relishes, sorbic acid is more effective than benzoic acid on account of the relatively high pH. The selective control of yeasts, moulds and putrefactive bacteria is the reason why sorbic acid is useful at levels of 500–2000 mg kg^{-1} in vegetable fermentations, where it allows the growth of the desired lactic acid-producing organisms. ***See*** Pickling

Fresh vegetables are generally dipped in sulphite solutions to prevent enzymatic browning though this also protects against microbial spoilage. It is feasible to preserve vegetarian burger-type products with sulphite. On the other hand, dehydrated vegetables are microbiologically stable but require sulphite as an antibrowning agent for the dehydration process and subsequent storage, at a final level of some 2500 mg SO_2 per kg of dried product.

Baked Products

Propionic acid is widely used in the making of bread and cakes to inhibit surface mould and *Bacillus subtilis*, which causes rope. The sodium salt is used chiefly in confectionery products whilst the calcium salt is used for bread. Levels of propionic acid up to 3000 mg kg^{-1} may be found in such applications. Sorbic acid is also an effective preservative for cereal products, at concentrations similar to those of propionic acid, and inhibits the mould *Trichosporon* found in rye bread and cakes. Unfortunately, sorbic acid also inhibits yeast. A potential solution could be to use sorboyl palmitate which has no antimicrobial activity but which degrades on heating, liberating sorbic acid. Mixtures of esters of *p*-hydroxybenzoic acid (methyl:propyl, 3:1) at 300–600 mg per kg are useful preservatives for cakes (particularly fruit cakes) but inhibit yeasts and cannot be used in bread. Some components of flour confectionery, e.g. icing, fillings and marzipan, can be preserved with benzoic acid, esters of *p*-hydroxybenzoic acid and sorbic acid.

Dairy Products

Apart from the small-scale use of hydrogen peroxide to preserve milk by reducing the severity of heat treatment required for sterilization, the major use of preservatives in dairy products is for cheese. Propionic acid is formed naturally during the ripening of certain cheeses, but may be used as an added preservative for surface treatment to prevent mould growth. Nitrite ion, which is formed from added nitrate in cheese, inhibits fermentations caused by clostridia and coliforms such as *Enterobacter* which are undesirable because they may give rise to the production of carbon dioxide with resultant blowing of the cheese. Nitrite ion is to be avoided in cheeses where the production of propionic acid is important because nitrite also inhibits the bacteria required for its formation. Nisin is a good antimicrobial agent for cheese because it is particularly effective against clostridia such as *Cl. tyrobutyricum* in cheeses with a higher pH. The egg white antimicrobial enzyme, lysozyme, is employed for a similar purpose. Cheese-making represents a major use of sorbic acid, where it serves two purposes: to inhibit surface mould growth, achieved by dipping, dusting (preferably with calcium sorbate) or incorporation of the preservative into wax coatings or packaging films, or to protect packs of processed cheese after opening, by incorporation of the preservative into the cheese at levels of 500–700 mg sorbic acid per kg. ***See*** Cheeses, Chemistry and Microbiology of Maturation; Nisin

Meat and Fish Products

The curing of meat is probably the best known use of nitrites and nitrates. They are used in conjunction with salt, which has the effect of reducing the water activity sufficiently to reduce spoilage by *Pseudomonas* and related organisms, whilst nitrite prevents growth of germinating spores. Curing is regarded as protection against food poisoning caused by *Clostridium botulinum*. Eventual spoilage in cured meat is caused by lactic acid bacteria if the salt content is low, or by micrococci and vibrios if it is high together with a variety of moulds and yeasts. Off-flavours are associated with hydrolysis of fat. The storage life of cured meat is extended by smoking as a result of additional antimicrobial agents from the smoke; again, when the meat ultimately spoils it is dominated by lactic acid bacteria. Typical concentrations of sodium nitrite and nitrate in cured meat are 50–200 mg and 500 mg per kg, respectively. ***See*** Meat, Preservation

Sulphites are added to a small number of comminuted meat products (sausages, burgers) at levels of 600 mg SO_2 per kg, to achieve a maximum of 450 mg SO_2 in the finished product. Whilst the total sulphite content of such a product falls slightly during storage, a large proportion becomes reversibly bound to carbonyl compounds of microbial origin. The success of sulphite in this application is the result of its inhibition of salmonellae whilst spoilage of sulphited meat is confined to Gram-positive microflora consisting of lactobacilli and *Brochothrix thermosphacta*. The result is that the spoilage is associated with a sour odour, unlike the spoilage of untreated meat which gives rise to a 'putrid' smell. The use of sulphites in meat products leads to the destruction of thiamin and is not permitted in important meat sources of the vitamin. Mould growth on the surface of dry sausage, during the drying period, may be controlled by dipping the casings in a solution of potassium sorbate prior to stuffing.

Fish may be preserved in acetic acid (10–30 g per kg) by the traditional process of marinating. Whilst this provides basic protection against pathogenic microorganisms, lactobacilli may grow and the use of acetic acid should be combined with a mild thermal process or other antimicrobial agents. Benzoic acid is slightly more effective than sorbic acid in this application (1000 mg per kg) and esters of *p*-hydroxybenzoic acid (300–600 mg per kg) may also be used. ***See*** Fish, Spoilage

Fat Products

Although benzoic acid has been used for some time for the preservation of margarine at concentrations of 800–1500 mg per kg, it is far from ideal for this application owing to the relatively high pH of the food. On the other hand, it is a good preservative for mayonnaise, though improved performance against acid-producing bacteria can be obtained by using mixtures of benzoic and sorbic acids. Margarine and low-fat spreads can be most effectively preserved by means of sorbic acid. ***See*** Dressings and Mayonnaise, Chemistry of the Products; Margarine, Methods of Manufacture

Miscellaneous Uses

Food ingredients which are normally manufactured in the form of solutions, or have a water activity sufficient to sustain microbial growth, may be treated with preservatives. Examples include flavouring preparations, gelatin, malt extracts, antifoam agents. ***See*** Water Activity, Effect on Food Stability

Carbon Dioxide

Carbon dioxide is not a preservative in the usual meaning of the word. However, in most of the modified atmosphere packaged foods, it does help to preserve. It specifically inhibits the growth of the typical fast-growing oxidative Gram-negative bacteria that quickly spoil e.g. meat and fish. Slower growing, less obnoxious microorganisms, such as lactic acid bacteria, often then take over, but shelf life can be substantially extended. The extent of such inhibition by CO_2 is greatly enhanced as temperature is reduced. It has been speculated that this results mainly from the increased solubility of the gas at low temperatures. Shelf life increases for meat and fish can be by 2–3 times, if chill control is effective.

Bibliography

Branen AL and Davidson PM (eds) (1983) *Antimicrobials in Food.* New York: Marcel Dekker.

Hayes PR (1985) *Food Microbiology and Hygiene.* London: Elsevier Applied Science.

Lueck E (1980) *Antimicrobial Food Additives. Characteristics, Uses, Effects.* Berlin: Springer-Verlag.

Wedzicha BL (1984) *Chemistry of Sulphur Dioxide in Foods.* London: Elsevier Applied Science.

B.L. Wedzicha
University of Leeds, Leeds, UK

Analysis

Food preservatives constitute a group of compounds of widely different molecular structures; they are organic and inorganic substances with different functional groups and tendencies to form ions. There are no procedures which are generally applicable to the analysis of preservatives as a class of food additive; the procedures are specific to the preservative being analysed. The lowest concentrations of commonly used preservatives are of the order of a few milligrams per kilogram of food and, with few exceptions, recommended or statutory methods of analysis are designed to give good accuracy at levels of 10 to >1000 mg preservative per kg food. The question of the lower limit of detection is rarely an issue unless it is desired to use small sample sizes, e.g. <1 g, or to determine whether or not a food or its ingredients had been treated with a preservative. For solid foods, small sample sizes often lead to nonrepresentative sampling and should be avoided. Not all the procedures described constitute official methods of analysis. Frequently, for routine analysis a food manufacturer would use a rapid or cheap analytical technique standardized against an official method. The official status of given procedures varies from country to country.

Organic and inorganic acid preservatives may be added in the form of the undissociated acid or a variety of salts. In food the ionic composition is determined largely by concentration and pH, but it is generally impossible to predict this accurately for any given situation. In order to avoid complications with the specification of the amount of preservative in a food, this is usually referred to as the weight for weight concentration of the undissociated acid, e.g. benzoic acid, sorbic acid or sulphur dioxide. Nitrite and nitrate levels are expressed in terms of the weight of the sodium salt.

There are, of course, a very large number of possible analytical procedures available for each preservative. Those given here represent a selection to illustrate the variety of methods recommended for use on food samples.

Carboxylic Acids and Esters

Extraction

All the carboxylic acid and ester preservatives are steam-volatile from acidified food samples, and steam distillation offers an effective means of separating them from the sample. In the case of sorbic acid, the use of an acid environment causes the decomposition of sorbic acid–thiol adducts which may be formed in some foods and therefore go undetected. The nonionic nature of these preservatives in acid solution also allows extraction with a variety of organic solvents, e.g. diethyl ether, chloroform, dichloromethane. Such organic solvents may also be used to isolate, or concentrate, preservatives obtained in large volumes of steam distillate. The subsequent detection and determination of preservatives depends on the substance to be analysed.

Analysis of Extracts and Distillates

Rapid screening of extracts and distillates for sorbic and benzoic acid and the esters of *p*-hydroxybenzoic acid is possible by TLC on silica gel. If separated on plates containing an F_{254} fluor, these acids may readily be visualized under UV light. Acetic and propionic acids can be identified by paper chromatography using an acid–base indicator for visualization of spots. ***See*** *Analysis*

Chromatography, Thin-layer Chromatography; Chromatography, Principles

The total acidity in steam distillates is an acceptable measure of the concentration of an acid preservative provided that only one such preservative is used and that a blank for the food sample is known. Propionic and acetic acids can be separated using a silica gel column with butanol–chloroform solvent and the eluate titrated for quantitation. Solvent extracts of food samples can only be subjected to direct spectrophotometric analysis with caution. For example, benzoic acid, extracted into diethyl ether from tomato products, jams, jellies and soft drinks, may be determined by measuring its absorbance at 267·5, 272 and 276·5 nm and using the average of the highest and lowest wavelength readings as the background value for the absorbance at 272 nm. In general, solvent extracts and steam distillates can be analysed directly by reversed-phase HPLC or by GLC after forming trimethylsilyl or methyl esters. Internal standards for chromatography are chosen to have extraction and chromatographic behaviour similar to that of the preservative being analysed. Standards are generally added to the food before extraction and any derivatization is carried out on both the analyte and the standard. Examples of internal standards include phenylacetic and caproic acids for gas chromatographic analysis of benzoic and sorbic acids, respectively. ***See*** Chromatography, High-performance Liquid Chromatography; Chromatography, Gas Chromatography

Esters of *p*-hydroxybenzoic acid are determined in steam distillates or extracts either by reversed-phase HPLC or by direct spectrophotometry as *p*-hydroxybenzoic acid after saponification.

Oxidation of sorbic acid to malonaldehyde by means of dichromate and subsequent reaction with thiobarbituric acid (TBA) offers a specific nonchromatographic assay of this preservative in steam distillates from all types of food. Frequently, crude solvent extracts can be analysed in the same way without undue interference from other TBA-reactive substances either present in food or formed during dichromate oxidation.

Enzymatic Methods

Specific enzymatic procedures for the determination of acetic and formic acids are available. Acetic acid is determined by conversion to pyruvate (using ATP and coenzyme A in the presence of acetylcoenzyme A synthetase), which reacts in turn with oxaloacetate (citrate synthetase) obtained by reduction of malate by NAD^+ (malate dehydrogenase). The rate of utilization of NAD^+ is a measure of acetic acid concentration. Formic acid is determined more simply by measuring the rate of utilization of NAD^+ during the oxidation of the acid to hydrogen carbonate (formate dehydrogenase). In general, sample preparation for enzymatic methods involves the preparation of aqueous extracts of solid foods, separating fat and precipitating protein (perchloric acid), decolorizing strongly coloured samples (charcoal) and adjusting to pH 7–8. ***See*** Enzymes, Use in Analysis

Sulphur Dioxide and Sulphites

Classification

Sulphur dioxide and sulphite are considered to exist in food in two forms: free and bound. The former includes all species of sulphur in oxidation state +4, i.e. SO_2, HSO_3^-, SO_3^{2-} and $S_2O_5^{2-}$. The latter term is used to describe that preservative which is in the form of hydroxysulphonate (carbonyl–HSO_3^-) adducts. Legislation requires that total (free + bound) preservative be determined. In food processing, the concentration of free additive determines its antimicrobial effect and is frequently measured. A specific problem which exists with this preservative is that its measured concentration falls with time as a result of reactions with food components, and as a result of autoxidation after packaged foods are opened and exposed to air. Samples must be analysed without delay and exposure to air minimized.

A broad classification of analytical methods for sulphur dioxide and sulphites is based on whether or not a separation procedure is involved. Methods are usually referred to as direct when a specific analytical reaction is applied to the whole food sample and indirect when the additive is recovered before determination.

Direct Methods of Analysis

The simplest analysis of sulphites in foods involves iodimetric titration after allowing for a blank representing the reactivity of the food sample itself towards iodine. This is normally found by adding an aldehyde or ketone (formaldehyde, acetaldehyde, acetone) to combine with free sulphite before the blank titration. The free sulphite content of a food is determined by acidifying the sample and titrating with iodine. The total (free + bound) sulphite may be determined by raising the pH of the sample to pH 10 to decompose any hydroxysulphonates before acidifying and titration. Iodimetric titration is applicable to beverages and aqueous extracts/homogenates of solid foods. Dark-coloured beverages may be titrated using electrometric detection of the endpoint.

Sulphites in solution may be determined spectrophotometrically by reaction with tetrachloromercurate

+pararosaniline+HCHO reagent, 5,5-dithiobis(2-nitrobenzoic acid) (DTNB) or malachite green. These reactions depend on the nucleophilic behaviour of sulphite ion and suffer interference from nucleophiles in foods, particularly thiols. Despite this, the pararosaniline method has been used widely for direct determination of the preservative in beverages, solutions of sugars, and aqueous extracts or homogenates of solid foods including fruits and vegetables, and has been adapted successfully for an 'autoanalyser' system. ***See*** Spectroscopy, Visible Spectroscopy and Colorimetry

Few chromatographic methods of analysis of sulphites in foods have gained acceptance, mainly because of the problems associated with detection of the species. Ion-exclusion chromatography of sulphite ion with amperometric detection is the best-established chromatographic method available and can be applied directly to beverages and aqueous extracts of foods.

Other direct analyses include the use of polarography and sulphur dioxide-sensitive electrodes but are not in widespread use. Enzymatic methods of analysis make use of the oxidation of sulphite to sulphate+H_2O_2 (sulphite oxidase) and assay of H_2O_2 with NADH (NADH-peroxidase). The rate of formation of NAD^+ is used to determine the sulphite concentration. Sulphite oxidase may also be immobilized at the tip of an oxygen-measuring electrode for a direct indication of the rate of utilization of oxygen by the enzyme in sulphite oxidation. The pH requirement for these enzymatic analyses is slightly alkaline (pH 7·5–8) and the method can only give the total additive concentration.

Indirect Methods of Analysis

All indirect methods of analysis of sulphur dioxide and sulphites in foods involve the conversion of the ionic forms to gaseous SO_2 as the method of separating the additive from the food. The Monier–Williams distillation technique, described in 1927, is still the standard by which other methods are evaluated. This method, and variants of it, involve the distillation, under reflux, of the food sample acidified with HCl, H_2SO_4 or H_3PO_4, in a gentle stream of an inert gas. SO_2 is trapped in H_2O_2 or iodine. Some variations of this technique include the codistillation of SO_2 with water with a downward sloping condenser, to speed up the procedure. The most serious interference arises from the presence of volatile compounds which are oxidized by H_2O_2 to acid, or react with iodine. Such interfering components are found in onions, leeks and cabbage. Small samples may be analysed by replacing the conventional trapping agents by pararosaniline reagent or DTNB, which allow spectrophotometric determination of SO_2 at much greater sensitivity than is possible by titration. Polarographic analysis of the distillate offers a good alternative. Desorption of SO_2 from cold acidified samples enables free preservative to be determined; prolonged boiling (e.g. 1·75 h) or pretreatment with alkali causes the decomposition of hydroxysulphonates and gives the total preservative present. In general, distillation methods do not require any sample preparation and are equally applicable to solid and liquid foods.

An automatic distillation unit ahead of an 'autoanalyser' with spectrophotometric detection provides an effective automatic indirect method of analysis for liquid samples and aqueous extracts. An ingenious approach to separating dissolved SO_2 from an acidified food sample is by diffusion through a PTFE membrane. This may be used in a flow injection arrangement where the diffusion cell carries streams of acidified liquid food or extract solution on one side of the membrane and a suitable reagent for spectrophotometric analysis on the other.

Analysis of gaseous SO_2 in the headspace above acidified liquid foods or homogenates of solid foods, or transfer of SO_2 through the headspace from the sample to a reagent for spectrophotometric analysis in a microdiffusion cell, offer alternative indirect methods of analysis.

Nitrites and Nitrates

The most important uses of nitrite and nitrate as food preservatives are in meat products and the majority of analytical procedures for food have been devised specially for this application. There are, however, numerous published procedures for the determination of the ions in water, and soil and those occurring naturally in plant material.

Since nitrites and nitrates are readily soluble in water, aqueous extracts from homogenates of the food at neutral or weakly alkaline pH are sufficient. A low extraction pH leads to the loss of nitrite. Apart from pH, important variables in extraction are time and the addition of heavy-metal salts (e.g. $BaCl_2$, $CdCl_2$, $HgCl_2$). A proportion of nitrite present in food may exist in bound form as nitrosothiols, nitrosylmyoglobin, etc.; methods of extraction which involve the heavy-metal ions and long extraction times (e.g. 1–4 h) cause the release of such bound additives. Techniques based on short extraction periods (e.g. 10 min) at neutral pH measure mainly the free nitrite. Nitrosyl haemoproteins may be decomposed to species convertible to nitrite if the extracting medium contains 80% acetone.

Extracts generally contain dissolved and suspended protein. This is removed and solutions are clarified by means of zinc acetate+potassium ferrocyanide, alumina cream, potash alum or zinc sulphate+borax. In general, the reagents allow recoveries of nitrite of >98% with standard deviations no greater than 0·75%, from meat samples, with the exception of the procedure involving the use of alumina cream, for which the recovery is slightly lower at 95%.

Nitrate may be determined directly using a reaction in which it is used as a nitrating agent for variety of organic substrates and particularly the xylenols (dimethylphenols); the nitro-compounds so formed may be determined spectrophotometrically. This reaction can also be used to determine the concentration of nitrite if it is first oxidized to nitrate with permanganate. Alternatively, nitrate may be reduced to nitrite by spongy cadmium or enzymatically with NADPH (nitrate reductase). The rate of formation of NADP in the enzymatic procedure may be used to measure nitrate concentration directly. The most widely used reaction for the determination of nitrite is that with an aromatic primary amine (usually sulphanilic acid) in acid solution to form a diazonium salt, which is coupled to an aromatic amine or hydroxy compound to give an azo dye. This can be determined spectrophotometrically. Typical coupling agents have included 1-naphthylamine and 1-naphthol which are carcinogenic, and 1-naphtholsulphonic acid and *N*-1-naphthylethylenediamine dihydrochloride are now frequently recommended. Both oxidizing and reducing agents interfere; ascorbic acid is a particular problem.

The sample blank can represent a significant error in analysis. There is no completely satisfactory way of preparing this; suggestions have included the passing of the deproteinized sample through an ion exchange resin to remove nitrate and nitrite before determining the blank.

Other Preservatives

Biphenyl

Steam distillation of citrus peel homogenates and extraction of distillate with heptane provides a sample which may be analysed for biphenyl by TLC on silica gel with heptane as developing solvent. Quantitative analysis is by extracting biphenyl from the TLC medium with ethanol and measuring the absorbance of resulting solutions at 248 and 300 nm. The absorbance at 300 nm is used to correct for background absorbance. GLC methods of analysis are also available.

Thiabendazole

Thiabendazole may be extracted from food with acid (HCl) in which it is subsequently reduced with zinc in 30% glycerol + phenylenediamine. Subsequent oxidation with ferric ions gives rise to a blue complex which is extracted into butanol for spectrophotometric measurement.

Lactic Acid

Standard methods for the determination of lactic acid involve its conversion to the Fe(III) complex for spectrophotometric measurement, or by GLC after complete methylation. In an enzymatic procedure, lactic acid is converted to pyruvate by NAD^+ (lactate dehydrogenase); the product is trapped by reacting with glutamate (glutamate–pyruvate transaminase) in order to displace the lactic acid–pyruvate equilibrium to the right. The rate of utilization of NAD^+ is used to measure lactic acid concentration. Sample preparation is generally as described for carboxylic acids above.

Hydrogen Peroxide

Spot tests for the presence of hydrogen peroxide in milk use either a solution of vanadium pentoxide in H_2SO_4 (pink or red colour) or KI/starch (blue colour).

Quantitative analysis is with peroxidase and a suitable substrate. The frequently recommended *o*-dianisidine is carcinogenic and substrates such as guaiacol or 2,2′-azinobis(3-ethylbenzthiazolidine-6-sulphonic acid) (ABTS) are preferred. Assays on milk samples are carried out after precipitation of the protein by adjusting to pH 4·5 with HCl.

Nisin

The complexity of procedures for isolating and chemically quantifying nisin has lead to the development of microbiological assays as normal analysis procedures. One International Unit of nisin activity is the extent of inhibition of a test microorganism caused by approximately 25 ng of pure nisin. A suitable test microorganism is *Streptococcus agalactiae*. A particular assay procedure is an adaptation of the methylene blue reduction test; the delay in decolorization of the dye is proportional to nisin concentration over a tenfold concentration range. A new method of analysis involves ELISA which is much more straightforward than the microbiological assay and should become popular. ***See*** Nisin

Bibliography

Branen AL and Davidson PM (eds) (1983) *Antimicrobials in Food.* New York: Marcel Dekker.

Furia TE (ed.) (1972) *Handbook of Food Additives*. Boca Raton, Florida: CRC Press.
Helrich K (ed.) (1990) *Official Methods of Analysis of the Association of Official Analytical Chemists*, 15th edn and First and Second Supplements (1990 and 1991). Arlington, Virginia: Association of Official Analytical Chemists.
Usher CD and Telling GM (1975) Analysis of nitrate and nitrite in foodstuffs: a critical review. *Journal of the Science of Food and Agriculture* 2: 1973–1805.
Wedzicha BL (1984) *Chemistry of Sulphur Dioxide in Foods*. London: Elsevier Applied Science.

B.L. Wedzicha
University of Leeds, Leeds, UK

PRESERVES

See Jams and Preserves

PRETERM INFANTS – NUTRITIONAL REQUIREMENTS AND MANAGEMENT

To say that the newly born is only a digestive tube is certainly to exaggerate. Nevertheless, the importance of the alimentary tract must not be underestimated.

Pierre Budin, 1907

The larger preterm babies (i.e. > 2000 g) usually present no particular feeding problems and can often feed from birth as effectively as a full-term infant. The group of babies who present most problems with feeding are the very smallest and the most immature of the preterm infants. Because of the many illnesses and complications that occur in this group of newborn babies, the provision of adequate nutrition is often forgotten, but unless careful attention is paid to the baby's intake of fluid, energy and nutrients, survival will be jeopardized, and at the very least complications will occur in the newborn period. The consequences of poor nutrition in this vulnerable period will be noticeable in infancy, later childhood and even adult life. ***See*** Infants, Nutritional Requirements

The smallest preterm babies have poor carbohydrate and fat stores and poor muscle bulk. In infants born at less than 26 weeks of gestation the function of the gastrointestinal tract is insufficiently mature to allow them to tolerate enteral feeding at the start. If healthy, they are able to adapt quickly to enteral feeding and maturation seems to occur much sooner than expected after birth.

By 25 weeks the fetus is thought to swallow as much amniotic fluid per kilogram as the equivalent volume of breast milk which will be taken by normal term infants. Interference with this continuum might deprive the baby of intestinal hormones, and prolonged parenteral nutrition may lead to intestinal mucosal atrophy. Early feeding leads to more mature intestinal motor function, increased gastrin and gastrin inhibitory peptide, a lower incidence of jaundice and improved growth rate.

The very-low-birth-weight (VLBW) or extremely-low-birth-weight (ELBW) baby may have a total endogenous energy store of only *c*. 500 kcal. Little of this will be available for immediate metabolic use because much of it is structural protein. The protein requirements of the ELBW and small-for-gestational-age (SGA) baby are particularly high.

Fetal Weight Gain

Weight-specific weight gain of the fetus (g kg^{-1} day^{-1}) decreases from 30 at 22 weeks of gestation to about 15 at 30 weeks and to 10 by 38 weeks. There is a reciprocal increase in weight gain when this expressed in absolute terms (day^{-1}) in the same period from 10 to 30 g day^{-1}.

Nutritional Requirements of the Preterm Baby

Illness and some treatments place extra demands on the preterm baby's nutritional status. Low ambient temperature, infection, fluid deprivation, the work of breathing and the catabolic effects of drugs such as corticosteroids provide additional nutritional stress. Careful attention must be paid to the need to prevent complications when possible or to compensate for their effects.

More energy and protein are required by SGA babies to achieve catch-up growth if their poor growth has arisen from nutritional deprivation before birth. Blood flow to the gut in the growth-retarded fetus is frequently reduced as part of the process of diversion of blood flow to the brain that occurs in response to fetal hypoxia. This reduction of intestinal blood flow might persist into neonatal life. Greater caution needs to be observed when feeding growth retarded preterm babies enterally because they are particularly predisposed to the risk of necrotizing enterocolitis (NEC).

Water

The high insensible loss of water from the skin and lungs (30–60 ml kg day^{-1}) and poor renal concentrating ability with losses in the urine (90–100 ml kg^{-1} day^{-1}) combine to make VLBW babies dependent on a high fluid intake. When taken together with the small amount of water to allow for growth (10–15 ml kg^{-1} day^{-1}) and for gastrointestinal loss (5–10 ml kg^{-1} day^{-1}), the total water requirement ranges from 130 to 185 ml $kg^{-1}day^{-1}$. In order to avoid a large calorie density of administered feed, it is necessary to prevent the use of hyperosomolar feeds and to keep the urine osmolality between 100 and 200 mosmol. Urine flow should be approximately 4–6 ml kg^{-1} h^{-1}. Water requirements will be increased by a high ambient temperature, radiant heat gain such as occurs from a radiant heater or from phototherapy, a low ambient humidity, and poor skin integrity such as that found in the smallest infants. Fluid intake may need to be raised to as much as 200 ml kg^{-1} day^{-1} or more to take account of increased water requirements. If the energy density of the milk is low, e.g. breast milk, large volumes may be needed to provide the required calories and other nutrients. An excessive water intake might lead to oedema, pulmonary oedema and a patent ductus arteriosus.

Energy

Energy requirements are estimated to vary between 110 and 165 kcal kg^{-1} day^{-1}. The lower limit will apply to those preterm babies nursed in optimal thermal conditions, who are well, are handled minimally, have good absorption of nutrients and have minimal activity. Conversely, a higher intake will be needed by those infants who are hypothermic, who have to endure a number of therapeutic procedures, whose absorption of nutrients is imperfect, or who have increased activity and increased handling, and who are ill. At an energy intake of 130 kcal $kg^{-1}day^{-1}$, and a feed volume of 150–200 ml kg^{-1} day^{-1}, the energy density of the feed will need to be 85–65 kcal per 100 ml. If energy intake is increased disproportionately to the protein intake, this could lead to excessive fat deposition and relative protein deprivation. ***See*** Energy, Measurement of Food Energy

Protein

The calculated requirement for protein in the low-birth-weight (LBW) infant is controversial but probably lies between 3·0 and 4·0 g kg^{-1} day^{-1}. This is higher than for the full-term infant but needs to supply the requirement for growth and to achieve a normal body composition. Too high an intake will result in acidosis, hyperaminoacidaemia and hyperammonaemia, and will also result in an increased osmotic load and a raised blood urea. If the total energy intake is inadequate and if protein intake exceeds the need for synthesis, excretion of protein is increased. Conversely a high energy intake without sufficient protein results in protein depletion, low plasma urea and a low plasma albumin level, and also causes oedema. A positive nitrogen balance will occur only if sufficient nonprotein calories are given simultaneously with protein (e.g. 60 nonprotein calories per kg with a protein intake of 2·5 g kg^{-1} day^{-1}). ***See*** Protein, Requirements; Protein, Digestion and Absorption of Protein and Nitrogen Balance

The sulphur-containing amino acid taurine is abundant in human milk, whereas it is absent in soya milk and deficient in unmodified cow's milk-derived formulae. In animals, visual problems and even blindness occur with taurine deficiency. Taurine deficiency may play a part in the cholestasis that so often accompanies parenteral nutrition.

Fat

Fat storage begins at about 26 weeks of gestation, when it increases from about 1% of body weight to 10–20% at term. Although there is some transfer of fatty acids across the placenta, most of the fat in the fetus arises from *de novo* synthesis. ***See*** Fats, Requirements; Fatty Acids, Metabolism

Fat absorption from the intestinal tract of the preterm infant is compromised and steatorrhoea is likely if too much fat is given. Absorption of fat is influenced by the chain length of fatty acids and by the proportion of fat that is saturated. Long-chain and saturated fatty acids such as those that predominate in cow's milk are absorbed less well than the mainly polyunsaturated fatty acids present in breast milk. Hydrolysis of triglycerides in the intestinal tract follows from the action of lipases. In the term infant, pancreatic lipase and intestinal lipase are the most important of these, acting in concert with bile salts. Both the enzyme and the bile salt pool are reduced in preterm infants. Lingual lipase, secreted in the saliva, may not be available to the preterm infant, who is usually tube-fed and thus deprived of salivary enzymes. Unheated human milk contains bile salt-stimulated lipase which is important in the digestion of fat in the upper bowel. Once absorbed, further hydrolysis of triglycerides is limited by the deficiency of lipoprotein lipase which occurs in the preterm infant. Carnitine is important in the transfer of long chain fatty acids and some medium chain fatty acids across the mitochondrial membrane. Synthesis is deficient in the extremely preterm infant who is dependent on exogenous carnitine, such as is available in human milk, and in most cow's milk based formulae.

Preterm infants are more vulnerable to essential fatty acid deficiency, particularly of linoleic and linolenic acids. Formulae should provide no more than 20% of total fat as fatty acids and linoleic acid should provide at least 4·5% of total calories and linolenic acid at least 0·5% of total calories.

At an average energy intake of 130 kcal kg^{-1} day^{-1}, fat intake might be expected to vary from 4·7 to 9 g kg^{-1} day^{-1}. Fat provides about 50% of the calories in milk.

Newborn VLBW infants cannot convert parent fatty acids of the omega-6 or omega-3 series to longer-chain, more unsaturated fatty acids. These fatty acids are important for neurological development. Consideration is now being given to providing longer-chain fatty acids in enterally administered preterm milk. Medium-chain triglycerides (MCTs) are absorbed more readily, metabolized completely and are mainly independent of carnitine for transfer into the mitochondria. They are also used completely to supply energy rather than contributing to fat storage and will improve nitrogen retention if given in high concentration. If in excess to requirement they will merely be metabolized for thermogenesis. Breast milk contains few MCTs but adapted formulae contain increased amounts. However, if given in overly excessive amounts, they will cause diarrhoea. ***See*** Thermogenesis

Carbohydrate

Preterm babies will normally require 9–11 g glucose per kg per day to maintain normoglycaemia. This is equivalent to 6–8 mg kg^{-1} min^{-1}. This will be partly derived from endogenous glucose production, but with deficient glycogen stores and poor ability for gluconeogenesis these infants are remarkably dependent on exogenous provision of carbohydrate. The carbohydrate in breast milk (lactose) is normally in a concentration of about 7·5 g per 100 ml (at 200 ml per kg this will provide 15·6 g kg^{-1} day^{-1}). Whereas in some milks carbohydrate might be partly in the form of glucose and starch hydrolysates, in most formulae lactose predominates as the main constituent of the carbohydrate fraction. Too high a level of glucose in enteral feeds will provoke osmotic diarrhoea. During fetal and neonatal life, glucose is the main fuel for the brain. In the preterm infant, carbohydrate provides 25–50% of calories in the enterally fed baby. Raising the carbohydrate content without considering the need for a proportional increase in protein might result in protein deficiency. ***See*** Carbohydrates, Requirements and Dietary Importance; Glucose, Maintenance of Blood Glucose Level; Glucose, Function and Metabolism

Hyperglycaemia (glucose greater than 6·9 mmol l^{-1}) is a common problem in ELBW babies. It leads to glycosuria, osmotic diuresis, dehydration and hyperosmolality. It is predisposed by intravenous glucose therapy and the administration of excessive amounts of glucose. Corticosteroid treatment, infection and the stress of illness are likely to precipitate this problem. If a baby is receiving an infusion of glucose this may need to be reduced, and in some cases insulin needs to be administered either by infusion or intermittent injection of 0·01–0·1 units per kg per hour.

Hypoglycaemia with glucose less than 2·6 mmol l^{-1} is especially likely to occur in SGA and VLBW babies. It may be asymptomatic or may also produce nonspecific neurological symptoms. The blood glucose of these babies needs to be carefully measured and intravenous glucose given if the levels are low or if there are any suggestive symptoms of hypoglycaemia. High glucose infusion rates can produce a fall in both phosphate and potassium and if continued for more than 24 hours serum phosphate and potassium levels must be carefully measured and maintenance potassium and phosphate given if necessary. ***See*** Hypoglycaemia for Nutrition

Minerals

Sodium. Initially many preterm babies have a high fractional sodium excretion and lose sodium excessively in the urine. Early sodium intake needs to be about 3 mmol $kg^{-1}day^{-1}$ and in some instances, especially in ELBW babies, up to 10 mmol kg^{-1} day^{-1} is needed to maintain a serum sodium concentration of greater than 130 mmol l^{-1}. The level of serum sodium needs to be carefully monitored because sodium losses are variable. Enteral feeds should provide 6–15 mmol l^{-1}. Breast milk is likely to be deficient in sodium. ***See*** Sodium, Physiology

Potassium and Chloride. These should be provided by feeds in an amount of 2–4 mmol kg^{-1} day^{-1} and are usually present in adequate amounts in both breast and formula feeds, though some breast milk has been found to be chloride deficient. ***See*** Potassium, Physiology

Calcium, Phosphorus and Magnesium. Calcium and phosphorus may be lost excessively in VLBW infants. Enteral phosphate absorption is usually complete but phosphate concentration in enteral feeds, especially breast milk, is likely to be deficient and consequent phosphorus depletion from urinary losses will then lead to hypophosphataemia (less than 1·5 mmol l^{-1}). This may be associated with a high serum alkaline phosphatase level and evidence of osteopenia with poor osseous mineralization particularly involving the skull and the remainder of the skeleton. Sometimes fractures will occur. The soft rib cage may lead to less effective respiratory mechanics and the soft skull to a flattened elongated skull shape. ***See*** Calcium, Physiology; Magnesium

Standard formulae and human breast feeds do not contain sufficient calcium and phosphorus to support extrauterine bone mineralization in VLBW infants. The calcium-to-phosphorus ratio in administered feeds should normally be kept at 1·5 to 2·0 and precipitation should be avoided. If calcium and phosphorus are added to feeds then phosphorus should be added first, when it will enter the milk fat globule, and calcium later so that precipitation is avoided. About 2–3 mmol of phosphate will usually be adequate and proportionately more of calcium. Calcium should never be given without phosphate because of the risk of renal calcification in the presence of urinary calcium loss.

Iron. The amount of iron in a 1-kg baby at birth is only sufficient to synthesize 18·0 g of haemoglobin (Hb). The infant will become iron deficient by the time it doubles its birthweight. After birth, erythropoietin normally falls; the haemoglobin falls to about 9·0 g per 100 ml by 4–6 weeks. If iron intake is inadequate, iron stores will be quickly exhausted and Hb will fall. In breast fed babies one needs to administer 2–3 mg ferrous iron per kilogram from about 6–8 weeks of age. If many blood transfusions were given in early life, the administration of iron can be delayed. ***See*** Iron, Physiology

Zinc and Copper. Absorption is poor in VLBW babies given heated banked milk, though absorption is better with commercial formula feeds adapted for the preterm infant in which a high proportion of fat is given as MCT. Absorption is better from human preterm milk. Deficient zinc intake has been reported in a baby whose mother had low levels in the breast milk. Deficiency of zinc and copper is only likely to occur in those babies recovering from bowel surgery who have additional losses and increased requirements. In these babies particularly, zinc and copper levels should be measured. ***See*** Copper, Physiology; Zinc, Physiology

Vitamins

Vitamin A. There is increasing evidence that the VLBW baby is deficient, since most vitamin A is transported to the fetus in the third trimester and hepatic storage of vitamin A increases with gestation. Vitamin A deficiency is incriminated as one factor associated with chronic lung disease (CLD) in babies treated with prolonged ventilation for the respiratory distress syndrome of prematurity, and babies with this complication have been found to have low levels of this vitamin. It is recommended that VLBW babies should be given 1000–3000 iu vitamin A per day (300–1000 μg per day). An excessively high dose may give rise to intracranial hypertension. Formulae for low-birth-weight infants are supplemented with vitamin A and this reduces the need for extra supplements of the vitamin. ***See*** Retinol, Physiology

Vitamin D. Infants fed human milk need to be given 15–30 μg (600–1200 iu) vitamin D per day. The dosage of vitamin D can usually be kept low if phosphate supplementation is provided. Supplementation can be reduced when the baby reaches term. Cholecalciferol or ergocalciferol (vitamin D_3) is usually used. No advantage has been shown for the use of 25-hydroxyvitamin D or 1-α-hydroxyvitamin D nor for 1,25-dihydroxyvitamin D_3. Artificially fed VLBW infants need less supplementation since the milk has already been enriched with this vitamin. ***See*** Cholecalciferol, Physiology

Vitamin E. Storage of this vitamin is poor, and in the preterm infant intake is usually deficient and requirements during growth are higher. Deficiency is associated with haemolytic anaemia. It has also been proposed that vitamin E has a role in the prevention of intraventricular haemorrhage, of CLD in ventilated babies, and of severe retinopathy of prematurity. These therapeutic effects are controversial. Excessive dosage has been associated with an increase risk of NEC.

Vitamin E is usually given in a dose of 5–20 mg per day. There are adequate amounts in breast milk and in fortified preterm milk formulae. ***See*** Tocopherols, Physiology

Vitamin K. Deficiency results in a haemorrhagic tendency which can manifest as serious bleeding. Human milk contains little vitamin K. Preterm infants are especially vulnerable and are given vitamin K at birth, in most cases intramuscularly, and if they are fed with human milk then repeated doses are recommended. There is a current controversy whether intramuscular vitamin K (but not orally administered vitamin K) might be associated with an increased risk of childhood cancer, though the evidence of this association is yet far from certain and many units are choosing to continue use of the vitamin until more definite proof appears. The usual dose is 0·5–1 mg vitamin K intramuscularly.

Vitamin C. Vitamin C is deficient in breast milk. The requirements of preterm babies for vitamin C are raised and it should be given in a dosage of 25–50 mg daily. It may be of particular value in the preterm baby for its action in stimulating the activity of tyrosine oxidase, an enzyme that might be important in preventing transient hypertyrosinaemia. The vitamin is destroyed by heating and is added to formula feeds. ***See*** Ascorbic Acid, Physiology

Vitamin B complex. Members of the vitamin B group are not normally deficient in the preterm infant but are included in multivitamin preparations. ***See*** individual vitamins

Folic Acid. Human milk contains low levels of folic acid and preterm babies given human milk should be supplemented with this vitamin. It is usually not present in multivitamin preparations and has to be given separately. ***See*** Folic Acid, Physiology

Feeding the Preterm Baby

Parenteral Nutrition – Partial or Total

Satisfactory tissue accretion can be achieved with parenteral nutrition. It can also produce normal weight gain and brain growth, and the risk of NEC and of aspiration from the stomach is low. In the smallest babies with poor intestinal function, and postoperatively, parenteral nutrition may be essential for prolonged periods but usually this method of feeding is only used for brief periods until enteral nutrition can be established. If parenteral nutrition needs to be prolonged, the insertion of a central venous line will usually be needed.

There are many complications associated with the use of parenteral nutrition in the preterm baby, including an increased risk of infection, the development of acidosis, hyperammonaemia and hyperaminoacidaemia consequent on amino acid infusion. Also there is the risk of the development of cholestasis. Lipid emulsions should be given in 20% concentration rather than 10%. This has been found to be an advantage resulting in lower plasma phospholipid-to-triglyceride ratios with less inhibition of lipoprotein lipase and improved triglyceride clearance. Hypertriglyceridaemia may lead to diminished leukocyte function and may interfere with gas exchange in the lung. In addition it also interferes with chemical analysis of the serum. Lipid cannot be used in the presence of significant hyperbilirubinaemia for fear of displacement of bilirubin from albumin binding sites.

Usually one aims to give lipid $1\ \mathrm{g\ kg^{-1}\ day^{-1}}$ increasing gradually to a maximum of $4\ \mathrm{g\ kg^{-1}\ day^{-1}}$, glucose $8{\cdot}6$–$11{\cdot}5\ \mathrm{g\ kg^{-1}\ day^{-1}}$ (6–$8\ \mathrm{mg\ kg^{-1}\ min^{-1}}$) increasing if required; protein $0{\cdot}5\ \mathrm{g\ kg^{-1}\ day^{-1}}$ increasing to a maximum of $2\ \mathrm{g\ kg^{-1}\ day^{-1}}$. Supplements of trace minerals and vitamins are needed if parenteral nutrition is given for more than a few days as the sole source of nutrition.

Enteral Feeds

Breast milk was the preferred feed for preterm babies until the 1940s, when it was first realized that weight gain was faster in babies who were fed a modified cow's milk feed. Protein intake was much lower in those babies given pooled breast milk (less than 2 g per 100 ml) than in those given cow's milk with its higher protein concentration (greater than 3 g per 100 ml) and these babies grew faster. However, excessive protein intake (greater than $4{\cdot}5\ \mathrm{g\ kg^{-1}\ day^{-1}}$) was noted to lead to the development of acidosis and a raised blood urea and fever. This was especially seen in those babies given cow's milk formulae which contained a high concentration of casein compared to whey. High protein intakes also lead to hyperammonaemia, hyperaminoacidaemia and hyperphenylalaninaemia and tyrosinaemia, which are known to be associated with sleepiness and later on with slow intellectual and motor development. ***See*** Infants, Breast- and Bottle-feeding

The energy density of breast milk is lower than that of modified cow's milk. To achieve an adequate energy intake and promote rates of growth similar to those occurring *in utero*, the volume needed becomes unacceptably high. Breast milk once again came into vogue after the 1950s but the inadequacy of donor milk in preterm nutrition was soon being realized. At present, although pooled donor breast milk is used in some neonatal units that have milk banks, the milk of the baby's own mother is given in many more centres. In some centres breast milk is not used at all but special formula milks which have been adapted for the low-birth-weight babies are preferred.

Milk Banking

In spite of concerns about the possible transmission of human immunodeficiency virus (HIV) via donor milk to preterm babies, some neonatal units continue to operate milk banks, though exercising the most stringent safeguards. Donors are screened with a suitable questionnaire and are tested for HIV, hepatitis and syphilis. Each milk sample is bacteriologically tested and after passing the safety standards for bacteriological contamination the donor milk is pooled so that it is more likely to be uniformally constituted. Nutrients of the pooled sample are often measured and thereafter the milk is carefully pasteurized usually using Holder pasteurization at 62°C for 30 min. This temperature will unfortunately destroy much of immunoglobulin present in the milk as well as hormones, enzymes and living cells.

'Preterm Milk'

The milk of the mother of the preterm baby is nutritionally advantageous. Compared to mature donor milk, it contains more protein, more energy and more sodium. It is usually administered without undergoing heat treatment and as a result its inherent immune functions are preserved and enzymes and hormones are intact. In some neonatal units this milk is bacteriologically tested before being given, and is deep frozen until the results of these tests are available. The major transmission of maternal antibody to the fetus takes place after 28 weeks' gestation. Extremely preterm babies are thus unlikely to have protective antibody even against the mother's own bacterial flora. Organisms present in her breast milk, some of which may be pathogenic, can therefore constitute a danger to her baby and if they are found, her milk is then subjected to heat treatment.

The nutritional advantage of 'preterm milk' is only present for a few weeks and thereafter it resembles mature donor breast milk.

Disadvantages of Breast Milk

Breast milk is unlikely to meet the nutritional demands of the growing VLBW baby. The energy, protein content and sodium, calcium and phosphorus concentrations in the milk are all deficient. In order to meet the baby's nutritional demands and because of its low energy density, unacceptably large volumes (greater than 200 ml per kg) of milk may be needed, with the attendant dangers of fluid overload. The fat in breast milk tends to adhere to syringes and feeding tubes and the energy content of the milk may thus be substantially reduced.

Hyponatraemia and poor bone mineralization (osteopenia) have been found in these small breast-fed babies. In addition, their growth is slower.

The destruction of enzymes in the milk by heat treatment reduces its nutritional value, and fat absorption might be halved.

The Advantages of Breast Milk for the Preterm Infant

Unheated breast milk contains a number of anti-infection systems, enzymes and hormones. The fluid also contains growth factors which have been shown in animal studies to have an effect on intestinal growth and maturation. Most breast milk fat is unsaturated and absorbed more easily. The long-chain polyunsaturated fatty acids in breast milk are more easily absorbed and have an important function in the synthesis of cell membranes. They are thus likely to be involved in myelination of neurones. Preterm babies given breast milk are less likely to suffer with NEC, even when given heated breast milk. The occurrence of confirmed NEC was at least six times more common if formula feed was used exclusively. Gastric emptying is faster with breast milk, vomiting is less frequent and the baby attains full enteral feeding more quickly than is the case with formula feeds. The babies are less constipated than with formula feeds and milk casts are hardly ever seen when breast milk is used. Recent work has shown that the intelligence of preterm babies when followed to the age of 8 years is significantly better if they received breast milk in the newborn period.

Nutritional Quality of Breast Milk

The nutritional quality of breast milk can be improved by encouraging mothers to express milk for administration to their own baby, and this can be given unheated if shown to be bacteriologically safe, so that the hormones and enzymes are preserved. If donor milk is used, hindmilk should be collected rather than foremilk, which has a lower energy content. Likewise mature milk rather than transitional milk should be collected. Pasteurization should be accurately controlled.

Care should be taken when administering the milk. Fat present in breast milk separates out on standing and adheres to plastic. Fat loss can be minimized, though not eliminated, by simple expedients such as administering small volumes using a syringe with an eccentric nozzle with the syringe propped up so that fat, as it separates, is less likely to adhere to the syringe barrel. The syringe should be agitated to keep the fat in suspension.

In addition to these measures, donor breast milk can be fortified in various ways:

(1) with added sodium to counter the hyponatraemia that is likely to occur;

(2) with added phosphorus and calcium to prevent osteopenia;

(3) with added energy in the form of dextrose polymer or medium-chain triglycerides or both.

Great care must be taken to avoid raising the osmolality of the feed unacceptably (greater than 300 mosmol) when making these additions.

(4) with commercially available breast-milk fortifiers that provide additional protein, carbohydrate and minerals, or by adding small amounts of preterm-adapted milk-formula to the breast milk. It is preferable to know the constituents in the milk before making these additions and essential that levels of minerals, glucose

and protein in the baby are carefully monitored to ensure that the dose chosen is appropriate;
(5) with breast-milk components. In a few neonatal units breast-milk fractions including fat and protein derived from the pooled supply are added to the deficient breast milk. Such 'lacto-engineering' is not practicable for most centres.

Adapted Formula for the Preterm Infant

At least 15 milks are now available commercially which have been adapted to fulfil the needs of the LBW newborn more effectively than conventional formulae which are designed for full-term infants. Protein content has been increased above 2 gm per 100 ml and the whey:casein ratio is standardized to 60:40. The fat concentration has also been increased to greater than 4 g per 100 ml using polyunsaturated/saturated fat mixtures and in some instances using MCT oil up to a concentration of 50%. One manufacturer has recently supplemented its preparation with long-chain polyunsaturated fatty acids which are missing in other formula feeds. Fat concentration in conventional milks lies between 2·5 and 3·8 g per 100 ml. Carbohydrate concentration of most milks is in the range of that found in conventional formulae. Lactose is used in most milks, sometimes combined with glucose or maltodextrin. Calorie concentration in all these adapted milks is above 70 kcal per 100 ml and in a few above 80 kcal per 100 ml. In conventional formulae and breast milk, calorie concentration usually lies between 60 and 70 kcal per 100 ml. Sodium concentration is increased in most of these milks to above 12·5 mmol l^{-1}, i.e. above the level found in conventional formulae and in breast milk. Of the minerals, calcium, zinc and magnesium are present in higher concentrations than in conventional formulae while phosphate concentration tends to be in the same range (7·5–17·5 mmol l^{-1}), though the levels are higher than those present in breast milk. The changes in composition of these feeds designed for LBW babies should therefore meet their demands more effectively than would other feeds. High energy density would allow an adequate calorie intake without the risk of fluid overload. The increased protein contact would ensure more physiological partitioning of weight gain than would occur if fat or carbohydrate had been independently increased. The increased sodium concentration would be more advantageous in the early weeks when most LBW infants are natriuretic but should only be used with careful surveillance of sodium levels. When sodium conservation matures in the baby, these milks may no longer be suitable. ***See*** Infant Foods, Milk Formulas

It has recently been shown that preterm babies of a few months of age, if given the opportunity to take as much feed as they wish, ingest a larger volume per kilogram of body weight than term infants. Also their weight gain is improved with an adapted milk that has a higher energy, protein and mineral content though with less sodium than provided in preterm formulae.

These milks that have been adapted for the LBW infant result in greater weight gain than breast milk or conventional formulae. There were, however, some methodological deficiencies inherent when these comparisons were made. Breast milk is not a homogeneous fluid and its nutritional efficiency is influenced by a number of factors which were not taken into account when comparisons were made. Donor milk was foremilk that was heated and no precautions were taken to prevent the separation of fat. When 'preterm milk' was compared with some of these more recent formulae, the formula still resulted in better weight gain and growth although the difference was much less marked.

Timing of the First Feed

Up to the 1960s the first feed was delayed, sometimes for several days, because of the fear of aspiration and pneumonia and of oedema, but from the 1960s it became clear that this delayed feeding policy was contributing to cerebral palsy, visual defects, mental impairment, hypoglycaemia and jaundice. This critical period of undernutrition was leading to later poor growth and was interfering with development. The benefit of early feeding was clear. At present there is still reluctance to begin enteral feeds early because of concerns about precipitating NEC. Also, there remain worries about giving excessive amounts of fluid and the risk of fluid overload leading to patent ductus arteriosus and perhaps even predisposing to chronic lung disease. There is however general agreement that cautious fluid and calorie administration should begin early within the first few hours of delivery. Enteral feeds are administered either by gastric tube or by transpyloric tube.

Oral Feeds

Sucking via a teat or on the breast should be promoted, especially in larger preterm babies or when the VLBW baby is gaining weight adequately and is well and robust. Normal oral feeding is more likely to stimulate intestinal hormonal output and to promote normal intestinal function and growth. The importance of fostering a normal mother–infant relationship by encouraging feeding early should not be underestimated. While tube feeding is in progress and the baby is vigorous, 'nonnutritive sucking' on a teat might be encouraged to promote intestinal hormonal and enzyme production and intestinal motility.

Feeding Plan

The smallest and most ill preterm infants are normally initially given intravenous fluid containing glucose and solute. After 2–3 days parenterally administered amino acids and lipids, in gradually increasing amounts, may be started and continued until the baby is thought to be well enough to tolerate enteral feeds.

Minimal enteral feeds via an indwelling gastric tube are started after a few days, beginning with 0·25–0·5 ml hourly and increasing by 0·25–0·5 ml 6–12-hourly if the aspirated residual gastric fluid volumes are not increasing. Constipation is treated by the insertion of glycerine suppositories and stasis of intestinal contents is thereby discouraged. Feeding is started using breast milk at least until the baby is receiving more than 150 ml kg^{-1} day^{-1}. The volume of feeds is only increased by a maximum of 20 ml kg^{-1} day^{-1}. Vitamin E may be started and phosphate added as sodium or potassium dihydrogenphosphate. Additional sodium may be required at this time. After about a week, a multivitamin preparation is given orally if this is tolerated, and this will provide vitamins A, D, C and B. Vitamin K may be given at weekly intervals, especially if the baby is receiving antibiotic therapy. After some weeks, an oral iron preparation and orally administered folic acid are begun. Breast milk is continued and the amount is increased up to a maximum of 200 ml kg^{-1} day^{-1}. The breast milk should be fortified as necessary. If weight gain is adequate (about 15 g kg^{-1} day^{-1} falling to 10 g kg^{-1} day^{-1} when the baby weighs about 2 kg) this schedule can be continued. Alternatively an adapted LBW formula could be given instead. In those centres where donor milk is not available or is felt to be inadvisable, such formulae are used instead of breast milk.

Objectives in the Nutrition of the Preterm Infant

1. To aim to achieve a postnatal growth rate comparable to that of the healthy fetus of equivalent gestational age.
2. To attain the same bodily habitus as would occur in the fetus at the same gestation, e.g. in the amount and distribution of protein and fat.
3. To maintain the normal chemical constitution of tissues and body fluids.
4. To allow for any additional demands that might be imposed by illness or stress and also to minimize the need for these by effective prevention and treatment.
5. To ensure that in trying to achieve these objectives no additional risk is imposed on the baby.
6. To bear in mind that any beneficial or detrimental effect of nutritional manipulation may only become apparent during infancy, childhood or even in adult life.

Bibliography

Hay WW (ed.) (1991) *Neonatal Nutrition and Metabolism*. St. Louis: Mosby Year Book Publishers.

Hay WW (1991) Nutritional needs of the extremely low birthweight infant. *Seminars in Perinatology* 15: 482–492.

Lucas A and Cole TJ (1990) Breast milk and neonatal necrotising enterocolitis. *Lancet* 336: 1519–1523.

Neu J (1992) Nutritional support of the high risk very low birthweight and premature infant. *Current Opinion in Pediatrics* 4: 212–216.

Neu J, Valentine C and Mestze W (1991) Scientifically based strategies for nutrition of the high risk low birthweight infant. *European Journal of Pediatrics* 15: 2–13.

Uauy R (ed.) (1989) Current concepts in newborn nutrition. *Seminars in Perinatology* 13 (2).

Wharton BA (ed.) (1987) *Nutrition and Feeding of Preterm Infants*. Oxford: Blackwell Scientific.

H.R. Gamsu
King's College Hospital, London, UK

PROBIOTICS

See Microflora of the Intestine

PROCESS CONTROL

See Plant Design, and Instrumentation and Process Control

PROCESSED CHEESE

See Cheeses

PROSTAGLANDINS AND LEUKOTRIENES

It has been over 60 years since the prostaglandins (PGs) were discovered and 30 years since they were first chemically isolated and identified in sheep seminal vesicles and renal medulla. Although together with thromboxanes (TXs) and leukotrienes (LTs) the PGs possess the most potent and most divergent biological activities of any naturally occurring compounds, their true physiological role in many instances remains unknown. From a pathophysiological viewpoint, an absolute or relative deficiency of PGs relative to TXs has been implicated in the aetiology of hypertension, thrombosis and atherogenesis for many years. Since PGs and TXs are derived from essential fatty acids the role of dietary intake of these fatty acids on PG and TX production, particularly in relation to the cardiovascular system, will be examined. The possible role of dietary essential fatty acids on LT and PG production in relation to inflammation, the immune system and gastric function will also be discussed.

Chemistry

In the 1930s it was reported that fresh human semen causes rhythmic contractions of human endometrium. These observations were later confirmed and the active substance called 'prostaglandin'. It was not until the 1960s, however, that prostaglandin was determined to be a mixture of biologically active compounds and isolated and identified PGE_{1-3} and PGF_{1-2} from sheep seminal vesicles. Also independently isolated and identified in the 1960s was PGA_2, PGE_2 and $PGF_{2\alpha}$ from extracts of rabbit kidney medulla. Later, in 1975 thromboxane A_2 (TXA_2) which is a potent vasoconstrictor and platelet aggregant was isolated, and also in 1976 prostacyclin (PGI_2) was found, which was synthesized mainly by endothelial cells, prevented platelet aggregation and was vasodilatory. Studies of the slow-reacting substance of anaphylaxis (SRS-A), found it to be composed of certain LTs which are products of leucocyte membrane arachidonate.

The PGs are composed of a basic 20-carbon fatty acid containing a cyclopentane ring, the so-called hypothetical prostanoic acid. The carbons are numbered 1 to 20 from the carboxyl to the terminal methyl group. The designations of PGE_1, PGE_2 and PGE_3 refer only to the number of double bonds in the aliphatic side-chains. The PG_2s are the most abundant naturally occurring class. For PG_1s, the precursor is 8,11,14-eicosatrienoic acid (dihomo-γ-linolenic acid), and for PG_2s the precursor is 5,8,11,14-eicosatetraenoic acid (arachidonic acid). The PG_3s can be formed from 5,8,11,14,17-eicosapentaenoic acid (EPA).

On the other hand, LTs were named because they were discovered in leucocytes and because the common structural feature is a conjugated triene. Various members of the group have been designated alphabetically and a subscript denotes the number of double bonds.

Analysis of PGs and LTs

Prostaglandins and leukotrienes are present in many different biological samples including plasma. However, eicosanoid quantification in the peripheral circulation does not accurately reflect *in vitro* biosynthesis and disposition because of rapid turnover, low concentration, metabolism by lung and liver, and sample processing problems. For example, plasma concentrations of PGE_2 and TXB_2 determined by gas chromatography have been reported to be 12 and several $pg\,ml^{-1}$ respectively; however, this does not accurately reflect platelet TXA_1 biosynthesis or production of PGs or LTs by neutrophils. This is best analysed by measuring the supernatant content. To evaluate *in vivo* PGs or LTs produced in any local or specific organ, urinary samples can be utilized for the kidneys or pleural fluid or bronchial lavages for pulmonary production. The most accurate reflection of local synthesis is obtained by incubations of slices or homogenates of specific cells and/or tissues with measurement of PG, TX or LT production *in vitro*.

The most precise method to determine PGs and LTs is undoubtedly gas chromatography linked to mass spectrometry. However, this type of analysis is not widely available; it requires expensive equipment and highly specialized analytical expertise and is thus not generally employed. Radioimmunoassay using a specific antibody against various PGs or LTs and ^{3}H- or ^{125}I-labelled tracer are available, highly specific, sensitive and relatively inexpensive and have been widely utilized for eicosanoid analysis. ***See*** Chromatography, Gas Chromatography; Mass Spectrometry, Principles and Instrumentation; Immunoassays, Radioimmunoassay and Enzyme Immunoassay

Prior to radioimmunoassay, extraction of arachidonic acid metabolites from plasma, urine or tissue homogenates may be required. The most common extraction method is to extract acidified aqueous solutions with organic solvent such as diethylether or ethyl acetate. Further purification by column chromatography with a Sep-pak C-18 cartridge, by silicic acid chromatography, or sometimes by high-performance liquid chromatography, may be necessary because of cross-reactivity of antibodies. Recently, a prepacked octadecylsilyl (ODS) silica column has become available to extract and separate PGs and LTs. ***See*** Chromatography, High-performance Liquid Chromatography

Biosynthesis of PGs and LTs

Arachidonic acid and its metabolite by-products (the arachidonic acid cascade) are the important mediators of a number of physiological phenomena, described later. As shown in Fig. 1, the rate-limiting step in the formation of the metabolites of arachidonic acid seems to be the initial step, i.e. the release of free arachidonic acid from the cell membrane phospholipid pool mediated through activation of phospholipase A_2. Phospholipase C may play another role in arachidonic acid release through liberating a diglyceride which is then hydrolysed by another lipase to yield arachidonic acid. Following the release of arachidonic acid, three enzymes are involved in its subsequent metabolism. In the first pathway, PG cyclooxygenase synthesizes the initial step, giving two PG endoperoxide substances – PGG_2 and PGH_2. These intermediates are labile and can be converted into stable PGs such as PGE_2, PGD_2 and $PGF_{2\alpha}$. In addition, unstable PGG_2 and PGH_2 can be metabolized to TXA_2 by thromboxane synthetase and PGI_2 by prostacyclin synthetase. These compounds are rapidly converted to stable but biologically inactive TXB_2 and 6-keto-$PGF_{1\alpha}$ respectively.

In the second pathway of arachidonic acid metabolism, 5-lipoxygenase forms a series of products named LTs, as illustrated in Fig. 2. 5-Hydroperoxyeicosatetraenoic acid (5-HPETE) may be enzymatically dehydrated to LTA_4 (5(*S*)-*trans*-oxido-7,9-*trans*-11,14-*cis*-eicosatetraenoic acid). Subsequent enzymatic hydrolysis of LTA_4 results in LTB_4 (5(*S*)-12(R)-dihydroxy-6-*cis*-8,10-*trans*-14-*cis*-eicosatetraenoic acid). Leukotriene A_4 also reacts with sulphydryl compounds by addition of gluthathione (Glu-Cys-Gly) by glutathione-*S*-transferase, yielding LTC_4 (5(*S*)-hydroxy-6(*S*)-glutathionyl-7,9-*trans*-11,14-*cis*-eicosatetraenoic acid). Leukotriene D_4 can be then produced from LTC_4 through elimination of glutamyl residue by a γ-glutamyl transpeptidase. The remaining peptide bond in LTD_4 is hydrolysed to give LTE_4 by a dipeptidase. ***See*** Fatty Acids, Metabolism

Physiological Action of PGs and LTs

The relative amounts of compound formed depend on the tissue or cell being studied. When platelets are activated by collagen or thrombin, unstable TXA_2 is predominant, resulting in a rapid and irreversible aggregation of platelets and contraction of vascular smooth muscle cells. On the other hand, vascular endothelial cells form unstable PGI_2 which has a potent vasodilating and antiaggregatory action similar to stable PGE_2. Neutrophils as well as monocytes or macrophages produce PGE_2, while PGD_2 is formed in mast cells and basophils. In kidneys, renal interstitial cells and collecting duct cells synthesize PGE_2 and the renovasculature endothelia synthesize PGI_2, while glomerular mesangial endothelia form $PGF_{2\alpha}$. In addition to PGE_2, other PGs are also produced in the gastrointestinal tract. Prostaglandin E_1, PGE_2 and their analogues, administered exogenously, suppress gastric acid secretion and possess the property of protecting the

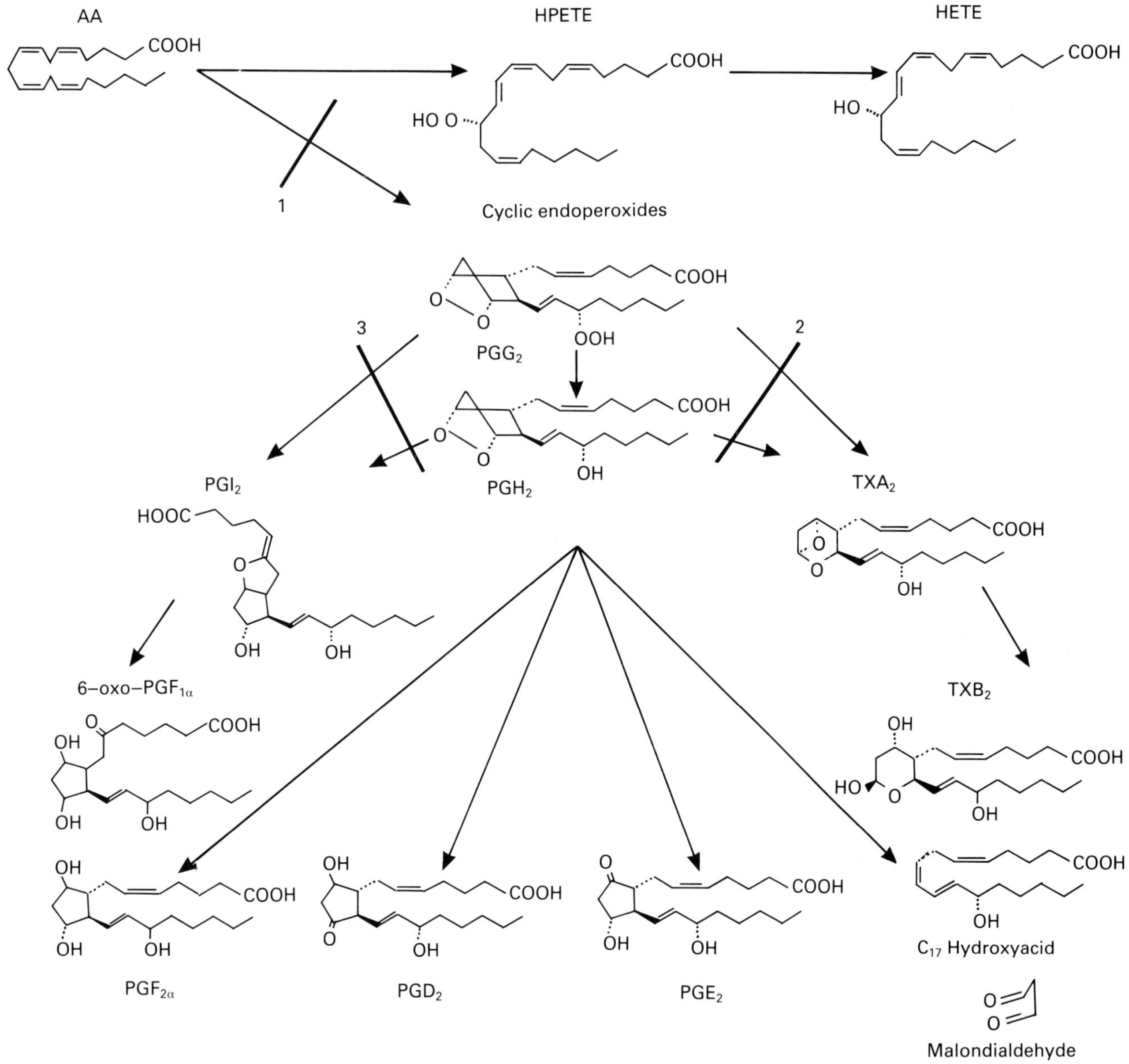

Fig. 1 Metabolic pathways of progstaglandin (PG) biosynthesis from arachidonic acid. Sites where cyclooxygenase inhibitors (aspirin-like drugs), thromboxane synthetase inhibitors (imidazole and 1-methyl imidazole) and prostacyclin synthetase inhibitors (15-hydroperoxyarachidonic acid and 13-hydroperoxylinoleic acid) act are indicated by the numerals 1, 2, and 3 respectively. AA, arachidonic acid; HETE, 12-hydroxyarachidonic acid; HPETE, 12-hydroxyperoxyarachidonic acid; PGI_2, prostacyclin; TXA_2, thromboxane A_2; TXB_2, thromboxane B_2. (From Moncada S and Vane JR (1978) Unstable metabolites of arachidonic acid and their role in haemostasis and thrombosis. *British Medical Bulletin* 34: 129–135. Reproduced with permission of the author.)

gastric mucosa against necrotizing agents such as absolute ethanol or hydrochloric acid (cytoprotection).

Prostaglandins play many physiological roles in human reproduction, e.g. menstrual regulation, pregnancy and induction of labour. Prostaglandin E_1, PGE_2 or their analogues are now widely used clinically for induction of labour or abortion.

Prostaglandin E_2 and PGI_2 produce or augment vasodilation and enhance bradykinin or serotonin-induced pain, resulting in redness, heat, swelling and pain in the inflammatory processes. Prostaglandins such as PGE_2 and PGI_2 have also been shown to increase during pyrogen-induced fever. In fact, PGE_2 administered into the hypothalamus increases body temperature, while PGD_2 deceases body temperature, indicating a possible role of PGs in human thermoregulation. Prostaglandins and thromboxane, and some monohydroxy derivatives of arachidonic acid (hydroxyeicosa-

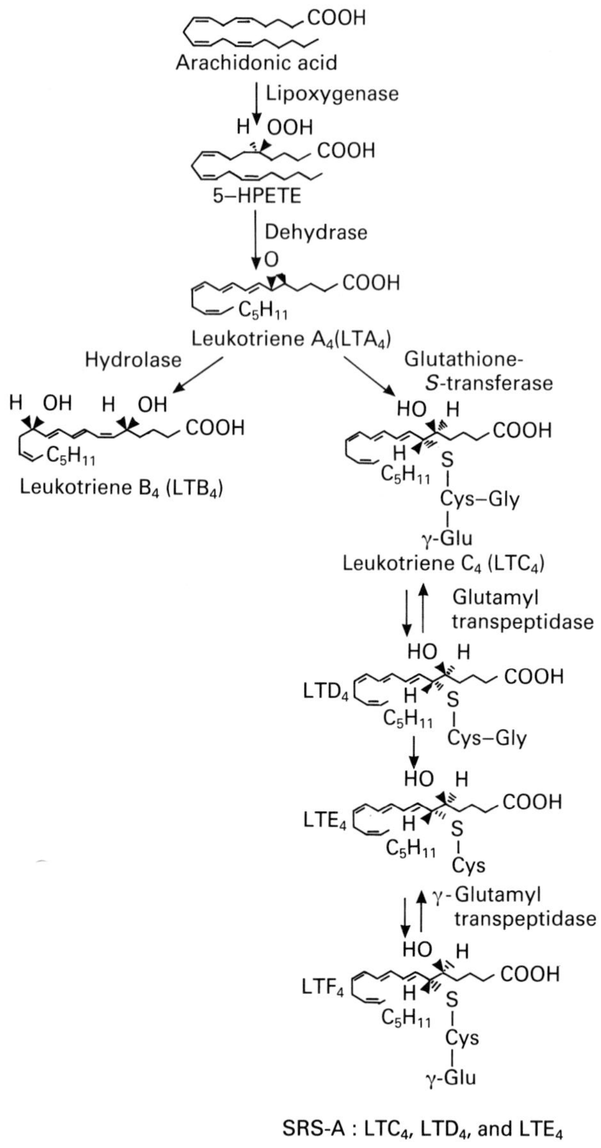

Fig. 2 Formation of leukotrienes from arachidonic acid by way of 5-lipoxygenase pathway. (From Samuelsson B (1983) Leukotrienes: Mediators of immediate hypersensitivity reaction and inflammation. *Science* 220: 568–575. Copyright by the Nobel Foundation, 1983. Reproduced with permission of the publisher).

tetraenoic acids) also have chemotactic effects on polymophonuclear leucocytes. Leukotriene B_4 is not only chemotactic for neutrophils and eosinophils but also for monocytic macrophages. These phenomena are considered to be the initial events of the inflammatory response. Polymorphonuclear leucocytes only survive for hours in the inflammatory area, while monocytes may stay for weeks and finally may be transformed to fibroblasts initiating the repair process of the wounds. Furthermore, monocytes can present antigens to cells capable of producing antibodies and can synthesize all the members of the arachidonic acid cascade. Monocytes are also capable of forming interleukin, interferon, complement and protease, all of which participate in tissue disruption. The lipoxygenase system is more important in subacute and chronic inflammation, whereas the cyclooxygenase system produces, modifies or modulates acute inflammation. The anti-inflammatory action of aspirin, indomethacin and other nonsteroidal anti-inflammatory agents is produced almost entirely by PG synthesis inhibition as shown by studies revealing inhibition of PG-cyclooxygenase with aspirin.

Leukotriene C_4, LTD_4, LTE_4 and their 11-*trans* isomers do not have any effects on chemotaxis, enzyme release or leucocyte aggregation. However, they possess potent biological activities, which were formerly attributed to SRS-A release from sensitized lungs treated with a specific antigen. These compounds may be important mediators in asthma and other acute hypersensitivity reactions. Contraction of guinea pig ileum and other smooth muscle by the LTs exhibits a slow onset and relaxation which is the basis of the original designation as SRS-A, distinguishing these substances from histamine, bradykinin and $PGF_{2\alpha}$. Leukotriene C_4 and LTD_4 are respectively about 200-fold and 20 000-fold more potent than histamine in promoting small airway contraction. In addition, these LTs cause rapid arteriolar contraction and promote plasma leakage in postcapillary venules. They also slow the rate of mucus clearance from the airways of patients with asthma after the inhalation of antigen, and increase the amount of mucus glycoprotein synthesis in the airways.

Effects of Diet on PG Formation

Polyunsaturated Fatty Acids

In 1974, it was first shown that dietary deficiency of essential fatty acids in animals results in hypertension which depends on a high sodium intake thus being very similar to renoprival hypertension or experimental "salt sensitive" hypertension. This was later confirmed in 1983 and the blood pressure response attributed to a dietary deficiency of linoleic acid. However, it would appear unlikely that this is a cause of human hypertension since selective deficiency of essential fatty acids in the human diet is rare. ***See*** Essential Fatty Acids, Physiology; Fatty Acids, Dietary Importance

A low prevalence of atherosclerosis and a low mortality from myocardial infarction despite a diet high in fat and cholesterol, as consumed by Danes and North Americans among Greenland Eskimos, suggested that dietary fish may have some properties that could prevent coronary artery disease. The Eskimos consume 5–10 g daily of the long-chain $n-3$ polyunsaturated fatty acids, EPA (C_{20}; $5n-3$) (Fig. 3) and docosahexaenoic acid (C_{22}; $6n-3$) which are present in fish oils. When

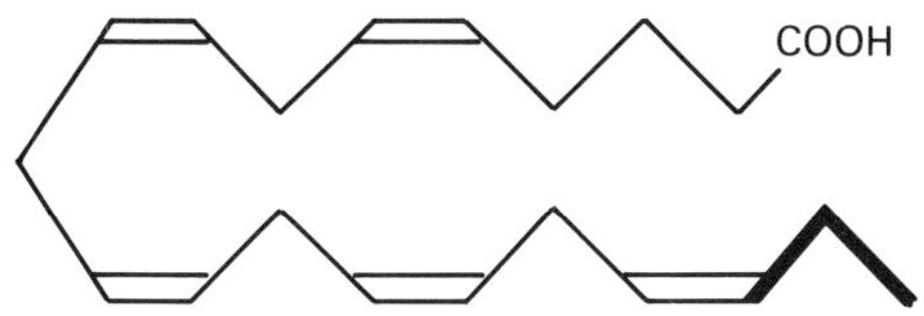

Fig. 3 Eicosapentaenoic acid (EPA).

n−3 fatty acids are included in the diet EPA inhibits conversion of linoleic acid to arachidonic acid and competes with arachidonic acid for the 2-acyl position in membrane phospholipids, reducing plasma and cellular levels of arachidonic acid. In addition, EPA competes with arachidonic acid as the substrate for cyclooxygenase. The platelets produce biologically inactive TXA_3 instead of TXA_2. However, PGI_2 synthesis in endothelial cells is not markedly inhibited and any newly synthesized PGI_3 has the same potency as PGI_2. Thus diet abundant in EPA may inhibit platelet aggregation and cause vascular dilatation, resulting in blood pressure reduction. In fact, high daily doses (15 g) of fish oils for 4 weeks have been shown to lower systolic and diastolic blood pressure in men with essential hypertension. However, the safety and long-term benefits of this treatment have been questioned because PGE_2 synthesis is inhibited and replaced by PGE_3 from EPA in the fish oils. Since renal blood flow is dependent on PGE_2 in renal disease and during volume contraction, any decrease in PGE_2 might result in more severely compromised renal function. *See* Atherosclerosis; Fish Oils, Dietary Importance

The other merit of dietary fish or fish oil supplements may be its plasma-lipid-lowering effect. The principal effect has been found to be a reduced production of triglycerides and very-low-density lipoprotein (VLDL) in the liver. Because VLDL is a precursor of low-density lipoprotein (LDL), a secondary reduction in LDL-cholesterol is also seen. Such conditions are considered to reduce the atherogenicity of lipoprotein particles. Furthermore, n−3 fatty acids have been reported to improve insulin sensitivity in skeletal muscle, possibly through a decrease in triglyceride accumulation. However, fish oil supplementation to diabetic patients resulted in no change in glycaemic control in one study and an actual deterioration in glycaemic control in another study.

Increased intake of polyunsaturated fatty acids has also been reported to result in reduced gastric acid secretion and an increase in gastric mucosal protection. Diets enriched in fish oil or dihomo-γ-linolenic acid also reduce LTB_4 formation in leucocytes or monocytes, suggesting a possible role for fish oil diets in suppressing inflammation in such disorders as rheumatoid arthritis or systemic lupus erythematosus. *See* Fatty Acids, Analysis

Proteins

Protein intake has a profound effect on renal haemodynamics and excretory function. High protein intake has been shown to have deleterious effects on the kidneys, especially in pre-existing renal disease, indicating a rationale for dietary protein restriction to lessen renal parenchymal damage. However, not all proteins are equipotent in relation to their renal effects. Lower glomerular filtration rate and renal plasma flow, as well as reduced excretion of albumin and some proteins have been reported in vegan and lactovegetarian individuals compared to omnivorous subjects. In fact, acute (more than 80 g) or chronic loading (more than 1 g per kg of bodyweight per day) with animal proteins resulted in an increase in renal plasma flow and glomerular filtration associated with a higher clearance of albumin than that on vegetable protein diets. Vasodilatory PGs such as PGI_2, which may be partly dependent on a rise in glucagon secretion from the pancreas, have been proposed as possible mediators of such renal effects of animal proteins since the administration of indomethacin was found to abolish the renal response to a meat meal. *See* Kidney, Structure and Function

Minerals

The isolation and identification of renal PGE_2 and PGA_2 in 1965–1967 provided biochemical support for the renal antihypertensive-endocrine function. Numerous studies in humans and animals thereafter showed that PGE_2 has a potent vasodilatory and natriuretic action and hence a pivotal role in sodium homeostasis. In fact, PGE_2 administration augments renal blood flow and urinary sodium excretion. In addition, PGE_2 stimulates renal renin secretion and inhibits vasopressin-mediated water reabsorption. During a low sodium intake the renin–angiotensin axis is activated by volume depletion resulting in antinatriuresis and vasoconstriction. Renal PGE_2 synthesis is stimulated by elevated circulating angiotensin II, which then counteracts the action of angiotensin II, leading to normalization of renal blood flow, blood pressure and sodium excretion (Fig. 4). This is the rationale for the beneficial effects of a low-sodium diet on these cardiovascular parameters. The same can be said for diuretic therapy since the renomedullary synthesis of PGE_2 from arachidonic acid is directly stimulated by furosemide. The natriurietic and antihypertensive effects of furosemide are mediated by this newly synthesized PGE_2 since both effects are inhibited by indomethacin. *See* Sodium, Physiology

Dietary potassium deficiency leads to hypokalaemia and impairment of maximal urinary concentrating ability. Since PGE_2 antagonizes vasopressin-induced water reabsorption, an increase in renomedullary PGE_2

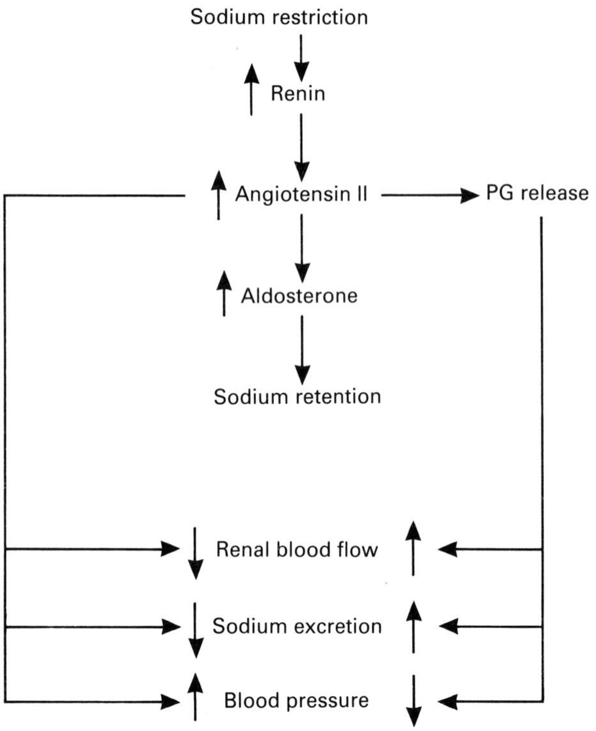

Fig. 4 Scheme whereby volume depletion secondary to dietary sodium restriction leads to an increase in PG release from the renal medulla and resultant physiological antagonism of the antinatriuretic and hypertensive actions of angiotensin II. (From Lee JB (1980) Prostaglandins and the renin-angiotensin axis. *Clinical Nephrology* 14: 159–163. Reproduced with permission of the publisher.)

production was postulated to underlie this concentrating defect. However, during chronic dietary potassium depletion secondary to dietary potassium deprivation in animals, PGE_2 synthesis by the renal medulla is markedly reduced and thus current evidence does not support a role for PGs in hypokalaemic polyuria. ***See*** Potassium, Physiology

Oral calcium supplementation has been reported to decrease blood pressure in some patients with essential hypertension and hypertensive rats. However, in other studies, calcium supplementation produced no effect on blood pressure in normotensive or hypertensive subjects. Although calcium supplementation has been shown to increase urinary PGE_2 excretion in normotensive women, there was no associated fall in blood pressure. ***See*** Calcium, Physiology; Hypertension, Hypertension and Diet

Spices

Spices have been consumed by many cultures for thousands of years. In fact, in classical Indian (Ayurvedism), Chinese or Graeco-Arabic systems of medicine some spice-containing plant materials were used to prevent or treat some diseases. Some spices have been shown to inhibit platelet aggregation. These include onion, garlic, ginger, cloves, omum, cumin and turmeric. Extracts from these spices were found to inhibit platelet cyclooxygenase, allowing less TXA_2 to be produced. As cyclooxygenase is inhibited, more substrate (arachidonic acid) becomes available to the lipoxygenase pathway, as shown by the observation that 12-HPETE increases in platelets treated with extracts from ginger, cloves, cumin and turmeric. On the other hand, onion extracts have been demonstrated to inhibit the 5- or 12-lipoxygenase in platelets. This might at least partially explain the observation that crude ethanolic extracts of onion prevented allergen-induced bronchoconstriction in animals and humans. Ginger, which is demonstrated to have antihistaminic and antioxidant action, also inhibits both the cyclooxygenase and 5-lipoxygenase pathway.

Conclusion

The only unequivocal dietary factor affecting PG synthesis and blood pressure is the level of sodium intake. Volume contraction secondary to lowered sodium intake leads to a lowering of blood pressure in many human hypertensives and experimentally hypertensive animals through an increased renal synthesis of PGE_2 from arachidonic acid. The converse is true for high sodium intake.

Although dietary fish oils containing EPA and/or $n-6$ unsaturated fatty acids have been shown to lower blood pressure, their safety and long-term effects have yet to be determined, as well as their effects on thrombosis and atherogenesis. Obviously, transfer of the high dietary fish intakes by Greenland Eskimos to other ethnic groups is a practical impossibility. Much further investigation is thus needed to clarify the interesting and provocative observations on dietary fish oil intake.

The role of calcium, potassium, protein and spices in the diet in relation to cardiovascular phenomena remains moot and presently unresolved.

Bibliography

Bergstrom S, Ryhage R, Samuelsson B and Sjovall J (1962) The structure of prostaglandin E_1, F_1 and F_2. *Acta Chemica Scandinavica* 16: 501–507.

Dusing R, Scherhag R, Glanzer K, Budde U and Kramer HJ (1983) Effect of changes in dietary prostaglandin precursor fatty acids on arterial blood pressure and vascular prostacyclin synthesis. In: (Samuelsson B, Paoletti R and Ramwell P, eds) *Advances in Prostaglandin, Thromboxane and Leukotriene Research*, vol. 12. New York: Raven Press.

Fioretto P, Trevisan R, Valerio A *et al.* (1990) Impaired renal response to a meat meal in insulin-dependent diabetes: role of glucagon and prostaglandins. *American Journal of Physiology* 258: F675–683.

Hamburg M, Svensson J and Samuelsson B (1976) Novel transformations of prostaglandin endoperoxides: formation of thromboxanes. In: (Samuelsson B and Paoletti R, eds) *Advances in Prostaglandin and Thromboxane Research.* New York: Raven Press.

Katayama S, Maruno A, Inaba M *et al.* (1991) Effect of dietary calcium on renal prostaglandins. *Prostaglandins, Leukotrienes and Essential Fatty Acids* 42: 197–200.

Knapp HR and Fitzgerald GA (1989) The antihypertensive effects of fish oil. *New England Journal of Medicine.* 320: 1037–1043.

Kontessis P, Jones S, Dodds R *et al.* (1990) Metabolic and hormonal responses to ingestion of animal and vegetable proteins. *Kidney International* 38: 136–144.

Kurzrok P and Lieb CC (1930) Depressor substance in seminal fluid. *Chemistry and Industry* 28: 268–272.

Lands WEM and Smith WL (eds) (1982) *Methods in Enzymology*, vol. 86. New York: Academic Press.

Leaf A and Weber P (1988) Cardiovascular effects of n−3 fatty acids. *New England Journal of Medicine* 318: 549–557.

Lee JB, Covino B, Takman BH and Smith ER (1965) Renomedullary depressor substance, medullin: Isolation, chemical characterization and physiological properties. *Circulation Research* 7: 57–77.

Lee JB (ed) (1982) *Prostglandins.* New York: Elsevier North Holland.

Lee JB and Katayama S (1985) Prostaglandins, thromboxanes and leukotrienes. In: (Wilson JD and Foster DW, eds) *Williams Textbook of Endocrinology.* Philadelphia: WB Saunders.

Lee JB, Cronshaw K, Takman BH and Attrep KA (1967) The identification of prostaglandin E_2, A_2 and F_2 from rabbit kidney medulla. *Biochemical Journal* 105: 1251–1260.

Moncada S, Gryglewski R, Bunting S and Vane JR (1976) An enzyme isolated from arteries transforms prostaglandin endoperoxides to an unstable substance that inhibits platelet aggregation. *Nature* 263: 633–665.

Oates JA, Fitzgerald GA, Branch RA *et al.* (1988) Clinical implications of prostaglandin and thromboxane A_2 formation. *New England Journal of Medicine* 319: 689–698 and 761–767.

Rosenthal J, Simone PG and Silbergleit A (1974) Effects of prostaglandin deficiency on natriuresis, diuresis and blood pressure. *Prostaglandins* 5: 435–441.

Samuelsson B (1983) Leukotrienes: Mediators of immediate hypersensitivity reaction and inflammation. *Science* 220: 568–575.

Smith WL and Willis AL (1971) Aspirin selectively inhibits prostaglandin production in human platelets. *Nature New Biology* 231: 235–237.

Srivastava KC and Mustafa T (1989) Antiplatelet activity and prostanoid metabolism. *Prostaglandins, Leukotrienes and Essential Fatty Acids* 38: 255–266.

Vane JR (1971) Inhibition of prostaglandin synthesis as a mechanism of action for aspirin-like drugs. *Nature New Biology* 231: 232–235.

von Euler US (1935) The specific blood pressure lowering substance in human prostate and seminal vesicle secretions. *Klinische Wochenschrift* 14: 1182–1183.

Shigehiro Katayama
Saitama Medical School, Saitama, Japan
James B. Lee
State University of New York School of Medicine and Biomedical Sciences, Erie County Medical Center, Buffalo, USA

PROTEIN

Contents

Chemistry

Like peptides, proteins are formed from amino acids through amide linkages. Covalently bound hetero constituents can also be incorporated into proteins. For example, phosphoproteins such as milk casein or phosvitin of egg yolk contain phosphoric acid esters of serine and threonine residues. Glycoproteins, such as κ-casein, various components of egg white and egg yolk, collagen from connective tissue, and serum proteins of some species of fish, contain one or more monosaccharide or oligosaccharide units bound *O*-glycosidically to serine, threonine or δ-hydroxylysine or *N*-glycosidically to asparagine.

The structure of a protein is dependent on the amino

acid sequence (the primary structure) which determines the molecular conformation (secondary and tertiary structures). Proteins sometimes occur as molecular aggregates which are arranged in an orderly geometric fashion (quaternary structure).

Amino Acid Sequence

Amino Acid Composition, Subunits

Sequence analysis can only be conducted on a pure protein. First, the amino acid composition is determined after acidic hydrolysis. The procedure (separation on a single cation exchange resin column and colour development with ninhydrin reagent) has been standardized and automated (amino acid analysers). It is also necessary to know the molecular weight of the protein. This is determined by gel column chromatography, ultracentrifugation or sodium dodecyl sulphate–polyacrylamide gel electrophoresis (SDS–PAGE). Furthermore, it is necessary to determine whether the protein is a single molecule or consists of a number of identical or different polypeptide chains (subunits) associated through disulphide bonds or noncovalent bonding. Dissociation into subunits can be accomplished by a change in pH, by chemical modification of the protein, such as by succinylation, or with denaturing agents (urea, guanidine hydrochloride, sodium dodecyl sulphate). Disulphide bonds, which are also found in proteins which consist of only one peptide chain, can be cleaved by oxidation of cystine to cysteic acid or by reduction to cysteine with subsequent alkylation of the thiol group to prevent reoxidation. Separation of subunits is achieved by chromatographic or electrophoretic methods. ***See*** Amino Acids, Properties and Occurrence; Amino Acids, Determination

Terminal Groups

N-Terminal amino acids can be determined by treating a protein with 1-fluoro-2,4-dinitrobenzene (Sanger's reagent) or 5-dimethylaminonaphthalene-1-sulphonyl chloride (dansyl chloride). Another possibility is the reaction with cyanate, followed by elimination of the *N*-terminal amino acid in the form of hydantoin, and separation and recovery of the amino acid by cleavage of hydantoin. The *N*-terminal amino acid (and the amino acid sequence close to the *N* terminus) is accessible by hydrolysis with aminopeptidase, in which case it should be remembered that the hydrolysis rate is dependent on amino acid side-chains and that proline residues are not cleaved. A special procedure is required when the *N*-terminal residue is acylated (*N*-formyl- or *N*-acetyl-amino acids, or pyroglutamic acid).

Determination of *C*-terminal amino acids is possible via the hydrazinolysis procedure. The *C*-terminal amino acid is separated from the amino acid hydrazides, e.g. by a cation exchange resin, and identified. The *C*-terminal amino acids can be removed enzymatically by carboxypeptidase A, which preferentially cleaves amino acids with aromatic and large aliphatic side-chains, carboxypeptidase B, which preferentially cleaves lysine, arginine and amino acids with neutral side-chains, or carboxypeptidase C, which cleaves with less specificity and cleaves proline.

Partial Hydrolysis

Longer peptide chains are usually fragmented. The fragments are then separated and analysed individually for amino acid sequences. Selective enzymatic cleavage of peptide bonds is accomplished primarily with trypsin, which only cleaves Lys—X and Arg—X bonds, and chymotrypsin, which cleaves peptide bonds with less specificity (Tyr—X, Phe—X, Trp—X and Leu—X). The enzymatic attack can be influenced by modification of the protein. For example, acylation of the ε-amino group of lysine limits tryptic hydrolysis to Arg—X, whereas substitution of the SH group of a cysteine residue with an aminoethyl group introduces a new cleavage position for trypsin into the molecule ('pseudolysine residue'). The most important chemical method for selective cleavage uses cyanogen bromide (BrCN) to attack Met—X linkages. Separation of peptide fragments is achieved by gel and ion exchange column chromatography using a volatile buffer as the eluent (e.g. pyridine, morpholine acetate) which can be removed by freeze drying of the fractions collected. Recently the separation of peptides and proteins by reversed-phase high-performance liquid chromatography (HPLC) has gained great importance, using volatile buffers mixed with organic, water-soluble solvents (e.g. acetonitrile) as the mobile phase. ***See*** Chromatography, High-performance Liquid Chromatography; Chromatography, Principles

The fragmentation of the protein is performed by different enzymatic and/or chemical techniques, at least by two enzymes of different specificity. The arrangement of the peptides obtained, in the same order as they occur in the intact protein, is accomplished with the aid of overlapping sequences. ***See*** Peptides

Sequence Analysis

The Edman degradation is by far the most important method in sequence analysis. It involves stepwise degradation of peptides with phenylisothiocyanate. The resultant phenylthiohydantoin is either identified directly or the amino acid is recovered. The stepwise reactions are performed in solution or on peptide bound to a carrier, i.e. to a solid phase. Both approaches have been automated (sequencer). Carriers used include resins containing amino groups (e.g. aminopolystyrene) or glass beads treated with aminoalkylsiloxane. The peptides are then attached to the carrier by carboxyl groups (activation with carbodiimide or carbonyl diimidazole, as in peptide synthesis) or by amino groups. Methods other than the Edman degradation can provide additional information. These include determination of terminal residues with amino- and carboxypeptidases, as already discussed, or mass spectrometric analysis of suitable volatile peptide derivatives.

Recently, amino acid sequences of proteins increasingly have been deduced from the nucleotide sequences of the corresponding genes. Examples of food proteins are gliadins and glutenins from wheat. If the gene responsible for the expression of a protein is known, the sequence analysis of the nucleic acid can be performed much easier than that of the protein itself.

Examples for amino acid sequences of food proteins are summarized in Tables 1 (milk proteins), 2 (collagen), 3 (monellin) and 4 (thaumatin).

Conformation

Information about conformation is available through X ray crystallographic analysis of protein crystals and through the determination of the H—H distances by proton nuclear magnetic resonance (NMR) spectroscopy in solution. In principle, the conformation of the protein in crystalline form can be assumed to be similar to that of the protein in solution. In 1960 the structure of myoglobin (17·8 kDa) was elucidated with a resolution of 0·2 nm. Individual atoms are well revealed at 0·11 nm. Such a resolution has not been achieved with proteins. Reliable localization of the C_α atoms of the peptide chain requires a resolution of less than 0·3 nm. ***See*** Spectroscopy, Nuclear Magnetic Resonance

Extended Peptide Chains

X Ray structural analysis and other physical measurements of a fully extended peptide chain reveal the lengths and angles of bonds. The peptide bonds have partial (40%) double bond character with π electrons shared between the C′—O and C′—N bonds. The resonance energy is about 83·6 kJ mol^{-1}. Normally the bond has a *trans* configuration, i.e. the oxygen of the carbonyl group and the hydrogen of the NH group are in the *trans* position; a *cis* configuration occurs only in exceptional cases (e.g. in small cyclic peptides or in proteins before proline residues). Because of the partial double bond character, six atoms of the peptide bonds, C^α_i, C'_i, O_i, N_{i+1} and H_{i+1}, lie in one plane. For a *trans* peptide bond, ω_i is 180°. The position of two neighbouring planes is determined by the numerical value of the angles ψ_i (rotational bond between a carbonyl carbon and an α-carbon) and ϕ_i (rotational bond between an amide nitrogen and an α-carbon). For an extended peptide chain $\psi_i = 180°$ and $\phi_i = 180°$. The position of side-chains can also be described by a series of angles χ_i^{1-n}.

Secondary Structure (Regular Structural Elements)

The primary structure gives the sequence of amino acids in a protein chain while the secondary structure reveals the arrangement of the chain in space. The peptide chains are not in an extended or unfolded form (ψ_i, $\phi_i \neq 180°$). It can be shown with models that ψ_i and ϕ_i, at a permissible minimum distance between atoms, can assume only particular angles. It has been shown for many proteins empirically that they have values of ψ_i, ϕ_i pairs within the permissible range. When a multitude of equal ψ_i, ϕ_i pairs occurs consecutively in a peptide chain, the chain acquires regular repeating structural elements. The types of structural elements are compiled in Table 5.

β Sheet

Three regular structural elements (pleated-sheet structures) have values in the range of $\phi = -120°C$ and $\psi = +120°$. The peptide chain is always lightly folded on the C_α atom, thus the R side-chains extend perpendicularly to the extension axis of the chain, i.e. the side-chains change their projections alternately from $+z$ to $-z$. Such a pleated structure is stabilized when more chains interact along the x axis by hydrogen bonding, thus providing the cross-linking required for stability. When adjacent chains run in the same direction, the peptide chains are parallel. This provides a stabilized, planar, parallel sheet structure. When the chains run in opposite directions, a planar, antiparallel sheet structure is stabilized. The lower free energy, twisted sheet structures, in which the main axes of the neighbouring chains are arranged at an angle of 25°, are more common than planar sheet structures.

Helical Structures

There are three regular structural elements in the range $\phi = -60°$ and $\psi = -60°$, in which the peptide chains are coiled like a threaded screw. These structures are stabilized by intrachain hydrogen bridges which extend

Table 1. Amino acid sequences of proteins from bovine milk

α_{s1}-Casein B																			
R	P	K	H	P	I	K	H	Q	G	L	P	Q	E	V	L	N	E	N	L
L	R	F	F	V	A	P	F	P	Q	V	F	G	K	E	K	V	N	E	L
S	K	D	I	G	S[a]	E	S[a]	T	E	D	Q	A	M	E	D	I	K	E	M
E	A	E	S[a]	I	S[a]	S[a]	S[a]	E	E	I	V	P	N	S[a]	V	E	Q	K	H
I	Q	K	E	D	V	P	S	E	R	Y	L	G	Y	L	E	Q	L	L	R
L	K	K	Y	K	V	P	Q	L	E	I	V	P	N	S[a]	A	E	E	R	L
H	S	M	K	Q	G	I	H	A	Q	Q	K	E	P	M	I	G	V	N	Q
E	L	A	Y	F	Y	P	E	L	F	R	Q	F	Y	Q	L	D	A	Y	P
S	G	A	W	Y	Y	V	P	L	G	T	Q	Y	T	D	A	P	S	F	S
D	I	P	N	P	I	G	S	E	N	S	E	K	T	T	M	P	L	W	
α_{S2}-Casein																			
K	N	T	M	E	H	V	S[a]	S[a]	S[a]	E	E	S	I	I	S[a]	Q	E	T	Y
K	Q	E	K	N	M	A	I	N	P	S	K	E	N	L	C	S	T	F	C
K	E	V	V	R	N	A	N	E	E	E	Y	S	I	G	S[a]	S[a]	S[a]	E	E
S[a]	A	E	V	A	T	E	E	V	K	I	T	V	D	D	K	H	Y	Q	K
A	L	N	E	I	N	E	F	Y	Q	K	F	P	Q	Y	L	Q	Y	L	Y
Q	G	P	I	V	L	N	P	W	N	Q	V	K	R	N	A	V	P	I	T
P	T	L	N	R	E	Q	L	S[a]	T	S[a]	E	E	N	S	K	K	T	V	D
M	E	S[a]	T	E	V	F	T	K	K	T	K	L	T	E	E	E	K	N	R
L	N	F	L	K	K	I	S	Q	R	Y	Q	K	F	A	L	P	Q	Y	L
K	T	V	Y	Q	H	Q	K	A	M	K	P	W	I	Q	P	K	T	K	V
I	P	Y	V	R	Y	L													
β-Casein A^2																			
R	E	L	E	E	L	N	V	P	G	E	I	V	E	S[a]	L	S[a]	S[a]	S[a]	E
E	S	I	T	R	I	N	K	K	I	E	K	F	Q	S[a]	E	E	Q	Q	Q
T	E	D	E	L	Q	D	K	I	H	P	F	A	Q	T	Q	S	L	V	Y
P	F	P	G	P	I	P	N	S	L	P	Q	N	I	P	P	L	T	Q	T
P	V	V	V	P	P	F	L	Q	P	E	V	M	G	V	S	K	V	K	E
A	M	A	P	K	H	K	E	M	P	F	P	K	Y	P	V	Q	P	F	T
E	S	Q	S	L	T	L	T	D	V	E	N	L	H	L	P	P	L	L	L
Q	S	W	M	H	Q	P	H	Q	P	L	P	P	T	V	M	F	P	P	Q
S	V	L	S	L	S	Q	S	K	V	L	P	V	P	E	K	A	V	P	Y
P	Q	R	D	M	P	I	Q	A	F	L	L	Y	Q	Q	P	V	L	G	P
V	R	G	P	F	P	I	I	V											
κ-Casein B																			
Z[d]	E	Q	N	Q	E	Q	P	I	R	C	E	K	D	E	R	F	F	S	D
K	I	A	K	Y	I	P	I	Q	Y	V	L	S	R	Y	P	S	Y	G	L
N	Y	Y	Q	Q	K	P	V	A	L	I	N	N	Q	F	L	P	Y	P	Y
Y	A	K	P	A	A	V	R	S	P	A	Q	I	L	Q	W	Q	V	L	S
D	T	V	P	A	K	S	C	Q	A	Q	P	T	T	M	A	R	H	P	H
P	H	L	S	F	M	A	I	P	P	K	K	N	Q	D	K	T	E	I	P
T	I	N	T	I	A	S	G	E	P	T	S	T[b]	P	T	I	E	A	V	E
S	T	V	A	T	L	E	A	S[a]	P	E	V	I	E	S	P	P	E	I	N
T	V	Q	V	T	S	T	A	V											
α-Lactalbumin B[c]																			
E	Q	L	T	K	C	E	V	F	R	E	L	K	D	L	K	G	Y	G	G
V	S	L	P	E	W	V	C	T	T	F	H	T	S	G	Y	D	T	E	A
I	V	E	N	N	Q	S	T	D	Y	G	L	F	Q	I	N	N	K	I	W
C	K	N	D	Q	D	P	H	S	S	N	I	C	N	I	S	C	D	K	F
L	N	N	D	L	T	N	N	I	M	C	V	K	K	I	L	D	K	V	G
I	N	Y	W	L	A	H	K	A	L	C	S	E	K	L	D	Q	W	L	C
E	K	L																	
β-Lactoglobulin B[e]																			
L	I	V	T	Q	T	M	K	G	L	D	I	Q	K	V	A	G	T	W	Y
S	L	A	M	A	A	S	D	I	S	L	L	D	A	Q	S	A	P	L	R
V	Y	V	E	E	L	K	P	T	P	E	G	D	L	E	I	L	L	Q	K
W	E	N	G	E	C	A	Q	K	K	I	I	A	E	K	T	K	I	P	A
V	F	K	I	D	A	L	N	E	N	K	V	L	V	L	D	T	D	Y	K
K	Y	L	L	F	C	M	E	N	S	A	E	P	E	Q	S	L	A	C	Q
C	L	V	R	T	P	E	V	D	D	E	A	L	E	K	F	D	K	A	L
K	A	L	P	M	H	I	R	L	S	F	N	P	T	Q	L	E	E	Q	C
H	I																		

[a] Phosphorylated serine.
[b] Glycosylated threonine.
[c] Disulphide bonds: 6–120, 28–111, 61–77, 73–91.
[d] 2-Pyrrolidone-5-carboxylic acid.
[e] Disulphide bonds: 66–160 and 106–119 or 106–121, respectively.

Table 2. Amino acid sequence of collagen from mammalian skin α^1 chain[a]

				Z*	M S Y	G Y D	E K S	A G V	S V P
G P M	G P S	G P R	G L P*	G P P*	G A P*	G P Q	G F Q	G P P*	G E P*
G E P*	G A S	G P M	G P R	G P P*	G P P*	G K N	G D D	G E A	G K P
G R P*	G Q R	G P P*	G P Q	G A R	G L P*	G T A	G L P*	G M K*	G H R
G F S	G L D	G A K	G N T	G P A	G P K	G E P*	G S P*	G Z B	G A P*
G Q M	G P R	G L P*	G E R	G R P*	G P P*	G S A	G A R	G D D	G A V
G A A	G P P*	G P T	G P T	G P P*	G F P*	G A A	G A K	G E A	G P Q
G A R	G S E	G P Q	G V R	G E P*	G P P*	G P A	G A A	G P A	G N P*
G A D	G Q P*	G A K	G A N	G A P*	G I A	G A P*	G F P*	G A R	G P S
G P Q	G P S	G A P*	G P K	G N S	G E P*	G A P*	G N K	G D T	G A K
G E P*	G P A	G V Q	G P P*	G P A	G E E	G K R	G A R	G E P*	G P S
G L P*	G P P*	G E R	G G P*	G S R	G F P*	G A D	G V A	G P K	G P A
G E R	G S P*	G P A	G P K	G S P*	G E A	G R P*	G E A	G L P*	G A K
G L T	G S P*	G S P*	G P D	G K T	G P P*	G P A	G Q D	G R P*	G P A
G P P*	G A R	G Q A	G V M	G F P*	G P K	G A A	G E P*	G K A	G E R
G V P*	G P P*	G A V	G P A	G K D	G E A	G A Q	G P P*	G P A	G P A
G E R	G E Q	G P A	G S P*	G F Q	G L P*	G P A	G P P*	G E A	G K P*
G E Q	G V P*	G D L	G A P*	G P S	G A R	G E R	G F P*	G E R	G V E
G P P*	G P A	G P R	G A N	G A P*	G N D	G A K	G D A	G A P*	G A P*
G S Q	G A P*	G L Q	G M P*	G E R	G A A	G L P*	G P K	G D R	G D A
G P K	G A D	G A P	G K D	G V R	G L T	G P I	G P P*	G P A	G A P*
G D K	G E A	G P S	G P A	G T R	G A P*	G D R	G E P*	G P P*	G P A
G F A	G P P*	G A D	G Q P*	G A K	G E P*	G D A	G A K	G D A	G P P*
G P A	G P A	G P P*	G P I	G N V	G A P*	G P K*	G A R	G S A	G P P*
G A T	G F P*	G A A	G R V	G P P*	G P S	G N A	G P P*	G P P*	G P A
G K E	G S K	G P R	G E T	G P A	G R P*	G E V	G P P*	G P P*	G P A
G E K	G A P*	G A D	G P A	G A P*	G T P	G P Q	G I A	G Q R	G V V
G L P*	G Q R	G E R	G F P*	G L P*	G P S	G E P*	G K Q	G P S	G A S
G E R	G P P*	G P M	G P P*	G L A	G P P*	G E S	G R E	G A P*	G A E
G S P*	G R D	G S P*	G A K	G D R	G E T	G P A	G A P*	G P P*	G A P*
G A P*	G P V	G P A	G K S	G D R	G E T	G P A	G P I	G P V	G P A
G A R	G P A	G P Q	G P R	G B K*	G Z T	G Z Z	G B R	G I K*	G H R
G F S	G L Q	G P P*	G P P*	G S P*	G E Q	G P S	G A S	G P A	G P R
G P P*	G S A	G S P*	G K D	G L N	G L P*	G P I	G P* P*	G P R	G R T
G D A	G P A	G P P*	G P P*	G P P*	G P P*	G P P	S G G	Y D L	S F L
P Q P	P Q Q	Z K A	H D G	G R Y	Y				

[a] The sequence has been deduced from sequences of different, but very similar, mammalian skin collagens.
Z*, 2-pyrrolidone-5-carboxylic acid; P*, 4-hydroxyproline, K*, 5-hydroxylysine.

Table 3. Amino acid sequences of the A and B chains of the sweet protein monellin (the residues Y-13 to D-16 are localized within a β-turn and are part of the structure, which is postulated to come into contact with the sweet receptor)

		5	10	15	20
A chain: F[a]	R E I	K G Y E Y	Q L Y V V	Y A S D K	L F R
	A D I	S E D Y K	T R G R K	L L R F N	G P
	V P P P				
		5	10	15	20
B chain: T[b]	G E W	E I I D I	G P F T Q	N L G K F	A V
	D E E	N K I G Q	Y G R L T	F N K V I	R P
	C M K	K T I Y E	E N		

[a] About 10% of the A chains additionally contain phenylalanine at the *N*-terminus (Phe-A chain).
[b] About 19% of the B chains additionally contain threonine at the *N*-terminus (Thr-B chain), and about 24% have no *N*-terminal Gly (Des-Gly[1] chain).

Table 4. Amino acid sequence of the sweet protein thaumatin I (disulphide bonds, 9–204, 56–66, 71–77, 121–193, 126–177, 134–145, 149–158, 159–164; residues Y-57 to D-59 are localized within a β turn and are part of the structure, which is postulated to come into contact with the sweet receptor)

					5					10					15					20					25					30					35					40
	A	T	F	E	I	V	N	R	C	S	Y	T	V	W	A	A	A	S	K	G	D	A	A	L	D	A	G	G	R	Q	L	N	S	G	E	S	W	T	I	N
41	V	E	P	G	T	N	G	G	K	I	W	A	R	T	D	C	Y	F	D	D	S	G	S	G	I	C	K	T	G	D	C	G	G	L	L	R	C	K	R	F
81	G	R	P	P	T	T	L	A	I	F	S	L	N	Q	Y	G	K	D	Y	I	D	I	S	N	I	K	G	F	N	V	P	M	N	F	S	P	T	T	R	G
121	C	R	G	V	R	C	A	A	D	I	V	G	Q	C	P	A	K	L	K	A	P	G	G	G	C	N	D	A	C	T	V	F	Q	T	S	E	Y	C	C	T
161	T	G	K	C	G	P	T	E	Y	S	R	F	F	K	R	L	C	P	D	A	F	S	Y	V	L	D	K	P	T	T	V	T	C	P	G	S	S	N	Y	R
201	V	T	F	C	P	T	A																																	

Table 5. Regular structure elements (secondary structure) present in protein

	K (°)	ψ (°)	n^a	d^b (Å)	r^c (Å)
β sheet, parallel	−119	+113	2·0	3·2	1·1
β sheet, antiparallel	−139	+135	2·0	3·4	0·9
Twisted sheet			2·3	3·3	1·0
3_{10} helix	−49	−26	3·0	2·0	1·9
α helix (right handed)	−57	−47	3·6	1·5	2·3
α helix (left handed)	+57	+47	3·6	1·5	2·3
π helix	−57	−70	4·3	1·1	2·8
Poly-L-proline helix I	−83	+158	3·3	1·9	
Poly-L-proline helix II	−78	+149	3·0	3·1	
Polyglycine helix II (left handed)	−80	+150	3·0	3·1	

[a] Amino acid residues per turn.
[b] The rise along the axis direction, per residue.
[c] The radius of the helix.

almost parallel to the chain axis, cross-linking the CO and NH groups, specifically the CO group of amino acid residue i with the NH group of residue $i+3$ (3_{10} helix), $i+4$ (α helix) or $i+5$ (π helix). The α helix, and for polypeptides from L-amino acids exclusively the right-handed α helix, is the most common. The 3_{10} helix is observed only at the ends of α helices, while the existence of a π helix is hypothetical. A helix is characterized by the angles ϕ and ψ, or by the parameters derived from these angles: n, the number of amino acid residues per turn; d, the rise along the main axis per amino acids residue; and r, the radius of the helix. Thus, the equation for the pitch, p, is $p=nd$.

Reverse Turns

An important conformational feature of globular proteins are the reverse turns, β turns or β bends. They occur at 'hairpin' corners, where the peptide chain changes direction abruptly. Such corners involve four amino acids residues, among them frequently proline. Glycine is favoured in the third position of β bends on the basis of energy considerations. Different types of β turns are known, for which different amino acids are allowed.

Super-secondary Structures

Analysis of known structures has demonstrated that regular elements can exist in combined forms. Examples are the coiled-coil α helix, chain segments with antiparallel β structures (β meander structure) and combinations of α helix and β structure (e.g. $\beta\alpha\beta\alpha$).

Tertiary and Quaternary Structure

Proteins can be divided into two large groups on the basis of conformation: (1) fibrillar (fibrous) or scleroproteins, and (2) folded or globular proteins.

Fibrous Proteins

The entire peptide chain is packed or arranged within a single regular structure for a variety of fibrous proteins. Examples are wool keratin (α helix), silk fibroin (β sheet) and collagen (a triple helix). Stabilization of these structures is achieved by intermolecular bonding (elec-

Table 6. Bond types in proteins

Type	Examples	Bond strength (kJ mol⁻¹)
Covalent bonds	—S—S—	−230
Electrostatic bonds[a]	$-COO^- \; H_3N^+-$	−21
	$>C=O \; O=C<$	+1
Hydrogen bonds	—O—H. . . O>	−17
	>N—H. . .O=C<	−13
Hydrophobic bonds	$-CH(CH_3)_2 \cdots (H_3C)_2CH-$	−0·1[b]
	—Ala. . .Ala—	−3
	—Val. . .Val—	−8
	—Leu. . .Leu—	−9
	—Phe. . .Phe—	−13
	—Trp. . .Trp—	−19

[a] For ε=4.
[b] Per square angstrom of surface area.

trostatic interaction and disulphide linkages, but primarily hydrogen bonds and hydrophobic interactions).

Globular Proteins

Regular structural elements are mixed with randomly extended chain segments (randomly coiled structures) in globular proteins. The proportion of regular structural elements is highly variable: 20–30% in casein, 45% in lysozyme and 75% in myoglobin. Five structural subgroups are known in this group of proteins: (1) α helices occur only; (2) β structures occur only; (3) α-helical and β-structural portions occur in separate segments on the peptide chain; (4) α helix and β structures alternate along the peptide chain; and (5) α helix and β structures do not exist. The process of peptide chain folding is not yet fully understood. It occurs spontaneously, probably arising from one centre or from several centres of high stability in larger proteins. Folding of the peptide chain packs it densely, by formation of a large number of intermolecular noncovalent bonds. Data about the nature of the bonds involved are provided in Table 6.

The hydrogen bonds formed between main chains, main and side-chains and side–side-chains are of particular importance for folding. The portion of polar groups involved in hydrogen bond build-up in proteins of >8·9 kDa appears to be fairly constant at about 50%. The hydrophobic interaction of the nonpolar regions of the peptide chains also plays an important role in protein folding. These interactions are responsible for the fact that nonpolar groups are folded to a great extent towards the interior of the protein globule. The surface areas accessible to water molecules have been calculated for both unfolded and native folded forms for a number of monomeric proteins with known conformations. The proportion of the accessible surface of a stretched state which tends to be fixed in the interior of the globule as a result of folding is a simple linear function of the molecular weight.

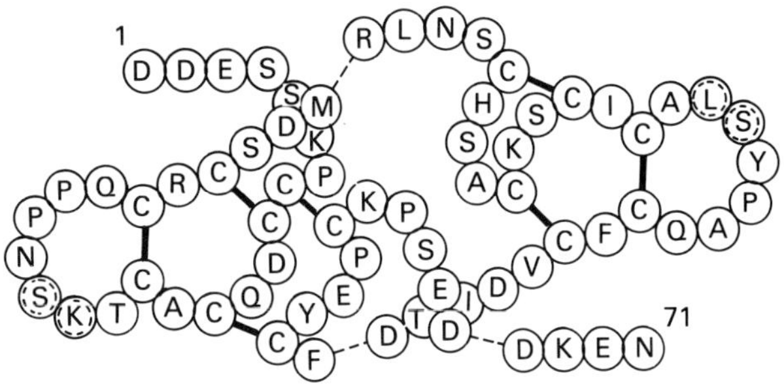

Fig. 1 Bowman–Birk inhibitor from soya beans (according to Ikenaka *et al.*, 1974). Amino acids are given in the IUPAC one-letter notation. Amino acids encircled by broken circles denote reactive sites; broken lines denote bonds split to separate the two domains.

Proteins with disulphide bonds fold at a significantly slower rate than those without disulphide bonds. Folding is not limited by the reaction rate of disulphide formation. Therefore the folding process of disulphide-containing proteins seems to proceed in a different way. The reverse process, protein unfolding, is very much slowed down by the presence of disulphide bridges which generally impart great stability to globular proteins. This stability is particularly effective against denaturation. An example is the Bowman–Birk inhibitor from soya beans, which inhibits the activity of trypsin and chymotrypsin. Its tertiary structure is stabilized by seven disulphide bridges. The reactive sites of inhibition are Lys16–Ser17 and Leu43–Ser44, i.e. both sites are located in relatively small rings, each of which consists of nine amino acid residues held in ring form by a disulphide bridge. The thermal stability of this inhibitor is high (Fig. 1).

Quaternary Structures

In addition to the free energy gain by folding of a single peptide chain, association of more than one peptide chain (subunit) can provide further gains in free energy. For example, haemoglobin (four associated peptide chains) $\Delta G^\circ = -46$ kJ mol^{-1} and the trypsin–trypsin inhibitor complex (association of two peptide chains) $\Delta G^\circ = -75{\cdot}2$ kJ mol^{-1}. In principle such associations correspond to the folding of a larger peptide chain around several structural domains without covalently binding the subunits.

Examples for protein conformations are given in Fig. 2, which shows the sweet proteins monellin and thaumatin.

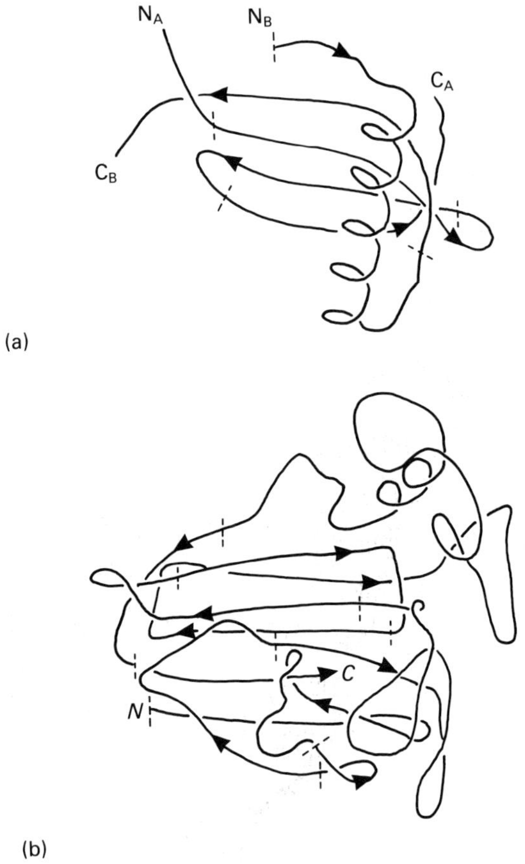

Fig. 2 Conformation of the peptide chains of the sweet proteins monellin (a) and thaumatin (b) (according to Ogata C *et al.*, 1987). β sheet ⟶; α helix; β turn; N_A, N_B, C_A, C_B denote *N* and *C* termini of the A and B chains.

Denaturation

The term 'denaturation' denotes a reversible or irreversible change of native conformation (tertiary structure) without cleavage of covalent bonds (except for disulphide bridges). Denaturation is possible with any treatment that cleaves hydrogen bridges, or ionic or hydrophobic bonds. This can be accomplished by changing the temperature, adjusting the pH, increasing the interfacial area, or adding organic solvents, salts, urea, guanidine hydrochloride or detergents such as sodium dodecyl sulphate. Denaturation is generally reversible when the peptide chain is stabilized in its unfolded state by the denaturing agent and the native conformation can be re-established after removal of the agent. Irreversible denaturation occurs when the unfolded peptide chain is stabilized by interaction with other chains (as occurs for instance with egg proteins during boiling). During unfolding, reactive groups, such as thiol groups, that were covered or blocked may be exposed. Their participation in the formation of disulphide bonds may also cause an irreversible denaturation. Denaturation of biologically active proteins is usually associated with loss of activity. The fact that denatured proteins are more readily digested by proteolytic enzymes is also of interest.

Physical Properties

Dissociation

Proteins, like amino acids, are amphoteric. Depending on pH, they can exist as polyvalent cations, anions or zwitterions. Since α-carboxyl and α-amino groups are linked together by peptide bonds, the uptake or release of protons is limited to free terminal groups, and mostly to side-chains. In contrast to free amino acids, the pK_a values fluctuate greatly for proteins since the dissociation is influenced by neighbouring groups in the macromolecule. For example, in lysozyme the γ-carboxyl group of Glu35 has a pK_a of 6–6·5, while the pK_a of the β-carboxyl group of Asp66 is 1·5–2, of Asp52 is 3–4·6 and of Asp101 is 4·2–4·7. The total charge of a protein, which is the absolute sum of all positive and negative charges, is differentiated from the so-called net charge which, depending on the pH, may be positive, zero or negative. By definition the net charge is zero and the total charge is maximal at the isoelectric point. Lowering or raising the pH tends to increase the net charge toward its maximum, while the total charge always becomes less than at the isolectric point.

Since proteins interact not only with protons but also with other ions, there is a further differentiation between an isoionic and an isoelectric point. The isoionic point is defined as the pH of a protein solution at infinite dilution, with no other ions present except for H^+ and OH^-. Such a protein solution can be acquired by extensive dialysis (or, better, electrodialysis) against water. The isoionic point is constant for a given substance while the isoelectric point is variable, depending on the ions present and their concentration. In the presence of salts, i.e. when binding of anions is stronger than that of cations, the isoelectric point is lower than the isoionic point. The reverse is true when cationic binding is dominant. In most cases the shift in pH is consistently positive, i.e. the protein binds more anions than cations. At its isoelectric point a protein is the least soluble and the most likely to precipitate ('isoelectric precipitation') and is at its maximal crystallization capacity. The viscosity of solubilized proteins and the swelling power of insoluble proteins are at a minimum at the isoelectric point.

Optical Activity

The optical activity of proteins is due not only to asymmetry of amino acids but also to the chirality resulting from the arrangement of the peptide chain. Information on the conformation of protein can be obtained from recording the optical rotatory dispersion (ORD) or the circular dichroism (CD), especially in the range of peptide bond absorption wavelengths (190–200 nm). The Cotton effect occurs in this range and reveals quantitative information on secondary structure. An α helix or a β structure gives a negative Cotton effect, with absorption maxima at 199 and 205 nm, while a randomly coiled conformation shifts the maximum to shorter wavelengths, i.e. results in a positive Cotton effect.

Solubility, Hydration and Swelling Power

Protein solubility is variable and is influenced by the number of polar and apolar groups and their arrangement along the molecule. Generally, proteins are soluble only in strongly polar solvents such as water, glycerol, formamide, dimethylformamide or formic acid. In a less polar solvent such as ethanol, few proteins have appreciable solubility (e.g. prolamins). The solubility in water is dependent on pH and on salt concentration. At low ionic strength, the solubility rises with increase in ionic strength and the solubility minimum (isoelectric point) is shifted to a somewhat lower pH. This shift is due to preferential binding of anions to the protein. As a rule, neutral salts have a twofold effect on protein solubility. At low concentrations they increase the solubility ('salting in' effect) by suppressing the electrostatic protein–protein interaction (binding forces). The logarithm of the solubility (S) is proportional to the ionic strength (μ) at low concentrations:

$$\log_{10} S = K\mu. \quad (1)$$

Protein solubility is decreased ('salting out' effect) at higher salt concentrations due to the ion hydration tendency of the salts. The following relationship applies (S_0, solubility at $\mu = 0$; K, salting out constant):

$$\log_{10} S = \log_{10} S_0 - K\mu. \quad (2)$$

Cations and anions in the presence of the same counter ion can be arranged in the following orders (Hofmeister series) based on their salting out effects: $K^+ > Rb^+ > Na^+ > Cs^+ > Li^+ > NH_4^+$; $SO_4^{2-} >$ citrate$^{2-} >$ tartrate$^{2-} >$ acetate$^- > Cl^- > NO_3^- > Br^- > I^- > CNS^-$. Multivalent anions are more effective than monovalent anions, while divalent cations are less effective than monovalent cations. Since proteins are polar substances, they are hydrated in water. The degree of hydration (grams of water of hydration per gram protein) is variable. It is 0·22 for ovalbumin (in ammonium sulphate), 0·06 for edestin (in ammonium sulphate), 0·8 for β-lactoglobulin and 0·3 for haemoglobin.

The swelling of insoluble proteins corresponds to the hydration of soluble proteins in that insertion of water between the peptide chains results in an increase in volume and other changes in the physical properties of the protein. The amount of water taken up during swelling can exceed the dry weight of the protein by several times. For example, muscle tissue contains 3·5–3·6 g of water per gram of protein (dry matter).

Chemical Reactions

The chemical modification of protein is of importance for a number of reasons. It provides derivatives suitable for sequence analysis, identifies the reactive groups in catalytically active sites of an enzyme, enables the binding of protein to a carrier (protein immobilization) and provides changes in protein properties which are important in food processing. In contrast to free amino acids and, except for the relatively small number of functional groups on the terminal amino acids, only the functional groups on protein side-chains are available for chemical reactions.

Lysine Residue

Reactions involving the lysine residue can be divided into several groups: (1) reactions leading to a positively charged derivative; (2) reactions eliminating the positive charge; (3) derivatizations introducing a negative charge; and (4) reversible reactions. The latter are of particular importance.

Reactions which Retain the Positive Charge

Alkylation of the free amino group of lysines with aldehydes and ketones is possible, with a simultaneous reduction step. A dimethyl derivative (Prot—$N(CH_3)_2$) can be obtained with formaldehyde.

Guanidination can be accomplished by using *O*-methylisourea as a reactant. α-Amino groups react at a much slower rate than ε-amino groups. This reaction is used analytically to assess the amount of biologically available ε-amino groups.

Derivatization with imido esters to amidines is also possible. The reactant is readily accessible from the corresponding nitriles. Proteins can be cross-bonded with the use of a bifunctional imido ester.

Treatment of the amino acid residue with amino acid carboxyanhydrides yields a polycondensation reaction product, made up of the protein as main chain and peptide side-chains. The length of the side-chains depends on reaction conditions. The carboxyanhydrides are readily accessible through interaction of the amino acid with phosgene.

Reactions Resulting in a Loss of Positive Change

Acetic anhydride reacts with lysine, cysteine, histidine, serine, threonine and tyrosine residues. Subsequent treatment of the protein with hydroxylamine leaves only the acetylated amino groups intact.

Carbamylation with cyanate attacks α- and ε-amino groups as well as cysteine and tyrosine residues. However, their derivatization is reversible under alkaline conditions.

Arylation is possible with 1-fluoro-2,4-dinitrobenzene (FDNB, Sanger's reagent) and trinitrobenzene sulphonic acid. FDNB also reacts with cysteine, histidine and tyrosine. 4-Fluoro-3-nitrobenzene sulphonic acid, a reactant which has good solubility in water, is also of interest for derivatization of proteins.

Deamination can be accomplished with nitrous acid. This reaction involves α- and ε-amino groups as well as tryptophan, tyrosine, cysteine and methionine residues.

Reactions Resulting in a Negative Charge

Acylation with dicarboxylic acid anhydrides, e.g. succinic acid anhydride, introduces additional carboxyl groups into the protein.

Introduction of a fluorescent acid group is possible by interaction of the protein with pyridoxal phosphate, followed by reduction of an intermediate Schiff base.

Reversible Reactions

N-Maleyl derivatives of proteins are obtained at alkaline pH by reaction with maleic anhydride. The acylated product is cleaved at pH <5, regenerating the protein. The half-life (τ) of ε-*N*-maleyl lysine is 11 h at pH 3·5 and 37°C. More rapid cleavage is observed with the 2-methylmaleyl derivative ($\tau < 3$ min at pH 3·5 and 20°C) and the 2,2,3,3-tetrafluorosuccinyl derivative (τ very low at pH 9·5 and 0°C). Cysteine binds maleic anhydride through an addition reaction. The *S*-succinyl derivative is quite stable. This side reaction is, however, avoided when protein derivatization is carried out with *exo-cis*-3,6-*endo*-hexahydrophthalic acid anhydride.

Acetoacetyl derivatives are obtained with diketene. This type of reaction also occurs with cysteine and tyrosine residues. The acyl group is readily split from tyrosine at pH 9·5. Complete release of protein from its derivatized form is possible by treatment with phenylhydrazine or hydroxylamine at pH 7.

Arginine Residue

The arginine residue of protein reacts with α- or β-dicarbonyl compounds, e.g. with nitromalondialdehyde, to form cyclic derivatives. The resulting nitropyrimidine absorbs at 335 nm. The arginyl bond of this derivative is not cleaved by trypsin but it is cleaved in its tetrahydro form, obtained by reduction with sodium borohydride. In the reaction with benzil, an iminoimidazolidone derivative is obtained after a benzilic acid rearrangement. Reaction of the arginine residue with 1,2-cyclohexanedione is highly selective and proceeds under mild conditions. Regeneration of the arginine residue is again possible with hydroxylamine.

Glutamic and Aspartic Acid Residues

These amino acid residues are usually esterified with methanolic hydrochloric acid. There can be side reactions, such as methanolysis of amide derivatives or *N*,*O*-acyl migration in serine or threonine residues.

Diazoacetamide reacts with a carboxyl group and also with the cysteine residue to carboxamidomethyl derivatives.

Amino acid esters or other similar nucleophilic compounds can be attached to a carboxyl group of a protein with the help of a carbodiimide.

Cystine Residue

Cleavage of cystine is possible by a nucleophilic attack. The nucleophilic reactivity of the reagents decreases in the following series: arsenite and phosphite > alkanethiol > aminoalkanethiol > thiophenol and cyanide > sulphite > OH^- > *p*-nitrophenol > thiosulphate > thiocyanate. Cleavage with sodium borohydride and with thiols is also possible. Complete cleavage with sulphite requires that oxidative agents (e.g. Cu^{2+}) are present and that the pH is higher than 7. The resultant *S*-sulpho derivative is quite stable in neutral and acidic media and is fairly soluble in water. The *S*-sulpho group can be eliminated with an excess of thiol reagent. Cleavage of cystine residues with cyanides (nitriles) is of interest since the thiocyanate formed in the reaction is cyclized into a 3-acyl-2-iminothiazolidine under cleavage of the *N*-acyl bond. This reaction can be utilized for the selective cleavage of peptide chains. Initially, all the disulphide bridges are reduced with dithiothreitol, and are then converted to mixed disulphides through reaction with 5,5′-dithio-bis(2-nitrobenzoic acid). These mixed disulphides are then cleaved by cyanide at pH 7.

Electrophilic cleavage occurs with Ag^+ and Hg^+ or Hg^{2+}. Electrophilic cleavage with H^+ is possible only in strong acids (e.g. 10 M hydrochloric acid). The sulphenium cation which is formed can catalyse a disulphide exchange reaction. In neutral and alkaline solutions a disulphide exchange reaction is catalysed by the thiolate anion.

Cysteine Residue

A number of alkylating agents, e.g. iodoacetic acid, iodoacetamide, ethylenimine and vinylpyridine, yield derivatives which are stable under the conditions for acidic hydrolysis of protein. The reaction with ethylenimine gives an *S*-aminoethyl derivative and, hence, an additional linkage position in the protein for hydrolysis by trypsin. Iodoacetic acid, depending on the pH, can react with cysteine, methionine, lysine and histidine residues. Maleic acid anhydride and methyl-*p*-nitrobenzene sulphonate are also alkylating agents. A number of reagents make it possible to measure thiol group content spectrophotometrically (azobenzene-2-sulphenylbromide, 5,5′-dithiobis(2-nitrobenzoic acid), *p*-mercuribenzoate, *N*-ethylmaleic imide).

Cysteine is readily converted to the corresponding disulphide, cystine, even under mild oxidative conditions, such as treatment with iodine or potassium hexacyanoferrate(III).

Stronger oxidation of cysteine, and also of cystine, e.g. with performic acid, yields the corresponding sulphonic acid, cysteic acid.

Methionine Residue

Methionine residues are oxidized to sulphoxides with hydrogen peroxide. The sulphoxide can be reduced, regenerating methionine, using an excess of thiol reagent.

α-Halogen carboxylic acids, β-propiolactone and alkyl halides convert methionine into sulphonium derivatives, from which methionine can be regenerated in an alkaline medium with an excess of thiol reagent.

Reaction with cyanogen bromide, which splits the peptide bond on the carboxyl side of the methionine molecule is used for selective cleavage of proteins.

Histidine Residue

Selective modification of histidine residues present on active sites of serine proteinases is possible. Substrate analogues such as halogenated methyl ketones inactive such enzymes (e.g. 1-chloro-3-tosylamido-7-aminoheptan-2-one inactivates trypsin and 1-chloro-3-tosylamido-4-phenylbutan-2-one inactivates chymotrypsin) by *N*-alkylation of the histidine residue.

Tryptophan Residue

N-Bromosuccinimide oxidizes the tryptophan side-chain and also tyrosine, histidine and cysteine. The reaction is used for the selective cleavage of peptide chains and the spectrophotometric determination of tryptophan.

Tyrosine Residue

Selective acylation of tyrosine can occur with acetylimidazole as a reagent.

Diazotized arsanilic acid reacts with tyrosine (*ortho* substitution) and with histidine, lysine, tryptophan and arginine.

Tetranitromethane introduces a nitro group into the *ortho* position.

Bifunctional Reagents

Bifunctional reagents enable intra- and intermolecular cross-linking of proteins. Examples are bifunctional imidoester, fluoronitrobenzene and isocyanate derivatives.

Bibliography

Creighton TE (1983) *Proteins. Structures and Molecular Properties.* New York: WH Freeman.

Croft LR (1980) *Introduction to Protein Sequence Analysis*, 2nd edn. Chichester: Wiley.

Edsall JT and Wyman J (1958) *Biophysical Chemistry*, vol. I. New York: Academic Press.

Glazer AN (1976) The chemical modification of proteins by group-specific and site-specific reagents. In: Neurath H, Hill RL and Boeder C-L (eds) *The Proteins*, 3rd edn, vol. II, p 1. New York: Academic Press.

Hudson BJF (ed.) (1982) *Developments in Food Proteins.* London: Applied Science.

Ikenaka T, Odani S and Koide T (1974) Chemical structure and inhibitory activities of soybean proteinase inhibitors. *Bayer-Symposium V "Proteinase Inhibitors"*, p 325. Berlin: Springer-Verlag.

Lottspeich F, Henschen A and Hupe K-P (eds) (1981) *High Performance Liquid Chromatography in Protein and Peptide Chemistry.* Berlin: Walter de Gruyter.

Needleman SB (ed.) (1970) *Protein Sequence Determination.* Berlin: Springer-Verlag.

Needleman SB (ed.) (1977) *Advanced Methods in Protein Sequence Determination.* Berlin: Springer-Verlag.

Ogata C, Hatada M, Tomlinson G, Shin WC and Kim S-H (1987) Crystal structure of the intensely sweet protein monellin. *Nature* 328: 739–742.

Schulz GE and Schirmer RH (1979) *Principles of Protein Structure.* Berlin: Springer-Verlag.

Sheppard RC (ed.) (1970–) *Amino-acids, Peptides and Proteins.* London: The Chemical Society.

Walton AG (1981) *Polypeptides and Protein Structure.* New York: Elsevier North Holland.

Wittmann-Liebold B, Salnikono J and Erdman VA (eds) (1986) *Advanced Methods in Protein Microsequence Analysis.* Berlin: Springer-Verlag.

H.-D. Belitz
Technical University Munich, Garching, FRG

Food Sources

Amino acids, peptides and proteins are important constituents of food. They supply the required building blocks for protein biosynthesis. In addition, they directly contribute to the flavour of food and are precursors for aroma compounds and colours formed during thermal or enzymatic reactions in production, processing and storage of food. Other food constituents, eg. carbohydrates, take part in such reactions. Proteins also contribute significantly to the physical properties of food through their ability to stabilize gels, foams, doughs, emulsions and fibrillar structures. ***See*** Physical Properties of Foods

Table 1 shows the most important protein sources and their contribution to worldwide production of protein. Besides plants and animals, algae (*Chlorella*, *Scenedesmus* and *Spirulina* spp.), yeasts and bacteria may be used for protein production (single-cell protein, SCP). Common carbon sources are glucose, molasses, starch, waste water, higher *n*-alkanes and methanol. Yeasts of the genus *Candida* can be grown, for example, on paraffins, and they deliver about 0·75 t of protein per t of alkane. Bacteria of the genus *Pseudomonas* deliver about 0·3 t of protein per t of alcohol in aqueous methanol. Because of the high nucleic acid content of yeasts and bacteria (6–17% of dry mass), isolation of the protein from biomass is necessary. The importance of SCP in the future will depend on the total protein market, especially on prices and functional properties of individual proteins. ***See*** Single-cell Protein, Algae; Single-cell Protein, Yeasts and Bacteria

Table 2 provides data on the average protein content and the nutrient density – ratio between the amount of protein (g) and the total energy value (MJ) of the digestible constituents – of selected foodstuffs. The protein content varies as follows: >20% (cheeses, meats, legumes, oil seeds); 10–20% (fishes, eggs); 5–10% (products from cereals); and <5% (milk, fruits, vegetables). ***See*** individual foods

The daily requirements of humans for essential amino acids, and the amino acid compositions of some important foodstuffs are shown in Table 3. The biological value of a protein (g protein formed in the body per 100 g food protein) depends on the absolute content of essential amino acids, on the relative proportions of essential amino acids, and on their ratios to nonessential amino acids. The biological value of a protein is generally limited by the following:

- Lysine: deficient in proteins of cereals and other plants,
- Methionine: deficient in proteins of bovine milk and meat,
- Threonine: deficient in wheat and rye,
- Tryptophan: deficient in casein, corn and rice. ***See*** Amino Acids, Metabolism

The biological values of some important food proteins are listed in Table 4. The highest value so far observed is for a blend of 35% egg and 65% potato proteins (one chicken egg and 500 g of potatoes).

Bibliography

Altschul AM (ed.) (1974–1985) *New Protein Foods*, vols 1–5. New York: Academic Press.

Bartholomew E, Biekert E, Hillmann H, By H, Weigert M and Weise E (eds) (1980) *Ullmanns Encyklopädie der technischen Chemie*, pp 491–557. Weinheim: Verlag Chemie.

München BB (ed) (1989) *Souci–Fachmann–Kraut Food Composition and Nutrition Tables 1989/90* 4th edn. Stuttgart: Wiss. Verlagsgesellschaft mbH.

Rehm H-J and Reed G (eds) (1983) *Biotechnology*, vol. 3. Weinheim: Verlag Chemie.

H.-D. Belitz
Technical University of Munich, Garching, FRG

Table 1. Worldwide protein production (1978–1979)

Source	Amount ($\times 10^6$ t)	Yield (kg ha^{-1})	Price (\$ kg^{-1})
Cereals	140	200–700	1
Oil seeds	40	500–1200	0·8
Legumes[a]	8·6	200–1000	1
Vegetables[b]	8·3		7
Meat	18	50–200	17
Fish	13		11
Milk	15	50–400	12
Eggs	3		10

[a] Without oil seeds.
[b] Roots and tubers.

Table 2. Protein content (%) of selected foodstuffs

No.[a]	Foodstuff	Average protein content: Raw product	Average protein content: Edible portion	Nutrient density (g MJ^{-1})
	Milk and dairy products			
1	Human milk	1·13	1·13	3·94
2	Cows' milk	3·33	3·33	11·98
3	Ewes' milk	5·27	5·27	13·10
4	Condensed milk (7·5% fat)	6·49	6·49	11·72
5	Whole dried milk	25·20	25·20	12·43
6	Yoghurt (1·5–1·8% fat)	3·55	3·55	16·44
7	Camembert cheese (40% fat dm)	22·50	22·50	19·47
8	Cheddar cheese (50% fat dm)	25·40	25·40	15·25
9	Edam cheese (40% fat dm)	24·80	26·10	19·68
10	Blue cheese (50% fat dm)	21·10	21·10	14·21
11	Emmental cheese (45% fat dm)	26·98	28·70	17·80
12	Feta cheese (45% fat dm)	17·00	17·00	16·97
	Egg			
13	Chicken egg	11·35	12·90	19·25
14	Yolk	16·10	16·10	10·71
15	White	11·10	11·10	50·00
	Meat			
16	Lamb (muscles only)	20·80	20·80	40·69
17	Veal (muscles only)	21·30	21·30	51·83
18	Beef (muscles only)	22·00	22·00	47·35
19	Pork (muscles only)	22·00	22·00	47·51
20	Sausage cervelat	16·56	16·90	8·74
21	Duck	14·48	18·10	18·61
22	Goose	9·89	15·70	10·82
23	Chicken	15·24	20·60	35·56
24	Turkey	14·75	20·20	21·79
	Fish			
25	Flounder	7·43	16·50	51·39
26	Halibut	16·08	20·10	45·13
27	Herring	12·74	18·20	18·25
28	Cod	13·28	17·70	53·45
29	Tuna	13·12	21·50	22·18
30	Trout	10·14	19·50	43·21
31	Salmon	12·74	19·90	22·90
	Cereals			
32	Oats (whole grain)	11·69	11·69	8·24
33	Rolled oats	12·53	12·53	8·43
34	Corn Flakes	7·15	7·15	4·91
35	Rice, polished	6·83	6·83	4·77
36	Rye (whole grain)	8·82	8·82	7·42
37	Rye flour, type 997	6·86	6·86	5·40
38	Wheat (whole grain)	11·73	11·73	9·26
39	Wheat flour, type 405	9·84	9·84	7·09
40	Wheat flour, type 1700	11·23	11·23	8·99
41	Wheat bread	7·61	7·61	7·83

continued

Table 2. *continued*

No.[a]	Foodstuff	Average protein content		Nutrient density (g MJ^{-1})
		Raw product	Edible portion	
	Roots and tubers			
42	Potato	1·63	2·04	7·15
43	Potato flakes	8·60	8·60	6·58
44	Beet root	1·19	1·53	9·25
45	Celeriac	1·13	1·55	23·06
46	Taro	1·68	2·00	4·61
47	Topinambur	1·68	2·44	22·26
48	Yam	1·68	2·00	4·97
	Leaves, stems and flowers			
49	Artichoke	1·15	2·40	31·08
50	Cauliflower	1·53	2·46	31·78
51	Broccoli	2·01	3·30	37·56
52	Brussel sprouts	3·47	4·45	36·96
53	Red cabbage	1·17	1·50	18·63
54	Chinese leaves	0·94	1·19	26·40
55	Soya bean sprouts	4·40	5·30	24·34
	Vegetable fruits			
56	Aubergine	1·03	1·24	19·54
57	String beans	2·25	2·39	18·15
58	Cucumber	0·44	0·60	12·56
59	Tomato	0·91	0·95	12·70
	Legumes and oil seeds			
60	Bean (dry)	21·09	21·30	18·28
61	Pea (dry)	22·67	22·90	17·19
62	Lentil (dry)	23·50	23·50	18·65
63	Peanut	25·25	25·25	11·20
64	Soya bean (dry)	28·00	33·73	27·52
	Fruits			
65	Apple	0·31	0·34	1·50
66	Pear	0·44	0·47	2·06
67	Apricot	0·82	0·90	4·90
68	Apricot (dry)	5·00	5·00	4·90
69	Cherry (morello)	0·80	0·90	3·99
70	Peach	0·70	0·76	4·41
71	Plum	0·56	0·60	2·92
72	Strawberry	0·80	0·82	6·14
73	Grape	0·65	0·68	2·35
74	Grape (dry, raisin)	2·46	2·46	2·12
75	Orange	0·72	1·00	5·72
76	Banana	0·77	1·15	3·02
77	Date (dry)	1·61	1·85	1·60
78	Fig	1·30	1·30	5·17
79	Fig (dry)	3·50	3·54	3·54
	Mushrooms			
80	Champignon	2·69	2·74	43·65
81	Chanterelle	0·93	1·52	34·66
82	Yellow boletus	2·22	2·77	41·75
83	Yellow boletus (dry)	19·70	19·70	40·21
84	Truffles	5·53	5·53	52·85

[a] These numbers are used in Table 3.
dm, Dry matter.

Table 3. Adult requirement (AR) for essential amino acids, and amino acid compositions of various foodstuffs

Amino acids	AR[a]	Foodstuff[b] 1	2	3	4	5	6	7	8	9	10	11	13	14	15	17	18	19
Alanine		0·056	0·13	0·22	0·27	0·91	0·16	0·89	0·76	0·89	0·73	0·92	0·89	1·03	0·83	1·64	1·69	1·53
Arginine		0·051	0·13	0·18	0·23	0·92	0·13	0·82	0·90	1·02	1·65	1·00	0·89	1·28	0·68	1·54	1·54	1·53
Aspartic acid		0·12	0·29	0·46	0·55	1·99	0·28	1·67	1·81	2·00	0·83	1·58	1·46	1·76	1·23	2·40	2·34	2·43
Cystine	11–14[c]	0·024	0·028	0·060	0·066	0·23	0·027	0·13	0·21	0·25	0·12	0·17	0·31	0·31	0·29	0·28	0·28	0·31
Glutamic acid		0·22	0·79	1·07	1·49	5·51	0·70	4·95	6·62	6·15	4·97	5·76	1·81	2·20	1·64	3·97	4·13	3·91
Glycine		0·036	0·076	0·11	0·18	0·56	0·086	0·46	0·47	0·55	0·51	0·51	0·53	0·62	0·50	1·34	1·56	1·42
Histidine		0·031	0·095	0·13	0·17	0·66	0·095	0·73	0·80	0·93	0·99	1·02	0·33	0·44	0·28	0·80	0·85	0·99
Isoleucine	10–11	0·077	0·22	0·28	0·45	1·61	0·22	1·52	1·81	1·48	1·19	1·73	0·93	1·09	0·74	1·29	1·25	1·27
Leucine	11–14	0·13	0·36	0·53	0·72	2·47	0·38	2·19	2·52	2·65	2·14	2·99	1·26	1·63	1·08	1·89	1·95	1·92
Lysine	9–12	0·086	0·28	0·46	0·52	1·96	0·28	1·65	2·07	2·37	2·38	2·39	0·89	1·30	0·74	2·05	2·31	2·20
Methionine	11–14[c]	0·024	0·090	0·14	0·17	0·62	0·092	0·59	0·77	0·78	0·52	0·79	0·45	0·47	0·47	0·60	0·65	0·72
Phenylalanine	13–14[d]	0·054	0·18	0·26	0·34	1·22	0·19	1·20	1·45	1·44	1·22	1·61	0·80	0·79	0·76	1·02	1·06	0·98
Proline		0·12	0·34	0·55	0·72	2·55	0·42	2·38	3·05	2·91	2·35	3·73	0·59	0·78	0·50	1·15	1·28	1·21
Serine		0·059	0·21	0·32	0·41	1·43	0·22	1·29	1·58	1·57	1·33	1·66	1·15	1·62	0·92	1·15	1·14	1·12
Threonine	6–7	0·063	0·16	0·24	0·33	1·16	0·16	0·84	0·98	1·13	0·92	1·14	0·71	1·01	0·58	1·13	1·15	1·25
Tryptophan	3	0·022	0·049	0·070	0·088	0·35	0·042	0·31	0·29	0·40	0·21	0·43	0·23	0·29	0·20	0·30	0·29	0·31
Tyrosine	13–14[d]	0·056	0·18	0·26	0·32	1·28	0·18	1·08	1·30	1·33	1·02	1·61	0·59	0·78	0·46	0·88	0·89	0·91
Valine	11–14	0·081	0·24	0·32	0·48	1·73	0·27	1·61	1·81	1·88	1·46	2·12	1·12	1·24	0·98	1·31	1·32	1·42

continued

Table 3. *continued*

Amino acids	AR[a]	Foodstuff[b] 21	22	23	24	25	26	27	28	29	30	31	32	33	34	35	36	37	38
Alanine		—	0·97	1·49	—	1·24	1·77	1·52	1·42	1·61	1·55	1·67	0·72	0·79	0·80	0·50	0·52	0·41	0·51
Arginine		1·10	0·98	1·44	1·21	1·16	1·37	1·18	1·21	1·25	1·40	1·33	0·85	0·87	0·24	0·57	0·49	0·33	0·62
Aspartic acid		—	1·41	2·35	—	2·07	2·15	2·31	2·01	2·88	2·36	2·22	1·11	1·29	0·54	0·78	0·68	0·61	0·70
Cystine	11–14[c]	—	—	0·31	0·27	0·15	0·24	0·24	0·25	0·29	0·17	0·29	0·32	0·39	0·16	0·11	0·19	0·14	0·29
Glutamic acid		—	2·34	3·82	—	3·18	3·01	3·23	3·13	3·52	3·33	3·23	2·90	3·08	1·86	1·58	2·57	2·05	4·08
Glycine		—	0·99	1·45	—	0·93	1·15	1·13	0·94	1·17	1·47	1·63	0·78	0·85	0·34	0·41	0·50	0·43	0·72
Histidine		0·41	0·44	0·63	0·52	0·45	0·48	0·52	0·52	1·09	0·57	0·66	0·27	0·30	0·24	0·17	0·19	0·18	0·28
Isoleucine	10–11	0·94	0·74	1·33	1·01	0·92	1·27	1·04	0·99	1·21	1·27	1·16	0·56	0·61	0·33	0·34	0·39	0·32	0·54
Leucine	11–14	1·40	1·32	1·84	1·47	1·60	1·94	1·75	1·69	2·17	1·78	1·77	1·02	1·13	1·24	0·66	0·67	0·54	0·92
Lysine	9–12	1·56	1·24	2·11	1·74	1·82	1·56	1·75	2·05	2·21	2·02	2·02	0·55	0·50	0·18	0·29	0·40	0·28	0·38
Methionine	11–14[c]	0·45	0·38	0·66	0·53	0·58	0·80	0·66	0·60	0·61	0·66	0·70	0·23	0·24	0·17	0·17	0·14	0·12	0·22
Phenylalanine	13–14[d]	0·71	0·66	0·94	0·77	0·70	0·68	0·75	0·84	1·05	0·92	0·91	0·70	0·78	0·43	0·39	0·47	0·36	0·64
Proline		—	0·76	1·09	—	0·60	0·81	0·84	0·82	0·88	0·85	1·00	0·87	0·84	0·97	0·42	1·25	0·84	1·56
Serine		—	0·62	0·95	—	0·86	1·27	1·05	0·99	1·05	0·97	1·01	0·74	0·71	0·47	0·41	0·45	0·34	0·71
Threonine	6–7	0·79	0·70	1·04	0·81	0·92	0·99	1·04	0·97	1·18	1·08	1·11	0·49	0·53	0·32	0·28	0·36	0·31	0·43
Tryptophan	3	—	0·20	0·29	0·16	0·21	0·26	0·21	0·24	0·30	0·24	0·26	0·19	0·19	0·050	0·090	0·11	0·070	0·15
Tyrosine	13–14[d]	—	0·51	0·79	0·28	0·64	0·68	0·67	0·71	0·97	0·68	0·72	0·45	0·57	0·27	0·26	0·23	0·22	0·41
Valine	11–14	0·87	0·97	1·22	0·95	1·06	1·30	1·21	1·09	1·42	1·25	1·39	0·79	0·81	0·44	0·49	0·53	0·41	0·62

continued

Table 3. *continued*

Amino acids	AR[a]	Foodstuff[b] 39	40	41	42	43	44	45	50	51	52	53	54	56	57	58	59	60	61
Alanine		0·37	0·49	0·24	0·11	0·35	—	0·086	—	—	—	—	—	—	—	—	0·026	0·74	0·48
Arginine		0·43	0·60	0·31	0·12	0·61	0·027	0·044	0·11	0·19	0·28	0·110	0·080	0·042	0·10	0·045	0·018	1·49	3·71
Aspartic acid		0·48	0·66	0·39	0·43	1·83	—	0·164	—	—	—	—	—	—	—	—	0·121	2·45	1·92
Cystine	11–14[c]	0·24	0·25	0·18	0·020	0·14	—	0·004	—	—	—	0·030	0·010	—	0·024	—	0·001	0·23	0·45
Glutamic acid		3·66	3·75	3·15	0·46	1·89	—	0·283	—	—	—	—	—	—	—	—	0·337	4·33	3·46
Glycine		0·42	0·63	0·29	0·12	0·34	—	0·047	—	—	—	—	—	—	—	—	0·018	0·95	0·59
Histidine		0·22	0·25	0·18	0·040	0·29	0·021	0·024	0·049	0·063	0·11	0·027	0·026	0·021	0·049	0·008	0·013	0·70	0·77
Isoleucine	10–11	0·46	0·52	0·38	0·10	0·33	0·049	0·048	0·11	0·13	0·21	0·043	—	0·063	0·11	0·019	0·023	1·49	1·88
Leucine	11–14	0·82	0·86	0·59	0·14	0·59	0·053	0·075	0·17	0·16	0·23	0·061	—	0·077	0·14	0·025	0·030	2·26	2·34
Lysine	9–12	0·24	0·35	0·20	0·13	0·69	0·082	0·074	0·14	0·15	0·25	0·071	0·058	0·034	0·14	0·026	0·029	1·87	2·13
Methionine	11–14[c]	0·17	0·21	0·13	0·030	0·080	0·005	0·018	0·048	0·050	0·040	0·014	0·032	0·007	0·034	0·006	0·007	0·26	0·35
Phenylalanine	13–14[c]	0·55	0·59	0·42	0·10	0·42	0·026	0·047	0·077	0·12	0·15	0·032	0·047	0·054	0·073	0·014	0·024	1·40	1·39
Proline		1·45	1·57	0·96	0·11	0·47	—	0·040	—	—	—	—	—	—	—	—	0·016	0·98	0·49
Serine		0·66	0·74	0·39	0·10	0·46	—	0·049	—	—	—	—	—	—	—	—	0·028	1·38	0·98
Threonine	6–7	0·32	0·39	0·25	0·090	0·36	0·033	0·044	0·11	0·12	0·16	0·042	0·052	0·043	0·093	0·016	0·023	1·15	1·57
Tryptophan	3	0·12	0·15	0·080	0·030	0·070	0·013	0·012	0·034	0·037	0·050	0·012	0·020	0·011	0·027	0·004	0·006	0·23	0·35
Tyrosine	13–14[d]	0·32	0·37	0·21	0·080	0·50	—	0·025	0·035	—	—	—	0·039	—	0·050	—	0·012	0·97	1·22
Valine	11–14	0·49	0·60	0·39	0·13	0·49	0·047	0·073	0·15	0·17	0·24	0·046	0·070	0·073	0·13	0·021	0·023	1·63	1·82

continued

Table 3. *continued*

Amino acids	AR[a]	Foodstuff[b] 62	63	64	65	70	72	74	75	76	77	80	81	82	83	84
Alanine		1·29	0·81	1·53	0·015	0·039	0·044	0·091	0·029	0·046	—	—	—	—	—	—
Arginine		2·24	3·46	2·36	0·008	0·017	0·037	0·305	0·073	0·054	0·040	0·200	0·090	0·26	—	0·65
Aspartic acid		3·16	3·31	3·99	0·101	0·090	0·191	0·087	0·122	0·115	—	—	—	—	—	—
Cystine	11–14[c]	0·25	0·43	0·59	0·001	0·009	0·007	0·006	0·003	0·002	—	0·014	0·120	0·29	2·08	0·15
Glutamic acid		4·49	5·63	6·49	0·025	0·139	0·126	0·118	0·066	0·105	—	—	—	—	—	—
Glycine		1·30	1·64	1·42	0·009	0·015	0·034	0·063	0·023	0·042	—	—	—	—	—	—
Histidine		0·71	0·71	0·83	0·006	0·017	0·016	0·051	0·012	0·077	0·040	0·057	0·030	0·22	1·59	0·10
Isoleucine	10–11	1·19	1·23	1·78	0·010	0·013	0·019	0·047	0·020	0·038	0·060	0·110	0·039	0·030	0·21	0·16
Leucine	11–14	2·11	2·03	2·84	0·016	0·028	0·044	0·075	0·032	0·085	0·062	0·120	0·110	0·12	0·84	0·40
Lysine	9–12	1·89	1·10	1·90	0·015	0·029	0·034	0·071	0·039	0·057	0·044	0·170	0·039	0·19	1·35	0·49
Methionine	11–14[c]	0·22	0·31	0·58	0·003	0·030	0·001	0·013	0·008	0·009	0·022	0·023	0·008	0·058	0·42	—
Phenylalanine	13–14[d]	1·40	1·54	1·97	0·009	0·018	0·025	0·047	0·020	0·034	0·051	0·074	0·089	0·10	0·73	0·19
Proline		1·22	1·43	1·82	0·010	0·027	0·027	0·157	0·189	0·040	—	—	—	—	—	—
Serine		1·51	1·83	1·69	0·012	0·033	0·033	0·051	0·043	0·049	—	—	—	—	—	—
Threonine	6–7	1·12	0·85	1·49	0·008	0·027	0·026	0·055	0·020	0·038	0·049	0·087	0·130	0·11	0·75	0·38
Tryptophan	3	0·25	0·32	0·45	0·002	0·005	0·015	0·005	0·007	0·018	0·049	0·024	0·048	0·21	1·46	0·020
Tyrosine	13–14[d]	0·84	1·19	1·25	0·005	0·020	0·029	0·010	0·013	0·021	—	0·066	0·085	0·12	0·86	0·18
Valine	11–14	1·39	1·45	1·76	0·012	0·039	0·025	0·071	0·033	0·057	0·076	0·090	0·062	0·078	0·56	0·25

[a] Daily requirement in mg per kg bodyweight.
[b] Numbering as in Table 2; units are g amino acid per 100 g edible portion.
[c] Cystine plus methionine.
[d] Phenylalanine plus tyrosine.

Table 4. Biological value of various food proteins

	Biological value (g per 100 g)	
Foodstuff	Adult rat	Young rat
Chicken egg	91	97
Cows' milk	74	90
Beef	67	76
Casein	72	69
Wheat flour	41	52
Lentil		41
Pea		32

Determination and Characterization

The important role of proteins as therapeutic and diagnostic agents has led to the development of many rapid analysis techniques. These vary in their selectivity and sensitivity, but can all be applied to food protein measurement. The main analysis methods used in the biochemical laboratory are shown in Table 1. These are the measurement of total protein using ultraviolet (UV) absorption and visible region spectrophotometry, and more selective approaches, involving chromatographic, electrophoretic and immunological techniques. It is important to note that all these analysis methods are, in the absence of known standards, semiquantitative at best, since there is no universal protein against which calibration can be made. Quantification is only achieved by comparison with a known, pure sample and is only accurate if the analyte itself is sufficiently resolved in the analysis. Amino acid analysis of protein hydrolysates is an important method of protein analysis and is discussed in detail in the context of amino acids. Prior to a description of these methods of analysis, certain aspects relating to food pretreatment must be considered. ***See*** Amino Acids, Determination

Table 1. Examples of the main techniques of protein analysis

Analysis technique	Examples
Total protein measurement	Total nitrogen UV absorption Reaction with a chromophore
Chromatography	Ion exchange Hydrophobic interaction Gel permeation
Electrophoresis	Sodium dodecyl sulphate–polyacrylamide gel electrophoresis (SDS–PAGE) Continuous zone electrophoresis (CZE) Isoelectric focusing (IEF)
Immunochemistry	Immunoelectrophoresis Radioimmunoassay Enzyme-linked immunosorbent assay (ELISA)

Food Pretreatment

Food material may contain hundreds, if not thousands, of different types of proteins. These may be associated with a variety of nonproteinaceous material such as carbohydrates, fats and polyphenolics, which may interfere with the analysis, or be enclosed within cells or subcellular organelles where their immediate analysis is not possible. It is therefore desirable to release entrapped protein and remove as much of the nonproteinaceous components of food as possible. Time spent in such pretreatment of the starting material is often worthwhile so as to increase the accuracy of subsequent analysis. Consequently, prior to considering alternative analysis methods, the following points should be noted:

(1) Although there is frequently little choice over the starting material in the analysis of protein, the fresher the food the better. Globular proteins are often labile and subject to degradation by a variety of means. Freezing of food is recommended for storage, with rapid thawing prior to analysis. If protein lability is suspected, denaturation, enzyme inactivation and proteolysis can be reduced significantly by including 2–5 mM EDTA, 1 mM phenylmethylsulphonylfluoride (PMSF) and 10^{-7} M pepstatin A in the extract.
(2) Depending on the food, the majority of the protein may be intracellular and require cell disruption prior to analysis. The technique used depends on the raw material, but a blade homogenizer is generally suitable for plant material and mincing or hand homogenizing for animal tissue. Neutral pH extraction buffer is used during disruption (e.g. 0·15–0·2 M phosphate, pH 7), with at least two volumes of extraction buffer to one volume of material.
(3) If the protein is insoluble or particle associated, a variety of solubilization techniques may be used, including alkaline hydrolysis (pH 8–11), ultrasonication, addition of detergents (e.g. 1–3% w/v Triton X-100) or solvent extraction. Where hydrolysis of fibrous proteins (e.g. connective tissue) is required to a give a total protein concentration, the protein is boiled in strong acid or alkali to yield the constituent amino acids in free form.
(4) Following protein extraction, the insoluble material

is removed by centrifugation. If the preparation is still very crude, further protein extraction can be achieved by dialysis or precipitation with ammonium sulphate (80% saturation will precipitate most protein), trichloroacetic acid (15% w/v) or solvent (55% v/v acetone).

The protocol outlined above is designed to produce a semipurified extract containing protein in a soluble form suitable for subsequent analysis. Having prepared the soluble protein extract it should be analysed within hours or stored for a limited time at 0–5°C.

Total Protein Analysis

The key methods of total protein analysis are the determination of total nitrogen, UV absorption and a variety of techniques involving protein reaction with a chromophore. All these techniques are subject to interference from nonprotein species and consequently the degree of accuracy will depend on the level of protein purity. The accuracy can be improved significantly by precipitation of protein prior to analysis. All methods are semiquantitative at best since there is no universal protein against which calibraton can be made.

Total Nitrogen Measurement

Total nitrogen is usually determined by the Kjeldahl method, to give an approximation of the protein content. Many variations on the Kjeldahl method exist, but essentially sample heating with sulphuric acid with a boiling aid (e.g. potassium sulphate) and a catalyst (e.g. selenium) converts covalent nitrogen to an ammonium salt which is then alkali treated to release ammonia which is measured titrimetrically following distillation, or photometrically after reaction with Nessler's reagent. The level of interference in this method may be significant due to the presence of amino acids, nucleic acids and low-molecular-weight nitrogenous compounds.

UV Absorption

Most proteins have an absorption maximum at 280 nm due to the presence of tyrosine and tryptophan residues. This is a quick and sensitive measurement of total protein but may be very inaccurate due to interference from nucleic acids and polyphenols. The pH should be controlled to near neutrality as the absorption can shift with pH. The absorption must be read against a suitable blank such as a buffer used in the initial extraction. If nucleic acids are present, their interference can be allowed for by measuring sample absorption at 280 and 260 nm and using the following empirical formula: protein (mg ml^{-1}) = 12·55 × (absorbance at 280 nm) − 0·76 × (absorbance at 260 nm).

Chromophoric Methods

These techniques all involve visible region spectrophotometry after reaction of the protein with a chromophore. All are semiquantitative unless the protein is highly pure and the assay calibrated using the respective prepurified protein. In most cases bovine serum albumin is used as a calibrating protein for heterogenous protein samples. The Biuret method involves the formation of copper chelates with the peptide linkages of the protein backbone and is consequently more accurate than the Lowry, bicinchoninic acid (BCA) and Bradford techniques which rely on the reduction of copper(II) to copper(I) by the protein and vary in sensitivity, depending on the protein aromatic amino acid composition. The BCA and Bradford methods are the most sensitive. ***See*** Spectroscopy, Visible Spectroscopy and Colorimetry

The determination of amino nitrogen following protein hydrolysis is a more accurate approach but results in the destruction of some amino acids, particularly tryptophan. Hydrolysis of protein in strong acid is usually followed by heating with ninhydrin (triketohydridene) to give a purple coloration (Ruhemann's purple). Although more accurate than the other chromophoric methods, the ninhydrin method is time-consuming and consequently less often used.

Chromatographic Methods

Chromatography is the fractionation of a mixture on the basis of differences in its partitioning between a stationary solid phase packed into a column and a mobile liquid phase passing through it. The rapid fractionations achievable using small-diameter wide-pore packings in high-performance liquid chromatography (HPLC) and fast protein liquid chromatography (FPLC) has enabled complex protein mixtures to be analysed using separation times of under an hour. The common chromatography techniques used in protein analysis (Table 2) separate on the basis of different protein properties. These include charge (ion exchange), hydrophobicity (reversed phase and hydrophobic inter-

Table 2. Examples of chromatography techniques used in protein analysis

Chromatographic technique	Basis of separation
Ion exchange	Protein charge
Hydrophobic interaction	Protein hydrophobicity
Size exclusion	Protein size or radius of gyration
Reversed phase	Protein hydrophobicity
Affinity	Biospecific interactions

action) and size (gel permeation). All techniques use a pump to apply the mobile phase to a column containing the solid phase. The sample is injected into the mobile phase stream prior to the column and the separation analysed using a detector (usually UV absorption at 254 or 280 nm) downstream from the column. ***See*** Chromatography, Principles; Chromatography, High-performance Liquid Chromatography

Ion Exchange Chromatography

This is the fractionation of proteins on the basis of their charge using cation exchange to fractionate positively charged proteins using a negatively charged solid phase, and, more commonly, anion exchange to fractionate negatively charged proteins using a postively charged solid phase. Most cation exchange separations use a negatively charged carboxymethyl group attached to the solid phase and a mobile phase buffered to pH 5–6 (i.e. below the isoelectric point of most proteins). Conversely, most anion exchange separations use a positively charged diethylaminoethyl group attached to a solid phase with a mobile phase buffered to pH 7–8 (i.e. above the isoelectric point of most proteins). Following sample injection on to the column, fractionation is normally achieved by applying a gradient of increasing salt concentration. As an example, anion exchange will typically use a mobile phase of 20 mM phosphate pH 7 or Tris–hydrochloric acid pH 8 with a gradient to 1 M sodium chloride in the same buffer. Altering the mobile phase pH can also be used to promote elution, but this is less common than the use of salt displacement.

Hydrophobic Interaction Chromatography

Here the protein is salted out on to the solid phase by addition of ammonium sulphate or sodium chloride. The support is derivatized with hydrophobic functional groups octyl and phenyl groups being most common. The column is pre-equilibrated in a high salt concentration buffer (e.g. 1·5–2·0 M ammonium sulphate in 0·1 M sodium phosphate pH 7) and the sample ionic strength similarly adjusted. Following sample application, elution is typically promoted by reduction of the ionic strength using a gradient of decreasing salt concentration. Proteins are fractionated according to their hydrophobicity, with the least hydrophobic eluting first. Very hydrophobic proteins which adsorb strongly can be eluted by applying a mobile phase of lower polarity by addition of ethylene glycol.

Size Exclusion Chromatography

Here proteins are separated on the basis of their size, or more correctly their radius of gyration. In contrast to the other techniques, no adsorption of the protein to the solid phase occurs. The matrix consists of beads of a porous gel fully permeable to the mobile phase, which is usually a neutral pH buffer (e.g. 20 mM sodium phosphate pH 7). If a mixture of proteins is applied to the column, smaller molecules will have full access to the internal structure of the gel beads since they can enter the pores. In contrast, very large protein molecules will have limited access to the gel pores. The volume of mobile phase available to the large proteins is therefore much smaller than the volume available to the small molecules. Consequently the large proteins will pass through the column much more quickly than the small molecules and be eluted from the column earlier. In between these two extremes, proteins of intermediate molecular weight will have varying degrees of access to the internal pores and will emerge from the column in order of decreasing radius to gyration. For globular proteins the radius of gyration will approximate to molecular weight. The degree of access that any protein will have to the internal structure of the beads will depend on the relative mean diameter of the pores and the radius of gyration to the protein. Matrices of different pore sizes are manufactured to allow fractionation of different molecular weight ranges.

Reversed-phase Chromatography

Reversed-phase chromatography is the most common form of high-performance separation, but has only been applied to protein separation in the last decade. It is capable of providing high-resolution separations, but will denature the protein due to the hydrophobicity of the matrix and the nonpolar solvents used. It is therefore most appropriate for peptide fractionation and for applications where retention of native protein structure is not essential. Reversed-phase chromatography is performed on silica derivatized with aliphatic functional groups. Octadecyl (C_{18}) bonded silica is most common, using a gradient of 20–60% v/v acetonitrile in water. Commonly, 0·1% trifluoroacetic acid (TFA) is added to assist in protein solubilization and increase protein affinity for the stationary phase. Since protein denaturation occurs, separation is based on the relative hydrophobicity of internally located amino acids, and protein mixtures which are normally difficult to resolve using other techniques can be fractionated successfully using reversed-phase chromatography.

Affinity Chromatography

Here an immobilized ligand with a high degree of affinity for a particular protein is used. Examples include the use of an antibody to purify an antigen or an

Table 3. Characteristics of commonly used electrophoretic techniques

Electrophoretic technique	Characteristics
Sodium dodecyl sulphate–polyacrylamide gel electrophoresis (SDS–PAGE)	Separates according to molecular weight, denatures proteins
Continuous zone	Uses native conditions to separate according to size and charge
Multiphasic zone	Uses native conditions and discontinuous buffers to improve resolution
Isoelectric focusing (IEF)	Separation according to isoelectric point using a continuous pH gradient
Two-dimensional	Separation on basis of size and charge using IEF followed by SDS–PAGE in a second dimension
Capillary	Separation in a carrier buffer within a hollow capillary

enzyme to purify its substrate or a substrate analogue. The attraction of the ligand to the target molecule is based on biospecific interactions and is highly specific. It is therefore useful for the purification of a single protein from a complex mixture and is used to a greater extent in protein purification than in analysis.

Electrophoretic Methods

Electrophoresis is the separation of molecules within an electric field on the basis of the sign and number of electric charges on the molecule. In proteins this charge is contributed by the side-chains of amino acids and charged carboxyl and amino terminal groups. In an electric field, proteins will migrate to the anode or cathode, depending on the overall charge. The isoelectric point of a protein is the pH at which the protein has zero net charge and reflects the relative number of acidic and basic amino acid side-chains. Electrophoresis is usually carried out in a matrix such as paper, cellulose acetate or a hydrophilic gel so as to slow down diffusion through the aqueous phase, counteract the effects of convection and facilitate subsequent protein immobilization. Most electrophoretic methods (with the exception of capillary electrophoresis) are qualitative unless a gel scanner is used and the protein bands are compared with purified proteins of known concentration. The most common forms of electrophoresis (Table 3) are based on the high-resolution capacity of polyacrylamide gels to provide protein separations on the basis of size, net charge and hydrophobicity. *See* Electrophoresis

Sodium Dodecyl Sulphate–Polyacrylamide Gel Electrophoresis (SDS–PAGE)

PAGE in the presence of the anionic detergent sodium dodecyl sulphate is the most popular form of electrophoresis. Here proteins are characterized according to the molecular size of their polypeptides. Proteins are separated under denaturing conditions, which resolves multimeric proteins into subunit polypeptide chains. The anionic detergent denatures and solubilizes proteins so as to mask their charge and give a net negative charge per unit of mass which is approximately constant for all proteins. Hence subsequent electrophoretic separation depends on the molecular radius of the peptide, which is roughly proportional to the molecular size. Sodium dodecyl sulphate (0·1%) will saturate peptide chains to give approximately one detergent molecule per two amino acid residues. The percentage of polyacrylamide in the gel will determine the mean pore size and the fractionation range of the gel. Hence with 5% w/v of the polymer the fractionation range is 20–350 kDa. If proteins with a wide range of molecular weights are present a gradient of polyacrylamide can be used (e.g. 3–30% for a fractionation range of 15–300 kDa). Separation is therefore by molecular sieving and a set of standard proteins of known molecular weight should be fractionated along with samples so as to determine protein molecular weight accurately. A discontinuous buffer system may be used to give protein band sharpening. Here, proteins are first concentrated in a stacking gel region prior to electrophoretic separation and then fractionated between a rapidly migrating lower buffer phase (e.g. chloride) and a slower migrating upper buffer phase (e.g. glycinate).

The protein concentrates between the buffer phases, reducing their diffusion so as to improve resolution. SDS–PAGE has the drawback that it cannot resolve proteins of the same molecular weight. Furthermore, anomalies may arise from proteins containing carbohydrate and lipid groups.

Continuous Zone Electrophoresis (CZE)

Electrophoresis under native conditions is useful if it is desirable to retain activity. However, the protein must be soluble with no tendency to aggregate or form a precipitate. CZE is a rapid and simple fractionation method of lower resolution than SDS–PAGE. A uniform polyacrylamide concentration and a single-phase buffer solution are used. Separation is on the basis of size and charge. The protein mixture is applied at the cathode, usually in buffer of up to 10 mM strength and of pH 8–9. Proteins will migrate to the anode and basic proteins will be lost. If basic proteins are of interest then a lower pH and migration from anode to cathode should be used. This technique is most suitable for concentrated protein samples, which should be applied to the gel as a narrow band with a concentration of at least 1 mg ml^{-1}.

Multiphasic Zone Electophoresis (MZE)

MZE is most appropriate for dilute samples. This technique uses discontinuous (multiphasic) buffers to concentrate the protein into a stack as in SDS–PAGE. Separation occurs under native conditions, typically using a glycine–phosphate buffer system. Nonionic detergents such as 0·5% Triton X-100, Tween 80 and Brij 35 can be added to sample and buffers in both MZE and CZE to improve protein solubility.

Isoelectric Focusing (IEF)

Here high resolution is provided by a continuous pH gradient. Proteins migrate through a flat gel bed until they reach a pH corresponding to their isoelectric point. Low polyacrylamide levels (3–5% w/v) are used to minimize molecular sieving effects. IEF is used in both native or denaturing conditions (e.g. with 8 M urea). Although sodium dodecyl sulphate cannot be added to IEF buffers, SDS–PAGE can be used in conjunction with IEF to give data on both size and charge heterogeneity. IEF is best for easily solubilized proteins, and, although it is a highly sensitive technique, it can prove a disadvantage if slight protein heterogeneity exists.

Two-dimensional Electrophoresis

Two-dimensional PAGE usually provides separation on the basis of size and charge. IEF in 8 M urea is typically used on a flat bed of gel in one dimension followed by SDS–PAGE at right angles. It is invaluable for the fractionation of complex protein mixtures.

Protein Fixing and Staining

Electrophoresis gels are usually fixed after use to precipitate and immobilize separated proteins in the gel and to remove nonprotein contaminants. Common fixatives are 20% w/v trichloracetic acid, methanolic acetic acid (3 parts methanol to 1 part acetic acid to 6 parts water) or 10–20% sulphosalicylic acid. Proteins are then visualized by dyeing. The most common dye is Coomassie brilliant blue R-250 at 0·1% w/v in 45% methanol, 10% acetic acid and 45% water. Staining takes around 2 h and is followed by destaining to remove dye from the gel, leaving stained protein. An effective destainer is the dye solution without Coomassie blue, with gel treatment for up to 24 h. Coomassie blue is sensitive down to 0·5 μg of protein per square centimetre of gel. Silver staining using silver nitrate is more sensitive and may be used to visualize trace impurities.

Capillary Electrophoresis

This is an emerging protein fractionation technique based on the separation of solutes within a capillary. Samples are injected by electrostatic or hydrostatic means into a hollow capillary in a moving carrier buffer (typically neutral pH phosphate buffer). A potential is applied across the capillary such that proteins will migrate in the carrier buffer within the capillary lumen to emerge according to protein charge density, with negative charged proteins emerging first, followed by neutral and then positively charged proteins. This technique has the advantages of short fractionation times and minimal sample and buffer volume requirements. Furthermore, it is more quantitative than gel-based electrophoresis techniques, provided protein standards of known concentration are used.

Immunological Methods

Immunological methods of protein analysis rely on the interaction between an antigen and an antibody to give rise to a number of secondary effects, including precipitation and agglutination. In contrast to the other methods of protein analysis discussed (with the exception of affinity chromatography), immunological techniques are typically used for the specific determination of the level of one protein type in a mixture. The levels and types of the remaining proteins in the same mixture remain unknown. The protein to be detected may be an antibody itself or, more commonly, an antigen, in which case the antibody must be available for the analysis method to be used. Immunological techniques (Table 4) are commonly based on protein measurement using precipitation or radioactive counting. ***See*** Immunoassays, Principles; Immunoassays, Radioimmunoassay and Enzyme Immunoassay

Precipitation Techniques

This requires a multivalent antigen (i.e. several sites available for antibody binding, as in a protein) and an excess of antibody. If antigen is present in excess no

Table 4. Characteristics of immunological techniques used for protein measurement

Immunotechnique	Characteristics
Ouchterlony assay	Diffusion of antigen and antibody from wells in a gel to form a precipitin arc
Single radial immunodiffusion	Antigen diffusion from a well into a gel containing antibody to form a precipitation ring
Immunoelectrophoresis	Electrophoresis of an antigen in a gel followed by antibody diffusion to form a precipitin arc
Rocket electrophoresis	Electrophoresis of an antigen into a gel containing antibody to form a 'rocket' of precipitation
Two-dimensional electrophoresis	Conventional electrophoresis followed by rocket electrophoresis
Radioactive binding for antibody measurement	Uses radioactive or immobilized antigen to bind to antibody. Followed by adding a second antibody or precipitation (Farr method)
Radioimmunoassay	Radiolabelled antigen competes with unlabelled antigen for binding to an antibody
Immunoradiometric assay	Antigen is bound to an immobilized antibody. Radiolabelled antibody is then added
Enzyme-linked immunosorbent assay (ELISA)	Antigen added to an immobilized antibody. A second antibody with an enzyme tag is then bound, substrate added and the level of product formed is measured

precipitation will result. All methods use immunoprecipitation in gels.

Double Diffusion (Ouchterlony) Assay

Here antigen and antibody are placed in different wells in a gel and diffuse towards each other. Opaque lines of a precipitate are formed where they meet at optimum levels. This allows identification of similarity between antigens.

Single Radial Immunodiffusion

In this technique an antigen diffuses from a well into an agar gel containing the antibody. Initially a high concentration of antigen exists, giving soluble complexes. Diffusion of antigen occurs until a precipitate forms, with higher concentrations of antigen producing greater ring diameters. This technique may be used to determine the concentration of immunoglobulin in milk, in which case standards of antigen can be used for calibration.

Immunoelectrophoresis

This is used to determine the identity of an antigen by its electrophoretic mobility. Electrophoresis of the antigen is performed in an agar gel, the antigen migrating to the anode. The current is then stopped and a trough cut in the agar which is then filled with antibody. This is allowed to diffuse into the gel to form a precipitin arc where antigen and antibody meet at optimum concentration. The arc is closest to the trough at the point where antigen is at its highest concentration.

Rocket Electrophoresis

This is the quantitative electrophoresis method which has been used for albumin measurement. Electrophoresis of an antigen into a gel containing antibody forms a precipitate in the shape of a rocket. The antigen must move to the anode; the length of the rocket is proportional to the antigen concentration.

Two-dimensional Immunoelectrophoresis

Here a preliminary electrophoretic separation of the antigen is carried out perpendicular to the final direction of rocket electrophoresis. In the second dimension the antigen is driven into a gel containing antiserum. It is used for more complex protein mixtures and, as with rocket electrophoresis, the area under the peak of precipitation is proportional to the antigen concentration.

Radioactive Binding Techniques for Measuring Antibody

These techniques determine the capacity of antiserum to complex with radioactive or immobilized antigen. Since antibodies have a range of binding affinities, a semi-quantitative measurement results. When using radioactive antigen, excess antigen (which will not form a precipitate), is added to antibody and the bound antigen measured either by precipitation with ammonium sulphate so that only free antigen stays in solution (the Farr method), or by adding anti-antibody to form a precipitate and leave only free antigen in solution (the anti-globulin coprecipitation method). Alternatively, antiserum may be added to immobilized antigen and radiolabelled anti-IgG added to bind to the antigen–IgG complex. Unbound radiolabelled anti-IgG is washed out and the radioactive count measured as an estimation of antibody.

Radioactive Binding Techniques for Measuring Antigen

Radioimmunoassay

Here a known concentration of radiolabelled antigen binding to a fixed concentration of antibody competes with unlabelled antigen. The reduction in the level of bound radiolabelled antigen is therefore a measure of the unlabelled antigen concentration.

Immunoradiometric Assay

In this technique the radiolabelled reagent (a second antibody) is added in excess. Antigen is first bound to an immobilized antibody. Following washing to remove unbound antigen, a radiolabelled antibody is added in excess to bind to the antigen. A different radiolabelled antibody to the immobilized antibody may be used to increase the sensitivity.

Nonradioactive Labelling Techniques

In these techniques enzymes are commonly used in the technique of enzyme-linked immunosorbent assay (ELISA). Antigen is again added to an immobilized antibody. However, the second antibody is now tagged with an enzyme such as a peroxidase or phosphatase. Following binding of the enzyme-linked antibody and washing to remove unbound protein, the substrate of the enzyme is added. Measurement of the level of product formed by action of the enzyme on the substrate typically used visible region absorbance, giving a highly sensitive measure of the concentration of antigen.

Bibliography

Clausen J (ed.) (1988) *Immunological Techniques for the Identification and Estimation of Macromolecules. Laboratory Techniques in Biochemistry and Molecular Biology*, 3rd edn. Amsterdam: Elsevier.

Hames BD and Rickwood D (eds) (1990) *Gel Electrophoresis of Proteins*. Oxford: IRL Press.

Harris ELV and Angal S (eds) (1989) *Protein Purification Methods. A Practical Approach*. Oxford: IRL Press.

Herbert D, Phipps PJ and Strange RE (1971) *Chemical analysis of microbial cells*. In: Norris JR and Ribbons DW (eds) *Methods in Microbiology*, vol. 5B, Chap 3. London: Academic Press.

Scopes R (1982) *Protein Purification. Principles and Practice*. New York: Springer-Verlag.

Simon Roe
AEA Technology, Harwell, UK

Requirements

The protein requirement is defined as the lowest level of dietary protein that will balance the losses of nitrogen (N) in persons maintaining energy balance. The body proteins are constantly undergoing breakdown and resynthesis at rates that vary from tissue to tissue. The amino acids released by tissue breakdown are mostly reused for synthesis but there is some loss – the obligatory nitrogen loss – by oxidative catabolism.

It is necessary to specify that the subject is in energy balance because utilization of dietary protein is influenced by energy intake. It has been shown that at any given level of dietary protein the addition of energy improves nitrogen balance until it reaches a plateau, which represents the limitation imposed by protein. Any change above or below the energy needs of the subject is likely to influence nitrogen balance. This is why the current international recommendations consider energy and protein requirements together. ***See*** Energy, Measurement of Food Energy; Energy, Energy Expenditure and Energy Balance

If more protein is consumed than is physiologically required, the excess is metabolized as a source of energy. However, there has been some concern that excessive protein intakes may contribute to demineralization of bone and deterioration of renal function in patients with renal disease; it is therefore suggested that the upper limit should be 1·5 g of protein per kg of bodyweight per day (Department of Health, 1991).

Earlier Recommendations

According to the perceived wisdom of the day, figures for requirements for protein have varied enormously (see Table 1). In the early days such figures were effectively pronouncements by leading physiologists and it was not until the discussions of the Food and

Table 1. History of recommended protein intakes (adult man)

Author	Recommended intake
Playfair (1865)	57–184 g (observed)
Voit (1881)	118 g (recommended)
Chittenden (1905)	50–55 g (observed and recommended)
Sherman (1920)	33–40 g at 70 kg bodyweight (0·59 g per kg)
League of Nations (1936)	1 g per kg bodyweight
FAO (1957)	0·35 g kg^{-1}
FAO (1965)	0·71 g kg^{-1}
FAO (1973)	0·57 g kg^{-1}
FAO (1985)	0·86 g kg^{-1}

Agriculture Organization of the United Nations (FAO) that figures were based on scientific evidence. At intervals of about 10 years, namely 1957, 1965, 1973 and 1985, revised FAO recommendations have been made in the light of increasing knowledge.

The first of these reports, in 1957, emphasized the proportions of the various amino acids required and they quantified dietary protein requirements in terms of a theoretical reference protein of perfect amino acid composition, i.e. satisfying the experimentally determined needs for each amino acid. The reason for terming this a theoretical protein is that dietary protein is most effectively assimilated only at low levels of intake (as used in laboratory assessment of protein quality) and efficiency falls off as the level rises. The efficiency of utilization of the theoretical protein is constant at all levels of intake.

National Recommended Dietary Allowances (RDAs)

While the FAO reports represent the conclusions of expert committees, each national authority has compiled its own figures based on the opinions of the committee involved and taking national habits into account. This has led to different figures in different countries. Thus the RDA for protein for adult males is 81 g per day in France, 64 g in the FRG, 70 g in The Netherlands, and 54 g in Spain. In practical food terms, the amount of protein one would recommend in the diet must depend on the total food (energy) intake; otherwise, heavy work would apparently call for provision of the extra energy from pure carbohydrate and fat. Two examples of the practical food approach that obviate this difficulty are the figure of 10–15% of the total energy intake as protein recommended by the Scandinavian countries, and 10% (69 g of protein on average) in the 1979 UK tables.

Requirements

Table 2. UK Dietary Reference Values for protein

Age	Weight (kg)	Estimated average requirement (g per day)	Reference intake (g per day)
0–3 months	5·9	—	12·5
4–6 months	7·7	10·6	12·7
7–9 months	8·8	11·0	13·7
10–12 months	9·7	11·2	14·9
1–3 years	12·5	11·7	14·5
4–6 years	17·8	14·8	19·7
7–10 years	28·3	22·8	28·3
Males			
11–14 years	43·0	33·8	42·1
15–18 years	64·5	46·1	55·2
19–50 years	74·0	44·4	55·5
50+ years	71·0	42·6	53·3
Females			
11–14 years	43·8	33·1	41·2
15–18 years	55·5	37·1	45·4
19–50 years	60·0	36·0	45·0
50+ years	62·0	37·2	46·5
Pregnancy			+6
Lactation			
0–6 months			+11
6+ months			+8

The 1991 UK revision (Table 2; Department of Health, 1991) adopted the FAO/WHO (World Health Organization)/UNU (United Nations University) figures (WHO, 1985). For adult males aged 19–50 years this is listed as 42·6 g of protein for males weighing 74 kg as the estimated average requirement, with 55·5 g as the Reference Nutrient Intake – the term introduced in the 1991 revision and which is equivalent to RDA. For women weighing 60 kg the corresponding figures are 36 and 45 g per day. The figures are based on milk or egg protein which are taken as completely digestible. For diets based on unrefined cereal grains and vegetables a correction factor for digestibility of 85% is applied; for diets based on refined cereals the correction factor is 95%. ***See*** Dietary Reference Values

Methods of Assessment of Needs

There are two methods of determining nitrogen requirements. The one used until the 1985 FAO report was the factorial method in which the losses of nitrogen from the body are measured to establish the amount that is required. The second method, adopted in the 1985 report, measures the intake of nitrogen needed to maintain nitrogen equilibrium.

Factorial Method

On a diet free from protein the body loses nitrogen in the urine as urea, creatinine, uric acid and ammonium salts together with small amounts of other compounds; this is termed the endogenous loss. There are also losses in the faeces from bacteria, mucosal cells shed from the lining of the intestine and residues of digestive juices that have not been reabsorbed; this is termed metabolic loss.

In addition, there are small losses from skin cells, hair, finger nails and body fluids.

These obligatory losses are measured on subjects fed a diet free from protein. On such a diet the urinary output of nitrogen does not attain a constant level but falls sharply for 4 to 6 days and then very slowly. The levels reached are 2–3 g per day or 30–45 mg per kg of bodyweight. Forty-five milligrams of nitrogen approximates to 2 mg of nitrogen per basal kcal. The two phases of this decline are taken to indicate two processes. The first rapid fall is loss from organs with a rapid protein turnover and is called labile body protein. The second is largely from muscles and skin. The obligatory nitrogen loss is taken as the figure at the end of the rapid fall. This relative inexactitude is one of the disadvantages of the factorial method.

The 1965 report took the figures of 46 mg of nitrogen per kg (2 mg per basal kcal) for urine loss, 20 mg per kg for faecal loss, and 20 mg for other losses. In the light of better evidence the 1973 report changed these findings to 37, 12 and 5 mg respectively. This accounts for the marked reduction in protein recommendations between 1965 ($0{\cdot}71\ g\ kg^{-1}$) and 1973 ($0{\cdot}57\ g\ kg^{-1}$). At first consideration obligatory losses should be balanced by an intake of the same amount. This was assumed in the 1965 report but it was later realized that when an individual is supplied with the same amount of nitrogen as his or her obligatory loss he or she remains in negative nitrogen balance. It was found necessary to increase that figure by 30% to achieve nitrogen equilibrium. An additional 30% was added to allow for individual variation. All measurements are made as nitrogen and then converted into protein by multiplying by the factor 6·25.

While the basic data show a small difference for nitrogen losses between males and females, figures are not sufficiently firm to allow any distinction. The difference between final protein figures is due to the different standard bodyweights used for males and females; in the 1973 report this was 0·71 g of protein per kg of bodyweight per day, which is 39 g for females and 46 g for males. This was called the practical allowance and based on full utilization of the protein in the diet, i.e. high-quality protein, termed reference protein.

An extra allowance is made for protein of lower quality, i.e. protein with net protein utilization (NPU) values of 70–80 in developed countries and 60–70 in less developed countries (with values as low as 50–60 in special conditions); this is stated in the 1973 report.

Apart from the difficulty of deciding the level of the endogenous loss (since there is no clear sharp break in the curve), most workers did not measure the cutaneous losses because of the difficulty of doing so, and they accepted published values which were derived from only a limited number of experiments.

Nitrogen Balance Method

Nitrogen balance is the difference between nitrogen excreted from the body and nitrogen ingested in the diet (of which the greater part by far is protein). During growth, pregnancy, lactation and recovery from convalescence the body is in positive nitrogen balance since it is retaining nitrogen for the purpose of synthesizing new protein tissues. During dietary deprivation, most illnesses and certain types of stress the body loses nitrogen and is in negative balance. The healthy adult is in nitrogen equilibrium. The basis of this method of determining nitrogen requirements is to feed the subjects a series of diets with different levels of protein while measuring nitrogen excretion, then to interpolate to nitrogen equilibrium (zero nitrogen balance).

In early studies the diets included very low levels of protein and a zero level but since the nitrogen balance response is not linear throughout the entire submaintenance range recent studies use levels around the expected range of requirements. An allowance has to be made for the variable losses of nitrogen in sweat, which are considerable in heavy work in a hot climate.

It takes some time at a given level of dietary protein to achieve a steady state, i.e. adjustments in urine output do not immediately follow changes in nitrogen intake, so that diets are fed for periods of 1–3 weeks at each intake level. These short-term nitrogen balance determinations do not take account of adaptation to low levels of dietary protein as experienced in developing countries. Long-term studies require several months, but these are usually limited to a single level of protein intake which would provide evidence that a particular diet is adequate.

It must be borne in mind that in these detailed consultations about protein (and energy) requirements and recommendations the FAO is largely concerned with the adequacy of diets in poorly fed populations of developing countries. In the well-fed Western world there is never a problem of protein shortage in healthy people as long as enough food is consumed to satisfy hunger. The fundamental physiological considerations, of course, still apply.

Direct nitrogen balance studies are the currently accepted method of determining protein requirements but they suffer from the lack of long-term balance studies, the absence of independent validation of an optimal state of protein nutrition, and lack of knowledge of the functional significance of the size of the total nitrogen pool and rate of turnover of tissue proteins. There are no functional indicators of protein inadequacy at a stage before clinically detectable changes occur.

The limitations of the data are indicated by the small numbers of studies at the time of the 1985 report – nine short-term studies of single protein sources on a total of 93 subjects, eight short-term studies on the typical mixed diets of eight countries on a total of 73 subjects, and six long-term studies, 24–89 days, five on egg and one on milk, on a total of 34 subjects.

The short-term studies provide an estimate of a mean daily requirement of 0·63 g of highly digestible good-quality protein – a figure slightly higher than the 1973 'safe level'. The long-term studies suggest that 0·58 g kg^{-1} is a reasonable estimate.

From all the results available it is suggested that 0·6 g per kg of bodyweight per day is the average requirement for good-quality proteins such as meat, milk, egg and fish.

The coefficient of variation was 12·5%, i.e. 2 SD equals 25%. This provides the currently accepted figure of 0·75 g of protein per kg of bodyweight per day as the safe level of intake for an adult.

Table 3. Values for the digestibility of protein in humans

Protein source	True digestibility (mean ± SD)	Digestibility relative to reference proteins
Egg	97±3	
Milk, cheese	95±3	100
Meat, fish	94±3	
Maize	85±6	89
Rice, polished	88±4	93
Wheat, whole	86±5	90
Wheat, refined	96±4	101
Oatmeal	86±7	90
Millet	79	83
Peas, mature	88	93
Peanut butter	95	100
Soya flour	86±7	90
Beans	78	82
Maize plus beans	78	82
Maize plus beans plus milk	84	88
Indian rice diet	77	81
Indian rice diet plus milk	87	92
Chinese mixed diet	96	98
Brazilian mixed diet	78	82
Fillipino mixed diet	88	93
American mixed diet	96	101
Indian rice plus beans diet	78	82

From FAO (1985).

Quality and Digestibility

Earlier recommendations for protein requirements were based on 'good-quality' protein or reference proteins such as eggs or milk, i.e. proteins with NPU close to 1·0 (100 in the older nomenclature). Thus the recommendation of 0·75 g protein per kg of bodyweight is for protein of this high quality and the amount has to be increased proportionately to compensate for lower quality. For example, 0·75 g of NPU 1·0 is equivalent to 1·07 g of quality 0·7, or 1·25 g of NPU 0·6.

In practice, the quality of the mixture of dietary proteins in underdeveloped countries (which rely very largely on a single staple food, usually a cereal), is about 0·7, and the addition of protein-rich foods such as meat and milk – cereals, meat and milk are the major sources of dietary protein in the developed countries – increases this to only 0·8. Consequently, the 1985 report lays less emphasis on the quality of the protein because the figure recommended is higher than the 1973 figure and digestibility correction alone is adequate. Digestibility was not mentioned at that time.

Only four amino acids are likely to limit the protein quality of mixed diets; they are lysine, the sulphur amino acids, threonine and tryptophan, and diets in developing countries generally meet the adult requirement for these amino acids. Quality can be taken care of by calculating the proportion of the most limiting amino acid compared with that in the requirement pattern and then increasing the protein figure in that ratio. ***See*** Amino Acids, Metabolism

Differences in digestibility arise from the nature of the cell walls of the food, the presence of dietary fibre and polyphenols, and from chemical reactions that alter the release of amino acids during digestion. Adjustments are therefore made to the amount of food protein to be ingested by comparing the digestibility of the protein food with that of reference proteins (Table 3).

The digestibility of a complete diet can be calculated by using the values for individual sources of protein and multiplying by the proportion of those foods in the diet.

Diets based on coarse, whole grain cereals and vegetables are generally about 85% digested, while those based on refined cereals are 95% digested; meat, milk and eggs are 100% digested.

Extra Allowances

Figures for protein requirements are based on measurements made on young adults with corrections for age, sex and bodyweight. They apply to healthy adults in nitrogen equilibrium. There are obviously greater needs for growing children, pregnant women and lactating mothers, and these are shown in Table 4 (the UK figure for lactation, in Table 2, has been modified). The extra needs for growth are small compared with the nitrogen required to maintain the tissues, even in young children, except for babies under one year of age. For example, the average requirement for growth of boys 4–5 years of age is 0·36 g of protein per day while for maintenance it is 12·6 g. It is accepted that different amounts of protein may be laid down from day to day as part of the normal process of growth. The body has very limited capacity for storing amino acids; it is therefore necessary to provide enough protein every day for possible extra demands of higher rates of growth. For this reason the extra demands for growth were estimated at 50% greater than the theoretical growth rate. ***See*** Children, Nutritional Requirements; Lactation; Pregnancy, Safe Diet

Table 4. Safe level of protein intake (good-quality protein)

Age	g per kg bodyweight	g per day
3–6 months	1·85	13
9–12 months	1·50	14
1–2 years	1·2	13·5
2–3 years	1·15	15·5
3–5 years	1·10	17·5
5–7 years	1·0	21
7–10 years	1·0	27
10–12 years		
Male	1·0	34
Female	1·0	36
12–14 years		
Male	1·0	43
Female	0·95	44
14–16 years		
Male	0·95	52
Female	0·9	46
16–18 years		
Male	0·9	56
Female	0·8	42
Adult		
Male	0·75	49 (65 kg bodyweight)
Female	0·75	41 (55 kg bodyweight)
Pregnancy		+6
Lactation		+17·5

From WHO (1985).

Trauma

Protein requirements for healthy adults take into account the 'normal stresses of everyday life' but there are increased losses of nitrogen from the body in pain, anxiety and psychological stresses, with very large losses in trauma, chronic disorders, parasitic infections, etc. Clearly, these will vary enormously with the severity and duration of the condition. Tables 5 and 6 show some examples of these losses which must be made good during convalescence. ***See*** Surgery, Nutrition in Elective and Emergency Surgery

Table 5. Protein losses related to surgery

Time of loss	Protein loss (g)
Before surgery	
Bleeding peptic ulcer	90 (over 5 days)
Long bone fracture	140–190 (over 10 weeks)
Limb burn	Up to 1000
During surgery (blood loss)	
Gastric surgery	9–18
Pneumonectomy	20–60
Thyroidectomy	3–12
Catabolic response after surgery	
Herniorraphy	18 (over 10 days)
Gastric resection	50–175 (5–10 days)
Appendectomy	50 (10 days)
Bed rest (inactivity), 3–4 weeks	300

Table 6. Approximate protein loss (g) during first 10 days following injury or operation

	Loss of tissue	Haemorrhage or exudate	Catabolic phase	Total
Simple fracture of femur	—	200	700	900–1100
Muscle wound	500–750	150–400	750	1350–1900
35% Burn	500	150–400	750	1400–1650
Gastrectomy	(20–180)	20–100	625–750	645–850
Untreated typhoid fever	—	—	675	675

Requirements

Bibliography

Department of Health (1991) *Dietary Reference Values for Food Energy and Nutrients for the United Kingdom.* Report on Health and Social Subjects 41. London: Her Majesty's Stationery Office.

FAO (1957) Protein requirements. *FAO Nutritional Studies.* 16. Rome: Food and Agriculture Organization.

FAO (1965) Protein requirements. Report of the FAO/WHO Expert Committee. *FAO Nutrition Report Series* 37. Rome: Food and Agriculture Organization.

FAO (1973) Energy and protein requirements. Report of the Joint FAO/WHO Expert Committee. *FAO Nutrition Report Series* 52. Rome: Food and Agriculture Organization.

WHO (1985) Energy and protein requirements. Report of the Joint FAO/WHO/UNU Expert Consultation. *WHO Technical Report Series* 724. Geneva: World Health Organization.

A. E. Bender
Leatherhead, Surrey, UK

Functional Properties

Standardization of food properties, in order to meet nutritional, physiological and toxicological demands and requirements of food processing operations, is a perennial endeavour. Food production is similar to a standard industrial fabrication process: on the one hand, there is the food commodity with all its required properties; on the other hand, there are the components of the product, each of which supplies a distinct part of the required properties. Such considerations have prompted investigations into the relationship in food between macroscopic physical and chemical properties, and the structure and reactions at the molecular level. Reliable understanding of such relationships is a fundamental prerequisite for the design and operation of a process, either to optimize the process or to modify the food components to meet the desired properties of the product.

Functional Properties

In common with other food constituents, proteins contribute significantly to the physical properties of foodstuffs, especially through their ability to build or stabilize gels, foams, doughs, emulsions and fibrillar structures. Table 1 shows some typical examples of functional properties of proteins in relation to important food systems. Foaming, gelling and emulsifying properties will be discussed in more detail. ***See*** Physical Properties of Foods

Foaming Properties

Proteins act as foam-forming and foam-stabilizing agents in various foodstuffs, e.g. in baked products, confectionery, desserts, and beer. The properties of various proteins are different: serum albumin is excellent and ovalbumin poor in foam formation. Mixtures of proteins, e.g. albumen, may be particularly effective. The globulins start the formation of foam; ovomucin is important for its stabilization, and ovalbumin and conalbumin are responsible for the heat-setting properties.

Foams are dispersions of gases in liquids. Proteins stabilize such systems by formation of flexible, cohesive films around the surface of the gas bubbles. During

Table 1. Typical functional properties performed by proteins in food systems

Functional property	Mode of action	Food system
Solubility	Protein solvation, pH-dependent	Beverages
Water absorption and binding	Hydrogen-bonding of water, entrapment of water (no drip)	Meats, sausages, breads, cakes
Viscosity	Thickening, water binding	Soups, gravies
Gelation	Protein matrix formation and setting	Meats, curds, cheese
Cohesion–adhesion	Protein acts as adhesive material	Meats, sausages, baked goods, pasta products
Elasticity	Hydrophobic bonding in gluten, disulphide links in gels (deformable)	Meats, bakery
Emulsification	Formation and stabilization of fat emulsions	Sausages (e.g. bologna), soup, cakes
Fat adsorption	Binding of free fat	Meat, sausages, doughnuts
Flavour binding	Adsorption, entrapment, release	Simulated meats, bakery, etc.
Foaming	Forms stable films to entrap gas	Whipped toppings, chiffon desserts, angel cakes

From Kinsella JE and Srinivasan D (1981).

whipping, the protein is adsorbed at the interface via hydrophobic areas, followed by partial unfolding (surface denaturation). The decrease of the surface tension, caused by protein adsorption, facilitates the formation of new interfaces and more bubbles. The denatured proteins interact under film formation. ***See*** Stabilizers, Types and Function

The foamability of a protein molecule depends on its diffusion rate and the ease with which it is denatured. These parameters in turn are dependent on the molecular mass, the surface hydrophobicity, and the stability of the conformation.

Foams break down, because large gas bubbles grow at the expense of smaller ones (disproportionation). The stability of a foam therefore depends on the stability of the protein film and its permeability to gas. The stability of the film in turn depends on the amount of adsorbed protein and on the ability of the molecules to interact. Usually the surface denaturation exposes additional amino acid side-chains, which can participate in intermolecular interactions. The stronger the cross-linking, the more stable is the film. The pH value of the system should be near the isoelectric points of the involved proteins because the association is favoured by a low net charge. ***See*** Amino Acids, Properties and Occurrence

To summarize, the ideal foaming protein is characterized by low molecular mass, high surface hydrophobicity, good solubility, low net charge at the pH of the foodstuff, and low conformational stability.

Foams are destroyed by lipids and organic solvents, e.g. higher alcohols, which, on account of their hydrophobicity, displace proteins from the surface of the gas bubbles without being able to form stable films. Egg yolk, for example, prevents the whipping of albumen, even at low concentrations. The disruption of protein association by the lecithins is responsible for this.

The foaming properties of proteins may be increased by chemical and physical modification. Partial enzymic hydrolysis produces smaller molecules of higher diffusion rate, better solubility, and higher surface hydrophobicity. A disadvantage is the decrease of film stability and the loss of heat coagulation properties. The introduction of charged or neutral groups, and the partial heat denaturation (e.g. of whey proteins), may also increase the desired properties. Recently, the addition of strongly basic proteins (e.g. clupeins) has been tested; this obviously increases the protein association within the films and allows the foaming of lipid-containing systems.

Gel Formation

Gels are dispersions made up of at least two components. The solid phase forms a cohesive network within the dispersing agent. Gels are characterized by their lack of fluidity and their elastic deformability. They are situated between solutions with repulsive forces between the molecules of the dispersed phase, and precipitates with strong intermolecular forces. A distinction can be made between two gel types, the *polymeric networks* and the *aggregated dispersions*. Intermediate stages are also known.

Examples of *polymeric networks* are gels formed by gelatin, agarose and carrageenan, respectively. The formation of a three-dimensional network occurs by aggregation of disordered fibrillar molecules via limited ordered structures, e.g. double helices. Gels of this type are characterized by low concentrations (*c.* 1%) of polymer, transparency and fine texture. Gel formation is started by adjustment to a suitable pH value, addition of suitable ions, and heating or cooling. Since the aggregation occurs mainly via hydrogen bridges, which are easily broken by heat, polymeric networks are *thermoreversible*, i.e. they are formed by cooling of a solution, and they melt on heating.

Examples of *aggregated dispersions* are the gels formed by globular proteins after denaturation by heat. The thermal unfolding of the protein exposes amino acid side-chains, which may take part in intermolecular interactions. The following association leads to small special aggregates, which form linear strands. The interaction of these strands delivers the three-dimensional gel network. Because of the unordered type of aggregation, the gel formation requires rather high protein concentrations (5–10%). To avoid the formation of coarse, less structured gels, especially in the range of the isoelectric point, the rate of aggregation should be smaller than that of unfolding. To start aggregation, a certain degree of unfolding is necessary, which seems to depend on the specific protein. Since the partial denaturation liberates mainly hydrophobic groups, the aggregation occurs predominantly via intermolecular hydrophobic bonds and delivers *thermoplastic* (*thermo-irreversible*) gels. This type of gel is not liquefied by heating, but weakening or shrinking may occur. Besides hydrophobic interactions, disulphide bonds, formed from exposed thiol groups, may contribute to cross-linking, as well as ionic bonds between proteins with different isoelectric points, e.g. in heterogeneous systems (e.g. albumen).

Addition of salts improves the gel formation: a moderate increase of the ionic strength promotes the interaction between charged molecules or molecule aggregates by shielding charges without precipitation. An example is the improvement of the heat coagulation of soya globulin by calcium ions (tofu). ***See*** Soya Cheeses

Emulsifying Properties

Emulsions are dispersions of two or more immiscible liquids. They are stabilized by emulsifiers, compounds, which form interfacial films and prevent the dispersed phase from coalescing. Proteins are able to stabilize emulsions (e.g. milk) on account of their amphiphatic nature. The emulsifying properties of a protein depend on the rate at which it diffuses to the interface, its adsorption there, and the deformability of its conformation by the surface tension (surface denaturation). The rate of diffusion depends on the temperature and the molecular mass, which in turn may be influenced by the pH and the ionic strength. The adsorption depends on the exposition of hydrophilic and hydrophobic groups, which is determined by the amino acid composition, and by the pH value, the ionic strength, and also the temperature. Finally, the stability of the conformation depends on the molecular mass, the amino acid profile and on intramolecular disulphide bonds.

Ideal emulsifying properties could be attributed to a protein of rather low molecular mass, balanced amino acid composition in respect to charged, polar and apolar side-chains, good solubility in water, marked surface hydrophobicity, and stable conformation. ***See*** Emulsifiers, Organic Emulsifiers

Modification of Functional Properties

Modification of proteins is still far from common in food processing, but is increasingly being recognized as essential, for two main reasons. First, proteins fulfil many functions in food, and some of these can be better served by modified than by native proteins. Second, persistent nutritional problems the world over necessitate the utilization of new raw materials. Modifying reactions can ensure that such new raw materials (e.g. proteins of plant or microbial origin) meet stringent standards of food safety, palatability, and acceptable biological value. A review is given here of several protein modifications that are being used or are being considered for use. They involve chemical or enzymatic methods or a combination of both. Examples have been selected to emphasize existing trends. Table 1 presents some protein properties which are of interest to food processing. These properties are related to the amino acid composition and sequence, and the conformation of proteins. Modification of the properties of proteins is possible by changing the amino acid composition or the size of the molecule, or by removing or inserting heteroconstituents. Such changes can be accomplished by chemical and/or enzymatic reactions.

From a food processing point of view, the aims of modification of protein are as follows:

1. To block the reactions involved in deterioration of food (e.g. the *Maillard* reaction). ***See*** Browning, Nonenzymatic
2. To improve some physical properties of proteins (e.g. texture, foam stability, whippability, solubility).
3. To improve the nutritional value (by increasing the extent of digestibility, by inactivation of toxic or other undesirable constituents, and by introducing essential ingredients such as some amino acids).

Table 2. Chemical reactions of proteins significant in food

Reactive group	Reaction	Product
—NH_2	Acylation	—NH—CO—R
—NH_2	Reductive alkylation with HCHO	—$N(CH_3)_2$
—$CONH_2$	Hydrolysis	—COOH
—COOH	Esterification	—COOR
—OH	Esterification	—O—CO—R
—SH	Oxidation	—S—S—
—S—S—	Reduction	—SH
—CO—NH—	Hydrolysis	—COOH + H_2N—

Chemical Modification

A selection of chemical reactions of proteins that are pertinent to and of current importance in food processing are shown in Table 2.

Acylation

Treatment with *succinic anhydride* generally improves the solubility of protein. For example, succinylated *wheat gluten* is quite soluble at pH 5. This effect is related to disaggregation of high-molecular-weight gluten fractions. In the case of succinylated *casein*, it is obvious that the modification shifts the isoelectric point of the protein (and thereby the solubility minimum) to a lower pH. Succinylation of *leaf proteins* improves the solubility as well as the flavour and emulsifying properties. Succinylated *yeast protein* has not only an increased solubility in the pH range of 4–6, but is more heat-stable above pH 5. It has better emulsifying properties, surpassing many other proteins, and has increased whippability.

Introduction of *aminoacyl groups* into protein can be achieved by reactions involving amino acid carboxy anhydrides, amino acids and carbodiimides, or *t*-butyloxycarbonyl (BOC)–amino acid hydroxysuccinimides with subsequent removal of the amino-protecting group (BOC). Feeding tests with *casein* with attached methionine, as produced by this method, have demonstrated a satisfactory availability of methionine. Such covalent attachment of essential amino acids to a protein may avoid the problems associated with food supplementation with free amino acids: losses in processing, development of undesired aroma due to methional, etc. With *β-casein* as an example, it was shown that the association of a protein is significantly affected by its acylation with fatty acids of various chain lengths. ***See*** Casein and Caseinates, Methods of Manufacture

Alkylation

Modification of protein by reductive methylation of amino groups with formaldehyde and sodium borohydride ($NaBH_4$) retards *Maillard reactions*. The resultant methyl derivative, depending on the degree of substitution, is less accessible to proteolysis. For this reason, its value from a nutritional and physiological point of view is under investigation.

Redox Reactions Involving Cysteine and Cystine

Disulphide bonds exert a strong influence on the properties of proteins. *Wheat gluten* can be modified by reduction of its disulphide bonds to sulphydryl groups, and subsequent reoxidation of these groups under various conditions. Reoxidation of a diluted suspension in the presence of urea results in a weak, soluble, adhesive product (gluten A), whereas reoxidation of a concentrated suspension in the presence of a higher concentration of urea yields an insoluble, stiff, cohesive product (gluten C). Additional viscosity data have shown that the disulphide bridges in gluten A are mostly intramolecular, while those in gluten C are preferentially intermolecular. ***See*** Wheat

Enzymatic Modification

Of the great number of possible enzymatic reactions with protein as a substrate, only a small number have so far been found to be suitable for use in food processing.

Dephosphorylation

With *β-casein* as an example, it was shown that the solubility of a phosphoprotein in the presence of calcium ions is greatly improved by enzymatic dephosphorylation.

Plastein Reaction

The plastein reaction enables peptide fragments of a hydrolysate to join enzymatically through peptide bonds, forming a larger polypeptide. The reaction rate is affected by, among other things, the nature of the amino acid residues. Hydrophobic amino acid residues are preferentially linked together. Incorporation of amino acid esters into protein is affected by the alkyl chain length of the ester. Short-chain alkyl esters have a low rate of incorporation, while the long-chain alkyl esters have a higher rate of incorporation. This is especially important for the incorporation of amino acids with a short side-chain, such as alanine. The plastein reaction can help to improve the biological value of a protein. An example is the plastein enrichment of zein with tryptophan, threonine and lysine.

Enrichment of a protein with selected amino acids can be achieved with the corresponding amino acid esters or, equally well, by using suitable partial hydrolysates of another protein. For example, the enrichment of *soya protein* with sulphur-containing amino acids is possible through 'adulteration' with the partial hydrolysate of *wool keratin*. The protein efficiency ratio (PER) values of such plastein products are significantly improved. In this way the production of plasteins with an amino acid profile very close to that recommended by the Food and Agriculture Organization and the World Health Organization (FAO/WHO) can be achieved from very diverse proteins, e.g. from *leaf*, *bacterial* and *algal protein*. ***See*** Novel Protein, Production and Uses of Leaf Protein; Novel Protein, Sources of Food-grade Protein

The plastein reaction also makes it possible to improve the *solubility* of a protein, e.g. by increasing the content of glutamic acid. A *soya protein* with 25% glutamic acid yields a plastein with 42% glutamic acid. Soya protein has a pronounced solubility minimum in the pH range of 3–6. The minimum is much less pronounced in the case of the unmodified plastein, whereas the glutamic-acid-enriched soya plastein has a satisfactory solubility over the whole pH range and is also resistant to thermal coagulation. Proteins with an increased content of glutamic acid show an interesting sensory effect: partial hydrolysis of modified plastein results not in a bitter taste, but in the generation of a pronounced 'meat broth' flavour. ***See*** Soya Beans, Processing for the Food Industry

Elimination of the bitter taste from a protein hydrolysate is also possible without incorporation of hydrophilic amino acids: bitter-tasting peptides, such as leucine–phenylalanine, which are released by partial hydrolysis of protein, react preferentially in the subsequent plastein reaction and are incorporated into higher-molecular-weight peptides with a neutral taste.

The versatility of the plastein reaction is also demonstrated when undesired amino acids are removed from a protein. A *phenylalanine-free diet*, which can be satisfied by mixing amino acids, is recommended for certain metabolic defects. However, the use of a phenylalanine-free, higher-molecular-weight peptide is more advantageous from a sensory or osmotic point of view. Such peptides can be prepared from protein by the plastein reaction. First, the protein is partially hydrolysed with pepsin. Treatment with pronase under suitable conditions then preferentially releases amino acids with long hydrophobic side-chains. The remaining peptides are separated by gel chromatography and subjected to the plastein reaction in the presence of added tyrosine and tryptophan. This yields a plastein which is practically phenylalanine-free and has a predetermined ratio of other amino acids, including tyrosine.

The plastein reaction can also be carried out as a *one-step process*, thus putting these reactions to economic, industrial-scale use.

Associations Involving Cross-linking

Cross-linking between protein molecules is achieved with peroxidase. The cross-linking occurs between tyrosine residues when a protein is incubated with a mixture of peroxidase and hydrogen peroxide (H_2O_2). Incubation of protein with a mixture of peroxidase, H_2O_2 and catechol also results in cross-linking. The reactions in this case are the oxidative deamination of lysine residues, followed by aldol and aldimine condensations, i.e. reactions analogous to those catalysed by lysyl oxidase in connective tissue.

Texturized Proteins

The world production of protein for nutrition is currently about 30% from animal sources and 70% from plant sources. The plant proteins are primarily from cereals (50%) and oil seed meal (20%). Some nonconventional sources of protein (single-cell proteins, leaves) have also acquired some importance. Proteins are responsible for the distinct physical structure of a number of foods, e.g. the fibrous structure of muscle tissue (meat, fish), the porous structure of bread, and the gel structure of some dairy and soya products.

Many plant proteins have a globular structure and, although available in large amounts, are used to only a limited extent in food processing. In the mid-1950s, in an attempt to broaden the use of such proteins, a number of processes were developed which confer a fibre-like structure to globular proteins. Suitable processes give products with cooking strength and a meat-like structure. They are marketed as meat extenders and meat analogues, and can be used whenever a lumpy structure is desired.

Starting Material

The following protein sources are suitable for the production of texturized products: soya; casein; wheat gluten; oil seed meals such as those from cottonseed, groundnut, sesame, sunflower, safflower or rapeseed; zein (corn protein); yeast; whey; blood plasma; or packing plant offal such as lungs or stomach tissue. The required protein content of the starting material varies and depends on the process used for texturization. The starting material is often a mixture, such as soya with lactalbumin, or protein plus acidic polysaccharide (alginate, carrageenan, or pectin).

The suitability of proteins for texturization is variable, but the molecular weight should be in the range of 10–50 kDa. Proteins of less than 10 kDa are weak fibre builders, while those higher than 50 kDa are disadvantageous owing to their high viscosity and their tendency to gel in the alkaline pH range. The proportion of amino acid residues with polar side-chains should be high in order to enhance intermolecular binding of chains. Bulky side-chains obstruct such interactions, so that the amounts of amino acids possessing these structures should be low.

Texturization

The globular protein is unfolded during texturization by breaking the intramolecular binding forces. The resultant extended protein chains are stabilized through interaction with neighbouring chains. In practice, texturization is achieved in one of two ways:

1. The starting protein is solubilized and the resultant viscous solution is extruded through a spinning nozzle into a coagulating bath (*spin process*).
2. The starting protein is moistened slightly and then, at high temperature and pressure, is extruded with shear force through the orifices of a die (*extrusion process*).

In the *spin process*, the starting material (protein content >90%, e.g. a soya protein isolate) is suspended in water and solubilized by the addition of alkali. The 20% solution is then aged at pH 11 with constant stirring. The viscosity rises during this time as the protein unfolds. The solution is then pressed through the orifices of a die (5000–15 000 orifices, each with a diameter of 0·01–0·08 mm) into a coagulating bath at pH 2–3. This bath contains an acid (citric, acetic, phosphoric, lactic or hydrochloric) and, usually, 10% sodium chloride. Spinning solutions of protein and acidic polysaccharide mixtures also contain earth alkali salts. The protein fibres are extended further (to about two to four times the original length) in a 'winding up' step, and are bundled into thicker fibres with diameters of 10–20 mm. The molecular interactions are enhanced during stretching of the fibres, thus increasing the mechanical strength of the fibre bundles.

The adherent coagulation solvent is then removed by pressing the fibres between rollers, then placing them in a neutralizing bath (sodium bicarbonate plus sodium chloride) of pH 5·5–6 and, occasionally, also in a hardening bath (concentrated sodium chloride). The fibre bundles may be combined into larger aggregates with diameters of 7–10 cm. Additional treatment involves passage of the bundles through a bath containing a binder and other additives (a protein which coagulates when heated, such as egg protein; modified starch or other polysaccharides; aroma compounds; lipids). This treatment produces bundles with improved thermal stability and aroma. A typical bath for fibres which are to be processed into a meat analogue might consist of 51% water, 15% ovalbumin, 10% wheat gluten, 8% soya flour, 7% onion powder, 2% protein hydrolysate, 1% sodium chloride, 0·15% monosodium glutamate and 0·5% pigments. Finally, the soaked fibre bundles are heated and sliced.

In the *extrusion process*, the moisture content of the starting material (protein content about 50%, e.g. soya flour) is adjusted to 30–40% and additives (salt, buffers, aroma compounds, pigments) are incorporated. Aroma compounds are added in fat as a carrier, when necessary, after the extrusion step to compensate for aroma losses. The protein mixture is fed into the extruder (a thermostatically controlled cylinder or conical body which contains a polished, rotating screw with a gradually decreasing pitch) which is heated to 120–180°C and develops a pressure of 30×10^5–40×10^5 Pa. Under these conditions, the mixture is transformed into a plastic, viscous state in which solids are dispersed in the molten protein. Hydration of the protein takes place after partial unfolding of the globular molecules and stretching and rearrangement of the protein strands along the direction of mass transfer. The process is affected by the rotation rate and shape of the screw, and by the heat transfer and viscosity of the extruded material and its residence time in the extruder. As the molten material exits from the extruder, the water vaporizes, leaving behind vacuoles in the ramified protein strands.

The extrusion process is more economical than the spinning process. However, it yields fibre-like particles rather than well-defined fibres. A great number and variety of extruders are now in operation. As with other food processes, there is a trend toward developing and utilizing high-temperature, short-time (HTST) extrusion cooking. ***See*** Extrusion Cooking, Principles and Practice

Bibliography

Aeschbach R, Amado R and Neukom H (1976) Formation of dityrosine cross-links in proteins by oxidation of tyrosine residues. *Biochimica Biophysica Acta* 439: 292.

Arai S, Yamashita M and Fujimaki M (1978) Nutritional improvement of food proteins by means of the plastein reaction and its novel modification. *Advances in Experimental Medicine and Biology* 105: 663.

Cherry JP (ed.) (1981) *Protein functionality in foods. ACS Symposium Series*, No. 147. Washington, DC: American Chemical Society.

Finot P-A, Mottu F, Bujard E and Mauron J (1978) *N*-Substituted lysines as sources of lysine in nutrition. *Advances in Experimental Medicine and Biology* 105: 549.

Grant DR (1973) The modification of wheat flour proteins with succinic anhydride. *Cereal Chemistry* 50: 417.

Kinsella JE (1976) Functional properties of proteins in foods: a survey. *Critical Review of Food Science and Nutrition* 7: 219.

Kinsella JE (1978) Texturized proteins: fabrication, flavoring, and nutrition. *Critical Review of Food Science and Nutrition* 10: 147.

Kinsella JE and Shetty KJ (1978) Yeast proteins: recovery and functional properties. *Advances in Experimental Medicine and Biology* 105: 797.

Kinsella JE and Srinivasan D (1981) Nutritional, chemical, and physical criteria affecting the use and acceptability of proteins in foods. In: Solms J and Hall RL (eds) *Criteria of Food Acceptance*. Zürich: Foster Verlag.

Mitchell JR (1986) Foaming and emulsifying properties of proteins. In: Hudson BJF (ed.) *Developments in Food Proteins* – 4, p 291. Essex: Elsevier Science Publishers.

Pomeranz Y (1991) *Functional Properties of Food Components*, 2nd edn. New York: Academic Press.

Poole S and Fry JC (1987) High performance protein foaming and gelation systems. In: Hudson BJF (ed.) *Developments in Food Proteins* – 5, p 257. Essex: Elsevier Science Publishers.

Puigserver AJ, Sen LC, Clifford AJ, Feeney RE and Whitaker JR (1978) A method for improving the nutritional value of food proteins: covalent attachment of amino acids. *Advances in Experimental Medicine and Biology* 105: 587.

Yamashita M, Arai S, Amano Y and Fujimaki M (1979) A novel one-step process for enzymatic incorporation of amino acids into proteins: application to soy protein and flour for enhancing their methionine levels. *Agricultural Biology and Chemistry* 43: 1065.

H.-D. Belitz
Technical University of Munich, Garching, FRG

Interactions and Reactions Involved in Food Processing

The nature and extent of the chemical changes induced in proteins by food processing depends on a number of parameters, e.g. composition of the food, and processing conditions, such as temperature, pH or the presence of oxygen. Consequences of such reactions may be desirable or undesirable. For example, the biological value of proteins may be decreased by destruction of essential amino acids, conversion of essential amino acids into derivatives which are not metabolizable or decrease in the digestibility of protein as a result of intra- or interchain cross-linking. Formation of toxic degradation products is also possible. The nutritional/physiological and toxicological assessment of changes induced by processing of food is a subject of some controversy and opposing opinions.

Reactions with Carbohydrates

Many foodstuffs contain reducing sugars and amino compounds, such as proteins, peptides, amino acids and

Table 1. Mutagenic compounds from pyrolysates of amino acids and proteins

Mutagenic compound	Abbreviation	Pyrolysed compound
3-Amino-1,4-dimethyl-5*H*-pyrido[4,3-*b*]indole	Trp-P-1	Tryptophan
3-Amino-1-methyl-5*H*-pyrido[4,3-*b*]indole	Trp-P-2	Tryptophan
2-Amino-6-methyldipyrido[1,2-α:3′2′-*d*]imidazole	Glu-P-1	Glutamic acid
2-Aminodipyrido[1,2-α:3′2′-*d*]imidazole	Glu-P-2	Glutamic acid
3,4-Cyclopentenopyrido[3,2-α]carbazole	Lys-P-1	Lysine
4-Amino-6-methyl-1*H*-2,5,10,10*b*-tetraazafluoranthene	Orn-P-1	Ornithine
2-Amino-5-phenylpyridine	Phe-P-1	Phenylalanine
2-Amino-9*H*-pyrido[2,3-*b*]indole	AαC	Soya globulin
2-Amino-3-methyl-9*H*-pyrido[2,3-*b*]indole	MeAαC	Soya globulin

amines. Reactions between these components are usually summarized under the term 'Maillard reaction'. They occur especially at higher temperature, low water activity and during long-term storage. Reactive sugars are glucose, fructose, maltose, lactose and, to a lesser extent, reducing pentoses. Reactive amino compounds are mainly primary amines, which are generally present in higher concentrations than secondary amines. An exception is, for example, malt, which contains high levels of proline. Proteins react mainly via the ε-amino groups of their lysine residues, but guanidino groups of arginine residues are also involved. These reactions may lead to:

(1) Brown pigments, which are called melanoidins. These contain nitrogen in varying amounts, and exhibit different molecular masses and different solubilities in water. Little is known about their structure. Their formation is desirable during baking and roasting, and undesirable in the case of foodstuffs which are only slightly coloured or have a distinct colour (evaporated milk, light-coloured dry soups, tomato soup).
(2) Volatile compounds, which often contribute to the aroma. Aroma formation via the Maillard reaction is desirable during cooking, baking and roasting. However, the development of off-flavours during storage of foodstuffs, especially in a dry state, or during pasteurization, sterilization and roasting is undesirable.
(3) Bitter compounds, which may be desirable (coffee), or may cause an off-flavour, e.g. in the case of grilled meat or fish.
(4) Reductones, which have strongly reducing properties, and contribute to the stabilization of foodstuffs against oxidative spoilage.
(5) Losses of essential amino acids (e.g. lysine, methionine).
(6) Mutagenic compounds.
(7) Cross-linking of proteins.

The Maillard reaction of the ε-amino group of lysine prevails in the presence of reducing sugars, e.g. lactose or glucose, which yield the Amadori compounds, namely protein-bound ε-*N*-desoxylactulosyl-1-lysine or ε-*N*-desoxyfructosyl-1-lysine, respectively. Lysine is not biologically available in these forms. Acidic hydrolysis of such primary reaction products yields lysine as well as the degradation products furosine and pyridosine in a constant ratio. A nonreducing sugar (e.g. sucrose) can also cause a loss of lysine when conditions for sugar hydrolysis are favourable, leading to the formation of the reactive sugars glucose and fructose. ***See*** Amines; Amino Acids, Properties and Occurrence; Browning, Nonenzymatic; Carbohydrates, Interactions with Other Food Components; Peptides

At the end of the 1970s it was shown that burnt parts of the surface of grilled fish and meat, as well as the condensates, isolated during grilling, exhibit strong mutagenic effects in microbial tests (*Salmonella typhimurium* strain TA 98). Model experiments showed that products of pyrolysis of amino acids and proteins were responsible for these effects. In Table 1 several mutagenic compounds (pyridoindoles, pyridoimidazoles and tetraazafluoranthenes) are listed, which were isolated from such pyrolysates.

Mutagenic compounds may be formed also at lower temperatures. The compounds listed in Table 2 were isolated from meat extract, deep fried meat, grilled fish, and model mixtures of creatinine, an amino acid (glycine, alanine or threonine) and glucose. The compounds, mainly imidazoloquinolines and imidazoloquinoxalines, are formed from creatinine, products of the Maillard reaction (pyridines, pyrimidines), and amino acids. Their toxicity depends on the heteroaromatic amino group. ***See*** Mutagens

Reactions with Lipid Oxidation Products

Products Formed From Hydroperoxides

Hydroperoxides formed thermally or enzymatically in food are usually degraded further. This degradation can

Table 2. Mutagenic compounds from various heated foodstuffs and model systems

Mutagenic compound	Abbreviation	Foodstuff/model system[a]
2-Amino-3-methylimidazo[4,5-*f*]chinoline	IQ	1,2,3
2-Amino-3,4-methylimidazo[4,5-*f*]chinoline	MeIQ	3
2-Amino-3,8-dimethylimidazo[4,5-*f*]chinoxaline	MeIQx	2,3
2-Amino-3,4,8-trimethylimidazo-[4,5-*f*]chinoxaline	4,8-DiMeIQx	2,3,5,6
2-Amino-3,7,8-trimethylimidazo-[4,5-*f*]chinoxaline	7,8-DiMeIQx	4
2-Amino-1-methyl-6-phenylimidazo[4,5-*b*]pyridine	PhIP	2

[a] 1, meat extract; 2, deep fried meat; 3, grilled fish; 4, model mixture of creatinine, glycine, glucose; 5, as 4, but with alanine; 6, as 4, but with threonine.

also be of a nonenzymatic nature. In nonspecific reactions involving heavy metal ions, haeme(in) compounds or proteins, hydroperoxides are transformed into oxo, epoxy, mono-, di- and trihydroxy acids. Unlike hydroperoxides, i.e. the primary products of autoxidation, some of these derivatives are characterized as having a bitter taste. Such compounds are detected in legumes and cereals. They may play a role in other foods rich in unsaturated fatty acids and proteins, such as fish and fish products.

Lipid–Protein Complexes

Studies of the interaction of hydroperoxides with proteins have shown that, in the absence of oxygen, linoleic acid 13-hydroperoxide reacts with *N*-acetylcysteine, yielding an adduct which is made up of several isomers. However, in the presence of oxygen, covalently bound amino acid–fatty acid adduct formation is significantly suppressed; instead, oxidized fatty acids are formed.

The difference in reaction products is explained by the different reaction pathways. The thiyl radical, derived from cysteine by abstraction of a hydrogen atom, is added to the epoxyallyllic radical only in the absence of oxygen. Otherwise, in the presence of oxygen, oxidation of cysteine to cysteine oxide and of fatty acids to their more oxidized forms both occur with a higher reaction rate than the previous reaction. As a consequence, a large portion of the oxidized lipid from a protein-containing food stored in air does have lipid–protein covalent bonds and, hence, is readily extracted with a lipid solvent such as chloroform/methanol (2:1, v/v). ***See*** Fatty Acids, Properties

Protein Changes

Some properties of proteins are changed when they react with hydroperoxides and their degradation products. This is reflected by changes in food texture, decreases in protein solubility (formation of cross-linked proteins), colour (browning) and changes in nutritive value (loss of essential amino acids).

The radicals generated from hydroperoxides can abstract hydrogen atoms from protein, preferentially from the amino acids tryptophan, lysine, tyrosine, arginine, histidine, cysteine and cystine, wherein the phenolic hydroxy-, sulphur- or nitrogen-containing group reacts.

Protein radicals combine with each other, resulting in the formation of a protein network. Malonaldehyde is generated under certain conditions during lipid peroxidation. As a bifunctional reagent, malonaldehyde can crosslink proteins through a Schiff base reaction with the ε-amino group of lysine.

The Schiff base adduct is a conjugated fluorochrome that has distinct spectral properties (λ_{max} excitation ~350 mm; λ_{max} emission 450 nm). Hence, it can be used for detecting lipid peroxidation and the reactions derived from it with the proteins present.

Reactions resulting in the formation of a protein network like that outlined above have practical implications, for example they are responsible for the decrease in solubility of fish protein during frozen storage. Also, the monocarbonyl compounds derived from autoxidation of unsaturated fatty acids readily condense with protein-free amino groups, forming Schiff bases that can provide brown polymers by repeated aldol condensations. The brown polymers are often nitrogen-free since the amino compound can be readily eliminated by hydrolysis. When hydrolysis occurs in the early stages of aldol condensations (after the first or second condensation) and the released aldehyde, which has a powerful odour, does not re-enter the reaction, the condensation process results not only in discoloration (browning) but also in a change in aroma.

Decomposition of Amino Acids

Studies of model systems have revealed that protein cleavage and degradation of side-chains, rather than the formation of protein networks, are the preferred reactions when the water content of protein–lipid mixtures decreases. Several examples of the extent of losses of amino acids in a protein in the presence of an oxidized

Table 3. Amino acid losses occurring in protein reaction with peroxidized lipids

Reaction system		Reaction conditions		
Protein	Lipid	Time	Temperature (°C)	Amino acid (loss, %)
Cytochrome *c*	Linolenic acid	5 h	37	His(59), Ser(55), Pro(53), Val(49), Arg(42), Met(38), Cys(35)[a]
Trypsin	Linoleic acid	40 min	37	Met(83), His(12)[a]
Lysozyme	Linoleic acid	8 days	37	Trp(56), His(42), Lys(17), Met(14), Arg(9)
Casein	Linoleic acid ethyl ester	4 days	60	Lys(50), Met(47), Ile(30), Phe(30), Arg(29), Asp(29), Gly(29), His(28), Thr(27), Ala(27), Tyr(27)[a,b]
Ovalbumin	Linoleic acid ethyl ester	24 h	55	Met(17), Ser(10), Lys(9), Ala(8), Leu(8)[a,b]

[a] Trp analysis was not performed.
[b] Cys analysis was omitted.

Table 4. Amino acid products formed in reaction with peroxidized lipid

Reaction system		Compounds formed from amino acids
Amino acid	Lipid	
Histidine	Methyl linoleate	Imidazolelactic acid, imidazoleacetic acid
Cystine	Ethyl arachidonate	Cystine, hydrogen sulphide, cysteic acid, alanine, cystine disulphoxide
Methionine	Methyl linoleate	Methionine sulphoxide
Lysine	Methyl linoleate	Diaminopentane, aspartic acid, glycine, alanine, α-aminoadipic acid, pipecolinic acid, 1,10-diamino-1,10-dicarboxydecane

lipid are presented in Table 3. The strong dependence of this loss on the nature of the protein and reaction conditions is obvious. Degradation products obtained in model systems of pure amino acids and oxidized lipids are described in Table 4.

Reactions under Alkaline Conditions

Losses of available lysine, cystine, serine, threonine, arginine and some other amino acids occur at high pH values. Hydrolysates of alkali-treated proteins often contain some unusual compounds, such as ornithine, β-aminoalanine, lysinoalanine, ornithinoalanine, lanthionine, methyllanthionine and D-alloisoleucine, as well as other D-amino acids. The formation of these compounds is based on the following reactions. 1,2-Elimination in the case of hydroxyamino acids and thioamino acids result in the formation of 2-aminoacrylic acid (dehydroalanine) and 2-aminocrotonic acid (dehydroaminobutyric acid), respectively. In the case of cystine, the eliminated thiolcysteine can form a second dehydroalanine residue. Alternatively, cleavage of the cystine disulphide bond can occur by nucleophilic attack on sulphur, yielding a dehydroalanine residue through thiol and sulphinate intermediates. Intra- and interchain cross-linking of proteins can occur in dehydroalanine reactions involving additions of amines and thiols. Ammonia may also react via an addition reaction. Acidic hydrolysis of such a cross-linked protein yields the unusual amino acids listed in Table 5. Ornithine is formed during cleavage of arginine. ***See*** pH – Principles and Measurement

Formation of D-amino acids occurs through abstraction of a proton via a C_2 carbanion. The reaction with L-isoleucine is particularly interesting. L-Isoleucine is isomerized to D-alloisoleucine, which, unlike other D-amino acids, is a diastereomer and so has a retention time different from L-isoleucine, making its determination possible directly by amino acid analysis.

Heating proteins in a dry state at neutral pH results in

Table 5. Formation of unusual amino acids by alkali treatment of protein

Name	Formula
3-N^6-Lysinoalanine (R = H) 3-N^6-Lysino-3-methylalanine (R = CH_3)	COOH | $CHNH_2$ | CHR—NH— $(CH_2)_4$—$CHNH_2$—COOH
3-N^5-Ornithinoalanine (R = H) 3-N^5-Ornithino-3-methylalanine (R = CH_3)	COOH | $CHNH_2$ | CHR—NH— $(CH_2)_3$—$CHNH_2$—COOH
Lanthionine (R = H) 3-Methyllanthionine (R = CH_3)	COOH COOH | | $CHNH_2$ $CHNH_2$ | | CHR —S — CH_2
3-Aminoalanine (R = H) 2,3-Diaminobutyric acid (R = CH_3)	COOH | $CHNH_2$ | $CHRNH_2$

the formation of isopeptide bonds between lysine residues and the β- or γ-carboxamide groups of asparagine and glutamine residues. These isopeptide bonds are cleaved during acidic hydrolysis of protein and, therefore, do not contribute to the occurrence of unusual amino acids. A more intensive heat treatment of proteins in the presence of water leads to more extensive degradation. The effect of alkaline treatment of a protein isolate of sunflower seeds shows that serine, threonine, arginine and isoleucine concentrations are markedly decreased with increasing concentrations of sodium hydroxide. New amino acids (ornithine and alloisoleucine) are formed. Initially, the lysine concentration decreases, but increases at higher concentrations of alkali. Lysinoalanine behaves in the opposite manner. The formation of lysinoalanine is influenced not only by pH but also by the protein source. An extensive reaction occurs in casein even at pH 5·0 due to the presence of phosphorylated serine residues, while noticeable reactions occur in gluten from wheat or in zein from corn only in the pH range 8–11. Table 6 lists the contents of lysinoalanine in food products processed industrially or prepared under 'usual household conditions'. The contents are obviously affected by the food type and by the processing conditions.

Reactions Under Oxidative Conditions

Oxidative changes in proteins involve methionine, which quite readily forms methionine sulphoxide. After *in vivo* reduction to methionine, protein-bound methionine sulphoxide is apparently biologically available. ***See*** Oxidation of Food Components

Bibliography

Aeschbacher HU (1986) Possible cancer risk of dietary heat reaction products. *Proceedings of Euro Food Tox II, Zürich*, pp 112–126.

Chen C, Pearson AM and Gray JI (1990) Meat mutagens. *Adv Food Nutr Research* 34: 387–449.

Eriksson C (ed.) (1981) *Maillard Reaction in Food. Chemical Physiological and Technological Aspects.* Oxford: Pergamon Press.

Fujimaki M, Namiki M and Kato H (eds) (1986) *Aminocarbonyl Reactions in Food and Biological Systems.* Amsterdam: Elsevier.

Gardner HW (1979) Lipid hydroperoxide reactivity with proteins and amino acids: a review. *J. Agric. Food Chem.* 27: 220–229.

Haagsma N and Slump P (1978) Evaluation of lysinoalanine determinations in food proteins. *Z. Lebensm. Unters. Forsch.* 167: 238–240.

Ledl F (1990) Chemical pathways of the Maillard reaction. In: Finot PA, Aeschbacher HU, Hurrell RF and Liardon R

Table 6. Lysinoalanine content of various foods

Food	Origin	Treatment	Lysinoalanine (mg kg^{-1} protein)
Frankfurter	CP[a]	Raw	0
		Cooked	50
		Roasted in oven	170
Chicken drums	CP	Raw	0
		Roasted in oven	110
		Roasted in microwave oven	200
Egg white, fluid	CP		15
Egg white		Boiled	
		(3 min)	140
		(10 min)	270
		(30 min)	370
		Baked	
		(10 min at 150°C)	350
		(30 min at 150°C)	1100
Dried egg white	CP		160–1820[b]
Condensed milk, sweetened	CP		360–540
Condensed milk, unsweetened	CP		590–860
Milk product for infants	CP		150–640
Infant food	CP		<55–150
Soya protein isolate	CP		0–370
Hydrolysed vegetable protein	CP		40–500
Cocoa powder	CP		130–190
Sodium caseinate	CP		45–560
Sodium caseinate	CP		430–6900
Calcium caseinate	CP		250–4320

[a] Commercial product.
[b] Variation range for different brand name products.

(eds) *The Maillard Reaction in Food Processing, Human Nutrition and Physiology*, p 19. Basel: Birkhäuser.

Masters PM and Friedman M (1979) Racemization of amino acids in alkali-treated food proteins. *J. Agric. Food Chem.* 27: 507–511.

Mauron J (1975) Ernährungsphysiologische Beurteilung bearbeiteter Eiweissstoffe. *Dtsch. Lebensm. Rundsch.* 71: 27–35.

Perkins EG (ed.) (1975) *Analysis of Lipids and Lipoproteins.* Champaign, IL: American Oil Chemists' Society.

Sternberg M and Kim CY (1977) Lysinoalanine formation in protein food ingredients. *Adv. Exp. Med. Biol.* 86B: 73–84.

Waller GR and Feather MS (eds) (1983) *The Maillard Reaction in Foods and Nutrition. ACS Symposium Series*, No. 215. Washington, DC: American Chemical Society.

Whitaker JR and Fujimaki M (eds) (1980) *Chemical Deterioration of Proteins. ACS Symposium Series*, No. 123. Washington, DC: American Chemical Society.

H.-D. Belitz
Technical University Munich, Garching, FRG

Quality

Quality

The quality of dietary protein is a measure of its usefulness in the body, i.e. to what extent the amino acids provided satisfy requirements.

This cannot be assessed by comparing the amino acid composition of the dietary protein with that of the body since the various tissues of the body are broken down and resynthesized at different rates. Thus the half-life (time taken to replace half of the protein) of the organs such as liver and heart is 10 days, whereas that of muscle is about 150 days and that of some enzymes is only a few hours or minutes.

Historically, protein quality (PQ) was being measured before all the amino acids had been discovered, so that it was many years before the two were correlated.

The earliest measures of PQ were in 1909, yet threonine was not discovered until 1935. It was not until

1948 that amino acid composition was correlated with PQ.

Until the early part of the nineteenth century, all proteins – then called albuminous substances – were considered to be the same, apart from their water content. In 1832 it was demonstrated that dogs could not survive when provided with gelatin as their sole source of protein, yet they flourished on meat. This was the first indication of differences between albuminous substances.

It was subsequently shown that gelatin is relatively deficient in tryptophan and methionine.

As far as satisfying the requirements of the body is concerned, the diet must supply enough of the essential amino acids, listed in Table 1, together with enough dietary nitrogen for synthesis of the nonessential amino acids (NEAAs). An essential amino acid (EAA) is defined as one that cannot be synthesized in the body or at least not in adequate amounts. Nonessential amino acids are not dietary essentials since they can be synthesized from EAA. *See* Amino Acids, Metabolism

In practice, with very few exceptions, dietary proteins contain a mixture of all 20 amino acids but in varying proportions, so that NEAAs are provided ready-made.

The quality of a protein is limited by that EAA present in least amount relative to requirements; occasionally there may be more than one with the same shortfall. Hence all measurements of PQ depend on the limiting EAA.

Methods in Use

There are two main methods of measuring PQ in experimental animals. One is based on the balance between the intake and output of nitrogen and the second is based on growth. Numerous modifications and indicators of quality have been suggested from time to time, including the following:

1. Rapid biological methods: growth of protozoan, *Tetrahymena* spp.; growth of beetles, *Tribolium* spp.; growth of bacteria, *Escherichia coli*, *Leuonostoc* spp.
2. Specific response: regeneration of plasma protein; regeneration of liver protein, liver enzymes.
3. Abbreviated indices: plasma amino acid index; blood urea concentration; urinary creatinine divided by total urine nitrogen; short-term nitrogen retention; pepsin-digest-residue.
4. Other methods: formol titration; protein quality index; dye-binding capacity; egg replacement value.

Table 1. Amino acids

Essential	Nonessential
Leucine	Alanine
Isoleucine	Glycine
Lysine	Serine
Methionine	Proline
Valine	Hydroxyproline
Threonine	Cystine
Tryptophan	Tyrosine
Phenylalanine	Glutamic acid
Plus, for children,	Aspartic acid
Arginine	Cysteine
Histidine	

Biological Value Determination

The biological value (BV) of a protein is a measure of its usefulness to the body and is measured as the proportion of the absorbed nitrogen that is retained in the body (R/A).

Protein is the only major dietary constituent that contains nitrogen (there is nitrogen in some of the vitamins and in minor constituents of foods); its progress through the body can therefore be traced by measuring intake and output of nitrogen in the faeces and the urine.

Faecal nitrogen is derived from undigested and unabsorbed dietary protein and its breakdown products and from nondietary sources. These are residues of unabsorbed digestive enzymes, bacteria and the cells shed regularly from the lining of the intestinal mucosa. Hence faecal nitrogen is not a correct measure of undigested protein.

If the faecal nitrogen (designated F) is subtracted from the nitrogen intake (I) and expressed as a proportion of the intake, $(I-F)/I$, the result is termed apparent digestibility. Only when allowance is made for metabolic faecal nitrogen (M) is the measure that of true digestibility, $[I-(F-M)]/I$.

This is expressed as a ratio but was formerly expressed as a percentage. Thus 0·6 is equal to 60% digestibility.

After the products of protein hydrolysis, i.e. the amino acids, have been absorbed they participate in various metabolic reactions in the body, including synthesis of protein tissues, synthesis of enzymes and hormones, etc. If the amino acids derived from the diet are not in the correct proportions for these purposes, then the unusable part is converted to urea and excreted in the urine. There are, however, other nitrogen compounds excreted in the urine which are the end products of various metabolic processes and are not the consequence of unusable dietary protein; these are creatine, creatinine, uric acid, ammonium salts, allantoin and traces of other compounds. This is analogous to the metabolic faecal nitrogen but is termed endogenous urinary nitrogen (arising from internal sources; exogenous nitrogen arises from dietary sources).

Thus urine nitrogen (U) must be corrected for

endogenous nitrogen (E) to determine how much absorbed nitrogen could not be used.

We therefore have two measures:

$$\text{Absorption, } A = I - (F - M)$$
$$\text{Retention, } R = I - (F - M) - (U - E)$$

The proportion of absorbed nitrogen that is retained for use is given by the following equation:

$$R/A = \frac{I - (F - M) - (U - E)}{I - (F - M)}$$

This is defined as biological value.

Biological value is a measure of the usefulness of that part of the dietary protein that has been digested and absorbed. If digestibility (D) is taken into account, i.e. if the measure is retention of ingested instead of absorbed protein, it is termed net protein utilization (NPU), which is equal to BV multiplied by D. At one time BV was measured on a percentage scale but in the SI system is given as a ratio.

The metabolic and endogenous nitrogen excretion was measured by feeding the animals a protein-free diet and assuming that the nitrogen excreted on this diet was the same as M and E when the experimental animals were subsequently fed on the protein diet under test. However, it is not certain that endogenous losses are the same on a protein diet as on a nonprotein diet. Furthermore, animals fed on a protein-free diet for several days (and several days are needed for an accurate measurement) lose weight and protein tissue.

This was rectified by feeding a low level of a perfect protein, i.e. one that, when fed at a low level, is completely used by the body and makes no contribution to the nitrogen excreted in the urine or the faeces. This consisted of egg protein fed at 4% of the weight of the diet while the test protein is fed at 10%.

The principle of the method is to measure how much of the dietary protein is usable. There is clearly a limit to the amount of protein that the body can use even if its amino acid composition matches requirements perfectly. The maximum growth rate (which determines the maximum amount of protein that can be used) of a young growing animal requires 15–20% protein in the diet but the efficiency of utilization decreases with increasing levels of protein. Accordingly the test is carried out at 10% dietary protein level so that it is efficiently used and the use limited only by the amount of the limiting EAA.

The procedure consists of feeding 12 rats on 4% egg protein diet for 10 days – 3 days to standardize the excretion and 7 days collection of urine and faeces to measure M and E. During the subsequent three 10-day periods three proteins can be tested. This is followed by a final period on 4% egg protein diet because endogenous output of nitrogen changes as the animals grow. The endogenous level corresponding to each of the test periods is derived by interpolation between the first and final egg periods. Each animal is treated separately, so that there are 12 measurements for each protein.

This procedure takes 50 days and numerous analyses of urine and faeces to determine the quality of three proteins. The method was abbreviated (carcass analysis method) by measuring the protein in the body directly rather than by the difference between intake and output. This measures NPU and also D, from which BV can be calculated.

Using four litters, each of eight rats, the procedure is to divide these into eight groups, each of four littermates. One group is fed the nonprotein diet for 10 days to determine endogenous losses and each of the seven groups is fed one of the seven test proteins.

So seven proteins can be assayed in 10 days (plus a duplicate set of analyses taking another 10 days).

Instead of analysing each carcass for total nitrogen, it is sufficient simply to determine body water since for a given colony, after the age of about 40 days, there is a constant relation between body water and body nitrogen. Each group of four is combined for feeding and analysis.

Since this method is so much shorter than the original it has become widely used.

Protein Efficiency Ratio (PER)

The other main method of measuring PQ is based on growth rate. Since the rate of growth will depend on the amount of food eaten, including protein, PER is measured as the rate of growth per gram of protein eaten. It ranges from zero to 4·4.

Originally, the test protein was fed at several levels and the highest PER value obtained was noted. The method was standardized by the Association of Official Agricultural Chemists of USA (AOAC) at 9% protein diet fed for 28 days. The results are standardized by comparison with a group fed casein at the same time, and the PER is corrected to a value of 2·3 for casein.

However, there are acknowledged drawbacks to the method. First, it assumes that the weight gain is all protein tissue, whereas the amount of fat laid down can vary to some extent with different foods. Second, the result varies with the amount of protein, i.e. total food, eaten. Third, proteins with BV less than 40, admittedly uncommon, cannot be measured since this corresponds to a PER of zero.

Other Methods

The animal assays are expensive, tedious and provide variable results and many attempts have been made to develop alternative procedures. Some of these have been listed above but few have been much used.

Amino Acids

The limitation on PQ is the amount of the EAA present in least amount relative to needs, i.e. the limiting amino acid.

The amount of the limiting amino acid was compared with the amount of the same amino acid in egg protein – the 'perfect' protein – and termed 'chemical score'. It was chemical score that correlated with BV.

Instead of egg protein the Food and Agriculture Organization (FAO) used the human amino acid requirements – referred to as the Provisional Amino Acid Pattern – for comparision with the test protein and termed the value 'protein score'.

Another method of using EAA composition is by calculating the geometric mean of all the EAAs compared with egg and termed the 'EAA index'.

These analytical methods have advantages over the relatively imprecise and laborious bioassays but there are two major problems.

First, part of some of the amino acids can combine with other ingredients of a food and within the protein itself to form bonds that are not hydrolysed by the digestive enzymes. However, these bonds are hydrolysed by the acid treatment that precedes chemical analysis and so leads to false information.

As far as lysine – the principal amino acid that can be rendered partly unavailable – is concerned, there are chemical tests for availability, although they are not always applicable to all foods. Little is known about the chemical reactions that render other EAAs unavailable, and no useful method of measuring these has been developed.

The second problem is that at low levels of PQ there is not a straight-line relationship between limiting EAA and BV. ***See*** Amino Acids, Determination

Complementation

Table 2 lists the BVs of a number of proteins. Those with the higher values are mostly animal proteins and those of lower values mostly plant proteins. At one time this gave rise to a division into first-class (animal) and second-class (plant) proteins but the term has been discarded since a mixture of proteins often has a higher BV than the average of the components. This is because a surplus of an EAA in one protein may be able to compensate for a shortfall in the limiting EAA of another protein. This is termed complementation.

An example is provided by a mixture of pea flour and maize meal. The BV of the former is 0·43 and is limited by methionine, with a surplus of lysine. Maize meal protein BV is 0·35, limited by lysine with a surplus of methionine. The BV of an equal mixture of these proteins should be the mean of the two BVs, i.e. 0·39. In practice, it is found to be 0·7. This means that the BV of single protein sources is not a matter of practical importance. It may be of value in measuring processing changes such as damage to a limiting amino acid or the liberation of a bound form; otherwise, the BV of individual protein sources is of limited practical value.

Table 2. Protein quality

Individual foods	Biological value	Protein efficiency ratio	Chemical score
Human milk	0·95–1·0	4·5	100
Hen's egg	0·95–1·0	4·5	100
Cows' milk	0·75	3·0	100
Fish	0·75	3·0	100
Meat	0·75	3·0	100
Soya beans (heat-treated)	0·7	2·5	63
Casein	0·7	2·5	60
Rice, white	0·6	1·0	60
Wheat grain	0·6	1·0	60
Bread, white	0·5	0·5	35
Peas and beans	0·35–0·50	0–0·5	25–40
Maize (corn)	0·4	0	40
Gelatin	0	—	0
Complete diets			
Industrialized countries (based on meat, milk, wheat)	0·8		
Some developing countries (based largely on cereals)	0·7		
Some developing countries (based on cassava, etc.)	0·6		

This is well illustrated by the BV of whole diets. In developing countries, where the diet is based largely on a single cereal, the BV is usually about 0·7 and rarely as low as 0·6. In industrialized countries, where the protein sources are largely cereal, milk and meat, it is not higher than about 0·8.

In general, cereal proteins are limited by lysine which is in surplus in milk and meat products; hence the last two foods serve to complement the cereals.

Except for young children the quality of the protein of the diet is rarely a matter of major importance. This is because the amount of protein in most diets usually exceeds the minimum requirements (as long as enough food is available to satisfy the appetite) and quantity compensates for quality. The requirements for protein are around 6–7% of the total energy intake and even cereals contain between 8% and 12% protein.

Protein quality becomes a matter of importance when the diet is inadequate in energy or the staple food is low in protein and that protein is of low quality.

The situation becomes serious in infants weaned from

breast milk, when the quality and quantity of the dietary protein is of major importance.

Recommended Method of Measuring Protein Quality

The many methods partly listed earlier in this article illustrate the perceived need to shorten the lengthy and expensive classical methods. However, the main drawback to methods involving experimental rats is that the results are not directly applicable to human beings. The FAO (1990) recommended a method consisting of amino acid scores corrected for true digestibility and/or bioavailability.

Amino acid score is calculated from a determination of the limiting EAA after hydrolysis compared with the amino acid requirements of a child aged 2–5 years, as suggested by the World Health Organization (WHO, 1985). Children of this age require 30% of their protein intake as EAA, while adults require only 15% but this amino acid pattern is intended for all ages except infants. When applied to adults the pattern ensures enough EAA to allow for any uncertainty in the figures of adult EAA needs – pending further research. (The WHO report of 1985 used different scoring patterns for different age groups.) For infants under 1 year of age the amino acid pattern of human milk is used as the standard.

The scores are corrected for true digestibility determined on rats or by *in vitro* enzymatic methods or, in the instance of established protein foods, taken from food composition tables.

Bibliography

Bender AE (1982) Nutritional value of proteins and its assessment. In: Fox PF and Condon JJ (eds) *Food Proteins*. Essex: Elsevier Science Publishers.

Bodwell CE (ed.) (1977) *Evaluation of Proteins for Humans*. Westport, Connecticut: AVI Publishing.

FAO (1990) *Report of the Joint FAO/WHO Expert Consultation on Protein Quality Evaluation*. Rome: Food and Agriculture Organization.

Pellet PL and Young VR (1980) *Nutritional Evaluation of Protein Foods*. UNU Food and Nutrition Bulletin Supplement. Tokyo, Japan: United Nations University.

Porter JWG and Rolls BA (eds) (1979) *Proteins in Human Nutrition*. New York: Academic Press.

WHO (1985) Energy and protein requirements. Report of the Joint FAO/WHO/UNU Expert Consultation. *WHO Technical Report Series* 724. Geneva: World Health Organization.

A. E. Bender
Leatherhead, Surrey, UK

Digestion and Absorption of Protein and Nitrogen Balance

The assimilation of protein into the body has classically been viewed as involving digestion of protein to amino acids in the gut, absorption of these amino acids through the gut mucosa, and transport of amino acids in the plasma for subsequent metabolism, including protein synthesis, in other tissues. However, it is now clear that substantial quantities of peptides containing up to five amino acid residues can also be absorbed into the intestinal mucosa. Here they are largely hydrolysed to free amino acids before being released into the circulation, so that the circulating concentration of peptides is normally very low, although many other tissues are capable of taking up and utilizing peptides. On the other hand, whole proteins are not normally absorbed from the gut in measurable amounts.

Digestion

Protein digestion begins in the stomach with the action of pepsins, hydrolytic enzymes which preferentially cleave peptide bonds adjacent to phenylalanine, tyrosine or leucine residues. The human stomach secretes at least three distinct pepsins with different pH optima ranging from almost neutral, corresponding to the condition of the stomach contents immediately after a meal, to pH 1·2, as would be found several hours later.

Further digestion takes place in the small intestine, under the influence of enzymes which originate in the acinar cells of the exocrine pancreas. The specificity of these enzymes is listed in Table 1.

Both pepsin and the pancreatic proteolytic enzymes are synthesized in the form of inactive proenzymes (also termed zymogens) containing one or more extra peptide segments which prevent the formation of the three-dimensional structure that confers catalytic activity. The extra peptide of pepsinogen (containing 42 amino acids) is hydrolysed by hydrochloric acid in the stomach. Trypsinogen is activated by the removal of a hexapeptide from the amino terminal by enteropeptidase (formerly known as enterokinase), an enzyme secreted by the small intestine. Trypsin in turn activates all the pancreatic proenzymes, including itself.

The final stage of protein digestion is accomplished by a wide array of aminopeptidases and di- and tripeptidases manufactured in the absorptive cells of the intestinal epithelium. Some of these act in the lumen of the gut, but most of the activity is located within the brush border membrane or within the cytoplasm of the cells.

Absorption

Absorption of the products of digestion involves active transport into the enterocyte, metabolism within the cell and diffusion down a concentration gradient into the portal circulation. Absorption occurs throughout the length of the small intestine, but is confined to the cells in the top third of the villus.

Amino acids and peptides are transported into cells by a number of active transport systems, most of which are linked to the simultaneous entry of sodium ions. Sodium enters the cell down a chemical concentration gradient; this is maintained by the membrane 'pump' which exchanges sodium for potassium ions and is fuelled by the hydrolysis of ATP. The specificity of the amino acid transport systems in the gut is broadly similar to that of the systems which are found in other tissues, particularly the brush border of the kidney tubule, and this is summarized in Table 2. Evidence from studies *in vitro* suggests that small peptides may be transported more rapidly than free amino acids, but these studies have often not been carried out under the conditions which occur *in vivo* after a normal meal.

After transport across the intestinal wall amino acids can be degraded, metabolized to other compounds, including other amino acids, incorporated into proteins, or released unaltered into the circulation. The rate of protein synthesis in the small intestinal mucosa is amongst the highest of any tissue in the body. This is no doubt related to the high rate of cell turnover and the requirement for secretion of digestive enzymes. These cells also utilize energy at a high rate, and the amino acid glutamine, supplied both from the lumen of the gut and from the circulation, is a major fuel. Some of the ammonia which is formed during the utilization of glutamine is released into the gut, from where it may subsequently be reabsorbed and transported to other tissues to be used for synthesis of nonessential amino acids. ***See*** Amino Acids, Metabolism

Most of the peptides which are absorbed into the mucosal cells are hydrolysed to single amino acids before being released into the portal circulation. However, some intact peptides have also been observed to enter the portal circulation, and this has raised some interest because of the possibility of peptides with hormonal activity being absorbed intact.

The circulating concentration of peptides is normally very low. This may suggest that only small quantities of peptides are released into the portal circulation. However, it has also been observed that when peptides are infused directly into the circulation they are rapidly metabolized, probably extracellularly, since the plasma concentration of the free amino acid constituents of the peptide rises almost immediately. This means that glutamine, an amino acid which is notoriously unstable during storage in solution, can be supplied in parenteral nutrition mixtures in the form of stable peptides which are readily available to the tissues.

Amino Acid Pools

Amino acids are transported in the plasma and may be taken up into all tissues by mechanisms based on the systems referred to in Table 2. Amino acids are also present in blood cells, but the interaction of these amino acids with other tissues may be quite different from that of plasma amino acids. For example, the plasma amino acid concentration increases as blood traverses the gastrointestinal tract after a meal, whereas the amino acid content of blood cells actually decreases.

The intracellular free amino acid content varies between different tissues, and is affected by hormonal and nutritional influences. However, the intracellular concentration is greater than the plasma concentration for most amino acids in most tissues. The distribution ratio (intracellular concentration divided by plasma concentration) is close to unity for many of the essential

Table 1. Pancreatic proteolytic enzymes

Enzyme	Proenzyme	Substrate	Catalytic function
Trypsin	Trypsinogen	Proteins and polypeptides	Cleaves peptide bonds adjacent to arginine or lysine
Chymotrypsins	Chymotrypsinogens	Proteins and polypeptides	Cleave peptide bonds between aromatic and neutral or basic amino acids
Elastase	Proelastase	Elastin and some similar proteins	Cleaves peptide bonds adjacent to glycine, alanine or serine
Carboxypeptidase A	Procarboxypeptidase A	Proteins and polypeptides	Cleaves carboxy-terminal amino acids with aromatic or branched side-chains
Carboxypeptidase B	Procarboxypeptidase B	Proteins and polypeptides	Cleaves carboxy-terminal amino acids with basic side-chains

Table 2. Amino acid transport systems

System	Sodium dependence	Preferred amino acids
A	Yes	Alanine, glycine, proline, serine, methionine
L	No	Leucine, isoleucine, valine, phenylalanine, tyrosine, tryptophan, histidine, methionine
ASCP	Yes	Alanine, serine, cysteine, proline
Ly^+	No	Lysine, histidine, arginine, ornithine
Dicarboxylate	Yes	Glutamate, aspartate
Beta	Yes	β-Alanine, taurine
N	Yes	Histidine, glutamine, asparagine

amino acids, particularly in muscle, but is several-fold higher for some of the nonessential acids, with a value as high as 70 having been reported for aspartic acid in liver. The interpretation of changes in amino acid concentrations in particular situations is complex. Changes in concentration may result from changes in outflow to or inflow from other tissues, or pools of other metabolites in the same or different tissues, including utilization for protein synthesis and supply from protein degradation.

Only about 2% of the amino acids in the body are present as free amino acids, the rest being present as protein. Again there is considerable variation between different amino acids and between different tissues in the ratio of the free to protein-bound amino acids, with much lower values generally being found for essential than nonessential amino acids. Given the currently accepted values for rates of protein synthesis it can be shown that in some tissues the entire free pool of some essential amino acids is incorporated into protein within seconds, emphasizing the rapidity with which free amino acids must be resupplied.

Nitrogen Balance

The state of protein metabolism in the body as a whole may be assessed by measuring nitrogen balance. This is because almost all the nitrogen in the body is in the form of protein, and the nitrogen content of a wide range of proteins is relatively constant at around 16%. Thus if the body is in positive nitrogen balance it must be laying down new protein, and this is normally associated with growth in children. Conversely, if the body is in negative nitrogen balance it must be losing tissue protein, either because the diet is inadequate or because of a pathological response to injury or illness. Nitrogen balance in an individual will fluctuate from day to day by a few grams either side of zero, but over a period of weeks a nongrowing adult will normally be in zero nitrogen balance.

Every protein in the body has a specific functional role. These roles include the catalytic function of enzymes, the contractile function of muscle proteins, the structural role of connective tissue proteins, roles in transport, immunological recognition, hormone receptors, and many others. There is no inert store of protein to be drawn on in times of need, as exists for fat (adipose tissue triglyceride) and carbohydrate (glycogen); any sustained period of negative nitrogen balance is likely to have functional consequences. Nevertheless, different tissues do lose proteins at different rates when the body goes into negative nitrogen balance, with muscle protein (including the smooth muscle of the gut) being lost in the early stages while the liver and the heart are relatively protected.

Measurement of Nitrogen Balance

Nitrogen balance is defined as the difference between intake and output, and may be formally represented by the following equation:

$$\text{Balance} = I - (U + F + M)$$

I is nitrogen intake, U is urinary nitrogen excretion, F is faecal nitrogen excretion, and M is the sum of all the other routes by which nitrogen is lost from the body.

Nitrogen intake is mainly in the form of protein, and can conveniently be measured by the Kjeldahl method, which measures virtually all nitrogen except that bonded to oxygen (e.g. nitrates and nitrites). The same method is thus also appropriate for measurement of the various components of nitrogen excretion. Accurate determination of nitrogen balance requires direct analysis of duplicate portions of diet, although a reasonable approximation can be made using a weighed food record and tables of food composition. An alternative approach is to feed the subjects a constant amount of a diet which has been specially formulated using known quantities of ingredients.

The urine is the major route for excretion of nitrogen, so that urinary nitrogen is the major determinant of nitrogen balance and reliable collection of 24-h urine specimens is required for measuring nitrogen balance. Most of the nitrogen in urine is in the form of urea, which is the end product of amino acid oxidation. Other nitrogenous compounds in the urine include ammonia, creatinine and uric acid, as well as small quantities of peptides, amino acids and other small molecules. When protein intake is low the proportion of urea in the urine will decrease, so that urinary urea content is not a reliable indicator of total urinary nitrogen.

Faecal nitrogen is composed mainly of bacterial cells from the large intestine, together with mucosal cells which have been sloughed off from the intestinal wall, and some remnants of undigested food proteins. On a normal mixed diet faecal nitrogen amounts to a fairly constant 8% of nitrogen intake, although on very-high-fibre diets and particularly those containing large quantities of legumes this proportion may increase. In the measurement of nitrogen balance it is conventional to use a non-absorbable marker to mark the beginning and the end of the faecal collection, or to use a continuous faecal marker.

The other routes by which nitrogen is lost from the body include the shedding of hair, nails and dead skin cells, sweat, saliva, semen, blood lost during menstruation or removed for clinical testing, and the ammonia exhaled in the breath. Dermal losses vary with the rate of sweating and with protein intake. Many of these losses are small and hard to measure, so that a single figure of 0·5 g of nitrogen per day may be taken to approximate the sum of these miscellaneous routes.

The most important use of nitrogen balance is in evaluating the adequacy of dietary intake. It is the major criterion which is used to assess quantitative protein requirements. Such studies have demonstrated that energy intake is also a major determinant of nitrogen balance. Measurement of nitrogen balance is also the basis of many biological methods for measuring protein quality.

Amino Acid Imbalance and Antagonism

It is normally the case that once the minimum requirements for total nitrogen and for each essential amino acid have been met, any surplus protein intake has no deleterious effect on nitrogen balance. However, there are reports of the addition of a very large quantity of a single essential amino acid causing a reduction in the biological value of an otherwise adequate protein. This has been called amino acid imbalance, and appears to occur only when the protein intake is marginally adequate. It can thus be overcome by increasing the intake either of total protein or of the limiting amino acid.

The related phenomenon of amino acid antagonism refers to a situation in which the addition of one amino acid affects the efficiency of utilization of another related amino acid. For example, the addition of one of the branched-chain amino acids (leucine, isoleucine or valine) will impair the utilization of dietary protein, and this can be overcome by adding the other two branched-chain amino acids. It is tempting to speculate that this is because of competition for membrane transport, or because of increased activity of the common pathway by which these amino acids are oxidized, but no such mechanism has yet been confirmed.

Bibliography

Alpers DH (1986) Uptake and fate of absorbed amino acids and peptides in the mammalian intestine. *Federation Proceedings* 45: 2261–2267.

Bender DA (1985) *Amino Acid Metabolism*, 2nd edn. Chichester: John Wiley.

Calloway DH, Odell ACF and Margen S (1971) Sweat and miscellaneous nitrogen losses in human balance studies. *Journal of Nutrition* 101: 775–786.

Gardner MLG (1984) Intestinal assimilation of intact peptides and proteins from the diet – a neglected field? *Biological Reviews* 59: 289–331.

Waterlow JC, Garlick PJ and Millward DJ (1978) *Protein Turnover in Mammalian Tissues and in the Whole Body*. Amsterdam: North Holland.

Webb KE (1986) Amino acid and peptide absorption from the gastrointestinal tract. *Federation Proceedings* 45: 2268–2271.

Yudilevitch DL and Boyd CAR (eds) (1987) *Amino Acid Transport in Animal Cells*. Manchester: Manchester University Press.

Peter W. Emery
King's College London

Synthesis and Turnover

It is self-evident that growth must involve the synthesis of new protein. Equally, there are situations, usually associated with illness or undernutrition, when the organism loses protein, indicating that a mechanism must also exist for degrading tissue proteins. However, not until the 1940s was it demonstrated that the processes of protein synthesis and degradation occur continuously even in nongrowing adult animals which are neither gaining nor losing weight. Tyrosine labelled with the stable isotope, ^{15}N, was fed to an adult rat, and after 10 days only half the label had been excreted in the urine while the rest had been incorporated into tissue proteins. Protein turnover includes both the synthesis of protein from amino acids and the breakdown of protein to amino acids. Much experimental work over the last 40 years has been devoted to measuring the rates at which these two processes occur and identifying the factors which modulate them. It is now estimated that adult man synthesizes and degrades about 300 g of protein each day, which is three or four times as much as is consumed in the diet.

Mechanism of Protein Synthesis

The major biochemical pathways by which protein is synthesized in cells are now well established. Information specifying the primary structure is carried in the sequence of base pairs in the DNA molecules in the nucleus. This information is transcribed by the formation of short-lived molecules of messenger ribonucleic acid (mRNA) with base sequences complementary to those of the deoxyribonucleic acid (DNA). The mRNA moves out of the nucleus into the cytosol, where it associates with ribosomes to allow translation of the message as a sequence of amino acids. Every three bases on the mRNA specify a single amino acid which then binds to the ribosome in association with a specific molecule of transfer RNA (tRNA). A peptide bond is formed to join the new amino acid to the growing polypeptide chain, and the deacylated tRNA is released from the ribosome. ***See*** Nucleic Acids, Physiology

Even when the completed polypeptide has been released from the ribosome it may undergo considerable post-translational modification. Some amino acid residues may be hydroxylated or methylated, while others may form covalent links with other parts of the molecule as the protein adopts a stable three-dimensional structure. Nonprotein molecules, particularly carbohydrates, may be added. Finally, parts of the protein may be removed altogether, possibly after being transported to a different compartment of the cell, to allow activation of an inactive precursor or to facilitate secretion of export proteins.

Regulation of the rate of protein synthesis within a given cell occurs at two levels. There is *specific* control of the types of protein synthesized, which occurs mainly by controlling the transcription of particular genes. There is also *general* control of the total amount of protein synthesized, which occurs mainly at the level of translation. This includes regulation of the number of ribosomes present in the cell, and hence its capacity for protein synthesis, as well as regulation of the efficiency of translation, which appears to be achieved largely by changes in the rate of initiation of peptide synthesis, although elongation and termination may also be modulated under some circumstances.

Protein Degradation

The mechanisms by which proteins are broken down are much less well understood. A large number of proteolytic enzymes have been identified, with differing substrate specificities and conditions for optimal function. This may reflect the complexity and variety of the substrates on which they operate, and may also indicate a degree of cooperative activity, since once a protein is committed to degradation it appears to be broken down to its constituent amino acids very rapidly.

Many of the proteolytic enzymes appear to operate within a specific organelle, the lysosome. This provides the acidic environment in which enzymes such as the cathepsins are optimally active. Lysosomes can engulf and degrade large structures such as whole mitochondria, but individual soluble proteins can also enter the lysosome from the cytosol.

Lysosomes are believed to be responsible for the majority of protein degradation, particularly in tissues such as the liver where the overall rate of protein turnover is high. However, there are also a number of proteinases which operate in the cytosol, and are active at neutral or even alkaline pH. These include the following: the calpains, which are activated by calcium ions and may be particularly active in muscle; a large multicatalytic proteinase sometimes called a proteosome; and a series of enzymes which commit proteins to degradation by binding them to a 76-amino-acid polypeptide called ubiquitin.

There also exist extracellular proteinases which degrade specific extracellular proteins such as collagen.

The control of protein degradation may be exercised at both a general and a specific level, as with protein synthesis. Some of the factors which affect the rate of protein degradation are intrinsic to the protein substrate itself, so that, for example, increasing size and acidic nature favour more rapid degradation. At least some protein degradation appears to require energy. There also appears to be a requirement for continuing protein synthesis, although this may simply reflect the fact that the enzymes which degrade proteins are, of course, proteins themselves. However, this association may at least partly explain the observation that rates of protein synthesis and degradation often change in the same direction in response to external influences.

Measurement of Protein Synthesis and Breakdown

Having identified the pathways by which proteins are synthesized and degraded it is desirable to measure the rates at which these processes occur. The ultimate aim is to be able to measure the turnover rates of individual proteins, but for technical reasons most current methods are only able to give reliable data for turnover rates in whole tissues or even the whole body. However, such data do provide a useful insight into the mechanisms by which nutritionally significant quantities of protein are gained or lost, and such measurements are being made increasingly frequently in both clinical and agricultural research.

One lesson which has been learned is that rates of protein synthesis and degradation measured in most *in vitro* preparations are often poor indicators of what happens *in vivo*. This is particularly true of tissues such as muscle, where factors such as tissue integrity, stretch and electrical stimulation appear to be as important as nutrient supply, oxygen delivery, hormone concentrations and locally produced growth factors in determining the rates of protein turnover.

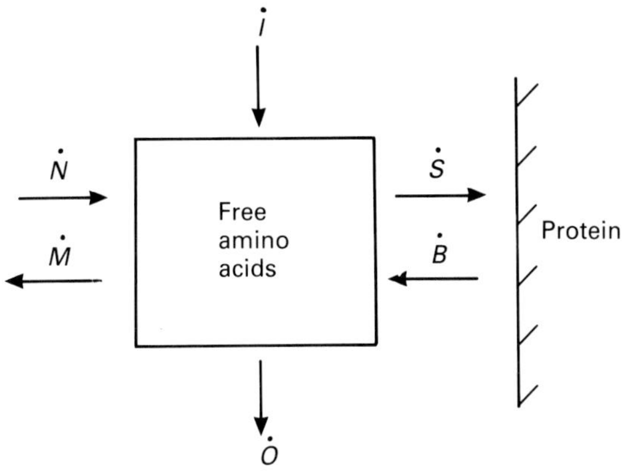

Fig. 1 Two-pool model for investigating whole-body protein turnover. $\dot{I}$, intake of amino acids from dietary protein; $\dot{S}$, protein synthesis; $\dot{B}$, protein breakdown; $\dot{O}$, amino acid oxidation; $\dot{N}$, synthesis of nonessential amino acids; $\dot{M}$, metabolism to other compounds.

Measurement of Whole-body Protein Turnover

The measurement of protein turnover necessitates a model which is simple enough to analyse fully but can be reasonably interpreted in terms of what happens in a living animal or person. The model used most commonly is the two-pool model shown in Fig. 1. All the protein in the body is represented by one pool, while the other pool represents the total amount of one particular free amino acid (both intracellular and extracellular). The flux of that amino acid ($\dot{Q}$) is defined as the sum of all the processes by which the amino acid leaves the free pool:

$$\dot{Q} = \dot{O} + \dot{S} + \dot{M}$$

In this equation $\dot{O}$ is amino acid oxidation, $\dot{S}$ is protein synthesis, and $\dot{M}$ is metabolism to other compounds. Measurements are normally made in a steady state, when the amino acid pool is neither expanding nor contracting, so that flux is also equal to the sum of all the processes by which the amino acid enters the pool:

$$\dot{Q} = \dot{I} + \dot{B} + \dot{N}$$

$\dot{I}$ is intake of amino acids from dietary protein, $\dot{B}$ is protein breakdown, and $\dot{N}$ is synthesis of nonessential amino acids. Judicious choice of amino acid allows these equations to be simplified: for essential amino acids, $\dot{N} = 0$, and several of the amino acids have no metabolic fates other than oxidation so that $\dot{M} = 0$. Dietary intake can be controlled during the period of measurement, and is often set to zero. Thus, if the rates of flux and oxidation can be measured, the rates of incorporation of amino acid into protein (i.e. protein synthesis) and release from protein (i.e. protein degradation) can be calculated. Knowledge of the proportion of that amino acid in tissue proteins then allows the overall rates of protein synthesis and degradation to be calculated. *See* Amino Acids, Metabolism

There are two commonly used methods for measuring amino acid flux. One involves the administration of a single dose (d) of amino acid (often glycine) labelled with ^{15}N and measurement of the cumulative urinary excretion of label in an end product of amino acid metabolism, usually either urea or ammonia, over a period of up to 60 h. The flux is calculated from the following equation:

$$\dot{Q} = d(E/e)$$

In this equation e is the cumulative excretion of the labelled end product and E is the total excretion rate of that end product. Oxidation rate can be estimated from the rate of excretion of urea nitrogen expressed as a proportion of the nitrogen flux. This method is noninvasive and therefore particularly suitable for certain clinical situations. However, from a theoretical point of view it has major disadvantages in that the long period of collection necessary to obtain a reasonably complete collection of label allows some isotope to re-enter the free amino acid pool after incorporation into rapidly turning over proteins. In addition, the fact that very different values for flux are obtained using different end products even in the same experiment indicates that the end products are not derived from the same pool as the precursors for protein synthesis. This therefore casts doubt on the validity of the method, particularly when comparing different clinical and metabolic conditions.

The other method for measuring protein turnover is to give a constant intravenous infusion of labelled amino acid (at a rate i) and to measure the labelling of the amino acid in the plasma when this reaches a steady state value (E_{max}), at which time isotope is entering and leaving the plasma pool at the same rate. The flux is then calculated from

$$\dot{Q} = i/E_{max}$$

E_{max} is expressed as the ratio of plasma labelling to infusate labelling. The most commonly used amino acid is leucine labelled with ^{13}C in the carboxyl group so that the oxidation rate can be obtained by measuring the rate at which ^{13}C-labelled carbon dioxide ($^{13}CO_2$) is excreted in the breath. By giving priming doses of labelled amino acid and bicarbonate, steady-state labelling can be reached very quickly and the whole procedure can be completed in 4 h, although the need to take repeated blood samples is considered a practical disadvantage. The major theoretical limitation is that the labelling of amino acids in the plasma may not be the same as that at the site of protein synthesis, which is likely to be intracellular and will thus vary between different tissues anyway. The labelling of the intracellular amino acid pool is normally lower than that of the plasma pool, so that this method tends to underestimate the true flux rate.

Synthesis and Turnover

Measurement of Protein Synthesis in Individual Tissues

The measurement of protein synthesis in individual tissues involves administration of a labelled amino acid and measurement of the amount of the labelled amino acid incorporated into protein over a given time. The labelling of the precursor pool for protein synthesis during the incorporation time must also be measured. One way of doing this is again to give a constant infusion of the labelled amino acid, so that the free amino acid labelling remains constant for a period of several hours. Since the practical approach is similar this can conveniently be combined with measuring whole-body protein turnover. The major limitation of this technique is again that the labelling of free amino acids reaches different steady state values in the plasma and within the cells, and it is still unclear which value best reflects the precursor pool for protein synthesis. Furthermore, in some tissues which contain proteins that turn over very rapidly there is the possibility of some re-entry of label from protein after several hours of infusion.

A method which largely overcomes these problems involves giving a very large single dose of labelled amino acid. This effectively floods all the free amino acid pools, so that they reach virtually the same specific activity, and this value remains constant, or declines slowly and linearly, for a period of minutes after which the tissue sample is taken.

The practical drawback to both these methods for measuring protein synthesis in tissues is the need to take a sample of the tissue. This can be achieved in animals which can be killed at the end of the experiment, but studies in humans are restricted to those tissues which can safely be sampled by percutaneous biopsy (such as muscle) or those which can be sampled during scheduled surgical operations.

Measurement of Protein Degradation

The measurement of protein degradation in tissues is much more problematical. Methods based on isotopic labelling are limited by the problem of reutilization of amino acids after release from protein. Most workers have concluded that the most reliable estimate of protein breakdown is obtained from the difference between protein synthesis and the net rate of change of the protein mass of the tissue. This involves sequentially killing closely matched groups of animals over a period of days, and thus cannot give any information about short-term fluctuations in protein degradation.

One approach to estimating protein degradation specifically in skeletal muscle involves 3-methylhistidine. This amino acid is formed by post-translational modification of certain histidine residues in the myofibrillar proteins, actin and myosin, and once it is released from degraded protein it cannot be reutilized for protein synthesis. 3-Methylhistidine is not metabolized, except for *N*-acetylation in certain species, including the rat; assessment of its production rate, from measurement of urinary excretion, should therefore indicate the rate of myofibrillar protein breakdown. Unfortunately, it is now known that many other tissues, including smooth muscle and skin, contain significant quantities of actin, and can make appreciable contributions to 3-methylhistidine excretion, and this has led to erroneous conclusions about muscle protein degradation in certain circumstances such as fasting. Some useful data have been derived from measurements of arteriovenous differences in 3-methylhistidine across muscles, but results based solely on measurement of urinary excretion of 3-methylhistidine should be interpreted with extreme caution.

Rates of Protein Synthesis

Whole-body protein turnover has been measured in a number of species. Values (per kg of bodyweight per day) reported for mature animals range from 40 g in the mouse to 3 g in the cow, and with humans at 4–5 g. Both synthesis and degradation are considerably increased in young growing animals, with values of 12 g per kg per day being reported for premature infants, falling to 7 g per kg per day in 1-year-old children.

Whole-body protein turnover represents the sum of the contributions from many different tissues. The contribution of each tissue depends on the fractional rate of protein synthesis in the tissue and on the proportion of the body's protein mass present in that tissue. Table 1 shows representative data for a number of tissues in the rat; isolated values are available for several other species, including humans, but these do not yet form a comprehensive picture.

It should be noted that protein turnover in different

Table 1. Contributions of tissues to whole-body protein synthesis (average values for 100-g rats)

Tissue	Protein content (% of whole body)	Fractional synthetic rate (% per day)	Absolute synthesis (% of whole body)
Whole body	100	34	100
Muscle	45	17	23
Skin	10	64	19
Liver	4	90	12
Gut	4	100	11
Bone	3	90	8
Remainder	34	26	27

tissues is controlled independently. Thus a simple measurement of whole-body protein turnover may fail to detect changes in opposite directions in two different tissues, such as muscle and liver.

Regulation of Protein Synthesis and Degradation

The rates of both protein synthesis and degradation change in response to nutritional and other environmental influences. For example, muscle protein synthesis increases rapidly after a meal, as the tissue goes temporarily into positive balance, then subsequently declines as amino acids are mobilized during the postabsorptive phase. It is possible that the concentrations of free amino acids within the cell have an influence on the rate of protein synthesis, although this view is hard to reconcile with the observation that the Michaelis constants (K_m) of the amino acyl tRNA ligases are generally well below the intracellular amino acid concentrations, so that the tRNAs are normally fully charged. Moreover, intracellular amino acid concentrations have been observed to increase in starvation, when the rate of protein synthesis is known to go down. Thus it is likely that hormones play a more important role in regulating protein synthesis.

However, it is still possible that the concentration of one or another amino acid may affect protein turnover. Tryptophan is the amino acid which is generally present in the lowest concentrations in intracellular free amino acid pools, at least in mammalian tissues, and there is evidence from *in vitro* studies using rat liver cells that changes in tryptophan supply can affect the rate of protein synthesis by affecting the aggregation of ribosomes into polysomes (the form in which they are active). However, the circumstances in which such large changes in tryptophan concentration would occur *in vivo* are not clear.

In vitro studies have also suggested that the branched-chain amino acids, or leucine alone, may regulate protein synthesis in skeletal muscle, and that the deaminated metabolite of leucine, α-ketoisocaproic acid, may regulate protein degradation. However, injection of a large quantity of leucine has been shown to have no effect on muscle protein synthesis *in vivo*.

Recent evidence has indicated that the rate of protein synthesis in muscle correlates with the intracellular concentration of glutamine, particularly in circumstances where both are reduced, including injury and infection. However, it has not yet been shown that increasing muscle glutamine concentration will restore the balance between protein synthesis and degradation.

Of the many hormones that are known to affect protein turnover, insulin has received perhaps the most attention. Studies in rats show that insulin increases protein synthesis, and this may be the major mechanism by which food intake affects protein metabolism. In contrast, insulin's main action in humans appears to be to decrease protein degradation. Glucocorticoids depress muscle protein synthesis, and thyroid hormones increase protein breakdown. Testosterone and its synthetic derivatives appear to exert their anabolic effects either by increasing protein synthesis or by decreasing protein degradation, depending on concentration. ***See*** Hormones, Adrenal Hormones; Hormones, Thyroid Hormones

Factors that Depress Protein Synthesis

One of the major reasons for studying protein turnover in relation to nutrition is to investigate the mechanisms by which growth rate is reduced or by which lean body mass is lost in situations of undernutrition or disease. It is perhaps not surprising that retardation or even cessation of growth is associated with a reduction in the rate of protein synthesis. There is now considerable evidence that the net loss of lean body mass in adults suffering from chronic diseases such as cancer as well as primary muscle wasting diseases is also caused by a decrease in protein synthesis. Protein breakdown is not increased, and may even be decreased as a secondary response to decreased protein synthesis. This has major implications for therapy, which should be directed towards augmenting protein synthesis rather than suppressing protein breakdown. ***See*** Cancer, Diet in Cancer Treatment

Several factors appear to be involved in the depression of protein synthesis. Patients with chronic diseases often have inadequate intakes of protein and energy, and this is known to depress protein turnover. Lack of physical activity, particularly in bedridden patients, is another important factor. Changes in conventional hormones are likely to be involved, as are cytokines such as the interleukins and tumour necrosis factor. These are peptides produced by the cells of the immune system in response to infection and injury which, apart from their immediate local action, now appear to have systemic effects including suppressing muscle protein synthesis. This presumably evolved as an adaptive response to acute injury or infection, when it would be useful to mobilize amino acids from muscle protein to provide substrates for such processes as gluconeogenesis and acute-phase protein synthesis, but is now seen as a maladaptive response to chronic disease. At the cellular level there is some evidence that these cytokines, and indeed some conventional hormones, may act by modulating the production of eicosanoids. In this context prostaglandin $F_{2\alpha}$ appears to increase the rate of protein synthesis while prostaglandin E_2 promotes protein breakdown. ***See*** Prostaglandins and Leukotrienes

Bibliography

Garrow JS and Halliday D (eds) (1984) *Substrate and Energy Metabolism in Man.* London: John Libbey.

Millward DJ (1990) The hormonal control of protein turnover. *Clinical Nutrition* 9: 115–126.

Rennie MJ, Edwards RHT, Emery PW *et al.* (1983) Hypothesis: depressed protein synthesis is the dominant characteristic of muscle wasting and cachexia. *Clinical Physiology* 3: 387–398.

Rennie MJ, Hundal HS, Babij P *et al.* (1986) Characteristics of a glutamine transporter in skeletal muscle may have important consequences for nitrogen loss in injury, infection and chronic disease. *Lancet* ii: 1008–1012.

Waterlow JC, Garlick PJ and Millward DJ (1978) *Protein Turnover in Mammalian Tissues and in the Whole Body.* Amsterdam: Elsevier Science Publishers BV.

Waterlow JC and Stephen JML (eds) (1981) *Nitrogen Metabolism in Man.* Essex: Elsevier Science Publishers.

Peter W. Emery
King's College, London

Deficiency

Protein deficiency indicates a lack of body protein or a relative deficiency of one or several essential amino acids. Thus protein deficiency is synonymous with a negative nitrogen balance. The deficiency can result from a protein-deficient diet or other events, such as diseases, and it must be distinguished from the multifactorial syndrome of kwashiorkor. However, the clinical features and physiological effects are generally similar to those of kwashiorkor.

Causes and Groups At Risk

Causes

Although the main cause of protein deficiency is a protein-deficient diet, frequently seen in children in developing countries, it can also occur in patients suffering from various diseases. In this sense, protein deficiency (as well as other nutritional deficiencies) is a secondary consequence of the particular disease. ***See*** Malnutrition, The Problem of Malnutrition

Secondary protein deficiencies can be ascribed to six causes:

1. Irregular food habits, e.g. in the case of chronic alcoholics. The diet of alcoholics can be severely deficient in protein.
2. Inability to digest and absorb the protein that is consumed; this occurs in patients with chronic gastrointestinal disorders, such as coeliac disease or enteritis. ***See*** Coeliac Disease

Table 1. Postoperative protein losses

Operation	Loss (g per day)
Stomach removal	110–112
Strumectomy	72
Cholecystectomy	71
Herniotomy	11
Amputation of the breast	9–18
Fractures of long bones	86–312
Skull injuries	67
Hip joint dislocation	17

From Welsch A (1986) *Krankenernährung* 5th edn, p 410. Stuttgart: Thieme Verlag.

3. A disturbed protein metabolism, which may exist in patients with cirrhosis of the liver but also in patients with hormonal disorders or in some cases of diabetes.
4. A continuous loss of protein; this predominates in patients with diseases such as chronic renal disease, bleeding or exudative gastroenteropathy. High losses of albumin into the urine are indicators of the nephrotic syndrome. ***See*** Kidney, Structure and Function
5. An increased protein turnover, which is characteristic in cases of infection, fever or gastroenteritis.
6. Enhanced catabolism of protein, with increased nitrogen losses, seen in patients with severe injuries, especially burns, or in post-operative stress. ***See*** Burns Patients – Nutritional Management; Surgery, Nutrition in Elective and Emergency Surgery

Protein losses after operations depend on the kind of surgery (Table 1). If patients are unable to eat normally, insufficient food intake worsens the protein deficiency.

Groups At Risk

The groups at risk can be classified according to the causes of protein deficiency (see above) or better according to socioeconomic parameters. According to socioeconomic parameters, by far the largest group is the poor population in developing countries, primarily children in the second year of life. This group is at risk from primary protein deficiency because of a protein-deficient diet, caused by economic, ecological and political factors. In developed societies, groups at risk are those who adhere to extreme diets or suffer from anorexia nervosa or bulimia nervosa. ***See*** Anorexia Nervosa; Bulimia Nervosa

Clinical Features

Fatty Liver

Protein deficiency contributes to fatty infiltration of the liver and results in hepatomegaly. This can be seen in healthy subjects receiving a low-protein diet as well as in protein-deficient animals fed adequate quantities of other nutrients. Because of a decreased synthesis of β-lipoproteins needed for the transport of triglycerides, the fat accumulates in small droplets within the cells, first in the periphery of the lobules and then spreads to the centre of the lobules. In spite of this, the liver functions are well maintained. ***See*** Fats, Digestion, Absorption and Transport

Muscle Wasting

Because of increased protein catabolism, skeletal muscle is often wasted, and subjects limit their physical activity. In many cases the parallel prevailing oedema conceals the wasting.

Oedema

Oedema is a frequent sign in protein deficiency. It generally causes swelling of the tissues, but usually appears in the feet and lower legs. The aetiology of oedema has not been completely clarified. Protein deficiency or, rather, hypoalbuminaemia is only one of the factors.

Other common features of protein deficiency are changes in the hair; it becomes easily plucked and there are changes in texture and colour. Also common are skin changes, with areas of desquamation, hypo- and hyperpigmentation. Protein deficiency also leads to mental changes, lethargy, fatigue, anorexia and some degree of anaemia and is frequently associated with infections. In children, severe protein deficiency leads to growth retardation.

Physiological Effects

Organs and systems can be changed in protein deficiency, and this may alter their physiological functions.

Cardiovascular System

Muscle wasting in the heart leads to reduced cardiac output and poor circulation. As a result, the pulse rate of many patients is low and even impalpable, and the extremities are cold and pale.

Renal Function

Structural abnormalities of the kidney have not been observed, but the glomerular filtration rate may be diminished. The concentrating power can also be impaired, which may be attributable to a low blood urea concentration or an accompanying potassium deficiency.

Pancreas

Atrophic changes and fibrosis of the pancreas have been observed both in children dying of protein malnutrition, and in animals fed low-protein diets. The atrophy affects the acinar cells, responsible for exocrine secretion, but not the islets, responsible for hormone (glucagon and insulin) production. A reduced number of secretory granules has also been observed. These changes result in impaired pancreatic function with a diminished secretion of pancreatic enzymes into the intestine. Glucagon and insulin secretion are also lowered in severe cases of protein deficiency.

In another abnormality of the pancreas in protein-deficient animals the protein content and phospholipid levels are reduced, and triglyceride and cholesteryl ester levels are increased. This altered pancreatic lipid composition in protein-deficient animals may lead to changes in the integrity of pancreatic membranes and can predispose to injury.

Gastrointestinal Tract

The mucosal epithelial cells of the small intestine make up the tissue with the highest turnover rate and are therefore very vulnerable to protein deficiency. The intestine wall becomes thin, the mucous membrane is smooth and atrophic, and the villi are frequently flattened. The enzyme-secreting cells and the pancreatic acinar cells may be affected in a similar way. As a result, the production of digestive enzymes, primarily the disaccharidases, is reduced and the absorptive and digestive capacity of nutrients is impaired.

A feature seen in biopsis after protein deficiency is the intense cellular infiltration in the mucosa and submucosa, especially with lymphocytes and plasma cells. The colon can be less affected, but the muscle coats may also be atrophic and the surface epithelium may show infiltration of plasma cells.

Endocrine System

The pituitary gland responds effectively to the stimulus of protein deficiency; thus the secretion of human growth hormone is normal or, possibly, supranormal and the thyroid-stimulating hormone is elevated. ***See*** Hormones, Thyroid Hormones; Hormones, Pituitary Hormones

The thyroid may be atrophic, with reduced colloid tissue, and the vascular cells are flattened and inactive. The thyroid function appears to be normal, but there are data suggesting a reduced function of the thyroid gland. The concentrations of thyroxine-binding proteins and of thyroxine are reduced.

The adrenal gland appears atrophic but, nevertheless, plasma cortisol levels are elevated and may lead to metabolic disturbances. A reduced insulin secretion, as seen after glucose tolerance tests, can also be predominant in protein deficiency. ***See*** Glucose, Glucose Tolerance and the Glycaemic Index; Hormones, Adrenal Hormones

Immunological System

In protein deficiency, the lymphoid tissue – primarily the thymus gland, but also lymph nodes and spleen – is atrophic. First, the production of thymus-dependent lymphocytes (T cells) is reduced and is responsible for the diminished cell-mediated immunity to infectious agents, as demonstrated in specific tuberculin skin tests. In addition, the phagocytic activity of neutrophils and antibody formation are reduced. The body therefore becomes more susceptible to infections which can be fatal. ***See*** Immunity and Nutrition

Treatment

The dietary treatment of protein deficiency depends on the cause of the deficiency. In patients with nephrotic syndrome, characterized by massive losses of albumin into urine, the intake of protein should be increased to 90–120 g per day for adults when hepatic synthesis of albumin can compensate in part the urinary losses. ***See*** Kwashiorkor; Marasmus

In cases of acute renal failure, the protein intake should be only 20 g per day. This reduces protein metabolism and the production of urea. In patients with liver cirrhosis the protein intake should also be reduced because a high-protein diet can precipitate hepatic encephalopathy.

Special Feeding Methods

Protein deficiency can be treated by tube feeding, intravenous feeding or supplementary feeding. Tube feeding is indicated in patients with severe malnutrition who are unable to eat. Parenteral nutrition is essential when the small intestine is unable to digest and absorb nutrients, or after severe injuries such as burns.

In mild protein deficiencies an increased oral intake of protein is sufficient to meet dietary needs. Several special preparations containing high levels of protein are available on the market.

Complications

Since protein deficiency is frequently accompanied by infections, dehydration and deprivation of vitamins and electrolytes, these conditions need special attention. In most cases of severe protein deficiency, therefore, treatment has to start with fluid replacement and, if indicated, with antibiotic therapy.

Adaptation to Low-protein Intakes

Waterlow (1986) described two possibilities of nutritional adaptation, which he called the first and second line of defence. The first line of defence is the ability to achieve nitrogen balance at various levels of protein intake. The second line of defence is necessary after the capacity to economize nitrogen is exhausted, resulting in a reduced lean body mass or reduced growth rate in children. In this section, only the first line of defence is discussed.

Nitrogen Balance

Nitrogen equilibrium is a state in which, for a given intake of nitrogen, an equivalent amount of nitrogen is lost from the body via urine, faeces, skin, sweat, etc. When protein intake is low, dietary protein is used more efficiently; this is exemplified by the excretion of nitrogen, particularly in the form of urea, which is reduced and so contributes to the restoration of nitrogen balance. In this situation the pathways of amino acid synthesis predominate.

The liver plays an important role in the adaptive process since it is the only organ which can transform the nitrogen from amino acids into urea. The alterations in the nitrogen pathway are mainly brought about by changes in the activity of liver enzymes. Experiments with animals have shown that in protein deficiency urea cycle enzymes, such as argininosuccinase, are low and the amino-acid-activating enzymes are high. Table 2 shows this finding in children with severe protein malnutrition, soon after admission to hospital and after recovery.

The metabolic activity of the gastrointestinal tract is important in the adaptive process. Normally, one third of the urea produced is passed into the bowel and can be hydrolysed by the gut microflora. Thus urea nitrogen becomes available for metabolic interaction, e.g. for the synthesis of nonessential and essential amino acids. However, the necessary carbon skeletons can be the

Table 2. Enzymatic activity of the livers of children with malnutrition

	Amino-acid-activating enzymes	Argininosuccinase
Soon after admission	1·44	1·06
1–2 Months later	0·91	1·46

The figures for the amino-acid-activating enzymes are the mean of 18 measurements and are expressed in μmol phosphorus exchanged per mg protein; for argininosuccinase the figures are the mean of 11 measurements and expressed in μmol urea per mg protein. The changes on recovery are statistically significant. From Passmore R and Eastwood MA (1986), p 52.

limiting factor in adaptation. In protein deficiency a greater proportion of urea is retained by the body. ***See*** Microflora of the Intestine, Role and Effects

Recent investigations have shown that the colon is permeable to urea and amino acids, thus supporting the view that hydrolysis of urea plays an important nutritional role.

Plasma Albumin

Plasma albumin reacts very sensitively to a reduced protein intake. Although the albumin turnover is reduced by about 36% in studies with adults, the plasma albumin concentration is reduced by only about 7%. This maintenance of intravascular circulating albumin mass is attributable to a reduced breakdown and a shift of albumin from the extravascular to the intravascular compartment.

Factors Affecting Adaptation

Among the factors that can affect the adaptation to low protein intake are infections, diarrhoeal diseases, and injuries. In infections, protein from muscle and skin is needed for the immune response, and this leads to a negative nitrogen balance. In diarrhoeal diseases the malabsorption has a negative influence on the adaptive process; in patients with injury the severe losses of nitrogen have the same effect.

Energy balance is also essential for nitrogen balance because of its nitrogen-sparing effect. Thus if protein deficiency is accompanied by energy deficiency, the adaptation to a low protein intake cannot be achieved completely.

Protein Reserves

The body of a human adult (65 kg) contains 12 kg of protein, about 50% of which is found in muscles. The amount of body protein depends on, among other things, the dietary protein and carbohydrate intake; if carbohydrate is lacking, the amino acids are utilized for gluconeogenesis. ***See*** Glucose, Function and Metabolism

Table 3. Relative losses of protein in different organs and tissues from rats over 7 days

Organ or tissue	Loss (percentage of primary content)
Liver	40
Prostate gland	29
Seminal vesicle	29
Gastrointestinal tract	28
Kidney	20
Blood plasma	20
Heart	18
Muscle, skin, skeleton	8
Brain	5
Eyes	0
Testicle	0
Adrenal gland	0

From Kraut H (1981), p 148.

Protein reserves are not comparable to special fat depots, and not all body proteins can serve as protein reserve. Reserves are primarily organs which contain labile body protein such as liver, plasma (with proteins such as albumin and enzymes) and the gastrointestinal tract which has an assessed protein content of about 700–800 g. Although the protein turnover rate in muscle is very slow, this tissue is a very important protein reserve owing to its large mass.

In protein deficiency the labile body proteins are metabolized first; when deficiency is long-term, all organs are affected to various extents (Table 3). The well-fed human adult can lose about 3 kg of protein without disturbances to his or her health.

Bibliography

Elmadfa I and Leitzmann C (1990) *Ernährung des Menschen* 2nd edn, pp 139–162. Stuttgart: Verlag Ulmer.

Jackson AA (1990) The aetiology of kwashiorkor. In: Harrison GA and Waterlow JC (eds) *Diet and Disease in Traditional and Developing Societies*, pp 76–113. Cambridge: Cambridge University Press.

Kraut H (ed.) (1981) *Der Nahrungsbedarf des Menschen: 1 – Stoffwechsel, Ernährung und Nahrungsbedarf, Energiebedarf, Proteinbedarf*, pp 140–153. Darmstadt: Steinkopff Verlag.

Latham MC (1990) Protein–energy malnutrition. In: Brown ML (ed.) *Present Knowledge in Nutrition* 6th edn, pp 39–46. Washington, DC: International Life Sciences Institute – Nutrition Foundation.

Passmore R and Eastwood MA (1986) *Human Nutrition and Dietetics* 8th edn, pp 40–53, 279–291, 403–407, 452, 482–501. Edinburgh: Churchill Livingstone.

Siegenthaler W, Kaufmann W, Hornbostel H and Waller HD (eds) (1984) *Lehrbuch der inneren Medizin*, pp 14.17–14.19. Stuttgart: Thieme Verlag.

Trowell HC, Davies JNP and Dean RFA (1982) *A Nutrition Foundation's Reprint of Kwashiorkor*, pp 122–160. New York: Academic Press.

Waterlow JC (1986) Metabolic adaptation to low intakes of energy and protein. *Annual Reviews of Nutrition* 6: 495–526.

C. Leitzmann
Justus-Liebig University, Giessen, FRG

Heat Treatment on Food Proteins

Virtually all foods are complex mixtures of biological macromolecules such as proteins, nucleic acids and polysaccharides as well as fats, vitamins, minerals and a host of other small molecules in trace amounts. While many are eaten raw, a very large number of foods and food ingredients are subjected to some form of processing and in the great majority of cases this processing involves the application of heat.

Typical Heating Processes

Heat treatments are sometimes used to break down plant or animal tissues to facilitate the extraction and preparation of a food or food ingredients, but more frequently they form part of a cooking or preservation process. *Cooking* is generally also intended to break down tissue matrices partially to improve texture, pallatability, digestibility and frequently flavour. There are many ways of cooking foods, with a wide range of temperatures and heating times, but in general the heat treatment is quite severe and may cause extensive changes in macromolecules. *Canning* likewise involves a relatively severe heat treatment, but most other heat preservation processes are less severe and, particularly if the intention is merely to reduce the microbiological cell count and hence to prolong storage life, the heating process is frequently termed *Pasteurization*. More severe heating to eliminate bacteria and fungi entirely is usually referred to as *sterilization*. Both these terms are very general and extremely imprecise in relation to the heat treatment given to the product. This is often true even when only a single product is being considered. For example the in-bottle sterilization of milk involves heating at about 115–120°C for 10–20 min, while UHT-processing is typically performed at 139–145°C for 2–4 s. Both processes give essentially sterile products with very long shelf lives, but since the amount of heat energy supplied is so different, it is not surprising that the amount of thermally-induced change in the various constituents differs. This is most immediately obvious in the more cooked flavour of the in-bottle sterilized milk and its darker colour. ***See*** Canning, Quality Changes During Canning; Heat Treatment, Chemical and Microbiological Changes; Pasteurization, Principles; Sterilization of Foods

With milk, fruit juices, soups and other liquid foods, owing to convection currents during processing the extent of heat damage to components is reasonably uniform throughout the product, but with solid foods the need to ensure adequate heating of the interior generally means that surface layers are subjected to a more prolonged and severe heating than the inside. This of course means a gradation of damage to the constituents and greatly complicates rigorous measurements and interpretation of the effects of heat on either nutritional quality or chemical composition.

Heat-induced Chemical Damage to Proteins

The Maillard Reaction in Protein–Carbohydrate Mixtures

The most obvious visual indication of heat damage to proteins in foods is the development of a brown coloration due to the Maillard reaction. This will occur to some extent in virtually all foods subjected to almost any heat treatment more severe than the mildest pasteurization conditions and, while it may be considered as damage in that it causes changes in the proteins, it is not necessarily undesirable. Indeed, in many foods such as coffee, beer, toasted bread, baked goods, many milk desserts and most snack foods and fried foods the occurrence of the Maillard reaction is essential for the correct flavour and appearance. In essence it is a heat-induced reaction between the reducing groups of sugars and amino groups of protein polypeptide chains. Typically both terminal α-NH_2 groups and the ε-NH_2 groups of lysine side chains are involved, and since many of the chemical reactions taking place have high activation energies they are greatly accelerated by heating. However, it is worth remembering that, like many other chemical reactions, these nonenzymic browning reactions approximately double in rate for every 10°C rise in temperature, and although they are therefore rapid at high temperatures they do proceed at a slow but finite rate even at low temperatures during storage of many foods. Indeed, the presence of glycosylated haemoglobin is readily detected in the blood of diabetic patients with high blood glucose levels, so at least the initial stages of reaction occur even under physiological conditions. The Maillard reaction begins with the condensa-

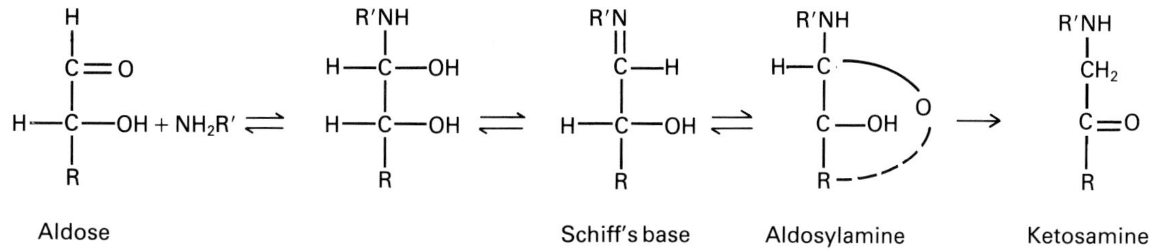

(1)

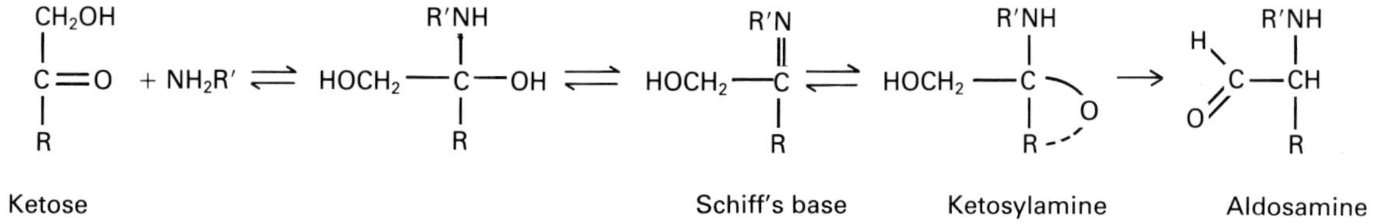

(2)

tion of reducing sugar molecules with nonionized amino groups, eliminating a water molecule, but these initial adducts are not very stable and either revert to the starting components or pass on to similarly unstable Schiff's bases which in turn rapidly isomerize to aldosylamines which then rearrange (the Amadori rearrangement reaction) into stable ketosamines (see eqn [1]). Similarly, ketoses give rise to aldosamine products (eqn [2]). ***See*** Browning, Nonenzymatic

These early stages of the Maillard reaction are largely reversible and many of the protein lysine groups can easily be recovered (e.g. following mild acid hydrolysis), but the ketosamines and aldosamines may then be subjected to further reactions, such as thermal fragmentations, condensations, oxidations or rearrangements, giving very complex reaction mixtures rich in carbonyl compounds which can react further with amino acids and amino acid derivatives. These in turn can give rise to polycarbonyl derivatives which, by further cleavage and polymerization reactions, lead ultimately to brown and black pigments termed melanoidins. Many of the lower-molecular-weight volatile carbonyl compounds are important flavour constituents of heated foods, while the high-molecular-weight melanoidins give colour to bread and baked products, for example.

Nutritionally, there is a loss of available lysine, which may be slight in mildly heated products where only limited reaction has occurred, and is about 50% at the ketosamine/aldosamine stage and total at the melanoidin stage. Also, in the later stages of the reaction there may be formation of both intermolecular and intramolecular covalent bonds between protein polypeptide chains which hinder hydrolysis by proteolytic enzymes and effectively diminish the digestibility of the protein as a whole. Melanoidins produced in heated protein–carbohydrate model systems have been shown to be mutagenic, but they are poorly absorbed through the intestinal wall, so this will minimize physiological effects. Like other aspects of the Maillard reaction, this is still the subject of considerable research. ***See*** Carbohydrates, Interactions with Other Food Components

Chemical Damage in Carbohydrate-free Foods

Because the majority of foods containing proteins also contain at least small amounts of carbohydrate, the Maillard reaction is probably the most important single route whereby chemical thermal damage to proteins occurs, but there are other routes not requiring the participation of carbohydrate. Reactions such as the desulphation of sulphur-containing amino acids, deamidation, oxidation and isomerization of amino acids can occur when proteins alone are heated under appropriate conditions. Above about 115°C, cysteine and cystine are partially destroyed with the formation of hydrogen sulphide and dimethyl sulphide, and oxidation to cysteic acid also occurs. Some of these and other fragments contribute to the flavour of heated fish, meat and milk. The loss of ammonia in deamidation reactions from the side-chains of asparagine, and rather more slowly from glutamine, to give aspartic and glutamic acids occurs at a rather lower temperature (about 100°C). While this may not be important in a nutritional context, the conversion of a neutral group into an acidic one may have significant conformational effects and modify functional properties. Heating in air accelerates the oxidation of amino acids such as cysteine, cystine and methionine, but tryptophan is also susceptible and is partially destroyed.

Racemization reactions via a β-elimination mechanism require much more severe heat treatments, generally in excess of 200°C, and the resulting mixtures of D and L isomers of amino acids will have reduced nutritional value, both because the D isomers have little nutritional value and because their presence will reduce the susceptibility of the protein polypeptide chain to digestion by proteolytic enzymes. There is also the possibility of toxicity if D amino acids or proteins containing them are consumed. At similar temperatures, cyclization reactions can also occur and may give rise to toxic or mutagenic derivatives, such as the α-, β- and γ-carbolines derived from tryptophan. Likewise, very severe heating of foods such as meat and fish which are low in carbohydrate may lead to the formation of isopeptide bonds, covalent cross-links such as ε-*N*-(γ-glutamyl)lysine or ε-*N*-(β-aspartyl)lysine between lysine and glutamic or aspartic acids, respectively, which may reduce the digestibility and hence nutritional value of the proteins. ***See*** Amino Acids, Properties and Occurrence

Many of the above reactions are pH dependent as well as temperature dependent, but there are other reactions that also occur at more extreme pH values. Extensive acid hydrolysis of proteins requires several hours of heating at 100–110°C in strong (e.g. 6 M) acid but not all peptide bonds are of equal stability and the Asp–Pro bond if it occurs in a protein is very much more readily hydrolysed, so that heating for just a few minutes at 100°C at a pH of 4–5 is quite sufficient to cause partial hydrolysis. Under alkaline conditions a rather more extensive range of heat-induced chemical changes can occur, and serine, threonine and lysine are all partially destroyed. Some of the changes are in fact similar to those seen at pH values closer to neutrality under more severe heating. Such changes include the formation of lysinoalanine, lanthionine and dehydroalanine from lysine, cysteine or phosphoserine residues and result in the formation of new cross-links between polypeptide chains, which reduces digestibility and, together with the loss in important amino acids, decreases nutritional values. ***See*** pH – Principles and Measurement

Heat-induced Physical Damage to Proteins

Conformational Effects

In all the effects of heat on proteins discussed so far only chemical changes to proteins that modify specific amino acids and thus alter the composition and primary structure of the polypeptide chains have been considered. While all of these changes can occur quite readily in foods, they only become significant in cooked, sterilized or other extensively heated products where the heat energy input is far higher than under the much more mild conditions that would be described as pasteurization. Even mild heat treatments may have dramatic effects on food proteins, but in this case by disrupting protein secondary, tertiary and perhaps quaternary structure; in other words by altering the conformation or shape of the molecules rather than by causing changes in the primary amino acid sequence. Severe heating destroys the native conformations of protein molecules as well as bringing about those changes already discussed, but mild heating may cause extensive conformational change with no accompanying chemical change that can be detected by compositional analysis.

In spite of this there may be pronounced nutritional consequences, since changing the shape of a protein molecule can have a very considerable impact on the ease with which individual peptide bonds can be cleaved by proteolytic enzymes and hence on the digestibility of the protein. A good example of the influence of molecular shape on digestibility is given by milk proteins, in which the whey proteins with typical globular protein structures, are as a generalization about three orders of magnitude less susceptible to hydrolysis by various different proteinases than the caseins, which all have much more open and randomized shapes. As would be expected, heat denaturation of globular proteins is nearly always accompanied by increased susceptibility to proteolysis.

Denaturation and Refolding

Secondary structural features of protein molecules, such as α-helices, β-sheets and β-turns, and the way in which the polypeptide chains fold into the correct globular protein three-dimensional shape (tertiary structure), as well as the way in which the subunits of multisubunit molecules are assembled (quaternary structure), are all stabilized by purely physical forces rather than by covalent bonds. These physical forces (hydrogen bonds, van der Waals forces, electrostatic bonds and hydrophobic bonds) vary considerably in strength, but can all be broken by purely physical means. Since all these interactions are reversible, as the protein is synthesized on the ribosome the polypeptide chains fold up in such a way as to minimize the overall energy of the system. Subsequent post-translational modifications, if they occur, may change the situation somewhat and mean that other states would have still lower energy, but the native conformation of any protein always represents an energy minimum even if not the lowest one. This of course means that for a molecule to change shape requires overcoming an energy barrier before falling into a new minimum.

Protein denaturing agents such as urea, detergents and organic solvents function by lowering the height of the energy barriers, but heat operates rather differently. Individual atoms and groups in a protein can rotate around carbon–carbon, carbon–nitrogen and carbon–oxygen bonds and bonds can flex and stretch so that there are many different types of movement within protein molecules. The application of thermal energy increases the kinetic energy of all these movements and makes it easier for the molecule as a whole to 'flip' from one conformation to another. Sometimes, as protein molecules are unfolded in this way, hydrophobic groups on different molecules become exposed and meet and form a region of strong hydrophobic bonding. Unlike the other physical bonds, hydrophobic bonds are strengthened by increasing temperature, so this process can lead to aggregation and precipitation. The formation of insoluble material on heating is often taken to be characteristic of protein denaturation but this is by no means always the case, and frequently proteins can lose all vestiges of native structure and hence be completely 'denatured' while still remaining soluble.

To add further complication, it is now appreciated that denaturation and any subsequent refolding process to regain the native conformation are not all-or-nothing processes and many intermediate partially denatured states can exist. The individual stages are often not totally independent, however, and there are a number of established cases of cooperativity in which, for example, the generation of an element of structure (e.g. the formation of an α-helix) during protein refolding acts as a nucleus and facilitates further regain of structure. Thus, while native protein structures can sometimes be 'unzipped' and then 'zipped-up' again when denaturing conditions are removed, this is not always true. Most frequently there are a large number of conformational energy minima, so that many different refolding pathways are possible, and then the percentage regain of native structure may be very small. In situations where there has been a post-translational modification after the protein has folded into its native conformation on the ribosome, as for example in the addition of carbohydrate to give glycoproteins, alternative refolding pathways may be much more favoured on energy grounds and essentially no native structure will be regenerated. Likewise, if post-translational modifications have involved cleavage of the polypeptide chain (as in chymotrypsin, plasmin, etc.) it is virtually impossible to reassemble the pieces of the molecules back into the correct conformation once they have been denatured.

The height of the energy barriers separating the native conformation of a protein molecule from neighbouring conformational states is very variable, as would be expected when so many different bonds and other factors are involved in stabilizing the molecules. At one extreme there are a number of proteins and enzymes that begin to unfold and lose activity at a temperature of 40°C or even less, while at the other extreme there are enzymes, many of them from thermophilic bacteria and archebacteria, that function perfectly at well over 100°C, in thermal springs and volcanic vents for example.

It is worth pointing out here that the term 'heat stable' is often misused, or at least used imprecisely, and may describe two quite different situations. The description heat stable is correctly used for enzymes, such as some amylases used in starch saccharification and enzymes from thermophiles which are active at high temperatures, but it is also sometimes applied to enzymes such as the proteinases and lipases from psychrotrophic bacteria (e.g. *Pseudomonas fluorescens*). These latter enzymes are in fact completely unfolded and inactivated at quite low temperatures (e.g. 50–60°C) but have the ability to refold spontaneously back to the native enzymatically active conformations in almost 100% yield on subsequent cooling. They can frequently emerge unscathed after heating for many minutes at temperatures in excess of 100°C, but of course they are not active at such temperatures and should therefore be termed heat-survivable rather than heat-stable. ***See*** Enzymes, Functions and Characteristics

The Consequences of Protein Denaturation in Foods

The activities of all enzymes, and virtually all other biological activities of proteins as well, rely upon the correct juxtaposition in three-dimensional space of specific amino acid side-chain groups; in other words, the maintenance of the native conformation. Likewise the physical properties of the protein, including functional properties such as solubility, foaming properties, emulsifying ability, surface tension, elasticity, etc., are also highly conformation dependent. Thus, partial or complete thermal denaturation will not only destroy any enzymic activity but will also change functional behaviour.

Protein or enzyme denaturation is put to good use in a very wide range of food applications. For example, it is the thermal denaturation and inactivation of metabolic enzymes in pathogenic bacteria, leading to their death, which is the basis of pasteurization as a means of increasing food safety. Higher sterilization temperatures similarly disrupt the metabolism of other microorganisms as well, leading to sterile long-life products. Intermediate temperature–time combinations are currently gaining popularity as a means of extending shelf-life in a wide range of cook–chill foods, which, while not sterile (and hence not long-life), avoid most of the flavour defects often associated with the chemical changes referred to above resulting from more severe heat treatments. Blanching of fruits and vegetables is a comparatively brief treatment in water or steam at about 100°C carried out prior to freezing, dehydration or irradiation and its principal purpose is to inactivate enzymes such as lipases, lipoxygenases, proteinases, polyphenoloxidases and glycohydrolases that can alter the flavour, colour or texture of the food during further processing or subsequent storage. There are a number of degradative enzymes released from plant cells when they are disrupted by ice-crystal growth during freezing which can cause extensive tissue breakdown and produce 'mushy' products, for example, if the blanching is omitted. *See* Freezing, Structural and Flavour Changes; Irradiation of Foods, Applications

A number of enzymes are themselves toxic or can give rise to toxic products if allowed to remain active in foods (e.g. myrosinase in rapeseed), but many other toxins and antinutritional factors in foods are also proteins. Sometimes these will be toxins from contaminating microorganisms and sometimes indigenous toxic proteins, such as the trypsin inhibitors (Kunitz and Bowman-Birk inhibitors) and chymotrypsin inhibitors present in many legumes such as soya beans, peanuts, peas and most other beans. Legumes also contain lectins (or phytohaemagglutinins) which form complexes with carbohydrates, including the oligosaccharide portions of cell-surface glycoproteins, and interfere with membrane transport processes or cause the cells to aggregate, often leading to toxic effects. In all these examples the biological activity is completely destroyed when the protein concerned is denatured, so heat treatments are an essential step in obtaining safe and palatable food products. *See* Plant Toxins, Trypsin Inhibitors; Plant Toxins, Haemagglutinins

It has already been noted that heat-induced unfolding of a protein molecule renders it more susceptible to proteolytic attack and hence increases digestibility and sometimes the nutritional value; this may be regarded as a change in a functional property, but other properties are also influenced by conformational state. For example, highly insoluble collagen molecules unfold, dissociate and then dissolve readily when heated in water to 65°C or more, while other proteins behave very differently (immunoglobulins and egg proteins, for example, unfold, aggregate and precipitate, so becoming insoluble) and still others, such as caseins, are little influenced by heating. Aggregation and precipitation are particularly likely to occur if heating occurs at a pH close to the isoelectric point of the protein, but the influence of pH can be complex and unexpected. For example, at pH 5–6, β-lactoglobulin and mixed whey protein fractions precipitate on heating, but treatment for 10–15 min at 50–80°C at either an acid (2–4) or basic (7–9) pH is used to improve functional properties, and the proteins remain soluble even if the pH is then readjusted to 5–6 and also show improved emulsifying, foaming, gelling and thickening properties when compared with the native protein. Insoluble precipitated aggregates of denatured proteins usually possess few useful functional properties except for high water binding. *See* Whey and Whey Powders, Protein Concentrates and Fractions

Unfortunately, when considering the effects of heat on food proteins, it is seldom possible to isolate the effects of heat alone, as even in simplified model systems (which have been studied extensively in a research context) many other factors can influence the outcome. For example, pH, water content, protein concentration, presence of anions and cations, the identity of anions and cations, presence of other solutes, substrates, inhibitors or cofactors all affect the extent and rate of reaction and even the type of reaction. Thus, heating at a pH close to the isoelectric point of a protein favours denaturation, aggregation and precipitation of dilute or quite concentrated solutions, but with highly concentrated solutions it may lead to gelation, both processes also being greatly influenced by the presence of Ca^{2+} ions (the amount of which may also be pH-dependent). Likewise the Maillard reaction is influenced by the concentration of both protein and reducing sugar reactants as well as time and temperature, while the water content of a protein solution greatly influences both the temperature and enthalpy of denaturation and hence all the enzymic, biological and functional properties resting on it.

It is difficult to be specific about the influence of heat on the proteins in any given, and often complex, food products but it is clear that heat treatments have many different both harmful and beneficial effects during food preparation and processing. Treatments have been developed and perfected over many millennia to optimize the beneficial effects for many different purposes, so that heat treatments are now an indispensable part of the preparation of not only human food but also frequently of processed animal feeds. It is only in the last few years of all this effort that we have begun to be able to explain the underlying science and to exploit heat treatments on a more rational basis.

Bibliography

Borgstrom G (1968) *Principles of Food Science, vol. 1, Food Technology*. New York: Macmillan.

Charalambous G (ed.) (1986) *Handbook of Food and Beverage Stability*. New York: Academic Press.

Eskin NAM (1990) *Biochemistry of Foods*, 2nd edn. New York: Academic Press.

Fennema OR (ed.) (1985) *Food Chemistry*, 2nd edn. New York: Marcel Dekker.

Gaman PM and Sherrington KB (1986) *The Science of Food*, 2nd edn. Oxford: Pergamon Press.

Høyem T and Kvåle O (eds) (1976) *Physical, Chemical and Biological Changes in Food Caused by Thermal Processing*. London: Applied Science Publishers.

Kinsella JE and Soucia WG (eds) (1989) *Food Proteins*. Champaign, Illinois: American Oil Chemists Society.

Mohsenin NN (1988) *Thermal Properties of Foods and Agricultural Materials*. New York: Gordon and Breach.

Anthony T. Andrews
University of Reading, Reading, UK

PROTEIN CONCENTRATES

See Whey and Whey Powders

PULSES

Pulses, or grain legumes, are those species of the plant family Leguminosae that may be consumed by humans or domestic animals as mature dry seeds. The two oleaginous crops also belonging to this group, peanuts (*Arachis hypogaea*) and soya beans (*Glycine max*) are normally classified as oilseeds and are dealt with elsewhere. Beans, peas and lentils, which are also referred to elsewhere, are, however, usually considered pulses. Moreover, the reader is warned of a great deal of confusion and disagreement regarding the nomenclature of different species of grain legumes; common and scientific names most frequently used in the food science literature have been adopted here. *See* Beans; Legumes, Legumes in the Diet; Peanuts; Peas and Lentils; Soya Beans, Properties and Analysis

Importance and Global Distribution

Pulses are characterized by a high average protein content (20–40%) that places them second in importance to cereals as source of protein. Pulses are a major part of the traditional foods in the diet of many less developed countries (LDC), most notably in India, Mexico and countries of Africa and Central America. Current world production of pulses according to the Food and Agriculture Organization (FAO) is about 58×10^6 t, up from an average of 41×10^6 t for the period 1979–1981 with almost 56% corresponding to dry peas and beans. More than half of the pulses are produced in the LDC, where they are sometimes referred to as 'the meat of poor people'. Food legumes have also been described as '. . . potentially the most valuable, yet probably the least developed, of the naturally occurring sources of food proteins'.

Botanical and common names of legume plants and pulses and locations where they are primarily consumed are listed in Table 1. Many less prominent species are cultivated at a subsistence level and consumed locally, failing to enter the statistics. For example, the bambara groundnut (*Voandzeia subterranea*) is reported to be the third most important leguminous crop in Africa south of the Sahara, superseded only by cowpeas and peanuts, but reliable data are not available.

Morphology of Seeds

Legume seeds are normally of medium to large size (0·3–2·5 cm in diameter), with a large embryo and a hard, usually smooth testa or seed coat. The seeds may be white, coloured or mottled, shiny or dull. The two major structural constituents are the seed coat (8–15% of the seed weight), and the cotyledons which provide nourishment for the developing plant. Other features include the epicotyl or embryonic stem tip, the radicle or embryonic root, the hilum, which is the scar left where the seed coat separates from the stalk, and the micropyle, a minute opening which functions in water absorption (Fig. 1).

Legume testae are composed of a thin cuticle overlying a layer of thick-walled cells called palisade cells. Hourglass cells and the mesophyll form the innermost layers which are not attached to the cotyledons, allowing for easy dehulling. The ultrastructure of cotyledon cells is characterized by starch granules (8–30 μm in diameter) embedded in a proteinaceous matrix consisting of protein bodies (3–8 μm in diameter) (Fig. 2). The cellulosic or primary cell walls of the cotyledons are separated by middle lamellae, composed of an intercellular cement based on pectic material, which softens after heating, thus contributing to the smooth textural characteristics of cooked legumes.

Chemical Composition and Nutritional Aspects

Composition of selected pulses is presented in Table 2. Total carbohydrates in dry legumes range from 24% to 68%. They include mono- and oligosaccharides, starch and other polysaccharides. Starch is the most abundant carbohydrate and varies from 24% in wrinkled peas to

Table 1. Pulses around the world

Common name	Botanical name	Distribution
Pigeon pea	*Cajanus cajan*	Southeast Asia, primarily India
Chick pea, Bengal gram	*Cicer arietinum*	India, Turkey, Pakistan
Horse gram	*Dolichos biflorus*	India
Lentil, split pea	*Lens culinaris*	India, Turkey
Lupin, tarwi	*Lupinus* spp.	Australia, USSR, Mediterranean basin
Velvet bean	*Macuna pruriens*	Africa, Southeast Asia
Tepary bean	*Phaseolus acutifolius* Gray	Uganda
Adzuki bean	*P. angularis*	Japan
Lima bean	*P. lunatus*	USA, Central and South America
Green gram, mung bean	*P.* or *Vigna radiatus*	India, Thailand, Philippines
Common bean, haricot	*P. vulgaris*	Worldwide
Pea	*Pisum sativum*	China, USA
Winged bean	*Psophocarpus tetragonolobus*	Papua, New Guinea
Broad bean, butter bean	*Vicia faba*	China, Italy
Cowpea	*Vigna unguiculata*	Nigeria, Central Africa
Black gram, urd	*V. mungo*	Southeast Asia, mainly India
Bambara groundnut	*Voandzeia subterranea*	Africa

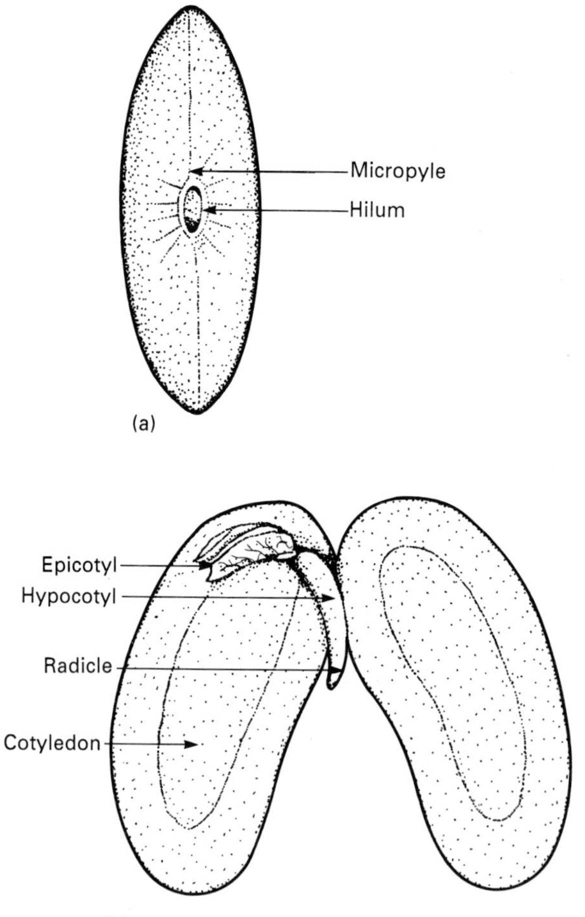

Fig. 1 Morphological characteristics of a typical legume seed. (a) External view; (b) internal view. Reproduced from Stanley and Aguilera (1985).

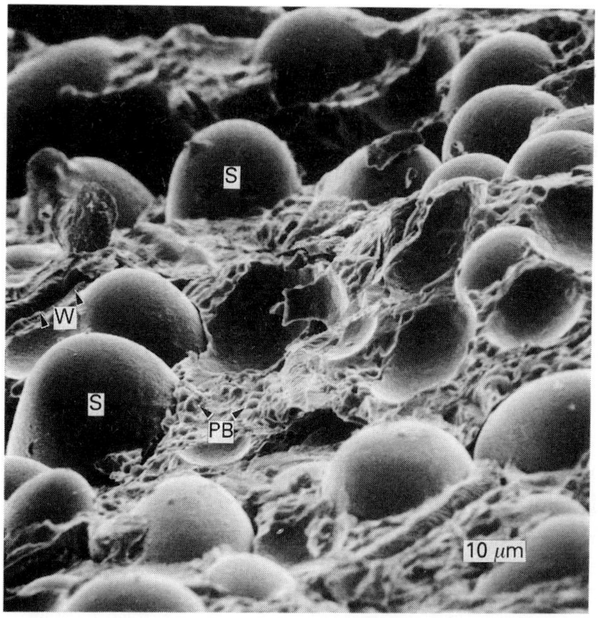

Fig. 2 Microstructural view of a cross-section of a legume cotyledon. S, starch granules; PB, protein bodies.

almost 60% in some *Phaseolus* beans. Lupins contain almost no starch. Oligosaccharides of the raffinose family (e.g. raffinose, verbascose and stachyose), present in mature legume seeds at levels of 1–6%, are thought to be responsible for flatulence. Stachyose tends to predominate in *Phaseolus* spp., cowpeas and lupins, while verbascose is higher in black and green grams, chickpeas and broad beans. Pulses contain appreciable amounts of crude fibre (1·2–13·5%) specifically in their seed coats and cell walls. Cellulose is the major component of crude fibre in peas and beans, while in other legumes (lupins,

Table 2. Main chemical components in selected pulses

Pulse	Protein (%)	Ether extract (%)	Fibre (%)	Ash (%)	Starch (%)
Cajanus cajan	22·0	2·1	7·2	3·8	35·7
Cicer arietinum	17·0	5	4·0	3·0	37·2–50·0
Vigna radiata	19·7–24·2	1·2	4·3	3·5	37·0–53·6
Vicia faba	26·0–33·0	2	8·0	4·0	30·7–42·3
Lupinus albus	39	10	10·5	3·0	<4·0
Phaseolus vulgaris	22	1·6	4·0	3·6	35·2
P. lunatus	19·7	1·1	4·4	3·4	—
Psophocarpus tetragonolobus	29·8–37·4	16·5	5·4	4·3	—
Vigna unguiculata	28·4	1·3	3·9	3·6	41·2–50·6
V. mungo	23·9	1·4	3·4	3·2	32·2–47·9

Adapted from Kay DE (1979).

Table 3. Essential amino acid composition (mg per g N) of some pulses

	FAO pattern	*Cajanus cajan*	*Cicer arietinum*	*Vigna radiata*	*Vicia faba*	*Lupinus albus*	*Phaseolus vulgaris*	*Phaseolus lunatus*	*Psophocarpus tetragonolobus*	*Vigna unguiculata*	*Vigna mungo*
Isoleucine	278	206	277	223	256	319	280	310	306	230	300
Leucine	305	450	468	441	472	531	490	509	538	440	544
Lysine	279	409	428	504	397	350	440	465	462	427	450
Methionine and cysteine	275	172	189	77	128	175	180	141	175	141	100
Phenylalanine and tyrosine	360	746	541	462	460	556	550	581	562	489	556
Threonine	180	219	235	209	222	256	240	261	281	225	219
Tryptophan	90	NR	NR	NR	53	56	NR	63	63	68	NR
Valine	270	243	284	259	284	312	340	322	356	283	362

Adapted from Arora SK (1983) and Kay DE (1979).
NR, not reported.

lentil, red and black gram) hemicellulose predominates. ***See*** Carbohydrates, Classification and Properties; Dietary Fibre, Properties and Sources

The special contribution of grain legumes to the human diet lies in the quantity and quality of the protein. Storage proteins in legumes (globulins) account for about 80% of the seed nitrogen, and are usually classified as legumins or vicilins. Legumins are larger, less soluble in salt solution and more heat-stable than vicilins. Pulse proteins are lacking in sulphur-containing amino acids (methionine and cysteine), tryptophan and in some cases in valine and isoleucine, but are rich in lysine, in which cereal grains are relatively deficient (Table 3). Thus there is a complementary effect between proteins from pulses and those in cereals, starchy roots or fruit that explains why most indigenous diets evolved around the combination of a pulse and corn (Latin America) or rice (the Orient), or a pulse, plantain and cassava (Africa). ***See*** Protein, Quality

The biological quality of legumes measured as the protein efficiency ratio (PER) varies between 0·2 and 2·0. As a result of the complementary amino acid effect, the PER of common beans (0·5) is raised to a value of 1·8 in a 50:50 bean:corn mixture. The digestibility of most legume grains varies between 68% and 83%, lower than for animal proteins.

It is remarkable that in spite of the presence of many so-called antinutritional or toxic factors, pulses have nevertheless provided humans over the centuries with a valuable source of nutrients. Protease inhibitors retard proteolytic enzyme activity; phytohaemagglutinins, or lectins, are capable of agglutinating red blood cells; lathyrogens (in *Lathyrus sativus*) cause nervous paralysis; phytic acid binds to trace metals; cyanogenic glycosides (in *Phaseolus lunatus*) are precursors of hydrogen cyanide (HCN); favism inducers (in *Vicia fava*) can lead to haemolytic anaemia; polyphenols interfere with protein metabolism and are among the most common antinutritional factors that may be present in pulses. Fortunately, most of these are almost completely destroyed after normal cooking. Water-soluble, bitter and poisonous alkaloids are present in most lupin varieties up to a level of about 4%, but since antiquity the extremely toxic *Lupinus mutabilis* has been a preferred protein source for the dwellers of the highlands of Peru and Bolivia who detoxify the grains in running water. ***See*** Anaemia, Other Nutritional Causes; Plant Toxins, Haemagglutinins; Phytic Acid, Nutritional Impact

Grading, Handling and Storage

Most grain legumes are harvested after a period of field drying, when pods and grains are exposed to attack by birds, rodents and insects. The harvested legumes may carry field infestation, mainly by bruchid beetle species which lay eggs on the maturing pods, to the storage sites. Seeds are graded according to the amount of broken, unripe, shrivelled, infested, dented, pitted, off-coloured and mouldy grains present. Normal impurities include parts of leaves and pods, weeds, other seeds and stones.

Reported losses within the post-harvest system in LDC caused by insect and mould infestation average between 10% and 20% of the crop. A most subtle form of post-harvest quality loss is the hard-to-cook (HTC) phenomenon, whereby seeds do not imbibe sufficient water and fail to soften upon cooking. This defect is accelerated when pulses are stored under high temperature (>30°C) and relative humidity (>80%) conditions, as is the case in tropical countries. Losses due to HTC involve extra energy required to cook hardened grains for longer times, reduced nutritional value of the protein and costly methods of disposal of extremely hard beans. Loss of edible beans by HTC in Central America alone is estimated at US\$15–16 × 10^6 per year. ***See*** Insect Pests, Problems Caused by Insects and Mites; Spoilage, Moulds in Food Spoilage; Storage Stability, Parameters Affecting Storage Stability

Processing and Food Uses

Food legumes have the advantage over many crops in the simplicity of their preparation and in the multiplicity of their edible forms. Tender green shoots and leaves, unripe whole pods and dry seeds are commonly eaten; some species (e.g. the winged bean) produce edible tubers as well. Most pulses are consumed at home with the only processing being soaking of the dry seeds and cooking in boiling water. In Central America the cooking broth is usually fed to infants. It must be emphasized that heating is a required step, not only to gelatinize the starchy cotyledons but also to inactivate antinutritional factors.

Substantial amounts of grain legumes are consumed, particularly in Africa and Asia, after milling to dehusk and split the cotyledons. In India, dhal is prepared by cleaning the seeds, tempering them with water until the seed coat has loosened. Black gram is processed into dhal by removing the seed coat, covering the seed with vegetable oil, and drying it in the sun. The dried seed is then passed through a roller. In parts of southern India, the whole seed is soaked in water and the softened seed coat is removed by passing it through a roller. The yield after preparation of dhal at the rural level may be as low as 65%. In East and West Africa, dry cowpeas are soaked, decorticated, ground to a paste and consumed in the form of steamed (*moin-moin*) or deep-fat fried (*akara*) foods.

Canning greatly increases the convenience of using beans in urban areas but the price is usually too high for poorer consumers. Cowpeas and chickpeas are usually canned in brine. Quick-cook legumes have been developed in an attempt to increase the convenience in preparation of dry legumes in the home. The process consists of presoaking the dry seeds in a solution of inorganic salts (sodium chloride, sodium tripolyphosphate, sodium bicarbonate and sodium carbonate), rinsing and air drying.

Special emphasis has been given in LDC to the development of effective industrial processes to remove or inactivate antinutritional or toxic components, to reduce losses during dehusking and milling, and to utilize off-grade or hardened seeds (e.g. by extrusion-cooking of flours). In other cultures, where flatulence (leading to discomfort and sometimes pain, particularly in infants) limits more extensive utilization of pulses, technological alternatives have been sought to reduce the content of flatulence-producing oligosaccharides. These efforts include breeding of oligosaccharide-reduced varieties, germination before cooking, and the use of hydrolytic enzymes.

Based on the compartmentalization of starch and protein at the cellular level (Fig. 2), fractionation methods have been developed that employ dehulling, fine grinding and air classification to separate the large and heavy starch granules from a shattered protein matrix fraction. High-protein (>50%), high-starch and high-fibre products, from field peas, green and yellow peas and common beans, have been commercially produced in Canada, the USA and Denmark. Less successful, however, have been industrial efforts to produce protein concentrates and isolates by the wet method developed for soya beans. The high-fibre by-product made from bean hulls provides a unique nutritional functionality by containing both water-insoluble and water-soluble dietary fibre.

Grain legumes are sometimes ground into a nutritious flour and used in bakery products and snacks. Seeds are occasionally utilized for the production of seed sprouts. Lesser known local uses of pulses include the preparation of confectionery products from Adzuki bean paste ('Ann') in Japan, puffed chickpeas prepared after soaking and roasting in hot sand in Asia, and cooked and salted lupins consumed as snacks in the Iberian peninsula.

Bibliography

Aguilera JM and Trier A (1978) The revival of the lupin. *Food Technology* 32(8): 70–76.

Arora SK (1983) *Chemistry and Biochemistry of Legumes.* London: Edward Arnold.

Daussant J, Mosse J and Vaughan J (1983) *Seed Proteins.* London: Academic Press.

Kay DE (1979) *Crop and Product Digest 3 – Food Legumes.* London: Tropical Products Institute.

Stanley DW and Aguilera JM (1985) A review of textural defects in cooked reconstituted legumes – the influence of structure and composition. *Food Biochemistry* 9: 277–323.

Jose M. Aguilera
Universidad Católica, Santiago, Chile

PYRIDOXINE

See Vitamin B_6

Q FEVER

See Zoonoses

QUALITY ASSURANCE AND QUALITY CONTROL

A principal factor in the performance of a food enterprise is the quality of its products. There is a worldwide trend towards more stringent customer expectations with regards to food quality and safety. Accompanying this trend has been a growing realization that continual improvements in quality are often necessary to achieve and sustain good economic performance.

Today, every customer demands quality from his or her supplier. In particular, the last customer in the chain of suppliers and customers – the consumer – more and more vociferously demands quality in the sense that (1) the future of the world will not be endangered by harm to the environment resulting from food production, and (2) that he or she will suffer no economic damage or injury to health as a result of consumption of food components.

Definition and Description of Food Quality and Safety

The quality of food can be defined as a composite of characteristics which affect the ability of foods to satisfy definite requirements and which determine its fitness for consumption.

Quality in this sense can be discussed under three headings:

1. Sensory quality, the basic characteristics of which are the product's colour, its appearance, texture, tenderness, juiciness, taste and smell; these characteristics, being easily perceptible, are the basis of evaluation and preference on the part of consumers. ***See*** Sensory Evaluation, Sensory Characteristics of Human Foods
2. Nutritive value, which determines the content and level of essential nutrients (biological value), some microbial components (e.g. desired lactic bacteria), energy content and dietary value, its digestibility and availability. ***See*** Bioavailability of Nutrients
3, Convenience and technological quality, characterizing the easy handling and functionality of a food product, to manufacturers and tradesmen and then to consumers; these qualities are based on characteristics such as shelf life, processing level, good price–quality relation, durability, ease of storage and transportation, dimensions, functionality, and usefulness in the production of definite food products. ***See*** Storage Stability, Mechanisms of Degradation; Storage Stability, Parameters Affecting Storage Stability; Storage Stability, Shelf-life Testing

Wholesomeness, also called food safety or food hygiene, may be defined as the result of all conditions and measures that are necessary during the production, processing, storage, distribution and preparation of food to ensure that it is safe, sound, wholesome and fit for human consumption. The aim of all these conditions and measures is to protect the health of people as consumers and to ensure fair practice in the food trade.

The wholesomeness of food determines its fitness for consumption, the classification of which should be based on specified sanitary criteria. The bases of these criteria are the assumptions that a food product must not (1) be harmful to human health, (2) have undergone deterioration (spoilage), (3) have been adulterated, and (4) have been produced, distributed and stored under improper sanitary or physical conditions.

Factors harmful to food consumers include the following:

1. Pathogenic microorganisms and their toxins.
2. Pathogenic parasites and products of their metabolism.
3. Foreign substances, or their degradation products, with toxicity to humans. ***See*** Parasites, Occurrence and Detection; Spoilage, Bacterial Spoilage; Spoilage, Moulds in Food Spoilage; Spoilage, Yeasts in Food Spoilage

In relation to pathogenic microorganisms and parasites

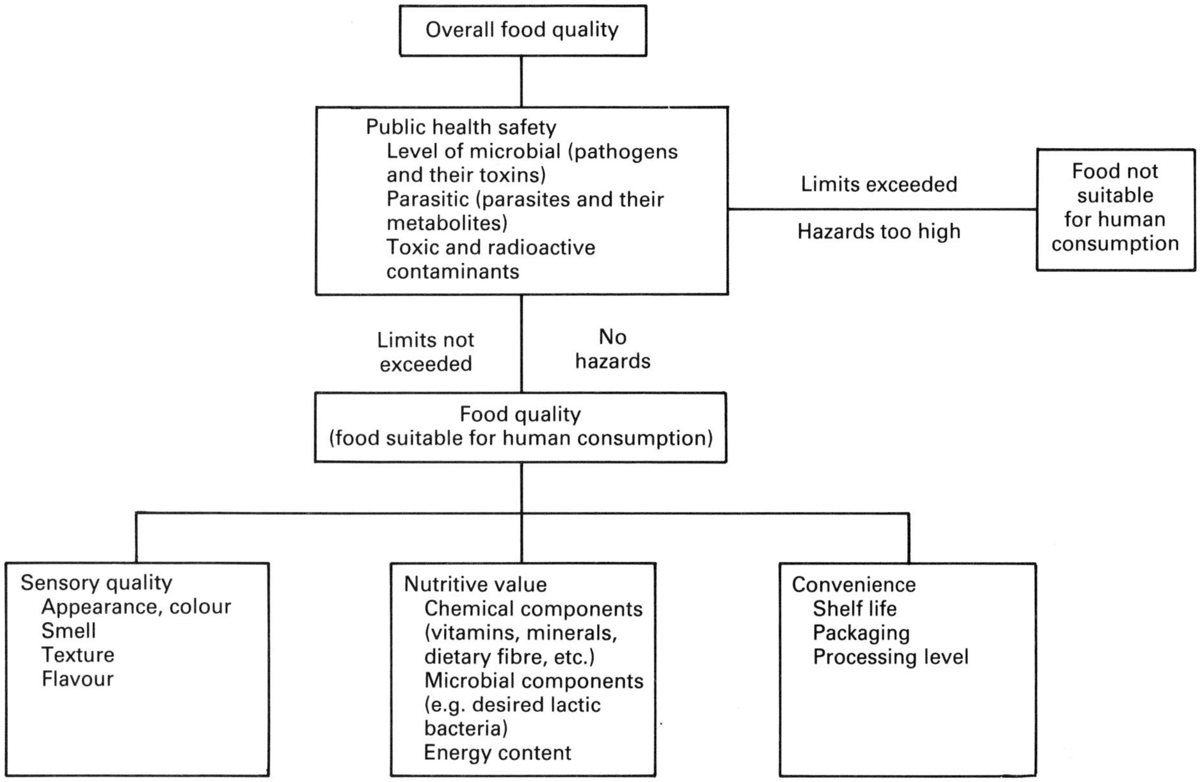

Fig. 1 Main components of food quality and safety.

there is an obligatory sanitory rule, which says that they must not occur in foods owing to their decisively negative influence on human health. Foreign substances may be differentiated as follows:

1. Food additives introduced intentionally to foods to improve their utilization quality, and
2. Technical and accidental contaminations present in food as residues resulting from technological processes or environmental pollution. These include toxic metals, pesticides, medicines, fodder supplements (premix) and growth stimulants. ***See*** Contamination, Types and Causes; Food Additives, Safety

The main components of food quality and safety are shown in Fig. 1.

Role of Quality Assurance and Quality Control

Quality assurance can be characterized by the quality loop for processed food (Fig. 2). Quality assurance is only consistent if the *origins* of lack of quality are eliminated or the emergence of faults is excluded from the beginning: problems result in, rather than originate from, exceeded limits of contamination, adulteration, etc. The quality assurance system is based on the following:

1. Product planning and development, including the results of marketing and market research.
2. Education and training for management and quality control staff as well as for each counterpart.
3. Application of statistical control method and sampling inspection plan in process control and for final inspection.
4. Responsibility for the quality in each step of the quality loop.
5. Using an information and motivation system on product quality.

Assisted by the internal quality partners of a company, if necessary, with additional commitment from the external quality partners (suppliers and customers) the quality assurance of a product is structured into the following steps:

1. Definition of requirements and responsibilities with regard to the quality of individual activities, processes and materials.
2. Definition of the desired properties and required parameters of the product.
3. Definition of materials, technologies, packagings, storage conditions, number and qualification of personnel, etc.
4. Analysis of possible sources of faults and factors affecting quality.
5. Definition of the quality assurance actions: informa-

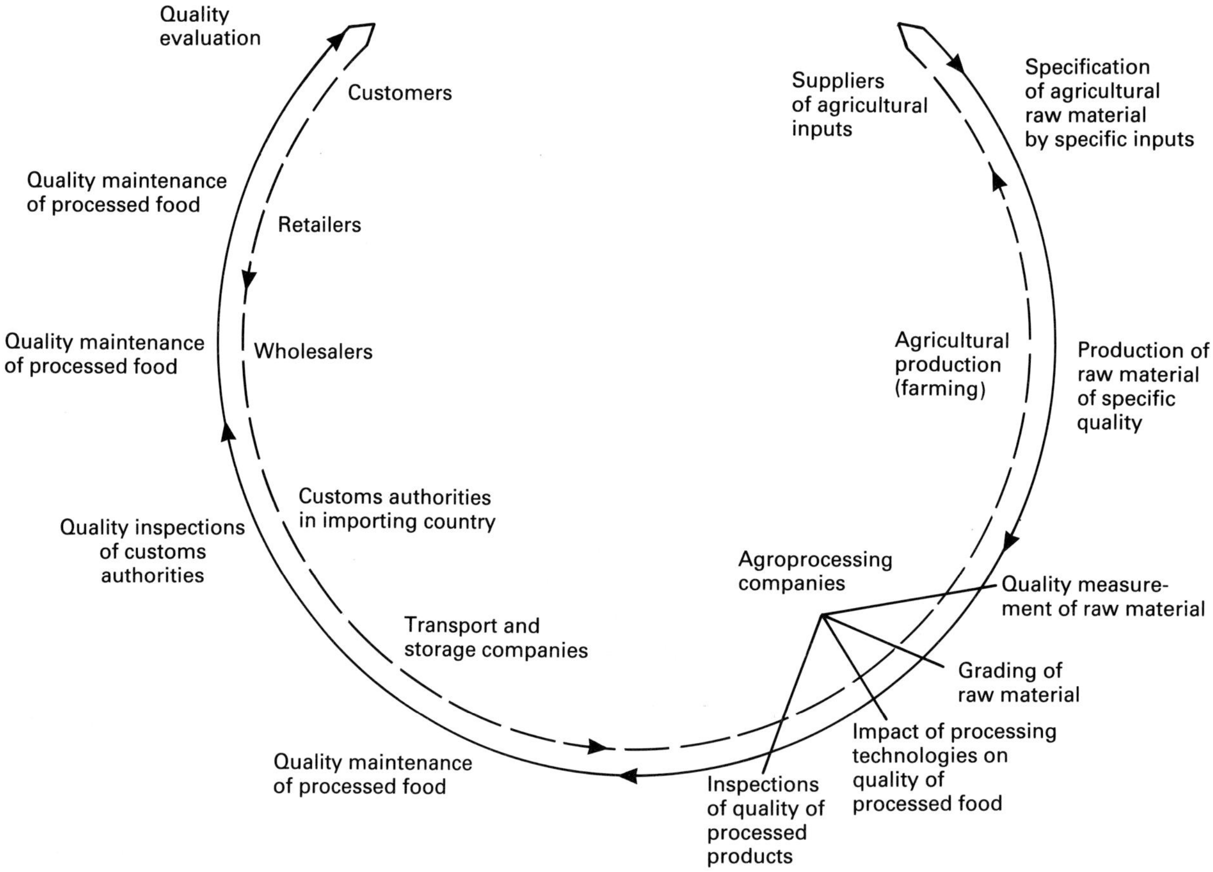

Fig. 2 Quality loop for processed food. —— Flow of products; - - - flow of information on quality requirements.

tion, education, training, motivation, parameters of investigations, methods of analysis and sampling documentation, etc.

6. Definition of responsibilities for certain quality assurance actions.

Appropriate quality control procedures should be part of the process to ensure product safety and quality and hence part of the company quality assurance system. Figure 3 shows the quality control of food manufacturing, including inspection and laboratory investigation in a general flow chart.

All plant operations should be controlled by written procedures and records of performance data. The range and complexity of procedures will vary, depending upon the size and scope of the operation. However, management personnel of even the smallest operation should be aware of the need for quality control consistent with the type of food products being manufactured, stored or distributed. In larger, more complex operations, the inspector should determine if written control procedures exist and should review records detailing results of quality control testing relating to product safety or other factors where regulatory requirements exist.

Ideally, quality control management should be separate and autonomous from production management. The quality control department analyses any planned production changes, whether in preparation or in equipment, and should have responsibility for final approval of all such changes. Final approval for production should not be given until all quality control records have been reviewed and properly endorsed by the quality control management team.

Laboratory personnel and equipment must be evaluated to determine whether or not they are able to carry out all their responsibilities under the firm's specifications and existing regulations. Laboratory inspection should include spot checks of in-process controls made by personnel on the production lines.

Implementation in Food Production (Good Manufacturing Practice)

The guidelines on good manufacturing practice (GMP), elaborated and published initially for drugs, were later modified for human foods and oriented mainly to food safety. The formulated criteria apply in determining whether or not the facilities, methods, practices and controls used in the manufacture, processing, packing or holding of food conform with or are operated or

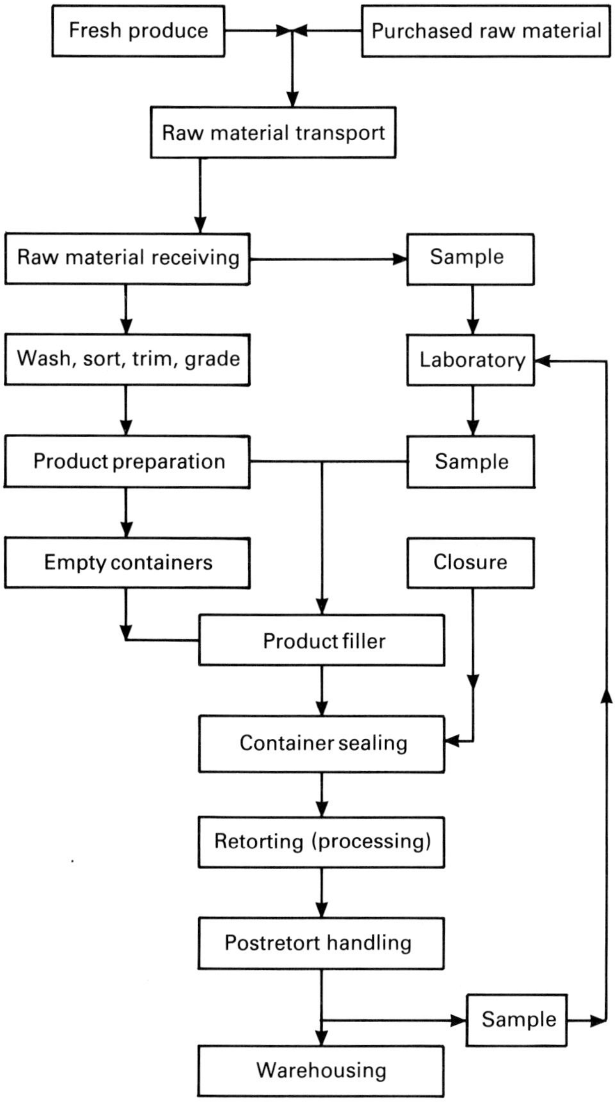

Fig. 3 General flow chart including quality control.

administered in line with good manufacturing practices to assure that food for human consumption is safe and has been prepared, packed and held under sanitary conditions.

More recently systematic, rule-based approaches to quality assurance have been widely applied, most notably that based on the hazard analysis and critical control point (HACCP) programme. Such techniques provide ways of applying our current knowledge of microbial ecology of food in a systematic manner to ensure that nothing is overlooked and that the risk of microbiological food contamination is minimized. ***See*** Hazard Analysis Critical Control Point

The HACCP programme includes assessment of potential hazards, prescribes for the elimination of avoidable hazards and sets tolerances for the hazards that cannot be eliminated in the processing of food. It defines the appropriate control measures, the tests to be carried out and the criteria for product acceptance. It gives a rational systematic documented procedure which can be used for organizing and implementing the entire quality assurance programme.

However, GMP includes not only food safety, but also, as a part of the quality assurance system, the application of product specifications. The specification for a product is usually the description of the requirements with which the product has to comply. These can have a regulatory, commercial or in-house basis. Specifications are an integral part of product development, manufacture, marketing and sales. From primary producer to the retail trade, each industry has its own specific requirements in the description of the following characteristics of their products:

- Name
- Product description
- Specific properties
- Sensory attributes
- Chemical parameters
- Physical data
- Microbiological parameters
- Shelf life
- Storage conditions
- Packaging and labelling
- Handling
- Proper sampling and test methods

From a company viewpoint, specifications can have an internal and external function. The internal function comprises purchasing criteria, incoming goods control, and recording of knowledge and information in a systematic and accessible way. The external function includes sales criteria, sampling procedures and quality control of finished products. It simplifies functional contacts between companies and can serve as a base for product comparisons. This has led to the development of quality system standards and guidelines that complement relevant product requirements given in the technical specifications.

The series of standards set by the International Standards Organization (ISO 9000, ISO 9002 and ISO 9004) embodies a rationalization of the many and various national approaches in this sphere which must be applied in adequate form under the circumstances of the food industry. According to the ISO 9004, more and more food companies have a quality assurance handbook or quality manual covering the following main topics:

1. General quality policy.
2. Description of the quality system and the responsibilities and interrelationships of all operating and staff functions.
3. General description of quality planning requirements, with specific details for each product category where appropriate.

4. Policy requirements for specific elements of the system (relating to GMP) and procedures which implement these policies.

The quality manual is of direct benefit not only to the producer (e.g. cost reduction, help in product liability cases or improved competitivity), but also to the consumer, who will perceive its effect as a high level of quality will be produced and maintained by the application of an integrated and documented quality assurance system.

Implementation in the Food Laboratory (Good Laboratory Practice)

Whereas quality assurance of the work done by laboratories was formerly based largely on confidence in the individual laboratory and in individual members of the staff, a more formalized requirement has recently developed for factual documentation of quality in the results of the investigations presented. In this respect, the work has proceeded largely along international lines where, for example, the ISO has been extremely active.

Incorrect results from a laboratory may have serious health or economic consequences as a result of an unsuitable food incorrectly being assessed as suitable, or by unnecessary rejection of an acceptable food. To ensure reliable analytical results and avoid mistakes, the laboratories must continuously check and improve the quality of analytical work.

The quality manual, which consists of general advice for 'quality assurance', should be supplemented with more detailed regulations for individual laboratories. Consequently, it is recommended that each laboratory prepares its own quality assurance directives in a quality manual as a supplement to the general principles. A quality manual is individual and therefore differs from laboratory to laboratory. Consequently, each laboratory needs to prepare its own manual. However, the following information is necessary in almost every case:

1, Requirements placed on the laboratory: staff (management, experts, chemists, physicists, microbiologists, etc.); premises (type and extent of the activities); apparatus (equipment, instruments etc.); glassware and plastic equipment; chemicals, gases and solvents.
2. Sampling (sampling and sampling staff do not necessarily belong to the laboratory).
3. Sample reception at the laboratory: marking and keeping records; storage of samples before and after analysis.
4. Sample preparation.
5. Selection of methodology.

Defining method (type I) is one which determines a value that can only be arrived at in terms of the method *per se* and serves for calibration purposes.

Reference method (type II) is designated where type I methods do not apply. It should be selected from type III methods, and should be recommended for use in cases of dispute and for calibration purposes.

Alternative approved method (type III) is one which meets the criteria required by the Codex Committee on Methods of Analysis and Sampling for methods that may be used for control, inspection or regulatory purposes.

Tentative method (type IV) is one which has been used traditionally or else has been recently introduced but for which the criteria required for acceptance by the Codex Committee on Methods of Analysis and Sampling have not yet been determined.

6. Documentation and reporting: registration; laboratory register; analytical documentation; use of computers and electronic data processing; method description.
7. Quality assurance of analytical results: acceptance criteria for an evaluation of analytical results; standard substances, recovery tests reference materials; repeated determinations; intercalibration and collaborative tests; control chart.

The intention with this quality manual is to provide a general survey of the factors that influence analytical reliability at the food laboratories. It is hoped that the implementation of good laboratory practice and the guidelines contained in the manual will be of value to all the staff categories working within the laboratory and will inspire the creation of efficiently operating quality assurance systems for laboratory work within the food companies. ***See*** Analysis of Food

Bibliography

ASQC (1986) *Food Processing Industry Quality Control System Guidelines.* Wisconsin: American Society for Quality Control.

FAO (1990) *Manuals of Food Quality Control: 5. Food Inspection.* Rome: Food and Agriculture Organization.

ISO (1983) ISO 9000 series, equivalent to EN 2900. International Standards Organization.

ISO (1983) *ISO 9000 Quality Management and Quality Assurance Standards: Guidelines for Selection and Use.* International Standards Organization.

ISO (1983) *ISO 9002 Quality System – Model for Quality Assurance in Production and Installation.* International Standards Organization.

ISO (1983) *ISO 9004 Quality Management and Quality System Elements – Guidelines.* International Standards Organization.

ISO (1983) ISO Guide 25, EN 45001. International Standards Organization.

Nordic Council of Ministers (1990) *Quality Assurance Principles for Chemical Food Laboratories.* Copenhagen: Nordic Council of Ministers.

P. J. Molnar
Central Food Research Institute, Budapest, Hungary

QUARG

***See* Cheeses**

QUINOA

Quinoa has been an important crop in the Andes for many centuries. Because it offers certain advantages over common cereal crops, it is now gaining popularity in other regions of the world. This article reviews these advantages as well as quinoa's global distribution, morphology, nutritional composition, processing and food applications. ***See*** Cereals, Contribution to the Diet

Global Distribution

Quinoa grows well at high altitudes and is currently grown in Argentina, Bolivia, Chile, Colombia, Ecuador and Peru.

Because it can produce high yields under such adverse conditions as drought, frost and poor soils, quinoa has potential agronomic importance elsewhere. It has been grown experimentally in other regions of the world where climatic conditions are similar to those of the Andes.

The grain grows at high elevations in Colorado, USA, where it is limited by temperature to locations between 2000 and 3500 m. The acclimatization of quinoa to Finland has been reported. In Jokionen, Finland, approximately 50 quinoa ecotypes produced mature seeds. The grain can also be grown in the Himalayas.

Morphology and Relationship

The quinoa plant has an erect stem, which may be branched or unbranched, bearing alternate, highly polymorphic leaves. Its height varies from 0·7 to 3·0 m. The fruit is a pale yellow colour with an occasional tinge of magenta pigment. The grain may be conical, cylindrical or ellipsoidal in shape with a diameter of 1·8–2·6 mm.

The grain is protected by a perianth consisting of loosely adhering cells, a pericarp and two seed coat layers. The perianth cells can be easily removed by washing. ***See*** Wheat, Grain Structure of Wheat and Wheat-based Products

Two other grain chenopods which originated in the tropical American highlands are closely related to quinoa (*Chenopodium quinoa* Willd.). These are huazontle (*C. nuttalliae*) and cañihua (*C. pallidicaule*). Four other varieties of grain chenopods, 'bathu, bithu, dhangar and taak' (*C. album* L.), are found in the Himalayas.

Quinoa Seed Composition

While its proximate composition varies with species, quinoa is generally higher in protein, fat, ash and fibre than are common cereal grains (Table 1). Quinoa's proximate composition ranges from 10% to 18% for protein, from 4·5% to 8·75% for crude fat, from 2·4% to 3·65% for ash, from 2·1% to 4·9% for crude fibre, and from 54·1% to 64·2% for carbohydrates. ***See*** Cereals, Dietary Importance

Lysine, the first limiting amino acid in many grains, including wheat and corn, is present in relatively higher amounts in quinoa. This makes its amino acid balance more favourable (Table 2).

Information on vitamin content is scarce and variable. Quinoa appears to be a good source of vitamin E. Levels vary from 46 to 59 ppm, depending on variety. This is somewhat higher than the amount reported for wheat (Table 3). Low levels of carotenoids and B vitamins have been reported. The level of riboflavin is higher, while levels of nicotinic acid and thiamin are lower than found in some common cereals such as wheat, rice, corn and oats (Table 3).

Little has been presented on the individual mineral content of the quinoa seed. Most available information is rather old. Phosphorus and potassium make up 65% of the ash content of quinoa. In relation to some cereal grains, quinoa appears to be higher in potassium, calcium, phosphorus, magnesium and iron. ***See*** individual minerals

Quinoa starch has a gelatinization temperature range

Table 1. Proximate composition (on a dry weight basis) of selected cereal grains

Component	Quinoa[a] (%)	Polished white rice[b] (%)	Oats[b] (%)	Hard red winter wheat[b] (%)	Corn[b] (%)
Protein	15·4	6·7	14·2	12·3	8·9
Fat	7·5	0·4	7·4	1·8	3·9
Ash	3·0	0·5	1·9	1·7	1·2
Crude fibre	2·4	0·3	1·2	2·3	2·0
Total carbohydrates	68·4	80·4	68·2	71·7	72·2

[a] DeBruin A (1964) Investigation of the food value of quinoa and cañihua seed. *Journal of Food Science* 29: 872–876.
[b] Anderson RA (1976) Wild rice: nutritional review. *Cereal Chemistry* 53: 948–955.

Table 2. Amino acid content (g per 16 g N) of quinoa[a] and four cereal grains[b]

Amino acid	Quinoa	Wheat	Rice	Corn	Oats
Isoleucine	3·6	3·3	3·8	3·7	3·8
Leucine	6·0	6·7	8·2	12·5	7·3
Lysine	5·6	2·9	3·8	2·7	3·7
Methionine	2·0	1·5	2·3	1·9	1·7
Cysteine	—	2·5	1·1	1·6	2·7
Phenylalanine	4·1	4·5	5·2	4·9	5·0
Tyrosine	2·8	3·0	3·5	3·8	3·3
Threonine	3·5	2·9	3·9	3·6	3·3
Valine	4·5	4·4	5·5	4·8	5·1
Arginine	7·0	4·6	8·3	4·2	6·3
Histidine	2·4	2·3	2·5	2·7	2·1
Alanine	4·7	3·6	6·0	7·5	4·5
Aspartic acid	7·3	4·9	10·3	6·3	7·7
Glutamic acid	11·9	29·9	20·6	18·9	20·9
Glycine	5·2	3·9	5·0	3·7	4·7
Proline	3·1	9·9	4·7	8·9	5·2
Serine	3·7	4·6	5·4	5·0	4·7

[a] Van Etten CH, Miller RW and Wolff IA (1963) Amino acid composition of seeds from 200 angiosperm plant species. *Journal of Agricultural and Food Chemistry* 11: 399–410.
[b] Food and Agriculture Organization (FAO) (1970) Amino acid content of foods and biological data on proteins. *FAO Nutritional Study* 24.

of 57–64°C. Although the starch loses its birefringence at about the same temperature as wheat starch, it has different pasting characteristics (changes in viscosity as determined on amylograph during various stages of the heating/cooling cycle). Quinoa starch produces a greater viscosity than wheat starch at equal concentrations. In contrast to most cereal starches, quinoa starch has a low amylose content of approximately 11%. ***See*** Starch, Sources and Processing

The colour and palatability of the quinoa seed is affected by the presence of saponins. The term saponin is applied to a group of glycosides which form soapy solutions in water. These bitter compounds, which may have antinutritional effects, are located in the seed coat layer of the grain. They must be eliminated through washing and/or friction prior to consumption. ***See*** Saponins

Protein Quality

A number of studies have been performed to determine the quality of the protein in quinoa. Growth response was used to determine the protein quality of a saponin-free quinoa, which contained approximately 11% protein. Quinoa produced a growth response which was better than that produced by white rice, corn or wheat. Improved growth rates were obtained with a combination of quinoa and casein, which provided a total dietary protein of 12·42%. From these results, it was concluded that, when judged by growth response, quinoa supplies a high-quality protein.

In another study, young rats were maintained for 50 days on a diet of whole quinoa, supplemented with vitamins A and D. The quinoa had been washed to remove saponins. Throughout the study, rats gained weight and maintained a healthy appearance. These results indicate that quinoa supplies adequate amounts of protein, calories, minerals and B vitamins, which support rat growth. Thus quinoa has a high food value.

Another experiment compared quinoa protein to milk protein. Saponins were removed from the quinoa through washing. Rats maintained on the quinoa protein diet gained significantly more weight than did those on the milk protein diet. No beneficial effects were produced by supplementing the quinoa diet with small amounts of milk protein. In the same investigation, a depletion–repletion study was performed. The quinoa protein diet was determined to be superior to the milk protein diet in replenishing the body stores of nitrogen.

Table 3. Vitamin and mineral composition (mg per 100 g) of selected cereal products[a]

Component	Quinoa	Wheat	Rice	Corn	Oats
Minerals					
Potassium	1040[b]	580	340	330	460
Calcium	71·2[c]	60	68	30	95
Phosphorus	377[c]	410	285	310	340
Magnesium	310[b]	180	90	140	140
Iron	9·1[c]	6	—	2	7
Manganese	4·3[b]	5·5	6	0·6	5
Copper	0·6[b]	0·8	0·9	0·2	4
Vitamins					
Thiamin	0·24[c]	0·55	0·33	0·44	0·77
Riboflavin	0·22[c]	0·13	0·09	0·13	0·18
Nicotinic acid	1·17[c]	6·4	4·9	2·6	1·8
Tocopherols	5·9	3·9	—	—	—

[a] Hoseney RC (1986) *Principles of Cereal Science and Technology.* St Paul, Minnesota: American Association of Cereal Chemists.
[b] DeBruin A (1979) Investigation of the food value of quinoa and cañihua seed. *Journal of Food Science* 29: 219–221.
[c] Ministerio de Previsión Social y Salud Publica (1979 *Tabla de Composición de Alimentos Bolivianos,* 1–8.

The protein quality of quinoa has also been investigated by studying nitrogen efficiency for growth (NEG) and protein efficiency ratio (PER). Uncooked quinoa and casein had similar NEG values. Cooking quinoa improved its NEG value, without changing the amino acid composition of the grain. By comparing PER values, the protein quality of cooked quinoa and casein were determined to be nearly identical. The quality of the protein in quinoa was again shown to be high. ***See*** Protein, Quality

Processing

Quinoa must be processed to remove bitter saponins, which decrease the palatability of the grain. Saponins can be removed by washing the grain vigorously in cold, running water followed by drying overnight at 60°C. Other methods include repeated washing of the grain in alkaline water, with pounding and rubbing to remove the pericarp. Cooking has also been reported to remove saponins.

Saponins can also be removed by first rinsing the grain with water, followed by soaking overnight at refrigerated temperatures. The following morning the grain is rinsed quickly in hot water. Saponin content is checked by placing the grain in a tube, adding water and shaking it vigorously for 30 s. If no foaming occurs, saponins are assumed to be removed. The grain must then be dried quickly to prevent germination.

In Colorado, a barley pearling machine has been modified for on-farm use. It removes the pericarp and saponins from quinoa. A similar piece of equipment, a tangential abrasive dehulling device (TADD), has been developed in Canada. Because of its ability to remove successive layers of seed, it has been used to remove saponins from quinoa. With this device, saponin content was effectively decreased by minimal abrasive dehulling. Processing the grain into a flour to an extraction level of 85·2–98·8% decreased the saponin content to a low level. This type of processing would be advantageous, because the grain would not have to be dried as it does after saponin removal by washing.

Food Applications

Quinoa may be consumed as a whole grain or it may be ground and used as a flour. The whole grain may be cooked and served in a way similar to rice or it may be incorporated into other foods, such as soups. The grain may be fermented to produce a beverage called 'chicha'. The flour is used to produce a coarse bread called 'kispina'. The flour is also used in combination with wheat flour to make a number of items. High protein cakes and biscuits can be made by combining 60% quinoa flour with wheat flour. Noodles have been prepared using up to 40% quinoa flour without adversely affecting the appearance or other characteristics of the final product. Many recipes that incorporate quinoa into biscuits, chowder, croquettes, salads and casseroles are available.

Although quinoa can be used in numerous ways, few scientific studies report its effects on the sensory, nutritional and functional qualities of breads, biscuits or pasta. In one of the few studies available, 20% quinoa was blended with wheat flour and baked into bread. Baking resulted in a slightly decreased lysine content and PER. No indication was given on the effect of the quinoa flour on the volume, texture or colour of the bread.

Quinoa was used in a composite flour blend for Bolivia. Levels of 5% and 10% quinoa combined with wheat flour resulted in decreased loaf volume. Addition of 20 or 40 ppm potassium bromate counteracted the volume decrease and resulted in breads similar to wheat bread with no dough additives. An 8% level of quinoa was used in a composite flour with wheat and in combinations with rice, soy and corn. No adverse physiological reactions were demonstrated in children fed the breads containing quinoa. Although it seemed promising, quinoa was eventually deleted from the flour blends due to its high cost and lack of assured production increases to guarantee sufficient quantities of flour.

Flour milled from germinated quinoa has been used to reduce the viscosity of starch foods. Supplementation with germinated quinoa flour may help to increase the palatability and the effective caloric density of food used for weaning children. This may be useful in Bolivia and other areas of the world, where children are weaned on starchy foods. ***See*** Flour, Analysis of Wheat Flours; Flour, Dietary Importance

Extrusion has been used to produce expanded and textured quinoa. The sensory quality of the products was acceptable and rats fed the products remained in good health. However, the PER for the extruded products was lower than that of cooked quinoa flour and casein.

Extrusion processing may be a potentially useful method for processing quinoa. Blends of corn grits and quinoa have been extruded successfully. Quinoa was combined with corn grits, at levels of 10%, 20% and 30%, and extruded with corn grits, at levels of 10%, 20% and 30%, and extruded under various conditions. Products prepared with an initial moisture content of 15% and extruded with a compression ratio of 3:1 were found to be the most acceptable. These products had greater expansion, lower density and lower shear strength than products extruded at 25% moisture or with a 1:1 compression ratio. Products containing quinoa were higher in protein, fibre, ash and some amino acids than were products containing 100% corn grits. Products with quinoa also had greater nitrogen solubility than those with only corn grits. Greater levels of quinoa resulted in products with increased density, reduced expansion and reduced shear strength. Quinoa addition also resulted in darker, less yellow products than corn grits alone. ***See*** Extrusion Cooking, Principles and Practice

Bibliography

Coulter L and Lorenz K (1990) Quinoa – composition, nutritional value, food applications. *Lebensmittel – Wissenschaft und Technologie* 23: 203–207.

Quiros-Perez F and Elvehjem CA (1957) Nutritive value of quinoa proteins. *Journal of Agricultural and Food Chemistry* 5: 538–541.

Risi CJ and Galwey NW (1984) The *Chenopodium* grains of the Andes: Inca crops for modern agriculture. *Advances in Applied Biology* 10: 145–216.

Simmonds NW (1964) The grain chenopods of the tropical American highlands. *Economic Botany* 19: 223–235.

White PL, Alvistur E, Dias C, Vinas E, White HS and Collazos C (1955) Nutrient content and protein quality of quinoa and cañihua, edible seed products of the Andes mountains. *Journal of Agricultural and Food Chemistry* 3: 531–534.

L. A. Coulter
Department of Food Science and Human Nutrition, Colorado State University, USA

RADIATION

See Drying, Irradiation of Foods and Legislation

RADIOACTIVITY IN FOOD

Principles

Radioactivity is the process of spontaneous disintegration of unstable atomic nuclei, a process associated with emission of ionizing radiation. Radioactive substances, of natural origin or man-made, are present in all foodstuffs. This article describes the radionuclides of interest and considers the internal radiation exposure of man due to ingested radionuclides and discusses the relative contribution of natural and man-made radionuclides to the resulting effective dose equivalent.

Naturally Occurring Radionuclides

Radioactive species of certain elements are continuously formed in the upper atmosphere by interaction of cosmic radiation with elements present in the atmosphere ('cosmogenic radionuclides'). Other radioactive species were formed in the nuclear processes associated with the earth's origin ('primordial radionuclides'). The major cosmogenic and primordial radionuclides and some of their properties are listed in Table 1. The becquerel (Bq) is the unit of radioactivity; one Bq corresponds to one nuclear disintegration per second.

Data concerning concentrations in foods and in the human body, and annual intake of the cosmogenic radionuclides and of potassium-40 (^{40}K) and rubidium-89 (^{89}Rb) are rather uniform in the whole world, whereas the radionuclides of the uranium and thorium series may be present at considerably higher levels (compared to Table 1), depending on local geological conditions. The population living along the Kerala coast of India, for instance, has an annual intake of 40 Bq of radium-226 (^{226}Ra) and 2000 Bq of radium-228 (^{228}Ra). Another case of elevated intake relates to aboriginals of uranium-rich areas of Western Australia, where the annual intake of lead-210 (^{210}Pb) from the carcasses and offal of sheep and kangaroos can be as high as 3000 Bq. Arctic lichen accumulate polonium-210 (^{210}Po) and reindeer graze mostly on lichens in the winter, so that a further accumulation occurs in reindeer meat. Laplanders consuming reindeer meat may have a tenfold higher intake of ^{210}Po than populations of the temperate latitudes. ***See*** individual foods

The occurrence of certain natural radionuclides plays a role in analytical methods. Fossil fuels do not contain carbon-14 (^{14}C) because their age is much greater than this radionuclide's half-life. (*Half-life*, also called physical half-life or radioactive half-life, is the time required for the disintegration of one-half of the atoms of a given radioactive substance.) This is of interest in food control because adulterations of alcoholic beverages with synthetic ethanol can be recognized by a lower content of ^{14}C. Similarly, synthetically produced acetic acid can be differentiated from natural vinegar. ***See*** Adulteration of Foods, Detection

Alpha radiation consists of particles containing two protons and two neutrons (i.e. helium nuclei) emitted by a radionuclide. *Beta radiation* consists of electrons or positrons emitted by a radionuclide. *Gamma radiation* is electromagnetic radiation emitted by a radionuclide. In contrast to α and β radiation, which cannot penetrate matter deeply, γ radiation has a high penetrating power. The ^{40}K in a body can thus be determined by measuring ^{40}K γ radiation outside this body. Whole-body counters for humans and animals are available that permit measurement of ^{40}K and other γ-emitting radionuclides. Since adipose tissue is more or less free of potassium, lean body mass can be determined in this way. The principle can also be used for nondestructive estimation of the fat content of meat.

Man-made Radionuclides

Consequences of Atmospheric Explosions of Nuclear Weapons

Since the first nuclear test explosion in the desert of New Mexico on 16 July 1945 large amounts of radioactive fission products have been released into the stratosphere

Table 1. Radionuclides naturally occurring in foods

Radionuclide	Radiation emitted	Half-life[a]	Concentration in foods	Annual intake (Bq)	Human body weighing 70 kg contains (Bq)
Cosmogenic					
Carbon-14	β	5730 y	220 Bq per kg C	22 000	2800
Tritium	β	12·3 y	0·4 Bq per kg water	350	22
Beryllium-7	γ	53·6 d	—	50	—
Sodium-22	β/γ	2·6 y	—	50	—
Primordial					
Potassium-40	β/γ	$1{\cdot}3 \times 10^9$ y	30·9 Bq per g K	33 000	4300
Rubidium-87	β	$4{\cdot}7 \times 10^{10}$ y	3 Bq per kg food	—	600
Uranium series					
Uranium-238	α/γ	$4{\cdot}5 \times 10^9$ y	<0·05 Bq per kg food	5	1
Thorium-234	β/γ	24·1 d	—	5	—
Uranium-234	α/γ	$2{\cdot}4 \times 10^5$ y	—	5	—
Radium-226	α/γ	1600 y	50 Bq per kg brazil nuts 2 Bq per kg cockles 0·5 Bq per kg eggs 0·7 Bq per kg peanuts <0·1 Bq per kg other foods	15	1
Lead-210	β/γ	22·3 y	20 Bq per kg shellfish <0·1 Bq per kg other foods	40	20
Polonium-210	α	138·4 d	200 Bq per kg mussels 10 Bq per kg reindeer meat 6 Bq per kg fish 0·1 Bq per kg eggs <0·1 Bq per kg other foods	40	6
Thorium series					
Radium-228	β	5·8 y	25 Bq per kg brazil nuts <0·1 Bq per kg other foods	15	1
Thorium-228	α/γ	1·9 y	40 Bq per kg brazil nuts 2 Bq per kg cereals <0·1 Bq per kg other foods	15	0·3

Source: United Nations (1982, pp 83–105), supplemented from other sources.
[a] y, year; d, day.

and have eventually settled on the earth's surface as 'nuclear fallout'. Short-lived radionuclides such as ^{131}I, a β- and γ-emitter with a half-life of 8 days, disappeared quickly, while long-lived species remained in the biosphere for many years, the most important ones being strontium-90 (^{90}Sr), a β emitter with a half-life of 29 years, and caesium-137 (^{137}Cs), a β- and γ-emitter with a half-life of 30 years. The total release of ^{90}Sr and ^{137}Cs by atmospheric tests was about 600 and 960 PBq, respectively (P, peta $= 10^{15}$). Fallout radioactivity in food reached a peak in 1964, when an agreement between the Soviet Union and the United States of America to stop atmospheric nuclear weapons tests became effective. This is demonstrated in Fig. 1 for ^{90}Sr in total diet samples collected in New York City. The majority of nuclear weapons tests were carried out in the northern hemisphere. The extent of radioactive contamination was therefore considerably lower in the southern hemisphere, as shown for Argentina.

Yearly averages of specific activity of radiocaesium in beef and pork produced in Germany are shown in Fig. 2. Results obtained during the 1960s and 1970s in other countries of the 30° to 70° north latitudes, which received the highest fallout deposits, were similar. After the peak of 1964, caesium activity decreased faster than strontium activity (shown in Fig. 1). This was a consequence of binding of caesium ions to clay minerals in the soil, which makes this element rather unavailable to the roots of plants, whereas the uptake of caesium ions deposited on the leaves of plants is quite rapid. Caesium activity in animals and in foods derived from animals closely followed the pattern of caesium activity in plants used for feed. In most soils, availability of strontium is much higher than that of caesium. Uptake

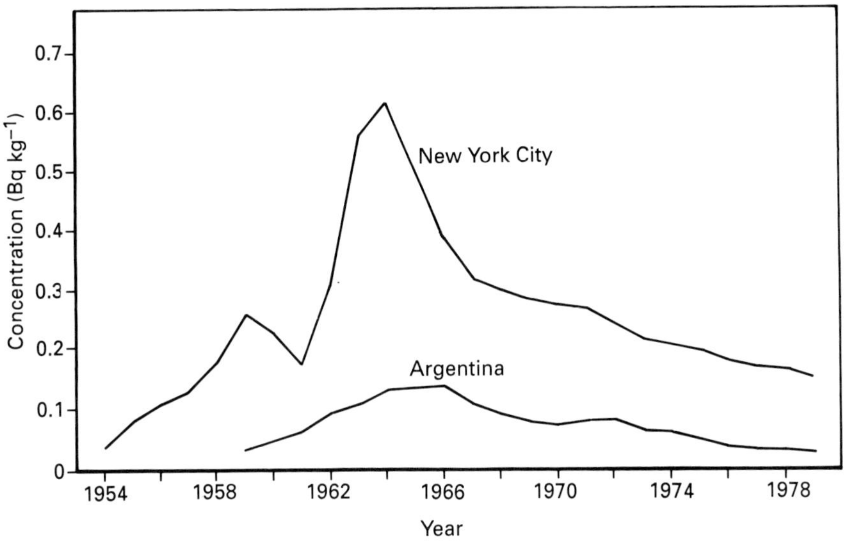

Fig. 1 Strontium-90 in total diet of Argentina and New York City, 1954–1979. Source: United Nations (1982, p. 217).

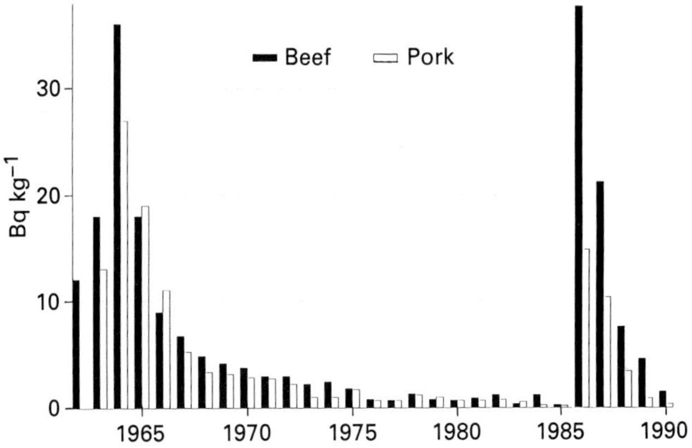

Fig. 2 Annual average of specific activity of radiocaesium (^{137}Cs until 1985, $^{134}Cs+^{137}Cs$ after 1985) in beef and pork in the Federal Republic of Germany, 1962–1990.

of strontium ions through plant roots continued long after deposition of fallout had decreased to below a measurable level. ***See*** Beef; Pork

Consequences of Nuclear Accidents

A severe accident occurred on 10 October 1957 at the Windscale Works, now known as Sellafield, in northwest England, when a graphite fire broke out in a uranium reactor primarily designed for production of plutonium. The accident released 740 TBq of ^{131}I, 22 TBq of ^{137}Cs, 0·07 TBq of ^{90}Sr and other radionuclides (T, tera$=10^{12}$). Authorities restricted the consumption of milk in an area of 500 km^2 for various periods of time. Environmental monitoring of radioactivity was not well developed at that time, and the extent of contamination remained largely unknown.

The accident at the Three Mile Island reactor in Pennsylvania, USA, on 28 March 1979 released less than 1 TBq of ^{131}I and no radioactive strontium or caesium. Radioactive iodine at barely detectable levels was found in some foods produced in the area.

According to information provided by the Soviet Government, the catastrophe of the Chernobyl reactor on 26 April 1986 released 260 PBq ^{131}I, 38 PBq ^{137}Cs, 18 PBq ^{134}Cs, 8 PBq ^{90}Sr, 5 Bq ^{241}Pu and large quantities of many other radionuclides. Shifting winds carried the radioactive clouds over various parts of the Soviet Union, Scandinavia, Central Europe, the Balkan countries and Turkey. Other parts of Europe were much less affected. A European Community Directive of 31 May 1986 established the following limits for radiocaesium (i.e. ^{137}Cs and ^{134}Cs): milk and milk products 370 Bq l^{-1}; baby food 370 Bq kg^{-1}; all other foods 600 Bq kg^{-1}.

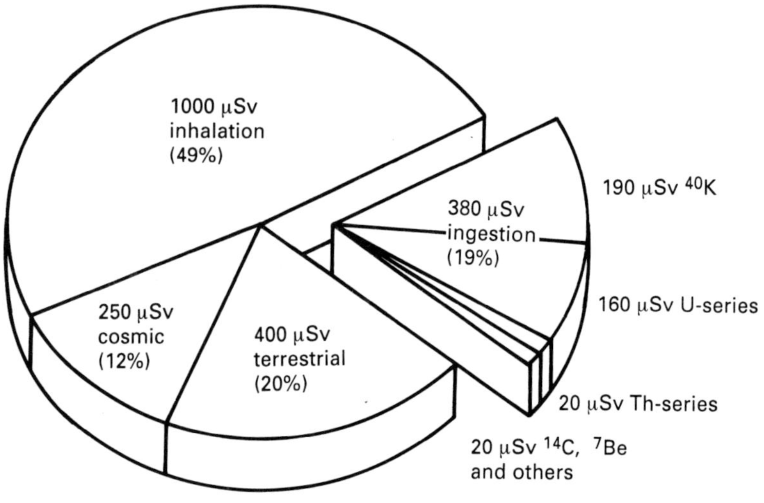

Fig. 3 Annual effective dose equivalent from natural sources of radiation. The dose from incorporated radionuclides contributes 19% of the total of 2030 μSv/person.

The results presented in Fig. 2 permit a comparison of the specific activity of radiocaesium in beef and pork after the Chernobyl accident with that observed at the time of atmospheric testing of nuclear weapons. It should be noted that the data of the pre-Chernobyl years represent ^{137}Cs alone, those of the post-Chernobyl years represent the sum of ^{137}Cs and ^{134}Cs. Measurable levels of ^{134}Cs were not present in the fallout from weapons' tests.

In countries reached by the radioactive plumes from Chernobyl, regional differences in the deposition of radioactivity correlated with rainfall during the critical days ('washout effect'). Levels of ^{90}Sr were little affected in Central Europe by the Chernobyl accident. Released radionuclides of low volatility, such as plutonium, were deposited mostly in the near vicinity of the accident site.

Information on the degree of radioactive contamination of foods produced in the more severely affected areas of Byelorussia, the Ukraine and the Russian Federation is conflicting. External radiation exposure due to radioactivity deposited on the ground appears to be a greater problem than the internal exposure from radioactivity ingested with food and drink. More than 135 000 people and thousands of farm animals were evacuated from a 30-km zone around the reactor site. Five years after the accident, restrictions to the production of milk and certain other foods still apply to certain areas outside the 30-km zone.

Radiation Exposure from Incorporated Radionuclides

Radioactivity alone, measured in becquerels, does not provide enough information to evaluate possible risks associated with the intake of radionuclides because one becquerel of one radionuclide does not usually have the same biological effect as one becquerel of another radionuclide. It is the amount of energy deposited in living tissues which primarily determines the possible biological effects of radiation. The radiation energy imparted per gram of tissue, the type of radiation (α, β, γ), the distribution of the particular radionuclide in the body and its rate of excretion, and the different vulnerability to radiation of different tissues and organs, must all be taken into account. Following procedures developed by ICRP (the International Commission for Radiological Protection), the *effective dose equivalent* can be calculated, taking all these factors into consideration. It is an approximate measure of the actual ('effective') risk of genetic and somatic radiation damage to man during the remaining life time. For brevity it is simply referred to as 'dose' in the following discussion – unless the full term is required for clarification. The dose unit is called the sievert (Sv), and one sievert is defined as one joule per kilogram.

Following the ICRP procedure, the annual dose caused by ingestion of various naturally occurring radionuclides has been estimated to be 380 μSv for adults, with ^{40}K making the largest single contribution (Fig. 3). It is obvious from comparison of Fig. 3 with Table 1 that the radionuclides of the uranium and thorium series contribute significantly to the ingestion dose in spite of their low radioactivity in the diet. This is partly due to the higher vulnerability of biological systems to α radiation as compared to β and γ radiation, but the long biological half-life of some of these α-emitters also plays a role.

It is very difficult to achieve a total ingestion dose of less than 380 μSv by resorting to special diets. Potassium (and with it ^{40}K), is an essential mineral, under close homeostatic control in the body. Carbon (and with it ^{14}C) is contained in all foods. The elements of the uranium and thorium series are present in most foods of plant or animal origin, although at very low concentrations. Populations living in areas where the soils are rich in uranium/thorium minerals or individuals regularly consuming reindeer meat, brazil nuts or seafood can reach annual exposure levels considerably higher than 380 μSv. ***See*** Potassium, Physiology

Ingested radionuclides are not the only source of radiation to which man is inevitably exposed. Inhalation of the natural radionuclides radon-222 (^{222}Rn) and radon-220 (^{220}Rn) (also called thoron) causes an annual dose of 1000 μSv (Fig. 3). This must be seen together with an annual external dose of 650 μSv, of which 250 μSv comes from cosmic radiation and 400 μSv from terrestrial radiation, for instance the γ-radiation emitted from ^{40}K present in the soil and in building materials. The dose from ingested natural radionuclides thus contributes 19% of the total dose of 2030 μSv from all natural sources.

The average dose values here presented apply for the adult population in Germany. The contribution from cosmic radiation increases with increasing altitude, that from terrestrial radiation and from inhalation fluctuates considerably with local geological conditions. In Switzerland, for instance, the average dose from cosmic radiation is 400, from terrestrial radiation 550 and from inhalation 1600 μSv. The internal dose from ingested natural radionuclides is 380 μSv, as in Germany; it contributes 13% to the total dose of 2930 μSv from all natural sources.

How does the exposure from incorporated man-made radionuclides compare with that from natural radionuclides? The average intake of ^{137}Cs from fallout in 1964 reached 9 Bq per day in Central Europe, or 3280 Bq in the whole year. For adult individuals this meant an effective dose equivalent to 46 μSv. To this must be added 14 μSv from ^{90}Sr. In all, the intake of fallout radioactivity up to 1985 in adults living in Central Europe resulted in a life-time dose of 240 μSv from ^{137}Cs and 160 μSv from ^{90}Sr.

After the Chernobyl accident, the amounts of ^{131}I, ^{137}Cs and ^{134}Cs ingested by adults between 1986 and 1990 have contributed 125 μSv to the life-time dose. The low levels of radioactive contamination still present in some foods may cause an additional contribution of 15 μSv in coming years, so that the total dose resulting from ingestion of radionuclides derived from the Chernobyl reactor should be around 140 μSv in Central Europe. In the United Kingdom a dose of less than 20 μSv may be expected. Populations living in Central Europe in areas that received particularly high radioactive depositions in May 1986 may receive twofold or threefold higher doses. Even these doses are small compared to an *annual* dose of 380 μSv from ingestion of natural radionuclides. They are also small compared to the ICRP *annual* dose limit of 1 mSv for the general public. With the exception of populations living in the most severely affected areas of the Soviet Union, it is most unlikely that the radioactive contamination resulting from the Chernobyl disaster will lead to a statistically detectable increase in health effects, even if it is taken into account that the dose from external radiation will be several times higher than the internal dose.

Bibliography

Carter MW (ed.) (1988) *Radionuclides in the Food Chain.* Berlin: Springer.

Diehl JF, Frindik O and Müller H (1989) Radioactivity in total diet before and after the Chernobyl reactor accident. *Zeitschrift für Lebensmittel-Untersuchung und -Forschung* 189: 36–38.

Eisenbud M (1987) *Environmental Radioactivity – From Natural, Industrial and Military Sources.* Orlando: Academic Press.

International Atomic Energy Agency (1989) *Measurements of Radionuclides in Food and the Environment. A Guidebook.* Vienna: IAEA.

Maul PR and O'Hara JP (1989) Background radioactivity in environmental materials. *Journal of Environmental Radioactivity* 9: 265–280.

United Nations (1982) *Ionizing Radiation – Sources and Biological Effects.* UN Committee on the Effects of Atomic Radiation (UNSCEAR), 1982 Report to the General Assembly. New York: UNO.

Walker MI, Walters B and Mondon KJ (1991) The assessment of radiocaesium intake from food using duplicate diet and whole-body monitoring techniques. *Food Additives and Contaminants* 8: 85–95.

J. F. Diehl
Federal Research Centre for Nutrition, Karlsruhe, FRG

RADIOIMMUNOASSAY

See Immunoassays

RAISING AGENTS

See Leavening Agents and Yeasts

RAISINS

See Grapes

RAMAN SPECTROSCOPY

See Spectroscopy

RASPBERRIES AND RELATED FRUITS*

Raspberries are a high-value crop due to their unique flavour, exacting climatic requirements, high costs of production, and perishability. This article describes the relationships among various raspberry species and relatives, discuss the commercial raspberry industry, describe the morphology, anatomy and chemical composition of fruit, review harvesting and handling techniques for fresh fruit, and present various food uses for the raspberry.

Taxonomy and Commercial Importance

Raspberries are a diverse group of flowering plants that are closely related to blackberries. Both raspberries and blackberries belong to the genus *Rubus*. Taxonomists recognize 12 subgenera within *Rubus*, but only the raspberries (*Idaeobatus*) and blackberries (*Eubatus*) have obtained commercial significance. The fruit of raspberries detaches from the receptacle when picked, leaving a white torus attached to the plant and a hollow fruit. This characteristic distinguishes them from blackberries in which fruit abscission occurs behind the receptacle.

The genus *Rubus* is a member of the rose family (Rosaceae) which also includes important fruit crops such as apples, pears, cherries, peaches, plums and strawberries. Wild raspberries occur on five continents, but are most abundant in the Northern Hemisphere. The temperate and subtropical region of eastern Asia is recognized as the centre of origin where the most diversity exists. More than 200 species have been identified, but only a few are important commercially. These include the European red raspberry (*R. idaeus* subsp. *vulgatus* Arrhen.), the North American red raspberry (*R. idaeus* subsp. *strigosus* Michx.) and the black raspberry (*R. occidentalis* L.). Hybrids between the red and black raspberry are commonly called purple raspberries because of the fruit and cane colour, and these were once given the specific rank of *R. neglectus*

* The colour plate section for this article appears between p. 3896 and p. 3897.

Peck. However, most taxonomists do not recognize hybrids as distinct species. Interspecific hybrids with blackberries have also been made; some of these are commercially important, such as Tayberry, Loganberry, Boysenberry and Youngberry.

Several other species within the genus *Rubus* have edible fruit, or have been used by plant breeders to improve cold hardiness and resistance to diseases and insects in raspberries. Examples are *R. glaucus* Benth., a South American tetraploid black raspberry that is probably a raspberry × blackberry hybrid, *R. leucodermis* Torr. and Gr. (the western North American black raspberry), and *R. spectabilis* Pursh. (salmonberry); the Asiatic species *R. coreanus* Miq., *R. phoenicolasius* Maxim. (Japanese wineberry), *R. parvifolius* Nutt. (trailing raspberry), *R. ellipticus* Sm. (golden evergreen), *R. illecebrosus* Focke (strawberry raspberry), *R. kuntzeanus* Hemsl. (Chinese raspberry), and *R. nivens* Thumb.; the Hawaiian species *R. macraei* Gray and *R. hawaiiensis* Gray (Akala berries); and the arctic raspberries of Europe (*R. arcticus* L.) and North America (*R. stellatus* Sm.).

Commercial Industry

Raspberries were first introduced into cultivation in Europe nearly 450 years ago. By the early nineteenth century, more than 20 cultivars of red raspberry were grown in both England and the USA. English cultivars were then exported to the USA, where crosses between them and North American seedlings gave improved cultivars. Red raspberries are the most widely grown, while black raspberries are popular only in certain regions of the eastern USA. The progeny of black and red raspberries have purple fruits and canes; these types are popular in eastern North America. Yellow-fruited *R. idaeus*, caused by a recessive mutation, is also grown on a limited scale for speciality markets. *R. occidentalis* genotypes with yellow fruit are not grown commercially.

The three major raspberry production regions are (1) the USSR, (2) Europe (mostly in Poland, Hungary, Yugoslavia, Germany, and the UK), and (3) the Pacific Coast of North America (British Columbia, Washington and Oregon). Much of the fruit produced in these regions is harvested mechanically and processed. In other production regions, such as eastern North America, nearly all the production is for the fresh market. Many other countries, such as Chile, New Zealand, and Australia, have small, but growing industries. World production is estimated at 375 000 t.

Patterns of production in North America shifted dramatically in the early 1900s. In 1920, New York State growers harvested more than 4000 ha. The systemic 'mosaic virus disease' infected most of the planting stock, and the processing raspberry industry collapsed in this region. With the development of tissue culture propagation techniques, virus-indexing of nursery stock, and breeding for resistance to the virus vector, the raspberry processing industry redeveloped on the West Coast.

Varieties

Two types of bearing habits are found in commercial red raspberries. The first type is called a 'summer-bearing' habit. Canes originate from either crown buds or adventitious root buds in early spring. Canes elongate during the growing season, forming fruit buds in the axils of leaves in the autumn when temperatures decrease and day lengths shorten. The plants become dormant for winter, then the buds on the cane grow the following spring. The lateral buds contain both leaves and flowers. Fruiting occurs in early summer, then the cane senesces. While second year canes (floricanes) are flowering, first year canes (primocanes) are growing from the crown or roots. More than 40 summer-bearing red raspberries are grown commercially, and these change with the release of new, improved varieties. Among the major varieties originating in North America are 'Boyne', 'Canby', 'Chilcotin', 'Newburgh', 'Skeena', 'Taylor', 'Titan' and 'Willamette'. The Glen and Malling series are important varieties from the UK.

The second type of growth habit is called 'fall-bearing'. In some varieties, fruiting laterals will develop from the top of first year primocanes when they reach a certain size. If the growing season is sufficiently long, fruit can be harvested from the tops of these canes in autumn. The lower portion of the cane will fruit the following summer, if it is allowed to remain in the field. The major variety is 'Heritage'; other important varieties are 'Amity' and 'Autumn Bliss'.

At least 13 varieties of black raspberries are grown in North America. The most successful are 'Alleghany', 'Allen', 'Bristol', and 'Jewel'. 'Royalty' is the most popular purple raspberry. Several yellow or golden raspberry varieties are grown on a small scale for speciality markets.

Morphology and Anatomy of Fruit

Each raspberry flower contains from 60 to 160 ovaries. Each ovary contains two ovules, but one usually aborts after differentiation. About one month after pollination, the ovaries ripen simultaneously to form the fruit.

Fruits ripen in three phases. Pollination is followed by a period of rapid cell division. In the second phase, cell division slows while the embryo develops and the seed coat hardens. Finally, very rapid growth occurs due to cell enlargement. Each phase lasts from 10 to 12 days.

Considerable variation in fruit size exists, with a range from 1 to 5 g.

As with most climacteric fruit, ethylene production in the raspberry begins in the receptacle when the fruit starts to colour, and peaks when fully ripe. Respiration, however, decreases as ripening proceeds. Growth stresses help to fracture the middle lamella and walls of cortical cells where the fruit attaches to the receptacle. Cell wall breakdown and disintegration are complete when the fruit is fully ripe. The detachment force is small, usually 25 g, when fruits are ripe. A 12-fold difference in detachment force can exist between underripe and overripe fruit.

The raspberry fruit is not a true berry, but rather an aggregate of many individual drupelets. Each drupelet is anatomically analogous to a cherry with a hard endocarpic seed (pyrene) surrounded by a fleshy mesocarp and an outer exocarp. The fleshy mesocarp is composed of thin-walled, turgid parenchymatous cells. Just below the exocarp is a thin layer of oval, collenchymatous cells. Large seeds are undesirable, but there is a relationship between seed size and drupelet size. Small seeds tend to be associated with small drupelets. An average seed has a mass of about 1 mg, and comprises between 4% and 5% of the total mass of a berry. A 100 g sample of raspberries may contain more than 4000 seeds.

The cohesion of individual drupelets results from the entanglement of unicellular, epidermal hairs that are most abundant on the sides and base of the drupelet. In some black raspberry varieties, fusion of the cuticle or wax also contributes to drupelet cohesion. Drupelets cannot normally be separated without tearing the exocarp. Considerable variation in fruit firmness and drupelet cohesion exists among commercial varieties. If the percentage of developed drupelets is low, then cohesion will be poor. Firmness is related to cell diameter and tissue compactness.

Chemical and Nutritional Composition

Fruit Composition

The main constituent of the raspberry fruit is water (*c.* 87%). Of the remaining solids, 9% are soluble and the rest insoluble. Pectins compose 0·1–1·0% of the soluble fraction, but this amount decreases with ripening due to hydrolysis.

The main sugars are glucose, fructose and, to some extent, sucrose. These compose the major soluble component of the juice. A typical ripe raspberry fruit will contain 5–6% sugar. Citric acid is the second largest component of the soluble fraction; raspberries contain very little malic acid and at least 10 other acids in trace amounts. The amount of acid in the fruit increases early in development, then decreases as the fruit begins to ripen. The balance between the sugars and acids is important for consumer acceptance. A fruit with a low sugar : acid ratio will taste tart; one with a high ratio will taste bland. A typical pH of a ripe raspberry fruit is 3·0–3·5; the ratio between sugars and acids (w/w) is approximately 1·0. Fruits grown under warm, dry summers are sweeter, less acid, more aromatic, and more highly coloured. Hot weather will reduce the aroma of the fruit, and wet weather will reduce the sugar content.

A large number of volatile compounds contribute to the flavour of raspberries; most are present at less than 10 ppm. Compounds include alcohols, acids, esters, carbonyls, ketones, and other hydrocarbons (naphthalene and related compounds). A particular ketone, 1-(*p*-hydroxyphenyl)-3-butanone, has an odour very much like raspberry.

Considerable changes occur in phenolic compounds as fruits mature, and these are linked with oxidative enzyme systems. Raspberry juice from ripe fruit contains 0·10–0·14% polyphenols, mainly as catechin and chlorogenic, ferulic and neochlorogenic acids.

Raspberry fruits contain small amounts of vitamins; only vitamin C is present at a significant level. Amino acids include alanine, serine, asparagine, glutamic acid, glutamine, aminobutyric acid, valine, leucine, and aspartic acid. Table 1 presents data on the vitamin and nutrient content of red raspberries.

Fruit Colour

The colour of raspberry fruit is imparted by anthocyanins. The anthocyanin molecule in raspberry consists of cyanidin and pelargonidin with glucose attached at the 3 position. Additional glucose, rhamnose or xylose sugars may be present in various combinations to give diglycosides or triglycosides. Fruits with a preponderance of pelargonidin glycosides have an orange-red colour, as opposed to a deep red colour with cyanidin glycosides.

The type and concentration of anthocyanins in the fruit are controlled by several genes. Fruits with a yellow colour are produced when certain genes suppress anthocyanin production. Molecules that are complexed with the anthocyanin also affect colour development, and the pH of the fruit has a small effect. During storage, raspberry fruit increases in anthocyanin, darkens, and becomes more blue.

Handling and Storage of Fresh Produce

Harvest Considerations

Raspberries have one of the highest respiration rates of any fruit. This, coupled with their thin skin and sugary

Table 1. Reported values or ranges of nutrient content in 100 g of fresh raspberries

Nutrient	Amount
Water (g)	84–87
Food energy (kcal)	31–49
Protein (g)	0·42–1·40
Fat (g)	0·20–0·55
Carbohydrate (g)	5·8–11·6
Fibre (g)	3·0–7·4
Ash (g)	0·40–0·51
Minerals (mg)	
Calcium	22–50
Iron	0·57–1·20
Magnesium	18–30
Phosphorus	12–50
Potassium	130–221
Sodium	0–2·5
Zinc	0·46
Copper	0·07–0·21
Manganese	1·01
Sulphur	17·3
Chlorine	22·3–22·8
Boron	71–125
Vitamins (mg)	
Carotene	0·05–0·08
Thiamin	0·01–0·03
Riboflavin	0·03–0·10
Pantothenic acid	0·24–0·30
Nicotinamide	0·20–1·00
Vitamin B_6	0·06–0·90
Vitamin C	13–38
Tocopherols	0·3–4·5

interior, makes them among the most perishable of all fruits. With any given variety, fruit will ripen over a period of several weeks. Harvesting the same planting frequently (once every two days) is critical. Fruit harvested before it is fully ripe will have a much longer shelf life than fully ripe or overripe fruit. The optimum stage of maturity for the raspberry occurs when the berry first becomes completely red, but before any darker hues develop.

Fruit quality for fresh market raspberries usually declines as the season progresses. Marketing channels must be open before the first berries ripen as these are likely to have the highest quality and largest size for the season. Berries should not be touched before harvest, and only undamaged berries with good appearance should be placed in the pack. The magnitude of injury caused by human pickers can be so great as to mask any other causes of deterioration.

Overripe or damaged berries should be harvested and discarded because they are susceptible to moulds. *Botrytis* is the most common pathogen of raspberry fruit. Once the mould growing on overripe berries sporulates, large amounts of inoculum will be present to infect other ripening fruit. Overripe berries also attract ants, wasps and other pests.

Containers holding approximately 150 g of raspberries should be used. Wide, shallow containers are preferable to deep containers; each should have no more than four layers of raspberries. Many different types of containers are available.

Post-harvest Considerations

The objective of post-harvest handling of raspberries is to slow the respiration and transpiration rate. Respiration and transpiration result in shrinkage and reduced sweetness. Conditions that slow the respiration process are low temperatures, high carbon dioxide levels, and low oxygen in the storage chamber. Transpiration is slowed by high humidity.

Temperature is the easiest environmental variable to modify for extended storage of raspberries. A 5°C reduction in temperature reduces the respiration rate by approximately 50%: at 0°C the respiration rate is 24 mg of carbon dioxide per kg of raspberry per hour; at 5°C the rate is 55; at 10°C the rate is 92; at 15°C the rate is 135; and at 20°C the rate is 200. Rapid movement of cold, humid air through the berries is essential during the first few hours after harvest. Large growers may have a separate precooling facility specifically designed for removing field heat. For every hour delay in cooling, shelf life is reduced by one day. Growers can take advantage of night cooling by harvesting fruit as early in the morning as possible.

Once the berries are cool, containers should be wrapped in plastic to prevent water loss from the fruit and condensation on the berries when they are removed from the cooler. The plastic should not be removed until the temperature of the berries warms to near the ambient temperature.

The storage room can be maintained as low as −1°C. Berries will not freeze at or above this temperature because the sugars in the fruit depress the freezing point.

When temperature is lowered, the amount of moisture in the air is reduced. For raspberries, it is critical to maintain a humid atmosphere (90–95%) simultaneously with low temperature to prevent water loss from the fruit. Special cooling units designed to maintain a high humidity are required for raspberries. At 25°C and 30% relative humidity, fruit loses water 35 times faster than at 0°C and 90% relative humidity.

A high carbon dioxide atmosphere (up to 40%) will reduce respiration and mould growth. *Botrytis*, *Rhizopus*, *Alternaria*, *Penicillium* and *Cladosporium* cause post-harvest fruit rot, depending on storage temperature and carbon dioxide level. High carbon dioxide atmospheres are used when raspberries are transported long distances. Special semipermeable wraps are also available that create a modified atmosphere within individual containers. Low-oxygen atmospheres will extend the shelf life of raspberries, but bad-tasting aldehydes and alcohols will accumulate in the raspberry fruit when oxygen is limited. Off-flavours also develop when raspberries are held under elevated carbon dioxide levels for an extended period.

The loss of raspberries from harvest to the consumer's table has been estimated at more than 40%. A 14% loss occurs from farmer to wholesaler, a 6% loss occurs from wholesaler to retailer, and 22% is lost between the retailer and consumer. Most of these losses are due to poor handling of berries after harvest.

Fruit Uses

In the early 1900s, black raspberry juice was extracted, concentrated, and used as an edible dye for foodstuffs, such as meat. Raspberries were also dehydrated for long-distance transport. Today, the majority of raspberry fruits are harvested by hand and eaten fresh, or machine-harvested and processed into purées, preserves, juice, jam, jelly, dessert topping, pie filling, or yoghurt. A limited amount of wine is made from raspberry juice.

Bibliography

Crandall PC and Daubeny HA (1990) Raspberry management. In: Galletta GJ and Himelrick DG (eds) *Small Fruit Crop Management*, pp 157–213. New Jersey: Prentice Hall.

Green A (1971) Soft fruits. In: Hulme AC (ed) *The Biochemistry of Fruits and their Products*, vol. 2, pp 375–410. New York: Academic Press.

Jennings DL (1988) *Raspberries and Blackberries: their Breeding, Diseases and Growth.* New York: Academic Press.

Ourecky DK (1975) Brambles. In: Janick J and Moore J (eds) *Advances in fruit breeding*, pp 98–129. Indiana: Purdue University Press.

Pritts M and Handley D (1989) *Bramble Production Guide.* Northeast Regional Agricultural Engineering Service Bulletin 35. Cornell University, New York.

M. P. Pritts
Department of Fruit and Vegetable Science, Cornell University, USA

RATIONING OF FOODS

Historical Introduction

The first Food Controller was appointed by the British Government in December 1916, not because of shortage of food, but because of high food prices and industrial unrest. The government then assumed almost unlimited control over food supplies and prices, despite the statement in the House of Commons by the President of the Board of Trade, as recently as 17 October 1916, that 'We want to avoid any rationing of our people in food'. When war clouds began to loom again in the 1930s the first suggestion that a food defence department should be set up sprang from a desire to forestall inflation, should war break out.

The Food (Defence Plans) Department of the Board of Trade was established in November 1936 and its first report was published in 1938 'to explain in broad outline a part of the preparations that are being made . . . for feeding the nation in the event of war'. One of the aims was 'to ensure that supplies of essential foodstuffs at controlled prices are available to meet the requirements of all types of consumers and in all parts of the country, if, and when, an emergency arises'. The report states that to secure this food control, the organization of supplies and the regulation of consumers' demands were essential. The report also states that food control has three aspects with which the consumer is vitally concerned:

1. Guarantee of regular supplies.
2. Limitation of prices and profits.
3. Equality of sacrifice in the event of shortage.

On these principles, plans for food rationing and distribution were made by the Board of Trade. They were adopted by the Ministry of Food (the second; the first was established when the first Food Controller was appointed in December 1916) instituted in September 1939, immediately after the outbreak of war.

Control by Ministry of Food

Legally and politically the Ministry of Food was brought into existence overnight, and the growth of its administration machine was rapid. Its headquarters staff numbered about 350 when war started, and increased 10-fold by March 1940. By the summer of that year the ministry had achieved a settled form – bearing the main features of the Food (Defence Plans) Department – that was to survive well into the postwar years.

The control of prices and distribution of foods extended beyond what might be regarded as essential foodstuffs, and the ministry's responsibilities towards nutritionally vulnerable groups, such as infants, children, expectant and nursing mothers and invalids, also went beyond what was probably originally foreseen. Nutritional considerations also became more important than they were when the initial plans were made.

The ministry's objectives were as follows: to ensure equal shares of the more scarce foods to all consumers; to make special provision of milk and other foods for the nutritionally vulnerable groups; to leave unrationed for as long as possible certain 'buffer' foods, in particular bread and potatoes; and to provide communal eating facilities so that industrial and other workers and school children could supplement their rations with meals away from home. These measures, taken as a whole, might be summarized as 'fair shares for all', a phrase which implies special provision for those sections of the community with special needs, e.g. high-energy foods for heavy manual workers and growth-promoting nutrients for children and child-bearing women. There was also a 'Points' rationing scheme, introduced in December 1941, whereby nonperishable foods, mainly canned and dried, too scarce to ration, could be bought against special coupons. This scheme was introduced to counter public criticism that these foods, previously unrationed, were being distributed unequally.

Domestic Rationing

Rationing of foods for domestic consumers began in January 1940 with butter, bacon and ham, and sugar. It was extended, as supplies of other foods became scarce, to include carcass meat, margarine, cooking fat, tea, and cheese. By the end of 1943, most foods were distributed by one or other of the different domestic rationing systems:

1. *Straight rationing*: carcass meat, bacon and ham, cheese, fats, sugar, preserves, tea (carcass meat was rationed by cash value; the others by weight).
2. *Points rationing*: canned meats, canned fish, canned peas and beans, canned fruit, dried fruit, rice, sago, tapioca, dried pulses, condensed milk, cereal breakfast foods, biscuits, oat flakes and rolled oats, syrup and treacle.
3. *Personal points rationing*: chocolate and sugar confectionery.
4. *Controlled distribution schemes*: liquid and dried milk, shell eggs and dried eggs, oranges and onions. Distribution of eggs, oranges and onions entailed systems of allocations which depended on supplies and other factors. Schemes for milk distribution are shown in Table 1 (see *Nutritional Aspects of Rationing*, below).
5. *Schemes for invalids*: special rations, either additional to or in substitution for ordinary rations, were supplied against a medical certificate to invalids suffering from certain specified complaints.

The only important foods not rationed in 1945 were bread, flour, oatmeal, potatoes, other fresh vegetables, fresh fruit (other than oranges) and fish. In 1940, the first year of rationing, 25% of household food expenditure went on rationed foods, whereas at the height of wartime rationing, in 1943, only 40% of food expenditure was on unrationed foods.

After the end of the war there was a severe world shortage of cereals, and flour; bread and flour confectionery had to be rationed for 2 years from July 1946. Because of the severe winter and late spring in 1947, the potato crop in that year was unusually small and the distribution of potatoes had to be controlled from November 1947 until April 1948.

Communal Feeding

An initial plan for the surrender in catering establishments of coupons for meals containing meat or bacon was abandoned when bacon was rationed in January 1940. Ultimately, the principle was accepted that occasional meals eaten in catering establishments should be additional to domestic rations, because most of such meals were eaten by people whose employment did not permit them to return home for a meal. Thus the supply of foods to the catering industry was strictly controlled. For residential establishments, food was allocated on the basis of the number of ration books held by residents; for nonresidential catering establishments, allocation was on the basis of the number of meals and hot beverages served, and the scale of allocation was devised to make catering and domestic meals roughly equivalent. For example, it was assumed that the fat ration would be spread over 4 meals a day or 28 a week. Thus the caterer was allocated one twenty-eighth of the domestic ration for each meal served. It was assumed that meat would be eaten at 2 main meals a day or 14 a week, and the caterer's allowance was one fourteenth of the domestic ration for each meal served.

For particular groups of the population, communal meals were used as a means of supplementing ordinary domestic rations in the following ways.

Industrial Canteens

From November 1940, the Ministry of Labour and National Service could require the establishment of a

Table 1. Nutritional measures employed during rationing in UK

Measure	Object	Class benefited
Steps to increase milk (milk solids) consumption: increased milk production; diversion of milk to liquid market; increased imports of skimmed dried milk and cheese National Milk Scheme (cheap milk for expectant mothers, infants and children up to 5 years) Milk-in-Schools Scheme (expansion) Milk Cocoa Drink Scheme (for adolescents) Milk Supply Scheme (to provide for differential rationing of milk)	To raise levels of intake of Animal protein Calcium Riboflavin	Primarily, the 'vulnerable' groups; secondarily, the whole population
Addition of vitamins A and D to margarine	To make margarine nutritionally similar to butter To raise intakes of vitamins A and D To compensate to some extent for the shortage of eggs, which are one of the few natural foods rich in Vitamin D	Whole population
Introduction of 85% extraction flour and National bread	To remove all risk of vitamin B_1 deficiency To improve iron consumption To improve riboflavin and nicotinic acid consumption	Whole population
Addition of calcium carbonate to flour	To raise calcium consumption To counteract any immobilizing influence on calcium of phytic acid in high extraction flour	Whole population
Increased production and consumption of carrots and green vegetables	To maintain supply of vitamin C to compensate for reduction of fruit imports To improve vitamin A consumption	Whole population
Vitamins Welfare Scheme: orange juice, cod liver oil, vitamin A and D tablets	To ensure adequacy in respect to vitamins A, C and D during pregnancy and early life of children	Pregnant women, infants and children up to 5 years

Modified from Drummond JC and Wilbraham A (1958), pp 451–453.

canteen at any factory employing more than 250 persons on munitions or other work on behalf of the Crown; this was extended to all kinds of industrial establishments in 1943. For all industrial canteens there were special scales of allowance which gave more rationed foods per meal than in restaurants for the general public.

School Meals

The School Meals Scheme for England and Wales, which started in 1906, was extended in 1941 to include Scotland. Its purpose was to ensure an adequate diet, particularly in animal protein, for schoolchildren. Especially generous scales of allowances of rationed foods, including double the ordinary catering meat allowance, were available for school meals.

British Restaurants

In 1940 the Ministry of Food encouraged local authorities to establish communal centres in which cheap and nourishing meals could be provided for people who were obliged to eat away from home and for whom no other catering facilities were available. No special allocation of rationed foods was available for these restaurants.

Table 2. Energy value and nutrients in domestic food consumption (percentages from rationed and controlled foods)

	Year	Rationed by: Weight or cash value	Points	Milk and eggs	Total from rationed and controlled foods	Total from unrationed foods
Energy value (%)	1945	29	7	11	47	53
	1949	27	5	11	43	57
Protein (%)	1945	16	7	19	42	58
	1949	14	4	19	37	63
Fat (%)	1945	56	6	16	78	22
	1949	58	4	16	78	22
Calcium (%)	1945	10	5	50	65	35
	1949	7	4	44	55	45
Iron (%)	1945	16	12	8	36	64
	1949	13	5	7	25	75
Vitamin A (%)	1945	29	2	19	50	50
	1949	30	2	18	50	50
Thiamin (%)	1945	12	7	12	31	69
	1949	7	2	13	22	78
Riboflavin (%)	1945	16	4	42	62	38
	1949	14	3	40	57	43
Nicotinic acid (%)	1945	23	7	3	33	67
	1949	19	3	3	25	75
Vitamin C (%)	1945	0	0	4	4	96
	1949	0	1	3	4	96
Vitamin D (%)	1945	42	21	13	76	24
	1949	50	7	12	69	31

Source: Ministry of Food (1951a).

Other Schemes

A special cheese ration was introduced in 1941, and the Rural Pie Scheme a year later, for agricultural workers unable to get meals in canteens. There were also special schemes for feeding people who had suffered from severe air raids.

Nutritional Aspects of Rationing

Although the nutritional state of the population aroused increasing attention in the 1930s, the prewar plans for food control took no particular account of the nutritional needs of the people. The government assumed that most foods would continue to be in adequate, albeit perhaps reduced, supply, and that the choice of diet could be left to the individual, subject to such rationing as shortages might necessitate.

Before the war, about 30% of the UK food supply (measured in terms of its energy value) was home-produced; the rest was imported. Early in the war it became clear that there would be competition for shipping space between imports of food and other war materials, and throughout 1941 many ships were sunk by enemy action. Thus imports of foods halved between the immediate prewar years and 1942–1944. Agricultural production was increased, but by 1943–1944 only about 40% of the food supply was produced at home. These events made it essential to produce detailed nutritional analyses and justification for the total programme of food consumption. The first comprehensive report on this by the Scientific Advisor to the Ministry of Food (JC Drummond) was prepared just before the collapse of France. Thereafter, the worsening food situation forced continual attention to the nutritional value of food supplies and the nutritional needs of the people. The Scientific Advisor set out from the start not only to maintain but to improve the nutritional value of the British diet. The psychological importance of the traditional foods to which people were accustomed was fully taken into account in planning food supplies and rationing. The nutritional measures that were planned and put into effect are summarized in Table 1. These were superimposed on the general system of rationing already described.

According to the records of the National Food Survey, nearly 50% of the energy value of the working class diet in the 1940s was derived from foods that were rationed by weight, cash value or points, or under controlled distribution, except in the years when flour and its products were rationed and potato distribution was controlled, when the proportion rose to over 80%. The contribution of rationed and controlled foods to urban working class consumption of energy and nutrients in 1945 and 1949 is shown in Table 2.

Administration of Rationing

Before the outbreak of war, each local authority was asked to appoint a Food Control Committee. By September 1939 about 1520 such committees had been set up, later falling to 1250. Most consisted of 16 to 18 members, representing local consumers and retailers and employees in local food shops. The powers and functions of the committees, assigned by the Minister of Food, included the enforcement of the minister's orders (including the rationing orders). The executive functions of the committees and much of the daily business of food control were performed by local food offices, of which there were about 1500 in 1939 and 1200 in 1949. As an economy measure some of the work of local food offices was transferred to district offices in 1949.

The administration was a great success and requisite rations were always available in the shops. This had an important effect on public morale.

End of Food Rationing

Tea was the first food to be taken off the ration (in October 1952); bacon, ham and carcass meat were the last. The only foods that were prematurely derationed (in April 1949) were chocolates and sweets. These had to be rationed again in August 1949 because postwar demand exceeded supplies which, on the basis of prewar consumption, had been thought to be sufficient. They were finally freed from rationing in February 1953.

All rationing came quietly to an end in July 1954.

Bibliography

Board of Trade (1938) *Report of the Food (Defence Plans) Department. For the Year ended 31st December, 1937.* London: Her Majesty's Stationery Office (HMSO).

Drummond JC and Wilbraham A (1958) *The Englishman's Food* 2nd edn (revised and with a new chapter by Dorothy Hollingsworth). London: Jonathan Cape.

Hammond RJ (1951) *Food*, vol. I. *The Growth of Policy*. London: HMSO and Longmans.

Hammond RJ (1956) *Food*, vol. II. *Studies in Administration and Control*. London: HMSO and Longmans.

Hammond RJ (1962) *Food*, vol. III. *Studies in Administration and Control*. London: HMSO and Longmans.

Ministry of Food (1946) *How Britain was fed in War Time. Food Control 1939–1945*. London: HMSO.

Ministry of Food (1951a) *The Urban Working-Class Diet 1940 to 1949. First Report of the National Food Survey Committee*. London: HMSO.

Ministry of Food (1951b) *ABC of Rationing in the United Kingdom*. London: Ministry of Food.

Dorothy F. Hollingsworth
Petts Wood, Orpington, UK

RECOMBINED AND FILLED MILKS

General Considerations

The manufacture of recombined milk and milk products is a technology which has emerged over the past three decades. During this time there has been a worldwide development in the establishment of dairy recombination plants, mainly in countries which have a limited indigenous dairy industry, or in regions suffering from seasonal deficiencies in milk supply. Recombined dairy products are therefore ensuring a year-round supply for many households in developing countries, and also enabling the extension of the local dairy industries. On the other hand, for the industrialized countries milk recombination offers the opportunity to transfer raw materials (milk powders, anhydrous milk fat, etc.) from surplus production areas to deficiency areas, in order to compensate for the above mentioned problems and to open up new markets. Hence it should be taken into consideration that world stock levels and prices of butter and milk powder are strongly affecting the development of recombination industry as well as the trend of prices of recombined dairy products.

Table 1. Typical composition of recombined milk products

Ingredients	Product formulations (kg t^{-1}) Recombined milk		Recombined cream	Recombined sweetened condensed milk
Skim-milk powder	80·5	85	30	229
Buttermilk powder	9·7	—	45	—
Anhydrous milk fat	33·3	35	190	88·5
Water	876·4	880	733·2	250
Carrageenan	—	—	0·3	—
Emulsifiers	—	—	1·5	—
Sucrose	—	—	—	432
Seed lactose	—	—	—	0·5

Definitions

In principle, most types of dairy products can also be produced in a reconstituted or recombined form. In the *Code of Principles Concerning Milk and Milk Products*, the Codex Alimentarius Commission of the Food and Agriculture Organization and World Health Organization (FAO/WHO) has considered the relation between conventional and recombined milk products: it was determined that all products must be based on the use of fresh milk, reconstituted milk or recombined milk. Recombined products can be subcategorized into seven main groups: milk, cream, evaporated milk, sweetened condensed milk, cultured milk products, butter, and cheese. According to definition, a *reconstituted milk product* is the product resulting from the addition of water to the dried or condensed form of the product in the amount necessary to re-establish the specified water:solids ratio. A *recombined milk product* is the product resulting from the combining of milk fat and milk solids-non-fat (SNF) in one or more of their various forms, with or without water. This recombination must be made in order to re-establish the product's specified fat:SNF ratio and solids:water ratio. Among these two fundamental categories of special milk products, the following designations are also commonly distinguished: *toned milk* is a product containing locally produced milk enriched with reconstituted skim-milk solids in order to obtain the average milk composition (e.g. high-fat milk, such as buffalo milk, is adapted to cows' milk composition; or local cows' milk showing adverse variations in composition is adjusted to normal composition). *Blended milk* is a product in which recombined milk is used in mixture with local fresh milk in order to restore the fluctuations in local milk production. *Filled milk* is a recombined product in which milk fat is partly or completely replaced by locally available vegetable oils.

Ingredients Used for Recombination

The quality of recombined dairy products is directly influenced by the quality of the ingredients used. Nutritional, technological, and sensory specifications of raw materials must be considered in order that the recombined products resemble the original products. Typical compositions of some selected products are shown in Table 1. In the following section the basic components and their properties are described.

Dried Milk and Milk Protein Powders

Skim-milk powder is generally the most usual component for providing milk proteins in the manufacture of recombined dairy products. *Whole-milk powder* (prevailingly 'instantized' types) can be used as a source for both SNF and milk fat, but oxidative changes in the fat phase, leading to sensory deterioration, are limiting its application. Certain proportions of *buttermilk powder* can be used more or less advantageously to enhance typical flavour characteristics. In addition, different types of *whey protein powders* and *caseinates*, and, recently, *retentate powders* (for the manufacture of cheese) made from ultrafiltered milk are coprocessed because of attributing certain functional properties to the final products. ***See*** Casein and Caseinates, Uses in the Food Industry; Whey and Whey Powders, Production and Uses

Reconstitutability, heat stability and viscosity properties are the most important parameters, limiting as well as enabling the utilization of milk powders in recombined products. These properties are determined mainly by the preheat treatment applied during milk powder production as well as by seasonal variations in raw milk composition. Because a number of milk proteins are heat-labile, the extent to which they are denatured reflects the heat treatment applied during

Table 2. Heat classification and applicability (*) of skim-milk powder to recombined dairy products

	Categories of skim-milk powder				
	Extra-low-heat	Low-heat	Medium-heat	Medium-high-heat	High-heat
Classification parameters					
Whey protein index (ADMI[a], IDF)	ND	≥6·0	5·9–4·5	4·4–1·5	≤1·4
Heat number (IDF)	ND	≤80	80·1–83·0	83·1–88·0	≥88·1
Cysteine number	26–33	34–41	42–49	50–60	≥60
Recombined products					
Pasteurized milk		*	*	*	
UHT milk			*	*	
Sterilized milk			*	*	
Evaporated milk					*[b]
Sweetened condensed milk		*	*		
Yoghurt		*	*		
Cheese	*	*			
Butter					*
Ice cream		*	*	*	*

[a] American Dairy Products Institute (formerly American Dry Milk Institute), Chicago, Illinois, USA.
[b] Specially manufactured, 'heat-stable', high-heat powder is used.
ND, no data available.

milk powder manufacture. This effect is also used as an indicator for its suitability to be applied in a diverse array of recombined products. The amount of undenatured whey protein (as characterized by the 'whey protein index' or the 'heat number' analysis) is usually taken as a measure for milk powder classification, but for certain purposes, dried milk is also classified based on other analytical indicators (see Table 2). Besides these parameters, moisture and fat content, solubility index, bulk density, flowability, wettability, scorched particles, rennetability, emulsification properties, titratable acidity, sensory aspects, and bacteriological requirements are also included in various milk powder specifications. Most of these properties can be influenced to a certain extent by applying a defined powder technology. However, some particular factors are additionally improved by using certain additives to the product. ***See*** Whey and Whey Powders, Protein Concentrates and Fractions

Milk powder is usually received in 25-kg multi-wall bags, with an inner layer consisting of polyethylene. Alternatively, large bulk bins with a capacity of 200–1000 kg are used.

Fats and Oils

Among the traditional dairy-based fat sources – *anhydrous milk fat* (AMF), *unsalted butter*, *anhydrous butter oil*, *butter oil* – AMF has been used most because of its better keeping quality. In order to maintain good storage stability (6–12 months) of AMF, suitable packaging in steel drums under an inert atmosphere must be applied. Recently, fresh frozen milk fat and soft-fat fraction (produced by fractional crystallization from milk fat) have also been successfully employed in recombined products. Because of their pronounced organoleptic properties, the use of vegetable oils for the manufacture of filled dairy products is restricted to a few oil sources. Only *palm oil*, *coconut oil*, *soya bean oil* and, to a certain extent, *maize oil* have proved to be suitable. For filled milks, highly refined, bleached vegetable oils are suggested. Some of the oils contain considerable amounts of natural antioxidants. However, all fat components have to be regarded as very sensitive to oxidative and lipolytic rancidity, leading to the formation of objectionable off-flavours. Specifications regarding fats or oils usually contain parameters such as fat and moisture content, fatty acid composition, maximum concentration of free fatty acids, rancidity, peroxide value, trace elements levels, and organoleptic properties. ***See*** Butter, The Product and its Manufacture; Vegetable Oils, Applications

Water

Water is the major constituent in many of the recombined dairy products. It is therefore essential that the water used fulfils several standards for *drinking water*, as laid down by the WHO. Besides sensory properties, physical, chemical, (micro)biological and radiological factors are important. It has been shown that some physicochemical parameters strongly co-influence deposit formation during thermal processing of recombined products. Today, many countries impose established limits and specifications for drinking water. According to the recommendations of the International Dairy Federation (IDF), water to be used for the recombination of dairy products should not exceed the following maximum salt concentrations: total hardness, 100 μg of calcium carbonate ($CaCO_3$) per g of water; chloride, 100 $\mu g\ g^{-1}$; sulphate, 100 $\mu g\ g^{-1}$; nitrate, 45 $\mu g\ g^{-1}$.

Additives

Additives can fulfil several targets in recombined products. Owing to certain losses during production and storage, the raw materials used for recombination may contain a lower vitamin content than conventionally produced milk products. In this case, the final vitamin content of the recombined product would be reduced. Fortification with *fat-* and *water-soluble vitamins*, either as single components or in readily prepared mixtures, is therefore applied frequently or is even obligatory. In general, it was noted at the last IDF seminar on milk recombination, in 1988, that recombined dairy products are of high nutritive value and comparable to conventional dairy products. ***See*** Vitamins, Overview

Antioxidants exhibit a very complex function in preventing the generation of oxidized flavour of the fat components in the product, Since some particular vitamins (tocopherols, ascorbic acid) are capable of inhibiting free radical chain reactions in the lipid phase, they are commonly added as antioxidants. ***See*** Antioxidants, Natural Antioxidants

Stabilizers and *emulsifiers* (carrageenan, alginates, gelatin, lecithin, glycerolmonostearate) are used to stabilize the fat phase and also to improve the texture and the mouth-feel of the products. ***See*** Emulsifiers, Uses in Processed Foods; Stabilizers, Types and Function

Different *salts* (sodium citrates, phosphates, calcium salts and sodium chloride) are used to assist the reconstitution properties, the heat stability, the coagulation, and to act as stabilizing reagents.

Besides *rennet*, the addition of *glucono-δ-lactone* helps to improve the coagulation properties of milk in the manufacture of recombined cheese. Pregastric *lipases* are applied to enhance the development of desired cheese flavour.

Sugar, as needed for recombined sweetened condensed milk, must fulfil high quality requirements: sucrose should be refined 'A1 granular quality' and of 'water white' colour; the lactose seed material must be finely ground with a maximum particle size of 10 μm.

Depending on the wishes of the consumers, various natural or synthetic *flavours* and *colours* of different origin are also added. ***See*** Colours, Properties and Determination of Natural Pigments; Colours, Properties and Determination of Synthetic Pigments; Flavour Compounds, Production Methods

Recombination Technology

The processes for recombination of dairy products range from rather simple (e.g. recombined pasteurized milk) to more advanced technologies, such as those applied in the manufacture of defined cheese varieties or sweetened condensed milk. Milk recombination plants are normally built with capacities of up to 15 000 l h^{-1}. Parallel lines are installed to achieve higher performances. In general, the recombination technique has been derived from the conventional processes of dairying, which were further adapted and supplemented by procedures such as weighing, dissolving, mixing and filtration. 'Tailor-made' recombination systems are frequently built in order to meet the individual requirements. The raw materials are processed using discontinuous weighing, and semicontinuous or continuous recombination systems. Typical equipment for batch-type milk recombination basically consists of the following parts: a jacketed *mixing vat*, which is equipped with a calibrated sight-glass or mounted on load cells so that its contents can be exactly measured, and is thermostatically controlled by circulation through a heat exchanger; a *powder–liquid blender* or a *dumping funnel* with a circulating pump for adding and dispersing powder; equipment for *melting milk fat*; a *clarifier* or *duplex filters*; a two-stage *homogenizer*; equipment for *pasteurization*; a *packaging* line; equipment for *refrigerated storage* of the products.

Modern, continuously working equipment for milk recombination with automatic fat dosage, as designed by Alfa Laval (Lund, Sweden), is shown in Fig. 1. It consists of *mixing tanks* (6), in which warm water at 40–50°C is metered. Skim-milk powder is supplied to the *funnel* (4) and transported automatically into the mixing tank by water flowing through a bypass line. The agitator in the mixing tank is started at the same time as the *circulation pump* (5). After completely dissolving the milk powder, AMF is added via a *measuring tank* (3) from the *fat storage* and *melting tanks* (1). Then the whole mixture is agitated until complete fat dispersion is obtained. The process is repeated in the next tank when all ingredients have been mixed in and added to one tank. The mixture is continuously drawn from the full mixing tank by a *pump* (7), which propels the mixture through *duplex filters* (8) which remove foreign particles. After preheating in the *heat exchanger* (11),

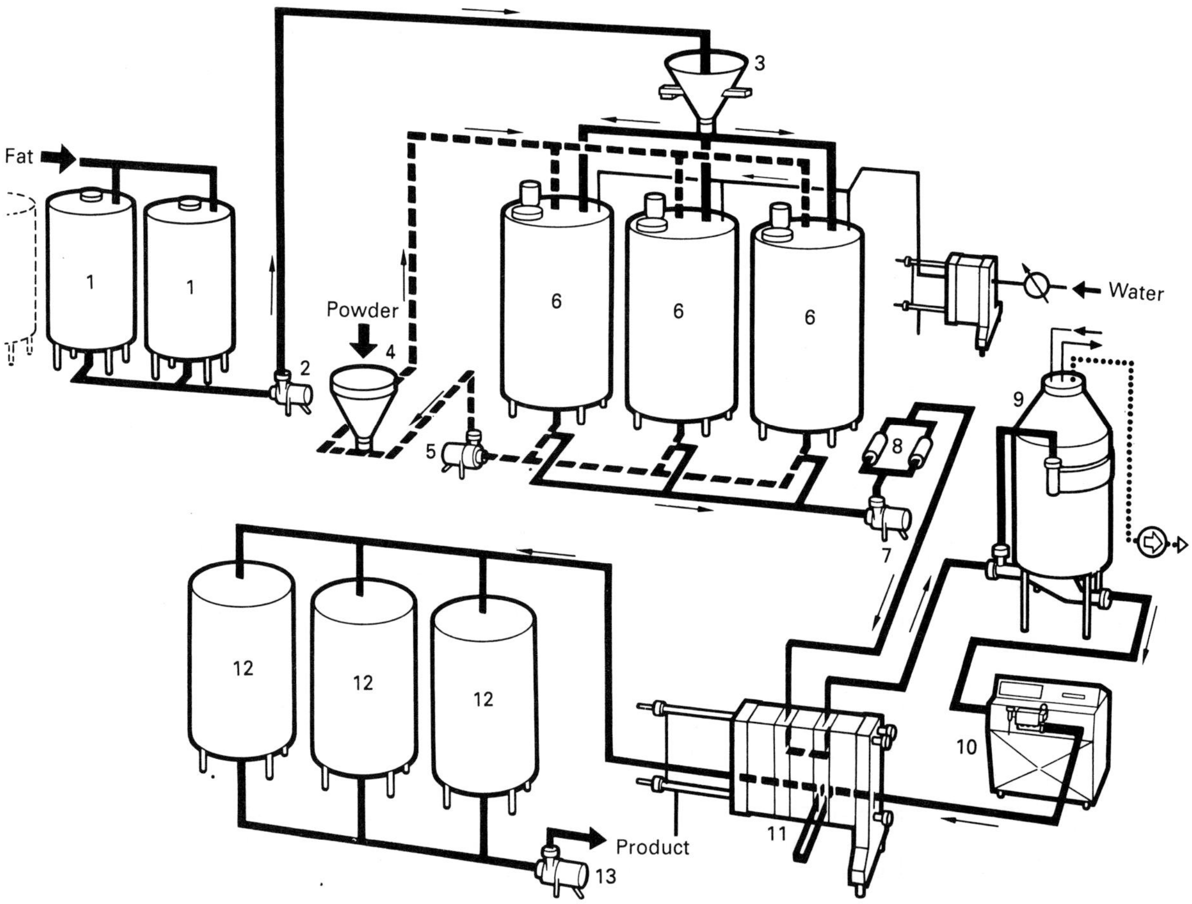

Fig. 1 Plant for the manufacture of recombined milk with fat supply to mixing tanks. (1) Tanks for fat; (2) pumps; (3) weighing tank; (4) powder-blending funnel; (5) circulation pump; (6) mixing tanks; (7) pump; (8) filters; (9) vacuum deaerator; (10) homogenizer; (11) heat exchanger; (12) storage tanks; (13) pump. For explanation, see text. (By courtesy of Alfa Laval Food Engineering AB, Lund, Sweden.)

the product is pumped through a *homogenizer* (10) where the fat dispersion is completed. A *vacuum deaerator vessel* (9) installed in the line before the homogenizer can be used to eliminate the air occluded in the powder particles as well as that picked up during powder mixing. The homogenized product is then cooled in the *plate exchanger* (11) before storage in buffer tanks or packaging.

Besides high-capacity recombination systems such as those shown in Fig. 1, other automatic processors can be used advantageously. The Primodan (Holbaek, Denmark) recombiner, for example, is equipped with an adjustable steam-heated butter melter and is capable of milk production capacity of up to 25 000 l h^{-1}. Another fully automated and energy-saving milk powder dissolver is available from Jongia (Leeuwarden, The Netherlands).

Mixing and Fat Melting

In smaller plants, the powder bags are mostly handled manually. They are unsacked directly into the mixer. In larger plants, automatic emptying systems are installed and the powder is transported pneumatically into storage silos, from which it is mechanically conveyed to the mixing system via a weighing hopper or a screw-feeder.

Milk powder is usually dissolved at high stirring or mixing rates. Simple mixing devices consist of a centrifugal pump attached to a tank filled with the defined amount of warm water, with a powder funnel mounted between. The pump is connected to the jacketed tank with a pipe enabling sufficient circulation. The milk powder is transported into the tank by the vacuum thereby obtained.

In general, the choice of mixer used for dissolving the powder and dispersing the fat depends on the viscosity and/or on the dry matter content of the product mix. For recombined products of low viscosity (pasteurized, sterilized and ultra-heat-treated (UHT) fluid milk, stirred yoghurts), the following systems are used: turbine or propellor mixers (e.g. Alfa Laval; APV Rosista, Horsens, Denmark; Scanima, Aalborg, Denmark) driven by an electric motor; Venturi jet mixers (e.g. Danish

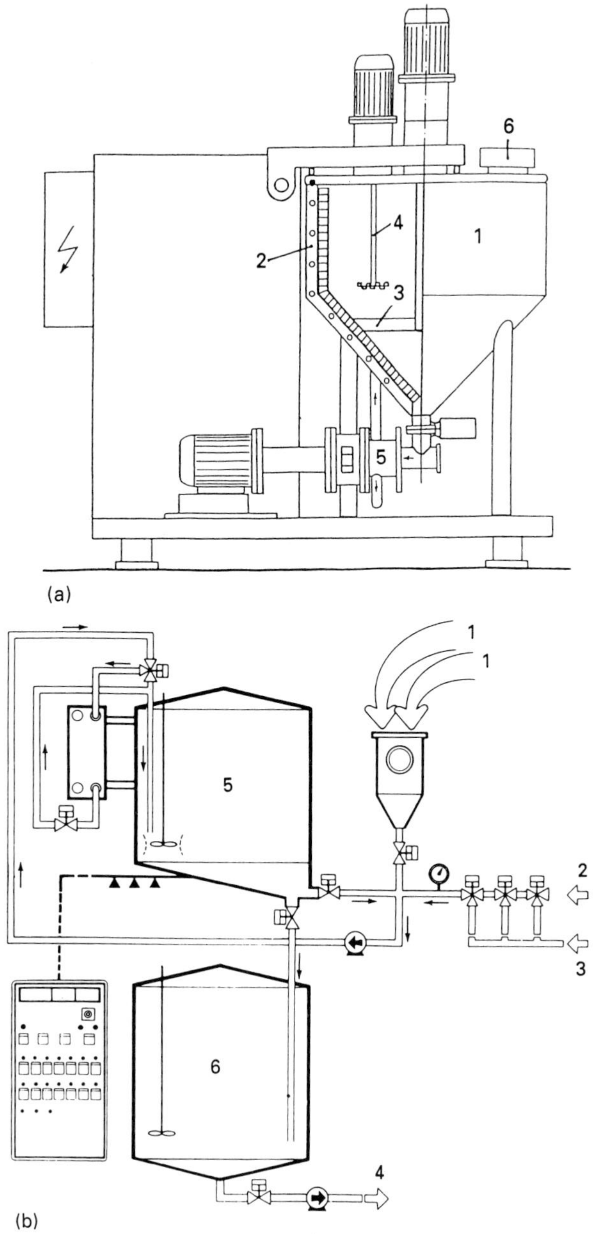

Fig. 2 Examples of special mixing units. (a) Disko-Emulgiermix (courtesy of Balik GmbH, Vienna, Austria): (1) vacuum mixing tank; (2) thermostatically controlled jacket; (3) anchor-type stirrer; (4) disc mill dissolver; (5) flow-through homogenizer; (6) milk powder intake. (b) Weighmatic (courtesy of Alfa Laval Food Engineering AB, Lund, Sweden): (1) powder intake; (2) liquid inlet; (3) cleaning-in-place (CIP) inlet; (4) product outlet; (5) weighing tank; (6) buffer tank.

Turnkey Dairies, Aarhus, Denmark); static mixers (Primodan, Jongia); high-speed blenders (TMP mixer, APV Rosista; Tri-Blender, Tri-Clover, Kenosha WI, USA). Best dispersion is achieved when applying a continuous recirculating motion in the mixing tanks. High-speed blenders are also suited for the preparation of high-viscosity recombined products (e.g. evaporated and sweetened condensed milk, ice cream). Another category of mixers are the disc mill mixers, which are also mainly used in the production of recombined sweetened condensed milk. The so-called Disko-Eulgiermix (Fig. 2a), capable of simultaneously homogenizing whilst recirculating, and of working under vacuum conditions, was introduced recently (Balik, Vienna, Austria). It is suitable for preparing products of different viscosity ranges. By applying a defined vacuum, foaming and picking up of air can be avoided during mixing. Another example for a sophisticated mixing unit is the Weighmatic (Fig. 2b) two-tank system (Alfa Laval), which has been specially designed for the manufacture of recombined ice cream and confectionery. The whole mixing process is performed with high accuracy under computer-controlled conditions.

Different systems are used for melting the milk fat. If the fat is packed in cans, the melting process is carried out by placing the cans in water at 80°C for 2–3 h. Drums containing AMF are either stored in rooms at 45–50°C for 24 h or rapidly treated in steam tunnels or in a hot-water-heated Primodan butter melter before use. The fat or oil is added into the mixing tank after complete dissolution of the milk powder has been achieved. Depending on the equipment, the melted fat is either pumped or discontinuously added into the mixing vessel.

Filtration, Homogenization and Pasteurization

From the mixing tank the product is transferred via a balance tank to a separation unit, consisting either of duplex filters, made from stainless steel with nylon nets, or of a clarifier, in order to remove extraneous matter and undissolved particles. Most recombined milk products are homogenized. Depending on the product type, different conditions of homogenization are employed. For most of the products, two-stage homogenization should be preferred. Typical pressures are 14 MPa plus 3·5 MPa for whole milk, 17 MPa plus 3·5 MPa for evaporated milk. *See* Filtration of Liquids

Pasteurization is carried out mainly following the conventional dairy technology, using continuous heat exchangers rather than batch heating. The latter system is only used in small-scale production units. Typical conditions are 73°C for 15 s for whole milk, 75–80°C for 20–30 s for evaporated milk, 86–92°C for 30 s for sweetened condensed milk. *See* Pasteurization, Principles

Normally, the recombined milk flows from the production line to the filling station, thereby passing buffer tanks. They should be of the aseptic type in the case of sterilized or UHT milk. *See* Heat Treatment, Ultra-high Temperature (UHT) Treatments; Sterilization of Foods

Future Perspectives

During the last few years, scientific progress has contributed considerably to the development of improved technologies for recombination of dairy products as well as to the successful use of special dairy-based ingredients. Examples such as the manufacture of recombined feta-type and other cheeses, using retentate or high-protein powders from ultrafiltration technology, confirm this trend. In this context, the functional properties of various components should be well recognized since they can be utilized specifically. Furthermore, in many of the countries with a recombination industry, long-life milk products would allow a more convenient distribution and storage. Following this aim, modern processes such as the UHT technology, cleaning-in-place (CIP) cleaning facilities and aseptic packaging lines must be considered as factors of growing importance. ***See*** Cheeses, White Brined Varieties

Bibliography

Al-Tahiri R (1987) Recombined and reconstituted milk products. *New Zealand Journal of Dairy Science and Technology* 22: 1–23.

International Dairy Federation (1982) *Milk for the Millions.* Proceedings of the IDF Seminar on Recombination of Milk and Milk Products, Singapore 1980. Bulletin No. 142.

International Dairy Federation (1990) *Recombination of Milk and Milk Products.* Proceedings of the International Seminar, Alexandria, Egypt, 1988. Special Issue No. 9001.

Kjaergaard Jensen G and Nielsen P (1982) Reviews of the progress of dairy science: milk powder and recombination of milk and milk products. *Journal of Dairy Research* 49: 515–544.

Mann EJ (1988) Recombined and reconstituted milk. *Dairy Industries International* 53: 15–16.

Pedersen PJ (1985) Plants for recombination. In: Hansen R (ed.) *Evaporation, Membrane Filtration and Spray Drying in Milk Powder and Cheese Production*, pp 373–383. Vanlose, Denmark: North European Dairy Journal.

Wolfgang Kneifel
Agricultural University, Vienna, Austria

RECOMMENDED DIETARY ALLOWANCES

See Dietary Reference Values

REFINING

See Sugar and Vegetable Oils

REFRIGERATION

See Chilled Storage

REFUGEES – NUTRITIONAL MANAGEMENT*

Characteristics: Refugees and Emergencies

The mass movement and displacement of large populations as a result of natural and man-made disasters is not a modern phenomenon, but has been recorded throughout history. Only recently, however, starting with a more systematic documentation of feeding operations during the Nigerian civil war in 1967, and during refugee emergencies in the 1980s, has a more scientific base of knowledge been recorded to assist in food and nutrition planning in emergencies.

Definition of Refugees

The strict definition of the term 'refugee', according to international legal instruments, refers to a person who has fled his or her country of nationality owing to a well-founded fear of persecution. The term 'displaced person', however, may refer to someone who is uprooted both inside or outside his or her country of nationality or origin, and is therefore displaced from his or her normal means of livelihood. In the early 1990s, the United Nations (UN) determined that some 15×10^6 refugees, and more than 20×10^6 internally displaced persons had been forced to leave their homes as a result of man-made or natural disasters. For the purpose of simplicity, the term refugee herein will imply both internally and externally displaced persons, as both situations lead to similar challenges with regards to nutritional risk and requirements.

Vulnerable Nutritional Status

The nutritional status of refugees is inherently vulnerable given a number of factors. Predisplacement conditions of food shortages prior to movement, scarcity of food and difficulties for food preparation during movement, and the conditions of relief camp after arrival contribute to poor nutritional status of most refugee groups.

Predisplacement Conditions

The ultimate decision by a refugee to flee his or her residence is most often based on the perception of a severe risk regarding personal safety and survival. In this context, the most influential factor in many situations centres around the inability to obtain sufficient food for survival. Modern-day warfare and civil strife situations commonly involve innocent civilians. A disruption of agricultural activities and local economies is often evident in these circumstances, as well as during natural disasters such as famine or floods. Interference with the delivery of relief food supplies to persons caught in civil strife situations is also a growing occurrence, and is sometimes carried out intentionally by warring factions. For these reasons, refugees are often found with marginal nutritional status even before displacement occurs.

Nutritional Vulnerability during Movement

As documented in recent years (Table 1), depending on security conditions and distances, the average duration a refugee must spend walking in transit to relief camps is normally between 15 and 20 days. During this period, refugees are often without a steady food supply and have difficulty preparing meals owing to a lack of water, cooking utensils, or threats to their security. Foraging and hunting of wild game is commonly reported. In the worst cases, such as in the Cambodian emergency of 1979 and the Sudan emergency of 1984, digging up roots and eating items such as inedible leaves and soil has been reported. Avoiding starvation is difficult during periods of transit, and for this reason refugees most often arrive at relief camps in an already deteriorated nutritional condition.

Nutritional Risks at Reception

The nutritional status of refugees is often further at risk owing to the difficult and harsh conditions awaiting them in the early stages of relief operations. Camps are often spontaneously created, and overcrowding, poor water and sanitation facilities, and exposure to extreme climatic variation as well as to previously non-encountered infectious diseases can result in epidemics such as measles, malaria and acute respiratory infections. This begins a vicious cycle of disease and malnutrition and most often results in increased mortality, especially among children.

To compound this situation, refugee situations by

* The colour plate section for this article appears between p. 3896 and p. 3897.

Table 1. Mass refugee movements (1979–1989) and duration in transit

Year(s)	Country of origin	Country of asylum	Approximate number[a]	Duration in transit (days)
1979	Cambodia	Thailand	300 000	20
1980	Ethiopia	Somalia	1 200 000	10
1981–1982	Afghanistan	Pakistan	1 800 000	15
1984–1985	Ethiopia	Sudan	300 000	25
1987–1988	Mozambique	Malawi	600 000	14
1988–1989	Sudan	Ethiopia	300 000	30

[a] Average of statistics from gross influx data, from the United Nations High Commissioner for Refugees (UNHCR), Programme and Technical Support Section, Geneva.

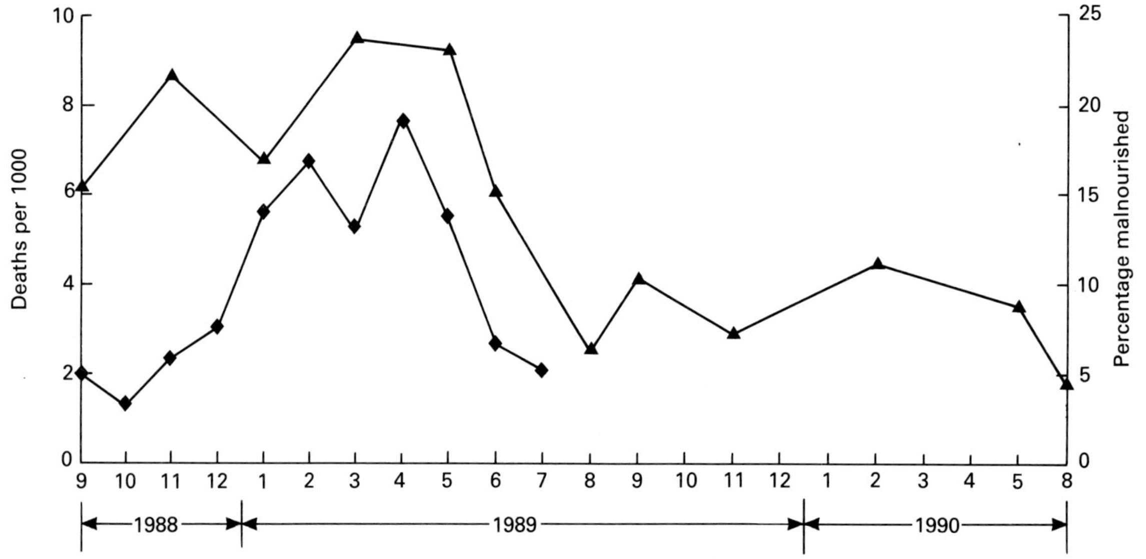

Fig. 1 Protein–energy malnutrition (PEM) prevalence (% weight-for-height <80% median; ▲—▲) among children under 5 years of age, 1988–1989, and crude mortality rates (deaths (all ages) per 1000 per month; ◆—◆), September 1988 to July 1989, Hartishek A Camp, eastern Ethiopia. From Toole M (1991) Case study: Somali refugees in Hartishek A Camp, eastern Ethiopia, health and nutritional profile, July 1988–June 1990. *Responding to the Nutrition Crisis Among Refugees: The Need for New Approaches.* Report of the International Symposium, 17–20 March 1991, Oxford.

their very nature are difficult to predict, and food stocks and logistical systems are often tenuous at best in the early relief stages of an emergency. The UN World Food Programme, which manages global food supplies and donations for most refugee and emergency situations, estimates that some 6–9 months are required to acquire international food aid and establish an adequate system for its delivery. The daily energy level of the first 3 months of food rations in the above-mentioned refugee emergencies (Table 1), was approximately 5880 kJ (1400 kcal) per person. This further results in vulnerability and often increased malnutrition and mortality.

The correlation between increased malnutrition and mortality in relief camps has been well documented in recent years (Fig. 1). ***See*** Malnutrition, The Problem of Malnutrition; Malnutrition, Malnutrition in Developing Countries

Because of the many common factors influencing the nutritional status of refugees before, during and after movement, it can be stated that mass refugee emergencies are usually nutritional emergencies, and proper nutritional management should form the cornerstone of all other relief efforts.

Principles of Management

The nutritional management of refugee situations is multi-dimensional, and involves all aspects of relief: site selection; logistics; registration; housing, water and sanitation, and health interventions; social services; income-generating schemes. Determinating an adequate food aid ration, or establishing special feeding programmes for malnourished children is not sufficient. All aspects of the relief programme ultimately affect the nutritional outcome of refugees. Adequate site selection, and proper management of health, water and sanitation programmes are necessary for support of proper utilization of nutritional inputs. In addition, a correct and accurate census and registration of the population forms the foundation of any relief feeding programme.

Integration

At the start of a refugee emergency, displaced persons normally move to a specific geographical location where there is a perceived availability of assistance or support, e.g. at a water source such as a river or at a border station or designated area where health services and food are available. In addition, they may move to a location designated by some authority, often immigration personnel, police or military personnel. These locations have a tendency to become permanent camp sites, and their location and characteristics can in turn dictate the future nutritional vulnerability as well as success of food relief efforts.

If the risk of dependency on external food aid and future nutritional problems among refugees are to be reduced, experience has shown that the establishment of relief camps should be avoided wherever possible. Camps are artificial social and economic structures which do not support self-determination with regard to household food security. Dependency on minimal and often unreliable food aid normally results in dietary imbalances. The creation of camp 'dependency syndrome' can deter from self-sufficiency if the situation is prolonged. Artificial power-structures are common, and those who can control the food distribution may repress and violate more vulnerable members of the population.

Although security conditions will often dictate the designated location of a refugee population, all efforts should be made to negotiate the possibility for refugees to integrate with local populations. This may require greater efforts with regard to registration and food distribution systems, but will offer more protection for the future nutritional status of the population. In this situation, a food relief programme which benefits locals as well as refugees is introduced to refugee-affected villages. Where relief camps are unavoidable, they should be located near towns and economic centres, where road access is assured and economic integration more likely.

Methods of Feeding

With the exception of supplementary feeding programmes for particularly vulnerable groups, central feeding of cooked meals or soup kitchens for prolonged periods should be avoided. They may be a convenient means of supplying food at the start of an emergency, or in transit centres, but after a few weeks they lead to nutritional problems. Allowing refugees to prepare their own meals on a family basis is most likely to foster an acceptable consumption level. In addition, contamination and food poisoning is a noted problem in large central kitchens where hygienic conditions are not assured. ***See*** Food Aid for Emergencies

In practice, a standard dry or uncooked food ration is distributed weekly or every 10 days to all individuals regardless of age, and refugees prepare this food in their homes or shelters. Special attention to the provision of cooking utensils, cooking fuel and other items necessary for meal preparation must be made. Where whole-grain cereal is provided, some means of grinding it, either in commercial mills or with traditional methods, must be provided.

The use of soup kitchens is advised for supplementary feeding programmes, where an additional 1260–2100 kJ (300–500 kcal) and 15–20 g of protein per day are provided to particularly vulnerable groups. Beneficiaries of such programmes can include all children under a specific age, e.g. 3 years, or selected malnourished children, as well as pregnant and lactating women and the elderly. Uncooked extra food rations can also be distributed to such groups on a weekly basis, if logistical or staff constraints do not permit the establishment of a kitchen.

Food Rations

According to more recent studies, a relief food ration should consist of the daily provision of 7980 kJ (1900 kcal) per refugee (regardless of age), in a population with average age and gender composition. This level, when shared among family members, should provide enough energy to sustain minimal activity in a healthy individual. An example of three acceptable food rations which supply this energy level is illustrated in Table 2.

Of great importance at the early stage of operations is a logistical system that is adequate to ensure the timely delivery and distribution of the food. If food deliveries are erratic, a balanced diet cannot be achieved. Use of experienced logistical personnel and regular meetings of food aid coordination committees, composed of relevant coordinating and technical personnel, are essential to secure a reliable supply of commodities.

Factors Affecting Food Requirements

The per capita daily target level of 7980 kJ is considered adequate for a healthy population, with minimal acti-

Table 2. Example of food rations containing at least 7980 kJ (1900 kcal)

	Amount per person per day[a]		
	1	2	3
Cereal flour (wheat flour, maize meal or rice)	400	400	400
Pulses (beans, peas or lentils)	60	20	40
Fat sources (oil or fats)	25	25	25
Fortified cereal blend[b]	—	30	—
Canned fish or meat	—	—	20
Sugar	15	20	20
Salt	5	5	5
Approximate food value[c]			
Energy (kJ)	8106	8106	8106
Protein	45	45	45
Fat	45	45	30

[a] Dry weight of commodities. All values given in g, except for energy.
[b] Specially fortified relief foods such as Corn Soya Blend, etc.
[c] Compiled from Katona-Apte J (1991) *1990 World Food Programme, Approximate Nutritional Values of Commodities Per 100-Gramme Portion.* Rome: World Food Programme.
Source: UNHCR (1991) *Provisional Guidelines for Calculating Food Rations for Refugees.* UNHCR Internal Instruction No. 68/91. Geneva: United Nations High Commissioner for Refugees.

Table 3. Summary of selected reports of nutritional diseases in refugee camps

Disease and location affected	Date	Prevalence (estimated number)	Main groups	Source
Scurvy				
Somalia	1982	—	—	Magan *et al.* (1983)[a]
Somalia	1984–1987	5·6–45·2%	Total population	Mursal and Hassan (1991)[b]
		1·0–28·8%	Children	
Sudan	1985	22% (18 000)	Women	WHO (1989)[c]
Xeropthalmia				
Sudan	1985	6·7% (10 000)	Children	Pizzarello (1986)[d]
Pellagra				
Zimbabwe	1988	(1544)	Adults	UNHCR/PTSS Internal Document
Malawi	1989	5·4% (23 000)	Adults	UNHCR/PTSS Internal Document
Berberi				
Thailand	1980	8% (6000)	Adults	US Government CDC Internal Document

UNHCR, United Nations High Commissioner for Refugees; PTSS (Programme Technical Support Section), CDC, Centers for Disease Control.
[a] Magan AM *et al.* (1983) An outbreak of scurvy in Somali refugee camps. *Disasters* 7: 94–97.
[b] Mursal H and Hassan B (1991) Anaemia and scurvy in refugee camps in Somalia: Refugee Health Unit Experience. (Unpublished.)
[c] WHO (1989) Scurvy and food aid among refugees in the Horn of Africa. *Weekly Epidemiology Record* 64(12): 85–92.
[d] Pizzarello LD (1986) Age specific xerophthalmia rates among displaced Ethiopians. *Archives of Disease in Childhood* 61:1101–1103.
Modified from Berry-Koch A, Moench R, Hakewill P and Dualeh M (1990) Alleviation of nutritional deficiency diseases in refugees. *Food and Nutrition Bulletin* 12(2).

vity levels, in mild climates, and where an average age and gender demographic composition exists. This food ration should be increased in situations where the refugee population is already malnourished or suffering ill-health, e.g. where mortality rates are above the normal level for that region or population. In addition, the food ration should be increased where refugees must perform heavy physical labour such as agricultural work, and/or where there is exposure to cold climates, and/or where there exists a skewed demographic composition with a greater proportion of adult males than is usual. In all of these circumstances the energy requirement is greatly increased and a larger food ration is needed.

The extent to which the food ration must meet all nutrient requirements of the population, in addition to energy, will depend on the extent of dependency on external supplies. This must be assessed through household food security and market surveys, and possible inequities of access to additional food supplies.

Nutritional Deficiency Diseases

Vitamin and mineral deficiency diseases have been a phenomenon in refugee camp situations of long duration. Mass outbreaks of scurvy and pellagra have been documented in camps in Africa throughout recent refugee emergencies (Table 3). *See* Scurvy

Mass distribution of vitamin and mineral supplements can prevent the outbreak of these disease states, but must be considered a short-term solution. Fortification of food intended for refugees is another possible solution to this problem, which has been attempted in various situations. *See* Food Fortification

Overall, a nutritionally adequate food supply depends on availability and purchasing power. This might include some barter or trade of the food ration, which should not be discouraged on a small and individual basis. It may be the only means for refugees to diversify and balance their diet, and permits, for example, access to spices and other culturally important items needed to make a food ration palatable.

Bibliography

Brown R and Berry AM (1987) Prevention of malnutrition and supplementary feeding programmes. In: Sandler RH and Jones TC (eds) *Medical Care of Refugees*, pp 113–124. New York: Oxford University Press.

de Ville de Goyet C, Seaman J and Geijer U (1978) *The Management of Nutritional Emergencies in Large Populations*. Geneva: World Health Organization.

Refugee Studies Programme, Oxford University (1991) *Responding to the Nutrition Crisis Among Refugees: The Need for New Approaches*. Report on the International Symposium, 17–20 March 1991, Oxford.

Rivers J (1988) The nutritional biology of famine. In: Harrison GA (ed.) *Famine*, pp 57–106. Oxford: Oxford University Press.

United Nations Administrative Committee for Coordination, Subcommittee on Nutrition (ACC/SCN) (1988) *Nutrition in Times of Disaster*. Report of an International Conference, 27–30 September 1988. Geneva: World Health Organization.

UNHCR (1982) *Handbook for Emergencies*. Geneva: United Nations High Commissioner for Refugees.

Angela Berry-Koch
UNHCR, Geneva, Switzerland

REGULATIONS

See Legislation

RELIGIOUS CUSTOMS AND NUTRITION

Many of the factors which shape eating habits are derived from religious laws. Since eating is an activity which must be carried out repeatedly to ensure survival, it provides a daily reminder for members of a religious group of their reliance on the bounty of their deity; the need for obedience to his will and their separateness from nonbelievers.

These factors may be restrictions or prohibitions which form part of religious doctrine or they may simply be strongly held beliefs common to the members of a particular religious sect, but which do not have any real doctrinal basis. For instance, the origins of some eating practices may lie in the commemoration of particular events in religious history or may have been instituted specifically as a way of establishing the 'separateness' of the religious group.

Thus some dietary restrictions can claim direct origins in the Holy texts of the religion concerned. 'Ye shall not eat the swine because it parts the hoof but does not chew the cud' (Leviticus 11:4) may be cited as the basis for the rejection of pork in Judaism. In contrast, the eating of matzohs at Passover by Jewish people commemorates the deliverance from Egypt, when their ancestors did not have time to allow the bread to rise. However, the taboo against horsemeat in much of the Western world has its origins in the order from Pope Gregory III to Boniface, undertaking missionary work among the Germans, forbidding the consumption of horseflesh by Christian converts to demonstrate their separateness from the pagan Vandals.

As well as demonstrating separateness, obeying food laws which are enshrined in religious dogma is a way of displaying devoutness and of expressing belief and respect for the religion's supreme being. It gives the adherent a sense of security and community with co-religionists and new converts to a religion are often the most zealous practitioners of its rules and regulations. Such dietary rules may include the designation of acceptable and forbidden foods; foods which should or should not be eaten on particular days of the year; the time of day at which food should be eaten and how to prepare food. In addition, fasting may be seen to have an important function in developing self-discipline among a religion's followers and be accompanied by regulations concerning who should fast, when and for how long.

Such food regulations may have a range of practical implications, including effects on the sustainability of natural resources, for example, through seasonal food taboos; on the management and distribution of food resources within a community or household and the nutritional status of individuals, based on beliefs about the types and quantities of foods which people in different categories and situations should eat.

Food prohibitions may be either permanent or temporary and this difference may be an important factor in determining whether they are detrimental to nutritional status and health. The permanent prohibitions, such as that of pork amongst Jews and Moslems, are unlikely to be harmful. However, temporary restrictions are often applied to groups who are potentially nutritionally vulnerable and these are consequently of greater concern: pregnant and lactating women; infants; during weaning; puberty and adolescence; during illness. *See* Adolescents, Nutritional Problems; Infants, Nutritional Requirements; Infants, Weaning; Lactation

It has been estimated that about 60% of the world's population are adherents of one of the five major religions – Christianity, Islam, Hinduism, Buddhism and Judaism. Consequently the eating habits of large numbers of people are affected by their regulations, which are considered in more detail below.

Christianity

Although it is a common practice among Christians to say a prayer before and/or after a meal to encourage an attitude of gratitude and reverence while eating; many branches of the Christian Church do not exercise any dietary regulations, instead following the guidance given by St Paul to the Christians at Corinth (1 Corinthians 10: 25–27) that they were free to eat anything sold in the meat market, since 'The Earth and everything in it belong to the Lord' and advising them to eat without question foods served by unbelievers.

However, other groups apply dietary laws, with varying degrees of strictness. Traditionally, Roman Catholics were required to abstain from eating meat on Fridays, in remembrance of the death of Christ, and fish was commonly substituted. But in 1966, this was amended so that meat was permitted except on Fridays during Lent. In contrast the Mormons, Seventh Day Adventists and Eastern Orthodox Church have far more extensive regulations, as described below.

The Mormons

Founded in 1830 in the United States by Joseph Smith Jnr, The Church of Jesus Christ of the Latter Day Saints, has spread worldwide. Healthy living is considered to be important for members, who are forbidden to smoke and encouraged to eat a well-balanced diet. Consumption of vegetables is emphasized and meat is used sparingly. Alcohol and caffeine-containing beverages like tea and coffee are forbidden. Because of these eating patterns, there has been considerable research interest in the prevalence of a number of degenerative diseases amongst Mormons. Such studies have tended to indicate lower rates of heart disease and some forms of cancer. *See* Cancer, Epidemiology

Like the Jews and Seventh Day Adventists, Mormons consider that the Sabbath is a day of rest when work, including cooking, should not be undertaken. Consequently foods must be prepared on the previous day. Those who are in good health are expected to take part in organized fasting each month as a religious discipline. This lasts for 24 hours, from Saturday through to Sunday evening and the money or food saved is contributed to help the needy.

Seventh Day Adventists

Another sect founded in nineteenth-century America, the central tenet of faith amongst Seventh Day Adventists is that the Second Coming of Christ on Earth is imminent. The name of this religious group reflects the dreams and visions of an early member of the church, Mrs Ellen White, which resulted in the Jewish Sabbath – Saturday – being adopted as the holy day of rest rather than Sunday. As a result food must be made ready on a Friday and the dishes washed on Sunday.

As in the Mormon Church, emphasis is placed on healthful living, based on the text 'Know ye not that ye are the temple of God and that the Spirit of God dwelleth within you. If any man defile the temple of God, him shall God destroy; for the temple of God is holy, which temple ye are' (Corinthians 3:16–17). As a result, members are expected to regulate their eating habits and take exercise and rest appropriately. Eating between meals is discouraged in the belief that the body needs time to digest and assimilate the food taken in at meal times. Tea, coffee, alcohol and tobacco are all avoided, as well as many spices which are regarded as stimulants. Adventists believe in a simple diet and are basically vegetarian. This is partly a response to Old Testament teaching, for example, on the uncleanness of swine, but in addition the writings of Mrs White reject meat-eating because it makes people more animalistic and insensitive to the needs of others. However, while basically vegetarian, most adventists do include milk and eggs in their diet, and like the Mormons have been the subject of several studies exploring the links between diet and health. ***See*** Vegetarian Diets

Eastern Orthodox Church

As the result of a number of doctrinal differences, the Eastern Orthodox Church split from the Church of Rome in 1054, when the Patriarch of Constantinople and the Bishop of Rome excommunicated each other.

One of the major characteristics of the practice of Eastern Orthodoxy is fasting. However, this does not mean a total abstinence from food, but that certain foods should not be eaten during periods of fasting while others continue to be acceptable (fasting of this type is also observed by Hindus). All animal products are forbidden and all fish is excluded from the diet except for shellfish. In the areas where Eastern Orthodoxy is practised, olive oil is widely used, so that the requirement that it be avoided during fasting is a sign of true sacrifice.

Two major fasts are observed during the Church year; the Lent fast in the 40 days preceding Easter, which commemorates Christ's 40 days in the wilderness, and the Advent fast in the 40 days leading up to Christmas. in addition there are two shorter periods of fasting in the summer months. Throughout the year Wednesday and Friday are designated as fast days, with the exception of the two before Ascension Day.

Preparation for the Lent fast involves the consumption or disposal of all meat and dairy products. (A similar origin is usually claimed to underlie the British tradition of eating pancakes on Shrove Tuesday, before the start of Lent on Ash Wednesday.) After that, no meat is eaten until Easter Day – although fish is allowed on Palm Sunday and on the Annunciation day of the Virgin Mary.

The foods which are eaten during Easter are rich with symbolism. On Good Friday lentil soup is eaten to symbolize the tears of the Virgin Mary and this soup is often flavoured with vinegar in memory of Christ's suffering on the cross. Lamb is a central feature of the resurrection celebrations, appearing in the lamb-based soup which is served to break the Easter fast after a midmight service on Easter Saturday (served with olives, bread and fruit). While traditionally spit-roast lamb forms the centrepiece of the Easter Sunday celebrations, hard-boiled eggs which have been dyed red (representing the blood of Christ) are cracked open on Easter morning, symbolizing the opening of the tomb. Coloured eggs like these are regarded as tokens of good luck and some are also baked on top of the specially prepared Easter bread.

In addition to the symbolic use of food during the Easter celebrations, ritual food is also important in other ceremonies. For example, a dish called Kolyva (boiled wholegrain wheat, mixed with pomegranate seeds, nuts and spices) is used in mourning and commemorating the death of a family member. The wholegrain wheat symbolizes everlasting life and sugar is sprinkled on the top of the Kolyva to represent the wish that the deceased will have a sweet life in heaven. Further decoration includes a cross and the initials of the dead person either in brown sugar or almonds. The Kolvya is blessed by the priest and later distributed to friends and family 3, 9 and 40 days after the death and also on the first and third anniversaries.

No adverse nutritional effects have been associated with the dietary practices associated with the Eastern Orthodox Church and to some extent these may contribute to the 'Mediterranean diet' which is generally accepted to have health benefits. In addition, historically, the tradition of fasting may have played a useful role in preventing the overexploitation and inequitable distribution of food resources.

Judaism

Instructions regulating every aspect of behaviour among orthodox Jews are found in the sacred writings of the Torah, i.e. the books of Genesis, Exodus, Leviticus, Numbers and Deuteronomy in the Bible. The detailed guide to dietary practices as described in the Torah regulates what foods can be eaten and how they are to be prepared. Foods which are permitted and have been prepared correctly are described as 'kosher'. Those which are not permitted or have not been prepared in a ritually correct way are referred to as 'trayf'.

Only certain animal foods are considered to be acceptable, others are considered as unclean and are therefore forbidden. These are described in a comprehensive listing sometimes referred to as the 'Abominations of Leviticus' (Leviticus 11). In summary, animals which have cloven hooves and also chew the cud are acceptable, this would include the cow, sheep, ox and goat. However, animals like the pig or the camel which do not fulfil both criteria are considered unclean. Even animals which are acceptable must have the blood removed before the meat can be eaten. Blood is taboo because it is considered to be the vital life of an animal. Since it is also forbidden to eat the sciatic nerve and abdominal and intestinal fat of an animal, only the forequarters of these animals are used. In addition, reptiles, creeping animals, birds of prey and most winged insects are also forbidden. Shellfish are considered unclean and only fish with fins and scales are acceptable.

Only meat which has been ritually slaughtered is 'kosher'. Any animal which dies of disease or natural causes is considered unclean. The process of ritual slaughter is undertaken by a 'shochet' and must be supervised by a rabbi. The animal's throat is cut with one deep slash; this method combined with further treatment ensures the complete removal of blood from the meat. It is soaked for thirty minutes then allowed to drain on a slatted board before it is sprinkled with salt. After a further hour it is washed again and then considered ready for cooking. ***See*** Meat, Slaughter

Separate sets of utensils are used by orthodox Jews to store, prepare and serve fish, meat and dairy products. The separation of meat and dairy products is particularly important and they should never be eaten at the same meal. Strict observance requires that six hours elapse after eating meat before dairy products can be consumed. Conversely, after taking milk or dairy products it is necessary to wait up to an hour before eating meat.

As with other religious groups who use the Bible as their source of divine reference, the Sabbath is a sacred day of rest in the Jewish faith and all food preparation should be carried out on a Friday.

Fasting also forms part of the Jewish tradition and may be undertaken by girls over the age of 12 years and one day and by boys aged over 13 years and one day on a number of occasions. This may be in connection with decrees from God or events referred to in the scriptures. For example, on the Day of Atonement everyone over the age of 13 should undertake a complete fast. Fasting may be undertaken in response to decrees by the rabbis; on the eve of the Passover it is customary for the first-born to fast, symbolizing the slaying of the first-born. Fasting is also undertaken to commemorate sad days in the history of the Jewish people. A series of designated fasts maybe undertaken as an expression of sorrow at the events surrounding the siege and fall of Jerusalem under Nebuchadnezzar.

Private fasts may also be undertaken to show sorrow or repentance; when facing temptations or tests and in need of divine guidance, or to commemorate the death of a parent.

Extensive food symbolism is also apparent at most major religious events. At Rosh Hashannah (Jewish New Year) the traditional bread, called challah, is decorated with birds or ladders to carry the prayers of the family to heaven. Bread and slices of apple are dipped in honey to symbolize the hopes for a year full of sweetness.

On the first and third nights of the Passover a 'seder' plate is prepared including various foods which are symbolic of events in Jewish history. A roasted egg recalls the burnt offerings made in the temple at Jerusalem; a roast shank bone commemorates the ancient sacrifice of the Paschal lamb; horseradish, celery and parsley represent the bitterness and poor food of the years of slavery in Egypt; haroseth (a mixture of chopped apple, cinnamon, nuts and wine) resembles the clay that was used by the Jews to make bricks during their stay in Egypt; three pieces of matzoh bread symbolize the three measures of meal that Abraham asked Sarah to prepare for the three angels who visited him on the night of the passover, and the middle matzoh is broken by the celebrant to symbolize the parting of the Red Sea. During the course of the meal, at least four cups of wine must be drunk in memory of the biblical promises of deliverance given to the ancient Israelites.

The extent to which regulations and traditional customs are still followed varies greatly between different Jewish Groups. Orthodox Jews strictly follow the dietary rules of Judaism while Reform Jews have abandoned many of the dietary restrictions and ritual practices.

Islam

Moslems believe that Mohammed was the last of God's prophets, so that, while they accept the divine inspiration of the Bible, they believe that this has been superseded by the instructions given to Mohammed by Allah and recorded in the Holy writings of the Qur'an. One of the chapters, or suras, in the Qur'an, known as the 'The Cow' deals specifically with dietary regulations.

In common with Judaism, certain foods are considered to be unclean and should not be eaten. This includes pork, carnivorous animals and birds, the domestic ass, carrion and anything over which the name of another God has been invoked. Blood is also avoided and alcohol is strictly forbidden. Animals which are to be eaten should undergo ritual slaughter by a blow to the head, while words of dedication to Allah are spoken. Any animal that dies of disease, strangulation or beating is considered unacceptable. One practical outcome of the anxiety to avoid foods which are not 'Halal' can be a narrowing of the variety of foods included in diets of Moslems living in Western countries where there is heavy reliance on processed and ready-prepared foods whose ingredients are unfamiliar. This may contribute to anaemia and poor growth of some infants who undergo prolonged milk-feeding and extensive reliance on nonmeat, principally dessert-type, commercial weaning foods.

Fasting forms a central religious practice for Moslems. It is one of the Five Pillars of Islam, together with faith, prayer, almsgiving and pilgrimage to Mecca. Mohammed said: 'Every good act that a man does shall receive from ten to seven hundred rewards, but the rewards of fasting are beyond bounds for fasting is for God alone and he will give its rewards'. Fast days are observed throughout the year. Very strict Moslems may fast on Monday and Thursday each week and on the 13th, 14th and 15th day of each month. The 10th day of Muhurram is a voluntary fast which is generally observed. But the most important fast of all is Ramadan which takes place in the ninth lunar month of the year and commemorates the first Quranic revelations to Mohammed. The fast lasts for the entire month, during which time no water or food should be taken between sunrise and sunset. All Moslems who have reached the 'age of responsibility' (12 years for girls and 15 years for boys) are expected to fast.

However, there are exemptions for certain groups. Elderly people in poor health, pregnant and nursing women are exempt but expected to fast for an equivalent number of days at another time during the year or to 'substitute fast' by feeding the poor. Menstruating women, the sick, travellers on a journey of more than three days and people involved in hard labour are not allowed to 'substitute fast' but must make up the lost days when they are able.

It must be recognized, however, that it is easier to fast when everyone else is doing it, than to undertake the fast alone at another time. For this reason pregnant women may choose not to take up this exemption and some concern has been expressed about the nutritional implications of this pattern of eating during pregnancy.

When Ramadan falls in the summer months it is particularly demanding on Moslems living at Northern latitudes, where day lengths can be extremely long. Fainting due to prolonged periods without food has been reported amongst Moslem schoolchildren in Britain.

Hinduism

The practice of Hinduism, which originated in India over 4000 years ago, involves the worship of a multitude of deities, all of whom are part of the one universal spirit, Brahman. The social structure embodied in the caste system and a central belief in reincarnation shape many aspects of life for followers, including eating habits and diets.

Dietary regulations are included in the Code of Manu which enshrines the sacred laws of Hinduism. Devout Hindus are vegetarians because of their belief in the sanctity of life and the belief that the soul of one's ancestors may be reincarnated in an animal. Although members of lower castes may sometimes eat other meats, beef is strictly prohibited because the cow is sacred. Fish is not generally eaten except in the Bengal region. Many also reject eggs because these are also potential lives. Other foods forbidden under the Code of Manu are domestic fowl, salted pork, onions, garlic, turnips and mushrooms. However, the vegetarian diet followed by Hindus is generally nutritionally adequate, when eaten in sufficient quantities, although the high esteem in which 'ghee' (clarified butter) is held as a pure food can help to contribute to overweight and related problems among the more affluent.

Fasting may be undertaken by Hindus for a variety of reasons associated with the family, the time of year, religious festivals, caste, age or sex. These fasts may be total, or partial. Older women may fast regularly on several days each week, restricting themselves to 'pure' foods.

The caste system is strongly reflected in eating practices. There are many rules concerning who may eat with whom or give food to whom. Ritual purity attaches to each caste, so that accepting food from a lower caste or accepting food or water from them defiles members of a higher caste. If a member of the highest caste, a Brahmin, touches food cooked by a person from a lower caste then he loses his ritual purity and his own caste status. Members of different castes must not eat together and any family member who ignores caste differences is likely to be excluded from meals with his family lest he pollute them.

Sikhism

Guru Nanak sought to unify Hinduism and Islam when he founded Sikhism in the fifteenth century. Sikhs worship one God and their dietary laws reflect a mixture of ideas from its antecedents. Alcohol is forbidden and Sikhs have a reverence for cattle similar to Hindus and will not eat beef. Some may be vegetarian. However, eating meat killed according to the rules of Islam is forbidden and some Sikhs may eat pork. In general

Sikhs are more flexible about the regulations governing their eating habits than either Hindus or Moslems.

Buddhism

Although once the official religion of India, Buddhism is now predominant in Sri Lanka, Burma, Thailand, Laos, Cambodia and Japan. Like Hindus, Buddhists believe in reincarnation and vow to abstain from killing or injuring any living creature. Nonetheless, while Buddhist monks are usually strictly vegetarian, meat is eaten by some Buddhists. In Thailand fish is commonly eaten, on the basis that the fish is not killed, it is merely removed from the water. Food is always offered to visitors in a Buddhist home as a demonstration of 'right action', one element of the 'Eight-Fold Path' by which Buddhists seek to achieve Nirvana or perfection. Buddhist monks are seen to embody that ideal and by supporting them with gifts of food lay Buddhists can acquire merit.

As discussed, foods are used in most of the major religions of the world for a variety of purposes. When food and nutrition programmes are being planned, there is always a temptation for outsiders to decry any practices with which they are unfamiliar as irrational and superstitious. However, people may have strong attachments to such eating habits, so that attempts to bring about change may well lead to resentment and resistance. Consequently, in encountering an example of a taboo which affects the eating habits of any population group it is important to recognize the nutritional outcome. A practice which is likely to improve nutritional status can be encouraged, even if the rationale is unfamiliar; a practice which is unlikely to have any nutritional effect should be left well alone; while a practice that puts health and nutrition at risk needs to be handled with sensitivity and care should be taken to discover how alternative, beneficial practices can be developed which accord with the cultural and religious regulations of the group concerned.

Bibliography

Douglas M (1978) *Purity and Danger: An Analysis of Concepts of Pollution and Taboo*. London: Routledge & Kegan Paul.

Farb P and Armelagos G (1980) *Consuming Passions: The Anthropology of Eating*. Boston: Houghton Mifflin.

Harris M (1986) *Riddles of Food and Culture*. London: George Allen and Unwin.

Jane Thomas
King's College London, UK

RENAL FUNCTION AND DISORDERS

Contents

Kidney Structure and Function

Role of the Kidney

The kidneys' functions are essential for life. They regulate the volume and composition of the body's fluids, eliminating in the urine waste products, including inactivated hormones, foreign substances and their derivatives, as well as surplus water and normal soluble constituents of the body. The kidneys also produce metabolites such as ammonia and hormones, including erythropoietin, active metabolites of vitamin D, renin and prostaglandins. Unless their functions are replaced by dialysis, or transplanted kidneys from another person, patients without kidneys survive barely 2 weeks, during which time nitrogenous waste products accumulate in their blood in increasing concentrations. However, it does scant justice to the kidneys to say that they merely excrete wastes in the urine and 'purify the blood'. Their most important function in an adult person is to secrete into the blood 180 l of extracellular fluid (ECF) – a 'mirror-image' of the urine, in which the body's cells live – per day. The renal tubules reabsorb 175 l each day. ***See*** Water, Physiology

Structure and Function of the Kidneys

What follows is a brief account of what most physiologists think various parts of the kidneys do. There is

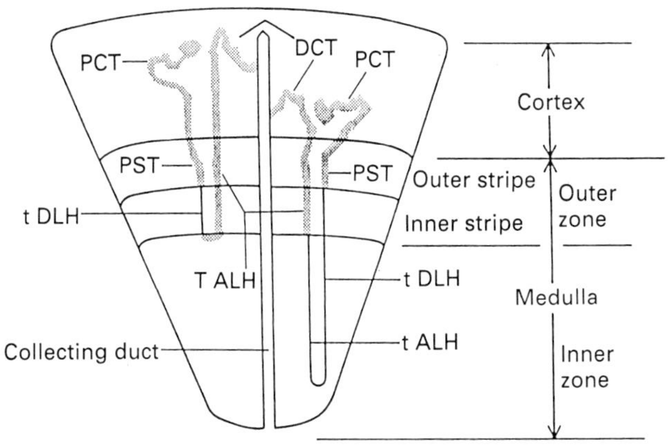

Fig. 1 The principal features of typical nephrons. PCT, proximal convoluted tubule; PST, proximal straight tubule; DCT, distal convoluted tubule; tDLH, thin descending limb of loop of Henle; tALH, thin ascending limb of loop of Henle; TALH, thick ascending limb of loop of Henle. The cortical nephron is shown on the left, and the juxtamedullary nephron on the right. (Reproduced from Robinson JR, 1988; p 3, Fig. 1.1.)

no space for theories about how they do it, or for the often contradictory evidence. These were discussed by Robinson (1988) and, often in more detail, by Brenner and Rector (1981) and by Seldin and Giebisch (1985). Descriptions of presumed human kidney functions are extrapolated from studies of intact kidneys, single nephrons *in situ*, isolated tubules and cells in culture media.

Each human kidney has about a million rather similar nephrons arranged in parallel between vascular glomeruli, in the outer cortex, and the renal pelvis in the centre which collects the urine. The various segments of the nephron and their arrangement are illustrated in Fig. 1.

The glomeruli contain tufts of capillaries perfused in parallel through afferent arterioles by blood at a pressure about half that in the aorta. The more superficial 'cortical' nephrons, with glomeruli in the cortex, have short loops of Henle which do not penetrate into the medulla beyond the outer zone. The more deeply placed 'juxtamedullary' nephrons, with glomeruli near the boundary between cortex and medulla, have long loops and hairpin bends deep in the medulla. Nephrons of both kinds end in distal convoluted tubules draining into collecting ducts that traverse the whole thickness of the medulla before discharging into the renal pelvis.

The renal tubules receive only blood that has first passed through glomeruli. Most of the blood that leaves the glomeruli in their efferent arterioles perfuses capillaries in the cortex amongst the proximal tubules; little more than 1% penetrates deep into the medulla, in capillary channels (vasa recta) among the loops of Henle and collecting ducts. These carry the only blood supply to the medulla; its sluggish circulation, confined to vessels that double back on their tracks, has the effect that solutes, especially sodium chloride and urea, reabsorbed in the medulla, are not immediately carried up to the cortex and away in the renal veins. Instead, they accumulate in increasing concentrations deep in the medulla, where they can extract water osmotically from urine in the collecting ducts, making it more concentrated and reducing its volume.

Glomerular Filtration

An adult person's two kidneys, about 300 g in all, receive daily about 1800 l of blood – about 25% of the resting cardiac output. In the 2×10^6 glomerular capillary tufts this blood is exposed to a pressure greater than the oncotic (colloid osmotic) pressure of the plasma proteins over an area, comparable with the whole external surface of the body; and about 180 l of protein-free 'glomerular filtrate' is expressed each day (125 ml min^{-1}) by ultrafiltration to enter the renal tubules, where more selective actions convert it into a litre or so of urine. Since the entire volume of blood plasma is submitted to filtration about 50 times every day, glomerular filtration alone could effectively rid the body of soluble waste products, but the rapid loss of ECF would make life extremely short.

The rate of glomerular filtration can be calculated by dividing the rate of excretion, UV (where U is the concentration in the urine and V the volume of urine produced each minute) of a substance such as the polyfructosan, inulin (M_r 5000), which enters the filtrate freely and then passes through the tubules without gain or loss, by its concentration, P, in the plasma. The result, UV/P, is the volume of plasma containing the amount excreted per minute. This is also called the renal plasma clearance because this volume of plasma could be cleared each minute at the observed rate of renal excretion. Adult human clearances of inulin average 125 ml min^{-1}. Clearances of other substances may be smaller or larger. Smaller clearances (if the substances are filtered) indicate reabsorption by the tubules. Larger clearances imply addition to the urine on the way through the tubules, i.e. tubular excretion.

Excretion Versus Conservation of Various Solutes

From the nutritional point of view the most important renal function is possibly the conservation of essential nutrients and metabolites. Table 1 shows how the tubules return essential solutes and water from the filtrate to the body. The tubular epithelial cells expend energy derived from their own metabolism to do this, whereas the energy for glomerular filtration is provided by the heart.

For all of the substances shown in Table 1, except

Table 1. Approximate daily amounts in round figures of some substances in an adult person's glomerular filtrate and urine

Substance	Amount in glomerular filtrate	Amount in urine
Sodium	25 000 mmol (575 g)	100 mmol (2·3 g)
Potassium	800 mmol (31 g)	80 mmol (3·1 g)
Calcium	250 mmol (10 g)	5 mmol (200 mg)
Glucose	900 mmol (160 g)	Nil
Amino acids	60 g	2 g
Urea	1000 mmol (60 g)	600 mmol (36 g)
Water	180 l	0·5–2 l

Modified (by adding nonmolar units) from Table 3.8, p 96 in Robinson JR (1988).

urea, the kidneys' primary activity is conservation. All the glucose and most of the amino acids and other solutes filtered by the glomeruli are reabsorbed. The urine contains only small residues that escape reabsorption.

Conservation of Sodium, Potassium, Calcium and Magnesium

Sodium

To keep the amount in the body constant the kidneys must excrete as much of any substance as is absorbed or produced in the body and not lost in other ways. Sodium may be lost in sweat under hot conditions, but other nonrenal routes are normally unimportant. About 50% of the filtered sodium is recovered from the proximal tubules, 35–40% from the loops of Henle, 5–8% from the distal tubules, and 2–3% from the collecting ducts. Remarkably small adjustments of tubular reabsorption can allow more sodium to escape in the urine when dietary intake increases. Thus the reabsorption of 24 900 instead of 25 000 mmol, a reduction of 0·4%, would double the rate of excretion.

The balance between filtration and reabsorption is equally sensitive to alterations in glomerular filtration rate (GFR), which is affected by cardiovascular adjustments to physical activity, posture, etc., and is reduced when plasma volume shrinks with depletion of sodium and water. Bulk reabsorption from the proximal tubules and loops parallels the rate of filtration. A region of closely packed cells, the macula densa, between the ascending limb of Henle and the distal tubule, appears to function as a sensor and somehow depress GFR in nephrons which have not reabsorbed most of the filtered sodium upstream. There is argument about the mechanism of this 'glomerular–tubular feedback', discovered by Thurau in Munich; but it serves to keep filtration adjusted to the capacity of each tubule to reabsorb sodium. When the body is deficient in sodium the macula also appears to be involved in causing cells near the vascular pole of the glomerulus (juxtaglomerular apparatus) to release renin into the blood. Renin converts angiotensin I to angiotensin II, which stimulates the adrenal glands to secrete aldosterone and enhance renal conservation of sodium, whilst promoting the excretion of hydrogen ions (H^+) and potassium. Sodium may be almost completely absent from the urine of depleted persons. *See* Sodium, Physiology

Potassium

A person in balance with a daily intake of 100 mmol (3·9 g) of potassium loses about 10 mmol in faeces and 90 mmol in the urine, but the urine may ultimately become almost free of potassium in depleted individuals. Most filtered potassium is reabsorbed routinely from the proximal tubules and loops of Henle, especially the thick ascending limbs, where active reabsorption of sodium makes fluid in the lumen 10 mV positive to the surroundings, promoting the passive reabsorption of other cations – potassium, calcium and magnesium. Most of the urinary potassium is secreted from cells in the distal tubules and collecting ducts, and is not residual potassium from the glomerular filtrate. *See* Potassium, Physiology

Calcium

Calcium in faeces has mostly missed being absorbed, and the kidneys are largely responsible for maintaining the body's calcium stores. Nearly 99% of filtered calcium is reabsorbed, about 50% of it along with sodium from the proximal tubules and 20–30% more from the loops. About 10% is reabsorbed from the distal tubules and cortical parts of collecting ducts, where the process is regulated, mainly by parathyroid hormone. *See* Calcium, Physiology

The Handling of some Organic Compounds

Glucose (Table 1) disappears half way along the proximal tubules and none reaches the urine. If the concentration in the blood is increased, the point at which glucose has all been recovered shifts along the tubule. Above twice the normal plasma concentration, when more is filtered than the tubules can reabsorb, glucose appears in the urine. The plasma concentration when this occurs is the 'threshold', and it is high enough to avoid losing glucose in the urine during moderate increases in concentration after meals. It is not a fixed concentration, however. If GFR is half normal, the concentration needed to saturate tubular reabsorption will be doubled;

if the reabsorbing capacity of the tubules is reduced, the threshold will be lowered in proportion. ***See*** Glucose, Maintenance of Blood Glucose Level

Urea is the only substance in Table 1 which is not conserved. About one third of filtered urea is lost from the proximal tubules by 'back-diffusion' as its concentration increases with reabsorption of water. When the urine is very scanty and highly concentrated, another third can diffuse back distally. But urea is never totally reabsorbed; it used to be called a 'no-threshold substance' because however little there is in the blood, it is always excreted. Increased metabolic production automatically increases the rate of excretion as the amount in the glomerular filtrate increases with greater concentrations in the blood.

Active Tubular Secretion into the Urine

The straight terminal portion of the proximal tubule actively transports some foreign anions such as *p*-aminohippurate (PAH) and penicillin from the blood to the tubular fluid against steep concentration gradients. (So little PAH remains in the renal veins that it can be used to estimate the flow of blood through the kidneys). Some physiological substances, e.g. creatinine, urate, oxalate and a variety of end products of metabolism are also excreted; some, such as urate, may be partially reabsorbed from later segments. Hydrogen ions and potassium are secreted by cells in distal tubules and collecting ducts.

The Kidney and Acid–Base Balance

The kidney helps to correct excess acidity or alkalinity of the body fluids by excreting acid or alkaline urine. The reaction (pH) of the urine depends upon the balance between the amount of H^+ secreted by the tubular epithelium and the amount of bicarbonate ions delivered in the glomerular filtrate. This is proportional to the concentration of bicarbonate in the plasma and to the GFR. The H^+ come from carbonic acid, formed from carbon dioxide and water in cells all along the nephron. The reaction is catalysed by the enzyme carbonic anhydrase, and the rate of production varies with the partial pressure of carbon dioxide in the blood.

Hydrogen ions enter proximal tubular fluid in exchange for reabsorbed sodium, and combine with filtered bicarbonate ions. Since one bicarbonate ion is left in the cell as a by-product of the generation of each hydrogen ion, and passes into the plasma with a sodium ion, the net effect is not to acidify the tubular fluid but to replace every filtered bicarbonate ion destroyed in the lumen by a new bicarbonate ion in the plasma, thus 'reabsorbing' filtered bicarbonate which would otherwise be lost.

Active excretion of H^+ into urine in the distal tubules and collecting ducts after all the bicarbonate has gone can lower the pH of the urine to 4·4, i.e. 1000 times more acid than plasma at pH 7·4. The H^+ excreted to protect the bland alkalinity of the body fluids (pH 6·8 is neutral at 37°C) are excreted in the forms of 'titratable acid', mostly dihydrogen phosphate, and ammonium, formed in tubular cells from glutamine and amino acids.

Acids ingested or produced in the body remove bicarbonate from the plasma so that less is filtered. More of the H^+ produced by the tubules then go to acidify the urine, and new bicarbonate, equivalent to the urinary titratable acid and ammonium, is restored to the plasma.

The urine becomes alkaline when more bicarbonate is filtered than available H^+ can 'reabsorb'. This may occur if large amounts are ingested; if a low partial pressure of carbon dioxide in the blood slows renal H^+ production, as at high altitudes where respiration driven by lack of oxygen eliminates carbon dioxide excessively; or when inhibitors of carbonic anhydrase depress the production of carbonic acid to yield H^+ and 'reabsorb' filtered bicarbonate.

The Kidney and Water Balance

The fluid entering the loop is always hypotonic in the absence of antidiuretic hormone (ADH). The distal tubules and collecting ducts continue to reabsorb sodium actively, but are virtually waterproof, and extraction of further sodium without water leaves the urine copious, up to 20 ml min^{-1}, and far more dilute than the plasma. Permanent failure to produce ADH leads to diabetes insipidus, in which the kidneys cannot concentrate the urine and conserve water. The patients live in a state of perpetual water diuresis, and have to drink up to 10 l per day to avoid thirst and dehydration. (NB Above 10 l per day they would receive treatment to reduce diuresis, e.g. Vasopressin, Desmopressin.)

Antidiuretic hormone from the neurohypophysis (formerly posterior pituitary gland) makes the distal tubules and collecting ducts more permeable, so that urine slowly traversing the deep medulla loses water by osmosis to the high concentration of solutes parked there; it also gains urea. During maximal antidiuresis, the day's urinary solutes are contained in about 600 ml of urine four times as concentrated as the plasma. However, such highly concentrated urine can only be produced in small quantities. In general, large amounts of unreabsorbed solutes overwhelm the concentrating process and the urine cannot be made hypertonic, but during antidiuresis the urine is made hypertonic compared to plasma. Persons in liferafts who drink sea water (more concentrated than the most concentrated urine) actually lose body water as they excrete the excess salt with far more water than they imbibe.

Epilogue

One common process throughout the nephron is active reabsorption of sodium by an (Na^+-K^+)-ATPase, which transports sodium (Na) out of the epithelial cells and at the same time takes in potassium (K), using energy in the form of adenosine triphosphate (ATP). This keeps cell [Na] low, cell [K] high, and the cytoplasm about 70 mV negative to its surroundings. The steep electrochemical gradient drives sodium in from the luminal fluid. In the proximal tubule, incoming sodium ions (Na^+) share specific membrane carriers with glucose and several classes of amino acids, and carry these with them (cotransport). Chloride, phosphate and foreign anions also enter by cotransport with Na^+, and other Na^+ are taken up in exchange for H^+ used to 'reabsorb' bicarbonate. Hence active transport of sodium largely drives the proximal bulk recovery of filtrate as well as contributing to more specialized functions in the distal nephron. The transport mechanisms were admirably reviewed (with over 200 references) by Rose (1991) and Kinne (1991).

Bibliography

Brenner BM and Rector FC (eds) (1981) *The Kidney*. Philadelphia: WB Saunders.

Kinne RKH (1991) Selectivity and direction: plasma membranes in renal transport. (Homer Smith Award Lecture). *American Journal of Physiology* 260: F153–F162.

Robinson JR (1988) *Reflections on Renal Function* 2nd edn. Oxford: Blackwell Scientific Publications.

Rose BD (1991) Diuretics. (Nephrology Forum.) *Kidney International* 39: 336–352.

Seldin DW and Giebsich G (eds) (1985) *The Kidney, Physiology and Pathophysiology*. New York: Raven Press.

James R. Robinson
Dunedin, New Zealand

Nutritional Management of Renal Disorders

The nutritional and metabolic state of the body is dependent upon the intake of adequate amounts of nutrients as food and drink and the ability of the body to utilize these substances. The kidney disposes of nitrogenous waste products from the breakdown of protein and excretes excess amounts of fluid, sodium, potassium, phosphate, hydrogen ions and organic acids. The kidney also has a synthetic function converting vitamin D_3 to its active metabolite, 1,25-dihydroxycholecalciferol and in producing the hormone erythropoietin (EPO), which is important in promoting the formation of red blood cells.

In many types of kidney disease the excretion of these substances is impaired. The rationale behind the dietary management of kidney disease is simply that reducing the intake of protein, phosphate and electrolytes reduces the need for excretion and thus slows down the accumulation of waste products in the blood. Improving or maintaining a good nutritional state will also prevent unwanted breakdown of body tissue (catabolism) which would otherwise increase the 'workload' on the kidney. Finally, dietary intervention may slow down the progressive deterioration of kidney function which can occur long after the initiating event has been dealt with.

Individual kidney diseases show many similarities in their clinical and pathological features. To aid clinical and dietetic management it is useful to describe the five main syndromes of kidney disease which most commonly require dietetic intervention: 1. chronic renal failure, 2. renal calculi, 3. nephrotic syndrome, 4. end-stage renal failure needing haemodialysis and continuous ambulatory peritoneal dialysis, 5. acute renal failure.

Chronic Renal Failure (CRF)

Giovanetti described CRF as a chronic reduction of glomerular filtration rate (GRF) below the normal range (*c.* 120 ml min^{-1}) irrespective of the underlying cause (Table 1). A state of uraemia (renal failure) then ensues.

The severity of uraemia is related to the accumulation of protein metabolites, the abnormal electrolyte pattern, acid–base imbalance, salt–water imbalance, and hormonal imbalance. A number of changes in plasma constituents are seen in CRF and these are discussed below.

Creatinine

The excretion of creatinine (from creatine and phosphocreatine in muscle tissue) occurs predominantly by filtration. A decrease in GFR below 50% of normal leads to a rise in plasma creatinine. Reciprocal plots of creatinine against time are commonly used to chart the progression of CRF. However, as creatinine is propor-

Table 1. Main causes of chronic renal failure

Glomerulonephritis	Hypertension
Chronic pyelonephritis	Polycystic disease of the kidney
Obstructive, reflux, analgesic nephropathies	Systemic disease, e.g. scleroderma, systemic lupus erythematosus
Diabetic nephropathy	Vasculitis
Renal transplant failure	Myeloma, amyloidosis

Table 2. Stage of renal failure related to creatinine level

Stage	Urinary creatinine concentration
Early	< 150 μmol l^{-1}
Moderate	< 300 μmol l^{-1}
Advanced	> 400 μmol l^{-1}
End-stage renal failure	> 1000 μmol l^{-1} (requires renal replacement therapy – dialysis or kidney transplantation)

Table 3. Uraemic manifestations

	Uraemic symptoms
Cardiovascular	High blood pressure; fluid overload; pericarditis; cardiomyopathy
Gastrointestinal	Nausea and vomiting; gastritis
Peripheral nervous system	Neuropathy; myopathy
Central nervous system	Confusion, drowsiness, coma; impaired higher mental function; insomnia
Blood	Anaemia
Skeletal	Renal osteodystrophy; bone pain; fractures
Skin	Pruritis; pigmentation; dystrophic calcification

tional to lean body mass, a loss of lean tissue during the course of CRF can give a false impression that the rate of decline of GFR is slowing down. Table 2 relates creatinine levels to stage of renal failure.

Urea

Urea is the main metabolite of protein. In CRF plasma urea increases with an increased breakdown of proteins from dietary sources or catabolism. A decrease in the urea : creatinine ratio after starting a low-protein diet may be used as a measure of compliance with the diet. Other factors, however, may raise urea, e.g. drugs (frusemide, steroids), gastrointestinal bleeding, hypercatabolic state.

Urea itself was thought to be nontoxic (except in very high concentration) but its elevation correlates with the rise in plasma levels of other more toxic compounds, e.g. guanidines, amines, phenols, oxalic acid, uric acid, middle molecules. Urea may therefore be used as a surrogate molecule for the other toxins. Uraemic symptoms such as drowsiness, glucose intolerance and nausea have been linked to high plasma concentrations of urea.

A reduction in protein intake in pre-dialysis patients will decrease the amount of nitrogenous waste, phosphate, potassium and acids that must be dealt with by the kidney. In addition, low-protein diets have also been associated with an improvement in plasma lipid levels, glucose intolerance, decreased proteinuria and improved albumin levels. The improved biochemical and metabolic indices are correlated with a reduction of uraemic symptoms (Table 3).

Phosphate and Calcium

Hyperphosphataemia and hypocalcaemia are characteristic of CRF. Decreased GFR causes an increased plasma phosphate. Simultaneously, low vitamin D levels (production of active vitamin D_3 is impaired in CRF) leads to a reduction of calcium absorption from the gut, hence lower plasma calcium. These two factors stimulate parathyroid hormone (PTH) release, which in turn causes bone resorption. The release of calcium from bone, driven by PTH, is an attempt to normalize plasma calcium levels at the expense of the integrity of the skeletal system. The extraskeletal effect of an increased phosphate × calcium product is the calcification of vascular and visceral tissues.

A low-phosphate diet (*c.* 18–25 mmol per day (550–750 mg per day)) combined with phosphate binders such as calcium carbonate and aluminium hydroxide is employed to keep plasma phosphate levels within the normal range, therefore preventing the metabolic cascade which leads to renal osteodystrophy (renal bone disease) and metastatic calcification.

The intake of calcium-containing foods is limited on a low-protein diet and supplementation of 1000–1500 mg calcium is required. The use of calcium carbonate as a phosphate binder will help to correct the hypocalcaemia as some of the calcium is absorbed. Great care must be taken with vitamin D_3 supplements and reference to blood biochemistry is necessary to avoid undue hypercalcaemia. If the plasma phosphate level is high at the time of vitamin D_3 supplementation, further calcification will result.

Potassium

Hyperkalaemia does not tend to develop until the GFR is ≤ 10 ml min^{-1} (i.e. endstage renal failure). Hyperkalaemia, however, may occur in moderate CRF under certain circumstances: increased catabolism, e.g. infection, trauma; acidosis; drugs – angiotensin converting enzyme (ACE) inhibitors, cyclosporin A, erythropoietin (EPO).

A plasma potassium > 6·5 mmol l^{-1} (26 mg dl^{-1}) can cause asystolic cardiac arrest. A dietary potassium restriction will be required if levels begin to rise above

5·5–6·0 mmol l^{-1} (22–24 mg per 100 ml). ***See*** Potassium, Physiology

Salt and Water

Kidney disease such as diabetic nephropathy and some forms of glomerulonephritis are characterized by retention of salt and water with resulting oedema and hypertension. On the other hand, interstitial nephritis, e.g. chronic pyelonephritis, may initially present with a failure to concentrate urine, leading to salt and water loss and eventual dehydration.

In nonrenal patients many hypertensive clinics are adopting a nonpharmacological approach as a first-line strategy in the treatment of hypertension. This includes salt restriction, weight control and control of the other risk factors of coronary heart disease. In renal patients salt restriction may potentiate the action of ACE inhibitors, diuretics and β-blockers. As the kidney loses its ability to respond to rapid changes in salt and fluid intake, any salt restriction should be implemented gradually. The level of restriction is not normally less than 80–100 mmol (1·8–2·3 g) sodium per day.

Other Metabolic Derangements of CRF which may Respond to Diet Therapy

Anaemia

The anaemia of CRF is due to: (1) decreased production of erythropoietin (EPO); (2) poor iron intake due to a poor appetite or a low-protein diet; (3) poor iron absorption due to drug interactions e.g. H_2 antagonists, $CaCO_3$, $Al(OH)_3$; (4) folate, vitamin B_{12} deficiency – due to a poor diet or drug interactions; (5) chronic blood loss.

Anaemia contributes to the extreme fatigue, loss of appetite and taste changes experienced in CRF. Plasma levels of ferritin, folate and vitamin B_{12} and other haematological parameters such as haemoglobin, and mean corpuscular volume are easily measured and may indicate the need for iron, folate and vitamin B_{12} supplements. ***See*** Anaemia, Iron Deficiency Anaemia; Anaemia, Other Nutritional Causes; Anaemia, Megaloblastic Anaemias

Hyperlipidaemia

Hyperlipidaemia is a common complication of renal failure. Table 4 shows lipid profiles in renal failure. Hyperlipidaemia is correlated with an increased risk of atherosclerosis and consequently cerebrovascular disease (CVD) and coronary heart disease (CHD). In the 15–34 year age group, patients with CRF have 250 times the risk of CVD and 180 times the risk of CHD compared with nonrenal patients. The association between a high saturated fat and high sugar intake with raised cholesterol and triglycerides is well documented. Yet in renal patients the intake of these foods used to be encouraged to make up the energy deficit of a low-protein diet. With the increase and improvement in dialysis facilities and transplantation, renal patients are living longer. A more holistic approach to dietetic treatment of CRF is required and consideration is now given to the types of fat and carbohydrate recommended in a low-protein diet. Trials involving the increased intake of polyunsaturated fats (PUFA) and/or fish oil supplements have been promising, as has the use of pharmacological agents such as the HMG CoA reductase inhibitor, simvastatin. Weight loss (if overweight) and increased exercise are also beneficial. ***See*** Hyperlipidaemia

Table 4. Lipid profile in renal failure

Renal state	Effect on blood lipids		
CRF	↑TG	↓HDL	sl↑LDL
Haemodialysis/peritoneal dialysis	↑TG	↓HDL	sl↑LDL
Nephrotic syndrome	↑TG	↓HDL	↑LDL
Transplantation	↑↓TG	↓HDL	↑LDL

↑, Increase; ↓, decrease; sl, slight increase.
CRF, chronic renal failure; TG, triglycerides; H(L)DL, high (low) density lipoprotein.

Glucose Intolerance

Glucose intolerance has been noted in over half the patients with CRF. Drug intervention is not normally required but it may be part of a wider syndrome which includes hyperinsulinaemia, peripheral resistance to insulin, hypertension, raised lipids, central obesity and heart disease. ***See*** Glucose, Glucose Tolerance and the Glycaemic Index

Progression of Renal Failure

Some dietary manipulations have been shown to have a protective effect on renal function. These include: (1) low-protein diet, by reducing hyperfiltration; (2) low-phosphate diet, through reduced calcification of renal tissue. Good control of hypertension and improving the lipid profile also prolongs kidney function, through decreased atherosclerosis and occlusion of renal arteries and arterioles.

Dietary Manipulations in Chronic Renal Failure

The previous paragraphs have described how dietary manipulation in CRF attempts to normalize or at least improve blood biochemistry and thus not only reduces the symptoms and consequences of uraemia but also may slow down the progression of kidney disease. The diet may involve altering the intake of one or more of the following components: protein/energy (vitamins/minerals); phosphate/calcium; potassium; sodium; fluid.

Protein Restriction

'Normal' protein intake in Western society is approximately 1–1·5 g kg^{-1} day^{-1}. Protein restriction aims to reduce intake to the minimum requirement. This has been calculated as 0·6 g per kg ideal body weight (IBW), where 70% of the protein is of high biological value (HBV). HBV proteins are proteins containing a high percentage of essential amino acids in proportions to allow efficient incorporation into body proteins. The protein intake from food can be reduced further, for example 0·3 kg per kg IBW if the diet is supplemented with essential amino acids or their ketoanalogues. The source of protein would be unrestricted but in practice would be mainly of low-biological-value (LBV) proteins, i.e. essentially a vegetarian or vegan diet. Table 5 describes the implementation of a low-protein diet. ***See*** Protein, Requirements; Protein, Quality; Vegan Diets; Vegetarian Diets

Energy Requirements

Calculation of energy requirements should be made depending upon the need for weight loss or weight gain. The source of calories should be from complex carbohydrates and mono/polyunsaturated fats where possible. A high intake of sugars and saturated fats may exacerbate the lipid abnormalities already present in CRF. ***See*** Energy, Measurement of Food Energy

Vitamins and Minerals

The reduction in intake of protein-containing foods has a deleterious effect on vitamin and mineral intake. A number of studies have shown that intakes of vitamins B_1, B_2, B_6, folate, iron, calcium and zinc can be affected. A potassium restriction will further affect folate and also vitamin C. More work needs to be done to correlate the perceived reduced intake of B vitamins with specific changes in enzyme activity, e.g. vitamin B_1 and erythrocyte transketolase. In the meantime it is wise to consider supplementation of low-protein diets with B vitamins. Large doses of vitamin C are not advisable because of its conversion to oxalic acid (see section on renal stones). Fat-soluble vitamins accumulate in CRF and supplements could have a toxic effect. Serum iron, ferritin, folate and vitamin B_{12} and red blood cell folate should be monitored regularly. The anaemia of CRF usually responds well to injections of EPO, providing the

Table 5. Procedure for prescribing a low-protein diet

1. *Medical history.* Primary diagnosis. Intercurrent illness. Blood biochemistry. Present symptoms. Anthropometry.
2. *Social history.* Family (level of commitments and support). Employment/financial situation. Cultural background.
3. *Diet history.* (Interview or 4–7 day diary). Assessment of current nutrient intake. Cultural/ethnic food habits. Meal pattern. Methods of cooking. Likes and dislikes.
4. Assess protein requirement for example, 70-kg man (no oedema), height 1·76 m, BMI 23.

Protein requirement:	70 × 0·6 = 42 g protein
	70% HBV = 28 g protein
	30% LBV = 14 g protein
Exchange system (Table 6):	28 g = 4 × 7 g exchanges
	14 g = 7 × 2 g exchanges

5. The 7 g and 2 g exchanges are fitted into a daily meal plan.
6. The importance of meeting energy requirements must be emphasized (despite the risk of atheroma it is difficult not to use the simple sugars and some saturated fats). Vegan and vegetarian diets may not require these to such an extent because they allow greater intake of complex carbohydrates and this may be one of the reasons behind the marked improvement in lipid profiles seen with these diets.
7. *Foods to avoid.* The exchange lists can be extended to include favourite foods. This leaves a few foods which are probably best avoided because small quantities contain significant amounts of protein, e.g. dried/evaporated/condensed milk, chocolate, instant desserts, ovaltine and malted drinks.
8. It is useful to give recipes, samples of prescribable products, a GP prescription letter and advice on obtaining a prescription 'season ticket' on the first visit.
9. The patient should be contacted at regular intervals to answer any queries and determine compliance to the diet. First follow-up should be within 2 weeks and then every 2–3 months, via telephone, diet diary, clinic appointment.
10. Monitoring blood biochemistry, anthropometry, i.e. BMI (height/weight^2) and mid arm circumference, regularly. ***See*** Nutritional Status, Anthropometry and Clinical Examination
11. There should be close liaison between the physician, dietitian and other specialists, e.g. renal nursing staff, diabetic liaison nurse and social worker. A number of studies have shown that regular visits to pre-dialysis clinics improves compliance with drug and diet therapy and also improves the quality of life of the patient. There may be a number of reasons to explain this, e.g. drug and and dietary compliance improves uraemic symptoms. The patient receives counselling which helps them to come to terms with the kidney disease and its treatment.

BMI, body mass index.
HBV, high biological value.
LBV, low biological value.

Table 6. Examples of protein exchanges

7-g Exchanges (each portion contains approximately 7 g of protein)	2-g Exchanges (each portion contains approximately 2 g of protein)
1 oz. meat (cooked), e.g. beef, lamb, pork, chicken	1 oz. bread (1 large thin slice)
1·5 oz. fish (cooked) e.g. haddock, cod, mackerel	4 oz. potato, yam, plantain
1 oz. cheese	1 oz. rice (uncooked) (=1 tblsp)
1 egg	0·5 oz. macaroni, pasta (uncooked)
0·33 pint milk	1 oz. biscuits (2–3 medium)
4 oz. yoghurt	1 oz. breakfast cereal
1 oz. nuts	1 small chapati (0·5 oz. flour)
1 oz. pulses (uncooked)	2 oz. sweetcorn, peas, dark green vegetables

1 oz. is approximately equivalent to 25 g. 1 pint equivalent to 570 ml.

Table 7. Phosphate (mg): protein (g) ratio of some common foods

Phosphate : protein ratio < 10
Beef, lamb, turkey, chicken, pork, cod, haddock, tuna, corned beef

Phosphate : protein ratio 10–15
Bacon, ham, meat paste, plaice, herring, mackerel, salmon, cottage cheese, white bread, white pasta

Phosphate : protein 15–20
Liver, pilchards, mussels, prawns, soya milk, peanuts, pulses, soya flour, white rice

Phosphate : protein ratio > 20
Milk, ice cream, cheese, cheese spread, yoghurt, paté, roe, sardines, whitebait, nuts, chocolate, instant desserts, wholemeal flour, brown rice, brown spaghetti, wholemeal bread, bran flakes, sponge, scones, marmite

ferritin stores are adequate ($>100\ \mu g\ l^{-1}$). ***See*** individual vitamins and minerals.

Phosphate/Calcium

A reduction of intake to *c.* 18–25 mmol phosphate per day is required to slow down the elevation of plasma phosphate levels. On a low-protein diet this is quite easy to achieve as phosphate is reduced simultaneously. Some renal units prefer to combine a moderately reduced protein diet (*c.* 0·8 g protein per kg IBW) with strict avoidance of foods with a high phosphate-to-protein ratio. The benefits of this diet are that a greater amount of LBV foods can be used, e.g. cereal products, which increases the intake of calories and other nutrients. The risk of malnutrition is not as great as with the stricter low-protein diets. The phosphate (mg): protein (g) ratio is a useful way to identify high-phosphate foods (Table 7). Despite the high phosphate content, the use of some wholemeal products is recommended because of the beneficial effect on bowel function.

A daily allowance for milk is usually set at 0·33–0·5 pint (190–280 ml). (Drug companies are beginning to market a low-protein, low-phosphate milk substitute.)

Potassium

Potassium is abundant in meat, fish and dairy products, as well as fruit, vegetables, nuts, pulses and cereals. As with phosphate, a low-protein diet will automatically reduce the intake of potassium. If the potassium levels remain elevated ($>6\ mmol\ l^{-1}$ (24 mg per 100 ml)) further restrictions will be necessary, aiming for approximately 0·6–1 mmol (24–40 mg) kg^{-1} IBW per day. This could be achieved by reducing the intake of particularly high potassium foods and/or by changing the cooking method to one which reduces the potassium content e.g. boiling. Alternatively, potassium binding agents such as calcium resonium can be used and may be preferable to further dietary restrictions.

Salt Restriction

For reasons discussed earlier salt restriction is not normally stricter than 'no-added-salt' (NAS), i.e. 80–100 mmol sodium per day.

On a low-protein diet a small amount of salt added in cooking and 1 oz. (25 g) 'salty' meat or fish would not increase the salt intake over 80 mmol. Salt should not be added 'at the table' and the use of most processed, packet or tinned foods are restricted. To increase palatability of the diet, a wide range of herbs, spices, pepper and vinegar can be used.

Renal Stone Disease

There are six main types of renal stone disease: calcium oxalate and calcium phosphate, calcium oxalate, magnesium ammonium phosphate, calcium phosphate, uric acid and cystine. Dietary intervention is mainly required in calcium oxalate stone formation. The formation of stones requires an increased urine concentration of stone-forming substances and decreased levels of natural inhibitors of stone formation. Dietary factors which influence these include the intake of calcium, oxalate, animal protein, sodium and fluid. Vitamin C, fibre and fat also have a role.

Useful biochemical measurements include urine creatinine, creatinine clearance and urine calcium and oxalate. It is necessary to establish whether hypercalcuria is due to: excessive intake of calcium and/or vitamin D; hyperabsorption of a low to normal intake of calcium; or failure of renal tubular reabsorption of calcium. An excess of vitamin D or calcium intake can be remedied. A diet history will reveal where the problem lies and the patient will be advised to decrease calcium intake to normal levels of < 1000 mg per day. If the problem is hyperabsorption, a low intake of calcium (400–600 mg per day) is advisable, although in the long term this may lead to calcium depletion and bone disease.

Hyperoxaluria has a much greater influence on stone formation than hypercalcuria and is significantly affected by oxalate intake. It is therefore recommended that high oxalate foods – tea (3–4 cups of weak tea per day are allowed), rhubarb, beetroot, chocolate, cocoa, nuts, spinach and strawberries – are avoided in conjunction with a reduced intake of calcium-rich foods. An allowance for milk is usually given as 190 ml (0·33 pint).

Protein Intake

A diet high in animal protein encourages stone formation by increased urinary excretion of calcium, oxalate and uric acid; increased acidity of the urine; and decreased citrate (inhibitor of stone formation). The prevalence of stone disease in vegetarians is 50% that of the general population which may be partly due to their lower protein intake. A moderate protein restriction of 1 g per kg IBW is therefore advised.

Sodium Intake

A high sodium intake has been associated with an increased renal secretion of calcium. Wendland in Canada suggested a moderate restriction of 100 mmol per day. *See* Sodium, Physiology

Fluid Intake

A high fluid intake of 2–3 litres per day is recommended in all forms of stone disease. An increased urine output leads to a decrease in concentration of 'stone-forming' factors.

Nephrotic Syndrome

Nephrotic syndrome is defined as the appearance of proteinuria > 3 g per day, hypoalbuminaemia, oedema and hyperlipidaemia. Nephrotic syndrome is not a disease in itself. It may appear in many different types of renal disease, some of which may proceed to end-stage renal failure (Table 8).

Table 8. Renal diseases which develop nephrotic syndrome

Glomerulonephritis
Diabetic nephropathy
Amyloid
Systemic lupus erythematosus
Vasculitis (PAN (polyarteritis nodosa), Wegener's granulomatosis)
Toxic glomerulopathy (gold, penicillamine, mercurials)

The treatment involves antihypertensive agents, immunosuppressants (steroids and others), diuretics, salt-poor albumin, and diet – protein, salt, and lipid lowering. In the past a high-protein diet was advised on the grounds that protein lost in the urine needed to be replaced. The protein intake could be as high as 150 g per day. Studies comparing low-protein diets and high-protein diets showed that proteinuria was reduced on a low-protein diet in diabetic and nondiabetic patients.

Studies using 0·3 g protein per kg IBW supplemented with amino acid/ketoacids also showed a decreased proteinuria and an increase in serum proteins. The mechanism could involve an improvement in permselectivity of the basement membrane in accordance with Brenner's hypothesis.

The current treatment for nephrotic syndrome ranges from 0·8 g protein per kg IBW with no extra allowance for protein losses to vegan diets with or without amino acid/ketoacid supplements. Nitrogen balance must be monitored regularly to ensure that protein depletion is not occurring. Reducing proteinuria also has a favourable effect on lipids perhaps because a lipid regulatory factor is no longer being lost into the urine. A moderate salt restriction (NAS) is advised when oedema is present and a fluid restriction is necessary.

Lipid Modification

Giovanetti has recommended vegetarian low-protein diets for their lipid-lowering action. Lipid abnormalities are part of the definition of nephrotic syndrome and there is a reported 85% increase in ischaemic heart disease in these patients. Lipid-lowering advice (35% energy from fat, increased PUFA:saturated fatty acid ratio) is recommended and in the future lipid-lowering agents may be used. *See* Fatty Acids, Metabolism

Haemodialysis and Continuous Ambulatory Peritoneal Dialysis (CAPD)

The nutritional problems of renal patients do not stop when renal replacement therapy is started. On the

Table 9. Haemodialysis diet

Nutrient (Recommended daily intake)	Rationale
Protein 1–1·2 g per kg IBW	8–12 g amino acids lost per dialysis. Dialysis may disrupt meal times 2–3 times a week. Protein exchanges are not necessary. Regular meals are encouraged.
Energy 35 kcal per kg IBW (or using Schofield equations)	Catabolic effects are induced by blood contact with dialysis membrane. 25 g glucose lost in each dialysis. Acidosis increases between each dialysis, leading to muscle and bone breakdown.
Phosphate 0·5 mmol kg^{-1} IBW	Not well dialysed, therefore restriction is still necessary, aided by phosphate binders.
Sodium 80–100 mmol Fluid 500 ml+PDUO	Urine output may be normal on starting dialysis but diminish with time. Advice on 'eking out' fluid allowance and avoiding a dry mouth will be necessary. Salt restriction will help to prevent thirst (as will good blood sugar control in diabetics). For practical purposes, 25–50 g of 'salty' tinned meat or fish, e.g. ham, corned beef, tuna, is 'allowed' in sandwiches.
Potassium 1 mmol kg^{-1}	Build-up of potassium can occur in the days between dialysis. Advice as per CRF.

IBW, ideal body weight.
PDUO, previous day's urine output.
CRF, chronic renal failure.

positive side, some of the symptoms of uraemia such as nausea, taste changes, gastrointestinal disturbances ('acidity', diarrhoea), lack of energy, should improve and therefore appetite should increase. On the negative side, coming to terms with a chronic illness, financial problems, sexual/partner problems, fluid restriction, reaction to dialysis (severe headaches, nausea), reaction to CAPD (stomach cramps, constipation), anaemia, peritonitis and other intercurrent infections all contribute to a poor appetite.

The haemodialysis and CAPD diet should be tailored to each patient. Regular blood tests are taken and the dietitian can monitor trends in changing urea and electrolyte levels and advise accordingly (Tables 9 and 10).

Patients are encouraged to take their full food allowances and to exercise as much as their condition allows to improve general health. As long as the patient is not underweight the 'healthy' eating principles of increased fibre, decreased saturated fat should be encouraged. Malnourished patients are given proprietary carbohydrate and protein supplements. Some renal units are administering intradialytic parenteral nutrition mixtures where oral supplements have failed to improve the nutritional state.

Acute Renal Failure (ARF)

Acute renal failure can be defined as an abrupt decline in renal function. Initially there is oliguria/anuria, retention of the nitrogenous end-products of protein metabolism, acidosis and electrolyte imbalance. As the kidney repairs itself and tubular function returns, there is polyuria to an extent where potassium and phosphate plasma levels drop below normal and salt and water depletion may occur.

ARF can be divided into three categories:

1. Noncatabolic, e.g. nontraumatic cause – interstitial nephritis from drugs, obstruction (prostate);
2. Catabolic, e.g. after surgery (transplant); rhabdomyolysis (massive muscle protein breakdown); haemolytic uraemic syndrome; postnatal;
3. Hypercatabolic, e.g. major trauma, burns and sepsis.

Noncatabolic ARF

As in CRF the aims of treatment are to prevent build-up of uraemic toxins and fluid overload in order to alleviate/prevent symptoms; and to maintain good nutritional status and prevent loss of flesh weight. The medical history is taken along with the diet history to identify any longstanding or current nutritional problems (nausea and lack of appetite is common).

If the patient is not symptomatic and the kidney condition is stable or improving, the standard hospital meals containing *c.* 1 g protein per kg IBW per day are adequate. Advice on salt intake (NAS), fluid restriction and potassium restriction may be necessary. In practice all that is needed is a simple diet sheet advising the patient and their relatives which snacks and drinks are preferable. The usual 'treats' which people receive in hospital – bowls of fruit, chocolate, fruit juice – are not acceptable. High phosphate levels are controlled with binders.

If the renal condition is worsening and the patient is symptomatic, a low-protein diet supplemented with high-calorie products may be beneficial to improve uraemia.

Catabolic and Hypercatabolic ARF

The patient will almost certainly require dialysis treatment. The catabolic state involves increased breakdown

Table 10. CAPD diet

Nutrient (Recommended daily intake)	Rationale
Protein 1·1–1·4 g per kg IBW	Albumin is lost across the peritoneal membrane into the dialysis fluid. Minimum *c.* 5 g per day. During peritonitis, losses may be >20 g per day; prescribable supplements will be required. Some vegetarians and people with a poor appetite, may need routine protein supplements.
Energy 25 kcal per kg IBW	70–80% of the glucose is absorbed from the CAPD fluid. This could amount to an energy intake of 200–500 kcal per day. Weight gain, increased triglycerides, increased blood sugars in diabetics are probable. Low-fat/low-sugar/high-fibre diet is encouraged.
Phosphate 0·5–0·6 mmol per kg IBW (15–18 mg kg^{-1})	Not well dialysed, therefore restriction is still necessary, aided by phosphate binders.
Salt 80–100 mmol (NAS) Fluid 500 ml+PDUO	Some patients ultrafiltrate (removal of fluid during dialysis) well. Fluid allowance can be adjusted accordingly. CAPD patients still have problems coping with fluid restriction and therefore salt should remain at NAS unless ultrafiltrating well.
Potassium 1 mmol per kg IBW	Dialysis is continual therefore there is not such a problem with accumulation. If consecutive blood results show a well-controlled potassium (<5·5 mmol l^{-1}) extra fruit and vegetables can be taken and/or cooking methods such as stir frying of vegetables can be used.
Vitamins (haemodialysis and CAPD)	Vitamin C has a high peritoneal clearance in CAPD. Vitamin B_6 plasma levels may decrease in the long term, vegetarians and people on H_2 antagonists are particularly at risk. Vitamin B_1, B_2 and folate losses may be increased. Fat-soluble vitamins are not lost in dialysis. B vitamin intake will increase with protein intake. Vitamin C supplement (approximately 100 mg) may be advisable and B vitamins for people who are not meeting their protein requirements.

IBW, ideal body weight.
NAS, no added salt.
PDUO, previous days urine output.

of muscle tissue and the release of urea, potassium, phosphoric acid and other organic acids. The patient may require nasogastric feeding or intravenous feeding (total parenteral nutrition).

Catabolic patients require 9–14 g nitrogen (56–87 g protein) per day. Hypercatabolic patients require 14–18 g nitrogen per day (87–112 g protein): they may be breaking down up to 40 g per day (235 g protein) but the liver cannot deaminate more than 20 g nitrogen per day (117 g protein), so this is the maximum that can be given.

The Schofield equation can be used to calculate the energy requirements on an individual basis. This calculates Basal Metabolic Rate, and additional calories are added according to a stress factor (level of injury), activity, specific dynamic action of food and body temperature. ***See*** Energy, Energy Expenditure and Energy Balance

Continual ultrafiltration is required in order to create a feeding space, otherwise nutrient intake would be severely limited by the fluid restriction: 500 ml plus previous day's urine output plus losses (fistula, diarrhoea, temperature); approximately 2–2·5 litres of feeding space is needed.

Nasogastric feeding. A feed containing a high calorie content (1·5–2 kcal ml^{-1}) is useful. It may be necessary to use a feed which is low in sodium, potassium and phosphate.

Total Parenteral Nutrition. Both essential and nonessential amino acids are required and the energy source should contain fat and carbohydrate. Additrace and Solvito will supply the trace elements and water-soluble vitamins, respectively. Fat-soluble vitamins will not be needed in the short term. As the patient enters the anabolic phase, potassium will be required (0·5–1·0 mmol kg^{-1} (20–40 mg kg^{-1})) as well as phosphate (10–30 mmol per day (300–900 mg per day)), although a phosphate-free regimen will help to control hyperphosphataemia in the early stages of ARF. A sodium-free regimen is not necessary if the patient is being dialysed and magnesium and calcium are provided in line with normal daily requirements. Regular monitoring of blood biochemistry will indicate any changes that have to be made to the TPN regimen.

Acknowledgements

I would like to thank Jonathan Kwan, Dr Peter Hart (Lecturers in Renal Medicine), Marianne Vennegoor, Gemma Bircher (Renal Dietitians) and Joanna Lee (Senior Secretary) for all their help.

Bibliography

British Dietetic Association Renal Dialysis Group. Lecture Notes, 1983–1992.

Fonque D and Laville M (1992) Controlled low protein diets in chronic renal failure. A meta-analysis. *British Medical Journal* 304: 216–220.

Giovannetti S (1989) *Nutritional Treatment of Chronic Renal Failure*. Kluwer Academic Press.

Goldstein DJ and Storm JA (1991) Intradialytic parenteral nutrition: Evolution and current concepts. *Journal of Renal Nutrition* 1: 9–22.

Hartley G and Forrest C (1989) Parenteral nutrition in acute renal failure. British Dietetic Association Renal Dialysis Group Meeting, Royal Victoria Infirmary, Newcastle, 1989.

Mitch WE and Walser M (1991) Nutritional therapy of the uraemic patient. In: Brenner BM and Rector FC (eds) *The Kidney*, 4th edn. Philadelphia: WB Saunders.

Paul AA and Southgate DAT (1991) *McCance and Widdowson's Composition of Foods*, 5th edn. London: MAFF and RSC.

Schofield WN (1985) Predicting BMR, new standards and review of previous work. *Human Nutrition: Clinical Nutrition* 39c (Supplement): 5–41.

Thomas B (1988) *Manual of Dietetic Practice*. Oxford: Blackwell Scientific.

Wendland BE (1990) Nutritional management of patients with urolithiasis. *European Dialysis and Transplant Nurses Association/European Renal Care Association*. 5–7.

Whitworth AJ and Lawrence RJ (1987) *Textbook of Renal Disease*. Hong Kong: Churchill Livingstone.

Barbara Engel
The Royal London Hospital, London, UK

RENNIN

See Cheeses

RESIDUE DETERMINATION

See Antibiotics and Drugs, Contamination, Fumigants, and Pesticides and Herbicides

RESISTANT STARCH

See Starch

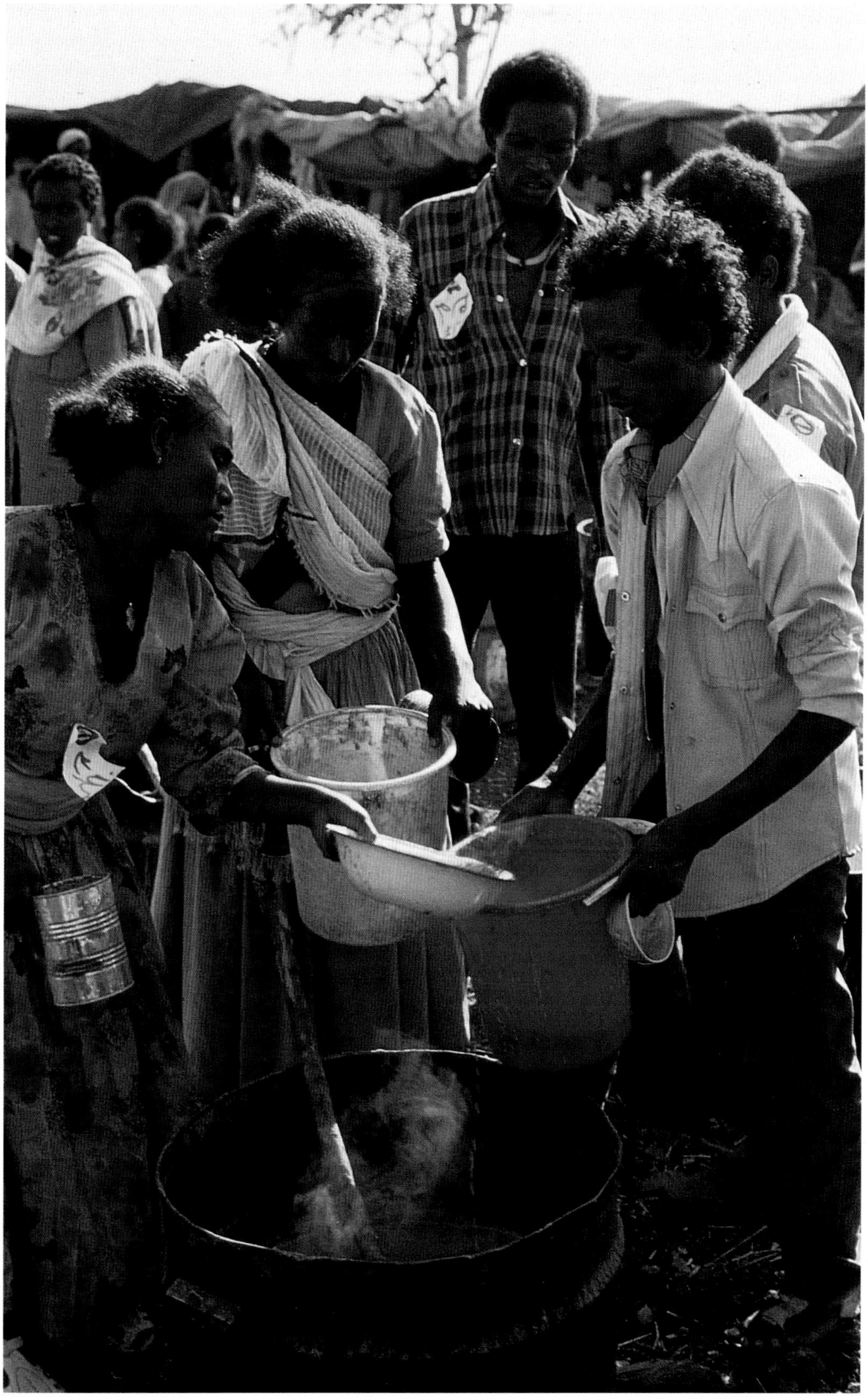

Plate 1 Food distribution to Ethiopian refugees in Sudan during February 1987. (T. Brevik, UNHCR.)

Plate 2 Sake bottles at the annual sake contest held at the National Research Institute of Brewing, Tokyo.

Plate 3 A sake tasting cup.

Plate 4 Poultry: Processed turkey products.

Plate 5 Red raspberry fruit ready for harvest.

Plate 6 Black raspberry fruit showing various stages of maturity.

Plate 7 Sensory evaluation: Preparation area in a development centre.

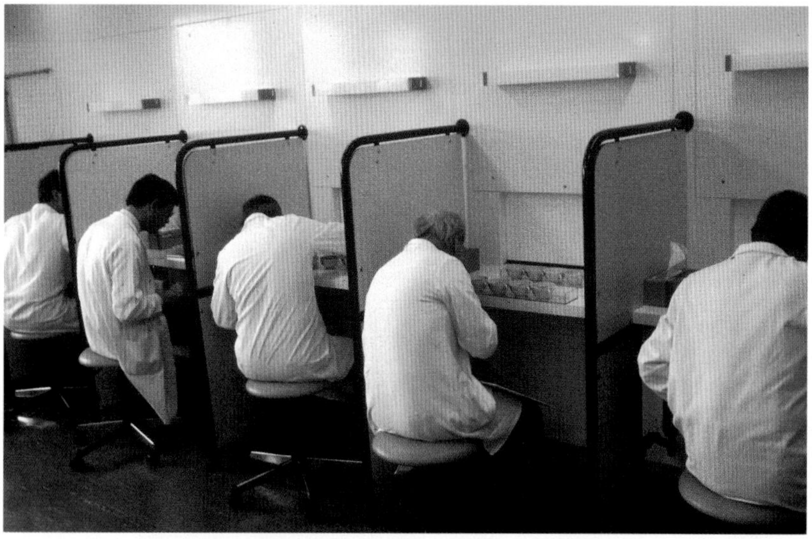

Plate 8 Sensory evaluation: Tasting booths in use in a development centre.

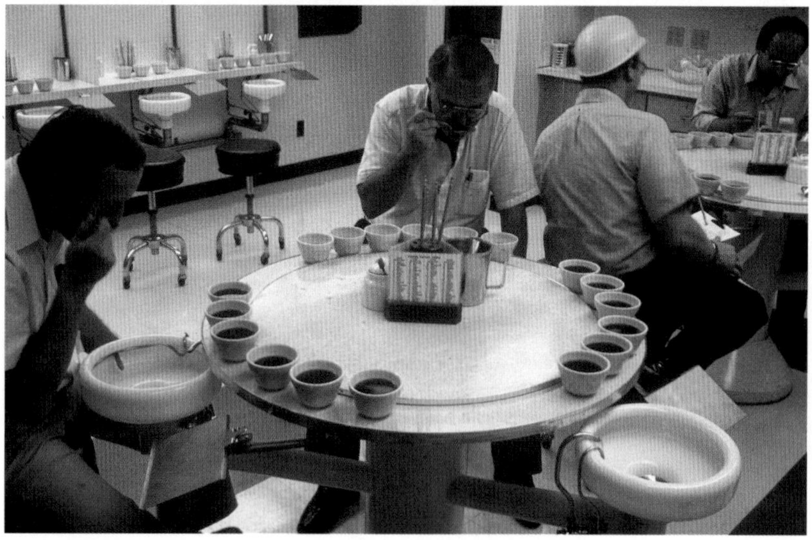

Plate 9 Tasting products in a factory around a revolving table.

RETAILING OF FOOD IN THE UK

In a relatively developed country such as the UK, the long food marketing chain from farmer to consumer contains many specialized activities, including manufacturing, wholesaling, retailing and catering. The elements in this chain contribute around 9% of gross domestic product (GDP), with a labour force of about 3 000 000, or 11% of those employed (Slater, 1988). In the UK, approximately 80% of food is processed to a greater or lesser degree, and all foods, other than those bought directly from farmers by consumers, undergo such marketing functions as transport and storage. Although catering plays an important and growing role at the consumer end of this chain, retailers remain the principal agents for the final distribution of food. Food wholesaling and retailing together contribute some 2·5% of GDP, and has over 1 000 000 employees, but the food retailing component accounts for more than two thirds of this value added and employment (Slater, 1988).

In 1988, about 80% of the £36 × 10⁹ food sales (including confectionery and soft drinks) was made through food retailers, the majority of the remainder being split roughly equally between 'drink, confectionery and tobacco retailers' and 'mixed retail businesses', such as Marks and Spencer, and Boots (calculated from BSO, 1991). Food retailers are usually divided into 'specialist food' and 'grocery' businesses, with the former classified by their main sales as greengrocers, dairymen, butchers, fishmongers or bread and flour confectioners, while the majority of the latter's sales consist of 'other food and soft drinks' (previously called 'groceries and provisions'). In addition, food was sold through retail outlets owned by businesses whose main activity was other than in retailing, such as food processors (particularly important for bread and for milk and dairy products), and directly to consumers from farmers' shops or pick-your-own (especially for fruit and vegetables).

Food retailers' sales have increasingly favoured the grocery trade and particularly the larger multiple chains. In 1988, the UK's 25 000 grocery businesses made up some 37% of the firms in food retailing, but their sales of over £33 × 10⁹ (including nonfoods) were more than 85% of the sector's total turnover. However, sales by the 69 'large grocery firms' exceeded 75% of the trade, with the top five alone accounting for more than 40% of retailed food (BSO, 1991). The power and importance of these major grocery retailing chains have been the most significant features in the development of food marketing over the past 20 years.

Economic Environment of Food Retailing

In the 1950s, the retailing of food appeared to be largely as it had before World War II. The generally fragmented structure consisted of more than 250 000 outlets in Britain, including numerous local 'corner stores' (BSO, 1975). These, and an extensive system of doorstep delivery, especially for milk and bread, provided the household with sources of frequently purchased food products. By the beginning of the 1950s, self-service began developing in the UK along the lines of the American model, and leading grocery retailers such as the Co-op and Sainsbury started to introduce supermarkets. Developments in the structure of food retailing and the rise of large-scale businesses were facilitated and promoted by a number of technical, economic and social changes. Packaging technology made self-service feasible in the grocery trade, and, along with transport advances such as multitemperature-controlled vehicles, enabled many improvements in the physical distribution of food. Higher standards of living, and acquisition of household consumer durables, allowed consumers with cars, refrigerators and freezers to undertake many of the storage and marketing functions previously performed by the corner shop and delivery van. The high proportion of women employed outside the home, along with the emphasis on leisure activities and more informal mealtimes, encouraged consumers to prefer foods that were increasingly convenient in buying, preparation and cooking. By the 1980s, food shopping on a large scale at a supermarket or superstore had become a weekly or fortnightly event for most consumers, with unanticipated or forgotten items obtained from local convenience stores.

Consumer expenditure on food alters only very slowly, and is much more stable than purchases of durable goods. In addition, since consumers in the UK generally have sufficient food to meet basic nutritional requirements, overall food consumption is almost static in energy terms, and expenditure tends to decline as a proportion of total consumer expenditure when real incomes increase. In 1961 household spending on food accounted for about a quarter of consumers' total expenditure, but by 1989 it was less than one eighth (Central Statistical Office, 1991). However, growth in real disposable incomes results in 'trading up', with better quality, more convenient foods replacing poorer, basic items in consumers' purchases. In addition, factors such as age, location, household size and family circumstances all give rise to differences in patterns of food

Table 1. Retailing of food: number of outlets

	1971	1980	1986	1988
Large grocers	105 283	11 906	8873	8328
Other grocers		48 871	32 942	26 992
Dairymen	3853	9003	9700	11 264
Butchers	33 939	22 795	20 721	18 215
Fishmongers	4678	2866	2572	2802
Greengrocers	23 318	17 778	16 805	14 263
Bakers	17 299	8381	7553	5895
All food retailers	188 370	121 600	99 751	87 758

Sources: BSO (1975, 1991).
Figures for 1980, 1986 and 1988 are on a revised basis.

expenditure and consumption, and to some degree in choice of retailer. Food retailing similarly presents a heterogeneous picture, although with a structure markedly different to that of 30 years ago.

Retail Structure

The structural changes that occurred in food retailing after World War II may be classified into two broad types: the decline of small grocery and specialist retailers, many of whom have gone out of business; and the rise of the large multiples who have grown internally or by merger, and in the process closed down small outlets and opened bigger shops and superstores. Table 1 indicates the effects of these developments on the numbers of food retail outlets over the past two decades; although the 1971 figures are not strictly comparable because of changed definitions, the general trend is clear. Simply comparing 1980 and 1988, a reduction in the number of outlets by over a quarter is apparent, with the largest falls occurring amongst grocers and bakers; dairymen are the only retailers to show an increase in numbers.

Although two thirds of the businesses in food retailing operate with a single outlet, the 'large grocery retailers' averaged over 120 outlets each in 1988. In the same year, the mean turnover of food retailing outlets was less than £500 000 per annum, but that of outlets owned by large grocers approached £3 500 000 per annum. Measured at constant 1985 prices, the average turnover per outlet of all food retailers increased by 50% during the period 1980–1988. However, outlets owned by large grocery retailers showed gains of 97%, while other grocery retailers' turnover per outlet rose by only 12%. (All calculations are from BSO, 1991).

From pre-World War II until the late 1950s, independent retailers accounted for about half of all grocery sales, with the rest of the market shared roughly equally between multiples and cooperative societies. The 1960s saw a rapid pace of structural change, and by the early 1970s multiple grocer's share of the grocery trade, as defined by Nielsen, exceeded that of the much more numerous independents. Figure 1 (based on data from Nielsen, AC) indicates that the position of the multiples continued to strengthen over the next two decades, while independents suffered continuous decline, and the cooperatives' fortunes waivered about a generally falling trend.

Within the multiple sector, the leading companies have established and enhanced their dominance in recent years. Five major grocery multiples – Sainsbury, Tesco, Argyll, Asda and Gateway (Isosceles) – are estimated to have accounted for about 50% of *total food and drink* sales in the UK in 1989 (Retail Business, 1990a). By comparison, the top five firms in 1985 held only 30% of the same market. (Nielsen's statistics indicate a five-firm concentration ratio of 60% of the more narrowly defined grocery trade in 1988.) Although the major multiples cover most of the UK, regional strengths vary, and in some cases companies outside the top five hold a leading position, e.g. Kwik Save in the northwest. In terms of two- or four-firm concentration ratios, regional concentration is generally higher than national (OFT, 1985; Burdus, 1988).

The two largest companies – Sainsbury and Tesco – have grown primarily by investment in new stores, and only marginally through the acquisition of existing competitors. By contrast, much of the rise of Argyll and Gateway can be attributed to takeovers, and Argyll has especially benefited from its acquisition of the UK part of Safeway in 1987. Asda, created in 1965 by Associated Dairies, was a pioneer in UK out-of-town superstores and has grown mostly by internal investment. However, in 1989 Asda took the opportunity to expand rapidly by purchasing 60 Gateway superstores from Isosceles plc (which had accquired the Gateway group earlier in the year).

Independents have attempted to secure some of the advantages of bulk buying and marketing by involvement in the voluntary 'Symbol' groups, which are sponsored by wholesalers. Spar, with about 2500 stores, is the largest in the UK. Many of its outlets have been converted to 'Eight til Late' convenience stores, and this format (with long opening hours and a limited range of products) has become more common amongst small food retailers.

Having once been pre-eminent in the grocery trade, the cooperative societies have tended to lag behind the leading multiples, but faired better than the other, smaller grocers in the late 1980s. The organization is dominated by the CRS (Cooperative Retail Society) with 20% of the groups' retail outlets and a quarter of the turnover, and the CWS (Cooperative Wholesale Society), which although primarily a manufacturer and wholesaler, also runs a substantial retail division. Some 80 separate societies existed in 1989, about 40% of the

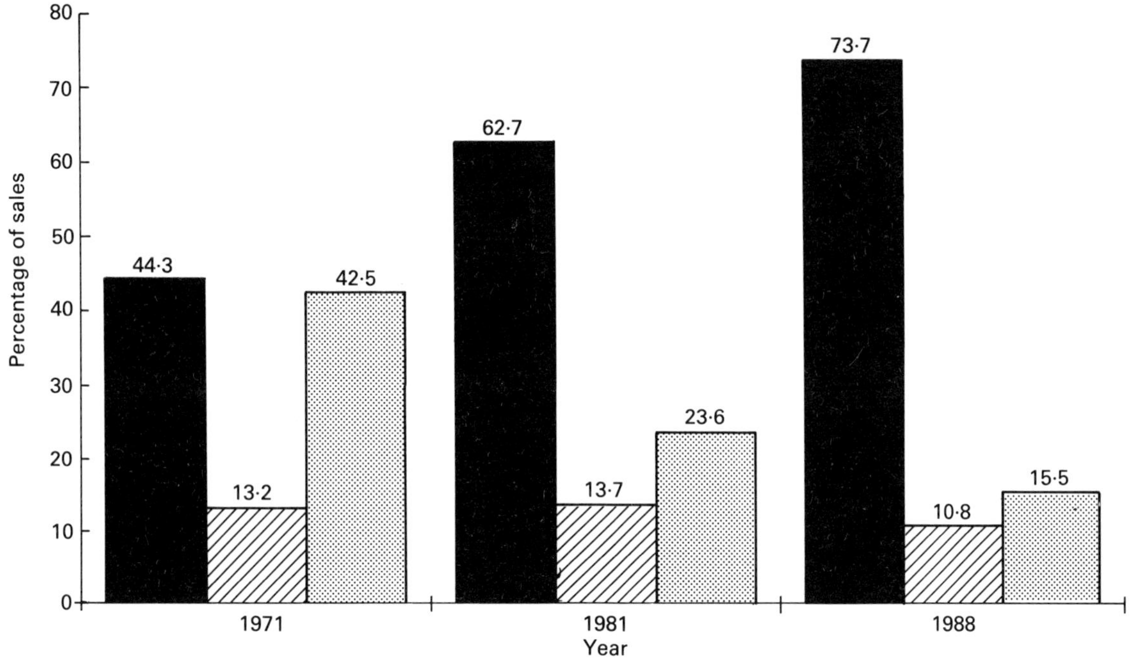

Fig. 1 Grocery retailing: market shares (■ multiples; ▨ cooperative societies; ▩ independents).

number at the beginning of the decade (Retail Business, 1990b).

Specialist food retailers are generally individual operators or small local multiples, although there are some significant groups. Amongst butchers, Union International had over 1000 butchers shops in 1990, operating under Dewhurst, Baxters, Matthews, and other names. The same company has major interests in meat processing and distribution. Vertical integration of manufacturing and retailing is also apparent in the dairy sector, with the Co-op, Unigate, Northern Dairies the leading roundsmen and processors, and amongst bakers, where the two main millers and bakers, Associated British Foods and Rank Hovis McDougall, also have significant retail interests. Doorstep milk delivery has tended to decline and, like other specialist areas, has lost market share to the grocery multiples (Retail Business, 1990c)

Competition

The process of development in food retailing follows from the economic environment and firms' competitive responses. Competition is not simply a matter of lower prices but may focus on advertising, product and store quality, product range, store location, or ancilliary services such as petrol and restaurant facilities. Consumer concerns over health and the environment are recent trends that retailers have utilized in their product choice and specifications, and more generally in their marketing and corporate image strategies.

Price competition has been a very significant feature of retailing over the last three decades. The effective removal of collective and individual resale price maintenance in 1956 and 1964, respectively, provided a catalyst to the growing practice of price discounting, albeit on a limited range of items. Consumers were attracted into shops on the basis of low prices on specific products but would then purchase other items at 'normal' prices. However, in the more price-conscious years of high inflation of the early 1970s, stores such as Asda and Kwik Save gained market share by more across-the-board price reductions. Other major multiples soon followed suit, particularly aware of the need to tempt consumers to the new, larger, out-of-town superstores and exploit their economies of scale. Once one-stop-shopping was established in the 1980s, most retailers had shifted their competitive emphasis to nonprice features, although Kwik Save and the German multiple Aldi (which arrived in the UK in 1990) continue to offer a limited line of manufacturers' brands at low prices.

Locational convenience may also be a significant factor in consumers' choice of shop. This is not simply a question of proximity to consumers' homes, since out-of-town or shopping centre stores may be especially convenient for the car-borne shopper. Town centre sites may also prove useful to those who shop during lunchbreaks, as Marks and Spencer's continued success in food retailing has shown. An associated aspect of convenience is consumers' opportunity to obtain the

Table 2. Commodity sales, 1988: percentage distribution between kinds of business

<table>
<tr><th>Commodity</th><th>Food retailers</th><th>Large grocers</th><th>Other grocers</th><th>Dairymen</th><th>Butchers</th><th>Fishmongers</th><th>Greengrocers</th><th>Bakers</th><th>CTNs</th></tr>
<tr><td>Fresh fruit and vegetables</td><td>89·7</td><td>50·2</td><td>6·3</td><td>0·5</td><td>1·1</td><td>0·1</td><td>31·5</td><td>0·1</td><td rowspan="6">3·0</td></tr>
<tr><td>Fresh milk and cream</td><td rowspan="3">84·2</td><td>25·5</td><td rowspan="3">2·9</td><td>40·2</td><td>0·2</td><td>0·0</td><td>0·4</td><td>0·2</td></tr>
<tr><td>Fresh meat</td><td>42·8</td><td>0·4</td><td>46·8</td><td>0·2</td><td>0·2</td><td>0·0</td></tr>
<tr><td>Fresh fish</td><td>20·1</td><td>0·4</td><td>1·7</td><td>53·4</td><td>3·2</td><td>0·0</td></tr>
<tr><td>Bakery products</td><td>64·9</td><td>37·7</td><td>9·1</td><td>1·4</td><td>1·6</td><td>0·0</td><td>0·8</td><td>14·2</td></tr>
<tr><td>Other foods and drinks</td><td>86·6</td><td>72·1</td><td>12·2</td><td>0·9</td><td>0·8</td><td>0·0</td><td>0·3</td><td>0·2</td></tr>
<tr><td>Alcoholic drinks</td><td>50·0</td><td>43·6</td><td>5·5</td><td>0·6</td><td>0·1</td><td>0·0</td><td>0·1</td><td>0·0</td><td>6·5</td></tr>
<tr><td>Chocolate and sugar confectionery</td><td>29·4</td><td>21·8</td><td>5·7</td><td>0·4</td><td>0·2</td><td>0·0</td><td>0·8</td><td>0·5</td><td>48·9</td></tr>
</table>

CTN: Confectioners, tobacconists and newsagents.

majority of requirements in one store (one-stop-shopping) and this has implied an increasingly wide range of food and nonfood items in larger grocery outlets. Multiples have expanded beyond their traditional sphere of packaged goods to encompass many of the specialist food retailers' areas. In-store bakeries, delicatessens, fresh meat and fresh fish counters, and free-flow fruit and vegetable displays show the multiples' ability to adapt their formats to compete directly with products and services offered by specialist retailers. In 1988, large grocers sold more than 50% of all fresh fruit and vegetables, over 40% of fresh meat and 33% of bakery products (Table 2).

A further aspect which links both price and product issues is the retailers' (or wholesalers') own-label goods. The proportion of packaged grocery items sold under this general heading now averages 30%, having been less than 25% at the beginning of the 1980s (IGD, 1990, based on data from Audits of Great Britain Ltd). Amongst food retailers, Sainsbury's proportion of own-label packaged grocery sales is highest, at over 50%, but in Waitrose (John Lewis Partnership), Tesco, Safeway (Argyll), Asda and the Co-op the own-label share of turnover is greater than the sector average. No longer seen as mere cheap alternatives to manufacturers' brands, own-labels are often of comparable standard and provide consumers with the retailers' assurance of quality. Own-labels enable retailers to react quickly to changing consumer requirements by altering specifications and introducing new products. They may also be used, in conjunction with advertising campaigns, by retailers promoting a corporate image of concern with issues such as healthy eating or conservation and the environment. Distributors' own-labels are frequently produced by major UK food manufacturers, either as a main enterprise, or using excess capacity from their own product lines. Since marketing costs are reduced by the saving in advertising and promotion, own-labels can prove to be attractive to consumers, manufacturers and retailers, as margins may be higher than for manufacturers' brands, yet retail prices may be lower.

Conclusions

Food retailing has undergone major developments over the past 20 years in structure, location, store size and product range. These reflect changing consumer incomes, lifestyles and tastes, the economic and technical environment, and the firms' competitive initiatives and responses. The largely static overall demand and falling share of the national market has created a highly competitive food retailing environment, in which successful businesses grow essentially by taking custom from their declining rivals. In food retailing this has promoted the long-term rationalization of multiple and cooperative firms and outlets and the decline of independents. Major retailing chains have become more aware of consumers' interests and behaviour, and are able to monitor shoppers' reactions to new products, price and merchandising changes through the information provided by electronic point-of-sale scanning equipment. This technology, coupled with retailers' own centralized warehousing, has improved stores' stock holding efficiency and lowered distribution costs.

More affluent consumers have chosen to undertake their main food shopping in larger stores and are still able to choose between a reasonable number of different retailers (Burdus, 1988). However, as the competition for prime sites intensifies, so retailers may locate superstores further from consumers. The range of products in these stores has increased and the choice of brands remains fairly wide. Poorer consumers, especially in the third of UK households without cars, may not be so well placed. Their local stores may offer neither the range nor low prices of the out-of-town superstores, especially if these local stores adopt the convenience store format. Specialist food retailers may prove to be more resilient, but their relative decline in comparison with the multiple chains appears likely to continue. The higher margins obtainable in specialist foods are likely to encourage multiples to continue to develop these areas.

The success of leading food multiples, as measured in growth, margins, profitability and return on capital, has occasionally given rise to concerns that exploitation of dominant positions has both depressed manufacturers' profits and led to unfair competitive advantages *vis-à-vis* smaller retailers. The Monopolies and Mergers Commission (MMC) report (1981) and the Office of Fair Trading's follow-up study (1985) both suggested that while major supermarket chains used buying power to gain significant discounts from their suppliers, these were passed to consumers in the form of lower prices. These reports indicated no anticompetitive abuses at the time, and although the MMC felt that structural changes in retailing should be kept under review, mergers and takeovers in this sector have continued. The argument has been to favour the strengthening of multiples aiming to compete with the two leaders. However, multiples are not all enjoying equal success, and if, as some commentators predict, the number of effective competitors were to fall in the future, there may be renewed calls for action by competition authorities in this area.

Bibliography

Burdus A (1988) Competition in the food distribution sector. In: Burns J and Swinbank A (eds) *Competition Policy in the Food Industries*, pp 68–102 University of Reading: Department of Agricultural Economics and Management.

BSO (Business Statistics Office) (1975) *Census of Distribution 1971*. London: Her Majesty's Stationery Office (HMSO).

BSO (1991) *Retailing 1988*. London: HMSO.

Central Statistical Office (1991) *Annual Abstract of Statistics*. London: HMSO.

IGD (1990) *Food Retailing*. Letchmore Heath, UK: Institute of Grocery Distribution.

Monopolies and Mergers Commission (1981) *Discounts to Retailers*, HC 311. London: HMSO.

OFT (1985) *Retailing and Competition*. London: Office of Fair Trading.

Retail Business (1990a) *Strategies in Grocery Retailing*. London: Economist Intelligence Unit.

Retail Business (1990b) *Co-operatives*. London: Economist Intelligence Unit.

Retail Business (1990c) *Specialist Food Retailers*. London: Economist Intelligence Unit.

Slater JM (1988) The Food Sector in the UK. In: Burns J and Swinbank A (eds) *Competition Policy in the Food Industries*, pp 1–27. University of Reading: Department of Agricultural Economics and Management.

J. A. Burns
University of Reading, Reading, UK

RETINOL

Contents

Properties and Determination

Physical Properties

Retinol (9,13-dimethyl-7-[1,1,5-trimethyl-6-cyclohexene-5-yl]-7,9,11,13-nonatetraene-15-ol) is a pale yellow crystalline powder or oily mass, depending on purity. It is insoluble in water but readily miscible with most organic solvents. Its formula, formula weight and certain other characteristic physical properties are summarized in Table 1, along with those of the two commercially significant ester derivatives.

Chemical Properties

All-*trans* retinol, considered the parent compound of the vitamin A group, is a complex unsaturated alcohol. Its trivial name originated with the historical recognition of its role in vision. This generic structure illustrated in Fig. 1, shares a common β-ionone ring, with attached conjugated isoprenoid side-chain.

Chemical and structural modifications to the ring, side-chain, or polar functional group generate many retinoids possessing a wide spectrum of properties. In nature only relatively few of these compounds exhibit significant vitamin A activity. These include the predominant all-*trans* forms of retinol (originally designated vitamin A_1) and its esters, retinal, retinoic acid

Table 1. Physical properties of retinol (all-*trans*) and its esters

Property	Retinol	Retinyl acetate	Retinyl palmitate
Formula	$C_{20}H_{30}O$	$C_{22}H_{32}O_2$	$C_{36}H_{60}O_2$
Formula weight	286·46	328·50	524·88
Melting point (°C)	63–64	57–59	28–29
UV max.[a] (nm)	325	326	326
Extinction coefficient, $E^{1\%}_{1\,cm}$	1820	1530	960
Molar absorptivity, ε	52 140	50 260	50 390
Fluorescence			
Excitation max. (nm)	325	325	325
Emission max. (nm)	470	470	470

[a] In 2-propanol. Spectral properties vary slightly between protic and aprotic solvents.

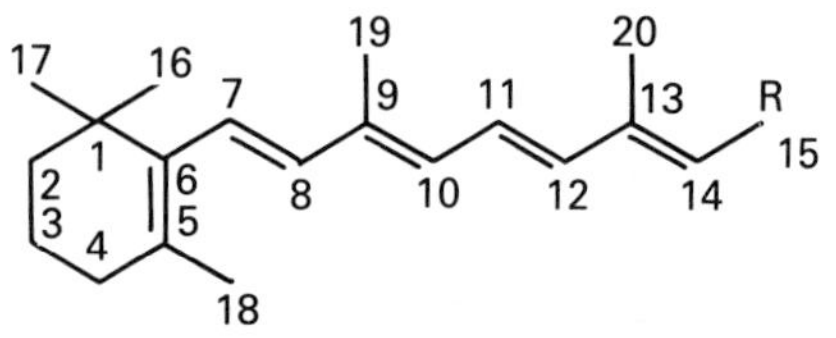

R = CH_2OH	all-*trans* retinol
CH_2OCOCH_3	all-*trans* retinyl acetate
$CH_2OCO(CH_2)_{14}CH_3$	all-*trans* retinyl palmitate
CHO	all-*trans* retinal
COOH	all-*trans* retinoic acid

Fig. 1 Structures of the basic retinoids. The *cis*-isomer positional variants are indicated by the side-chain numbering system.

and their associated 9-, 11- and 13-*cis* isomers. The *cis*-isomers, are selectively interconvertible with the *trans*-forms in the body and thereby express fractional biological activities of approximately 0·25, 0·50 and 0·75, respectively, relative to the all-*trans* vitamer. Some marine and freshwater fish contain significant quantities of the cyclic diene 3-dehydroretinol (originally designated vitamin A_2), which possesses approximately 40% of the bioactivity of all-*trans* retinol.

Vitamin A activity is commonly expressed in International Units (one iu is equivalent to 0·300 μg all-*trans* retinol, 0·344 μg retinyl acetate or 0·549 μg retinyl palmitate), USP Units, or more recently, Retinol Equivalents (one RE is equivalent to 1·0 μg all-*trans* retinol). Such units have been useful in nutritional studies where contributions from numerous vitamin A active congeners may be combined to yield a single value. For the food scientist, it is probably more appropriate to express vitamin A content as the summation of each retinoid, in absolute mass units (usually micrograms).

The dominant structural feature of retinol is the extensive conjugated double bond system, to which many of the physicochemical and biological properties may be attributed. It is also the principal factor in the lability of vitamin A. Thus, retinol and its derivatives are particularly sensitive to oxidizing conditions and are rapidly destroyed by heat, light and acids when in solution. Isomerization, oxidation and, ultimately, molecular cleavage may all occur concurrently, depending on the extent of environmental stress. Cleavage may result in the formation of volatile β-ionone fragmentation products which have importance in the development of off-flavour in some foods. Retinol is, however, relatively resistant towards alkali, and most degradation processes are minimized if it is maintained under inert gas at low temperature and in the absence of short-wavelength light.

In solution, several fat-soluble antioxidants offer protection, while in foods and food extracts, accompanying lipids and endogenous antioxidants such as phospholipids, tocopherols and carotenes act to stabilize retinol until they are sacrificially depleted. Both processing and storage expose vitamin A in foods to the risks of considerable loss (5–40%), although published data are often highly variable. ***See*** Antioxidants, Natural Antioxidants

Retinyl esters are considerably more stable than the parent alcohol, a feature exploited in the principal commercial formulations available to the food industry. Although minor differences in absorption efficiency have been reported, dietary retinyl esters are considered to exhibit identical biological activities as a result of their conversion to retinol in the intestinal wall.

Occurrence and Forms in Foods

All forms of vitamin A found in foods ultimately derive from the provitamin A carotenoids, which are ubiquitous in the higher plants and lower animal organisms. Humans obtain preformed vitamin exclusively from animal sources, while carotenoids are gained from foods of both plant and animal origin.

Table 2. Preformed vitamin A content of selected foods, expressed as retinol

Food	Retinol content (range) of edible portion (μg per 100 g)[a]
Milk	32–45
Butter	800–1000
Eggs	140–250
Beef	2–5
Liver (lamb)	7000–10 000
Mackerel	25–50
Cod liver oil	15 000–30 000

[a] To convert to iu per 100 g, multiply by 3·33.

Retinol and its derivatives are not widely distributed in foods. Fish liver oils are by far the most concentrated natural source, while animal liver, milk (and dairy products) and eggs contain significant quantities. The average Western diet is generally assessed as satisfying the recommended daily intake level of 1000 RE total vitamin A, with about 25% supplied by β-carotene. ***See*** Carotenoids, Properties and Determination; Carotenoids, Physiology

Retinoids in foods occur mainly as mixtures of retinyl esters, with lesser contributions from retinol itself. The predominant forms are long-chain esters, notably palmitate, with contributions from stearate and oleate. However, eggs are an exception where unesterified retinol is the major form. Certain foods also contain contributions from retinal (eggs and fish roe) and *cis*-isomers (mainly 13-*cis*), the latter particularly in foods subjected to processing.

Vitamin A deficiency is one of the prevalent diet-related issues and has received global attention. Consequently, nutritional tables are replete with information regarding retinol distribution in foods. Table 2 lists the vitamin A content of a few representative foods. Carotenoid-rich vegetable sources are not included. ***See*** individual foods

Use in Food Fortification

Increasing the vitamin A intake amongst populations at risk of retinol deficiency is a simple expediency and has been employed for several decades, both prophylactically and therapeutically. Many of the specific symptoms of severe deficiency (e.g. xerophthalmia, keratomalacia) are usually reversible, provided other nutritional criteria are similarly satisfied. In the developed nations, potential risks of subclinical deficiency are also avoided through food enrichment, thereby ensuring satisfactory intake. ***See*** Food Fortification

Fat-based foods such as margarine, milk and infant formula provide an ideal carrier in which to supply supplemental vitamin A. Dried milk, by reason of cost and convenience, is globally, the preferred foodstuff and additionally offers a high-quality protein and mineral enrichment medium appropriate to nutritional rehabilitation programmes. An increasing range of alternative food items are being used as carriers, ranging from specialized dietary beverages to breakfast cereals.

In most cases, supplementation is performed with synthetic all-*trans* retinyl acetate or palmitate. The esters in the case of margarine are usually added directly in an oil carrier, often in the presence of appropriate antioxidants. Where additional stability is required, or when dried food products are involved, it is common practice to use a powder preparation in which vitamin A (and stabilizer) is deposited in a suitable carrier, such as gelatine/carbohydrate, although alternatives (e.g. acacia gum) have recently been developed. However, uneven vitamin distribution is a common problem when dry-blending food products.

Water-miscible vitamin A powders have been developed containing emulsifiers that facilitate reconstitution into aqueous solution. These solutions can then be processed directly into fluids such as milk prior to drying. Using the wet-blending technique, the protective environment of the additive is hopefully substituted by intimate interaction with the lipid phase of the finished food. The principal advantage of this supplementation route is an enhanced vitamin homogeneity in the finished product. However, the added ingredients are generally found to be less stable than their endogenous counterparts. ***See*** Emulsifiers, Uses in Processed Foods

During food storage there are significant advantages in using packaging which provides effective oxygen and light protection. The use of cans allows nitrogen purging of the headspace, thereby greatly enhancing oxidative stability of the food and extending its shelf life.

Extraction and Clean-up

During analytical procedures, precautions are mandatory to exclude exposure of vitamin A to UV radiation, thermal, and oxidative stresses. Protective antioxidants and exclusion of air are essential to the success of the analysis. Purity of vulnerable retinoid standards is important during quantitative investigations, and spectrophotometric procedures are generally necessary to ascertain accurately their purity concentration.

Alkaline digestion (saponification) is most commonly employed as the first stage during analysis of retinol in foods. A representative sample is homogenized and digested in ethanolic potassium hydroxide or similar media. Saponification has the threefold effect of eliminating the bulk of lipid materials, releasing the vitamin from within the sample and converting the various esters into free retinol. The procedure can be performed under reflux temperatures, or at ambient temperature (for longer periods), the details of which are determined by the sample matrix. In some food samples where the vitamin A concentration may be low, a preliminary fat extraction step will assist in achieving the required assay sensitivity.

Retinol is partitioned from the digest into an immiscible nonpolar solvent, commonly consisting of hexane and ether. The solution is washed with water, dried and usually concentrated by evaporation, to provide a crude extract. Direct spectrophotometric or fluorometric assay of retinol may be impractical in poorly characterized foods, owing to interfering coextractives, and further clean-up is often necessary. Traditionally this has been achieved by open-column or thin-layer chromatography, using silica or alumina stationary phases, while gel-permeation chromatography has also been advocated. The laborious nature of these clean-up techniques has recently been improved through the growing use of prepacked, disposable chromatographic cartridges.

As the majority of native and supplementary retinol in foods exists in esterified form, there are analytical advantages to be gained through avoiding preliminary saponification, providing the nature of the food allows such simplified methodology. The total lipid fraction of such foods may therefore be directly extracted into organic solvent, which is then ready for analysis of the intact esters. Advantages include a reduction in the number of manipulative steps and the decreased risk of analyte loss through retinol degradation.

Chromatographic and Other Methods of Determination

The detection of retinol is advantaged by its extensive conjugated double bond functionality, which provides an intense and relatively specific chromophore in the high-UV region. The sensitivity and selectivity of spectral analysis is consequently enhanced, irrespective of the separation technique used. In addition, retinol exhibits a strong fluorescence which offers selectivity advantages and often reduces the need for prior purification. In situations involving well-characterized foods, and where discrimination between retinol and related compounds is unnecessary, spectrophotometric or fluorometric measurement of the crude extracts can often supply a reliable estimate, providing the limitations of such strategies are recognized. ***See*** Spectroscopy, Fluorescence

The traditional Carr–Price method for determination of retinol in foods has exploited the reaction with antimony trichloride to produce a blue complex in direct proportion to vitamin A concentration. Although the colour is transient and difficult to control, the reaction is still employed in laboratories with limited access to modern chromatographic instrumentation. Other Lewis acids, notably trifluoroacetic acid, will react in a similar way and offer some manipulative advantages over the Carr–Price approach.

There has been increasing recognition that a reliable estimation of vitamin A in foods is best achieved through exploiting chromatography to separate coextracted substances and, in particular, to differentiate the active retinoid species. Most variants of chromatography have been used during attempts to separate and quantify the retinoids. Thin-layer chromatography (TLC) and low-pressure liquid chromatography, while reasonably successful, have traditionally lacked the resolving power for reliable quantitative measurements, although high-performance TLC shows potential. Gas–liquid chromatography has not been widely utilized, as retinol (and its esters) degrade at elevated temperatures. The use of high-performance liquid chromatography (HPLC) has dominated vitamin A measurement over the last 20 years and has been largely responsible for the rapid proliferation of knowledge regarding this vitamin. This nondestructive technique facilitates a rapid analysis with minimum sample preparation. ***See*** Chromatography, High-performance Liquid Chromatography; Chromatography, Thin-layer Chromatography; Chromatography, Gas Chromatography

Since isomerization can be a common phenomenon during food production, *cis* and *trans* differentiation may be required, with the application of appropriate factors to account for their selective bioactivities. In addition, oxidation products such as oxyretinoids, epoxyretinoids and their fragmentation products, as well as varous retroretinoids, can exist in many food samples. These have poorly defined vitamin A activity and may distort a nonchromatographic measurement. Published literature is available for the determination of retinol isomers and esters, 3-dehydroretinol, retinal and retinoic acid in a wide variety of foods. These analyses are accomplished using normal- or reversed-phase HPLC, often with simple isocratic solvent systems. Normal-phase is generally the more efficient mechanism for isomer separations, while esters are best resolved under reversed-phase conditions. Figure 2 illustrates the isomeric distribution in a saponified sample of cod liver oil utilizing normal-phase conditions.

The estimation of retinal and retinoic acid in foods is not frequently carried out. They exist in some foods as metabolites of the live animal but their contribution to the vitamin A pool is negligible compared to that of retinol. Despite the occurrence of 3-dehydroretinol in fish and particularly fish liver oils, retinol remains the primary source of vitamin A owing to its higher biopotency. Thus retinol, and its esters, are generally the retinoids of major interest during routine food analysis.

Fig. 2 A 1-g sample of commercially refined cod liver oil was saponified with alcoholic potassium hydroxide and extracted into hydrocarbon. A portion was analysed by normal-phase HPLC using a Waters Radial-PAK silica column with a mobile phase of hexane/2-propanol (96:4, v/v) at 1·5 ml min^{-1}. Measurement was by fluorescence detection at 325 nm (excitation) and 470 nm (emission). Peak identification: **1**=11,13-di-*cis*-retinol; **2**=11-*cis*-retinol; **3**=13-*cis*-retinol; **4**=9,13-di-*cis*-retinol; **5**=9-*cis*-retinol; **6**=all-*trans*-retinol.

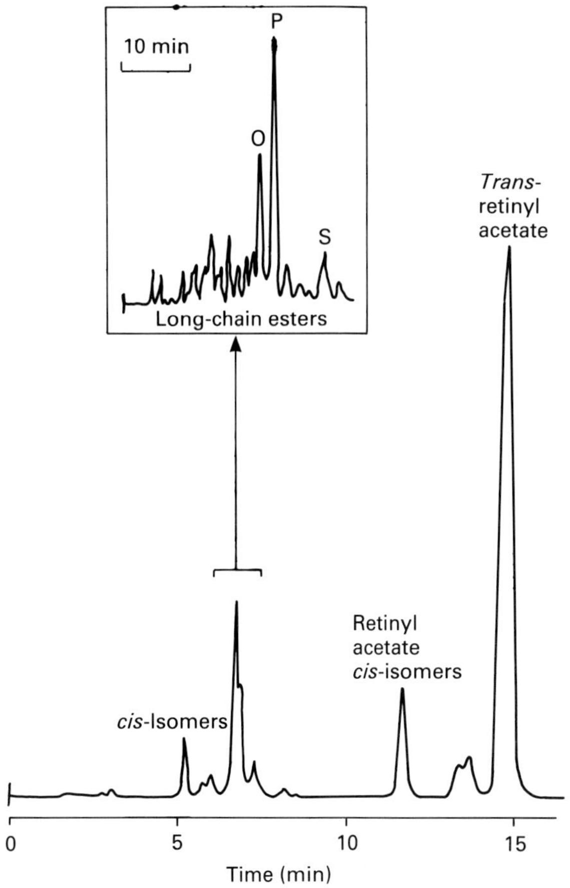

Fig. 3 The lipids from 2 g of milk-based infant formula (22% fat) were extracted without saponification into 5 ml of organic solvent. 100 μl was injected directly into an HPLC equipped with a silica column and a fluorescence detector (325 nm excitation, 470 nm emission). Mobile phase composition was hexane/2-propanol (99·92:0·08 v/v). *Insert*: The endogenous esters were collected and separated by nonaqueous reversed-phase HPLC on a C18 column with an isocratic mobile phase of acetonitrile dichloromethane (80:20). The major identified long chain esters are retinyl oleate (O), retinyl palmitate (P), retinyl sterate (S).

The specific details of HPLC procedures used for retinol analysis are too numerous to cover comprehensively. However, there are several features common to most schemes, notably the almost universal use of UV or fluorescence detection. (Electrochemical detection has been reported recently, but is not in widespread use.) Normal-phase separations are generally performed on 5–10 μm particulate silica columns, replacing the earlier use of alternative adsorbents (alumina, keiselguhr, etc.). Amino-, cyano- and diol-bonded silica is sometimes advocated to circumvent the problems associated with moisture on underivatized silica. The primary mobile phase component is usually a hydrocarbon in binary combination with a polar organic modifier, although ternary systems are sometimes recommended. Reversed-phase separations are usually confined to C18 bonded-phase columns, although other less hydrophobic derivatives are sometimes used (C2, C8 and phenyl). Mobile phases are often a simple binary combination of methanol (or acetonitrile) and water, and in some cases may be entirely nonaqueous.

Many foods, including poorly characterized or unknown samples, will usually be assayed subsequent to saponification, which offers the option of employing either chromatographic mode, and permits analytical simplification through conversion of multiple esters to a single retinol form. Milks, dairy products and fortified cereals, may, alternatively be tested by direct injection of a total lipid extract. Figure 3 illustrates a typical chromatogram obtained for a milk-based infant formula by this procedure. While retinyl acetate is the

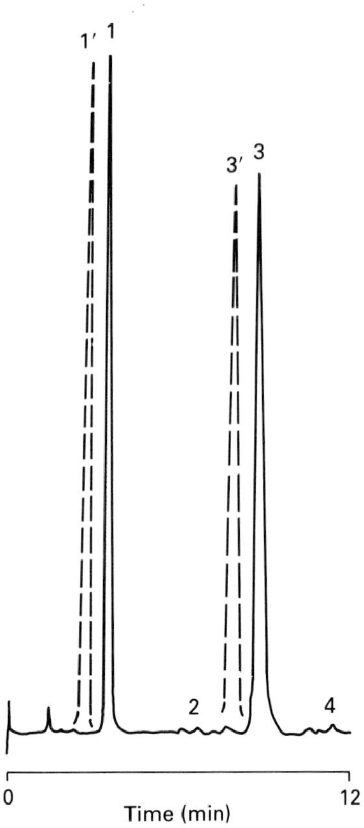

Fig. 4 Retinyl acetate was extracted from a multivitamin formulation without saponification, using DMSO. HPLC analysis was achieved on a Waters C18 Radial-Pak column with a methanol mobile phase flowing at 2 ml min^{-1}. Detection was achieved by UV at 280 nm, 0·3 AUFS. Vitamins D, E and K can be assayed concurrently. Peak identification: **1**=retinyl acetate (vitamin A); **2**=cholecalciferol (vitamin D_3); **3**=tocopheryl acetate (vitamin E); **4**=phylloquinone (vitamin K_1). The elution positions of retinol (1′) and tocopherol (3′) are indicated (broken curves).

additive, the long-chain endogenous esters are visible as a composite peak. (The resolution of the esters by nonaqueous, reversed-phase chromatography is illustrated in the insert.)

In fats and oil products retinyl esters can be assayed by direct dissolution in hydrocarbon, followed by clarification and injection into the chromatographic system. Analytical success with these strategies relies on the selective nature of spectral detection modes, the narrow elution window of the various esters, and their extinction equivalence. Limitations of these approaches include the requirement for normal-phase conditions in order to avoid triglyceride column fouling, and anhydrous or isohydric eluents to obtain reproducible retention. In addition, any nonesterified retinol present in the sample will not be concurrently viewed owing to its tenacious retention on silica relative to the esters.

High-potency pharmaceutical preparations and vitamin premix formulations can also be successfully assayed without the need for saponification. Thus, providing the carrier materials are free from significant fat, the retinyl esters can be extracted into a suitable solvent (e.g., dimethylsulphoxide), ensuring release from encapsulation, diluted into ethanol, clarified and injected onto a reversed-phase column. Such products usually contain vitamins D, E and K, all of which can be concurrently assayed using both wavelength and sensitivity attenuation. However, under these conditions, it is difficult to analyse for retinyl acetate and palmitate concurrently, as the latter is highly retained by nonpolar columns. This problem can be circumvented by performing the analyses subsequent to saponification and thereby recovering total vitamin A as retinol. Figure 4 illustrates a typical chromatogram of retinyl acetate in a pharmaceutical product. The elution position of retinol is also indicated.

Bibliography

Ball GFM (1988) *Fat-Soluble Vitamin Assays in Food Analysis. A Comprehensive Review*. London: Elsevier Applied Science.

Cashell K, English R and Lewis J (1989) *Composition of Foods. Australia*. Canberra: Australian Government Publishing Service.

Friedrich W (1988) *Vitamins*, pp 63–140. de Gruyter: Berlin.

Groenendijk GWT *et al.* (1980) Analysis of geometrically isomeric vitamin A compounds. In: McCormick DB and Wright LD (eds) *Methods of Enzymology, vol. 67, Vitamins and Coenzymes*, part F, pp 203–220. London: Academic Press.

Klaui HM, Hausheer W and Huske G (1970) Technological aspects of the use of fat-soluble vitamins and carotenoids and of the development of stabilised marketable forms. In: Morton RA (ed.) *Fat Soluble Vitamins; International Encyclopaedia of Food and Nutrition*, vol. 9, pp. 113–159. Oxford: Pergamon Press.

Lambert WE, Nelis HJ, De Ruyter MGM and De Leenheer AP (1985) Vitamin A: Retinol, carotenoids and related compounds. *Modern Chromatographic Analysis of the Vitamins*, Chromatographic Series, vol. 30, pp. 1–72. New York: Marcel Dekker.

Olson JA (1991) Vitamin A. In: Machlin LJ (ed.) *Handbook of Vitamins*, pp 1–57, New York: Marcel Dekker.

Pitt GAJ (1985) Vitamin A. In: Diplock AT (ed.) *Fat-Soluble Vitamins. Their Biochemistry and Applications*, pp 1–75. London: Heinemann.

Thompson JN (1982) Fat-soluble vitamins. *Trace Analysis*, Vol. 2, pp 3–43. London: Academic Press.

Thompson JN (1986) Review: Official methods for measurement of vitamin A. *Journal of the Association of Official Analytical Chemists* 69(5): 727–738.

Wagner AF and Folkers K (1975) *Vitamins and Coenzymes*, pp 280–307. New York: Robert E. Krieger Publishing Co.

Wyss R (1990) Chromatography of retinoids. *Journal of Chromatography* 531: 481–505.

D.C. Woollard
Ministry of Agriculture and Fisheries, Auckland, New Zealand
H.E. Indyk
Anchor Products, Waitoa, New Zealand

Physiology

Absorption, Bioavailability, Transport and Distribution

'Vitamin A' is the collective term for compounds that show the biological properties of retinol, including maintenance of epithelial tissue and visual function. This classification includes retinol, retinyl esters and retinal (vitamin A aldehyde); retinoic acid is included even through it does not sustain visual function. These are isoprenoid compounds, having in common an 11-carbon polyene chain attached to a trimethyl-substituted cyclohexenyl ring (Fig. 1). The term 'retinoids' refers to all compounds, natural or synthetic, that show some biological activity typical of vitamin A, such as promoting differentiation of cells in culture; not all retinoids can support all the functions of vitamin A, e.g. some are unable to contribute to vision. Vitamin A compounds are not found as such in plant tissues, but rather are characteristic of the animal kingdom; the notable exception is 13-*cis*-retinal, which serves as a chromophore in the purple membrane of certain halobacteria.

Dietary vitamin A comes from two sources: preformed vitamin A, and provitamin A carotenoids. Preformed vitamin A (mostly as esters of retinol with long-chain fatty acids) comes from animal products or from dietary supplements; retinyl esters, e.g. retinyl acetate and retinyl palmitate, are more stable chemically than is free retinol. Provitamin A carotenoids arise

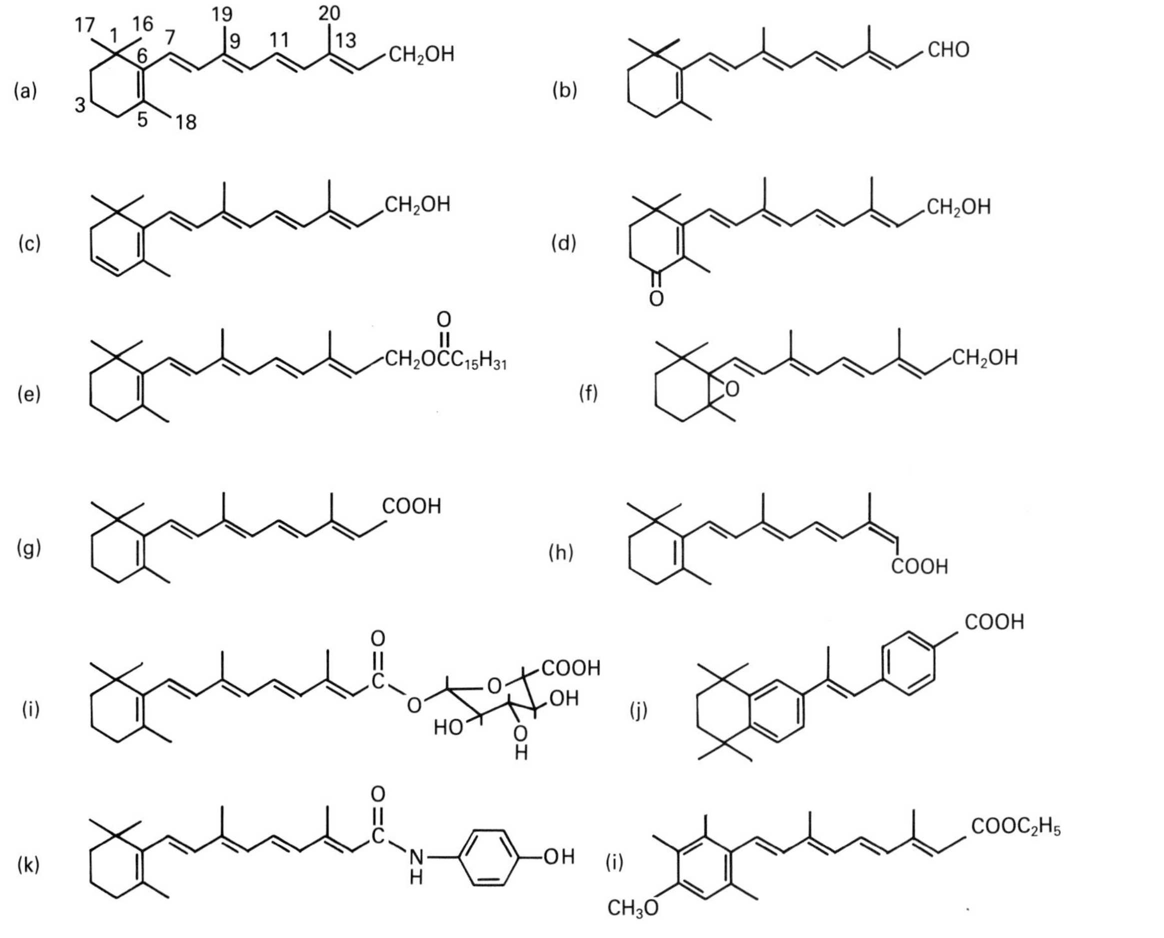

Fig. 1 Some naturally occurring retinoids: (a) all-*trans*-retinol (vitamin A alcohol), showing the conventional numbering system for the carbon atoms; (b) all-*trans*-retinal (vitamin A aldehyde); (c) all-*trans*-3,4-didehydroretinol (vitamin A_2 alcohol); (d) all-*trans*-4-oxoretinol (also called 4-keto-retinol), a metabolite of vitamin A; (e) all-*trans*-retinyl palmitate (vitamin A palmitate), a major storage form of vitamin A; (f) all-*trans*-5,6-epoxy retinol, a metabolite of vitamin A; (g) all-*trans*-retinoic acid (vitamin A acid, tretinoin); (h) 13-*cis*-retinoic acid (also called isotretinoin, Accutane, Ro 4-3780); (i) all-*trans*-retinoyl *β*-glucuronide, a naturally occurring metabolite. Some synthetic retinoids tested for dermatological or anticancer uses: (j) tetrahydrotetramethylnaphthalenylpropenylbenzoic acid, abbreviated TTNPB and trivially termed an 'arotinoid'; (k) 4-hydroxyphenylretinamide (4-HPR, *N*-retinoyl-4-aminophenol); (l) 'Acetretin', a trimethylmethoxyphenyl analogue of ethyl retinoate (also called Etretinate, Tigason, Ro 10-9359).

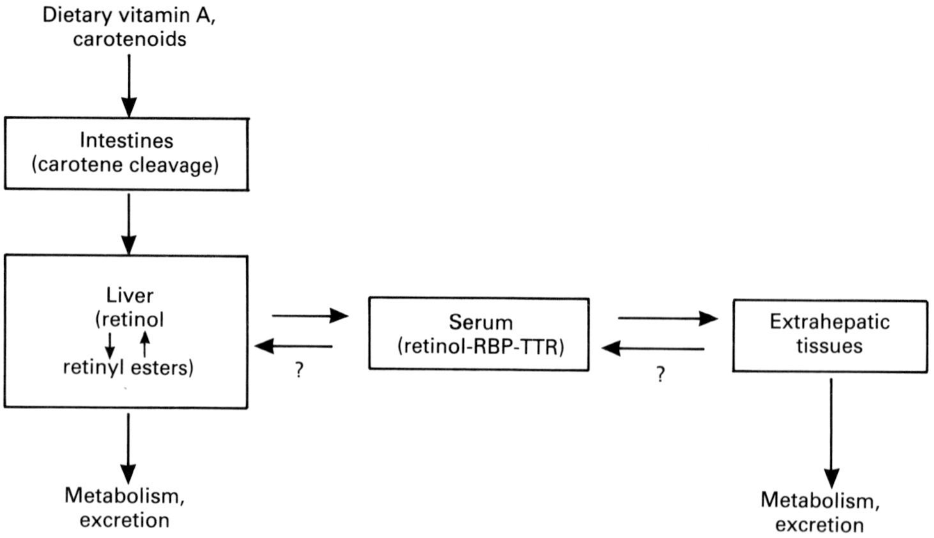

Fig. 2 Outline of vitamin A metabolism. RBP, retinol-binding protein; TTR, transthyretin, formerly called prealbumin.

mostly from plant products: β-carotene is the most active and is widely distributed in plants, but other carotenoids (such as α-carotene and β-cryptoxanthin and β-apocarotenals) can be important sources of vitamin A from particular foods. The relative importance of these sources of vitamin A is very dependent on diet. Other carotenoids, such as lycopene and xanthophylls, are major carotenoids in some foods and may have other important physiological functions, but they have no provitamin A activity. *See* Carotenoids, Properties and Determination; Carotenoids, Physiology

Typical estimates of dietary vitamin A absorption efficiency are 70–85%. Estimates of carotenoid absorption are usually much lower, but are confounded by slow intestinal absorption and rapid metabolism; there is considerable species variability in absorption efficiency and in metabolism of carotenoids. Animal feeding studies show that the biological matrix of food carotenoids has profound effects on their bioavailability.

Since both vitamin A and carotenoids are lipids, intestinal micelle formation with bile acids is essential for their absorption. Human subjects with impaired bile acid formation or flow (e.g. with biliary atresia) may require intramuscular supplementation with vitamin A and the other fat-soluble vitamins. Within the intestinal lumen, vitamin A esters (retinyl esters) are hydrolysed to free retinol and are absorbed as such; this free retinol is promptly re-esterified within the intestinal cells (Fig. 2). Provitamin A carotenoids, such as β-carotene, are often cleaved within intestinal cells; whether this metabolism is by central cleavage (mediated by the enzyme carotene 15,15′ dioxygenase) or by asymmetric cleavage followed by chain shortening is uncertain. Retinal (vitamin A aldehyde), the final product of carotenoid cleavage, is enzymatically reduced to retinol and is then esterified. The physiological ligand for this esterification process seems to be retinol bound to an intracellular retinol-binding protein (CRBP II, M_r approximately 14 600 Da, one of several small cellular retinoid-binding proteins); the primary retinyl-ester-synthesizing activity transfers fatty acid from phosphatidyl choline (lecithin-retinol acyltransferase; LRAT), although an acyl-coenzyme-A-dependent esterifying activity (acylCoA-retinol acyltransferase; ARAT) is also present.

Regardless of their dietary source, the retinyl esters are incorporated in the core of chylomicra and transported in the lymph. After removal of triacylglycerols by lipoprotein lipase as the lipoprotein particle circulates through peripheral tissues, the chylomicron remnants are rapidly taken up by the liver, and the vitamin A esters are hydrolysed by retinyl ester hydrolase there. The resulting retinol is then either re-esterified (primarily by LRAT, although ARAT activity may be important at high retinol concentrations) and stored in the liver, or released into the plasma as a complex with plasma retinol-binding protein.

Because most forms of vitamin A are hydrophobic, the transport, metabolism and function of vitamin A are dependent on a series of binding proteins, each specific for its ligand and tissue. The M_r of plasma retinol-binding protein (RBP) is typically approximately 21 000 Da in mammalian species; the complete amino acid composition has been determined for several species, and the gene from some species has been cloned. Retinol-binding protein binds retinol with 1:1 stoichiometry. The hydrophobic retinol molecule fits into a 'barrel' within the protein, shielded from interactions with the aqueous environment. In turn, Holo-RBP binds transthyretin (TTR; formerly called prealbumin) in plasma; it seems that a TTR tetramer can bind up to four molecules of RBP. Usual concentrations of human plasma RBP are 1·9–2·4 $\mu mol\ l^{-1}$ (40–50 $\mu g\ ml^{-1}$); typically, total circulating RBP is 80–90% saturated with retinol ligand. Although retinol, retinal and retinoic acid bind to RBP with similar affinity, retinol is present in highest concentrations and is the predominant ligand. Retinyl esters have much less affinity for RBP and are transported by serum lipoproteins instead.

The release of holo-RBP from liver is carefully controlled to maintain levels of circulating retinol within narrow limits, but the mechanism of this control is not yet clear. In the absence of adequate vitamin A, apo RBP accumulates in the liver, ready to be released as soon as vitamin A is available (this is the basis of the relative dose–response assay for vitamin A deficiency, as discussed below). Clearly, this mechanism has developed because of the dichotomy of vitamin A action: vitamin A is essential, in small amounts, for proper differentiation and maintenance of epithelial cells, but excesses of vitamin A are toxic and must be controlled by the organism.

Metabolism, Storage and Excretion

Vitamin A in excess of immediate requirements is stored in the liver as esters of long-chain fatty acids (Fig. 2); the ratio of liver retinol to retinyl esters decreases as total liver vitamin A increases, but 95% of total liver vitamin A is typically present as the esters. Retinyl palmitate is the primary ester in liver of the human, rat, cow, sheep, rabbit, cat, frog, trout and polar bear, although significant amounts of other esters are found (especially oleate and stearate), and liver retinyl ester composition can be affected by dietary fatty acid composition. (Remarkably, the predominant vitamin A ester in rat adrenal cortex is retinyl stearate.) Liver frequently contains as much as 90% of total body vitamin A, although other organs, such as kidney, testes and adrenal glands, also contain detectable retinyl esters. The efficiency of liver storage of vitamin A is particularly noteworthy in the polar bear, where concentrations as great as 36 μmol per g of liver tissue (10 380 $\mu g\ g^{-1}$) have been reported. Typical values for human liver vitamin A (autopsy specimens) are 0·44–0·74 $\mu mol\ g^{-1}$ (126–211 $\mu g\ g^{-1}$) in the USA.

The two cell types involved in liver vitamin A metabolism are the hepatocytes (parenchymal cells) and the lipocytes (also called stellate cells, Ito cells, or fat-storing cells). The hepatocytes are the major site of RBP synthesis and retinol-RBP release; some retinyl esters are also found here. The lipocytes store retinyl esters in cytoplasmic lipid droplets which also contain triacylglycerols and some cholesteryl esters within a phospholipid coat. It has been suggested that RBP transfers vitamin A, as retinol, between lipocytes and hepatocytes.

Kinetic studies carried out in rats have shown that there is extensive cycling of liver vitamin A stores, and extensive recycling of vitamin A between liver and other tissues. It is certain that retinol-RBP is the major form of transport from liver to peripheral tissues; it is not yet clear whether the transport of vitamin A from other tissues back to the liver is via retinyl esters transported by lipoproteins (normally present at about 5–10% of the serum concentration of retinol-RBP) or via retinol on RBP synthesized in the outlying tissues (messenger ribonucleic acid, or mRNA, for plasma RBP has been detected in a number of other tissues in addition to liver). In vitamin-A-deficient rats the recycling is even more extensive, and catabolism of vitamin A is markedly diminished.

Catabolism of vitamin A in the rat is by oxidation of the cyclohexenyl ring (particularly at the 4-position), epoxide formation at the 5,6-positions, hydroxylation of the ring methyl groups, and chain shortening (some examples of metabolites are shown in Fig. 1). The resulting metabolites are generally inactive, although some may have a little biological activity. These more polar retinoids are excreted in the urine and in the bile.

Retinol can be reversibly oxidized to retinal (vitamin A aldehyde); retinal can be oxidized to retinoic acid but retinoic acid cannot be converted back to the other forms. Thus retinol, retinyl esters, and retinal have equal biological activity because they are freely interconverted; retinoic acid fills some (but not all) of the functions of retinol, but it and its derivatives are not stored. Typical human serum concentrations of retinoic acid are 5–10 $nmol\ l^{-1}$ (1·5–3 $ng\ ml^{-1}$), compared with typical retinol concentrations of 1–2 $\mu mol\ l^{-1}$. Retinoyl β-glucuronide and retinyl β-glucuronide (formed in the liver from retinoic acid and retinol, respectively) are secreted into the bile but can be hydrolysed and reabsorbed in the intestine (enterohepatic circulation); both these compounds have vitamin A activity in a variety of tests; they are found in human blood, and may have regulatory roles in vitamin A function.

Roles in the Body

The major roles of vitamin A are in vision, differentiation of epithelial tissues, and in the immune system. Metabolism of vitamin A in the retina of the eye is unique, in keeping with the unusual role of vitamin A in that tissue. Vitamin A is stored in the retinal pigment epithelium as retinyl esters. All-*trans*-retinyl esters are simultaneously hydrolysed and isomerized to 11-*cis*-retinol, a compound unique to the eye. 11-*cis*-Retinol is then oxidized to 11-*cis*-retinal. 11-*cis*-Retinal is transferred from the pigment epithelium to the rod cells by interstitial retinoid binding protein (IRBP), a distinct binding protein (140 000 Da). In the rod cells, 11-*cis*-retinal binds (as a Schiff base with the ε-amino group of a specific lysine residue) to the protein opsin to form the visual pigment, rhodopsin. When a photon of light is absorbed by a rhodopsin complex, the 11-*cis*-retinal is isomerized to all-*trans*-retinal and released from the protein complex; the resulting conformation change of the protein initiates a cascade of reactions, resulting in a neural signal to the brain. The protein opsin is then available to bind another molecule of 11-*cis*-retinal for another round of the visual cycle. The all-*trans*-retinal which was released from the protein complex is transferred back to the pigment epithelium via IRBP, enzymatically reduced to all-*trans*-retinol, and esterified again. In contrast to the high turnover rates of vitamin A in other tissues, vitamin A in the eye is highly conserved, with little leakage back to the liver. Prolonged vitamin A deficiency, however, leads to reduced sensitivity to light, usually first noted as impaired vision at night (night blindness). These effects of vitamin A deficiency are generally reversible by subsequent vitamin A supplementation.

In a very different role, the cornea of the eye depends on vitamin A for proper cell differentiation and for secretion of protective glycoproteins. In vitamin A deficiency, these tissues are susceptible to attack by opportunistic bacteria; such attack may not be reversible and, especially on the corneal surface of the eye, may result in permanent scarring and permanent vision loss. These effects of vitamin A deficiency, unlike those in the retina, may not be reversible by subsequent vitamin A supplementation. Such vitamin-A-dependent corneal degeneration (given the general name 'xerophthalmia') accounts for an estimated 500 000 new cases of blindness in preschool children in the world each year.

The action of retinoids in differentiation is manifest in various systems, including maintenance of epithelial tissue (e.g. the lung, intestines and skin, and the cornea of the eye). In the absence of adequate vitamin A, cells of these tissues do not differentiate normally, but change structure (becoming stratified and cornified) and lose the ability to secrete glycoproteins. The common mechanism underlying these roles of retinoids in diverse tissues seem to involve the binding of retinoic acid (and perhaps retinol) to specific proteins associated with nuclear deoxyribonucleic acid (DNA). These nuclear retinoic-acid-receptor proteins (RARs), which are distinct from the cytoplasmic 'cellular retinoic-acid-binding proteins' (CRABPs), can then bind to specific regions of DNA, either promoting or inhibiting transcription of specific genes. A number of RARs have now been identified, perhaps explaining in part the variety of effects shown by retinoids on different cell types. The RARs identified to date all belong to a superfamily of nuclear proteins that includes steroid-binding receptors and thyroxine-binding receptors; all act by a common mechanism, but differ in specific DNA binding.

It has been argued, without conclusive proof as yet, that retinoic acid is the active form of vitamin A required for cellular differentiation. Retinoic acid is an endogenous metabolite of vitamin A. Animals maintained on retinoic acid as sole source of vitamin A seem to grow normally and maintain good health, but become blind (because retinoic acid cannot be converted to retinal); some but not all species also show loss of testicular function. Retinoic acid (all-*trans*- and 13-*cis*-isomers) and other acidic retinoid analogues have been found to ameliorate various skin disorders, including acne. Synthetic retinoids (Fig. 1) have been developed to reduce detrimental side-effects, which include irritation of the skin, increased serum cholesterol concentrations, and teratogenic consequences.

Retinoids have been shown to inhibit cancer development in a variety of tissues. Again, this seems to be through the role of vitamin A in promoting differentiation of epithelial tissues, in as much as cancer may be thought of as the proliferation of undifferentiated cells. ***See*** Antioxidants, Role of Antioxidant Nutrients in Defence Systems; Cancer, Diet in Cancer Prevention

Although animal studies have long shown a necessity for vitamin A in immune function, the molecular action of retinoids is still unknown. It also seems that some carotenoids function as such in immune function, and perhaps additionally as precursors of vitamin A. ***See*** Immunity and Nutrition

Requirements and Recommended Intakes

Because of confusion among the various units for presenting vitamin A values, the concept of the *Retinol Equivalent* (RE) has been proposed: 1 RE is equal to 1 μg of all-*trans*-retinol, either free or as the retinyl component of a retinyl ester; 1 RE is also equal to 3·33 iu (international units) of vitamin A, or 3·5 nmol retinol or retinyl ester. Although the exact vitamin A value of carotenoids depends on several factors, the RE has been defined as 6 μg of all-*trans*-β-carotene or 12 μg of other provitamin A carotenoids (10 iu of provitamin A carotenoid).

Liver concentrations provide the best appraisal of vitamin A status. Human liver specimens are difficult to obtain, but analysis of autopsy specimens can be useful in evaluating the vitamin A status of a population. As indicated above, serum retinol levels are maintained nearly constant and are not generally useful in assessing vitamin A status, except when liver vitamin A reserves fall well below 0·07 μmol per g of liver (20 $\mu g\ g^{-1}$). Serum (or plasma) retinol levels are normally 1–2 $\mu mol\ l^{-1}$ (290–570 $ng\ ml^{-1}$) across a wide range of mammalian species.

Because liver concentrations of vitamin A are not readily measured and serum retinol values are not an adequate indicator of an individual's vitamin A status, several indirect methods have been developed.

Conjunctival impression cytology (CIC) evaluates the development of squamous metaplasia (enlarged epithelial cells) and loss of goblet cells from the conjunctiva of the eye by histological examination of cells transferred from the cornea to filter paper.

The *relative dose–response* (RDR) depends on the release of retinol-RBP from liver into serum after a large oral dose of vitamin A (typically, 450 μg of retinyl acetate in human studies); as described above, apo RBP accumulates in liver in vitamin A depletion. The RDR is calculated as follows:

$$RDR = (A_5 - A_0/A_5) \times 100\%$$

A_5 represents serum retinol concentration at 5 h after the oral dose, and A_0 represents fasting serum retinol at the time of the dose. Studies in both humans and rats indicate that an RDR value greater than 20% indicates liver vitamin A reserves less than 0·07 μmol per g of liver (<20 μg g^{-1}), i.e. inadequate vitamin A status. The RDR assay requires, of course, that the oral dose be normally absorbed (an intramuscular injection has been used in human subjects with biliary atresia), and that protein metabolism is normal: protein deficiency or zinc deficiency or liver cirrhosis impairs the dose response.

An alternative approach, the *modified relative dose–response* (MRDR), uses the vitamin A analogue 3,4-didehydroretinol (vitamin A_2, a form found in some freshwater fish and found in very low levels in human skin). The ratio of vitamin A_2 to vitamin A_1 in serum retinol at 5 h after an oral dose is used as a measure of vitamin A status, with high values (ratio $>0{\cdot}03$ after an oral dose of 100 μg per kg of bodyweight) indicating poor vitamin A status. Chlorinated vitamin A analogues have been used in similar fashion.

The RDR and MRDR, as well as CIC, have been used successfully in human population studies. Isotope dilution of tracer-labelled vitamin A (radioactive or stable-isotope labelling) has been used to estimate human vitamin A status, but technical difficulties have so far prevented general use of the technique.

Nutritional requirements for vitamin A have not been well defined because of the diversity of vitamin A functions. In animal studies, daily intakes of 3–8 RE (10–28 nmol) per kg of bodyweight cure deficiency symptoms, and daily intakes of 30–60 RE (100–210 nmol) per kg of bodyweight produce optimal growth. Kinetic studies of vitamin A metabolism in rats show that irreversible loss of vitamin A is decreased on low vitamin A intakes. Studies on vitamin A requirements have not yet addressed functional criteria such as immune function and possible anticancer effects.

To provide an *adequate and safe human intake*, current WHO/FAO (World Health Organization, Food and Agriculture Organization) dietary recommendations (1988) and suggested Recommended Dietary Intakes (RDIs) for adults are based on intakes of 9·3 RE (33 nmol of vitamin A) per kg of bodyweight per day. In contrast, the Recommended Dietary Allowances (RDAs) of the US National Research Council (1989) attempt to set an *optimal intake level*, and so recommend a higher human male intake of 1000 RE per day, 800 RE for women (because of lower bodyweight).

Toxicity

Over 600 individual cases of human vitamin A toxicity have been reported, attributable either to acute (single or a few large doses ingested over a brief period of time) or chronic intake (moderately high doses taken frequently for periods of months or years). Acute toxicity in human adults is reported from doses of 300 000 to 10 000 000 RE; chronic doses of 15 000–300 000 RE have produced hypervitaminosis A. Symptoms include headache, vomiting, diplopia, alopecia, dryness of mucous membranes, desquamation, bone abnormalities, and liver damage. On the other hand, single oral doses of 60 000 RE in oil have been successfully used in vitamin A intervention programmes for preschool children, with transient toxic symptoms observed in no more than 3% of subjects. Massive doses of β-carotene are not toxic, but may be less efficiently absorbed and used than vitamin A itself. Rodent animals have been extremely valuable in elucidating vitamin A requirements and metabolism, but the rat seems to be much less susceptible to hypervitaminosis A than is the human. ***See*** Hypovitaminosis A

The toxic effects of high vitamin A intakes are mediated by serum retinyl esters; retinol-RBP concentrations are maintained at normal levels in hypervitaminosis A, but serum retinyl esters are markedly elevated, bypassing the normal homeostatic controls on vitamin A transport. Some carnivores, including the dog, are unusual in having high fasting levels of serum retinyl esters, presumably reflecting differences in lipoprotein metabolism in these species; the implications of this for vitamin A metabolism and for resistance to hypervitaminosis A are not clear.

The most tragic consequences of excessive vitamin A intake are teratogenicity (malformations of the cranium, face, heart, thymus, and central nervous system) and embryotoxicity. Acidic retinoids, such as those that have been used in dermatology, are particularly potent, as they can attain high serum levels and readily pass the placental barrier. Although large intakes of vitamin A itself (>7500 RE, or 26 μmol, per day early in human pregnancy) can cause birth defects (possibly as a result of metabolism of retinol to retinoic acid), serum concentrations of retinol and retinyl esters are normally maintained at moderate levels during pregnancy. It is assumed that these teratogenic effects are related to the important role of retinoids in differentiation of cells and that these effects are mediated via the nuclear receptor proteins. (Interestingly, the retinoid β-glucuronides are less teratogenic than retinoic acid.) In view of these teratogenic effects occurring at less than 10 times the recommended daily intakes, the consensus of several professional organizations is that women should avoid vitamin A supplements during the first trimester of pregnancy, and that subsequent supplements, if taken at all, should be limited to 1500–3000 RE (5·5–11 μmol) per day.

Continuing challenges in vitamin A research include the following: (1) the development and confirmation of indirect indices of vitamin A status; (2) elucidation of the mechanism of control of serum retinol-RBP concentrations; (3) more exact determination of vitamin A requirements for specific functions (not only growth and prevention of blindness, but also immune function and cell differentiation in individual tissues); (4) definition of the role of vitamin A in differentiation in specific tissues, perhaps leading to chemical design of distinctive retinoids to prevent specific cancers and dermatological diseases.

Bibliography

Blomhoff R, Green MH, Berg T and Norum KR (1990) Transport and storage of vitamin A. *Science* 250: 399–404.

Ganguly J (1989) *Biochemistry of Vitamin A*. Boca Raton, Florida: CRC Press.

Olson JA (1991) Vitamin A. In Machlin LJ (ed.) *Handbook of Vitamins* 2nd edn, pp 1–57. New York: Marcel Dekker.

Sporn MB, Roberts AB and Goodman DS (eds) (1984) *The Retinoids*, vols 1 and 2. New York: Academic Press.

Harold C. Furr
University of Connecticut, Storrs, USA

REVERSE OSMOSIS

See Membrane Techniques

RHEOLOGY

See Physical Properties of Foods

RHEOLOGY OF LIQUIDS

Food materials are multiphase, multicomponent systems whose rheological characteristics frequently present process engineers with difficult handling and pumping problems.

The food industry has, almost since its inception, been concerned with the physical properties of its products, both in terms of their processing and their eating characteristics. Consequently, considerable care must be taken in designing plant and process if the texture, consistency and quality of the final product is not to be compromised. The selection of appropriate pumps and conveying and mixing systems is especially germane to food materials which may be shear-dependent and/or abrasive in nature. ***See*** Physical Properties of Foods

Although it is a relatively easy matter to describe mathematically the flow behaviour of 'simple' liquids, food solid/liquid mixtures present unique problems in describing their flow behaviour.

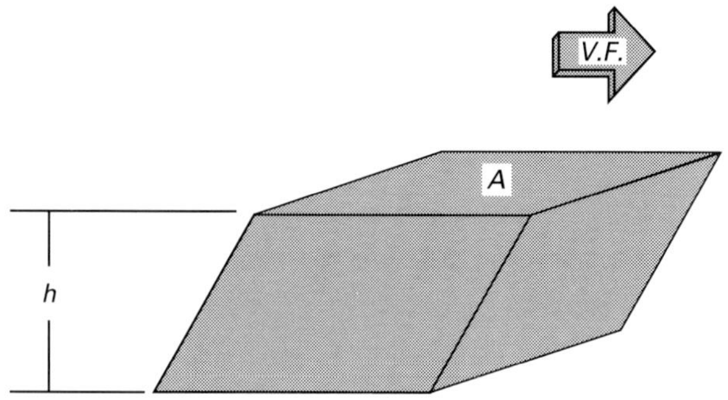

Fig. 1 Classical shear viscosity.

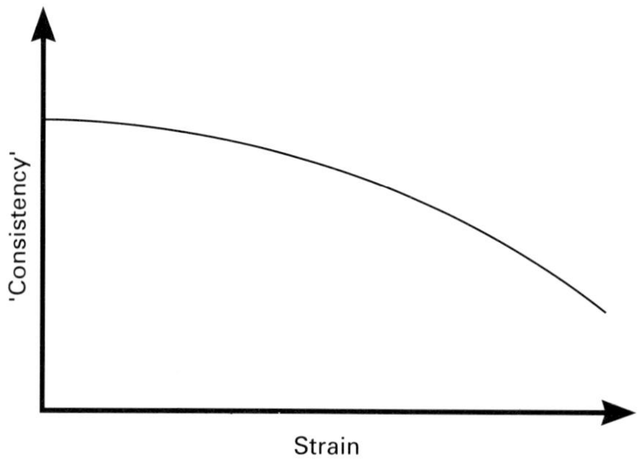

Fig. 2 Effect of increasing strain.

Liquid Behaviour

When assessing the behaviour of a liquid system, it is usual to investigate the stresses generated when the sample is sheared. That is to say, the force per unit area produced when the 'layers' of liquid move over each other in the direction of flow. This gives rise to the classical definition of a shear viscosity (Fig. 1).

Shear rate, $\gamma = V/h$ (s^{-1})
Shear stress, $\sigma = F/A$ (Pa)
Shear viscosity, $\eta = \sigma/\gamma$ (Pa s)

where V is the velocity, F is the force and A is the area over which it acts.

The range of shear rates observed in the food industries is vast, from values as low as 10^{-2} s^{-1}, for the self-levelling of a liquid due to surface tension, to 10^{6} s^{-1} seen in the spray drying of slurries. (The chewing and swallowing of foods generates shear values of about 10^{1} to 10^{2} s^{-1}.) This presents us with considerable difficulties with regard to most food systems, as very few behave in an ideal or Newtonian manner, with the shear stress proportional to the shear rate over all of the possible shear rate values.

Since viscosity is essentially related to shear rate and stress it is possible to operate a viscometer in one or two modes: (1) by applying a rotation and measuring the torque generated (controlled stain), or (2) by applying a fixed torque and measuring the speed generated (controlled stress). These two methods have their advantages and disadvantages; for example, the controlled-strain instruments are relatively cheap, robust and well understood but usually have a limited shear range (1–1000 s^{-1}). Controlled-stress instruments, on the other hand, have large shear rate ranges and can carry out other forms of testing (creep, oscillatory, etc.), but do tend to be somewhat expensive.

Working Liquids

When we work a liquid system we apply energy to it. This 'force' has a number of characteristics that we can define: (1) the area over which it is applied, (2) the distance over which it acts, and (3) the rate at which it is applied. We may then define the stresses involved (force/area) and the strains produced (relative deformation). If a more complex analysis is required, these forces may be further resolved into three components, the tensile or normal forces, perpendicular to the plane of the applied force and two shear or tangential forces parallel to it.

The strain (distance moved) when processing or measuring a sample has a considerable effect on its mechanical behaviour/flow properties. If a sample is moved such that its molecular domains or structural elements no longer interact/overlap the network may become 'overloaded' and ultimately break down (Fig. 2). For example, a sample of an emulsion such as mayonnaise which is stabilized by electrostatic or steric means may have a very low value of 'critical strain' after which structural damage to the emulsion will occur, and it is very important that testing and processing be carried out in such a way so as not to exceed this point. ***See*** Colloids and Emulsions

While, in general, increasing the strain tends to reduce interactions, increasing the shear rate may give rise to either increases or decreases in the apparent degrees of interaction. If we consider the effect of increasing the shear rate on entangled long-chain polymer systems (e.g. a long-chain polysaccharide), at zero shear rate these chains will be entangled by Brownian motion and will have a degree of interaction based on physical constraints. On moving one of the chains slowly with respect to the others, very little resistance would be encountered, as the molecules would have sufficient time to allow any rearrangements needed to take place. However, if the same process were carried out at a higher rate then the entanglement would have insufficient time for rearrangement and would resist such movement. This would produce a material which appeared to increase its degree of interaction at higher shear rates. Such a

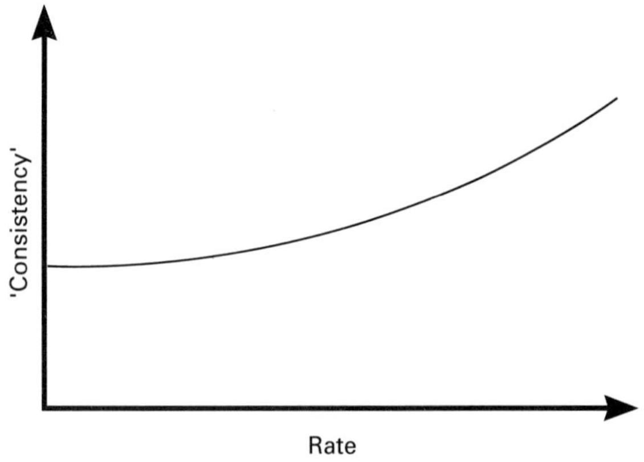

Fig. 3 Increasing interaction with shear rate.

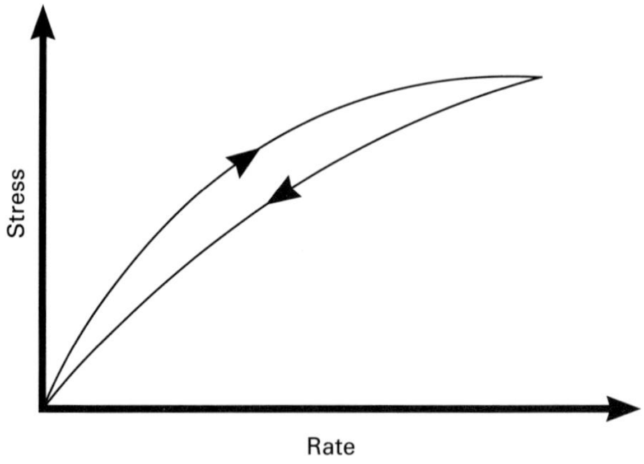

Fig. 5 Plastic/pseudoplastic behaviour.

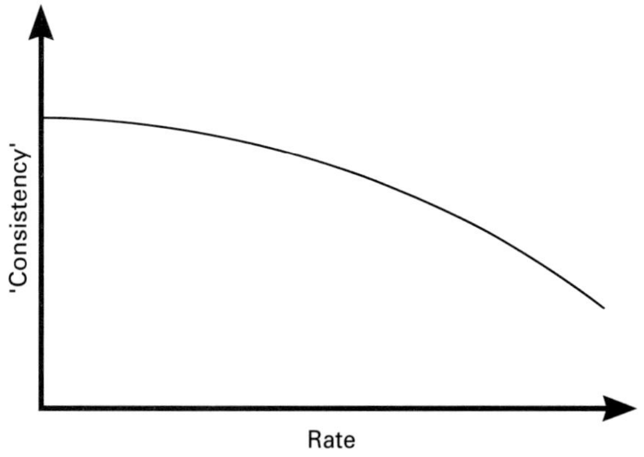

Fig. 4 Decreasing interaction with shear rate.

system may become unworkable if, for example, you try to pump it too quickly (Fig. 3). ***See*** Carbohydrates, Classification and Properties

However, if a system were stabilized by, for example, some form of hydrogen bonding, an increase in shear rate might reduce the effectiveness of the interactions. This would lead to a fall in the apparent consistency (Fig. 4).

The two parameters, strain and rate, when combined give an overall picture of the behaviour of the liquid. This rheological 'plane' may be very flat (basic Newtonian behaviour) or more commonly steeply curving (nonideal behaviour). The consequences of not having assessed the physical characteristics over a suitable testing regime may prove disastrous in terms of predicting processing behaviour.

Behaviour as a Function of Time

In the description of the different types of non-Newtonian behaviour it was implied that although the viscosity of a fluid might vary with shear rate, it was independent of the length of time that the shear rate was applied and also that replicate determinations at the same shear rate would always produce the same viscosity. This must be considered as the ideal situation, since most non-Newtonian food materials are colloidal in nature and as such the flowing elements, whether particles or macromolecules, may not adapt immediately to the new conditions. Therefore, when such a material is subjected to a particular shear rate, the shear stress and consequently the viscosity will decrease with time. Furthermore, once the shear stress has been removed, even if the structure which has been broken down is reversible, it may not return to its original structure (rheological ground state) instantly. The common feature of all these materials is that if they are subjected to a gradually increasing shear rate followed immediately by a shear rate decreasing to zero, then the downcurve will be displaced with regard to the upcurve and the rheogram will exhibit a hysteresis loop.

In the case of plastic and pseudoplastic materials the downcurve will be displaced to below the upcurve (Fig. 5), whereas for dilatant substances the reverse will be true (Fig. 6).

The presence of the hysteresis loop indicates that a 'breakdown' in structure has occurred and the area with the loop may be used as an index of the degree of 'breakdown'.

The term which is used to describe one such type of behaviour is *thixotropy*, which means 'to change by touch'. Strictly this term should only be applied to an isothermal sol–gel transformation. It has, however, become common to describe as thixotropic any material which exhibits a reversible time-dependent decrease in apparent viscosity. Thixotropic systems are usually composed of asymmetric particles or macromolecules which are capable of interacting by numerous secondary bonds to produce a loose three-dimensional struc-

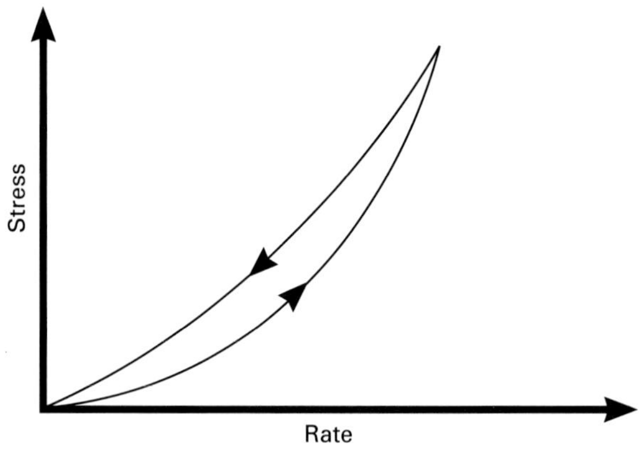

Fig. 6 Dilatant behaviour.

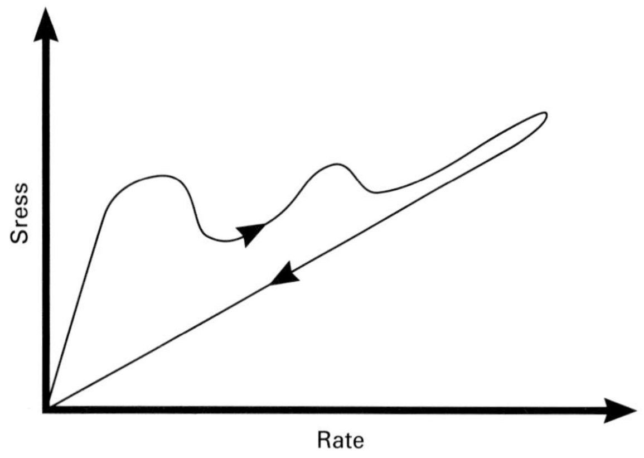

Fig. 7 Complex flow behaviour.

ture, so that the material is gel-like when unsheared. The energy which is imparted during shearing disrupts these bonds so that the flowing elements become aligned and the viscosity falls since a gel–sol transformation has occurred. When the shear stress is eventually removed, the structure will tend to reform, although the process is not immediate and will increase with time as the molecules return to the original state under the influence of Brownian motion. Furthermore, the time taken for recovery, which can vary from minutes to days depending upon the system, will be directly related to the length of time that the material was subjected to the shear stress, since this will affect the degree of breakdown.

In some cases the structure that has been destroyed is never recovered no matter how long the system is left. Repeat determinations of the flow curve will then only produce the downcurve that was obtained in the experiment that resulted in the material's destruction. Such behaviour should be referred to as 'shear destruction' rather than thixotropy, which, as will be appreciated from above, is a misnomer.

An example of such behaviour are the gels produced by high-molecular-weight polysaccharides which are stabilized by large numbers of secondary bonds. Such systems undergo extensive reorganization during shearing such that the three-dimensional structure is reduced: the gel-like nature of the original is then never recovered.

The occurrence of such complex behaviour creates problems in quantitative classification because not only will the apparent viscosity change with shear rate but there will also be two 'viscosities' that can be calculated for any given shear rate (i.e. from the upcurve and the downcurve). It is usual to attempt to calculate one viscosity for the upcurve and another for the downcurve. This must assume of course that each of the curves achieves linearity over some of its length, otherwise a defined shear rate must be used: only the former situation is truly satisfactory. Each of the lines used to derive the viscosity may be extrapolated to the shear stress axis to give an associated yield value. However, only the one derived from the upcurve has any significance, since that derived from the downcurve will relate to the broken-down system. Consequently the most useful index of thixotropy can be obtained by integration of the area contained within the loop. This will not of course take into account the shape of the up and down curves and consequently two materials may produce loops of similar area but have completely different shapes representing totally different flow behaviours. In order to prevent confusion it is best to adopt a method whereby an estimate of area is accompanied by yield value(s). This is of particular importance with flow curves which exhibit complex upcurves.

This situation is typical of the type of loop obtained with, for example, some samples of white soft paraffin, where the upcurve exhibits a number of bulges (Fig. 7). Those at lower shear rate are thought to be associated with the initial loss of three-dimensional structure, while the smoother deviations occurring at the higher shear rates are associated with molecular reorientation. Such behaviour is common in food and pharmaceutical systems and is one of the major causes of difficulties in their evaluation.

With such a wide range of rheological behaviour it is extremely important to carry out measurements that will produce meaningful results. It is crucial, therefore, not to use a determination of viscosity at one shear rate (such as would be acceptable for a Newtonian fluid) since it could lead to completely erroneous comparative results.

Figure 8 shows rheograms which are an example of four different types of flow behaviour all of which intersect at point A which is equivalent to a shear rate of 100 s^{-1}. Therefore, if a measurement were made at this one shear rate then all four materials would be

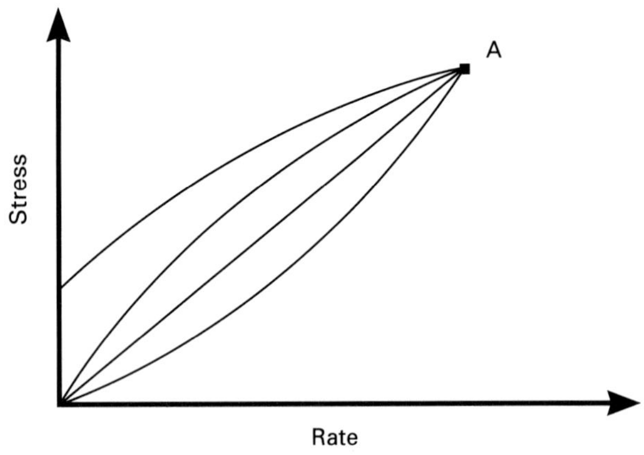

Fig. 8 Convergence of flow behaviour at a single rate.

shown to have the same viscosity although they all possess different properties and behaviour. Single-point determinations are probably an extreme example but are used to emphasize the importance of properly designed experiments.

It should be noted that while more complex rheological measurements may give rise to parameters other than viscosity (for example the storage and loss moduli), the arguments given here still apply, and time and shear dependence remain major problems when measuring changing food systems.

Bibliography

Ferry JD (1980) *Viscoelastic Properties of Polymers*, 3rd edn. New York: Wiley.

Lewis MJ (1987) *The Physical Properties of Foods and Food Processing Systems*. Chichester: Ellis Horwood.

A.E. Bell
University of Reading, Reading, UK

RIBOFLAVIN

Contents

Properties and Determination

Riboflavin, a member of the water-soluble B complex has been also called vitamin B_2, vitamin G, ovoflavin, lactoflavin, hepatoflavin and verdoflavin. The International Union of Nutrition Sciences Committee on Nomenclature and the Commission of Biochemical Nomenclature (IUPAC-IUB) have designated riboflavin as the name for this vitamin. It was isolated from milk and egg white and its structure elucidated. It was synthesized in 1934. Warburg's yellow enzyme, necessary for the oxidation of glucose 6-phosphate, was shown to contain riboflavin, and this elucidated the role of the vitamin in biological oxidation and reduction.

This article discusses those physical and chemical properties of riboflavin which are relevant to the effects of processing and cooking on the naturally occurring vitamin in foods. The sources and forms found in foods, uses in food fortification, and methods of determination will also be included.

Physical and Chemical Properties

Riboflavin has a molecular mass of 376·4 Da with an empirical formula of $C_{17}H_{20}N_4O_6$. Its elemental composition is 54·25% carbon, 5·36% hydrogen and 14·89% nitrogen. In tissues and foods it exists as riboflavin (7,8-dimethyl-10-(1′-D-ribityl)isoalloxazine), flavin mononucleotide (FMN; D-riboflavin 5′-phosphate) and flavin adenine dinucleotide (FAD; the 5′-(adenosine 5′-pyrophosphoryl) derivative of the vitamin). The molecular weight for FMN is 456·4 and for FAD is 785·6. Riboflavin is sparingly soluble in water, 10–13 mg per 100 ml at room temperature. The sodium salt of FMN is very soluble in water and is used in pharmaceutical preparations. The decomposition temperature of riboflavin is 278°C and darkening occurs at 240°C. Powdered riboflavin is orange in colour and solutions are greenish yellow with a high fluorescence at 565 nm. It is amphoteric in solutions with dissociation constants at $6{\cdot}3\times10^{-12}$ for K_a and $0{\cdot}5\times10^{-5}$ for K_b. It is light-sensitive and under acidic conditions it is rapidly converted to lumichrome (7,8-dimethylalloxazine). In alkaline solutions it is converted by light to lumiflavin (7,8,10-trimethylalloxazine). Riboflavin is readily reduced by sodium dithionite to colourless dihydroriboflavin which, upon shaking in air, is reconverted into riboflavin. The vitamin and its naturally occurring forms are relatively stable to heat and, during the early studies of the vitamins, this distinguished it from thiamin (vitamin B_1) which was heat-labile. However, riboflavin, FMN and FAD are unstable to heating in alkaline solution. ***See*** Thiamin, Properties and Determination

Occurrence in Foods

Milk and Cheese

The consumption of milk and dairy products provide about one-half the riboflavin intake of the average European and American. Raw cow's milk contains most of its activity in the form of riboflavin and FAD in the ratio of about 2:1. After pasteurization very little total riboflavin is lost. However, the heat process converts some of the FAD to riboflavin, increasing the ratio of riboflavin to FAD to 5–6. This contrasts with the riboflavin content of human milk where there is twice as much FAD as riboflavin.

In view of the fact that riboflavin is highly sensitive to destruction by light, the packaging of pasteurized milk becomes an important consideration. Milk in glass bottles exposed to light for several hours will lose most of its riboflavin content. During this reaction, singlet oxygen is produced which causes the destruction of vitamin C (ascorbic acid). The use of opaque cardboard containers for milk prevents loss of these vitamins. It is of interest that FAD is considerably less sensitive to light destruction than either riboflavin or FMN. ***See*** Ascorbic Acid, Properties and Determination; Milk, Dietary Importance

During the production of evaporated and powdered milk little riboflavin content is lost, attesting to the heat stability of this vitamin.

During the manufacture of hard cheeses such as Cheddar, most of the riboflavin remains in the whey rather than the curd. However, there is no further loss of riboflavin content during the ripening process. ***See*** Cheeses, Dietary Importance

Animal and Legume Sources

Commercial processing of meat, fish or poultry by canning, freezing, dehydration and irradiation have little effect on the riboflavin content. The vitamin is heat-stable and, while heat processing may convert FAD to riboflavin, the total assayable content is not changed. Sun drying of fish or meat will result in considerable loss of riboflavin owing to its light sensitivity. ***See*** Heat Treatment, Chemical and Microbiological Changes

Although preparation of meat, fish or poultry by dry heat results in little loss, the water-soluble properties of riboflavin will cause it to leach out in meat juices and water used in preparations such as stews. In the latter case, there will be a partition between the content in the meat and in the liquid with about three-quarters in the meat and the rest in the juices. To gain the maximum benefit of the vitamin, the liquid should be recycled for gravies and sauces.

Eggs in storage for a year lose only a small portion of the riboflavin content. Frying eggs in a pan open to light can cause destruction of one-half of the vitamin content. Little loss occurs in eggs hard boiled in the shell. ***See*** Eggs, Dietary Importance

As riboflavin is unstable in alkaline conditions the addition of sodium bicarbonate to the water of fresh peas to retain the green colour will result in considerable destruction of the vitamin. Its use to soften dried beans will also destroy much of the riboflavin content. ***See*** Cooking, Domestic Techniques

Fruits and Vegetables

Fruits and vegetables contribute about 10% of the total intake of riboflavin in the diet of the western world. Plants are richest in this vitamin during the rapid growth phase. Blanching, freezing or boiling of vegetables results in little loss if the content of the vegetable and water are both considered. Here again, the use of sodium bicarbonate in cooking water to prevent the loss of magnesium from chlorophyll to preserve the green colour of vegetables will result in the loss of riboflavin content.

Cereal Grains

In cereal grains, riboflavin is found primarily in the germ and bran. It is necessary to eat the whole grain in order to maximize intake of the riboflavin. In the case of wheat, most people prefer products such as bread, spaghetti and macaroni made from white flour. In the milling process to produce flour, most of the germ and bran are lost so that only about one-third of the riboflavin content remains in the flour. During World War II, the US government mandated that all white flour be enriched with thiamin, riboflavin, nicotinic acid and iron. Although the enrichment is no longer mandatory, most white flour millers still enrich with the vitamins and iron, and several other countries have adopted this procedure. ***See*** Cereals, Dietary Importance; Food Fortification; Iron, Physiology; Niacin, Physiology; Thiamin, Physiology

Millers obtain wafers for enrichment from the pharmaceutical industry and add it to the flour in a rather simple manipulation. Enriched flour contains more riboflavin than the whole wheat kernel. This was not the intention of the regulation, but arose due to an error in the method of assay which overestimated the riboflavin content of whole wheat. The amount used was never changed. This has benefited that portion of the population which consumes few dairy or meat products.

Rice is the staple grain for a considerable portion of the world population, and the majority consume polished rice. This is owing to the fact that the germ contains unsaturated fatty acids and, in tropical climates where refrigeration is not widely available, rancidity will occur. Since most of the vitamins are contained in the germ, vitamin deficiencies were endemic in those areas where polished rice was the main constituent of the diet. The exceptions were those areas such as India where parboiling of whole rice was practised. In this process, rice is steeped in water, steamed, dried and then milled to remove bran and germ, resulting in the water-soluble vitamins being carried into the starchy part of the endosperm. In Western countries, steaming under pressure and vacuum drying is used. The rice is called converted rice. ***See*** Rice

Enrichment of rice presents a difficulty not encountered in the enrichment of wheat flour. If the water-soluble vitamins are painted on the kernels of polished rice, they would be lost during the washing occurring prior to cooking. This has been solved by adding the vitamins to the rice which are then sprayed with an alcoholic solution of zein, the protein of maize. Since zein is insoluble in cold water, the rice can be washed without loss of vitamins. Presently, rice is enriched with thiamin, nicotinic acid and iron. However, no riboflavin is added because of the yellow colour of the vitamin. The enriched rice is added to the batch of rice to be enriched and mixed. If riboflavin were added, consumers would pick out the yellow grains and discard them.

Maize grits and meal which are degermed lose about one-half of the riboflavin content of the whole kernel. These are enriched with riboflavin and this is acceptable to consumers.

In the foregoing, it can be seen that the stability of riboflavin during processing and cooking, except during exposure to light and heat treatment in alkaline conditions, results from its chemical properties of resistance to oxidation and heat stability. However, in common with other water-soluble vitamins, cooking in water will result in partial extraction of the vitamin into the fluid. ***See*** Vitamins, Overview

Nature and Content of Riboflavin in Foods

Riboflavin occurs in food primarily as FAD, FMN and riboflavin. FAD usually predominates, cow's milk being the main exception. During cooking, a portion of FAD is broken down to riboflavin. However, there are hydrolytic enzymes in the small intestine capable of hydrolysing both FMN and FAD to riboflavin. Based on their riboflavin content, FAD and FMN both serve equally as sources of the vitamin.

The coenzyme forms in foods are converted to riboflavin before assay for riboflavin content and food tables give total riboflavin values. This is illustrated in Table 1 where the total riboflavin content of foods is given. Representative samples of dairy products, animal products and legumes, cereal grains and fruits and vegetables are listed. It is apparent that dairy products and animal products provide about 75% of the riboflavin intake of Western diets. A litre of milk, for example, will supply a full day's requirement of the vitamin. Fruits and vegetables are poorer sources and the riboflavin in them is not as efficiently absorbed as from animal sources. ***See*** Coenzymes

Use in Food Fortification

White flour (70% extraction) and maize grits and meal are enriched in many countries with a vitamin mixture containing riboflavin. Dry breakfast cereals are also often enriched with vitamin and minerals including riboflavin. However, for the reason stated above, polished rice is not enriched with this vitamin. ***See*** Cereals, Breakfast Cereals

When a normal individual ingests a vitamin pill or a food containing a high concentration of riboflavin, a large portion of the dose is excreted rapidly in the urine with a peak occurring about 2 h later. The urine colour changes from a straw colour to an orange-yellow colour, and the colour is readily discernable. Physicians often take advantage of this phenomenon by prescribing drugs that should be taken regularly for an extended period of time with 3–5 mg of riboflavin. The patient is asked to collect a specimen of urine 2 h after taking the drug the day before he is to visit the physician. From the colour of the specimen the physician can ascertain whether the drug was taken. This procedure can also be used in experimental studies as a marker for food ingestion.

Riboflavin is relatively insoluble compared to other B vitamins and more is absorbed if taken with food.

Isolation and Clean-up

Solvent extraction, adsorption and elution from columns and recrystallization has been the method to isolate riboflavin from natural materials such as liver and milk. Solvents such as acetone or alcohols, including methanol, ethanol and butanol, will both free

Table 1. Riboflavin content of foods

Food	Riboflavin content (mg per 100 g)	Food	Riboflavin content (mg per 100 g)
Milk and dairy products		*Vegetables*	
Whole milk	0·16	Asparagus (cooked)	0·12
Evaporated milk	0·32	Broccoli (cooked)	0·11
Skim milk	0·18	Cabbage (cooked)	0·06
Cheddar cheese	0·38	Carrots (cooked)	0·06
Cottage cheese	0·14	Potatoes (baked)	0·03
US cheese	0·35	Onions	0·02
Ice cream	0·19	Spinach	0·24
Yoghurt	0·23	Tomato	0·04
Meats, fish, poultry, eggs and legumes		*Cereal grains*	
		Whole wheat bread	0·11
Liver (cooked)	4·10	Enriched white bread	0·20
Beef (cooked)	0·21	White rice (cooked)	0·02
Pork (cooked)	0·34	Oatmeal (cooked)	0·02
Ham (cured and cooked)	0·28		
Lamb (cooked)	0·28	*Fruits*	
Veal (cooked)	0·27	Apple	0·01
Mackerel (cooked)	0·41	Banana	0·10
Salmon (canned)	0·19	Orange	0·04
Sardines (canned)	0·23	Peach	0·04
Herring (cooked)	0·30	Pear	0·04
Egg (chicken)	0·44	Strawberry	0·07
Chicken (cooked)	0·18	Watermelon	0·01
Kidney beans (cooked)	0·06		
Baked beans	0·06		
Peanuts (cooked)	0·06		
Soya beans (cooked)	0·29		

riboflavin from binders and dissolve it. These solutions have been adsorbed into columns of Florisil, Fuller's earth in an acid solution, Floridin XXF or Frankonit in a neutral solution. An eluent containing pyridine diluted with aqueous methanol or ethanol is most often used. Crystalline riboflavin is obtained from the eluate by extraction with an aqueous acetone–petroleum ether mixture.

Commercially, riboflavin is obtained in a broth from bacterial synthesis. Advantage is taken of the insolubility of reduced riboflavin. Sodium dithionite is added to the broth, resulting in the precipitation of reduced riboflavin, which is then filtered off, purified and oxidized to crystalline riboflavin.

Chromatographic and Other Methods of Assay

Several methods have been used to determine the riboflavin content of foods. The earliest method devised in 1931 was a rat bioassay. This was followed by fluorometric and microbiological methods. More recently, a high-performance liquid chromatography (HPLC) method has been devised.

The rat bioassay is seldom used since it is costly, time-consuming and lacks sensitivity.

Fluorometric methods can be completed within one working day and are more sensitive than rat bioassays. Since riboflavin exists in most foods as FAD bound to proteins, it is necessary first to free the FAD from protein and hydrolyse it to FMN. This is accomplished by treating finely divided samples with dilute trichloroacetic acid, which results in the conversion of bound FAD to free FMN. In most foods there is very little free riboflavin so that this treatment is sufficient to estimate total riboflavin content. FMN and riboflavin have similar fluorescence on a molar basis while that of FAD is only 14% of this.

The extract is then passed through a Florisil column and adsorbed riboflavins eluted with a solution of pyridine in acetic acid. A fluorescein solution which has similar absorption and fluorescent properties to riboflavin is used to set the fluorometer. The samples to be assayed are then placed in a cuvette in the fluorometer and a reading is taken before and after the addition of

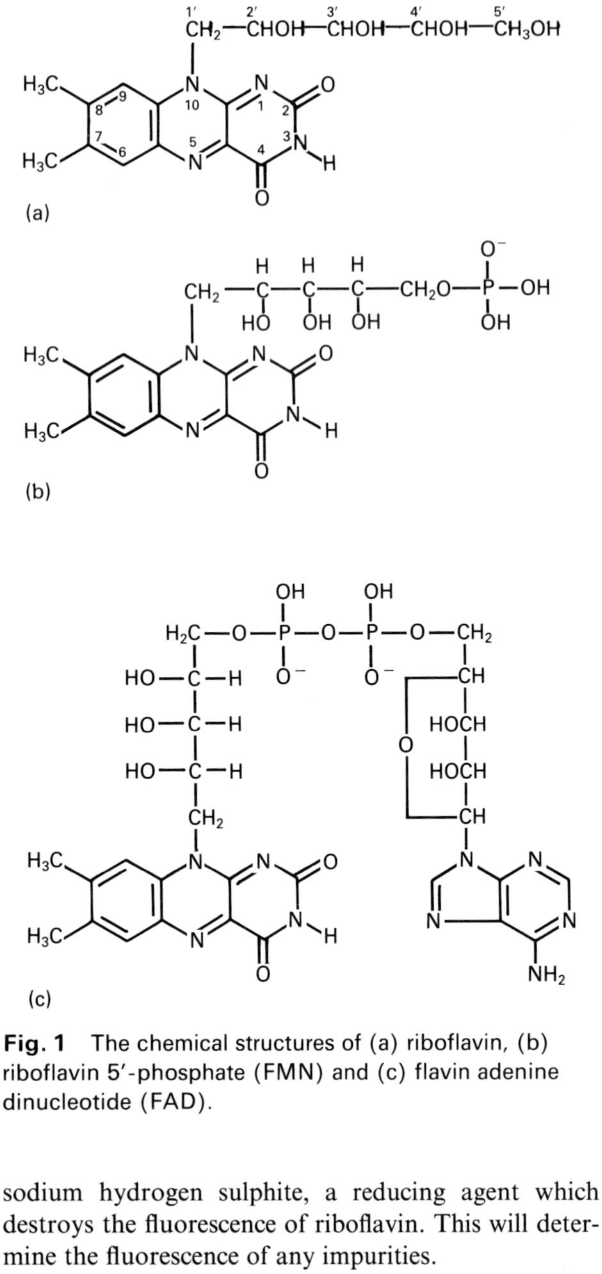

Fig. 1 The chemical structures of (a) riboflavin, (b) riboflavin 5′-phosphate (FMN) and (c) flavin adenine dinucleotide (FAD).

sodium hydrogen sulphite, a reducing agent which destroys the fluorescence of riboflavin. This will determine the fluorescence of any impurities.

If free riboflavin in the original sample is to be determined, the first extract is shaken with benzyl alcohol, in which riboflavin is very soluble, but FMN and FAD are not. This extract is then processed through the method described in the foregoing paragraph.

The microbiological assay, though more time consuming than the fluorometric assay, has the advantage of being more sensitive and more specific. It will measure only the biologically active form and is stereospecific. In this method, a finely homogenized sample of food in 0·1 N hydrochloric acid is autoclaved at 121°C. This liberates free riboflavin in the solution which is then neutralized. Treatment of the sample with an enzyme preparation such as Clarase can also be used to obtain free riboflavin in solution. In either case, the solution is then filtered to remove insoluble particles and fatty acids which inhibit the growth of the test microorganism, *Lactobacillus casei*. The solutions are diluted further with water to the range of riboflavin concentration the bacteria require for linear growth. A standard solution similarly diluted of pure crystalline riboflavin is used to create a standard curve. An assay medium, commercially available, containing all the nutrients with the exception of riboflavin is used. After the test solutions are added to the medium in test tubes, they are autoclaved for 15 min at 121°C to sterilize the solutions and, after cooling, an inoculum of the bacteria suspended in sterile saline is added. After incubation at 37°C for 16–24 h, the turbidity owing to growth of the bacteria is read in a colorimeter, and the turbidity in the test samples compared to that of the standard riboflavin solution from the standard curve. ***See*** Spectroscopy, Visible Spectroscopy and Colorimetry

The HPLC method requires more expensive equipment, but has the advantage of determining all the different forms of riboflavin in a food or tissue. In this method, a sample is homogenized in water and treated with trichloroacetic acid to remove the proteins. After filtering, the solution is neutralized and saturated with ammonium sulphate. After centrifuging, the supernatant is separated and shaken with 80% aqueous phenol. The upper phenol layer contains the riboflavins. After the addition of an equal volume of distilled water, the solution is extracted with water-saturated diethyl ether to remove phenol and the aqueous solution used for HPLC using specific columns. The eluents are ammonium acetate buffer (pH 6·0) and methanol, and the eluates are detected by fluorescence in a fluorometer. Through this method, riboflavin, FMN, FAD and other riboflavin derivatives can be separated and determined quantitatively. For estimation of the various forms of riboflavin this is the method of choice.

A number of alternative extraction procedures use dilute mineral acids, and the resulting extracts may be purified using solid phase extraction cartridges prior to HPLC.

Bibliography

Cooperman JM and Lopez R (1990) Riboflavin. In: Machlin LJ (ed.) *Handbook of Vitamins*, 2nd edn. New York: Marcel Dekker.

Rivlin RS (ed.) (1975) *Riboflavin*. New York: Plenum Press.

Roughead ZK and McCormick DB (1990) Flavin composition of human milk. *American Journal of Clinical Nutrition* 52: 854–857.

USDA (1984–1989) *Composition of Foods. Agriculture Handbooks* 8-1, 8-5, 8-9, 8-10, 8-11, 8-15, 8-16, 8-17, 8-20. Washington, DC: Human Nutrition Information Service, US Department of Agriculture.

J. M. Cooperman
New York Medical College, Valhalla, USA
R. Lopez
Our Lady of Mercy Medical Center, New York, USA

Physiology

Following the earlier identification of riboflavin and the two most common flavocoenzymes, the more recent recognition of the greater diversity of natural flavins has led to a broader appreciation of the multiple functions and metabolic processing of these important compounds. Much of the progress in this area and detail of function of the flavins and flavoproteins have been given in the periodic symposia held on this subject. This article reviews the principal features of the physiological handling of riboflavin and its natural derivatives in the mammalian body and, where known, in the human.

Digestion and Bioavailability

Riboflavin and lesser amounts of natural derivatives are released by digestion of complexes, mostly flavoproteins, contained within foods. Coenzyme forms of the vitamin, mainly flavin adenine dinucleotide (FAD) and flavin mononucleotide (FMN), are released from non-covalent attachment to proteins as a consequence of gastric acidification. Non-specific hydrolyses of the coenzyme forms by pyrophosphatase and phosphatase occur in the upper small intestine. By such actions, FAD is converted to FMN, which is further converted to riboflavin. Several per cent of 8α-(amino acid)riboflavins originally in covalent attachment as 8α-FAD linked to certain enzymes, notably of mitochondrial origin, are also released by such hydrolases that function together with proteolysis of the attached protein chains, which begins in the stomach with pepsin and continues in the small intestine with trypsin, chymotrypsin and exopeptidases. Traces of other ring and side-chain substituted flavins are similarly released by combinations of the above actions on non-covalently and covalently bound flavins. Riboflavin 5′-glycosides, for example, are cleaved by glycosidases present in the succus entericus. The digestive processes and locale for release of flavins from ingested material is shown in Fig. 1. ***See*** Coenzymes

While the efficiency of release of riboflavin from non-covalently bound forms is essentially complete in the normal gastro-intestinal tract, the vitamin is not recovered intact from flavin covalently linked to protein. Since the latter is less than 10% of total flavin within the diverse foods ingested by people and most mammals, the average bioavailability of riboflavin is fairly high. ***See*** Bioavailability of Nutrients

Absorption and Transport

Riboflavin and a fraction of flavin metabolites, including ring-altered forms, e.g. 8α-(amino acid)riboflavins, and side-chain derivatives, e.g. 7,8-dimethyl-10-formylmethylisoalloxazine, are absorbed primarily in the proximal small intestine by a saturable transport system that is rapid and approximately proportional to dose before levelling off. This saturation level is achieved with about 20 mg of the vitamin given in a single bolus to adult humans. Bile salts appear to facilitate the uptake, and a modest amount of flavin circulates via the enterohepatic system. The initial uptake of riboflavin by enterocytes is Na^+ dependent and reflects an adenosine 5′-triphosphatase (ATPase)-involved active cotransport system. Metabolic trapping by conversion to FMN and FAD occurs before release of the vitamin to circulation

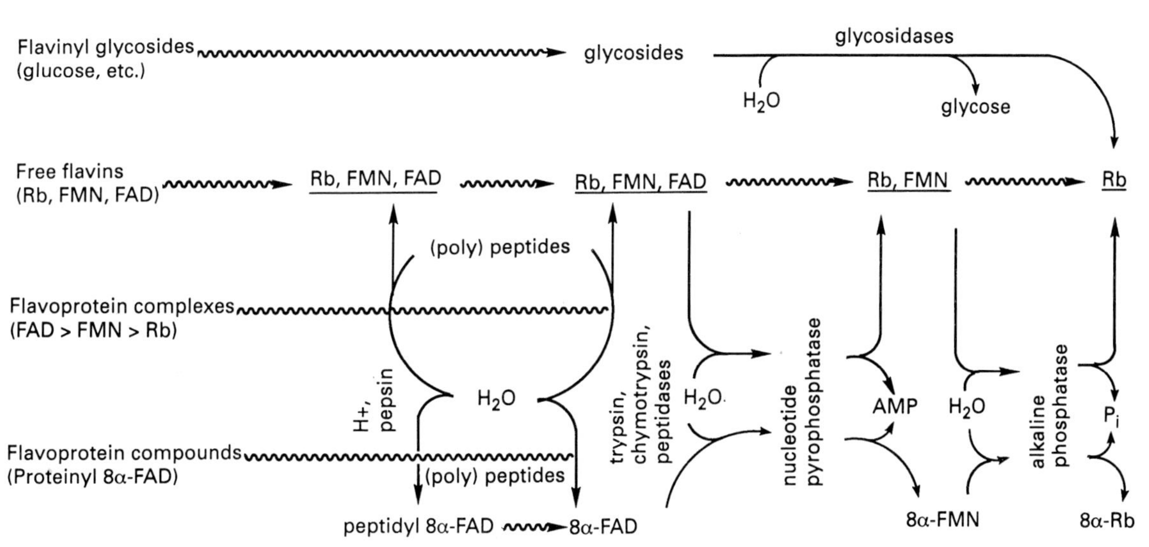

Fig. 1 Digestion of flavins in monogastric mammals. Rb, riboflavin; Pi, inorganic phosphate.

by non-specific pyrophosphatase and phosphatase. ***See*** Bile

Circulatory transport of flavin involves loose association with albumin and tight associations with some globulins. A subfraction of immunoglobulin G (IgG) has been found to bind avidly a small portion of the total free flavin in blood, and several immunoglobulins contribute significantly to plasma transport of the vitamin. Some riboflavin-binding proteins in plasma are pregnancy specific, including the classic case of the oestrogen-induced egg-white protein. These proteins have at least some portion of the binding domain in common and are essential for fetal development. Placental transfer of riboflavin in the human and other mammals involves binding proteins that help vector the vitamin and enhance supply to the fetus.

Uptake processes for flavins by mammalian cells have some characteristics in common, but there are both qualitative and quantitative differences among different cell types. Entry of riboflavin appears to be carrier mediated (facilitated) at physiological concentrations of the vitamin, since there is relative specificity to a saturable component that is responsible for initial rapid uptake. A riboflavin-binding protein has even been isolated from the plasma membrane of rat liver cells. The non-epithelial hepatocyte does not depend on Na^+ for riboflavin import, as do bipolar epithelial types such as the enterocyte or renal proximal tubular cell. Slower passive diffusion becomes more evident when the facilitating transporter is exceeded by pharmacological levels of the vitamin. In all cases, metabolic trapping of riboflavin by phosphorylation (dependent upon cytosolic flavokinase) follows passage of the vitamin through the plasma membrane. Release of riboflavin from cells requires hydrolysis of FMN by non-specific phosphatases.

Cellular Interconversions

The metabolic interconversions of riboflavin and flavocoenzymes are summarized in Fig. 2.

Conversion of riboflavin to coenzymes occurs within the cellular cytoplasm of most tissues, but particularly in the small intestine, liver, heart, kidney, and brain. The obligatory first step is the adenosine 5′-triphosphate (ATP)-dependent phosphorylation of the vitamin catalysed by flavokinase, which utilizes Zn^{2+}. The FMN product can be complexed with specific apoenzymes to form several functional flavoproteins, but the major portion is further converted to FAD in a second ATP-dependent reaction catalysed by FAD synthetase, which utilizes Mg^{2+}. It is clear that the biosynthesis of flavocoenzymes is regulated by supply of riboflavin (flavin status), competition for ATP (energy status), and hormonal balances. Thyroxine and triiodothyronine stimulate FMN and FAD synthesis in mammalian systems. This seems to involve a hormone-mediated increase in an active form of flavokinase. As product of the synthetase, FAD is also an effective inhibitor at this second step and may help regulate its own formation. FAD is the predominant flavocoenzyme present in tissues where it is mainly complexed with numerous flavoprotein dehydrogenases and oxidases. Several per cent of the FAD also becomes covalently attached to specific amino acid residues of a few important apoenzymes. Examples for the human include the 8α-N^3-histidyl FAD within the mitochondrial dehydrogenases for succinate, dimethylglycine and sarcosine, and the 8α-*S*-cysteinyl FAD within monoamine oxidase, also of mitochondrial localization. ***See*** Hormones, Thyroid Hormones

Turnover of covalently attached flavocoenzymes requires intracellular proteolysis, and further degradation of the coenzymes involves non-specific pyrophosphatase and specific 5′-nucleotidase cleavage of FAD to AMP (adenosine monophosphate) and FMN and action by non-specific phosphatases on the latter.

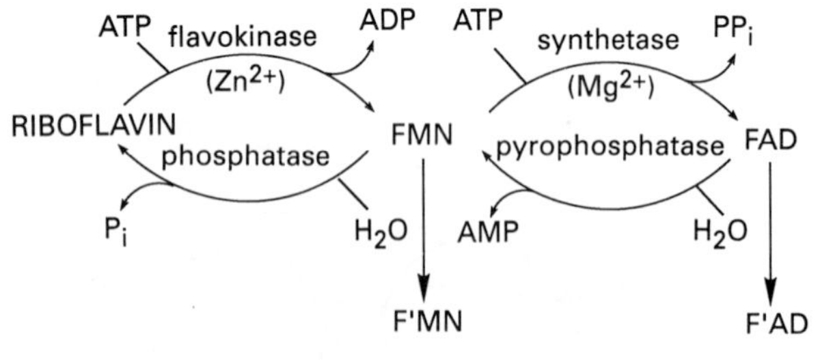

Fig. 2 Metabolic interconversions of riboflavin and flavocoenzymes.

Storage and Catabolism

There is little storage of riboflavin as such, since most exists within flavocoenzymes which are in relatively tight associations in holoenzymatic systems. During such severe deficiency of the vitamin as leads to death of experimental animals, there is a reduction in the level of extractable flavin that can approach about half of that found in an optimally supplemented control. Hence, there is moderately effective retention of riboflavin by its metabolic commitment to bound forms; however, even modest deficit of the vitamin is reflected in the decrease in function of certain flavoproteins well before full blown symptoms of deficiency.

Though certain bacteria, especially of the *Pseudomonas* genus, can extensively degrade both the ring system and side-chains of flavins, mammals are more limited in their abilities to catabolize the vitamin. The considerable diversity of flavin metabolites in and from mammals reflects the composite of reactions of photochemi-

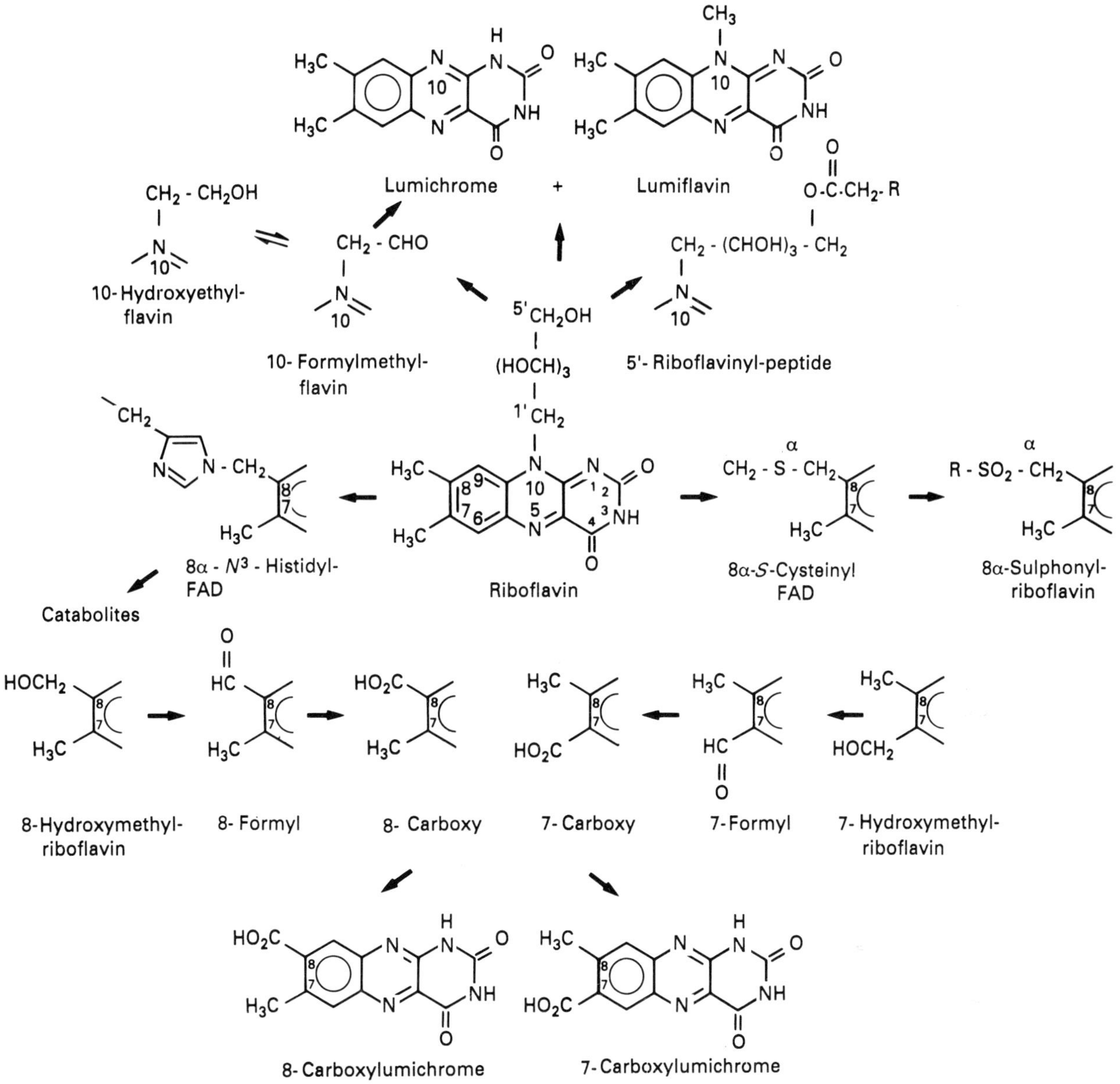

Fig. 3 Photochemical, microfloral, and cellular catabolism of riboflavin within mammals.

cal processes on the skin, microfloral activities in the gastrointestinal tract, as well as somatic actions both directly on flavin and on derivatives presented to cells by circulatory recovery from dermal tissue and by enterohepatic retrieval from the gut. The diverse flavin-related products identified from humans and other mammals are summarized in Fig. 3.

Cleavage of the D-ribityl side-chain at position 10 is mainly, if not entirely, attributable to light and intestinal microflora. Both can lead to partial fragmentation to form the 10-formylmethylflavin. This can be oxidized by alimentary bacteria of the ruminant and human to form the 10-carboxymethylflavin, and a further fraction of the formylmethyl compound is interconverted with the 10-hydroxymethylflavin as a result of pyridine-nucleotide-dependent dehydrogenase in tissue. Lumichrome-level compounds not only can result from complete removal of the side-chain by microflora, which can be decreased by antibiotic administration, but can accompany lumiflavin as a photoproduct from action of light on flavin within the dermal tissue. Catabolites of riboflavin that primarily derive from oxidations within tissues are the 7- and 8-hydroxymethylriboflavins (7α- and 8α-hydroxyriboflavins). These and products from further oxidation of the hydroxymethyl functions to formyl and carboxyl groups reflect microsomal mixed-function oxidase activity. Other flavin catabolites include those from 8α-(amino acid)riboflavins released from covalently bonded FAD. An 8α-sulphonylriboflavin may derive from the 8α-cysteinyl-FAD of monoamine oxidase. A peptide ester and a glucoside, both linked to the 5′-hydroxymethyl terminus of the vitamin, have also been found.

Excretion and Secretion

Since no isoalloxazine (flavin) can be biosynthesized within the cells of mammals which lack riboflavin synthetase, excretion and secretion reflect dietary intake and catabolic and photodegradative events. Essentially all known catabolites of riboflavin have been detected in urine; many of the lumichrome-level compounds are also in faeces. For normal adults eating varied diets, riboflavin comprises 60–90% of urinary flavin, 7-hydroxymethylriboflavin 3–7%, 8α-sulphonylriboflavin 2–15%, 8-hydroxymethylriboflavin 1–8%, 10-hydroxyethylflavin 1–7%, riboflavinyl peptide ester up to 5%, with traces of lumiflavin and, sometimes, the 10-formylmethyl- and carboxymethylflavins.

The presence in milk of 'lactoflavin', an early name for riboflavin, led to the recognition of this food as a good source of the vitamin. For milk from both cows and humans, the flavin in highest concentration other than the free vitamin is FAD, which can comprise over a third of total flavin. Much of this is hydrolysed to FMN by pasteurization. Fairly significant quantities of the 10-(2′-hydroxyethyl)flavin are notable, since this catabolite has antivitamin activities as reflected in competitive inhibition both of cellular uptake and subsequent flavokinase-catalysed phosphorylation of riboflavin. Hence, this catabolite, which may reach 10–12% of flavin in cow's milk, modestly subtracts from the biological activity of this food. Several per cent of both 7- and 8-hydroxymethylriboflavins are also present, with more of the former. Smaller amounts of other catabolites, including the 10-formylmethylflavin and lumichrome, comprise most of the rest. *See* Milk, Dietary Importance

Biochemical Functions

In bound coenzymic forms, riboflavin participates in oxidation–reduction reactions in numerous metabolic pathways and in energy production via the respiratory chain. A variety of chemical reactions are catalysed by flavoproteins. The redox functions of a flavocoenzyme (Fig. 4) include one-electron transfers, during which the neutral, oxidized (quinone) level of flavin is half reduced to the radical semiquinone, which can exist within natural pH ranges as neutral or anionic species. A further electron transfer can lead to a fully reduced hydroquinone. In addition, a single-step, two-electron transfer from substrate to flavin can occur with hydride ion transfer, e.g. from reduced pyridine nucleotide, or by base abstraction of a substrate proton together with carbanion addition.

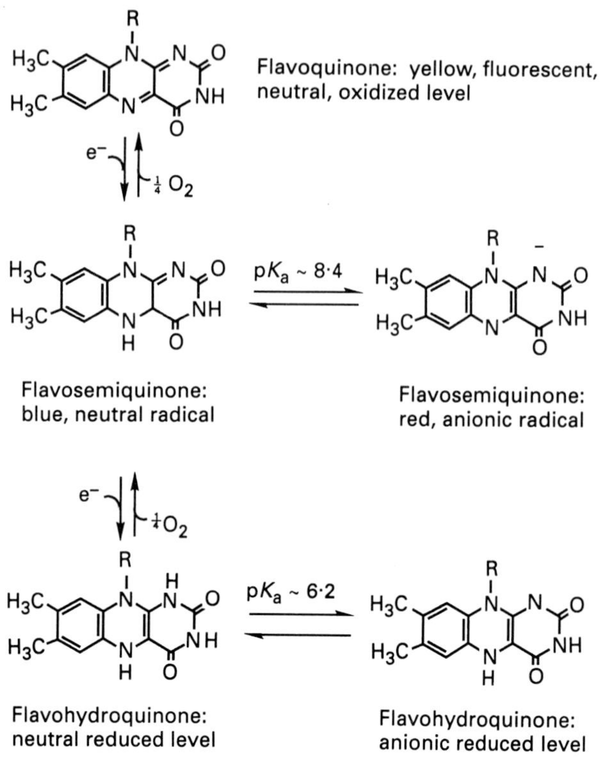

Fig. 4 Oxidation–reduction states of flavocoenzymes functioning physiologically.

There are flavoprotein-catalysed dehydrogenations that are both pyridine-nucleotide dependent and independent, reactions with sulphur-containing compounds, hydroxylations, oxidative decarboxylations, dioxygenations, and reduction of oxygen to hydrogen peroxide. The intrinsic abilities of flavins – to be varyingly potentiated as redox carriers upon differential binding to proteins, to participate in both one- and two-electron transfers, and to react in reduced (1,5-dihydro) form with oxygen – permit wide scope in their operation.

Requirements and Intakes

The requirement levels for riboflavin, in contrast to those for thiamin, are not raised when energy utilization is increased. Because of the interdependence of protein, energy intake, and metabolic body size, however, allowances calculated on these three bases do not differ significantly. Clinical signs of deficiency in adults can be prevented with intakes of riboflavin above 0·4 mg per 1000 kcal but over 0·5 mg per 1000 kcal may be required to maintain tissue reserves in adults and children as reflected in urinary excretion, red cell riboflavin, and erythrocyte glutathione reductase. From these considerations, the riboflavin allowances are now computed as 0·6 mg per 1000 kcal for people of all ages. This leads to US Recommended Dietary Allowances (RDAs) ranging from 0·4 mg per day for early infants to 1·7 mg per day for young adult males. However, for elderly people and others whose daily calorie intake may be less than 2000 kcal, a minimum of 1·2 mg per day is recommended in the USA. Since pregnancy imposes extra demands, reflected by decreased excretion and an elevated FAD stimulation of erythrocyte glutathione activity, an additional 0·3 mg per day is recommended. The lactating woman secretes approximately 35 μg per 100 ml of milk for an output of about 0·26 mg per day (750 ml) during the first six months and 0·21 mg per day (600 ml) during the second six months. Since the utilization of the additional riboflavin for milk production is assumed to be 70%, an additional intake of 0·5 mg is recommended for the first six months and 0·4 mg for the second. ***See*** Lactation

Small amounts of riboflavin, largely as digestible coenzymes, are present in most plant and animal tissue. Especially good sources are eggs, lean meats, milk, broccoli, and enriched breads and cereals. Such losses as occur during cooking are largely attributable to leaching of the heat-stable but light-sensitive flavins into water.

When supplementation or therapy with riboflavin is warranted, oral administration of 5 to 10 times the RDA is usually satisfactory.

Deficiency Causes and Symptoms

Pure, uncomplicated riboflavin deficiency is probably never encountered in patients, but is accompanied by multiple nutrient deficiencies. Ariboflavinosis can result from such primary and secondary factors as commonly affect supply or utilization of other nutrients as well. Inadequate dietary intake most commonly related to limited availability of food, but sometimes exacerbated by poor storage or processing, remains the major cause. In addition, anorexic persons rarely ingest adequate amounts of riboflavin and other nutrients.

Decreased assimilation results from abnormal digestion, absorption, or both. Lactose intolerance as a result of lactase insufficiency, mostly encountered among Blacks and Asians, argues against such people consuming non-lactase-treated milk, which is a good source of the vitamin. Malabsorption can occur as a result of tropical sprue, coeliac disease, malignancy and resection of the small bowel, and gastrointestinal and biliary obstruction. Poor absorption also results from disorders that increase motility and decrease gastrointestinal passage time, such as diarrhoea, infectious enteritis and irritable bowel syndrome. ***See*** Food Intolerance, Lactose Intolerances

Rather rarely encountered, but usually significantly improved by therapeutic treatment with riboflavin, are certain inborn errors where the genetic defect is in formation of a normal flavoprotein. Cases in this category include fatty acid desaturases in which specific defects have been found for the mitochondrial FAD-dependent dehydrogenases for short-chain, long-chain, and multi-chain acyl-CoAs (acyl coenzyme As). The young patients have a lipid storage myopathy, often accompanied by carnitine insufficiency, and exhibit glutaric aciduria. A low, FMN-dependent pyridoxine 5′-phosphate oxidase activity due to an erythrocyte deficiency of FMN, confirmed by response to oral riboflavin, was reported in the majority of subjects with D-glucose 6-phosphate dehydrogenase deficiency. Such cases seem to have an accelerated conversion of FMN to FAD so that glutathione reductase is saturated. This contrasts with heterozygous β-thalassaemia, in which there is an inherited slow erythrocyte conversion of riboflavin to FMN, a decrease in subsequent FAD, and a high stimulation of the erythrocyte glutathione reductase by extraneous FAD.

Defective utilization can result from disturbances in hormonal production, certainly relating to thyroid hormone, but less likely as a result of taking oral contraceptives. Phenothiazine derivatives appear to impair use of riboflavin.

Increased destruction of riboflavin occurs during treatment of neonatal jaundice with phototherapy. In this case, the side chain of the vitamin is photochemically destroyed, as it is involved in the photosensitized oxidation of bilirubin to more polar, excretable compounds.

The finding that phenobarbital induces microsomal oxidation of the 7-methyl function of the vitamin lends credence to the belief that long-time use of barbiturates may jeopardize flavin status.

Enhanced excretion of riboflavin occurs in catabolic patients undergoing nitrogen loss. The relationship of the vitamin to protein status has long been recognized. Also, certain antibiotics and phenothiazine drugs increase excretion of riboflavin. ***See*** Drug–Nutrient Interactions

Increased requirements can, of course, be the consequence of one or more of the above-mentioned factors. For example, protein–calorie malnutrition commonly accompanies a diminution in both absorption and utilization of riboflavin. Systemic infections, even without gastrointestinal involvement, sometimes lead to increased requirements that can result from decreased intake, defective absorption, poor utilization and increased excretion. ***See*** Protein, Deficiency

Clinical deficiency of riboflavin has been reduced by feeding a riboflavin-deficient diet and/or by the administration of an antagonist such as galactoflavin. The deficiency syndrome is characterized by sore throat, hyperaemia and oedema of the pharyngeal and oral mucous membranes, cheilosis, angular stomatitis, glossitis (magenta tongue), seborrhoeic dermatitis, and normochromic, normocytic anaemia associated with pure red cell cytoplasia of the bone marrow. As noted above, some of these symptoms, e.g. glossitis and dermatitis, when encountered in the field, may have resulted from other complicating deficiencies. Severe riboflavin deficiency can also affect the conversion of vitamin B_6 to its coenzyme and even curtail conversion of tryptophan to niacin. ***See*** Niacin, Physiology; Vitamin B_6, Properties and Determination

Toxicity

Toxicity from ingestion of excess riboflavin by experimental animals or humans is doubtful. The capacity of the human gastrointestinal tract to absorb orally administered riboflavin may be less than 25 mg in a single dose. The limited solubility and absorptivity of this vitamin as encountered in multivitamin preparations and natural foodstuffs, and its ready excretion as typical of water-soluble vitamins, normally precludes a health risk. There is one report of EEG (electroencephalogram) abnormalities in two patients during long-term treatment with riboflavin and niacin.

Bibliography

Edmondson DE and McCormick DB (eds) (1987) *Flavins and Flavoproteins*. Berlin/New York: Walter de Gruyter.

McCormick DB (1988) Riboflavin. In: Shils ME and Young VR (eds), *Modern Nutrition in Health and Disease*, pp. 362–369. Philadelphia: Lea & Febiger.

McCormick DB (1989) Two interconnected B vitamins: riboflavin and pyridoxine. *Physiological Reviews* 69, 1170–1198.

McCormick DB (1990) Riboflavin. In: Brown ML (ed), *Present Knowledge in Nutrition*, pp. 146–154. Washington DC: International Life Sciences Institute-Nutrition Foundation.

McCormick DB (1991) Coenzymes, Biochemistry. In: Dulbecco R (ed.), *Encyclopedia of Human Biology*, Vol. 2, pp. 1–19. San Diego; Academic Press.

Merrill AH, Jr, Lambeth JD, Edmondson DE and McCormick DB (1981) Formation and mode of action of flavoproteins. *Annual Review of Nutrition* 1: 281–317.

Müller F (ed) (1991) *Chemistry and Biochemistry of Flavoenzymes*, Vol. I-III. Boca Raton, Florida: CRC Press.

Donald B. McCormick
Emory University, Atlanta, USA

RICE

Global Distribution, Varieties and Commercial Importance

Rice is the most important crop in the world in terms of total developing-world production (481×10^6 t) and the number of consumers (2·5 billion) dependent on it as their staple food. Rice is widely grown in over 100 countries in every continent (except Antarctica), from 53°N to 40°S, and from sea level to an altitude of 3 km. Total production in 1989 was 507×10^6 t of rough rice. Asia accounts for 131 of the 146×10^6 ha world area that is used for rice cultivation. Mean rough rice yield was 3·47 t per ha in 1989. About 90% of the world's rice is grown and consumed in Asia. Major rice producers in 1988 were China, India, Indonesia, Bangladesh and Thailand. In terms of water regime, 49% of total rice area was irrigated in 1985, 29% rainfed lowland, 13% upland and 9% as deepwater or tidal wetlands, corresponding to total rough rice production of 72% from irrigated, 19% of rainfed lowland, 5% from upland and 4% from deepwater or tidal wetlands. Since the rice-growing area is shrinking, rice production must keep up with the 2·1% per year increase in population in tropical Asia through increased yield.

There are estimated to be about 100 000 rice varieties; only a small proportion is actually widely cultivated. They vary in grain weight, degree of dormancy, longevity and seed vigour, and some have red to purple-black pigments. More than 60% of the world's rice area is now planted with varieties of improved semidwarf plant type with erect leaves. The newer, improved varieties have similar yield potential as the first variety (IR8) but have better resistance or tolerance to biotic and abiotic stresses. Only about 4% of the world's rice production enters the international trade. The major exporters in 1988 were Thailand, the USA and Pakistan, but Vietnam was the third largest exporter in 1989. Major importers in 1988 were Iraq, the USSR, Hong Kong, Saudi Arabia, Malaysia, Singapore, Sri Lanka, Nigeria and Brazil.

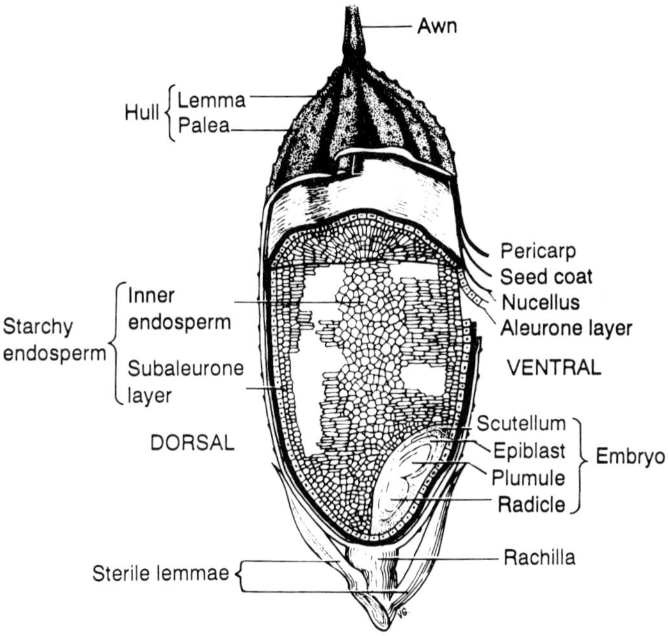

Fig. 1 Longitudinal section of the rice grain. Juliano BO (1984) In Whistler RL, BeMiller JN and Paschall EF (eds) *Starch Chemistry and Technology* 2nd edn, p 509. Orlando, Florida: Academic Press.

Morphology and Anatomy of the Spikelet or Grain

The rice grain (rough rice or paddy) consists of an outer protective covering, the hull (husk), and the edible rice caryopsis of fruit (brown, cargo, dehulled, or dehusked rice) (Fig. 1). Brown rice consists of the outer layers of pericarp, seed coat and nucellus, the germ or embryo, which are maternal tissues, and the endosperm. The endosperm consists of the aleurone layer, the subaleurone layer and the starchy or inner endosperm. Pigment is confined to the pericarp. The aleurone layer encloses the embryo. *See* Wheat, Grain Structure of Wheat and Wheat-based Products

The inedible hull constitutes 16–28% (mean 20%) of the rough rice weight. Brown rice consists of 1–2% pericarp, 4–6% aleurone plus nucellus and seed coat, 1% embryo, 2% scutellum and 90–91% endosperm. The aleurone and embryo cells are rich in protein bodies containing globoids or phytate bodies and in lipid bodies (spherosomes).

The endosperm cells are thin-walled and packed with amyloplasts containing polyhedral compound starch granules about 3–9 μm in size. Protein occurs mainly in the form of large (1–2 μm) and small (0·5–0·8 μm) spherical protein bodies and crystalline protein bodies (2–4 μm). Spherical protein bodies (PB I) are rich in prolamin (alcohol-soluble protein) and crystalline protein bodies (PB II) are rich in glutelin (alkali-soluble protein). Spherosomes are present in the subaleurone or two outermost cell layers.

Chemical and Nutritional Composition

Rice has one of the lowest protein contents (7%) among the cereals. The bran layers and embryo are richer in nonstarch constituents than the milled rice (Table 1). The major nutritional advantage of brown rice over milled rice is its higher content of B vitamins. Although higher in minerals, phytin in the aleurone forms complexes with minerals, reducing their bioavailability. The dietary fibre level of brown rice is higher than that of milled rice. The energy content of brown rice is higher than in milled rice due to the higher fat content. Rice has no vitamin A, C or D. *See* Cereals, Contribution to the Diet; Cereals, Dietary Importance

Starch varies in apparent amylose content (by iodine colorimetry): waxy, 1–2%; very low amylose, 2–10%; low, 10–20%; intermediate, 20–25; and high, 25–33%. The soluble fractions of protein are about 15% of albumin-globulin (water- and salt-soluble), 5% prolamin and 80% glutelin in milled rice. Bran proteins are 66–98% albumins. *See* Starch, Structure, Properties and Determination

Although cereal proteins are deficient in lysine, rice protein has one of the highest lysine contents among them, corresponding to amino acid scores of 55–69% based on the FAO/WHO/UNU (Food and Agriculture Organization/World Health Organization/United Nations University) amino acid pattern of 5·8% lysine (Table 2). Energy digestibility is higher in milled rice than in brown rice due to a lower dietary fibre level. Protein digestibility of milled rice is also higher, but the

Table 1. Comparison of nutrient composition of brown rice, milled rice and rice bran

	Amount per 100 g		
Property	Brown rice	Milled rice	Rice bran
Moisture (g)	14·0	14·0	14·0
Crude protein (g N × 5·95)	7·1–8·3	6·3–7·1	11·3–14·9
Crude fat (g)	1·6–2·8	0·3–0·5	15·0–19·7
Crude fibre (g)	0·6–1·0	0·2–0·5	7·0–11·4
Crude ash (g)	1·0–1·5	0·3–0·8	6·6–9·9
Available carbohydrates (g)	73–87	77–89	34–62
Total dietary fibre (g)	2·9–4·0	0·7–2·3	17–29
Water-insoluble fibre (g)	2·7	0·5	15–27
Thiamin (mg)	0·3–0·6	0·02–0·11	1·2–2·4
Riboflavin (mg)	0·04–0·14	0·02–0·06	0·18–0·43
Nicotinic acid (mg)	0·35–0·53	0·13–0·24	26·7–49·9
Vitamin E (mg)	0·8–2·5	tr[a]–0·30	3–15
Calcium (mg)	10–50	10–30	30–120
Phosphorus (g)	0·17–0·43	0·08–0·15	1·1–2·5
Phytin P (g)	0·13–0·27	0·02–0·07	0·9–2·2
Iron (mg)	0·2–5·2	0·2–2·8	8·6–43·0
Zinc (mg)	0·6–2·8	0·6–2·3	4·3–25·8
Energy content (kJ)	1520–1610	1460–1560	1670–1990
Energy content (kcal)	363–385	349–373	399–476

[a] tr=trace or less than 0·01 mg.

Table 2. Amino acid profile and energy and nitrogen balance in growing rats of brown rice, milled rice and rice bran

Property	Brown rice	Milled rice	Rice bran
Histidine (g per 16 g N)	2·3, 2·5	2·2–2·6	2·7–3·3
Isoleucine (g per 16 g N)	3·4–4·4	3·5–4·6	2·7–4·1
Leucine (g per 16 g N)	7·9–8·5	8·0–8·2	6·9–7·6
Lysine (g per 16 g N)	3·7, 4·1	3·2–4·0	4·8–5·4
Methionine + cystine (g per 16 g N)	4·4–4·6	4·3–5·0	4·2–4·8
Phenylalanine + tyrosine (g per 16 g N)	8·6–9·3	9·3–10·4	7·7–8·0
Threonine (g per 16 g N)	3·7–3·8	3·5–3·7	3·8–4·2
Tryptophan (g per 16 g N)	1·2–1·4	1·2, 1·7[f]	0·6, 1·2[f]
Valine (g per 16 g N)	4·8–6·3	4·7–6·5	4·9–6·0
Amino acid score[a] (%)	64, 71[f]	*55–69*	*83–93*
Digestibility energy[b] (% of total)	94·3	96·6	67·4
True digestibility[c] (% of diet N)	96·9	98·4	78·8
Biological value[d] (% of digested N)	68·9	67·5	86·6
Net protein utilization[e] (% of diet N)	66·7	66·4	68·3

[a] Based on 5·8 g per 16 g N as 100%.
[b] Mean values significantly different at the 5% level.
[c] Mean values significantly different at the 5% level.
[d] Mean values for Brown rice and Milled rice are *not* significantly different at the 5% level, but significantly lower than mean value for Rice bran at the 5% level.
[e] Mean values for Brown rice and Milled rice are *not* significantly different at the 5% level; mean values for Milled rice and Rice bran are significantly different at 5% level.
[f] Only two values reported.

biological value is lower, resulting in similar net protein utilization (NPU). Black or purple rices have lower brown rice NPUs than red and nonpigmented rices, but their milled rices have identical NPUs in growing rats. Rice complements legumes in amino acid composition for human diets. ***See*** Protein, Quality; Protein, Interactions and Reactions Involved in Food Processing

Cooking and parboiling reduces protein digestibility by 10–15%, with a corresponding increase in biological value but little change in NPU, in growing rats. However, lysine digestibility remains close to 100%. The fraction that remains in the faeces as faecal protein particles represents the lipid-rich core of large PB I, with less than 1% lysine but rich in cystine. PB II is readily digested.

Grading, Handling and Storage

There is no international standard for milled rice size and shape. IRRI (International Rice Research Institute) uses the following scale for size: extra long, > 7·50 mm; long, 6·61–7·50 mm; medium, 5·51–6·60 mm; and short, < 5·50 mm. For grain shape based on length:width ratio, the following scale is used: slender, > 3·0; medium, 2·1–3·0; bold, 1·1–2·0; and round, < 1·0. The FAO/WHO Codex Alimentarius Committee proposed size classification based on length:width ratio as long, ⩾3·0; medium, 2·0–2·9 and short, ⩽1·9. Grades are based on degree of milling, percentage head or whole-grain milled rice, immature grains, damaged (discoloured) and heat-damaged grains (chalky grains, red grains and red-streaked grains), and organic and inorganic extraneous matter.

The most important property of harvested grain is its moisture content: 14% on a wet weight basis is considered a safe storage value; grains become susceptible to fissuring from moisture adsorption stress at 12–16%, depending on variety. Japanese mills have shifted to milling at > 14% moisture to minimize grain breakage. Ageing for 3–4 months improves milling yields and makes milled rice expand more during cooking and become more flaky. ***See*** Cereals, Handling of Grain for Storage; Cereals, Bulk Storage of Grain

Processing and Food Uses

The per capita supply of milled rice is 85 kg per year in 1986–1988 in Asia as compared to 4–32 kg in other continents. It is mainly consumed as milled rice. About 20% of rice is consumed as parboiled rice. Parboiling consists of boiling or steaming steeped rough rice until the hull starts to open and then cooling and drying the gelatinized grain. Diffusion of bran B vitamins into the endosperm occurs during parboiling. Milling involves dehulling followed by removal of the outer 7–10% of the brown rice, either by friction or abrasion, removing most of the pericarp, seed coat, nucellus, aleurone layer, and the germ. In the Engelberg mills, hull and bran are removed together in one step with high grain breakage. Milling is done in several steps in modern cone mills, with tempering in between to minimize breakage.

Amylose content correlates positively with water absorption and volume expansion during cooking, and with hardness of boiled rice. Various rice products are prepared for which specific amylose types are preferred. Parboiled rices are preferably high and intermediate amylose, while extruded and flat noodles use mainly aged, high-amylose rices. Freshly and well-milled rice is preferred for rice products to minimize fat content and rancidity. Rice with a low starch gelatinization is preferred in rice puddings, breads and cakes, and beer adjuncts; this allows starch gelatinization at lower processing temperatures, particularly in the presence of sucrose. Waxy and low amylose rices are preferred for rice wines (for higher ethanol yield) and in frozen sauces, desserts and sweets because of the slow staling rate. Most of the waxy rices are consumed as desserts and snacks; only in Laos and north and northeast Thailand is steamed waxy rice consumed as a staple. Parboiled rice is preferred for 'idli' and 'dosa' with fermented rice : black gram pudding or cake at 3 : 1 weight ratio. ***See*** Fermented Foods, Fermentations of the Far East

There are various ways of cooking milled rices. In tropical Asia, they are prewashed to remove dirt, but this results in losses of B vitamins and fat. It may be cooked in the amount of water which it will absorb, or boiled in excess water and the cooking liquor discarded. A steaming step is used in Indonesia and also for waxy rices. Oil or ghee may be added in the Middle East, and this reduces surface cohesion. Enriched rice premixes containing iron and B vitamins resistant to washing have been developed for rice, but have not been popular due to additional expense; the enriched premix can be readily distinguished from ordinary grain.

Use of rice bran in cereal products increased in recent years due to the hypocholesterolaemic effect in man of the factor(s) in its oil-unsaponifiable fraction. Total rice bran oil production in 1986–1988 was 598 000 t per year, mainly in India, China, Japan and Vietnam, according to the FAO. This represents only 13% of potential bran oil production, because of the problem of potent lipase in rice bran requiring immediate oil extraction or stabilization, and the hull contamination of rice bran produced by Engelberg mills. Essential fatty acid content of rice oil is 29–42% $C_{18:2}$ and 0·8–1·0% $C_{18:3}$.

Bibliography

Araullo EV, de Padua DB and Graham M (eds) (1976) *Rice Postharvest Technology*. Ottawa: International Development Research Centre.

Barker R, Herdt RW and Rose B (1985) *The Rice Economy of Asia.* Resources for the Future, Washington DC, and International Rice Research Institute, Manila.
De Datta SK (1981) *Principles and Practices of Rice Production.* New York: John Wiley.
Food and Agriculture Organization (FAO) (1985) Rice Processing Industries. Proceedings FAO/UNDP Regional Workshop, Jakarta, 15–20 July 1985. Bangkok: FAO Regional Office for Asia and the Pacific.
Juliano BO (1991) *Rice in Human Nutrition.* Rome: Food and Agriculture Organization (In press).
Juliano BO (ed.) (1985) *Rice: Chemistry and Technology* 2nd edn. St Paul, Minnesota: American Association of Cereal Chemists.
Luh BS (ed.) (1980) *Rice: Production and Utilization.* Westport, Connecticut: AVI Publishing Co.
Pillaiyar P (1988) *Rice Production Manual.* New Delhi: Wiley Eastern Ltd.

Bienvenido O. Juliano
International Rice Research Institute, Manila, Philippines

RICKETS AND OSTEOMALACIA

The terms rickets and osteomalacia refer to the histological and radiological abnormalities seen in a variety of vitamin D deficiency conditions. These two deficiency diseases remain significant examples of undernutrition in many parts of the world today, even though they are largely preventable with improved calcium and vitamin D intakes. The clinical spectra of these two disorders are very similar, but rickets has more severely adverse effects.

Primary Causes and Abnormalities

Human vitamin D deficiency, which may result from inadequate dietary intake and/or skin biosynthesis of vitamin D (Fig. 1) is a primary cause of rickets. The active hormonal form of vitamin D, 1,25-dihydroxyvitamin D, exerts important roles in intestinal absorption of calcium and phosphate, bone metabolism and mineralization, and calcium homeostasis. Other roles are not yet well understood and include conservation of calcium and phosphate by the kidney, and interaction of the hormone with the immune system, skin and skeletal muscle. Rickets may also be caused by inadequate calcium intake alone, as in the case of children living in Nigeria. It is characterized by several clinical features and laboratory measurements of blood (Table 1). ***See*** Calcium, Physiology; Cholecalciferol, Physiology

Rickets describes the disordered growth and mineralization of the growth plate of the long bones. The resulting abnormalities, widening of the growth plate, delayed conversion of the growth plate cartilage to bone, and irregularly arranged bone trabeculae, are only seen when the vitamin D deficiency occurs during childhood or adolescence, i.e. before the growth plates close. ***See*** Bone

Osteomalacia describes abnormalities resulting from delayed and reduced mineralization of mature bone, whether trabecular or cortical. This mineralization defect leads to a widened area of unmineralized bone

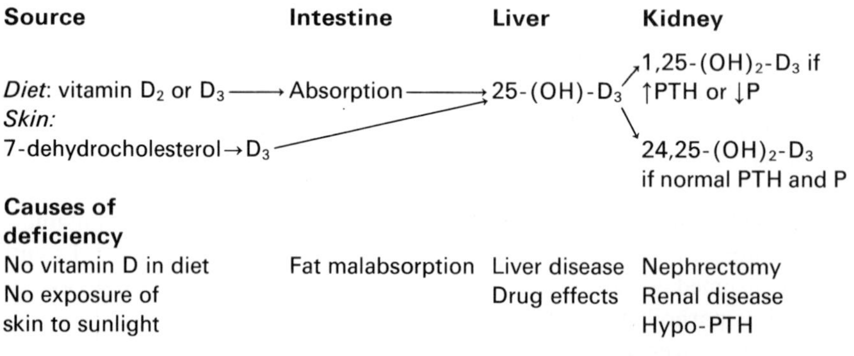

Fig. 1 Biotransformation of vitamin D and causes of deficiency. 25(-OH)-D_3=25-hydroxyvitamin-D; 1,25-$(OH)_2$-D_3=1,25-dihydroxyvitamin D; PTH=parathyroid hormone; P=phosphorus; 24,25-$(OH)_2$-D_3=24,25-dihydroxyvitamin D.

Table 1. Characteristic clinical features and blood serum measurements in rickets and osteomalacia

Age	Clinical features	Blood serum measurements
Children	Skeletal deformations (rickets) Impaired growth Undermineralized bone	Hypocalcaemia Hypophosphataemia Secondary hyperparathyroidism Low 25-Hydroxyvitamin D Elevated alkaline phosphatase
Adults	Undermineralized bone (Osteomalacia) Fractures	Hypocalcaemia Hypophosphataemia Low 25-Hydroxyvitamin D Elevated alkaline phosphatase Elevated osteocalcin[a]

[a]This finding has not been consistently reported (see Vieth R, 1990).

matrix on the bone surface, called osteoid. Osteomalacia is observed at all ages.

Infants and young children suffer from growth retardation primarily because of the stunted growth of the long bones, and may become bowlegged because of the reduced mechanical strength of these bones. Skeletal muscle weakness and increased urinary excretion of amino acids may also occur.

Radiographic evidence of rickets includes widened growth plate of the long bones of both arms and legs, and reduced bone density. In osteomalacia of the adult, undermineralization of the bones may manifest as pseudofractures, which are radiolucent lines on the radiographs that mimic stress fractures. However, the pseudofractures are usually bilateral and often involve non-weight-bearing bones such as the scapula.

Biochemical Changes

Biochemical changes in most cases of rickets and osteomalacia include decreased serum calcium and phosphate levels because of insufficient intestinal absorption of these ions, increased serum parathyroid hormone levels, as a result of the reduced serum calcium concentration, and elevated serum alkaline phosphatase concentration. In cases of inadequate supply of vitamin D or in liver disease, levels of serum 25-hydroxyvitamin D (25-(OH)-D, a metabolite) are invariably low, but the levels of 1,25-dihydroxyvitamin D (1,25-$(OH)_2$-D), the active hormone may still be normal, depending on the severity of the condition. Kidney disease will invariably lead to low serum 1,25-$(OH)_2$-D levels, while the serum 25-(OH)-D may be normal.

Secondary Causes

Causes of rickets and osteomalacia include not only insufficient dietary intake of vitamin D or inadequate exposure to sunlight, but also diseases that may interfere with absorption and biotransformations of vitamin D_3 or D_2 and conditions leading to decreased tissue responsiveness to the hormone and to renal loss of phosphate. In fact, relatively few cases of rickets in the Western world today are due simply to inadequacy of supply of vitamin D, and most of these are due to strict dietary or religious practices that involve vegetarian diets devoid of sources of vitamin D or dress codes that prevent exposure of the skin to sunlight. ***See*** Vegetarian Diets

Patients suffering from malabsorption of fats may have inadequate absorption of vitamin D from the small intestine. Liver disease may interfere with the conversion of vitamin D to 25-(OH)-D. Patients whose kidneys have been removed or are severely diseased will have decreased vitamin D hormone production and may develop osteomalacia.

Finally, three inborn errors of metabolism also lead to rickets. Vitamin D-dependent rickets type 1 is caused by a defect in the kidney enzyme 25-(OH)-D-1-hydroxylase, which is responsible for the synthesis of 1,25-$(OH)_2$-D; in vitamin D-dependent rickets type 2 the receptor for 1,25-$(OH)_2$-D is defective and cannot mediate the genomic action of the hormone. In the third condition, called X-linked hypophosphataemia, defects occur in the renal reabsorption of phosphate, leading to low circulating phosphate levels. Rickets and osteomalacia are present in this condition despite the presence of normal adult plasma 1,25-$(OH)_2$-D concentrations. ***See*** Inborn Errors of Metabolism

Public Health Consequences of Vitamin D Deficiency

Vitamin D, derived either from the diet or skin production, has become an increasingly important molecule in human function because of the extended longevity of human populations in many nations.

Much greater efforts by health organizations are needed to reduce the incidence and prevalence of rickets and osteomalacia in both the developing and the developed nations of the world. Assurance of adequate calcium intakes must be the first priority, and then an emphasis needs to be placed on daily sunlight exposure for infants and children, and especially for institutionalized elderly, in the prevention of these nutritional deficiency diseases.

In recent years, supplements of vitamin D (400 iu or greater) have been increasingly recommended for elderly people, especially those who have become institutionalized or shut-in because of limited mobility or confining conditions and diseases. Since the elderly have declining capabilities of skin biosynthesis and vitamin D conversion, larger doses of vitamin D supplements are being recommended for this section of the population. This recommendation is supported by findings of low circulating blood concentrations of 25-(OH)-D or 1,25-$(OH)_2$-D. Such low blood levels of vitamin D metabolites are now being defined as characteristic of vitamin D insufficiency rather than frank deficiency.

Oral therapy with 1-α-hydroxycholecalciferol has become fairly widespread in the United States among elderly subjects with low bone mass or fractures. The 1-α form must be converted in the liver to 1,25-$(OH)_2$-D by 25-hydroxylation. Dosages are generally low because of concerns about toxic effects. The efficacy of this therapy has not yet been fully evaluated.

Life-cycle Changes in Vitamin D Production and Metabolism

The skin production of vitamin D does not remain constant throughout the life-cycle, and it depends on the changes in the skin tissue itself as well as the environmental conditions in which people live. The skin of older individuals has considerably less capacity, for example, to produce vitamin D from its precursor. In the elderly, the concentration of 7-dehydrocholesterol is diminished, and the rate of conversion to vitamin D_3 is also reduced. When one considers that the elderly usually have reduced sun exposure as well, it becomes clear that they are at great risk of developing vitamin D insufficiency unless they have an adequate dietary supply of vitamin D.

Large numbers of elderly residents in the northern parts of the United States and of Europe have been shown to have depressed circulating levels of 25-(OH)-D and, in some reports, also low serum concentrations of 1,25-$(OH)_2$-D. Seasonal differences have become well established in these geographic regions, but 1,25-$(OH)_2$-D insufficiency without symptoms of a full-scale deficiency have only recently been documented. Because of the increasing lifespans of so many elderly, and because so many of them reside in nursing homes or similar facilities which permit little exposure of residents to sunlight, frank deficiency of vitamin D is likely to become more common and might even reach epidemic proportions. The public health consequences of osteomalacic bone tissue and increased fracture incidence could be enormous. Accordingly, some investigators have raised questions about the adequacy of the current allowances of vitamin D for the elderly. ***See*** Elderly, Nutritionally Related Problems

Alterations in the vitamin D endocrine system in old age, such as a decrease in the circulating 1,25-$(OH)_2$-D level, has been implicated in the pathogenesis of postmenopausal osteoporosis, because reduced serum 1,25-$(OH)_2$-D levels have been found in some women afflicted with this condition. However, there is no evidence indicating a general age-related decrease in serum 1,25-$(OH)_2$-D concentration. A recent study of a large group of healthy Caucasian adults ranging in age from 20 to 94 years revealed no decline in serum 25-(OH)-D or 1,25-$(OH)_2$-D concentrations in either sex. In another study, involving women between 26 and 88 years, serum 1,25-$(OH)_2$-D levels were actually higher after age 65 than in the youngest subjects, but the rate of intestinal calcium absorption was unaltered, suggesting decreased responsiveness to the hormone with increased age followed by a compensatory increase in hormone synthesis. ***See*** Hormones, Thyroid Hormones; Osteoporosis

Bibliography

Bouillon RA, Auwerx JH, Lissens WD and Peleman WK (1987) Vitamin D status in the elderly: seasonal substrate deficiency causes 1,25-dihydroxycholecalciferol deficiency. *American Journal of Clinical Nutrition* 45: 755–763.

Dagnelie PC, Vergote FJVRA, vanStaveren WA, *et al.* (1990) High prevalence of rickets in infants on macrobiotic diets. *American Journal of Clinical Nutrition* 51: 202–208.

DeLuca HF (1986) The metabolism and functions of vitamin D. *Steroid Hormone Resistance: Mechanisms and Clinical Aspects. Advances in Experimental Medicine and Biology* 196: 361–377.

Eastell R, Yeargey A, Veiera N, Cedel S, Kumar R and Riggs B (1991) Interrelationship among Vitamin D metabolism, true calcium absorption, parathyroid function, and age in women: Evidence of an age-related intestinal resistance to 1,25-dihydroxyvitamin D action. *Journal of Bone and Mineral Research* 6: 125–132.

Egsmose C, Lund B, McNair P, Lund B, Storm T and Sørensen O (1987) Low serum levels of 25-hydroxyvitamin D and 1,25-dihydroxyvitamin D in institutionalized old people: Influence of solar exposure and vitamin D supplementation. *Age and Ageing* 16: 35–40.

Glorieux FH (ed.) (1991) *Rickets.* Nestle Nutrition Workshop Series, vol.21. New York: Raven Press.

Gloth F III, Tobin J, Sherman S and Hollis B (1991) Is the recommended daily allowance for vitamin D too low for the homebound elderly? *Journal of the American Geriatric Society* 39: 137–141.

Holick MF (1991) Vitamin D: cutaneous production and therapeutic effect in psoriasis. In: Norman AW (ed) *Eighth Workshop on Vitamin D*, Paris, p 263.

Lawson DEM (1980) Metabolism of vitamin D. In: Norman AW (ed) *Vitamin D; Molecular Biology and Clinical Nutrition*, pp 93–126. New York: Marcel Dekker.

Meunier PJ, Chapuy MC, Aulot ME, *et al.* (1991) Effects of a calcium and vitamin D_3 supplement on non-vertebral fracture rate, femoral bone density and parathyroid function in elderly women: A prospective placebo-controlled study. *Journal of Bone and Mineral Research* 6 (Supplement 1): s135.

Okonofua F, Gill DS, Alabi ZO, *et al.* (1991) Rickets in Nigerian children: A consequence of calcium malnutrition. *Metabolism* 40(2): 209–213.

Omdahl JL, Garry PJ, Hunsaker LA, Hunt WC and Goodwin JS (1982) Nutritional status in a healthy population: vitamin D. *American Journal of Clinical Nutrition* 36: 1225–1233.

Pitt MJ (1991) Rickets and osteomalacia are still around. *Radiologic Clinics of North America* 29: 97–119.

Sherman S, Hollis B and Tobin J (1990) Vitamin D status and related parameters in a healthy population: The effects of age, sex, and season. *Journal of Clinical Endocrinology and Metabolism* 71: 405–413.

Stryd RP, Gilbertson TJ and Brunden MN (1979) A seasonal variation study of 25-hydroxyvitamin D_3 serum levels in normal humans. *Journal of Clinical Endocrinology and Metabolism* 48: 771–775.

Subcommittee on Dietary Allowances (1989) National Research Council, *Recommended Dietary Allowances*, 10th edn. Washington DC: National Academy Press.

Tjellesen L and Christiansen C (1983). Vitamin D metabolites in normal subjects during one year. A longitudinal study. *Scandinavian Journal of Clinical and Laboratory Investigation* 43: 85–89.

US Department of Agriculture (USDA), Human Information Service (1985) Dietary levels: Households in the United States, Spring 1977. *1977–78 Nationwide Food Consumption Survey*. Washington, DC: US Government Printing Office.

Vieth, Reinhold (1990) The mechanisms of vitamin D toxicity. *Bone and Mineral* 11: 267–272.

Webb AR, Kline L and Holick MF (1988) Influence of season and latitude on the cutaneous synthesis of vitamin D_3: Exposure to winter sunlight in Boston and Edmonton will not promote vitamin D_3 synthesis in human skin. *Journal of Clinical Endocrinology and Metabolism* 67(2): 373–378.

J. J. B. Anderson and S. U. Toverud
University of North Carolina, Chapel Hill, USA

RIPENING OF FRUIT

Most fleshy fruits attain physiological maturity at an unripe state unsuitable for immediate consumption. The ripening process that follows, either on the tree or after picking, is the result of a number of physiological and biochemical processes which are revealed in a sequence of changes in colour, texture, aroma and taste, leading eventually to a physiological state at which the fruit is commercially considered as edible. However, in the context of sensorial changes the meaning of ripening is rather ambiguous as it is based on a subjective appreciation of a very intricate process that involves a number of metabolic changes. Table 1 illustrates some of the most important organoleptic and physiological changes that occur during ripening and the biochemical processes most likely to be responsible for them.

The ripening process is mainly concerned with changes in components formed during fruit growth and development. So, many fruits are able to ripen after being detached from the mother plant, provided physiological maturity has been attained before picking.

Respiratory Patterns – Climacteric and Nonclimacteric Fruits

Even after detachment from the plant, fruits continue to display for some time the metabolic activities characteristic of living organisms. As changes in respiratory rates are good indicators of alterations in general metabolism, fruit respiration is useful as a reference to physiological state, which determines the potential storage life of a fruit. The first systematic studies of fruit respiration were carried out by Kidd and West in the 1920s with Bramley Seedling apples. Fruits, picked at different stages of development and stored at various temperatures, exhibited a typical respiratory pattern characterized by a decline in the respiration rate after picking which, sooner or later, depending on the storage temperature, was followed by a temporary rise in respiration (Fig. 1). The visible changes normally associated with ripening (colour, texture and flavour) began at or shortly after the maximum rate of respiration. The sudden change in the respiration rate was recognized by Kidd and West as a critical phase in the life of the fruit, denoting the transition from maturity to senescence,

Table 1. Organoleptic and metabolic changes that occur during ripening

Organoleptic and physiological changes	Biochemical changes
Texture	Rise in water-soluble pectins and decreasing degree of esterification through pectolytic activity. Partial cellulose breakdown
Colour	Degradation of chlorophyll and unmasking of underlying pigments. Synthesis of new pigments (carotenoids or anthocyanins)
Aroma and taste	Qualitative and quantitative changes in carbohydrates and organic acids. Synthesis of alcohols, esters and other volatile compounds
Rise in respiration	Autocatalytic synthesis of ethylene. Increased protein synthesis and altered metabolic control of respiratory pathways

and was named 'climacteric'. The same authors showed later that a similar increase in respiration also occurred in apples attached to the tree. In this situation the rise in respiration takes place at a slower rate but, eventually, reaches higher peaks than those attained by the detached fruits. Subsequent work indicated that many other ripening fruits exhibit a similar pattern of respiration, most of them, mainly tropical and subtropical fruits, displaying higher climacteric peaks than apples (Fig. 2(a)). On the other hand, some fruits show a continuous decline in respiration throughout maturation and ripening (Fig. 2(b)) and have been termed 'nonclimacteric'. A list of commercially important climacteric and nonclimacteric fruits is given in Table 2.

Climacteric Behaviour and Ethylene Production

It was recognized early in this century that ethylene can cause physiological responses in various plant tissues. The suggestion that fruit tissues themselves release gaseous substances that may affect the ripening behaviour of nearby fruits was made by Coussins, who reported, in 1910, that gases emanating from oranges caused premature ripening of bananas. It was later showed that the active substance released by ripening fruit was ethylene and that in most climacteric fruits the rise in ethylene production occurred at or prior to the beginning of the respiratory upsurge. So the gas was considered responsible for the initiation of the climacteric and the related ripening processes and was named the 'ripening hormone'.

The climacteric or nonclimacteric behaviour of fruits has been associated with the capacity of their tissues to synthesize ethylene. In climacteric fruits the respiratory increment is always accompanied, and often preceded, by an increase in ethylene levels within their tissues. This increase in endogenous ethylene induces an autocatalytic process of ethylene synthesis that, apparently, triggers off the respiratory upsurge of climacteric fruits. On the other hand, in nonclimacteric fruits the ethylene levels are always low and usually tend to decrease slowly during ripening at a rate approaching the rate of diminishing respiration. Climacteric and nonclimacteric fruits also differ in their responses to ethylene applied exogenously. The autocatalytic response of climacteric fruits to applied ethylene leads to the onset of climacteric respiration which, once initiated, cannot be stopped. Even if the exogenous gas is removed, the endogenous levels of ethylene will continue to increase once its catalytic synthesis starts. Thus, the respiratory rate at the climacteric peak does not depend upon the concentration of ethylene initially present. The respiratory rate of nonclimacteric fruits also responds to the application of exogenous ethylene, but does not exhibit an autocatalytic effect. Thus, in nonclimacteric fruits, the magnitude of the response varies proportionally to the concentration of the gas applied and the process can be reversed; that is, the respiration rate decreases to basal levels upon removal of ethylene, and can be reproduced by renewed applications of the gas. Nevertheless, the application of exogenous ethylene accelerates the ripening in nonclimacteric fruits as it does in climacteric fruits when furnished in the preclimacteric phase.

The belief that ethylene is the only substance directly responsible for the climacteric rise in respiration and the ensuing ripening has been, recently, under close reexamination. The existence of a strong link between ethylene and respiration is well documented in the literature, being observed in many other plant tissues besides fruits. However, other volatile substances that are produced during ripening when applied exogenously may mimic the respiratory changes caused by ethylene. On the other hand, the onset of increased ethylene synthesis does not always precede the respiratory upsurge in climacteric fruits. The literature reports a variety of fruits in which the increase in ethylene production either coincides with (e.g. Cox's Orange Pippin apples and Anjou pears) or follows (e.g. avocado var. Fuerte and plum var. Wickson) the rise in respiration. It has been suggested that, in these fruits, ethylene is not the only factor that determines the initiation of the climacteric and of ripening. The minimum ethylene production rate that triggers off the rise in respiration, although varying between species and culti-

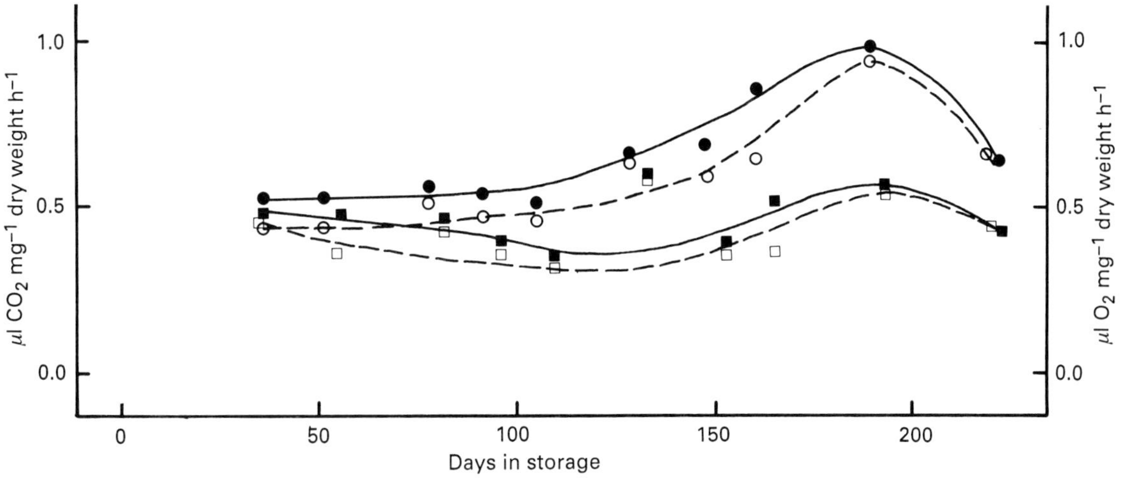

Fig. 1 Climacteric rise in respiration of 'Rocha' pears. Carbon dioxide evolution in fruits stored at 0°C (●——●) and −1°C (○——○) and oxygen absorption at 0°C (■ – – ■) and −1°C (□ – – □). From Teixeira AR, Carmona MA, Barreiro MJ, Silva MJ and Cabral ML (1978) A counservaçáo frigorifica de pera Rocha. *Agronomia Lusitana* 39: 57–84, with permission.

vars, is always very low (frequently less than 1 μl kg $^{-1}$ h^{-1}, which is approximately equivalent to an internal concentration of 2 ppm). So, the accurate determination of the relative timing of initiation of the rises in the rates of ethylene synthesis and respiration is rather difficult, particularly if, at this stage, other substances are being formed whose effects add up to those of ethylene. Notwithstanding the evidence suggesting that the stimulation of respiratory activity in fruits and other plant tissues may be induced by a number of factors, in the majority of climacteric fruits the autocatalytic ethylene synthesis begins before the respiratory upsurge and in most nonclimacteric plant tissues respiratory activity is tightly coupled to ethylene concentrations. Apparently, ethylene stimulates metabolism in general and may alter the control mechanism of existing metabolic pathways and induce changes in gene expression.

Climacteric Behaviour and Ripening

If the link between ethylene and climacteric respiration is partially resolved, the role of the climacteric respiratory rise in ripening remains obscure. It is now evident that many of the physiological changes associated with ripening (e.g. softening, colour changes and ethylene synthesis) can be, with some manipulation, separated from the respiratory climacteric. Thus, some fruits can initiate softening and pigment synthesis without an increase in respiration and the application of certain plant hormones and of protein synthesis inhibitors may prevent ripening changes without obviating the climacteric rise in respiration. Such observations do not support the opinion that the increased rate of respiration supplies the extra adenosine triphosphate (ATP) required as the energy source for physiological transformations that occur during ripening, such as synthesis of proteins, nucleic acids and pigments. In fact, the estimated energy demands of ripening are much lower than those provided by climacteric respiration; apparently, the energy generated by the basal respiratory rate is sufficient to drive the ripening events.

The actual evidence does not uphold the existence of an integrated relationship between the respiratory climacteric and ripening, but rather suggests that the rise in respiration during ripening may simply be a facet of ethylene action. The suggestion that ethylene may act by lowering the organizational resistance of the cell, mainly by increasing membrane permeability, has some experimental support. On the assumption that fruit tissues, like other living tissues, react to any injurious situation by changing their metabolism in order to maintain a homeostatic condition and that such changes are driven by respiration, it has been suggested that the respiratory climacteric is largely a homeostatic response, in an attempt to counterbalance the injury imposed on mitochondria by the initiation of senescence. That is, the degradative effects of incipient cellular senescence, probably accelerated by ethylene, impose a stress on mitochondria which leads to an increased respiratory rate. Such a suggestion is supported by the fact that mitochondrial structure and function, including coupling, are kept intact throughout the ripening process.

Effects on Texture

The texture of a fruit depends on the turgor of the cells as well as on the presence of supporting structures like cell

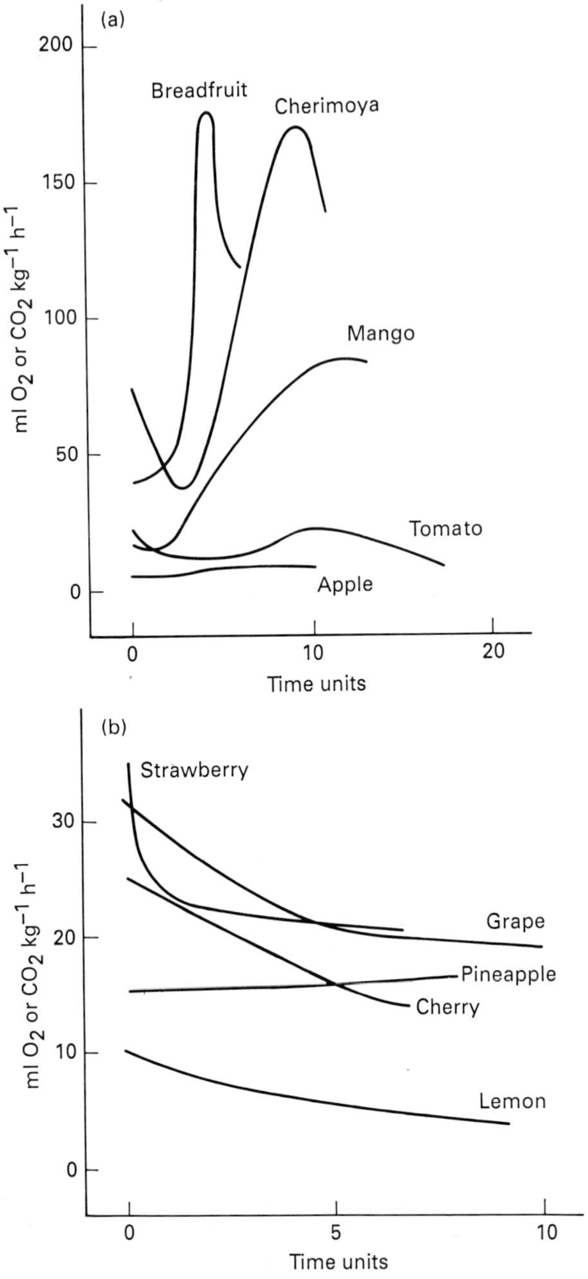

Fig. 2 Respiration patterns of some (a) climacteric and (b) nonclimacteric fruits. From Biale JB and Young RE (1982) Respiration and ripening in fruits – retrospect and prospect. In: Friend J and Rhodes MJC (eds) *Recent Advances in the Biochemistry of Fruits and Vegetables*, pp 1–39. London: Academic Press, with permission.

Table 2. Fruits with climacteric and nonclimacteric patterns of respiration

Climacteric	Nonclimacteric
Apple	Blueberry
Apricot	Cacao
Avocado	Cherry
Banana	Cucumber
Mango	Grape
Papaya	Grapefruit
Peach	Lemon
Pear	Lychee
Plum	Olive
Tomato	Orange
Watermelon	Pineapple
	Strawberry

walls and the cohesiveness of the cells. The latter property is conferred by the middle lamella, the intercellular structure that cements together the primary walls of contiguous cells and that is composed mainly of pectic material. Structurally, fruit cell walls are similar to primary cell walls of other plant tissues whose main components are polysaccharides, namely cellulose, hemicelluloses and pectins. Essential for the structural integrity of cell walls is the presence of calcium. Calcium ions are particularly important in cell-to-cell adhesion, the cementing effect being primarily attributed to the formation of a calcium–pectate complex in the middle lamella. The role of calcium in the maintenance of cell wall structure is particularly important in fruits in which the middle lamella gives a larger contribution as a cell wall component than in other plant tissues. The extensive softening that takes place during the ripening of many fruits is mainly attributed to changes occurring in cell walls, including the middle lamellae. ***See*** Cellulose; Hemicelluloses

In ripe fruit, cell walls and middle lamellae are partly dissolved as a result of the action of a number of enzymes that break down pectins and cellulose, of which polygalacturonase, pectinesterase and cellulase are the most extensively studied and whose activities are often correlated with the rate at which softening occurs during ripening. It is apparent from such studies that the major changes occurring in cell walls during ripening are associated with depolymerization and solubilization of the pectic substances due to the action of polygalacturonase and pectinesterase, respectively. The solubilization of pectin during the ripening of apples has been attributed to diminished levels of calcium ions in the cell walls and there is evidence that the softening of the apple fruit is associated with the transfer of divalent cations, particularly calcium, from the cell wall into storage compartments inside the cell. Postharvest calcium treatments that are carried out to prevent some physiological disorders, such as bitter-pit in apples, confer greater firmness to the fruits during storage.

The role of cellulase in tissue softening is not clear. Though it seems likely that hydrolysis of cellulose would weaken the cell wall structure, the experimental evidence is rather inconclusive as to its importance as a primary cause of fruit softening.

Effects on Sweetness

In fresh fruits the main compounds responsible for sweetness are soluble carbohydrates, i.e. sugars. It is apparent from the degree of sweetness of different ripe fruits that their sugar contents vary widely. Even within a particular species the sugar concentration in fruits may depend on the variety and on the environmental conditions and agricultural practices to which the mother plant has been subjected, prior to harvesting. As most climacteric fruits are harvested before ripening, considerable changes in sugar content may occur during storage. The average total sugar contents of most ripe fruits lie in the range 5–10% of fresh weight. Outside this range are, typically, lime and lemon with lower sugar contents (0·5–3·0%), and grape, cider apple and other fruits used for alcoholic fermentation purposes with higher contents (13–20%). ***See*** Carbohydrates, Classification and Properties

The main individual sugars responsible for sweetness in fruits are the monosaccharides glucose and fructose and the disaccharide sucrose. Further monosaccharides may also be present in fruits, but their contribution to sweetness is negligible. The relative proportions of glucose and fructose vary with species but in many fruits glucose levels exceed those of fructose. This does not necessarily mean a higher contribution of glucose for sweetness since the sweetening power of glucose is 2–3 times lower than that of fructose. Two important exceptions are apples and pears in which fructose may be present in concentrations up to three times those of glucose. In grapes, berries and oranges both sugars are often present in similar amounts. ***See*** Fructose; Sucrose, Properties and Determination

Sucrose is an important product of photosynthesis and the main form in which carbon is translocated from the leaves to other parts of the plant, including fruits. So, it is not surprising that in most fruits sucrose is the main respiratory substrate utilized for the provision of energy and intermediary metabolites for biosynthesis. The first step in the metabolism of sucrose, catalysed by β-fructofuranosidase (invertase), is the hydrolysis of the disaccharide to glucose and fructose. These facts explain the predominance of glucose, fructose and sucrose in most fruit tissues. The sucrose concentration varies from fruit to fruit but in most fruits the total hexose content exceeds that of sucrose, and in some fruits (e.g. cherry, grape and tomato) sucrose is almost absent.

A significant fraction of the sugar translocated to the growing fruit is metabolized and used in various biosynthetic processes, another fraction being stored after conversion into starch, the most common reserve carbohydrate in plants. Starch breakdown is one of the most important biochemical changes that occur during the ripening of many fruits. Some fruits may contain, when mature but unripe, large amounts of starch, the levels of which decrease dramatically during ripening. For instance, bananas may contain up to 20% starch at the mature green stage and less than 1% when fully ripened. Large reductions in starch content also occur during the ripening of other fruits, notably mango, apple and pear. The increased levels of sucrose, glucose and fructose present in such fruits when ripened result primarily from the enzymatic hydrolysis of starch. However, some fruits do not accumulate starch during their development, attaining their mature unripe state with little or no starch at all (e.g. melon, pineapple, plum and grape). In these fruits, carbohydrates are stored in the form of soluble sugars, their sweetness depending, largely, on the transport of sucrose from other parts of the plant, leaves being usually the main source. Consequently, the optimum sugar content of these fruits cannot be reached if they are harvested before ripening. ***See*** Starch, Structure, Properties and Determination

Effects on Aroma and Taste

Aroma and taste are considered to be the two major components of flavour, a rather subjective attribute to which all the other senses contribute in an integrated manner. Most of the changes in aroma and taste, together with changes in colour and texture, that are responsible for the commercial acceptability of fruits take place during the ripening process. ***See*** Flavour Compounds, Structures and Characteristics

In spite of the important advances in flavour chemistry made in the last decades, our understanding concerning the metabolism of the compounds involved and the way they interact in the perception of aroma and taste is very incomplete. Aroma, being the sensation associated with smelling, requires the presence of volatile relatively hydrophobic compounds. On the other hand, substances responsible for taste and other nonspecific saporous sensations are water-soluble and most nonvolatile, usually being present at higher concentrations than those responsible for aromas.

Despite the relatively low lipid contents of most fruits, lipid metabolism plays an important role in a number of ontogenic events, including ripening and senescence. Many of the most important aliphatic fruit volatiles are formed through an oxidative degradation of unsaturated fatty acids such as linoleic and linolenic acids. For instance, the action of lipoxygenase and a lyase on linolenic acid leads to the formation of 2-*trans*-hexenal or 2-*trans*-6-*cis*-nonadienal (depending on the site specificity of the enzymatic attack), these being important aroma constituents of tomatoes and cucumbers, re-

Table 3. Volatile esters associated with the aroma of individual fruits

Ester	Fruit
Benzyl benzoate	Cranberry
Ethyl butanoate	Orange
Ethyl 2-methylbutanoate	Melon
Ethyl 3-methylbutanoate	Blueberry
Isopentyl acetate	Banana
Methyl anthranilate	Concord grapes
Ethyl 2-methylbutanoate	Apple
Methyl (3-methylthio)propanoate	Pineapple
Methyl and ethyl derivatives of *trans*-2-*cis*-4-decadienoate	Bartlett pear

spectively. The aldehydes and ketones generated by lipoxygenase action can be converted further, by the action of dehydrogenases, to the corresponding alcohols, which usually have stronger aromas than the parent carbonyl compounds. For example, the dehydrogenation of 2-*trans*-6-*cis*-nonadienol gives 2-*trans*-6-*cis*-nonadienal, important in the aroma of cucumbers and melons.

The class of compounds that most frequently makes the major contribution to the pleasant fruity aromas developed during ripening is the esters of carboxylic acids. These are generated by esterification of the alcohols formed by reduction of the aldehydes derived either from oxidation of long-chain fatty acids or from amino acids. The mechanism of ester formation has not been satisfactorily explained but it is thought that esterification of the alcohols occurs with acyl-coenzyme A derivatives of fatty acids acting as acyl donors. While all the esters present may contribute to the characteristic aroma, certain individual esters have been associated with the aroma of specific fruits (Table 3). Some exceptions are known. For example, the aroma of peaches is attributed to the presence of lactones and that of raspberries to the ketone 1-(*p*-hydroxyphenyl)-3-butanone.

Volatile terpenoids are largely responsible for the characteristic aroma of citrus fruits. The monoterpenes citral and limonene exhibit distinct aromas of lemons and limes, respectively. Sesquiterpenes are also characteristic aroma compounds, such as β-sinensal in oranges and nootkatone in grapefruits.

Of the four basic taste sensations (sweet, sour, bitter and salt), the dominant changes that occur during fruit ripening involve an increase in sweetness and a decrease in sourness (mainly due to acidity). However, changes in bitterness (due to terpenoids) are important in citrus fruits as are changes in astringency (a taste sensation due to the association of phenolic substances, including tannins, with proteins in the saliva) that occur during the ripening of bananas, grapes and other fruits. The changes in the concentration of acids during fruit development and ripening differ with the type of fruit. The optimal acidity is rather high for citrus fruits, somewhat less for pome fruits and even less for tomatoes. Most fruit tissues accumulate excess acid in the preripening stage of development but, when ripe, each fruit has a range of acid:sugar ratio corresponding to optimum taste. In most fruits, citric and malic acids are the major contributors to the desired degree of acidity. *See* Acids, Properties and Determination; Phenolic Compounds

In postharvest fruits, as in other plant living tissues, organic acids are in a constant state of flux. They may arise or be used up through the operation of glycolysis, the citric acid cycle and, possibly, the glyoxylate cycle. Much of the loss of organic acids observed during ripening is attributed to their oxidation in respiratory metabolism as suggested by the increased respiratory quotient (RQ, the ratio of carbon dioxide evolution to oxygen absorption) during the climacteric. The RQ is approximately 1·0 when sugars are the respiratory substrates, increasing to about 1·3 when malate or citrate is the substrate, and further increasing to 1·6 when tartrate is respired. The rise in RQ during the climacteric is often accompanied by an increased activity of certain decarboxylation enzymes, such as malic enzyme and pyruvate decarboxylase. There is evidence that malic enzyme is synthesized *de novo* during the climacteric of pome fruits.

Changes in Colour

During the ripening of almost all fruits there occurs a disruption in the organization of chloroplasts and their reorganization into chromoplasts. As the photosynthetic apparatus is dismantled and the chlorophyll degraded, the colour of existing carotenoids is unmasked. This phenomenon may be responsible for the yellow to red colours of ripe fruits, as in tomatoes. Although carotenoids are normally synthesized in green tissues, it has been shown that additional β-carotene and lycopene is formed during ripening of tomatoes. Carotenoid synthesis during postharvest storage is affected by the surrounding atmosphere (decreased oxygen and increased carbon dioxide or nitrogen concentrations inhibit pigment formation in tomatoes), by light (it has been claimed that tomatoes synthesize more pigment in the dark) and by temperature (tomatoes, but not watermelons, synthesize less lycopene above 30°C). *See* Colours, Properties and Determination of Natural Pigments

The colour of many ripe fruits, including pomes, several types of berries, grapes, oranges, cherries, peaches, plums, bananas and figs, is due to the production of anthocyanins. These are conspicuous water-soluble plant pigments of phenolic character that accumulate in the vacuoles and are responsible for the pink, red, violet and blue colours of fruits, flowers and vegetables. The colours of anthocyanins are strongly affected by pH and can be intensified and stabilized by intermolecular associations with a number of metal ions, such as those of aluminium, iron, manganese and copper. The anthocyanins are typically located in the epidermal layers of the fruits but, like carotenoids, may also be present in the flesh. Anthocyanin synthesis during ripening is strongly stimulated by light. Apparently, this stimulatory effect has at least two mechanisms. By increasing the photosynthetic rate, light increases the availability of the carbohydrate needed to supply the energy and building blocks for anthocyanin biosynthesis. The second photoreaction involves the activation of phytochrome, a photoreceptor pigment that mediates several physiological processes in plant cells, including the synthesis of anthocyanins. Preharvest applications of some growth regulators, such as Alar (*N*-dimethylaminosuccinamic acid), can induce earlier formation of anthocyanins in fruits.

Postharvest Storage of Fruits

After picking, the ripening of fruits can be controlled by temperature adjustment or by changing the concentration of gases (usually carbon dioxide and oxygen) in the surrounding atmosphere. These two major methods of fruit storage are usually referred to as 'air refrigeration' and 'controlled-atmosphere storage', respectively, the latter being normally combined with some degree of refrigeration. Although effective in retarding ripening, both methods of environmental control impose stresses on fruits that may lead to different types of physiological disorders.

Low-temperature Storage

Lowering of the storage temperature is one of the commonest methods for extending the shelf life of fruits. With the exception of tropical and subtropical fruits, the lowest temperature that, in theory, still permits a normal metabolism is near the freezing point of a fruit, which in most cases lies between 0 and $-2°C$. The reduction of storage temperature decreases the rate of respiration and of metabolism in general by its direct effect on the rate of the enzyme-catalysed reactions. In climacteric fruits, low-temperature storage also delays the onset of ripening by retarding the autocatalytic production of ethylene and so extending the preclimacteric phase of ripening. Lowering the temperature also reduces the degree of response to exogenous ethylene.

Controlled-atmosphere Storage

A decreased rate of respiration and of other metabolic reactions, including ethylene synthesis, can also be achieved by altering the composition of the atmosphere in which the fruits are stored, usually by increasing the carbon dioxide and decreasing the oxygen concentrations. In controlled-atmosphere storage the proportion of the gases is carefully controlled, usually within $\pm 1\%$ of the desired value. When the control of the gas concentrations is less accurate, the practice is named modified-atmosphere storage. For example, in a confined atmosphere the carbon dioxide : oxygen concentration ratio may be increased simply as a consequence of the normal respiratory process or by sublimation of solid carbon dioxide (dry ice). Another type of controlled-atmosphere storage is hypobaric storage in which the fruits are kept in partial vacuum (0·1–0·3 atm). The decreased partial pressure of oxygen and the reduction of internal ethylene levels due to the increased diffusibility of the gases formed within the tissues retard the ripening of the fruits. ***See*** Controlled Atmosphere Storage, Applications for Bulk Storage of Foodstuffs

The efficacy of controlled-atmosphere techniques in extending the storage life has been well demonstrated for a number of fruits. Nevertheless, their commercial use has been somewhat limited. Controlled-atmosphere techniques are expensive and some of the more profitable fruits either do not respond favourably to atmosphere regulation or have such a rapid market turnover that the need for an improved storage procedure is curtailed. Nevertheless, conventional controlled-atmosphere storage is currently used for apples and pears and the practice of modified-atmosphere consumer sealed packs is becoming more popular due to the discovery of new packaging materials having a selective gas permeability that allows the retention of adequate gas concentrations when equilibrium is reached. Hypobaric storage is successfully employed in the storage of cut flowers but, to our knowledge, it has not been utilized for the commercial storage of fruits. ***See*** Chilled Storage, Use of Modified Atmosphere Packaging

Physiological Disorders

Different fruit species or even different cultivars of the same species may exhibit different tolerances to low temperatures and to modified atmospheres. Exposure of a particular fruit either to a temperature or to an atmosphere with a carbon dioxide : oxygen concentration ratio outside the recommended range usually causes injuries that lead to a decrease in both quality and storage life. The extent of such injuries depends strongly on the duration of exposure.

The mechanism by which physiological disorders are

induced is not well understood. Chilling injuries are generally attributed to a deregulation in metabolism leading to the underproduction of some essential metabolites and the overproduction of substances whose accumulation has toxic effects. For example, accumulation of ethanol, acetaldehyde and oxaloacetic acid has been associated with low-temperature breakdown of certain fruits, such as apples. Visible effects of cold injury include tissue necrosis, surface pitting, internal browning and failure to ripen. Some fruits (e.g. pears) normally exhibit a low susceptibility to cold injury, but for most fruits there is a critical temperature below which the fruits may be damaged, particularly if submitted to prolonged exposure. As recovery may often take place after short exposures to potentially damaging temperatures, it is believed that a minimum time of continuous exposure is required for injury to occur. In fact, interrupted exposures are often effective in diminishing damage. The lower temperature limit tolerated by fruits varies with their origin. Many fruits from temperate regions (e.g. apples) may tolerate temperatures in the range 0–4°C, whilst the banana becomes susceptible at temperatures below 12°C. Subtropical fruits, such as pineapples, avocados and citrus fruits, show intermediary limits, often below 8°C.

Identical metabolic imbalances are induced by extremes in atmospheric composition. Toxic levels of ethanol, acetaldehyde and succinic acid have been shown to accumulate in certain fruits prior to injury symptoms. Some fruits, such as citrus and pome fruits, are rather susceptible to high carbon dioxide levels and the use of low carbon dioxide concentrations (1–5%) are recommended. For more tolerant fruits, such as cherries, grapes, plums and strawberries, higher concentrations (15–30%) are usually advised. During storage, a minimum concentration of oxygen is required to support respiration. Below this level, anaerobic metabolism takes over and ethanol is produced. Although many fruits appear to tolerate oxygen concentrations below 5%, the exact limit for normal ripening is variable, depending on carbon dioxide concentrations, temperature and duration of exposure.

Ionizing Radiation

Alhough remaining an isolated rather than an extensive method of food preservation, the application of ionizing radiation is effective in delaying the ripening of fruits. Depending on the fruit, doses of 0·3–3·0 kGy (kilogray) can prevent, temporarily or permanently, the onset of fruit ripening. However, the minimum dose required to achieve this effect often exceeds the maximum tolerable dose. Even when successful in prolonging storage life, fruit irradiation is frequently hampered by a rapid softening of the tissues, possibly due to a direct or indirect activation of pectic enzymes. Actually, the mechanisms responsible for the observed effects of ionizing radiations on fruits are far from clear, but this is not surprising if one considers how numerous are the doubts over the biochemical and physiological control of fruit ripening even under normal conditions. ***See*** Irradiation of Foods, Applications

Bibliography

Frenkel C (1984) Factors regulating the respiratory upsurge in developing storage tissues. In: Palmer JM (ed.) *The Physiology and Biochemistry of Plant Respiration*, pp 33–46. Cambridge: Cambridge University Press.

Friend J and Rhodes MJC (eds) (1981) *Recent Advances in the Biochemistry of Fruits and Vegetables*. London: Academic Press.

Hulme AC (ed.) (1970) *The Biochemistry of Fruits and their Products*, vol. 1. London: Academic Press.

Lieberman M (ed.) (1983) *Post-Harvest Physiology and Crop Preservation*. New York: Plenum Press.

Salunkhe DK and Desai BB (1984) *Postharvest Biotechnology of Fruits*, vol. 1. Boca Raton: CRC Press.

Solomos T (1988) Respiration in senescing plant organs: its nature, regulation and physiological significance. In: Noodin LD and Leopold AC (eds) *Senescence and Aging in Plants*, pp 111–145. San Diego: Academic Press.

Tucker GA and Grierson D (1987) Fruit ripening. In: Davies DD (ed.) *The Biochemistry of Plants*, vol. 12, pp 265–318. San Diego: Academic Press.

A.R.N. Teixeira and R.M.B. Ferreira
Technical University of Lisbon, Lisbon, Portugal

ROLLER MILLING OPERATIONS

See Flour

ROOT VEGETABLES

See Cassava, Potatoes and Related Crops, Vegetables of Temperate Climates and Vegetables of Tropical Climates

ROUGHAGE

See Dietary Fibre

RUM

Rum is a spirit produced from various raw materials (juice, concentrate, molasses which are derived from sugar cane *Saccharum officinarum* L). It is an alcoholic beverage composed mainly of water and ethanol, but containing, in solution, more than 400 compounds, in very small quantities, giving the aroma and taste which differentiate rum from other spirits.

The traditional area of production was centred on the West Indies but, more recently, production has become established in other countries with a climate suitable for the cultivation of sugar cane, e.g. Australia, South Africa and the Philippines. Indeed, since the eighteenth century, rum production has been seen as a natural way of utilizing the by-products of sugar production.

Rum is manufactured by the distillation of mashes fermented by alcoholigenous yeasts. The raw distillate has a pungent taste, and a coarse, often unpleasant bouquet. Before sale, it has to be submitted to various operations to obtain the retail product:

- Initial maturation to develop and enhance the latent qualities.
- Blending.
- Dilution to the commercial alcoholic strength.
- Coloration, usually with caramel, or discoloration after maturation.
- Clarification if needed.

The distillation of fermented mashes derived from sugar cane is also carried out for ethanol production. The aim is pure alcohol without congeneric compounds, whereas in rum production the objective is ethanol, together with desirable congeneric compounds and some of the water.

Raw Materials

The Sugar Cane

It is generally agreed that Oceania, where it enjoys extensive use as a food, is the most plausible region of origin of sugar cane. The ancient status of the crop led Linnaeus to name all sugar canes that came into his possession as *Saccharum officinarum*, prior to 1753, but this binominal now refers only to the cultivated types. The production of sugar for commercial purposes began on Santo Domingo Island in 1509, and the crop has always been the most important source of sugar in the world.

Sugar cane is a member of the Gramineae, and sugars are stored in the stalk, which achieves a height of about 4 m and a diameter of 4–6 cm. It is formed of a succession of nodes and internodes, and the upper part of the stem is low in sugar. It grows easily in a tropical or subtropical climate, and while generally an annual, it is biennial in Hawaii. ***See*** Sugar, Sugar Cane

The main criterion for varietal selection of sugar cane for sugar manufacture is that the plant should show high juice purity, i.e. a minimum of nonsugar components (see Table 1), but, for alcoholic fermentations, the presence of minerals, vitamins and other nonsugar substances is important for growth and metabolism of the yeast.

The Juice

At the end of its growth cycle, the sugar cane has the highest content of sugar, while the nonsugar compo-

Table 1. Average percentages of components of the raw materials used in the distillery

Component	Cane	Juice	Syrup	HTM	Molasses
Water	69·0	78·0	29·0	16·0	20·0
Sugar	16·0	20·0	66·0	77·0	62·0
Nonsugar	3·0	2·0	5·0	7·0	18·0
Fibre	12·0	—	—	—	—

HTM, high test molasses.

nents are, in relation to dry matter, at their lowest. At this point, the stalks are cut, headed-down, and conveyed to the mills. Here the stalks are crushed and processed to extract as much sucrose as possible, and extraction rates of 98% are possible. The juice issuing from this stage is very dilute, two or three times more so than the original sap in the stems, and it is this juice which is used in the distillery or for processing to sugar (see Fig. 1). However, it is only in the French West Indies (Guadeloupe, Martinique) and Haiti that this basic sugar cane juice is used directly to manufacture rum on an industrial scale; it is known as 'agricultural rum' (rhum agricole) as it is derived from an agricultural material. The Martinique was also, at one time the source of a type of rum made from heated and clarified cane juice. The spirit produced in this way had some special, and much appreciated, features, in that the harmonious blend of the different components gave rise to a complex and stable aroma of great amplitude and penetration.

The Molasses

A by-product of the sugar industry, molasses is the most important raw material for rum manufacture, and is derived as follows. After extraction, the raw juice is purified to eliminate as many of the nonsugar elements as possible, concentrated by evaporation and, finally, centrifuged to extract the crystallized sucrose from the so-called 'mass-cuite'. After removal of the sucrose, there remains a brown, viscous liquid – molasses, and this can be used to produce so-called 'industrial rum' (rhum industriel). Its composition is complex and variable, and its quality depends on both the phytosanitary state of the cane and the efficiency of the factory process.

However, there is also on the market high test molasses (HTM), which is a syrup derived from sugar cane juice that has been clarified, partially inverted to avoid crystallization of the sucrose, and evaporated to 85°Brix. It has a high content of fermentable sugars (75–79%) and a low ash content (2·0–2·2%). This product is used for ethanol production in the USA, Japan and the UK, and Porto Rico has recently begun to produce rum from HTM.

Other Substrates

Cane sugar syrup, which is a defecated and evaporated juice (not inverted and desalted like HTM), was at one time used in the West Indies for rum production, but is now only used on a small scale in Haitian countries.

Some use is also made of the 'slop', which is waste stream from the distillation of the fermented liquor. Its utilization to dilute the sugar-rich raw materials gives rise to a rum which is coarse and highly aromatic. 'Aged' slop is composed of nitrogen compounds, minerals and organic acids, and these materials enhance both the degree of esterification and the growth of *Schizosaccharomyces* yeasts in the fermentation medium. The slop from molasses-based fermentations is richer in organic compounds than that from cane juice processes, and is widely used for the mashing operations.

Water

The mineral composition and bacteriological quality of the water used for dilution during the mashing operation can affect the fermentation stage.

Mashing Operations

Mashing operations consist of mixing proportions of sugar-rich raw materials with water, or sometimes slop, to a predetermined sugar level which permits fermentation without an excessive rise in temperature of the mash (<34°C). The pH of the medium is generally adjusted with sulphuric acid to around 4·5 to limit bacterial growth and activity. Nitrogen and phosphorus salts, needed by the yeasts for fermentation, are also added to overcome the deficiency of the sugar cane juice.

Cane Juice

The mashes made up from cane juice generally have a sugar content of around 10%. At the end of fermentation, ethanol concentrations range from 3·5 to 6·0% (v/v). Fluoride salts are sometimes added for additional antibacterial action. The normal fermentation time is 20–30 h.

Molasses

In traditional rum manufacture, the molasses is diluted with water to give a sugar concentration in the range of 9–25%, giving rise to ethanol concentrations in the fermented liquor of 5–12% (v/v). For 'white' rum, or rum with 'light' features, the molasses is sometimes pasteurized and clarified by centrifugation to remove any colloidal matter. These processes enhance the

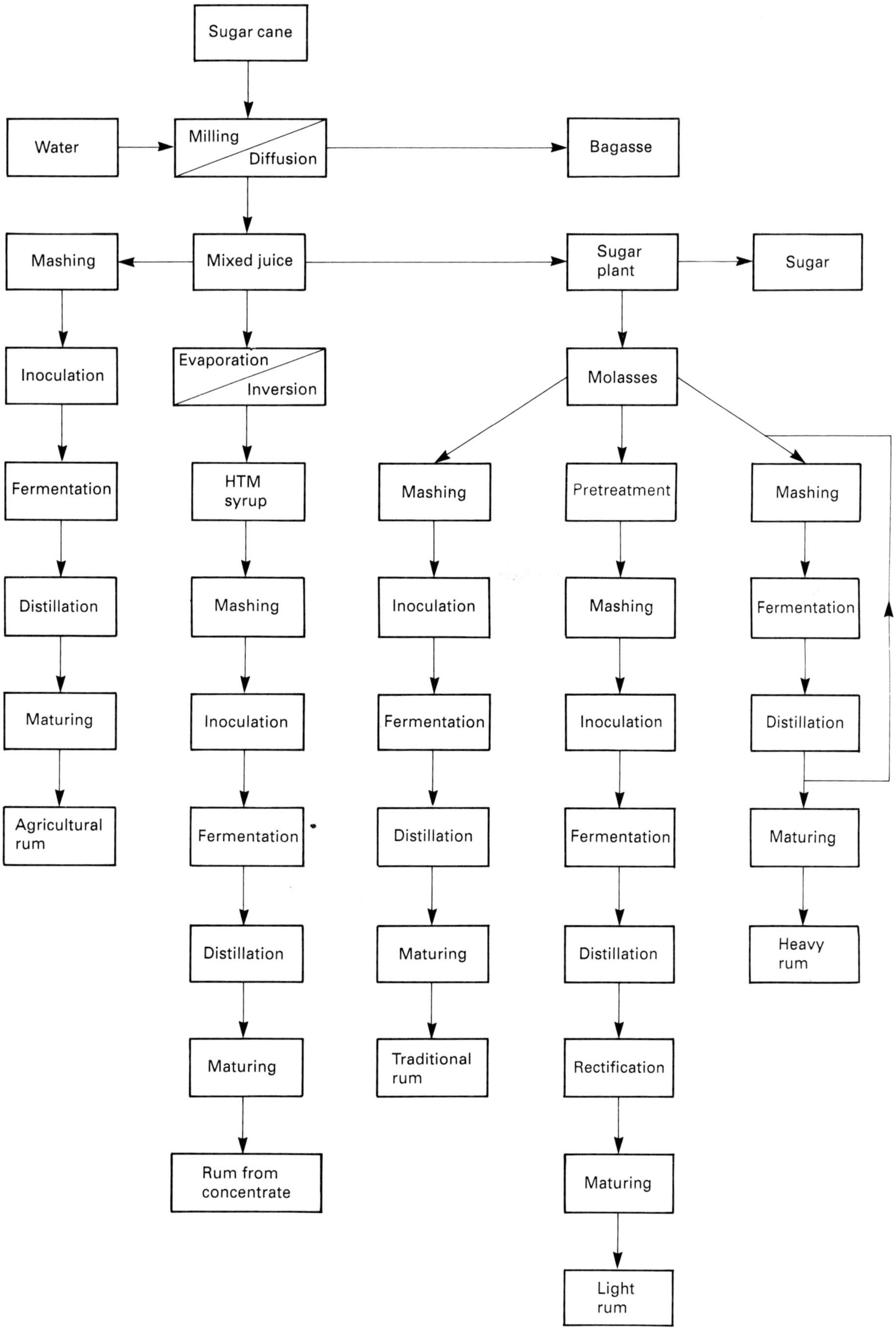

Fig. 1 Flow diagram of the manufacture of different types of rum.

fermentability of the molasses and, in addition, the fermented liquor is less likely to clog the distillation columns. In particular, the mashes are of good microbiological quality, well supplied with nutrients and, provided that the correct amount of antiseptic materials are added, e.g. sulphuric acid or fluoride salts, fast fermentations occur with little growth of bacterial spores. The yeast inoculation levels can remain consistent. The process for light rum production was described by Arroyo (1945).

Molasses is also the main raw material for the production of 'heavy' rum but, in this case, the dilution stage involves slop rather than water. Acidification of the mash and the addition of nutrients may not be needed.

Problems can arise from physicochemical factors, e.g. excessive temperature, or from overdevelopment of the bacterial flora as a result of low acidity of the mash.

Rum Fermentation

Microbiology

Mashes that are not inoculated with yeast are subject to spontaneous alcoholic fermentation by yeasts originating in the raw materials or on the surfaces of plant and equipment. Simultaneously with, and/or subsequent to, the alcoholic fermentation, there are sundry secondary fermentations (acetic, butyric, lactic) owing to bacteria in the raw materials. Such secondary fermentations were common until the end of the last century.

Nowadays, secondary fermentations are less important. In light rum production, pasteurization of the raw materials, inoculation with a selected yeast and the utilization of water of good microbiological quality means that bacterial problems have been largely eliminated. However, in the production of heavy rum, bacteria play an important part in the progress of the fermentation and in the formation of aroma compounds. Between these extremes, there are many fermentations in which the bacterial flora is present and active, but its precise role is not clear.

Yeasts in Rum Fermentation

The original yeast associated with rum fermentations belongs to the genus, *Schizosaccharomyces*. This genus is dominant in the media used for the production of heavy rum, and in the mash for Haitian rum based on cane syrup. The substrates for both these fermentations are peculiar in that the raw materials are often diluted with aged slop and, as a consequence, the mashes possess a high acidity, coming from bacterial fermentations that occurred while the slop was stored. In addition, the osmotic pressure of the medium is higher in mashes containing slop, so that it is the osmophilic yeasts, such as *Schizosaccharomyces*, that prevail. As a result, the fermentation time for producing the base liquor for a heavy rum can be quite lengthy (10–12 days), the more so as indigenous species of *Saccharomyces* grow only slowly at high osmotic pressures.

During the early days of rum production, all mashes tended to be made from molasses and slop, but, as the demand for light rum increased, so the need to improve yields and productivity arose. As a result, inoculation of the medium with yeast became normal, yeast selection and pure yeast fermentations, as well as the production of starter cultures, became possible. In the latter case, yeast propagation takes place in a mash with a low sugar content (5–8%) and well supplied with air. Under good operating conditions, the volume of starter culture will be equal to one third of the volume of the main fermentation. Yeasts specially selected for rum production are not available on the commercial market, although some big companies claim that they possess special strains. More often, any active dry yeast (baker's yeast or an oenologic yeast) available on the market is used. ***See*** Starter Cultures

Bacterial Flora

Cane juice issuing from stalks in a good phytosanitary state has a total bacterial count of around 1×10^6 colony-forming units (cfu) per ml, consisting mainly of the genera, *Lactobacillus, Clostridium, Streptococcus,* and *Corynebacterium*. Other common members of the flora, namely *Enterobacter, Erwinia,* and *Escherichia*, have little influence on the alcoholic fermentation.

The sanitary state of the stalks is diminished by the sugar cane borer (*Diatraea saccharalis*), while stalks that have been standing in the field for 2 days in bad conditions, or have been harvested in rainy weather, also show a marked deterioration in bacterial quality. *Lactobacillus* and *Corynebacterium*, in particular, increase under these latter conditions, and the total flora in the juice can reach 1×10^8 cfu ml^{-1}. Consequently, during the fermentation, there is an excessive build-up of volatile acidity and the risk of off-flavours. ***See*** Spoilage, Bacterial Spoilage

Molasses contains mainly spore-forming and thermoresistant bacteria (*Bacillus, Clostridium*), along with a superficial flora of lactobacillae. The total bacterial count is usually around 1×10^6 cfu g^{-1}. Under bad storage conditions, this latter count can reach 1×10^9 cfu g^{-1}, and this excessive flora can give rise to increases in volatile acidity and cause sluggish fermentations. Depending upon the precise conditions, the duration of molasses fermentations ranges from 15 h for light rum to 30 h for traditional rum.

Contamination can also arise from the water used for

dilution, e.g. water from a river, pond or underground source, and such water may contain facultative anaerobes, pathogenic organisms (coliforms or faecal streptococci), clostridia and sulphate-reducing bacteria. During rainy weather, surface water may contain up to 1×10^3 cfu ml^{-1} of sulphate-reducing or lactate-oxidizing bacteria, while underground waters tend to be of poor microbial quality during drought or semidrought conditions.

Distillation

Distillation is an important stage in the production of a spirit, and its purpose is the separation from the fermented mash of ethanol, together with those desirable congeneric compounds that give the aroma and taste typical of rum. There are two systems available.

The *pot still* is generally made of copper, and is operated batchwise with a series of two or more distillations. It is the simplest form of still, and is now obsolete at commercial level, except for the production of small volumes of very-high-quality product. A variant of this system is the *batch still*, provided with rectifying and condensing devices, which is employed to give rise directly to potable spirits. It is still used for heavy rum production in Jamaica.

Continuous distillation replaced the pot still during the second half of the nineteenth century, and is now widely used for large-scale production. The continuous system consists of one to three columns which contain perforated plates. Vapour condenses on the plates and a volume of condensate, depending on the operating conditions, returns to the lower level. The choice of the number of columns for a distillation system depends on the alcoholic strength desired in the distillate and the level of nonalcoholic components that is being sought. Light rums are produced in multiple-column devices, but traditional rums are distilled without rectification, i.e. there is no removal of any of the volatile fractions. Some sections of the still are made of copper, because copper catalyses certain chemical reactions that are important for product quality; for example, the volatility of sulphur compounds is lowered.

Maturation

Rum maturation is an operation designed to give a product of maximum consumer acceptability. It is not a process of change or transformation, but rather seeks to develop and enhance the latent qualities already existing in the raw distillate and inherited from previous stages.

Diverse changes take place during maturation, some of which are time-dependent, such as oxidation, and some of which are not, e.g. blending, coloration and clarification.

In some cases, distillates are consumed after merely diluting to normal trade strength (40–50% v/v), and this happens with locally consumed, agricultural rums in the French West Indies and with the Haitian 'Clairin'. If matured for a few weeks in stainless steel tanks and stirred to remove some of the volatile compounds, the raw distillates emerge as white rums, but if stored in wooden vessels the distillate tends to become coloured.

Ageing

For products of better quality, the distillate will be stored in oak casks of 150–450 l for a period of 3 years. During storage, physicochemical changes take place, including evaporation, changes in the alcoholic strength, and extraction of tannins from the wood, as well as various oxidation, polymerization and esterification reactions. All these transformations contribute to the maturation of the raw distillate.

Alternatively, maturation may be carried out over a period of months in large casks of several thousand litres capacity. The distillate then acquires a light-straw colour, and the organoleptic properties are half-way between white rum and aged rum. ***See*** Barrels

Finishing

Coloration

Rum is delivered colourless (white rum, Clairin, Ron Carta Blanca) or lightly tinted. Dark rums obtain their colour from compounds extracted from the wood of storage casks or from added caramel. ***See*** Colours, Properties and Determination of Natural Pigments

Dilution

Raw and matured distillates generally possess an alcoholic strength that requires dilution before bottling. The water used for this purpose must be as pure as possible and free from minerals and taints or flavours.

Blending

This operation is carried out so that any given product or brand name will have the same quality year after year. The process involves mixing, as appropriate, several manufactured lots in order to obtain one batch with the organoleptic properties requested by the consumers.

Clarification

Clarification is carried out by filtration through a range of materials, sometimes after treatment with charcoal, to give a bright product. White rums from aged distillates have to be treated in the same way. ***See*** Filtration of Liquids

Types of Product

In commercial circles, rum is usually classified according to the land of manufacture. However, the final product results from the interaction of the raw material, the biotechnical processes (fermentation, distillation) and the maturation procedure, and hence rum can be classified, taking these points into consideration, as follows:

1. Agricultural rum based on cane juice is produced on a cottage-industry basis, independent of the sugar manufacturing process. The product is usually consumed locally, although some small plants have the capacity to produce for export. Agricultural rums can be white, coloured or aged with 'light' features, and their organoleptic properties depend a great deal on the quality of the original cane juice.
2. Traditional rum, based on molasses, is produced for export in the West Indies.
3. Heavy rum is again based on molasses but, in this case, the dilution of the mash is achieved with slop (often aged). The product is not consumed directly but its aromatic strength makes it ideal as a cooking ingredient, or as an improver of blended rums. It is the typical product of Jamaica.
4. Light rum covers a series of molasses-based products with low contents of aromatic compounds. It is excellent as a base for long drinks or cocktails.

Rums manufactured from cane syrups were, at one time, appreciated for their bouquet, which is close to molasses rums, but more uniform, suave and delicate. The product has now disappeared from the trade. However, the product from HTM comes quite close in quality to the product from cane syrup.

Variants of the above types can be obtained by altering the maturation conditions (see Fig. 1), e.g. white, straw-coloured, aged, amber-coloured, or coloured.

Bibliography

Arroyo R (1945) *Studies on Rum. Research Bulletin No. 5.* University of Puerto Rico.

Fahrasmane L, Parfait A, Jouret C and Galzy P (1985) Production of higher alcohols and short chain fatty acids by different yeasts used in rum fermentations. *Journal of Food Science* 50: 1427–1436.

Guillaume J (1946) *Le Rhum, sa Fabrication et sa Chimie.* Annecy: Depollier and Cie.

Kervegant D (1946) *Rhums et Eaux-de-Vie de Canne.* Vannes: Editions du Golf.

Louis Fahrasmane
INRA, Point-à-Pitre, France

RYE

Rye is a cereal crop, the grain of which is used for food and feed similar to wheat and barley. Like all cereals, rye is a member of the grass family, *Gramineae*, and the genus *Secale*. Its species name is *S. cereale*. The species comprises many genotypes (varieties or cultivars). In terms of chromosomal composition, rye is a diploid ($2n = 14$) and contains seven pairs of chromosomes. Unlike wheat, rye tends to cross-pollinate extensively, pure varieties of rye are therefore difficult to maintain.

The centre of origin of rye is considered to be southwestern Asia. From there it moved gradually to eastern and northern Europe, initially as a weed in wheat and barley. It was domesticated much later than wheat. Rye was brought to North America in the sixteenth and the seventeenth centuries and later introduced into South America, South Africa and Australia.

Global Distribution, Varieties and Commercial Importance

Rye is typically a cool-temperature-zone crop but it can also grow in the semiarid regions and at high altitudes. It is more productive than wheat on acidic soils of low fertility. On good-quality soils, rye produces higher yields than wheat.

Most of the cultivated rye is grown as a winter crop, sown in early autumn and harvested in early summer. Rye has excellent winter-hardiness and can be grown at latitudes that are too cold for other cereals. Spring rye is grown in regions where the winters are too severe for winter rye.

The area of cultivated land devoted to rye (Fig. 1) decreased sharply during the 1960s and the 1970s but has stabilized in the 1980s at approximately 16×10^6 ha. Yields vary widely from region to region depending on soil conditions, climate and agronomic practices. Yields have generally increased during the past 25 years due to improved varieties and introduction of more effective agronomic practices. Accordingly, world production has not declined as much as the area.

In terms of world production, rye ranks last among the eight cereals (Table 1), the position it has occupied during the last three decades. The USSR leads in rye production, while Poland and Germany have main-

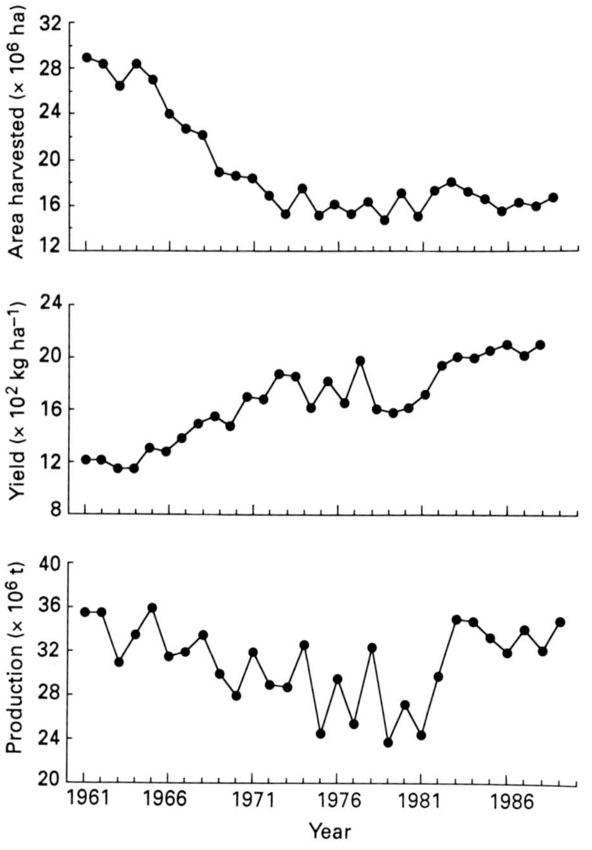

Fig. 1 World area, yield and production of rye.

Table 1. World production of cereal grains

Cereal	Production ($\times 10^3$ t)
Wheat	537 062
Rice, rough	504 367
Maize	460 238
Barley	168 734
Sorghum	55 836
Millet[a]	43 000
Oats	41 882
Rye	37 350

Source: Statistical Handbook 90, Canada Grains Council.
[a] Millet data from Food and Agriculture Organization (FAO) statistics.

tained second and third positions (Table 2). In Poland, some of the rye area is being used for the production of triticale, a new synthetic cereal species developed by the hybridization of wheat and rye. Rye is an important crop in the Scandinavian countries, especially Denmark, Sweden and Finland. ***See*** Cereals, Contribution to the Diet

Table 2. World rye production in 1989

Country	Area ($\times 10^3$ ha)	Yield (kg ha^{-1})	Production ($\times 10^3$ t)
USSR	10 598	1774	18 800
Poland	2275	2733	6216
Germany	1080	3611	3900
Canada	364	2294	835
Czechoslovakia	175	4054	708
Denmark	101	4802	485
Austria	91	4188	381
USA	194	1767	343
Spain	227	1485	337
Sweden	67	4716	316
France	74	3523	262
Hungary	97	2691	261
Finland	69	2945	202
Turkey	165	1109	183
All others	1078	1543	1664
Total	16 655	—	34 893

Source: FAO statistics.

Most of the rye grain is consumed domestically but considerable amounts are traded in the world grain markets. Canada and Germany are the major exporters, and the biggest importer is Japan, where the grain is used as animal feed.

Plant and Seed Morphology

The rye plant has a slender, tough, fibrous stem (straw) and elongated leaves. Plant heights vary widely from about 30 cm to over 2 m. The spike (inflorescence) is long and slender with stiff beards. Rye grain (caryopsis) is arranged in pairs alternately on a zig-zag shaped rachis. The grain is covered with a glume which is normally awned (bearded). Like wheat, rye is free-threshing; the grain separates freely from the glume (chaff) during threshing. ***See*** Wheat, Grain Structure of Wheat and Wheat-based Products

Rye grain is more slender and elongated than wheat varying in length from 4·5 to 10 mm and in width from 1·5 to 3·5 mm. The kernels are normally of greyish-yellow colour but the colour can vary widely depending on variety, region of production, and harvesting conditions. A crease extends the full length of the ventral side of the kernel. The surface of the kernels usually has a shrivelled appearance. Kernel weights are about 20 mg.

The rye kernel comprises three distinct morphological parts (Fig. 2): the starchy endosperm (86·5%); the bran, comprising the pericarp and the testa (10%); and the germ which includes the embryo and the scutellum (3·5%). In milling rye into high-quality bread flour, the bran and germ are separated from the endosperm which is then ground into flour.

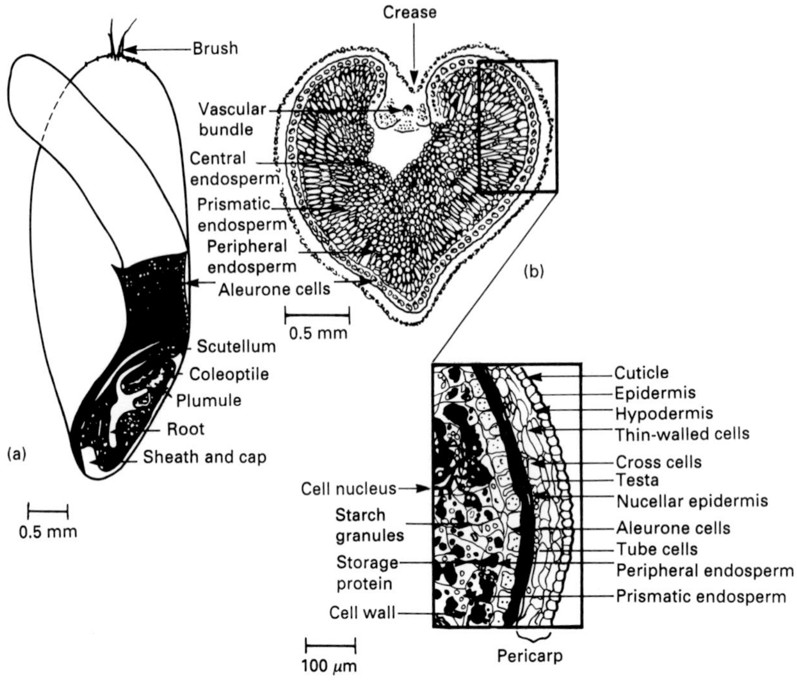

Fig. 2 Diagrammatic view of rye in longitudinal section (a) and in transverse midsection (b). From: Simmonds DH and Campbell WP (1976) Morphology and chemistry of the rye grain. In: Bushuk W (ed.) *Rye: Production, Chemistry, and Technology*, pp 63–110. St Paul, Minnesota: American Association of Cereal Chemists.

Table 3. Proximate composition of rye compared with other cereal grains[a]

				Barley			Oats		Rice	
	Rye (%)	Triticale (%)	Wheat (%)	Whole grain (%)	Kernel only (%)	Corn (%)	Whole grain (%)	Kernel only (%)	Whole grain (%)	Kernel only (%)
Protein	13·4	14·8	14·3	13·1	14·5	10·4	13·0	17·0	8·2	9·4
Ether extract	1·8	1·5	1·9	2·1	2·1	4·5	5·5	7·7	2·2	1·8
Crude fibre	2·6	3·1	2·9	6·0	2·1	2·4	11·8	1·6	10·1	0·9
Ash	2·1	2·0	2·0	3·1	2·3	1·5	3·7	2·0	5·7	1·1
Nitrogen-free extract	80·1	78·6	78·9	75·7	79·0	81·2	66·0	71·6	73·8	86·8

[a] Simmonds DH and Campbell WP (1976) Morphology and chemistry of the rye grain. In: Bushuk W (ed.) *Rye: Production, Chemistry, and Technology*, pp 63–110. St Paul, Minnesota: American Association of Cereal Chemists.

Chemical Composition and Nutritional Properties

The chemical composition of rye (Table 3) is typical of all cereals with carbohydrate (nitrogen-free extract) the major constitutent forming about 80% of the total kernel. Starch is the major carbohydrate component. Rye contains considerably more hemicellulose (pentosans) than wheat, 6–9% compared with about 2% for wheat. The hemicellulose contributes significantly to the functional properties of rye flour in breadmaking, where it interferes with the aggregation of flour proteins into gluten required for the viscoelastic properties of doughs. ***See*** Carbohydrates, Classification and Properties

Rye has some advantages over wheat when used for human nutrition because of its higher lysine content. Tryptophan is the first limiting amino acid in rye. Because rye flours are generally milled to higher extraction, they normally have a higher dietary fibre content than wheat flours. ***See*** Dietary Fibre, Properties and Sources; Protein, Quality

The mineral and vitamin content and composition of rye are similar to those of other cereals (Tables 4 and 5). Rye is considered a good source of thiamin, nicotinic

Table 4. Mineral composition of rye and other cereal grains (mg per 100 g, dry weight basis)[a]

			Barley			Oats		Rice	
	Rye	Wheat	Whole grain	Kernel only	Corn	Whole grain	Kernel only	Whole grain	Kernel only
Phosphorus	380	410	470	400	310	340	400	285	290
Potassium	520	580	630	600	330	460	380	340	120
Calcium	70	60	90	80	30	95	66	68	67
Magnesium	130	180	140	130	140	140	120	90	47
Iron	9	6	6	—	2	7	4	—	6
Copper	0·9	0·8	0·9	—	0·2	4	5	0·3	0·4
Manganese	7·5	5·5	1·8	—	0·6	5	4	6	2

[a] Simmonds DH and Campbell WP (1976) Morphology and chemistry of the rye grain. In: Bushuk W (ed.) *Rye: Production, Chemistry, and Technology*, pp 63–110. St Paul, Minnesota: American Association of Cereal Chemists.

Table 5. Composition of B group vitamins and carotene in rye and other cereal grains (mg per 100 g, dry weight basis)[a]

	Rye	Wheat	Barley (whole)	Corn	Oats (whole)	Rice (brown)
Thiamin	0·44	0·55	0·57	0·44	0·70	0·33
Riboflavin	0·18	0·13	0·22	0·13	0·18	0·09
Nicotinic acid	1·5	6·4	6·4	2·6	1·8	4·9
Pantothenic acid	0·77	1·36	0·73	0·70	1·4	1·2
Pyridoxine	0·33	0·53	0·33	0·57	0·13	0·79
Carotene	0	0	0·04	0·40	0	0·013

[a] Simmonds DH and Campbell WP (1976) Morphology and chemistry of the rye grain. In: Bushuk W (ed.) *Rye: Production, Chemistry, and Technology*, pp 63–110. St Paul, Minnesota: American Association of Cereal Chemists.

acid, riboflavin, pyridoxine, pantothenic acid, and tocopherol. Since these constitutents are mainly present in the germ and the aleurone layer of the endosperm, they are mostly removed during milling. ***See*** individual minerals and vitamins

Rye contains some constituents which have antinutritional properties, especially in animal nutrition where the total grain is used. Among these constituents are the alkyl resorcinols; the soluble pentosans which interfere with digestion of rye-based diets in monogastric animals; phytic acid which binds mineral elements such as calcium and zinc; and trypsin inhibitor. Rye samples often contain ergot bodies, and these contain the alkaloid ergotamine which, if consumed, can cause abortions in animals. The antinutritional constituents are of little concern in human nutrition because they are either removed during milling or destroyed during baking. ***See*** Cereals, Dietary Importance

Grading, Handling and Storage

Official grades have been established for rye in all of the major rye-producing countries. By way of example, grade standards for Germany are given in Table 6.

Table 6. German standards for milling rye

Moisture content	Max. 16%
Broken kernels	Max. 2%
Grain besatz (shrunken kernels, other grains, insect-damaged kernels)	Max. 1·5%
Sprouted kernels	Max. 1%
Black besatz (wheat seed, ergot, unsound grain, chaff, impurities)	Max. 0·5%
Hectolitre weight	Min. 71 kg

In general, the grades are based on the physical characteristics of the grain itself and on the presence of foreign materials. In some countries, α-amylase activity, an index of sprouting damage, is used to subclassify food grades of rye. In the Canadian system, this aspect of rye quality is taken into account through the specification of the maximum percentage of sprouted kernels that is permitted in the higher grades. Official moisture limits for the 'straight' grades vary among countries; in Canada this limit is 14%. Grain with moisture content between 14·1% and 17% and ⩾17·1% is graded tough and damp, respectively.

Table 7. German types of rye flour

Type of flour	Ash content (dry weight basis) (min. %)	(max. %)
610	0·58	0·65
815	0·79	0·87
997	0·95	1·07
1150	1·10	1·25
1370	1·30	1·45
1590	1·53	1·63
1740	1·64	1·84
1800 (meal)	1·65	2·00

Rye is usually handled in bulk like other major cereals. If necessary, the grain can be artificially dried before it is placed into storage if its moisture is above the safe limit. Care must be exercised to avoid damage to the functional or nutritional properties of the rye constituents by excessive drying temperature.

Rye grain of low moisture and free of live insects can be stored safely under cool, dry conditions over long periods. It is important to ensure that during storage the grain is not infested by microorganisms, insects, rodents or birds. *See* Cereals, Bulk Storage of Grain

Processing and Food Uses

On a global basis, more than half of the rye grain produced is used for animal feed. For food uses, the grain is usually milled into flour by dry milling processes similar to those used for wheat.

Prior to milling, the grain is cleaned using milling separators, disc separators, gravity separators and magnets. Each machine is designed to remove specific impurities with high efficiency. The clean grain is tempered to approximately 15% moisture. Milling is achieved with standard break-and-reduction roller-grinders, all of which are corrugated. Separations and purification of milling stocks are achieved by plansifters. Similar milling systems are employed in North America and in Europe although the European mills tend to produce a wider range of rye flour grades. *See* Flour, Roller Milling Operations

The full range of rye flours produced by a typical German mill is given in Table 7. The flours are graded according to ash content (colour), but may also carry specifications for other characteristics such as α-amylase activity (Falling Number Value) and swelling power which is strongly influenced by the content and nature of the pentosan constituent. Some rye is processed into meals, ranging in particle size, by crushing or cutting machines. *See* Flour, Analysis of Wheat Flours

Rye and rye products are used to produce a large variety of baked goods. In Europe, especially in Germany, many different types of bread are made from rye flours; the ash content of the flour determines the actual type. In addition, a wide variety of breads are produced from blends of rye and wheat flours containing from 15% to 70% rye flour. European bread types produced from rye are classified into the following groups:

1. Rye bread is produced from 100% rye flour, ash content 0·8–1·6%.
2. Rye–wheat mixed bread is produced from a mixture of rye and wheat flours containing at least 50% of rye flour.
3. Wheat–rye mixed bread is produced from a mixture of flours containing at least 50% wheat flour and at least 10% rye flour.
4. Wholemeal rye bread is produced from wholemeal rye flour.

In Canada and the USA, rye breads are usually made from blends of wheat and rye flours. Bread from 100% rye flour is available at speciality bread shops. *See* Flour, Dietary Importance

Another popular rye baked product, first introduced in Scandinavia but now available worldwide, is the so-called crispbread. It has excellent shelf life and its sensory qualities are preferred by many consumers. Crispbread is made by a special process and requires rye flour of defined particle size and low α-amylase activity.

Substantial quantities of rye grain are used as the source of fermentable starch in the alcoholic fermentation industry. The alcohol is used as a beverage and in the pharmaceutical industry. Small quantities of rye are used in the production of speciality breakfast cereals, e.g. muesli, and various snack items. Rye will continue to be an important, albeit small crop because of its unique agronomic and organoleptic properties. *See* Beers, Ales and Stouts, Preparation of Wort; Cereals, Breakfast Cereals

Bibliography

Bushuk W (1976) *Rye: Production, Chemistry, and Technology*. St Paul, Minnesota: American Association of Cereal Chemists.

Kent NL (1983) *Technology of Cereals*. Oxford: Pergamon Press.

Lorenz KJ (1982) Rye: utilization and processing. In: Wolff IA (ed.) *Handbook of Processing and Utilization in Agriculture*, vol. II, Part 1, pp 243–275. Boca Raton, Florida: CRC Press

Lorenz KJ (1991) Rye. In: Lorenz KJ and Kulp K (eds) *Handbook of Cereal Science and Technology*, pp 331–371. New York: Marcel Dekker.

Walter Bushuk
University of Manitoba, Winnipeg, Canada

SACCHARIN

Saccharin has been in use as an artifical sweetener for about 100 years. It was discovered accidentally by researchers at Johns Hopkins University in 1879 during studies on the oxidation of *o*-toluenesulphonamide. Its structure is shown in Fig. 1. It is not metabolized but is excreted unchanged by the human body and thus contributes no calories to the diet. The safety of the substance has been the subject of debate for many years and the controversy continues with arguments both for and against its use as an artificial sweetener.

Sweetness

Saccharin is about 300–500 times as sweet as sucrose, depending upon its concentration and the type of food medium in which it is used. Compared to sucrose it has a slow onset of sweetness which increases to a maximum and then persists. The major drawback of saccharin is its bitter metallic aftertaste which is particularly evident at higher concentrations. Because of this, efforts have been made to mask the aftertaste by adding substances such as cream of tartar, lemon flavour, pectin, ribonucleotides, glycine, gentian root and artificial sweeteners such as aspartame or cyclamate. Combinations of saccharin with aspartame or cyclamate have proved to be very successful for many applications. In these cases it has been found that saccharin exerts a substantial synergistic effect on the sweetness of either aspartame or cyclamate. For example a mixture of 20 mg of aspartame and 4 mg of saccharin is equal in sweetness to either 45 mg of aspartame or 35 mg of saccharin if used alone in a cup of coffee. Not only does saccharin exhibit a synergistic effect with aspartame, it apparently improves the stability of the latter in acidic soft drinks, enabling extended storage of such products. Saccharin/cyclamate mixtures were widely accepted in the United States until the ban on cyclamate in 1969. At present, saccharin/aspartame concentrations are commonly used in the United States in diet soft drinks. In some European countries saccharin/cyclamate combinations are still permitted. ***See*** Aspartame; Carbohydrates, Sensory Properties; Cyclamates; Sweeteners – Intense

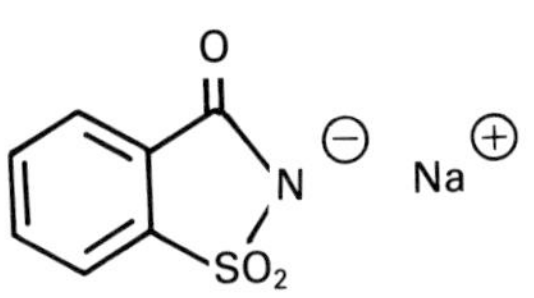

Fig. 1 Structure of saccharin (sodium salt).

Production, Physical and Chemical Properties

Commercial production of saccharin has been carried out by two different chemical processes. The Remsen–Fahlberg method involves reaction of toluene with chlorosulphonic acid to produce *o*- and *p*-toluene sulphonyl chlorides. The *o*-isomer is separated and then treated with ammonia to form *o*-toluene sulphonamide. The last compound is oxidized to *o*-sulphamoyl benzoic acid which upon heating cyclizes to saccharin.

The Maumee process involves diazotization of methyl anthranylate with sodium nitrite in the presence of hydrochloric acid to form 2-carbomethoxybenzenediazonium chloride. Sulphonation followed by treatment with chlorine yields 2-carbomethoxybenzenesulphonyl chloride which is converted to saccharin after amidation and acidification.

The calcium or sodium salts of saccharin are produced by reaction of the respective hydroxides with saccharin.

Saccharin and its calcium or sodium salts are white crystalline powders with intensely sweet tastes. A dilute aqueous solution of saccharin is about 500 times sweeter than a solution with an equal concentration of sucrose, while the salts are about 300 times sweeter than a solution containing an equivalent weight of sucrose. The melting point of saccharin is 228·8–229·7°C, while the salts decompose above 300°C. The excellent stability of saccharin and its salts under normal food-processing makes it ideally suited technically as a noncaloric sweetener in many different products. It is unaffected, for example, by the heat processing required for jams, jellies and canned fruits. The salts of saccharin are much more water soluble than saccharin (1 g salts per 1·5 ml water versus 1 g saccharin per 290 ml water). Aqueous solutions of saccharin are slightly acidic while the calcium and sodium salts are neutral or slightly basic. They are very stable in aqueous solution over a wide pH range, although below pH 2·5 at elevated temperatures saccharin has a tendency to hydrolyse slowly to *o*-sulphamoyl benzoic acid and *o*-sulphobenzoic acid.

Food Uses

Because of its stability, saccharin (usually as its calcium or sodium salt) has been used in a wide variety of food products. These include carbonated and noncarbonated soft drink beverages, table-top sweeteners, dry beverage bases, canned fruits, gelatin desserts, cooked and instant puddings, salad dressings, jams, jellies, preserves, chewing gum and baked goods. Typical amounts of sodium saccharin used in food products are, for example, 13·5 mg per teaspoon of table-top sweetener, 9·5 mg per fluid ounce of carbonated soft drinks, 5·5 mg per fluid ounce of noncarbonated soft drinks, 4·5 mg per teaspoon of jams and jellies and 2·2 mg per stick of chewing gum.

Nonfood uses of saccharin include applications in cosmetics such as toothpastes, mouthwashes and lipstick; pharmaceuticals such as coatings on pills; in cattle feed; in electroplating; and, in Japan, as an intermediate in the production of the rice blast fungicide, probenazole.

World commercial production of saccharin in 1900 was estimated to be about 190 tonnes. This increased more or less steadily up to the late 1970s when world annual production reached approximately 5000 tonnes. Since then, concerns about its safety have led to partial or complete bans on the use of saccharin in food products by a number of countries.

Metabolism and Safety

Many feeding studies with a variety of animal species have shown that saccharin is excreted unchanged. In rats, saccharin is excreted rapidly. However, there is some accumulation in the bladder, but the sweetener is completely cleared in 3 days after removal of saccharin from the diet. No metabolites have ever been detected. Studies have shown that the compound is not metabolized by liver microsomal enzymes and it does bind to DNA after oral administration. The metabolic profiles of saccharin fed to dogs, rabbits, guinea-pigs and hamsters are similar.

The greatest concern relating to metabolic effects of saccharin is the uncertainty about its carcinogenicity to humans. Several carcinogenicity studies in rats which included *in utero* exposure have indicated that feeding of saccharin may lead to bladder tumours. However, epidemiological studies in humans have provided no clear evidence that saccharin causes urinary bladder cancer. ***See*** Carcinogens, Carcinogenic Substances in Food

Regulatory Status

In 1972, because of the uncertainty of the safety of saccharin, the US Food and Drug Administration delisted saccharin and in 1977 proposed to ban its use in foods and beverages. However, public protest led congress to impose a moratorium on the ban which has been extended up to the present. In Canada, saccharin was banned in 1978 for use in foods. However, it is permitted to be sold in pharmacies as a table-top sweetener. In many other countries throughout the world, saccharin is permitted although its use is restricted to varying degrees. ***See*** Legislation, Additives

Analysis

A number of methods for the determination of saccharin in foods and pharmaceutical preparations have been reported. These include gas chromatography with flame ionization or electron capture detection, thin-layer chromatography, ultraviolet absorption spectrophotometry and high-performance liquid chromatography (HPLC) with ultraviolet absorbance detection. The last technique is the most commonly used method at present, especially for the determination of saccharin in foods or beverages. For liquid samples, minimal sample preparation is required. The samples are filtered and perhaps diluted before being analysed directly by HPLC. Solid foods are usually mixed with water, which extracts the saccharin. An aliquot of the aqueous extract is then directly analysed by HPLC. The sensitivity of the technique is more than adequate for the concentrations of saccharin normally encountered in foods. Saccharin has been incorporated into a multisweetener HPLC method (see Lawrence and Charbonneau, 1988). ***See*** Chromatography, Gas Chromatography; Chromatography, High-performance Liquid Chromatography; Chromatography, Thin-layer Chromatography

Bibliography

Franta R and Beck B (1986) Alternatives to cane and beet sugar. *Food Technology* 40: 116–128.

International Agency for Research on Cancer (1990) *IARC Monographs on the Evaluation of the Carcinogenic Risk of Chemicals to Humans. Some Non-Nutritive Sweetening Agents*, vol. 22. Lyon: IARC.

Lawrence JF and Charbonneau CF (1988) Determination of seven artificial sweeteners in diet food preparations by reverse-phase liquid chromatography with absorbance detection. *Journal of the Association of Official Analytical Chemists* 71: 934–937.

Shaw JH and Roussos GG (1978) *Sweeteners and Dental Caries*. Washington, DC: Information Retrieval Inc.

James F. Lawrence
Sir FG Banting Research Centre, Ottawa, Canada

SACCHAROSE

See Sucrose

SAFETY OF FOOD

See Cleaning Procedures in the Factory, Food Poisoning, Hazard Analysis Critical Control Point, Laboratory Management, Antiseptic Products for Personal Hygiene and specific Pathogens

SAGO PALMS

Sago has been used as a food by indigenous peoples of New Guinea, New Britain and the Moluccas since time immemorial. In spite of the usefulness of sago starch and other plant parts, it has benefited little from research, probably because it is a smallholders' crop, and because of its long gestation period. There is growing global interest in sago as an important tropical starch source. Since 1976 four international symposia have been devoted to the sago palm. Commercial production of sago starch for export is only undertaken in Malaysia and Indonesia. In Sarawak, Malaysia, the first sago plantation was established in 1987. This article discusses the nature of the plant and its uses.

Nature of the Plant

Taxonomy

The sago palm belongs to the order, Spadicifloreae, family Palmae, subfamily Lepidocaryoid. It is a species of the genus, *Metroxylon* (derived from the Greek words 'metra', meaning heart, and 'xylon', meaning xylem or wood). Two species differentiated by the presence or absence of spines are currently identified: *M. sagu* Rottb. (spineless) and *M. rumphii* Mart. (spiny). The validity of such species distinction is often questioned as the two are often found together with intermediates in wild populations and appear to cross freely. The true sago can be distinguished from other species by its fruit, which has less than 18 vertical series of scales.

Distribution and Gene Centre

Sago palms grow naturally in the South Pacific Islands, extending westward through Melanesia to Indonesia, Malaysia and Thailand, and from the Kai-Aru islands in the south to Mindanao in the Philippines.

The centre of diversity is in the Moluccas. The gene centre could also include New Guinea, where extensive natural stands are found. At least 14 cultivars of *M. sagu* and *M. rumphii* have been identified in Papua New Guinea. It is not easy to distinguish between wild and cultivated sago because of the low maintenance requirement.

In Sarawak, Malaysia tissue culture research is in progress to supply elite palms of short maturity period and high starch concentration to the commercial plantations.

Habitat

Sago is considered to belong to the hot, humid, tropical swamp environment. Natural stands occur in freshwater swamps in coastal regions and flood plains of rivers from sea level to 1000 m altitude. Best yielders are on lowlands below 400 m. They are mainly found between longitudes 90° and 180° east and between latitudes 10° north and 10° south, where the annual rainfall is higher than 2000 mm. Under natural conditions, temporary inundation suits the palm best as other plant species are more competitive in wetter or drier situations. The relative humidity should be high (72–97%). Although experiments have shown the sago palms to be tolerant of salinity (up to 5 μmhos cm^{-1}), in practice it does not compete successfully with the nipa palm (*Nypa fruticans*) under conditions of daily tidal inundation.

Temperature and moisture requirements are the two parameters which define the conditions for the growth of sago. The optimal temperature range is 25–30°C, while the minimum tolerable is 15°C. A favourable rainfall is perhaps 2000–4000 mm, evenly distributed throughout the year. Prolonged soil moisture deficit in the root zone, lasting more than 1 month, cannot be tolerated. Temporary moisture stress is generally tolerated by well-established palms because the thick woody bark, acting as an excellent insulator, makes the whole plant relatively desiccation-resistant. High insolation is favourable.

Although the sago palm can grow on a variety of soils, the best growth is observed on clay soils with high organic matter content. It is tolerant of low soil pH, high aluminium, iron and manganese, soil pans, heavy impervious clays, and peaty conditions. On mineral soils, it matures in 9–11 years, while on peat soils it reaches maturity in 11–14 years.

Sago stands seem to have a favourable climatic effect when they occur over an extended area. The cover that sago palm provides is 10–17 m high and has a stabilizing effect on the climate.

Morphology and Anatomy

The palm is a hapaxanthic (single-flowering) and soboliferous (bearing suckers) perennial. The stem acts as a sink site in which carbohydrates are stored for sucker production and the final flowering and fruiting effort.

Sucker production commences in the first year of planting, drawing food reserves from the mother palm until the leaf and root systems are well developed. Suckering results in a clump of plants.

The sago palm possesses a large number of superficial primary roots from which pneumatophores (erect roots, with aerenchyma) sprout above the ground water level. The trunk may also be covered with small roots. On peat soils the roots are mainly concentrated in the top 1·0 m layer and the root density is three to nine times greater than in mineral soils, indicating that the palm can compensate for the inherent low nutrient status of the peat by forming a denser root system.

The well-grown palm crown consists of 8–23 fronds, each composed of a leaf sheath, petiole, and 50 pairs of leaflets 70–180 cm long and 5–9 cm wide. The old fronds undergoing senescence drop off, leaving leaf scars on the trunk. The phyllotaxy of the sago palm is such that the fronds gyrate along a single generative spiral, ascending helically around the trunk axis in either a right- or left-handed direction. The divergence angle is 110·4°. Trunk formation starts around the fourth year. The trunk may reach 7–15 m in length. The stem diameter at the trunk base is usually 35–50 cm which gradually increases with height to 50–60 cm and tapers again towards the terminal end.

At approximately 9–14 years of age, the sago palm reaches maturity and flowers. The trunk tapers off, the frond size decreases and grades into bracts, and a large terminal inflorescence, resembling a much branched antler, is put forth. The panicle, a continuation of the main terminal axis, consists of a primary axis, dividing into secondary and tertiary axes. The tertiary axis bears small flowers in pairs, a male and a hermaphrodite. Both have three sepals, three petals, and six stamens, the male flower having a short pistillode of three tiny knobs and the complete flower, a shallow trioculary ovary with one chamber and three ovules. Flowering is protandrous (male flowers open and shed their pollen before complete flowers open). The stamens of the hermaphrodite flower are nonfunctional. The sago palm is therefore an obligatory cross-pollinator. The chromosome number of sago palm is 26.

The development of the inflorescence to the production of mature fruit takes 2–3 years, exhausting the carbohydrate reserve in the stem which turns woody and fibrous. On average, 2500 fruits may be formed. The fruits are globose, covered with 9–10 rows of spiral scales, and about 5 cm in diameter with a flattened apex. They are green when immature and straw-coloured when ripe.

The transverse section of the pith shows the xylem vessels spread irregularly over the section. The largest vessel, of 1·4 mm diameter, is visible to the naked eye. The surrounding tissues forming the pith are cells filled with starch granules 20–60 μm in size. Intercellular spaces are 200 μm in size. The base of the trunk is more fibrous and woody.

The cortex or bark of the trunk is a distinct layer, 2–3 cm thick, containing no starch. It is hard and woody.

Extraction of Sago Starch

In the processing of sago starch it is important to complete the whole process within the shortest time possible in order to avoid fermentation. Fermentation causes inferior quality and starch losses because part of the starch, on fermentation, is converted into soluble sugar and lost during processing.

Disintegration

The first stage in the extraction of sago starch is to separate the bark from the pith. This is performed manually by stripping the bark with an axe or 'parang'. The debarked log is then split into manageable pieces, or billets in preparation for rasping (disintegration). Rasping may be carried out using various tools, such as the traditional, manually operated rasper, the modern disc rasper, cylindrical rasper, chipping machine, and the hammer mill. The manually operated rasper consists of a long, narrow, wooden board with nails mounted on one face. The assembly is moved across a debarked log section in a sawing motion by either one or two operators. The modern disc rasper and cylindrical rasper are mechanically driven. The rasping action is executed by feeding the sago billets against a rotating element provided with sharp-pointed metal nails or rows of saw-like blades. The rasped pith, called 'repos', has a granular texture similar to that of coarse sawdust. The starch yield is dependent on efficient rasping. Contact with iron is detrimental to the quality of starch, and iron parts are therefore excluded from the processing equipment.

Extraction

In the extraction stage, the starch is separated from the cellulose. Traditionally, the repos is placed on a woven reed mat and, while adding copious quantities of water, a woman shuffles her feet in the repos, trampling it to express the starch. However, in commercial sago processing, the basic equipment in extraction is a revolving screen. A wide variety of designs have been used, differing in the complexity of their construction and their efficiency of operation. These include the shaking screen, rotating screen, vibrating screen sieves, vertical and horizontal centrifugal sieve, hydrocyclone, decanter and the nozzle separator. For efficient separation, the starch slurry is passed through a series of extractors with sieves of increasing fineness. The starch slurry leaving the first extractor, which has holes 125 μm in diameter, is pumped to another extractor of the same type. This second extractor consists of one or two stages and is equipped with stainless sieve gauze with holes of 80 μm. The operation requires that wash-water be sprayed on the pulp countercurrent to the flow along the screen. In some processes, fine screens are added in order to remove the fine fibres. The pulp and fine fibres are led off into storage as wet waste. On leaving the extractors, the starch milk is stored in a crude milk tank and may be passed through a sand cyclone to remove sand and dirt particles. ***See*** Starch, Sources and Processing

Starch Refining

In the refining stage, the pure starch is separated from small, insoluble cellulose particles. The method used determines the quality of the product obtained. In the traditional method, the starch slurry is concentrated through a series of settling tanks. The problem with this method is the time required to complete settling. A time frame of 4–24 h may be required, allowing contamination and fermentation by contact between the starch and the fruit water. In the fruit water, which is rather rich in sugars and other nutrients, microorganisms will readily develop, resulting in fermentation. Another modified traditional method which reduces the time required is the settling table. This is basically a flat wooden trough about 2·5–5 m wide and 15–30 m long, which may be horizontal or gently sloping from the end where the starch enters. The liquid flows in at a controlled rate, and the larger particles of starch separate initially, so that the starch is graded along the length of the table. As the starch settles, it builds up and increases the slope of the incline. The starch must therefore be periodically removed when a layer up to 15–20 cm thick has formed. This is performed either by shovelling or by washing off with a hose.

In the modern system, a high-speed centrifugal separator is used in order to perform the separation and refining as rapidly as possible. In case the liquid contains contaminants, the fine screened raw starch slurry having a concentration of about 3° Be′ (54 kg dry starch per m^3) is pumped through a safety strainer and a sand cyclone in order to protect the separators. The sludge is intermittently discharged from the strainer and the cyclone. To ensure a high-quality starch, two separators in series are normally installed. Both separators are of the same type, but as the second separator works with more concentrated feed, it can sometimes be of a smaller size. The concentration of the refined starch is 20–22° Be′.

Starch Dewatering and Drying

On completion of separation and refining, the concentrated starch slurry is ready for drying. A preliminary dewatering is generally undertaken to reduce the moisture content to 35–40%. This is achieved by a centrifugal filter or a vacuum filter. Drying of starch is achieved

Table 1. Properties of sago and other starches

	Origin of starch					
	Sago	Maize	Potato	Rice	Cassava	Wheat
Granule shape	Oval	Round, polygonal	Oval	Polygonal	Oval with indentation	Round
Granule size (μm)	20–60	15	15–100	3–8	5–35	2–10 and 20–35
Amylose (%)	27	26	24	17	17	25
Swelling power (%)	97	24	> 1000[a]	19	71	21
Initial pasting temperature[b] (°C)	69	62	56	66	58	65

[a] Swelling is greatly influenced by phosphate ester usually present in potato starch.
[b] Estimated from graphs by Cecil JE *et al.* (1982).

Table 2. Utilization of sago starch

Food industry	*Biotechnological industry*	*Animal feed*	*Nonfood industries*
Confectionery	High-fructose syrup		Biodegradable plastic
Sago pearl	Glucose syrup		Textile
Breadmaking	Dextrose monohydrate		Paper
Desserts	Caramel		Adhesive
Noodles	Maltose		Plywood
Crackers	Maltodextrin		
Modified starch	Sweeteners		
	Monosodium glutamate		
	Sorbitol		

either by sun-drying, which requires longer periods of time, or by the continuous processing system by flash drier using heated air. The starch is dried to a moisture content of 10–13%.

Nutritive Value of Sago Starch

Sago starch is an oval or egg-shape granule with sizes varying between 20 and 60 μm. Under polarized light, sago starch granules produce irregular black crosses, indicating some degree of crystalline structure. Compared to tuber starches (Table 1), sago starch has a higher pasting temperature (72–90°C), a higher ratio of amylose to amylopectin (27:73), and a much greater swelling power (97%). Results from proximate analysis show that the refined sago starch contains about 12–15% moisture, 0·1–1·0% protein, 0·1–0·3% fat, 0·1–0·5% fibre, 0·1–0·8% ash and 87·7% carbohydrate. The chemical composition of sago starch is not greatly influenced by the palm species, age, habitat, or agronomic practices; the main variability in the chemical composition of the commercial product originates from the processing system. ***See*** Starch, Structure, Properties and Determination

Utilization of Sago Starch

The starch extracted from the sago pith has a variety of uses, many of which are common to starches derived from cassava, maize and wheat. Starch is basically divided into food and industrial grades. The former is used by the food processing industries, the latter in a wide range of industries. Sago starch has to compete in terms of price and quality with the other starches. ***See*** Cassava, Use as a Raw Material; Wheat

Sago starch can be used in both food and nonfood industries. In the Malaysian food industry, sago starch is used in various traditional food products, such as sago pearls, biscuits (e.g. 'tabaloi' and 'kuih bangkit'), noodles (e.g. 'kuih teow' and 'beehoon') and desserts (e.g. 'cendol'). In Irian Jaya and Papua New Guinea, sago starch is used as a staple food. On a commercial scale, sago starch is used in the production of vermicelli and fish crackers.

Table 2 shows the full range of products made totally or partially from sago starch. Among the list of products, glucose syrup, dextrose monohydrate, caramel, monosodium glutamate, vermicelli, and fish crackers are the main consumers of sago starch.

Glucose is produced from the enzymatic breakdown

of sago starch. Glucose syrup is used in the production of confectionery, beverages, bakery products, and desserts. Dextrose monohydrate, which is the dry crystalline form of glucose, is used as a source of immediate energy in various soft drinks, tonic drinks, and medical products.

Caramel, which is a chemical conversion of sago starch, is used as a colouring and flavouring agent in beverages such as Coca Cola and Sarsi, and in sauces such as soya sauce. It is also used as a colouring agent in various bakery and biscuit products. ***See*** Colours, Properties and Determination of Natural Pigments

Monosodium glutamate (MSG) is a flavour enhancer made from the bioconversion of sago starch using microbes. In this process, sago starch serves as the raw material for the growth of the microbes. During this fermentation process, glutamic acid is produced and is then isolated and purified to produce MSG. Ajinomoto (M) Sdn. Bhd., the main producer of MSG in the ASEAN (Association of Southeast Asian Nations) region, uses a mixture of cassava and sago starches to produce MSG.

With advances in biotechnology, bioconversion products from sago starch such as maltodextrin, maltose and high-fructose syrup can be manufactured by employing established procedures. Chemical modification of sago starch to obtain value-added products is another area of utilization of sago starch presently being exploited by industry.

Sago starch can also be used as a substitute to corn and cassava starches in animal feed formulations. Field trials have indicated that the sago pith is a good source of raw material for animal feed. Using appropriate protein supplementation through the use of fish meal and other feed components, sago pith can be pelletized to produce good-quality animal feed.

In the nonfood industries, sago starch is used in the production of adhesives and glazing material for the paper industry. The production of glue from sago starch is currently being practised. Modification of sago starch can produce various tailor-made speciality starches for industrial use.

In addition to sago starch, other parts of the palm are also used for food. The fruit may be eaten but is regarded as not particularly palatable. The young shoot or palm cabbage is widely eaten as a vegetable. The larvae of the palm weevil (*Rhyncophorus schach* Oliv.), which breeds in the pith, are considered a delicacy by some people in Sarawak. On a dry-matter basis it contains 63·7% fat, 14·3% protein and 22·0% water.

The pith has a nutritional composition of 92·5% carbohydrates, 1·5% protein, 0·5% fat, 1·5% fibre, and 4·0% ash. The vitamin and mineral content is negligible. It is used as an animal feed component. ***See*** Carbohydrates, Classification and Properties; Dietary Fibre, Properties and Sources; Fats, Digestion, Absorption and Transport; Protein, Food Sources

Bibliography

Cecil JE, Lau G, Heng SH and Ku CK (1982) *The sago starch industry; a technical profile based on a preliminary study made in Sarawak*. London: Tropical Product Institute.

Kueh HS, Elone R, Tie YL, Ung CM and Jaman HO (1987) *Mukah Sago Plantation Feasibility Study, February 1987*. Sarawak: Land Custody and Development Authority.

Lim ETK (1991) *A comparative study on the performance of* Metroxylon sagu *and* Metroxylon rumphii *grown on organic and gleyed mineral soils*. MSc thesis, Malaysian Agricultural University, Serdang.

Ng TT, Tie YL and Kueh HS (eds) (1990) *Proceedings of the Fourth International Sago Symposium, 6–9 August 1990, Kuching, Sarawak, Malaysia*. Available from: Ministry of Agriculture and Community Development, Sarawak, Malaysia.

Stanton WR and Flach M (eds) (1990) *Proceedings of Second International Sago Symposium 15–17 September 1979, Kuala Lumpur, Malaysia*. The Hague: Martinus Nijhoff.

Tan HT (1982) Sago palm – a review. *Abstracts on Tropical Agriculture* 8(No. 9): 9–23.

Tan K (ed.) (1977) *Proceedings of First International Sago Symposium, Sago-76, 5–7 July 1976, Kuching, Sarawak, Malaysia*. Kuala Lumpur: Kemajuan Kanji Sdn. Bhd.

Yamada N and Kainuma K (eds) (1986) *Proceedings of the Third International Sago Symposium, 20–23 May 1985, Tokyo, Japan*. Available from: Tropical Agriculture Research Centre, Tsukuba, Japan.

Kueh Hong-Siong and Kelvin Lim Eng-Tian
Department of Agriculture, Kuching, Sarawak, Malaysia

SAKE*

Sake is produced in Japan in about 2000 breweries, and is distributed to all regions of Japan. Since sake brewing requires a cold temperature in winter, good rice and good water, the main places of production are located in districts with the above three conditions, such as Nada (Hyogo Pref.), Fushimi (Kyoto Pref.), Saijo (Hiroshima Pref.), Nigata (Nigata Pref.), Akita (Akita Pref.), etc. The product (1131·5 million litres, 1989) is consumed mainly in Japan, but recently the export (6·8 million litres, 1989) is increasing. In the U.S.A., where sake is consumed in some quantity (2·5 million litres, 1989), there are seven sake breweries established by Japanese sake companies, which are estimated to produce about 2·5 million litres in total per year.

Method of Manufacture

Water

The raw materials used in sake brewing are water and rice (see Fig. 1). Since water accounts for about 80% (v/v) of the end product, the ideal water for sake brewing is colourless, tasteless and odourless. The minerals in water are classified into effective and undesirable components. Components such as potassium, magnesium and phosphate promote yeast growth and alcoholic fermentation. Components such as calcium and chloride are thought to promote the elution of some enzymes from rice koji. The undesirable components are iron, manganese and copper. Iron is the most undesirable component because it gives the resulting sake a reddish brown colour by reaction with the cyclic polypeptide deferriferrichrome produced by koji mould. The maximum allowable iron content in the brewing water is 0·02 ppm. Therefore, there are good underground supplies of water in the main areas of sake production, e.g. 'Miya-mizu' water in the Nada district.

Rice

Rice selected for sake brewing tends to have large grains. One typical rice for sake, called 'yamadanishiki', weighs 28 g per thousand grains, while rice for eating weighs only 21–22 g. The peripheral layers of brown rice contain large amounts of inorganic components, vitamins, fats and proteins that are undesirable for sake brewing. Brown rice is polished in the mill to remove these materials. In sake brewing, the polishing percentage rate is defined as the weight percentage of white rice obtained from brown rice. In general, cleaned rice at a 60–70% polishing rate is used, but rice at a 35–50% rate is used in brewing superior sake ('ginjo-shu') for sake contests. More polishing thus results in a cleaner sake. In the past, polished rice at a 90% polishing rate, which is eaten for food in Japan today, had been used for sake brewing.

The polished rice is washed mechanically to remove the bran on the surface of the grains, and then steeped in water for several hours to absorb 28–30% water. After steeping, the excess water is drained off from the grains for about 4–8 h before steaming. The drained rice is transferred into a tank, called a 'koshiki', which has a hole in the bottom to admit steam. The koshiki is placed on the top of a large kettle filled with boiling water and the rice is steamed for about 1 h under atmospheric pressure. Recently, a machine for continuous rice steaming has been devised to replace the older koshiki. The steaming procedure increases the moisture content of the starch granules within the kernel by about 10%. *See* Rice

Koji

For koji preparation, 20% of the total rice is used and the rest is used for sake brewing (Table 1). The steamed rice is cooled down to about 36°C, and transferred to the koji-making room (koji-muro), where temperature and humidity can be controlled. Tane-koji (spores of *Aspergillus oryzae*) is scattered over the surface of the rice containing approximately 35% moisture at an inocula-

Table 1. Raw materials for sake brewing (junmai-shu)[a]

Addition	Amount (kg) of rice for Steaming	Amount (kg) of rice for koji	Water (litres)	Mash temperature (°C)
Moto	140	70	230	20
Soe (1st)	320	130	440	15
Naka (2nd)	700	200	1050	9
Tome (3rd)	1140	300	2000	7
Total	2500	700	4000	

[a] In the case of honjozo-shu, about 1200 litres of brewer's alcohol (30% solution) was added before filtration of the mash.

* The colour plate section for this article appears between p. 3896 and p. 3897

tion rate of 1 g of tane-koji preparation per kilogram of raw rice, at about 30°C, then mixed thoroughly and covered with cloths. After 10–12 h, the rice is mixed again and heaped on a table. After the growth of mould mycelium on the grains for about 20 h, small white spots, visible to the naked eye, appear and the temperature of the heaped grains begins to rise. At this stage, the developing koji is transferred into many shallow wooden trays (koji-buta) that are stacked on shelves and covered with cloths. The koji mould begins to grow rapidly and vigorously and the temperature of the koji rice rises. It is stirred twice every 4 h so that the temperature of the rice does not become too high (40–42°C). After 40–45 h from the time of spore scattering, the rice koji in the trays is moved out of the warm koji-muro, so that the low temperature (about 5°C) outside stops the growth. The resulting rice koji is white and has an odour like roasted chestnuts.

Recently, many sake factories have adopted the use of a koji-making machine. Koji prepared by the above method is white in colour, since the growth stage is halted prior to sporulation of the mould. The mycelial growth is observed not only on the surface of the grain, but also in the centre of the kernel. Growth of the mould within the kernel is considered desirable for the slow and progressive release of α-amylase and glucoamylase in the 'multiple parallel fermentation' of sake brewing.

Moto (Seed Mash)

About 7% of the total rice is used for moto preparation (Table 1), and in 2–4 weeks it is ready for use as a starter for the main mash. Moto plays an important role as a starter culture in carrying out the fermentation of moromi (main mash). Koji is mixed with water, steamed rice, and sake yeast (*Saccharomyces cerevisiae*) in the prefermentation tanks. This seed mash is a concentrated culture of pure and healthy living cells of the sake yeast.

Moromi (Main Mash)

On the 1st day, a mixture of steamed rice, water, koji and moto is put into the main fermentation tank (moromi preparation temperature, 15°C). On the 3rd (9°C) and 4th (7°C) days, additional volumes of steamed rice, water and koji are added to the tank in order to keep the yeast cell number at a level of about 10^8 ml^{-1} in the mash. During the fermentation of moromi, the temperature of which is kept about 10–15°C, the starch in the rice is liquefied and saccharified by amylases in the koji, and this converted mixture is fermented into ethanol by the action of the sake yeast. Both processes, saccharification and alcohol fermentation, proceed simultaneously in a well-balanced manner. This unique, complex method, called 'multiple parallel fermentation', contributes about 20% (v/v) of the alcohol content, higher than any other naturally fermented beverage.

Filtration of the Mash

When the fermentation is finished (20–30 days), the mash is press filtered to separate the sake from the solids. The mash residue remaining on the filtering cloth is called sake cake.

Pasteurization and Storage

The fresh sake thus obtained still contains yeast cells and turbid materials, such as fibre, starch and protein. After settling for 5–10 days, the supernatant is filtered with activated carbon (200–500 g per 1000 litres), cotton fibre, and celite through a filter press.

After blending, the fresh sake is heated to 60–65°C by passing through a heat exchanger. This pasteurization inactivates the enzymes (amylase, protease, etc.) and kills the so-called hiochi bacteria, harmful microorganisms that spoil sake during storage. Immediately after pasteurization, the hot sake is transferred to a sealed vessel for storage.

Recently, fresh sake filtered with a membrane filter (pore size about 0·45 μm) has also been produced; this process is employed instead of pasteurization.

Microbiology

Koji Making

Tane-koji (spores of *A. oryzae*) is available from about 10 tane-koji manufacturers in Japan who provide the product to sake companies. *A. oryzae* is also used for shoyu (soy sauce) and miso (soya bean paste) making. However, the strain for sake brewing is different from the others with respect to its enzymatic activity. In sake brewing, the following mycological characteristics have been found empirically to be of major importance in *A. oryzae*:

(1) rapid growth on and into the kernel of the steamed rice;
(2) production of abundant amylase (α-amylase and glucoamylase), a little acid carboxypeptidase and less tyrosinase;
(3) low production of coloured substances, such as deferriferrichrome, flavines, etc.

Sake Fermentation

The Brewing Society of Japan (Nihon Jozo Kyokai) provides the sake yeast strains (*S. cerevisiae*) used for

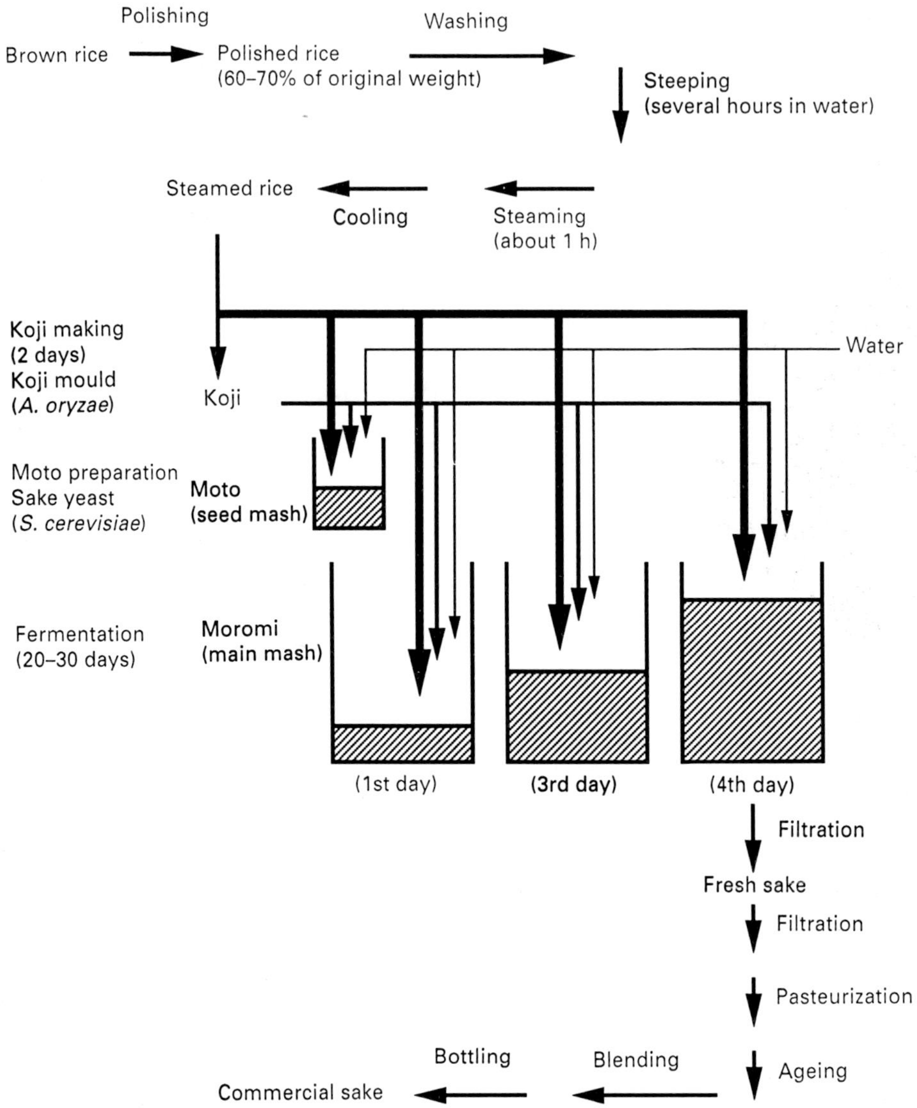

Fig. 1 Process flow chart for sake brewing.

sake brewing, named kyokai No. 6, No. 7, No. 9, No. 10, etc. Kyokai No. 6 and 7 strains are used for brewing the usual types of sake. Kyokai Nos. 9 and 10 are used for brewing ginjo-shu, because of their property of high aroma formation. Recently, breweries have used their own sake yeast strains for part of the brewing process. The characteristics needed for sake yeast are both high fermentation activity at low temperature (5–15°C), and resistance to high alcohol contents (about 20%).

Moto (Seed Mash)

There are several types of moto used for sake brewing. They are divided into two types. One is sokujo-moto, which is the most popular and does not take too long (about 2 weeks) before being able to be used in the main fermentation. The other is traditional moto (ki-moto, yamahai-moto), in which many bacteria contribute, such as nitrate-reducing bacteria (*Pseudomonas* spp.) and lactic acid bacteria (*Leuconostoc mesenteroides*, *Lactobacillus sake*). The latter lactic acid bacteria play an important role in the essential acidification of moto that is necessary for the pure culture of sake yeast. Since the production of lactic acid by these bacteria takes about 2 weeks, traditional moto takes longer (about 4 weeks). For shortening the time for making sokujo-moto, lactic acid is added. Acidification with lactic acid is one of the important procedures for ensuring a pure culture of the added sake yeast in spite of the open fermentation. The other procedure is the brewing method itself, which involves a step-by-step scale-up (Table 1). ***See*** Lactic Acid Bacteria

Other Microorganisms

It is known that there are microorganisms harmful to sake brewing. They are lactic acid bacteria (*Leuconostoc*

Table 2. Names for several types of sake

Name	Ingredients	Polishing rate	Remarks
Ginjo-shu	Rice, rice koji, brewer's alcohol	<60%	Characteristic flavour, taste and good for colour by ginjo brewing
Daiginjo-shu	Rice, rice koji, brewer's alcohol	<50%	Characteristic flavour, taste and good for colour by ginjo brewing
Junmai-shu	Rice, rice koji	<70%	Good for flavour, taste and colour
Junmaiginjo-shu	Rice, rice koji	<60%	Characteristic flavour, taste and good for colour by ginjo brewing
Junmaidaiginjo-shu	Rice, rice koji	<50%	Characteristic flavour, taste and good for colour by ginjo brewing
Honjozo-shu	Rice, rice koji, brewer's alcohol	<70%	Good for flavour, taste and colour

casei, L. leichmannii), and wild yeasts which spoil the moromi mash by their growth instead of the cultured sake yeast. Finished sake contains about 20% alcohol, so that only the hiochi bacteria (*Lactobacillus heterohiochii, Lac. homohiochii*) can grow in it. The sake in storage tanks and bottles spoiled by the bacteria changes, become turbid and sour, and have an off-flavour. To prevent spoilage, pasteurization is carried out and the history of pasteurization of sake began in the 16th century, long before Pasteur's discovery in 1868. ***See*** Spoilage, Bacterial Spoilage; Spoilage, Yeasts in Food Spoilage

Types of Sake

The most popular type of sake is produced by the new techniques of adding alcohol, sugar, organic acids and amino acids during the production process. These new techniques were developed after 1945, so as to ensure a stable production of sake from limited raw materials. Recently, the amounts of brewer's alcohol and other ingredients added to the moromi mash has decreased every year and, in 1990, a law controlling the types of sake was established (Table 2). This system is similar to the French Appellation Contrôlèe. In 1989 the production of honjozo-shu, junmai-shu and ginjo-shu sakes were 198·9 million, 54·6 million and 32·4 million litres, respectively, and these amounts correspond to about 25% of the total production.

Junmai-shu

This is 'pure rice' sake in which only rice with a less than 70% polishing rate, koji rice, and water are used as ingredients, and there are no further additions of other substances such as alcohol or sugar. In terms of ingredients, junmai-shu is closest to the traditional sakes of the Edo period. However, since the polishing rate of the rice used was more than 90%, the taste and flavour were assumed to be different from the modern type. Although junmai-shu tends to have a heavier taste than the usual types of sake; it is becoming popular.

Honjozo-shu

Less than 10% (w/w) of alcohol (95% alcohol) to white rice may be used, and no other ingredients may be added during the brewing process. Rice with a less than 70% polishing rate must also be used, and the raw alcohol must be added to the moromi, where it interacts with and brings out the flavour elements in the mash.

Ginjo-shu

This is a special type of junmai-shu or honjozo-shu, and has a fruity flavour like an apple. All the rice employed in brewing ginjo-shu must be polished to at least 60% of its original size. Recently, most of the products presented at sake contests held by the National Research Institute of Brewing of Japan are those brewed with a 35–40% polishing rate.

Others

Taru-zake or cask sake is aged in wooden casks (18–72 litres) so that the colour, aroma and flavour elements from the wood are absorbed into the sake. Today, taru-zake is sold both in casks of various sizes and in glass bottles.

Koshu is literally 'old sake'. This sake is aged for 2–3 years before it is bottled and put on the market. Most sakes are aged for less than 1 year before bottling, and should be drunk while they are young. Some high-quality sakes, however, such as ginjo-shu, may improve with age, taking on a distinctive mellow flavour.

Sake

Table 3. Organic acid composition of Junmai-shu

Acid	mg l^{-1}	%
Lactic acid	475·0	29·2
Acetic acid	40·8	2·5
Pyruvic acid	17·4	1·1
Malic acid	315·3	19·4
Citric acid	78·0	4·8
Succinic acid	698·8	43·0
Total	1625·3	100

Nigori-sake is 'cloudy sake' or 'roughly filtered sake'. This sake is filtered with open-weave sacks that leave some of the solid particles of the rice and koji rice suspended in the liquid, giving the sake a white, cloudy appearance.

Kijo-shu is produced by replacing half of the water used in brewing with sake. It is extremely heavy and sweet, and is usually served as an aperitif.

Akai-sake is 'red sake'. There are two methods for making akai-sake. One is produced using a red-coloured koji instead of the usual yellow type, and the other is fermented using an adenine auxotrophic mutant of the normal sake yeast.

Table 4. Amino acid composition of junmai-shu

Amino acid	mg l^{-1}	%
Aspartic acid	99·8	3·1
Threonine	243·5	7·5
Serine	137·1	4·2
Glutamic acid	281·6	8·6
Glycine	179·0	5·5
Alanine	291·8	8·9
Cysteine	73·4	2·3
Valine	177·7	5·4
Methionine	59·4	1·8
Isoleucine	112·7	3·5
Leucine	240·8	7·4
Tyrosine	198·6	6·1
Phenylalanine	150·8	4·6
Ornithine	37·7	1·2
Lysine	147·6	4·5
Histidine	84·5	2·6
Tryptophan	14·2	0·4
Arginine	407·9	12·5
Proline	191·7	5·9
γ-Aminobutyric acid	36·1	1·1
Ethanolamine	66·3	2·0
Ammonia	28·5	0·9
Total	3261·0	100

Components

Typical commercial sake contains about 16% alcohol, 3% total sugar (as glucose) and has a titratable acidity of 1·4 (the volume in millilitres of 0·1 N sodium hydroxide which titrates 10 ml of sake) and an amino acidity of 1·5 (the volume in millilitres of 0·1 N sodium hydroxide which titrates formol nitrogen in 10 ml of sake). Sugars other than glucose are contained in sake: maltose is not detected, but small amounts of isomaltose and ethylglycoside are present. The organic acid and amino acid compositions are shown in Tables 3 and 4. The nitrogen compounds in sake consist mainly of amino acids or peptides. These compounds are important for a mild taste. Of the aroma components, esters and higher alcohols are present, especially isoamyl acetate (3–5 ppm) and ethyl caproate (1–2 ppm), which are the most important components in ginjo-shu, which has a fruity flavour.

Bibliography

Hara S, Kitamoto K and Gomi K (1991) New developments in fermented beverages and foods with *Aspergillus*. In: Bennett JW and Klich MA (eds) Aspergillus: *The Biology and Industrial Application*. Stoneham, MA: Butterworths.

Kitamoto K, Oda K, Gomi K and Takahashi K (1991) Genetic engineering of a sake yeast producing no urea by successive disruption of arginase gene. *Applied and Environmental Microbiology* 57: 301–306.

Kondo H (ed.) *Sake: A Drinker's Guide*. Tokyo: Kodansha International.

Nunokawa Y (1972) Chemistry and Technology. In: Houston DF (ed.) *Rice*, pp 449–487. St Paul: American Association of Cereal Chemists.

Ohba T and Sato M (1986) Sake. In: Charalambous G (ed.) *Handbook of Food and Beverage Stability*, pp 773–799. London: Academic Press.

K. Kitamoto
National Research Institute of Brewing, Tokyo, Japan

SALAD CROPS

Contents

Dietary Importance

Vegetables that are normally consumed raw or uncooked are known as salad crops. They are seasonal and are attractive in appearance and valued for their appetizing succulence, bulk and nutritional value. With improved packaging and transportation, the demand for salad vegetables, especially celery and lettuce, is increasing.

Types of Plant in the Salad Category

Since there are a vast number of salad vegetables grown and consumed in the world, it is essential that they be grouped or classified so that information on them can be readily obtained. They may be classified according to botanical traits, season of growth, edible part used, and soil acidity (see Tables 1–4). As salad vegetables are consumed fresh, it may be particularly pertinent to classify them on the basis of edible portion consumed (Table 1). ***See*** Onions and Related Crops; Tomatoes

Nutritional and Dietary Significance

Salad vegetables are protective foods and are rich in macro- and micronutrients and fibre, and play a very important role in human health. In general, salad crops are not expensive. They are popular, easy to grow, have a long period of harvest, ship well, and have a relatively long storage life. They are important in neutralizing the acid substances produced during the course of digestion of meats, cheese and other foods. They are of value as roughage, which promotes digestion and helps to prevent constipation. A certain amount of bulky food is necessary for good health. Most vegetables, particularly the leafy ones like celery, cabbage, spinach and lettuce, are characterized by a high water content and a relatively high percentage of cellulose or fibre because of their succulence and large bulk. The leafy salad crops and root crops probably aid the digestion of the more concentrated foods. ***See*** Cellulose; Dietary Fibre, Physiological Effects; Dietary Fibre, Effects of Fibre on Absorption

The carbohydrates of vegetables comprise starches and sugars – mainly sucrose, glucose and fructose – which are easily available to the body. Vegetables also provide unsaturated lipids containing essential fatty acids such as linoleic, linolenic and arachidonic acid which are required for normal growth and maintenance of the body. Vegetables provide a variety of proteins. Proteins from the leafy parts of vegetables are comparable to animal proteins in terms of their amino acid make-up. This is due to the fact that the leaf is an actively metabolizing organ, and most of the proteins consist of enzymes; these proteins are more complete than the storage proteins derived from seeds. Thus, the consumption of leafy salads could make a major contribution towards maintaining optimum health of an individual. ***See*** Amino Acids, Metabolism; Carbohydrates, Requirements and Dietary Importance; Fatty Acids, Dietary Importance; Protein, Requirements

Spinach, kale, mustard, green turnip 'tops', spring onions, parsley leaves, chives, chicory and endive are excellent sources of vitamins A and C. Vitamin A ranges from 2500 to 14 000 iu and vitamin C from 30 to 180 mg per 100 g of the edible part (see Table 7). The plain leaf endive (escarole) and the dandelion (leaves) are much higher in vitamin A (14 000 iu) and contain 100 mg and 35 mg vitamin C per 100 g, respectively. ***See*** Ascorbic Acid, Physiology; Retinol, Physiology

Carrot roots contribute a significant proportion of the daily requirement for vitamin A in US diets for adults, and in addition to vitamins and minerals provide important dietary fibre. Beet crops and carrots are considered primarily as a source of carbohydrates, especially sugars.

Parsley and watercress are good sources of iron, calcium, vitamins A and C and riboflavin. Turnip greens (tops) are considered valuable in the diet primarily because of the mineral (calcium and iron) and vitamin A and C contents. Leafy, green and yellow salad vegetables are important sources of vitamin A (33%) and vitamin C (25%), and contain appreciable quantities of thiamin, niacin and folic acid. Salad crops ranking highest in vitamin A are carrot, turnip greens, spinach,

Table 1. Classification of salad crops on the basis of edible part and botanical family[a]

Edible part	Botanical name	Family
(a) **Underground portion**		
Enlarged tap root		
Beet	*Beta vulgaris*	Chenopodiaceae
Carrot	*Daucus carota* var. *sativus*	Umbelliferae (Apiaceae)
Celeriac	*Apium graveolens* var. *rapeceum*	Umbelliferae
Radish	*Raphanus sativus*	Cruciferae (Brassicaceae)
Turnip	*Brassica compestris* var. *rapifera*	Cruciferae
Parsnip	*Pastinaca sativa*	Umbelliferae
Tuber		
Jerusalem artichoke	*Helianthus tuberosus*	Compositae (Asteraceae)
Bulb		
Florence	*Foeniculum vulgare* var. *azoricum*	Umbelliferae
Onion	*Allium cepa*	Alliaceae
Leek	*Allium ampeloprasum* var. *porum.*	Alliaceae
(b) **Above-ground portion**		
Stem		
Asparagus	*Asparagus officinalis*	Liliaceae
Kohlrabi	*Brassica oleracea gongylodes* (group)	Cruciferae
Petiole		
Rhubarb	*Rheum rhabarbarum*	Polygonaceae
Celery	*Apium graveolens* var. *dulce*	Umbelliferae
Leaf		
Cabbage	*Brassica oleracea Capitata* group	Cruciferae
Chicory	*Cichorium intybus*	Compositae
Cress	*Lepidium sativum* L.	Cruciferae
Dandelion	*Taraxicum officinale*	Compositae
Garlic	*Allium sativum*	Alliaceae
Lettuce	*Lactuca sativa*	Compositae
Mustard	*Brassica juncea*	Cruciferae
Onion	*Allium cepa* group	Alliaceae
Endive	*Cichorium endiva*	Compositae
Parsley	*Petroselinum crispum*[b] var. *neopolitanum*	Umbelliferae
Watercress	*Rorippa nasturtium* var. *aquaticum.*	Cruciferae
Flowers		
Cauliflower	*Brassica oleracea botrytis* (group)	Cruciferae
Artichoke (globe)	*Cynara scolymus*	Compositae
Fruits		
Immature		
Cucumber	*Cucumis sativus* L.	Cucurbitaceae
Mature		
Tomato	*Lycopersicum esculentum*	Solanaceae

[a] All the salad vegetables belong to the Angiospermae and Dicotyledonae except onion, which belongs to subclass Monocotyledonae.
[b] Recently assigned to the genus, *Carum*; hence variety now recognized as *Carum petroselenum* var. *neopolitanum*

Dietary Importance

Table 2. Classification based on life cycle and optimum temperature (season)

Cycle/vegetable	Season
Perennial	**Cool season crops**
Artichoke	*Hardy*
Garlic	Asparagus
Tomato	Leek
Asparagus	Parsley
Dandelion	Rhubarb
Watercress	Turnip
Chicory	Cabbage
Onion	Dandelion
Rhubarb	Watercress
Chive	Garlic
	Onion
Biennial	Kohlrabi
Beet	Radish
Cabbage	
Celery	*Half-hardy*
Leek	Carrot
Parsley	Celeriac
Carrot	Fennel
Celeriac	Lettuce
Parsnip	Artichoke
Cauliflower	Beet
Florence fennel	Celery
Kohlrabi	Chicory
Turnip	Endive
Annual	**Warm Season**
Cucumber	Tender – tomato
Lettuce	Very tender – cucumber
Radish	
Endive	
Mustard	

Table 3. Classification based on tolerance to soil acidity[a]

Slightly tolerant (pH 6·8–6·0)	Moderately tolerant (pH 6·8–5·5)	Very tolerant (pH 6·8–5·0)
Asparagus	Carrot	Chicory
Beet	Cucumber	Endive
Cabbage	Garlic	Rhubarb
Cauliflower	Kohlrabi	Fennel
Celery	Radish	Dandelion
Leek	Tomato	
Onion	Turnip	
Lettuce	Mustard	
Cress		
Watercress		

[a] From Lorenz OA and Maynard DN (eds) (1982) *Knott's Handbook for Vegetable Growers*. p 840. New York: Wiley.

Table 4. Classification of salad vegetables by rooting depth[a]

Shallow (45–60 cm)	Moderately deep (90–120 cm)	Deep (>120 cm)
Cabbage	Beet	Artichoke
Celery	Carrot	Asparagus
Endive	Cucumber	Parsnip
Garlic	Mustard	Tomato
Leek	Turnip	
Lettuce		
Onion		
Parsley		

[a] From Lorenz OA and Maynard, DN (eds) (1980) *Knott's Handbook for Vegetable Growers*, p 139. New York: Wiley.

mustard and dandelion (Tables 5–7). By levels of vitamins and minerals, artichokes rank seventh among the top vegetables. Spinach, kale and collard contain high levels of vitamins and minerals, particularly vitamin A, calcium, phosphorus and iron. One serving meets the adult daily requirement of vitamin A and C, and 13% of the calcium requirement. Sprouting broccoli is most nutritious amongst the cole crops, especially for vitamins, calcium and iron. ***See*** individual vitamins and minerals

Kholrabi is superior to turnip and rutabaga, but is similar to red cabbage. Celery is a good source of calcium, phosphorus and iron and is also rich in carotene, a precursor of vitamin A, and in riboflavin and niacin. Spinach is one of the best vegetable sources of vitamin A, providing 8100 iu 100 g. This amount also provides 51 mg of vitamin C and 93 mg of calcium, making spinach amongst the more highly nutritious of the salad vegetables. Spinach is also a good source of iron (3·1 mg per 100 g), phosphorus (51 mg) and potassium (470 mg). The quality of spinach is related to its bright green colour, tenderness and flavour. ***See*** individual vitamins and minerals

Industrial Use

The continuous supply of fresh and processed vegetables on the markets of United States indicates the tremendous economic importance of the vegetable industry, and the contribution of vegetables to the dietary requirement of the people. With the expected increases in population, it is safe to predict that vegetable crops will continue to comprise a major segment of the agricultural economy.

The dehydration of onions is a very important industry in the United States, especially in California. Asparagus, tomatoes, lettuce, cucumber and carrot are

Table 5. Nutritional constituents of some of salad crops (per 100 g edible portion).[a]

	Water (g)	Energy (kJ(kcal))	Protein (g)	Fat (g)	Carbo-hydrate (g)	Vitamins A (iu)	C (mg)	Thia-min (mg)	Niacin (mg)	Minerals Ca (mg)	P (mg)	Fe (mg)	K (mg)
Asparagus	91·7	100 (26)	2·5	0·2	5·0	900	33	0·18	1·5	22	62	10·0	278
Artichoke	85·0	84 (20)	2·7	0·2	2·3	160	11	0·06	0·8	53	78	15·0	430
Rhubarb	95·0	67 (16)	0·6	0·1	3·7	100	9	0·03	0·3	96	18	0·8	251
Spinach	91·0	100 (26)	3·2	0·3	4·3	8100	51	0·10	0·6	93	51	3·1	470
Kale	83·0	220 (53)	6·0	0·8	9·0	10000	186	0·16	2·1	249	93	2·7	378
Mustard	90·0	130 (31)	3·0	0·5	5·6	7000	97	0·11	0·8	183	50	3·0	377
Turnip greens	90·0	117 (28)	3·0	0·3	5·0	7600	139	0·21	0·8	246	58	1·8	–
Beet greens	91·0	100 (24)	2·2	0·3	4·6	6100	30	0·10	0·4	119	40	3·3	570
Beet roots	88·0	180 (43)	1·6	0·1	9·9	20	10	0·03	0·4	16	33	0·7	335

[a] From Howard FD *et al.* (1962); USDA National Food Review, 1978.
In some cases, the analytical data relating to a given crop will appear to vary between Articles. These differences may arise because of contrasts in experimental techniques and/or raw materials employed by the original authors, and hence no attempt has been made to standardize the available information.

Table 6. Nutritional constituents of some salad crops (per 100 g edible portion)[a]

	Energy (kJ(kcal))	Macronutrient Water (g)	Protein (g)	Fat (g)	Carbo-hydrate (g)	Vitamins A (iu)	B_1 (mg)	B_2 (mg)	Niacin (mg)	C (mg)	Minerals Ca (mg)	Fe (mg)	Mg (mg)	P (mg)
Mustard greens	63 (15)	91	2·7	0·2	0·1	5300	0·08	0·11	0·8	70	140	2·0	48	45
Turnip (tops and roots)	75 (18)	92	1·8	0·2	0·2	2700	0·05	0·07	0·5	70	125	1·5	45	45
Greens	58 (14)	91	1·5	0·3	0·1	3400	0·07	0·10	0·6	60	90	1·1	31	42
Radish (Chinese winter Icicle, Scarlet globe	54–58 (13–14)	94	0·6–0·1	0·1	0·1–0·2	0	0·03	0·02	0·3	21	27	0·8	22	27
Carrot (roots)	130 (31)	89	0·8	0·2	6·6	1300	0·05	0·04	0·3	6	30	1·4	17	36
Celeriac roots	84 (20)	88	1·5	0·3	3·5	0	0·05	0·06	0·7	8	43	0·7	20	115
Celery														
Self-blanching (petiole)	29 (7)	96	0·7	0·1	1·2	90	0·03	0·02	0·3	7	25	0·3	10	27
Green (petiole)	34 (8)	95	0·9	0·1	1·2	120	0·03	0·04	0·3	10	70	0·5	14	34
Florence fennel	63 (15)	93	1·1	0·1	2·6	100	0·04	0·02	0·4	9	44	0·8	23	38
Parsley leaves	67 (16)	90	2·2	0·3	1·3	5200	0·08	0·11	0·7	90	125	2·0	79	40
Tomato fruit	80 (19)	94	0·9	0·1	3·7	1700	0·10	0·02	0·6	21	6	0·3	10	16
Cucumber fruit (immature)	50 (12)	96	0·6	0·1	2·2	45	0·03	0·02	0·3	12	12	0·3	15	24

[a] From Howard FD *et al.* (1962); USDA National Food Review, 1978.
In some cases, the analytical data relating to a given crop will appear to vary between Articles. These differences may arise because of contrasts in experimental techniques and/or raw materials employed by the original authors, and hence no attempt has been made to standardize the available information.

produced commercially and processed into different food products on an industrial scale. Sauerkraut is the main processed product from cabbage. Some cauliflower, cabbage, rat-tailed radish, turnip and carrot are stored, canned and shipped. Commercial quantities of chives are sold to processors, and chopped leaves are mixed with cottage cheese or cream cheese for the retail market. The commercial production of kale and spinach is concentrated along the east coast and California respectively. Chinese cabbage is fermented to produce

Dietary Importance

Table 7. Nutritional constituents of some salad crops (per 100 g edible portion)[a]

Crop	Edible part	Energy (kcal)	Macronutrient				Vitamin					Minerals			
			Water (g)	Protein (g)	Fat (g)	Carbo-hydrate (g)	A (iu)	B_1 (mg)	B_2 (mg)	Niacin (mg)	C (mg)	Ca (mg)	Fe (mg)	Mg (mg)	P (mg)
Chives	Leaves	20	92	2·8	0·6	1·1	6400	0·10	0·18	0·7	79	81	1·6	55	51
Onion, green	Bulb	21	90	1·3	0·2	4·0	330	0·06	0·05	0·3	32	62	0·5	25	43
	Leaves	19	92	2·0	0·2	3·4	5000	0·07	0·14	0·2	45	80	1·0	24	30
Artichoke, globe		20	83	2·7	2·3	2·3	160	0·08	0·06	0·8	11	53	1·5	48	78
Celtuce	Leaves	12	94	1·1	0·4	1·3	3500	0·09	0·12	0·5	33	59	0·8	38	34
Celtuce	Stalks	12	95	0·6	0·2	2·4	70	0·02	0·02	0·6	6	18	0·3	17	43
Chicory	Leaves	13	92	1·7	0·3	1·1	4000	0·06	0·10	0·5	24	100	0·9	30	47
Chicory	Roots	23	80	1·4	0·2	4·6	0	0·04	0·03	0·4	5	41	0·8	22	61
Endive	Leaves	11	95	1·3	0·2	1·2	2500	0·07	0·08	0·4	8	42	2·0	20	30
Endive	Curly	20	93	1·7	0·1	4·1	3300	0·07	0·14	0·5	10	81	1·7	—	54
Lettuce leaves															
Butter head		11	96	1·2	0·2	1·2	1200	0·07	0·07	0·4	9	40	1·1	16	31
Cos (Romaine)		16	94	1·6	0·2	2·1	2600	0·10	0·10	0·5	24	36	1·1	6	45
Crisp head (great lakes)		11	95	0·8	0·1	2·3	300	0·07	0·03	0·3	5	13	1·5	7	25
Cabbage (White, Red, Savoy),		20	92	1·2–2·0	0·1	0·4	200–1000	0·06	0·03	0·3	31–60	35–51	0·4–0·7	17–28	34–42
Cauliflower (Snowball)		22	90	2·2	0·3	0·9	40	0·08	0·06	0·6	72	30	0·5	12	60

[a] From Howard FD *et al.* (1962); USDA National Food Review, 1978.

In some cases, the analytical data relating to a given crop will appear to vary between Articles. These differences may arise because of contrasts in experimental techniques and/or raw materials employed by the original authors, and hence no attempt has been made to standardize the available information.

the Korean dish, 'kimchi', which includes either fermented chinese cabbage or fermented radish along with hot peppers, garlic and other ingredients. Lettuce leaves are sometimes used as a substitute for tobacco in cigarettes.

Use in the Home as Raw or Industrially Processed Food

Freshness is the key to crisp, tasty salads: salad crops are normally consumed uncooked and in a raw state. They are attractive and have value for their appetizing succulence, bulk and nutrients. They are also processed to obtain more tasty food products. Turnip, radish, carrot, cucumber, cabbage and cauliflower are used for preparing pickles in addition to their use in fresh salads. Celery, lettuce, tomatoes and onions are the most important salad crops available throughout the year, but carrots, onion, garlic, rhubarb, cabbage, watercress, mustard, spinach and lettuce have been used for medical purposes. The mustard plant is used as medicine for diseases of the liver and spleen, and also as a tonic in India. In 350 BC, cabbage was used as a treatment for gout, stomach problems, deafness, headache and hangovers. *See* Pickling

Celery, chervil, leeks and tomatoes are used in soups and stews. Rhubarb and tomatoes are also used for preparing sauces. Fringed-leaved endive, parsley, chervil, chicory, leeks and fennel leaves are also used for flavouring and for garnishing. Cucumber, cabbage, turnip, radish and lettuce are industrially fermented to dill pickles, sour pickles, salt stock, sauerkraut, sauerriipen, lettuce kraut and pawtsay. These fermented foods are more flavourful than the raw materials. Chicory roots are dried, roasted and ground for use as a coffee substitute, or in blends with coffee. Endive and chicory leaves are used as pot-herbs to some extent. *See* Chicory Beverages

Globe artichoke and celery are used as luxury vegetables. Chives are used in salads or in omelettes and other dishes. The early use of rhubarb was as a medicine. Rhubarb or 'pie plant' is prized for use in pies, tarts and sauces. It is also used for making wines and tanning hides; rhubarb leaves are toxic to humans. Globe artichokes have been reported to have medicinal benefits, including reduced blood clotting and capillary resistance, neutralization of some toxic substances and amelioration of gastrointestinal processes. A relative of the globe artichoke, a leafy plant known as cardoon, is used for its edible roots, or more often its petioles which may be cooked or eaten raw in salads.

Onions have been prized for their medicinal value since Gerarde, in the *Great Herbal* (1596), stated 'The juice of onions snuffed up into the nose purgeth the head . . . Stamped with salt, rue and honey . . . they are good against the bitting of mad dogs . . . annointed upon a bald head in the sun bringeth the hair again very speedily.'

Bibliography

Howard FD, MacGillivray JH and Yamaguchi M (1962) *Bulletin No. 788*. Berkley: California Agricultural Experimental Station.

Lincoln CP (1987) *Vegetables: Characteristics, Production and Marketing*. New York: Wiley.

Ryder EJ (1979) *Leafy Salad Vegetables*. Westport: AVI.

Salunkhe DK and Desai BB (1984) *Post-harvest Biotechnology of Vegetables*, vol. II. Boca Raton: CRC Press.

Wood R (1979) *Salad Crops*. London: Marshall Cavendish.

Yamaguchi M (1983) *World Vegetables: Principles, Production and Nutritive Values*. Westport: AVI.

K. Gupta, G.R. Singh, J.L. Mangal and D.S. Wagle
Haryana Agricultural University, Haryana, India

Leaf-types

Leafy salad vegetables are amongst the most important and universally used vegetable crops grown throughout the world, and are also available throughout the year. They are attractive in appearance and have value for their appetizing succulence, bulk and nutrients.

Leafy salad vegetables seem to be a natural crop grouping have mostly crisp, tasty and tender edible leaves, and their prominence as important crops has been heightened over several decades through an awareness on the part of consumers of the nutritional and other obvious values they offer to the diet as 'fresh greens'. These have become a daily table staple, at least in the United States. As a result, the acreage planted to leafy salad vegetable crop has expanded to a remarkable degree over the past few decades, making these crops an important segment of the agricultural and marketing industries.

Nearly all vegetables can be consumed raw in salads, but few of them are commonly referred to as salad crops and, of these, only a few are logically considered as leafy salad vegetables based on the criterion that their leaves alone are consumed. Leafy salad crops have been defined as these crops from which leaves and associated parts are harvested for use as raw vegetables. Mostly leafy salad crops are considered 'high risk' vegetables. Their succulent leaves and petioles are susceptible to damage by pests, by poor production and handling systems, and by environmental extremes over which a grower may have little or no control.

There are four major leafy salad crops – lettuce (*Lactuca sativa* L.), celery (*Apium graveolens*), endive

(*Cichorium endivia* L.) and chicory (*Cichorium intybus*) and a number of lesser vegetables used primarily for salads, flavouring or garnishing foods. The latter include watercress (*Rorippa nasturtium aquaticum*), garden cress (*Lepidium sativum*), parsley (*Carum petroselinum*), chervil (*Anthriscus cerefolium*), dill (*Anethum graveolens*), mustard (*Brassica juncea*), Chinese cabbage (*Brassica pekinensis*) and spinach (*Spinacia oleracea*). Most salad crops thrive best during cooler and moist seasons, and do not tolerate hot weather. Leafy salad vegetables are nutritionally important sources of protein, fibre, minerals and vitamins.

Global Distribution, Commercial Importance and Varieties

Lettuce

Lettuce, a most popular leafy salad crop, is grown throughout temperate, tropical and subtropical climates and is of great commercial importance. Lettuce is known as 'green gold' in California, as befits its place as the major salad crop grown in the most important vegetable-producing state. Lettuce (*Lactuca sativa* L.) is an annual, belonging to the family Compositae. Both cultivated and wild lettuce species are native to Europe and Asia, and have been in cultivation for the last 2500 years. In India, it was introduced by the Portuguese and the British during the 16th century.

Its production and consumption has increased several-fold over the last two decades throughout temperate countries. However, in the tropical and subtropical world, its popularity still rests with urban populations. Lettuce ranks sixth in acreage and third in value among 22 principal vegetables. It is widely used in all salad preparations, and has a somewhat bizarre use as a substitute for tobacco.

Thompson and Kelly (1957) stated that four types of lettuce are grown in the United States and each is recognized as a subspecies or botanical cultivar: (1) head type, *capitata*; (2) cutting or leaf type, *crispa*; (3) Cos or Romaine, *longifolia*; and (4) Asparagus or stem lettuce type, *asparagina*.

There are six morphological types of commercial lettuce cultivars: crisphead, butterhead, cos, leaf, stem and latin. The crisphead type is most common in the United States. About 70% of US crisphead lettuce is produced commercially in California, 16% in Arizona, and the rest in several other states. Crisphead lettuce is often called Iceberg lettuce. Butterhead and Cos (Romaine) are popular in northern and southern Europe.

Tracy (1904) and Thompson (1951) report 150 distinct varieties, of which 20–25 are commercially important. They have classified these varieties into three general classes, with subclasses:

1. *Butter varieties*
   (I) Cabbage heading varieties
   (II) Bunching varieties
2. *Crisp varieties*
   (I) Cabbage heading varieties
   (II) Bunching varieties
3. *Cos varieties*
   (I) Spatulate-leaved varieties
   (II) Lanceolate-leaved varieties
   (III) Lobed-leaved varieties

Most of the important commercial varieties are in the 'crisphead' class. The best known names are New York 515 and Imperial 44, 152, 456, 615 and 847; Great Lakes and Pennlake are other important crisphead varieties; there are several strains of Great Lakes. Big Boston, White Boston, May King, Salamandar and Wayahead are the best known varieties of butterhead type, but are of relatively little importance commercially. Black-seeded Simpson, Early-curled Simpson, Grand Rapids and Slowbolt are the main varieties of leaf or bunching lettuce. Paris white (White Cos) and Dark Green varieties are of the Cos type.

Ryder (1981) and Tigchelaar (1980) report on the recently released cultivars Sea Green, Empress Calsweet, Green Ice, Cabrillo, Cal K-60, Calmaria, Chaparral, Mesaverde, Morangold, Morgangourd and Royal Oak leaf. Great Lakes (headtype) and Chinese yellow (loose leaf type) cultivars of lettuce are commonly grown in India. Imperial 859 and Slow Bolt are also improved cultivars of lettuce, and all four cultivars have been released by the Indian Agricultural Research Institute, New Delhi.

Endive and Chicory

Endive and chicory are especially popular salad vegetables in Europe, and their popularity has also increased in the United States. The two plants are closely related. Endive is believed to be a native of Egypt or India. During the 13th century, it reached northern Europe. It was noted in the United States in 1806. Chicory is also known as French endive, witloof, witloof chicory and succory. The first mention of chicory culture was in 1616 in Germany. Cultivation began in England in 1686 and in France in 1826. Endive is an important market garden crop in Florida, USA. Endive is classified into two general groups: (1) the curled or fringe-leaved cultivars and (2) the broad-leaved cultivars. The green curled Ruffic, Deep Heart Fringed, Green Curled Pancalier and White Curled are curled or fringe-leaved types. Varieties representing the broad-leaved class are Broad-leaved Batavian, Full-heart, Batavian or Escarole and Florida Deep Heart.

Radichetta and Rosso di verona are the principal chicory cultivars in United States and Italy, respectively.

Penninck, Christaens, Mueninck and Van Espen are Belgian cultivars, while Normato, Malina, Bubbal Blank and Slusia Meilof are Dutch cultivars. Flambor, Bergere and Zoom are chicory hybrids.

Parsley

Parsley (*Carum petroselinum*) is native to the Mediterranean region and Europe. It was popular in early times among the Greeks and Romans. Its value was reputed to be in aiding digestion, and in suppressing odours of onion and wine. It has been in cultivation for over 2000 years. Parsley is produced commercially in Texas, California, New Jersey, Florida and New York. Parsley was introduced into England from Sardinia in 1548, and in to America by 1806.

Five types of parsley cultivars are available: plain-leaved, celery-leaved or Neapolitan, curled, fern-leaved, and turnip-rooted type. The curled type is common in the United States. Plain and turnip-rooted parsley are also grown in the United States. The best-known varieties are Moss, Extra Double Curled, Fern-leaved, Curled dwarf, Evergreen, Extra Triple Curled, Plain and Dark Green Italian. Morgo was developed in Sweden, and Hamburg is the only turnip-rooted cultivar.

Watercress

Watercress is unique among the vegetables, as it commonly grows in flowing water as a wild plant or as a cultivated crop. Watercress is believed to be native to the eastern Mediterranean and Europe. Watercress grows wild in streams in the cool regions of the world, but is also found growing wild above 1500 m in the Himalayas of Nepal and hilly areas of the United States. Most production is in Virginia, Maryland and Pennsylvania. It is also grown in California, the United Kingdom and Europe. It was cultivated in England for the first time in 1808, and in France in 1811. It enjoys worldwide distribution, and when grown commercially outdoors it is planted in beds covered with slowly flowing, clean water. There are two main cultivars, namely Green cress which is a diploid and fertile, and Brown cress or winter cress, which is triploid and sterile.

Cress and Chervil

Cress is also known as Peppergrass or garden cress, and is grown as a salad crop all over the world. It is native to Iran and Europe. From Persia it spread to India, Egypt and Greece. It was mentioned by Xenophon in 400 BC, and has been cultivated in England since 1548. It grows wild and cultivated in the United States.

Four types of cress are cultivated: common, curled, broad-leaved and golden cultivars. Chervil is native to Europe and has finally divided pinnate leaves. It was cultivated in England in 1597 and in America by 1806.

Spinach

Spinach (*Spinacia oleracea*) is the most important 'potherb' used raw in salads, albeit on a limited scale, and it is grown worldwide. Spinach is native to Central Asia, most probably Persia (Iran) and was cultivated by the Persians 2000 years ago. Cultivation of spinach developed during the Greek and Roman civilizations, and it was introduced into China in AD 647. The crop was known in Germany in the 13th century, in North Africa to Spain in 1100, and in America in 1806.

The major, fresh market, commercial areas are California, mostly in Ventura county, the Winter Garden region of the lower Rio Grande Valley of Texas, Maryland, New Jersey and Colorado. California is responsible for one-third of the total produced for industrial processing.

Jones and Rosa (1928) tested 121 varieties obtained from American and European sources and concluded that only five distinct varietal types exist. They classified the varieties into two classes or groups: flat-leaved types and savoy-leaved types, instead of the original four groups – Norfolk or Bloomsdale, round-leaved, thick-leaved and prickly-leaved. Juliana, Virginia Savoy, Old Dominion, Hollandia, Viroflag, Amsterdam Giant, King of Denmark and Nobel are important varieties of the two classes. Beltsville Arlington (Va.), Beltsville (Md), Winter Haven (Tex.) and Davis (Calif.) are longstanding varieties in different countries, and many hybrids are available.

Mustard

There are several species of mustard used for spice, flavouring, oil seeds, fodder, manures and 'greens' or salad. All the Brassicae are native to central Asia and the Himalayas, with secondary centres in India and eastern China. White mustard (*Sinapsis alba*), Brown mustard (*Brassica nigra*) and Ethiopian mustard (*Brassica corinata*) are commercially important in many parts of the world, and grown all over the world for their various uses. They are particularly common in the United States, Canada, Great Britain, most countries of Europe, India, China and the southeast Asian countries. In the United States, mustard for salads and greens is grown in Texas, California, Florida, Georgia, Louisiana and Mississippi. Mustard leaves are very strongly flavoured and pungent. The inner, younger leaves are milder and are best suited for salad use. There are two principal salad or 'greens' cultivars, namely Florida (broad leaf) and Southern (giant curled leaf) which are grown in the United States.

Chinese Cabbage

Cabbage and Chinese cabbage are grown as potherbs and also as salad vegetables and, a native to China, they have been cultivated since the 5th century and were introduced to the United States in the late 19th century. Numerous types of Chinese cabbage are extremely important in the diets of Asians, used raw in salads, cooked as 'greens' or fermented to preserve it for later use. The Korean dish 'kimchi' includes fermented Chinese cabbage or fermented radish along with hot pepper, garlic and other ingredients. In the United States, Chinese cabbage is grown throughout the year in California, Florida and Hawaii. Chinese cabbage is increasing in importance in European countries, notably Britain, France and Germany. The most popular varieties are chihili and Wong Bok, grown in the United States. Two more or less distinct species are grown; Petsai (*Brassica pekinensis*) and Pakchoi (*B. chinensis*). Shantung cabbage, chefoo cabbage, Pekin cabbage, chou de chine and celery cabbage are known varieties of Chinese cabbage.

Morphology and Anatomy of the Edible Portion

The cultivated lettuce (*Lactuca sativa* L.) is related to wild lettuce (*L. scariola*), a common weed in the United States. It forms a rosette of large, longish leaves which are somewhat spoon-shaped, and more or less undulated/toothed at the edges. From the centre of the rosette springs a nearly cylindrical stem, which narrows very rapidly and branches at about one-third of its height. It is furnished with clasping leaves, which encircle the stem and become narrower as they approach the top. There are six morphological types of lettuce; crisphead, butterhead, cos, leaf, stem and latin. Head-type lettuces have unbranched stems which generally remain less than 30 cm long, owing to the growth in length being arrested at an early stage. As the growing point continues to form leaf primordia, a rosette of sessile leaves arises. The first leaf unfolds normally, and can reach a length of over 30 cm. After some time, other leaves are produced which unfold only partially and form a 'skin', embracing the laterally formed leaves which do not expand. By this continuous segregation and growth of young leaves, a head is developed. Crisphead cultivars are usually large, heavy, brittle-textured and tightly folded, with green outer and white or yellowish inner leaves. Leafy lettuces primarily produce a rosette of leaves and have no heading tendency. Cos (Romaine) lettuces have elongated heads of long leaves with heavy midribs. The outer leaves are coarse in appearance and dark green, but the inner leaves are fine-textured and light green. Stem lettuce is sold as 'celtuce' in the United States; the stems are peeled and used as raw or cooked vegetables. Latin lettuce leaves are somewhat elongated, but more leathery than Romaine. Butterhead is characterized by its heads of crumpled leaves, which have a very soft, buttery texture; veins and midribs are less prominent than in crisphead types. The variety Imperial has light to medium green leaves with serrated or wavy edges, and is of relatively soft texture. Great Lakes leaves are brittle with ruffled margins, and are bright green to yellow with prominent ribs. Empire leaves are light green, deeply serrated and crisp, and the heads are often slightly elongate or conical.

Vangourd is dull green with a yellow interior, and the leaves have a softer texture than Great Lakes, with scalloped margins and flat ribs.

Endive grows as a loose head of leaves, which are usually strongly ruffled and serrated. The outside leaves are green and bitter, but the inner leaves are light green to whitish. In the case of Radichetta chicory, the leaves are dark green, narrow and notched, and the Italian cultivar is a red, nonforcing type. Endive is a plant with numerous radical leaves, which are smooth, lobed, more or less deeply cut and spreading into a rosette; the stem is hollow, from 50 cm to over 1 m high, ribbed and branching.

Parsley is a rosette of divided leaves on a short stem. The most common and attractive is the curled-leaf type, which contains three subtypes – doubled-curled (moss curled), evergreen and triple-curled. These subtypes are distinguished by the degree of leaf curling, coarseness of the leaf and plant growth habit. The plain leaf parsley (Dark Green Italian) has deeply cut leaves but no curling or fringing. Moss curled (double-curled) has a stem 30·5 cm tall, and vigorous, compact and very dark green leaves, which are finely cut, deeply curled and frost resistant. Evergreen leaves are coarsely cut, while extra triple-curled leaves are finely cut and very closely curled; the triple-curled variety has slightly shorter leaves which are closely curled. The leaves of Paramount are tall, very uniform, triple-curled and very dark green colour. Plain-leaved types include plain (singles) and have flat leaves. The leaves of the Dark Green Italian cultivar are heavy and glossy green.

Watercress has three sections: projecting above the water is an aerial stem portion with an apex, leaves but no roots; a stem portion beneath the flowing water which has foliage and a system of adventitious roots which arise from the leaf axils and remain free in water; and normal roots. Watercress leaves are pinnately compound with 3–11 leaflets, rounded or oblong, slightly fleshy and with smooth margins.

Table 1. Nutritional constituents of the major salad crops (data per 100 g basis)

						Vitamins					Minerals		
Crop	Water (g)	Energy (kJ (kcal))	Protein (g)	Fat (g)	Carbohydrate (g)	A (iu)	C (mg)	B_1 (mg)	B_2 (mg)	Niacin (mg)	Ca (mg)	P (mg)	Fe (mg)
Lettuce	96	54 (13)	0·9	0·1	2·9	330	66	0·60	0·06	0·3	20	22	0·5
Chicory	95	63 (15)	1·0	0·1	3·2	Trace	—	—	—	—	18	21	0·5
Endive (curly)	93	84 (20)	1·7	0·1	4·1	3300	10	0·07	0·14	0·5	81	54	1·7
Parsley	85	184 (44)	3·6	0·6	8·5	8500	172	0·12	0·26	1·2	203	63	6·2
Spinach	91	109 (26)	3·2	0·3	4·3	8100	51	0·10	0·20	0·6	93	51	3·1
Mustard	90	130 (31)	3·0	0·5	5·6	7000	97	0·11	0·22	0·8	183	50	3·0
Chinese cabbage	95	58 (14)	1·2	0·1	3·0	150	25	0·05	0·04	0·6	43	40	0·6
Garden cress	82	280 (67)	5·8	1·0	8·7	—	—	0·15	—	—	360	110	28·6
Watercress	89	138 (33)	2·9	0·2	4·9	2803	13	0·12	0·38	0·8	290	140	4·6

From USDA (1978) *National Food Reviews*: Gopalan C, Ramasastri, BV and Balasubramanian, SC (1971) *Nutritive values of Indian Foods.* Hyderabad: National Institute of Nutrition.
In some cases, the analytical data relating to a given crop will appear to vary between Articles. These differences may arise because of contrasts in experimental techniques and/or raw materials employed in experimental techniques and/or raw materials employed by the original authors, and hence no attempt has been made to standardize the available information.

Cress (*Lepidium sativum* L. Cruciferae) is a much-branched annual with linear leaves. The moss curled (curled) variety is branching, dark green and with a mild, but distinct, flavour.

In general, spinach produces a rosette of fleshy leaves, which may be crinkled (Savoy) or smooth in the vegetative stage. The leaves are ovate, rounded or triangular, succulent and borne on a short stem. In the second stage of growth, the stem elongates, producing a seed stalk with narrow, pointed leaves. The shape, size and colour may differ and be characteristic of a varietal group. Viroflay, Fruremona and Wiremona are used for processing, and have large, smooth, dark green/medium green leaves, and a semi-erect or spreading habit. Most common names are similar in various languages.

The basal rosette leaves of *mustard* vary in form, while the stem leaves are mostly entire. Leaves are large, broadening towards the apex, and have smooth or curled margins. Florida Broad Leaf is a vigorous, large and erect plant, with large, thick, broad, oval leaves with toothed margins. Southern Giant (curled leaf) is large and upright, with large, wide leaves and curly leaf margins.

An improved 'chihili' called Michihili is the most common variety of *Chinese cabbage*. It is uniform in size and heading. The heads are usually 46 cm long and 10 cm thick, cylindrical in shape with a tapered tip. The outer foliage is dark green, the inner leaves blanched, and the midrib is white and broad. Won Bok is a very old Chinese cultivar with light green leaves. The Pe-tsai resembles cos lettuce, but produces a much larger, elongated and compact head; the leaves are slightly wrinkled, green, thin and much-veined. The midrib is broad and light in colour. The Pakchoi variety resembles Swiss chard in its habit of growth. The leaves are long, dark green and oblong or oval, and do not form a solid head.

Chemical and Nutritional Composition

Lettuce, on the basis of overall consumption, is rated fourth, behind only tomato, citrus and potato, and 26th on the basis of nutritional value. Butterhead lettuce is more nutritious than the crisphead type owing to the large number of green leaves. Still more nutritious are cos and leaf lettuces, mainly because of their high vitamin A (1900 iu per 100 g) and vitamin C (18 mg per 100 g) values. Lettuce is also a good source of calcium and phosphorus. By contrast, crisphead lettuce supplies moderate amounts of ascorbic acid (6 mg per 100 g), vitamin A (330 iu per 100 g) and calcium (20 mg per 100 g), whilst butterhead lettuce (per 100 g) contributes 8 mg ascorbic acid, 970 iu of vitamin A and 35 mg of calcium. Although lettuce has a very high water content (94–95%), it can still prove a useful source of minerals. ***See*** individual nutrients

Fringed-leaved endive is more ornamental and more popular as a salad vegetable than is the broad-leaved type. These are used as potherbs as well as in salads. Broad-leaved types are used mainly unblanched in stews and soups. Curly endive (per 100 g) provides 81 mg calcium, 3300 iu of vitamin A and 10 mg of vitamin C, as well as moderate amounts of phosphorus, potassium and iron. Escarole has 14 000 iu of vitamin A per 100 g and 100 mg of vitamin C. All endives contain moderate amounts of minerals (Table 1).

Chicory is an important salad vegetable. Witloof chicory has a delicate flavour, and is used in salads in France, Belgium and Holland. The roots dried, roasted and ground are also used as a coffee substitute, or in blends with coffee. ***See*** Chicory Beverages

Parsley is popular as a garnish for salads, sandwiches and cooked dishes, as a flavouring for soups and pasta and as a salad ingredient – only in small amounts owing to its pungent flavour. The Hamburg type is used mainly in stews and soups. The swollen root of turnip-rooted parsley is eaten as a cooked vegetable. Parsely is one of the best vegetable sources of vitamins A and C, but its contribution to the diet is negligible.

Parsley contains (per 100 g) 8500 iu of vitamin A, 172 mg of vitamin C and 203 mg of calcium. It is also a good source of potassium, and a moderate source of iron, sodium and phosphorus.

Watercress is used in mixed salads, in sandwiches, as a garnish and in soups, and has a pungent taste. It was used as a medicinal plant in AD 77, and it is antiscorbutic. Raw leaves (100 g) and stews supply 79 mg of ascorbic acid (vitamin C), 4900 iu of vitamin A, 151 mg of calcium and moderate to large amounts of other elements. Cress leaves are harvested for garnishing, add pungency to salads, and resemble radish in flavour. Seedlings with the hypocotyl and green cotyledons are cut and used in Britain for salads and sandwiches. It has a pungent taste. Chervil may also be used in salads, as a garnish, in soups or as a cooked vegetable.

Spinach is an important potherb or green vegetable, and is used raw in salads or mixed with other salad vegetables. It is also processed and canned. Spinach is the best vegetable source of vitamin A, providing 8100 iu per 100 g, and it also contains 51 mg vitamin C and calcium (93 mg). Spinach is a good source of iron (3·1 mg), phosphorus (51 mg), and potassium (470 mg). However, the calcium is said to be unavailable owing to the presence of oxalic acid, which forms calcium oxalate. It may also contain appreciable amounts of nitrate and nitrite, which are injurious to health.

The young, inner leaves of mustard are tender and mild in taste, and are consumed in salads; they are best mixed with bland leaves like lettuce. The crop is also a source of flavouring, spice, oil seed and green fodder. Mustard greens are highly nutritious. A 100-g serving of cooked mustard contains 48 mg of vitamin C, 5800 iu of vitamin A, 138 mg of calcium and moderate amounts of other elements (Watt and Merrill, 1963).

Cabbage and chinese cabbage are important as potherbs and as salad crops. Chinese cabbage provides 25 mg of vitamin C and 240 iu vitamin A and moderate amounts of calcium, phosphorus, iron and potassium in 100 g of raw product (Watt and Merrill, 1963).

Handling and Storage (Fresh Produce Only)

The harvesting of lettuce begins as soon as the plant reaches acceptable size and firmness. The lettuce grown for market is allowed to grow to full size and develop a solid head; for home use it is often harvested before the head is well developed. The heads are hand-cut just below the lowest leaf. Loose and damaged leaves are removed, and the product is washed prior to packaging. There are three general types of market preparation:

1. *Naked pack*: The heads are cut, trimmed and packed in two tiers (12 heads each) in cardboard boxes.
2. *Source wrapped*: After being cut and trimmed, the heads are wrapped, heat sealed and placed 24 per carton.
3. *Bulk pack*: Cut heads are loaded into bulk containers and transported to a central packing shed where they are cored, shredded, washed and cooled. Cooling is predominantly by the vacuum method, particularly for head lettuce shipped to distant markets. Hydrocooling is common for market gardens. Most lettuce is shipped by refrigerated trucks. ***See*** Chilled Storage, Packaging Under Vacuum

Most commonly, lettuce is packed in cartons and vacuum cooled immediately after harvest. Lettuce intended for long storage should be packed in perforated plastic films – used as individual head wraps, or in carton or crates with liners to maintain a high humidity around the lettuce. It is advisable to trim two 'wrapper' leaves before packing, rather than keep all 5–6 leaves, to save on space and weight. The extra wrapper leaves are not needed to maintain quality. Lettuce can be stored for 2–3 weeks at 0°C (32°F); high relative humidity (95%) should be maintained to keep the lettuce fresh. It will keep about twice as long at 0°C (32°F) as at 3–4°C (38°F). The rate of respiration in lettuce increases greatly with an increase in storage temperature, and leaf lettuces respire twice as fast as head lettuces. Lettuce is not tolerant to carbon dioxide, and is injured by a 4–5% concentration of CO_2. Problems such as russet spotting, bacterial soft rot and carbon dioxide injury can cause serious losses. ***See*** Chilled Storage, Use of Modified Atmosphere Packaging; Chilled Storage, Effect of Modified Atmosphere Packaging on Food Quality; Controlled Atmosphere Storage, Applications for Bulk Storage of Foodstuffs

Endive and green chicory are harvested when the heads are compact, bright-coloured and well-sized; the plants are cut at the base, and damaged or diseased leaves are trimmed. The trimmed heads are washed and cooled by vacuum cooling or hydrocooling to 0°C (32°F), the heads are then packed in wire-bound crates or cartons or baskets for transit. They are stored at 0°C (32°F) and 90–95% relative humidity for 2–3 weeks.

They can be stored for a longer period with crushed ice in and around the package.

Chicory is harvested by cutting off the head at the base, removing the outside leaves, and packing in baskets or in layers in crates. The smaller package is desirable.

Parsley can be harvested over a long period of time, cutting outer and larger leaves only, and tying in bunches for market. The leaves are packed loose or bunched, and washed thoroughly before packing in crates or bushel baskets. Crushed ice is usually placed in the package, and shipment is made under refrigeration to distant markets. Much of the crop goes into the institutional trade (restaurants and hotels) and for dehydration. Under storage at 0°C (32°F) and high humidity, parsley can be held for up to 2 months.

Watercress is a highly perishable crop and needs prompt handling and refrigeration after harvesting. Watercress will lose its colour and the leaves become slimy when improperly handled. It should be kept cool during shipment by spraying with cold water and/or storage in ice, and stored at 0°C (32°F) with a high relative humidity. The use of polyethylene film liners may be justified to prevent moisture loss under some circumstances.

Spinach is harvested at maturity, and for market the plants are cut just above the soil surface; for processing, the plants are cut 5 cm above the soil surface to allow for regrowth. Diseased or yellow leaves are removed, and the crop must be handled carefully to prevent bruising or breakage of leaves and stems. Washing in tanks usually results in injury to the leaves, so spraying with water on a moving belt is advantageous. Spinach is packed in various containers, including round bushel baskets, hampers and crates. For long hauls, crushed ice is placed in the containers, and shipment is made in refrigerator cars or trucks. Spinach is stored in bags of transparent film.

In the case of mustard, small bundles are made after discarding old and diseased leaves. It is handled like other greens at 0°C (32°F). It is commonly shipped with in-package and top ice to maintain freshness. Vitamin content and quality are retained so long as wilting is prevented.

Chinese cabbage is harvested by hand, at the base, trimmed of outer diseased or injured leaves and packed individually or in cartons or wooden crates and vacuum cooled. Spacing in storage should be allowed for air circulation. It can be stored for 1–2 months at 0°C (32°F) with 90–95% relative humidity.

Bibliography

Jones HA and Rosa JT (1928) *Truck Crop Plants*. New York: McGraw-Hill.

Pierce LC (1987) *Vegetables: Characteristics, Production and Marketing*, pp. 163–268. New York: Wiley.

Ryder EJ (1979) *Leafy Salad Vegetables*. Westport: AVI.

Ryder EJ and Whitaker TW (1980) The lettuce industry in California. A quarter century of change, 1954–79. *Horticultural Reviews* 2: 164–207.

Ryder EJ (1981) 'Sea Green' lettuce. *Horticultural Science* 16(4): 571.

Salunkhe DK and Desai BB (1984) *Post Harvest Biotechnology of Vegetables*, vol. II. Boca Raton: CRC Press.

Thompson HC and Kelly WC (1957) *Vegetable Crops*, 5th edn. New York: McGraw-Hill.

Thompson RC (1951) *Lettuce Varieties and Culture*. USDA Farmers Bulletin No. 1953.

Tigchelaar EC (ed.) (1980) New vegetable varieties List XXI. The Garden Research Committee, American Seed Trade Association. *Horticultural Science* 15(5): 565.

Tracy WW (1904) American varieties of lettuce. USDA Business Periodicals Index Bulletin No. 69.

Watt BK and Merrill AL (1963) *Composition of Food*. USDA Agricultural Handbook No. 8.

Wood R (1979) *Salad Crops*. London: Marshall Cavendish.

Yamaguchi M (1983) *World Vegetables: Principles, Production and Nutritive Values*. Westport: AVI.

K. Gupta, J. L. Mangal, G. R. Singh and D. S. Wagle
Haryana Agricultural University, Haryana, India

Leaf Stem Crops

Among leaf-stem types of salad crops, celery (*Apium graveolens*) is the most important. The other crops of this group are American winter green (*Gaultheria procumbens* L.), florence fennel (*Foeniculum officinale* D.C.) and dandelion (*Taraxacium officinale* Wig). These crops are consumed in different forms, especially in temperate countries.

Celery is important in Europe and America. It is native to Mediterranean regions, and its habitat extends from Sweden southwards to Algeria, Egypt and Abyssinia, and in Asia from the Caucasus, Buluchistan to the mountains of India. Wild celery can be found in damp or marshy areas of different European countries, Egypt and parts of Asia. It was first used as a medicinal plant, and as food in AD 1500. Homer's *Odyssey* mentions salinon (celery) grown in 850 BC, but this early form, later called 'smollage', was a leafy plant, quite pungent and bitter, used exclusively for medicines.

Fresh use of celery began in the 18th century, and its use as a vegetable in the United States was recorded in 1806. It is used in mixed salads or as an appetizer, in which case it is often stuffed with cream cheese or peanut butter. The petioles are sometimes cut up and served as cooked vegetables, although this use is declining. Some celery is grown for processing in soups or pickled vegetable combinations. In the United States, a large part of the celery crop is consumed in the raw state, but considerable quantities are used in vegetable juices, soups, stews and as cooked vegetables. In England, some celery is also canned.

American winter green (*Gaultheria procumbens* L.) is commonly used as a salad crop for its tender leaves and stem. They are consumed in America and other Western countries, as well as Asia to Australia. It belongs to the family, Ericaceae, and *Gaultheria* was named after Dr Gaultier, a physician in Quebec, about 1750.

Florence fennel is also a leaf-stem type salad crop. *Foeniculum officinale* is native of southern Europe, and resembles wild fennel. The famous 'Carosella', so extensively used in Naples and scarcely known in any other place, is referred by authors to *Foeniculum piperitum*, D.C., a species very closely related to *F. officinale*. The plant is used while running to bloom. The stems, fresh and tender, are served up raw and still enclosed in the expanded leaf stalks; it is available only from the end of March until June. Italian fennel (*F. dulce*), also called finnochio, is a short-lived perennial cultivated as an annual. The base of the leaves becomes enlarged and bulb-like, and when blanched has the texture of celery. It is used primarily for flavouring.

The word 'dandelion' comes from the French 'Dent de Lion' meaning 'lion's tooth' and reflects the jagged shape of the leaf. It is a great favourite in Europe, and to a lesser extent in the United States. Previously it was considered a noxious weed, but now cultivated species are available. The cultivated dandelion has been developed from the wild species, *Taraxacum officinale*, a member of the Compositae family. Its leaves are used raw in salads or cooked as greens. The flowers can be fermented to produce wine. The roots may be used as a coffee adulterant similar to chicory roots. ***See*** Coffee, Analysis of Coffee Products

Morphology and Anatomy of the Vegetables

Celery is a member of the family Apiaceae (Umbelliferae). The most common form bears 5–12 thick petioles in a tight bunch or head. A less popular form is turnip-rooted celery, known as celeriac (var. *rapeceum*). It is a biennial, cool season crop, but it is sensitive to prolonged cold temperature. Growth through the first year normally produces a tight cluster of petioles and leaves attached to a very compressed stem. Celery leaves are divided, pinnatified, smooth with almost triangular, toothed leaflets which are dark green in colour. The leaf stalk is rather broad, furrowed and concave on the inside. The stem, which appears in the second year, is furrowed and branching. Early vegetative growth is quite spreading, but leaves arising from the apex of the short stem form compact, elongated heads. The root system of celery is spreading and fibrous with many feeder roots close to the soil surface. The plant produces compound clusters, or umbels, of small, white, perfect flowers with five petals and five stamens. The seeds (actually fruits formed from two compressed carpels enclosing the actual seed) are very small.

American winter green plants are erect, or prostrate with creeping stems giving rise to erect branches (15 cm high and leafy at the top). The lower leaves are alternate, serrate and short-petioled, but the top leaves are oval, 2·5–5 cm long, apiculate, glaucous and shining above. The flowers are solitary or in racemes or panicles, and scarlet coloured.

Florence fennel is an annual, very distinct, low growing and thick-set plant, with a very stout stem, which has joints very close together towards the base. The leaves are very finely cut, and of a glaucous, green colour. The leaf stalks are very broad, of a whitish-green hue and overlap one another at the base of the stem, the whole forming a kind of head or enlargement that varies in size; the inside is white and sweet. The greatest height of the plant, even when grown for seed, does not exceed 60–75 cm. The flowers are greenish and in a broad umbel of stout, stiff rays, and the seed, twice as long as wide, is flat on one side and convex on the other, and traversed by five, thick, yellowish ribs which occupy almost the entire surface of the skin.

Dandelion produces leaves in a basal tuft. The leaves are deeply-toothed, and a single, composite, yellow flower is produced on a hollow stalk. Cultivated forms of dandelion are larger and more leafy than the weedy forms, and improved cultivars, mostly of French origin, are larger, more tender, less bitter and slightly greener than the wild types.

Chemical and Nutritional Composition

Having a low content of fat (0·6 g per 100 g of edible part), celery is quite popular in the Western world; of 100 g of celery, around 71% is edible, comprising moisture content (88%), 6·3 g protein, 2·1 g minerals, 1·4 g fibre and 1·6 g carbohydrates. It has 155 kJ (37 kcal) energy, 230 mg calcium, 140 mg phosphorus, 6·3 mg iron, 3990 μg carotene, 0·9 mg thiamin, 0·62 mg vitamin C, 1·2 mg niacin, and 0·11 mg riboflavin. It has 240 iu of vitamin A and 126 mg of sodium. The composition varies at different stages of maturity, and reducing and total sugars show a marked decline in the leaf blades from harvest time to the end of storage. This decrease is accompanied by an increased concentration in the petiole. Soluble nitrogen increases in both the leaf blades and the petioles towards the end of storage. ***See*** individual nutrients

Hardly any information is available with respect to nutritional/chemical composition of florence fennel and American winter green vegetables. However, it is reported that dandelion leaves are an excellent source of vitamin A (14 000 iu per 100 g sample), vitamin C (35 mg per 100 g) and calcium (187 mg per 100 g).

Handling and Storage

There is no definite stage of maturity at which celery should be harvested for satisfactory quality. Celery plants are cut off below the soil surface. Stripping to remove suckers and diseased and damaged leaves, washing, grading, packing and precooling are done for marketing. Celery in storge absorbs foreign flavours, so it should be kept away from the odours of other products. Various methods of storage are in use, including trenching in the field, storing in pits and storing in cold storage and warehouses. Ripening changes takes place during storage. In California, celery is generally vacuum cooled. In Florida, most celery, packed in wire-bound crates, is hydrocooled. Celery can be stored for 2–4 weeks at 0°C (32°F) and 95% relative humidity. Dandelion plants are trimmed, washed and stored in a container at low temperature (0°C) and high humidity (90–95%) to preserve their quality. Similar conditions will also be used for other crops. ***See*** Controlled Atmosphere Storage, Applications for Bulk Storage of Foodstuffs

Bibliography

Lincoln CP (1987) *Vegetables: Characteristics, Production and Marketing*, pp. 163–268. New York: Wiley.

Ryder EJ (1979) *Leafy Salad Vegetables*. Westport: AVI.

Salunkhe DK and Desia BB (1984) *Post Harvest Biotechnology of Vegetables*, vol. II. Boca Raton: CRC Press.

Wood R (1979) *Salad Crops*. London: Marshall Cavendish.

Yamaguchi M (1983) *World Vegetables: Principles, Production and Nutritive Values*. Westport: AVI.

K. Gupta, J.L. Mangal, G.R. Singh and D.S. Wagle
Haryana Agricultural University, Haryana, India

Root Crops

The traditional salad root crops are carrots, bunching onions, garlic and radishes. Added to this list are potatoes, rutabaga/turnips, leeks, chicory (witloof), Jerusalem artichoke and salsify. Like many plant foods, some of these species are only relatively recently considered as salad crops, while the others are of long standing. The *true root* crops are carrots, radishes, chicory and salsify; other crops are *bulbous* stems such as onions, garlic, and leeks; others are *hypocotyl* swollen stems such as beets, rutabagas and turnips; and, finally, some are *tuberous*, i.e. they are developed from rhizomes (underground stems), such as the potato and the Jerusalem artichoke. Salsify and chicory are also grown for their leafy qualities but for this discussion they are considered for the stored root qualities (salsify), or are stored (chicory) and forced in the dark to develop a tender enlarged bud known as a chicon. They all share the similarities of salad crops in that they are harvested at their peak quality, and prepared for consumption mostly raw, though on occasion they are cooked, as for beets, leeks, chicons, and potatoes, but are always served cold either singly or in concert with other salad crops, together with the appropriate dressings or condiments.

Nutritive Composition

All vegetable crops provide varying amounts of food value, especially minerals, carbohydrates and vitamins, which in most instances are always higher in the raw state than when cooked. Table 1 presents the physical constituents of the salad root crops. All raw vegetables offer high water content, varying small amounts of protein, little or no fat and some fibre and are a good source of carbohydrates. In addition, as in Table 2, raw salad crops are a good source of minerals and vitamins. Salad root crops are not generally high in vitamin A, except for carrots, bunching onions and rutabagas. Most salad root crops (with the exception of dry onion) offer a good source of vitamin C. On average the daily human intake of minerals and vitamins is not large; however, natural sources such as the salad root crops are important in a daily diet of balanced nutrients and are within the reach of most consumers. Neither animal or cereal sources of food can match the mineral and vitamin content of raw vegetables. Animal and cereal food sources, unlike vegetables, are rarely consumed in their living state either unprocessed or raw. ***See*** individual nutrients

Description of Salad Root Crops

Carrot: *Daucus carota* L. var *sativus*

Belonging to the family Umbelliferae, this is a biennial plant normally grown for its first-year root. Where winters are too severe, carrot roots are placed in temperature/humidity controlled storage. A relatively ancient vegetable out of central Asia, it is grown worldwide in all climatic regions. Carrots are grown either for the speciality trade as small finger-size types, or to full-size, high-quality shipping and storing types. Carrot growth can be regulated by precision seeding spacing. In Europe, the demand for carrot quality overrides the convenience of mechanical harvesting prevalent in North America, where the demand for long, slender, fresh market, smooth, pointed roots with strong stem attachment permits mechanical harvesting and topping. In Europe, the varieties centre around the high-

Table 1. Physical constituents of salad root crops (Approximate values per 100-g raw edible portion)

Crop	Water (%)	Energy kcal g^{-1}	Protein (g)	Fat (g)	Carbohydrate (g)	Fibre (g)
Beetroot	87	44	1·5	0·1	10·0	0·8
Carrot	88	43	1·0	0·2	10·1	1·0
Chicory	95	15	1·0	0·1	3·2	—
Garlic	59	149	6·4	0·5	33·1	1·5
Jerusalem artichoke	79	67	2·0	0·1	7·0	1·2
Leek	83	35	1·5	0·3	14·0	1·5
Onion, green bunching	92	25	1·7	0·1	5·6	0·8
Onion, dry bulb	91	34	1·2	0·3	7·3	0·4
Potato	79	79	2·1	0·1	18·0	0·4
Radish	95	17	0·6	0·5	3·6	0·5
Rutabaga	90	36	1·2	0·2	8·1	1·1
Salsify	82	34	3·3	10·2	5·2	—
Turnip root	92	27	0·9	0·1	6·2	0·9

Adapted from USDA *Agricultural Handbook* 8 (1963, 1982).

Table 2. Mineral and vitamin content of salad root crops (Approximate values per 100-g raw edible portions)

Crop	Ca (mg)	P (mg)	Fe (mg)	Na (mg)	K (mg)	Vit. A. (iu)	Vit. B_1 (Thiamin) (mg)	Vit. B_2 (Riboflavin) (mg)	Niacin (mg)	Vit. C (Ascorbic acid) (mg)
Beetroot	16	48	0·9	72	324	20	0·05	0·02	0·40	11·0
Carrot	27	44	0·5	35	323	28 129	0·10	0·06	0·93	9·3
Chicory	—	21	0·5	7	182	0	0·07	0·14	0·50	10·0
Garlic	181	153	1·7	17	401	0	0·20	0·11	0·70	31·2
Jerusalem artichoke	14	78	3·4	—	—	0	0·20	0·06	1·30	4·0
Leek	59	35	2·1	20	180	95	0·06	0·03	0·40	12·0
Onion, green bunching	60	33	1·9	4	257	5000	0·07	0·14	0·20	45·0
Potato	7	46	0·8	6	543	0	0·09	0·04	1·48	19·7
Radish	21	18	0·3	24	232	8	0·01	0·05	0·30	22·8
Rutabaga	47	58	0·5	20	337	580[a]	0·08	0·22	0·50	
Salsify	60	75	0·7	—	380	0	0·08	0·22	0·50	8·0
Turnip, root	30	27	0·3	67	191	0	0·04	0·03	0·40	21·0

Adapted from USDA *Agricultural Handbook* 8 (1963, 1982).
[a] Source: *National Food Review*, USDA (1978).

quality coreless Nantes types. Other quality carrots are the miniatures (Amsterdam Coreless; 55 days) and Nantes types with blunt or rounded tips. For prime maturity, carrots are sown according to the time it takes to reach optimum root growth. If seeded during periods of continuous cool temperatures (7–10°C), young carrot plants develop a tendency to bolt (go to seed) prematurely (1–2%). The darker the root colour, the higher the carotene content: some areas of Europe prefer white carrots, others blood red to purple. The bulk of carrots in commerce are either red-orange or orange. Roots with small steles, or centre cores, are of greater culinary quality than those with well developed cores. The core or water-carrying tissue (xylem) increases in size as carrots mature according to varietal variations. Carrots grown on muck or peat soils readily develop longer, straighter roots than when grown on upland mineral soils. The older (and larger) a carrot becomes, the more woody is its texture. Carrot shape may be long and pointed (Imperator types), medium long and blunt-ended (Nantes types) or relatively short and broad at the top (Chanentay or Danvers types). As a salad root crop the crisp, colourful, sweet carrot root served as sticks, diced, sliced or shredded adds both flavour and colour to any combination salad or singly as carrot dips. The high vitamin A (carotene) content is a valuable contribution to the food system. The current interest in health foods has generated wide demand for carrots in restaurants and fast food outlets. *See* Controlled Atmosphere Storage, Effect on Fruit and Vegetables

Onion: *Allium cepa* L., *Allium fistulosum* L.

The dry bulb, and the green bunching onion also known as the Welsh or Japanese bunching onion, belong to the family Liliaceae. The onion has been in cultivation since very ancient times and, like the carrot, originated in central Asia. Tomb paintings in Egypt attest to the presence of the onion thousands of years ago. It is a perennial plant where winter freezes do not destroy the plants. Salad onions are primarily the short-term green bunching onion, whereas the full-season (dry bulb) onions are either of the Spanish (mild flavour) types or the more pungent full-season storage types. Like carrots, onion seed can be spaced to control bulb shape and size using precision seeding methods. Since the dry bulb onion pushes upwards as it grows, muck soils are preferred for onion production, but mineral soil types will produce excellent onions using sound cultural practices. From seeding to harvest the onion plant is very subject to frequent disease and insect attacks. In addition, onions are poor weed competitors. These factors must therefore be taken into account in the pursuit of viable onion crops. The major onion cultivars today are hybrids bred for both long- and short-day bulbing, attaining high levels of uniformity of production. At harvest, dry bulb onions are undercut and left to field dry before being topped and transferred to storage either in pallet boxes or bulk temperature-controlled storages. High-quality onions, like carrots, are available on a year-round basis. The bunching onion provides nonbulbing, long slender plants that are hand-pulled and tied in the field. They are an integral part of vetetable salads. Welsh onions likewise provide a continuous flow of green bunching salad onions. The dry bulb onion may be white-, yellow-, brown- or red-skinned, and for salads is sliced, diced or shredded. The mild-flavoured dry bulb onion, the mild Spanish onion, or the mild short-day dry bulb southern winter onions is also in great demand for salads. Onions are utilized in all types of fresh or cooked salads. Thinly sliced they are also served with the appropriate condiments and dressings in buns together with meats and other salad vegetables. *See* Onions and Related Crops

Beet: *Beta vulgaris*

Beet belongs to the family Chenopodiaceae and is also called beetroot or garden beet. It has been a food crop for several thousand years, originating in the Mediterranean area. This root (hypocotyl) crop is popular in salads, for both its young, tender leaves and for its young, tender root. The beet is a true annual, completing its life cycle in one season. The edible parts are consumed at the prime vegetative stage. Like other 'root' crops the tissue becomes more fibrous and woody as the root enlarges with age. As a salad crop, the beetroot is optimum at 2·5–4·5 cm in diameter. Beets are bunched or topped according to market preferences. Beet seed is sown in rows and planted at regular intervals to ensure a constant supply of tender, sweet roots. Occasionally beets develop dark, hard areas known as black spot, indicative of boron deficiency, which can be prevented or corrected by applying a boron compound to the leaves or the soil. Beets perform best at temperatures of 16–18°C. At higher temperatures, the photosynthates dissipate rapidly, creating the familiar dark-and-pale ring pattern indicative of lower quality as the rings enlarge. Beet shapes may be globular, top-shaped, flattened or elongated. The elongated beet is useful for uniform slicing. Beets are usually cooked and served cold or warm either alone or in conjugation with leafy salads. Beets are added to fish and meat salads to add colour and flavour. As a pickled product, beets are also used in various salad combinations.

Garlic: *Allium sativum* L.

Garlic, also called Clowns Treacle, belongs to the family Liliaceae and has been known since ancient times, originating in central Asia. The Romans knew of garlic, which the elite did not like to eat owing to its flavour, though they fed it to the working people. Garlic is not as a rule propagated by seeds as are onions. As the garlic basal stem develops it forms a segmented structure which gives rise to individual cloves. The cloves may have a sheath-like enclosing membrane, or this may be absent. The unsheathed-clove types are the less pungent. Garlic is slow-growing, requiring a long season to produce a full crop. Garlic differs from the onion in that its underground stem structure remains below the soil surface during its growing cycle. Maturity is indicated by the withering of the leaves in the fall. Warmer climates permit longer more abundant growth owing to long frost-free winter conditions. The pungency of garlic has traditionally been a background flavouring agent in salad making. However, in current cuisines it is also finely crushed and incorporated as a regular salad ingredient.

Radish: *Raphanus sativus* L.

Radish belongs to the family Cruciferae, originating from a Chinese plant. The type of radish root mostly used for salads belongs to the group of fast-growing, relatively small-sized and spherical to slower-growing somewhat elongated types. Radishes may be all red or scarlet or have the upper part red and the lower white, with the flesh being white. The elongated and fully white radishes are generally stronger-flavoured. The smaller-sized radish is harvested before the inner flesh becomes pithy and spongy. For the most part these radishes are optimum at 2·0–4 cm in diameter. Radish seeds are sized and precision-seeded to ensure greater root uniformity. Because the widely used spherical types germinate quickly and uniformly when sown in muck soil, the seed-to-harvest cycle is completed in 23–26 days. This rapid production cycle bypasses serious weed problems. Radishes are harvested in a single digging and topping operation. Market gardeners sometimes bunch radishes, leaving the tops intact. Radishes, if unharvested beyond optimum size, together with lengthening spring days, rapidly initiate the flowering cycle, depleting the stored carbohydrates in the roots and making them undesirable for consumption. Radishes are served whole, sprinkled with salt or dipped in dressing, or are sliced or cubed and incorporated in combination with other salads. The smooth, colourful radish root is decorative in salads, especially when the root is partially sliced to form a rosette of the root interior.

Potato: *Solanum tuberosum* L.

Potatoes belong to the family Solanaceae, whose origin lies in the northern Andes mountains of South America. The potato was brought to Europe after the discovery of the Americas. The potato produces underground stems (rhizomes) which terminate in an enlarged structure (tuber) made up of stored photosynthates. Potato tubers come in all sizes, shapes and colours, with regional preferences having developed over time. Tubers may have deep eyes (points of regrowth) or shallow eyes that are conspicuous or innocuous respectively, due to the pigmentation around them. The skin may be white, red, yellow, bronze or purple; the flesh may be white, yellow or purple. The preferences for a particular type of potato are as varied as are the places people come from. The potato is a perennial but mainly it is grown as an annual. The tubers are the source of each new crop. As a clone, it is a replica of the parent plant, and it is therefore the source for maintaining the type. Usually ‘seed’ tubers are quartered, balancing the ‘eyes’ on each segment. Early-planted potato pieces require suberization before planting to prevent ready entry of soil pathogens. One or more hillings will prevent the developing tubers from turning green on exposure to light. Spacing of potato pieces helps to control tuber sizing, eliminating the development of the ‘hollow heart’ typical of oversized tubers. The potato is almost a complete food, producing more food value per hectare than any other crop. Taste preferences run from raw to fried, to boiled, to baked, to salads where the potato is cooked and cut up for potato salad in combination with many other vegetable ingredients. ***See*** Potatoes and Related Crops, Fruits of the *Solanaceae*; Potatoes and Related Crops, Processing Potato Tubers

Leek: *Allium ampeloprasum* L. *porrum*

The leek belongs to the family Liliaceae and has a long association with food use, especially in Europe. The leek has been in continuous cultivation for over 4000 years, originating in the Mediterranean region. A biennial, the leek is much hardier and more cold-resistant than the onion. As with other members of the *Allium* group, this plant is a stem/root type, which means that, in order to extend the white elongated parts above the stem plate, leeks are planted in a 12–15 cm trench which, as the crop develops, is filled to produce blanching and tenderizing. The leek does not form a distinct bulb. The leaf sheaths form an elongated series of wrappings, with the youngest growth on the inside originating from the intercalary or growth-generating region of the stembase. The leek, like the onion, is classed as a root crop mainly because of the management of its growth. When harvested, leeks must be thoroughly washed before cooking to removed trapped grit in the folds of the leaves. By cutting the leek in half along the vertical, washing is facilitated. As a salad the leek is boiled and chilled and may be served alone or together with other crops, with appropriate condiments and dressings. In this manner the leek is considered an appetizer with the main salad offering. Leeks are also served as a hot appetizer, usually in conjunction with whipped butter.

Rutabaga: *Brassica napus napobrassica*. Turnip: *Brassica rapa* L. *rapifera*

These belong to the family Cruciferae. Other common names for rutabaga are Swedish turnip, Swede and turnip-rooted cabbage. Owing to their adaptation to the cooler parts of Eurasia, these crops have played a long role in providing food. The rutabaga differs from the turnip in several ways. Possibly it is less a true hypocotyl than the turnip because it has a more elongated stem connection. The turnip, which has no clear stem base formation, consists of numerous eyes or growth points on the top of the hypocotyl. The two differ in flesh colour of the root/hypocotyl; the rutabaga is distinctly yellowish, containing carotene in contrast to the white-fleshed, carotene-free turnip. The rutabaga is much slower-growing than the turnip and is milder in flavour. Neither vegetable is difficult to grow. Precision seeding and spacing ensures uniform growth control. Both are readily storable and produce seed in the second year from overwintered roots. Rutabaga/turnips are perhaps not usual candidates as salad root crops in that traditionally they are chopped into pieces, boiled, mashed and served hot. However, offered as a fresh 'stick' with salads and the appropriate dressing and condiments, these crops are salad items. In addition, if the roots are stored in a darkened warm room, they develop crisp tasty shoots for fresh salads.

Jerusalem artichoke: *Helianthus tuberosus* L.

This belongs to the family Compositae of North American origin, it is widely accepted in Europe. A perennial like the potato, it is sometimes compared with the potato because of its tuber development. The similarity stops there; the Jerusalem artichoke provides an entirely different culinary response. Like the potato, this crop is propagated by tubers or tuber pieces. It matures in about 100 days under much cooler and more difficult growing conditions than the potato. As a root salad crop, the knobby tubers are used raw, sliced thinly and incorporated in various vegetable salad combinations. The Jerusalem artichoke consists of little or no carbohydrates or starches, but its insulin content make it a valued health food, particularly for diabetics. This crop has probably been neglected owing to lack of understanding of its unique differences on the one hand, and to unfounded comparisons with the potato on the other.

Chicory: *Chichorium intybus* L.

Also known as Belgian witloof, French endive, or Barbe-de-capucion, chicory belongs to the family Compositae. It is a perennial grown as an annual for its unique stored root, which is forced in the dark to develop a compact, blanched bud head known as a chicon. Chicory is comparatively recent in its utilization as a salad crop. There seems to be no mention of it prior to the 1600s. It may be debatable whether chicory should be classed as a root or a leafy crop. Since the root, with part of the stem attached, becomes the main source of the new growing point, it is classed as a salad root crop. The chicon is a popular salad item enjoying great favour in Europe and parts of North America. Whether chicory is developed as a leaf crop similar to endive or grown as a chicon, it requires blanching so as to suppress the strong flavours characteristic of the unblanched daylight-grown types. Chicory or witloof today is largely grown in a dark controlled-temperature area, with the roots fed hydroponically rather than being placed in soil and covered. Quality chicons should not be washed prior to reaching the consumer, hence the change to the nutrient solution method of growing which overcomes this problem. The chicons may be served as a component of a fresh salad or cooked and handled as an appetizer with meats and appropriate dressings and condiments.

Salsify: *Tragopogon porrifolius* L.

Also known as the vegetable oyster or the oyster plant, this belongs to the family Compositae and is a comparative newcomer, originating in the Mediterranean basin. Salsify, a biennial is grown somewhat like carrots for its long, slender root. It is very hardy under moderately temperate conditions and can be left in the soil well into the autumn (fall). As with carrots, the seedbed should be deep and friable to ensure straight, smooth roots. The stored roots are washed and cleaned and served raw in longitudinal strips, diced or cubed. The attraction is its distinctive flavour, akin to the oyster, hence the name, and it is served raw as a separate dish and dipped in the appropriate dressing and condiments. This vegetable has never gained any degree of popularity in North America, but is enjoyed in Europe.

Bibliography

Howard FD, Maggillivray JH, Yamaguchi M *et al.* (1962) Nutrient composition of fresh California grown vegetables. *California Agricultural Experimental Station Bulletin.* 788.

Lorenz OA and Maynard DN (1988) *Knott's Handbook for Vegetable Growers*, 3rd edn. New York: Wiley.

Nicholson BE, Harrison SG, Masefield GB and Wallis M (1985) *The Illustrated Book of Food Plants*, 2nd edn. London, W1: Peerage Books.

Nonnecke IL (1989) *Vegetable Production.* New York: Van Nostrand Reinhold.

Pierce LC (1987) *Vegetable – Characteristics, Production and Marketing.* New York: Wiley.

USDA (1984) *Agricultural Handbook* 8.

When the Good Cooks Garden. (1974). San Francisco: Ortho Books.

Ib L. Nonnecke
Guelph, Ontario, Canada

SALAD DRESSINGS AND OILS

See Dressings and Mayonnaise

SALAMI

See Meat

SALMON

See Fish

SALMONELLA

Contents

Properties and Occurrence

Salmonella is a genus within the family Enterobacteriaceae. The organisms are rod-shaped bacteria that are usually motile, growing both in air and in its absence. Some biochemical properties are given in Table 1. Salmonellae are pathogenic for man and animals causing enteric and/or more generalized infections that vary widely in severity. Approximately 2200 serotypes are recognized, some being adapted to specific hosts and largely restricted to them, e.g. *S. typhi* in man, *S. dublin* in cattle, *S. cholerae-suis* in pigs. The majority, however, are host nonspecific and include those that are often associated with human food poisoning; these nonspecific serotypes are generally widespread. ***See*** Enterobacteriaceae, Occurrence of *Escherichia coli*

Traditionally, salmonella serotypes are named as if they were separate species but, because of their genetic similarity, a single species, *S. enterica*, has been proposed, with food-poisoning serotypes mostly classified within a subspecies, also named *enterica*. Thus, the serotype known as *S. typhimurium* becomes *S. enterica* subsp. *enterica* serotype *typhimurium*. Although the newer nomenclature is more appropriate scientifically, it has yet to be widely recognized and therefore is not used here.

The present article considers the occurrence and transmission of salmonellae, their sources and behaviour as food contaminants and the effects of process-

Table 1. Common properties of salmonellae

Fermentation of[a]:			
Glucose	+	Production of H_2S	+
Maltose	+	Lysine decarboxylase	+
Mannitol	+	Ornithine decarboxylase	+
Sorbitol	+	Phenylanaline deaminase	–
Sucrose	–	Urea hydrolysis	–
Salicin	–	Methyl red reaction	+
Adonitol	–	Voges–Proskauer reaction	–
Production of:			
Indole	–	Utilization of citrate	+

[a] Positive reactions usually result in production of acid and gas.

ing. The role of the organisms in food poisoning is also considered.

Occurrence and Transmission

Salmonellae are primarily intestinal parasites of vertebrate animals in which the infections are mainly asymptomatic and confined to the alimentary tract. Animals, both wild and domestic, play an important part in salmonella transmission and there are well established cycles of infection involving man, animals, their foods and the general environment. Humans may act as passive carriers on footwear and clothing or as intestinal carriers, transmitting the organisms mainly via sewage pollution. Only in special circumstances, e.g. in some hospital wards, is person-to-person contact significant in spreading the infection.

Among domestic livestock, the major reservoirs are poultry, pigs and cattle and, in all cases, dissemination of salmonellae is favoured by modern conditions of intensive rearing. This is especially so for poultry because thousands of birds are usually reared in the same house, whether for egg or meat production. Under these conditions, salmonellae may be acquired (1) by transmission from parent to progeny via contamination in the hatchery, (2) from contaminated feed, or (3) from environmental sources such as small-animal vectors. Within a poultry flock, infection may be maintained and spread among the birds by ingestion of contaminated droppings. Salmonella carriage in the live animal often leads to ultimate contamination of eggs or meat, as discussed below. ***See*** Beef; Pork; Poultry, Chicken; Poultry, Ducks and Geese; Poultry, Turkey

Animal feed has been considered a major source of salmonellae, and the manufacturing process involves use of contaminated ingredients such as meat and bone meal, fish meal and other types of protein, both animal and vegetable. Most feeds now receive some degree of heat treatment that aims to destroy pathogens. There is a legal requirement in the United Kingdom to test processed animal proteins for salmonellae, although testing is necessarily limited and may fail to reveal low levels of salmonella contamination.

Animals other than domestic livestock that can become infected with salmonellae and facilitate their dissemination include wild birds, rodents, insects and pets. Gulls are an important source of salmonellae owing to their feeding habits and frequent presence at sewage works, rubbish tips, etc. However, infections in rodents have been found to increase in the vicinity of farms and slaughterhouses, thus demonstrating that salmonella transmission is not solely in the direction of meat animals. ***See*** Insect Pests, Problems Caused by Insects and Mites

Salmonellae are frequently found in sewage, polluted waters and soil, where they may survive for weeks, months or even years if conditions are favourable. The spreading of manure on agricultural land can also lead to environmental contamination when holding periods prior to spreading are inadequate and salmonellae in the manure remain viable.

Mode of Introduction into Foods

Contamination of foods with salmonellae is often derived, either directly or indirectly, from infections of livestock. Virtually any kind of raw food can become contaminated, including fruits and vegetables, but the foods most often implicated in human salmonellosis are poultry, eggs, red meats, milk, cream, seafood, certain types of dessert, e.g. trifles, and some bakery products.

Poultry meat in particular shows a high incidence of salmonella contamination and, on average, about 50% of carcasses carry salmonellae, albeit usually in low numbers. While the problem clearly originates in the live bird, the process used to prepare oven-ready carcasses in commercial slaughterhouses is particularly conducive to the spread of minority organisms such as salmonellae. The stages in processing that are most likely to cause cross-contamination are scalding, defeathering and evisceration. Because of the need to avoid damaging the skin during defeathering, birds intended for sale in the chilled state are scalded by immersing them in water at temperatures as low as 49–50°C. This is sufficient to loosen the feathers, but leads to an accumulation of microorganisms in the scald tank and considerable opportunity for cross-contamination. Defeathering machines not only create aerosols that disperse microorganisms but provide conditions for microbial colonization of the equipment and further contamination of carcasses. The problem with automatic evisceration machinery is its tendency to cause gut breakage and spread of faecal bacteria. Effective washing and prompt chilling of eviscerated carcasses are essential to minimize levels of microbial contamination in general, but do little to reduce the number of birds that carry salmonellae.

During the 1980s, *S. enteritidis* emerged as a major cause of foodborne disease in Europe and North America. For the first time, table eggs were implicated in numerous outbreaks associated with this serotype. The organism can invade the bloodstream of the bird, causing infection of the oviduct or ovaries and contamination of egg contents. Penetration of the shell from external contamination can also occur, although salmonellae on the shell may be more important as a source of contamination for bulk liquid egg, when eggs are broken out commercially. Studies have shown that about 0·5% of eggs from a naturally infected flock may contain low numbers of salmonellae (in the region of 20 per egg). These are usually located in the albumen, presumably as a result of oviduct infection, and are unlikely to multiply while the egg remains fresh and its natural defences are intact. Nevertheless, food poisoning has arisen from a wide range of egg-containing products, especially home-made food that includes mayonnaise, indicating the need for greater care in preparing and handling these items and chill storage of eggs after purchase.

Although red meats tend to be less frequently contaminated with salmonellae than poultry (Table 2), the situation is again influenced by the carrier rate in the live animal. This in turn is affected by husbandry practices and by transportation conditions. The stress of being transported may be exacerbated by mixing the animals with others in unfamiliar surroundings, sometimes on more than one journey. The result is a greater faecal load and increased opportunity for cross-contamination. In pigs, pre-slaughter stress has been shown to increase the proportion of salmonella shedders from 1% to 50%. Similar problems can occur if animals are held under bad conditions at the abattoir while awaiting slaughter. ***See*** Meat, Slaughter; Meat, Preservation

With red-meat animals, the skill and care of the slaughterman are of paramount importance if clean carcasses are to be produced. Careful removal of the hide or fleece and the gut can help to minimize carcass contamination and reduce the spread of any salmonellae (Table 3).

Cows, like other domestic animals, can become infected with salmonellae from feed or other sources of contamination on the farm. If the organisms are being shed in the faeces, the udder or hide may become contaminated and ultimately salmonellae will appear in the milk. Milkborne salmonellosis is common in those parts of the world where milk is neither boiled nor pasteurized, but also occurs, much less frequently, in developed countries where the main products implicated are pasteurized milk, powdered milk and certain cheeses. ***See*** Milk, Processing of Liquid Milk

Seafoods are among those prone to environmental contamination since they may be obtained from inshore waters that receive discharges of human or animal waste. The most susceptible types include oysters, mussels and clams, which are filter feeders and concentrate microorganisms from the water. These animals require special cleansing procedures. ***See*** Shellfish, Contamination and Spoilage of Molluscs and Crustacea

In the case of vegetable crops, a close relationship has been observed between the salmonella serotypes isolated and those found in water used for irrigation. Such contamination can be hazardous in salad crops, which are not normally cooked before consumption.

Foods such as trifles and custards invariably have milk and egg constituents and readily support microbial growth. These are usually eaten cold and there is often ample opportunity for growth of salmonellae in the period between preparation and consumption if the foods are not properly refrigerated. Items of this kind should not be stored in close proximity to raw foods

Table 2. Incidence of salmonellae in meat animal tissues: results of some surveys over the period 1973–1990

Type of animal	Sample	No. of samples	Percentage positive
Sheep	Carcass surface	127	0
Cattle	Carcass surface	277	0·4
Cattle	Various tissues	674	2
Calves	Various tissues	270	4·3
Pigs	Lymph nodes	20 000	0·02
Pigs	Lymph nodes	407	19·2
Poultry[a]	Various tissues	292	48
Poultry[a]	Various tissues	204	59

Adapted from Mackey BM (1989) The incidence of food poisoning bacteria on red meat and poultry in the United Kingdom. *Food Science and Technology Today* 3: 246–249.
[a] Data from surveys of the UK Public Health Laboratory Service.

Table 3. Influence of slaughter procedures on salmonella contamination of pig carcasses

Procedures	No. of carcasses	No. positive	Percentage
Normal	35	16	46
Extra care in singeing and evisceration	30	2	7

Data of Oosterom J and Notermans S (1983) Further research into the possibility of *Salmonella*-free fattening and slaughter of pigs. *Journal of Hygiene, Cambridge* 91: 59–69.

Table 4. Minimum pH values for growth of salmonellae[a] in tryptone–yeast extract–glucose medium containing different acids

Type of acid	pH value
Hydrochloric	4·05
Citric	4·05
Tartaric	4·10
Lactic	4·40
Succinic	4·60
Acetic	5·40
Propionic	5·50

Adapted from Chung KC and Goepfert JM (1970) Growth of *Salmonella* at low pH. *Journal of Food Science* 35: 326–328.
[a] Inoculum: 10^4 per ml of *S. anatum, S. senftenberg* or *S. tennessee.*

such as meats because of the risk of cross-contamination.

Factors Affecting Growth in Foods

The growth range for salmonellae is 5–47°C, although most are unable to multiply below 7°C and even up to 10°C growth is slow. Optimum growth occurs at 35–37°C. The organisms can utilize simple carbon compounds as sources of carbon and energy and a wide range of nitrogenous compounds to satisfy their nitrogen requirements.

As with other microorganisms, growth of salmonellae in foods is affected by several factors, e.g. temperature, pH, salt concentration, which may act in combination to produce conditions that are more inhibitory to the organisms than any individual factor acting in isolation. Under conditions that are otherwise optimal, salmonellae grow at pH values between 4·0 and 9·0, with an optimum range of 6·5–7·5. In practice, however, the minimum pH value for initiation of growth varies with the type of acid used to control pH (Table 4). In this respect, acetic and propionic acids are much more inhibitory than either hydrochloric or citric acid. Below pH 4·0, the organisms tend to die out. ***See*** pH – Principles and Measurement

Salmonellae are not particularly salt-tolerant, although growth can occur in the presence of 4% sodium chloride and around 350 ppm sodium nitrite. The lower limit of water activity (a_w) permitting growth is 0·93. Despite their ability to grow in the absence of air, salmonellae are inhibited at oxidation–reduction potential (E_h) values below 30 mV. ***See*** Water Activity, Effect on Food Stability

Properties and Occurrence

Fate of Salmonellae in Food Processing

In high a_w foods, salmonellae are rapidly destroyed by heat pasteurization at temperatures in the region of 70°C. At 60°C decimal reduction times (time to effect a tenfold reduction) for most strains vary between 0·2 and 6·5 min when the organisms are tested in laboratory media. Certain strains of *S. senftenberg* are among the most heat-resistant. Some serotypes are more heat-resistant at pH 5·5 than pH 8·5, whereas heat-resistance in *S. enteritidis* at pH 7·0 has been increased by brief exposure to pH 9·2. Prior 'heat shock' at 42–48°C also increases heat-resistance, as do conditions of steadily rising temperature that might occur in long, slow cooking of some foods. However, only under marginal conditions of destruction is survival likely to be enhanced by pre-heating.

The importance of both temperature and time of heating can be appreciated in relation to commercially prepared beef roasts, for which a 'rare' (undercooked) appearance is often required. For beef to remain 'rare', the internal temperature must be kept below 58°C. At 57·2°C, the time required to eliminate 10^7 salmonellae (the usual objective of pasteurization processes) is 3 min, while at 54·4°C, which allows more latitude to safeguard meat colour, it is 30–60 min.

Salmonellae are relatively sensitive to ionizing radiation and radiation treatment would reduce their incidence in contaminated foods to suitably low levels. For a 10^7-fold reduction in frozen whole egg, a treatment dose of 3·6–5·4 kGy is necessary. With chilled poultry, the maximum dose that would avoid undesirable sensory changes is 2·5 kGy and this would be expected to reduce the proportion of salmonella-contaminated carcasses to about 1%. A decimal reduction dose for *S. typhimurium* in chicken mince has been calculated at 0·50–0·62 kGy, depending on the type of packaging used.

Most salmonellae are rather sensitive to freezing and frozen storage, and are more so at temperatures close to the freezing point of the substrate, although even repeated freezing and thawing have failed to effect complete elimination. *S. hadar* is reported to be more resistant to frozen storage than other serotypes tested.

Human Salmonellosis

Over more than 20 years, human salmonellosis has continued to increase throughout much of Western Europe and North America. In the United Kingdom some 12 000 cases were reported in 1982, increasing to over 27 000 by 1988 and comprising 69% of all reports of food poisoning in England and Wales. The real incidence is thought to be at least ten times higher. Since 1985, most of the increase has been due to the upsurge of one particular serotype, *S. enteritidis*, and this has occurred in several countries. It remains to be seen how long the organism will continue to predominate and why such a phenomenal rise due to a single serotype could have occurred simultaneously in different parts of the world. ***See*** Food Poisoning, Statistics

Bibliography

Bryan FL (1983) Epidemiology of milk-borne diseases. *Journal of Food Protection* 46: 637–649.

Humphrey TJ (1990) Public health implications of the infection of egg-laying hens with *Salmonella enteritidis* phage type 4. *World's Poultry Science Journal* 46: 5–13.

Mead GC (1989) Hygiene problems and control of process contamination. In: Mead GC (ed.) *Processing of Poultry*, pp 183–220. London: Elsevier Applied Science.

Old DC (1990) Salmonella. In: Parker MT and Duerden BL (eds.) *Topley & Wilson's Principles of Bacteriology, Virology and Immunity*, 8th edn, vol. 2, *Systematic Bacteriology*, pp 469–493. London: Edward Arnold.

Roberts D (1982) Bacteria of public health significance. In: Brown MH (ed.) *Meat Microbiology*, pp 319–386. London: Applied Science.

World Health Organization (1988) Salmonellosis control: the role of animal and product hygiene. *Technical Report Series 774*, Geneva: World Health Organization.

G.C. Mead
The Royal Veterinary College, London, UK

Detection

In the developed world, the majority of cases of human salmonellosis are zoonotic and result from the colonization/infection of food animals and resultant contamination of products derived from them. Salmonellae are also ubiquitous, and a diverse range of foodstuffs – including dried milk, fermented sausage, bean sprouts, chocolate and yeast-based flavouring – have been implicated as vehicles for human salmonellosis in the United Kingdom in the last few years. Natural water systems, particularly those receiving human and animal waste, are also frequently contaminated.

The ubiquity of salmonellae and the wide range of materials with which they have been associated has meant that numerous media and techniques have been developed for their isolation. It is not the intention of this paper to go into detail on the possible virtues or disadvantages of individual methods. Attention is focused instead on the fundamental aspects of the isolation and detection of salmonellae, particularly from food and environmental samples. Special emphasis is given to sublethal injury and its impact on recovery of the salmonellae and the possible inhibition by other organisms. The application of rapid methods is also discussed. ***See*** Food Poisoning, Tracing Origins and Testing

Sublethal Injury and its Impact on Detection

Salmonellae are primarily enteric organisms and, whether in the gut or on contaminated foodstuffs, they are usually present as part of a mixed microbial population. Exceptions to this are seen with invasive, host-specific strains such as *Salmonella typhi* in humans and *S. enteritidis* in poultry. The latter organism, for example, can be isolated in pure culture from both the tissues of infected chickens and the contents of intact shell eggs. It is generally necessary, however, to apply techniques which selectively encourage the growth of salmonellae and to use agars which permit the organism to produce characteristic colonies. The similarity in antibiotic resistance patterns between salmonellae and other enterobacteria has largely precluded the use of antibiotics and selection has relied upon the use of chemicals such as selenite, tetrathionate and deoxycholate. ***See*** Enterobacteriaceae, Detection of *Escherichia coli*

When microorganisms, particularly Gram-negative bacteria, are exposed to unfavourable conditions such as high or low temperature, they may become sublethally injured. This can be defined as 'a sensitivity to either selective agents or conditions to which normal cells are resistant'. In salmonellae, injury can be manifested by, for example, lowered resistance to the chemicals mentioned above. The organisms may also be less able to grow, particularly in selective media, at elevated incubation temperatures, such as 43°C, which are used to improve the selectivity of isolation media when samples like sewage are examined.

Techniques used routinely for clinical specimens when large numbers of the target pathogen may be present and where it is relatively unaffected by its environment are not suitable for use with either food or environmental samples. These might have been exposed to conditions damaging to bacteria and they may also contain only low numbers of salmonellae. The direct inoculation of food or environmental samples into selenite broth or another selective medium may lead to false negative results.

The principal site of injury in the bacterial cell is the outer membrane, and alterations in permeability allow the ingress of selective agents which are normally excluded. If cells are exposed to more extreme conditions, damage can also be caused to the chromosome and ribosomes. Much work remains to be done on the response of salmonellae to adverse conditions and on the full impact of the exposure of these bacteria to

Table 1. The influence of period of pre-enrichment and the addition of novobiocin and cefsulodin on the isolation of salmonellae from samples of liquid raw egg

Culture medium	No. of salmonella-positive samples[a] after incubation for 24 h	48 h
Egg/BPW only	32	14
Egg/BPW plus novobiocin (5 μg ml^{-1}), cefsulodin (10 μg ml^{-1})	74	70

Data from Humphrey TJ and Whitehead A (1992).
[a] 110 samples were examined. Rappaport Vassiliadis broth was used as the selective medium.

different environments and their subsequent ability to grow in culture media. Salmonellae, like other bacterial cells, are able to repair the lesions causing sublethal injury, either partially or completely, if they are incubated in a nonselective medium at 37°C. Two media, buffered peptone water (BPW) and lactose broth, are the most commonly used, with the buffering capacity of BPW giving it an advantage over the other medium – particularly when foods containing fermentable substrates are examined.

Studies have demonstrated that the isolation rate from selective broth cultures is greatly improved if they are inoculated with an actively growing population of salmonellae. It is thus essential that initial incubation in BPW allows the organism to both overcome the effects of sublethal injury and to multiply.

The period of recovery or repair is often referred to as pre-enrichment. Incubation is usually for 18–24 hours, although some workers have reported improved isolation rates when pre-enrichment is extended to 72 hours. This is dependent on the material under investigation. Salmonellae do not appear to be able to compete well with some other bacteria. Recovery media, by necessity, are nonselective and permit the growth of the majority of aerobic and facultative organisms present in the sample. Where this encourages the proliferation of either *Proteus* or *Pseudomonas* spp. – as with liquid raw egg and chicken meat – or lactobacilli – as with dairy products – salmonellae may be inhibited. This can be particularly pronounced when incubation is prolonged (Table 1).

The addition to BPW of novobiocin and cefsulodin, which inhibit *Proteus* or *Pseudomonas* spp., respectively, but have no effect on the growth of salmonellae, was found to improve significantly the isolation rate of salmonellae from naturally contaminated samples of raw liquid egg (Table 1). The recovery of salmonellae from dried milk can also be improved if, after 2–4 hours' initial incubation, Brilliant green at a final concentration of 0·0002% is added to the pre-enrichment culture.

The presence of any number of salmonellae in a processed or cooked food is regarded as being potentially hazardous and most microbiological guidelines suggest an absence of salmonellae in 25 g of product. There is also evidence that with some foodstuffs, particularly those rich in fat, and with certain vulnerable individuals such as the very young or the elderly, the infective doses of salmonellae may be low. Recent studies have also demonstrated that sublethally injured cells of *Campylobacter jejuni* are capable of repair in the intestinal tract. Salmonellae may also be able to do this and it is important that sensitive techniques are used for their detection in the laboratory. In this respect, the importance of effective pre-enrichment cannot be overemphasized. As a general rule, it is prudent, when examining food and environmental samples from any source, to carry out primary incubation in BPW at 37°C for a full 24 hours. The ratio of sample to medium should be at least 1:10 and, where appropriate, the growth of competing flora may be suppressed by the addition of antibiotics.

Selective Culture

The necessity for selective culture techniques in microbiology has been recognized for over 100 years and much of the early work was centred upon salmonellae. Selective isolation is usually in two stages: growth in an enrichment broth followed by plating on a selective agar.

Media make use of either dyes or chemicals. A formulation which is still used extensively is based on selenite. It is believed that this compound inhibits bacteria either by reaction with sulphydryl groups or by the formation of analogues of sulphur amino acids. Selenite broth has been used successfully for many years and modifications to improve the isolation rate have included the addition of cystine and the incorporation of fermentable sugars such as either mannitol or lactose which prevent increases in the pH of the medium. Selenite is less selective at alkaline pH values.

Selenite can be used for direct inoculation with specimens such as faeces or used following pre-enrichment. It is probably less suitable, however, for the latter when compared to other media. The medium can inhibit many bacterial types, but two important exceptions – *Proteus* and *Pseudomonas* spp. – can cause problems, particularly as they are capable of preventing the growth of salmonellae and can also resemble them on selective agars.

Salmonellae have been shown to possess the enzyme tetrathionate reductase, which is absent in many other bacteria. This provides a selective advantage and has been used in the development of another commonly used enrichment medium, tetrathionate broth. Tetrathionate is usually employed in combination with Brilliant green and bile salts, the medium being known as Müller–Kauffman tetrathionate broth. It is probably more selective than selenite, but more care may be required in the control of incubation temperatures.

Malachite green, in combination with high concentrations (4%) of magnesium chloride and a medium pH of 5·0–5·2, has been shown to be effective for the isolation of salmonellae. The medium was originally developed for use with faeces but modifications in the last 10–15 years, which have seen a reduction in the malachite green concentration from 0·012% to 0·004% and use of soya peptone rather than tryptone, have meant that the formulation, usually known as Rappaport–Vassiliadis (RV) broth, is more suitable for selective culture following pre-enrichment. This medium is probably the one of choice in the examination of nonclinical samples and numerous studies, on many different materials, have confirmed the superiority of RV broth over both selenite and tetrathionate.

The selective nature of the above broths is often finely balanced and can be disturbed by either a large inoculum, relative to broth volume, or overlong incubation. A change in incubation temperature can also profoundly affect the toxicity of the medium.

An inoculum to broth ratio of 1:10 is recommended for both selenite and tetrathionate broths, although following pre-enrichment culture there would appear to be no reduction in sensitivity if smaller volumes of BPW are used. The size of the inoculum is critical with RV broth and its ratio to the medium should be at least 1:100. With this medium, studies have indicated that more samples are salmonella-positive after 48 hours' rather than 24 hours' incubation, although the differences are often not significant. In contrast, extended incubation significantly reduces the effectiveness of tetrathionate and selenite broths and, with the latter, overgrowth of competing flora occurs frequently. These media can also become more toxic to the target organism and salmonellae may die if incubation is continued much beyond 24 hours. Care also has to be taken in controlling incubation temperature. Selenite and tetrathionate can be used at either 37°C or 43°C and are more selective at the higher temperature. Some studies have shown, however, that salmonellae will not always survive in these media at 43°C. Initial studies on RV broth recommended 43°C but it is now accepted that incubation at 40±0·5°C increases sensitivity.

Studies with particular samples or in particular situations have demonstrated that one selective medium may be superior to another. No one medium, however, can be relied upon to always maximize the isolation rate with all samples. There may thus be advantages in using two selective broths.

Selective agars should be so formulated that the competing flora are suppressed while the target organism is able to form discrete, characteristic colonies. A degree of selectivity is achieved by the addition of either dyes like Brilliant green or chemicals such as sodium deoxycholate. The resistance of salmonellae to selective agents is similar to that of many other bacteria and this can create difficulties in media formulation. Thus, the detection of salmonellae on selective agars relies heavily on diagnostic aspects. Principal among these are the inability of most strains to ferment lactose and the production of hydrogen sulphide from thiosulphate at a neutral pH (leading to black colonies).

As with enrichment broths, many selective agars are available, but none has been shown definitively to be suitable for all applications and the best course of action would be to use two agars in parallel. The most commonly used media appear to be deoxycholate citrate (DCA), xylose lysine deoxycholate (XLD) and Brilliant green (BG) agars.

Many workers have attempted to improve selectivity by the incorporation of antibiotics and/or other selective agents. Of these, novobiocin, which prevents the growth of *Proteus* spp., seems to be the most useful.

Confirmatory Tests

Other organisms, particularly members of the family Enterobacteriaceae, can resemble salmonellae on selective agars and confirmation of 'salmonella-like' colonies is necessary. Tests take the form of biochemical confirmation which usually involves assessment of urease and lysine decarboxylase activity, fermentation of dulcitol, indole production, growth in the presence of potassium cyanide, and utilization of sodium malonate. These reactions, in combination with serological tests, are usually sufficient for identification, but additional tests may sometimes may be necessary. If these are to form part of a laboratory routine, the use of commercial kits may be cost-effective. Antibodies against somatic (O) and flagella (H) antigens are used to confirm/identify salmonella-like isolates. Somatic antigens are composed of polysaccharide while those from the flagella are proteinaceous. Testing will usually comprise slide agglutinations with polyvalent 'O' and 'H' antisera followed by the use of sera raised against specific antigens.

Rapid Methods

The time taken to confirm or deny the presence of salmonellae, especially in food and environmental samples, can create difficulties and, with perishable foods, can mean that products may be distributed and consumed before test results are known. It is not surprising that considerable attention has been paid to the development of rapid methods of detection.

Studies on this topic have examined the feasibility of either reducing the incubation times with conventional culture methods or the use of new technologies based on either immunology or DNA–DNA hybridization.

At first sight, it would seem possible to reduce the length of either pre-enrichment or selective broth culture. This approach has largely been unsuccessful because much has yet to be learnt concerning the ability of damaged salmonellae to grow and recover from sublethal injury. A number of publications have stated that recovery from injury is complete within 8 hours at 37°C. This is dependent on the culture medium used and the degree of cellular damage. In some systems, lag times of up to 72 hours have been reported, and other studies have indicated that repair of damaged cells is not complete until some hours after growth has begun. Thus, subculture of pre-enrichment broths after 6 rather than 24 hours' incubation was found to reduce the isolation rate. Until additional information becomes available it would be wise to incubate primary cultures for at least 18 hours before subculture. Attempts to shorten the time of selective culture have proved similarly disappointing and a number of workers have demonstrated that isolation rates after 6 hours' incubation are markedly lower than those obtained after 24 hours.

'Rapid' in terms of food microbiology has been defined as a test which takes between 4 and 12 hours to complete while 'very rapid' tests can be completed in less than one hour. New methods for the detection of salmonellae fall into the former category and most take considerably longer than 12 hours. They also rely, in the main, on the initial use of traditional culture systems and are thus governed by the constraints outlined above.

Several DNA probes have been developed for commercial application but the most widely used is the Gene-Trak assay. The system uses two probes against 23SrRNA and specificity was found to be 100% against 239 different salmonellae. Studies with naturally contaminated chickens would indicate that the sensitivity of DNA probes is equal to that of culture. Best results were obtained when samples were cultured in selenite broth followed by Gram-negative broth before DNA probe assay.

A variety of salmonella immunoassays is available. These include Bio-Enza bead, a monoclonal ELISA; Tecra, a polyclonal sandwich ELISA; Spectate, a group-specific polyclonal latex agglutination test and the 1-2 Test which is an immunoband diffusion test using polyclonal antibodies. ***See*** Immunoassays, Radioimmunoassay and Enzyme Immunoassay; Immunoassays, Principles

A recent comparative study using naturally contaminated chicken carcasses demonstrated that 'rapid' methods based on either immunoassay or DNA hybridization were as sensitive as culture systems for the detection of salmonellae. The 'rapid' methods, which gave results within 48–53 hours, all required pre-enrichment and selective culture, but time was saved because selective plating and biochemical and/or serological confirmatory tests were not necessary. 'Rapid' tests would appear to offer advantages over traditional culture systems. It is likely that improvements will be made in both speed and sensitivity. This may encourage a wider use of the new technologies. If this is the case, it is to be hoped that people continue to bear in mind the vital roles of proper sample handling and the importance of using initial culture systems which facilitate the recovery and growth of injured organisms.

Bibliography

Bailey JS, Cox NA and Blankenship LC (1991) Comparison of an enzyme immunoassay, DNA hybridization, antibody immobilization and conventional methods for recovery of naturally occurring salmonellae from processed broiler carcases. *Journal of Food Protection* 54: 354–356.

Candlish AAG (1991) Immunological methods in food microbiology. *Food Microbiology* 8: 1–14.

Humphrey TJ and Whitehead A (1992) Techniques for the isolation of salmonellas from eggs. *British Poultry Science*, in press.

Patel P (1989) Diagnostic tests for particular bacteria and bacterial toxins. *Food Manufacture* 64: 70–74.

Speck ML (1984) *Compendium of Methods for the Microbiological Examination of Foods*. Compiled by the APHA Technical Committee on Microbiological Methods for Foods. Washington, DC: American Public Health Association.

T. J. Humphrey
Public Health Laboratory, Heavitree, UK

Salmonellosis

Salmonellosis is the general term for all human illnesses caused by members of the genus *Salmonella* and salmonellosis is one of the major forms of foodborne disease. Every year, in excess of 45 000 cases are reported in the United States alone, and it is likely that this figure represents no more than 10% of the actual infections; an illness only enters the statistics when a patient visits a physician and the organism is identified. This same point also highlights the fact that different serotypes (species) of *Salmonella* can bring about a spectrum of reactions in human beings ranging from mild inconvenience to fatality, but with the majority of patients never needing to seek medical advice. Obviously, each individual patient will differ slightly in their response to infection but, as the main differences are dependent upon the serotypes of *Salmonella* involved, it may be convenient to discuss the topic of disease on the same basis.

Typhoid (Enteric) Fever

This acute infectious fever is caused by *Salmonella typhi* and, along with the related disease paratyphoid fever (caused by *Salmonella paratyphi* A, *S. paratyphi* B (sometimes also referred to as *S. schottmuelleri*) and *S. paratyphi* C (sometimes also known as *S. hirschfeldii*)), is endemic in many parts of the world. It is potentially fatal for young children, particularly during the first year, and for the elderly, but under normal circumstances, many populations in developing countries appear to have acquired a degree of natural immunity. However, if the risks of infection and/or the levels of bacteria in the environment increase, then typhoid is still a major problem. The advent of floods is a case in point, for the spread of raw sewage to watercourses exposes many people to levels of bacteria well beyond the capacity of their immune systems.

This link between poor sanitation and disease is long established because, unlike many species of *Salmonella* which occur widely in birds and animals, *S. typhi* and *S. paratyphi* can only infect humans and certain apes. Consequently, infection is always from one human being to another, either directly or more usually through accidental contamination of water or food consumed by the disease-free individual, and the history of the disease confirms this pattern. During the Boer War, for example, the insanitary conditions of the battlefront resulted in over 10% of the troops being plagued by typhoid/paratyphoid, and the Union and Confederate Forces in the American Civil War suffered in the same way. This latter conflagration also indicated the dangers posed by symptomless carriers, for many soldiers who recovered from the disease retained high population levels of *S. typhi* in their bodies, and hence continued to pass contaminated faeces for weeks or even months. Consequently, men returning to their home towns or villages spread the disease across the country, and typhoid and paratyphoid reached almost epidemic proportions. Similar patterns can be observed today, in that incidents of paratyphoid A and C in the United Kingdom, for example, are invariably the result of a symptomless carrier returning from a foreign holiday; only type B is normally present in the local population.

Clinically, paratyphoid tends to be less severe than typhoid, but infection by either species causes inflammation and/or ulceration of the small intestine, septicaemia, enlargement of the spleen and, on occasions, infection of other organs leading, for example, to meningitis, nephritis or pneumonia. The incubation period is around 10 days, though it can be longer, and the symptoms experienced by the patient tend to be both vague and variable. Headaches tend to be common, as does a general feeling of lethargy, along with intermittent abdominal pain/discomfort and, at least initially, constipation is more usual than diarrhoea; as is to be expected, the body temperature rises slowly. Fatalities usually arise from intestinal haemorrhaging, perforation of the intestine and/or pneumonia, but otherwise the disease runs a slow course over 5–6 weeks and then leaves the patient to a longer period of gradual recuperation. However, even when the patient is clinically fit, bacteria can remain in the intestine and/or bloodstream, and such carriers can continue to excrete viable organisms in faeces, or even urine, for many months. As mentioned earlier, individuals of this type can provide a major source of infection via water supplies or shellfish living in contaminated water, or more directly if the carrier happens to be a handler of retail food items such as ice cream; flies are also known to be vectors of the bacteria to foods exposed to the atmosphere. It is for these reasons that even countries with advanced standards of hygiene can still suffer from outbreaks of typhoid or paratyphoid.

Prevention through improved standards of hygiene offers the best hope for avoidance, and the education of food handlers at all stages of production and preparation is essential. Legislation covering the display and storage of foods provides some useful guidelines, but without adequate expenditure on the training and monitoring services the impact may be more political than real. Personal hygiene is obviously important for an individual at risk from infection, but there is increasing evidence that diet may influence the course of infection as well.

In a healthy individual, the lower end of the small intestine – the site of initial infection by *S. typhi*, for example – is colonized by lactobacilli and, in particular *Lactobacillus acidophilus*, and high population levels of lactobacilli can prevent the salmonellae from becoming established. Similarly, the duration of the carrier state can be reduced by stimulation of the natural intestinal flora through the consumption of fermented milks, especially those that contain *L. acidophilus* alongside a normal yoghurt culture. ***See*** Microflora of the Intestine, Role and Effects

However, while simple preventive measures have much to recommend them, vaccination is advisable for visitors to areas where the disease is endemic. The usual vaccine, known as TAB as it offers protection against typhoid and paratyphoid fevers A and B, grants a high degree of immunity for around 2 years. Immunity is not, of course, guaranteed, but it is of note that the incidence of typhoid amongst troops serving in the First World War is reported to have fallen to around 0·2% – from the level that would have been anticipated of over 10% – solely as the result of vaccination. Even when the disease is contracted, the severity of the symptoms is much reduced following vaccination. If avoidance does fail, then antibiotics are able to eradicate an infection in most cases, although the carrier state can persist if the bacterium has spread beyond the intestine. This initial survival of medication – more than one course of antibiotic treatment may be necessary – may arise because the cells of *S. typhi* can find protection within the lymphatic system of the host. More specifically, according to Doyle and Cliver (1990), the macrophages of the human ingest salmonellae phagocytically just as they would any other invasive bacterium, but whereas most bacteria are killed by this defensive mechanism, *S. typhi* can survive and grow within the macrophage. Eventually, the macrophage bursts and releases the salmonellae into the bloodstream; if this release happens to coincide with the termination of the antibiotic treatment, i.e. the level of antibiotic in the blood is declining, the bacterium and the infection can persist. ***See*** Immunity and Nutrition

Gastroenteritis

In industrialized countries, salmonellae are much more likely to be encountered as causes of 'foodborne infection' and over the years several hundred serotypes have been implicated in incidents of food or waterborne disease. Often the serotype (species) name indicates where the oubreak occurred (e.g. *S. newport* or *S. panama*) but others have definite 'species' names like *S. typhimurium* or *S. enteritidis*. Despite the array of serotypes, many are of extremely local occurrence and, in most countries, disease outbreaks tend to be confined to a relatively few serotypes. In the United States, for example, *S. typhimurium* tends to head the list of isolates with, in descending order, *S. enteritidis*, *S. heidelberg* and *S. newport* being the next most important; the importance of other serotypes tends to vary from year to year.

The reason for this plethora of serotypes associated with food poisoning is that all food animals such as cattle or pigs, as well as domestic fowl such as turkeys, chickens and ducks, have intestinal floras that include salmonellae, and hence passage to food is all too easy. Domestic pets such as cat and dogs can be equally important sources of infection, as can reptiles such as terrapins, or casual invaders such as rats or mice. Consequently, any lapses in food-handling practice or personal hygiene can lead to contamination of a retail item and, if growth of the salmonellae can occur during storage and distribution, the unfortunate consumer will be faced with an infective dose of the serotype concerned.

Exactly what constitutes an infective dose probably differs from individual to individual and from serotype to serotype, and reported figures range from the generally accepted level of 10^9 viable cells to produce disease, through 11×10^3 for *S. typhimurium* in ice cream to 50 cells of *S. napoli* in chocolate. This extraordinary range of values implies that any food capable of supporting the growth and survival of a specific serotype may be suspect, and if the consumer is susceptible then infection will follow. The symptoms usually appear some 12–24 hours after ingestion and include diarrhoea, vomiting, abdominal pain, fever and headache, with the severity of each reaction varying with the resistance of the host. The emphasis on the intestinal problems is a reflection of the fact that the salmoneliae associated with gastroenteritis cause only localized infection. Thus, although members of both this group and the enteric fever group penetrate the epithelial cells at the lower end of the small intestine, only *S. typhi* and *S. paratyphi* can progress into the lymphatic system. By contrast, *S. typhimurium*, *S. enteritidis* and other serotypes give rise to local areas of inflammation but no septicaemia or infection of other organs.

In general, gastroenteritis resulting from a *Salmonella* infection is usually regarded as inconvenient rather than life-threatening, and hence no antibiotic or other prescribed therapy is recommended. The self-administration of drugs to combat diarrhoea is probably the common response of most patients, for salmonelloses of this type are rarely dangerous except to the very young or the elderly; in these vulnerable groups, the mortality rate is roughly 1 in 1000 cases. The symptoms of the disease usually last for 2–3 days but of more potential significance to the food industry is the fact that individuals may continue to excrete viable salmonellae for several weeks after the end of the illness. One month is a common time-span for the carrier/excretion state, and around 15% of patients may continue to pass salmonellae in their faeces for up to 2 months. The potential consequences of this situation for consumers is self-evident and, as with the enteric fevers, it may well be relevant that the length of the carrier state can be reduced if the natural intestinal microflora is activated by the ingestion of fermented milks.

Future Trends

Following the introduction of vaccination and other measures, typhoid and paratyphoid probably account for less than 5% of reported cases of salmonellosis in most industrialized countries, and hence it is gastroenteritic serotypes that are of most concern. Thus, aside from the human misery, the impact on a national economy in terms of absenteeism from work or the recall and destruction of contaminated products, for example, can lead to financial losses running into millions of dollars. If the food in question, or even some of its ingredients, is imported, then the financial implications will affect the exporting country as well, so that avoidance of infection offers considerable benefits all round.

Yet in spite of the obvious incentives for improvement, national statistics for reported cases of salmonellosis continue to increase. In the United States, for example, 20 000 cases of disease involving salmonellae were recorded in 1963 as against a current figure of around 50 000 and, bearing in mind the earlier comment concerning the likely ratio of 'reported to unreported' cases, it is clear that salmonellae pose considerable problems for the food industry. One of the reasons for this is that there are some 2300 serotypes of *Salmonella* currently recognized, of which 140–160 have been implicated in human disease. The status of the remaining types is not known for certain but, given that most of the serotypes are naturally present in the intestinal tracts of a variety of animals, the occasional involvement of any serotype with human disease cannot be ruled out; less than 1% of salmonellae appear to be truly host-specific. Consequently, there is an extensive reservoir of serotypes of potential importance, for although some serotypes like *S. typhimurium* are of major importance at the present time, the situation could change in the future; just as *S. enteritidis* has risen to prominence over the last 10 years, so other serotypes could emerge as potential threats at some time in the future. ***See*** Food Poisoning, Statistics

It may well be that certain serotypes were in the past restricted to small geographical areas, but the rapidly changing nature of society has altered this situation for the worse. Ayres *et al.* (1980) summarized some of the relevant factors as: expanding populations with potential for contaminating the environment; a much more mobile society; food establishments producing millions of units for retail purchase; proliferation of nonsterile convenience foods, many of which are susceptible to infection by salmonellae; more opportunities for the employment of carriers; little change in the levels of training provided for food handlers, especially outside of the major food companies; and increasing contamination of agricultural land and waterways. It is against this background that the food industry, environmental and public health officers and the medical profession have to operate in an attempt to contain, or if possible to reduce, the incidence of salmonellosis. With allocations of financial resources to these groups declining in real terms, and the research/education base to support these services being eroded in the interests of 'economy', it is hardly surprising that the incidence of gastroenteritic salmonellosis continues to rise in developed and developing countries alike.

Bibliography

Ayres JC, Mundt JO and Sandine WE (1980) *Microbiology of Foods*. San Francisco: WH Freeman.

D'Aoust J-Y (1989) Salmonella. In: Doyle MP (ed.) *Food-borne Bacterial Pathogens*. New York: Marcel Dekker.

Doyle MP and Cliver DO (1990) In: Cliver DO (ed.) *Food-borne Diseases*. New York: Academic Press.

Hobbs BC and Gilbert RJ (1987) *Food Poisoning and Food Hygiene*. London: Edward Arnold.

Jay JM (1992) *Modern Food Microbiology*. London: Van Nostrand.

Robinson RK (1991) *Therapeutic Properties of Fermented Milks*. London: Elsevier Science Publishers.

Rothwell J (1990) Microbiology of ice cream and related products. In: Robinson RK (ed.) *Dairy Microbiology*. London: Elsevier Science Publishers.

R.K. Robinson,
University of Reading, Reading, UK

SALTING

See Curing, Smoked Foods and Preservation of Food

SAMNA

Although similar to ghee and other concentrated milk fats, samna is of especial importance to the dairy industry in Egypt. This article deals specifically, therefore, with the production and properties of samna as traditionally manufactured in the rural areas of Egypt, and other milk fats are dealt with elsewhere. ***See*** Ghee

Samna is pure, clarified milk fat, produced in Egypt. It is generally prepared from buffaloes milk or cows' milk which constitute 63·5% and 35% respectively of total milk output in Egypt, while small amounts of samna are prepared from sheeps' milk and goats' milk which constitute 1% and 0·5%, respectively, of the total milk output. ***See*** Buffalo, Milk; Milk, Dietary Importance; Sheep, Milk

The main object of the primitive dairy industry in the rural districts of Egypt is to separate milk fats for making butter, and to make the remainder into products, which are consumed as such or, after storage, throughout the year. In lower Egypt, farmers put fresh milk in shallow or deep earthenware pots, 'matrad' or 'shalia', and leave it undisturbed in a warm, dark place till the cream rises and the milk coagulates. The cream layer is removed and beaten into butter, which is converted into samna. The presence of earthenware pots, such as used in cream-making, in the tomb of King Horaha of the first dynasty (3200 BC), indicate that the art of samnamaking was known to the Ancient Egyptians.

Samnamaking by the Traditional Method

The traditional process of samnamaking depends primarily upon the properties of the butter with respect to its acid degree value, taste, odour and weight. The latter parameter is important for determining both the yield of the samna, and the amount of salt to be added to the butter prior to the making process.

Liquefaction of Butter and Addition of Salt

Butter is placed in a stainless steel or aluminium vat; iron or copper pans should not be used, for if the samna is contaminated with heavy metals, oxidation of the fat is accelerated. The butter is then heated with continous stirring until liquefied (50–60°C), and the salt is added at a rate of about 2–4% of butter weight. ***See*** Heavy Metal Toxicology; Oxidation of Food Components

The addition of salt leads to the following: (1) a decrease in the water content of the resultant samna because the salt increases the boiling point of the water in the butter; (2) it helps in the fat separation that arises from the difference in density between fat and nonfat phases; (3) it increases the quantity and shelf life of the by-product ('morta'); (4) it plays an important role in the precipitation of the proteins in butter during boiling. In addition, some makers of samna believe that the addition of high levels of salt during the making of samna is necessary in order to store samna for a long time. This latter view is misguided for the following reasons: (1) the salt is not a fat-soluble material, and hence does not have a preservative effect; (2) contamination of the salt with heavy metals, e.g. iron or copper, accelerates fat deterioration; (3) the salt is a good absorber of water, and thus the presence of salt in samna can increase the moisture level and hence risk of rancidity. In general, after the butter has been heated and salted, it is filtered through a cloth, such as cheesecloth, to remove any foreign matter mixed with the butter. In the case of a good-quality butter, this step is omitted.

Butter Boiling and Ripening

Butter boiling and ripening are the most important parts of the process of samnamaking with respect to the quality of the resultant product. They are performed as follows. The liquefied butter is heated further to 90–96°C and, in the course of heating, a foam is produced which is known as the boiling foam. At this point, the level of heating should be reduced. After the disappearance of the foam, regular boiling or simmering of the fat starts. After a certain time, floating particles appear, which are principally composed of proteins and phospholipids. ***See*** Phospholipids, Properties and Occurrence

Some makers prefer to remove the floating particles, but others do not. In general, at 103–107°C most particles precipitate to the bottom of the vat. At 107–110°C, the amount of nonfat particles increases. At 110–115°C the all-nonfat suspended particles are precipitated to form 'morta', which has the same colour as samna, i.e. bright yellow. At 115–125°C, the colour of morta becomes dark and the characteristic cooked odour of samna starts to appear. Moreover, a foam suddenly appears, at which point the heating should be stopped (this foam is called the 'ripening foam'). Excessive heating after this stage leads to the following:

(1) production of a darker product; (2) the precipitated nonfat particles start to disperse and suspend again, and will be difficult to separate; (3) the samna acquires an undesirable flavour; (4) the keeping quality of resultant samna is very poor owing to the weak body derived from the crystallization of the fat.

The quality of samna depends on the control of the boiling and ripening processes and the expertise of the maker. For example, continuous stirring during boiling is very important to avoid any overcooking defects, as is the time at which the heating and stirring should be stopped. All these aspects must be controlled by the samnamaker.

Decantation and Filtration

Decantation and filtration are carried out when the samna is slightly hot. The clear fat layer is decanted by pouring it into another vat until just above the level of precipitated particles (morta). The latter portion of samna is passed through cheesecloth, at least twice, to retain any precipitated particles, and the clear fat is then added to the rest.

Packaging of Samna

The packing of samna is considered a vital process. Samna should be filled hot (at 50°C) into the storage cans in order to (1) expel air bubbles inside the cans, and (2) act as a sterilization process.

Many different containers can be used for packaging samna, e.g. jars of dark-coloured glass to avoid light-induced oxidation, tin-cans free of rust, or there are special pots (which are known as 'barany' and mostly used in Egypt) made from clay with a glazed inner-surface.

The proper conditions for the storage of samna are moderate temperatures and exclusion of direct light.

Samnamaking by Nontraditional Methods

Mechanical Separation

The mechanical separation method depends mainly upon the separation of fat from other nonfat matter by the use of special separators. The product has a higher fat content than samna made traditionally. The product of this method is known as butteroil.

Factors Affecting the Keeping Quality of Samna

1. An increase in the acidity of the butter or cream decreases the keeping quality of samna.
2. The heat treatment of the fat during the making process should be not less 110°C and not more than 125°C.
3. The presence of traces of heavy metals, e.g. iron and copper.
4. The presence of air or oxygen inside pots or cans.
5. The storage temperature.
6. The exposure of samna to light during storage.
7. A high level of moisture and/or nonfat matter in the samna.
8. The presence or absence of natural antioxidants.

Characteristics of Good-quality Samna

1. Samna produced from cows' milk fat has a golden-yellow colour (owing to the high content of β-carotene), but in the case of buffalo milk fat, the product has a white, slightly greenish colour.
2. The samna should have a slightly cooked, slightly sweet flavour, and it should be free of rancidity.
3. It should have a gritty texture (at 10–20°C).
4. The fat content should be not less than 99·5%, the moisture not more than 0·3%; the expected chemical properties of the fat(s) are shown in Table 1.

Antioxidants used in Samnamaking

The most common cause of deterioration in samna is oxidative rancidity resulting from exposure to the light. To prohibit this phenomenon, antioxidants are added. These are categorized as follows:

1. Natural antioxidants, which are present in samna, derived from milk, e.g. phospholipids and vitamin E (tocopherol). *See* Antioxidants, Natural Antioxidants
2. Antioxidants derived from nonfat matter (proteins) during the heat treatment process of samnamaking e.g. sulphydryl groups (−SH groups). *See* Antioxidants, Synthetic Antioxidants

Table 1. The fat constants of samna made from both cows' milk and buffaloes' milk

Fat constant	Cows' milk samna	Buffaloes' milk samna
Reichert–Meissl value	⩾22	⩾25
Polenske value	⩽2·7	⩽2·7
Kirschner value	⩾19	⩾22
Saponification value	⩾220	⩾222
Butyrorefractometer reading	40–44	40–43

Sometimes hydrogenated oils or other fats are added to samna, and, for detecting this kind of adulteration, milk fat constants should be measured, e.g. Reichert–Meissl, Polenske, Kirschner, and saponfication values.

3. Some natural materials that may be added to certain brands of samna during packaging, e.g. soya bean and wheat flours, fenugreek grains and safflower at a rate of 0·5–1%.
4. Chemical substances, e.g. propylgallate at a rate of 0·01–0·003%, but these materials are rarely used.

Nutritive Value of Samna

Samna is composed mainly of pure fat, and thus it is considered a good source of energy. Moreover, samna provides significant quantities of fat-soluble vitamins, e.g. A, D and E. In addition, the use of samna in cooking improves the taste of a meal, and hence overall intake may be increased. ***See*** Cholecalciferol, Physiology; Fats, Requirements; Retinol, Physiology; Tocopherols, Physiology

The By-product (Morta) of Samnamaking

Morta is a by-product of samnamaking, and if butter rather than cream is used, the amount of morta is usually equivalent to 6% of the weight of butter.

The composition of morta depends primarily upon the steps in, and conditions of, samnamaking but it is generally as follows:

- Water, 10–18%.
- Fat, 40–65%.
- Nonfat solids, 10–25%.
- Ash and salt, 10–15%.

The nutritive value of morta is mainly attributable to the high content of phopholipids. Morta is consumed as is, or used in making 'mish cheese'.

Bibliography

Abou-Donia SA (1984) Egyptian fresh fermented milk products. *New Zealand Journal of Dairy Science and Technology* 19: 7–18.
Lambert LM (1975) *Modern Dairy Products*, 3rd edn, p 307. London: Food Trade Press.
Warner JN (1976) *Principles of Dairy Processing*, p 253. New Delhi: Wiley Eastern Ltd.

S. A. Abou-Donia and S. I. El-Agamy
University of Alexandria, Alexandria, Egypt

SANITIZATION

Principles

Sanitization, or disinfection, is the elimination of microorganisms such that they may remain only at levels that are not harmful to health. This elimination may be by removal (i.e. cleaning), by killing (such as by chemicals or heat), or by a combination of these methods. The level of microorganisms designated as 'safe' will vary according to their location: more bacteria could be present on a 'safe' floor than on a 'safe' chopping board. It will also depend on the type of microorganism present: for a pathogen such a shigella, where very small numbers can give rise to infection, only a total absence is safe; for nonpathogenic (non-disease-causing) spoilage organisms, low numbers present no threat to health or immediate threat to quality. The process of disinfection does not normally include the killing of bacterial spores. These are highly resistant dormant forms in the life cycle of members of the genera bacillus and clostridia (such as *Bacillus cereus* and *Clostridium botulinum*). These are normally only eliminated by the more stringent process of sterilization, that is the total elimination of all microbial life. Some chemical disinfectants are capable of killing bacterial spores and are termed 'sporicides'. ***See*** *Bacillus*, Occurrence; Cleaning Procedures in the Factory, Types of Disinfectant; *Clostridium*, Occurrence of *Clostridium botulinum*; *Shigella*; Sterilization of Foods

Mechanisms and Kinetics

Chemical disinfectants kill their target microorganisms by reaction with one or more components vital to the internal integrity and function of that cell. This reaction can be chemical, physicochemical or both. A chemical reaction is one where covalent bonds within molecules are broken or formed or where the charge on an ion is altered. A physicochemical reaction is one where noncovalent bonds, i.e. areas of hydrophobic affinity or of noncovalent polarized attraction, are disrupted. These reactions follow closely the kinetics of chemical reactions in general, thus the killing of microorganisms by disinfectants can be similarly categorized. As the end effect, the death of microbes, may be the result of a single reaction or, more likely, of a combination of many reactions, the kinetics of microbial death can become somewhat more complex than those of more simple chemical reactions. The death of microorganisms in these circumstances approximates to the kinetics of a first-order chemical reaction. In essence, rates of microbial death occur in a logarithmic manner. To illustrate this: if 1 000 000 bacteria are acted on by a lethal chemical agent, they will lose a proportion, rather than a set number, of their population with each unit of time. If a lethal agent kills nine out ten of a particular bacterium every minute (i.e. one logarithmic reduction per minute), then 1 000 000 bacteria at time zero (i.e. the start of the reaction) will be reduced to 100 000 after 1 minute; to 10 000 in 2 minutes, and so on. After 6 minutes there will be one bacterium left. After 7 minutes there will (in theory) be 0·1 bacterium left. In statistical terms, this means one surviving bacterium in every ten samples. This continues so that at 12 minutes from the start there will be statistically one millionth of a bacterium left: in effect, one bacterium in every million samples. Thus, the concept of disinfection, as with the concept of safety, takes on a statistical element.

Microbial Resistance to Chemical Disinfectants

Microbial resistance to chemical disinfectants is a less well-defined phenomenon than the better-known analogy of microbial resistance to antibiotics. With antibiotics, the concentration of an agent achievable in the body will determine the so-called 'break point'. This provides a naturally set level of antibiotic concentration against which to judge resistance. If an organism is killed or inhibited below this break point, it is considered sensitive; if it can grow at or above the break point, it is resistant. Attainable concentrations of chemical disinfectants are not similarly fixed and can usually be varied within any given use situation. Thus, 'resistance' to chemical disinfectants has to be considered within these constraints, but can still in practice be a useful concept.

There are several possible mechanisms by which resistance to chemical disinfection exists.

Innate Resistance. This implies that there is either no susceptible target within the microbial cell or that the microbial cell is in some way impermeable to the agent. An example of the former is the resistance of viruses which comprise just nucleic acid and protein (such as the picornaviruses, a group which includes a number of viruses which infect by ingestion, for example the poliovirus) to disinfectants such as phenolics and quaternary ammonium compounds. An example of the latter is the high resistance of bacterial spores to most disinfectants. Bacterial spores are a 'survival' form of certain bacterial genera that show extreme resistance to chemical and physical inactivation. Only a few chemical disinfectants ('sporicides') are capable of inactivation of bacterial spores. ***See*** Viruses

Acquired Resistance. In this form of resistance, certain bacteria can adapt to grow or survive in disinfectant solutions. This can be through a process of 'training', whereby a bacterium is exposed to a sublethal concentration of a disinfectant and can then bring in to play a variety of phenotypic adaptations, such as capsule production, that can enable it to survive increasing concentrations of the disinfectant. Another mechanism is by acquisition of transmissible genetic elements that confer resistance to lethal agents. Although this is better known with bacterial resistance to antibiotics, it also exists with disinfectants.

Reasons for Failure of Disinfection

It is not only genuine resistances that can result in failures of a disinfection process. Other factors can affect the performance of a disinfectant and lead to a suboptimal disinfection result.

Neutralization. Neutralization of a chemical disinfectant can occur by chemical or physical reaction. Disinfectants, by their very nature, will react with molecules in a microbial cell, covalently or by charge polarization, and thus disrupt its vital functions. They will similarly react with nonliving organic matter, which removes disinfectant molecules from effective solution. If there is an excess of organic matter and the disinfectant has a high affinity for it, this can lead to a disinfection failure owing to neutralization of the disinfectant. The disinfectant will have reacted with nonliving organic matter, leaving insufficient for microbicidal purposes. Although excess organic matter is the most commonly encountered disinfectant neutralizer, others exist. There can be reactions between disinfectant and neutralizing compounds, converting the disinfectant compound to a nonmicrobicidal reaction product. These are specific chemical reactions and tend to occur in unusual circumstances. The other main source of disinfectant neutralization is from physical reaction with surfactants – charged or polarized large molecules – again in effect taking disinfectant molecules out of active solution. Disinfectants that are especially susceptible to this are those whose activity depends on charge or polarization within their molecule. This both makes them prone to these charge-based interactions and also means that, once their charge or polarization has been disrupted, their microbicidal capacity is neutralized.

Lack of Contact. Failure to make contact with the target organism is a common source of ineffective disinfection. Organic matter, as well as neutralizing disinfectants (see above), can form a barrier to disinfectants, shielding microorganisms within it and facilitating their survival. In this way microorganisms can withstand a disinfection procedure that, to the operator, should have been effective; i.e. involving the correct concentration of disinfectant applied for the right length of time to act against a microorganism innately susceptible to that disinfectant. Some disinfectants, mainly those with innate or additional surface active capabilities, are better able to overcome shielding barriers whilst others, especially those which can coagulate proteinaceous matter, are particularly prone to this shortcoming. However, even those disinfectants with surfactant properties cannot be expected to penetrate thick layers of viscous or solid organic matter. A clean surface is an essential prerequisite for efficient disinfection.

As with organic matter, organisms within crevices in surfaces are usually protected from contact with disinfectants. This protection is enhanced by the presence of organic matter. In situations when it may be *difficult* for a disinfectant to penetrate a crevice, it will probably be *impossible* for it to penetrate that crevice when it is also filled and protected by organic matter. So not only is absence of soiling needed, but also a crevice-free surface. Crevices can be innate, as in the grain of a wooden chopping board; a result of poor design, as in machinery or pipework; or a by-product of use, as in knife scours in a plastic chopping board.

Another situation where disinfectants do not work due to a failure to make contact with their target is that of poor-quality application such that the agent does not make contact with all the surfaces intended for disinfection. On an accessible surface, this could be a result of human error – due to inept or badly instructed staff – or to equipment inadequate to apply disinfectant to areas that are difficult to reach; or to failure to remove extraneous objects that may protect a surface from a directly applied or sprayed disinfectant. Inaccessible surfaces, such as those within pipework or other enclosed vessels, present a more substantial problem. The application of disinfectants should, where necessary, be considered at the design stage, either in the form of a modification allowing free access to all surfaces or 'cleaning-in-place' dispensing nozzles are permanent internal features. ***See*** Plant Design, Designing for Hygienic Operation

Concentration. The concentration of a disinfectant will also determine microbicidal efficacy. Disinfectants are formulated to work at a specific concentration. Sometimes different concentrations for different situations are stipulated, for example in the presence or absence of organic soiling. Should too low a concentration be used, inefficient microbicidal action will result; should too high a concentration be used, the consequences could be wasted resources, corrosion problems or taint and toxicity problems. Disinfectants should be made up for use accurately. A 'splash in a bucket' approach cannot ensure accuracy. Similarly, if a disinfectant is going to become substantially diluted during its use, it should be formulated to be near its designated use concentration after rather than before this dilution.

Time of Exposure. All disinfectants need adequate time in which to work. Exposure time of a microorganism to a disinfectant can be determined by the method of application; for example, it is very convenient to use a volatile surface disinfectant, such as alcohol, in the form of a wipe. The surface to be disinfected will be dry and useable within seconds. The very convenience of this form of application can also make it a less effective method.

Wetting Ability. Another factor that will determine efficacy of surface disinfection is the wetting ability of a disinfectant. When a disinfectant without a detergent or other surface-active agent (i.e. one which has no wetting ability) is spread on a surface, the film that it forms will rapidly turn into discrete droplets with dry areas between the droplets. No substantial disinfection will occur in these dry areas, which will comprise the majority of the area to which liquid was initially applied. A wetting agent, usually a detergent, will allow the disinfectant to remain as a continuous film on the surface and exert its full action until it is taken out of effective solution by drying, so becoming unavailable to microbial cells. Wetting ability also enhances a disinfectant's ability to penetrate or remove layers of organic matter.

Temperature. As with chemical reactions in general, the temperature at which a disinfectant acts will affect the speed of its activity. Unless otherwise specified, disinfectants are formulated to work around normal ambient temperatures. There are applications where they might be expected to work at low temperatures; for instance, in a refrigerated food production unit or out of doors in a cold climate. For a disinfectant to work effectively in such situations, it will need either a higher concentration, a longer exposure time, or both. From the same principles, if the temperature is higher than normal ambient, a more rapid and efficient disinfection will occur. Temperature itself starts to become lethal to vegetative bacteria (not bacterial spores) and other microbes around 60–65°C, at which temperatures the time needed for disinfection is the range of several minutes to hours. At 80°C disinfection has become far more rapid, needing only seconds for susceptible microbes to be killed at this temperature.

Other Factors to be Considered in Disinfectant Choice

Toxicity. Chemical disinfectants have a high reactivity with biological systems, this being an integral part of their microbicidal mechanism. As living systems share many biochemical similarities, it is not unusual for disinfectants to have a toxic effect on humans. Toxicity may occur through one or more routes of contact, such as through the skin or by ingestion, inhalation or absorption into mucous membranes, for example of the eyes or nose. There are two separate considerations in the use of chemical disinfectants in catering and the food industry: risk to user and risk to consumer.

The major toxic risk is to the users of chemical disinfectants. It is they who will handle concentrated disinfectant solutions or solids, where any toxicity will be many times that of the diluted solution; they who will make up the use dilution, involving splash and skin contact risks during this process; they who will apply the disinfectant with attendant risks of skin contact and inhalation; and they who will dispose of the disinfectant, with more contact and inhalation risks. Chemical disinfectants, especially in their undiluted form, must always be assessed for hazard before use. Any hazard must be minimized by handling procedures that ensure minimal contact (especially uncontrolled contact such as splashing), and use of effective and appropriate personal protective equipment (gloves, eye protection, etc.) where contact cannot be ruled out. As a general rule, prevention of operator contact with toxic agents by containment of the agent is preferable to use of personal protective equipment.

Consumer toxicity, whilst much rarer, is a serious consideration in terms of number of people affected as well as commercial considerations. Dilution factors of disinfectants that find their way from food processing areas to consumers via a food product will be immense. Nevertheless, the consequences of such an occurrence are equally immense. Control is achieved by segregation of significantly toxic chemicals from food-handling areas or routes into those areas.

Taint. Taint is a problem allied in many ways to toxicity in terms of origin and consequences. The problem here is that certain disinfectants can impart characteristic and undesirable taints to foods. Taint can be caused by extremely low concentrations of these chemicals, usually in terms of a few parts per billion of a contaminant in food. The disinfectants most implicated in taint problems are those with phenolic-derived compounds in them. These should be regarded with suspicion unless an assurance of lack of taint can be given. As with toxic agents, such disinfectants are best completely excluded from use in food-handling areas.

Corrosion. Some disinfectants, particularly those of an oxidative nature, can cause or accelerate corrosion of a variety of metals. Whilst this is not directly connected to microbiocidal issues, it is a factor that must be considered in the wider context of the practicalities of disinfectant use. Particularly implicated in this are hypochlorite disinfectants, which can start rusting of carbon steels in minutes. Corrosion is dependent on a combination of factors, mainly the constituents and concentration of the disinfectant, the materials which are in contact with the disinfectant, and the contact time and temperature. ***See*** Corrosion Chemistry

Characteristics of Commonly Used Disinfectants

Hypochlorites. These are probably the most useful general group of disinfectants in food-associated use. Advantages are low toxicity, low taint, fast action, wide microbicidal spectrum, and cheapness. Disadvantages are corrosivity to some metals (good quality stainless steels are usually fairly resistant) and ready inactivation of low concentrations by organic matter. They are available either as sodium hypochlorite, a liquid, or as sodium dichloroisocyanurate ('NaDCC') and chloramines, both soluble solids.

Iodine. Advantages are low toxicity, low taint and wide microbicidal spectrum. Disadvantages are some corrosion to metals (less so than with hypochlorites), ready inactivation of low concentrations by organic matter, and higher cost than hypochlorites. It is available as a complex ('iodophor') either with polyvinylpyrrolidone (PVP) or with detergents. These hold most of the iodine in complexed form with a small amount free in solution; more is released from the complex as iodine is used up. Iodine in potassium iodide or alcohol solutions are available for medical use but have no role in food hygiene.

Quaternary Ammonium Disinfectants (*QACs*). Advantages are low toxicity, low taint, cheapness, noncorrosivity. They all possess some cleaning ability and are convenient to use. Disadvantages are ready inactivation by a wide range of materials and an often incomplete microbicidal spectrum. They are useful as general hygiene agents but must be used with care if disinfection is a critical step. Benzylkonium chloride (**1**) is a commonly used quaternary ammonium disinfectant.

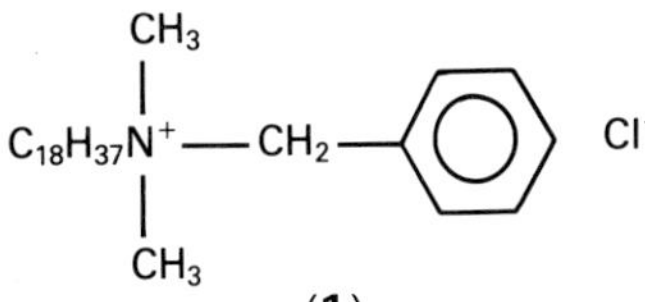

(**1**)

Bibliography

Ayliffe GAJ, Coates D and Hoffman PN (1984) *Chemical Disinfection in Hospitals*. London: Public Health Laboratory Service.

Block SS (ed) (1991) *Disinfection, Sterilization and Preservation*, 4th edn. Philadelphia: Lea and Febiger.
Gardner JF and Peel MM (1991) *Introduction to Sterilization and Disinfection*, 2nd edn. Edinburgh: Churchill Livingstone.
Hobbs BC and Roberts D (eds) (1987) *Food Hygiene and Food Poisoning*, 5th edn. London: Edward Arnold.
Russell AD, Hugo WB and Ayliffe GAJ (1982) *Principles and Practice of Disinfection, Preservation and Sterilisation*. Oxford: Blackwell Scientific.

P. N. Hoffman
Central Public Health Laboratory, London, UK

SAPONINS

Saponins are a heterogeneous group of glycosides which are widely distributed in plants of agricultural importance, particularly legumes. Many of these legumes are staple items of the human diet. Foods particularly rich in saponins are soya beans (*Glycine max*), chickpeas (*Cicer arietinum*) and beans derived from *Phaseolus vulgaris*. ***See*** Beans; Cereals, Dietary Importance; Legumes, Legumes in the Diet; Pulses; Soya Beans, Properties and Analysis

When saponins are agitated in water they form a soapy lather. Other properties generally ascribed to this wide group of compounds are haemolytic effects on red blood cells, cholesterol-binding properties, and a bitter taste. These properties characterize particular types of saponins and are not necessarily shared by all members of the group. From a biological point of view, some of these properties are beneficial while others are considered to be adverse.

Of particular interest is the observation that dietary saponins reduce plasma cholesterol levels in primates, thus having the potential to lower the risk of coronary heart disease in humans.

Chemical and Physical Properties of Saponins

Saponins consist of an aglycone unit linked to one or more carbohydrate chains (Fig. 1). The aglycone or sapogenin unit consists of either a sterol or the more common triterpene unit. In both the steroid and triterpenoid saponins the carbohydrate side-chain is usually attached to the 3 carbon of the sapogenin.

Saponins possess surface-active or detergent properties because the carbohydrate portion of the molecule is water-soluble while the sapogenin is fat-soluble. The stability and strength of forage saponin foams are affected by pH and this may have an effect on the development of bloat in ruminants. Saponins are remarkably stable to heat processing, and their biological activity is not reduced by normal cooking.

Isolation of saponins from plant material involves extraction with a polar solvent after removal of lipids, with petroleum ether or chloroform, followed by various purification techniques. A number of chromatographic procedures have been used to separate individual saponins. ***See*** Chromatography, Principles

The analysis of saponins is complex and potentially subject to considerable errors during their isolation, separation and quantification stages. Thus many early reports of the saponin content of food plants and processed food should be treated with caution. The data presented in Table 1 were obtained using rigorous methodology and are the most reliable data available at present. More recent studies have shown that many legume seeds contain several saponins, e.g. five different saponins have been separated from soya beans.

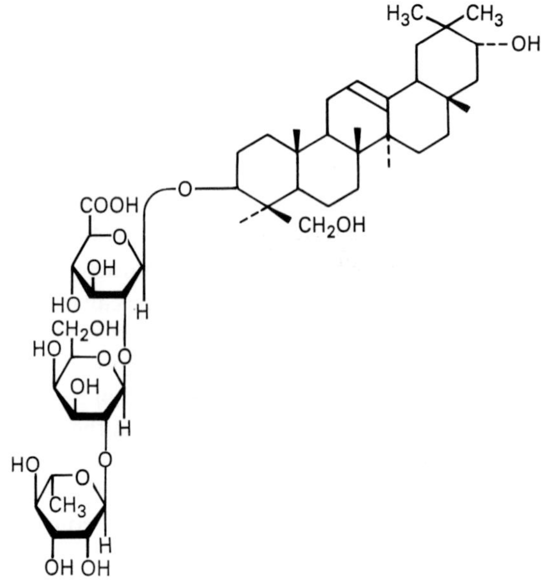

Fig. 1 The structure of a typical saponin (from soya beans).

Table 1. Saponin content of some legume seeds

Legume	Saponin content (g per kg of dry matter)
Chickpeas (*Cicer arietinum* L.)	2·3
Green pea (*Pisum sativum*)	1·8
Haricot bean (*Phaseolus vulgaris*)	4·1
Kidney beans (*Phaseolus vulgaris*)	3·5
Lentils (*Lens culinaris* Medik.)	1·1
Mung bean (*Vigna radiata* L.)	0·5
Runner bean (*Phaseolus coccineus* L.)	3·4
Soya beans (*Glycine max* L. Merrill)	6·5
Yellow split pea (*Pisum sativum*)	1·1

Biological Effects of Saponins

Ingested saponins can influence animal performance and metabolism in a number of ways.

Sensory Properties of Saponins

Recent studies have shown that a bitter or astringent taste is related to amounts of soya saponin I isolated from pea and soya flour. It is possible that saponins are a contributing factor to the undesirable organoleptic properties that humans associate with some legumes and legume products. The bitter taste of saponins may be responsible for the low palatability of lucerne to ruminants and may explain the reduced feed intake observed in many experiments when using this material as a forage. Many stock will discriminate against feeds containing high levels of lucerne meal.

Erythrocyte Haemolysis

Saponins have pronounced haemolytic properties when given intravenously, the degree of effect on the red blood cells varying among different mammalian species. The release of haemoglobin from red blood cells is the direct result of the interaction of saponins with membrane-bound sterols which causes an increase in the permeability of the plasma membrane, bringing about the destruction of the cell. ***See*** Sensory Evaluation, Taste

The haemolytic activity of saponins has been widely used as a means of detecting and assaying saponins in plant materials. The potentially toxic effects of intravenous injection of saponin extracts have resulted in this class of compounds being regarded as antinutritive factors in foods. Their low oral toxicity and the potentially useful nutritive value in foods have only recently been appreciated.

Effects on Blood and Tissue Cholesterol Levels

Since many legume saponins form insoluble addition complexes with cholesterol the effects of dietary saponin on cholesterol might be expected. Saponin in the diet of chicks has been reported to reduce plasma cholesterol in cholesterol-fed animals. Saponin extracted from *Quillaia saponaria* has been shown to reduce liver (but not plasma) cholesterol levels in chicks. An exhaustive series of experiments, with a variety of saponins, have consistently demonstrated cholesterol-lowering effects. These include the feeding of lucerne saponins to rats, rabbits and monkeys. Some workers have suggested that dietary saponins do not have any effect on plasma cholesterol concentrations. This is perhaps to be expected in experiments with rats which received no dietary cholesterol.

Experiments on humans have produced variable results. Different saponin levels in a dietary supplement based on soy flour had no effect on plasma cholesterol levels in hypercholesterolaemic men. As the experiment was conducted on free-living subjects it cannot be guaranteed that the subjects actually consumed the experimental diets.

In a more closely supervised trial, with subjects having normal blood cholesterol levels, it was found that a dietary supplement containing saponins did not have a significant effect on plasma cholesterol level. Increased faecal excretion of bile acids and neutral sterols was observed. Foods containing saponins, or diets supplemented with saponins, have been shown to reduce blood cholesterol levels in humans under conditions which would be expected to induce high levels of blood cholesterol. ***See*** Cholesterol, Role of Cholesterol in Heart Disease; Fats, Digestion, Absorption and Transport

Mode of Action

Saponins remain within the gastrointestinal tract. Some interact directly with dietary cholesterol, producing an insoluble complex which prevents the cholesterol being absorbed. Others appear to affect cholesterol metabolism indirectly by interacting with bile acids. An increased faecal excretion of bile acids is observed in response to feeding additional saponins in the diet. Bile acids thus diverted from the enterohepatic cycle would be replaced by hepatic synthesis from cholesterol. ***See*** Bile

In the digestive tract saponins form mixed micelles with cholesterol and bile salts; their hydrophobic triterpene or steroid groups stack together like small piles of coins. These micelles are then too large to pass through the intestinal wall.

Bile salts normally pass through the wall of the small intestine both by passive diffusion and active transport. Passive diffusion takes place along the entire length of the ileum and jejunum; active transport is confined to the terminal ileum. Saponins can interact with cell membranes, as shown by their haemolytic activity. It is possible that they may also have an effect on the cell membranes of the intestinal mucosa. However, the effects of saponins on both passive absorption and active transport can be explained simply as being due to the reduction in the concentration of free bile acid, because bile acids assist lipid absorption. Low concentrations of free bile acids would also impair the efficiency of lipid absorption and presumably affect the absorption of fat soluble vitamins. Thus there may also be a significant metabolic effect within the animal as well as in the digestive tract.

Effects on Growth

When lucerne saponin was added at high levels to the diets of monogastric animals, reduced feed efficiency and growth rates were observed. It is clear that species differ considerably in their responses to dietary saponin, e.g. poultry are more sensitive than rats. In contrast, soya bean saponins had little effect on the growth rate of experimental animals. Some saponins increase the permeability of the small intestinal mucosal cells, thereby inhibiting the active transport of some nutrients, but at the same time they facilitate the uptake of materials to which the gut would normally be impermeable.

The biochemical mechanism which accounts for the growth depressing effects has not been fully identified. In at least one experiment, the addition of 1% cholesterol to the diets of chicks completely overcame growth depression produced by 0·3% saponin.

It is possible that, in addition to their effects on lipid absorption, saponins also affect chymotrypsin and trypsin activity, which would affect the absorption of protein. ***See*** Protein, Digestion and Absorption of Protein and Nitrogen Balance

In addition to observing reduced intestinal uptake of cholesterol in rat intestinal perfusates on the addition of soy saponin extracts, some workers have also observed significant reduction in cholate and glucose uptakes. It was observed that the saponins could be washed out of the intestinal lumen, suggesting that inhibition, at least in the short term, was not caused by modification of, or damage to, the intestinal mucosa.

Metabolism of Saponins

Toxicity studies indicate that only very low levels of saponin absorption occur. Saponins are between 10 and 1000 times more toxic when administered intravenously than when given orally. Destruction of saponins in the digestive tract of both ruminants and monogastric animals has been observed. Saponins are degraded by rumen bacteria and by microflora found in the caecum of rats, mice and chicks. Since saponin in the caecum is past the major sites of absorption, the release of sapogenins and sugars is considered to be insignificant.

Conclusions

The substantial evidence that saponins from a number of plant species can reduce plasma cholesterol levels in man is likely to encourage further interest in these plant foods. It is generally recognized that the overall nutritional value of many Western type diets would be considered improved if more legumes or legume-based products were consumed regularly. The acceptance of more legumes in Western type diets is limited by undesirable taste characteristics, some of which may be due to the higher levels of saponin found in many legume seeds.

There is little doubt that saponins can be incorporated into human diets at levels which can give a beneficial effect and which would not entail a risk of acute toxicity. The fact that saponins can increase the permeability of intestinal mucosa raises the possibility of interesting nutritional and pharmacological uses.

Bibliography

Anderson JO (1957) Effect of alfalfa saponin on the performance of chicks and laying hens. *Poultry Science* 36: 873–876.

Birk Y (1969) Saponins. In: Liener IE (ed.) *Toxic Constituents of Plant Food Stuffs*, pp 169–210. New York: Academic Press.

Birk Y, Bondi A, Gestetner B and Ishaaya I (1963) A thermostable haemolytic factor in soybeans. *Nature* 197: 1089–1090.

Calvert GD, Bligh L, Illman RJ, Topping DL and Potter JD (1981) A trial of the effects of soya bean flour and soya-bean saponins on plasma lipids, faecal bile acids and neutral sterols in hypercholesterolaemic men. *British Journal of Nutrition* 45: 277–281.

Cheeke PR (1971) Nutritional and physiological implications of saponins: a review. *Canadian Journal of Animal Science* 51: 621–632.

Dietschy JM (1968) Mechanisms for the intestinal absorption of bile acids. *Journal of Lipid Research* 9: 297–309.

Fenwick DE and Oakenfull D (1983) Saponin content of food plants and some prepared foods. *Journal of the Science of Food and Agriculture* 34: 186–191.

Gestetner B, Birk Y and Tencer Y (1968) Soybean saponins. Fate of ingested soybean saponins and the physiological aspects of their hemolytic activity. *Journal of Agricultural and Food Chemistry* 16: 1031–1035.

Gibney MJ, Pathirana C and Smith L (1982) Saponins and fibre – lack of interactive effects on serum and liver cholesterol in rats and hamsters. *Atherosclerosis* 45: 365–367.

Griminger P and Fisher H (1958) Dietary saponin and plasma cholesterol in the chicken. *Proceedings of the Society for Experimental Biology and Medicine* 99: 424–426.

Heaton KW (1972) *Bile Salts in Health and Disease.* Edinburgh: Churchill Livingstone.

Ishaaga I and Birk Y (1965) Soybean saponins. IV The effect

of proteins on the inhibitory activity of soybean saponins on certain enzymes. *Journal of Food Science* 30: 118–120.

Johnson IT, Gee JM, Price K, Curl C and Fenwick GR (1986) Influence of saponins on gut permeability and active nutrient transport *in vitro*. *Journal of Nutrition* 116: 2270–2277.

Malinow MR, McLaughlin P, Kohler GO and Livingston AL (1977) Prevention of elevated cholesterolemia in monkeys by alfalfa saponins. *Steroids* 29: 105–110.

Malinow MR, McLaughlin P, Papworth L *et al.* (1977b) Effect of alfalfa saponins on intestinal cholesterol absorption in rats. *American Journal of Clinical Nutrition* 30: 2061–2067.

Malinow MR, McLaughlin P, Stafford C *et al.* (1979) Comparative effects of alfalfa saponins and alfalfa fiber on cholesterol adsorption in rats. *American Journal of Clinical Nutrition* 32: 1810–1812.

Malinow MR, McLaughlin P, Stafford C, Livingston AL and Kohler GO (1980) Alfalfa saponin and alfalfa seeds – dietary effects in cholesterol-fed rabbits. *Atherosclerosis* 37: 433–438.

Malinow MR, Connor WE, McLaughlin P, *et al.* (1981) Cholesterol and bile acid balance in *Macaca fascicularis* – effects of alfalfa saponins. *Journal of Clinical Investigation* 67: 156–162.

Mathur KS, Khan MA and Sharma RD (1968) Hypocholesterolaemic effect of Bengal Gram – a long-term study in man. *British Medical Journal* I: 30–31.

Newman HAI, Kummerow FA and Scott HM (1958) Dietary saponin, a factor which may reduce liver and serum cholesterol levels. *Poultry Science* 37: 42–46.

Oakenfull D (1981) Saponins in food – a review. *Food Chemistry* 6: 19–39.

Oakenfull DG, Fenwick DE, Hood RL *et al.* (1979) Effect of saponins on bile acids and plasma lipids in the rat. *British Journal of Nutrition* 42: 209–216.

Oakenfull DG, Topping DL, Illman RJ and Fenwick DE (1984) Prevention of dietary hypercholesterolaemia in the rat by soya bean and quillaja saponins. *Nutrition Reports International* 29: 1039–1046.

Pathirana C, Gibney MJ and Taylor TG (1980) Effects of soy protein and saponins on serum, and liver cholesterol in rats. *Atherosclerosis* 36: 595–599.

Potter JD, Illman RJ, Calvert GD, Oakenfull DG and Tapping DL (1980) Soya saponins, plasma lipids, lipoprotein and fecal bile acids. *Nutrition Reports International* 22: 521–528.

Price KR, Griffiths NM, Curl CL and Fenwick GR (1985) Undesirable sensory properties of the dried pea (*Pisum sativum*). The role of saponins. *Food Chemistry* 17: 105–115.

Price KR, Curl CL and Fenwick GR (1986) The saponin content and sapogenol composition of the seed of 13 varieties of legume. *Journal of the Science of Food and Agriculture* 37: 1185–1191.

Sautier C, Doucet C, Flamant C and Lemonnier D (1979) Effects of soy protein and saponins on serum, tissue and faeces steroids in rat. *Atherosclerosis* 34: 233–241.

Sidhu GS and Oakenfull DG (1986) A mechanism for hypocholesterolaemic activity of saponins. *British Journal of Nutrition* 55: 643–649.

Sidhu GS, Upson B and Malinow MR (1987) Effects of soy saponins and tigogenin cellobioside on intestinal uptake of cholesterol, cholate and glucose. *Nutrition Reports International* 35: 615–623.

Sirtori CR, Agradi E, Conti F, Mantero O and Gatti E (1977) Soybean protein diet in the treatment of type II hypercholesterolaemia. *Lancet i*: 275–277.

G. P. Savage
Lincoln University, Canterbury, New Zealand

SARDINES

See Fish

SATIETY AND APPETITE

Contents

The Role of Satiety in Nutrition

Satiety, a transient lack of interest in further ingestion, is a dispositional state of the individual. This state may limit how much food and drink is consumed on one occasion. It may also delay the next occasion of consumption. The satiating effect of a food is thought to contribute to that food's acceptability, by serving as part of the satisfaction to be gained from eating it. However, there are few relevant hard data, and thinking on this matter is confused.

Scientific Concept of Satiety

Satiety is specifically the inhibitory effect of dietary consumption on appetite. The decrease in hunger or thirst must, by definition, have been caused by some consequence of ingestion. In studies of satiety it is not sufficient to record satiety ratings alone. These may be graded expressions of the sensation of abdominal fullness, the wish to eat only a small amount, an awareness of how much was eaten and how recently, or some other measure of the disposition to refuse food. The origins of the lack of appetite must be shown to include an effect of eating for the rating to be a true measure of satiety.

Scientific analysis of satiation requires identification of the influence(s) from eating that reduce appetite. Satiety is not a response but the operation of a type of mechanism.

What Role does Satiety have in Food Acceptability?

It is reasonable to expect that short-term satisfaction of hunger by particular foods increases the acceptability of those foods. Another attraction of a food or meal could be a longer postponement of the need to eat. Understanding satiety mechanisms could be considered to have broad relevance to the formulation and marketing of foods.

However, there has so far been little scientific investigation in humans of the role of satiety in palatability and choice of foods. A considerable body of data in experimental animals has shown that the satiating effect of eating food carbohydrate induces sensory preferences for foods eaten shortly before absorption of the carbohydrate begins. Sensory preferences are also acquired by association with the effects of absorption of amino acids and of digested fat, under conditions where it is less clear that these nutrients are contributing to satiety. Experiments in humans on the acquisition of sensory preference by association with carbohydrate or protein also indicate that satiety may influence the learning of palatability. ***See*** Sucrose, Dietary Importance

What Role does Satiety have in Weight Control?

Satiety is widely assumed to play a major role in weight control. However, there is a dearth of data on the mechanisms by which satiety can inhibit food consumption in the short term, or on how such effects influence energy intake and long-term energy balance. Research methods are not generally attuned to the concept of satiety as a disposition integrated from sensory, physiological and cultural mechanisms, or to daily energy intake as an epiphenomenon of diverse sequences of complex dietary selection decisions. ***See*** Obesity, Treatment

As a result, satiety mechanisms are not effectively exploited in established ways of attempting weight control. This is a potential explanation for the facts that obesity has not been prevented, and generally reliable procedures have not been found to reduce unhealthy overweight. There are sound reasons for expecting satiety to have some part to play in weight control. The satiating effects of foods or meals could operate within the framework of certain patterns of eating in ways that help to reduce habitual daily intake, making excess energy intake over expenditure less likely.

Understanding satiety mechanisms is crucial for extrapolating to common eating patterns in order to identify foods and drinks that support successful long-term weight control. Unfortunately, most frequent research on human satiety has attempted only to test foods and food components for effects on appetite ratings and nutrient intake, without measuring the foods' mechanisms of action or their effects on eating habits. As a result, the role of satiety in weight control remains largely a matter of speculation.

Food-specific Satiety Mechanisms

Satiety is essentially sensory: it is a disposition to reject foods or to accept them in limited amounts. In principle, inhibition of eating could be similarly strong for all foods. Alternatively, it is conceivable that the strength of inhibition may vary with the sensed identity of the food. Such satiety with a degree of specificity to a food is called sensory-specific satiety. The Greek form of this term was applied to a taste-selective satiety that is claimed to depend on the nutrient consumed: negative alliesthesia (literally, a change in sensation – negative in the sense that the food is less attractive). The sensory impression does not necessarily change; the loss of interest is not because the taste seems weaker or different in gustatory quality. The change is in the reaction to the taste (or other sensory characteristics of the foodstuff).

There is evidence from experimental animals that hypertonic sugar solutions (which can be nauseating in people) cause a temporary loss of the sensual pleasure obtained from sweetness. In humans, however, suppression of a pleasurable reaction to foods has yet to be distinguished from a reduced disposition to consume them. Whether eating is motivated only by pleasures and pains, or by other things as well, is an empirical issue for cognitive psychology.

There are at least two sorts of food-specific satiety. One is caused by immediately preceding exposure to food having the consequently satiated sensory characteristics. The other is a learned reaction to those sensory characteristics in conjunction with a particular physiological signal, such as a partly filled stomach: it is not necessary for the particular food to have been ingested previously in that meal; the reaction is acquired from experience at one or more earlier meals.

Habituation Satiety

Satiation for a food induced by recent exposure to that food meets the basic criteria for the behavioural change known as habituation. This is a decline in a specific response (eating) during and shortly after repeated or continuous exposure to a specific stimulus (the food). The decline is not fatigue in the sensory pathways or in the eating movement circuits but involves a decline in effectiveness of central connections between these two neural systems. At the sensory end the rated strength of the sensation of, for example, flavour does not decrease, at least not by as much as the decrease in the rated pleasantness of eating the food.

Introduction of a new stimulus disrupts habituation and the original response re-emerges, an effect called dishabituation. In the case of satiety, this disruption is known as the variety effect. Presenting a food that was not in the preceding meal can reactivate eating and its rated pleasantness. Switching foods can provoke eating a larger meal than switches between samples of the same food, in experiments with rats and with human subjects.

If the larger meal is compensated by omission of a subsequent snack or by a smaller following meal, bodyweight should not be affected. There is no conclusive evidence that marketing a wide variety of flavour variants contributes to obesity.

The mental processes involved in habituation satiety remain to be elucidated. Boredom with a food seems to be more complex than mere fatigue in a central neuronal connection; yet it could still be based on a subconscious decline in a particular sensory facilitation of eating.

Habituation in invertebrates is mediated by a decline in effectiveness of synapses on the motor neurons from the repeatedly stimulated sensory pathway. Similarly, taste-specific satiation in monkeys involves no changes in the taste pathways through to the cerebral cortex: changes in line with the satiated behaviour occur only beyond the sensory cortex, towards the food-selection pathways.

Habituation also has a long-term component. The short-term habituated response recovers completely within an hour or so of the end of repeated stimulation. When the same stimulus is repeatedly presented on a later occasion, the decline in response is more rapid. Long-term habituation (i.e. more rapid rehabituation) may contribute to boredom with the same meal presented day after day.

There are indications of the more complex processes in the attractions of long-term variety. While the main item that characterizes a course in a meal might have to be varied, the same staple food (such as potatoes or bread) can be eaten day after day. In addition, drinks seem to show less item-specific satiation than many foods. Conventions about menus or personal dietary habits may contribute to cessation of eating one food or to starting to eat a different food.

Famine relief may provide only two or three foodstuffs, albeit in ample quantities. This diet might become tedious even to the chronically hungry who are being offered culturally appropriate staples. There is some evidence that intake increases when a variety of flavourings is provided with the basic commodities.

Conditioned Satiety

The second known type of food-specific satiety has been learned from experiences in the past. Satiety learning processes have the basic form of associative conditioning – stimuli elicit a reaction that they have not previously evoked, because they have been associated with consequent stimuli. Repetition is not enough; the stimuli must be paired with conditioning sequences.

The Russian behavioural physiologist Pavlov demonstrated conditioning of appetite when the ringing of a bell in hungry dogs was paired with feeding or a food taste; ringing the dinner bell when the dogs had empty stomachs elicited salivation and orientation to the food-provider or food bowl before the food was presented. The bell and the empty stomach predict the taste of food. The dog associated this cue or conditioned stimulus (CS) with the consequence or unconditioned stimulus (US).

The conditioning of satiety follows the same principles. A food stimulus and a partly filled stomach (the CS) are paired with subsequent strong satiation, such as a mildly bloated sensation (the US). When the food or flavour is presented again later, while the stomach is just as full, the food flavour has become less attractive or even slightly repulsive. To the extent that this loss of interest comes to consciousness, it will be expected that particular foods or meals will be satisfying.

A key difference from habituation satiety is that conditioned satiety does not depend on the sensory characteristics of the food that has just been eaten. The loss of interest can apply to a food that had not been presented before on that particular occasion, because the satiety had been learned on earlier occasions. All foods to which satiety has been previously conditioned will become less attractive when the stomach begins to fill.

Integration of Satiety Processes

Conditioned elicitation of loss of interest in eating by cues from food and from the digestive tract suggests that satiety in familiar situations is a learned response. If internal signals can become part of the food-specific satiating pattern, external cues such as emptied plates may also come to suppress appetite.

Experiments on satiety risk missing the normal mechanisms of satiety when they impose unfamiliar or extreme conditions. Eating would stop when visceral or social discomfort is strong enough, independently of prior learning. Therefore, to elucidate postingestional and social satieties, the role of any one satiety mechanism must be examined in its normal range and within the learned complex a person has established for his or her dietary habits.

Postingestional Satiety Mechanisms

The effects of food arise in sequence as the ingested food passes down the digestive tract, from sight and aroma, through texture, temperature, irritation, taste and retronasal aroma in the mouth, to stimulation of the gut wall, and the stimulation of tissues such as the liver and the brain itself as the food is absorbed. The traditional assumption in the physiology of ingestion has been that inhibitory stimuli subtract from facilitatory stimuli. Intake of sugar solutions in rats has an intestinal effect which subtracts from the palatability of the sweetness; however, these fluids are not familiar parts of the subjects' diet.

Linear or additive interactions are not the norm in neurophysiological studies, or in experiments on the interactions of mechanisms in the ingestion or rating of ordinary foods and drinks. There is no convincing evidence for interactions among satiety mechanisms that justify the use of the term 'cascade'. The evidence is consistent with satiety arising from recognition of the accustomed pattern of eating, postingestional stimulation and social constraints, with reduced satiety in the absence of any one element in that learned pattern.

Cultural and Interpersonal Satiety Mechanisms

Appropriate proportions of food items on a plate (e.g. meat and vegetables) and the total amount (e.g. for a man or at the price) are subject to cultural variation, although these in turn may be influenced by average physiological requirements.

The process of eating in company can influence food intake. For example, it may inhibit eating. This social satiety for food is likely to have been integrated with physiological and sensory satiety, at least in situations about which sufficient has been learned from personal experience.

Satiety Disruption in Dieting

Reducing dietary intake may evoke desires for more food and, possibly, cravings for forbidden foods. Cognitive bases for food cravings are more likely than hormonal imbalances or nutrient deficiencies.

Small meals are less likely to create satiety and will allow the hunger pattern in the viscera and the memory of that meal to form sooner than after a larger meal. Sufficiently severe restriction of food at meals, even when the gut and the tissues adapt, will activate innate mechanisms of distress, readily labelled hunger by the dieter. The temptation to look for food will be increased, as will the temptation to eat when food is available.

Satiety Manipulation in Weight Control

Permanently successful weight control will depend on keeping satiety as normal as possible in the new life style required to maintain target weight.

The Role of Satiety in Nutrition

Satiety Synapses and Antiobesity Drugs

The pathways in the brain over which the inhibitory effects of food on eating are transferred are poorly understood. Satiety signals, such as distension of the stomach, chemical stimulation of the upper intestine and oxidation of energy substrates in the liver, have been investigated in isolation from stimuli from food. Reactions to these food stimuli constitute the disposition to eat that satiety suppresses. Neural signals from the viscera and sensors of the circulation at the surface of the brain have their first relay at the back of the brain stem. It is not known where or how they interact there and higher in the brain with the sensory control of eating in the inhibition of appetite.

Visceral signals are relayed to the hypothalamus at the head of the brain stem; this has complex connections with the forebrain, where incoming and past information from different sources is put together. For example, gastric distension affects neuronal activity in the ventromedial hypothalamic nucleus. Yet this is not evidence that this region mediates satiety, because the hypothalamus has many autonomic and endocrine functions to perform. These are liable to affect eating behaviour indirectly (via neural circuitry elsewhere in the brain). In fact, it has been found that the ventromedial hypothalamic region is not crucial to any of the appetite-suppressant effects of food that have been tested.

These tests were carried out to evaluate the theory that the obesity seen after destruction of this region of the hypothalamus arises from a defect in satiety resulting from destruction of 'the satiety centre'. Since satiety mechanisms operate normally in rats made fat by lesions in this area, the hypothalamus is unlikely to contain any such centre.

The effects of drugs on the sizes of meals in experimental animals have inspired theories that particular monoamine neurotransmitters operate in the pathways on the brain that mediate satiety. A region of the hypothalamus forward from and above the originally supposed satiety and appetite centres appears to contain some of the synapses at which transmission alters meal sizes.

Noradrenalin appears to be the transmitter at synapses in this region that increases meal sizes and even restarts eating in a satiated animal. Analysis of the different possible mechanisms implicates a disruption of conditioned satiety.

However, it may be a localized effect of the arousal system that has noradrenalin synapses in many brain regions. This may be experimental isolation of the effect of distraction or excitement on normal moderation of eating. Conceivably, therefore, disruption of learned satiety is related to emotional eating. This is amenable to psychological management and does not need drug treatment.

Drugs that activate synapses using serotonin (5-hydroxytryptamine; 5-HT) decrease the size of the first meal that a rat takes after it has been deprived of food for a while. This has been interpreted as an augmentation of satiety. However, starved rats tend to eat fast and long at this initial meal. Serotonin is a transmitter in the brain-stem circuits that organize chewing and other tactually guided movements. Hence these drugs may simply be slowing movement or making chewing and swallowing more difficult; they have been shown to alter the effects of food textures on preference and intakes in rats. Serotonin is also important in sleep circuits; indeed, when these drugs are having their greatest impact on people's eating, they are also having their strongest sedative effect. The classic meal-size decrease in these circumstances may have nothing to do with satiety.

Serotonergic drugs that have been used as aids in rapid weight reduction may exert part of their effect peripherally. Serotonin is one of the most important transmitters in local neural control of the contractions that pass food down the digestive tract. These and other drugs therefore slow gastric emptying under some conditions, prolonging the satiating action of food and reducing the temptation to snack. They may even postpone the next meal.

Satiety in Dietary Treatment and Advice

There is a medical consensus that energy from starchy foods should not be decreased. Where energy is needed to replace that removed by reduction in fat intake, starch intake should be increased. Readily assimilated carbohydrate, particularly in the earlier stages of meals, is likely to play a major role in immediate satisfaction following the meal. This function and its training of expectations and palatability are important for acceptability of the diet.

In addition, intake of low-energy bulky foods, such as pulses, grains, and vegetables and fruit that provide nonstarch polysaccharides, is recommended. The satiating effect of normal amounts of fibre is probably mediated by learning by association with intestinal and metabolic effects such as those of glucose and fructose from the starch and sugar present in high-fibre foods. Therefore the sugar that is often required to make high-fibre foods edible also has a role in their satiating effects. Alternatively, some of the starch may be rapidly assimilated. However, much of the starch will be in slowly assimilable forms, potentially delaying the return of hunger. ***See*** Dietary Fibre, Fibre and Disease Prevention

Enthusiasm for a nutrient, a diet product or a dietary regimen by a professional or in another dieter can convey itself to the dieter or at least be tried in desperation. For example, it seems obvious that high-fibre foods are filling, and these foods have acquired a nutritious image. This sensory satiety can suggest visceral satiety, and these expectations may be fulfilled as long as sufficient energy is included with the fibre. The assumption is that sufficiently high doses of fibre augment satiety by postingestional mechanisms such as slower gastric emptying and slower absorption of nutrients through soluble fibre gels in the small intestine. These theories remain to be critically tested, by studies of normal eating with sensory factors dissociated out.

Satiety in Psychological Weight Therapy

Eating slowly is a standard part of behaviour modification for weight reduction, although its effectiveness has not been evaluated.

Part of the rationale has been to 'satisfy the taste buds'. Perhaps savouring all the food that is eaten helps to ward off the sense of restriction that makes forbidden eating more tempting. Another idea is that more prolonged sensing would augment food-specific satiety.

A different rationale for slow eating has been that it would fill the stomach more effectively and so satiate with less food. However, there is no evidence that gastric distension plays such a precise or dominant role in meal termination. Furthermore, if the eating were as slow as gastric emptying, the stomach would not begin to fill.

A variety of other suggestions have been made as to how to strengthen the influence of satiety but none has yet been taken up seriously, let alone tested for efficacy.

Bibliography

Bennett GA (1986) Behavior therapy for obesity: a quantitative review of the effects of selected treatment characteristics on outcome. *Behavior Therapy* 17: 554–562.

Blair AJ (1991) When are calories most fattening? *Appetite* 17: 161.

Bolles RC (ed.) (1991) *The Hedonics of Taste*. Hillsdale, New Jersey: Erlbaum.

Booth DA, Rodin J and Blackburn GL (eds) (1988) *Sweeteners, Appetite and Obesity* (supplement to *Appetite*) London: Academic Press.

Craighead LW, Stunkard AJ and O'Brien RM (1981) Behavior therapy and pharmacotherapy for obesity. *Archives of General Psychiatry* 38: 763–768.

Darga LL, Carroll-Michals L, Botsford SJ and Lucas CP (1991) Fluoxetine's effect on weight loss in obese subjects. *American Journal of Clinical Nutrition* 54: 321–325.

Dobbing J (ed.) (1989) *Dietary Starches and Sugars in Man: A Comparison*. London: Springer-Verlag.

Stunkard AJ (ed.) (1980) *Obesity*. Philadelphia: WB Saunders.

Wadden TA, Sternberg JA, Letizia KA, Stunkard AJ and Foster GD (1989) Treatment of obesity by very low calorie diet, behaviour therapy, and their combination: a five-year perspective. *International Journal of Obesity* 13(supplement 2): 39–46.

D. A. Booth
University of Birmingham, Birmingham, UK

Food, Nutrition and Appetite

Appetite for food and drink is the momentary disposition of an individual to seek and ingest edible or potable materials. The concept of having an appetite for food and drink has been widely misunderstood. Appetite has often been assumed to be a subjective phenomenon, private to a person's consciousness. The view has been held that appetite can be measured by a rating procedure, such as marking a visual analogue scale (a line labelled at each end, e.g. with the phrases 'extremely hungry' and 'not at all hungry'), or magnitude estimation (assigning numbers to the strengths of a named aspect of awareness).

A measure of appetite, whether based on ratings, behavioural patterns or intakes, is an objective estimate of the strength of influence of specified current determinants on the disposition to seek and consume food. Appetite is a publicly observable relationship between the tendency to ingest and the sensed food composition, the culturally interpreted context and the bodily state of the eater or drinker.

Appetite has also been restricted to facilitatory influences that come from food, or at least from outside the skin. In such terminology, facilitation from under the skin is often called hunger (and only the inhibition of eating from truly physiological sources is called satiety). Alternatively, appetite is in the mind, and hunger in the body.

However, appetite is not one sort of influence on eating behaviour – external, mental or other. Ingestive appetite is the causal structure by which all influences are momentarily affecting eating and drinking. It is sometimes necessary to distinguish appetite for food (hunger) from appetite for drink (thirst). The subjective experience of hunger can be referred to external sources, as in a food craving or the desire to eat at the usual time, not just to bodily sensations such as the epigastric pang.

Aspects of Appetite

The immediate causes of eating and drinking might be crudely divided into three categories, which can be dubbed sensory, somatic and social. This subdivision is not entirely coherent, however, because appetite is an integral whole. Not only do sensed food factors interact among themselves in the consumer's mind but also the sensory configuration interacts with physiological factors and economic, cultural and interpersonal factors, in determining a momentary decision to accept an item of food or drink.

It is therefore scientifically and practically unwise to study one factor by itself, especially out of the normal context of eating (or of food purchase or preparation).

Moreover, it cannot be assumed that sensory, social and somatic groups of factors independently add into the disposition to eat or purchase a food. Thus neither theory nor application is sound if based solely on economics or physiological or sensory research. ***See*** Food Acceptability, Affective Methods

Sensory Aspects of Appetite

The influences on eating or drinking that arise immediately from the sensing of the food or drink have traditionally been called palatability. However, incorrect assumptions are made in many uses of this term; one such assumption is that palatability is an invariant property of the food or drink itself. Palatability is not inherent to the food; it is an effect of the food on the eater. This effect can vary, not just between people, but within a person in different contexts of eating.

The ordinary person regards foods as having constant palatabilities, even if varying from person to person. It is natural to deny that one stopped eating because the food became less palatable; cessation of eating is generally attributed to feeling full or to having had an appropriate amount to eat.

Many scientific investigations and theories have also treated palatability as a constant sensory effect. In physiological psychology, for example, the standard account of the control of meal size is that accumulating satiety subtracts from constant palatability until insufficient facilitation of eating exists to continue the meal. It is more likely, however, that the sensory facilitation differs between the contexts of starting and ending a meal. When the sensory facilitation of eating is actually measured during or from before to after meals, it typically decreases to a minimum at the end of eating and for some while afterwards. Moreover, at the anecdotal level, a savoury course is not expected to be as palatable after a sweet dessert as it would be before the dessert.

Measurement of Sensory Preferences

Any measure of the sensory aspect of appetite must compare the responses to two or more foodstuffs differing in known sensed characteristics, unconfounded by any other differences that might affect the responses. Whether the responses are concrete (such as selection among the foods) or symbolic (such as numerical or line rating of liking, pleasantness, or likelihood of choice), and whether the food samples are presented simultaneously (e.g. triadic test) or successively (i.e. monadically), the relative acceptances give an estimate of the sensory preference of that assessor in the context of testing.

Attention must be given to the basic psychological mechanisms operative in sensory effects on food choice. The most preferred version of a food for a consumer in a given context is always a particular physicochemical configuration of its sensed characteristics. In fact, the individual's ideal point is a more precise sensory level than a descriptive verbal anchor (such as 'extremely strong') and is often no worse than a familiar physical standard. Taken with a second sensory anchor, such as apparent absence of the characteristic or perhaps a presence so weak (or so strong) as to be unacceptable, the ideal point defines a sensory scale as objectively and precisely as any purely descriptive rating. The crucial requirements are that each tested sensory level is described by the assessor or identified by the investigator as above or below the ideal level and that the preference-anchored responses are scored according to sensory level, not for degree of preference. ***See*** Sensory Evaluation, Sensory Characteristics of Human Foods; Sensory Evaluation, Food Acceptability and Sensory Evaluation

The raw preference sensory scores and the slopes of plots of these scores against physical levels are as meaningful as conventional descriptive scores and slopes (or psychophysical power function exponents). They can be averaged across a panel and tested statistically for differences between samples. This amounts merely to data analysis, however; it is not scientific measurement of the strength of influences on food perception or appetite. The most direct step towards efficient design of palatable foods, rather, is to use the effect of levels on score to measure causal strength: the sensitivity of each assessor's preference to a physically measured factor should be calculated from the slope and residual variance of a linear plot of responses against sensory levels, using general psychological theory of the mechanisms by which different responses are given to different signals (Fig. 1). ***See*** Sensory Evaluation, Sensory Rating and Scoring Methods

A great advantage of this approach is that sensitivity of perception or preference to each investigated factor can be estimated when the other foods, or the social and somatic contexts of the test, depart moderately from the combination in which the tested food is normally eaten or purchased. If tolerably low and high levels of a factor are well and equally represented among the responses, the ideal level and the sensitivity can be interpolated with fair accuracy. Moderate inappropriateness in the context of testing creates a Euclidean discontinuity which is virtually confined to the mid-range of the theoretical linear function (Fig. 1).

The sensitivities of a preference response to different factors can be used to identify how those factors interact in the individual's mind in the test situation, potentially enabling understanding of how a food works in dietary habits. Preference characteristics can also be aggregated

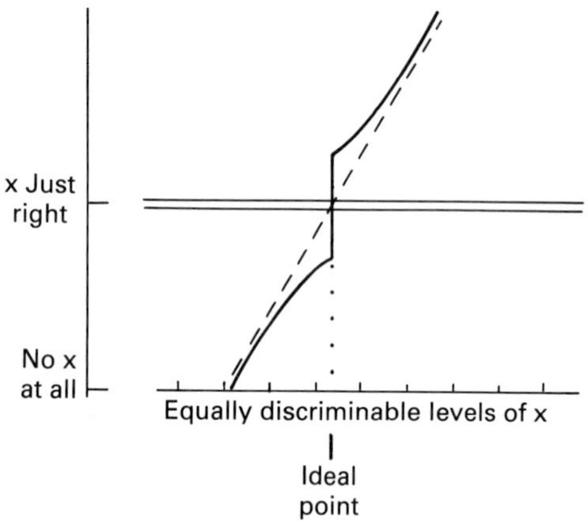

Fig. 1 Effect of equally discriminable differences in level of a described sensory or other influence on appetite on the strength of that factor, rated from too weak to be recognized to ideal strength, in a context where other factors are at inappropriate levels. The response line is in the Euclidean distance of the factor level from ideal, i.e. the square root of the sum of the square of the difference of a test sample from the influence's ideal level and a constant that represents the square of the size of the contextual defect; its discontinuity around the ideal point is the truncation of the preference peak by this defect.

across a panel of representative consumers. This could be used to give a physically and behaviourally objective market-response surface from which one or more popular products could be designed.

Somatic Aspects of Appetite

Somatic influences on appetite include physiological signals of hunger, thirst and satiety from the digestive tract or from tissues such as the liver or the brain itself, as well as these and other postingestional effects of foods that induce sensory preference or aversion.

As already indicated, even a physiological signal of hunger or satiety (such as gastric motility or distension) is likely to operate through particular sensory configurations when the eating situation is familiar. There is evidence that appetite from an empty stomach is stronger for an entrée than a dessert, and vice versa for a partly filled stomach. Similarly, we should expect conventions about foods that are appropriate to particular times of day (such as breakfast) to affect hunger from an empty upper intestine or the satisfaction of appetite arising from stimulation of the intestinal wall or hepatic metabolism, although such hypotheses remain to be tested physiologically. Investigation of behaviour at ordinary meals is necessary to elucidate the normal operation of somatic factors in appetite. Evidence for particular physiological signals has been obtained with unusual foods at imposed times but that leaves open the role of a signal in everyday eating and drinking.

Those physiological factors which alter food preferences by associative conditioning or other forms of learning are interacting with sensory factors. It must be noted that they are interacting quite differently from the contemporaneous mixing of factors considered elsewhere in this article. Indeed, a preference acquired by the association of certain sensory characteristics with a physiological consequence might be unaffected if the conditioning physiological event recurs when the conditioned food stimuli are also present. In psychology, this distinction in mechanisms is referred to as the difference between reinforcement and motivation, or between learning and performance. The distinction is crucial to the understanding of learned behaviour such as normal appetite.

The same metabolic signals that inhibit appetite may also induce preferences for food flavours and textures presented shortly beforehand. This is the main basis for the desire for familiar foods, ranging from the appetite for a staple food, such as potatoes or bread, to the craving for an indulgence item such as chocolate. Association with these reinforcers enables bodily states (such as a partly full stomach) and external situations (such as lunchtime, or a party atmosphere) to acquire a learned power to motivate eating or drinking.

The temporal pattern of appetite suppression following the consumption of a particular type of menu also appears to be learnt. Such expectations of satisfaction may be at least as important as physiological signals in the sensation of fullness and in the rise of appetite before the next meal.

Additional physiologically mediated psychological effects of foods are suspected to condition sensory preferences or to create functional expectations of sensorily identified foodstuffs or beverages, although conclusive evidence is lacking. These appetite-reinforcing somatic factors include the stimulant action of caffeine and the sedating action of a heavy meal or of alcohol. However, cultural stereotypes and personal experience can create stimulant, soothing or cheering effects from consumption of a drink or food, without any specific physiological basis.

A somatic factor in appetite can only be identified if the study disconfounds variations in the physiological effect from variations in both social and sensory effects. Neglect of this point has led to ill-founded claims of measuring physiologically specific aspects of appetite, such as carbohydrate craving, protein selection, protein- and fibre-induced postingestional satiety signals and gastric distension satiety. The technical difficulties are daunting, but there is no gain in knowledge from neglecting these variations.

All relevant sensory and social factors, as well as somatic factors, must be specifically manipulated and/or measured as perceived by the individual, and shown to be uncorrelated with the physiological effect(s) of interest. Only then can the sensitivity of appetite to that somatic factor in that context be estimated, or indeed even a qualitative effect of the factor be established.

Social Aspects of Appetite

Interpersonal, cultural and economic factors in appetite appear to influence food choices by modulating the effects of sensory factors. It may be feasible to investigate a range of foods or drinks for general effects of factors such as price, declared sweetener type, or use by children, on appetite at the point of purchase, serving or ingestion. However, effects may be specific to a food.

Social influences on appetite are often present in a symbolic form, e.g. the category of eating occasion or type of company at the meal, a brand logo, an advertising picture, compositional information, or the price label attached to a food item. Social factors do not differ from rated sensations generated by foods or in the body: descriptive scores that relate to physical factors can be used to model sensory and somatic aspects of appetite; hence the procedure for scaling a social factor and estimating its strength of influence on appetite is the same in principle as that for other factors.

Psychology, Society and Nature

Appetite is a psychological phenomenon, an aspect of the mental processes organizing an individual's behaviour. Hence the cognitive approach to food consumption should be capable of yielding general explanations of the anthropology of cuisine and the micro-economics of the food market. The constructs about cooking and eating held in common among individuals from a social group are evidence of their shared culinary culture. The effects of price, pack label, shelf display, etc. on the personal disposition to purchase a food item will aggregate across consumers into sales of that brand in the shops. The performance of selective eating behaviour provides challenges to scientific explanation that go beyond the physics and chemistry of foods and their biochemical and physiological effects. Each individual has his or her own causally structured behaviour beyond the food materials, the nutritional physiology and the socioeconomic systems.

Development of the quantitative behavioural science of food choice and intake is needed. With fundamental understanding of the determinants of eating habits at the level of the causal processes in the mind of the consumer, we would be able to bring together knowledge of the socioeconomic, physicochemical and biochemical–physiological processes into a coherent science and effective practice of food design and dietary health.

Bibliography

Axelson ML and Brinberg D (1989) *A Social-Psychological Perspective on Food-Related Behavior*. New York: Springer-Verlag.
Barker LM (ed.) (1982) *The Psychobiology of Human Food Selection*. Westport, Connecticut: AVI Publishing.
Bennett GA (1988) *Eating Matters: Why We Eat What We Eat*. London: Heinemann Kingswood.
Boakes RA, Burton MJ and Popplewell DA (eds) (1987) *Eating Habits: Food, Physiology and Learned Behaviour*. Chichester: John Wiley.
Bolles RC (1980) Historical note on the term 'appetite'. *Appetite* 1: 3–6.
Booth DA (ed.) (1978) *Hunger Models: Computable Theory of Feeding Control*. London: Academic Press.
Booth DA (1992) *Food and Drink: the Psychology of Nutrition*. Basingstoke: Falmer Press.
Chivers DJ, Wood BA and Bilsborough A (eds) (1984) *Food Acquisition and Processing in Primates*. New York: Plenum Press.
Dobbing J (ed.) (1987) *Sweetness*. London: Springer-Verlag.
Farb P and Armelagos G (1980) *Consuming Passions: the Anthropology of Eating*. Boston: Houghton Mifflin.
Fieldhouse P (1986) *Food & Nutrition: Customs & Culture*. London: Croom Helm.
Logue A (1991) *The Psychology of Eating and Drinking* 2nd edn. New York: WH Freeman & Co.
Murcott A (ed.) (1983) *The Sociology of Food and Eating*. Cardiff: Gower Press.
Ramsay DJ and Booth DA (eds) (1991) *Thirst: Physiological and Psychological Aspects*. London: Springer-Verlag.
Solms J and Hall RL (eds) (1981) *Criteria of Food Acceptance: How Man Chooses What He Eats*. Zurich: Forster.
Solms J, Booth DA, Pangborn RM and Raunhardt O (eds) (1987) *Food Acceptance and Nutrition*. London: Academic Press.

D. A. Booth
University of Birmingham, Birmingham, UK

SAUSAGES

See Meat

SCALLOPS

See Shellfish

SCANNING ELECTRON MICROSCOPY

See Microscopy

SCURVY

Etymology

'Scurvy' is the current word in the English language for what has at various times in history been referred to as skurvie, scurvie, skirvye, scurvey, scurby, skyrby, scorbie and scorby. The same word, in varied forms, has also been used to refer to scurf, i.e. dandruff. Confusion of meaning has also arisen on account of purpura (small bleeding points in the skin), a feature of both scurvy and many haematological disorders, being referred to loosely and quite incorrectly as scurvy. In addition, many of the early references to scurvy are almost certainly erroneous as only later was it possible to differentiate this clinical disorder from other diseases of close or superficial resemblance. Further confusion and lack of clarity was caused by reference to patients with mixed disorders as having scurvy, when the disease might in reality have been just a portion of the whole clinical presentation. Scurvy is a clinical syndrome caused by vitamin C deficiency which is associated with a derangement of collagenous protein synthesis; this is rapidly reversed with vitamin C supplementation.

History

The Ebers Papyrus from Ancient Egypt (*c.* 1500 BC) contains a reference to what could well be scurvy. In Ancient Greece Hippocrates referred to a disease occurring in soldiers and comprising pain in the legs, gangrene of the gums and the consequential loss of teeth. During the Middle Ages 'land scurvy' was recognized to be both commonplace and seasonal throughout the whole of Europe, usually being seen in the winter months long after the cessation of availability of summer fruits and vegetables.

'Land scurvy' became 'sea scurvy' during the fourteenth and fifteenth centuries at the time of a rising spirit of adventure, catalysed by new technical developments in ship design and advancements in navigational instrumentation, and compounded by growing and powerful commercial interests in the trading of silk and spices. These factors cleared the way for longer sea voyages of weeks or months in duration, without the opportunity for the crew to land at ports where fresh vegetables and fruits would be available. The diaries of Commodore George Anson's voyage of 1740–1744, during which he attempted unsuccessfully to circumnavigate the world and on which 1051 of his 1955 men died, the majority from scurvy, referred to 'the scars of old wounds, healed for many years, were forced open again', and stated 'many of our people, though confined to their hammocks, ate and drank heartily and were cheerful, yet having resolved to get out of their hammock, died before they could well reach the deck'.

Vasco da Gama sailed from Lisbon on 9 July 1497 with about 140 Portuguese sailors. He reached the southeastern coast of Africa 7 months later; the records indicate 'many of our men fell ill here, their feet and hands swelling and their gums grown over their teeth so that they could not eat'. On 6 April 1498 came the opportunity to purchase oranges from Moorish traders. Just 6 days later it is stated that 'all our sick recovered their health for the air of this place is very good'.

In the winter of 1536, on the frozen Saint Lawrence River, the North American Indians taught the French explorer Jacques Cartier and his men the value of boiling the bark and leaves of the white cedar tree in water and drinking the juice and dregs. Three doses on alternate days were sufficient to 'miraculously' cure their loss of strength and their swollen and inflamed legs.

The introduction of the potato to Europe from South America, where it had been cultivated for nearly 1500 years, was effected in the second half of the sixteenth century by the Spaniards who invaded that country. It is now seen to have provided a ready means of combating the problem of low vitamin C status in the winter months.

In 1617 John Woodall, Surgeon-General of the British East India Company, wrote *The Surgions Mate* which contains a 23-page chapter on the subject of scurvy. He emphasized at that time the necessity to provide ships with the juice of oranges, lemons or limes. He added a recommendation of two to three spoonfuls of lemon juice as a medicine against scurvy, and as a preventive too, if enough could be spared. This suggestion preceded the work and publications of James Lind which did not appear until the next century.

James Lind was born in Edinburgh in 1716 and eventually became Physician-in-charge of the 2000-bed Haslar Hospital near Portsmouth, the largest and newest of the naval hospitals. There he had the opportunity to study 300–400 cases of scurvy at a time. The famous experiment of his earlier days, in 1747, involved six groups of two men each, treated as follows: group 1 with cider, group 2 with elixir vitriol, group 3 with vinegar, group 4 with sea water, group 5 with two oranges and one lemon each day over 6 days, and group 6 with garlic, mustard seed, balsam of Peru, dried radish root and gum myrrh, together with barley water, tamarinds and cremor tartar. All received the same diet apart from the above 'medicines'. The best response was from group 5 by the sixth day, and group 1 came second at 2 weeks; the other 'remedies' proved to be of no value. Lind published his famous book *A Treatise of the Scurvy* in 1753, in Edinburgh. Nevertheless, it was not until 1804 that the Royal Navy decreed that lemon or lime juice must be provided daily; limes were later substituted for lemons as they were cheaper and could be obtained from the new West Indian colonies, although they are now known to contain less vitamin C. Hence the origin of the term 'limeys' for British sailors. In retrospect, one can now say that Lind made one major error in preparing his 'rob of oranges' by evaporating juice down to 10% of its original volume as heat is now known to destroy vitamin C; moreover, further storage following the preparative procedures permits yet greater deterioration in the vitamin C content.

The history of vitamin C is strewn with anecdotes such as these. Others should be briefly mentioned; they include Captain James Cook's 1768 voyage to the South Pacific in which his crew avoided scurvy by his insistence on the consumption of various fresh vegetables, sauerkraut in particular. An outbreak of scurvy at the National Penitentiary at Millbank in London in February 1825, following drastic reduction of the diet of the inmates, responded to the simple prescription of three oranges a day; however, neither the kitchen staff nor the prison officers suffered. The Great Potato Famine of 1845–1848 involved not only the UK and Ireland but also France and Belgium; many cases of scurvy were being reported in the British medical literature from 1847 onwards. Scurvy, not surprisingly, was seen in the general population and in the prisons and hospitals, too. The Royal Navy's Arctic expedition, which returned in 1876, suffered badly from scurvy in spite of the availability of lime juice. However, lime juice had not been supplied on sledging expeditions, and even when it had been taken the bottles had frozen solid, causing them to break on thawing over the evening fire.

In 1928 Albert Szent-Györgyi isolated a sugar from cabbage and lemon juice, and also from the adrenal gland. He was not permitted by the editor of the *Biochemical Journal* to call it 'ignose' or 'godnose' and had to call it 'hexuronic acid' instead. In 1932 CG King and WA Waugh identified the antiscorbutic factor from lemon juice to be the same as hexuronic acid; the name ascorbic acid was adopted. In 1933 Sir Norman Haworth and his colleagues in Birmingham, UK, established the chemical structure of ascorbic acid. Also in 1933 both Sir Norman Hawarths' team and the team led by Tadeus Reichstein in Switzerland succeeded in synthesizing vitamin C.

During the latter part of World War II, at the time when many of those captured in the Far East were in Japanese prisoner of war camps, it was accepted that deficiency of vitamins A and B was commonly seen; however, deficiency of vitamin C did not occur. The reason for this was that the prisoners themselves prepared and drank 'jungle juice', a mixture of the green leaf tops of sweet potatoes, kang kong (a vegetable eaten by the local natives in Borneo), leaves of the tapioca tree, and onion leaves for flavouring, all boiled up quickly in water (personal communication, Dr KH Trigg). Grass, which was in abundance, and which is now known to contain weight for weight more vitamin C than citrus fruits, was never used as it was thought to have no nutritional value.

Species Sensitivity

Most animals do not develop scurvy since they are able to synthesize vitamin C from glucose by the reactions in eqns [**1**] and [**2**].

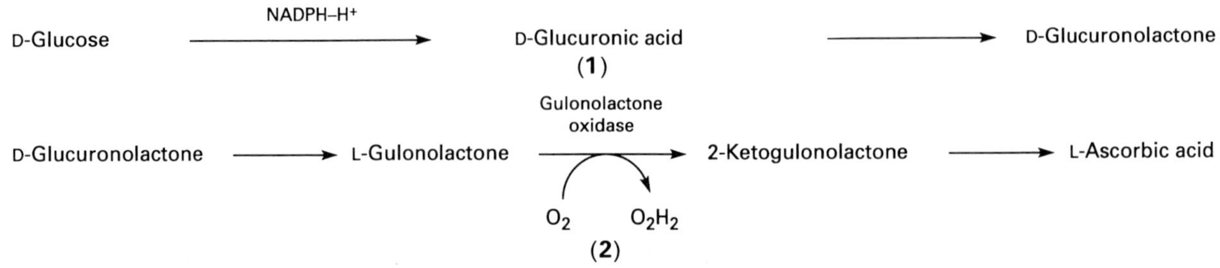

The only known species unable to synthesize vitamin C are the primates (including humans), guinea pigs, the red-vented bulbul, the fruit eating bat (*Pteropus medius*), the rainbow trout and the coho salmon. They lack the enzyme, gulonolactone oxidase, which is necessary for the conversion of L-gulonolactone to L-ascorbic acid. Hence, these species are dependent on dietary sources (Table 1) for the vitamin. ***See*** Ascorbic Acid, Physiology

Pathophysiology

Vitamin C deficiency leads to failure to maintain the cellular structure of the supporting tissues of mesenchymal origin, such as bone, dentine, cartilage and connective tissues. As a consequence, vitamin C deficiency is characterized by weakness and fatigue, hyperkeratosis of the hair follicles, perifollicular heamorrhages, petechiae and ecchymoses, swollen bleeding gums, delayed wound healing and the easy fracturing of bone. These are the characteristic clinical feature of scurvy.

The clinical manifestations of scurvy are the result of complete vitamin C deprivation of 100–160 days' duration (Table 2). During deficiency the vitamin C pool of the body is depleted at a daily rate of about 2·6% of the existing pool; 92% is lost after 100 days and symptoms of mild scurvy are evident when the pool is less than about 300 mg.

Fibroblasts play a central role in the wound healing process. In vitamin C deficiency the fibroblasts proliferate but in an unimpeded manner. These proliferating cells generally remain immature and fail to synthesize collagen molecules, which are the 'building blocks' of tissue repair. The impaired fibroblastic activity is essentially responsible for the delayed wound healing in vitamin C deficiency. Moreover, in vitamin C deficiency the cartilage cells of the epiphyseal plate at the diaphyseal end of the long bones continue to proliferate and line up in rows. The cartilage between the rows is calcified where osteoblasts do not migrate, resulting in compressed and brittle bone.

Ultrastructural studies in scurvy show distinct alterations of the ribosomal and polyribosomal pattern. The endoplasmic reticulum shows considerable dilation and loss of ribosomal granules. The polyribosomes are disaggregated, and unbeaded fibrillar material is seen in the extracellular space.

Table 1. Vitamin C content of foods and plants

Food or plant	Vitamin C content (mg per100 g or 100 ml)
Fruits	
Apples	4
Banana with peel	7
Citrus fruits	25–60
Gooseberries, fresh	60–65
Strawberries	58
Vegetables	
Broccoli	90
Cabbage	
Fresh	45–60
Boiled	24–30
Peas	
Sprouting	25–50
Dried	0
Potatoes	
Uncooked	10–30
Baked in skin	15
Boiled	5–15
Boiled and reheated	2–8
Sauerkraut	
Stored 1 month	10–15
Tomatoes	20
Animal origin	
Adrenal glands	100–200
Eggs	Trace
Meat or fish, well cooked	Trace
Milk	
Cows', pasteurized	1
Human breast milk	30–55
Botanical	
Scurvy grass	66–100
Spruce pine needles	65–200
Gramineae (grasses)	140–173
Liquids	
Apple cider, fresh unpasteurized	4–5
Blackcurrant syrup	60
Lemon juice, fresh	45
Lime juice, fresh	30
Orange juice, fresh	48
Rosehip syrup	150–200

Table 2. The consequences of withdrawing vitamin C from the diet

	Plasma ascorbic acid		Buffy coat ascorbic acid in leucocytes (WC)		Body pool of ascorbic acid		
Day	($mg\,l^{-1}$)	($\mu mol\,l^{-1}$)	(μg per 10^8 WC)	(nmol per 10^8 WC)	(g)	(nmol)	Clinical state
0	8–15	45–85	21–57	119–323	0·6–1·5	3·4–8·5	Adults
20	3	17	10–38	57–216			
40	1–3	6–17	2–10	11–57	0·3–0·6	1·7–3·4	Subclinical deficiency
60	< 1	< 6	< 5	< 28	0·3–0·6	1·7–3·4	
80	< 1	< 6	< 5	< 28	0·3–0·6	1·7–3·4	
100	< 1	< 6	< 5	< 28	0·3–0·6	1·7–3·4	
120	< 1	< 6	< 2	< 11	< 0·3	< 1·7	Perifollicular hyperkeratosis
140	< 1	< 6	< 2	< 11	< 0·3	< 1·7	
160	< 1	< 6	< 2	< 11	< 0·3	< 1·7	Petechiae and ecchymoses of the skin, and failure of wounds to heal
180	< 1	< 6	< 2	< 11	< 0·1–0·3	0·6–1·7	Gingival changes
200	< 1	< 6	< 2	< 11	< 0·1	< 0·6	Dyspnoea, oedema and very rapid progression

Biochemical Aspects

The characteristic feature of scurvy is an inability of the supporting tissues to maintain the synthesis of collagen, the intercellular substance. The collagen molecule consists of three linked polypeptide chains of amino acid residues. For the collagen molecule to aggregate into its triple helix configuration, the proline and lysine residues of newly synthesized collagen must be hydroxylated. Formation of the triple helix is important because it is in this this configuration that the procollagen is secreted from the fibroblast. Hydroxylation of both proline and lysine occurs after the amino acids have been incorporated into the peptide chain; vitamin C is required in these reactions. The essential process mediated by vitamin C is the formation of hydroxyproline from proline and of hydroxylysine from lysine as shown in the following reactions:

$$\text{Proline} + \text{Ascorbate} + O_2 \xrightarrow{\text{Proline hydroxylase}} \text{OH-Proline} + \text{Dehydro-ascorbate} + H_2O$$

$$\text{Lysine} + \text{Ascorbate} + O_2 \xrightarrow{\text{Lysine hydroxylase}} \text{OH-Lysine} + \text{Dehydro-ascorbate} + H_2O$$

Vitamin C appears to act as a cofactor for peptidyl proline and lysine hydroxylases, possibly by keeping copper and iron in a reduced state. Defective hydroxylation within the synthesizing fibroblast gives rise to the formation of an abnormal collagen precursor called protocollagen. Unlike the normal precursor (tropocollagen), this substance may not be extruded from the cell. If it is extruded it polymerizes into an abnormal collagen which lacks tensile strength. In addition to its requirement for collagen synthesis, vitamin C is believed to be an essential cofactor in forming the intercellular cement substance which binds the lining cells of blood vessels not only to their basement membrane but also to each other. ***See*** Coenzymes

Occurrence of Scurvy

Scurvy is now a rare condition but isolated cases can still be seen in certain population groups. The 'at-risk' groups are often the elderly (not so much in North America), food faddists, alcoholics, those living in institutions and patients with psychiatric disorders. Outbreaks of scurvy occur in poor nomadic populations in arid or semidesert districts when there is a threat of famine or a long-standing drought.

Bibliography

Basu TK and Schorah CJ (1982) *Vitamin C in Health and Disease*. London: Routledge.

Carpenter KJ (1986) *The History of Scurvy and Vitamin C*. Cambridge: Cambridge University Press.

Irwin MA (1976) A conspectus of research on vitamin C requirements of man. *Journal of Nutrition* 106: 823–879.

Lind J (1753) *A Treatise on the Scurvy*. Republished (1953) by Edinburgh University Press, Edinburgh.

Vilter RW (1978) Nutrient deficiencies in man: vitamin C. *CRC Handbook Series (Nutrition and Food)* 3: 91–103.

T. K. Basu
Department of Foods and Nutrition, University of Alberta, Edmonton, Canada
D. Donaldson
Department of Chemical Pathology, East Surrey Hospital, Redhill, UK and Crawley Hospital, Crawley, UK

SEAFOOD

See ***Fish, Shellfish and Marine Foods***

SEAWEED

See ***Marine Foods***

SELENIUM

Contents

Properties and Determination

The chemical properties of the trace element selenium are similar to those of sulphur; however, unlike sulphur, which tends to be oxidized in biological systems, selenium tends to undergo reduction in the tissues of plants and animals. Selenium is present in most biological tissues at very low concentrations and bound to proteins and as analogues of sulphur-containing amino acids. Several methods are available for the analysis of selenium; the most commonly used ones are chemical analysis based on the formation of a piazoselenol derivative and electrothermal atomic absorption spectrophotometry.

Chemical Forms of Selenium

Selenium is in group VIA of the periodic table of elements. This group includes the nonmetals sulphur and oxygen in the periods above selenium, and the metals tellurium and polonium in the periods below. By period, selenium lies between the group VA metal arsenic and the group VIIA nonmetal bromine. Thus, selenium is considered a metalloid, having both metallic and nonmetallic properties; its atomic properties are listed in Table 1.

Elemental selenium shows allotropy; that is, it can exist in either an amorphous state or one of three crystalline states. Amorphous selenium is a hard, brittle glass at temperatures below 31°C, is vitreous at 31–230°C, and is a free-flowing liquid above 230°C. The increased viscosity of amorphous selenium at temperatures less than 230°C results from the formation of polymeric chains at lower temperatures. Crystalline selenium can take the form of flat hexagonal and polygonal crystals (called α-monoclinic or 'red'), prismatic or needle-like crystals (called β-monoclinic or 'dark red'), or spiral polyatomic chains of the form Se_n (variously called hexagonal, trigonal, metallic, 'grey' or 'black'). The hexagonal crystalline form of selenium is the most stable; both monoclinic

Table 1. Atomic properties of selenium

Atomic number	34
Atomic weight	78·96
Electronic configuration	$Ar3d^{10}4s^{2}2p^{4}$
Atomic radius	1·40 Å
Covalent radius	1·16 Å
Ionic radius	1·98 Å
Common oxidation states	−2, 0, +4, +6
M–M bond energy	44 kcal mol^{-1}
M–H bond energy	67 kcal mol^{-1}
Ionization potential	9·75 eV
Electron affinity	−4·21 eV
Electronegativity	2·55 (Pauling's scale)

forms convert to the hexagonal form at temperatures above 110°C, and amorphous selenium converts spontaneously to the hexagonal crystalline form at 70–120°C.

Elemental selenium can be reduced to the −2 oxidation state (selenide), or oxidized to the +4 (selenite) or +6 (selenate) oxidation states. Hydrogen selenide (H_2Se) is a fairly strong acid with a pK_a of 3·8 in aqueous systems. The gas is colourless, has an unpleasant odour, and is highly toxic (the LC_{50} for 30 min exposure for guinea-pigs is 6 ppm). It can be produced by heating elemental selenium above 400°C in air, but it decomposes in air to form the element and water. Hydrogen selenide is fairly soluble in water (270 ml per 100 ml of water at 22·5°C). It reacts directly with most metals to form metal selenides, but these compounds are practically insoluble in water. Organic selenides are ready electron donors to their surrounding environments, thus oxidizing these forms of selenium to higher oxidation states.

In the +4 oxidation state, selenium can exist as selenium dioxide (SeO_2), as selenious acid (H_2SeO_3) or as selenite (SeO_3^{2-}). Selenium dioxide is formed by burning elemental selenium in air or reacting it with nitric acid. It is readily reduced to the elemental state by ammonia, hydroxylamine, sulphur dioxide and several organic compounds. It is soluble in water (38·4 g per 100 ml at 14°C) and forms selenious acid (H_2SeO_3) when dissolved in hot water. Selenious acid is weakly dibasic; with a pK_a of 2·6, dissolved selenite salts exist as biselenite ions in aqueous solutions at pH 3·5–9. In contrast to the organic selenide, selenites readily accept electrons from their environments, their selenium being easily reduced. At low pH, selenite is readily reduced to the elemental state by mild reducing agents such as ascorbic acid and sulphur dioxide. Selenites in soils are strongly bound by hydrous oxides of iron, forming insoluble complexes at moderate pH.

In its highest oxidation state (+6), selenium can exist as selenic acid (H_2SeO_4) or as selenate (SeO_4^{2-}) salts. Selenic acid is a strong acid. It is formed by the oxidation of selenious acid or elemental selenium with strong oxidizing agents in the presence of water. Selenic acid is very soluble in water; most selenate salts are soluble in water, in contrast to the corresponding selenite salts and metal selenides. Selenates tend to be rather inert and are very resistant to reduction.

Six stable isotopes of selenium exist in nature: ^{74}Se (natural abundance 0·815 mass %), ^{76}Se (natural abundance 8·66 mass %), ^{77}Se (natural abundance 7·31 mass %), ^{78}Se (natural abundance 23·21 mass %), ^{80}Se (natural abundance 50·65 mass %) and ^{82}Se (natural abundance 9·35 mass %). These isotopes have been employed in the study of the biological utilization of selenium in foods in which their quantification has been achieved by neutron activation (^{74}Se, ^{76}Se and ^{80}Se), or mass spectrometry (^{76}Se, ^{80}Se and ^{82}Se). More than two dozen radioisotopes of selenium can be produced by neutron activation or by radionuclear decay. These include such short-lived species as ^{77m}Se (half-life 17·5 s) and ^{87}Se (half-life 16 s), and such long-lived species as ^{75}Se (half-life 120 days). Of these, ^{75}Se and ^{77m}Se have been found to be suitable for the measurement of the element in biological materials by neutron activation analysis. Due to its emission of γ radiation and to its relatively long half-life of about 120 days, ^{75}Se has been widely employed in biological experimentation and in medical diagnostic work. ***See*** Mass Spectrometry, Principles and Instrumentation

Selenium is a semiconductor and exhibits photoconductivity (excitation with electromagnetic radiation can markedly increase its conductivity). This property has made selenium compounds useful in photocells and in xerography. As a result of utility conferred by these properties, the literature on the chemistry of inorganic and organic selenium compounds is large.

The selenium compounds of greatest interest in nutrition are presented in Table 2. It should be noted that, whereas forms of selenium available as supplements of foods and feeds are primarily compounds of the higher oxidation states, the metabolites of chief concern in nutrition and biochemistry are compounds in which selenium occurs in the reduced state.

Table 2. Selenium compounds important in nutrition

Oxidation state	Compound
Se^{2-}	H_2Se
	Na_2Se
	$(CH_3)_2Se$
	$(CH_3)_3Se^+$
	Selenomethionine
	Selenocysteine
	Se-methylselenocysteine
	Selenocystathionine
	Selenotaurine
Se^0	Selenodiglutathione
	Amorphous selenium
	Red selenium (α-monoclinic)
	Dark red selenium (β-monoclinic)
	Grey selenium (hexagonal)
Se^{4+}	H_2SeO_3
	Na_2SeO_3
Se^{6+}	Na_2SeO_4

Chemical Properties of Selenium

The chemical and physical properties of selenium are very similar to those of sulphur. The two elements have similar outer valence shell electronic configurations and atomic sizes (in both covalent and ionic states); and their bond energies, ionization potentials, electron affinities, electronegativities and polarizabilities are virtually the same. Despite these similarities, the chemistry of selenium and sulphur differ in two respects which distinguish them in biological systems.

The first important difference in the chemistry of selenium and sulphur is in the ease of reduction of their oxyanions. Quadrivalent selenium in selenite tends to undergo reduction, but quadrivalent sulphur in sulphite tends to undergo oxidation. This difference is demonstrated by the following reaction: $H_2SeO_3 + 2H_2SO_3 \rightarrow Se + 2H_2SO_4 + H_2O$. Thus, in biological systems, selenium compounds tend to be metabolized to reduced states, and sulphur compounds tend to be metabolized to more oxidized states.

The second important difference in the chemical behaviour of these elements is in the acid strengths of their hydrides. Although the analogous oxyacids of selenium and sulphur have comparable acid strengths ($SeO(OH)_2$, pK_a 2·6 versus $SO(OH)_2$, pK_a 1·9; $SeO_2(OH)_2$, pK_a 3 versus $SO_2(OH)_2$, pK_a 3), the hydride H_2Se (pK_a 3·8) is much more acidic than is H_2S (pK_a 7·0). This difference in acidic strengths is reflected in the dissociation behaviours of the selenohydryl group (–SeH) of selenocysteine (pK_a 5·24) and the sulphydryl (–SH) group on cysteine (pK_a 8·25). Thus, while thiols such as cysteine are mainly protonated at physiological pH, the selenohydryl groups of selenols such as selenocysteine are predominantly dissociated under the same conditions.

Selenite can react with nonprotein thiols and with protein sulphydryl groups to undergo reduction of Se^{4+} to SeO, with the concomitant formation of 1,3-dithio-2-selane products of the form RS–Se–SR, called 'selenotrisulphides'. The reaction is represented as follows: $4RSH + H_2SeO_3 \rightarrow RS{-}Se{-}SR + 3H_2O$. In this reaction, four sulphydryl sulphur atoms (-2 oxidation state) are oxidized to the -1 oxidation state of disulphides. This is balanced by the concomitant reduction of a single selenium atom from the selenite oxidation state of $+4$ to the 0 oxidation state. However, because the electronegativities of selenium and sulphur are very similar, the -2 charge may be distributed across the selenotrisulphide bridge, yielding an effective oxidation number of $-\frac{2}{3}$ for each of its members. A similar reaction can occur between selenite and the free sulphydryl groups of proteins to yield selenotrisulphide types of products. Thus, whereas the chemical and physical properties of selenium and sulphur are similar, important differences in their chemistries result in their having very different behaviours in biological systems.

Analysis of Selenium

Several procedures that have been employed for the analysis of selenium for industrial purposes do not lend themselves to biological applications due to low sensitivity. These methods include: (1) gravimetric measurement after reduction and quantitative precipitation with acid; (2) gravimetric measurement after electrolytic deposition with copper; (3) colorimetric titration with oxidizing agents after reduction with thiocyanate or other reducing agents; (4) colorimetric measurement of selenium hydrosols after reduction by hydrazine, tin chloride or ascorbic acid, and stabilization of the hydrosols with hydroxylamine hydrochloride, gum arabic or gelatin; (5) colorimetric measurement of azocompounds formed by the reaction of aromatic amines with diazonium salts, the latter being produced by the oxidation of organic compounds by Se^{4+}; (6) colorimetric measurement of complexes of Se^{2-} with phenyl-substituted thiocarbazides or semicarbazide after reduction of selenium to the Se^{2-} state. The limits of detection of selenium by these procedures are generally in excess of 0·5 ppm and, for the gravimetric methods, can be several parts per million. These procedures are not free of interferences by elements that can coprecipitate in the case of gravimetric methods, or by oxidizing agents in the case of the chemical methods. Therefore, they are generally not suitable for analysis of selenium in biological materials. *See* Spectroscopy, Visible Spectroscopy and Colorimetry

Other methods have been found to be useful for the determination of selenium in plant and animal specimens. Of these procedures, the fluorometric method using diaminonaphthalene (DAN) has been the most popular. This method involves oxidation of the sample to Se^{4+}, and reaction with DAN to form benzopiazoselenol. The product fluoresces intensely at 520 nm when excited at 390 nm and can be quantified using a fluorometer. The chief advantages of the DAN procedure are its high sensitivity (about 0·002 ppm) and its relatively low cost. Nevertheless, the method has two potential pitfalls.

The first involves the loss of selenium during the acid digestion of samples containing large amounts of organic materials. Adequate acid digestion of selenium in biological materials requires the complete conversion of the native forms of the mineral to Se^{4+} and/or Se^{6+}, and the subsequent reduction of any Se^{6+} formed in the process to Se^{4+} without loss. Inorganic selenium can be volatilized to an appreciable extent under the conditions of acidic digestion in the presence of such large amounts of organic materials that charring occurs, especially when sulphuric acid is used as an oxidant. The volatilized selenium, probably in the form of H_2Se, can result in significant errors in the analysis of fatty materials, such as egg yolks or adipose tissues. Because it is volatilized from acid solutions by reducing agents, this loss can be avoided by maintaining strongly oxidizing conditions during digestion and by using low heat such that the oxidation of Se^{4+} to Se^{6+} proceeds relatively slowly by gradually raising the temperature of the perchloric acid solution to 210°C. When the nitric–perchloric acid digestion is controlled and carefully attended, it produces satisfactory conversion to Se^{4+} even of such forms as trimethylselenonium (a major urinary metabolite) which are resistant to oxidation by nitric acid alone. Comparisons of the nitric–perchloric acid digestion method with direct combustion in an oxygen environment have shown that both yield comparable results.

The second pitfall of the DAN method involves interfering fluorescence due to apparent degradation products of DAN itself, which can produce fluorescence. These can be avoided by purifying the DAN reagent by recrystallization from water in the presence of sodium sulphite and activated charcoal, or by stabilizing the DAN reagent with hydrochloric acid and extraction with hexane. Several investigators have incorporated these procedures into methods using DAN which are very convenient for use in the routine analysis of selenium in biological materials. ***See*** Spectroscopy, Fluorescence

Conventional atomic absorption spectroscopy (AAS) has not been suitable for the determination of selenium in biological samples due to the generally high limit of detection (about 0·1 ppm). Variant AAS methods, however, have been developed with sensitivities adequate for biological use. One such method involves hydride generation of sample selenium followed by quantitative detection by AAS. This method requires only small sample sizes (e.g. 0·1 ml of serum), has adequate sensitivity (about 0·01 ppm), and the hydride generation step may be automated. However, it suffers from possible interferences due to other elements which can form hydrides (e.g. copper, astatine, tin). Of these, the most serious interference is due to copper; steps must be taken to remove copper by the use of hydrochloric acid, tellurite or thiourea. Better sensitivity has been obtained using electrothermal AAS. This method avoids the problems associated with wet digestion by employing high-temperature oxidation in a graphite furnace. This reduces interferences due to nonspecific absorption of organic compounds and nonselenium salts, but introduces the problem of volatility of selenium under such conditions. This problem can be avoided by using palladium for thermal stabilization. In practice, electrothermal AAS with the use of a Zeeman effect background correction system can achieve sensitivities approaching 0·05 ppm.

Plasma atomic emission spectrometry (PAES) has not been used widely for the analysis of selenium in biological materials. Although very good sensitivity (about 0·001 ppm) has been reported using inductively coupled PAES, matrix effects present such a high level of interference that most laboratories are not able to obtain reasonable sensitivity by this method. Direct current PAES has not had adequate sensitivity for biological use.

Instrumental neutron activation analysis (INAA) of selenium offers the advantages of applicability to small sample sizes and relative ease of sample preparation. Although the greatest sensitivity (about 0·02 ppm) by this method is obtained by measuring ^{75}Se, its use necessitates lengthy irradiation (100 h), and long periods of postirradiation holding (60 days) and counting (2 h). Greater economy by increased sample throughput has been achieved, at the expense of sensitivity, by the use of the short-lived (half-life 17·38 s) isotope ^{77m}Se. This isotope can be irradiated (5 s), decayed (15 s) and counted (25 s) very quickly in an automated system. Due to the ease of this procedure as well as its nondestructive nature, some investigators with access to research reactors have found instrumental neutron activation analysis useful for the measurement of selenium. Nevertheless, the utility of the 'fast' method is limited at the present time by its relatively low sensitivity, rendering it unsuitable for accurate quantification of such low concentrations as are found in tissues of animals chronically deficient in the element.

The measurement of selenium by proton-induced X ray emission (PIXE), offers the potential advantage of simultaneous elemental analysis of biological materials. This method involves proton bombardment of target atoms (the sample) to cause loss of inner shell electrons and their consequent replacement by electrons from the outer shell. The X rays emitted during that transition are characteristic of the energy differences between electron shells and are, therefore, identifiable and quantifiable. At the present time, the sensitivity of this procedure for the determination of selenium (about 0·01 ppm) makes it useful for many biological purposes, especially when simultaneous elemental analysis may be needed; however, it is not sensitive enough for the accurate determination of very low tissue levels. X ray fluorescence spectrometry offers another nondestructive technique for multielement analysis; however, its sensitivity for selenium does not compare favourably with other methods available for biological use.

A procedure for determining selenium by double-isotope dilution has been developed. This method involves the use of two stable isotopes of selenium as tracer (^{76}Se) and internal standard (^{82}Se). Samples spiked with a known quantity of the internal standard are digested in a nitric–phosphoric acid mixture, undigested lipids are removed with chloroform, and hydrochloric acid is used to reduce any Se^{6+} to Se^{4+}. Selenite is reacted with 4-nitro-*o*-phenylenediamine to form 5-nitropiazoselenol, and the nitropiazoselenonium ion cluster is determined by combined gas–liquid chromatography/mass spectrometry. The native selenium in the sample is calculated from the measured isotope ratios, using the ^{80}Se naturally present in the sample. This technique has a reported sensitivity of less than 0·001 ppm. ***See*** Chromatography, Gas Chromatography

An interlaboratory (12 site) comparison of the more widely used methods for the determination of selenium in clinical materials found statistically significant differences among the mean concentrations reported for levels in lyophilized human serum analysed by either (1) acid digestion/DAN-fluorometry, (2) electrothermal AAS, (3) acid digestion/hydride generation AAS, or (4) acid digestion/isotope dilution mass spectrometry, with slightly higher values reported by the first procedure. The four methods compared very favourably for the analysis of pooled lyophilized urine samples. However, only the DAN fluorometric method showed homogeneity of variance among laboratories.

Chemical Speciation of Selenium

The predominant amount of selenium naturally present in biological materials appears to be in the reduced state (Se^{2-}); however, little information is available concerning the distribution of the various chemical forms in biological materials. The selenium in wheat and corn grown on selenium-rich soils is predominantly protein bound; in wheat half or more occurs as the selenium analogue of the amino acid methionine, selenomethionine. Selenomethionine is thought to be the chief form of the element in other plant tissues; however, little analytical work has been conducted to address that point. In animal tissues, it is essentially all protein bound. Several selenium-containing proteins have been identified in the rat; it is present in each protein only as the selenium amino acid selenocysteine. Therefore, the limited information currently available indicates that selenium in food occurs in proteins, probably chiefly as analogues of amino acids. In plant-derived foods, the major form may be selenomethionine, but in animal products it appears to be selenocysteine. This means that the utilization of food selenium may be affected by the digestibility of food protein and by the methionine and/or cysteine contents of the diet.

Bibliography

Janghorbani M, Ting BTG and Young V (1981) Use of stable isotopes of selenium in human metabolic studies: development of analytical methodology. *American Journal of Clinical Nutrition* 34: 2816.

Klayman DL and Gunther WHH (1973) *Organic Selenium Compounds: Their Chemistry and Biology*. New York: Wiley.

Kumpalainen J, Raittila AM, Lehto J and Koivistoinen P (1983) Electrothermal atomic absorption spectrometric determination of selenium in foods and diets. *Journal of the Association of Official Analytical Chemists* 66: 1129.

Levander OA (1976) Selected aspects of the comparative metabolism and biochemistry of selenium and sulfur. *Trace Elements in Human Health* 2: 135.

McKown DM and Morris JS (1978) Rapid measurement of selenium in biological samples using instrumental neutron activation analysis. *Journal of Radioanalytical Chemistry* 43: 411.

Nazarenko II and Ermakov AN (1972) *Analytical Chemistry of Selenium and Tellurium*, New York: Wiley.

Reamer DC and Veillon C (1983) A double isotope dilution method for using stable selenium isotopes in metabolic tracer studies: analysis by gas chromatography/mass spectrometry (GC/MS). *Journal of Nutrition*. 113: 786.

Rosenfeld I and Beath OA (1964) *Selenium: Geobotany, Biochemistry, Toxicity and Nutrition*. New York: Academic Press.

Watkinson JH (1966) Fluorometric determination of selenium in biological materials with 2,3-diaminonaphthalene. *Analytical Chemistry* 32: 981.

Gerald F. Combs
Cornell University, Ithaca, USA

Physiology

Selenium (Se) first attracted interest in the 1930s as a toxic trace element that caused 'alkali disease' in livestock consuming high-selenium plants. In 1957 selenium was shown to be essential for animals when traces of this element prevented liver necrosis in vitamin-E-deficient rats, and later to prevent a variety of economically important diseases such as white muscle disease in cattle and sheep, hepatosis dietica in swine, and exudative diathesis in poultry. The demonstration in 1973 of a biochemical function for selenium as a constituent of the enzyme glutathione peroxidase helped to explain the interrelationship between selenium and vitamin E. The importance of selenium in human nutrition was highlighted in reports in 1979 of selenium deficiency in a patient on total parenteral nutrition and of the selenium-responsive condition, Keshan disease, in China. Considerable research during the last two decades has provided information on the metabolism and importance of selenium in human nutrition, leading to the recent establishment of recommended dietary intakes. ***See*** Tocopherols, Physiology

Metabolism of Selenium

Selenoamino acids are the main dietary forms of selenium: selenomethionine is derived from plants and selenocysteine from animals. Inorganic forms of selenium are also often used in experimental diets and as supplements. The metabolism of selenium, including absorption, transport, distribution, excretion, retention and transformation to the active form, is very much dependent on the chemical form and amount of selenium ingested, and on interacting dietary factors. There is considerable species variation in many aspects of selenium metabolism.

Absorption

Selenium is absorbed mainly from the duodenum both in monogastric animals and ruminants. Selenomethionine and methionine share the same active transport mechanism, but little is known about the transport of selenocysteine. Absorption of inorganic forms of selenium such as selenite and selenate is via a passive mechanism.

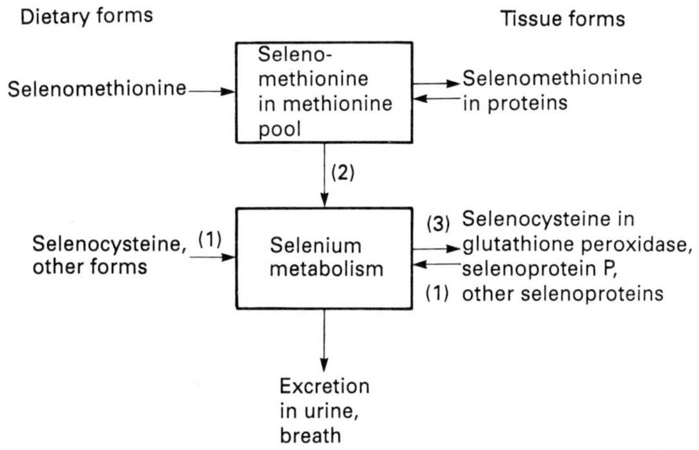

Fig. 1 Outline of selenium metabolism: (1) indicates selenocysteine-β-lysase which catabolizes selenocysteine, and (2) indicates processes that catabolize selenomethionine; both processes make selenium available to the organism; (3) indicates incorporation of selenium into serine during translation with formation of selenocysteine in selenoproteins. Modified from Levander OA and Burk RF (1990) Selenium. In: Brown ML (ed.) *Present Knowledge of Nutrition* 6th edn, pp 268–273. Washington DC: International Life Sciences Institute–Nutrition Foundation.

While the absorption of selenium is generally high in human subjects, probably about 80% from food selenium, selenomethionine appears to be better absorbed than selenite. Absorption of selenium is unaffected by selenium status and there appears to be no homeostatic regulation of absorption.

Transport

Little is known about the transport of selenium in the body, although it appears to be transported bound to plasma proteins, albumin in mice and β lipoproteins in humans. A selenium-containing protein called selenoprotein-P isolated from rat plasma has also been suggested as a transport protein.

Metabolism and Distribution

An outline of selenium metabolism is shown in Fig. 1. In animal tissues selenium occurs in association with protein, and is present in two main compartments or forms. The first is selenocysteine which is present as the active form of selenium in selenoproteins, including the selenoenzyme glutathione peroxidase and other selenoproteins. The second is selenomethionine, which is incorporated in place of methionine in a variety of proteins, unregulated by the selenium status of the animal.

Selenium levels in tissues are influenced by the dietary selenium intake; this is reflected in the wide variation in blood selenium levels of residents of countries with differing soil selenium levels (Fig. 2). Retention of selenium is also profoundly influenced by the form administered, with selenomethionine much more effective in raising blood selenium levels than sodium selenite or selenate. Selenomethionine appears to follow the same metabolic pathway as methionine and is incorporated into protein in place of methionine. This contributes to tissue selenium but has no physiological function and is not available for synthesis of functional forms of selenium until it is catabolized. Selenate is reduced to selenite, then selenide, and in this oxidation state (−2) is introduced into selenocysteine, the form in the active site of glutathione peroxidase and in other selenoproteins. The selenium from inorganic selenium, or from catabolism of selenocysteine or selenomethionine, is incorporated by replacement of oxygen into serine to form selenocysteine while serine is attached to a unique tRNA (transfer ribonucleic acid). Selenocysteine is then incorporated into a selenoprotein such as glutathione peroxidase.

Excretion

Urine is the principal route of excretion of selenium, followed by faeces, which mainly contains unabsorbed selenium. Homeostasis of selenium is achieved by regulation of its excretion. Daily urinary excretion of selenium is closely associated with plasma selenium and dietary selenium intake. Measurement of plasma renal clearance of selenium, which expresses its rate of excretion in the urine in terms of the amount contained in a unit volume of plasma, shows that the kidneys of residents of the low-selenium country, New Zealand, excrete selenium more sparingly than those of North Americans, and Chinese from low-selenium areas even more sparingly; this indicates possible adaptation of New Zealand and Chinese residents to their low selenium status. ***See*** Kidney, Structure and Function

Trimethylselenonium ion, only one of several urinary

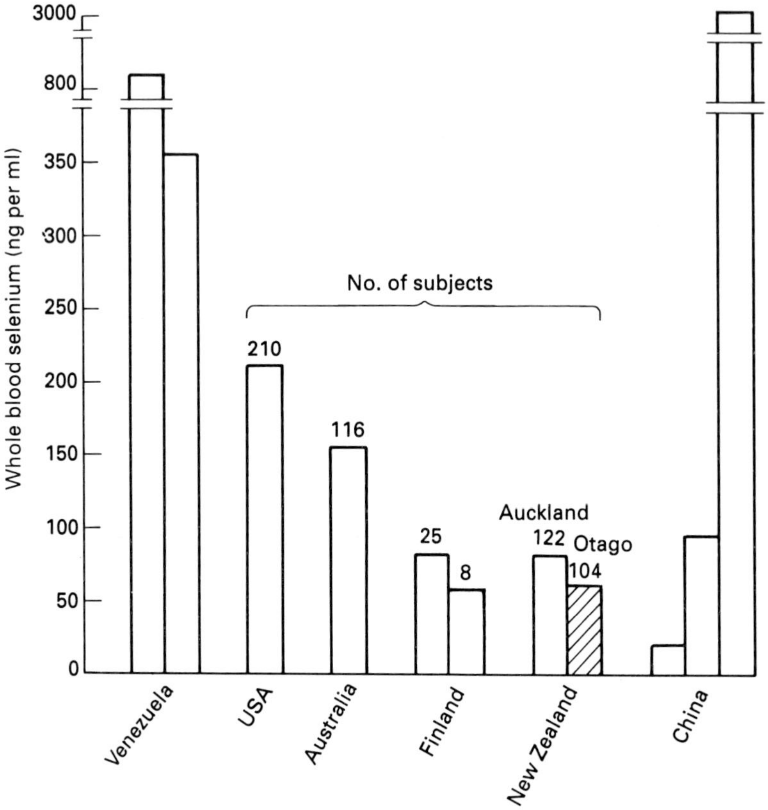

Fig. 2 Blood selenium concentrations of healthy adults in Venezuela, the USA, Australia, Finland, New Zealand and China (low- and high-selenium areas). (To convert ng Se per ml to μmol per litre, multiply by 0·0127.)

metabolites, has been identified, but appears to be a minor metabolite in humans. Small losses of selenium occur through the skin or hair or, at high intakes, in expired air as volatile dimethylselenide.

Bioavailability

As well as absorption, utilization of a nutrient may also include transformation to a biochemically active form, which for selenium is assessed by monitoring changes in tissue glutathione peroxidase. Few studies have been made in human subjects of the bioavailability of selenium, but animal studies show a wide variation in the availability of selenium from different foods. In rats the bioavailability of selenium from mushrooms, tuna, wheat, beef kidney and Brazil nuts is 5%, 57%, 83%, 97% and 124%, respectively, in comparison to sodium selenite. Human studies also show differences among various forms of selenium such as selenate, wheat and yeast, but this also depends upon the criterion of measurement used for availability, indicating the need to consider several variables, including short-term changes in glutathione peroxidase activity, long-term retention of tissue selenium, and metabolic conversion to biologically active forms. ***See*** Bioavailability of Nutrients

Role of Selenium in the Body

The only firmly established role of selenium in the animal body is as a defence against oxidant stress. Selenium exerts its biological effect as a constituent of several selenoproteins, of which three have been purified and studied.

The only well-characterized selenoprotein is the selenoenzyme, glutathione peroxidase, which consists of four identical subunits, each containing one selenocysteine at the active site. Activity of this enzyme can be reduced to less than 1% in the tissues of selenium-deficient animals. Glutathione peroxidase, present in two different forms in cells (including erythrocytes) and plasma, may function *in vivo* to remove hydrogen peroxide, thereby preventing the initiation of peroxidation of membranes and oxidative damage. However, the significance of this function in the body is uncertain, and it seems more likely that the oxidant defence role for selenium is exerted through other selenoproteins. Glutathione peroxidase may have more specific functions in arachidonic acid metabolism in platelets, microbiocidal activity in leucocytes, and in the immune response mechanism.

A second selenium-containing protein has been isolated which is different from the classic selenium-dependent glutathione peroxidase in that it can metabolize fatty acid hydroperoxides that remain esterified to phospholipids in cell membranes. Named phospholipid hydroperoxide glutathione peroxidase, this enzyme can inhibit microsomal lipid peroxidation and thus may account for some of the antioxidant activities of selenium. ***See*** Antioxidants, Natural Antioxidants

In addition to the selenium-containing peroxidases, a rat plasma protein designated selenoprotein P, with no glutathione peroxidase activity, has been purified and characterized. Selenoprotein P is a glycoprotein containing selenium as selenocysteine, and its concentration falls to less than 10% of control in selenium deficiency in rats. The function of selenoprotein P is unknown, but it has been suggested that it is an oxidant defence protein, or a transport protein.

Several selenium-containing enzymes have been identified in microorganisms, and other selenoproteins have been found in animal tissues, suggesting other functions for selenium. The recent discovery that hepatic iodothyronine 5′-deiodinase is a selenoenzyme indicates a role for selenium in iodine metabolism. Furthermore, selenium has strong interactions with heavy metals such as cadmium, silver and mercury and may protect against toxic effects of these metals. ***See*** Cadmium, Toxicology; Mercury, Toxicology

Dietary Intake

Food is the major source of selenium; drinking water contributes little. Dietary intake varies with the geographic source of the foods and the eating habits of the people. Plant food selenium levels reflect the selenium content of the soils and its availability for uptake by plants; cereals and grains grown in soils poor or rich in selenium may vary over 100-fold in selenium content. Animal foods vary less. Fish and organ meats are the richest sources, followed by muscle meats, cereals and grains, then dairy products, with fruits and vegetables mostly poor sources. Average dietary intakes are 10–20 μg of selenium per day in the selenium-poor soil areas of China where Keshan disease is endemic, about 30 μg per day in New Zealand with selenium-poor soils, and over 200 μg per day in seleniferous areas in Venezuela. Intakes of 100 μg per day and 70 μg per day for US men and women, respectively, are within the range of 50–200 μg per day, the estimated safe and adequate intake range proposed by the USA in 1980. Fertilizing certain crops with selenium raised the Finnish intake from 40 μg per day to close to 100 μg per day. These differing intakes are reflected in the wide range of blood selenium concentrations found in residents of these countries (Fig. 2).

Assessment of Selenium Status

Blood selenium concentration is generally considered a useful measure of both selenium status and selenium intake, but several other tissues are usually assessed as well. Plasma selenium reflects short-term selenium status and erythrocyte selenium long-term status; toenails are often used but selenium-containing shampoos restrict the use of hair selenium. The close relationship between blood or red cell glutathione peroxidase activity and selenium concentrations (Fig. 3) is useful for people with relatively low selenium status, but becomes useless once the maximum activity of the enzyme is reached at blood selenium concentrations above 100 ng per ml (1·27 μmol per litre), as may occur in countries like USA. The growing emphasis on functional rather than static indices poses difficulties for selenium status because it is not always clear whether selenium is involved directly or as a limiting factor in, for example, platelet aggregation or ethane/pentane production. It seems that selenium does not function on its own, so that the situation is further complicated by a long list of interacting factors:

- Methionine, cysteine, protein
- Polyunsaturated fatty acids and dietary fat
- Vitamins E, C, A, B_6, and B_2
- Iron, copper, zinc, manganese, molybdenum and arsenic
- Heavy metals: mercury, cadmium and lead
- Iodine
- Oxidant stressors
- Xenobiotics and carcinogens
- Viral infections

Biochemical techniques are being sought as sensitive indicators of selenium overexposure, which at present is diagnosed from hair loss and nail changes.

Requirements and Recommended Dietary Allowances (RDAs)

Many countries have recently proposed RDAs based upon estimates of minimum requirements from Chinese intakes for endemic and nonendemic Keshan disease areas as well as intakes at which maximum glutathione peroxidase activity occurs. RDAs are 85 and 70 μg selenium per day for Australian men and women, 15 μg per day more than the US RDAs for men and women (70 and 55 μg per day), and more than for adults in Canada (50 μg per day), Scandinavian countries (30–60 μg per day) and the UK (60–75 μg per day). Each set is available from habitual diets in each country. However, whether or not optimal health depends upon maximum glutathione peroxidase activity has yet to be resolved; if not, RDAs may approach intakes in countries with naturally low selenium status such as New Zealand. ***See*** Dietary Requirements of Adults

Toxicity

It is well established that the margin between an adequate and toxic intake is quite narrow. Overexposure, or sele-

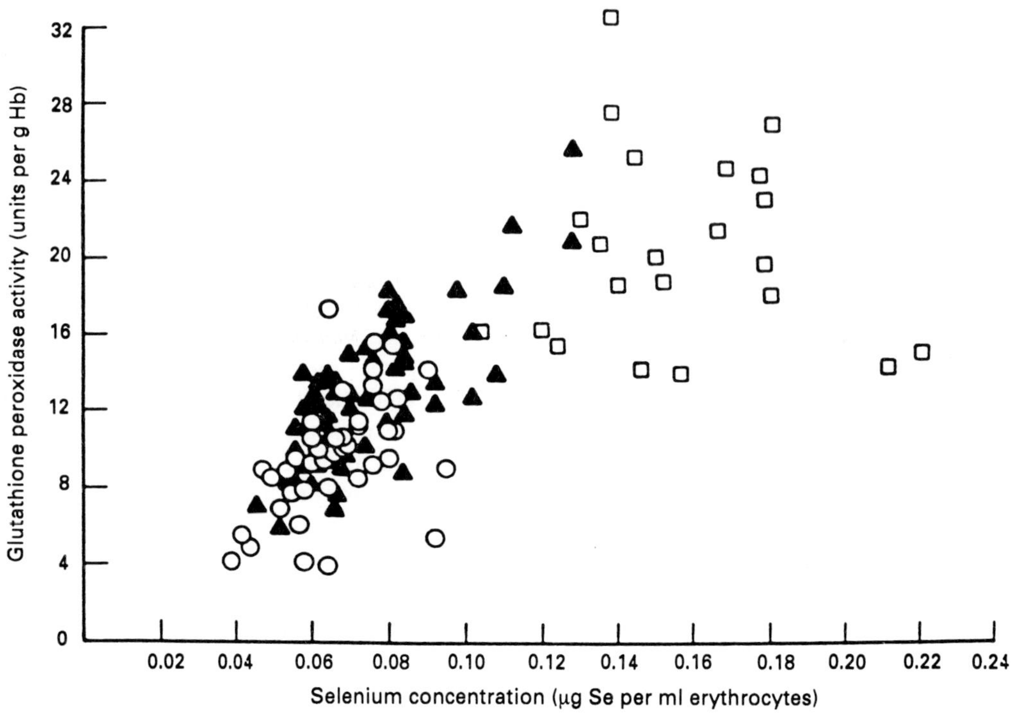

Fig. 3 Relationship between selenium concentration of erythrocytes and glutathione peroxidase activities for New Zealand residents: Otago patients (○); Otago blood donors (▲) and overseas subjects (□). (To convert μg Se per ml to μmol per litre, multiply by 12·66.) (Courtesy of Cambridge University Press, Cambridge.)

nosis may occur from consuming high-selenium foods grown in seleniferous areas. Moreover, people following long-term liberal megadosing can also attain an undesirably high selenium status. Threshold toxicity intakes for selenium have yet to be established.

Deficiency

For a deficiency to be established, a low selenium status needs to be linked with an impaired function or with a selenium-responsive condition or disease. Although residents in many low-selenium areas have both low selenium status and low levels of glutathione peroxidase activity, there is as yet no evidence that these are suboptimal or have resulted in noticeable oxidative damage or changes in other defence mechanisms. Moreover, people are not in noticeably better health when glutathione peroxidase activity is maximized by supplementation with selenium. Whether any of the newer functions of selenium are suboptimal in persons with low selenium status has yet to be investigated. However, an exception to this general rule may be found in the problem of Keshan disease in China.

Selenium-responsive Diseases in Humans

Immense selenium supplementation studies of over a million subjects in China were highly successful in controlling Keshan disease, an endemic cardiomyopathy that primarily affects children and women of childbearing age. However, features of Keshan disease cannot be explained entirely on the basis of selenium deficiency, suggesting that there may be some secondary factors such as a virus or environmental toxins. Kashin–Beck disease, an endemic osteoarthritis which occurs during preadolescent or adolescent years, is another disease that has been linked to low selenium status in China, and again interacting factors are being sought.

Selenium deficiency in some patients on hyperalimentation has been associated with a cardiomyopathy, muscle pain and muscular weakness but such clinical symptoms were not seen in all patients with an extremely low selenium status, indicating that there may be other interacting factors.

It is not clear whether syndromes in animals are related to human diseases such as the selenium-responsive muscular dystrophy observed in sheep. Although persistent anecdotal reports from New Zealand farmers in low selenium areas indicate their conviction that selenium relieves the farmers' muscular aches and pains, double-blind trials have failed to give a clear-cut answer.

Selenium and Chronic Disease

Many attempts have been made to link other diseases with a poor selenium status, such as cancer and heart disease. A number of epidemiological studies have suggested an association for cancer, but this has not been confirmed by others. It should be noted that two of the early proponents of the low-selenium high-cancer risk hypothesis concluded that the protective effect of selenium is still not proven and that the use of dietary supplements should not be recommended. Similarly, evaluation of prospective epidemiological studies failed to show a causal relationship between selenium status and the risk of ischaemic heart disease. Further studies are necessary to clarify any possible role for selenium in these degenerative diseases.

Bibliography

Burk RF (1989) Recent developments in trace element metabolism and function: newer roles for selenium in nutrition. *Journal of Nutrition* 119: 1051–1054.
Combs GF and Combs SB (1986) *The Role of Selenium in Nutrition*. Florida: Academic Press.
Levander OA (1987) A global view of nutrition. *Annual Reviews in Nutrition* 7: 227–250.
Robinson MF (1989) Selenium in human nutrition in New Zealand. *Nutrition Reviews* 47: 99–107.
Robinson MF and Thomson CD (1983) The role of selenium in the diet. *Nutrition Abstracts and Reviews* 53: 3–26.

Christine Thomson and Marion Robinson
Otago University, Dunedin, New Zealand

SENSORY EVALUATION

Contents

Sensory Characteristics of Human Foods

Sensory phenomena related to foods are an interaction between the physical or chemical stimulus and the person; thus they are not a simple mapping of physical and chemical properties of the food. When the reasons for the choice of foods are considered the sensory attributes are not the only factors likely to be influential. In surveys, taste, flavour and other sensory attributes are generally cited as among the most important influences, if not the most important. However, attempts to link sensory responses to particular compounds in foods such as salt and sugar with the intake of these have shown far less clear relationships.

The Biological Utility of Sensory Perception

Foods are recognized by their taste, flavour and appearance as well as their origin. Humans are omnivores and are able to eat many different types of foods. In their development the chances of survival of humans would have been increased by consuming a wide variety of foods which would be more likely to provide an appropriate mix of nutrients than would reliance on a limited number of foods. Thus the ability to perceive sensory attributes is biologically extremely important. For animals with relatively restricted diets there is the possibility of fixed genetic programming of sensory perception and of preferences, but in the case of omnivores this is impossible since there are no unique sensory features which characterize an adequate nutrient supply. There is also increased utility in being able to adapt to changing circumstances by eating new foods. Thus the problem for omnivores is to eat a variety of foods, but at the same time to avoid dangerous and poisonous materials.

Although there are no unique features of good food sources there are some general rules and some preferences may be in part genetically determined. There is evidence based upon facial expressions that new-born babies like sweet tastes and dislike bitter and sour tastes. There is even evidence that the human fetus will increase the rate of sucking-type responses with an injection of saccharin into the amniotic fluid. Even in these cases there is the possibility that experiences *in utero* could account for such preferences, but they are more likely to be genetic. These responses are adaptive since naturally sweet substances are usually good, safe sources of energy, while bitter substances are often poisonous.

However, the overwhelming evidence is that most food preferences and aversions are learned through experience. There are powerful learning mechanisms associating the eating of a food with subsequent illness, and this leads to aversions both in animals and in humans. Such aversions are rapid (often following just one occasion), very strong, difficult to remove and result in a strong dislike of the foods; this is not the same as people avoiding foods which they see as fattening or bad for health, and it involves a dramatic change in sensory preference. This occurs even when the person knows that the food was not the cause of the illness. People will develop aversions to foods eaten before cancer chemotherapy treatment because the therapy causes nausea and vomiting, even though they know the food was not the cause. This mechanism is biologically useful in directing the animal away from poisonous and harmful materials.

The converse learning of positive preferences for sensory attributes is much more difficult to demonstrate experimentally, but seems likely to be a major factor in the acquisition of preferences. Again, this learning can involve the association between sensory attributes of foods and positive postingestional effects, e.g. reducing hunger or making the person feel better. Although there is good evidence for the increased liking of high-energy foods, which reduce hunger, there is less evidence to support such a mechanism for other nutrients. However, there is speculation that an animal deficient in a particular nutrient will feel better after consuming a source of that nutrient, and hence come to like the food. Certainly, the development of learned preferences is less clear than the development of aversions and there is no increased liking for the flavours put into medicines even though the medicines will make people feel very much better. In this case, people seem able to dissociate the effects from the flavour in a way that they cannot with sickness and aversions. In humans, much of the learning associated with foods takes place within a social context and this will interact with the phenomena described above.

Thus sensory perception is potentially useful in our interactions with foods and is likely to have some fairly fundamental biological antecedents.

Sensory Responses and Physical or Chemical Characteristics

When foods are evaluated by human assessors there are clear relationships between the concentrations of some chemicals in the foods and the responses of the assessors. Thus with higher concentrations of sodium chloride there will be higher ratings of saltiness, and similar relationships exist for responses to more complex flavour stimuli or variations in physical properties. There is often a temptation to ascribe these sensory phenomena to the food and chemical components themselves. However, the sensory attributes exist only in the interaction between the food and the person assessing it. *See* Flavour Compounds, Structures and Characteristics

In psychology, this type of relationship has been described as the stimulus–response, or more accurately the stimulus–organism–response, relationship. The foods containing components at certain concentrations form the stimuli, and the responses are the actions of assessors, whether that is a rating on a verbally anchored scale, a choice between alternatives, saying whether a taste is present, or describing the nature of a taste or flavour. Although it might appear that there is a simple transduction from stimulus to response, this is far from the case. Figure 1 shows one simplified schematic

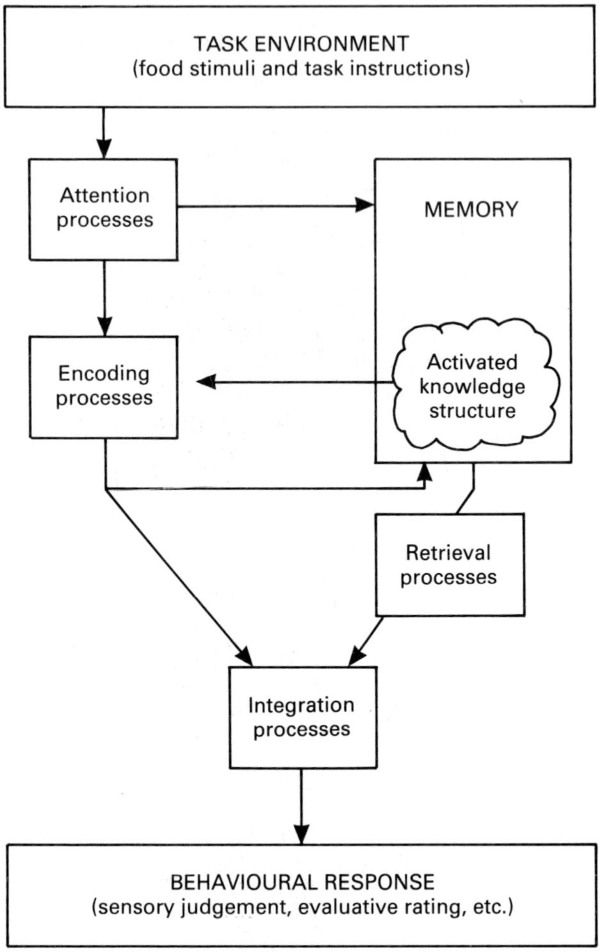

Fig. 1 Schematic representation of stages involved in an assessor making sensory responses. Adapted, with the publishers permission, from Olson JC (1991) The importance of cognitive processes and existing knowledge structures for understanding food acceptance. In: Solms J and Hall RL (eds) *Criteria of Food Acceptance*, pp 69–81. Zurich: Springer-Verlag.

representation of some of the processes that are involved when assessors rate samples. First, there is a stimulus-input phase, during which chemical stimuli trigger receptors, but these are only one type of input among a mass of signals both from the food (which is of interest) and from everything else in the environment. The person has to allocate attention to the signals which are deemed to be important. Incoming information is compared to internal memories, followed by integration of information and decision-making. Only following this can some overt response be made. It is clear that this is not a peripheral function of the human subject, and it involves complex cognitive processing, no matter how seemingly simple the response.

It follows from this that there will be differences not only in the physiology and biology of the person making the assessment, but also in the psychological processes which the person must go through in order to decide upon the response to give. Responses will be influenced by many factors in addition to the physical and chemical characteristics of the foods presented, and the exact experimental procedures, response formats, sample orders, physiological state, learning, experience and expectations of the assessor will all influence the responses. Assessors are not purely analytical and will be influenced by factors other than the chemical and physical characteristics of the foods.

Sensory Responses and Food Choice

Although some variations in chemical composition are perceived, there remains the question of whether or not, and if so by how much, this influences the choice of foods. Here the evidence that liking for particular sensory attributes influences choice is far less clear. Attempts to relate sensory responses to particular tastes and intake of foods have largely concerned salt and sugar and, to a lesser extent, fat.

In the case of sugar intake, much of the early work centred on attempts to relate sugar preferences to bodyweight and obesity, on the assumption that a higher preference for sugar would lead to a greater intake of sugar and hence a higher bodyweight. Similar thinking with salt led to attempts to relate sensory responses to blood pressure and hypertension, on the grounds that higher preferences for salt would lead to higher intake and hence higher blood pressure. There are some strong assumptions here – that sugar intake will lead to a higher bodyweight and that salt intake will raise blood pressure – and these assumptions are not entirely borne out by the evidence. Many of these studies did not actually include a measure of intake, although it would often be assumed to be the intervening variable. More recent studies have included measures of intake in addition to sensory measures, but in many cases have not provided strong evidence for specific sensory attributes relating to intake. ***See*** Carbohydrates, Sensory Properties

The measures used have typically been threshold sensitivity, above-threshold sensitivity, or some measure of preference or liking for different concentrations of tastant. Measures of preference are usually based upon varying the concentration of sugar or salt in a food, and assessing the level at which a person maximally likes the food. The argument that preference should relate to intake seems obvious; if a person likes a high level of a tastant, such as salt or sugar, in the food then that person would be likely to have a high intake of that compound. The argument for sensitivity relating to intake is less clear. Despite some early work with salt, when it was argued that less sensitive people would have a lower intake, almost all of the studies in this area have followed the hypothesis that those people who are less sensitive will have a higher intake in order to achieve an equivalent taste experience.

Threshold sensitivity is a measure of the lowest concentration that an individual can detect or recognize. Although it is a simple measure to make, and appears relatively objective, its importance in determining consumption is more doubtful. Few tastants or odorants are at threshold levels in real foods and there is little evidence that sensitivity at threshold relates directly to sensitivity at above-threshold levels. For this reason, more recent studies have included measures of sensitivity to changes in concentrations above threshold levels. A number of the studies have included measures of sensitivity to tastants in solution, although this does not necessarily relate to sensitivity to the same tastants in foods. Moreover, the use of solutions for hedonic or preference studies is problematic since liking will be food-specific and the degree of liking of, for example, salt in water has little significance for consumption of salt in real foods.

The evidence for a relationship between measures of sensitivity or preference for particular sensory attributes and intake is poor. Although a number of studies have shown the expected relationships, there are a number of studies failing to show any relationship. Even where there are statistically significant relationships these are often small and may be one significant finding from among many nonsignificant findings.

There are a number of possible reasons for this lack of a clear relationship. As always, there are methodological problems in many of the studies, including poor measurement of both sensory responses and intake. In particular, intake has often been assessed using short, unvalidated questionnaires which may give a poor indication of usual intake. Even so, there are well-designed and well-executed studies which show only slight relationships between sensitivity and preference on the one hand and intake on the other. One reason for the lack of a clear relationship is that foods form

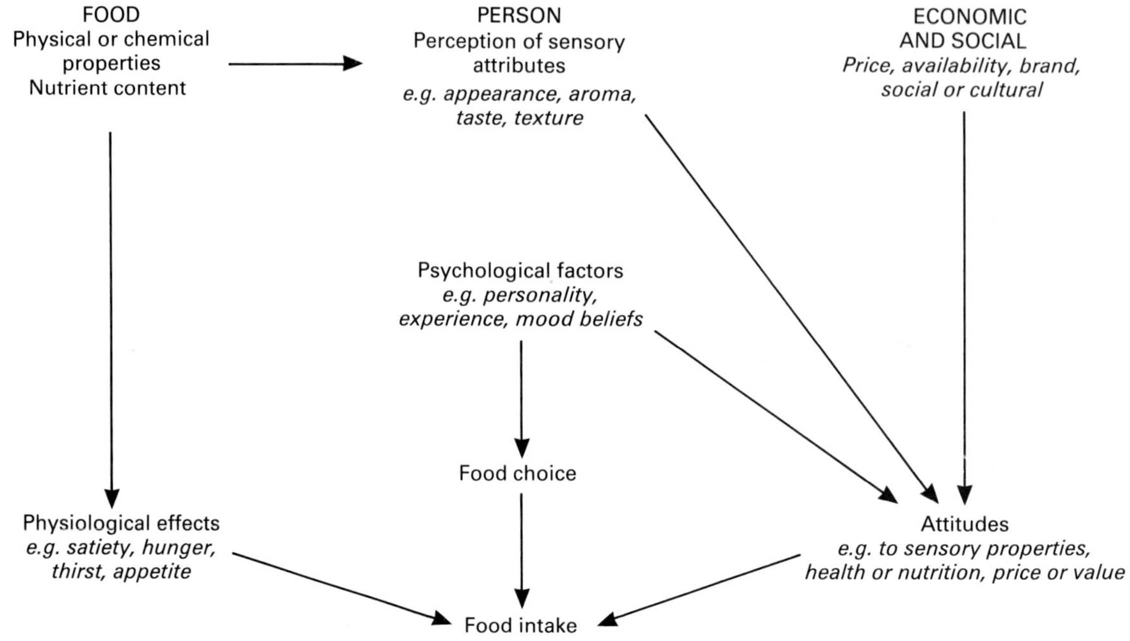

Fig. 2 Possible factors influencing the choice of foods.

complex sensory stimuli, varying in many sensory attributes – not only the sweetness or saltiness studied. There are also a large number of potential influences on food choice and intake other than sensory attributes. ***See*** Food Acceptability, Affective Methods

Other Influences on Food Choice

Figure 2 shows some of the potentially large number of factors influencing choice. In addition to liking for the sensory attributes, there are the post-ingestional effects, described earlier, but also a range of factors related to the person making the choice and to the economic and social environment within which the choice is made. Choices may be influenced by beliefs about the nutritional and health benefits of the foods or by issues of availability, advertising and convenience. Thus sensory factors are not the sole determinants of food choice.

Although the studies of the relationship between specific sensory responses and intake do not offer clear evidence, sensory factors do appear to be important. In surveys where people are asked to rate the importance of factors influencing food choice, sensory and taste factors generally come out as one of the most important influences. This is also true in more detailed studies where beliefs and attitudes have been quantitatively related to choice; beliefs concerning the liking of the sensory experiences of eating the food are generally the most closely related to choice and consumption, and are found to be more important than beliefs about the nutritional and health benefits of the foods or matters of price or convenience.

It is also the case that although the liking for specific sensory attributes may be difficult to relate to choice there is a good correspondence between liking or preference for the foods themselves and their consumption. If a person rates tomatoes as highly liked then that person's consumption of tomatoes is likely to be higher than that of a person rating tomatoes as disliked. However, it does not always follow that a high rating of liking will mean the food will be eaten more often. Someone may, for example, rate lobster as very highly liked and bread as only moderately liked but still eat far more bread than lobster. Some of the other factors in Fig. 2 are likely to be playing a more major role here. The exact nature of the relationship between the liking for individual sensory attributes and total liking or preference for the food is not clear and in some cases measures of overall liking may also incorporate elements of factors other than purely sensory ones.

While sensory factors and liking are important in food choice they need to be understood within this wider context. Working within this wider framework appropriate sensory evaluation offers a potential means to bridge the gap between the physical or chemical properties of foods and consumer choice.

Bibliography

Khan MA (1981) Evaluation of food selection patterns and preferences. *CRC Critical Reviews in Food Science and Nutrition* 15: 129–153.

Randall E and Sanjur D (1981) Food preferences – their conceptualization and relationship to consumption. *Ecology of Food and Nutrition* 11: 151–161.

Shepherd R (1988) Sensory influences on salt, sugar and fat intake. *Nutrition Research Reviews* 1: 125–144.

Sensory Characteristics of Human Foods

Shepherd R (ed.) (1989) *Handbook of Psychophysiology of Human Eating*. Chichester: John Wiley.
Solms J, Booth DA, Pangborn RM and Raunhardt O (eds) (1987) *Food Acceptance and Nutrition*. London: Academic Press.

R. Shepherd
AFRC Institute of Food Research, Reading, UK

Food Acceptability and Sensory Evaluation

The area of food acceptability and food choice is complex, with many factors influencing an individual's choice of one food item in preference to others available. While the sensory characteristics of a product are important, they are only a small part of the food choice equation. Social, cultural and psychological factors also influence selection. This article illustrates the importance of sensory evaluation in consumer and market research, and indicates how data from consumers can be related to data from trained sensory panels to provide a better understanding of food acceptability.

Food Choice and Food Acceptability

The terms 'food choice' and 'food acceptability' are often used interchangeably, but in fact they refer to two different aspects of consumer behaviour. Food choice, in general terms, is defined as the selection of foods for consumption, and encompasses a whole library of factors in addition to those relating directly to acceptance. These are covered elsewhere in the Encyclopaedia. ***See*** Food Acceptability, Affective Methods

Food acceptability, or acceptance, relates directly to the reaction of an individual to a particular food item at a particular time. In a sensory context, this can be measured as how much an individual likes or dislikes a product, how much an individual prefers one product to another, how different sensory characteristics influence enjoyment of a product, and so on. Within this definition, an individual may find a food acceptable, but may not choose it for a variety of reasons; these may include marketplace influences, situational influences and personal factors such as concerns about weight, health, etc.

Sensory Characteristics and Food Acceptability

When presented with a food or drink the eyes, nose, mouth, ears and tactile senses receive messages, generated through interaction with the product. Signals resulting from these messages are transmitted to the brain, which consequently modifies, sorts, classifies and interprets the information to provide a judgement. Past experiences and memories enable the brain to decipher this information. If a particular food or drink product has not previously been experienced, then an individual is likely to be more alert, using the senses to gather as much information as possible, and recalling any relevant past experiences. This, in conjunction with current experiences, will produce an acceptability judgement. When an item is familiar or recognized, personal expectations about the sensory characteristics will already exist. Evaluation of acceptability will then be in relation to these expectations, either consciously or subconsciously. If the flavour, texture or other sensory characteristics are unusual or unexpected. e.g. when a food product has picked up a taint, then messages to the brain will require a judgement to be made on whether or not the item should be eaten.

New sensations are continually experienced by individuals, as more varieties of food and drink are consumed. These, together with new situations which are encountered, provide the brain with more information for processing. This can have an immediate effect on consumption. For example, a food may be liked initially, but on continued eating, it may be considered too sweet and sickly to be desirable. This type of information is processed by the brain and stored, so that it can be used to evaluate similar foods in the future. Another example might result from more indirect sources, such as media coverage on health properties of a particular product. Although not preferred when compared to other products, concern about health may lead to actual choice of one product in preference to others available.

The appearance characteristics of colour, size, shape and surface texture are all evaluated by the eye, and individuals rely greatly upon visual recognition for evaluating the acceptability of a food. Other factors, such as the flavour or texture, become more important on eating food, and will be more influential in subsequent selections. The importance of initial appearance can be illustrated by a plate of food, on which attention is drawn to a large blemish on a potato. The impact of this will affect initial reactions, as the blemish will either need to be cut away, or the potato rejected as unacceptable. Had the blemish been removed prior to serving, then the judgement would be different.

Colour is particularly important in quality judgements of fresh produce, where colour changes may be associated with an accompanied undesirable change in texture or flavour. Many colourings are added to food products by industry to make them look more appealing. Through experience, colour associations are learnt and individuals can evaluate the appropriateness of colours to particular food items. Certain colours are immediately associated with certain flavours. For example, a yellow-coloured sweet is assumed to be lemon-flavoured. Uncertainty of the flavour of a product has been illustrated because the colour was not characteristic, or flavour ratings have varied as the amount of characteristic colour has altered. Memory is important in this aspect of food selection. Based on past experience, items may be chosen because they look good. An orange may be chosen because it looks sweet and juicy, or a cake may be rejected because it looks dry, stale and tasteless. *See* Colours, Properties and Determination of Natural Pigments; Colours, Properties and Determination of Synthetic Pigments

Odour can indicate the food type, and smell is often used to evaluate the quality or freshness of a product; for example, cream and milk may be smelt before they are used. Undesirable odours may develop in foods as a result of inadequate or improper storage conditions. In some products, desirable characteristics develop on storage; these can be detected by smell, and are important criteria for acceptability. For example, a quality wine may be said to have a 'good nose'. During cooking, certain volatiles may be produced and in some supermarkets, in-store bakeries use the smell of baking bread to attract potential purchasers.

When food is eaten, flavour and texture have their largest influence on acceptability of the food. Flavour of food is determined by a combination of taste, detected mainly by the tongue, the volatiles released from the food when in the mouth, and sensations such as hotness (for a curry) and coolness (for menthol). Flavour is often quoted as important in the acceptability of food, as a product may be selected or rejected on its expected flavour. In some products, texture is the dominant characteristic relating to acceptability, and a particular food may be added to a recipe dish to give it texture, e.g. water chestnuts added to a Chinese meal. Again, certain texture characteristics are associated with certain foods, and if a product does not satisfy these criteria it may be rejected. Slimy textures in the mouth in certain products may render otherwise acceptable products unacceptable. *See* Flavour Compounds, Structures and Characteristics

The human senses are continually collecting and processing information from the surroundings, but individuals are generally unaware of this as the brain filters out the irrelevant sensations, perceiving only those relevant at a particular time. For example, we are unaware of our arms and legs when walking, and similarly, unless studying a plate of food carefully, we are unlikely to notice the usual, probably summarizing the appearance as 'good'. However, if something is not as expected then the brain will register this fact to enable action to be taken. Take, for example, the plate of food referred to previously, in which the potato had a large blemish. By registering this fact, it was possible to cut the blemish away, and therefore avoid eating it.

In investigating food acceptability it is obvious that some foods are liked by some people, but the same foods may be strongly disliked by others. Food preferences and acceptability are individual and influenced largely by previous experiences, current trends and knowledge. Sensitivities vary not only between individuals, but also within an individual during a lifetime and at different times of the day. In addition to this, two foods may be perceptually different in sensory characteristics, but may be equally liked. In studying food acceptability, it is therefore necessary to look for patterns between groups of individuals in their acceptability behaviour and to segment the sample population on this basis, targeting the product accordingly.

The Importance of a Sensory Input

The importance of a product having the right sensory characteristics is reflected by the large research efforts within food companies directed towards improving and optimizing formulations, or developing new products with different sensory characteristics. For a product to be successful it is necessary to understand the consumer reactions and the role of the sensory characteristics in the overall product image.

Consumers often find it difficult to verbalize the sensations they are experiencing when eating a food, describing the food as 'nice' or claiming to dislike the flavour, and so on. For this reason, it is often more effective to ask a trained sensory panel to describe and quantify the sensory characteristics of a product, and relate these data to information collected from consumers on product acceptability. Individuals selected and trained as sensory assessors are able to give detailed descriptions, identifying the similarities and differences between products. As they are trained to characterize products in detail, they are sensitive to small differences. Sensory assessors are not representative of the consumer and should not be used to give acceptability or preference judgements.

Two types of panel can therefore be defined: the trained sensory panel who describe and discriminate between products, and the consumer or untrained panel who give subjective information about product acceptability or state which product they prefer. Information collected from these two panels can be related to optimize the sensory characteristics of a product.

Designing the Experiment

Correct design of an experiment will ensure that all objectives are met, and enable the influence of individual variables within the formulation to be examined. As large-scale acceptability studies are expensive and time-consuming, it is often only practical to have a few of the total number of samples evaluated by each consumer, necessitating an appropriate design which ensures adequate distribution of samples among respondents.

The recruitment of a representative sample of consumers to take part in the test is therefore important. This sample needs to include all the relevant views of the population from which it was drawn, so that reliable conclusions can be reached.

If acceptability information is to be segmented or related to other data sets, such as that from a trained sensory panel, then it is desirable to collect information on all the products in the test from all the respondents, particularly in the acceptability trial. If this is not the case, variations in individual preferences will not be recorded and this may affect segmentation procedures.

Measuring Consumer Responses

Data which measures the liking or disliking of samples can be collected from consumers in a variety of different situations: at home, in a street interview, questionnaire with an interviewer present, or a questionnaire completed by the respondent themselves. Many food companies conduct acceptability studies on their own products among their staff, to give an indication of likely consumer acceptability, or to select a few samples from a larger range of samples for a consumer test. Such people are not likely to be representative of the consuming public, as they form only a small, often limited section of the population. For example, the majority may be between 20 and 50 years old. These people are also likely to have considerable knowledge about the company's product, and may be able to pick it out in a blind test. Employees may also know about the development stages of a product, and would therefore have information which could bias their evaluations. To test the validity of an employee panel in determining the acceptability of a product, one or more of the products allocated a low acceptability score should be included in the main consumer trial.

The type of test chosen, e.g. a paired preference or hedonic scale, will depend on the objectives of the study. Acceptability testing measures how much a product is actually liked or disliked, and the results from different samples can be compared. In preference testing the samples are directly compared and judged on the basis of preference for one sample over another. A person may prefer one sample, but not like either of them very much.

The instructions given and the format of the test are important in consumer testing. If the respondent has only a small amount of the product to taste, this will allow an immediate response, but no evaluation of the product on continued eating will be determined.

The situation of the test can also influence the results. For example, in a testing centre in a town hall, respondents are unlikely to have the time to spend evaluating a product in detail and are not in familiar surroundings, or perhaps not particularly comfortable ones. Concentration is therefore likely to be minimal, with many respondents wanting to finish the test quickly and resume their original tasks. At home the consumer rarely compares two products side by side, but relies on memory of similar products when making a comparative judgement with their normal brand. A paired comparison test or a test where the respondent is able to directly compare the test formulation with their usual brand is therefore likely to be more stringent. The time of testing can often be determined by the respondent in a home test situation, but in a street interview this is not controllable, and foods may be tested out of context with normal eating behaviours. For example, cream cakes may receive a low rating by a respondent because they were tested early in the morning, a time when they would not normally be consumed.

Analysing Consumer Data

Data collected from consumer acceptability studies can be summarized and analysed in traditional ways. For example, histograms can be plotted, sample means calculated, and a number of significance tests (e.g. chi-square test, Wilcoxon signed rank test, Mann Whitney U test) can be applied in order to investigate more formally differences in acceptability for a number of samples. However, these methods give only limited information about actual product acceptability, as they do not identify segments within the test group. As previously mentioned, people are individual in their likes and dislikes, and thorough examination of acceptability data, using segmentation procedures, is therefore often a more realistic approach. This may be illustrated by examining the data from two distinct groups of respondent, where no significant preference was illustrated overall. The first group liked the products that the second group disliked. Averaging data to look at the mean acceptability rating of each sample in this instance is meaningless as it does not represent either view. However, applying internal preference mapping would reveal these differences.

Internal preference mapping is a multivariate technique which takes the acceptability data from individual consumers to derive a multidimensional space in which each of the samples is located. A line can be drawn from the origin of the plot to each individual consumer's position on the preference space (Fig. 1). From these preference vectors, it is possible to derive the order of sample acceptability for each consumer. This is determined by drawing perpendicular lines from the vector to each sample, as illustrated in Fig. 1 for consumer 8. In this example, sample A was most liked by consumer 8, sample I was second most liked and sample H was least

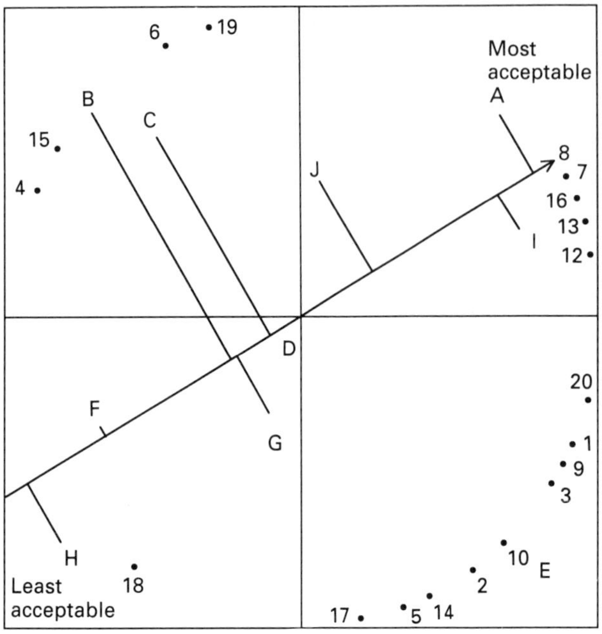

Fig. 1 Example of an internal preference mapping plot. The direction of preference for one consumer is indicated by an arrow. A–J, samples; 1–20, consumers.

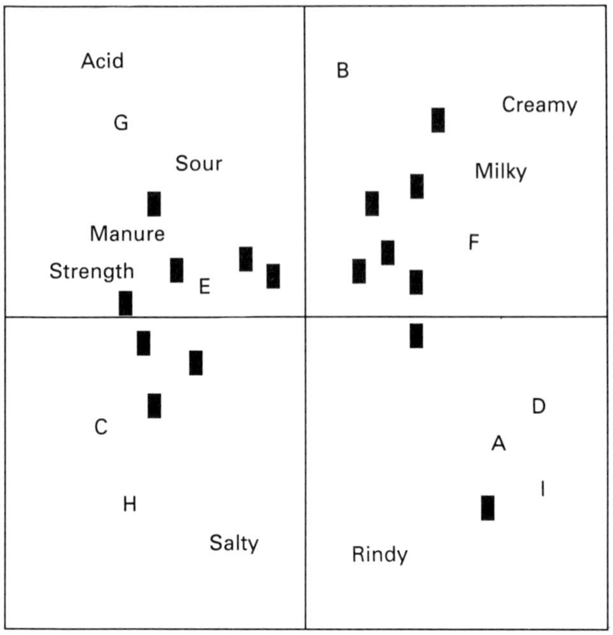

Fig. 2 Example of an external preference mapping plot illustrating the ideal-point model. A–I, samples; ■ ideal point for an individual consumer.

liked. Clusters of vectors are used to derive possible segments of consumers, identified as having similar acceptability behaviours. It is recommended that no fewer than five samples are used for this type of analysis.

Internal preference mapping examines the relationship between samples with respect to acceptability, but it does not say which sensory characteristics are important in determining the acceptability for each of the samples. Quantitative information on the sensory characteristics of the samples, obtained from a trained sensory panel, can be related to the acceptability data in several ways to give this information.

One method is to obtain the average panel ratings, of each sample, for each of the sensory characteristics showing discrimination, and correlate these values with each individual's acceptability data on the same samples. This identifies the sensory attributes related to acceptability judgements for each individual. However, this can be quite time-consuming to calculate and interpret. By segmenting the consumers into groups, the sample averages for each group can be correlated with the sensory characteristic data. This is usually a more efficient and meaningful approach.

External preference mapping is an approach which can be used to map consumer acceptability information onto a sample space derived from analysis of sensory profile data by principal component analysis or generalized Procrustes analysis. As with the above approach, the same set of samples need to be included in both the sensory profile and consumer acceptability tests. There are a number of different regression models used in this approach, and these can be considered under two headings: the vector model and the ideal-point model. The applicability of one of these models to a set of data can be determined by evaluating goodness-of-fit criteria.

The vector model represents the acceptability data of each consumer as a direction of increasing preference on the profile sample space. In principle, the interpretation is similar to that for internal preference mapping. The only difference is that external preference mapping uses a previously determined sample space, such as the one from a trained sensory panel profile analysis. The resulting vector model assumes that none of the samples examined has a sufficient level of the sensory characteristics influencing acceptability. This is illustrated by the directional vectors pointing outwards. Therefore this model does not indicate the most acceptable level of each of these characteristics, assuming that the more the characteristic increases, the greater the acceptability. In practice, this model is obviously limited in its application.

The ideal-point model takes into account all acceptability judgements from one person and locates the position where the ideal sample would lie on the profile space. An ideal point is determined for each consumer in this way, and the model assumes that the levels of each sensory characteristic, responsible for influencing acceptability, lie within the profile space determined by the trained sensory panel. Figure 2 shows a combined sample-and-attribute space, derived from principal component analysis, onto which the ideal points have been plotted. The rectangles represent ideal points for the 16 consumers whose ideal product was within the profile space. By using the attribute information it is possible to determine the levels of each characteristic which will give rise to the ideal sample for any one consumer. Segmentation is also possible from this technique, and is a realistic objective for proposing a marketing strategy for a product. The main drawback of this technique is that only some of the consumer's ideal points will fit within the sample space generated through the profile analysis.

In practice, the use of external preference mapping will depend on selection of an appropriate range of samples. Successful implementation also depends on having a sufficient number of samples, each of which must be evaluated by each person; at least six for the vector model and seven for the ideal-point model. These numbers are based on realistic considerations for most sensory experiments where acceptability data are also being collected. At least 20 samples are desirable before such a method can be used for serious predictive work. For these reasons, preference mapping methods are exploratory.

Further Approaches to Determining Sensory Acceptability

The sensory characteristics are important in product acceptability, as has been illustrated above, but the techniques to identify and optimize these characteristics are limited and still require development. Statistical models are one option, but although it has been shown that a trained sensory panel can reliably characterize the products, they may not in reality identify the important characteristics for product acceptability. Other techniques, such as the repertory grid method, are being investigated to assist consumers in describing sensory perceptions.

Bibliography

Lawless HT and Klein BP (ed.) (1991) *Sensory Science Theory and Applications in Foods*. New York: Marcel Dekker.

McBride RL and MacFie HJH (eds) (1990) *Psychological Basis of Sensory Evaluation*. Essex: Elsevier Publishers.

MacFie HJH and Thomson DMH (1988) Preference mapping and multidimensional scaling. In: Piggott JR (ed.) *Sensory Analysis of Foods*, pp 381–409. Essex: Elsevier Science Publishers.

Peryam DR and Pilgrim FJ (1957) Hedonic scale method of measuring food preferences. *Food Technology* 11(9): 9–14.

Thomson DMH (ed.) (1988) *Food Acceptability*. Essex: Elsevier Science Publishers.

Janet S. Colwill
Campden Food and Drink Research Association, Chipping Campden, UK

Practical Considerations*

Sensory analysis consists of methodically examining sensory perception elicited by products to decide whether to continue development or production, or to ascertain quality for other purposes. This decision should be taken in the light of clear information on the characteristics of the products, their level of perception and importance to the consumer. The information required depends on the type of decision to be taken:

1. Information related to consumer acceptance is obtained from subjective tests performed outside the laboratory towards large panels of untrained consumers. This corresponds to *sensory evaluation*. It allows the creation or launch of new products responding to consumers' expectations or needs.
2. Information on product characteristics involves objective *sensory analysis* and requires specific sensors. The detecting device is a panel of assessors, who must perform in such a way as to be sensitive and provide reliable results. To obtain a good measuring tool, the individual assessor must be selected and carefully trained. It is important, during sensory analysis, that external effects are reduced by providing a quiet and neutral tasting space, where interindividual exchanges and influence from witnessing the preparation of samples are excluded.

Sensory analysis is often performed in a research and development context to understand effects of technical parameters and to pilot production efficiently, to find out the impact of the ingredients on the final product and to optimize new formulae.

Sensory analysis is also performed at factory level to guarantee uniformity of production on a daily basis, for checking raw materials, intermediate and finished products, and storage stability.

In academia, sensory analysis can be carried out to increase knowledge on perception and human behaviour towards food.

In institutes or government laboratories, sensory analysis can be used for recognition of the quality level in grading products, evaluating the merit of different producers or brands. In each case, sensory equipment and groups of assessors are adapted to meet these different requirements.

The sensory tool used also depends on the product type. Conditions are not the same for beverages such as wine, coffee or tea, for culinary, bakery or dairy products, fresh fruits and vegetables, spices, aromas, cosmetics, medicines, or cigarettes. Each of these product ranges presents its own difficulties and has its specific methodology, although they do have some

* The colour plate section for this article appears between p. 3896 and p. 3897.

common features in the selection and training of assessors, as well as in the laboratory set-up where equipment is arranged to achieve optimum conditions for the preparation and the presentation of samples. ***See*** individual foods

Assessors

Product analysis is performed by laboratory panellists who should not be confused with experts; both types of judges are selected and trained to provide assessments independent of their affective feelings.

Laboratory panels are asked to assess and express perceived differences between samples; experts go further and are able to associate perceived characters with what causes them, and evaluate their significance in terms of prices. Assessors are often recruited from inside firms or institutes.

Selection and Characterization of Assessors

With the support of the management, a call for volunteers for objective panels is made. Availability, interest, state of health, and smoking habits are immediately checked. During the first meetings, the role and usefulness of tasting are stressed. Explanations on sensory receptors, perception of food and fundamentals of liking are given to the group of subjects. Above all, it is necessary to define the subjects' ability to perceive, to differentiate, to evaluate intensity of perception and to name it. Sensitivity, however, is not the limiting factor in determining whether or not a subject will be able to evaluate the products. Most experts had no special acuity as beginners but became proficient in their respective fields by developing concentration and applying good sensory techniques. Subjects should then become conscious of their perception by introspection, and be able to express it.

Ability to Perceive

Procedures to establish individuals' perception are described in national and international standards, or in the literature (ISO 6658-1985 and ASTM STP 758).

Taste

Exercise starts on what are usually recognized as basic tastes: sweet, bitter, salty and acid (sour). Possible taste blindness or agueusias are thus detected. For screening purposes a subject can at several sessions be presented with series of threshold concentrations of these basic tastes and asked to identify them. If a correct answer is given each time, the subject possesses average or above-average sensitivity. If he or she fails to give a correct answer, he or she is less sensitive than the average population. For introspection training, the subject can be presented with clearly perceived concentrations of the four basic tastes. The subject is asked to mark, on a drawing of the tongue, the exact spot where he or she perceives the specific tastes, concentrating on the inside of the mouth to determine these perceptions.

Odours

The subject's ability to identify odours, and possible odour blindness or anosmia are examined. A range of odorants, usually chosen within the domain of odours related to the type of product to be studied, is presented. Correct and incorrect answers are recorded. Perception of flavour components is explained by presenting a flask containing a concentrated syrup covered with a tight membrane and with a short straw inserted. First, the subject sips the air through the straw and tries to identify the aroma from his or her retronasal perception. Then, after opening the flask, the subject sniffs air and compares the odour perceived through the direct nasal pathway with his or her first perception. Thus the subject clearly distinguishes the odorant component of a food aroma. An untrained person would confuse it with the taste.

Other

If necessary, subjects' ability to perceive and name colours are tested. Subjects are also tested on their ability to distinguish consistencies. This aspect is usually developed during training rather than during the characterization step. Information about a subject is shared with him or her and kept in a data base in order to create a pool of potential and defined tasters, and for easy consultation. This information, given voluntarily by the subject, is absolutely confidential. Selection is made in relation to individual abilities and according to specificities of the products to be tested.

Training of Assessors

Laboratory Panels

The amount of training needed depends on the type of test to be carried out. For simple differentiation it may involve as little as 1 or 2 h; for complex profiling it may take weeks or months. Unless the assessors need to familiarize on solutions representing basic tastes for physiological or chemical studies, training takes place with the product.

Training is carried out in order to achieve the following goals:

1. Apply good procedures in product handling for

Table 1. Tasting procedures for beverages

Features observed	Wine (cold)	Coffee (Hot)
Visual approach: aspect and colour	The taster examines the appearance by bringing the glass to eye-level, then stirs the wine in the glass to observe the way it flows on the sides	The taster examines the coffee in the cup
Aroma		
The first nose	The taster sniffs above the glass to perceive the aroma	The taster sniffs above the cup to perceive the aroma
The second nose	The taster twists the wine and puts his or her nose in the glass to sniff the released aroma	The taster stirs the coffee with a spoon and sniffs the released aroma
Flavour	The taster sips a mouthful of wine and tastes in-sucking, warming and absorbing some air in the mouth. The internal overpressure pushes some flavour-saturated air through the retronasal pathway	The taster takes a spoonful of hot coffee and sips it rapidly and firmly in order to obtain a spray in the mouth; this allows the most volatile components to reach the nose through the retronasal pathway with the overpressure applied
Taste	While sucking the wine, the taster analyses the taste	Having the coffee in his or her mouth, the taster analyses the taste
Body	The taster examines the apparent consistency and the mouth-feel elicited by the wine	The taster examines the mouth-feel provoked by the strength and the astringency of the coffee
Persistency	After swallowing or spitting out, the taster examines the impression which remains in his or her mouth	After swallowing or spitting out, the taster examines the impression which remains in his or her mouth

tasting, concentrate on product characteristics, and develop an appropriate language (vocabulary).

2. Discover the product varieties and their technological treatments. Develop ability to differentiate, quantify intensity, and become confident in one's own judgement.

3. Become familiar with the test itself and the specific questionnaire, and try to reach a consensus with all the other panellists in scoring intensities of one or several descriptors.

Handling and Strategy of Observation

If the type of product does not correspond to well-established knowledge, the trainee is asked to describe samples of the product using his or her own terminology. This is followed by group discussions on the main features perceived in these products. Individual comments and strategy for observation are discussed. The importance of the sniffing approach is emphasized.

For solid substances, the texture is important. The perception of the first bite, the evolution during chewing, the swallowing consistency and the lingering perceptions (e.g. aftertaste) must be carefully studied in their natural sequence.

When the product has a traditional approach to its sensory analysis, e.g. a wine or a coffee, a procedure is exercised (see Table 1).

Ability to Differentiate Varieties and Treatments

After the presentation of the products and the description of their specificities, origins and varieties are commented on with the trainee. The ability to differentiate characteristics is exercised using triangle tests. Sometimes it is necessary to highlight one characteristic by masking another one.

The product characteristics provoked by technical treatments are then introduced to the trainee who is asked to differentiate them again using triangle tests.

Typical defects associated with growth, crop conditions or technical treatments are presented at different levels of intensity. The ability to quantify intensity is trained by ranking and scoring series of four or five concentrations of each characteristic.

At each step of the training, exercises are decoded with the trainee who is asked to retaste knowing the results to help memorization. Coefficients of correlations between given concentration and individual scores reflect the subject's proficiency in these exercises.

Training with the Questionnaire

The questionnaire is used to accustom trainees to the sequence of the questions and the structure of the scales. This is particularly important with questionnaires for profiling. Exercises are repeated in order to check reproducibility of response, but training should not be prolonged too much in order to prevent the participants becoming bored with a job which is not his or her main task.

A second selection, taking account of subject's performances, can be made at the end of the training.

Guidelines for training panellists are presented in Fig. 1.

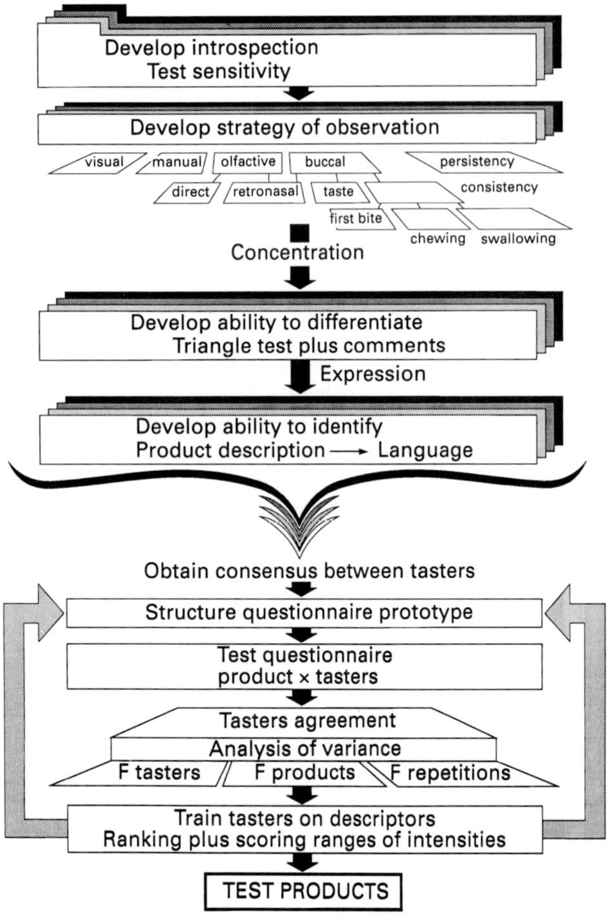

Fig. 1 Guidelines for training panellists.

Laboratory panellists are periodically submitted to checks, to increase their knowledge, to remind them of the references' specific notes and to avoid drifts.

Experts

Experts are usually trained individually by coaching with a senior expert who explains each step following a similar line. They become able to rate samples on quality, as: excellent, good, fair, poor and bad, according to the magnitude and significance of particular defects. Thus they develop an idea of the product value which they can express in terms of price.

Laboratory Design

Tasting is very much affected by surrounding effects, such as smells and noise. A quiet and neutral environment must be ensured. It is recommended that a sufficiently large area, in relation to the number of products to be treated daily, is allocated to this activity. This area must be divided into sections: preparation and storage area, tasting area, meeting area, and office.

Preparation and Storage Area

A section for the preparation of the samples should be equipped with counters and tables, sinks with hot and cold water, washing-up machines, cooking apparatus, microwave oven (if necessary), refrigerators and freezer. Cupboards are necessary to store glass or plastic utensils, the latter only if they do not release any flavour or interact with the product. Cups and glasses of standardized shape to taste tea and wine have been created (ISO 3591 (1977)–ISO 5494 (1978)). It is also necessary to keep questionnaire forms ready for use as well as standard laboratory equipment.

An adjacent room to store samples and for shelf-life studies might be necessary.

Tasting Area

Tasters must not see the preparation of samples before tasting. Therefore the preparation room should be close to the tasting area but separated from it. The individual panel booths should be slightly overpressured, to prevent outside odours from entering, and well ventilated. They should be air-conditioned to guarantee a constant temperature, particularly for tasting samples in which texture is very sensitive. Lighting of 323–538 lx on the surface of the table should be as near daylight as possible, diffusing an even light and not casting any shadow on the samples presented to the assessors. The use of coloured lights should be allowed to mask differences in appearance which may interfere with the perception of the characteristics in question. Talking of other activities should not be allowed in this area.

Tables and stools in the tasting room must be comfortable so that the taster is able to concentrate on his task. A cuspidor with running water is recommended for spitting out samples after examination. A dedicated computerized system is very useful for data acquisition, further decoding and on-line treatment of results. The decoded information might be fed back to the tasters for training purposes while their samples are still in front of them for retasting. Such equipment must not encumber the space or distract the assessors from their main task, which is the assessment of the samples (Fig. 2).

For some categories of products, appropriate equipment exists; for example, in coffee tasting, tasters sit around rotating tables on which hot samples are presented in series (Fig. 3). Tasters take a spoonful from each bowl, taste, spit it out and turn to the next one. In this way each sample passes in front of each taster who has to perform at regular intervals. This type of equipment is frequently used for quality control in production or for evaluation prior to purchase.

In some multiproduct plants, daily produced samples are presented on large tables. Tasters walk around and examine the series; each one is presented with its

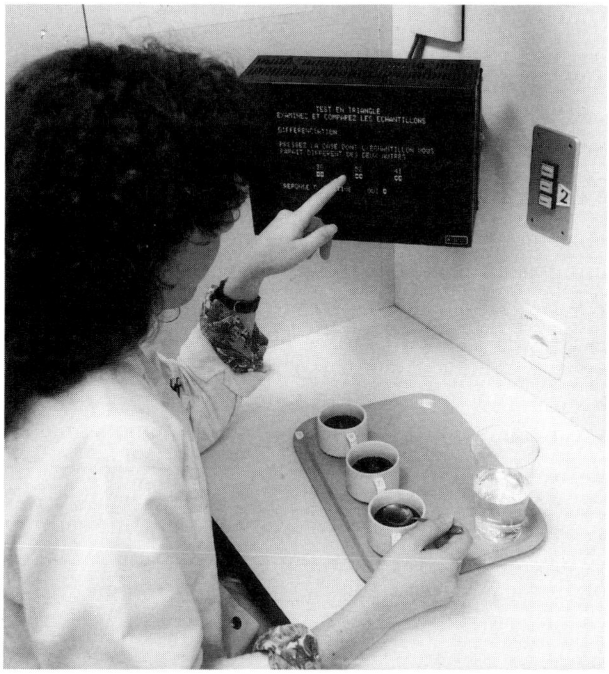

Fig. 2 Taster responding to a triangle test on the screen of a data acquisition system in a tasting booth.

Fig. 3 Rotating table for coffee tasting.

reference sample which represents the relative specifications for production.

Meeting Area

Beside the testing area, a meeting area should be available for group activities with the panel.

Office

An office close to the tasting and preparation areas is necessary for the panel leader who has to conduct interviews with the project leaders requesting sensory analysis and to discuss the final reports.

Floor plan standard specifications exist at national and international levels (ISO 5395-1983 and ASTM STP 913). An example of a floor plan is given in Fig. 4.

Preparation and Presentation of Samples for Assessment

Preparation and presentation of samples are also related to the type of products and purpose of the evaluation. In general, samples should be prepared in sufficient quantities, served in equal amount, in similar appearance and at an appropriate controlled temperature. They should be clearly coded and presented with the scoresheet and a glass of rinsing water, usually on an individual tray. Some specific procedures concerning tea, wine, spices, etc. are laid out in national and international standards. Tastings in a factory or in a control laboratory are not usually served under the same controlled conditions as, for example, in a development and research laboratory. Tasters can walk around the tables to examine uncoded samples.

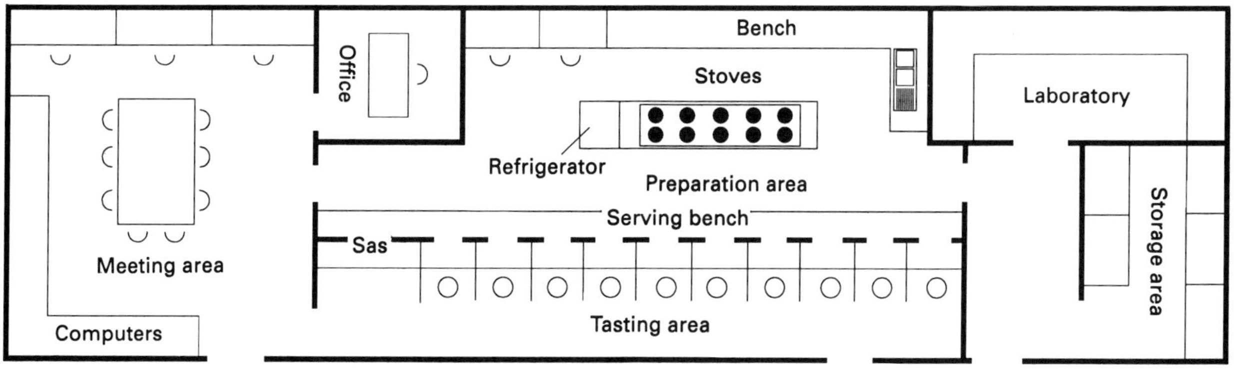

Fig. 4 Layout of sensory analysis laboratory (Sas, airlock).

Bibliography

Amerine MA, Pangborn RM and Roessler EB (1965) *Principles of Sensory Evaluation of Food*, p 602. New York: Academic Press.

Bodyfelt FW, Tobias J and Trout GM (1988) Sensory testing panel, an overview. *The Sensory Evaluation of Dairy Products*, pp 489–526. New York: Van Nostrand Reinhold.

Daget N, Voirol E, Resenterra P and Cabi-Akman R (1984) A system of data acquisition for sensory testing: the tactile plasma screen. *Development in Food Science: 10 Progress in Flavour Research*. Proceedings of the 4th Weurman Flavour Research Symposium, Dourdan. Amsterdam: Elsevier Science Publishers BV.

Gatchalian M (1981) Requirements for sensory analysis. In: Diliman (ed.) *Sensory Evaluation Methods with Statistical Analysis*, pp 98–142. Quezon City: College of Home Economics, University of the Philippines.

Jellinek G (1985) *Sensory Evaluation of Food: Theory and Practice*, pp 23–34, 308–331. Chichester: Ellis Horwood.

Meilgaard M, Civille GV and Thomas C (1989) Selection and training of panel members. *Sensory Evaluation Techniques*, pp 163–179. Boca Raton, Florida: CRC Press.

Peynaud E (1983) La formation des dégustateurs in le goût du vin. *Le Grand Livre de la Dégustation*, p 188–197. Paris: Dunod.

Société Scientifique d'Hygiène Alimentaire, Institut Scientifique d'Hygiene Alimentaire (1990) *L'Organisation Pratique de la Mesure Sensorielle en Evaluation Sensorielle, Manuel Méthodologique*, pp 44–87. Paris: Technique et Documentation Lavoisier.

Stone H and Sidel JL (1985) *Sensory Evaluation Practices*. New York: Academic Press.

Nicole Daget
Nestlé Research Centre, Vers-chez-les-Blanc, Switzerland

Sensory Difference Testing

Sensory difference tests are used to determine whether or not two (or more) samples which are of the same basic type, but known to be physically different in some way, are likely to be sensorially discriminable. Difference tests are widely used in the food industry for quality control/assurance and product-development purposes. Quality control/assurance functions include routine comparison of incoming ingredients against a purchaser's benchmark and routine monitoring of finished goods against 'standard' product. In product development, difference tests are used to monitor the effects of ingredient or process changes. *See* Quality Assurance and Quality Control

Methods in Brief

The three main types of difference test used to make comparisons between two samples are the Duo (also known as the paired or paired comparison test), Duo-trio and Triangle tests.

Duo Test

The sensory assessor is presented with two samples (A and B), side-by-side and asked which has the most (or least) of a specified sensory characteristic, for example *Which is the sweetest?* Half of the assessors should be instructed to try A first then B, and vice versa for the other half. The samples are identified to the assessor using 'neutral' codes, such as 3-digit random numbers. With only two samples to choose from, the probability that a sensory assessor will select one sample or the other by chance alone is 1 in 2 (i.e. $p=0{\cdot}5$). This is probably the least practical and most abused of the three methods because it assumes that the technologist conducting the test knows *a priori* which sensory characteristic will differ between the two samples (otherwise it cannot be specified). It must also be assumed that this sensory characteristic is the only source of difference and that all of the sensory assessors involved will recognize that characteristic.

Duo-trio Test

This also compares two samples (A and B). The assessor is first presented with a sample marked 'Standard'. This is used as a reference and can be either A or B. The assessor is then presented with a pair of coded samples, one of which will be A and the other B, and instructed to '*Select the sample which is* most different *from the Standard*', or alternatively, '*Select the sample which is* most similar *to the Standard*'. The assessor must not be instructed to select the sample which is different from (or the same as) Standard – see 'Basic Principles of Sensory Response Behaviour' below.

A and B should be designated as the 'Standard' an equal number of times. Also, the samples following the 'Standard' can be presented in two possible orders; A followed by B, or B followed by A. This means that there are in fact four possible combinations, each of whch should occur an equal number of times:

Standard = A, followed by A then B
Standard = A, followed by B then A
Standard = B, followed by A then B
Standard = B, followed by B then A

Again, with only two samples to choose from, the probability that a sensory assessor will select one sample or the other by chance alone is 1 in 2 (i.e. $p=0{\cdot}5$).

Triangle Test

The assessor is presented with two samples of one type and one of the other (i.e. AAB or BBA) and should be instructed to select either '*The two samples which are most similar*' or alternatively, '*The sample which is most different*'. The assessors must never be asked to identify the two samples which are 'the same' or the 'odd one out' – see 'Basic Principles of Sensory Response Behaviour' below.

There are six possible sample combinations which must occur an equal number of times, in order to avoid bias:

AAB, ABA, BAA, BBA, BAB, ABB

With three samples to choose from, the probability that a sensory assessor will select any one sample by chance alone is 1 in 3 (i.e. $p = 0{\cdot}333$).

Other Methods

There are several other so-called multiple sample tests which follow the same principle as the triangle test. For example, the Two-from-five test involves two samples of one type and three of the other, where the objective for the assessor is to group the two samples of one type and three of the other. The probability of achieving this by chance is 1 in 20 (i.e. $p = 0{\cdot}2 \times 0{\cdot}25 = 0{\cdot}05$).

Which Method to Use?

Difference tests *do not* compare samples, they compare perceptions (mental images) produced by the samples. With foods, this generally means that the assessor must try one sample, then another and then compare the perception of the second sample with their memory of the first. Visual comparison and some forms of tactile comparison are obvious exceptions. When the difference test method involves more than two samples, the 'memory problem' becomes increasingly acute. The Duo Test facilitates the most straightforward comparison, but this method is not generally applicable for the reasons described above. Duo-trio is the next simplest psychological task, then the Triangle test, then Two-from-five. With the Duo-trio test, the assessor can try the 'Standard' then the first sample specified and make a comparison. The assessor can then re-taste the 'Standard' and try the second specified sample. So, at any one time, the assessor is only comparing one perception with the memory of the other. With the triangle test, the assessor will probably try to compare the perception of the most recently tasted sample with the memory of the other two (depending on how the assessor chooses to make the comparison).

Basic Principles of Sensory Response Behaviour

In the foregoing description of the Duo-trio and Triangle tests, firm guidelines were given as to how the sensory assessor should be instructed to perform the sensory task. For example, with the Triangle test assessors should *not* be instructed to, identify the '*odd one out*' or the '*two samples which are the same*'; the proper instructions are to identify the sample which is '*most different*' or the '*two samples which are most similar*'. Two very basic principles underly the reason for this.

(1) *The same stimulus can evoke a variety of sensory responses on different occasions.* For example, if a sensory assessor is repeatedly (but unknowingly) presented with a particular sugar solution, a variable sweetness response is likely to be obtained. This is partly due to the fact that the assessor will become adapted to the sweet stimulus and may perceive it as less sweet on successive presentations. Clearly, this has important implications for the Duo-trio and Triangle tests. These practical problems are ameliorated by making sure that the order of sample presentation is rotated and by instructing the assessors correctly (see above).

(2) *Two physically different stimuli can evoke two identical sensory responses.* If an assessor tastes two sugar solutions which differ in concentration, he or she might perceive them to be identical. In other words, the assessor cannot discriminate on that occasion. If this is extended to the Triangle test, the assessor would probably not identify the physically different sample as 'most different'.

Forced Choice

What happens when a sensory assessor cannot identify a sample which is '*most different from standard*' in a Duo-trio test or cannot identify a sample which is '*most different*' in a Triangle test? There are two schools of thought: (1) The assessor should be forced to make a decision, even if he or she ultimately guesses. This is known as the *forced choice* option. (2) The assessor should be given the option of declaring 'no decision', i.e. that he or she cannot discriminate amongst the samples.

The forced choice option should be adopted in *all* cases, simply because the 'no-decision' option provides the assessor with an easy way out and does not encourage discrimination. This unequivocally diminishes the discriminating power of the test and therefore should be avoided. The statistical procedures used to interpret the outcome of a difference test were specifically designed to account for the fact that some of the 'correct answers' are guesses.

How Many Assessors?

The sensory professional is regularly faced with the decision of how many people should be involved in a difference test. In the practical reality of the food industry, the question is usually posed as 'How few can I get away with?' The answer to both questions is governed by one guiding principle: The more people involved, the less is the risk of making a wrong decision based on the outcome of the difference test.

Those involved in the decision making process must decide what are acceptable levels of risk and then calculate the total number involved using a specific formula. This is covered in detail in the 'Hypothesis Testing' section. If there is a lot at stake, then compromising on numbers is unwise.

General Design Considerations

Imagine a scenario in which a coffee manufacturer wishes to change his production process in some way, for reasons of economy. It is critical to establish whether loyal customers are likely to notice a difference between the existing and revised products as any difference may be construed as a defect. How should this difference test be conducted? There are two schools of thought.

(1) Conventional wisdom suggests that this type of difference test should be conducted using screened and trained sensory assessors working under rigorously controlled conditions in a sensory testing laboratory. This means that preparation of the coffees (e.g. method of brewing, concentration of coffee, amount of milk and/or sugar, if any), order of presentation, tasting temperature, etc., can all be regulated to remove extraneous sources of variability. Sensory assessors taste the coffee in a specified manner; sometimes they will 'sip and spit' or alternatively they may instructed to swallow, but they will never drink more than a few mouthfuls of each coffee. This philosophy is adopted in the belief that if skilled sensory assessors cannot find a difference under carefully controlled conditions, then ordinary consumers are unlikely to perceive a difference.

(2) The alternative school of thought discounts the above 'laboratory-based' method as unrealistic and therefore invalid, for several compelling reasons. First it is argued that coffee is never consumed under such sterile and regulated conditions. Ordinary consumers make and drink coffee in many different ways. The particular set of conditions selected for the laboratory-based difference test may mask differences which are apparent to consumers who prepare and drink their coffee in a totally different way. Secondly, it is argued that differences may not become apparent when only relatively small sips of coffee are tasted. For example, consumers might become aware of a lingering, bitter aftertaste after having consumed several mugs of a coffee over a period of several hours. This aftertaste is unlikely to be apparent if only one or two sips are tasted and then spat out. Thirdly, it is argued that a consumer who has consistently purchased a brand over several years will have a very clear 'taste expectation', so that any slight departure from that expectation will quickly be picked up. In other words, they may well be more 'expert' in that product than a trained sensory assessor.

Consumer based methodologies have now been developed to emulate difference tests. For example, a consumer is given three coded jars of coffee to use in-home. Two of the coffees are physically identical and different from the third. The consumer is instructed to use the coffees in the normal way, over a period of several weeks perhaps, and to identify the two coffee samples which are most similar to each other (or the one which is most different). This is directly analogous to the laboratory-based triangle test.

The main drawback of using this type of procedure is that it has much higher inherent variability, so more people need to take part to counteract the noise in the data. However, the main philosophical point is that this noise must not be construed as error, but as natural or real variability. Clearly, this type of difference test is more time consuming, more difficult to organize and very much more expensive than any laboratory-based method. However, the proponents of this technique claim that increased validity is essential when high-risk decisions are to be based on the outcome of a difference test, so that the increased cost is usually justified.

Alternatively, it is also possible to conduct difference tests with consumers in a central location (e.g. a hall or some other central facility that is readily accessible to the public). This combines some of the benefits of using consumers but allows the technologist to exercise control over sample preparation and tasting conditions.

The decision as to who should participate in a difference test and how and where it should be conducted ultimately depends on the objective of the test.

Preference Testing

The fundamental principles of difference testing can also be extended to preference judgements. Perhaps the most common example is the Paired Preference Test, where the respondent decides which of two samples he or she prefers.

Bona fide consumers must be used for preference tests. Consumers (often referred to as 'respondents' by market researchers) should be selected to be representative of the section of the population at whom the product(s) is directed. There are *no* circumstances where it is valid to use experts or selected and trained sensory assessors to make preference judgements that can be generalized.

Paired preference tests are most valid when conducted under realistic circumstances of consumption. ***See*** Food Acceptability, Affective Methods

The Meaning of 'Difference'

Imagine a Duo-trio test where 65 out of 100 sensory assessors have identified the sample or product that is known *a priori* by the technologist to be physically different from the Standard, as being '*Most different from Standard*'. On this occasion, 35 from 100 people did not discriminate between the two samples. Let us assume that these are genuine non-discriminators. Of the remaining 65 some unknown proportion (x) probably guessed correctly and therefore did not genuinely discriminate between the samples. The remainder $(65-x)$ are genuine discriminators.

What does this mean? Clearly, some people can discriminate between the two samples and others cannot, so there is no logical basis for declaring the samples to be either 'the same' or 'different' without qualifying the decision using an estimate of probability. Had only 45 out of 100 assessors given the expected answer in the above Duo-trio test, then there are likely to be fewer genuine discriminators. Conversely, if the result had been 85 from 100, there would probably be more genuine discriminators. This leads to two assumptions which are fundamental to understanding difference testing:

1. The larger the number of people who give the expected response in a difference test, the larger the number of genuine discriminators.
2. The more genuine discriminators, the larger the perceptual difference between the two physically different samples.

From a practical point of view, the technologist will need to make decisions based on whether or not the samples under test are likely to be different. Hypothesis tests are used to assist this decision-making process.

Hypothesis Testing

The starting point for any difference test *must* be the hypothesis test which will ultimately be used to decide whether or not the samples or products under scrutiny are likely to be discriminable.

Setting up a Hypothesis Test

A hypothesis test is formulated by first of all establishing the null hypothesis (H_0) and then the alternative hypothesis (H_a). The two hypotheses are mutually exclusive. In sensory difference testing, the hypotheses would be formulated as follows:

H_0 The physically different samples are perceptually *similar*.

H_a The physically different samples are perceptually *different*.

The object of the difference test is to provide evidence which will convince the sensory professional to select one hypothesis or the other.

There are two types of risk associated with this decision-making process. The first type of risk, known rather unimaginatively as 'error of the first kind', is the risk of wrongly rejecting the null hypothesis (i.e. generalizing that there is a perceptual difference when there probably is not). The probability of making this type of error is usually referred to as 'alpha' (α). Values of α range from 1 (i.e. error is certain) to infinitely small (i.e. error is very unlikely). The second type of risk, known as 'error of the second kind', is the risk of *not* rejecting the null hypothesis when it is probably incorrect (i.e. concluding that there is no perceptual difference when there probably is). The probability of making this type of error is known as 'beta' (β).

Alpha and beta are inversely related (although not directly), so if α is minimized, β will tend to increase, and vice versa. However, low levels of both α and β are possible if the total number of assessors (n) is increased, hence the previous assertion that the more people involved, the less is the risk of making a wrong decision based on the outcome of the difference test.

Which Type of Error is most Important?

Consider the following three scenarios:

1. A manufacturer of breakfast cereal wishes to substitute one ingredient with another to decrease the cost price of the product and hence improve the profit margin. However, it is critical that regular purchasers of this product do not perceive a difference as this may be construed as a defect, which might ultimately reduce overall sales and hence reduce total profitability. This is one of the most common uses of difference testing.

In this situation, it is absolutely critical that the manufacturer does not conclude that there is *no difference* when there probably is, so β must be minimized. All too often, this is not realized and the sensory professional will seek to minimize α. With a fixed number of respondents (usually too few), this has the effect of increasing β, often far beyond acceptable levels of risk.

2. A manufacturer of instant coffee has established that a slight increase in 'roasted' character will improve general acceptance of the product. He needs to be sure that his manufacturing process can consistently produce coffee which has higher levels of roasted notes, or the attempted improvement will be futile. In short, he needs to be sure that a difference really exists in the modified product.

With this scenario, it is important that the manufacturer does not conclude that there is a difference when in reality there is none. This means that he needs to minimize the risk of wrongly rejecting the null hypothesis, i.e. to minimize α.

3. A manufacturer of milk chocolate has shown that a change of cocoa butter improves his product. However, the new type of cocoa butter is more expensive, so he needs to be sure that the increased cost will confer real benefits in preference. In this case, the manufacturer would use a paired preference test and must seek to minimize α.

The most logical way of proceeding is to decide which type of error is most important and set the probability at an appropriate level. The probability of the other kind of error is then set at a slightly higher level. Ultimately, the probability levels for both types of error must depend on the degree of risk (i.e. what might be lost if the wrong decision is made).

Null and Alternative Hypotheses Distributions

Imagine a Duo-trio test conducted on two samples (A and B) which are physically similar. (This would of course be a thoroughly pointless exercise but it serves as a good illustration.) The probability of identifying A or B as the '*Most different from Standard*' is one chance in two or $p_0 = 0{\cdot}5$. This does not mean that if 100 people were involved in such a test that exactly 50/100 would associate A (or B) with the Standard. However, if the same test were repeated many times with different groups of 100 people, then on average, the number of people associating either A or B and Standard would tend towards 50. This happens because there is nothing other than chance causing the assessor to associate A or B with Standard.

If, on the other hand, A and B are physically different, and this difference has some effect on what is actually perceived, the average number of times A or B would be correctly associated with Standard would tend to increase. In other words, the number of correct associations of one sample or other with Standard would increase beyond 50/100 (i.e. $p_a > 0{\cdot}5$). This means that there is something other than chance guiding assessor judgements – a genuine perceptible difference which is evident to some but not all of the assessors. As this physical difference between A and B increases, the number of people correctly associating A or B with the Standard will increase (i.e. $p_a >> 0{\cdot}5$), because more and more of the assessors are guided by genuine discrimination. If there is something more than chance guiding assessor judgements, then the alternative hypothesis should prevail. In the Duo-trio test this means that the probability level will be greater than 0·5 (i.e. $p_a > 0{\cdot}5$).

However, there is one unsolvable problem. If for example, 70/100 assessors correctly associated either A or B with Standard (in a Duo-trio test where A and B are known to be physically different), how can the number of genuine discriminators be established? (Some of the 70 will be genuine discriminators while others will have guessed.) In other words, at what point should it be decided that A and B are probably generally discriminable? In statistical terms, this means that p_a is known to be greater than 0·5, but the precise value of p_a is unknown.

In spite of the fact that p_a is never known for sure, it is possible to make some important generalizations which help in the construction of the hypothesis test:

1. The greater the perceptual difference between two samples (A and B), the greater the value of p_a (i.e. p_a will tend towards 1).
2. The smaller the difference assumed between p_a and p_0, the larger the number of assessors that will be required to participate in a difference test at specified levels of α and β. This is because A and B are effectively less easily discriminated.

For most purposes, it can be assumed that $p_a - p_0 = 0{\cdot}2$. (This is based on accumulated empirical evidence.) So for the Duo-trio test, $p_a = 0{\cdot}7$ is normally assumed.

Calculating the Number of Assessors

The total number of assessors required to participate in a difference test is dependent on α, β and $p_a - p_0$. (N.B: p_0 is the only parameter which is fixed – the other three must be judged, principally on the extent of the loss which will be incurred if an incorrect decision is made on the basis of the outcome of the difference test.) When the levels of all four parameters have been decided, it is possible to calculate the total number of assessors using eqn [1].

$$n = \left[\frac{Z_\alpha[p_0(1-p_0)]^{1/2} + Z_\beta[p_a(1-p_a)]^{1/2}}{p_a - p_0}\right]^2 \qquad (1)$$

where:

α = probability of Type 1 error (Z_α from normal deviate tables)

β = probability of Type 2 error (Z_β from normal deviate tables)

p_0 = mean probability of H_0 distribution ($p_0 = \frac{1}{2}$ for Duo-trio)

p_a = mean probability of H_a distribution ($p_a = 0{\cdot}7$; unknown but usually estimated as $p_0 + 20\%$ of total probability)

The total number of assessors required for various combinations of α, β and $p_a - p_0$ is shown in Table 1. (N.B: this refers to a 1-tail test only – see later.)

Usually, if α is set at 0·05 then β is set at 0·10 or vice versa. Table 1 shows that at these levels of α and β, the minimum number of judgements is calculated as 48 or 49 (assuming $p_a = 0{\cdot}7$). There are some circumstances when it would be permissible to use fewer than this number of assessors, but great caution is required.

In order to conclude that there is likely to be a

Table 1. Number of independent judgements required in a discrimination test (one-tailed, i.e. *a priori*, expectation that either $p_a > p_0$ or $p_a < p_0$) at specific levels of p_a, α and β.

	$\alpha=0{\cdot}05$, $\beta=0{\cdot}10$	$\alpha=0{\cdot}05$, $\beta=0{\cdot}05$	$\alpha=0{\cdot}05$, $\beta=0{\cdot}01$	$\alpha=0{\cdot}10$, $\beta=0{\cdot}05$
$p_a - p_0$	n	n	n	n
0·10	209	263	383	208
0·15	91	114	165	90
0·20	49	62	89	48
0·25	30	37	53	29
0·30	20	25	34	19

perceptual difference between samples A and B (i.e. to reject the null hypothesis), a minimum proportion and hence a minimum number of 'correct' results is required. This is calculated using eqns [2]–[4].

$$Z_\alpha = (p_1 - p_0)/\sigma_0 \qquad (2)$$

which can be re-expressed as

$$p = (Z_\alpha\, \sigma_0) + p_0 \qquad (3)$$

where

p_1 = proportion of 'correct' responses (unknown)
p_0 = mean probability of H_0 distribution ($p_0 = \frac{1}{2}$)
α = probability of Type 1 error (0·05)
Z_α = number of standard deviations (σ_0) between p_1 and p_0
σ_0 = standard deviation of p_0

and

$$\sigma_0 = [p_0(1 - p_0)]^{1/2}/n \qquad (4)$$

where n = number of assessors (90). Therefore

$$p_1 = 0{\cdot}59.$$

Minimum number of 'correct' responses $= p_1 \times n = 0{\cdot}59 \times 90 = 53$

Duo-trio Test – Worked Example

Imagine the scenario of the breakfast cereal manufacturer described above. A Duo-trio test was used to establish whether or not the existing product and the modified product are discriminable. If α is set at 0·05 and β at 0·01 (i.e. Type 2 error is most important), with $p_a - p_0 = 0{\cdot}2$, the minimum number of assessors required is 89 (eqn [1]/Table 1). The minimum number of 'correct' responses is 53/89 (eqn [3]). This means that if 53 or more assessors gave the 'correct' answer in the Duo-trio test, the null hypothesis would be rejected and it would effectively be decided that the products were generally discriminable (at the accepted levels of risk).

Using Fewer Assessors

Sensory difference tests are regularly conducted with fewer than the recommended number of assessors. This is always dangerous, because it means that Type 1 and 2 error (α and β) will probably be unacceptably large. However, these risks can be offset to some extent by other factors. For example, if a group of assessors is very experienced in working with a product, they should be aware of its various sensory nuances and therefore they should be aware of relatively small departures from normality. So by being confident in the abilities of the assessors the sensory professional may be willing to accept higher levels of α and β.

Another option is to use repeat judgements; for example, 25 assessors performing the task twice, rather than 50 independent judgements. This places great emphasis on the abilities of the 25 people involved. If an assessor is insensitive, or complacent in performing the sensory task, then the problem is amplified by the duplication. Clearly, replication carries very grave risks, so it is absolutely essential that the assessors are carefully screened and trained; they must know the product well and their performances must be routinely monitored.

One-tail and Two-tail Tests

In any Duo-trio test, the test sample which is physically different from the Standard would be expected *a priori* to be identified as '*Most different from Standard*'. Similarly, in a Triangle test, the physically different sample would be expected a priori to be '*The most different*' (not withstanding comments made under 'Basic Principles of Sensory Response Behaviour'). In both cases, it is expected that the fact that the sample is physically different would aid discrimination to some extent (or not hinder it at least). This effectively means that the proportion of 'correct' responses would on average be either the same or larger than that expected by chance. It is not anticipated that the proportion would on average be smaller. The statistical test need only check whether or not the proportion of correct judgements is larger than expected. This is known as a one-sided or one-tail test. If, in a Duo-trio test, 20/50 assessors got the 'correct' answer, there would be no need to conduct any form of statistical analysis; it is obvious that the samples are not generally discriminable.

Paired preference tests require two-tailed statistical tests. This is because there can be no *a priori* expectation which sample (A or B) is preferred, so the proportion of respondents who prefer A over B could logically be more, the same as, or less than chance expectation ($p = 0{\cdot}5$).

Paired Preference Test – Worked Example

Imagine the scenario of the milk chocolate manufacturer mentioned above. He must be sure that the more expensive cocoa butter will make the chocolate taste better, so he conducted a paired preference test to compare the two chocolates (A versus B). Since a great deal was at risk, he wisely set $\alpha=0{\cdot}05$, $\beta=0{\cdot}10$ and $p_a=0{\cdot}6$ (i.e. $p_a-p_o=0{\cdot}1$). Using eqn [1] he calculated that the total number of respondents required was $n=259$. (Because this is a two-tail test, α (0·05) is shared between the two tails of the distribution. This means that α in either direction is $0{\cdot}05/2=0{\cdot}025$, which effectively increases the required number of respondents from 208 (one-tail test, Table 1) to 259.)

In this case, 156 respondents preferred A (the new recipe) while 103 preferred B (the current recipe). Should he reject the null hypothesis and conclude that A is generally preferred to B? To do this, the minimum number of preferences for A is calculated using eqn [3]. In this case, a minimum of 145 preferences for A is required (and by implication a maximum of 114 preferences for B.) The result from the preference test (preferences for A=156; preferences for B=103) is outside the range (114–145), so the decision is made to reject the null hypothesis and it is concluded that A is generally preferred to B at the specified levels of error.

Practical Issues

This description of difference testing aims to provide the reader with a good working knowledge of principles and methods. Detailed practical issues such as methods of sample preparation and coding are described in the references listed below.

Bibliography

Amerine MA, Pangborn RM and Roessler EB (1965) *Principles of Sensory Evaluation of Food.* London: Academic Press.

Fritjers JER (1988) Sensory difference testing and the measurement of sensory discriminability. In: Piggott JR (ed.) *Sensory Analysis of Foods*, pp 131–154. London: Elsevier Applied Science.

International Standards Organization (1985) *Methods for Sensory Analysis of Foods: Part 1 – General Guide to Methodology* (ISO 6658).

Jellinek G (1985) *Sensory Evaluation of Food.* Chichester: Ellis Horwood.

Stone H and Sidel JL (1985) *Sensory Evaluation Practices.* London: Academic Press.

David M.H. Thomson
Mathematical Market Research Ltd, Reading, UK

Sensory Rating and Scoring Methods

Sensory Rating and Scoring Methods

Purpose of Rating and Scoring Methods

Rating and scoring methods provide the basis for quantification of much sensory information. Although these two terms are sometimes used interchangeably by sensory analysts, they have different meanings.

Rating refers to the quantification of information by the use of ordinal categories, while scoring is a more defined form of rating since it uses a numerical interval or a ratio scale, of which the properties are known. A scale can be defined as a measurement continuum, divided into successive units according to the properties associated with it.

There are many different rating and scoring methods used in sensory analysis, as illustrated both here and in other articles. In each case, these scales are physical measurement tools used to measure some sensory phenomenon perceived by individuals. Thus, implicit in using rating and scoring is that these scales provide meaningful representations of some psychological process or processes.

When considering rating and scoring methods, the reader should be aware that the experimental design considerations given to sensory analysis procedures should be observed. In this article, example forms are given with the illustrations of different scales to aid the reader in designing appropriate questionnaires.

Type of Scale

Four types of scale can be used to collect data: nominal scales, ordinal scales, interval scales and ratio scales.

Nominal Scale

A nominal scale is one where data collected are categorized by a name or a number. Each observation collected using these scales must fall within one of the categories. For example, 'canned', 'frozen', 'dried', 'chilled' and 'fresh' are five categories used to describe methods of food preservation. These categories have no logical ordering and, thus, the key point about nominal scales is that the different categories have no quantitative relationship.

Ordinal Scale

An ordinal scale is one which allows observations to be ordered according to whether they have more or less of a particular attribute. Successive numbers or words are used to indicate more (or less) of the attribute being measured. Ordinal scales do not allow the amount of difference between observations to be quantified. The nine-point hedonic scale (described later) is ordinal, as is ranked data.

Interval Scale

An interval scale is one where the distance between points on the scale is quantifiable. In many instances, the distance or intervals between points on a scale will represent an equal perceptual distance. For example, if the perceptual distance between 1 and 2 on a seven-point scale of sweetness was the same perceptual distance as between 2 and 3, 3 and 4, and so on, then this scale would have interval properties.

Ratio Scale

A ratio scale is one where the observations collected can be expressed as a percentage or ratio of each other. For example, a person eating 100 g of chocolate a day eats twice as much as a person eating 50 g per day. An example of a ratio scale in sensory analysis is magnitude estimation, which will be discussed later. The main difference between interval and ratio scales is that the latter has a true zero, whereas the zero point of an interval scale is arbitrary.

Data Collection Methods and Data Analysis

Nominal Data

Nominal data can be collected in a number of ways (e.g. Fig. 1), but common to all nominal data is that each observation can only fall into one category. The most logical first step in analysis and interpretation of the data, therefore, is to produce a histogram indicating the frequency of occurrence of each category. The next step is to determine whether more observations fall in one category than another, or if the distribution of counts for one sample is the same as for another. In either case, the usual method of analysis is by the χ^2 test.

A more advanced method for treating nominal data is the technique of multiple correspondence analysis. This method produces a multidimensional spatial representation of the relationship between samples and attributes, which is often a useful summary of the data.

Rating Methods

Rating methods involve the quantification of perceived sensations by the use of scales.

Category Scales

Category scales are widely used in sensory analysis, both for objective assessment and affective (related to preference or liking) response. Category scales for objective sensory measurement are most often unipolar, as it is the amount of a particular attribute which is being measured. An example of a unipolar scale for firmness is given in Fig. 2. A category scale may also be constructed such that each category has a verbal label attached to it, as illustrated in Fig. 3. Bipolar scales can be used to measure attributes such as texture, e.g. substitute the two anchors of Fig. 2 with 'soft' and 'firm', respectively. While these scales give the appearance of having interval properties, care should be taken in making this assumption without experimental evidence. Generally speaking, category scales provide ratings and not scores; thus, category scales provide ordinal data.

A well-known scale for affective measurement is the nine-point hedonic scale (Fig. 4). Variations of this rating scale exist, comprising fewer categories and the absence of the 'middle' category.

Continuous Line Scales

Continuous line or visual analogue scales take the form of an unstructured line as illustrated in Fig. 5. Such

Please taste the sample coded 457, and identify which of the four basic tastes you perceive.

Sweet ____________

Sour ____________

Salt ____________

Bitter ____________

Unable to identify ____________

Fig. 1 Taste identification using a nominal scale.

Please taste the sample coded 658, and indicate how firm it is by placing a tick in the appropriate box below.

Not firm — Very firm

0	1	2	3	4	5	6	7	8	9

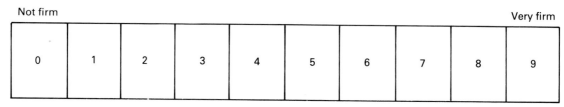

Fig. 2 Texture evaluation using a numeric category scale.

Please taste the sample coded 943, and rate the sweetness intensity by placing a tick in the appropriate box below.

Not sweet	Slightly sweet	Moderately sweet	Very sweet	Extremely sweet

Fig. 3 Flavour evaluation using a verbal category scale.

Please taste the sample coded 183, and indicate how much you like or dislike the flavour by placing a tick by the appropriate descriptor.

________ Like extremely

________ Like very much

________ Like moderately

________ Like slightly

________ Neither like nor dislike

________ Dislike slightly

________ Dislike moderately

________ Dislike very much

________ Dislike extremely

Fig. 4 Hedonic assessment using a nine-point verbal category scale.

Please taste the sample coded 563, and indicate how bitter it is by placing a vertical mark through the line below.

Not — Very

Fig. 5 Flavour evaluation using a continuous line scale.

scales are usually unipolar and measure the two extremes of an attribute, the extremes of which are represented as anchor points at the left and right of the scale. The length of the line is usually 100 mm, although this may be longer or shorter according to need. However, research conducted both by psychologists and market researchers has indicated that 100 mm is satisfactory. A line which is too short may decrease the discrimination achievable by the assessors, while beyond a particular length it is unlikely that an assessor will provide more precise information. Continuous line scales are often used with trained sensory panels as they allow the assessor more discrimination ability than the category scale. In spite of the fact that data collected on rating scales are measured by an interval ruler, strictly speaking the data should be considered ordinal, since it is usually impossible to prove the linear (interval) properties of every continuous line scale used for sensory measurement.

With a trained sensory panel, continuous line scales are usually assumed to be linear with the understanding that the extremes of the scale are slightly curved due to end-effects. If continuous line scales were truly interval, then the data would be scores rather than ratings. It would be unwise, however, to make this assumption with an untrained panel or a consumer panel without justification.

Data Analysis

Data which have interval properties can be analysed using parametric methods, providing certain assumptions are satisfied. These methods assume that the data are normally distributed, which implies a symmetric distribution (i.e. the mean, median and mode are equal) and that the data have interval or ratio properties. Parametric methods of analysis are powerful methods for sensory interval data as they enable more precise interpretation to be made of the results. There are a wide range of statistical methods available for interval data, depending on the question being asked. The simplest form of analysis is to calculate means and standard deviations, which measure the location and spread of the data, respectively. This can be taken one step further by calculation of a standard error and confidence interval for the mean. This allows the sensory analyst to make inferences about the mean of a sample with respect to the attribute being measured, or to test a hypothesis that the mean is equal to a particular value. t Tests can be used to test the hypothesis that two samples have the same mean values for a particular attribute, while analysis of variance will test whether more than two samples have the same mean values.

When several attributes are being evaluated, multivariate analysis procedures are often adopted. Techniques which can be used include principal component analysis, generalized Procrustes analysis, factor analysis, discriminant analysis, canonical variate analysis and cluster analysis.

Whilst data collected by rating methods are mostly ordinal, it is a common practice to use the parametric statistical methods discussed. However, these methods assume that the data are derived from an underlying normal distribution, which assumes that the data are continuous, i.e. they have interval or ratio properties. Clearly, by definition, ordinal data cannot be continuous. In fact, for each of the above-mentioned parametric statistics there is a nonparametric equivalent. These are medians (means), interquartile ranges (standard deviations), Mann–Whitney U test and Wilcoxon paired test (t tests), and the Kruskal–Wallis and Friedman rank test (analysis of variance).

Multivariate analysis on ordinal data can be achieved through correspondence analysis, which reduces the dimensionality (number of measured attributes) to a smaller number of dimensions which effectively describes the correspondence between samples and attributes.

Magnitude Estimation

Method

Magnitude estimation is a form of ratio scaling, where the perception of a specified attribute in one sample is measured as a ratio of the perception of that attribute in another sample. This method was introduced by Stevens in 1953 and one of its first applications in sensory analysis was to evaluate pleasantness of odours.

There are two main methods of collecting ratio data by this method. In the first method, one sample is designated the standard and allocated a whole positive (not zero) number (e.g. 100) to represent the perception of a specified attribute. The perception of this attribute in subsequent samples is represented as a fraction or multiple of the standard number. For example, if the sample evaluated is twice as sweet as the standard then it is allocated the value 200, if it is a quarter as sweet then it is allocated the value 25. It is often recommended that the standard sample should be one-third of the way along in the range of samples to be used in the experiment.

In the second method, the assessor allocates a positive (not zero) number to the first sample, and evaluates the perception of that attribute in subsequent samples as a ratio of the first sample. Whichever approach is adopted, magnitude estimation usually requires more training than the other methods of data collection. An example of a magnitude estimation form is given in Fig. 6.

Data Analysis

Data collected on a ratio scale, by magnitude estimation, require additional thought as they have to be transformed before analysis. In considering data analysis, it is important to recall that the data are ratios for each individual. Further, as each individual has used his own range within the ratio scale, the distribution of the data will differ from subject to subject. It is also likely that if an individual's data are plotted, the shape of the distribution of data about the standard sample will differ for each subject. Thus, not only are there scaling problems to contend with, but also inconsistency in the distributional plot of the data. For these reasons,

Please taste the reference sample (R), and allocate it a value of 100 to represent the sweetness intensity perceived.

Please taste each of the three samples in the order indicated. You are required to allocate each sample a number (not zero) indicating the degree to which it is more or less sweet than the reference sample.

R

100

749 ____ 381 ____ 526 ____

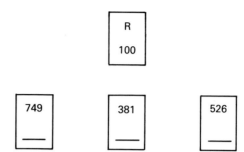

Fig. 6 Flavour assessment by magnitude estimation.

Please taste the four samples of chocolate, and rank them according to increasing intensity of cocoa flavour.

Rank	Code	
1	________	Least cocoa
2	________	
3	________	
4	________	Most cocoa

Fig. 7 Flavour assessment by ranking.

individual data cannot be assumed to come from a normal distribution, and further direct averaging of data would be misleading due to the scaling problems. Thus, data are transformed to ensure that each individual has in effect used the same scale. Several methods can be used to rescale magnitude estimation data: modulus equalization, modulus normalization and external calibration.

Modulus equalization requires that the ratings provided by each assessor are multiplied by a constant, with the aim of ensuring that the geometric mean of each individual is the same as the geometric mean of the grouped data. Modulus normalization is used if many samples are evaluated over a number of sessions, and the experimenter has built in common samples which appear at each tasting session. The information from the common samples is then used to rescale the data based on the geometric mean. Further details can be found in Moskowitz (1977) for this and the other two methods mentioned.

Averaging magnitude estimation data is not as straightforward as interval data. However, these data usually have a log-normal distribution and, hence, by taking the logarithms the distribution of the data becomes symmetric and hence more like a normal distribution. It is often easier to analyse logarithmic data. Parametric statistics, as described above, can then be used to analyse transformed magnitude estimation data if it is certain that the transformed data are normally distributed, otherwise nonparametic methods must be used.

Ranking

Ranking is neither a rating nor a scoring method, but as it is widely used in sensory analysis and provides ordinal data, it is included here for completeness. The method of ranking is used to order a number of samples according to increasing or decreasing perception of a specified attribute. Samples to be ranked need to be presented at the same time, and this limits the total number of samples that can be used in such a test, due to sensory fatigue. Ranking of five or six samples is often possible, although this number will depend largely on the nature of the samples; for example, chocolate has mouth-coating properties which make it difficult to rank many samples. An example of a ranking exercise is illustrated in Fig. 7. Data are analysed using nonparametric statistical methods, as described earlier in this article.

Bibliography

Chatfield C and Collins AJ (1980) *Introduction to Multivariate Analysis*. London: Chapman and Hall.

Land DG and Shepherd R (1988) Scaling and ranking methods. In: Piggott JR (ed.) *Sensory Analysis of Foods*. London: Elsevier.
Lebart L, Morineau A and Warwick KM (1984) *Multivariate Descriptive Analysis: Correspondence Analysis and Related Techniques for Large Matrices*. New York: Wiley.
Moskowitz HR (1977) Magnitude estimation. Notes on what, when and why to use it. *Journal of Food Quality* 3: 195–228.
O'Mahony M (1986) *Sensory Evaluation of Foods: Statistical Methods and Procedures*. New York: Marcel Dekker.
Peryam DR and Giradot NF (1952) Advanced taste-test method. *Food Engineering* 24: 58–61, 194.
Peryam DR and Pilgrim FJ (1957) Hedonic scale method for measuring food preferences. *Food Technology* 11(9): 9–14.
Piggott JR (1986) *Statistical Procedures in Food Research*. London: Elsevier.
Stevens SS (1953) On the brightness of lights and the loudness of sounds. *Science* 118: 576.

J. A. McEwan and D. H. Lyon
Campden Food and Drink Research Assoction, Chipping Campden, UK

Descriptive Analysis

In the past decade considerable attention has been given to descriptive analysis; more companies are learning about the usefulness of the methodology, and new methods have been proposed, offering the sensory professional a variety of options. As a methodology, descriptive analysis is the most sophisticated source of product information available in the context of providing a complete and quantitative description of a product's sensory properties. Such information can be considered in the same way as one obtains chemical or biological analyses of products. It is particularly important because it provides a focus for development efforts, it provides a basis for measuring the effects of a process or of ingredients, and it is essential for correlating consumer response behaviour and for identifying those product attributes that are most important to consumer preferences. Before discussing specific methods and their relative merits, it is necessary to define what is meant here by descriptive analysis.

Descriptive analysis is a sensory methodology that provides quantitative descriptions of products based on the perceptions of a group of qualified subjects. It is usually a total sensory description, taking into account all the sensations that are perceived – visual, auditory, gustatory, olfactory, kinaesthetic – when the product is evaluated. The evaluation could include product handling and use, and, in that sense, it is a total experience. The evaluation could also focus on one component of a product, such as aroma; however, there are risks in such a narrow focus, and these risks will be discussed later in this article.

The earliest practitioners of descriptive analysis were the brewmasters, perfumers, flavourists and other product specialists. The value of their information was appreciated by their employers because they not only described products and made recommendations about the purchase of specific raw materials, they also evaluated the effect of process variables on product quality (as they determined product quality), especially the determination that a particular product met their criteria (and that of their company) for manufacture and sale to the consumer. These activities served as the basis for the foundation of sensory evaluation as a science, although at the time it was not considered within that context. It was possible for the expert to be reasonably successful as long as the marketplace was less competitive. However, all this changed with the scientific and technological developments in most consumer products industries and, more specifically, the food and beverage industry. This explosion of technology, the use of more refined raw materials, and related developments made it increasingly difficult for experts to be as effective as in the past, and certainly not effective in a global sense. The expert's functions were further challenged by the increasingly competitive and international nature of the market-place, the rapid introduction and proliferation of new products, and the sophistication of the consumer's palate. At this same time period, sensory evaluation began its development, providing specialized test methods, scales for measurement, and statistical procedures for the analysis of results. Formal descriptive analysis (and the separation of the expert from sensory evaluation) received its major impetus from 'flavour profile', an approach which demonstrated that it was possible to select and train individuals to describe the sensory properties of a product in some agreed sequence, leading to actionable results without dependence on the individual expert. The method was also distinctive in that no direct judgement was made concerning consumer acceptance of the product, although most investigators assumed consumer acceptance based on a result. The method attracted considerable interest as well as controversy; however, there was no question as to its importance to the development of the field. Since then, other methods have been developed.

Background

In this discussion, it is useful first to consider the more fundamental issues on which all descriptive methods are based. In particular, it is necessary to consider the following: the basis on which subjects are selected to participate; the training they are given after selection, including the development of a descriptive language for the array of products being evaluated; the idea of subjects providing quantitative judgements like an instrument; and the analysis of the data to provide actionable recommendations.

Such a discussion is particularly important because there are numerous decisions made by the sensory professional (the panel leader) in the course of the organization and use of a descriptive panel to evaluate products. These decisions and actions derive from that person's perspective and understanding of (1) the perceptual process in general and (2) the specific descriptive process. Unlike discrimination and acceptance tests, where subjects exhibit choice behaviour in a global sense, i.e. all perceptions are taken into account to yield a single judgement, the descriptive test requires each subject to provide numerous judgements for each product.

In this discussion three methods are contrasted: quantitative descriptive analysis (QDA), spectrum analysis and free-choice profiling; these methods are used to a much greater extent than others, and therefore warrant most attention.

Screening

A descriptive test involves relatively few subjects (as few as 10 to as many as 20) and there must be good evidence that the specific differences obtained are reliable and valid, and not the result of spurious responses from one or two insensitive or more variable subjects. This ability to rely on a limited number of subjects derives from the knowledge that the subjects are qualified to participate based on their sensory skills, i.e. they have been screened. This means that an individual has demonstrated his or her ability to perceive differences at better than chance among an array of products from the type of products that will be tested. Failure to carry out adequate screening of each individual raises serious questions as to that person's ability to describe differences or score the intensities for an array of products with any degree of confidence and, of course, this will have a direct impact on the test results. It enables one to use the lowest number of subjects without loss of information.

There are very basic differences in the approaches that are used for the qualification of subjects, and panel leaders must decide what procedure(s) they are willing to use in their descriptive testing. It has been recommended that screening is based on sensitivity to particular chemicals (e.g. sweet, sour, salt and bitter stimuli or a selection of odorants) or on various personality tests. It is surprising that sensitivity tests continue to be recommended (flavour profile and similar methods) in spite of earlier evidence that it provided no indication of a subject's subsequent performance. In Spectrum Analysis, a combination of threshold testing followed by discrimination with actual products is recommended. Alternatively, Free-choice Profiling requires no subject screening and no measures of subject variability are provided.

In QDA, the discrimination methodology with products of the type to be tested has been and continues to be the most effective procedure for identifying subjects who can, and especially those who cannot, perceive differences (after first determining that they are regular users of the specific product category). It is also the most parsimonious in terms of time required to identify qualified subjects. Based on results of 20 to 30 discrimination trials completed over 3 or 4 days (sessions last about 90 min), one can select a pool of discriminators from amongst a group of totally naive individuals. Working with qualified subjects, screening tests are not needed unless one is using a different product category. The screening procedure, at this stage, is intended primarily to eliminate the nondiscriminator and, secondarily, to familiarize all subjects with the sensory properties of the products. Only after training and data collection can one empirically determine the effectiveness of the screening and subsequent training activities.

This screening procedure is intended to be product-category-specific, as is the subsequent training effort. For example, if one were testing Cheddar cheeses, then the products for screening would be Cheddar cheeses, including those for the test, and not other cheese types such as Brie. In this way one minimizes the risk of including insensitive and/or unreliable subjects in a test for a particular product category. This does not mean that one is confronted with the task of having to requalify subjects every time a different set of products within that category is evaluated. Most subjects are capable of meeting the qualifying criteria for many different categories of products. Experienced descriptive subjects (e.g. those who have participated in at least two tests) will have learned how to use their senses and how to evaluate products within a category being evaluated without experiencing a substantial loss of sensitivity, and are therefore very likely to be able to evaluate products beyond those for which they were screened.

Screening subjects for more than one type of product makes sense because it will eliminate any additional screening before each new category of products. This does not guarantee that a subject will stay qualified, nor does it eliminate the need for the other stages of training. In addition, it does not excuse the panel leader from monitoring subject performance before a test. Infrequent use of subjects requires that they be rescreened, in which case much effort will be wasted. A potential problem with screening subjects on several products is the assumption that these individuals will be willing to participate in future tests. If used too frequently (e.g. daily) there is every likelihood that the subjects will lose interest and thus not perform satisfactorily. It may be better to develop a larger pool of qualified subjects and reduce reliance on a small subset of people for all testing, or qualify small groups for each category depending on the anticipated need for the product information. Panel leaders must develop guidelines for subject use based on their company experiences.

In summary, it is necessary that subjects for a descriptive test demonstrate their ability to perceive differences at better than chance among the products that they will be testing. This will be ensured if products used are representative of the products (that will be tested). For inexperienced individuals, this skill can take as many as 20 to 30 trials to demonstrate, and it is very likely that about 30% of those who volunteer will fail to meet the chance probability requirement.

Training

Once screening is completed, training is initiated and it is here that one also encounters a diversity of approaches. While the training process is usually focused on the language (or attributes) used as a basis for scoring the products, there are other important activities, including the following: grouping attributes by modality; ordering them by occurrence within a modality; developing definitions for each attribute; familiarizing subjects with scoring the products; and identifying references that are helpful. If subjects are inexperienced and there is no list of attributes (developed and used by another panel), then the training effort will require the maximum amount of time, usually about 7–10 h. For these same inexperienced subjects, a period of about 7 h is usually required to familiarize themselves with an existing list of attributes. When experienced subjects are presented with an entirely new product category, about 5–7 h are required for training. These training times are intended solely as a guide, for the products and the skill of the panel leader also will have an effect. Regardless of the situation, the subjects work, at times individually but mostly as a group, to ensure that the attributes are sufficiently understood by one another, and that all of the product's characteristics have been fully accounted for. All of this is done under the direction of a panel leader who does not participate in developing the language. The QDA methodology was the first methodology and remains one of the few methods that excludes the panel leader from directly participating in the language development. This is true whether one is working with an experienced panel or with a panel that has never participated before. The panel leader's primary responsibility is to facilitate communications among subjects and organize results of their discussion. The panel leader who participates as a subject (as is allowed with the Spectrum method) is a biased subject because of his or her awareness of product differences in advance and the test objectives. Not only does this action communicate the wrong message to the subjects (subjects will tend to defer to the panel leader, whom they assume has the correct answer), but also the end result is more likely to be a group judgement rather than a group of judgements.

Developing a sensory language or using one that already exists is an interesting process and certainly one that is essential for the successful outcome of a test. For some, the language assumes almost mystical importance, such that a considerable body of literature has been developed in which lists of words are published for specific types of products. It appears that most of these lists were compiled by product technologists. Unfortunately, this is not very different from the efforts of product specialists or experts of 50 or more years ago when they developed quality scales and corresponding descriptions for products as part of an effort to establish food quality standards. Besides their interest in evaluation of their respective company's products, their technical and trade associations often formed committees for the express purpose of developing a common language for describing the flavour (or odour) of the particular category of products. For purposes of this discussion we chose the publication by Clapperton *et al.* (1975) on beer flavour terminology as an example of the efforts (of the association of brewers) to develop an internationally agreed terminology. These authors stated the purpose as follows: 'to allow flavor impressions to be described objectively and in precise terms'. Various literature and professional sources were screened for words (or descriptors) describing the flavour of beer, and, after reviews, the committee arrived at an agreed list. As the authors noted, the issue of chemical names compared with general descriptive terms was resolved by the inclusion of both.

From a technical viewpoint, the use of chemical names was appealing because it was believed that they could be related to specific chemicals in the beer. An example would be the term 'diacetyl', which could be ascribed to a specific chemical, as compared with 'buttery', a less precise term that might be ascribed to several chemicals. By including both types of terms it was stated that the terminology would be of value to the flavour specialist (the expert) and the layman (the consumer). While the concept of an agreed terminology should have considerable appeal, careful consideration of this approach reveals significant limitations, especially in relation to its application beyond that technical panel. It is risky to decide *a priori* what words subjects should use to describe a particular sensation. The fact that product changes – formulation, process, or both – rarely yield a single sensation means that, for example, a flavour change would not be totally represented. To restrict subjects to specific words assumes that the language is unchanging, as are the meanings assigned to it. The frame of reference for an attribute is unique to each subject, and attributes requiring a very different frame of reference are difficult for inexperienced and/or nontechnical subjects to understand. While chemical terminology is supposed to have specific meaning to an expert, it is unlikely to have the same meaning to a consumer or to a trained subject (and often to another expert).

It is interesting to note that use of technical terminology extends training time and this appears to be related to the complex nature of this terminology, especially for subjects with no technical training. The Spectrum method provides the subjects with attributes, as well as references and designated intensity scores for those references, all of which are intended to enhance the testing process. The use of standard attributes and references implies that product variables will not produce unique sensations, and forces the subjects to limit the value of their perceptions relative to what has been perceived in the past by others. The training effort is extended over a period of 3 months; this raises questions as to its responsiveness and overlooks the inherent variability in the subjects (over time) and the references that are being used (that also change over time). Regardless of the source, a language that does not provide for subject input is unlikely to yield uncomplicated sensory responses. Subjects are influenced by the information given to them, and are much less likely to question it, because of its source. While it can be argued that such approaches are merely intended to help panel leaders or the subjects, the temptation is very strong to use this approach rather than allowing subjects to use their own terminology.

To the student of the history of psychology, descriptive analysis can be considered as a type of introspection, a methodology used by the school of psychology known as structuralism in its study of the human experience. Structuralism required the use of highly trained observers in a controlled situation verbalizing their conscious experience (the method of introspection). Of particular interest to us is the use of the method of introspection as an integral part of the descriptive analysis process and, specifically, the language development component.

However, the obvious difference is that products are included (in descriptive analysis) so that the subject's responses are perceptual and not conceptual. In addition, the subjects are encouraged to use any words they want, provided that they use common, everyday words, and that they define the meaning of each word-sensation experience, if for no other reason than to ensure that they will be able to score the products using a terminology with which they are familiar. While each subject begins with his or her own set of words, they work as a group, under the direction of a panel leader, to come to agreement as to the meaning of those words, i.e. the definitions or explanations for each word-sensation experience, and also when they (the sensations) occur. In addition, a standard evaluation procedure is defined. All of these activities require time; in the QDA methodology, there can be as many as four or five sessions, each lasting about 90 min. This amount of time is essential if the sensory language is to be developed and understood, and the subjects are capable of using it (and the scale) to differentiate the products. These sessions help to identify attributes that could be misunderstood, and also enable the subjects to practise scoring products and discussing results, on an attribute-by-attribute basis, under direction of the panel leader. Of course, all this effort cannot make subjects equally sensitive to all attributes. In fact, when evaluating products, subjects rarely, if ever, achieve complete agreement for all attributes, nor are they all equally sensitive, nor is it ever expected (if this did occur one could rely on the *n* of 1). The effectiveness of this effort can be determined only after a test has been completed; each subject has scored the products on a repeated trial basis and appropriate analyses have been done. The idea that there should be a specific number of attributes is, at best, questionable, as is the issue of whether or not one has the correct attributes. How does one know that all the attributes have been developed or that they are the right ones? The answer to the former is empirical (and in part answered in much the same way as one determines the number of angels that can occupy the head of a pin). The answer to the latter is also empirical, i.e. given a set of variables, do some attributes exhibit systematic changes as a function of those variables.

Alternatively, Free-choice Profiling claims that no training is required; subjects can use any words they want and the results are collated for analysis. However, the 'no training' is not necessarily correct in the sense that there is time and effort required by the subjects to develop definitions for the words; in more recent publications on use of the method, as many as 10 training sessions are used. Clearly, there are wide differences in what is defined as subject training insofar as concerns Free-choice Profiling.

The words used to represent sensations are nothing more than labels that provide a common basis for judging an array of products. There is no reason to assume that these words represent anything beyond that. Although, it has been suggested that the words represent concepts, and that for a descriptive panel to be effective, concepts must be aligned, i.e. subjects must agree on all the sensations (or attributes that represent those sensations) to be included in a concept if the results are to be useful. The process by which subjects discuss their judgements for an attribute, and the definition for an attribute, could be considered as concept alignment, which is an integral part of the QDA training. Whether this, in fact, is concept alignment remains to be demonstrated; however, it is clear that subjects can reach agreement on attributes and can reliably differentiate amongst products, after completion of training. How attributes are formulated in the brain and the true meaning of those attributes are issues that go well beyond descriptive analysis and sensory evaluation, in general.

In addition, it should be noted that subjects will not agree on all the sensations to be included, any more than there is agreement on all the attributes. The sensations are themselves interactive, leading to multiple words to represent them. The individuality of each subject (sensitivity, motivation, personality) further complicates or adds to the complexity of the process. As a result, a descriptive panel typically develops many more attributes (30 or 40 or more) than will be necessary to describe an array of products fully. However, the fact that there are many more attributes than are needed should not be unexpected or of concern.

In addition to a descriptive language and definitions, it may be useful to have references available for training or retraining subjects. Here, too, one finds different opinions as to the types of references and how they are to be used. For example, a comprehensive list of references and how they are to be used may be presented, including their respective intensity scores for scale extremes. Unfortunately, these references are based on commercially available products, all of which are variable in the normal course of production, in addition to the intended changes based on changing technologies, ingredients and/or market considerations. The author then advised the reader that these references and their intensities may change for the aforementioned reasons, in which case the original references and their intensities will have to be redetermined. What, then, is the value of such references? References have a role to play in helping subjects to relate to a particular sensation that is not easily detected or not easily described. However, references should not introduce any additional sensory interaction or fatigue, or significantly increase training time. In most training (or retraining) situations, the most helpful references are usually a product's raw materials. Of course, there will be situations in which totally unrelated materials will prove helpful to some subjects, and it is the panel leader's responsibility to obtain such materials. There will also be situations in which no reference can be found within a reasonable time period. A panel leader should not delay training just because a reference cannot be found. While most professionals agree that references are helpful, there is no evidence that without them a panel cannot function or that results are unreliable and/or invalid. We have observed that so-called expert languages usually require numerous references, and subjects take considerably longer to learn (this language) than they do a language developed by themselves. This should not be surprising, if one thinks about it. After all, references are themselves a source of variability; they introduce other attributes unrelated to their purpose, and increase the potential for sensory interactions. The panel leader must therefore consider their use with appreciation for their value as well as for their limitations, and must decide when and what references will be used. In our experience they are of limited value, for use in the language development and training or retraining activities. In retraining, or when adding new subjects to a panel, they are helpful in enabling these individuals to experience what the other subjects are talking about and possibly to add their comments to the language.

The Spectrum method makes the greatest use of references, while the QDA and Free-choice Profiling use a more *ad hoc* process, in which references are used only where they are helpful to the subjects.

Scoring

All three methods are quantitative and use various types of scales. The QDA methodology makes use of a line scale (a graphic rating scale) and the concept of functional measurement to obtain intensity judgements. The scale is a 15-cm line with two vertical lines, each placed 1·27 cm from the scale ends. Above each of the vertical lines are words that designate scale direction and intensity for that attribute. For the scale to be used effectively, the subjects must be provided with ample opportunity to practise scoring products and become comfortable with using the scale to differentiate products. Successful use of the scale also depends on being able to provide the subjects with the extremes of product differences as part of the training effort. However, no attempt is made to require the subjects to use the same part of a scale, only that each subject be consistent with him or herself. Because each product is scored on a repeated trials basis by each subject, it makes no difference what part of the scale is used to differentiate the products. This use of a single scale is in contrast with the spectrum method, in which more than one type of scale may be used; for example, a scale will be used in which specific product standards are provided with numerical anchors and products will be scored relative to those standards. Free-choice profiling uses a line scale and, like QDA, the subjects practise scoring products to familiarize themselves with the testing process.

Design and Analysis

The design and the analysis of a descriptive test are equally important for a successful test. A descriptive test yields a large sensory data base (in comparison with a discrimination or an acceptance test) including both univariate and multivariate components and, as such, it permits a wide range of statistical analyses to be performed. One of the main features of the QDA methodology was the use of a comprehensive statistical analysis of the data, which represented a significant development for sensory evaluation. With the availability of statistical packages and of personal computers, panel leaders have unlimited and low-cost resources, providing an on-line capability for obtaining means, variance measures, ranks and pairwise correlations, and for factor analysis, multiple regression, cluster analysis, discriminant analysis, etc.

All three descriptive methods incorporate replication (or repeated trials) into their test designs. However, the QDA method is the only one that specifies a minimum number; however, this too is a decision that the panel leader makes based on the products, the expected degree of difficulty and past experience. The QDA recommendation of four to six was derived from empirical observations that variability exhibited a decline as the number of replicates increased from one to four. However, there was a decreasing return as the number of replicates increased versus the amount of time required for the subject to complete all the evaluations.

When using the analysis of variance to analyse descriptive data, the repeated trials design is a more sensitive model, and thus more likely to yield significant product differences. In addition, this design provides a basis for measuring the performance of each subject on an individual attribute basis. Neither the Spectrum nor the Free-choice methods specify these kinds of analyses, and the latter method relies primarily on the use of a specific factor analysis technique – Procrustes analysis – as a means of describing the results. However, it has been observed that the indiscriminate use of this analysis could lead to misleading conclusions, and that using random numbers yielded a better separation of products than did the obtained responses. The problem, in this example, was related to two issues: the first was the wide variation among subjects and the second was the experimental nature of the analysis, which the author noted 'will always produce an order even when no such order is present in the ingoing data'.

All sensory data need to be analysed in some detail as a basis on which to reach conclusions about product differences. As noted before, the descriptive test uses a limited number of subjects and one must be very confident about their behaviour before one initiates any extrapolations or investigations about the underlying structure of the data. Specifically, the use of replication is essential, regardless of the skill of the subject; it is a cost that cannot be forfeited without serious compromises in any conclusions that are reached about results.

In the QDA methodology, the one-way analysis of variance is used to identify subjects and attributes for which sensitivity is reduced and/or there is a significant interaction. This information is then used before the next test to help these subjects and those attributes. In this way one can identify attributes and subjects that are not contributing to product differences and could be dropped from subsequent tests. However, the attribute decision is a panel responsibility; the panel leader can only make suggestions.

As with any readily available resource, statistics are often misunderstood and/or are misused, particularly when responses are highly variable or when the panel leader confuses use of some of the multivariate procedures as evidence for the validity of results. In other instances, investigators use factor analysis and/or clustering techniques as a basis for excluding subjects who are not in agreement with the panel or to eliminate attributes that they believe are not being used to differentiate products or are used as substitutes for the same sensation. An example of this is MANOVA (multivariate analysis of variance) and factor analysis which may be used during training as a means of reducing the number of attributes on the scorecard (from 45 to 12). One must be careful about using procedures that, *a priori*, decide which subjects or attributes will be used. After all, the subjects are still in training and to have them score products and use the results as a basis for reducing either, or both, may be premature. This approach can lead to a group of subjects who are more likely to agree with each other, but how does one know that those who agree are correct. How does one differentiate subjects who are discriminators from nondiscriminators if, for example, the nondiscrimination is of 30% or 40% of the attributes? One of several objectives of training is to enable subjects to learn how to evaluate products using the attributes that they determined were helpful to them (at that stage of the training). To proceed from one session to another and substantially reduce the list of attributes is communicating a message that there is a correct list of words and, with a little patience, it will be provided. Using factor analysis to reduce the number of attributes during training is troublesome because it assumes that attributes highly correlated to a factor are measuring the same component of the product and, therefore, one or two attributes can represent the others. However, attributes correlated to a factor are only an indication of some common basis which assigns those attributes to that factor; it does not imply causality. It is interesting to note that one goes to considerable effort to encourage subjects to contribute words to use in a scorecard and then to devise procedures to eliminate about 75–80% of them before the actual test is performed, thus sending the message to subjects that there is a correct list. As mentioned previously, not all subjects will use all the attributes to the same extent in differentiating products, and eliminating too many of them substantially increases the likelihood of reducing sensitivity and overlooking product differences (an example of type 2 error).

After a test one can use univariate and multivariate procedures to help to identify subjects who are experiencing difficulty with specific attributes and/or with use of the scale. This information is then used prior to the start of the next test so as to further clarify the testing process. If one intends to identify which attributes are most important and, in turn, the extent to which one can reduce the attribute list for a specific application, e.g. quality control.

Conclusions

Descriptive methods are an established methodology for all sensory programmes. Results from a descriptive test provide precise quantitative descriptions of products, and this information has many applications that range from product development to advertising claims.

In the past decade much attention has been directed to development of methods that provide the sensory professional with alternatives on how the test is organized and fielded, and how the results are analysed. There are currently three very different methods described in the literature that concern the four main components of a descriptive test: subject selection; training; data collection; analysis of results.

For subject selection the Spectrum method relies on a combination of flavour profile followed by product screening, which is the primary method used with the QDA methodology. The former takes several weeks, while the latter requires about 5 days. Alternatively, Free-choice Profiling requires no screening and claims that anyone can participate (implying no sensory skill required).

For training, the QDA methodology requires that subjects develop the list of words and their ordering by modality, the definitions, and the evaluation procedure. No standards are used in this training. The time required to complete these activities is about 6–8 h for a totally new panel and half or less this time for an experienced panel. The spectrum method trains subjects for about 3 months, and provides an extensive language along with standards and intensity values. Free-choice profiling claims that no training is required, but current descriptions of the method indicate that training is used.

For data collection, QDA recommends at least three replicates, while the Spectrum and Free-choice methods indicate that replication is appropriate without further discussion as to how much or in what way the data will be used. It is clear that those who have used these methodologies and those who will use them in the future will make their own modifications.

Data analysis is particularly important for determining the quality of the basic information. While current interest is very high in the use of clustering techniques and other multivariate procedures, such analyses should be considered only after the quality of the data base has been assessed. Both the QDA and the Free-choice methods describe specific analyses of their respective data, while the Spectrum method leaves the specific choices to the panel leader.

For the sensory professional there is much to choose from when considering a descriptive test. Whatever methodology is chosen, there are many aspects that require decisions from the professional. It is the authors' contention that these decisions should be influenced by the following considerations:

1. All subjects should be screened for their sensory skill using products from the category being tested.
2. About 30% of the people who volunteer to participate in these screening tests will not meet minimum requirements of at least 50% correct matches (in a discrimination test).
3. Subjects should use words that are derived from common, everyday language to describe products.
4. Subjects should practise scoring products using the aforementioned words as part of the training effort.
5. Subjects should develop definitions or, if using a previously developed scorecard, they should have the opportunity to modify the definitions.
6. Subjects should score each product on a repeated trials basis with a minimum of three replications.
7. Data analysis must provide measures of subject reliability on an attribute basis, means and variance measures for products, and tests for significance.
8. Multivariate procedures should be used where the data warrant the effort.

Sensory evaluation has achieved considerable recognition as a source of unique product information. Much of this derives from the use of descriptive analysis methods. This discussion is intended as a perspective of the development and use of descriptive analysis, as well as a review of the main features of methods currently in use.

Bibliography

Boring EG (1950) *A History of Experimental Psychology* 2nd edn. New York: Appleton.

Cairncross WE and Sjöström LB (1950) Flavor profile – a new approach to flavor problems. *Food Technology* 4: 308–311.

Caul JR (1957) The profile method of flavor analysis. *Advances in Food Research* 7: 1–40.

Civille GV and Lawless HJ (1986) The importance of language in describing perceptions. *Journal of Sensory Studies* 1: 203–215.

Clapperton JR, Dalgliesh CE and Meilgaard MC (1975) Progress towards an international system of beer flavor terminology. *Master Brewers Association of America, Technical Quarterly* 12: 273–280.

Harper RM (1972) *Human Senses in Action*. Edinburgh: Churchill Livingstone.

Huitson A (1989) Problems with Procrustes analysis. *Journal of Applied Statistics* 16: 39–45.
Ishii R and O'Mahony M (1990) Group taste concept measurement: verbal and physical definition of the umami taste concept for Japanese and Americans. *Journal of Sensory Studies* 4: 215–227.
Johnson PB and Civille GV (1986) A standard lexicon of meat WOF descriptions. *Journal of Sensory Studies* 1: 99–104.
Jones FN (1958) Prerequisites for test environment. In: Little AD (ed.) *Flavor Research and Food Acceptance*, pp 107–111. New York: Van Nostrand Reinhold.
Lyon BG (1987) Development of chicken flavor descriptive attribute terms aided by multivariate statistical procedures. *Journal of Sensory Studies* 1: 99–104.
Mackey AO and Jones P (1954) Selection of members of a food tasting panel: discernment of primary tastes in water solution compared with judging ability for foods. *Food Technology* 8: 527–530.
Marshall RJ and Kirby SPJ (1988) Sensory measurement of food texture by free-choice profiling. *Journal of Sensory Studies* 3: 63–80.
Marx MH and Hillix WA (1963) *Systems and Theories in Psychology*. New York: McGraw-Hill.
Meilgaard M, Civille GV and Carr BT (1991) *Sensory Evaluation Techniques*. Boca Raton, Florida: CRC Press.
Muñoz AM (1986) Development and application of texture reference scales. *Journal of Sensory Studies* 1: 55–83.
O'Mahony M, Rothman L, Ellison T, Shaw D and Buteau L (1990) Taste descriptive analysis: concept formation, alignment and appropriateness. *Journal of Sensory Studies* 5: 71–103.
Powers JJ (1988) Current practices and applications of descriptive methods. In: Piggott JR (ed.) *Sensory Analysis of Foods*, 2nd edn, pp 187–266. London: Elsevier Applied Science.
Rainey BA (1986) Importance of reference standards in training panelists. *Journal of Sensory Studies* 1: 149–154.
Smith GL (1988) Statistical analysis of sensory data. In: Piggott JR (ed.) *Sensory Analysis of Foods* 2nd edn, pp 335–379. Essex: Elsevier Science Publishers.
Stone H and Sidel JL (1992) *Sensory Evaluation Practices*, 2nd edn. Orlando, Florida: Academic Press.
Stone H, Sidel J, Oliver S, Woolsey A and Singleton RC (1974) Sensory evaluation by quantitative descriptive analysis. *Food Technology* 28: 24, 26, 28, 29, 32, 34.
Williams AA and Arnold GM (1985) A comparison of the aromas of six coffees characterized by conventional profiling, free-choice profiling and similarity scaling methods. *Journal of the Science of Food and Agriculture* 36: 204–214.
Williams AA and Langron SP (1984) The use of free-choice profiling for the evaluation of commercial ports. *Journal of the Science of Food and Agriculture* 35: 558–568.

Herbert Stone and Joel L. Sidel
Tragon Corporation, Redwood City, USA

Appearance

Flavour and aroma are important in sensory evaluation but appearance is a vital attribute because it greatly influences the overall impression and acceptability of a product. Appearance analysis is essential to food product development, quality assessment, and the maintenance of product quality throughout the distribution system. Evaluation of appearance involves analysis of both geometric attributes and colour attributes. Foods vary greatly in their visual characteristics depending on their physical composition and optical qualities. The colour perceived is determined by the proportion of wavelengths reflected or transmitted. Colour systems and colour measurement devices have been developed in an attempt to qualify colour.

Product appearance is psychologically related to the potential sensory satisfaction and perceived value of that product. Purchasing decisions are frequently based on appearance. The observer determines quickly and unconsciously whether a product is uniform or irregular, shiny or dull, yellow or brown. Deviations from what is normally encountered are associated with deterioration of quality, lack of maturity, or a disproportionate quantity of ingredients.

Visual perception involves three elements: object, light source, and observer. The appearance of any given object involves all aspects, both physical and psychological, which characterize that product: colour, gloss, size, contour, brightness, clarity, translucency.

Visual perception results from an intricate combination of factors, including how an object modifies the light striking it and how the eye interprets the light reaching it from the object. Light entering the lens of the eye is focused on the retina. Rods and cones within the eye then convert the light to a neural impulse which travels to the brain via the optic nerve.

The eye–brain mechanism is extremely sensitive and discriminating. For example, humans can identify about 300 different colours but can discriminate among 5–10×10^6. However, the human eye has its physiological limitations. Visual acuity is affected by brightness and adaptation, and improved with illumination.

The visual qualities of an object can be divided into two categories: (1) geometric attributes and (2) chromatic or colour attributes.

Geometric attributes include the size and shape of an item. Length, thickness, conformation, width, particle size, volume, height, shape, and distribution of pieces contribute to the overall appearance of a product. With some products, beverages for example, degree of effervescence observed on pouring is a means of visually evaluating carbonation.

Geometric attributes also include the spatial aspects that cause light perception to vary from point to point over a surface of uniform colour. Visual attributes associated with the spatial distribution of light include gloss (sheen), haze and turbidity, opacity, transparency, and translucency.

Foods vary greatly in appearance depending on their optical qualities. Light is either absorbed, reflected or transmitted through the food depending on the physical structure and chemical nature of its components. Reflected light leaves the sample from the same side as that which is illuminated. The amount of light reflected depends on surface texture, the refractive index of the material, and the angle at which the beam strikes the surface. When all light bounces off the surface, the object is seen as very shiny. Transmitted light passes through the sample and is viewed from the exit side. Thus reflected and transmitted light are the visual stimuli received by the eye. Light distribution can be further classified as follows:

1. Specular reflection. Objects that have an optically smooth surface, such as a sheet of metal or uncrumpled foil, appear shiny because light is reflected specularly. The light is highly directional rather than diffuse; it strikes the sample and rebounds from the illuminated surface at right angles.
2. Diffuse reflection. Light can be diffused or reflected in many directions and at odd angles from the surface. Most foods reflect light diffusely because they are opaque with irregular surfaces. Fibres and pigments are responsible for diffusion. Flour, cheese, cornflakes, cooked roast beef, and mashed potatoes are seen by diffuse reflection. When light is scattered in all directions the product has little gloss.
3. Specular transmission. In some foods, such as wine, jelly, apple juice and vegetable oil, light penetrates the sample without deflection. Transparent foods allow the specular transmission of light.
4. Diffuse transmission. A hazy or turbid beverage contains particles that diffuse or impede the transmission of light. Light penetrates the product, scatters and exits in many directions.

Most foods possess a combination of optical characteristics so that light reacts in more than one type of distribution pattern. A shiny apple appears glossy because it has both smooth and irregular surfaces, causing light to reflect both specularly (at right angles) and diffusely (at odd angles). Translucent foods, such as orange juice, apple sauce and fruit jam, are seen by both reflected and transmitted light.

The appearance of a product is affected by both the spectral conditions of light and geometric conditions of viewing; the conditions of observation must therefore be standardized and reproducible. Conditions to be controlled when evaluating product appearance include the following:

1. Intensity of light source.
2. Type and wavelength of light (incandescent, fluorescent, daylight).
3. Angular size of light source.
4. Angle of incidence, i.e. the direction from which the light strikes the sample (direction of illumination).
5. Angle of viewing.
6. Background.

Physical Evaluation of Colour

Colour, a primary aspect of appearance, serves as an indicator of quality, maturity and degree of cooking, and plays a major role in whether a food is accepted or rejected. Certain foods are acceptable only if they fall within a specific range of colours. Vivid hues, colour surprises such as yellow-fleshed watermelon, and unnatural colours are seldom acceptable. Age, culture, socioeconomic background and previous experiences can influence one's reaction to a specific food colour. White corn is considered a delicacy by some people, but others know and prefer only the yellow variety. Colour influences one's ability to identify flavour. For example, a lime-flavoured beverage coloured red is deceiving and may not be perceived as citrus-flavoured because of the uncharacteristic appearance.

Colour is a psychological phenomenon that exists in the mind of the observer. It is not a physical property but a perception resulting from the effects of light waves rebounding from or passing through the item. The concept of colour results from the interaction of the light source, object being viewed, eye and brain. Without light one sees no colour; colours are merely terms that describe various blends of electromagnetic energy.

The spectrum of visible radiant energy referred to as light is fairly narrow, with wavelengths ranging from 380 to 770 nm. Wavelengths of 380–400 nm are violet; those from 400 to 500 nm produce the effect known as blue. Wavelengths of 500–600 nm are perceived as green and yellow, while those longer than 610 nm are red. Light waves differ from other types of radiant energy in that they are visible. Wavelengths shorter than 380 nm or longer than 760 nm do not initiate a visual response.

The spectral composition of light leaving any sample is determined by the optical properties and pigments of that product. Pigments and other materials absorb light. Wavelengths reflected or transmitted, i.e. those not absorbed, are visible and thus determine the colour brought to the brain from the light receptors in the eyes. ***See*** Colours, Properties and Determination of Natural Pigments; Colours, Properties and Determination of Synthetic Pigments

Consider a glass beaker containing vegetable oil. The oil appears transparent because no particles in the oil or glass diffuse the penetrating light. The oil looks yellow because it selectively reflects lightwaves in a particular,

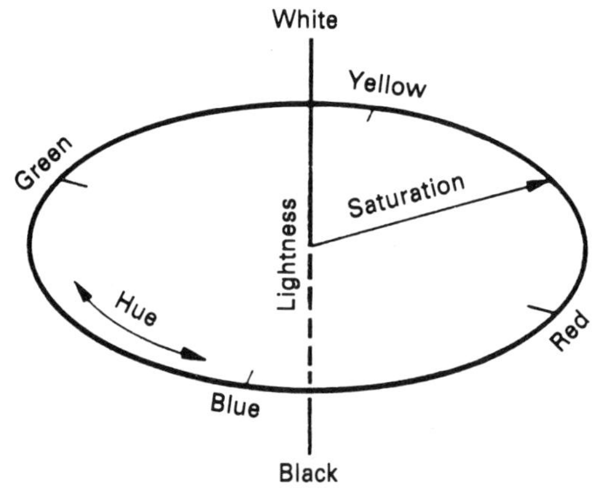

Fig. 1 Three-dimensional colour system.

narrow colour range while absorbing all others. Likewise, a tomato appears red because it absorbs all light except red.

Corn starch is white because it scatters light by multiple reflection; all wavelengths are reflected equally with none being absorbed. When reflection is minimal and absorption is the dominant process, dark colours result. If all wavelengths are absorbed, black results.

Colour of any given food sample will be affected by preparation and presentation. Texture, surface moisture, sample thickness, and exposure to light and air may influence colour perception. Samples that are representative of the product should be evaluated.

Colour can be thought of as a three-dimensional characteristic (Fig. 1) composed of two chromatic attributes and one luminous factor. To define colour precisely one must specify these dimensions.

1. Hue (usually known as 'colour') refers to whether a sample is red, orange, yellow, green, blue, or violet. The proportion of various wavelengths determines the hue perceived.
2. Colorimetric purity or saturation is the amount of colour present. Terms used to express depth, strength of hue or vividness refer to purity. Saturation reflects how different the colour is from grey. A pastel colour is less saturated than an intense colour. In the Munsell system, described below, purity is referred to as chroma.
3. Luminous intensity. The lightness or brightness of an object refers to its capacity to reflect or transmit light. Lightness is called value in the Munsell system.

The three-dimensional concept of colour can be clarified if one arranges the colours of the rainbow in a hue circle. Saturation is factored in if the centre of the circle is considered neutral grey and the most saturated colours are at the outside edge, furthest from the centre. However, a two-dimensional description of colour is inadequate. Consider prepared mustard and a fresh lemon. They are both the same hue and both are highly saturated but they differ in 'colour'. A third dimension, lightness, is needed to complete the description.

Analysis of small colour differences is important in food products. The effect of modifying ingredients or processes can be determined by measuring colour differences. Monitoring colour changes resulting from exposure to light, heat and frozen storage is an important aspect of quality control for many products.

Following are some aspects of colour perception to be considered when evaluating food:

1. Panelists often score the colour of two samples similarly even when filters are used to mask differences because filters mask hue but not necessarily brightness or purity.
2. Adjacent or background colour affects visual perception. Food placed against a colour background can appear tinged with that colour.
3. Colour perception is influenced by surface characteristics such as gloss, moisture and texture. High gloss tends to mask colour differences.
4. One's ability to differentiate colour is reduced under low-intensity light.
5. Visual acuity differs among individuals; some are highly sensitive and can discern subtle differences in colour.
6. Colour is best observed from directly above the product whereas gloss should be evaluated from an angle.
7. Humans are generally more sensitive to small differences in hue than to small differences in saturation.

Colour-measurement Systems

The human eye is unable to make reproducible, quantitative judgements about colour. The eye merely judges composite visual appearance without differentiating between light diffusion or absorption. However, the physical nature of colour can be measured. Light reflected or transmitted can be measured, stated in terms of numerical values, and compared to a standard. Colour-measuring instruments are best used for determining colour differences between samples, rather than the absolute colour of a product.

Instruments designed to measure appearance can be divided into two categories: those that measure geometric attributes (i.e. gloss) and those that measure chromatic attributes. Different instruments are necessary for measuring each optical type of food (translucent, transparent, opaque).

As with human evaluation, the conditions of observation must be standardized. Placement of the specimen, type and intensity of light source, and conditions of illumination must be controlled.

The colour-measurement devices described below are

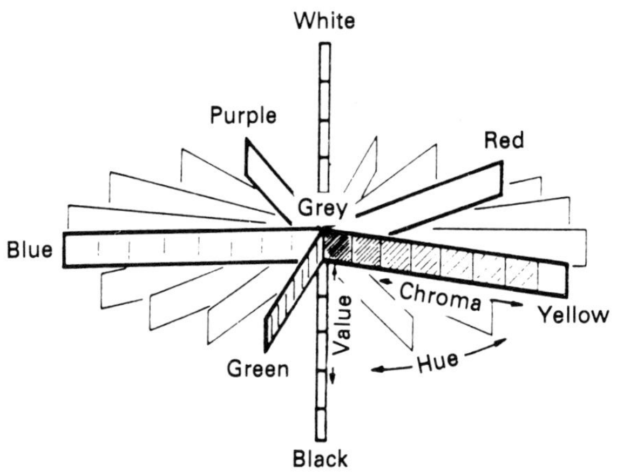

Fig. 2 Munsell colour system.

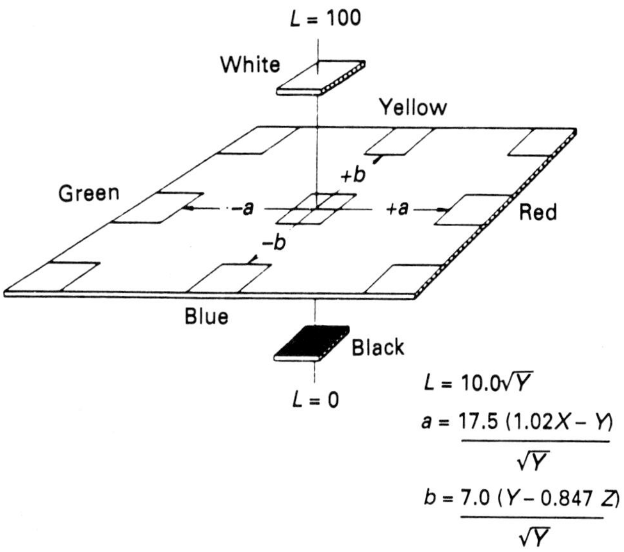

Fig. 3 Hunter *L,a,b* colour solid based on the opponent colour system.

designed to provide meaningful correlation with visual perception. They are based on colour systems developed in an attempt to quantify colour. Colour values in one system can be converted to those of another system by using special equations and numerical tables. Three widely used systems are described below.

CIE System

The Commission Internationale de l'Eclairage (CIE) system depends on the three primary colours – red, blue, and yellow. Colour is defined by mathematical computations relating percentage reflectance of the specimen to reflectance values for three filters that simulate the response of a standard observer (the average human eye). The filters are X (amber), Y (green) and Z (blue).

This system of specifying colour by X, Y, Z (tristimulus values) is based on experiments with a random sample of individuals with normal colour vision. The physical stimuli causing each colour to be perceived by the brain were quantified. In this system colour is characterized in terms of X, Y, Z for all three stimulus modes: incandescent light, noon sunlight and overcast daylight. Colour designations in other systems can be converted to CIE values by references to published tables.

Munsell System

The Munsell system and the CIE system differ in the way that colour is described. However, Munsell numbers can be converted to CIE values.

In the Munsell system (Fig. 2), hues are arranged around a vertical axis. The vertical axis is value (lightness) ranging from black (0) to white (10). Chroma (saturation) is described in units radiating outward from the central axis. Vivid, highly saturated colours are on the outside. Mint green is closer to the neutral point of grey (in the centre of the circle) than to green. Red is closer to the outer edge than pink.

The perceived dimensions of colour are expressed in numerical terms and written as hue-value/chroma. A green apple, for example, might be described as 5GY8/4. The fruit is greenish-yellow in hue (5GY), light in value (8/) and moderately weak (/4) in chroma. A wine red plum might be defined as 9RP3/6, i.e. reddish purple in hue (9RP), dark in value (3/) and moderate in chroma (/6).

Colours are determined by visually comparing the test sample with standardized colour chips or papers or by using a disc colorimeter. With this device discs of coloured paper are rotated to create an additive effect in the eye. This blend of reflected light can be compared to the food sample.

Opponent-colour System (*L,a,b*-type)

In the opponent-colour system the dimensions are black to white along a vertical axis, blue to yellow (going from front to back) and red to green (going from right to left). Uniform colour scales based on this system have been developed. On the widely accepted Hunter scale (Fig. 3), L measures darkness–lightness, a represents redness if positive, greenness if negative and b corresponds to yellowness if positive and blueness if negative. In other words, greenness or redness can be specified by $-a$ or $+a$, respectively. Instruments such as the Hunter colour and colour-difference meters calculate L, a and b functions automatically so that direct readouts are obtained. Opponent-colour dimensions correspond most closely to the actual visual signals transmitted to the brain from the eye.

Measuring Colour

Instruments for measuring the chromatic attributes of a product differ in the methods of wavelength selection and type of evaluation device. The instruments can be divided into two basic categories:

1. Spectrophotometers measure light reflectance or transmittance over a range of wavelengths.
2. Tristimulus instruments use filters that simulate the standard functions of the eye and give values for colour in terms of *X*, *Y*, *Z*, or *L*, *a*, *b*.

Tristimulus colorimeters can be thought of as psychophysical analysis instruments because they give measurements that correlate with impressions of the brain and eye. Spectrophotometers measure the physical properties of light distributed by the sample but give little information about how the observer perceives colour. ***See*** Spectroscopy, Visible Spectroscopy and Colorimetry

Spectrophotometers measure the spectral distribution of light by a food sample. Either reflectance or transmittance can be measured. Spectrophotometers are used in chemical analysis and ingredient identification. Food colour of transparent juices, colour extracts in solution, translucent syrups, and gels can be measured with the spectrophotometer. One limitation of this instrument is that the viewing area is quite small.

Operation of the spectrophotometer is simple. Monochromatic light is directed against the surface of a sample. Light from the sample exits through another opening and strikes a photo tube. Readings are taken at various wavelengths. The spectral response obtained can be compared to a standard, or percentage reflectance at each wavelength can be converted mathematically to *X*, *Y*, and *Z* (red, green and blue respectively). These values are used to calculate chromaticity coordinates *x* and *y* which are then plotted on a CIE *x*–*y* chromaticity diagram. Colour is further defined with *Y* (lightness). Although precise, this method is tedious unless direct readout is available.

Tristimulus instruments have light sources, filters and photodetectors, and a device that computes the chromatic dimensions of colour (*a* and *b*, or *x* and *y*). Tristimulus instruments have a larger viewing area than the spectrophotometer. Mirrors and lenses project the light onto the specimen. Light from the sample is simultaneously directed to three filters, and produces three signals which are detected by photosensors and converted to *x*, *y* and *z* values.

Specifying colour in tristimulus values is equivalent to locating that colour in three-dimensional space. Any colour can be defined by a statement of *x*, *y*, and percentage of *Y*. The spectral colours are plotted on *x*,*y* coordinates to obtain a CIE *x*–*y* chromaticity diagram. Brightness (luminosity) can be plotted as a percentage of *Y*.

The Hunter colour and colour-difference meters are tristimulus instruments widely used in food colour analysis. Either *Rd* (luminous reflectance) or *L* (visual lightness) measurements can be taken: *Rd* is related to the CIE system; *L* is related to the Munsell system.

Operation of the Hunter colour-difference meter is relatively simple. An intense, controlled source of light is focused against the sample. Light reflected from the sample is diffused within a white-coated enclosure. The reflected light is measured as *Rd* or *L*. Colour filters are located around the enclosure. Photocells in the back of the filters generate *a* and *b* readings.

Readings of *Rd* (or *L*), *a* and *b* are given for each test sample. These values can be reported as observed or as relative values (*a*/*b*), or they can be converted into CIE designations. Colour differences are determined by comparing the readout with previous readings or with a standard.

If instrumental evaluation of colour is impractical or impossible, other approaches can be implemented. Food can be visually compared to standardized coloured objects or guides, such as those used by government inspectors when grading butter, vegetables and fruit.

Single-number colour scales have been developed for rating colour quality of specific products, such as apple sauce, raw tomatoes, tomato juice, orange juice, and citrus fruits. Colour specifications, in terms of *X*, *Y*, *Z* or *L*, *a*, *b*, are converted to numbers which correspond to a particular quality or grade.

Unique methods may be developed for certain situations in the laboratory. Extracts, juices, artificially coloured liquids, or dried or frozen samples could serve as the standard reference. Photographs representing degrees of browning of cut fruit or baked products could also be used.

Analysing all appearance attributes, geometric and chromatic, is seldom necessary or practical. In most cases, it is appropriate to measure only those attributes relative to the problem or product.

Bibliography

Clydesdale FM (1976) Instrumental techniques for color measurement of foods. *Food Technology* 30(10): 52–59.

Farrel KT, Wagner JR, Peterson MS and Mackinney G (eds) (1954) *Color in Foods, A Symposium*. Washington, DC: National Academy of Sciences, National Research Council.

Francis FJ and Clydesdale FM (1975) *Food Colorimetry: Theory and Applications*. Westport, Connecticut: AVI Publishing.

Hunter RS (1975) *The Measurement of Appearance*. New York: John Wiley.

Kramer A (1976) Use of color measurement in quality control of foods. *Food Technology* 30(10): 62–71.

Kramer A and Twigg BA (1962) *Fundamentals of Quality Control for the Food Industry*. Westport, Connecticut: AVI Publishing.

Appearance

Little A (1976) Physical measurements as predictors of visual appearance. *Food Technology* 30(10): 74–82.
Mackinney G and Little AC (1962) *Color of Foods*. Westport, Connecticut: AVI Publishing.

Patricia A. Redlinger
Iowa State University, Ames, USA

Texture

A Quality Factor

Texture is a sensory property of foods which, together with appearance, aroma and basic taste, has a profound effect on consumer acceptance of foods. Each of these properties is composed of a number of notes. Texture lies between taste and aroma in this respect, several dozen different texture notes being detectable in foods.

Textural perception occurs directly through the tactile (touch) and kinaesthetic (movement) senses, and indirectly through the senses of vision and hearing. In contrast to colour and flavour, there are no specific sensory receptors for texture. Texture also has aspects related to the absence of defects and to the satisfaction and pleasure of eating. It is an important quality attribute in almost all foods and most important in foods that are bland in flavour, or have the characteristics of crispness or crunchiness.

Texture may be defined as 'that group of physical characteristics that arise from the structural elements of the food, are sensed primarily by the feeling of touch, are related to the deformation, disintegration and flow of the food under a force, and are measured objectively by functions of mass, time, and length'. This definition teaches that texture has its roots in structure (molecular, microscopic, macroscopic) and the manner in which this structure reacts to applied forces. It also emphasizes that texture is a multidimensional property comprising a number of sensory characteristics.

A large number of terms are popularly used to describe textural sensations. Table 1 organizes many of these terms into a manageable system that facilitates understanding their interrelationships. It classifies textural properties into *mechanical characteristics* (reaction of the food to stress), *geometrical characteristics* (the feeling of the size, shape and arrangements of particles in the food, sometimes called 'particulate properties'), and *other characteristics* (relating to the sensations of moisture, fat and oil in the mouth). Table 2 provides physical and sensory definitions of the mechanical characteristics.

Sensory Evaluation

Since, by definition, texture is a sensory property, the most logical approach to its description and quantification is by sensory evaluation. In the early days, panels with various degrees of training were used to score specific textural characteristics or 'texture' in general. The scoring methods used were either numerical intensity scales (frequently 0–7, with 0 denoting absence and 7 a very high intensity of a specific characteristic) or hedonic scales (ranging from 'dislike extremely' to 'like extremely'). The latter should not be used when the objective is to quantify the intensity of the characteristic present.

The sensory perception of texture is a dynamic process which involves the rate and magnitude of the

Table 1. Classification of textural characteristics and their relationship to popular nomenclature

Characteristics	Primary parameters	Secondary parameter	Popular terms
Mechanical	Hardness	—	Soft–firm–hard
	Cohesiveness	Brittleness	Crumbly–crunchy–brittle
		Chewiness	Tender–chewy–tough
		Gumminess	Short–mealy–pasty–gummy
	Viscosity	—	Thin–thick
	Springiness	—	Plastic–elastic
	Adhesiveness	—	Sticky–tacky–gooey
Geometrical	*Class*		*Examples*
	Particle size and shape	—	Gritty, grainy, coarse, etc.
	Particle shape and orientation	—	Fibrous, cellular, crystalline, etc.
Other	Moisture content	—	Dry–moist–wet–juicy
	Fat content	Oiliness	Oily
		Greasiness	Greasy

Adapted from Szczesniak AS (1963) Classification of textural characteristics. *Journal of Food Science* 28: 385–389.

Table 2. Definitions of the mechanical parameters of texture

	Physical	Sensory
Hardness	Force necessary to attain a given deformation	Force required to compress a substance between molar teeth (solids) or the tongue and palate (semisolids)
Cohesiveness	Strength of internal bonds	Amount of sample deformation before rupture when biting with molars
Fracturability	Forces necessary to fracture the material	Force with which the material crumbles, cracks or shatters
Chewiness	Energy required to disintegrate a solid food to a state ready for swallowing	Number of chews required to masticate a sample at one chew per s and constant rate of force application to reduce it to a consistency suitable for swallowing
Gumminess	Energy required to disintegrate a semisolid food to a state ready for swallowing	Denseness that persists throughout mastication of a semisolid food
Viscosity	Rate of flow per unit force	Force required to draw a liquid from a spoon over the tongue
Springiness	Rate at which a deformed material goes back to its undeformed condition following removal of the deforming force	Degree and speed with which the material returns to its original height following partial compression with molar teeth
Adhesiveness	Work necessary to overcome the attractive forces between the surface of the food and other surfaces with which the food comes in contact	Force of the tongue required to remove the material that adheres to the mouth (generally the palate, but also lips, teeth, etc.) during the normal eating process

applied forces, and also the effects of temperature, saliva and time. The time element includes the repeated application of destructive forces in the masticatory process, and the duration of the food's contact with saliva and mouth temperature.

The multiparameter nature of texture and the process dynamics of its sensory perception during mastication form the basis for the sensory texture profile, the schematic for which is shown in Fig. 1. The method is used to define the texture characteristics present, the intensity of each, the order in which they appear and all the changes that occur from the first bite through completion of mastication. Texture profile analysis is presently the preferred sensory method for texture characterization because it is the only method that provides a complete analysis of all the textural properties of a food. Using reference samples and standard scales for specific parameters, highly trained panels provide a descriptive and quantitative 'fingerprint' of the product's texture. The training and maintenance of the panel may be tedious and expensive, but the quality of the generated data usually compensates for it. The basic principles can be adapted to different products in different situations including untrained consumer panels.

Two examples of sensory texture profiling are shown in Table 3. Most other sensory methods can be regarded as partial texture profiles or modifications of the basic procedure.

Instrumental Evaluation

Although sensory methods are the final arbiter of textural quality, instrumental methods are the most frequently used for texture measurement because they are cheaper and less time-consuming. To be successful, any instrumental measurement must correlate highly with sensory evaluation of textural quality. Furthermore, it needs to be recognized that instrumental methods measure one or more textural properties of a food before it is placed in the mouth, but cannot follow the changes that occur during mastication. Most instrumental methods are 'one-point' measurements, i.e. they measure only one dimension of the texture, albeit usually a dominant textural property. Because there is a very wide range of types of foods and textures, people use a wide variety of methods to manipulate foods during mastication. A large number of texture-testing instruments have been described in the literature and almost 100 are available commercially. Table 4 classifies objective methods of food texture measurement on the

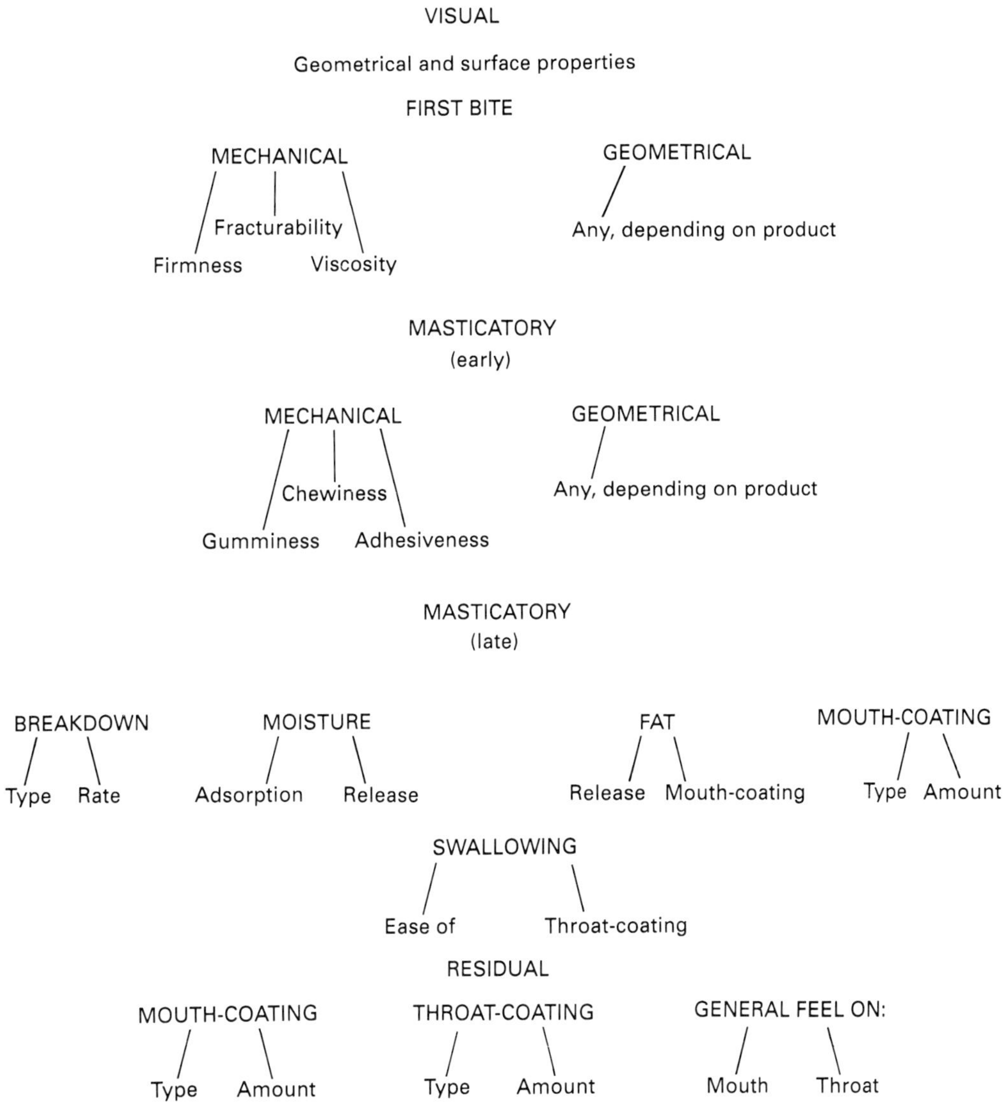

Fig. 1 Sensory texture profiling technique.

basis of the variable(s) measured and the principle of the test. Force-measuring instruments are the most common, but other principles, such as distance, time and energy, are also used. There are a few examples where a chemical analysis correlates well with a textural property. Finally, the sounds that are generated are an important dimension of the textural quality of crisp and crunchy foods.

The key element that differentiates force-measuring instruments is the geometry of the test cell which holds the sample and applies a force to it (cutting, puncture, compression, extrusion etc.). Recognition of the above fact has led to the widespread use of strength-of-materials-testing machines that provide a sophisticated driving mechanism, force sensor, and recording of the force–time relationship. They are more expensive than most of the simpler instruments but are widely used for research purposes.

The following criteria are recommended as a guide to selecting a texture testing instrument:

- Purpose – research or quality assurance.
- Nature of product – rheological type, heterogeneity.
- Required accuracy – high variability inherent in unprocessed foods often necessitates many replications.
- Cost – including operation and maintenance.
- Time – routine use requires rapid testing.
- Location – ability to withstand adverse environmental conditions when placed in factory surroundings.
- Nature of sensory evaluation method used by people (squeeze in hand, cut with incisors, crush between molars, roll with the tongue against the hard palate, etc.)

Table 3. Basic sensory texture profile ballot for meatballs and soda crackers

Texture notes	Meatballs	Soda crackers
Initial		
Mechanical		
Hardness (9-point scale)	3·4	4·0
Fracturability (7-point scale)	0·7	2·5
Viscosity (8-point scale)	Not applicable	Not applicable
Geometrical	Lumps, with a grainy surface	Flaky and puffy
Other	Moist, uncut surface is slippery and cut surface is not slippery	Dry
Masticatory		
Mechanical		
Gumminess (6-point scale)	1·2	0
Chewiness	17·7 chews	16 chews
Adhesiveness (5-point scale)	1·2	0·7
Geometrical	Coarse, grainy; some fibrous particles are present	Flaky
Other	Moist	Dry
Residual		
Rate of breakdown	Large lumps break down fast; grains break down at a medium rate	High
Type of breakdown	Lumps turn into a nonhomogeneous, grainy paste, and grain size decreases. Some stringy fibrous grains are present; they become more noticeable towards the end and require more effort to chew	It breaks down into little rough sheets, then changes into a smooth dough
Moisture absorption	Initially moist. Saliva mixes easily with slurry and the bolus becomes progressively more moist. Residual grains feel dry	It absorbs a lot of saliva slowly and changes into a moist dough
Mouth-coating	Slight residual oiliness. A few particles stick between the teeth and around the mouth	Little pieces stick to the mouth and gums

From Bourne MC (1982), pp 262, 263.

This should narrow the field to the most promising two or three test principles. The final candidates should be tested over the full range of textures normally encountered with the food and correlated with sensory evaluation. A statistical analysis of the results should identify which principle and instrument are the best for each particular application. The final step is to establish the test conditions that give the strongest resolution between different samples and then standardize them. These include sample size, test cell dimensions, force range, speed of travel of moving parts, chart speed, temperature, and perhaps other factors.

Texture Profile Analysis (TPA)

The methodology of TPA brings instrumental texture evaluation a step closer to sensory testing. It involves compressing a bite-size piece of food two or more times in a reciprocating motion that simulates the action of the jaw, and quantifying from the resulting force–time curve a number of textural parameters that correlate well with sensory evaluation. The method, originally developed for the General Foods Texturometer, has been adapted to universal testing machines. Figure 2 shows a generalized TPA force–time curve. *Hardness* is defined as the peak force on the first compression cycle (first bite); *fracturability* (originally called brittleness) is defined as the first significant break in the curve on the first bite. The areas under the curve during the first bite and the second bite are a measure of the work that is done in the compression. The ratio of these two areas (A2/A1) is defined as *cohesiveness*. The negative force peak in the first decompression is defined as *adhesive force* and the negative areas as *adhesive work*. The distance that the

Table 4. Objective methods for measuring food texture

Principle	Measured variable	Dimensional units	Examples
Force	Force (F)	mlt^{-2}	
Puncture	F	mlt^{-2}	Fruit pressure testers
Extrusion	F	mlt^{-2}	Shear Press, Tenderometer
Shear	F	mlt^{-2}	Warner-Bratzler Shear
Crushing	F	mlt^{-2}	—
Tensile	F	mlt^{-2}	Extensograph
Torque	F	mlt^{-2}	Rotary viscometers, Farinograph, Struct-o-Graph
Snapping	F	mlt^{-2}	
Deformation	F	mlt^{-2}	
Distance			
	Length	l	Penetrometers, Bostwick Consistometer
	Area	l^2	USDA Consistometer
	Volume	l^3	Bread volume, Succulometer
Time	Time (T)	t	Ostwald Viscometer, Biscuit Texture Meter
Energy	Work ($F \times D$)	$ml^2 t^{-2}$	Area under force–distance curves
Ratio	F or D or T measured twice	Dimensionless	Specific Gravity
Multiple	F and D and T	mlt^{-2}, l, t	Instron, Lloyd, Zwick
Chemical analysis	Concentration	Dimensionless (%)	Alcohol-insoluble solids
Miscellaneous	Anything	Anything	Optical density, crushing sounds

Adapted from Bourne MC (1966). A classification of objective methods for measuring texture and consistency of foods. *Journal of Food Science* 31: 1011–1015.

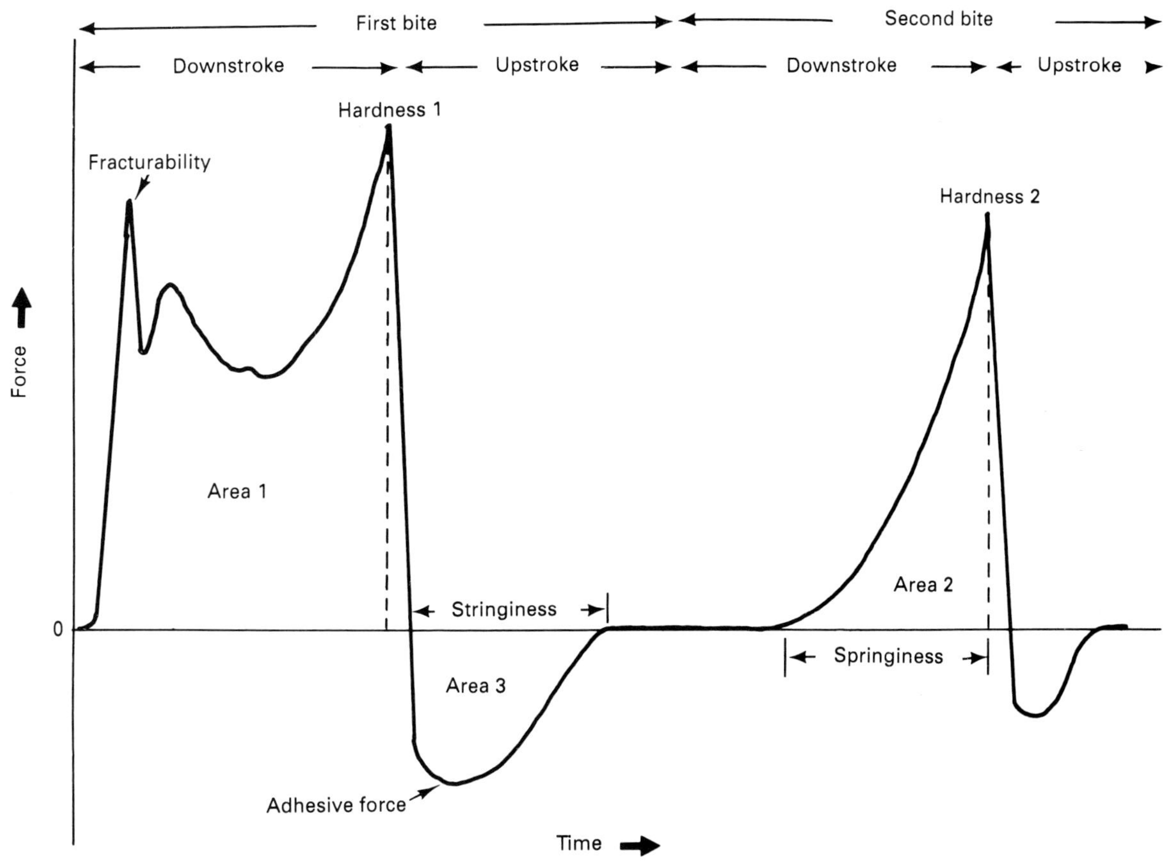

Fig. 2 A generalized texture profile analysis curve. (From Bourne MC (1978) Texture profile analysis. *Food Technology* 32(7):63 (copyright, Institute of Food Technologists).)

product extends in the decompression is defined as *stringiness*, and the distance that the product recovers its height between the first and second compressions is defined as *springiness*. Two other parameters are derived by calculation: *gumminess* is defined as the product of hardness and cohesiveness; *chewiness* is defined as the product of gumminess and springiness (which equals hardness multiplied by cohesiveness multiplied by springiness).

In the early days, instrumental texture profile analysis was an academic curiosity because of the long time required to extract the information from the force–time curves. With computer retrieval and analysis of the data, the time requirement has been reduced to the point where it can become a simple routine test. It will probably become the instrumental measurement of choice.

Rheological Measurements

Historically, rheology has been defined as the study of the deformation and flow of matter or the response of materials to stress. The science of rheology, proven so useful in the field of high polymers, has many applications to food, including raw materials (e.g. grains, meats, fruits), intermediate products in the manufacturing process (e.g. bread dough, cheese curd, sausage emulsions), and finished products (all foods). However, rheology does not cover all aspects of texture. The size reduction that occurs during mastication is not rheology, neither are the sensations of oiliness, moistness and particle size and shape.

Most people believe that there is a sharp distinction between solids (which do not flow) and liquids (which flow). In fact, the distinction between solids and liquids is far from clear because many liquids possess some of the properties of solids, and many solids possess some of the qualities of liquids. The science of rheology specializes in the study of these complex materials (of which there are many examples in foods) that are partly solid and partly liquid.

The flow of liquids may be divided into several broad classes:

1. *Newtonian flow*, in which the shear rate* is directly proportional to the shear stress†. Examples are edible oils, sugar syrups, milk and honey. The measurement of the properties of these foods is straightforward, since viscosity‡ is independent of shear rate.
2. *Plastic or Bingham flow*, in which a minimum shear stress, known as the 'yield stress', must be exceeded before flow begins. Examples are tomato ketchup, whipped egg white, mayonnaise, margarine and butter.
3. *Pseudoplastic flow*, in which an increasing shear force gives a more than proportionate increase in shear rate, i.e. apparent viscosity decreases with increasing shear rate. Salad dressings exemplify this type of flow.
4. *Dilatant flow*, in which equal increments in shear stress give less than equal increments in shear rate, i.e. apparent viscosity increases with increasing shear rate. This type of flow is rare in foods but is found in high-solids suspensions of raw starch and some chocolate syrups. ***See*** individual foods

* Shear rate (denoted by the symbol $\dot{\gamma}$ and expressed in s^{-1}) is the velocity gradient established in a fluid that results from the application of a shear stress.
† Shear stress (denoted by σ and expressed in Pa) is the force per unit area applied tangential to the plane on which the force acts.
‡ *Viscosity* (denoted by η and expressed in Pa s) is the internal friction of a fluid or its tendency to resist flow ($\eta = \sigma/\dot{\gamma}$); it should only be used for Newtonian fluids. *Apparent viscosity* (denoted by η_a) is the viscosity of a non-Newtonian fluid expressed as the viscosity of a Newtonian fluid at a specified shear rate (e.g. $\eta_a = \sigma/\dot{\gamma}_{50}$ is the apparent viscosity of a non-Newtonian fluid at a shear rate of 50 s^{-1}).

The differences between these types of flow are shown in Fig. 3. The shear-stress-versus-shear-rate plot (Fig. 3a) is the more common manner of presentation; another that is sometimes used is shown in Fig. 3b, in which the axes are interchanged.

Time Dependency

For some fluids the shear stress is a function of both the shear rate and the time it is subjected to shear. For *thixotropic* products the apparent viscosity decreases with time of shearing. This condition is frequently found in food systems such as gum solutions and starch pastes. *Rheopectic* products increase in apparent viscosity with time of shearing. This type of behaviour is rare in foods.

Viscoelasticity

As discussed above, most foods combine some of the properties of ideal liquids, which exhibit only viscosity (flow) and ideal solids which exhibit only elasticity (deformation). These are called viscoelastic foods. In characterizing these systems rheologically it is necessary to measure both the viscous component (loss modulus, G″) and the elastic component (storage modulus, G′). A food with high G′ and low G″ behaves more like a solid than a liquid, while a food with low G″ and high G′ behaves more like a liquid than a solid. An elastic solid such as rock candy will have a G″ value of zero, while a Newtonian liquid such as sucrose syrup will have a G′ value of zero.

Traditionally, viscoelasticity was measured by performing creep tests in which a weight was placed on the test material and the change in specimen height was monitored over a period of time. Presently, viscoelasticity is usually measured by an oscillation test in which a

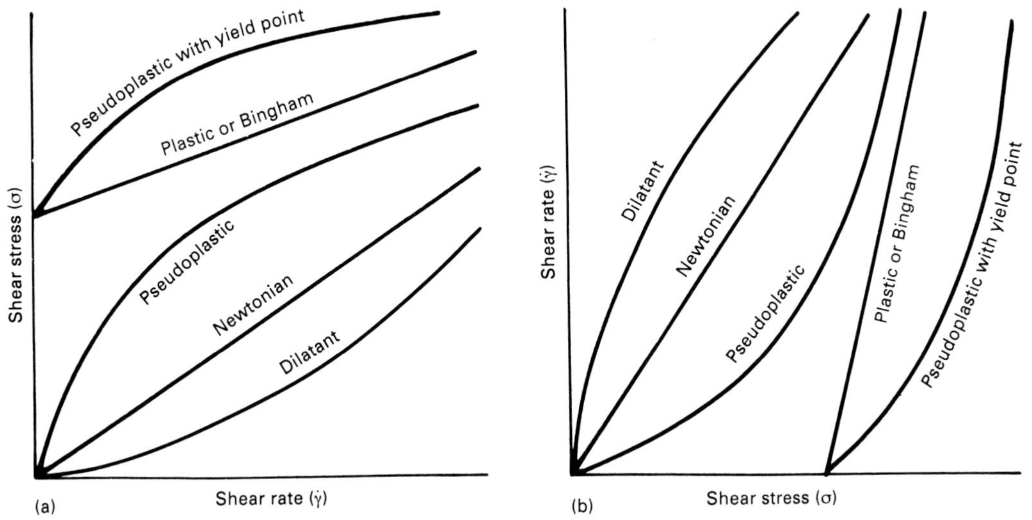

Fig. 3 (a) Shear-stress versus shear-rate plots for various types of flow. (b) Shear-rate versus shear-stress plots for the same types of flow. From Bourne MC (1982), p 215; by courtesy of Academic Press, New York.

sample of known dimensions (if solid) or filled into a cup with standard dimensions (if liquid) is subjected to repeated small sinusoidal deformations that do not fracture the sample. Analysis of the resulting shear stress versus time curves yields the numerical values of G′ and G″. Modern instrumentation provides a computerized system for quantifying these moduli. The rate and degree of deformation may be varied to provide information on the internal structure and mechanical behaviour of the material. ***See*** Food Acceptability, Affective Methods

Bibliography

Aguilera JM and Stanley DW (1990) *Microstructural Principles of Food Processing and Engineering*. Essex: Elsevier Science Publishers.

Bourne MC (1982) *Food Texture and Viscosity: Concept and Measurement*. New York: Academic Press.

Brennan JG (1980) Food texture measurement. In: King RD (ed.) *Development in Food Analysis Techniques*, vol. 2, pp 1–78. Essex: Elsevier Science Publishers.

Christensen CM (1984) Food texture perception. *Advances in Food Research* 29: 159–199.

De Man JM, Voisey PW, Rasper VF and Stanley DW (eds) (1976) *Rheology and Texture in Food Quality*. Westport, Connecticut: AVI, Publishing.

Melgaard M, Civille GV and Carr BT (1987) *Sensory Evaluation Techniques*, vol. 2. Boca Raton, Florida: CRC Press.

Mohsenin NN (1970) Application of engineering techniques to evaluation of texture of solid food materials. *Journal of Texture Studies* 1: 133–154.

Moskowitz HR (ed.) (1987) *Food Texture: Instrumental and Sensory Measurement*. New York: Marcel Dekker.

Muñoz AM and Civille GV (1987) Factors affecting perception and acceptance of food texture by American consumers. *Food Reviews International* 3(3): 285–322.

Sherman P (ed.) (1979) *Food Texture and Rheology*. New York: Academic Press.

Szczesniak AS (1987) Correlating sensory with instrumental texture measurements – an overview of recent developments. *Journal of Texture Studies* 18: 1–15.

Szczesniak AS (1991) Textural perceptions and food quality. *Journal of Food Quality* 14: 75–85.

Malcolm C. Bourne
Cornell University, Geneva, New York, USA
Alina S. Szczesniak
Mt Vernon, New York, USA

Aroma

The sense of smell (olfaction) is one of the two chemical senses in humans, the other being the sense of taste. The term ‘odour’ has been defined by the British Standards Institute as the quality of the sensation perceived, via the olfactory organ situated in the nasal cavity, from certain volatile substances; and aroma is an odour with a pleasant connotation. Since the olfactory organ is extremely sensitive, the concentrations of volatile compounds in foods contributing to aroma may be very low. Analysis of aroma compounds is concerned with the extraction, separation and characterization of complex mixtures of volatile compounds present in foods and beverages.

Olfaction

The olfactory receptors are sited in the olfactory epithelia, two small patches of mucous membrane in the recesses of the nasal passages (Fig. 1). The response of these receptors to airborne molecules determines the odour of a substance. Odorant molecules entering the nose find their way to the olfactory epithelia, where they come into contact with cilia (hairs) on the receptor cells. These produce an electrical signal that travels along nerve fibres to the olfactory bulb and thence to the brain. There are approximately 10^7 olfactory receptor cells in the human nose, while dogs may have 100 times more receptors and hence have a far more sensitive olfactory system.

Olfaction plays a primary role in the perception of flavour in foods and beverages, although the chemical compounds which stimulate odour perception (odorants) are quantitatively only very minor constituents of those commodities. When food is eaten, taste and smell act in concert to produce the integrated sensation of flavour. The aroma of the food is not just perceived in inhaled air; when food is in the mouth, odorants can reach the olfactory receptors by the retronasal route, entering the nasal cavity the same route as expired air.

Odour Sensitivity

The odour perception process is considerably more sensitive than the perception of taste, by factors of 10^6 to 10^9. Furthermore, a very large number of odorants can be differentiated, and thus aroma plays the major role in delineating the characteristic flavour of a particular food. The inability to differentiate flavours is usually associated with anosmia (lacking a sense of smell) rather than ageusia (lacking a sense of taste).

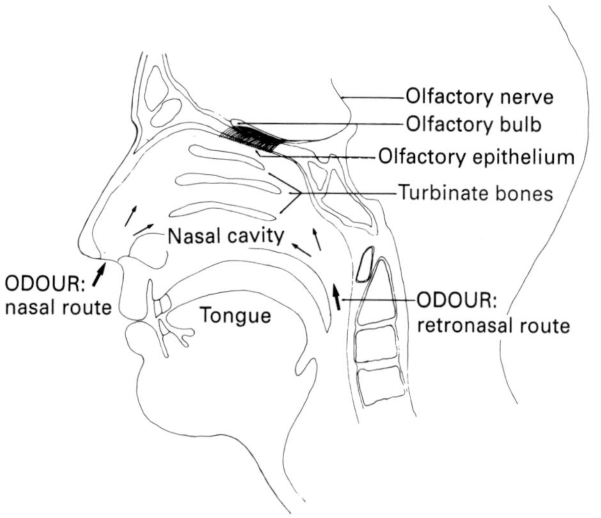

Fig. 1 A representation of a section through the nasal and oral cavities showing the olfactory organ. Reproduced, with permission, from Thomson DMH (1986) The meaning of flavour. In: Birch GG and Lindley MG (eds) *Developments in Food Flavours*, pp 1–21. Essex: Elsevier Science Publishers.

Table 1. Odour threshold of some aroma compounds in aqueous solution

Compound	Threshold parts per 10^9 ($\mu g\ l^{-1}$)
Ethanol	100 000
2,5-Dimethylpyrazine	1800
Butyric acid	250
Limonene	10
Hexanal	5
Ethyl 2-methylbutyrate	1
Methanethiol	2×10^{-1}
β-Ionone	7×10^{-3}
2-Isobutyl-3-methoxypyrazine	2×10^{-3}
bis-(2-Methyl-3-furyl) disulphide	2×10^{-5}
2,3,6-Trichloroanisole	3×10^{-7}

It has been suggested that the triggering of one human olfactory neuron by a powerful odorant may require only 8 molecules, and as few as 40 molecules can produce an identifiable sensation. If it is assumed that only 1 in 1000 molecules taken into the nasal cavity actually reaches a receptor site, the detection limit of olfaction is of the order of 40 000 molecules or 10^{-19} mol, which is several levels of magnitude greater in sensitivity than present-day analytical instruments.

The range of odour thresholds exhibited by odorants extends over at least 10 orders of magnitude (Table 1). A compound with one of the lowest reported odour thresholds is bis-(2-methyl-3-furyl) disulphide, which has a meaty aroma and can be detected at concentrations of 2 parts in 10^{14} parts of water. This concentration is equivalent to about 20 μg in the water of a 50-m swimming pool of 1000 m^3 capacity! At the other end of the range, at least 1 mg of ethanol needs to be present in 10 ml of water before the ethanol can be detected by smelling the solution.

Chemical Nature of Aroma Compounds in Foods

The range of chemical classes which contribute to food flavours is diverse in both chemical composition and physical properties, and includes aliphatic, alicyclic, aromatic and heterocyclic compounds. All aroma compounds have some volatility, but this ranges from permanent gases to compounds with very little vapour pressure, and molecular weight up to 300. Analysis of the volatile components associated with foods and beverages shows that most commodities contain mixtures of many different volatile compounds which usually have one or more functional groups. Examination of the literature shows a total of over 6000

Table 2. Stages in the analysis of aroma volatiles

Step	Analytical techniques
Isolation and concentration	Headspace analysis
	Distillation
	Extraction
	Adsorption
Separation	GC
	Liquid chromatography
Identification	Chromatographic retention
	Mass spectrometry
	Infrared spectroscopy
	Other instrumental methods
	Chemical synthesis
Sensory characterization	GC odour-port evaluation
	Sensory panel

GC, gas chromatography.

different volatile compounds in foods and beverages, and the numbers associated with complex cooked foods such as coffee and meat exceed 1000. Many of these compounds are common to a number of different foods, and the contribution of any compound to the characteristic aroma of a particular food will depend on the number of factors, including the following:

- Odour character.
- Concentration in product.
- Odour threshold.
- Vapour pressure.
- Adsorption on food matrix.
- Interaction with other components.
- Synergism with other volatiles.

Some volatiles, such as aliphatic hydrocarbons, have little aroma and do not contribute to food flavour, while other compounds define the aromas of specific foods (e.g. benzaldehyde – almonds; 2-isobutyl-3-methoxypyrazine – bell peppers; citral – lemons). However, for many foods, the aroma depends on contributions from a complex mixture of volatiles comprising different chemical classes. ***See*** Essential Oils, Properties and Uses; Flavour Compounds, Structures and Characteristics

Analysis of Aroma Volatiles

In attempting to understand the nature of those compounds which characterize different food aromas, flavour scientists have analysed the volatile composition of many different foods. The approach adopted usually involves the extraction of volatiles from the food matrix, their concentration, and the separation and identification of the individual components (Table 2). As far as possible the methods employed need to provide an extract which contains all the volatile components without the introduction of artefacts. If the relative contribution of the different volatiles to the overall aroma is to be evaluated, the extract needs to retain the volatiles in the same relative proportions as in the original food. The wide range of odour thresholds can result in quantitatively minor components making a major contribution to the aroma, while some large components will have no sensory significance. Successful isolation and identification of minor components with aroma significance is one of the major challenges of aroma analysis.

Isolation and Concentration of Aroma Volatiles

The components of food which are responsible for aroma are present in extremely small quantities compared with the major constituents, of which water is usually the most abundant. The first step in aroma analysis is to obtain a volatile extract of the food in sufficient quantity to be able to separate and identify the components of aroma significance, while maintaining the characteristic aroma properties of the food. A number of isolation techniques have been developed, all based on utilizing the volatile nature of the aroma compounds to separate them from the food matrix.

Headspace Analysis

The concentration of volatiles in the headspace vapour above a food or beverage will be very low. However, it will contain a representative mixture of those compounds which contribute to the aroma of the product and, therefore, analysis of that headspace will provide the best method of obtaining a sample representative of the food aroma. In its simplest form, headspace analysis involves the direct introduction of a given volume of the vapours above a food sample onto a gas chromatographic column. This technique is rarely used, however, because the volatiles are not present in sufficiently high concentration and because water from the sample will often cause interference. Methods have been described for the concentration of headspace volatiles by purging the headspace with a stream of an inert gas (nitrogen or helium) and condensing the volatiles in a series of cold traps cooled in ice-water, solid carbon dioxide or liquid nitrogen. Extraction of the condensate with a small amount of a suitable solvent provides an aroma extract suitable for chromatographic analysis. Instead of using cold traps, headspace volatiles can be collected on suitable adsorbents. ***See*** Chromatography, Gas Chromatography

Adsorption Methods

The ability of certain solid surfaces to adsorb volatile organic molecules is widely used in the analysis of aroma compounds in foods and beverages. The methods also find application in environmental analysis. Activated charcoal was the first to be used, but certain porous polymers (e.g. Chromosorb, Porapak and Tenax), which were originally developed for gas–solid chromatography, are now frequently employed. Like charcoal, all these materials readily adsorb volatiles while having little affinity for water and low-molecular-weight alcohols, making them particularly useful in the analysis of samples with a high water content.

In a typical headspace collection (entrainment) the volatiles are purged from a flask containing the sample using a flow of a purified inert gas (helium or nitrogen) which carries the volatiles to a small tube containing the adsorbent. The quantity of adsorbent may vary from 10 to 200 mg of porous polymer, while smaller quantities of charcoal can be used because of its greater adsorbing capacity. Collection times may range from a few minutes to several hours, but the volatile profile will change with the collection time and the nature of the adsorbent. A closed-loop system, involving a pump which circulates the headspace continuously through a trap (usually charcoal), was developed for water analysis but has also found application for food volatiles.

Removal of the adsorbed volatiles for chromatographic analysis can be achieved thermally or by extraction with solvents. Like charcoal, the porous polymers exhibit high thermal stability, some to temperatures above 300°C, and adsorbed volatiles can be desorbed by heating the polymer under a gas flow and collecting in a cooled tube for subsequent analysis. More conveniently, the volatiles can be desorbed directly onto the gas chromatographic column by placing the adsorbent trap in a specially modified injection port, thus avoiding the loss of components or unnecessary dilution. Cooling of the front of the column (cryofocusing) with solid carbon dioxide or liquid nitrogen during this desorption will avoid any loss in chromatographic resolution. Solvent extraction is effected simply by passing a small amount of a solvent through the adsorbent in the trap. Concentration of the collected solution by careful evaporation of the solvent provides a flavour extract for chromatographic analysis.

Distillation

The aim of the isolation step in aroma analysis is the separation of the volatile components from the nonvolatile food matrix; hence distillation is a commonly employed method. Steam distillation finds application in the analyses of volatiles from beverages and high-water-content foods, although it is less applicable to fats and oils. It has the disadvantage that the large quantities of aqueous distillate require further extraction with a solvent in order to separate the volatiles from the water. Concentration of the extract is then necessary. The formation of artefacts may also be a problem. Vacuum distillation finds wide application in the extraction of volatiles from fats and oils, and, provided that the temperature is kept low, artefact formation can be minimized. A number of different designs of equipment have been utilized, but all incorporate cold traps for the collection of the distillate. In molecular distillation, a high vacuum ($< 10^{-3}$ mmHg) is used and the volatiles have a relatively short path to travel from the surface of a liquid oil sample to a cold surface where they condense. Efficient collection of high-boiling-point components is achieved because at high vacuum the distance between the sample surface and the cold condenser (10–20 mm) is less than the mean free path of the molecules.

Extraction

Organic solvents may be used to extract volatiles from the aqueous distillate which results from steam distillation of the food. Initial concentration of the aqueous phase by freeze concentration is sometimes used to reduce the volumes requiring extraction; continuous liquid–liquid extraction equipment may also be used. Solvents are chosen on the basis of their selectivity for the volatile compounds of interest, and on boiling point. Commonly used solvents are diethyl ether, pentane, isopentane and some chlorofluorocarbons. After extraction, most of the solvent must be removed to provide an aroma concentrate suitable for chromatographic separation. Removal of traces of water, using a drying agent (sodium or calcium sulphate) or by freezing out at −20°C is essential before careful removal of solvent by distillation. Final concentration to a volume of 0·1–1 ml is often achieved by purging with a gentle stream of nitrogen.

Direct extraction of an aqueous food fraction with an organic solvent is of limited use because the extract will contain much nonvolatile matter. However, in recent years supercritical fluid carbon dioxide has been utilized for the extraction of flavours on both a laboratory and a commercial scale. It is a highly selective solvent, exhibiting strong affinity for most aroma constituents, whilst salts, sugars, lipids and many acids are insoluble. The ease of removal of the solvent, after extraction, to give a concentrated aroma extract is another attractive feature of supercritical carbon dioxide extraction.

One of the most widely used techniques in aroma analysis combines steam distillation and solvent extraction in the Likens–Nickerson apparatus (Fig. 2). Since the technique was first reported in 1964 for the extraction of hop oil, many variations to the original design have been suggested. The essential feature is the simultaneous condensation of the steam distillate with an immiscible extraction solvent. A simple U-tube, with appropriate side-arms, allows the return of water to the

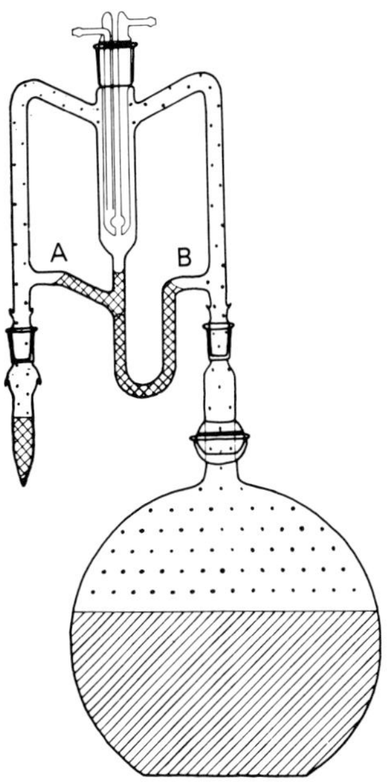

Fig. 2 Likens–Nickerson apparatus for continuous steam distillation and solvent extraction. Reproduced, with permission, from Nickerson GB and Likens ST (1966) Hop oil components in beer. *Journal of Chromatography* 21: 1–5.

steam distillation flask, and the return of the solvent, containing the extracted volatiles, to a reservoir flask where it redistils for further extraction. The technique provides a simple, rapid method involving only small volumes of solvent. Thermal degradation of labile components can be diminished by carrying out the distillation and extraction under reduced pressure.

Separation of Aroma Components

To determine the sensorially important compounds in an aroma isolate, the complex mixture needs to be separated into its components. The amount of isolate available is usually small, containing very many compounds of diverse chemical structures, varying greatly in concentration, and important components are often present in extremely low amounts. The success of any aroma analysis depends very heavily upon the efficiency of separation and the sensitivity of the detection. Gas chromatography (GC) is the most widely used separation method in aroma analysis because it best meets these criteria of separation efficiency and sensitivity. High-performance liquid chromatography (HPLC) offers some advantages over GC, principally in the analysis of thermally labile compounds. However, larger quantities of components are generally required, the separation efficiency is lower, and subsequent identification by mass spectrometry is not as easy; consequently, the use of HPLC in aroma analysis is usually limited to specific applications and to the fractionation of isolates prior to GC analysis. ***See*** Chromatography, High-performance Liquid Chromatography

The GC columns used for aroma analysis must show high resolution but must retain that resolution at relatively high sample loadings so that detectable amounts of trace components can be separated from major components present at concentrations many thousands times higher. As well as being able to separate complex mixtures, the column must not cause adsorption, degradation or chemical rearrangements since many aroma compounds are thermally labile, unstable or readily oxidized. The advantages of capillary columns over conventional packed columns have long been recognized in aroma research, and they have been used in flavour research laboratories since 1960, very soon after Golay's original demonstration of this type of column. Progress in aroma research has been closely linked with the development of capillary columns. Bonded-phase fused silica columns now provide the high resolution and long-term stability required for aroma analysis, as well as an inertness to adsorption and degradation which cannot be matched by columns made from glass or stainless steel. Phases commonly used include polar phases, such as Carbowax 20M, and nonpolar and semipolar silicones. Retention data obtained by analysing an aroma isolate on two different phases can be helpful in the identification of components. The stereochemistry of compounds can influence flavour, and capillary columns coated with chiral phases can be used to separate enantiomers.

Detection of Components of Sensory Significance

An essential step in the analysis of aroma volatiles is the determination of those components in the complex mixture which contribute to the aroma of the food or beverage. Sensory profiling of the whole product can give some indication of the important aroma characters present in the product, and such information should be used when assessing the contribution of the individual components of an aroma isolate to the overall flavour quality. A widely used technique for determining components which contribute to aroma, is odour-port smelling (or sniffing). The column effluent is split between a conventional GC detector and a vent to the outside of the oven where the odours emerging can be smelled and described. Thus chromatogram peaks which correspond to specific aromas can be identified, and 'aromagrams' can be compiled to complement the chromatogram. It should be noted that components with characteristic odours often have low threshold values and may not give detectable GC peaks.

Identification of Components

Structure elucidation for the chromatographically separated components is the next step in the analysis of an aroma isolate. Modern instrumental techniques, such as mass spectrometry (MS), nuclear magnetic resonance (NMR), infrared (IR) and ultraviolet (UV) spectroscopy, provide the most efficient means of volatile identification. Chemical derivatization and other reactions on components trapped from the chromatograph column have been employed in aroma analysis, but the quantity of compound required limits their use to situations in which instrumental techniques alone have been unable to provide an identity. Most instrumental techniques require compounds to be trapped as they elute from the chromatograph, but coupled GC–MS and, more recently, GC–IR allow direct analysis of the separated components.

Mass Spectrometry

Most flavour chemists would consider a mass spectrometer to be an essential instrument in a flavour laboratory. It provides excellent structural information for organic compounds at a level of sensitivity several factors better than other instrumental techniques. The availability of coupled GC–MS systems in the late 1960s was one of the most important developments for flavour research, and this was followed a decade later by the introduction of efficient computerized data-handling systems. ***See*** Mass Spectrometry, Principles and Instrumentation

The requirements of a modern coupled GC–MS for volatile analysis are fast pumping of the ion source to allow the capillary column to be connected directly to the ion source, and computer control for data acquisition and processing to permit automatic repetitive scanning throughout the GC–MS analysis with a scan time of 1 s or faster. Sensitivity is usually such that a full spectrum can be readily obtained on 1 ng of a single component in a complex mixture injected onto the GC column, and in many cases an identifiable spectrum can be obtained from as little as 10 pg. Higher sensitivity is possible using selected ion monitoring to analyse for the presence of known compounds. Quadrupole mass spectrometers offer good sensitivity and rapid scan rates, and are significantly cheaper than magnetic sector instruments, but only provide nominal mass data. The double-focusing magnetic sector instrument offers one main advantage over the simple quadrupole instrument, in that the high resolution capability permits the acquisition of accurate mass data, allowing the calculation of empirical formula for ions in a spectrum, which can be a great asset in the identification of unknown compounds.

The characterization of mass spectra is greatly facilitated by comparing with known spectra in compiled libraries incorporated into the MS data system. Searching programs in the computer provide matching of sample spectra with library spectra. Current data bases contain over 120 000 spectra of organic compounds, but special compilations containing only volatile compounds have also been prepared and at least one is available in computer-readable format. Confirmation of the identity of compounds should always be carried out, if possible by comparing mass spectra and retention times from authentic samples.

Other Instrumental Methods

Both NMR and IR spectroscopy are extremely valuable techniques in the identification of organic compounds. However, they require larger sample sizes than MS and are not as widely used in aroma analysis. They are primarily used to aid structure elucidation of unknown compounds. Samples for NMR must be collected, as they elute from the GC column, in a cooled glass capillary tube placed at a collection port at the end of the column. Addition of a few microlitres of deuterated solvent to the tube will provide a sample suitable for analysis in a microprobe of a Fourier transform instrument. An adequate proton spectrum can be obtained in a few hours with 1–10 μg of sample. A much larger sample is needed for an adequate ^{13}C spectrum. ***See*** Spectroscopy, Infrared and Raman; Spectroscopy, Nuclear Magnetic Resonance

Similar preparative GC separation can provide samples for IR spectroscopy, when up to 100 μg of the compound may be required. Fourier transform IR instruments directly coupled to the GC have been developed and are much more sensitive. They give vapour-phase spectra and are capable of rapid, repetitive scanning, allowing the acquisition of GC–IR chromatograms. Such IR detectors for GC can give adequate spectra on 10–100 ng of compounds. By offering functional group information IR is, in general, complementary and supportive of MS. Because IR is non-destructive, it has been possible to develop tandem GC–IR–MS systems, providing a powerful tool for aroma analysis. ***See*** Food Acceptability, Affective Methods

Bibliography

Maarse H (ed.) (1991) *Volatile Compounds in Foods and Beverages*. New York: Marcel Dekker.

Maarse H and Belz R (eds) (1981) *Isolation, Separation and Identification of Volatile Compounds in Aroma Research*. Berlin: Akademie-Verlag.

Morton ID and MacLeod AJ (eds) (1982) *Food Flavours*. Amsterdam: Elsevier Science Publishers BV.

Piggott JR (ed.) (1988) *Sensory Analysis of Foods* 2nd edn. Essex: Elsevier Science Publishers.

Teranishi R, Flath RA and Sugisawa H (eds) (1981) *Flavour Research – Recent Advances*. New York: Marcel Dekker.

Theimer ET (ed.) (1982) *Fragrance Chemistry. The Science of the Sense of Smell.* New York: Academic Press.

D. S. Mottram
University of Reading, Reading, UK

Taste

Taste (gustation) may be defined as the range of sensations elicited by the interaction of selected water-soluble compounds ('tastants') with specialized cells in the oral cavity.

Anatomy and Physiology of Taste Perception

Tastants are generally transported to taste receptor cells via saliva or other fluids in the mouth. Clusters of these cells, with supporting tissues, make up the 'taste buds', which are primarily found in varying numbers on the large, easily visible papillae of the tongue. A small number of functional taste buds may also be found on the soft palate and elsewhere in the oral cavity and pharynx. There is extremely wide interindividual variation in taste bud number and regional distribution, and these differences may, to some extent, translate into differences in taste perception.

At the molecular level, a range of different mechanisms have been proposed for the initial interaction between taste cells and different classes of tastants. Channels selectively responsive to hydrogen ion concentration (pH), and specific halide salts have been proposed as the site of origin for sour and salty tastes, respectively, although there is also evidence of receptor-mediated mechanisms for the latter. A number of biochemical and psychophysical studies provide strong support for the existence of at least one (and probably more than one) type of sweet taste receptor in humans. Mechanisms for bitterness remain the most elusive, and both receptor-mediated and electrostatic interactions have been proposed as basic mechanisms for initiation of bitter taste sensations.

It is commonly stated that different tastes are perceived solely or most intensely in certain regions of the tongue (e.g. bitter in the back of the tongue, and sweet in the front); however, this is an oversimplification. Most taste buds, regardless of location, appear to be receptive to multiple taste qualities, and most areas of the tongue can detect all taste qualities over a wide range of concentrations. While it does appear that certain regions may be most sensitive to very low concentrations of particular taste qualities, the same or another region may be more responsive to high concentrations of the same quality.

Taste Qualities and Taste–Smell Confusions

The classical view is that taste is limited to mixtures of four 'basic' qualities – sweet, sour, salty, and bitter – and this notion is supported by a number of lines of physiological and behavioural evidence. Nevertheless, authorities dating back to Aristotle have proposed other schemes with varying numbers of taste qualities, and the debate continues today. While it appears that most human languages provide names for a limited number of taste qualities corresponding to some or all of the basic four, other schemes exist. For example, the savoury quality associated with monosodium glutamate (MSG) and ribonucleotides is generally accepted as a basic taste (termed 'umami') in Japanese science and culture, and the inclusion of umami as a basic taste has many proponents. Other work suggests that basic tastes may merely be easily recognized points in a continuum, or the result of limitations in semantics or cultural exposure. Indeed, one could well argue that there may be different numbers and categories of basic tastes, depending upon whether the judgment is made on physiological, perceptual, or linguistic criteria.

For the purpose of descriptive evaluations of foods, it should be noted that even though the specific names of the basic four tastes are seemingly well established and familiar, they are often misapplied. Experiments have shown that a high proportion of adults fail to apply correctly the appropriate names to representative tastant solutions, with misuse of the terms 'bitter' and 'sour' being particularly common.

Some of the difficulty in assessing the existence or nature of basic tastes is attributable to confusions in the meaning and function of 'taste' and 'smell', common even in the scientific literature. Furthermore, 'taste' is frequently equated with 'flavour' in colloquial speech. It is clear that the peripheral taste and smell systems are completely separate and distinct in anatomy and physiology, as well as in the sensations they mediate. Whereas many different foods may share common tastes (sweet, salty, etc.), it is clearly in their volatile flavours that they are recognized as unique.

Although it seems obvious upon reflection, it is not commonly appreciated that the volatile flavours of ingested foods and beverages are largely perceived retronasally, i.e. during the outward flow of odorous compounds through the nasal passageways upon exhalation. There is no tactile experience in the olfactory sensory areas, but the foods producing those odours are present and felt in the mouth. Hence there is a natural but incorrect inclination to refer many volatile flavour sensations to the mouth, and to perceive them as being partially or totally 'taste'. Thus, while colds or other nasal blockages may lead to complaints of a loss of taste, it is in fact the ability to smell that is temporarily lost; the sense of taste is usually unaffected. Indeed, many if not most patients undergoing clinical evaluation for reported loss of taste and smell are diagnosed as having only loss of smell.

Tastants in Foods

From the point of view of isolation and analysis, taste, as opposed to odour, presents a somewhat simpler challenge to the food technologist. As suggested above, one important step is to ensure that it is indeed taste that is of interest, i.e. that the material is water-soluble and sensed in the mouth. Simply pinching the nose largely stops the flow of air carrying the unique volatile flavours of foods from the oral cavity to the olfactory receptor areas, and can be used to differentiate taste and smell. (A more sophisticated method is to blow a continuous gentle stream of clean air into the nostrils.) When this approach is applied to examine many common 'taste' problems in foods, such as soapy and metallic notes, they are often revealed to be odours.

The range of taste qualities and types of chemical compounds imparting them are to some degree limited. Along with a handful of synthetic sweeteners, sweetness is invariably attributable to mono- and disaccharides, either pre-existing in the food or possibly resulting from the action of salivary amylases upon polysaccharides. There may be other sweet compounds found in foods, such as particular amino acids, and some rarer compounds derived from specific plants sources. Salty tastes in foods are generally attributable to metal salts, particularly those of sodium and potassium. However, other than sodium chloride and lithium chloride (which is highly toxic), no compounds are known to impart a 'pure' salty taste. Materials such as potassium chloride are generally described as salty-bitter or by other combinations of terms. The savoury taste of MSG is also predominantly described as salty by Western consumers. Sour tastes are reliably associated with pH and organic acids. ***See*** Flavour Compounds, Structures and Characteristics

In contrast to other taste qualities, bitter taste in foods is associated with a much wider range of compounds, examples of which may differ substantially in origin and structure. In addition, the threshold for detection of many bitter compounds is extremely low, with notable bitterness imparted by concentrations in the micromolar range. For these reasons, bitter off-flavours in foods can be particularly troublesome to isolate and analyse.

Evaluation of Taste Function

The gustatory function of individuals and the characteristics of specific tastants may be assessed by a number of means. The parameters most commonly evaluated include threshold sensitivity, suprathreshold sensitivity or perceived intensity, and hedonics. Numerous variations and combinations of these types of methods have been employed, with the choice of one particular procedure over another often appearing to reflect simply the personal preferences of the investigator, although there may be more valid reasons to use certain methods. There is little question that identical methods should be used if it is important to relate results, in an absolute sense, to previous studies. ***See*** Food Acceptability, Affective Methods

Threshold Sensitivity

Thresholds are a measure of the lower limit of ability to detect the presence of a tastant above background (detection threshold) or to correctly identify its quality (recognition threshold), usually assessed as the level of stimulus which an individual can detect or correctly identify 50% (or other specified percentage) of the time. It is important to ensure, for recognition threshold testing, that subjects are sufficiently familiar with the terminology for naming the taste qualities of interest.

Taste thresholds are generally simple, but time- and labour-intensive to determine. Stimuli for classical taste threshold testing are almost always solutions of a single tastant in a neutral vehicle, usually deionized water, although there may be situations where it is of interest to determine thresholds (e.g. for an off-flavour or taint) in a real food product, in which case the food or other complex matrix may serve as the vehicle. Most procedures require preparation of a very wide range of tastant concentrations, usually in half-log or smaller steps.

There are numerous variations on several basic techniques for determination of thresholds. The approach in many types of tests is to repeatedly present single samples of taste stimuli (for recognition threshold) or pairs of stimuli (one sample of tastant, one sample of vehicle, for detection thresholds). The task is to identify the quality of the sample (recognition) or determine which sample in a pair contains the stimulus (detection) in a number of ascending and descending trials.

Other methods may determine thresholds solely from an ascending sequence, with the threshold taken as the lowest concentration at which a correct response is given on a predetermined number of repeated presentations. Alternatively, subjects may be given a mixed set of samples, and asked to separate them into those containing tastant and those containing only the solvent (e.g. deionized water).

Individual threshold values for specific taste stimuli are typically unrelated to gustatory function as determined by other measures. Furthermore, individuals with isolated decrements in threshold sensitivity generally do not report a consistent feeling of loss of taste function. Because real foods contain a variety of sensory stimuli at levels well above threshold and in a complex matrix, the measure of taste sensitivity to one or more compounds in a simple vehicle by itself may reveal little about individual perceptions of foods.

Despite these drawbacks, there are suitable applications of threshold measures in assessing gustatory function. They can be sensitive indices of the functionality of the sensory system itself, and may be more responsive than other measures to the initial stages of sensory loss. For food processors, knowledge of taste threshold values may find use in quality control, in that the level of human sensitivity to the taste of desirable or undesirable compounds may provide guidelines against which product formulation and processing procedures may be assessed.

Perceived Intensity

Concentrations of tastants at suprathreshold levels provide the most salient sensations in our everyday experiences. Testing of stimuli at this level generally involves an assessment of the relationship between stimulus concentration and perceived intensity. For taste, in common with other sensory systems, this relationship appears to be best described as a power function of the following form:

$$I = aX^{\beta}$$

where I is perceived intensity, a is a proportionality constant, X is physical concentration, and β is an exponent characteristic of the stimulus and testing conditions. In order to determine these variables, intensity ratings are commonly made using an open-ended ratio scaling system, such as magnitude estimation, where subjects basically assign their own numerical values to stimuli in proportion to the intensity of the sensation they perceive. When the log of intensity judgments are plotted against the log of tastant concentrations, the result is typically a linear relationship with slope β.

The values of β can be used to compare individuals and groups, but also reflects the nature of the tastant itself. When $\beta > 1$, perceived intensity more than doubles with a doubling of physical stimulus concentration. Although a wide range of values have appeared from different laboratories, the value of β is generally found to be close to or slightly greater than unity for substances such as sucrose and sodium chloride. When $\beta < 1$, perceived intensity increases at a rate proportionally lower than the increases in physical stimulus concentrations. This tends to be the case for many bitter and, to some degree, sour tastants. The appropriate use of different scaling procedures and corrections for the way in which subjects use scaling systems is critical to proper interpretation of data, particularly when contrasting the responses of individuals.

The most common alternative to magnitude estimation scaling for intensity is some form of line or category scale. Such scales are inherently easy to design and use, simple to explain, and require minimal data handling. However, categories are limited in range, are not necessarily equally spaced intervals, and do not offer the theoretical mathematical advantages of ratio scales. Nevertheless, several studies indicate that category scales can generate similar results to, and share common response characteristics with, magnitude estimation data. The relationships between category ratings and tastant concentrations are best described by semilogarithmic function of the following form:

$$I = a + \beta(\log X)$$

In practice, category scales may offer useful advantages in reliability and use by novice judges. Where subject groups of sets of samples are the focus of comparisons within a study, category and line scales are wholly suitable and easy to implement. However, where the study requires derivation of accurate (i.e. absolute, externally valid) values of the power function variables, such as in clinical classification of individuals with regard to gustatory function, or specific comparisons to published work, then magnitude estimation is presently the standard method.

Marked deficits in perceived intensity of stimuli at the suprathreshold level are likely to be associated with subjectively noticeable losses in sensory function. Such deficits may occur with or without altered thresholds, as shown in Fig. 1. The different stimulus–intensity relationships demonstrate why thresholds alone, or comparisons of responses to tastants at a single concentration or over a very narrow range, may provide an invalid assessment of sensory function.

The preceding methods consider intensity only at a single time-point after exposure. In reality, however, perceptions of chemosensory stimuli have a temporal component, generally incorporating a time-lag between exposure and sensation, then a rise in intensity followed by a decay to extinction. The sequence may take only a few seconds or several minutes, depending upon the stimulus and sample matrix. A number of different methods for assessing time–intensity (T–I) responses have been used, and with increasing availability of computerized data collection and improved statistical methodology for analysis of T–I curves, use of this technique will find many applications.

Hedonics

While threshold sensitivity and perceived intensity largely define in many ways the physiological capabili-

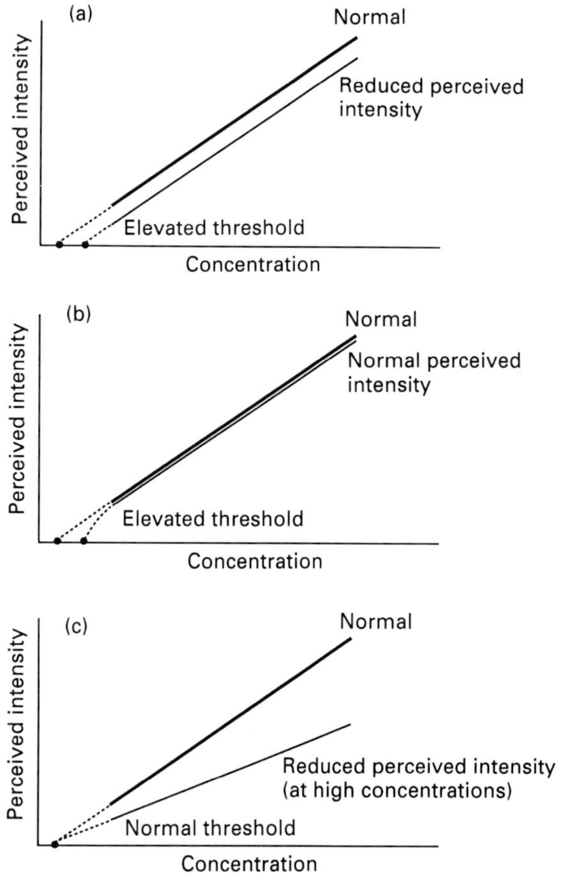

Fig. 1 Three different patterns of abnormal sensory response. (a) A relatively constant decrement in perceived intensity at all concentrations. (b) An elevated threshold with normal perceived intensity. (c) A change in the stimulus–intensity relationship reflected as relative decrement in perceived intensity at higher concentrations. Modified from Bartoshuk LM, Gent J, Catalanotto FA and Goodspeed RB (1983) Clinical evaluation of taste. *American Journal of Otolaryngology* 4: 257–260.

ties of sensory systems, taste sensations may elicit a robust affective response. The goal in hedonic testing for research or clinical purposes is usually to contrast preferences among individuals or populations. This is considerably different from similar work in food processing, where the objective is product optimization. In hedonic assessments, subjects are typically asked to rate stimuli on simple category or line scales for overall liking (e.g. ranging from 'extremely unpleasant' to 'extremely pleasant'), liking of specific taste attributes (e.g. 'not salty enough' to 'just right' to 'far too salty'), or to compare two or more stimuli and rank them for preference. While the test stimuli for determining thresholds or intensity functions are usually purified substances (e.g. a single tastant in water), stimuli for hedonic tests are often samples of real or modified foods varying in their content of one or more compounds of interest. In addition to this type of sensory testing, data on 'liking' could be derived from laboratory or home consumption trials, surveys of preference or preferred frequency of consumption of food items, habitual dietary intake, or combinations of these. However, sensory testing can reveal the nature and shape of the relationship between tastant concentration and liking, and the tastant concentration most preferred. These can then be used to compare or categorize populations, and to assess the effect of time or specific treatments on preferences of individuals.

Hedonic measures show closer associations with actual dietary intakes than assessments of sensory function. However, it is clear that sensory evaluation accesses only one dimension of food acceptance. Other tests, such as actual consumption, can reveal information about acceptability which may or may not coincide with sensory testing. Simple extrapolation from sensory data on taste hedonics to food preferences or selection or overall dietary intake, without actually measuring these, must be viewed with caution. Laboratory testing situations, while allowing a great deal of control, may also restrict the expression of behaviours and affect exhibited in more common eating circumstances, leading to highly reliable but invalid results.

While individual taste preferences undoubtedly have a significant idiosyncratic component, they may be strongly influenced by genetics and physiological state. Sweet-tasting stimuli are readily accepted and liked by humans at birth, while sour- and most bitter-tasting materials appear to elicit aversive responses. A favourable response to salt appears to develop at 4–6 months postnatal, at which time it is apparently liked. Putative teleological explanations have been assigned to this small core of innate sensory preferences, i.e. proposing that they would serve to direct humans toward safe sources of energy and selected nutrients, and allow recognition and avoidance of many naturally occurring (bitter and sour) toxins.

Preferred concentrations of sweetness and saltiness in foods appear to be greatest in young children, declining into adulthood, although it is difficult to distinguish the different contributions of genetics and experience on the developmental changes in taste preferences. It is clear that learned responses to taste qualities in specific foods are highly robust, and can ultimately supersede innate preferences as determinants of hedonic response. For example, it seems reasonable to ascribe the acquisition of taste preferences for (the innately disliked) bitterness in caffeinated beverages and beer to the rewarding psychobiological effects of consumption of these foods.

Genetic Differences in Taste Perception

Taste thresholds for common tastants often indicate wide variation between individuals, particularly in bitter

Table 1. Ability to taste phenylthiocarbamide (PTC) among selected populations

Population	Percentage of PTC 'tasters'
Northern European	69
Spanish and Portuguese	76
Japanese	93
Black African	97
Brazilian Indian	99

Data abstracted from Allison AC and Blumberg BS (1959) Ability to taste phenylthiocarbamide among Alaskan eskimos and other populations. *Human Biology* 31: 352–359.

taste sensitivity. One well-characterized interindividual difference in sensory function is the genetically determined sensitivity to the bitter taste of thioureas such as phenylthiocarbamide (PTC) or 6-*n*-propylthiouracil (PROP). These and related compounds have also been of nutritional interest as they are goitrogens which may be naturally occurring in selected cruciferous vegetables.

'Tasters' perceive weak concentrations of PTC or PROP as intensely bitter, while 'nontasters' generally perceive no taste at all except at very high levels. The proportion of tasters has been found to vary from 60% to almost 100% among different racial and ethnic groups (Table 1). There is further evidence suggesting that thiourea taster status may also relate to differences in the perception of several unrelated bitter and, possibly, some sweet tastants, at concentrations found in existing food products. However, evidence that PTC taster status is a meaningful determinant of food selection or acceptance is limited. Similar extreme variation in taste perceptions of other compounds, such as mannose, creatinine, and sodium benzoate, have also been described but are not as well documented. Thus it seems likely that humans differ not only in their acceptance of different taste qualities, but also in the actual sensation that many tastant materials produce.

Bibliography

ASTM (1968) *Manual of Sensory Testing Methods*. ASTM Special Technical Publication 434. Philadelphia: American Society for Testing and Materials.

Carterette EC and Friedman MP (eds) (1978) *Handbook of Perception. VIA: Tasting and Smelling*. London: Academic Press.

Finger TE and Silver WL (eds) (1978) *Neurobiology of Taste and Smell*. New York: John Wiley.

McBride RL and MacFie HJH (eds) (1990) *Psychological Basis of Sensory Evaluation*. Essex: Science Publishers.

McBurney DH and Collings VB (1977) *Introduction to Sensation/Perception*. Englewood Cliffs, New Jersey: Prentice Hall.

Meiselman HL and Rivlin RS (eds) (1986) *Clinical Measurement of Taste and Smell*. New York: Macmillan.

D. J. Mela
AFRC Institute of Food Research, Reading, UK

SEPARATION AND CLARIFICATION

Principles of Separation

Sedimentation

The natural process of sedimentation by gravity was one of the techniques used for separating a dilute slurry into a clear liquid and a deposit of denser material before the invention of centrifugal separators, which were first used in the 1880s. Sedimentation was used, for example, in the dairy industry, where fat globules that aggregated on the surface of milk could be skimmed off by hand.

A suspension of slurry to be treated in this way must be a mixture of two or more phases; the phases to be separated must be insoluble in each other and must be of different densities (denoted by ρ), so that the less dense phase will float on the surface of the denser phase. The point at which the slurry separates completely into a layer containing settled solids and a layer of clear liquid, with a distinct interface between the two, is called the *critical settling point*. From this point onwards, the solids continue to settle at a very slow rate; as they compress, more liquid is forced up into the liquid layer. The velocity with which a solid particle or liquid droplet moves through a viscous fluid under the influence of gravity is called the sedimentation velocity (v_g). This velocity is determined by the following physical parameters:

Particle diameter d (m)
Particle density ρ_P (kg m^{-3})
Density of the fluid medium ρ_L (kg m^{-3})
Viscosity of the fluid medium η (kg m^{-1} s^{-1})
Acceleration due to gravity g (9·81 m s^{-2})

If the values of these parameters are known, then the sedimentation velocity of the particle or droplet can be calculated using a formula based on Stokes' law:

$$v_g = [d^2(\rho_P - \rho_L)/18\eta]g \quad (1)$$

Thus it can be seen that sedimentation velocity increases with the square of the particle diameter (so that a particle of $d = 2$ mm will settle four times faster than a particle of $d = 1$ mm). The sedimentation velocity will also increase with increasing differential density between the phases, and with diminishing viscosity of the fluid medium (a particle will move more quickly through a freely flowing liquid).

Centrifugation Theory

Centrifuges use rotational acceleration to separate two phases, and the acceleration (a) increases with distance from the axis of rotation (radius r) and with the speed of rotation, or angular velocity (ω) according to the following formula:

$$a = r\omega^2$$

The force acting on a particle (which is directly proportional to its acceleration) therefore increases with increasing distance between a particle and the axis of rotation, and with the speed of rotation. A particle at any position is considered to move at the terminal velocity characteristic of its position. Substituting centrifugal acceleration ($r\omega^2$) for the acceleration due to gravity (g) in eqn [1], results in the equation:

$$v = [d^2(\rho_P - \rho_L)/18\eta]r\omega^2 \quad (2)$$

where v = terminal falling velocity of spherical particles of diameter d at radius r in a centrifugal field rotating at rate ω. From this formula, the sedimentation velocity of every particle in the centrifuge can be calculated. When the sedimentation velocity in a centrifuge is divided by the sedimentation velocity in gravity (eqn [1]), a measure of the efficiency of the centrifuge is obtained.

Clarification

The separation of solid particles from a liquid is called *clarification*. A centrifuge used for this purpose contains a series of baffle inserts, each a conical disc, which together form a disc stack (see Fig. 3). Radial strips known as caulks keep the discs the correct distance apart so that the separation channels formed are of the optimum width.

The fluid to be clarified normally enters the channel toward the outer edge and leaves at the inner edge. While the fluid moves through the channel the denser particles settle outwards towards the disc forming the upper boundary. The velocity of the fluid varies across the width of the channel; close to the discs it is near zero, and in the centre of the channel it is at its maximum. These particles in the fluid move both in the direction of the fluid and radially outwards (Fig. 1). These two movements result in the vector velocity (v_P). The velocity of the fluid on the surface of the disc is so low that any settled particles are no longer carried along with the fluid. They slide outwards along the underside of the disc until they are deposited on the wall of the centrifuge.

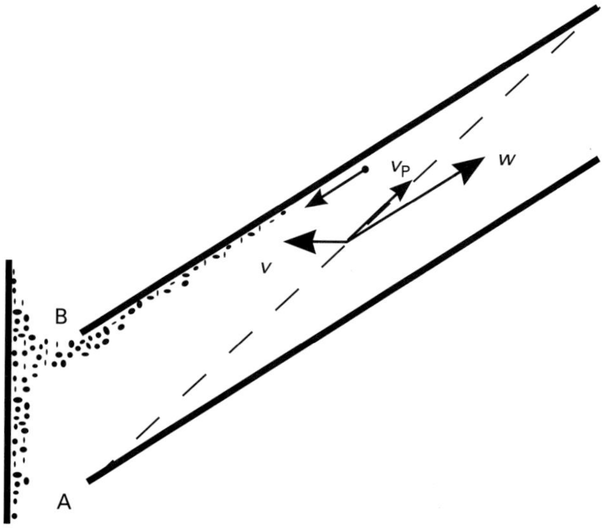

Fig. 1 If a particle is introduced at point A, it will be carried with the liquid phase in the direction w and towards the upper disc B by the sedimentation velocity v; the particle therefore follows a path representing the sum of the these two forces – the vector velocity (v_P). After impaction against the upper disc, the force w is negligible; the centrifugal force carries the particles along the underside of disc B as shown.

The *limit particle* is the smallest size of particle which can enter the centrifuge at the least favourable point in the separation channel (i.e. close to the upper surface of the lower disc) and still reach the underside of the upper disc by the time it has moved to the inner edge of the disc stack. Any particles larger than the limit particle will be separated. Particles smaller than the limit particle can also be separated if they enter the centrifuge closer to the top of the separation channel (i.e. closer to the underside of the disc on which they will settle). The smaller the particle is, the closer to the top of the separation channel it must enter in order to be separated. Therefore, for particles smaller than the limit particle, the proportion separated diminishes with size of particle.

Centrifuge Types and Applications

Centrifuges are used widely in the food industry for both liquid–liquid and solid–liquid separations. Low-speed centrifuges generate at least 1200g ($g = 9{\cdot}81$ m s^{-2}), and small tubular centrifuges can generate up to 63 000g. Several types of centrifuge are used depending on the application.

The solid bowl disc centrifuge is a liquid–liquid separator which generally operates at between 5000 and 10 000g, and, in the dairy industry for example, can handle throughput rates of up to 35 000 l h^{-1}. A nozzle

Fig. 2 A centrifugal separator capable of handling some 25 000 l of milk per hour for the production of skim-milk and market cream; the relative volumes of the output streams will depend on the fat contents of both the incoming milk and the cream.

version of this centrifuge separates solids as a pumpable slurry which is continuously discharged through a nozzle at the top of the centrifuge. Solids output rates as high as $5000\,l\,h^{-1}$ are possible. Disc centrifuges have been used for diverse applications in the food industry such as cream separation and the refining of fish oils, and also for catalyst separation and dehydration of marine lubrication oils. ***See*** Cream, Types of Cream; Fish Oils, Production

When the process demands a dry solid phase, and the fluid contains a large volume of solids, e.g. the separation of sugar from the mother syrup, a solid bowl decanter centrifuge can be used. At forces of 1000–3000*g*, throughput rates can be up to $26\,000\,l\,h^{-1}$, which, in the sugar industry, is equivalent to $20\,t\,h^{-1}$. High-speed decanters (5000–10 000*g*) can be used but with lower rates. ***See*** Sugar, Handling of Sugar Beet and Sugar Cane

Tubular centrifuges can be used for liquid–liquid separations but are unsuitable for solid–liquid separations since they have to be disassembled to retrieve the solids.

Many advances in centrifuge design have been made which increase both speed and capacity. In the dairy industry, centrifuges are used for the removal of solid impurities from milk before pasteurization, for removing bacteria from milk, for separating cream from milk (Fig. 2), for separating fat from whey, for separating quarg curd from whey and for purifying butter oil. ***See*** Milk, Processing of Liquid Milk

Example of Application: Continuous Separation of Milk

In a milk-skimming centrifugal separator, the milk enters through vertically aligned channelling holes in the

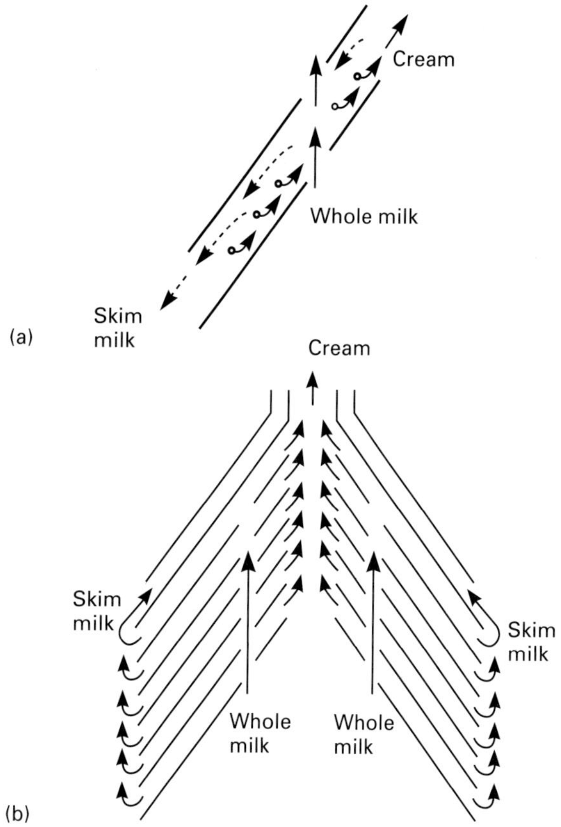

Fig. 3 (a) The flow of cream and skim-milk in the space between of a pair of discs in centrifugal separator; (b) a typical disc stack. (Redrawn after Towler C (1986).)

discs (Fig. 3). The particles settle radially outwards or inwards in the separation channels depending on their relative density. High-density solid impurities move towards the periphery of the separator and settle in the sediment space. Fat globules (cream) are less dense than the skim milk and so move inwards towards the axis of rotation. They then pass to an axial outlet. The skim milk moves outwards and enters a channel at the top of the disc stack from which it leaves the separator.

Efficiency of separation depends on the centrifuge design and on the flow rate of the milk through the separator. The smallest fat globules cannot be separated, leaving a residual fat content in the milk of between 0·05% and 0·07%. If the flow rate through the separator is reduced, the velocity of the fluid through the separation channels is reduced, so that smaller particles have time to sediment, thus increasing the efficiency of the separation.

The contaminating material in milk, e.g. dirt particles, blood cells and bacteria, comprises about 1 kg per 10 000 litres; this increases if sour milk is being separated. Older style separators necessitated manual cleaning to remove the solid sediment; now, separators may be used which automatically eject solids at designated intervals (normally hourly). The centrifuge has a bowl with peripheral discharge slots covered by a movable sliding bowl bottom which is held in place by the pressure of water underneath it. To eject the solids, the water pressure is momentarily released and the sliding bowl bottom moves downwards to expose the slots for about 0·15 s, just sufficient for the solids contents to be discharged. The ejected sediment is deposited in an annular receiver which circulates at the same high angular velocity as the periphery of the bowl. This motion carries the solids through a discharge duct to a cyclone where the kinetic energy of the sediment is absorbed. The solids are then discharged by gravity to a collecting tank.

Bibliography

Anon. (1975) *Cream Processing Manual.* Middlesex: Society of Dairy Technology.

Earle RL (1983) *Unit Operations in Food Processing.* Oxford: Pergamon Press.

Foust AS, Wenzel LA, Clump CW, Maus L and Andersen LB (1960) *Principles of Unit Operations.* New York: Wiley.

Towler C (1986) Developments in cream separation and processing. In: Robinson RK (ed.) *Modern Dairy Technology*, vol. I. London: Elsevier Applied Science.

Wiseman A (ed.) (1983) *Principles of Biotechnology.* Surrey: Surrey University Press.

S.C. Robertson and R.K. Robinson
University of Reading, Reading, UK

SHARK

See Fish

SHEEP

Contents

Meat

Lamb is the meat from young sheep. It should be distinguished from mutton, which is from adult animals. The two categories together are referred to as sheep meat.

Sheep, raised for both meat and wool production, form a significant part of the economy of many countries. The current world production of mutton and lamb is 6·5 million tonnes (FAO, 1990). Sheep are farmed in all the main regions of the world, but it is notable that 17% occur in Australia and New Zealand and that exports from these countries are the core of world trade.

Sheep farming is possibly the most diverse branch of animal production. There are hundreds of breeds, each with its particular environmental niche and economic function. There is a large variation between the major sheep-producing countries in climatic conditions, in management procedures and in the breed types available. At one end of the climatic scale there are the fat-tailed desert sheep which survive and thrive where other sheep would perish; at the other extreme are mountain breeds like the Scottish Blackface of Great Britain, which can stand harsh conditions, high rainfall, snow and indifferent herbage. Most sheep feed outside continuously with only limited supplementation of available pasture, whereas a few, mainly in Europe and North America, are housed from birth and fed carefully formulated diets. There is little mechanization or intensification.

In some countries, sheep-farming systems can cover wide ranges of climatic condition, as in Britain where there is a stratification system involving the crossing of breeds adapted to particular environments: crosses from hill and upland areas are the female breeding stock for lowland areas where they are mated with sire breeds selected specifically for meat production.

Genetic improvement in the world's sheep population has been very uneven. On the one hand there are pioneering examples of application of genetics, particularly in Norway, Australasia and the UK. Published information on sheep breeding is much more a review of the technology about to be applied, rather than a record of achievement.

Progress is slow because environmental, technical and economic conditions prevailing in areas where sheep are kept are frequently difficult. Breeding objectives are often not clear: indeed, there may be direct conflicts between adaptation for survival and for productivity. Relatively low reproduction rates and long generation intervals do not help. The multipurpose use of sheep for meat, wool and milk also reduces selection pressure for meat characteristics under some circumstances.

Sheep meat production is dominated by two factors; first there are the biological and agricultural considerations of resource use. Secondly there are the marketing considerations, in particular the ability of farmers to adapt output to conform more closely with changing market demand. In developed countries, this demand is for an increasingly lean product.

Carcass Quality

The value of sheep carcasses depends on several factors, namely weight, conformation (carcass shape), proportion of the main tissues (muscle, fat and bone), distribution of these tissues through the carcass, muscle thickness and meat quality. ***See*** Meat, Structure

The weight and size of the carcass has a major influence, not only on the quantity of the various tissues, but also on the size of the muscles exposed on cutting and of the individual joints prepared from it. This is of importance particularly in relation to a retailer's ability to provide cuts of suitable size for customer requirements. Over generations, wholesalers and retailers from different regions of countries and different parts of the world have become accustomed to handle certain weight ranges of carcasses; abattoir practices and cutting methods have been developed accordingly. Most wholesalers state desired weight ranges in buying schedules and apply discounts to carcasses outside these ranges. In some cases, these discounts are severe and are a constraint on the use of improved breeds and production systems. There has, for example, historically been a demand in Britain for small legs suitable for the Sunday roast, which has slowed the movement to heavier carcasses that are likely to be more efficient to produce and process. This demand has also had a major effect on the type of carcass produced in New Zealand for the British market. ***See*** Meat, Slaughter

Table 1. Composition of carcasses in different fat classes of the British classification scheme

	Fat class[a]					
	1	2	3L	3H	4	5
Lean meat in carcass (%)	62·4	58·7	56·1	54·3	51·7	48·1
Separable fat in carcass (%)	16·5	21·7	25·6	28·2	32·1	37·3
Total-protein in carcass (%)	14·2	13·3	12·7	12·2	11·4	10·2
Lipid in lean (%)[b]	3·6	5·1	6·2	6·9	8·1	9·6

[a] Most carcasses fall into fat class 3L. Fat classes 4 and 5 would be too fat for most requirements.
[b] Intramuscular fat.
Data from Kempster AJ *et al.* (1986).

Although conformation (the shape of the carcass) provides a poor indication of carcass composition, it is still regarded by many in the meat industry as valuable in this respect. Carcasses with good conformation normally command higher prices and most national classification and grading schemes include conformation as a factor. It is particularly important in some European markets.

Among carcasses of similar weight, the percentage formed by each tissue varies considerably depending on breed type and level of feeding. The proportion of lean meat in the carcass is of major importance since this is the principal determinant of yield and commercial value in many countries. Taken as a generalized ideal, the best carcasses should have an optimum level of fatness, sufficient to ensure that carcasses do not dry out and to ensure good eating quality, and minimum bone.

The general trend is now towards the production of leaner carcasses because of the consumers' increasing demand for meat with a low fat content. Table 1 gives the typical composition for carcasses in different fat classes of the National Carcass Classification Scheme operated in Britain (this can be linked directly to the developing EEC scheme). ***See*** Meat, Dietary Importance

Definition of Retail Cuts

Traditionally, lamb is sold by butchers largely bone-in; so about 90% of the carcass is 'saleable meat' and goes across the counter.

The lamb carcass is typically separated into the sides by cutting it lengthwise through the spine. The main cuts (using British terminology) are as follows.

The *forequarter* is a large cut which includes the neck, shoulder and part of the breast. It makes an economical family roast but is more difficult to carve than a leg. Boned and rolled or stuffed, it is easier to carve.

The *shoulder* is smaller than the leg. It is more easily carved after boning. Shoulder chops or cubed shoulder meat should be well trimmed to remove excess fat before grilling or stewing.

Neck chops may be stewed, braised or casseroled.

Breast is an economy cut. The ends of the rib bones can be cut out to simplify carving or rolling.

The *best end of neck* or *rack* is made up of six or seven rib chops. When it is to be cooked as whole, it may be boned and rolled, or chined (backbone removed) for easy carving. The rib chops are trimmed to make cutlets.

The *loin* can be cut into about seven or eight meaty chops, each with a short T bone; or it can be left in one piece and rolled; or trimmed, boned, rolled and cut into noisettes before cooking.

Chump chops (or *leg chops*) are lean, meaty chops cut from the end of the leg nearest the loin.

The *leg* is a large, lean roast which may be cooked with the bone in, or boned and stuffed. Meaty steaks can be cut from the thick part of the leg.

Consumer attitudes indicate that lamb is often regarded as fatty, not versatile, and difficult to carve. Major developments have, therefore, taken place in new cutting methods. This work produces boneless cuts which are convenient in size, easy to carve and lead to minimal waste on the plate.

The new techniques only work satisfactorily on lambs which are quite lean at the outset, since it is difficult to trim the intermuscular fat without adopting a complete muscle seaming technique. If there is increasing use of these boneless techniques it will create further pressure towards the production of heavier and leaner lambs.

Lamb and mutton are used less in processed form than other meats. One of the main reasons is the high bone-to-meat content of the carcass and the small size of the carcass. These two factors make it more labour-intensive for boning-out and processing and therefore more costly relative to other meats. Lamb fat is also more saturated and less suitable for processing.

The main area of development in Britain at the moment is in kitchen-ready products which offer variety, convenience and novelty to consumers. The range of products now being introduced in Britain, particularly by independent butchers, includes lamb burgers, marinated chops and steaks, and more exotic dishes such as lamb bengali, lamb pasanda and lamb italliene. Lamb is the basis for many spiced dishes on a worldwide basis, and quite a number of these are being introduced to Europe by the major supermarket chains.

Recently there has been renewed interest in the curing of lamb because of its cheapness in some countries relative to pigmeat. ***See*** Meat, Preservation

Bibliography

Food and Agriculture Organization of the United Nations (1990) *Production Yearbook, 1989*, vol. 43. Rome: FAO.

Kempster AJ, Cook GL and Grantley-Smith M (1986) National estimates of body composition of British cattle, sheep and pigs with special reference to trends in fatness. A review. *Meat Science* 17: 107–138.

A. J. Kempster
Meat and Livestock Commission, Milton Keynes, UK

Milk

The milking of sheep dates back to the domestication of animals. Although the cow has replaced the sheep as the major milk producer in many countries, sheep and goats remain important as sources of milk where land is too poor to support cows. With the search for alternative farm incomes there has been a rekindling of interest in sheep dairying in northern Europe, Australia and the USA. Attractions are low capital investment, limited land requirement, high profit margins per litre and lack of quota limits on production.

Cheese and yoghurt are the main products of sheep's milk worldwide. Smaller markets exist for liquid milk and other products. Liquid milk may be pasteurized, frozen or spray-dried. The main differences between the milk of sheep and that of cows and goats is that it contains almost double the solids, being rich in fat, protein, minerals and vitamins. However, the methods of production and processing and the potential microbiological problems are similar. ***See*** Milk, Dietary Importance

Production of Sheep's Milk

In 1989, 8 778 000 t of sheep's milk was produced worldwide. Asia and Europe are the continents with the highest levels of production, the main producers being concentrated in the North Mediterranean Basin, the Black Sea area, Iran and Central Asia. In many of these countries the normal practice is to milk the ewes after the lambs have been weaned, the milk being an essential food in the diet of the local population. However, sheep can only be considered as dairy animals when they are milked for the greater part of their lactation, the lambs only suckling for 1–2 months. When this criterion is applied the concentration of dairy production is at present mainly restricted to Europe and Israel.

Breeds of Dairy Sheep

All breeds of sheep can be used for milk production and in many localities strains of sheep have been selected for their ability to produce significant quantities of milk. In the Mediterranean at least 48 breeds can be classified primarily as milk producers; worldwide the number would be much higher. The breeds most used for milk production in large commercial flocks are the Friesland and the Texel, although the Lacaune is favoured in the Roquefort region of France and the Awassi in Israel. Many flockmasters cross a milk breed with a local breed to introduce hardiness, or use breeds such as the Finn Dorset cross to encourage prolificacy.

Yields of Milk

Yields vary depending on factors such as length of lactation, time of year of lambing, level of nutrition and housing. However, yields of 120 to 1000 l per lactation are quoted, the normal yield being 200–300 l per lactation.

To maintain a constant supply of milk, out-of-season lambing can be practised. Each ewe may then produce three lamb crops in 2 years, so increasing profitability. An alternative method is to keep two flocks, one lambing at the traditional season and the other out-of-season. Both methods require a high level of husbandry.

Milking

Milking is carried out mechanically in commercial flocks. Milking machines are similar to those used for dairy cows except that the vacuum level is set at 10–11 lb (4·5–5 kg) and the clusters are specialized units each with two lightweight teatcups (Fig. 1).

Milking is normally carried out in a parlour, which may be of the static or rotary design. The milk is then transferred either to milk churns or to a bulk milk tank depending on the size of the enterprise. The milk must be cooled to below 5°C as soon as possible after milking. In the case of churns this can be achieved by passing the milk over a surface cooler before it enters the churn, or by using an in-churn cooler. Churns should then be stored in a refrigerator. With a bulk tank, cooling is automatically carried out. *See* Milk, Liquid Milk for the Consumer

Hygiene in Milk Production

High levels of hygiene must be maintained at every stage of production and handling of the milk, paying particular attention to the following:

1. Sheep should be kept as clean as is practicable and the hair and wool clipped away from the area around the udder.
2. Udders should be cleaned prior to milking, particular attention being paid to teats and teat orifices.

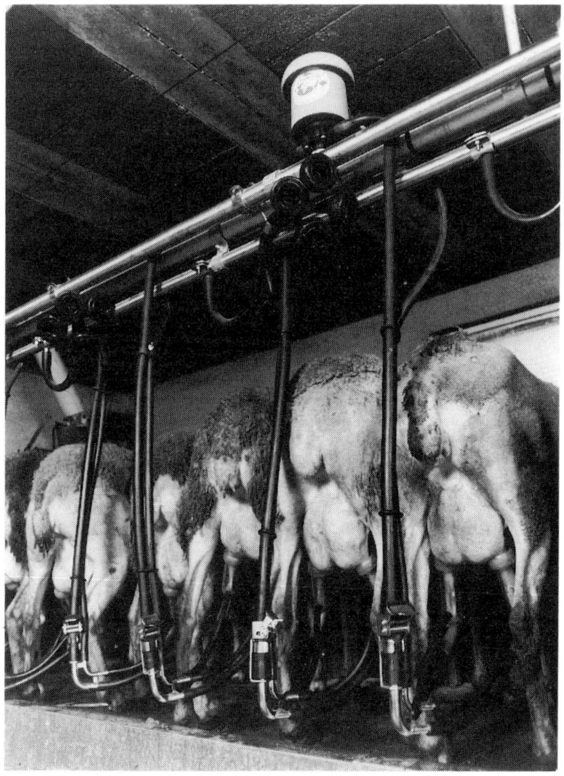

Fig. 1 Sheep milked in parlour. (Courtesy of RJ Fullwood and Bland Ltd.)

Table 1. Composition of ewes', goats' and cows' milk: range of levels reported in the literature

	Ewe[a] (%)	Goat[a] (%)	Cow (%)
Total solids	15·8–23·4	11·33–18·68	12·4
Fat	4·54–12·6	2·84–7·78	3·8
Protein	4·3–6·77	2·64–5·3	3·3
Lactose	4·19–5·25	3·91–6·3	4·7

[a] Data from Juarez M and Ramos M (1984) Dairy Products from ewe's and goat's milk. *Dairy Industries International* 49(7): 20–24. Other data from Holland B, Welch A, Unwin ID *et al.* (1991) *McCance and Widdowson's The Composition of Foods* 5th edn. London: Her Majesty's Stationery Office.

3. Cleaning and sterilization of the milking equipment must be carried out thoroughly after each milking.
4. Milk must be cooled promptly and kept at less than 5°C until used. *See* Milk, Processing of Liquid Milk

Composition and Nutritional Significance

Composition

Information on the composition of sheep's milk is sparse. However, it is generally richer in fat, protein and total solids than goats' or cows' milk (Table 1).

Table 2. Mineral constituents of ewes' and cows' milk: range of levels reported in the literature

	Ewe[a]	Cow
Sodium ($g\,l^{-1}$)	1·41–1·32	0·3–0·9
Potassium ($g\,l^{-1}$)	1·01–1·52	1·1–1·7
Calcium ($g\,l^{-1}$)	1·62–2·59	1·1–1·3
Magnesium ($g\,l^{-1}$)	0·14–0·19	0·09–0·1
Phosphorus ($g\,l^{-1}$)	0·82–1·83	0·9–1·0
Iron ($mg\,l^{-1}$)	0·29–1·39	0·3–0·6
Copper ($mg\,l^{-1}$)	0·19–1·3	0·1–0·3
Zinc ($mg\,l^{-1}$)	4·66–11·8	2·0–6·0
Manganese ($mg\,l^{-1}$)	0·032–0·13	

[a] Data from Juarez M and Ramos M (1984) Dairy Products from ewe's and goat's milk. *Dairy Industries International* 49(7): 20–24. Other data from Holland B, Welch A, Unwin ID *et al.* (1991) *McCance and Widdowson's The Composition of Foods* 5th edn. London: Her Majesty's Stationery Office.

Table 3. B vitamin content of ewes' and cows' milk: range of levels reported in the literature

	Ewe[a]	Cow
Riboflavin ($mg\,l^{-1}$)	4·3–5·0	1·5–2·3
Thiamin ($mg\,l^{-1}$)	0·6–1·2	0·3–0·6
Nicotinic acid ($mg\,l^{-1}$)	3·9–5·4	0·6–1·3
Pantothenic acid ($mg\,l^{-1}$)	3·5–5·3	3·5
Vitamin B_6 ($mg\,l^{-1}$)	0·7	0·4
Folic acid ($\mu g\,l^{-1}$)	54·0	50·0
Vitamin B_{12} ($\mu g\,l^{-1}$)	1·4–9·8	3·0
Biotin ($\mu g\,l^{-1}$)	50·0–90·0	20·0

[a] Data from Williams AP, Bishop DR, Cockburn JE and Scott KJ (1976) Composition of ewe's milk. *Journal of Dairy Research* 43: 325–329.
Other data from Holland B, Welch A, Unwin ID *et al.* (1991) *McCance and Widdowson's The Composition of Foods* 5th edn. London: Her Majesty's Stationery Office.

As with other milks, variations in composition with breed, stage of lactation, diet, season, milking times and procedure have been recorded. Levels of fat and total solids tend to increase with length of lactation, and milk at the end of lactation may have a fat content greater than 10% w/v. Fat content may vary with breed. The essential amino acids are present at concentrations similar to those in cows' milk, although isoleucine, leucine, serine, glutamic acid and tyrosine may be slightly lower. Mineral content is similar to or higher than that of cows' milk, levels of calcium being particularly high (Table 2). Levels of vitamins are again similar to or higher than those in cows' milk, sheep's milk being particularly rich in some of the B vitamins (Table 3). ***See*** individual nutrients

The very low level of carotene in sheep's milk, like goats' milk, means that it is white rather than cream in colour. Like goats' milk, sheep's milk contains a high proportion of small fat globules (86% 4·5 μm diameter or less), making it a 'naturally homogenized' milk. This has advantages in cheesemaking, where the small globules remain trapped in the curd so that less fat is lost in the whey. It also means that the milk can be frozen.

Nutritional Significance

From the compositional data it can be seen that sheep's milk has a higher nutritive and energy value than cows' and goats' milk. Advantages to the processor are that yoghurt can be made without increasing the levels of solids not fat, and cheese yields are high. The small size of the fat globules may enhance digestibility of the milk and products. However, the high levels of proteins and minerals make the milk unsuitable for infants under 12 months old.

Allergy sufferers who are intolerant of cows' milk may tolerate sheep's milk. However, as yet this is supported by anecdotal rather than scientific evidence.

Processing of Liquid Milk

Liquid milk may be frozen (before or after pasteurization), or spray-dried.

Freezing

Milk should be frozen rapidly and as soon as possible after milking to prevent separation of solids. Thin blocks (up to 3·5 cm thick) of up to 5 l of milk can be handled and frozen successfully to −25°C or lower. In this form it will be suitable for processing for 6 months or longer. Shorter-term storage is possible in a domestic freezer (−12°C to −18°C), where deterioration takes place after 1–3 months. Slow thawing at refrigeration temperatures avoids problems with separation, although more rapid thawing may still be satisfactory for product making. Microbiological quality generally remains unaltered by freezing, although slow freezing and storage at higher freezer temperatures may have some bactericidal effect. ***See*** Freezing, Principles

Pasteurization

Sheep's milk can be satisfactorily pasteurized by batch or high-temperature, short-time (HTST) processes, although milk pasteurized by the batch method may develop an unacceptable flavour. Alkaline phosphatase activity is higher in sheep's milk than in cows' milk, and both the Aschaffenburg and Mullen test and the Scharer test have been used to assess effectiveness of pasteurization. These tests rely on the detection of phenol compounds liberated from phosphate substrates (disodium phenyl phosphate – Scharer; and *p*-nitrophenyl phosphate – Aschaffenburg and Mullen) by the alkaline phosphatase enzyme. However, results may not always be reliable with sheep's milk. ***See*** Pasteurization, Principles

Spray-drying

Sheep's milk can be spray-dried, either in full-fat or skimmed form, in the same way as cows' milk. ***See*** Drying, Spray Drying

Production of Processed Foods

The high level of the solid components in sheep's milk makes it particularly suitable for manufacture into yoghurt, cheese, butter and ice cream.

Yoghurt

The uniform small fat globules in sheep's milk ensure that there is little or no separation of the cream in the yoghurt, so that the milk does not require homogenization. The method of manufacture is the same as for cows' milk; the set type is normally manufactured, although a small amount of the strained variety is made, particularly in Greece. ***See*** Yoghurt, The Product and its Manufacture; Yoghurt, Dietary Importance

Cheese

Sheep's milk can be successfully made into cheese by any of the recipes used for cows' milk, although the resultant cheese is unlikely to resemble the corresponding cows' milk cheese. A wide range of cheeses (fresh, soft, brined, hard, and blue-veined varieties) have been made traditionally, many of which are produced in one locality. Roquefort, feta, pecorino and ricotta are probably the best known. Until the introduction of modern technology, feta cheese – which should be white – could only be made successfully from sheep's milk as there is too much natural colour in cows' milk. Pecorino is the hardest of the sheep's milk cheeses and is associated with the production of ricotta. Ricotta, which is not technically a cheese as it contains no casein, is produced by heating the whey of any cheese which has not been salted to at least 85°C. The resultant white foam is skimmed off and allowed to drain overnight in a 'ricotta' basket or strainer. ***See*** Cheeses, White Brined Varieties; Cheeses, Dietary Importance

In recent years a wide range of 'new' cheeses has been developed to suit modern tastes.

Butter

Production of butter from sheep's milk is not a commercial proposition, but it is in some localities for domestic use and is sometimes clarified to produce ghee. ***See*** Ghee

Ice Cream

Sheep's milk ice cream is not normally a commercially viable product. However, some manufacturers make small quantities which they sell from their own premises to encourage visitors to their retail outlet. ***See*** Ice Cream, Methods of Manufacture

Microbiology and Contamination

Relatively little information exists on the microbiology of sheep's milk. However, such milk and its products are subject to similar public health and quality problems as are milks from other animals.

General Microbiology

Milk drawn aseptically from sheep can be practically sterile. However, it easily becomes contaminated during handling, and studies on freshly produced bulk milk (hand- and machine-milked) give a generally poor picture, with average total viable counts of 10^5–10^6 bacteria per ml often reported. This is partly due to the preponderance of production in developing countries and rural areas where conditions are primitive and understanding of hygiene lacking. Milk produced under hygienic conditions can be of excellent microbiological quality. However, contamination from milking equipment can be a problem as the high fat content of the milk makes surfaces difficult to clean.

Mastitis

The commonest cause of clinical mastitis and therefore the most likely contaminant of milk from this source is *Staphylococcus aureus*. However, *Pasteurella haemolytica* may be as or more important in some flocks. Mastitic streptococci and a variety of other bacteria may also be present in the milk owing to clinical or subclinical udder infections.

Public Health Risks

The only report of disease of animal origin transmitted via sheep's milk or products in the UK has been of staphylococcal intoxication from raw milk cheese. Many strains of mastitic *S. aureus* have been shown to be enterotoxigenic, so this seems to be the greatest public health risk. *Listeria monocytogenes* causes circling disease in sheep and can be excreted in the milk. This poses a potential health risk since infection is not usually associated with severe mastitis or organoleptic changes in the milk. Sheep can also become infected with salmonellas and campylobacters, and a carriage pattern of campylobacter in healthy sheep similar to that in cows has been found in the UK, creating a potential risk. *Chlamydia psittaci* infects sheep but does not seem to be excreted into the milk. Other pathogens which could be transmitted to humans through sheep's milk and products are *Mycobacterium tuberculosis* (rare in sheep), *Brucella melitensis/abortus*, *Coxiella burnettii* and *Toxoplasma gondii*. ***See** Listeria*, Listeriosis; Mycobacteria

Infections can also be transmitted via sheep's milk and products from milkers, food handlers, or equipment.

Postpasteurization contamination of milk is thought to have been the cause of an outbreak of *Salmonella enteritidis* from cheese in Spain. Contamination of human origin has caused outbreaks of *Staphylococcus aureus* intoxication and shigella infection. The occurrence of other food poisoning organisms in sheep's milk or products has not been recorded. ***See** Staphylococcus*, Food Poisoning

Bibliography

IDF (1981) *The Composition of Ewe's and Goat's Milk*. IDF Bulletin Document 140. Belgium: International Dairy Federation.

MAFF (1989) *The Hygienic Production of Ewes' Milk*. London: Ministry of Agriculture, Fisheries and Food.

Mills O (1982) *Practical Sheep Dairying* 2nd edn. Wellingborough, UK: Thorsons.

Alison J. Stubbs
ADAS, Reading, UK
Susan C. Morgan-Jones
Scottish Office Agriculture and Fisheries Department, Edinburgh, UK

SHELF LIFE

See Chilled Storage, Controlled Atmosphere Storage, Packaging, Retailing of Food in the UK, Spoilage and Storage Stability

SHELLFISH

Contents

Characteristics of Crustacea

As shellfish or shelled aquatic invertebrates, the Crustacea share important features, such as bilateral symmetry, body segmentation and the possession of a well-developed exoskeleton, with the other groups within the phylum Arthropoda. However, they may be separated at the subphylum level from the Cheliceriformes (scorpions, spiders, etc.), Trilobitomorpha (fossil trilobites) and Uniramia (insects and myriapods) by the features listed below:

1. Appendages are uniramous or biramous.

2. Brain is tripartite (with deutocerebrum).
3. Body is divided into cephalon and trunk, or latter subdivided into thorax and abdomen.
4. There are five pairs of cephalic appendages: preoral first antennae and four pairs postoral appendages – second antennae (which migrate to preoral position in adults), mandibles, maxillules, and maxillae.
5. Mandible is gnathobasic (mandible arises from limb base); endopod and exopod are reduced in adults.

There are at least 39 000 species of crustaceans, ranging in size from less than 1 mm in length to spider crabs with leg spans of over 4 m, and these are subdivided into 6 classes and 38 orders. Although most of these are marine, small groups have exploited the freshwater and terrestrial environments, so that crustaceans can be found from mountain tops and deserts to abyssal depths. Despite the large number of species, many are either too small or too dispersed to be exploited by man. The forms most commonly exploited commercially are crabs, lobsters and shrimps. Planktonic copepods (10 000 species), Cladocera (450 species), planktonic and benthic Ostracoda (5000 species), and mainly benthic Amphipoda and Isopoda (10 000 species) often form an essential link in food webs which support fisheries, but such forms are only rarely themselves the subject of fisheries. However, a key problem in the management of crustacean fisheries concerns euphausiids (krill) which play a central role in the Antarctic Ocean food web as food for many whales, seals and birds, yet are also increasingly being fished directly as food for man and domestic animals.

Taxonomy

A brief outline classification of crustaceans utilized directly by man is given in Table 1; for more detailed descriptions refer to the bibliography. It is clear that apart from larger planktonic species such as euphausiids, mysids and sergestids, which swarm in sufficient concentration to be harvested economically, and gourmet species such as stalked barnacles (*Pollicipes*) and stomatopods, the majority of edible crustaceans are found in Order Decapoda which comprises some 10 000 species. The possession of branched or dendrobranch gills, external fertilization and release of eggs into the sea separates the penaeid and sergestid shrimp from the rest of the Decapoda.

The Pleocyemeta have unbranched gill filaments and brood their eggs, which hatch at a later stage than the nauplius produced on hatching by the Dendrobranchiata. This group contains the majority of the familiar shrimps or prawns (the terms are now synonymous), crayfish, lobsters, squat lobsters and crabs. Caridean shrimps (*Palaemon, Macrobrachium*) possess phyllobranchiate or flattened, plate-like gills, separating them from the larger and more robust Astacidea (which possess trichobranchiate or unbranched, tubular gill filaments). All the freshwater crayfish (Astacidea) and marine lobsters (*Homarus* and *Nephrops*) have greatly enlarged chelae or claws on the first pair of walking legs, distinguishing them from the Palinura (spiny and slipper lobsters), which have similar gill structures, but lack the enlarged claws. This marine group tends to be more mobile and possesses a flattened abdomen and large tail fan used in swimming.

The Anomura contain forms in which the abdomen is either soft and twisted asymmetrically to fit into gastro-

Table 1. Pennant (1771) classification of Phylum, Subphylum or Superclass Crustacea utilized directly for human nutrition

Class Maxillopoda	
Subclass Cirripedia	
Order Thoracica	Stalked barnacles, e.g. *Pollicipes*
Subclass Copepoda	
Order Calanoida	Copepods, e.g. *Calanus plumchrus*
Class Malacostraca	
Subclass Hoplocarida	
Order Stomatopoda	Mantis shrimp (*Squilla mantis*)
Subclass Eumalacostraca	
Superorder Peracarida	
Order Mysidacea	Possum shrimp (*Neomysis intermedia*)
Superorder Eucarida	
Order Euphausiacea	Krill, e.g. *Euphausia superba*
Order Decapoda	
Suborder Dendrobranchiata	Penaeid and sergestid shrimps, e.g. *Penaeus, Sergestes*
Suborder Pleocyemata	
Infraorder Caridea	Caridean and procaridean shrimps, e.g. *Macrobrachium, Palaemon*
Infraorder Astacidea	Crayfish and chelate lobsters, e.g. *Astacus, Homarus, Nephrops*
Infraorder Palinura	Palinurid, spiny and slipper lobsters, e.g. *Panulirus, Palinurus, Thenus, Scyllarides*
Infraorder Anomura	Galatheid crabs, king crabs, e.g. *Paralithodes, Pleuroncodes*
Infraorder Brachyura	Crabs, e.g. *Cancer, Scylla, Callinectes, Maia*

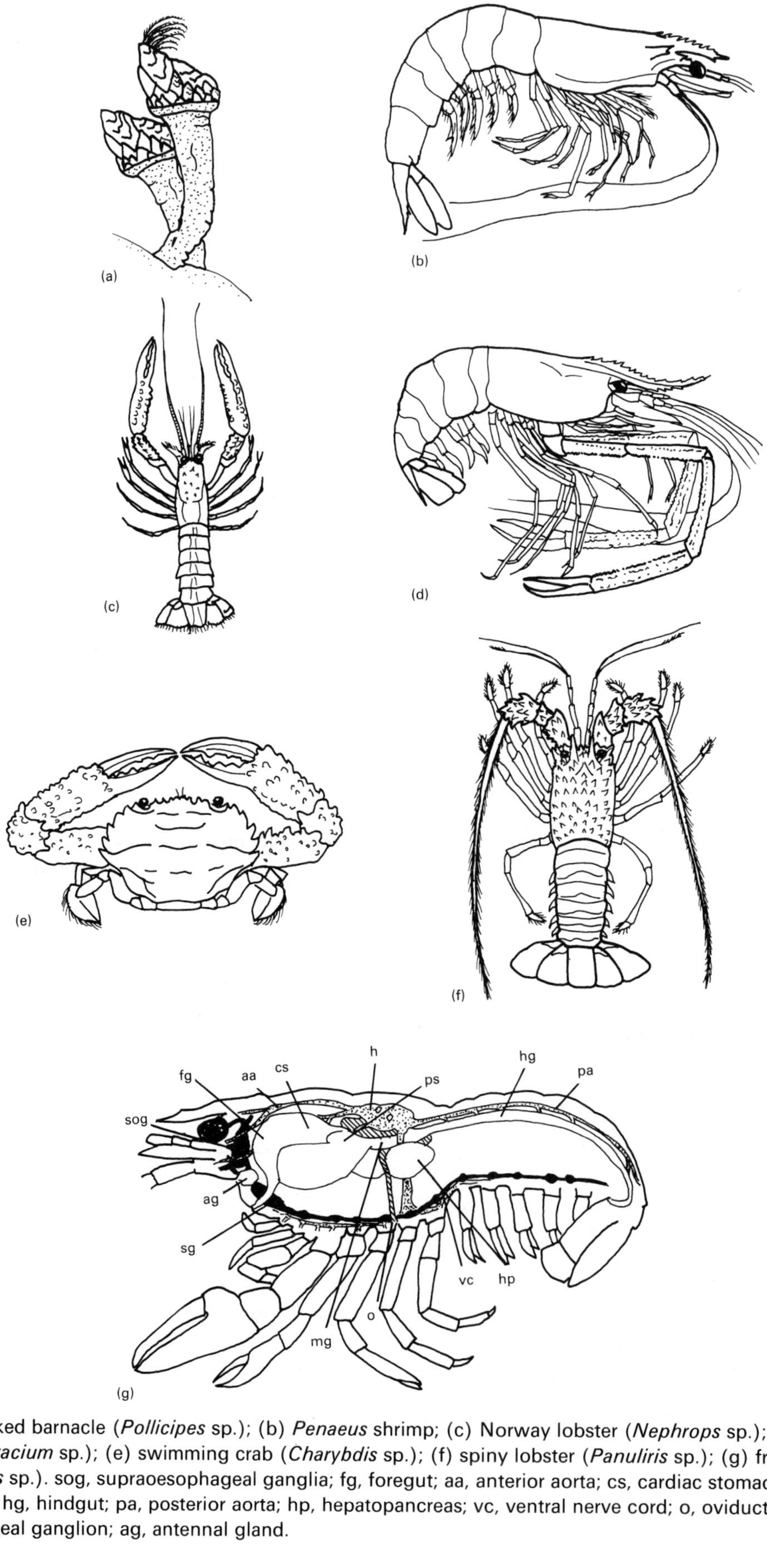

Fig. 1 (a) Stalked barnacle (*Pollicipes* sp.); (b) *Penaeus* shrimp; (c) Norway lobster (*Nephrops* sp.); (d) caridean shrimp (*Macrobracium* sp.); (e) swimming crab (*Charybdis* sp.); (f) spiny lobster (*Panuliris* sp.); (g) freshwater crayfish (*Astacus* sp.). sog, supraoesophageal ganglia; fg, foregut; aa, anterior aorta; cs, cardiac stomach; h, heart; ps, pyloric stomach; hg, hindgut; pa, posterior aorta; hp, hepatopancreas; vc, ventral nerve cord; o, oviduct; mg, midgut; sg, suboesophageal ganglion; ag, antennal gland.

Characteristics of Crustacea

Table 2. Composition of commercial crustaceans (all figures are per 100 g of raw material, except *Homarus*, which is boiled)

	Energy (kcal)	Carbohydrate (g)	Protein (g)	Total fat (g)	Fatty acids: Saturated (g)	Mono unsaturated (g)	Poly unsaturated (g)	$n-3$ (g)	Cholesterol (mg)
Crab (mixed)	74–95	0–2·2	15–18	0·8–1·9	0·16–0·17	0·22	0·4–0·5	0·38–0·44	60–78
Penaeid shrimp	87–100	0–2·7	17–22	0·4–0·8	0·11–0·2	0·05–0·15	0·09–0·49	0·07–0·34	96
Panulirid lobster	100	1·7	19·2	1·2	0·14	0·14	0·59	0·27	106
Homarid lobster	93	5·4	20·5	0·6	0·08	0·13	0·07	0·06	72

Data modified from various sources.

pod shells, or flexed beneath the cephalothorax as in galatheid lobsters. The coconut crabs (*Birgus*) and lithodids represent extreme groups in which the asymmetrical abdomen is closely applied to the underside of the thorax, giving the appearance of true crabs or Brachyura. This final group is distinguished by the lateral expansion of the cephalothorax, and reduction of the abdomen to form a symmetrical flap, lacking uropods, which is flexed under the thorax.

Crustacean Structure and Function

As a predominantly aquatic group, crustaceans have been able to exploit the arthropod chitinized exoskeleton fully without the weight restrictions imposed by such a system on land. In the classes under consideration (Table 1), fusion of the head and one to three thoracic segments has occurred to produce a carapace, or protective shield, which often extends forwards in the form of a rostrum and posteriorly to protect the cephalothorax (Fig. 1). The abdomen usually retains flexibility by means of soft cuticular joints between each segment. Even barnacles (Fig. 1) retain most of these features, although much modified for their sedentary existence.

Each segment typically bears a pair of jointed appendages, anteriorly. The two pairs of antennae are often elongated and mobile, and carry aesthetascs which are chemosensory hairs. Other head appendages usually form mouthparts concerned with feeding and include the mandibles or jaws, maxillules and maxillae, and, in malacostracans, additional maxillipeds.

The rest of the thoracic appendages are modified into pereopods (Fig. 1) specialized for walking, swimming, respiration, feeding or defence. The abdominal segments (pleonites) typically bear biramous pleopods, paddlelike appendages, used for swimming in malacostracans. In this group the terminal somite bearing the anus forms a flattened telson, and the last pair of abdominal segments are modified to form uropods; together with the telson, these form a tail fan used in swimming.

The gut is divided into a chitin-lined fore- and hindgut and a midgut lined with endoderm. The foregut oesophagus leads to a stomach which is often subdivided into cardiac and pyloric regions in malacostracans. The midgut forms an intestine of variable length and bears the digestive caecae or hepatopancreas, emptying into the pyloric chamber of the stomach. The hindgut is usually short, absorptive in function, and leads to the anus. In crabs and lobsters, mechanical breakdown of food in the gastric chambers of the stomach is performed by heavily sclerotized teeth which form a grinding gastric mill; in shrimps this structure may be absent and breakdown is solely by enzymes secreted from the hepatopancreas. The products of digestion are absorbed by cells in the fine tubules of the hepatopancreas, or cells lining the midgut trunk, where further intracellular digestion occurs. In some groups, expelled faecal material is sheathed in a peritrophic membrane.

The circulatory system is comprised of a dorsal muscular heart with ostia or pores to draw in blood from the pericardial cavity. In advanced malacostracans the heart has a series of blood vessels ensuring that blood flows to body organs and to the gills (Fig. 1). Return to the heart is via pools or the haemocoel, although active crustaceans may have a primitive venous system to return the blood to the pericardial cavity. Gaseous exchange in large, advanced crustaceans is via gills, which arise as branches from the base of the thoracic limbs. The gills are modified in various ways to provide a large surface area of thin, permeable cuticle, and in decapods are protected under the carapace in branchial chambers through which ventilating currents of water are drawn.

Excretion is in the form of ammonia, which is released across gill surfaces and via nephridia in maxillary or, in most malacostracans, antennal glands (Fig. 1). These glands are also active in osmoregulation, as are the gill surfaces. The crustacean cuticle, unlike the waxy insect

Table 3. Habitats and global distribution of crustaceans utilized for human nutrition

Group	Habitat	Distribution (commercial fisheries)
	Marine	
Copepods	Pelagic	Norway, Canada, Japan
Cirripedes	Rocky, coastal	Portugal
Mysids	Pelagic, coastal, estuarine	Japan, Southeast Asia, China, Korea
Euphausiids	Pelagic, offshore	Antarctica, Canada, Norway, Mediterranean
Sergestids	Pelagic, coastal, estuarine	China, Southeast Asia, Japan, East Africa, India, Brazil, Surinam, Philippines
Penaeid shrimp (60 species)	Benthic, soft-substrate, nutrient-rich, estuarine, coastal	Worldwide between 40°N and 40°S
Plesiopenaeus	Benthic, soft-substrate, deep-water	Atlantic, Australia, South Africa
Pleoticus	Soft-substrate, deep-water	Southwest Atlantic
Caridean shrimp		
Crangon	Benthic, soft-substrate, coastal	Europe, USSR, Algeria
Pandalus	Benthic, coastal to 1500 m	North Pacific, Atlantic
Palaemon	Benthic, rocky, coastal	Europe, Algeria
Stomatopoda	Benthic, rocky, coastal	Tropical to Mediterranean
Lobsters		
Homarus	Benthic, rocky-soft, coastal to 700 m	North Atlantic, Mediterranean
Nephrops	Benthic, soft substrate, 15–800 m	Northwest Atlantic, Mediterranean
Panulirids	Benthic, rocky, coastal, 700 m	30°N–50°S worldwide
Syllaridae	Benthic, soft-rocky, coastal	Mediterranean, Japan, Indian Ocean
Anomurans		
Galatheids	Benthic, rocky, coastal	Mediterrannean, Japan, Western USA
Lithodes	Benthic, rocky-soft, coastal	Southwest Atlantic
Paralithodes	Benthic, rocky-soft, coastal	Northwest Pacific
Crabs		
Chionecetes	Benthic, rocky, coastal	Northwest Atlantic, Northwest Pacific, East Central Atlantic
Maia	Benthic, rocky, seaweeds, coastal	Mediterranean
Cancer	Benthic, rocky-soft, coastal	Africa, Northeast and Central Atlantic, Mediterranean, Northeast and -west Pacific, East Central Pacific
Portunids	Benthic, rocky-soft, coastal	Northeast Atlantic, Asia, West Pacific
Callinectes	Benthic, rocky-soft, coastal, shelf edge	West and Northwest Atlantic
Scylla serratus	Benthic, mangal, coastal	Asia, India, West Central Pacific
Geryon	Benthic, soft, 300–1500 m	Northwest Atlantic
	Freshwater	
Caridean Shrimp		
Macrobrachium	Estuaries, rivers, benthic, soft-substrate	Tropical, introduced worldwide
Palaemonids	Estuaries, rivers, lakes, soft, benthic	Tropical
Astacidea		
Crayfish	Rivers, lakes, streams, rocky-soft, vegetated	Temperate to tropical, worldwide

cuticle, is largely permeable, and imposes severe constraints upon ionic regulation; hence few crustaceans are found away from water.

The crustacean brain is composed of three fused ganglia – two anterior dorsal supraoesophageal ganglia and a third which forms a pair of circumenteric connectives extending round the oesophagus to a suboesophageal ganglion linked to the ventral nerve cord. This cord bears paired segmental body ganglia. The optic and antennulary nerves run to the supraoesophageal ganglia, and the suboesophageal ganglia often become a large, fused mass serving the nerves to the mandibles, maxillules, maxillae and maxillipeds. In crabs all the thoracic ganglia fuse to form a large ventral nerve plate. Sensory systems are well developed, despite the exoskeleton, and take the form of innervated setae responding to touch or currents, whilst others, such as aesthetascs, detect chemicals or gradients in attractants emanating from food.

Crustaceans are well adapted to detect light, and photoreceptors range from the simple larval naupliar eye, responding to light direction and intensity, to the

Characteristics of Crustacea

stalked, multifaceted compound eye found in decapods. This is capable of discerning shapes, patterns, movement and at least some have colour vision. Moulting, chromatophore activity, tidal and daily locomotor rhythms, and aspects of reproduction are under hormonal and neurosecretory control, but an understanding of the mechanisms of this control is still at an early stage.

Sexes are separate in most crustaceans, but even some advanced malacostracans such as *Pandalus* may be protandrous hermaphrodites (maturing first as males, then later changing sex). Gonads are paired structures which are found in various regions of the trunk and empty via genital pores usually on a trunk sternite. In male decapods an anterior pair of pleopods is modified for sperm transfer. Sperm is deposited directly into the oviduct or into a seminal receptacle, where it may be stored for some time. Crustaceans may brood fertilized eggs, usually in an external pouch (mysids) or attached to pleopods (most decapods), but exceptionally (penaeids) may release their eggs freely into the sea.

Typically crustacean eggs hatch into planktonic larval forms, although these are suppressed, for example, in the Mysidacea, Amphipoda, Isopoda and freshwater Astacidea, where direct development occurs. The most primitive larva is the nauplius, with a single median simple eye and three pairs of biramous limbs; this is followed by the zoea and megalopa, gaining additional limbs and segments at each moult. A wide variation in numbers of larval stages and duration is seen, penaeids having 12 stages extending over 13 days, homarid lobsters 4 stages over 15 days, whilst the planktonic life of palinurids may extend up to 12 months.

Table 2 gives the composition of crustaceans commonly consumed by man and reveals that all are low in saturated fats, high in polyunsaturates, particularly the highly unsaturated $n-3$ series, and contain medium levels of cholesterol. ***See*** Carbohydrates, Requirements and Dietary Importance; Energy, Measurement of Food Energy; Fats, Requirements; Protein, Requirements

Several crustacean groups (copepods, branchiurans, cirripedes and isopods) have given rise to parasites which may cause serious infestations in edible fish and shellfish.

Habitats and Distribution

Apart from the pelagic zooplanktonic copepods, mysids, euphausiids and sergestids, which are mainly fished commercially in colder productive waters (Table 3), most other crustaceans only occur in commercial densities in shallow coastal seas. The fast-growing, high-value penaeid shrimp are restricted to warmer waters, as many have life cycles associated with estuarine mangals, and many burrow in sands and muds. In contrast, the slower-growing caridean shrimps are centred in the northern boreal region, where the shallow-sea pandalids make up the bulk of the fisheries.

The slow-growing homarid lobsters also have a cold-to-warm temperate distribution and are replaced in warmer seas by the panulirid and scyllarid lobsters. The most productive crab fisheries used to be those for king crabs (*Paralithodes*) and snow crabs (*Chionecetes*) from the North Pacific and Alaska, but blue crabs (*Callinectes*) from the western Atlantic have recently produced the highest landings. Although most lobsters and crabs show a preference for rocky habitats, both *Cancer* and *Homarus* will construct burrows in soft substrates.

Many crustacean groups have invaded freshwater habitats, but only caridean shrimp, notably *Macrobrachium* and crayfish, attain commercial fishery sizes. The former are restricted to tropical waters, whilst the latter occur worldwide. European crayfish populations declined as a result of disease, but recent introduction of expatriate species has led to these becoming pests, particularly in Africa and Southern Europe.

Bibliography

Bowman TE and Abele LG (1982) Classification of the recent crustacea. In: Bliss DE (ed. in chief) *The Biology of Crustacea*, vol. 1, pp 1–25. New York: Academic Press.

Brusca RC and Brusca GJ (1990) *Invertebrates*, pp 1–922. Sunderland, Massachusetts: Sinaver Associates.

Provenzano AJ (1985) Economic aspects: fisheries and culture. In: Bliss DE (ed. in chief) *The Biology of Crustacea*, vol. 10, pp 1–331. New York: Academic Press.

D. A. Jones
School of Ocean Sciences, University of Wales, Gwynedd, UK

Commercially Important Crustacea

The commercially important crustacea are decapods and include the crabs, lobsters, shrimps, prawns, crawfish, crayfish and krill. Total recorded global landings were $4{\cdot}5 \times 10^6$ t in 1989. Within each group numerous species are utilized commercially on a global basis. For example, 400 species of shrimps and prawns are listed in one publication as being of commercial importance. Shrimps and prawns are also the most important group in terms of landings (approximately $2{\cdot}5 \times 10^6$ t in 1989), being derived from both wild and farmed sources, the latter from tropical waters. The only other marine group which is farmed, albeit on a relatively minor scale, are the lobsters. In the USA, *Homarus americanus* has been farmed on a limited scale but attempts to farm the

Table 1. Some commercially important crabs

Species	Common name	Occurrence
Callinectes sapidus	Blue crab	Atlantic, North America
Cancer magister	Dungeness crab	Pacific
Cancer pagurus	European edible or brown crab	Europe
Maia squinado	Spider crab	Europe
Portunus or *Liocarcinus puber*	Swimming crab	Europe
Paralithodes camchaticus	King crab	Pacific, North America, Japan, USSR
Geryon quinquedens	Red crab	USA
Chionoecetes tanneri	Tanner, snow or queen crab	Pacific, Japan, USA

European lobster, *H. gammarus*, have, to date, failed on economic grounds. Attempts to enhance wild stocks by the seeding of fishing grounds, a process known as ranching, are currently in progress in Europe and show signs of promise. In contrast, freshwater crayfish are successfully farmed, especially in the USA, Australia and, to a lesser extent, in Europe. This article gives details of the major species utilized, and how they are handled and utilized. However, with such a wide range of species available in most of the groups, a big variation in product type occurs and only a brief survey is possible here.

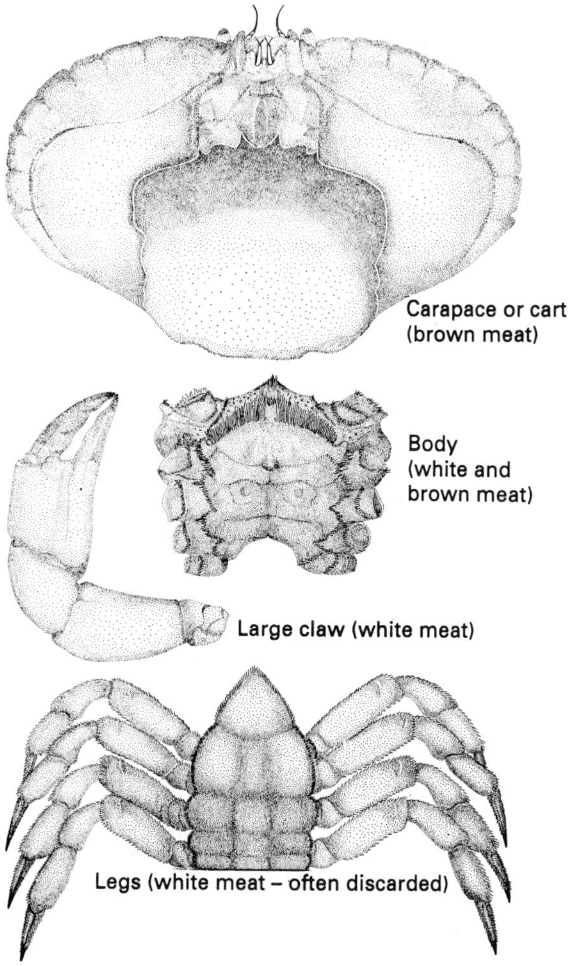

Fig. 1 Parts of the edible crab (*Cancer pagurus*).

Crabs

Globally, there are around 20 main species of crab of commercial importance. These vary considerably in their morphological characteristics, thus giving a wide variation in handling procedures and product types. The most important species utilized are shown in Table 1.

Handling, Utilization and Storage

A discussion of the utilization of crab is complicated by the variations in the parts of the animal which are eaten. Firstly, not only is muscle meat consumed but also, in the case of some species, the hepatopancreas (liver) and gonads are utilized. Clearly, with the latter, differences exist between the sexes but, with both liver and gonad, there are some very large seasonal variations. Even with muscle meat, owing to the utilization of different parts of the animal, variations in meat quality occur within a species. For example, in *Cancer pagurus*, the European edible crab, the most sought after meat is claw meat, while leg meat and body meat are regarded as second-grade and may, in many cases, be discarded. However, in the case of *Callinectes sapidus*, the blue crab, the claws are less important and the most prized meat is body muscle meat (see Fig. 1).

Much crab is either sold live for the final customer to cook and prepare, or merely cooked for sale in the shell. Live presentation has in recent years increased in importance owing to advances in so-called 'vivier' transportation, either by boat or road transport. In the latter case crabs are held in seawater which is often chilled and aerated so that the crabs can be maintained for several days prior to holding in shore-based tanks close to the market. These markets are, in the main, in continental Europe.

In spite of the importance of these methods of utilization, factory processing is still important, particularly in the UK, the USA and various countries in the Far East. This processed crab meat is marketed fresh (chilled at around 3°C), frozen, or sold in hermetically sealed cans. It should be noted that the freezing of uncooked crab results in a poor-quality end product and low yields. ***See*** Preservation of Food

The meat is picked after cooking by a variety of means, most often simply by hand, but mechanical picking machines do exist; the variation in meat type has led to the development of numerous picking devices. These range from machines which give a product indistinguishable from the hand-picked meat, where the morphology of the crab allows, to devices which merely recover the meat as a second-grade intermediate product for incorporation in soups, pâtés, etc., where the meat is very difficult to extract. In North America, liver–gonad is not utilized but it is in Europe, particularly from *Cancer pagurus* which has a well-developed hepatopancreas. This material is usually picked by hand, and is either utilized directly for 'dressed' crab or is incorporated in pâté or paste products.

The processing of crab is difficult and, owing to its morphology, the meat is relatively fragmented. Thus gross microbiological contamination can occur during processing and, in a precooked food which is usually not reheated prior to consumption, can result in food poisoning. Microbiological standards and guidelines exist to assist processors, and strict personal hygiene by operatives and thorough factory cleaning procedures are therefore necessary, all backed up by careful quality control of the final product.

Lobsters

The commercially important lobsters are somewhat simpler to identify, there being just two true lobsters: *Homarus gammarus*, the European lobster, and *Homarus americanus*, the American lobster. However, to these should be added *Nephrops norvegicus*, the Norway lobster or Dublin Bay prawn. Other related species, such as *Metanephrops* spp., which are now available on a global basis and which are used for similar products, should also be included in this group.

Handling, Utilization and Storage

In comparison with the crabs, the utilization of the true lobsters is relatively simple. The main parts utilized are the tail white meat and the large claw meat; the hepatopancreas–gonad (tomali) is utilized but, since very few lobsters are factory-processed, this is usually extracted at the domestic level.

Very many European and North American lobsters are sold to the final customer in the live state. The handling and transportation of live lobsters is relatively easy as the animal is robust compared with crabs. Lobsters are often held for several weeks in holding tanks, awaiting seasonal upturns in the market, e.g. around Christmas in continental Europe. Nevertheless, there is a healthy market for cooked whole lobsters, which are often sold frozen; a recent development has been the sale of frozen blanched lobsters, which are said to closely resemble freshly cooked material when finally cooked by the consumer or caterer. Like crab, lobsters cannot be successfully frozen raw as the meat sticks to the shell, resulting in low yields and poor quality.

Very few lobsters are cooked at the factory level and then picked. This only occurs as a last resort when animals are moribund, disfigured or sometimes if they are excessively large; there is a market requirement for lobsters up to around 1 kg, and larger ones are difficult to sell.

The utilization of *Nephrops* requires a separate discussion. The main part of the animal of commercial interest is the tail meat but, as with true lobsters, the hepatopancreas–gonad is utilized at the domestic level. Small amounts are utilized for speciality canned soups. As with lobsters, live storage and transportation, particularly to continental European markets, is practised, but *Nephrops* is not a robust animal and high losses have been sustained until recent advances in vivier transportation. A very major outlet for *Nephrops* is in the whole frozen form, once again in continental Europe. *Nephrops* will freeze satisfactorily in the raw form, and cooked presentation is rare.

The commercial handling of *Nephrops* and its presentation as 'scampi' in the UK is very different. The UK market, as with many other products, demands convenience foods. The catering trade in this context is very important too. Thus the factory shelling of *Nephrops* started at an early stage in the utilization of this species (1950s). This was initially done by blowing out the tail meat from the shell with compressed air but in more recent years, for health and safety reasons (the protein-containing aerosol can cause asthma), the industry has moved to water pressure for this process. The heads, if not discarded at sea, are discarded prior to meat removal from the tail shell. Initially, the tail meats were individually quick frozen and sold in this form. However, more recently, as fish size has decreased owing to heavy fishing, extrusion techniques to give more constant and larger sized pieces have been employed. Battering and breading of such so-called 'scampi' (named after a Mediterranean food, 'scampo') then takes place. Polyphosphate is sometimes used in the formulation of this product. An industry code of practice exists to attempt the control of product quality, particularly the percentage of batter and breading coating; control of the core composition where there is scope for adulteration with added water, or even with other, lower-value proteins, remains a problem. Legislation is planned in the UK to remove abuse in this area.

Commercially Important Crustacea

Table 2. Some commercially important prawns and shrimps

Species	Common name	Occurrence
Prawn		
Pandalus borealis	Deep or cold-water prawn	North Atlantic or north Pacific
Penaeus japonicus	Kuruma prawn	Mediterranean, Atlantic, Indo-Pacific
Penaeus monodon	Giant tiger prawn	Indo-Pacific
Penaeus esculentas	Common tiger prawn	Indo-Pacific
Penaeus indicus	Indian prawn	Indo-Pacific
Penaeus merguiensis	Banana prawn	Indo-Pacific
Macrobrachium carcinus	Freshwater prawn	Freshwater (estuarine Atlantic or Pacific)
Palaemon serratus	Common prawn	Atlantic, Mediterranean
Shrimp		
Crangon crangon or *C. vulgaris*	Brown shrimp	Northeast Atlantic, Mediterranean
Pandalus montagui	Pink shrimp	Northeast Atlantic, Mediterranean
Crangon septemspinosus	Sand shrimp	Atlantic (North America)
Crangon franciscorum, *C. nigricauda* and *C. nigromaculata*	Bay shrimp	Pacific (North America)

Shrimps and Prawns

The terms 'shrimp' and 'prawn' are often used synonymously in many parts of the world. In some countries all such species are termed shrimp, e.g. in the USA; in others, e.g. India, the term 'prawn' is used for all species. The UK differentiates legally on the basis of size, such that the term 'prawn' can only be used for the larger animals. This is specified by a count per unit weight. In general terms throughout the world, prawn usually refers to members of the families *Pandalidae, Penaeidae* and *Palaemonidae* and shrimp to members of the family *Crangonidae*, although some members of the above three families may also be called shrimp.

Some commercially important prawn and shrimp species are shown in Table 2.

Handling, Utilization and Storage

As with *Nephrops* species, only the tail meats of shrimps and prawns are utilized where processed commercially, although at the domestic level the hepatopancreas and gonad may be used in the case of the larger species. Handling at the commercial level can range from the simple presentation of whole raw prawn, either chilled or frozen, to fully cooked and peeled tail meats, fresh, frozen or otherwise preserved such as by canning. A major method of processing for tropical shrimp is for the shelling to take place by hand in the Far East, and for the tails to be then frozen in bulk (10-kg blocks are common). These blocks are then thawed and cooked in Europe or the USA, prior to individual quick-freezing. Such shrimp are ideal for a variety of purposes but are often labelled as being suitable for incorporation in cooked dishes, such as curries.

The northern cold-water prawn (*Pandalus borealis*) may be landed fresh or frozen, and is then processed through a highly automated line which cooks, peels by means of counter-rotating rollers, and then grades them, prior to individual quick-freezing and packaging. This is the species very commonly used in shrimp cocktails.

Traditionally, *Crangon crangon*, the brown shrimp has been utilized in northern Europe for a variety of presentations; in the UK it is 'potted' in spiced butter. This shrimp is cooked at sea and then hand-peeled, very often by the fisherman or his family at home prior to sale, usually to a cooperative for incorporation in the final product. Such a procedure is now judged both uneconomic and unsatisfactory on hygiene grounds, and this industry has regressed as repeated attempts to develop a peeling machine have failed. Very recently, a Dutch invention has proved satisfactory and the industry could expand again. New requirements under the Food Safety Act, 1990, however, could result in changes to the handling procedure, including the banning of cooking at sea.

Several of the larger *Pandalid* and *Penaeid* species are machine-processed in the USA where different styles of product are offered, e.g. 'fan tail' presentation. Like scampi in the UK, they are coated with batter and breading either raw or cooked for frozen or chilled presentation. Many other forms of presentation are possible, e.g. dried, smoked, pasteurized and semipreserved.

As shrimps and prawns are often precooked and not cooked again by the final customer prior to consumption, they can cause food poisoning. As with crab,

Table 3. Some commercially important crawfish

Species	Common name	Occurrence
Panulirus argus	Spiny lobster	Caribbean
Panulirus regius	Spiny lobster	Japan
Panulirus vulgaris	Spiny lobster	Europe
Jasus lalandii	Rock lobster	South Africa (Atlantic or Indian Ocean)
Jasus edwardii	Spiny rock lobster	New Zealand
Jasus verreauxi	Eastern rock lobster	Australia

careful hygienic handling is necessary and microbiological guidelines and specifications exist to control the final product. The irradiation of shrimps and prawns by pasteurization doses is permitted in some countries to allow the presentation of a microbiologically safer final product.

Crawfish and Crayfish

The term 'crawfish' refers to marine species belonging to the families *Palinurus, Panulirus* and *Jasus*. They are sometimes called lobsters in certain parts of the world. There is often confusion with 'crayfish' which refers to freshwater species belonging to the families *Cambarus* and *Astacus*. The most important of this group are the crawfish which are categorized in two main divisions, the spiny lobsters and the rock lobsters (see Table 3).

Handling, Utilization and Storage

The utilization of crawfish is very similar to that of the *Homarus* species of lobsters. Apart from live presentation to the final customer, which is common in continental Europe, crawfish are presented as cooked whole fish or tails, fresh (chilled) or frozen, or the tail meats are shelled raw and frozen for sale. Shelling is entirely manual.

Crayfish, on the other hand, are much smaller and are presented alive, fresh or frozen, usually in the entire form.

Krill

The term 'krill' covers many crustacean species, all of which look very similar to small shrimp. They occur very widely but predominantly in Arctic and Antarctic waters. Important species are *Meganyctiphanes norvegica, Thysadoessa inermis* and *Euphausia superba* (Antarctic krill).

The krill resource has attracted much attention from fishery scientists over many years, but particularly since the mid-1970s. In spite of extensive research, there is still no agreed estimate of the size of the resource, and hence no information on a sustainable level of exploitation. Several countries fish for krill, particularly Japan and eastern European countries (Russia, Poland, eastern Germany) and it is these countries which have studied both the resource and the technology of utilization.

Handling and Utilization

Krill is mainly utilized, albeit on a limited scale, for direct human consumption and for animal feed. Its use for so-called technological products, e.g. utilization of the chitin in the shell, has not progressed, just as the utilization of crustacean waste generally has largely failed to give economically viable products. The main problem here has been that whilst such applications are technically possible, synthetic products are cheaper.

Krill is used directly for human consumption, either as it is, or as a paste or spread or in soups. The characteristic shrimp-like flavour is useful in such products. A basic requirement for all these products is a means of shelling, and amongst several approaches a modification of the roller device used for shrimps, such as *Pandalus borealis*, was found effective by both Polish and Japanese workers, giving throughputs of up to 500 kg per hour and a yield of up to 25%. Fluorine levels can give rise to concern in krill products, but the roller shelling method gives satisfactory results acceptable to regulatory authorities, e.g. the US Food and Drug Administration.

As animal feed, krill meal is valuable. Production of meal is not difficult except that low yields are obtained (14–15%), rather than the theoretical 22–23%. The chemical composition of the meal is variable, reflecting intrinsic variation and variation in the handling of the raw material. Fluorine levels are unacceptably high, which means that the final feed must be compounded with a limited amount of krill meal. Nevertheless, krill is used as a feed in salmon farming, the amino acid composition being similar to other fish meals, particularly so far as lysine is concerned. ***See*** Fish Meal

Until more reliable information on the size of the resource, and therefore the sustainable yield, is available, attempts to increase the landings of krill would be folly. Concern has been voiced because krill is at the base of many food chains supporting many economically important species, already utilized in commercial fisheries.

Chemical and Nutritional Composition

Table 4 gives information on the chemical composition of crustaceans, including, where available, information

Table 4. Proximate composition and vitamin content (per 100 g) of various crustacean shellfish

Species	Moisture (g)	Protein (g)	Fat (g)	Ash (g)	Carbohydrate (g)	Energy (kJ)	Vitamins: A (iu)	Thiamin (μg)	Riboflavin (μg)	Nicotinic acid (mg)	Ascorbic acid (mg)	Vitamin B_{12} (μg)
Crab												
Cancer pagurus	77·4	17·6	1·4	3·5	3·6	407	90	135	247	2·80	3·5	
(muscle, raw)	75·3–79·5	16·0–19·2	1·3–1·5						145–350			
Cancer magister	80·0	17·3	1·3	1·5	0	340		176	20			
(muscle, raw)	78·1–82·3	13·8–23·4	0·8–3·0	1·2–2·0								
Crawfish (Spiny)												
Panulirus argus	70·0	19·2	7·5	1·6	1·7	634						
(muscle, raw)	64·3–75·6	15·3–23·1	0·3–14·6	1·5–1·7								
Lobster (European)												
Homarus gammarus	75·1	19·9	0·6	1·8	2·6	399	65					
(muscle, raw)												
Shrimp (deep or cold-water prawn)												
Pandalus borealis	78·0	15·0	3·3	2·8	0·9	389		76	194			
(muscle, raw)								57–90	142–250			
Shrimp (Jumbo)												
Penaeus monodon	77·0	19·2	0·6	1·4	1·8	374		17	54	3·97		2·99
(muscle, raw)	75·6–78·5	17·6–20·8	0·4–0·7	1·3–1·5								

From Sidwell VD (1981).

on the vitamin content. It should be noted that crustacean shellfish can vary widely in their composition as a result of their method of growth, whereby moisture is taken up to expand the new shell (exoskeleton) when it is still soft. ***See*** individual nutrients

Bibliography

Early JC and Stroud GD (1982) Shellfish. In: Aitken A, Mackie IM, Merritt JM and Windsor ML (eds) *Fish Handling and Processing* 2nd edn, pp 126–137. Edinburgh: Her Majesty's Stationery Office.

FAO (1989) *Fishery Statistics Yearbook: Catches and Landings.* vol. 68. Rome: Food and Agriculture Organization.

Holthuis LB (1980) *FAO Species Catalogue*, vol. 1: *Shrimp and Prawns of the World.* FAO Fisheries Synopsis No. 125 (FIR/S125 Vol 1). Rome: Food and Agriculture Organization.

OECD (1990) *Multilingual Dictionary of Fish and Fish Products.* Oxford: Fishing News Books.

Sidwell VD (1981) *Chemical and Nutritional Composition of Fin Fishes, Whales, Crustaceans and Molluscs and their Products.* NOAA Technical Memorandum, NMFS F/SEC II. Washington, DC: US Department of Commerce.

J. C. Early
Torry Research Station, Aberdeen, UK

Characteristics of Molluscs

Molluscs constitute a unique phylum of animals, the Mollusca, which are characterized by a combination of morphological and anatomical features separating them from all other invertebrate organisms. Many common names such as snails, clams and squids apply to representatives of these animals, which number fewer than 50 000 living species. Molluscs are known to have diversified early in the fossil record, namely the Cambrian of the Paleozoic over a half a billion years ago; the antecedent or ancestral mollusc is presumed to have evolved in the Precambrian and many different lineages have radiated into the vast array of ecological niches of the world's biotopes.

Molluscs are widespread in marine environments, living from the shore to the greatest abyssal depths and occurring in pelagic or oceanic realms as well as benthically, both on and in all kinds of substrates. In some parts of the world, they dominate the intertidal zone, and in the recently discovered aphotic deep-water vents and seeps as well as other oxygen-depleted environments, molluscs, especially mussels and certain clams, are particularly conspicuous, frequently utilizing endosymbiotic chemoautotrophic bacteria. Groups of representative snails and bivalves have also successfully adapted to terrestrial and freshwater habitats. Finally, some species have become specialized endoparasites.

In adult size, molluscs range from tiny gastropods and bivalves less than 1 mm in diameter or length to the giant squid which may be over 15 m long and upwards of 1000 kg in weight.

Typical Morphological and Anatomical Features

Molluscs are protostomous coelomates, exhibiting spiral, determinate cleavage and schizocoely as well as having trochophore larvae and the blastopore forming the mouth of the adult. They usually possess all or a combination of the following features: (1) a reduced coelom and vestiges of metamerism; (2) a mantle or fleshy epidermis of the dorsal body wall which has glands capable of secreting calcium carbonate to form an exoskeletal shell or shelly parts, such as plates, spines, and spicules; (3) a mantle cavity or an invagination of the mantle which contains a pair or more of specialized respiratory structures, the ctenidia or gills, and into which the digestive, metanephridial excretory, and reproductive systems debouch their products; (4) the ventral body surface modified into a pedal groove or muscularized foot for progression or locomotion; (5) a special chitinized, rasp-like tongue or radula; (6) an open haemocoelic circulatory system with a chambered heart having auricles and ventricles. Additionally, the nervous system has variously paired ganglionic portions, particularly cerebral, pedal and visceral ones, as well as ventral anteroposterior cord-like connective elements; specialized sensory structures were evolved for olfaction, vision, balance and tactile stimulation.

Primitively, these animals were gonochoristic, that is, having the male and female sexes in separate individuals; fertilization was external and the eggs developed into pelagic larvae; hermaphroditism with both sexes in the same individual, brooding of eggs, and ovoviviparity are a few among many of the modifications in the reproductive strategies of these animals.

Taxonomy of the Group

Currently seven classes of living molluscs are recognized. The entirely marine, shell-less Aplacophora, with at least 250 species, are vermiform animals which have calcareous spicules, scales or plates embedded in the mantle and which are separable into two principal groups, sometimes considered as independent classes. The more or less cylindrical, gonochoristic Caudofoveata (Fig. 1A), having the body somewhat separated into anterior, medial and posterior sections and possessing an anteroventral pedal shield, attain lengths of 140 mm and live mostly as infaunal burrowers. Fewer than 100 species are known, and they feed mainly on

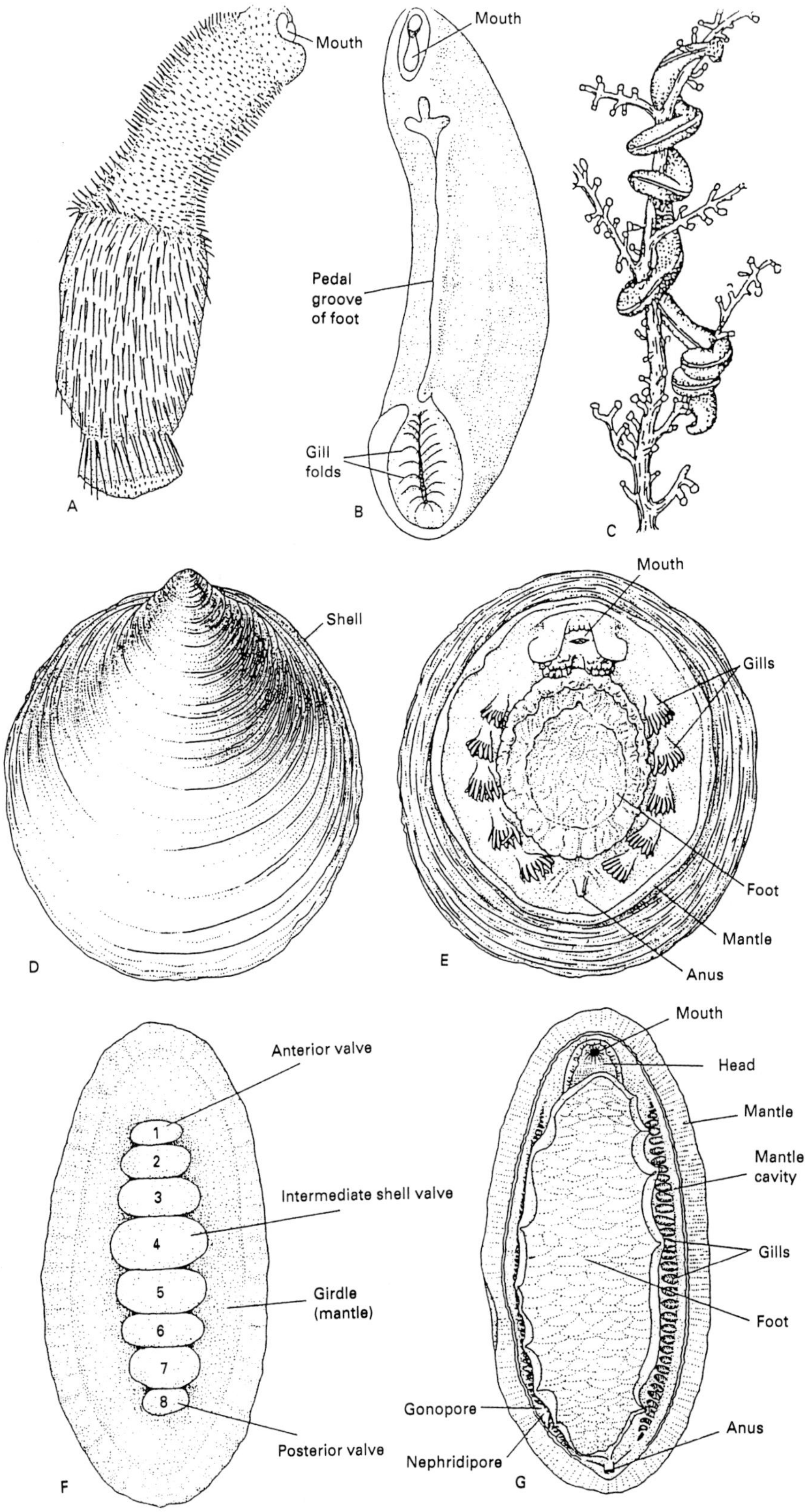

Fig. 1 (A) Aplacophoran caudofoveate showing tripartite body form and spicules. (B) Aplacophoran solenogastre showing the ventral pedal groove. (C) Aplacophoran solenogastre on a cnidarian. (D) Dorsal view of monoplacophoran shell. (E) Ventral view of monoplacophoran showing gross anatomical detail. (F) Dorsal view of polyplacophoran with the shell plates and girdle. (G) Ventral view of polyplacophoran showing gross anatomical detail.

Characteristics of Molluscs

microorganisms and detritus. The free-living, more or less uniformly worm-like, elongate and laterally compressed Solenogastres (Fig. 1B, C), with over 200 recognized species, are from 1 to 300 mm in length. Hermaphroditic and predacious, they have a distinct pedal groove, posterior gill folds and live either epibiotically on sediments or epizoically on cnidarians which usually constitute their principal food.

The Monoplacophora (Fig. 1D, E), with fewer than 20 living species, have a controversial fossil history dating from the Early Cambrian; they are benthic, mainly deep-sea animals having a single cap-shaped or limpet-like shell and variously paired organ systems such as the pedal retractor muscles, gills or ctenidia, nephridia or kidneys and gonads, reflecting a primitive metamerism. In monoplacophorans, the foot is large with a flattened, creeping sole, and the pallial mantle cavity is peripheral along the sides of the animal.

The chitons or so-called coat-of-mail shells, separated as the Class Polyplacophora (Fig. 1F, G), also have a broadened, flat muscular foot and a mantle cavity containing numerous pairs of ctenidia. Consisting of about 600 species of marine benthic epifaunal animals with dorsoventrally depressed bodies, they are unique in having eight calcareous shell plates held together by a strong, peripheral girdle. The body outline is elongate to ovate, and adult individuals range in size from 3 to almost 400 mm in length. Although some chitons are known to occur at depths beyond 7000 metres, most live in shallower waters and are mainly grazing herbivores, with a radula having 17 teeth in transverse rows and cusps reinforced with magnetite. The anterior portions of the alimentary canal have pairs of glands for the digestion of carbohydrates, and the broad, large foot is used to creep over, or to attach by suction to, the substrate. Special photosensory structures, the aesthetes, are found on the dorsal surface of the valves and may, along with the iron-containing radula, facilitate homing behaviour.

The Gastropoda or snails (Fig. 2A–C), the largest and most diverse class with as many as 40 000 species, have representatives in most of the world's biotopes and are known from the Early Cambrian. Although most species live in the marine environment, many kinds occur both on land and in fresh water; a few taxa have become specialized internal parasites. Principally, gastropods are univalves usually with a coiled, apically closed calcareous shell, which also exhibits a great many different shapes and forms, being sometimes merely cap-shaped, greatly reduced or even altogether lost. The very largest of living snails has a shell over 500 mm in length, the smallest less than 1 mm.

Anatomically, snails and their relatives are basically divisible into head, foot, and visceral mass; they are all characterized by a unique process called torsion which occurs during ontogeny. Primitively, the mantle cavity, into which the alimentary canal, excretory structures and gonads of the visceral mass empty, contains a pair of laterally disposed respiratory ctenidia or gills and is located posteriorly in the early larval gastropod (Fig. 2A). During development, the entire mantle cavity and its contents are twisted or torted through an arc of 180° to the right, bringing the cavity itself into an anterior position above and behind the neck and head of the snail (Fig. 2B). Internally, torsion causes a peculiar crossing (or streptoneury) of the laterally disposed, paired nervous connections between the anteriorly placed cerebral ganglia and the posteriorly located visceral ganglia.

The head of a gastropod usually has at least one pair of cephalic tentacles, and eyes are often associated with them; however, both structures can be lost. The foot has a creeping sole, and primitively, there are lateral grooves between it and the mantle, forming a so-called epipodium with sensory tentaculate and integumentary organs. The sole of the foot may be subdivided and variously modified for swimming and in the case of internal parasites, it is entirely lost. The distinctive visceral mass, a dorsal hump, is covered by the mantle which secretes the shell, and contains the internal organs or viscera, including the heart, kidneys, gonads and alimentary canal. Generally speaking, the jaws and radula assist the mouth in bringing food into the digestive tract; one or more pairs of salivary glands or specialized pouches may contribute enzymes to the buccal cavity or oesophagus; the stomach leads through the intestine to the terminal anus and may have a crystalline style, style pouch, typhlosole, digestive diverticula, and a triturating gizzard associated with it. All these structures, the head, foot, visceral mass, mantle and shell, are different and variously altered in the major subdivisions of the Gastropoda.

Traditionally, the snails are subdivided into three major subclasses: the Prosobranchia, Opisthobranchia, and Pulmonata. Recently, with the advent of cladistic analysis and the discovery of unusual forms in the deep sea, authorities have begun to dismantle the more conservative classification system long in use by elaborating complex split taxonomic systems incorporating numerous new higher-level taxa; but at least 350 families constitute the diversity of gastropods, about 150 in the prosobranchs and 100 each in pulmonates and opisthobranchs, and if one includes the fossil taxa, there are about 8000 different genera though many more than this have been named.

Prosobranchs (Fig. 2C), with the gills anterior to the heart, form the most primitive and most numerous of gastropods with more than 20 000 species; the shell is almost always present and usually spirally coiled, although it may sometimes be patelliform, tubular or lost in the adult; usually the foot with its creeping plantar sole bears a calcareous or corneous operculum,

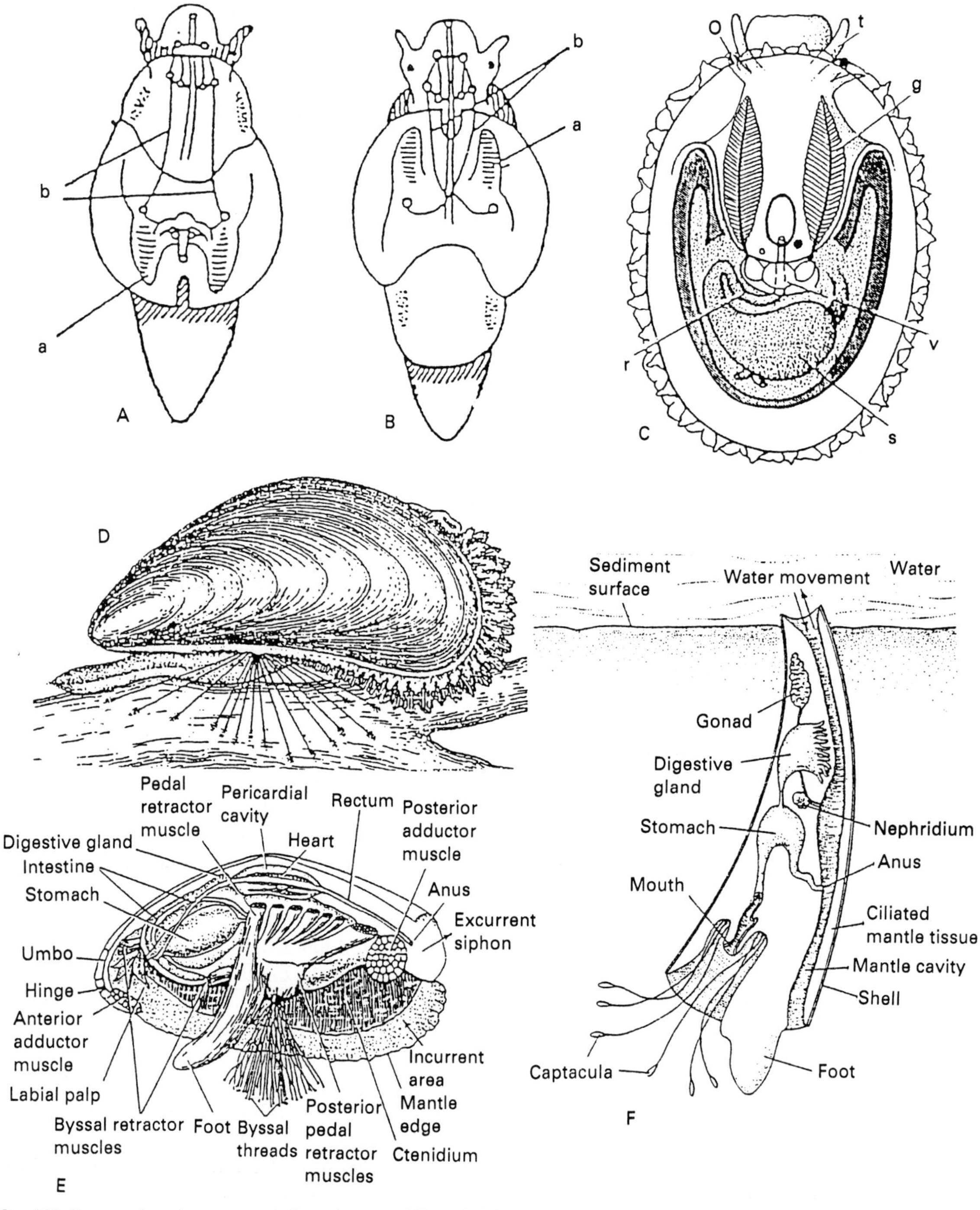

Fig. 2 (A) Pre-torsional gastropod showing ctenidium (a) in posterior mantle cavity and uncrossed cerebrovisceral connective nerves (b). (B) Post-torsional gastropod showing ctenidium (a) in anterior mantle cavity and crossed cerebrovisceral connective nerves (b), or streptoneury. (C) Primitive prosobranch gastropod showing: g, ctenidium; o, eye; r, rectum; s, stomach; t, tentacle; v, ventricle of heart traversed by rectum. (D) External view of a bivalve mussel with its byssal attachment. (E) Internal view of a bivalve mussel with gross anatomical detail. (F) Internal view of a scaphopod with gross anatomical detail.

a structure capable of covering the aperture of the shell when the animal is retracted. As a rule, the mantle cavity is anteriorly directed and it contains, in the most primitive prosobranchs, a pair of respiratory structures (the ctenidia), a pair of chemosensory organs (the osphradia), and a pair of mucous-secreting structures (the hypobranchial glands – sometimes also a source of purple pigments that are a problem in commercially-important species); the alimentary canal as well as the excretory and reproductive tracts empty into the mantle cavity. The ventricle of the heart is, in the most primitive forms, traversed by the rectum; there are also paired lateral auricles. In more advanced prosobranchs, for example the so-called neogastropods, the mantle cavity becomes somewhat detorted and shifted toward the right; the paired structures, e.g. the gills, osphradia, hypobranchial glands, auricles, and kidneys, are reduced to single ones; siphonal structures, elaborated by folds of the mantle, direct the flow of water into and out of the mantle cavity and thus facilitate respiration as well as the removal of excretory and alimentary waste products. The nervous system is streptoneurous, with the cerebrovisceral connectives crossed, and the animals are usually gonochoristic. Prosobranchs primitively are herbivorous grazers or detritus feeders, using the radula to scrape edible material off the substrate, but different evolutionary lines have culminated in many highly specialized carnivores, frequently accompanied by changes in, or even loss of, the radula. According to the kind of preferred diet, great differences obtain in the nature of the alimentary canal with its radula and various associated digestive glands.

On more than one occasion, prosobranchs have given rise to lineages which have invaded freshwater and terrestrial habitats, and, of course, concomitant alterations, especially in the respiratory and reproductive structures have resulted. Land prosobranchs tend to be more tropical or subtropical in their distribution while the freshwater groups are widely successful in both the tropics and temperate zones. In the marine realm, prosobranchs are the predominant snails from the intertidal areas over the continental shelf and slope down to the greatest depths.

The opisthobranchs, with the gills, if present, usually behind the heart, are a numerically much smaller group than either prosobranchs or pulmonates, with fewer than 2000 species; their hallmarks include the progressive loss of the shell and operculum in adults, further detorsion and loss of streptoneury, wherein the cerebrovisceral connectives are not crossed (euthyneury), and finally, hermaphroditism.

With only a few freshwater representatives, this subclass of snails is virtually entirely marine and includes herbivores, vigorous, rapacious carnivores, and floating mucous-net suspension feeders. Probably their most spectacular representatives are the nudibranchs, which, as adults, lack the shell and mantle cavity with its ctenidium; instead, they have sometimes elaborated separate respiratory structures, such as cerata and anal or lateral gills, which along with the mantle, may be brilliantly coloured.

The third subdivision of gastropods, the Pulmonata, with about 15 000 species, are with a few exceptions, entirely freshwater and terrestrial; these animals are hermaphroditic and detorted, euthyneurous, lack an operculum, ctenidium or osphradium, and have transformed the pallial cavity into a special, highly vascularized pulmonary or lung-like cavity which facilitates gaseous respiration and opens to the exterior by means of a pneumostome, a narrowed, contractile aperture. Generally, a shell is present, spirally coiled and variously shaped or reduced to a patelliform or flattened disk; sometimes, the shell is wholly enveloped by the mantle or totally lost, an evolutionary phenomenon adopted by several different lines of descent and leading to a slug-like condition as an adult. In contrast to the prosobranchs, the pulmonate radula shows a remarkable degree of uniformity, mostly being adaptations for herbivory, although some lineages of these snails have become carnivores.

Freshwater pulmonates, though occurring in tropical and subtropical waters, are particularly successful in exploiting marginal, sometimes ephemeral habitats; they are widespread but more dominating in the cool and temperate regions of the Paleoarctic. Terrestrial lunged snails are cosmopolitan in distribution, with some taxa found on the most remote islands or in the most rigorous deserts. Although largely living on the ground or in leaf-litter, some representatives of these snails have become arboreal, adapting to conditions on the bark, branches, or leafy canopy of trees. In these taxa, often the shells are brilliantly coloured and exhibit great polymorphism.

A sister-group of the Gastropods, the Cephalopoda, including squids and octopuses, with some 600 living species are also entirely marine organisms which have almost exclusively reduced, lost or internalized the shell, reorganized the body, and adopted a predatory mode of feeding either in the open ocean or along the bottom of the sea; these animals are discussed elsewhere. ***See*** Marine Foods, Edible Animals Found in the Sea

The Bivalvia (Fig. 2D, E), comprising familiar organisms such as clams and mussels, with fewer than 8000 species and over 100 families, are aquatic, marine and freshwater, epifaunal or infaunal, predominantly deposit or filter-feeding animals characterized by having bilaterally disposed shells or valves which can be tightly closed by adductor muscles. In size, living bivalves are known to range from 1–2 mm up to over 1000 mm such as the giant clams of the reefs of the Pacific Ocean. Formed by specialized distal portions of the flap-like mantle, the valves are connected dorsally by a hinge which may consist of a ligament and so-called dental elements. The body is generally laterally compressed; the spatulate or lanceolate foot is variously developed, adapted for creeping, burrowing or cleansing, and, in attached or cemented forms such as oysters, it may be greatly reduced or lost. A pedally secreted byssus normally occurs in bivalves though sometimes only in the young embryonic stages. In the reduction of a distinct cephalic region, these animals lost the radula and evolved specialized palps on either side of the mouth which assist in sorting the filtered, particulate food. Their nervous system is decentralized and more specialized posteriorly in conjunction with the posterior development of mantle openings or siphons.

In bivalves, the mantle cavity shows marked expansion and deepening, occupying much of the space between the shells, and the ctenidia exhibit a significant elaboration in most taxa. These form greatly enlarged structures consisting of lateral pairs of plates or sheets of filaments, in which there are blood vessels for respiration and on the surfaces of which are elaborated a complex ciliation, facilitating the filtering of edible materials from plankton suspended in the water column or microorganisms from the detritus or bottom sediments. Mainly gonochoristic with external fertilization, bivalves may also be hermaphroditic and sometimes protandrous, with free-swimming or parasitic larva; retention of the young in the mantle cavity or in special incubatory marsupia in the gills occurs in many species.

Living benthically both in fresh and saline waters, clams and their relatives have exploited many of the biomes of aquatic habitats; they occur from shallow intertidal waters to the greatest abyssal depths of the oceans; they utilize both soft sediments and hard substrates, attaching, burrowing, and even boring into limestones, other shells, or wood. Freshwater bivalves include the tiny so-called fingernail clams which almost become terrestrial, sometimes living in moist litter, and the pearly mussels, widely distributed on continental areas and even oceanic islands. One group of small marine epizoic and commensal clams has evolved as internal parasites in echinoderms.

Closely related to the bivalves, the Scaphopoda (Fig. 2F), commonly known as tusk or tooth shells, with about 350 species, first occur in the Middle Ordovician over 450 million years ago. With a bilaterally symmetrical body occupying an elongate, gently curved, tubular shell which is open at both ends, scaphopods have fused their lateral mantle flaps, reduced their cephalic region, and lost their ctenidia, replacing these respiratory structures, with crenulations or folds in the mantle cavity. Calcareous and composed of three layers, the shell is attenuated into a narrow posterior opening; the external shell surface is variously smooth or sculpted, and in size the animals range from 2 to 150 mm in length in adults, though one fossil species at the end of the Paleozic attained nearly 300 mm. Located on a protrusive proboscis, the mouth is surrounded by long filiform, tentacular organs called captacula which probe the sediment and collect, by adhesive glands and ciliary tracts, tiny interstitial microorganisms such as benthic foraminiferans, which constitute their food.

Entirely marine and relatively conservative in their radiation, scaphopods are shallow infaunal burrowers found throughout the world's oceans from the sublittoral to deep abyssal zones below 6000 m. Preferring soft substrates, they flourish mainly in muddy, sand bottoms with the posterior portion of the shell with its aperture projecting from the substrate.

Bibliography

Boss, KJ (1982) Mollusca and classification of Mollusca. In: Parker SP (ed.) *Synopsis and Classification of Living Organisms*, vol. 1, pp 945–1166; vol. 2, pp 1092–1096. New York: McGraw-Hill.

Pojeta J, Runnegar B, Peel JS and Mackenzie G Jr (1987) Phylum Mollusca. In: Boardman RS, Cheetham AH and Rowell AJ (eds) *Fossil Invertebrates*, pp 270–435. Oxford: Blackwell Scientific.

Wilbur KM (ed.-in-chief) (1983–1988) *The Mollusca*, 12 vols. New York: Academic Press.

K. J. Boss
Harvard University, Cambridge, USA

Commercially Important Molluscs

The types of molluscan shellfish important in commerce are grouped into three classes: the Gastropoda, the Lamellibranchiata or bivalves, and the Cephalopoda. On a global basis there are numerous mollusc species of commercial importance within each class. This article discusses the Gastropoda and bivalves; the Cephalopoda are covered elsewhere. ***See*** Marine Foods, Edible Animals Found in the Sea

The recorded global landings of marine molluscan shellfish were approximately $7{\cdot}5 \times 10^6$ t in 1988 from farmed and wild sources; fresh waters yielded some 250 000 t. Unlike crustacean shellfish, farming has long been of major importance for certain species, particularly oysters and mussels. This article gives details of how they are farmed and discusses handling, utilization and storage. It also gives information on their chemical composition.

Bivalves

The bivalve molluscs, which are all filter feeders, provide the major source of raw material for commercial exploitation and include the clams, cockles, mussels, oysters and scallops. As filter feeders, any pollutant or toxic substance ingested is concentrated, giving potential problems for public health. Microbiological problems and the means of overcoming them are covered in this article under the headings *Cockles* and *Mussels*, although similar problems potentially exist for all the bivalves, if eaten raw or inadequately cooked. The

Table 1. Common marine and freshwater clams

Species	Common name	Occurrence
Merceniara (or *Venus*) *Mercenaria*	Hard clam or hard-shell clam or quahog (quahaug)	Atlantic, North America and Europe
Tapes or *Venerupis* spp.	Carpet shells	Europe, North America
Sandomus giganteus	Butter clam	Atlantic or Pacific, North America
Mactra sachalinensis	Hen clam	Japan
Mya arenaria	Soft (-shell) clam; also, confusingly, hard clam (see above)	Atlantic or Pacific, North America
Titaria cordata	Gulf clam	Gulf of Mexico
Spisula solidissima	Surf clam	Atlantic, North America
Corbicula spp.	Freshwater clam	Japan

problems caused by the ingestion of algal toxins by bivalves, such as paralytic shellfish poisoning (PSP) and diarrhoetic shellfish poisoning (DSP), are covered elsewhere.

Clams

The vernacular term 'clam' causes difficulty since it is used in certain parts of the world for other bivalve molluscs, such as scallops (see below), and clams are sometimes called cockles. It is generally used, however, for the common marine and freshwater bivalve species shown in Table 1.

Whilst the majority of clams are from wild sources, clam farming is practised in continental Europe and in Japan.

Handling, Utilization and Storage

Many clam species are sold in the shell, either alive or fresh. They may be presented frozen in the shell too. Shelling on a commercial basis for sale fresh or frozen is also common and, in some cases, is highly mechanized. For example, the qhahog-processing industry in North America is very well developed and involves shucking or shelling by means of gas flames to denature the muscle attachment and to thereby open the shell, prior to removal of the meats by vigorous mechanical agitation. This is followed by grading and packaging.

Table 2. Commercially utilized cockles

Species	Common name	Occurrence
Cardium edule or *Cerastoderma edule*	Common cockle	Cosmopolitan
Cardium corbis	Common cockle	Pacific, North America
Cardium aculeatum	Spiny cockle	Atlantic or Mediterranean
Cardium tuberculatum	Knotted cockle	Atlantic or Mediterranean

Apart from fresh or frozen presentation, the meats may also be smoked, dried or canned. Soups (chowder) or clam liquor itself are commonly sold canned, especially in the USA. ***See*** Preservation of Food

Cockles

The true cockles are all members of the family Cardidae, of which four species are commonly used commercially (see Table 2).

Note that cockles may be called winkles in North America and certain clams are called cockles in New Zealand (see *Clams*, above).

Handling, Utilization and Storage

The factory processing of cockles has a very long history in northern Europe, particularly in the UK, where cockles were, until recent years, the most important shellfish landing in terms of weight, although not value.

On landing, cockles are boiled or steamed to release the meats from the shell and then riddled to separate the meats, which fall through the riddle into vats of water. Here, repeated washing takes place to remove grit. Traditionally, batch cooking in coal-fired boilers was used, with all the subsequent stages, riddling and washing, being manual. Gradually, batch cooking, using steam at atmospheric pressure, and mechanical riddling were employed, but hand washing continued. In the 1970s continuous steam cookers, so-called 'monoblock' cookers, largely replaced batch steamers.

The most important exploited stocks of cockles in the UK and in northern Europe are found in estuarine or coastal waters, the Thames Estuary and The Wash being the most important in the UK. These waters are sometimes heavily polluted and thus cockles have featured repeatedly in food poisoning outbreaks. This was in spite of close supervision by health authorities and the existence of bylaws governing the heat treatment to be applied, made under the Public Health (Shellfish) Regulations 1934, as amended by subsequent Food Acts. Initially, bacterial contamination was suspected and has undoubtedly given rise to illness in the past, as a result of either undercooking or cross-contamination between raw and cooked material, in the relatively primitive plants which were often operated by a single person.

Improvements were made progressively, initially, to give a better heat treatment and to reduce the risk of cross-contamination. However, illness still occurred and, in the late 1970s, epidemiological studies indicated that at least a proportion of the outbreaks was attributable to viruses (the entero-viruses or small, round-structured viruses; SRSVs) which had not been inactivated during the cooking process. *See* Viruses

Studies in the UK at the Central Public Health Laboratory indicated that the minimum heat treatment, measured at the centre of a cockle, required to inactivate hepatitis A virus (HAV, an SRSV) is 85°C, held for 1 min. This allowed the specification of a new process to give a safe product. This process, which is continuous, involves the use of hot water at 95°C, into which the shell-on cockles are fed in a thin layer (10 cm). The requirement for the process was increased so that a centre temperature of 90°C held for 1·5 min is provided to give a margin of safety and to inactivate other enteroviruses which are suspected to be more heat-resistant than HAV. Should the temperature not be met, for whatever reason (inadequate heat input or excessively cold (frozen) cockles being used), the conveyor belt is designed to stop so that undercooked material is not fed to the later stages of the process.

The specification also includes other matters, such as the general layout of the factory and the procedures to be adopted for riddling and degritting, to reduce the chance of cross-contamination. The bylaws governing the process to be used for cockle processing in the Thames Estuary region now require the system to be installed; the process is also being applied elsewhere in the UK, after recommendations from the health authorities. Since the introduction of the specification, judged on the basis of epidemiological evidence, the safety of cockle meats has improved.

Cockles are marketed alive in the shell, or as boiled shucked meats, fresh, frozen or salted. They are also bulk-packed in acetic acid for later sale in malt vinegar as a marinaded product.

Mussels

All the commercially important mussel species are members of the Mytilidae. The important species are listed in Table 3. Apart from the natural stocks, which occur as shown in Table 3 or as a result of historical introductions, the blue mussel is cultivated on a worldwide basis. Simple farming operations involve relaying young spat from areas where spatfalls are very successful to fattening grounds; at the other end of the scale, highly sophisticated operations involve rafts to which ropes are attached and upon which the mussels grow following a spatfall. The Spanish mussel industry based in Galicia uses this technique which has resulted in a very large and successful industry. Less sophisticated is the use of ropes, suspended from horizontal ropes (longlines), which are anchored at each end. This technique is employed on the west coast of Scotland but, nevertheless, gives high-quality mussels with all the advantages of growth off the sea-bed, including more rapid attainment of a marketable product and grit-free meats.

Table 3. Commercially important mussels

Species	Common name	Occurrence
Mytilus edulis	Blue mussel	North Atlantic, Europe, Pacific, New Zealand
Mytilus californianus	Common mussel	Pacific, North America
Modiolus modiolus	Horse mussel	Europe
Modiolus barbatus	Bearded horse mussel	Atlantic or Mediterranean
Mytilus smaragdinus	Green mussel	Southeast Asia
Perna canaliculus	Green-lipped mussel or Perna	New Zealand

Handling, Utilization and Storage

After harvesting (which may be by hand raking if taken from the littoral zone, or by dredge if sublittoral) from polluted waters, mussels must be cleansed because they are filter feeders and concentrate any microbiological pollutants in the gut. Alternatively, they must be cooked by an approved process.

Currently, provisions for cleansing or depuration are made by orders under the Public Health (Shellfish) Regulations 1934, as amended, and they involve analysis of the flesh of the mollusc for the number of *Escherichia coli*. In future (after 1 January 1993) under an EEC directive, the actual growing waters will be classified on the basis of the bacteriological quality of their shellfish (again using *E. coli*) and the subsequent handling will be determined on this basis, although the final check will be, once again, on the basis of the *E. coli* content and absence of *Salmonella* in the flesh and intravalvular fluid. The bacterial specification for raw consumption, whether taken directly from the sea or following cleansing, is <230 *E. coli* per 100 g and absence of *Salmonella* in 25 g. *See* Enterobacteriaceae, Occurrence of *Escherichia coli*

Cleansing is either by relaying in unpolluted water

Table 4. Commercially important oysters

Species	Common name	Occurrence
Ostrea edulis	Common or flat oyster	Europe
Ostrea lurida	Western oyster	Pacific, North America
Ostrea laperousei	—	Japan
Ostrea lutaria	Dredge oyster	New Zealand
Crassostrea angulata	Portuguese oyster	Europe
Crassostrea virginica	Blue point oyster	Atlantic, USA
Crassostrea gigas	Pacific oyster	Pacific, North America, Europe

over a period of time or by treatment in tanks through which clean seawater is circulated. If recirculation takes place, sterilization of the seawater with ultraviolet light or chlorine is employed in the UK. This method is very effective for bacteria, but viral contamination requires longer depuration. Relaying of shellfish from more polluted waters for a minimum of 2 months in clean waters reduces the viruses to an acceptable level.

Many mussels are dispatched to the market in the raw form and the only other treatment, in addition to cleansing, is cleaning and declumping followed by debyssing, i.e. the removal of the fine proteinaceous filaments which attach the mussels to the substrate. However, when the presentation of shucked meat is required, factory processing by heating to remove the shell is necessary; the heat treatment is closely controlled, particularly for mussels from polluted waters. A similar heat treatment to that described for cockles (i.e. one which raises the centre temperature of the meat to 90°C for a period of 1·5 min) is often required by the authorities in the UK these days. The length of the process will depend on the design of the cooker and each type of equipment must be tested to check on the total process necessary for safety. Complex processing lines exist for mussel processing, particularly in countries where large-scale farming operations are established, e.g. Spain. Such processing lines are often linked to a cannery.

Final sale of the meats can be in a variety of forms – frozen or otherwise preserved, e.g. by canning in brine or oil. A popular treatment is to smoke the mussels prior to canning in oil. Another form of presentation is to marinade in vinegar; this gives what is known as a semipreserve with a more limited shelf life.

Oysters

The commercially important oysters are members of the Ostreidae, those listed in Table 4 being of particular importance.

Oysters have been fished for many centuries, indeed since prehistoric times. They formed the diet of working people in the UK in the nineteenth century, but the stocks have since declined owing to overfishing, pollution, disease and severe winters. As a result, they are now regarded as a luxury food. Cultivation of the common and Portuguese oyster, and, more recently, the Pacific oyster in the UK, has improved the stock situation but oysters still remain very much a luxury. The Pacific oyster does not breed successfully in cool northern waters and must be raised in hatcheries where heated water is used. The seed is transferred later to the growing area. Oyster farms, based on hatchery seed, are now established on the south and southeast coasts of England and west coast of Scotland, as well as Wales and Ireland.

Handling, Utilization and Storage

As with mussels, any live oysters taken from polluted waters must be cleansed before being sold for consumption raw, and the same procedures are applied. The legal provisions to protect public health are currently under the Public Health (Shellfish) Regulations of 1934, as amended by subsequent Food Acts, and from 1 January 1993 the EEC directive on live bivalve molluscs will also apply.

The vast majority of oysters are either sold live, with the shell intact, or on the so-called half-shell. The factory shucking of oysters is extensively practised in the USA. Shucked meats may be frozen or fresh (raw). Factory-prepared products of oyster meats have been presented in Europe but have not found an extensive market compared with the traditional product in the shell, or half-shell.

Scallops

Around 20 species of scallops, also known as escallops, all members of the Pectinidae, are listed as being of commercial importance although those in Table 5 are the most important.

Apart from wild fisheries, scallops are also cultivated, particularly in Japan, where the overfishing of wild stocks has occurred. The experimental cultivation of scallops in UK waters has proved the feasibility of similar operations here, although only a few commercial operations have been established, mainly in Scotland.

Handling, Utilization and Storage

As with other bivalve molluscs, scallops are sold whole in the shell, either fresh or frozen, or they may be shucked, in which case the adductor muscle, with or without the gonad or roe (see Fig. 1), is separated from

Table 5. Scallops of greatest commercial importance

Species	Common name	Occurrence
Pecten maximus	Scallop, great scallop, Coquille St Jacques	Northeast Atlantic
Pecten yessoensis	Common scallop	Japan
Pecten (or *Placopecten*) *magellanicus*	Sea scallop	North America
Pecten jacobaeus	Great scallop	Atlantic or Mediterranean
Aequipecten gibbus	Calico scallop	Atlantic or North America
Chlamys opercularis	Queen scallop	Atlantic
Argopecten irradians	Bay scallop	Western Atlantic

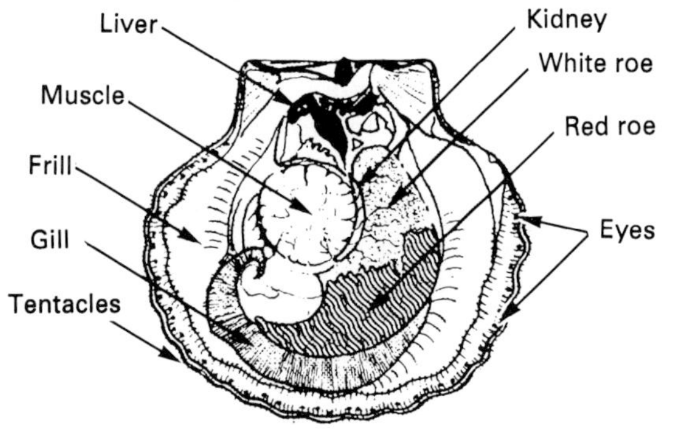

Fig. 1 Scallop with the flat shell removed.

the rest of the viscera for presentation fresh or frozen. The market determines the form of presentation; for example, certain markets in continental Europe demand the shell-on material since everything within the shell is consumed, just as with oysters and mussels. If shucked, continental European markets require the adductor muscle with gonad, whereas in the USA only the adductor muscle is consumed.

The larger species are hand-processed by separating the shells and cutting the muscle from the shell with a sharp knife. The viscera is then separated from the muscle, which is left with or without the roe.

Smaller species may be similarly treated by hand, particularly if the roe is required, but are usually machine-processed. This involves the severance of the mussel from the shell by immersion in hot water which denatures the attachment. The shell contents are then separated from the shell by a riddle, and the adductor muscle is then cleaned of the viscera, including the roe,

Table 6. Major commercially important abalones

Species	Common name	Occurrence
Haliotis refescens	Red abalone	Pacific, North America
Haliotis corrugata	Pink abalone	Pacific, North America
Haliotis gigantea	—	Japan
Haliotis tuberculata	Ormer	Atlantic, northern France, Channel Islands
Haliotis australis	Yellowfoot paua	New Zealand
Notohaliotis ruber	Red abalone	Australia
Schismotis laevigata	White abalone	Australia

by machine. The majority of shucked scallop meats are presented frozen.

Hand-processed meats absorb considerable amounts of processing (washing) water. The hot-water treatment used in machine shucking denatures the adductor muscle sufficiently to reduce the problem. In the case of hand processing, therefore, there is considerable scope for the fraudulent adulteration of the scallop meats during processing, and efforts to control this have been made, especially in France. Here, the measurement of the moisture content and protein, when expressed as a ratio (the WP ratio), must not exceed a value of 5. This is a useful approach but it is very difficult to achieve a ratio of less than 5 for small scallops, where the surface area to volume ratio is large, particularly as they are often trawled and therefore very dirty, so requiring vigorous washing which alone can cause an unacceptably high increase in water content.

Gastropods

The commercially important gastropod molluscs include abalones, conchs and winkles, which feed by grazing, and whelks, which are carnivores.

Abalones

Abalones are members of the Haliotidae. They occur widely off rocky coasts but are mainly found off the coasts of Australia and New Zealand, Japan and the west coast of North America. The main species are listed in Table 6.

In addition to wild stocks, abalones are cultivated most extensively in Japan, where they are grown in cylindrical nets suspended on rafts.

Table 7. Proximate composition and vitamin content (per 100 g) of various molluscan shellfish

							Vitamins					
Species	Moisture (g)	Protein (g)	Fat (g)	Ash (g)	Carbo-hydrate (g)	Energy (kJ)	A (iu)	Thiamin (μg)	Riboflavin (μg)	Nicotinic acid (mg)	Ascorbic acid (mg)	Vitamin B_{12} (μg)
Abalone	78·3	15·2	0·4	1·4	0	269		240	60	1·6		0·73
Haliotis gigantea (muscle, raw)	75·7–82·3	10·4–20·1	0·2–0·5	1·0–1·6								
Clam	84·6	10·7	1·2	1·4	2·1	260	300	10·9	136	1·75		85·51
Mya arenaria (whole, raw)	83·3–85·5	9·5–12·5	1·3–1·4	1·2–1·7				0·8–79·00	95–243	1·13–2·53		7·10–109
Cockle	81·4	10·6	1·0	2·6	4·4	290		69[a]	156[a]			
Cardium edule (whole, raw)	73·4–92·0	9·2–14·1	0·6–2·7	2·1–3·1								
Mussel	80·0	11·6	1·7	2·2	4·5	336		162	212			12·00
Mytilus edulis (whole, raw)	76·2–90·4	9·9–15·4	1·2–3·1	1·4–3·2					175–249			
Oyster	81·0	10·3	1·6	2·2	4·9	315					3·00	4·80
Ostrea edulis (whole, raw)	79·3–82·7	9·6–10·7	1·3–1·8	1·4–3·3								
Scallop	77·3	16·2	0·8	1·5	4·2	374	50[b]	70[b]	190[b]	1·3[b]	2·00	
Pecten maximus (muscle, raw)	72·5–80·7	12·5–18·0		1·3–1·9								

From Sidwell VD (1981).
[a] Data for *Cardium corbis.*
[b] Data for *Pecten yessoensis.*

Handling, Utilization and Storage

Several methods of fishing are employed for wild abalones. Two important methods are diving and the use of a long pole equipped with a hook used from small boats. Whilst some of the catch is sold whole, either fresh or frozen, there are well-established processing industries near to the main fisheries. Here, on landing, the animal is removed from the shell by hand and the edible fleshy foot and muscle separated from the viscera. After trimming, around one third of the original shell-on weight remains and this is then either sold as it is, fresh or frozen, or subjected to further processing. Since abalone meat is very tough, a common procedure is to cut it into 1-cm-thick slices and then to tenderize by beating, as with steak. A procedure followed for fleshy trimmings and sometimes whole meats is to cut or mince and recompound using a mould into tubes 8 cm in diameter. These are then frozen, and the resulting frozen cylinder of meat cut up into slices approximately 1 cm thick. These slices are battered and breaded for presentation in the frozen state. Another form of presentation is as sticks, similar to fish fingers.

Other methods of processing include drying, either naturally in the sun or in ovens. This product is then sometimes ground to a powder for use in soups or chowders. Abalone meats are also canned in brine.

Conchs

Conchs are economically important in certain parts of the world, most notably the Caribbean where they are fished by diving, or by poles, similar to those used to catch abalones. Numerous species exist, but the most important is *Strombus gigas* which can reach a size of 350 mm. The meats are extracted by hand and used on a local basis for a variety of culinary purposes, either directly or for the preparation of soups and chowders. The shells, which are very attractive, are perhaps equally important economically since an extensive souvenir trade is based on them.

Periwinkles

The common periwinkle, *Littorina littorea*, is found extensively in northern Europe and on the east coast of North America, where it is locally called a conch, although it should be emphasized that the winkle is much smaller and has a maximum length of 20 mm. It is robust and can be kept alive for weeks in tanks prior to sale which is normally in the live state, although it is sometimes precooked. The main markets are in continental Europe; for example, there is an extensive trade from Scotland to France.

Whelks

The vernacular term 'whelk' is used globally for several types of mollusc, but most commonly in Europe for *Buccinum undatum*. As a carnivore it can be fished successfully by baited pots in coastal waters.

The catch is marketed fresh or cooked in the shell; extracted meats are sometimes frozen. As the meat of the larger fish can be very tough, the medium-sized fish are in greater demand. In recent years a market for whelk meats from Europe has developed in Japan. These fish are cooked by steaming and are then mechanically crushed. The meats are then separated from the shell by brine flotation prior to freezing and packaging.

Whelk meats may also be smoked, canned or presented as a semipreserve in vinegar.

Chemical and Nutritional Composition

Table 7 gives information on the chemical composition of molluscan shellfish, including, where available, information on the vitamin content. Although their composition does not vary as extensively as crustacean shellfish, there is, nevertheless, considerable variation in composition from season to season and in association with the reproductive cycle. ***See*** individual nutrients

Bibliography

Early JC and Nicholson FJ (1991) Making shellfish products safe to eat. *Food Technology International Europe* 45–47.

Early JC and Stroud GD (1982) Shellfish. In: Aitken A, Mackie IM, Merritt JM and Windsor ML (eds) *Fish Handling and Processing* 2nd edn, pp 126–137. Edinburgh: Her Majesty's Stationery Office.

FAO (1989) *Fishery Statistics Yearbook: Catches and Landings*. vol. 68. Rome: Food and Agriculture Organization.

OECD (1990) *Multilingual Dictionary of Fish and Fish Products*. Oxford: Fishing News Books.

Sidwell VD (1981) *Chemical and Nutritional Composition of Fin Fishes, Whales, Crustaceans and Molluscs and their Products*. NOAA Technical Memorandum, NMFS F/SEC II. Washington, DC: US Department of Commerce.

J. C. Early
Torry Research Station, Aberdeen, UK

Commercially Important Molluscs

Contamination and Spoilage of Molluscs and Crustacea

Numerous species of molluscs and crustaceans are harvested worldwide for both commercial supply and private food sources. The bivalve molluscs (clams, oysters, mussels, scallops) and species of crab, shrimp, prawns, crayfish and lobster are economically important in many countries. Species of gastropods, i.e. whelks and abalone, are also harvested and are considered by some to be a delicacy. Many of these products are exported by countries to many parts of the world.

As with many food products, these seafoods are considered to be highly perishable. It is essential that these products are handled adequately after harvest to maintain wholesomeness prior to consumption. Proper handling to prevent or delay spoilage, to increase shelf life and to prevent cross-contamination by pathogenic organisms is a necessity. Preservation of these seafoods using cold temperatures (freezing, chilling), heat processing, and other preservation techniques, such as canning, are used. These seafoods can become contaminated, toxigenic, or infectious to humans by one or more of several routes. They can acquire microorganisms and toxins that are naturally present in the aquatic environment. They may be subjected to human sewage, animal faecal pollution, and chemical pollution of their habitat. During processing, they may be exposed to contamination by the processing environment and equipment, by the plant workers, and also during preparation at food service establishments. ***See*** Preservation of Food

Species for human consumption are normally harvested from waters that are generally considered to be nonpolluted. Areas that are subject to industrial pollution or near sewage treatment outfalls are usually classified as prohibited for harvesting. Bacterial monitoring programs have been established in many countries to limit harvest in suspected polluted areas owing to public health concerns. Toxins, enteric viruses, and human bacterial pathogens have been the principal causes of reported illnesses from these seafoods.

The bivalve molluscs are filter feeders, accumulating bacteria and other suspended particles that are present in the water in which they grow. These species are sessile and are therefore incapable of movement to cleaner waters. Several of these species are consumed raw, 'on the half-shell'. Whatever contaminants the animal has accumulated are then passed on to the consumer. Many documented outbreaks of human illnesses have occurred from consumption of raw molluscan shellfish harvested from polluted waters. Crustaceans, in contrast, are generally not consumed raw, so the microbiological picture may be much different.

Procedures have been developed to attempt to 'purify' molluscs prior to marketing. One process is to 'relay' molluscs from one growing area to another that has better water quality. After a period of 'cleansing', the product is then collected for market. A second procedure, depuration, involves placing the shellfish in man-made tanks with flowing seawater, sterilized usually by ultraviolet (UV) light. This procedure, over time, is effective in the purging of certain bacterial species accumulated by the molluscs. Both of these procedures add handling steps which can also effect the price of the product when it reaches the market. Live-holding tanks are gaining popularity for display of crustaceans and molluscs at retail markets and restaurants. Live species can be exhibited in large tanks which use circulating artificial seawater. Often mixed species are displayed, which allows for bacteriological cross-contamination among species. Concerns regarding adequate sanitation of these tanks, maintenance of water quality during recirculation, and the mixing of species have been raised. A recent study found that *Salmonella* and *Vibrio cholerae* (non-01) may be transmitted to consumers by oysters displayed in these tanks. Whether the oysters were the source of introduction of these bacteria to the display system was not determined.

Indigenous Microorganisms

There are numerous bacterial species that are indigenous to molluscs and crustaceans and are present in these seafoods at the time of harvest and subsequently involved in spoilage during storage. Several bacterial species have been reported to be pathogenic to humans.

Vibrio

In the past few decades, bacterial illnesses caused by species of *Vibrio* have increased in numbers. Many *Vibrio* species are indigenous in coastal areas, particularly brackish waters. Several mesophilic species are pathogenic to humans. Vibrios are involved in chitin degradation in the environment; they overwinter in the sediment and 'bloom' during warmer months of the year. As the water temperature warms (above 15°C) these organisms become more abundant in the water column and are then accumulated by molluscs during feeding. Crustaceans are also affected by the presence of vibrios in the water. Some *Vibrio* species actually attach to the shell of crustaceans.

The most notable species from a public health standpoint are *V. parahaemolyticus*, *V. cholerae* and *V. vulnificus*. These three account for the majority of human infections associated with mollusc and crustacea consumption. The predominance of reported illnesses occur during summer months, June to September, in temperate regions when estuarine water temperatures have warmed sufficiently.

V. parahaemolyticus is a very common species in coastal and estuarine areas. Fortunately, not all strains of this marine species cause human gastroenteritis. From previous studies, approximately 5% of strains isolated from the environment were found to produce a thermostable direct haemolysin (TDH), called 'Kanagawa phenomenon positive'. However, strains producing TDH are recovered from more than 95% of patients who have contracted this self-limiting disease. The exact mechanism of virulence to humans is as yet unknown. Consumption of raw molluscs has caused a number of localized outbreaks, while post-processing contamination of crab and shrimp has caused outbreaks involving large numbers of cases. ***See*** Vibrios, *Vibrio parahaemolyticus*

V. cholerae has been known for centuries, causing epidemics involving large numbers of people and many deaths. Generally, this pathogen is endemic in areas where there is poor sanitation and drinking water sources become contaminated. Cholera is a disease caused by two biotypes of serogroup 01 (somatic antigen 1) of *V. cholerae* that produce cholera toxin. The faecal–oral route is a common method of contamination, although various vectors have been involved, including seafoods. Cholera can be a severe, life-threatening disease, but is easily treated and is preventable by good hygiene and proper treatment of sewage and drinking water. In early 1991, epidemic cholera appeared in Peru and has subsequently spread to other South American and Western Hemisphere countries. This epidemic is of concern since estuaries have become contaminated, and crustaceans became a vector of infection of individuals in the United States in one instance. As of August 1991, nearly 275 000 cases had been reported in the Western Hemisphere with close to 3000 deaths.

Non-01 serogroups of *V. cholerae* (of the nearly 100 other somatic antigen designations) are widely distributed in the estuarine environment. Some strains produce cholera toxin (CT) or a cholera-like (CT-like) toxin and have been responsible for human illness. The disease produced by these strains is thought to be less severe than classical cholera. Molluscs and crustaceans were vectors in many of the reported cases. Some non-01 strains have also been reported to produce a rare septicaemic condition in humans. Several deaths have resulted from this type of infection. These strains did not produce CT or CT-like toxins; thus the virulence mechanism(s) is unknown and still being investigated. ***See*** Vibrios, *Vibrio cholerae*

The recently identified *V. vulnificus* is a particularly nasty human pathogen. This marine species can manifest a primary septicaemic condition with a mortality rate exceeding 50% of cases. Certain individuals are at greatest risk to infection: those with hepatic disorders (i.e. cirrhosis), the immunocompromised, and those with other medical disorders. Individuals with these conditions have been cautioned to avoid eating raw or undercooked seafoods. The majority of reported illnesses have been associated with the consumption of raw oysters. Most often, the oysters have been traced to harvest areas from Gulf Coast estuaries of the United States. However, *V. vulnificus* resides in many other coastal areas of the world. There are virulent and avirulent strains in the environment, as determined by animal bioassay. The mechanism(s) of virulence is under investigation and has not as yet been fully elucidated. ***See*** Vibrios, *Vibrio vulnificus*

There are several other species of *Vibrio* that are worthy of mention. These species are found in many coastal areas and have been isolated from seafoods. *V. mimicus*, *V. hollisae*, *V. fluvialis*, *V. furnissii*, *V. metschnikovii* are all recently identified species that have been isolated from stools of diarrhoeal patients. Food histories from some of these patients point to molluscan shellfish as the probable vector. These illnesses occur seasonally and are rare. Quite possibly, these species are opportunistic, attacking individuals with other underlying medical conditions. Mechanisms of virulence remain unclear, although some species produce protein toxins and proteases which may be part of the puzzle.

Aeromonas, Plesiomonas

Two other genera of indigenous bacteria in the family Vibrionaceae have been implicated in diarrhoeal illness associated with mollusc consumption. *Aeromonas hydrophila*, *A. caviae* and *A. sobria* and *Plesiomonas shigelloides* have been isolated from human stools although, as yet, these species have not gained acceptance as enteric pathogens. They appear to be opportunistic organisms, seemingly attacking hosts with other underlying medical problems. Virulence mechanisms have not been elucidated. These organisms are widely distributed in the environment and can be recovered from fresh and brackish water and shellfish. *Aeromonas* species have also been recovered from soil and animal sources. Strains of *Aeromonas* can grow at refrigeration temperatures and may be involved in spoilage of seafood products during storage. ***See*** Aeromonas

There is much yet to learn regarding the control of these indigenous bacteria to prevent human illness. Harvesting of molluscs only during certain seasons has been suggested to minimize the levels of these organisms. Similarly, indicator organisms within the taxonomic family have been suggested to estimate levels of the species of concern. All of these bacteria are susceptible to mild heating and would thus be eliminated from adequately cooked seafoods.

Nonindigenous Microorganisms

Bacteria

There are a number of bacterial species that are considered nonindigenous to molluscs and crustaceans. They are of concern in that they have caused human illness in the past mainly as a result of contamination of the product. Insanitary practices during processing are a major contributor of these enteric pathogens. Direct faecal transmission from human or animal reservoirs is another factor in contamination.

Salmonellosis associated with consumption of contaminated molluscs occurred as early as the 1900s in the United States when these shellfish were harvested from waters with human faecal pollution. Control measures for harvesting shellfish are based on these early outbreaks of illness. *Salmonella* is a major cause of foodborne illness in the world and contamination of a food with this organism should be taken seriously. There are many reservoirs of *Salmonella* besides humans, including mammals, birds and reptiles. The faecal–oral route is often the method of contamination of foods by food-handlers. ***See*** Food Poisoning, Economic Implications

The *Shigella* group of enterics also contaminate foods via faeces. They are primarily thought of as waterborne pathogens involved in foodborne disease owing to poor sanitation. Human waste in estuaries has resulted in contamination of shellfish and subsequent outbreaks of shigellosis. Carriers of the organism working in a food-processing environment are a major concern. Good personal hygiene habits and sanitation practices are essential in preventing the spread of *Shigella*. ***See*** *Shigella*

Two psychrotolerant bacteria that should be mentioned are *Yersinia enterocolitica* and *Listeria monocytogenes*. These organisms are widely distributed in the environment, are associated with many animal species, and have been recovered from estuarine waters and shellfish. These two species are capable of growth at refrigeration temperatures. Not much is known about the association between seafoods and the diseases yersiniosis and listeriosis; however, consumption of raw molluscs was noted in the food history of cases. *L. monocytogenes* has been recovered from several seafood commodities, including frozen crustaceans, and the processing environment. ***See*** *Listeria*, Properties and Occurrence; *Listeria*, Detection; *Listeria*, Listeriosis; *Yersinia enterocolitica*, Detection and Treatment; *Yersinia enterocolitica*, Properties and Occurrence

There are increasing reports of molluscan-borne human illness caused by *Campylobacter* species, an enteric pathogen with many animal reservoirs. The suspected low infectious dose of this pathogen elicits concern. There is recent evidence that points to estuarine contamination by waterfowl and terrestrial run-off and subsequent survival in the environment. Studies likewise indicate this pathogen may survive in shellfish after harvest, even when they are handled adequately during storage. ***See*** Campylobacter, Properties and Occurrence; Campylobacter, Detection; Campylobacter, Campylobacteriosis

Enterotoxin-producing strains of *Staphylococcus aureus* have been responsible for outbreaks of intoxication with shellfish as the vehicle. Food handlers are the primary source of contamination, although the equipment and food contact surfaces can be fomites. This pathogen is also widely distributed, with human and animal reservoirs. Seafoods receiving a heat treatment and then further handling, such as hand picking or sorting, have been vehicles, since competitive microorganisms have been destroyed allowing *S. aureus* to predominate. Temperature abuse of the product allows for growth and toxin production. Enterotoxins produced by this organism are resistant to heat and may persist in the product even though the vegetative cells have been destroyed. ***See*** *Staphylococcus*, Properties and Occurrence; *Staphylococcus*, Detection; *Staphylococcus*, Food Poisoning

Spore-forming bacteria, *Clostridium botulinum*, *C. perfringens* and *Bacillus cereus*, have rarely been implicated in seafood-borne intoxications. These organisms are widespread in the environment, particularly in soil and not infrequently in the intestinal tract of animals. The potent neurotoxin produced by *C. botulinum* and the disease botulism probably elicit the greatest reaction from health officials. Canning processes, vacuum packaging, modified-atmosphere storage, and the recently popular minimally processed procedure *sous vide* for shellfish elicit safety concerns. Heat processing must be adequate to destroy spores of the organism prior to storage under anaerobic conditions to prevent germination and subsequent toxin production. *C. botulinum*, type E, one of seven identified, is primarily of marine origin. Several strains of *C. botulinum* grow and produce toxin at a temperature as low as 3°C. The protein toxins can be inactivated by heating at 60°C for 5 minutes. However, this means a product with preformed toxin must receive a heating step prior to consumption. ***See*** *Bacillus*, Occurrence; *Bacillus*, Detection; *Bacillus*, Food Poisoning; *Clostridium*, Occurrence of *Clostridium perfringens*; *Clostridium*, Detection of *Clostridium perfringens*; *Clostridium*, Food Poisoning by *Clostridium perfringens*; *Clostridium*, Occurrence of *Clostridium botulinum*; *Clostridium*, Botulism

C. perfringens, since it is commonly found in the intestinal tract of warm-blooded animals, has been suggested for use as an indicator of faecal pollution in the estuarine environment. This organism, along with *Bacillus cereus*, must reach high numbers in the food such that the enterotoxins produced can cause diarrhoeal illness. Proper storage of shellfish at refrigeration temperatures will prevent the growth of these organisms. Shellfish-borne illness from these two organisms is rare compared to that from other aetiological agents.

Viruses

More than 100 enteric viruses can be found in human faeces. Of concern is the very low numbers of these viruses needed to cause human infection. Since these viruses may be passed on to the environment in sewage effluents, seafoods may become contaminated. Norwalk and hepatitis A viruses are documented to have been transmitted to humans by contaminated molluscs. In the United States all reported cases of seafood-associated viral infections have been from the consumption of raw or inadequately cooked molluscs. Only a few of the enteric viruses have been shown to be transmitted, including hepatitis A, nonhepatitis A and B, Norwalk, Snow Mountain agent, astroviruses, caliciviruses and small round viruses. Lack of detection methods are thought to be one reason that the list is not larger. Some enteric viruses can survive in the marine environment longer than the bacterial species used for monitoring water quality. Water temperature ($< 10°C$) appears to be most important for survival. Fortunately, no multiplication of these viruses has been demonstrated in molluscs, although some survived refrigerated storage for extended periods. Crustaceans can take up these viruses in experimental procedures, but field studies are lacking for natural conditions. Enteric viruses were found in crabs sampled at a sewage sludge dump. ***See*** Viruses

The amount of heat processing of shellfish is a factor in prevention of transmission of viruses. Although a reduction in numbers of surviving enteric viruses can be expected by commonly used cooking techniques (boiling, baking, frying, steaming), low-level survival is of concern owing to the low infectious dose of many of these viruses. Boiling of molluscs for at least 20 minutes has been suggested. The best preventative measure to limit transmission is to assure that shellfish are harvested from nonpolluted waters, as difficult a task as that might be.

Transmission of viruses via shellfish after contamination by infected food handlers again stresses the need for good hygiene and sanitation practices. Hepatitis A contamination by food handlers is the only virus demonstrated to cause illness from seafoods, besides the consumption of raw molluscs. Thus far in the United States there are no documented cases of transmission of the other human enteric viruses by seafoods contaminated during processing, distribution or food handling.

Spoilage

Since these seafoods are highly perishable, immediately after harvest they must be subjected to reduced temperature, throughout processing and distribution, to extend shelf life. Control of temperature is critical to delay or retard decomposition by bacterial and autolytic enzymes, and the oxidation and hydrolysis of fats. A general rule is that for each degree increase in temperature from 0°C shelf life is reduced by 2 days. Storage under or in ice provides maximum shelf life for products that are not stored frozen. Storage temperatures below 7°C inhibit growth of most human enteric bacteria and prevent toxin production. ***See*** Storage Stability, Mechanisms of Degradation; Storage Stability, Parameters Affecting Storage Stability

Some species, particularly bivalve molluscs, are transported as live shipments. A cool temperature and a damp environment must be maintained to keep these animals alive, which for some species can be approximately 2 weeks. Generally, problems with decomposition do not occur as long as the animal remains alive. Shipment and storage of raw products must be separate from processed product to prevent cross-contamination. Some countries have strict regulations regarding shipment and sale of certain live species.

The microflora of harvesting waters are highly variable and thus the microflora of the shellfish vary. For example, warm-water crustaceans harbour mainly Gram-positive bacteria such as *Micrococcus* or Coryneforms, while cold water species carry primarily Gram-negative organisms, including *Pseudomonas*, *Moraxella* and *Flavobacterium*. Warm-water shellfish generally have higher bacterial levels (mesophilic organisms) at harvest than cold-water shellfish. However, it has been suggested that cold-water species spoil more quickly since they already contain bacteria that are capable of growth at refrigeration temperatures. Studies have shown certain organisms during chilled storage to become predominant as spoilage progresses. In some cases, shifts in the type of bacteria occur. Species of *Pseudomonas*, *Flavobacterium*, *Acinetobacter* and *Moraxella* seem to be most commonly involved in spoilage, although other genera have also been associated with processes of decomposition. Yeasts can also be involved to some extent in spoilage. For example, *Rhodotorula*, *Candida* and *Torulopsis* species can result in discoloration of the product during storage. ***See*** Spoilage, Bacterial Spoilage; Spoilage, Yeasts in Food Spoilage

Growth of bioluminescent marine bacteria can result in products that 'glow in the dark', particularly noticeable in processed shrimp and crab. These bacteria survive the processing techniques or recontaminate the product and grow during chilled storage to levels where the luminescence is visible to the human eye. Bioluminescent organisms include species of *Photobacterium*,

Xenorhabdus, *Altermonas* and *Vibrio*. These bacteria are generally thought to be nonpathogenic to humans; however, a few species of *Vibrio* may prove otherwise.

The bacteria normally involved in spoilage are susceptible to heat and normal cooking processes would eliminate them. Temperature abuse at some point between harvest and consumption is the primary cause of rapid spoilage and also for allowing pathogenic organisms to grow.

Chemical Contamination

Marine Toxins

There are several natural or preformed toxins that occur in shellfish, primarily the bivalve molluscs owing to their filtering ability and concentration of the toxins. These toxins are quite potent and may be present in minute quantities. Some of these toxins may undergo transformations within the shellfish that can increase the toxicity to humans; however, the amount of accumulated toxin will not increase after harvest. The production of these toxins is a natural process and cannot be controlled. Monitoring programs, most using animal bioassays, that have been established have aided effectively in reducing human illness by closing harvesting areas when toxins reach certain levels. These toxins are far more stable to heat than most bacterial toxins and therefore cooking of shellfish is not a reliable measure of protection. No process can be relied upon to remove or destroy these toxins in shellfish. ***See*** Contamination, Types and Causes

Paralytic shellfish poisoning (PSP) is a potentially fatal syndrome resulting from ingestion of a family of potent neurotoxins called saxitoxins. PSP is generally associated with consumption of bivalve molluscs although crabs and whelks have been implicated. Seventeen different saxitoxins are recognized at present, derivatives of the parent saxitoxin molecule. These water-soluble toxins are produced by a group of dinoflagellates, frequently by species of *Alexandrium*. Low doses result in a tingling or burning sensation of the lips, with higher doses causing paralysis of extremities, loss of motor coordination and even death by respiratory paralysis. No antidote is known.

The 'red tide' of the Gulf of Mexico is from a bloom of *Ptychodiscus brevis* with production of brevetoxins which can cause neurotoxic shellfish poisoning (NSP) in humans. NSP is generally found along the west coast of Florida, and other Gulf States, occasionally extending up the Atlantic Coast of the United States. Brevetoxins, of which nine are identified, consist of cyclic polyether backbones. They are lipophilic in nature, causing nausea and neurological symptoms. This syndrome usually lasts a few days and no deaths of humans have been reported.

Okadaic acid and its derivatives have caused diarrhoetic shellfish poisoning (DSP), recognized in Japan and Europe. Bivalve molluscs are the known vectors. Acute diarrhoea is the primary symptom, with onset more rapid than from bacterial intoxications. DSP toxins have been isolated from molluscs implicated in illnesses and from two genera of dinoflagellates.

Amnesic shellfish poisoning (ASP) is a recently identified syndrome which seems to be localized in an area of Eastern Canada, Prince Edward Island. Consumption of mussels containing domoic acid have caused this rare form of poisoning, resulting in three deaths. Symptoms include nausea, loss of equilibrium and neural involvement including memory loss. Thus far, it appears that brain damage caused by ASP may be irreversible, resulting in permanent loss of memory.

Consumption of whelks, edible marine snails of the genus *Neptunea*, has caused rare instances of tetramine poisoning. Symptoms caused by tetramethylammonium ion appear approximately 30 minutes after ingestion. Recovery is usually complete within a few hours. These symptoms include headache, dizziness, and short periods of impaired vision. This toxin is considerably less toxic than most marine toxins. There have been no documented cases of whelk-derived human mortality.

Chemical Contaminants

Industrial and agricultural chemicals that can contaminate areas of shellfish harvest are difficult to assess from a risk-evaluation standpoint. The effects to health that might be suspected from these contaminants are not readily obvious in the form of distinctive or acute illnesses. The effects of long-term exposure to low levels of these contaminants, both to shellfish and to those who consume them as food, are areas that need to be studied further to determine actual risks.

Concentrations of these contaminants can vary by area and be distributed unevenly in the environment, thus being difficult to control. Organic contaminants (polychlorinated biphenyls, dioxin), chlorinated hydrocarbon insecticides (DDT, Endrin, Chlordane), petroleum hydrocarbons, along with inorganics, antimony, arsenic, cadmium, lead, mercury and selenium, to name only a few, may find their way into aquatic environments and utimately into shellfish. These chemicals and others are known to cause serious health effects in humans. Prohibiting the harvest of shellfish from areas that are subject to industrial pollution or excessive terrestrial run-offs may be the most effective way of reducing risks to human health. Future needs are to strengthen and then enforce regulations to minimize chemical contamination of the environment. ***See*** Heavy Metal Toxicology; Pesticides and Herbicides, Toxicology

Bibliography

Ahmed FE (ed.) (1991) *Seafood Safety*. Committee on Evaluation of the Safety of Fishery Products, Food and Nutrition Board, Institute of Medicine. Washington, DC: National Academy Press.

Anderson DM, White AW and Baden EG (eds) (1985) *Toxic Dinoflagellates*. New York: Elsevier Science Publishing.

Anonymous (1991) Update: Cholera – Western Hemisphere, and recommendations for treatment of cholera. *Morbidity and Mortality Weekly Report* 40: 562–565.

Colwell RR (ed.) (1984) *Vibrios in the Environment*. New York: Wiley.

Gerba CP (1988) Viral disease transmission by seafoods. *Food Technology* 42: 99–103.

International Committee on Microbiological Specifications for Food (1988) *Microorganisms in Foods 4. Application of the Hazard Analysis Critical Control Point (HACCP) System to Ensure Microbiological Safety and Quality*. Oxford: Blackwell Scientific.

Liston J (1990) Microbial hazards of seafood consumption. *Food Technology* 44(12): 56–62.

Tison DL and Kelly MT (1984) *Vibrio* species of medical importance. *Diagnostic Microbiology Infectious Diseases* 2: 263–276.

Ward DR and Hackney CR (eds.) (1991) *Microbiology of Marine Food Products*. New York: Van Nostrand Reinhold.

C. A. Kaysner
FDA – Seattle District, Washington, USA

Ranching of Commercially Important Molluscs and Crustacea

The aquaculture or rearing of aquatic organisms, unlike fisheries management as applied to wild populations, requires control of the crop. The term 'ranching' is sometimes used to describe the process of allowing free-ranging stocks to utilize natural pastures (e.g. salmon ranching, in which hatchery-reared smolts are released to the sea to complete their growth before returning to shore to be harvested). In this section we will use the term in the broader sense of cultivation, with or without restrictions on movement of the stocks. ***See*** Fish Farming

Molluscs

Types Suitable for Farming

To be suitable for farming, organisms ought to possess, in addition to such features as market acceptability and high value, biological characteristics that enhance their culture in captivity. Among these should be good food conversion efficiency or low trophic position, adaptability to crowding, rapid growth, tolerance to changes in water quality, and ease of reproduction in captivity.

Virtually all molluscs commercially cultured today are herbivorous throughout their life cycle. Most are filter-feeders, using phytoplankton as their primary source of nutrition, at least as larvae, and in most instances also as adults. Examples include oysters, clams, scallops and mussels. After the larval stage, a few use other kinds of plants as food. *Strombus gigas*, the Caribbean Queen Conch, feeds on benthic microalgae, while the abalones (primitive gastropods) use macroalgae. Predatory molluscs, such as cephalopods, are seldom cultured for food because of the expense of providing live animals for them to eat. The giant clams, *Tridacna gigas* and relatives, which live in tropical phytoplankton-poor coral reef environments, cultivate algae within their own tissues as their source of nutrition.

Most cultured molluscs are sessile inhabitants of the benthos after the planktonic larval stages. Important exceptions include the scallops which, because they can swim off the bottom, must be confined during grow-out. Although mussels and clams can move short distances on the bottom, such limited movements are not serious obstacles to their cultivation. Because of their low trophic position and lack of intraspecific aggressive behaviour, the commercially cultivated molluscs can be grown in relatively high densities, as long as adequate water currents bring them food and oxygen, and take away wastes.

Most cultured molluscs reach marketability in two or three years, but some mussels and tropical oysters attain this size in less than a year. Slower growth may be acceptable when the value of the end product is exceptionally high.

Since most cultivated species of molluscs are from estuarine or other inshore habitats, where fluctuations in environmental conditions are common, they are relatively tolerant of such changes.

The majority of commercially cultivated molluscs have very similar general requirements for their larval stages; thus similar hatchery methods are used for a variety of species. Typically, adults are conditioned by heavy feeding at temperatures lower than those required for spawning, and then are shocked by thermal or chemical stimuli to initiate release of gametes. Fertilization is usually external (e.g. *Crassostrea* spp.) but some species (e.g. *Ostrea* spp.) brood their embryos for a few days before liberating them into the water column to feed.

Free-swimming larvae are fed with appropriate species of phytoplankton. Two major approaches to the feeding of mollusc larvae in hatcheries have been developed. A natural mixture of phytoplankton species in nutrient-enriched sea water may be encouraged to bloom in large hatchery vessels and used as food. The other method is based on primarily monospecific, laboratory-reared cultures of desirable algal species. Use of such cultures requires careful timing so that adequate supplies of fresh algae will be available when needed. Techniques for concentrating and freezing phytoplankton permit the production of algae in advance of use. The concentrated cultured algae are then diluted prior to being fed to mollusc larvae, with results comparable to those obtained with freshly cultured algae. ***See*** Single-cell Protein, Algae

After a few days or weeks, depending upon the species, metamorphosis of the molluscan larvae takes place and the post-larvae settle to the bottom of the culture vessel. At metamorphosis, oysters seek hard substrates to which they attach themselves permanently. Preferred substrates, such as old oyster shells, may be supplied by the culturist. Molluscs, which in the field may show preferences for sediments or other substrates, will metamorphose in the absence of those stimuli under hatchery conditions. After a variable period in the nursery, the young juveniles of filter-feeding molluscs are planted in the field. Abalone may be stocked in rocky, natural habitats but can be grown in tanks where they can be provided with macroalgae or formulated rations and protected from enemies. Most filter-feeders require pumping of large volumes of plankton-rich water, which makes rearing them to market in tanks not economical. Survival rates of field-planted juvenile shellfish are strongly influenced by size at planting. Generally, the larger the juvenile molluscs are at planting, the better they are able to survive predation.

Practical and Environmental Considerations

Molluscs are nutritious, delicious, easily digested, high in protein, low in fats, and popular in many countries. Because natural stocks are frequently inadequate to meet demand, cultivation has become a major source of supply. Culture of filter-feeding molluscs to market size is feasible only under field conditions. As water is taken in by the molluscs, it is passed over gill structures which effectively remove suspended food particles. This filtering process can result in the concentration within the shellfish of organisms such as viruses or bacteria harmful to man. Serious diseases, such as hepatitis and cholera, can be contracted from eating raw shellfish from contaminated areas. Phytotoxins derived from certain dinoflagellate algae known as 'red tides' also can cause severe illness in humans without affecting the shellfish. Therefore, strict controls over growing and harvesting, including the monitoring of water quality, are necessary to assure safe product. Growers and their grounds are licensed and the fresh product carries a certification tag through distribution to the end user. In many areas where short-term variation in water quality occurs, health officials may require depuration of shellfish before sale. ***See*** Fats, Digestion, Absorption and Transport; Protein, Food Sources

A number of serious diseases to which shellfish are subject have had catastrophic effects on oyster and other shellfish populations. These diseases may not harm the human consumer directly, but affect the industry by reducing stocks, increasing production costs and creating market shortages. Many of these diseases have been spread from widely separated geographical areas to new areas through man's deliberate and uncontrolled transplantation of shellfish stocks. These same transplantation efforts have resulted in the spread of a number of shellfish predators and competitors.

Feeding for Maximum Yield

All commercially cultured molluscs are by nature bottom-dwellers. The filter-feeders, by far the most abundant category, depend upon currents to bring them their food and oxygen and to carry away their wastes. Although commercially formulated feeds are often used for other cultured aquatic species, such feeds are not yet available for filter-feeding molluscs. Growth is generally better in areas of good water circulation, and because wave action and water movement are greater near the surface than on or near the bottom, oysters, mussels and scallops may grow as much as two or three times faster if they are kept off the bottom. This is accomplished with floating rafts or other devices from which hang ropes with attached molluscs. By using the three-dimensional water column, up to 50 t per ha per year of some species can be produced under ideal conditions.

Handling and Marketing

Although molluscs are marketed in many forms, the highest prices are usually obtained for live product. Quality of all seafood begins to deteriorate immediately upon death. Thus, at harvest, the live shellfish must be kept moist and cool and moved as quickly as possible to market. Live filter-feeding molluscs are almost never kept in water during transportation to markets. Most species are capable of living for days or weeks out of water and this is generally sufficient time for the product to reach the consumer. Highest prices for oyster and clams are obtained when they are served on the half shell in the raw bar trade. There is a substantial market for fresh, shucked product as well as various processed forms of shellfish meats. Processing of molluscs for other markets involves shucking the meats out of the shells, then canning, cooking or freezing them. The entire body of most shellfish is eaten, but in North America only the adductor muscle of scallops is consumed, the remainder of the flesh being discarded. However, adductor muscles with attached roe is sought by Europeans and Australians, and the Asian practice of consuming all soft parts of the scallop is beginning to be adopted by gourmets.

The shells of commercial species often have commercial value as well. Shells may be planted on oyster beds as culch to encourage the setting of postlarval oysters which generally prefer old shells, especially of their own species, rather than other hard substrates. Shells may be ground for use as grit in poultry production. In former times, oyster shells were used for road beds. They are still used in some areas for the production of lime. Shells of some species, such as abalones and conchs, are used in jewellery-making or as curios in the tourist trade.

Potential for the Future

The demand for seafood in general, the high yields possible with filter-feeding molluscs, and the spread of hatchery technology are encouraging the expansion of mollusc culture in many parts of the world. Increasing pollution resulting from population pressure near traditional production grounds continues to reduce the natural areas suitable for mollusc culture along many coastlines. The devastating collapse of the oyster industry in Chesapeake Bay, one of the world's largest estuaries, is attributed largely to historical overfishing and the spread of oyster diseases during the last four decades. Some scientists believe that deterioration in water quality may be a major factor in reducing the disease resistance of these populations. Others maintain that reduction in shellfish stocks contributes to reduced water quality by removing a major filtration process from the estuarine environment.

Depuration is the practice of holding live shellfish in tanks of clean or continuously sterilized water to permit purging of potentially dangerous microorganisms from shellfish before sale. Ultraviolet radiation, chlorination, and ozonation of filtered water in recirculating systems are used for this process. Depuration serves to increase consumer confidence in the safety of raw shellfish. While shellfish from polluted or potentially polluted waters can be made safe with the process, depuration may not be required if the shellfish come from certified growing grounds. In some countries, depuration is required of all live shellfish. The trend towards universal use of this process will be driven by the general decrease in availability of pathogen-free rearing waters. In the USA, where the practice is not yet mandated for all shellfish, some government agencies are beginning to recommend that persons with compromised immune systems refrain from eating any raw shellfish, to avoid risk from possibly illegal or contraband product. ***See*** Plant Toxins, Detoxification of Naturally Occurring Toxicants of Plant Origin

Several factors may help increase production in the years ahead. The use of polyploidy to produce sterile oysters can result in better meat quality over the entire year, since the shellfish do not divert energy to reproduction during the normal spawning season. As finfish culture begins to move towards offshore waters, especially in Europe and Asia, the opportunity exists for offshore mollusc culture in association with the physical structures of these floating fish farms.

Crustaceans

Types Suitable for Farming

Virtually all crustaceans cultured for food are members of the order Decapoda, which consists of the prawns, crayfish, crabs, and lobsters. These crustaceans offer a set of problems very different from those of molluscs. They generally do not feed as low on the food chain as do molluscs, and they are much less amenable to crowding because of their need for substrate and their tendency towards cannibalism. To a considerable degree, their high unit value compensates for these disadvantages.

The most widely cultivated crustaceans are the shrimps or prawns, most of which are tropical or subtropical in distribution and capable of reaching market size in a few months. Food of most prawns consists of detritus, microorganisms and small animals associated with sediments; hence prawns are easier to feed than predaceous crabs and lobsters. Prawns are also much less cannibalistic than crabs and lobsters. Although temperature, salinity and other requirements vary somewhat among species, all important prawns are inshore or estuarine species capable of tolerating a range of environmental conditions.

There are significant differences in reproductive biology of the various commercially important crustacean groups. The penaeid prawns, by far the most important of all cultured crustaceans, are unique among the decapods in their reproductive pattern. The fertilized eggs of penaeids are broadcast into the water column, whereas the females of caridean prawns and all other decapods after spawning carry their eggs attached to abdominal pleopods during embryological development. The swimming larval stage which hatches from the free-floating penaeid egg is the non-feeding nauplius, a stage passed within the egg of other decapods. From the time the nauplius moults into the protozoal stage it must be provided with food, typically phytoplankton and other microorganisms. Throughout the sequential larval stages, the young prawn shifts its dietary preference increasingly towards animal matter such as zooplankton. In culture conditions, this food is provided commonly in the form of the newly hatched nauplii of the brine shrimp, *Artemia*. Recent advances in hatchery technology now permit the feeding of prepared diets to these larval stages. Upon completion of the larval series, the postlarva is capable of assuming a benthic existence, although it may continue to swim for a time. The sequence from hatching to metamorphosis varies by species and rearing conditions but typically may require 10–40 days. Postlarvae often are kept in the hatchery or nursery area for an additional few weeks to gain size and strength before being stocked into ponds.

In each of the other major groups of cultured crustaceans, special features of reproduction or feeding biology may have an impact on the rearing process. The two major groups of marine lobsters differ markedly in their larval development, social behaviour, growth rates and other characteristics. The clawed lobsters hatch in a relatively advanced stage and pass through only three moults over a period of 10–30 days, depending upon temperature, before attaining the benthic postlarval stage. They may then require 5 years or more to reach market size under ambient conditions. Even in heated water with ideal feeding, market size may require more than 2 years. The spiny or rock lobsters hatch as delicate phyllosoma larvae, which float in the plankton for up to a year before metamorphosis. Under culture conditions market size for some spiny lobsters can be achieved in less than two years from metamorphosis. In contrast, the freshwater crayfishes hatch directly as miniature adults, bypassing the free larval stages. Thus, their culture does not require a complex hatchery technology. Crayfish feed on detritus, much as do penaeid prawns, and do not require the kind or quantity of animal protein needed by marine lobsters.

Practical and Environmental Considerations

Because the commercially important prawns are all warm-water species requiring two-dimensional living space, as opposed to being able to use the water column as fish do, their culture is concentrated in tropical regions where climate is suitable and coastal land is available. Freshwater prawns, the larvae of which require brackish water, can be reared away from the coast. Artificial sea water suffices for the hatchery phase. The rapid expansion of penaeid prawn culture in the last decade has created a demand for coastal sites in many tropical countries. The construction of thousands of hectares of ponds has resulted in the destruction of ecologically valuable mangrove and other ecosystems.

Proper site selection is a major factor in establishing a successful prawn farm. Soil type, abundant clean sea water, good ground water supplies for adjusting salinity within the ponds, and other physical characteristics determine the environmental suitability for a prawn farm. Social, political and economic factors such as structure of the labour pool, ease of securing government permits, tax relief, repatriation of profits to foreign investors, and land costs are also important.

Most early prawn operations, and many current farms in areas of inexpensive land and labour, have relied on extensive low-density techniques which, while yielding only modest harvests, have proved to be profitable and relatively low risk. Such farms rely heavily upon natural foods produced within the ponds, occasionally supplemented by fertilizers to enhance pond productivity. More recently, the trend has been towards semi-intensive to very intensive culture, requiring more reliance on artificial rations, since the densities of prawns in such operations are greater than natural pond productivity can sustain. Because of the rapid growth which can be obtained under culture conditions, two or more crops of prawns can be grown in a year, with yields ranging from a few hundred kilograms to several tens of thousands of kilograms per hectare per annum. Obtaining the higher yields requires much more intensive management, including aeration, water changes, heavy feeding and, in the most advanced methods, the removal of suspended and settleable wastes from the ponds by specially designed hydraulics.

The discharge of wastes from culture ponds can be an environmental problem in areas of heavy concentration of prawn ponds or hatcheries. In the late 1980s, prawn culture in Taiwan reached such intense levels that many farms were forced to use sea water contaminated with effluents from adjacent farms. Ultimately, this led to severe outbreaks of disease and a crash in the industry. Similar problems exist in Ecuador where hatcheries are concentrated along the coast. Heavy and sometimes indiscriminate use of antibiotics is thought to be a factor in outbreaks of disease in that area also. Another environmental effect of discharge from culture ponds may be deterioration of water quality in receiving streams, a concern which is being increasingly addressed by regulatory agencies in many countries. ***See*** Effluents from Food Processing, Disposal of Waste Water

Crayfish farming is practised in Europe, Australia, China and, to a lesser extent, Africa and South America, but the largest production of crayfish is in the southern USA particularly in southern Louisiana. Of the six or eight major species of crayfish cultured around the world, the most widely used is probably the red swamp crayfish, *Procambarus clarkii*, which is the major contributor to the Louisiana production. For better or worse, this species is also the most widely transplanted and established exotic crayfish in the world. The traditional culture method for this species is based upon use of seasonally flooded fields in which rice or other vegetation is sown or allowed to grow as forage. An initial stocking with crayfish breeders is followed by slow draining of the ponds to encourage the breeders to dig burrows in which reproduction takes place. After a 'dry' season, during which the forage vegetation develops, the fields are reflooded, the young emerge from the burrows and feeding by the new generation begins. After the initial stocking, populations are self-sustaining, even with heavy fishing pressure. Traditionally, harvesting has been by trapping, an expensive and relatively inefficient technique which can cost up to 60–80% of the value of the crop. Production levels of the order of a few hundred to as many as several thousands of kilograms per hectare per annum can be achieved. Oxygen depletion is a major problem for producers, particularly when too much young vegetation decays in warm weather following the autumn flooding. Flushing the ponds with new water maintains oxygen levels. Methods known to improve production, such as aeration, feeding of prepared rations, keeping ponds flooded year-round, and batch harvesting are being adopted.

A major problem for the crayfish industry in Europe has been the decimation of native stocks by a fungal disease introduced to Italy from North America in the middle of the nineteenth century. Enormous economic and ecological damage was done as the disease extended west to the British Isles, north to Sweden, east to Turkey and south to Spain. North American crayfish are relatively resistant to the disease but are carriers capable of infecting new areas into which they are introduced. All non-North-American crayfish, including those native to the southern hemisphere, are thought to be susceptible to this 'plague'. The introduction of tolerant species such as *Procambarus clarkii* to areas outside North America precludes or threatens the culture of susceptible species. In an effort to compensate for the loss of native crayfish, particularly the noble crayfish, *Astacus astacus*, European culturists have introduced the somewhat similar North American signal crayfish, *Pacifastacus leniusculus*, which in many areas has completely replaced the remaining stocks of the native species.

Other environmental problems related to crayfish culture include damage to irrigation structures by burrowing species, competition between introduced and native species, and the introduction of disease by indiscriminate transfer of wild stocks. In Kenya the proliferation of introduced *Procambarus clarkii* caused economic damage to gill net fisheries.

Lobster culture is still in a relatively undeveloped state. The clawed lobsters, *Homarus* spp., are very aggressive and require physical isolation or large areas for growth after the larval stages. Furthermore, although they can be bred in captivity and reared from hatching to adult size, the time required for completing this cycle is rather long. Large-scale rearing of clawed lobsters to commercial size is not yet practical. Stocking of hatchery-reared *Homarus* for enhancement of fisheries has been practised to a limited extent. The most successful application of aquaculture techniques has been in 'pounding', the practice of short-term holding of market size lobsters to permit postmoult fattening or to keep harvested stocks off the market during periods of low prices.

Spiny or rock lobsters of the family Palinuridae are much less aggressive and somewhat faster growing after the larval period. The rearing of juvenile spiny lobsters to market size poses few problems as the animals tend to be more social and grow faster than their clawed counterparts. Unfortunately for culturists, the spiny lobsters have a larval type which has resisted large-scale laboratory or hatchery rearing to metamorphosis. This larva, the phyllosoma, is physically delicate and morphologically very different from the larva of most other decapods. It is adapted for long-distance, long-duration ocean transport. Its nutritional and other ecological requirements are not well understood.

Crab culture is limited to a very few species. Mass culture techniques for several commercially valuable species have been developed, but most crabs are predaceous and highly cannibalistic, which makes their culture in high densities to market size very difficult. Wild-caught juveniles of the large and highly esteemed Indo-Pacific mangrove or mud crab, *Scylla serrata*, are stocked in low densities in brackish ponds used for oyster and finfish culture. Postlarvae of the Chinese mitten crab, *Eriocheir sinensis*, are produced in hatcheries in very large numbers for stocking into natural waters.

A special, short-term form of crab culture is the production of highly esteemed soft-shelled crabs. The holding of premoult crabs until they shed their shells requires only a few days. The animals are not fed during this time. The entire soft crab can be eaten. The market value of soft-shelled crabs may be three or more times the value of a hard crab of the same species. Soft-shelled crabs (*Carcinus mediterraneus*) have been produced in Italy for many years, but the world's largest soft crab industry is based on the blue crab, *Callinectes sapidus*, the swimming crab of the Chesapeake Bay and other coastal waters of the eastern USA.

Production of soft crabs requires a convenient source of premoult crabs, so that shedding operations need to be located in the vicinity of a crab fishery. Premoult crabs are held in shallow wooden trays until they moult and are then removed while still very soft. Water may be pumped from the nearby bay or creek through the holding trays and returned without treatment to the natural environment; alternatively, closed, recirculating systems may be used. The advantages of the latter include no need for access to waterfront property, freedom from fluctuating water quality of the natural environment, and some degree of control over temperature and other variables. Crabs collected by traps, dredges or other methods are sorted by moult stage using colour indicators and other clues. Those of similar physiological condition are kept together. Crabs close to moulting do not feed, but those several days away from shedding are extremely cannibalistic, so that careful sorting is required. The blue crab may moult at any time of day or night. A crab left in the water for more than a few hours after shedding begins to toughen its new shell, thereby losing value quickly. Sorting and removal of soft crabs must be done continuously as long as there are crabs in the system.

Feeding for Maximum Yield

As population densities of prawns are increased from low to very high levels in ponds, natural food supplies quickly become inadequate for sustaining growth. At moderate densities supplemental feeding with nutritionally incomplete rations may be acceptable because prawns can derive the small quantities of necessary vitamins and other trace materials from natural foods, but as density increases, a nutritionally complete diet becomes essential for adequate growth and survival. Development of such diets led to a major expansion of the prawn farming industry in Taiwan in the early 1970s. As knowledge of dietary requirements has grown, many commercial rations have been developed. Conversion rates of better than two units of dry feed for each unit of live weight of prawn have been achieved.

The physical nature of the artificial diet is critical since, unlike fish which are able to swallow large particles, prawns are nibblers, taking small amounts over an extended time. The artificial ration, usually extruded or in the form of pellets, must retain its integrity for hours, yet must also retain its attractiveness by slowly leaching out odours or scents until the particle is found by the prawn. Because most prawns are nocturnally active, feeding is most effective in late afternoon or early evening to minimize the time between provision of the diet and its consumption by the prawns.

Crayfish also are basically detritivores. Natural vegetation, animal matter and associated microorganisms suffice as food in low-density culture. Supplemental feeding with natural or formulated feeds can greatly improve growth. Supplemental feeding has not been widely practised in Louisiana, where wild and cultivated crayfish production is high and product value low, but can be practical where prices are several times higher. Opening of new European and other markets has stimulated development of methods for production of larger crayfish and may justify supplemental feeding. Commercial rations are already available.

Handling and Marketing

As with all seafood, prawns and other crustaceans are at their best when they reach market alive or with minimal time between harvest and processing. Most cultured prawns are chilled at or near harvest and 'heads' removed before freezing. This extends shelf life by removal of sources of enzymes and other materials which would accelerate the deterioration of the stored product. Exceptions include prawns intended for the live market, such as those shipped to Japan and some European markets where the taste value of the 'head' of the prawn is appreciated.

Lobsters may be shipped alive, or the tails may be removed and frozen for shipment.

Soft crabs are shipped live or frozen after removal of some body parts such as eyes, gills and the abdomen. Best prices for all crustaceans are obtained for live or fresh product.

Potential for the Future

Prawn farming already accounts for at least 25% of all prawn supplies and the rate of expansion of prawn farming suggests that within a decade as much as 50% of the world's prawn supply may come from farms. The very large market demand, relatively low production costs and strong market prices make prawn farming the fastest growing aquaculture activity in the world. At present the largest producers are China, Indonesia, Taiwan, Ecuador and India, but many other countries are entering the competition.

It is unlikely that significant prawn production will be developed in the temperate zones, despite promising intensive culture methods that may make possible even indoor culture close to major markets. The lower costs of production in tropical, less developed countries, coupled with relatively inexpensive shipping costs even to markets on opposite sides of the globe, will probably continue to keep the major production in those areas.

Lobster farming will make slower progress because of much higher production costs relative to market value when compared with those for prawn farming. The market value of a kilogram of prawns is not very different from the value of a kilogram of lobsters, yet prawn farmers, with simpler hatchery technology in hand, can produce four to six times the number of crops of prawns in the time required to produce a single crop of lobsters.

In contrast to lobster farming, crayfish farming will certainly expand considerably from its current base. The several advantages for the crayfish farmer include less expensive feed, no requirement for sea water, coastal property or elaborate hatchery technology, and a relatively favourable market price for all forms of the product. Many species of crayfish not yet cultured on a large scale may come to dominate the world market. Some of the larger Australian species, such as *Cherax tenuimanus* (the marron) and *Cherax quadricarinatus* (the red claw), grow to sizes comparable to lobsters, with none of the disadvantages of their marine counterparts. The meat yield of those large crayfish is much greater on a percentage basis than that of smaller species such as *Procambarus clarkii*. Techniques for the production of soft-shelled crayfish, similar to those needed for soft crab production, have been developed.

Production of soft-shelled crustaceans is likely to expand. Traditional markets for the blue crab have been US coastal cities such as Baltimore and New York, but in recent years exports to Japan and Europe have become significant. As familiarity with this type of delicacy increases, soft crab production of other species is beginning in nontraditional areas. The large Caribbean spider crab, *Mithrax spinosissimus*, is scarcely available in large enough natural populations to support fisheries, yet because it has a rapidly developing nonfeeding larval stage, and eats a largely vegetarian diet, it is easily cultured, and is already being used for soft-shelled crab production. Since the entire soft-shelled crustacean can be eaten, eliminating much waste, and since the anterior portion of the body of prawns (the 'head') contains much flavour, it is likely that soft-shelled prawns will soon be tested in the market place.

Bibliography

Hahn KO (ed.) (1989) *CRC Handbook of Culture of Abalone and Other Marine Gastropods*. Boca Raton, Florida: CRC Press.

Huner JV and Brown EE (eds) (1985) *Crustacean and Mollusc Culture in the United States*. Westport, Connecticut: AVI Publishing Co.

Manzi JJ and Castagna M (eds) (1989) *Clam Mariculture in North America*. Developments in Aquaculture and Fisheries Science, 19. Amsterdam: Elsevier.

McVey JP (ed.) (1983) *CRC Handbook of Mariculture*, vol. 1, *Crustacean Aquaculture*. Florida: CRC Press, Boca Raton.

Morse DE, Chew KK and Mann R (eds) (1984) *Recent Innovations in Cultivation of Pacific Molluscs*. Developments in Aquaculture and Fisheries Science, 14. Amsterdam: Elsevier.

New MB (ed.) (1982) *Giant Prawn Farming*. Developments in Aquaculture and Fisheries Science, 10. Amsterdam: Elsevier.

Provenzano AJ (ed.) (1985) *The Biology of Crustacea*, vol. 10, *Economic Aspects: Fisheries and Culture*. New York: Academic Press.

A. J. Provenzano, Jr
Old Dominion University, Virginia, USA

SHELLFISH – DIETARY IMPORTANCE

See Fish, Dietary Importance of Fish and Shellfish

SHERRY

Contents

The Product and its Manufacture

Definition

Sherry is the name given to a number of related types of dessert wine originally developed in the area around Jerez-de-la-Frontera in the south of Spain. Until recently, the name was widely appropriated by the producers of white fortified wines in other areas for products that, in some cases, did not even attempt to imitate the qualities of the original Jerez wine. Because the quality and style of the final product is dependent on the nature and quality of the grapes, and this itself depends on the climate and viticulture methods used in raising these grapes, the use of geographical names to describe wines from other areas is now strongly discouraged on the international market, though Jerez producers have been less successful than many in having such usage outlawed. In this article the term 'sherry' refers only to the Jerez product, though technical procedures developed in other areas to imitate this wine are described when this seems appropriate.

The Vineyards

The best vineyards are situated in low hills composed of a rather chalky soil known locally as albariza, most of which occurs in a broad area immediately to the east of Jerez.

Rainfall is moderate (about 60 cm) but quite variable. Very little falls in the summer months from May to August. Rain during the vintage period, that is, September, is not uncommon. The rainfall is predominantly from November to February, when it may be prolonged and quite heavy.

The preferred soil is one containing from 30 to 80% calcium carbonate with the remainder consisting of sand, clay and humus. The advantages of this soil seem to lie in its physical structure rather than its chemical composition. It has the virtue of retaining water from winter rainfall even at the end of the very dry summer.

Planting, Training and Pruning

Traditionally, vines were cultivated very low and supported on small wooden props to keep the fruit off the ground. Almost all are now supported on wires but are still trained low with one wire at 50 cm or so and another at approximately 1 m above the ground.

The vines are always grafted onto hybrid rootstocks as the soil is contaminated with insects of the genus *Phylloxera*, which attack the roots of *Vitis vinifera*. The choice of rootstock is made after soil analysis has been carried out as not all hybrid rootstocks are suitable for growing in soil of high calcium carbonate content.

Virus-free rootstocks are now available, but could not be obtained when many of the existing vineyards were planted.

The traditional pruning requires that the vine is pruned to short spurs some 30–40 cm above soil level; one branch, the fruiting branch (the vara), is restricted to 8–10 buds, the other, the replacement spur, to one or two buds.

Grape Varieties

At one time half a dozen white grape varieties were grown for sherry, but currently only three may be planted, the Palamino, the Pedro Ximenes and the Moscatel Gordo Blanco. Of these, little, if any, Moscatel is incorporated in sherry by the leading producers, Pedro Ximenes is used only for sweetening wine and, for that purpose, it may be, and is, obtained from outside the delimited area. For all practical purposes sherry vineyards are planted with either the Palamino basto (the Palamino de Jerez) or, more often, the higher-yielding Palamino fina, both locally isolated subvarieties. As mentioned earlier, virus-free stock was not available when the majority of the vines were planted but, in spite of this and the very large area under a single variety, the vines seem very healthy.

The Vintage

One unfortunate result of the use of a single variety of grape, the Palamino, on similar soils in a compact area is that all the grapes ripen virtually simultaneously, usually in the first 10 days in September. To secure the crop in optimum condition for wine-making the harvest must be completed very rapidly within 3 weeks, a large labour force for this purpose must be assembled and a great deal of equipment is required to cope with the throughput; much of this will stand idle for the other 49 weeks of the year.

Wine-making

The first part of the sherry-making process is the production of a normal dry white wine. Standard methods are used so we need not go into detail. Grapes will normally be crushed through rollers within 4 h of picking, though those harvested late in the afternoon may be safely stored overnight in plastic boxes when temperatures are lower.

Pressing follows immediately after crushing. Horizontal batch presses of the type manufactured by Vaslin or Wilmes are most often used in smaller installations. Larger installations often employ a continuous pressing line with perhaps a continuous pre-dejuicer operating without pressure and two dejuicers operating at 1–2 atm. Whatever method is used, the free run juice and early pressings, which can reach tannin levels of up to some 200 ppm, are segregated from later pressings which reach levels of about 500 ppm. The remainder of the juice is extracted with a hydraulic press; this is very high in tannin and is not used to make sherry, but is sent for distillation.

As in other wine-making areas, must containing about 1% solids is preferred. The must is often cooled and settled in tanks for at least 8 h. This reduces the load on the centrifuges, which are a major item of capital expenditure.

Formerly, all fermentation was carried out in 500–600 litre oak butts, later used for maturing the wine. Due to the small capacity the lack of temperature control was not a disaster. Nowadays most wine is fermented in tanks, but a limited amount of fermentation is still carried out in the traditional way because (1) there is some demand for wine so produced, which has a characteristic flavour useful in some blends, (2) it is the best way to condition new casks about to enter the maturing system and (3) there is a lively demand from Scotch whisky producers for sherry-matured butts. ***See*** Wines, Production of Table Wines

In the larger fermentation tanks, temperature control is essential, and most wine-makers regard 25°C as about optimal. Currently, the majority of sherry makers do not use cultured yeast: natural yeasts are plentiful. During the fermentation the yeast flora goes through a succession similar to that described in other areas. Initial fermentation is by apiculate yeasts, *Kloeckera* and/or *Hanseniaspora* species, while later stages are conducted by a variety of *Saccharomyces* species, though *S. cerevisiae* almost always predominates.

The fermentation is more or less complete in 5 days but the wine continues to ferment slowly until it is completely dry.

Fortification

As soon as fermentation is complete (usually January) the wine clarifies spontaneously and is racked off the lees.

The alcoholic strength of the wine is increased to approximately 15·5% by volume by the addition of fortifying brandy. This alcohol is prepared by the continuous distillation of wine, which is carried to 95–96% by volume. At this strength the brandy is almost but not quite neutral in flavour. It is not distilled locally but is usually brought in from the La Mancha district of central Spain. Before the brandy is used it is mixed with an equal quantity of wine and allowed to rest and settle for about 3 days. The fortifying mixture so produced causes much less clouding than would straight brandy when it is added to the wine. ***See*** Brandy and Cognac, Brandy and its Manufacture

Maturation

Essentially, the wine destined to become sherry is matured in one of two ways: either a layer of yeast cells is allowed to develop on the surface of the liquid so that reducing conditions are established (the flor process), or further fortification prevents the development of yeast so that conditions are essentially oxidative (maturation without flor). A preliminary assessment of the wine is made by the tasters to determine which type of maturation would be most suitable for each batch. In either case, maturation is conducted in oak casks of 500–600 litres organized to facilitate fractional blending.

Maturation Under Flor (Fino Wines)

The wine should be low in tannin, preferably less than 250 ppm. Those wine-makers having centrifuges at their disposal may choose to allow musts destined for this sort of wine to oxidize rather freely as this helps to diminish the tannin levels, and contributes to a rather neutral flavour. It is said that the subvariety, Palamino fino, produces must particularly suitable for finos. ***See*** Tannins and Polyphenols

The wine is passed serially through casks organized in a series of discrete steps. These casks are laid out in cool, well-ventilated buildings, known as bodegas. A typical system for flor maturation might consist of casks arranged in six groups of approximately equal size with perhaps 500 casks in each group. The casks are on their sides and kept 80% full so there is a large surface area of wine exposed to air relative to the volume of wine. A film of flor yeast grows at the air–wine interface.

Locally, five of the six sets of casks are called criaderas, the other the solera (the whole assembly of casks is known as a solera system). A fraction of the contents of each cask in the solera is removed, with the minimum of disturbance of the flor film, when a supply of matured wine is required, and is replaced by a similar amount of homogenized wine from each cask in the first criadera, this is replaced with wine from the second criadera, and so on. Evidently, the wine in the fifth criadera must be replaced by wine which has not been in this maturation system. It will be apparent that the maturity of the wine in each set of casks will increase progressively from the fifth criadera to the solera. The wine may take anything from 2 to 8 years to pass through such a system, depending on the style and quality required. Typically, the wine might pass through the system in 3 years, a result which could theoretically be achieved by drawing off 25% from the solera eight times a year. A certain amount of pipework may be permanently installed to facilitate the frequent movement of wine through the system; nevertheless, the procedure remains labour intensive and contributes substantially to the cost of the wine. However, experience has shown that the flor yeast does not flourish if larger volumes of wine are moved at less frequent intervals. This is due, in part at least, to the need for atmospheric oxygen; tests demonstrate that all the oxygen dissolved in the wine and that in the headspace in the casks is used within a few days of a movement. In practice, such regularity cannot be achieved, if only because there is a greater demand for sherry in the months leading up to Christmas. However, once the system has reached equilibrium considerable inequalities in its operation are possible without variation in the flavour and quality of the product.

The Flor Yeast

Happily for the Spanish sherry producers, yeasts suitable for the production of flor films develop spontaneously in any cask of suitable Jerez wine maintained under the right conditions. The same was not originally true of other areas, such as Australia, where the process has been attempted. However, it is not difficult to culture suitable yeasts and establish them on low-tannin white wines, providing temperatures are not too high and the wine is fortified to some 15% to put film-forming acetic acid bacteria at a disadvantage. The Jerez flor yeasts have been studied and described under the names of *Saccharomyces beticus*, *S. cheresiensis* and *S. montuliensis*. They are believed to be strains of *S. cerevisiae* and/or *S. rouxii* which has become adapted to this environment. Their development is quite sensitive to oscillations in temperature. In Jerez itself, the thickness of the film falls off in summer in spite of attempts to regulate the temperature and humidity of the bodegas by ventilation and damping down of the flor. The optimum temperature for the development of flor is between 15 and 18°C, though some growth occurs between 12 and 25°C. The casks are stacked one on another, four high. In summer, those in the highest layer reach a temperature above the optimum for the growth of flor, so this position is reserved for non-flor wines.

Submerged Flor – An Alternative System

It is possible to induce changes in flavour somewhat similar to that produced by maturation under a film of flor yeast by inoculating a wine very heavily with a suitable yeast culture and then injecting either air or oxygen under pressure through a suitable sintered diffuser. Depending on the population of yeast used and the mass of oxygen injected, the process takes from 36 h to a month or more. Often the cultured yeasts used are similar to those used for maturation under a film, but a somewhat broader range of strains of *S. cerevisiae* and related species has been found to be suitable. The predominant effects are reduction of glycerol, ethanol and volatile acids, especially acetic acid, and increase in the levels of acetaldehyde. Wines so treated have been marketed straight or blended with untreated wines in the USA, Australia and the UK. The process has been tried in Jerez, but the product was not to the taste of the local wine blenders.

Maturation Without Flor (Oloroso Wines)

All the second press wine, and some of the free run juice, is matured after it has been further fortified with grape alcohol to bring the alcoholic strength to 18·5% by volume. This amount of alcohol is sufficient to prevent the growth of flor yeast, so the wine is exposed to air and undergoes oxidative reactions. These casks are commonly maintained about 95% full as this gives sufficient exposure to oxidation to maintain the reactions at the required rate. As with flor wines, maturation takes at least 3 years, but it is sufficient to move the wine perhaps every 6 months through perhaps three criaderas and a solera; under these conditions a little less than 50% of each cask might be moved on each occasion. The wines with the highest tannin levels are initially somewhat coarse and benefit from exposure to higher temperatures. Often they are stacked outdoors exposed to the summer sun for 6 months or so. This is a cheap source of heat though losses by evaporation are substantial.

Wines produced in this way are known as olorosos. It is said that Palamino basto produces must particularly suitable for olorosos.

Compound Maturation (Amontillado Wines)

Some wines initially matured under flor yeasts for several years, are racked, fortified to 17% and given a further maturation of some years in a oxidative solera. Such wines, known as amontillados, are inevitably expensive but develop a fine flavour, much sought after for consumption straight or for improving complex blends.

Sweetening Materials

The most characteristic material is PX wine made by drying Pedro Ximenes grapes in the sun to increase the specific gravity about two and a half times, pressing out the very sweet juice and, immediately at the onset of fermentation, fortifying with grape alcohol. Much of this wine is subject to oxidative maturation in a solera. Both the immature product and the matured wine have characteristic flavours valuable in blended sweet sherry. Very little PX wine is made in the Jerez district but, by special arrangement, it is imported from the district of Montilla, which is hotter and more suited to the production of such grapes.

PX wine is dark in colour, so to produce pale, sweet sherries hydrolysed beet sugar syrups were used at one time, but this has fallen out of favour and rectified concentrated grape must is now used.

Colouring Materials

Certain food caramels are permitted in sherry, but quality producers prefer to use 'colour wine', which imparts its own desirable flavour to the blend. It is made by boiling down grape must until the concentrate is heavily caramelized and then adding this concentrate to fermenting palomino must in quantities just insufficient to stop the fermentation. It is aged in an oxidative solera. ***See*** Caramel, Properties and Analysis

Commercial Styles of Sherry

Flor sherries are usually marketed unsweetened at strengths between 15 and 17% after between 2 and 8 years of maturation.

Wines matured without flor (olorosos) are normally sold at 17% or sometimes at 19·5% alcohol, as are amontillados; the finer and more aged examples of the latter may be consumed unsweetened, but the majority are slightly sweetened to soften the impact on the palate.

Those wines marked as medium dry, medium, sweet, cream or pale cream sherries, depending on the amount of sweetening material added, usually contain predominantly oloroso wine with a variable amount of fino to lighten the colour and flavour, together with a certain, usually rather small proportion of amontillado to confer fine character. The PX and colour wines in the darker blends confer additional complexity of flavour.

Bibliography

Criddle WJ, Goswell RW and Williams MAW (1981) The chemistry of sherry maturation. *American Journal of Enology and Viticulture* 32: 262–267.

Gonzalez-Gordon M (1970) *Jerez-Xeris-Scheris*. Jerez-de-la-Frontera, Spain: Jerez Graficas.

Gonzalez-Gordon M (1972) *Sherry: The Noble Wine*, London: Cassell.

Goswell RW (1986) Microbiology of winemaking. 1. Fortified wines. In: Robinson RK (ed.) *Developments in Food Microbiology*, vol 2, London: Elsevier.

Jeffs J (1982) *Sherry*. London: Faber and Faber.

Perez L (1982) Consideraciones yecnicas en la elaboracion del Jerez. *Proceeding ll Jornadas Universitarias sobre el Jerez*, pp 167–197. Cadiz: Universidad de Cadiz.

Robin W. Goswell
John Harvey and Sons Ltd., Bristol, UK

Composition and Analysis

There are three main types of sherry, made by three different ageing techniques. Sherry can be matured under flor yeasts, giving fino styles, without flor yeasts for the olorosos or by a combination of flor followed by a period of ageing without flor resulting in amontillados. In this article, variations in composition with the origin of the grape, the influence of processing and the chemical and organoleptic changes during maturation will be described.

Organoleptic Assessment

Gonzalez Gordon describes the nomenclature used by sherry tasters in Jerez. The set of descriptive terms appears to be related to alcoholic strength, total acidity, the time the wine has matured under flor and the concentration of acetaldehyde which has developed. Other terms also used are clean, dirty, soft, hard, full, empty and dull.

With the exception of some special products used for sweetening all the wine is made from one grape variety named palamino. Wine made from palamino grapes is fairly neutral, lacking in acidity and without any distinct varietal character. This neutral base forms an excellent background for the delicate flavours produced as a result of the maturation procedure. The appearance, bouquet and taste of the young wine in each cask is examined and the sherries are classified according to their quality. The end products have very complex sensory characteristics. Initially, all sherries are fermented to dryness, but some of them are sweetened for certain markets. Fino has a pale straw colour, is very dry but without acidity. Its bouquet is delicate yet pungent and the alcoholic strength usually lies between 15·5 and 17% (v/v). Manzanilla produced in the coastal town Sanlúcar de Barrameda is a regional variation of the fino style, which has a delicate and highly individual aroma. It is very dry, with a clean and slightly bitter aftertaste, being slightly less full bodied than fino. Manzanilla has a pale straw colour much like a fino, and an alcoholic strength between 15·5 and 16·5% (v/v). Amontillado is very dry and clean, with a pungent aroma reminiscent of fino but nuttier and fuller bodied. The colour is amber, and becomes darker with increasing age. The alcoholic strength is 17–18% (v/v). Oloroso has a strong bouquet, being less pungent than fino or amontillado, but with more body, sometimes referred to as 'fatness'. It is dry, but has a sweet aftertaste. Oloroso has the darkest colour, best described as dark gold, and its colour intensity increases with age. *See* Wines, Production of Table Wines; Wines, Wine Tasting

Variations due to Processing or Origin

All the grapes are grown in a relatively small, well-defined area. Although the location of the vineyard, such as distance from the sea and the direction it faces, and the structure of the soil has some influence on the end product, the predominant effect is exerted by the treatments during maturation.

Grapes, Musts and Wines

The specific gravity of the must is between 1·085 and 1·095, with a tannin content of 300–600 mg l^{-1} and a total acidity between 3·5 and 4·5 g l^{-1}, expressed as tartaric acid. Often the acidity is too low and is adjusted. Traditionally, calcium sulphate was added to the grapes at crushing, precipitating calcium tartrate and, apparently, liberating free tartaric acid. Nowadays, it is becoming more common to adjust the total acidity by addition of tartaric acid to the must.

Grapes affected by botrytis will result in wines containing increased concentrations of glycerol and gluconic acid; the latter is present in young wines in similar concentrations as in the must and can be used as an indicator of grape quality.

During crushing, sulphur dioxide may be added. However, this is not always considered desirable, since it tends to increase the amount of tannin in the wine. Sulphur dioxide inhibits the action of oxidative enzymes and increases the extraction of phenols from the grapes into the wine and prevents oxidation and precipitation of the phenols onto the grape solids. The concentration present during fermentation is also believed to influence the future wine aroma, since sulphur dioxide forms complexes with carbonyls, such as acetaldehyde, 2-ketoglutarate and 4-ketobutyrate. When these complexes dissociate during maturation, the carbonyls become available for the formation of aroma compounds. *See* Phenolic Compounds; Tannins and Polyphenols

Flor develops quite naturally on those base sherries which have a low tannin content, a pH between 3·1 and 3·4 and contain less than 100 mg l^{-1} of sulphur dioxide. The alcohol content is fortified to 15% (v/v) using a neutral spirit and the sherries are stored in 500 litre oak casks kept 80% full, at temperatures between 15 and 20°C. The flor protects the sherry from the uptake of oxygen, and prevents browning to which the wine is very susceptible. In order to produce a good oloroso the neutral base wine needs to contain sufficient oxidizable phenols. These sherries are fortified to between 18 and 19% (v/v) alcohol to avoid the growth of any flor and are stored in casks which are kept about 95% full.

Changes in Nonvolatile Compounds During Maturation

The changes in composition as a result of maturation are not easy to study. The solera systems contain blends of sherries of different ages and vintages. The blending procedure to which the solera system is subjected is very complex. In addition, the young wine with which the solera system is topped up will show differences in composition from year to year. One study overcame this problem by setting up small-scale solera systems in glass containers, and using sherry from one production year only to replenish the system.

It has been reported that, especially in olorosos and amontillados, there is a considerable loss in volume during maturation and usually an increase in alcoholic strength, resulting in a higher concentration of nonvolatiles. In finos the flor uses alcohol as a carbon source for growth, hence the net effect is often a decrease in alcohol concentration. The higher alcohol concentration in olorosos and amontillados during maturation, often combined with a higher storage temperature, may lead to increased extraction of phenols from the wood, thus explaining the higher concentrations of phenols in these sherries. In contrast to olorosos and amontillados, which mature under oxidative conditions, finos mature under nonoxidative conditions. The growing flor on top of the fino will consume any dissolved oxygen in the sherry and protects it from absorbing oxygen from the headspace in the cask. This lack of oxygen prevents the oxidation of phenols, hence finos maintain a pale yellow colour. In contrast, the amber and dark golden colour of the amontillados and the olorosos can be attributed to oxidation of phenols. *See* Barrels

Of research on sherries, most of the work has been directed towards understanding the aroma formation in fino under flor, while much less research has been directed towards studying changes in volatile compounds in other sherries during maturation, and only limited information is available about other changes in analytical composition. The available information is discussed under individual headings below; it was obtained mostly from primary publications since there were no reviews available.

Phenolic Acids and Aldehydes

The phenolic compounds in sherry can form a substrate for oxidative reactions, influencing the organoleptic properties of the product. The changes of three cinnamic acids (*p*-coumaric, caffeic and ferulic acids), six benzoic acids (gallic, protocatechuic, *p*-hydroxybenzoic, vanillic, syringic and gentisic acids), four phenolic aldehydes (*p*-hydroxybenzoic and 3,4-dihydroxybenzoic aldehydes, vanillin and syringaldehyde) and two coumarins (esculetin and scopoletin) were studied during the maturation of finos, amontillados and olorosos in two or three solera systems. It is likely that the presence of some of these compounds is due to extraction from the oak vats during storage, especially in olorosos and amontillados, which are matured much longer in oak vats than finos. For example, in whisky a number of these aromatic alcohols, acids and aldehydes are reported to be breakdown products from lignin, which is extracted from the oak vats.

The cinnamic acid concentration in finos (8–60 mg l^{-1}, mean 16 mg l^{-1}) is generally lower than in olorosos (12–138 mg l^{-1}, mean 52 mg l^{-1}) or amontillados (19–82 mg l^{-1}, mean 47 mg l^{-1}). Although there are some changes in concentrations of these compounds during maturation, there are no obvious trends. Curiously, in table wines, cinnamic acids occur in higher concentrations as tartrate esters, but these have not been reported in sherry; possibly, this indicates that some of these compounds are derived from the oak vats. Of the benzoic acids, gallic acid increases during maturation of finos, but this has not been clearly observed in amontillados or olorosos. However, there is generally an increase in the concentration of benzoic acids in all three types of wine. The concentration of benzoic acids ranges from 6 to 242 mg l^{-1} (mean 79·6 mg l^{-1}) in finos, from 34 to 293 mg l^{-1} (mean 137·9 mg l^{-1}) in amontillados and from 32 to 392 mg l^{-1} (mean 121·8 mg l^{-1}) in olorosos.

Olorosos contain the highest concentrations of *p*-hydroxybenzaldehyde, vanillin and syringaldehyde (11–147 mg l^{-1}, mean 76·2 mg l^{-1}), followed by amontillados (10–123 mg l^{-1}, mean 54·9 mg l^{-1}). During maturation the concentration of these aldehydes increases in olorosos and amontillados, presumably due to extraction from the oak vats. Finos contain lower concentrations of these aldehydes (4–45·6 mg l^{-1}, mean 21·8 mg l^{-1}) while no significant changes are observed during maturation.

The concentrations of coumarins are about the same in all three sherry types (finos, 9–94·2 mg l^{-1}, mean 43·2 mg l^{-1}; amontillados, 14·4–115 mg l^{-1}, mean 53·3 mg l^{-1}; olorosos, 15·2–112 mg l^{-1}, mean 54·5 mg l^{-1}) and there is no clear trend in changes occurring during maturation.

Flavonols

Quercetin, kaempferol and isorhamnetin have been identified in maturing finos, but not in maturing amontillados and olorosos. It has been postulated that the absence of these compounds in amontillados and olorosos could be due to polymerization reactions whereby, under the oxidative conditions prevalent during maturation, they may contribute to the formation of coloured compounds. In fino solera systems the concentration of all flavonols decreases with increasing age of the sherries. There are large differences between the initial concentrations of these compounds in the youngest of these, presumably due to the composition of the sherries used to feed the solera system. Quercetin is present in the highest concentration (105–328 $\mu g\ l^{-1}$, mean 182 $\mu g\ l^{-1}$), followed by kaempferol (13–48 $\mu g\ l^{-1}$, mean 25 $\mu g\ l^{-1}$) and isorhamnetin (9–19 $\mu g\ l^{-1}$, mean 12 $\mu g\ l^{-1}$).

Polyalcohols

The content of six polyalcohols (erythritol, arabitol, xylitol, mannitol, sorbitol and inositol) have been studied in fino, oloroso and amontillado solera systems. Generally, inositol is present in the greatest concentration (13–850 mg l^{-1}), followed by erythritol (100–240 mg l^{-1}). There is no trend in the changes during maturation, indicating that the differences between the young sherries are greater than possible changes occurring during maturation. All three types of sherry contain similar concentrations of erythritol, arabitol and xylitol, but olorosos are more abundant in mannitol, sorbitol and inositol. *See* Alcohol, Properties and Determination

Dry Extract

A total of 14 different parameters have been studied in five stages of fino, manzanilla, oloroso and amontillado solera systems. The final average age was 4 years for the flor sherries, while the olorosos and amontillados reached an average age of 12 years. Glycerol, tartaric acid and proteins and, to a lesser extent, polyphenols, potassium, sulphates and orthophosphate account for 60% of the dry extract determined by hydrometer.

The dry extract in finos and manzanillas decreases from 20·3 to 11·7 g l^{-1}, due to glycerol being used as a carbon source by the growing flor. Consequently, during maturation of finos and manzanillas the glycerol concentration is reduced from 7 to 0·2 g l^{-1}. Sherry made from grapes affected by botrytis contains an increased concentration of glycerol, most of it being metabolized during maturation under flor. The precipitation of potassium bitartrate during the storage period, often with low temperatures during the winter, reducing the tartrate concentration from 3·30 to 2·20 g l^{-1}, is a major cause of the reduction of total acidity from 5·3 to 4·1 g l^{-1}, also resulting in a slight increase in pH. As a further result of the growing flor, the amount of protein decreases during the early part of maturation (from 2·1 to 1·9 g l^{-1}), after which it remains relatively constant.

The dry extract in olorosos increases from 18 to 24 g l^{-1} and, in amontillados, from 10·8 to 20·9 g l^{-1}. The initial dry extract in amontillados is much lower, since this wine has first gone through maturation under flor before being fortified and matured further without flor. The sherries show increases in total acidity (olorosos from 4·8 to 5·6 g l^{-1}, amontillados from 3·9 to 5·8 g l^{-1}), proteins (olorosos from 1·8 to 2·1 g l^{-1}, amontillados from 1·7 to 2·1 g l^{-1}) and polyphenols (olorosos from 0·31 to 0·46 g l^{-1}, amontillados from 0·34 to 0·51 g l^{-1}), but a decrease in tartaric acid (olorosos from 1·96 to 0·85 g l^{-1}, amontillados from 2·2 to 1·09 g l^{-1}). The last is due to precipitation of tartrates. The annual evaporation of water, estimated to be between 3 and 5%, explains the increased concentrations of the other components.

Amino Acids

Amino acids are important as nutrients for flor and as precursors for the formation of volatiles. There is also interest in amino acids from a safety point of view, since some of the amino acids are thought to be precursors in the formation of ethylcarbamate, a compound with carcinogenic properties which can be formed naturally in alcoholic beverages. *See* Amino Acids, Properties and Occurrence

In very young sherries, proline is the predominant amino acid (4·5–6·2 mmol l^{-1}), accounting for about 70% of the amino acid content. During maturation under flor the proline concentration decreases continuously to about 1 mmol l^{-1}; the yeasts of growing flor presumably use proline as a nitrogen source. In ageing olorosos an increase in proline content during maturation is observed (5·5–6·6 mmol l^{-1}), attributed to the concentration of the sherries as a result of evaporation.

Volatile Compounds

Many valuable contributions to the understanding of flavour in sherry have been made by Webb and his coworkers. At present, 307 volatile compounds have been identified in sherries. The compounds consist of 28 alcohols, one hydrocarbon, 19 carbonyls, 47 acids, 65 esters, 16 lactones, 67 bases, four sulphur compounds, 13 acetals, one ether, 14 amides, 17 phenols, five furans, three coumarins, four dioxolanes and two dioxanes. Many of these compounds have also been identified in table wines, and are not especially typical for sherries. The volatile compounds in sherries can be derived from the grapes, and are formed during fermentation or maturation. It is during this last stage, especially when maturation under flor is used, that the aroma typical for the style of sherry is formed, and hence forms the most important stage for the formation of volatiles typical of sherry. However, there are only a few publications reporting changes in volatiles during flor maturation. During this process there appears to be a decrease in alcohols such as 2-phenylethanol, 2-methyl-1-propanol, 2- and 3-methylbutanols, 3-methyl-1-pentanol and 1-heptanol, but an increase in alcohols such as 1-propanol and 1-butanol. There is an increase in esters such as ethyl lactate and diethyl succinate, but a decrease in esters such as ethyl butanoate, ethyl hexanoate, 1-hexyl acetate and ethyl octanoate. In general, there is an increase in aliphatic and aromatic acids, terpenes and carbonyls. Important compounds contributing to the typical sherry aroma are the lactones and associated hydroxy and oxo compounds, formed during maturation under flor. Sotolon (4,5-dimethyl-3-hydroxy-2(5*H*)-furanone) has been identified in flor sherries only; it is present in sufficient concentrations (range 22–72 μg l^{-1}, mean 41·5 μg l^{-1}) to lend its nutty odour to the characteristic aroma of flor sherry. Solerone (4-hydroxy-5-ketohexanoic acid lactone) has been reported to contribute an aroma of ‘premium quality Pinot Noir wine’. Oak lactone (*trans*-3-methyl-4-hydroxyoctanoate acid lactone) has also been found in sherries; in brandy and whisky it is extracted from the wood during maturation.

Composition and Analysis

See Flavour Compounds, Structures and Characteristics

Changes in Major Volatiles

Changes occurring in the major volatile compounds have been studied in five stages of fino, manzanilla, oloroso and amontillado solera systems. The final average age was 4 years for the flor sherries, while the olorosos and amontillados reached an average age of 12 years.

During maturation under flor the most important changes in the concentrations of the major volatiles are a decrease in volatile acidity from 0·39 to 0·24 g l^{-1} (expressed as tartaric acid) and a reduction of 1·5% (v/v) in the ethanol content, which can be explained by the growing flor using acetic acid and ethanol as carbon sources, while the flor is also responsible for the considerable increase in the acetaldehyde concentration from 2·2 to 6·5 mmol l^{-1}. This increase tends to be greater during the early part of the maturation, while the final concentration in finos is higher than in manzanillas (6·5 and 5·3 mmol l^{-1}, respectively). Acetaldehyde makes an important contribution to the typical aroma of flor sherries. The isoamyl alcohol concentration is the same in both flor sherries at the beginning of their maturation period (2·73 mmol l^{-1}) but, at the end, finos contain a higher concentration (2·92 mmol l^{-1}) than manzanillas (2·28 mmol l^{-1}). The cooler and more humid climate at the coast, where manzanillas are matured, is presumed to influence the yeast metabolism, thus resulting in these differences between manzanillas and finos. Changes due to evaporation also occur, but these have less influence than the changes as a result of the flor.

Due to the longer maturation period of the olorosos and amontillados, the effect of concentration is much more pronounced, resulting in an increase in concentration of most compounds in these sherries. For instance the alcohol concentration increases in oloroso by about 1·5% (v/v). In addition, changes also occur as a result of oxidative processes, in contrast to sherries maturing under flor, which are protected from oxidative changes. For example, the acetaldehyde concentration decreases in amontillados from 6·5 to 4·2 mmol l^{-1}, but remains more or less constant in olorosos at 3·6 mmol l^{-1}; acetaldehyde is slowly being oxidized to acetic acid, thus partly explaining the increase in volatile acidity (in olorosos from 0·7 to 1·1 g l^{-1} and in amontillados from 0·2 to 0·7 g l^{-1}). There is an increase in the ethyl acetate concentration (in olorosos from 1·9 to 3·2 mmol l^{-1} and in amontillados from 0·7 to 2·0 mmol 1 l^{-1}).

Bibliography

Botella MA, Perez-Rodriguez L, Domecq B and Valpuesta V (1990) Amino acid content of fino and oloroso sherry wines. *American Journal of Enology and Viticulture* 41: 12–15.

Brock ML, Kepner RE and Webb AD (1984) Comparison of volatiles in palomino wine and submerged culture flor sherry. *American Journal of Enology and Viticulture* 35: 151–155.

Estrella MI, Hernandez MT and Olano A (1986) Changes in polyalcohol and phenol compound contents in the aging of Sherry wines. *Food Chemistry* 20: 137–152.

Estrella MI, Alonso E and Revilla E (1987) Presence of flavonol aglycones in sherry wines and changes in their content during ageing. *Zeitschrift für Lebensmitteluntersuchung und Forschung* 184: 27–29.

Gonzalez-Gordon M (1972) The sherry viniculture. *Sherry, The Noble Wine*, pp 10–136. London: Cassell.

Goswell RW (1986) Microbiology of fortified sherries. In: Robinson RK (ed.) *Developments in Food Microbiology*, vol. 2, pp 1–19. London: Elsevier.

Goswell RW and Kunkee RE (1977) Fortified sherries. In: (ed.) Rose AH *Alcoholic Beverages. Economic Microbiology*, vol. 1, pp 477–535. London: Academic Press.

Maarse H, Visscher CA, Willemsens LC and Boelens MH (eds) (1989) *Volatile Compounds in Food, Qualitative and Quantitative Data*, 6th edn, vol. II. Zeist, Netherlands: TNO-CIVO Food Analysis Institute.

Martin B, Etievant PX and Henry RN (1990) The chemistry of sotolon: a key parameter for the study of a key component of flor sherry wines. In: Bessiere Y and Thomas AF (eds) *Flavour Science and Technology*, pp 53–56. Chichester: Wiley.

Martinez de la Ossa E, Caro I, Bonat M, Perez L and Domecq B (1987) Dry extract in sherry and its evolution in the aging process. *American Journal of Enology and Viticulture* 38: 321–325.

Martinez de la Ossa E, Perez L and Caro I (1987) Variations of the major volatiles through the aging of Sherry. *American Journal of Enology and Viticulture* 38: 293–297.

Perez L, Valcarcel MJ, Gonzalez P and Domecq B (1991) Influence of Botrytis infection of the grapes on the biological ageing process of fino sherry. *American Journal of Enology and Viticulture* 42: 58–62.

Webb AD and Noble AC (1976) Aroma of sherry wines. *Biotechnology and Bioengineering* XVIII: pp 939–1052.

Jokie Bakker
AFRC Institute of Food Research, Reading, UK

SHIGELLA

Members of the bacterial genus *Shigella* cause the disease shigellosis (bacillary dysentery) which affects only humans and higher primates. The infection is usually self-limiting, but can be life-threatening to infants, the elderly and the malnourished. The microorganisms are transmitted through food or water that has been contaminated with faecal matter from infected individuals. Among persons with poor personal hygiene, the organisms can be spread by person-to-person contact, particularly under crowded conditions as found in custodial institutions and day-care centres. In industrialized nations the organisms are transmitted via food and to a lesser extent by water, but in nonindustrialized nations where chlorination may be lacking the organisms are mainly waterborne. Conditions such as wars or natural disasters which lead to crowding and breakdown of sanitation can give rise to epidemics of shigellosis.

It has been estimated that there are approximately 5·5 million cases of microbial foodborne disease per year in the United States, with 170 deaths and a yearly cost of approximately 7000 million dollars. Foodborne shigellosis is estimated at 163 000 cases per year with 3 deaths and a cost of 63 million dollars. Similarly, Canada has an estimated 1 million microbial foodborne poisoning cases per year with 29 deaths. Canadian citizens annually pay 1200 million dollars for these microbial foodborne outbreaks. Canada has an estimated 19 000 cases of foodborne shigellosis per year with 0·5 deaths at an annual cost of 7·5 million dollars. Thus, foodborne shigellosis is not a major disease in terms of number of foodborne disease cases (3% of the total in the United States and 2% in Canada), and in terms of cost the disease comprises less than 1% of the total microbial foodborne disease cost in both countries (*Journal of Food Protection* (1989) 52: 586–594; 595–601). ***See*** Food Poisoning, Statistics

The Organism

The genus *Shigella* is located in the family Enterobacteriaceae. The organisms are small straight rods, Gram-negative, nonmotile and facultatively anaerobic. With a few exceptions, sugars are fermented without gas production. On the basis of DNA homology, *Shigella* is quite closely related to *Escherichia coli* and it can be difficult to separate the two genera. Shigellae have been described as metabolically inactive biotypes of *E. coli.* While *E. coli* and *Shigella* are closely related genetically, most *Shigella* species cause shigellosis, whereas most strains of *E. coli* do not (however, enterohaemorrhagic *E. coli* secretes shiga-like toxin and enteroinvasive *E. coli* produces symptoms similar to shigellosis). The genus *Shigella* contains four species based on immunological and biochemical differences: *S. dysenteriae*, *S. flexneri*, *S. sonnei* and *S. boydii.* All shigella species except *S. sonnei* are further subdivided into serotypes. ***See*** Enterobacteriaceae, Food Poisioning by *Escherichia coli*

Shigella do not compete well with other microbial species, but they are known to survive for several weeks on inanimate objects stored at $<10°C$ and in foods stored at $\leqslant 25°C$. They can survive in 25% NaCl solutions for at least 9 days at $\leqslant 25°C$. Growth is inhibited at pH $\leqslant 6·0$ when pH adjustment is made with organic acids though they can survive in refrigerated fermented milks with pH values of 4·0–4·2. The temperature range for growth is 7–46°C but shigellae do not survive heating at 63°C for more than 5 minutes.

The Disease

The incubation period for shigellosis is usually 12–50 hours (range <1–7 days) and the illness may persist for approximately 2 weeks. Ingestion of as few as 10–100 bacterial cells can lead to infection. The disease is caused by the microorganisms attaching to the surface of the intestinal epithelium of the colon, invading the epithelial cells, multiplying intracellularly, and killing the host cells. The shigellae then spread to adjacent cells and connective tissue. The overall process is generally limited to the mucosal surface and results in a violent inflammatory reaction. Abscesses and ulcerations develop which may be due to shiga toxin activity; however, the exact role of toxin in pathogenesis of *Shigella* has not been fully determined. Shigellosis may vary from asymptomatic infections or a mild diarrhoea to dysentery with bloody stools, mucus secretion, dehydration, fever, chills, toxaemia and tenesmus. The disease in children is generally more severe, with high fever, neurological symptoms, convulsions, headaches, delirium and lethargy. Sequelae to shigellosis may include haemolytic uraemic syndrome and chronic rheumatoid diseases such as reactive arthritis and Reiter's syndrome. In industrialized countries, mortality is low – in the order of 0·1–0·2%. Mortality is higher in underdeveloped countries, particularly among children since *Shigella* is endemic and is a major cause of diarrhoea in those countries.

Oral rehydration appears to be the best treatment for shigellosis, particularly in malnourished individuals. Antibiotics and antibiotic treatment may be undesirable since resistance to antibiotics is widespread in the genus and appears to be increasing. Thus, antibiotics, whether used as a therapeutic treatment or as a prophylactic measure, may be ineffective.

Epidemiology

Most of the cases of shigellosis are spread person-to-person, but food and water can also be vehicles of transmission of shigellae. Person-to-person contact is common in day-care centres and other institutions where people are under custodial care.

In the United States it has been estimated that approximately 300 000 cases of shigellosis occur each year (foodborne, waterborne and by person-to-person contact) but only a few of these are actually reported. In a recent US Centers for Disease Control report on foodborne disease for the years 1983–1987 (*Journal of Food Protection* (1990) 53: 711), there were 37 reported outbreaks of foodborne shigellosis with 9971 cases and 2 deaths. In 45% of the outbreaks the food was eaten in restaurants, and in 9% the food was eaten at home. The major contributing factors to the outbreaks was poor personal hygiene of a food handler and improper holding temperature of foods. The major foods implicated were salads (potato, egg, meat, poultry, tossed, etc). The salads contained either uncooked items or cooked items that were put together by hand and the salads were not reheated before serving. Since the infective dose is low (10–100 bacterial cells), growth of *Shigella* is not necessary. Shell fish, raw vegetables and raw fruit were also implicated as vehicles for shigellosis.

Therefore, a *Shigella*-infected food handler with poor personal hygiene who prepares foods that are eaten raw or are not reheated after being prepared from cooked ingredients can be a primary contributor to foodborne shigellosis.

Two recent shigellosis outbreaks due to *S. sonnei* occurred in the United States; both outbreaks were open-air gatherings at camp sites and illustrate what happens when standard hygienic procedures for the prevention of enteric diseases are not observed. In the 1987 outbreak (*Journal of Infectious Diseases* (1990), 162: 1324), more than 3000 persons were affected and more than 7000 were infected in the 1988 outbreak (*American Journal of Epidemiology* 133: 608). Lack of appropriate sanitary and food-handling facilities and allowing untrained (and infected) individuals to prepare communal meals led to rapid dissemination of the disease in both outbreaks. Both outbreaks demonstrated that large gatherings, particularly at camp-sites, must be carefully planned in coordination with public health authorities to prevent the possibility of enteric disease.

In the United States (and probably in other industrialized countries), shigellosis is disproportionately common among migrant workers and lower socioeconomic groups, since both of these groups are subjected to crowded living conditions, inadequate water supplies and poor sanitary facilities. Shigellosis is also common in individuals with poor personal hygiene such as young children in day-care centres and people in institutions for the mentally retarded.

Of the *Shigella* species isolated from cases of diarrhoea in the United States in 1984, the majority of the isolates were *S. sonnei* (65%) followed by *S. flexneri* (31%). *S. dysenteriae* only accounted for 2% of the isolates. In industrialized countries *S. sonnei* is responsible for most cases of shigellosis, whereas in less-developed countries *S. flexneri* is usually seen. *S. dysenteriae* is the major cause of epidemic shigellosis.

Virulence

The virulence of *Shigella* species may be studied using animal models such as monkeys or opiated guinea pigs. However, more convenient methods include: (1) the rabbit ileal loop technique to measure both invasiveness of the organism and fluid accumulation induced by shiga toxin; (2) the Sereny test to determine the ability of *Shigella* to cause keratoconjunctivitis (involving invasion of corneal epithelial cells) in guinea-pigs, rabbits or mice; (3) the use of tissue culture to measure epithelial cell invasiveness; and (4) enzyme-linked immunosorbent assay (ELISA) to measure toxin production, and DNA probes for determination of virulence-associated plasmid and chromosomal genes. ***See*** Immunoassays, Radioimmunoassay and Enzyme Immunoassay

Surface polysaccharides (LPS) produced by the bacterial cells are necessary for virulence and are chromosomally determined in *S. flexneri* and *S. boydii* but are plasmid-mediated in *S. sonnei* and *S. dysenteriae*. The role of LPS is to protect the bacterial cells from the host defence systems.

Genetic control of invasiveness involves both plasmid and chromosomal genes. A deletion in or loss of a particular high-molecular-weight plasmid (120 MDa in *S. sonnei* and 140 MDa in the other species) leads to a decrease in or loss of the ability to invade epithelial cells. The ability of *Shigella* species to absorb Congo red from agar media containing the dye is associated with harbouring the large virulence plasmid also. Therefore, absorption of Congo red correlates with virulence in wild-type shigellae. However, it is possible to produce mutants that lack Congo red binding capability but are virulent.

Expression of virulence is dependent on temperature.

Shigella grown at 37°C invade tissue culture epithelial cells, produce keratoconjunctivitis, and absorb Congo red. However, cells grown at 30°C do not express these characteristics even though their virulence plasmid profile is the same as that of 37°C grown cells. A short incubation of 30°C-grown cells at 37°C leads to expression of virulence but chloramphenicol prevents the appearance of the virulence characteristics, indicating that protein synthesis is necessary during the shift-up phase. It appears that shigellae do not need to express virulence at low temperatures, but when the organisms sense the host environment the temperature of 37°C acts as a trigger for the expression of the virulence genes.

The classical shiga toxin is an exotoxin produced by *S. dysenteriae*. The toxin is cell associated and appears to be a periplasmic space protein. The toxin is released into the environment during the stationary stage of growth either by bacterial autolysis or by leakage from the periplasmic space. Shiga toxin production *in vitro* is iron dependent; addition of ferric ion to iron-deficient media results in decreased toxin production. The toxin has a molecular weight of 70 kDa and consists of two polypeptides: one A subunit combined with five B subunits. While structurally similar to cholera toxin and heat-labile toxin from enterotoxigenic *E. coli* strains, shiga toxin is immunologically distinct from those toxins. The other *Shigella* species also produce a toxin immunologically similar to shiga toxin; however, less toxin is produced. Enteropathogenic *E. coli* and enterohaemorrhagic *E. coli* are noninvasive microorganisms that produce shiga-like toxins. The shiga toxin genes in *Shigella* are located chromosomally, whereas bacteriophage conversion is responsible for production of shiga-like toxin in *E. coli* strains. Shiga-like toxins have been found in strains of *Vibrio*, *Salmonella* and *Campylobacter* also.

The role of toxin in shigellosis has not been completely elucidated. Several activities, all produced by the same protein, are associated with the toxin and these include: neurotoxicity, cytotoxicity, enterotoxicity, neuronotoxicity and inhibition of protein synthesis. Shiga toxin inhibits protein synthesis by acting on the 60S ribosomal unit and preventing peptide elongation. Toxicity (cell death) elicited by the toxin is due primarily to inhibition of protein synthesis.

Iron-binding in relation to virulence in *Shigella* has received some study. *S. dysenteriae* and *S. sonnei* produce the siderophore (iron-binding compound) enterobactin, whereas *S. boydii* and *S. flexneri* produce aerobactin, while some strains produce both. Aerobactin-deficient *S. flexneri* mutants behaved similarly to the wild type in their ability to invade tissue culture cells and to produce keratoconjunctivitis. Thus, the siderophore does not behave as a virulence factor. Siderophores in other Enterobacteriaceae have been shown to be involved in virulence, but the limited evidence accumulated to date suggests that siderophores are not virulence factors in *Shigella*.

Antibiotic resistance can be considered an indirect virulence factor in *Shigella* species since resistant shigellae can have a selective advantage over competitors. The use of broad-spectrum antibiotics eliminates sensitive organisms that might compete with shigellae, selects for those microorganisms with antibiotic resistance plasmids (R-plasmids) and selects for transfer of resistant plasmids to antibiotic-sensitive strains. Strains that are multiply resistant to a number of antimicrobial agents are common in shigellosis and make treatment of severe cases difficult.

Isolation and Detection of *Shigella* Species

The diagnosis of shigellosis is accomplished by isolation of the organism from faeces and from food or water. Unfortunately, there is no growth medium which is selective or specific for *Shigella* species. Freshly voided faecal samples should immediately be plated directly onto XLD, SS or MacConkey selective agars. If that is not possible, the specimens should be shipped to the laboratory refrigerated in buffered glycerol saline or Cary Blair transport medium. However, transportation of the faecal sample is not recommended. For isolation of shigellae from suspect foods, the food sample should be inoculated into both Hajna GN and selenite cystine broths. Overnight growth from the broths is streaked onto at least two (or more) of the following selective agars: MacConkey, XLD, deoxycholate citrate, EMB, tergitol 7, SS, or Hektoen enteric. Various other members of Enterobacteriaceae, such as *Klebsiella*, *Proteus* or *E. coli* inhibit the growth of *Shigella* and the organism may not always be found on the selective agars if they are present. ***See*** Food Poisoning, Tracing Origins and Testing

Non-lactose-fermenting colonies picked from the selective agars are streaked onto the surface of triple sugar iron slants as well as stabbed into the slant butt. Tubes showing a red streak and yellow butt with no gas or hydrogen sulphide production are further subjected to biochemical screening, serology, phage typing and colicin typing in order to speciate the isolates.

Selective media used for the isolation of *Shigella* may contain bile salts, deoxycholate or other inhibitors which interfere with the detection of injured cells and thereby give false-negative results. Injured cells resulting from sublethal stress (heat, acid, sanitizers or other stresses introduced by processing conditions) are a real possibility in underprocessed foods. Injured cells cannot repair and grow in the presence of toxic selective agents. In order to detect injury, the cells must be allowed to undergo a period of repair in the absence of selective agents. Detection of injured cells can be achieved by inoculating shigella-containing samples into a nonselective broth for a short period (6–8 hours) to allow repair, followed by the addition of double-strength enrichment broth. Broth growth is then streaked onto selective agars. Direct plating of injured cells can be accomplished by plating dilutions of the sample onto nonselective agar. After a period of repair, the nonselective agar is overlain with a selective agar and further incubated. Colonies found on the selective agars are then tested to determine whether they are *Shigella*.

Newer techniques that are available for detection of *Shigella* include subjection of colonies from selective media to DNA probes for detection of shigella virulence determinants. The virulence marker antigen (VMA) present in virulent *Shigella* and enteroinvasive *E. coli* can be detected by using ELISA. The presence of VMA is correlated with the presence of the large virulence plasmid and with epithelial cell invasiveness. Shiga toxin can be detected in faecal samples by the use of ELISA also.

The polymerase chain reaction (PCR), which amplifies selected DNA sequences, may be useful for detection of *Shigella* species. A DNA sequence common to *Shigella* species can be amplified by PCR and the detection of the pathogen can be made directly from foods, water or faeces without the necessity of using bacteriological media. The large virulence plasmid in shigellae is relatively unstable (particularly so in *S. sonnei*) and may be lost during enrichment and isolation, thereby preventing the determination of virulence of the isolates. The advantage of the PCR technique is that the organism can be determined directly in the sample without using techniques which may lead to loss of the virulence plasmid. Injured cells will be detected by PCR and a repair period is not necessary. Interpretation of PCR results may be difficult because it will amplify the DNA from dead cells also.

Control and Prevention of Shigellosis

The finding of *Shigella* in foods or water indicates human faecal contamination. Therefore, good sanitation is important. Water must be chlorinated if there is any possibility that it will be used by humans or will enter the food chain at some point. There must be proper treatment and disposal of sewage and human faeces should not be used as fertilizer. Since insects and rodents can be passive carriers of human faecal matter, their presence must be eliminated in places where foods are prepared and served. ***See*** Water Supplies, Water Treatment

Community-wide outbreaks are difficult to control because of the ease of person-to-person transmission in young children and other individuals with poor personal hygiene. Handwashing with soap and running water after defaecation and before eating may be the single most important preventive measure to break the cycle or transmission of shigellosis. Therefore, training of children and other individuals with poor personal hygiene in proper handwashing techniques is critical.

Individuals, particularly those who handle foods, must have good personal hygiene and practise proper handwashing and nail-cleaning procedures. Persons ill from shigellosis (or for that matter, from any infectious disease) should not be allowed to prepare or serve foods.

Ideally, vaccination of individuals and groups at risk (infants, children, elderly people, malnourished individuals, immunocompromised individuals, military personnel, laboratory and clinical personnel) should be an effective measure against the spread of shigellosis. However, at the present time, an effective vaccine against *Shigella*-borne disease is not available.

The initiation of good sanitation practices and use of an aggressive education programme, with emphasis on what constitutes good personal hygiene, will do much to alleviate the occurrence and spread of shigellosis.

Bibliography

Cantey JR (1985) Shiga toxin – an expanding role in the pathogenesis of infectious diseases. *Journal of Infectious Diseases* 151: 766–771.

Formal SB, Hale TL and Kapfer C (1989) *Shigella* vaccines. *Reviews of Infectious Diseases* 11(supplement 3): S547–S551.

Hale TL (1991) Genetic basis of virulence in *Shigella* species. *Microbiological Reviews* 55: 206–224.

Hale TL and Formal SB (1987) Pathogenesis of shigella infections. *Pathology and Immunopathology Research* 6: 117–127.

Maurelli AT and Sansonetti PJ (1988) Genetic determinants of *Shigella* pathogenicity. *Annual Reviews of Microbiology* 42: 127–150.

O'Brien AD and Holmes RK (1987) Shiga and shiga-like toxins. *Microbiological Reviews* 51: 206–220.

Smith JL (1987) *Shigella* as a foodborne pathogen. *Journal of Food Protection* 50: 788–801.

Wachsmuth K and Morris GK (1989) *Shigella*. In: Doyle MP (ed.) *Foodborne Bacterial Pathogens*, pp 447–462. Marcel Dekker: New York.

James L. Smith
Agricultural Research Service, USDA, Wyndmoor, USA

SHRIMPS

See ***Shellfish***

SINGLE-CELL PROTEIN

Contents

Algae

Single-cell protein (SCP) is a generic term for crude or refined protein which originates from bacteria, yeasts, moulds, or algae. The production and utilization of algae is discussed in this article. ***See*** Myco-protein

Suitable Organisms

Algae have been used for SCP production from carbon dioxide and light energy (autotrophic growth). By means of chlorophyll, the algal cells are capable of hydrolysing water as a source of reducing power that fuels carbon dioxide fixation.

The name 'algae' is generically given to photosynthetic organisms, either microscopic or macroscopic, living largely in water habitats, but growing with undifferentiated or slightly differentiated tissues. The term is applied to taxonomically unrelated species. Moreover, in some cases it is incorrectly given, e.g. in the case of cyanobacteria or *Euglena*.

The Cyanobacteria is a group included among the prokaryotes; the members of this group are therefore more closely related to bacteria than to 'true' algae. However, they are known as blue-green algae, and for SCP purposes they are usually considered as such. *Euglena* is a genus of microorganism belonging to the Protozoa; it is a strange case of an unicellular animal with chlorophyll.

'True' algae belong to the plant kingdom, being the most simple plants. There are unicellular and multicellular organisms, some of them reaching huge sizes.

Many algae have long been used as food; from single-cell organisms up to the multicellular seaweeds which have an important position in the diet of coastal communities. Table 1 shows the main algae used as food.

The most widely consumed seaweed is *Porphyra*, particularly *P. tenera*, which has a widespread distribution. It is mainly consumed in Japan, but also in the Philippines, Wales and New Zealand. Among the most important species of seaweed used as food is *Ulva lactuca* ('sea lettuce'), which is used as a salad ingredient in western Europe. *Enteromorpha* is consumed in Hawaii and the Philippines in salads or as a flavour enhancer for fish dishes. *Caulerpa* is also consumed in the Philippines.

Macroscopic algae hardly fit the SCP definition, owing to their multicellular nature, and the low protein content of the final product (from 6% to 30% on a dry-weight basis). They are mostly collected from the sea, and, when cultivated, the production resembles more a 'farming' facility than a biotechnological process.

In strict SCP terms, both 'true' unicellular algae and cyanobacteria are used, and they have been consumed for centuries. The practice dates back to the Aztecs, long before the discovery of the New World, when *Spirulina maxima* was harvested from natural habitats for human consumption. A similar strain is still used in Lake Chad in Central Africa for the same purpose. Other ancient cultures have used algae as food, but on a smaller scale. For example, *Nostoc* is consumed in Mongolia, China, Thailand and Peru, while *Oedogonium* and *Spirogyra* are consumed in Burma, Thailand, Vietnam and India, *Chlorella* is produced in Japan and Taiwan, and *Scenedesmus* in China. Industrial and experimental efforts in other countries depend on the construction of large, shallow ponds in areas of high insolation.

Table 1. Main groups of algae used as food

Genera	Growth form	Cellular structure	Kingdom	Group
Spirulina				
Arthrospira				
Nostoc		Prokaryote	Prokariotae	Cyanobacteria (Cyanophyceae)
Anabaena				
Tolypothrix	Unicellular			
Chlorella				
Scenedesmus				
Oedogonium				
Spirogyra				
Coelastrum		Eukaryote	Plantae	Chlorophyceae
Ulva				
Enteromorpha	Multicellular			
Caulerpa				
Porphyra				Rhodophyceae

Production

Single-cell algae are produced by a variety of methods, ranging from lakes or earthen ditches or ponds to highly sophisticated fermenters or bioreactors. Intensive cultivation was initiated in the 1940s in Germany, and is currently performed in many countries around the world, albeit on a modest scale.

Substrate Requirements

Algae are autotrophic organisms with the distinct advantage of using carbon dioxide as a carbon source. Carbon dioxide represents the cheapest and probably the most abundant carbon source for microbial growth. Owing to such inexpensive requirements, these organisms are very convenient for SCP production. Some species can also grow heterotrophically using organic carbon sources.

The process is limited by the carbon source; the concentration of dissolved carbon dioxide is rather low, as its solubility is low in aqueous solution. Some additional source of carbon can enhance cell growth, be it carbon dioxide injected from combustion gases or cheap organic materials such as manure, molasses, or industrial wastes. The added organic compounds are rapidly degraded by bacteria to make the carbon dioxide needed for algal reproduction. Lake Texcoco in Mexico contains high concentrations of carbonate and bicarbonate which are efficiently consumed by *Spirulina maxima*.

Nitrogen sources are generally nitrates, nitrites, ammonia, or urea. Oxidized nitrogen compounds require energy to be reduced; ammonia is therefore a more convenient form. Many cyanobacteria are able to fix atmospheric nitrogen; species of the genus *Anabaena* are particularly active in fixing nitrogen.

Another important nutrient is phosphorus that can be incorporated in the inorganic form. Micronutrients are needed in minimal amounts.

The use of waters polluted with organic wastes as an input to algal ponds has the advantages of using the cheap raw material and promoting decontamination, while obtaining a good source of SCP. The macronutrients needed (ammonium, phosphate) are normally present in domestic sewage, animal wastes and food industry residual waters.

Mass-culture Systems

Outdoor cultivation may be in open or closed systems. Growth occurs not more than 0·5 m from the water surface, and is controlled by light penetration. An example of a typical mass-culture open system is that conducted by the Sosa Texcoco Company in Mexico, in which an alkaline (pH 9–11) lake with a surface area of 900 ha produces about 400 t of *Spirulina maxima* per year.

Open systems show low cell densities with large variations in productivity, the propagation of mixed populations, and frequent problems of contamination by bacteria, fungi, protozoa and invertebrates. On the other hand, the facilities are constructed at low cost and large areas are available. Under such nonsterile, mixed-culture conditions some algal species tend to predominate, depending on water properties and environmental conditions. For example, *S. maxima* predominates on Lake Texcoco because of the high alkalinity of the water.

For open systems, lakes, ponds and ditches are used. They can have an earthen floor or be lined with concrete or plastic. The surface of the water in some ponds or ditches used for mass production is covered with polyethylene or other 'plastic' material to reduce the risk of infections. One method of avoiding contamination is the common practice of seeding a large inoculum to dominate the culture, at least during the first growth phase. Other attempts to promote a single-species culture have been evaluated. For example, the use of nitrogen-fixing species of the genus *Anabaena* in media without other nitrogen sources has proved useful. Similarly, the selection of thermophilic species has shown some preliminary potential.

A gentle agitation is very important to achieve high productivity. It prevents sedimentation, and allows a more homogeneous exposure of algal cells to light, and the reduction of nutrient and temperature gradients along the depth of the culture. To this end, several designs have been implemented in ponds and ditches, including paddlewheels, gravity flow and pump recycle, combined with special slope designs in oval ponds and horizontal raceways.

Productivities rarely exceed 30 g per m^2 per day and cell densities of 2 g l^{-1} which are much lower than the values for other industrial fermentation processes. Some experimental improvements have been achieved by optimizing the use of wastes through the addition of nitrogen sources, adding aeration ports, and the inoculation of selected bacteria that efficiently degrade the diluted organic materials. By such means, the dry-weight productivities reach about four and three times the average figures obtained for maize and soya bean on a yearly basis. With intensive modes of cultivation, algal cultures can produce up to 20–35 times more protein than soya bean for the same area of land.

The most advanced system developed so far is the high rate algal pond (HRAP), which combines the treatment of sewage with the simultaneous massive production of algae. The project was developed by the Technion Research Center at Haifa, Israel, and current facilities consist of shallow canals that add up to a maximum of 1000 m^2, equipped with systems for gentle agitation and aeration. The process is operated continuously, with retention times varying between 2 and 6 days depending on the season. A steady multiculture is established in the system within a few days of operation and includes bacteria that degrade organic compounds, and well-defined algal species, predominantly *Euglena*, *Chlorella* and *Scenedesmus*. The maximum productivity reported at times of maximum solar radiation is 30 g m^{-2}, with an average annual production of 7 kg of algae per m^2. To recover the cells, aluminium sulphate is added as a flocculant; the float is then dewatered by centrifugation and dried in a drum dryer to reach a final moisture content of 10%. The final product has been shown to be of excellent nutritional quality, containing 574 g of crude protein per kg, and with an amino acid profile superior to the average for soya bean protein. It has been used to replace at least 25% of the diet for fish and 10% of a poultry diet with no toxic effects. The resulting effluent can be used directly for crop irrigation.

Photobioreactors

In 'clean-culture' systems, a single species is inoculated and can be maintained over extended time periods. These systems are closed photobioreactors operating either outdoors or indoors.

Large systems operating outdoors consist of flat tubes, covering large areas exposed to sunlight, and can be operated either in batches or continuously. They have been used for the production of *Chlorella*, *Arthrospira* and *Spirulina*. Another innovative design, operated in the USA, is based on the use of oval plastic bags floating on thermal waters.

Photobioreactors operating indoors necessarily have smaller sizes because artificial light is needed. Designs can be either plastic tubular systems or stainless steel fermenter-like reactors with internal illumination to allow maximum light incidence. Their use is rather limited for SCP production because of low throughputs; however, they are quite adequate for the production of high-value-added metabolites, such as polysaccharides, carotenes, etc.

Table 2. Comparative composition of algae and soya bean

	Spirulina maxima (g kg^{-1})	*Scenedesmus obliquus* (g kg^{-1})	Soya bean (g kg^{-1})
Crude protein (N × 6·25)	600–710	500–560	400
Lipids	60–70	120–140	200
Carbohydrates	130–160	100–170	350
Minerals	40–90	40–90	50

Table 3. Nutritional parameters of protein from some genera of microalgae

Product	NPU	PER	BV
Spirulina	65	1·80	75
Spirulina plus methionine	73	—	82
Chlorella	66	—	72
Chlorella plus methionine	78	—	91
Scenedesmus	67	1·93	81
Casein	83	2·50	88

NPU, net protein utilization; PER, protein efficiency ratio; BV, biological value.

Harvesting

The recovery of microalgal biomass after production is rather difficult, particularly in lakes of large surface area, or when low concentrations occur. Some species, such as *Arthrospira platensis* (formerly *Spirulina platensis*), *S. maxima* and *Coelastrum probiscideum*, are easily skimmed off or harvested by filtration through cloths or screens. Filter presses can also be used. Owing to their small cell size (10 μm), other species need to be harvested by centrifugation or flocculation, i.e. by adding flocculants such as lime, alum or polyelectrolytes.

After harvesting, the algal biomass must be dewatered by centrifugation, and/or dried. Operations to dehydrate biomass normally involve drum-drying, sun-drying or spray-drying; the former is the most widely preferred (see below).

Nutritional Value and Human Consumption

The nutritional value of algal SCP is similar to those of other SCP sources. The chemical composition of some algae compared to soya beans is shown on Table 2. The protein content of *Spirulina maxima* and *Scenedesmus obliquus* is notably higher than the value for soya bean, whereas low amounts of lipids are present in the former. The mineral content is comparable for all products.

The amino acid content of *S. maxima* shows an adequate balance, except, as with any other microbial biomass, for the sulphur-containing amino acids, methionine and cystine. It is rich in vitamins, especially the water-soluble ones, and rich in essential fatty acids. However, the contents are highly dependent on cultivation and processing conditions. Table 3 shows some nutritional parameters for examples of algae. Although inferior in nutritional quality to casein, the protein efficiency ratio (PER), net protein utilization (NPU), and biological value (BV) show excellent levels for algae. Nutritional tests have shown promise when algae supplemented with methionine and cystine are fed to broilers. However, monogastrics experience problems digesting the whole cells, and some processing is therefore necessary. ***See*** Essential Fatty Acids, Physiology; Protein, Chemistry; Protein, Quality; Vitamins, Overview

Other uses include the cultivation of daphnid and similar species that thrive on plankton as a food source in aquaculture.

The method of drying affects product bioavailability; drum-drying algae compared to air-drying can result in increases of around 100% of NPU, and 60% in digestibility. This phenomenon may be caused by the rupture of the algal cell walls when water is removed under controlled conditions.

Algal SCP has been mainly used for the preparation of tablets and other products to be sold as health foods, as protein and vitamin supplements, or as aids to weight loss. However, very few in-depth studies, if any, have been conducted to evaluate the nutritional or health-associated benefits of algal SCP. ***See*** Health Foods, Dietary Supplements

There is also some experience related to the supplementation of cereal foods with algal SCP. Mixtures with doughs for baked goods and pasta have been suggested. In Mexico, *S. maxima* has been used as a supplement for biscuits produced by a state company as part of a national breakfast programme for schoolchildren.

Concerning acceptability, the major problem of algal biomass is the presence of dark green pigments that are difficult to mask. In addition to chlorophylls, other pigments, such as carotenes, xanthins and phytocyanin, are present in varying amounts. Flavour and colour may be improved if the algal biomass is treated during downstream processing for removal of undesirable components. ***See*** Colours, Properties and Determination of Natural Pigments

Toxicological Problems

A common problem for SCP from any microorganism is the high content of nucleic acids present in microbial cells. Consumption by humans of nucleic acids in amounts higher than 2 g per day can lead to accumulation of uric acid which develops into gout and kidney stones. The concentration of nucleic acids in algal biomass depends on several factors, such as species and growth conditions. Cyanobacteria have a nucleic acid content of 40–50 g kg^{-1}, while microscopic plant algae have 10–170 g kg^{-1}. These amounts are higher than in most other foodstuffs. ***See*** Nucleic Acids, Physiology

In order to reduce the nucleic acid content, protein concentrates can be prepared by cell disruption and protein separation. However, this procedure increases the production costs of the product. In general, when algal biomass is consumed by humans it is in small amounts, so that the consumption of nucleic acids is without risk.

The cell wall of microalgal biomass represents about 10% (w/w). It is mainly composed of indigestible carbohydrates and some other compounds, e.g. murein in cyanobacteria. The bioavailability of protein from whole cells is therefore very low. The preparation of protein concentrates or isolates is convenient to obtain products with a high nutritional value and free of undesirable pigments, although it represents a costly alternative.

Algae, in common with many other water plants such as hyacinth, have the ability to remove heavy metals from polluted waters. Similar physiological phenomena account for the accumulation of pesticides and organochlorinated compounds. This represents an objection when algae is intended for use as SCP. However, the causes are well identified and some actions can be implemented to maintain the final product composition within safety levels. ***See*** Heavy Metal Toxicology; Pesticides and Herbicides, Toxicology

The origin of the water used in cultivation ponds determines the need for pretreatments. In general, wastes generated in food industries carry low amounts of the contaminants mentioned. However, urban sewage and runoffs show high variability in the content of heavy metals and other toxicants. When these waste streams are subjected to standard secondary treatments, most organics are degraded, whereas metals remain associated with the activated sludge, rendering the water safe as an input for algal growth. Furthermore, some studies have demonstrated that the biological absorption of metals is a rather slow process, requiring more time than the usual retention time of the water within the bioreactor. It appears that fears concerning the presence of recalcitrant toxicants in algae might be excessive, although more research is needed to obtain the full picture.

Another problem to consider is the possibility of contamination by pathogenic microorganisms. The downstream processes of SCP products are designed to destroy most of the viable forms present, although some could survive in the product. Recommendations have been established for microbiological standards of SCP products for use in animal feeds.

Bibliography

Goldberg I (1985) *Single-Cell Protein*. Berlin: Springer Verlag.

Litchfield JH (1983) Single-cell proteins. *Science* 219: 740–746.

Moo-Young M and Gregory K (1986) *Microbial Biomass Proteins*. Essex: Elsevier Science Publishers.

Robinson RK and Toerien DF (1982) The algae – a future source of protein. In: Hudson BJF (ed.) *Developments in Food Proteins – 1*, pp 289–325. Essex: Elsevier Science Publishers.

Rose AH (ed.) (1979) *Microbial Biomass. Economic Microbiology*, vol. 4. London: Academic Press.

Switzer L (1980) *Spirulina. The Whole Food Revolution*. Berkeley: Proteus Corporation.

Wood A, Toerien DF and Robinson RK (1991) The algae – recent developments in cultivation and utilization. In: Hudson BJF (ed.) *Developments in Food Proteins – 7*, pp 79–123. Essex: Elsevier Science Publishers.

Mariano García-Garibay and Lorena Gómez-Ruiz
Universidad Autónoma Metropolitana, Iztapalapa, Mexico City, Mexico
Eduardo Bárzana
Universidad Nacional Autónoma de México, Mexico City, Mexico

Yeasts and Bacteria

Yeasts and bacteria biomass have been consumed by the human race since ancient times in fermented foods; however, the single-cell protein (SCP) concept is applied to the massive growth of microorganisms for human or animal consumption. Single-cell protein is a generic term for crude or refined protein which originates from bacteria, yeasts, moulds, or algae. This article discusses the production and utilization of bacteria and yeasts. ***See*** Myco-protein

Suitable Organisms and Substrates

Historical Developments

The first developments for SCP production were achieved during times of war when conventional foods were in short supply. During World War I, *Saccharomyces cerevisiae* was massively produced in Germany from molasses to replace up to 60% of imported protein. A similar experience was repeated during World War II for the mass production of *Candida utilis* on sulphite liquor from paper manufacturing wastes. After the war, several plants were built in the USA and Europe, mainly for *C. utilis* production.

Accelerated industrial development and general welfare expectancy led to a renewed interest in SCP as a means of alleviating food shortage resulting from a growing imbalance between food production and demand by the world's population, mainly in developing countries.

An important breakthrough took place when the production of SCP from hydrocarbons was demonstrated by several petroleum companies during the 1950s and 1960s; during the 1970s considerable research efforts resulted in the use of methanol and ethanol derived from petroleum as convenient substrates. However, concern about substrate safety and the increase in petroleum prices shifted the interest back to the utilization of renewal sources, mainly food and agriculture by-products such as molasses and whey, or industrial wastes rich in starch, cellulose and hemicellulose.

To date, an enormous number of reports about SCP production have appeared in the scientific literature. Two main approaches have been followed: the utilization of convenient substrates, and the use of waste materials, which makes SCP production an instrument of pollution control.

Many industrial processes have been developed worldwide; the most important are given in Tables 1 and 2. The country with the largest capacity for SCP production is the USSR, where at least 86 plants are in operation using different substrates. The situation is similar in some countries in Eastern Europe, such as Czechoslovakia. Other countries with important industrial outputs are the USA, the UK and France.

Microbial Strains

Yeasts and Bacteria

The most frequently used yeasts are the following: *Saccharomyces cerevisiae* (and related synonymous forms such as *S. carlsbergensis*, *S. uvarum*, etc.), *Kluyveromyces marxianus* (including synonymous, subspecies and asexual forms, such as *K. fragilis*, *K. lactis*, *K. bulgaricus*, *C. kefyr* and *C. pseudotropicalis*), *C. utilis* (and its sexual form, *Hansenula jadinii*) and *Yarrowia lipolytica* (formerly known as *C. lipolytica* and *Saccharomycopsis lipolytica*).

All these yeasts have been widely used in the manufacture of human foods; *S. cerevisiae*, *K. marxianus* and *C. utilis* have been classified as generally recognized as safe (GRAS) for human consumption by the Food and Drugs Administration (FDA) in the USA. *Saccharomyces cerevisiae* is also available as spent yeast from breweries and other alcohol industries, but it is not commonly used for SCP production because it can easily deviate into an alcoholic fermentation. *Kluyveromyces marxianus* and related species are widely used owing to their capacity to assimilate lactose, the carbohydrate present in cheese whey, but these species can also grow on inulin, a fructose polymer found in some plants. *Candida utilis* is used for a wide variety of substrates such as sucrose, ethanol and sulphite-spent liquor. It can also grow on wood hydrolysates because of its ability to assimilate pentoses.

Starchy solids or water streams from the potato and maize industries require previous hydrolysis for yeast growth, such as *C. utilis*, or as in the Symba process, the utilization of an amylolytic yeast (*S. fibuligera*). The following yeasts are able to assimilate hydrocarbons: *Y. lipolytica*, *C. tropicalis*, *C. rugosa* and *C. guilliermondii*, which can also be produced on lipids. Methanol is the preferred alcohol utilized as a substrate by *Pichia* species (*P. pastoris*, *P. methanolica*, etc.), *Hansenula polymorpha*, *H. capsulata* and *C. boidinii*. The most important processes developed for yeast SCP production are shown in Table 1.

Bacteria have been mostly used for the production of animal feed. The most commercially important species utilize methane and/or methanol as substrates, as shown in Table 2. Methanol is usually preferred over methane because it is water-soluble and less explosive. The Ministry of Agriculture, Fisheries and Food (MAFF) in the UK allows the use of the Imperial Chemical Industries (ICI) product, 'Pruteen', in animal feed. To produce bacterial SCP from whey, several lactic acid and propionic bacteria have been investigated, frequently in mixed cultures with yeasts, as in the Kiel process.

In general, yeasts have been preferred for SCP production over bacteria, especially for human consumption. It seems that yeasts are more accepted because they are more familiar to humans through age-old foods such as bread or beer. However, bacteria have some advantages over yeasts such as higher protein content, higher yields (carbon source to protein conversion) and faster growth rate, although a higher nucleic acid content limits the uptake of protein in the diet.

Production Processes

Since SCP must have a competitive price in the protein market, especially protein of vegetable origin, it is essential to guarantee efficiency at all stages of the process. In particular, the carbon source accounts for up to about 60% of the operation costs. It follows that high yields of substrate conversion are required, preferably in a continuous culture which offers the highest productivity, and using a cheap but easily assimilated carbon source. This explains the generalized use of molasses, whey, or industrial residues, depending on local availability, and the attempts to implement fossil fuels as substrates.

In order to maximize carbon assimilation, the nutrients must be balanced. Sources of nitrogen, minor

Table 1. Main yeast SCP industrial and pilot developments

Substrate yeast	Process/product	Country
Sulphite liquor		
Candida utilis	St Regis Paper	USA
C. utilis	Boise Cascade	USA
C. utilis	State industry	USSR
C. utilis	Attisholz	Switzerland
Hydrocarbons		
Yarrowia lipolytica	British Petroleum/Toprina	UK
C. tropicalis	British Petroleum	France
Y. lipolytica	Liquichimica	Italy
Y. lipolytica	State industry	USSR
Y. lipolytica	Swedt	FRG
Y. lipolytica	Roniprot	Rumania
Ethanol		
C. utilis	Amoco/Torutein	USA
Candida sp.	Mitsubichi	Japan
C. utilis	Pilot	Czechoslovakia
Methanol		
Pichia sp.	Mitsubichi	Japan
P. pastoris	Phillipps Petroleum/Provesteen	USA
Starch		
Saccharomycopsis fibuligera and *C. utilis*	Symba	Sweden
Saccharomyces cerevisiae	Brewers	Several
Molasses		
C. utilis	Industrial	Cuba
C. utilis	Industrial	Taiwan
C. utilis	CINVESTAV IPN	Mexico
Liquid sucrose		
Hansenula jadinii	Phillipps Petroleum/Provesta	USA
Whey		
Kluyveromyces marxianus	Wheast-Knudsen	USA
K. marxianus	Amber Lab. Universal	USA
K. marxianus, *K. lactis*, and *C. pintolopessii*	Fromageries Bel	France
K. marxianus	SAV	France
C. intermedia	Viena	Austria
C. utilis	Waldhof	USA
Confectionery effluent		
C. utilis	Bassett	UK

Table 2. Main bacterial SCP industrial and pilot developments

Substrate bacteria	Process/product	Country
Methane		
Methylococcus capsulatus	Shell Oil	UK
Methanol		
Methylophilus methylotrophus	ICI/Pruteen	UK
Methylomonas sp.	Hoechst-Uhde/Probion	FRG
Acinetobacter calcoaceticus	Exxon-Nestlé	
Cellulose		
Cellulomonas sp. and *Alcaligenes* sp.	Louisiana State University	USA
Whey		
Lactobacillus bulgaricus and *Candida krusei*	Kiel	FRG

Table 3. Nutritional parameters of SCP products

	Kluyveromyces marxianus	*Saccharomyces cerevisiae*	*Candida utilis*	*Methylophilus methylotrophus*	Whole egg	Soya meal
Protein (g kg^{-1} dry wt)						
crude (N × 6·25)	450–580	480	420–570	720–880	123	440–500
True protein	400–420	360	—	640	—	480
Essential amino acids (g per 16 g N)						
Isoleucine	4·0–5·1	4·6–5·5	4·3–5·3	5·2–5·4	5·6	5·4
Leucine	7·0–8·1	7·0–8·1	7·0	8·2–8·4	8·3	7·7
Phenylalanine	3·4–5·1	4·1–4·5	3·7–4·3	4·3–6·5	5·1	5·1
Tyrosine	2·5–4·6	4·9	3·3	3·5–3·8	4·0	2·7
Threonine	4·1–5·8	4·8–5·2	4·7–5·5	5·7–6·5	5·1	4·0
Tryptophan	0·9–1·7	1·0–1·2	1·2	1·1–1·6	1·8	1·5
Valine	5·4–5·9	5·3–6·7	5·3–6·3	6·3–6·5	7·5	5·0
Arginine	4·8–7·4	5·0–5·3	5·4–7·2	4·3–5·6	6·1	7·7
Histidine	1·9–4·0	3·1–4·0	1·9–2·1	2·2–2·3	2·4	2·4
Lysine	6·9–11·1	7·7–8·4	6·7–7·2	4·1–7·3	6·2	6·5
Cystine	1·7–1·9	1·6	0·6–0·7	0·8	1·8	1·4
Methionine	1·3–1·6	1·6–2·5	1·0–1·2	1·4–3·0	3·2	1·4
PER	1·8	2·0	1·7	—	2·6	1·4–2·2
NPU	67	—	—	84	98	64
Vitamins (μg g^{-1})						
Thiamin	24–26	104–250	8–9·5	—	0·9	9·0
Riboflavin	36–51	25–80	44–45	—	4·7	3·6
Pyridoxine	14	23–40	79–83	—	1·1	6·8
Nicotinic acid	136–280	300–627	450–550	—	0·7	24·0
Folic acid	6	19–30	4–21	—	0·3	—
Pantothenic acid	67	72–86	94–189	—	18·0	21·0
Biotin	2	1	0·4–0·8	—	0·3	—

PER, protein efficiency ratio; NPU, net protein utilization.

elements (phosphorus, potassium, sulphur, magnesium, etc.) and trace elements (vitamins and minerals) are adjusted according to the general composition of the carbon source. This in turn is highly dependent on the strain of organism used. In general, simple nitrogen sources, such as urea, ammonia and nitrate, are used to keep costs down. Phosphate is supplied as phosphoric acid or as soluble phosphate salts.

Process variables, such as dilution rate, temperature, pH, ionic strength and level of oxygenation, exert a strong influence in cellular yield. In particular, an abundant supply of oxygen promotes aerobic metabolism and higher growth rates. However, owing to the low solubility of oxygen in aqueous media, the cost of aeration, through air sparging and agitation, increases rapidly with the scale of operation, resulting in an important technical challenge.

In general, when yeast biomass is produced, alcohol accumulates owing to oxygen limitation. Some alternatives proposed for SCP from whey are the production of alcohol as a by-product from fermentation, or the use of *Kluyveromyces* in a mixed culture with *C. pintolopessii*, in which the latter consumes the alcohol produced by the former. The first approach has been followed by Amber Laboratories (Universal Foods), while the second one is favoured by the Fromageries Bel in France.

In order to sustain high oxygen transfer rates, large air volumes have to be supplied with high agitation rates. Alternative fermenter designs include the air-lift type, which achieves maximum oxygenation with minimum power requirements. In fact, the largest fermenter ever operated is an air-lift (3000 m^3) for the aerobic production of *Methylophilus methylotrophus* by ICI. Currently, the high cell density fermentation designs pioneered by Phillipps Petroleum allow users to obtain up to 160 g l^{-1} of yeast biomass, while traditional fermentation techniques reach, at most, 30 g l^{-1}. These systems possess very efficient means for heat removal and oxygen transfer.

Once obtained, the microbial biomass is recovered by filtration or centrifugation. The resulting cell suspension can be either spray-dried, or the cells are broken to obtain extracts, hydrolysates or autolysates. Finally, the protein can be concentrated or isolated.

Nutritional Value

The main nutritional contribution of SCP, either in human food or in animal feed, is its high protein content. Bacteria have a protein concentration ranging from 50% to 83% and yeasts from 45% to 55%. Protein quality is also quite acceptable, as compared to vegetable proteins. Single-cell protein from any source is generally limited in sulphur-containing amino acids. Some parameters of protein quality are given in Table 3. When methionine is added to SCP, the protein quality is increased considerably, reaching values similar to those of casein. ***See*** Protein, Quality

Single-cell protein is also an important source of vitamins; brewer's yeast has long been used as a vitamin supplement. Vitamin contents are also shown in Table 3. ***See*** individual vitamins

Toxicological Aspects

Safety of both microorganism and substrate are important considerations. Microorganisms used for SCP production have to be subjected to extensive toxicological clearance. Those microorganisms normally present in fermented foods, such as *Saccharomyces cerevisiae*, are free from suspicion and can be used without concern. The main concern has been linked to the use of petroleum derivatives; for those processes residual alkanes must be removed with solvents. However, some residual hydrocarbons may still be present, and several reports have noted the presence of unusually high amounts of odd-chain fatty acids and paraffins in tissues from animals fed with SCP from alkanes. These fatty acids, particularly unsaturated C17, could result in toxic effects. These arguments have been used against SCP plants based on hydrocarbons.

The high content of nucleic acids present in microbial cells also has some consequences. Human consumption of nucleic acids in amounts higher than 2 g per day could lead to the accumulation of uric acid, resulting in gout and kidney stones. The concentration of nucleic acids in the biomass depends on several factors. The first is the nature, species and strain of the microorganism; bacteria normally have higher concentrations than yeasts. Another factor is the set of growth conditions: the higher the growth rate, the higher the nucleic acids content in the cell. For *K. marxianus* grown on whey, the concentration of nucleic acids reached its peak at the middle of the exponential phase. Table 4 shows the nucleic acids content of several SCP products. ***See*** Nucleic Acids, Physiology

In order to reduce the nucleic acid content, two approaches have been followed: (1) produce the biomass at low growth rates, or (2) isolate the protein and thus eliminate undesirable compounds. The second method is usually applied, not only to eliminate nucleic acids but also the cell wall.

Table 4. Nucleic acids content of SCP products and other foodstuffs (dry wt.)

	Nucleic acids content	
	g per kg biomass	g per kg protein
Kluyveromyces fragilis		
Whole cell	570–870	110–200
Protein isolate	140	10–70
Saccharomyces cerevisiae	160	330
Candida utilis	100	240
Cereals	1·4–4·0	11–40
Liver	—	40

Yeast and bacterial cell walls are difficult to digest, leading to poor bioavailability of the proteins, flatulence, allergic responses, and diarrhoea.

Nucleic acid content in SCP for animal feed is not a problem, because animals possess the enzyme, uricase, which prevents uric acid accumulation. However, cell wall digestibility in monogastric animals is poor.

Utilization for Human Food

Other considerations, beyond toxicological concern and nutritional quality, are organoleptic acceptability and functional properties.

Single-cell protein has been used as a protein supplement in baked goods, biscuits, snacks, soups, and special foods such as geriatric or baby meals. Its use as an extender in sausages and other meat products is important, mainly in eastern Europe. Despite the justified production of SCP in terms of the world's protein shortage and widespread malnutrition, a real demand for a protein is based on its ability to compete in terms of its functional properties, e.g. solubility, water-binding ability, emulsification capacity, gelation, whippability, and foam stability. The successful supplementation of existing products and the replacement of traditional foods with SCP depend on the availability of proteins capable of matching functionality, low price, and organoleptic acceptability. ***See*** Protein, Functional Properties

Better opportunities to meet the needs of the food industry will be provided by protein concentrates or isolates. Moreover, SCP isolates can favourably compete with soya isolates from the functional point of view. However, isolation or concentration increases production costs dramatically.

Processes such as texturization by spinning and extrusion, and enzymatic or chemical modification, can improve the functionality of SCP. For example, protein fibres obtained by spinning can form textured protein products such as meat extenders.

Enzymatic modification includes partial proteolysis to improve solubility, emulsification capability, and whippability, or the reverse reaction, known as 'plastein' (peptide bond formation), to improve nutritional value through addition of limiting amino acids. Promising chemical modifications include acetylation, which improves thermal stability, or succinylation or phosphorylation to increase solubility, emulsification and foaming capacities; however, such modifications tend to reduce the nutritive value of the proteins.

In spite of the fact that dried whole cells have limited functional properties, they are frequently used as flavour-carrying agents and food binders. Dried yeast cells can act as oil-in-water emulsion stabilizers.

The major market for microbial biomass is as a flavour enhancer for meat products, soups, gravies, sauces, salad dressings, seasonings, and any food with an important note of the fifth basic flavour, called umami, can benefit from the increased savoury, cheesy or meaty flavours. In fact, yeast protein hydrolysates, autolysates and extracts have long been used as food flavourings.

Cell Disruption and Protein Isolation

Several techniques are applied for cell disruption. A common one is autolysis, in which the biomass is exposed to a heat shock, or to chemical compounds such as low-molecular-weight thiols. Further incubation induces activation of cellular enzymes, leading to the complete lysis of the cells. The process also activates endogenous ribonucleases that reduce nucleic acids. Lysis can also be facilitated by addition of exogenous enzymes, such as proteases, β-glucanases or lysozyme. Disadvantages of these techniques are extensive proteolysis, which reduces yield and functional properties of proteins, and the high cost involved. Chemical treatments with alkalis, organic solvents, or salts which weaken cell walls are also used. Alkaline treatment may result in undesirable side-reactions, forming compounds such as lysinoalanine, and off-flavours. ***See*** Enzymes, Uses in Food Processing

Physical methods to break cell walls are the most widely used. High shear rates are achieved by means of homogenizers or colloidal mills, and these have been used extensively for SCP processes.

Once the cells have been broken, the protein is extracted using water or alkaline conditions; the cell wall debris are removed by centrifugation and the protein is further precipitated with acid, salt or heat treatment, while the nucleic acids remain in the supernatant. The protein isolate is then obtained. Some chemical modifications during protein extraction include phosphorylation or succinylation, which facilitate protein separation from nucleic acids and improve its functional properties.

Prospects

Single-cell protein has to compete with other protein sources, such as soya bean, fish meal and milk proteins. Production and isolation of protein from microbial biomass is rather expensive because the processes are capital- and energy-intensive. The broad utilization of SCP has been limited for economic reasons. However, recent biotechnological advances, such as high-cell-density fermentations, more efficient downstream operations, and the possibility of genetically improved microorganisms, could lead to a re-evaluation of SCP.

High-cell-density fermenters have made possible a considerable reduction of equipment size, energy savings, very high productivities and cheaper downstream processing. For example, direct spray-drying from the fermenter is possible. These kinds of improvement may bring SCP to a competitive level. Phillipps Petroleum is currently producing Provesta and Provesteen, trade marks for SCP from different species of yeast, using this process (see Table 1).

Genetic engineering has focused on the possibility of improving substrate utilization. The first modified microorganism utilized in an industrial process, and the largest-scale application of genetic engineering, is a strain of *Methylophilus methylotrophus* developed by ICI in 1977. The improved strain, grown on methanol, is able to utilize ammonium ions as a nitrogen source more efficiently than the wild strain, saving one mole of adenosine triphosphate (ATP) per mole of ammonium assimilated; an increase in efficiency of 4–7% was achieved.

Considerable research based on genetic engineering has been carried out to obtain yeasts that are able to utilize a broader range of carbon sources, such as lactose, starch, cellulose, xylose, and chitin, in order to use cheaper and more available substrates.

Some other interesting possibilities are the selection of mutants with weaker cell walls, or the introduction of genes coding for lytic enzymes to facilitate cell disruption. In addition, an enhancement of the nutritional value by modification of the amino acid content, or the attainment of proteins with better functional properties, look promising. All these possibilities have been investigated but the practical results will take a long time to be implemented.

Another approach to improve the economics of SCP would be to produce it as a low-cost by-product in multiproduct microbial processes, e.g. during the processing of food wastes to reduce biological oxygen demand. Another possibility is the recovery of nucleic acids from bacterial or yeast biomass to produce 5′ nucleotides which can be used as flavour enhancers.

Bibliography

Batt CA and Sinskey AJ (1987) Single-cell protein: production modification and utilization. In: Knorr D (ed.) *Food Biotechnology*, pp 347–362. New York: Marcel Dekker.

Goldberg I (1985) *Single-Cell Protein*. Berlin: Springer Verlag.

Guzmán-Juarez M (1983) Yeast Protein. In: Hudson BJF (ed.) *Developments in Food Proteins – 2*, pp 263–291. Essex: Elsevier Science Publishers.

Litchfield JH (1983) Single-cell proteins. *Science* 219: 740–746.

Moo-Young M and Gregory K (1986) *Microbial Biomass Proteins*. Essex: Elsevier Science Publishers.

Reed G (1982) Microbial biomass, single cell protein, and other microbial products. In: Reed G (ed.) *Prescott & Dunn's Industrial Microbiology* 4th edn, pp 541–592. Westport Connecticut: AVI Publishing.

Rose AH (ed.) (1979) *Microbial Biomass. Economic Microbiology*, vol. 4. London: Academic Press.

Schlingmann M, Faust U and Scharf U (1984) Bacterial Proteins. In: Hudson BJF (ed.) *Developments in Food Proteins – 3*, pp 139–173. Essex: Elsevier Science Publishers.

Mariano García-Garibay and Lorena Gómez-Ruiz
Universidad Autónoma Metropolitana, Iztapalapa, Mexico City, Mexico
Eduardo Bárzana
Universidad Nacional Autónoma de México, Mexico City, Mexico

SLAUGHTER

***See* Meat and individual Animals**

SLICING

***See* Comminution of Foods**

SLIMMING

Contents

Slimming Diets

Definition

A slimming diet is a low-energy diet designed to enable overweight people to lose body fat until they achieve a healthy weight for their height, i.e. a body mass index (BMI; the bodyweight, in kg, divided by the square of the height, in m) between 20 and 25.

Principles

Energy balance in the body depends on energy input and output. If energy intake from the diet exactly balances energy output, the energy content of the body remains stable. When energy input exceeds output, the excess is stored in the body as fat, and bodyweight increases. Conversely, body energy stores are reduced when energy output is greater than input. The loss of body fat is reflected in a lower bodyweight. ***See*** Energy, Energy Expenditure and Energy Balance

Clearly, overweight people (except particularly muscular individuals) have at some time in their lives eaten more energy than their body required. To lose weight, such individuals must go into negative energy balance. In practice, a negative energy balance is easier to achieve by reducing energy intake than by increasing energy expenditure through physical activity. A programme of dietary restriction and exercise is the optimum combination for effective weight reduction. ***See*** Obesity, Treatment

There are no short cuts to losing weight. Although other treatments, including drug therapy (anorectic, diuretic or thermogenic), jaw wiring or surgical procedures, are employed to expedite weight loss, especially in the severely obese, individuals must learn to modify their eating habits to prevent weight regain.

One kilogram of body fat is equivalent to about 32 MJ (7700 kcal). A daily energy deficit of 4·6 MJ (1100 kcal) will therefore result in a weekly weight loss of approximately 1 kg, equivalent to an individual who maintains energy balance on 8·8 MJ (2100 kcals) reducing to 4·2 MJ (1000 kcal) per day. Realistic rates of weight loss lie between 0·5 and 1·0 kg per week for most people following slimming diets that provide 4·2–6·3 MJ (1000–1500 kcal) per day.

Theoretically, increasing energy restriction results in faster weight loss, but there are complicating factors. The composition of the weight loss varies. With moderate energy restriction 75% can be accounted for by fat and the remainder by nitrogen losses from lean tissue. The ratio of lean tissue to fat loss increases with increasing energy restriction such that, on complete starvation, half the weight loss is attributed to lean body tissue. This is clearly undesirable and ultimately fatal. Effective slimming diets are those which impose restrictions on energy intake compatible with the maximum of fat to minimum of nitrogen losses.

Consideration must also be given to the macronutrient composition of a slimming diet. Apart from the need to provide adequate protein to replace nitrogen losses, there are differences in energy utilization between diets. The increase in metabolic rate which automatically occurs postprandially is higher with a high-carbohydrate diet than with an isocaloric high-fat diet.

Resting metabolic rate, the energy required by the postabsorptive body at rest, accounts for two thirds of the total energy requirement of sedentary people. On energy-restricted diets the body adapts by lowering resting metabolic rate – by less than 10% on diets providing over 4·2 MJ (1000 kcal), and by more than 20% on very-low-energy diets. In effect, the total energy expenditure is reduced and the energy deficit incurred by following a slimming diet is not as great as may be assumed from the difference between prediet and slimming diet energy intakes, especially on the more restrictive regimes.

Initial weight losses on a slimming diet can be higher than predicted by energy deficit alone. In addition to fat, the body also has limited stores of carbohydrate as glycogen in the liver and muscle. The glycogen is readily available and is metabolized under conditions of energy restriction, concomitantly releasing its component water. Thus the greater proportion of the initial weight loss is made up of water and this rate of weight loss cannot be sustained in subsequent weeks. Equally, when reverting to normal energy intakes after a period of energy restriction, weight gain may be observed as glycogen is stored along with its component water. ***See*** Glycogen

Types of Slimming Diets

The characteristics of a good slimming diet are as follows:

1. It effectively reduces energy intake to ensure a steady weight loss of 0·5–1·0 kg per week.
2. It is nutritionally sound.
3. It ensures that compliance is possible over long periods of time.
4. It is compatible with individual life styles and eating patterns, e.g. it allows for ‘eating out’ and other social occasions.
5. It teaches slimmers how to modify their diet so that weight loss can be maintained.

Slimming diets fall into two categories: conventional food-based diets, and nutritionally complete formula preparations.

Conventional Slimming Diets

It is difficult to devise a nutritionally adequate diet, made up from normal everyday foods, the energy content of which is less than 4·2 MJ (1000 kcal) per day. Consequently, most diets of this type are designed to provide between 4·2 and 6·3 MJ (1000–1500 kcal) per day. Occasionally, diets providing 3·3 MJ (800 kcal) are advocated for short periods. There are many different types of dietary approaches within this category and all can be effective in achieving weight loss.

Set Diets

The slimmer is provided with a diet sheet from which daily menus can be selected to provide the required amount of energy. The number of meals can vary but most tend to be based on a three-meals-a-day eating pattern. The main advantage of these diets is that slimmers do not have to count calories. These diets can be designed to meet specific needs (e.g. vegetarian) and to provide alternative choices within meals. Some offer only a limited choice of food and may not result in weight loss, especially if the description of portion sizes is inadequate.

Calorie Counting Diets

The slimmer is allowed to choose foods up to a given calorie level. Although allowing slimmers considerable freedom of choice, it nevertheless requires them to select and weigh foods of known calorie value and to keep a running total of calories consumed. This type of regime requires considerable application and numerical ability on the part of the slimmer. The widespread availability of comprehensive guides on the calorie content of foods and nutritional labelling on food packs has facilitated this approach to slimming in recent years. There is the danger that the slimmer will consume a nutritionally unbalanced diet if calorie content is the only criterion for food choice.

No Counting Diets

No counting diets are based on the 'traffic light' system, whereby certain foods are allowed in unlimited quantities, others in restricted amounts, and a further group of foods are forbidden altogether. Forbidden foods include those high in fat and sugar – sugar, cakes, biscuits, chocolate, confectionery, pastries, fried foods, cream, and alcohol. Foods allowed in moderation include bread, pasta, rice, pulses, potatoes, milk, dairy products, fats, and oils. Lean meat, fish, fruit, and vegetables are allowed *ad libitum* since intake of these foods are considered to be self-limiting. The principle of such a diet is to modify food choices to those that are prepared and cooked very simply; highly processed foods are eschewed. The resulting diet is low in carbohydrate, high in protein, moderate in fat with adequate levels of micronutrients. The restriction on cereal food consumption runs counter to current recommendations for the role of cereal fibre for normal bowel function. The general guidelines of the diet will work well for individuals whose normal diets are largely made up of foods from the forbidden list and who are willing to make basic changes to their eating habits – ideal, perhaps, for first time dieters. Seasoned slimmers who have already made changes in their eating habits as a result of previous attempts to slim may find that they can consume in excess of their requirements while still following the dietary guidelines.

Low-carbohydrate Diets

The principle of the low-carbohydrate diet is to limit the intake of total carbohydrate, not because carbohydrate-rich foods are more fattening than other foods, but because on such diets voluntary energy intakes fall below 6·3 MJ (1500 kcal) per day while the overall nutrient content remains satisfactory. Carbohydrate foods tend to act as carriers for other foods – bread with butter and jam, pasta with sauce, cream or custard with pudding – so that a restriction in intake of carbohydrate foods will automatically reduce the intake of associated high-energy foods. Low-carbohydrate diets were popular and effective, and they generated the myth, still widely held, that bread and potatoes are particularly fattening. These diets were more successful under the traditional three-meals-a-day eating pattern, before the widespread introduction of low-energy versions of standard foods. Even so, it was still possible to achieve a high energy intake without exceeding the carbohydrate allowance by eating cheese, for example, which is carbohydrate-free but high in fat and energy. Conversely, the carbohydrate limit could easily be exceeded merely by consuming three or four pieces of fruit of modest calorie value. One immediate benefit perceived by the slimmer on a low-carbohydrate diet is a rapid initial weight loss caused by diuresis as glycogen stores in the body are utilized. This may be followed by ketosis when carbohydrate intakes are particularly low. The ketotic state may also exert an anorectic effect, thus facilitating compliance to the diet. The concept of advocating low-carbohydrate diets runs contrary to current nutritional advice, which recommends low-fat, high-fibre diets. *See* Carbohydrates, Requirements and Dietary Importance

Low-fat Diets

The principle of the low-fat diet is that limiting the intake of foods high in fat will reduce total energy intakes. Fat contains over twice the energy of proteins or carbohydrate, therefore high-fat foods are high in energy. Voluntary energy intakes on low-fat slimming diets providing about 30 g fat, are reduced to around 5 MJ (1200 kcal) per day, a level which will induce weight loss. These diets are in line with current nutritional guidelines to reduce total dietary fat intakes and are the basis of many current popular and successful slimming regimes. Compliance to a low-fat diet is aided by the widespread availability of low-fat versions of standard foods. *See* Fats, Requirements

High-fibre Diets

The aim of high-fibre diets is to achieve a daily fibre intake of between 30 and 50 g per day on daily intakes of 4·2–6·3 MJ (1000–1500 kcal). These levels of fibre are double customary intakes. Foods rich in dietary fibre are beneficial in a slimming diet as they provide additional bulk for minimal increase in energy, and the high satiety value of these diets improves compliance. High-fibre diets prevent constipation, a common side-effect of reducing diets, although flatulence can be a problem in some individuals. High-fibre diets are high in total carbohydrate and low in fat. The evidence for increased faecal energy losses on high-fibre diets is poor. Current nutritional guidelines for healthy eating advocate diets based on wholegrain cereals, fruits and vegetables – foods rich in fibre. Thus high-fibre slimming diets are to be recommended, especially where the fibre is derived from natural foods rather than from supplements. ***See*** Dietary Fibre, Fibre and Disease Prevention; Dietary Fibre, Physiological Effects

Meal Replacements

Meal replacements are preparations designed to be eaten instead of meals. Their main advantage is that they offer perfect portion control of known calorie content; many are supplemented with vitamins and minerals. Some are designed as total diet replacements, and provide all the necessary nutrients for a given calorie cost; others are intended to substitute for one or two meals a day, the remaining calories to be eaten as a conventional low-calorie meal. The original meal replacements were available as biscuits; the range now include milk shakes, cereal bars, meals, soups, and sweet or savoury liquid drinks. Meal replacements can be convenient for the busy slimmer, although improved nutritional labelling on all food packs, including sandwiches and ready meals, has reduced their traditional advantage.

Formula Diets

Formula diets provide a specific amount of energy in the form of nutritionally complete formula, usually a dried powder that is mixed with water and serves as the sole source of nourishment for long periods of time. Such diets remove the dieter from all contact with food and so theoretically improve compliance. However, these diets are monotonous and boring even if a variety of flavours are available. On returning to a normal diet the slimmer is ill-equipped to choose an appropriate diet to maintain any weight loss achieved. Formula diets of varying composition and energy value are available over the counter. Those providing less than 2·5 MJ (600 kcal) per day, the so-called very-low-calorie diets (VLCDs) have undergone particular scrutiny.

Very-low-calorie Diets

Launched commercially in the early 1980s, VLCDs reached the peak of their popularity in the latter half of the decade. The 1987 UK Committee on Medical Aspects of Food Policy (COMA) report recommends that these diets should provide a minimum of 1·7 MJ (400 kcal) and 2·1 MJ (500 kcal) per day for women and men respectively, 40–50 g of high-quality protein, and should meet Recommended Dietary Allowances (RDAs) for vitamins and minerals. Very-low-calorie diets are not recommended as a first-choice slimming method, nor should they be continued for longer than the stipulated 3–4 weeks before returning to a mixed diet. People with a BMI below 25 should not use VLCDs as a sole source of food and VLCDs are contraindicated for certain other groups. The companies marketing the most popular brands of VLCDs have initiated programmes to rehabilitate successful slimmers onto maintenance diets.

These VLCDs should not be confused with the notorious 'liquid protein sparing diets', widely marketed in the 1970s. These diets were prepared from hydrolysed collagen and gelatin, and were banned after several deaths were reported. The diets were deficient in some essential amino acids and minerals.

Success Rates

Numerous trials have demonstrated the effectiveness of many different types of slimming diets in inducing weight loss in overweight subjects. Even resistant slimmers lose weight when fed calorie-restricted diets under strictly controlled conditions. The vast majority of people try to lose weight under their own initiative. It would appear that many are striving to achieve the impossible for despite the fact that at any one time, one in four adults is trying to slim, the prevalence of obesity continues to increase. Clearly, most people find it difficult to resist food in a social environment which encourages 'grazing' rather than discrete meal occasions, and where palatable foods are readily available. No wonder the public are susceptible to the miracle cures peddled by quacks and charlatans.

In many respects, the prospects for a successful outcome to a slimming diet have never been better. The choice and variety of low-calorie foods continues to increase, and the slimming food market, worth 25 million pounds in the UK in 1965, is currently worth over 1000 million pounds per annum. Merely substituting low-calorie versions of standard foods (e.g. dairy products, fats and sugar) into the typical UK diet could result in a daily saving of around 1·5 MJ (370 kcal), equivalent to a weekly weight loss of 0·3 kg a week. Slimming diets are readily available in specialist and other magazines and newspapers, and books on slimming are frequently best sellers.

Anyone attempting to slim cannot but help be

aware – through the medical profession, media and advertising – that it is not going to be easy. If overweight people expect not to succeed, then it is not surprising that the majority live up to their worst expectations and fail.

There is a need for a fundamental change in the approach to slimming. Implicit in the term of 'going on' a slimming diet is 'coming off'. Overweight people should be seeking to make permanent changes to their diet, so that when they have lost weight they continue with the same diet but simply eat more of it. It has been shown that slimmers who integrate their diet into the family eating pattern are more successful in achieving and maintaining weight loss.

Fortunately, the diet advocated for successful slimming and for general health in the population concurs – a diet low in fat and high in complex carbohydrates. Thus a better understanding of the nutritional value of foods and information on which foods to choose for a healthy diet should improve the current situation. Successful weight loss clearly requires more than a good slimming diet. Overweight people also need to learn how to modify their eating behaviour and to increase activity levels if they are to ensure long-term success.

The Role of Exercise

Increased physical exercise is a positive aid to weight reduction. Not only does it increase energy expenditure but it may also build up lean body tissue which will maintain total energy requirement at a higher level. Numerous studies have shown that a programme of diet and exercise results in better weight losses than diet alone. Moreover, some studies have shown that individuals continue to do more voluntary exercise on completion of the exercise programme. ***See*** Exercise, Metabolic Requirements

Any aerobic exercise, such as walking, swimming, cycling or running, is suitable. Walking is the obvious first-choice exercise for the heavily overweight and the unfit, and a minimum of three 30 min exercise sessions a week are recommended. Regular exercise will improve fitness, cardiovascular function, and muscle tone, and will also promote a sense of 'wellbeing' which in turn will reinforce motivation to comply with the diet.

The Role of Slimming Clubs and Groups

Group therapy has been shown to be of positive benefit to overweight people trying to lose weight. Surveys have shown that most people who have tried a slimming club found it their most successful method for losing weight, and the weight losses achieved compare favourably with clinically based treatments. There are many such groups, administered by both lay and commercial companies, and the leading weight loss organizations run slimming clubs nationwide. The groups meet on a weekly basis and are usually run by leaders who have successfully lost weight, and maintained the loss, themselves and have a genuine interest in helping others to lose weight. At each meeting, the members are weighed and learn through talks and group work about different aspects of slimming and behaviour modification. Members pay a weekly fee, for which they receive ongoing support, advice, motivation and inspiration from fellow members and the group leader. The diets are nutritionally sound, follow healthy eating guidelines and provide at least 4·2 MJ (1000 kcal) per day. The leading organizations run weight maintenance programmes for those who have reached their target weight, and offer free membership as long as weight is maintained within a set limit.

Recently, commercial weight loss centres have opened in this country which offer one-to-one counselling, behaviour modification, and small group therapy. In addition to the substantial initial fee, slimmers must buy the special dietary foods and ready meals supplied at the centres. The cost to the slimmer is much greater than in the commercial weight loss clubs.

Bibliography

Ashwell MA (1973) Commercial weight loss groups. In: Bray G (ed.) *Recent Advances in Obesity Research: II*, pp 266–276. London: Newman Publishing.

COMA (1987) *The Use of Very Low Calorie Diets in Obesity*. Report on Health and Social Studies 31. London: Her Majesty's Stationery Office.

Evans E and Miller DS (1978) Slimming aids. *Journal of Human Nutrition* 32: 433–439.

Consumers' Association (1978) *Which? Way to Slim*. London: Consumers' Association.

Fisher RB (1986) *A Dictionary of Diets, Slimming and Nutrition*. London: Paladin.

Garrow JS (1988) *Obesity and Related Diseases*. London: Churchill Livingstone.

Miller DS and Parsonage SA (1975) Resistance to slimming: adaptation or illusion? *Lancet* 1: 773–778.

Royal College of Physicians Report (1983) Obesity. *Journal of the Royal College of Physicians of London* 17: 4–58.

Sanders TAB, Woolf R and Rantzen E (1990) Controlled evaluation of slimming diets: use of television for recruitment. *Lancet* 333: 918–920.

Elizabeth Evans
Slimming Magazine Clubs, London, UK

Metabolic Consequences of Slimming Diets and Weight Maintenance

In many societies today, a plethora of individuals pursue weight-reduction diets for a variety of reasons, ranging from social pressures to medical concerns. Dieters usually fail to perceive that the phenomena underlying weight gain and weight loss may be quite different in complexity. Weight gain, as a consequence of excess energy intake, is a relatively simple process whereby fat stores are increased. To a large extent, it is independent of the quality of food eaten. Thus, whenever calories are ingested in excess of calories expended, be they in the form of sugars, proteins, lipids or alcohol, the balance is stored in the most compact, weight-efficient form, as fat. The only exceptions to this rule are growth, pregnancy, muscle building, and rehabilitation following malnutrition, where excess calories, with adequate amounts of protein, can also be retained as protein.

Weight loss, as a consequence of insufficient energy intake, has a greater variety of effects which depend upon nutritional status, energy content of the diet and balance of nutrients. Thus, an excess daily intake of 1250 kJ (300 kcal) in the form of either 75 g of sugars, or 75 g of protein or 35 g of lipids, will result, in most individuals, in a gain of 35 g of body fat. A deficient daily intake of energy of 1250 kJ may induce a considerable range of changes in bodyweight and body composition. In appropriately chosen conditions, the subject will lose 35 g of body fat, but electrolytes and protein shifts will often alter the bodyweight response so that not all of the weight lost will be in the form of fat. ***See*** Energy, Energy Expenditure and Energy Balance

Regimens for Weight Control

The ideal diet should deplete body fat stores, which are excessive, without reducing the protein pool out of proportion to the small increment always associated with the obese state itself. For this to occur, the chosen diet should meet stringent qualitative and quantitative requirements. Brain metabolism requires 150 g of glucose per day: this mandatory need cannot be met by transformation of lipids, whether supplied by the diet or mobilized from body reserves. Glucose can be generated from protein, from the glycogen pool, or provided in the diet. Consequently, diets which provide at least 2500 kJ (600 kcal) per day in the form of carbohydrate, and a substantial portion of protein, will spare the waste of endogenous protein and mobilization of the glycogen pool. Many slimming diets do not satisfy the brain's need for glucose, and therefore deplete glycogen and protein stores. These diets may be grouped into (1) total fasting and certain very-low-calorie diets, or 'modified fasts', which do not contain enough glucose and protein precursors, and (2) diets which supply adequate calories but mostly in the form of fat. With both groups of diets, body proteins are mobilized and converted by the liver to glucose, which is then oxidized by the brain. The loss of body protein entails a loss of weight which is 10 times that of adipose tissue for an equivalent calorie content. A total fast will induce a protein loss not too dissimilar to that of a 5000 kJ-per-day diet, composed mainly of fat. In both instances, the weight loss will be very important: the high-weight–low-calorie protein waste will overwhelm the low-weight–high-calorie changes of fat stores. Concurrent major fluid losses will compound the phenomenon. ***See*** Carbohydrates, Digestion, Absorption and Metabolism; Carbohydrates, Requirements and Dietary Importance; Fats, Digestion, Absorption and Transport; Fats, Requirements

Dietary strategies other than a reduction in calorie content have not been thoroughly evaluated for the treatment of obesity. There is some evidence that macronutrients may act directly, by their nature rather than energy, on mechanisms which regulate bodyweight. Furthermore, the potential of micronutrients to affect fat storage has not been explored.

Behaviour modification targeted at overeating is also advocated to help reduce bodyweight. Some obese individuals may suffer from specific behavioural syndromes characterized by excessive consumption of certain foods, at particular times, e.g. carbohydrate cravings have been associated with seasonal affective disorders.

The major obstacle to dietary and behavioural treatment of obesity is that the voluntary control of energy intake is difficult and requires strong motivation. The success rate of dieting can be disappointing: studies published in the medical literature indicate that weight loss after 1 year is usually less than 10% of entry weight. However, most people who determine that they should control their weight, or are advised to do so, are never part of medical surveys. Thus the real success rate for populations are not known.

Metabolic Consequences of Weight-reducing Diets

Negative Nitrogen Balance

Whenever a diet, irrespective of its calorie content, does not have sufficient protein to maintain required rates of protein synthesis, or a sum of protein and carbohydrate to meet the brain's energy requirements, it will induce a negative nitrogen balance, with protein tissue waste. Obese individuals have an increased lean body mass and a higher metabolic rate than control individuals of 'ideal' bodyweight. Therefore, protein waste will be tolerated to some extent during dieting. Unfortunately, there is no evidence that protein is lost from those tissues where it accumulated during weight gain, i.e. from adipose tissue as a consequence of cell hyperplasia and hypertrophy, and from muscle and supporting tissues as a response to increased gravity. On the contrary, the protein loss may occur from all tissues. The digestive tract, and possibly the heart, are particularly sensitive to functional alterations during weight loss with excessive negative nitrogen balance. *See* Protein, Requirements

Water and Electrolytes Losses

Elimination of excess water is a common feature of weight loss, particularly during the early stages. A negative calorie balance reduces insulin secretion and has a diuretic effect which enhances weight loss beyond that expected from endogenous fat oxidation. The diuretic effect varies with the quantity and quality of energy consumed. In general, the lower the energy content as carbohydrate, the greater the diuretic effect. Concurrent loss of sodium and potassium in the urine reflects the contraction of the extracellular and intracellular water compartments. Cardiac arrhythmias, including ventricular fibrillation and death, have been reported, probably as a consequence of electrolyte changes in the serum and cells. The use of diuretics alone or in combination with diets, especially when markedly carbohydrate-restricted, is therefore dangerous or, at least, superfluous in the treatment of uncomplicated obesity. Before counteractive adaptive mechanisms are set in motion, water loss at the initiation of low-calorie, low-carbohydrate diets may induce an extra weight loss of 0·5 kg per day, or even more. *See* Water, Physiology

Water loss will confound changes in body fat weight. In extreme situations, such as those created by regimens very high in calories and fat, protein waste and dehydration will induce weight loss while fat stores are actually increased. Other diets, along the same principles, will also reduce weight with little fat loss.

Lowering of Blood Pressure

Overeating stimulates the release of catecholamines and insulin. Consequently, renal reabsorption of sodium is increased, and higher blood volume and cardiac output may then raise blood pressure. Hypocaloric diets will reverse this sequence of events and frequently reduce the hypertension associated with obesity, without medication or sodium restriction. *See* Hypertension, Physiology

Weight-reducing diets, in particular those low in carbohydrate, induce relaxation of arteriolar tonicity and lowering of diastolic blood pressure. This, in combination with fluid volume contraction mentioned above, is responsible for orthostatic hypotension, a frequent side-effect, characterized by the inability to redistribute blood from the lower limbs when standing. Salt supplementation can diminish such an effect, but not abolish it, as a new steady state is reached in which the extra sodium is simply excreted. **See** Hypertension, Hypertension and Diet

Lowering of Blood Lipids and Glucose

Although most obese individuals are free of metabolic complications such as dyslipoproteinaemias and non-insulin-dependent diabetes mellitus, many patients with these metabolic anomalies are obese. When energy and carbohydrate intake are excessive, high levels of triglycerides and low-density lipoprotein cholesterol may accumulate in plasma. Obesity is also often associated with a diminution of circulating high-density lipoproteins. These changes in plasma lipid composition will contribute to the development of atherosclerosis and increase the risks of cardiovascular disease. Hypocaloric diets can rapidly blunt or eliminate hypertriglyceridaemia, by the plasma-clearing action of lipoprotein lipase, and hypercholesterolaemia, by reducing both the endogenous and exogenous flow of atherogenic lipoproteins. Concomitantly, the level of high-density lipoproteins in the plasma increases. Weight reduction can therefore often help to control dyslipoproteinaemia without the use of pharmacological agents. *See* Atherosclerosis; Lipoproteins

Blood glucose abnormalities in obese subjects are frequent. They range from mild glucose intolerance to overt noninsulin-dependent diabetes mellitus. These anomalies of glucose homeostasis are characterized by the presence of normal-to-high levels of circulating insulin which are insufficient to overcome the resistance by target tissues to its biological effects. Excess plasma insulin is a recognized risk factor for atherosclerosis. One of the most impressive effects of energy-restricted diets is the improvement or normalization of the altered glucose homeostasis associated with obesity. In a matter of days, even without significant losses of bodyweight, the obese subject can be switched to a normal glucose metabolism. A negative calorie balance can rapidly restore insulin sensitivity in a dieting individual. The mechanisms of this spectacular metabolic regulation are not known. The treatment of diabetes in the obese requires hypoglycaemic agents only when energy restricted diets have proved unsatisfactory.

Reduced Resting Metabolic Rate and Postprandial Thermogenesis

With the loss of significant amounts of bodyweight, oxidative metabolism decreases as does energy expenditure. This occurs both at rest and following meals. At rest, most of the energy produced is utilized to maintain body temperature and vital processes. Following the ingestion of a meal, additional oxygen is required for mechanical work associated with digestion, and for chemical reactions involved in processing the nutrients. The energy cost of disposing of food in the body varies with the caloric content of the meal, its composition and sapidity. It represents only a small percentage of the overall resting metabolic rate. The main factors responsible for slowing of the oxidative metabolism are hormonal adaptive changes. Faced with reduced energy intake, the body responds by lowering the circulating levels of hormones which stimulate energy expenditure. Thus, the activity of the sympathetic nervous system is diminished and catecholamine production is lowered. Furthermore, the conversion of the thyroid hormone T_4 into the more powerful hormone T_3 is also reduced. These changes combine their effects in reducing mitochondrial oxidations and, therefore, energy output. This adaptive sparing of bodyweight varies with the energy restriction and the composition of the diet. It may represent approximately 850 kJ (200 kcal) per day after 2 to 3 weeks of fasting, but less if the diet has energy, particularly in the form of carbohydrates. In addition to implementing energy expenditure, exercise is able to blunt the drop in metabolic rate observed with adaptation to a negative energy balance, thereby enabling weight loss to continue unmitigated. ***See*** Hormones, Thyroid Hormones; Thermogenesis

A more chronic effect of weight-reducing diets is the loss of significant amounts of bodyweight when lean tissue is also wasted. The ensuing cell atrophy will contribute to lowered oxygen consumption.

Patterns of Weight Loss

After the initiation of a hypocaloric diet, a combination of behavioural and metabolic mechanisms takes place wherein the rate of weight loss decreases with time, despite a constant dietary intake.

On the behavioural side, many dieters tend to reduce their energy expenditure as a reflex in the face of scarce energy intake. Probably the most common cause for plateaus in weight loss is that, even with adherence to appropriate pattern and composition of the diet, portion sizes tend to increase.

On the metabolic side, a number of predictable adaptive reactions slow the rate of weight loss. First, water and electrolyte losses are rapidly blunted by sensitive hormonal counterregulatory responses. These may involve, depending upon the degree of water loss, the liberation of antidiuretic hormone by the hypothalamus or the stimulation of the renin–angiotensin–aldosterone axis. These responses are more decisive when diuretics are prescribed with the diet. Second, the body adapts by shifting its oxidative metabolism from glucose to fat and ketone-acids. The reduced need for glucose spares the energy-diluted protein tissue at the expense of the energy-dense fat stores. Thus, the rate of bodyweight loss declines to very near the theoretical level expected from conversion to fat as the primary source of energy utilization. Third, the oxidative metabolism diminishes *pari passu* with the activity of the thyroid gland and sympathetic nervous system. Fourth, the loss of lean body mass, with time, lowers the resting metabolic rate, thus reducing the energy deficit.

Occasional interventions in the course of dieting may also slow down the rate of weight loss. Rotating diets, depending on their sequence, may induce rapid shifts in water and protein pools. For example, when a 3500 kJ per day ketogenic diet is followed by a diet of the same calorie content but richer in carbohydrate and protein, the weight loss may be temporarily reduced and weight may even increase as a consequence of water and protein gains. This same diet would have the opposite predictable effects if administered *de novo*. Thus, the same hypocaloric diet is capable of inducing either weight and protein gain, or weight and protein loss in the same individual, depending on the previous metabolic balance at any given total bodyweight. These considerations emphasize the extent to which the dependence of the dieter upon bodyweight measurement as a source of feedback information is misleading.

A less common but interesting situation is created when dieting individuals embark on exercise programmes. This leads to loss of body energy, without the expected concurrent loss of weight. It is possible, for example, to lose 1 kg of body fat as a consequence of a negative calorie balance, and to gain 1 kg of tissue protein as muscle is built up. The bodyweight has not changed, but the individual has lost in the process 30 000 kJ of body energy. Recent studies have shown that a similar protein-sparing effect, with negative calorie balance, can be achieved by administration of appropriate doses of insulin or growth hormone. The usefulness of these hormonal manipulations in the dieting individual has not yet been established.

Maintenance of Weight Loss

Whenever the stability of bodyweight is disrupted, the hypothalamic counterregulation will tend to restore weight to its initial value. This phenomenon applies in general to any level of weight equilibrium, whether the subject is thin, of 'ideal' weight, or obese; it is also operative for whichever direction the change of weight takes place, be it a gain or a loss. It is therefore recommended, in order to overcome the constraints of the 'ponderostat', that slimming strategies aim for a moderate rather than a rapid rate of weight loss, interspersed with periods of weight stabilization.

Physical exercise has the ability, within a wide range of intensity, to maintain and possibly re-establish the regulatory functions of the brain concerning energy balance. Exercise programmes should therefore complement the slimming diet, with the reasonable expectation that they will help the dieter to maintain the weight loss.

The temporary use of anorectic drugs has been advocated to promote weight maintenance between episodes of weight loss. The administration of thyroid hormones, to offset their lowered secretion during dieting, has also been suggested. There is not sufficient evidence to recommend these pharmacological interventions.

Finally, therapies intended to improve the will to lose weight are an essential part of slimming programmes. Group support therapy, as provided by associations such as Weight Watchers, or individual psychotherapy, may be successful in some individuals, while knowledge of nutrition principles seems to improve the incentive to lose weight and maintain the loss.

Benefits of Exercise

Exercise has many beneficial effects for weight maintenance. The most acknowledged effect, albeit not the most important, is that exercise increases calorie expenditure above the resting metabolic rate. The energy cost of exercise is often overestimated by dieters and easily annulled by food self-reward. For example, a brisk, one-hour walk has approximately the equivalent energy of three small slices of bread.

Exercise, of some intensity and frequency, has anorectic properties which are probably mediated by changes in hypothalamic neurotransmitters and sex hormone production. With more moderate ranges of physical activity, as previously stated, the hypothalamic regulation of energy balance seems to better able to adjust energy intake to energy expenditure on a day-to-day basis, thereby improving the chances of weight maintenance.

Exercise has powerful effects on fuel utilization in general, and glucose metabolism in particular. Acute exercise accelerates the net rate of glucose utilization through the actions of humoral and hormonal factors on adipose tissue, muscle and liver. The blood-glucose-lowering effect of exercise may be rapid and important. It mimics the effect of intravenous injections of significant amounts of insulin, with the added characteristic that it manifests itself in most instances, whether the subject is insulin-resistant or not. Furthermore, chronic exercise counteracts insulin resistance associated with obesity, hence improving glucose and lipid metabolism. The metabolic properties of acute and chronic exercise should therefore be advantageously used in the treatment of obesity. ***See*** Anorexia Nervosa; Bulimia Nervosa; Obesity, Aetiology and Assessment; Obesity, Fat Distribution; Obesity, Treatment

Bibliography

Bray GA (1976) *Major Problems in Internal Medicine*, vol. 9. Philadelphia: WB Saunders.

Cowley DK and Sizer FS (1987) Fad diets, fact and fiction? *Nutrition Clinics* 2.

Garrow JS (1978) *Energy Balance and Obesity in Man*. Amsterdam: Elsevier.

Howard AN, Bray GA, Novin D and Bjorntorp P (eds) (1981) Proceedings of a symposium on obesity and hypertension. *International Journal of Obesity*, vol. 5, supplement 1.

Olson RE (ed.) (1986) Diet and behaviour: a multidisciplinary evaluation. *Nutrition Reviews* 44.

Roncari DAK (1986) Obesity and lipid metabolism. In: Spittel Jr JA and Volpe R (eds) *Clinical Medicine*, vol. 9 Philadelphia: Harper & Row.

Wurtman RJ and Wurtman JJ (eds) (1987) Human obesity. *Annals of the New York Academy of Sciences* 499.

E. Rasio
Université de Montréal, Quebec, Canada

Metabolic Consequences of Slimming Diets and Weight Maintenance

SMOKED FOODS

Contents

Principles

Smoking is an ancient process but its origins are obscure. The foods most commonly smoked are meats, fish, shellfish and cheese, although some snack foods and nuts are smoke flavoured. Meats or fish grilled over an open fire are incidentally smoked, but are not normally considered to be smoked foods as such. In addition, while outside the scope of this article, it should be noted that some alcoholic beverages are matured in charred wooden casks or incorporate ingredients that have been exposed to peat smoke. ***See*** Whisky, Whiskey and Bourbon, Products and Manufacture

Smoke is produced by substantially raising the temperature of wood, and limiting the air supply so as to prevent combustion but allow destructive distillation. Historically this was no doubt achieved by burning whatever pieces of wood were available, but it has become traditional to use small pieces, i.e. chips, shavings or sawdust. Alternatively, smoke may be generated by friction when the end grain of blocks of wood is pressed against a rapidly moving rough surface (e.g. a rotating carbide disc).

Smoke flavours are prepared by condensation of smoke obtained by pyrolysis of wood either in a limited supply of air or in the absence of air, when the product is known as pyroligneous acid. The initial condensate separates into an aqueous phase and a tarry phase. The smoke condensate may be separated into fractions by physical separation techniques or by solvent extraction. These fractions may be further purified to remove hazardous substances known to be present in smoke. Smoke flavourings include smoke condensates, fractions thereof, and mixtures of such fractions.

The temperature at the centre of a heap of smouldering sawdust may be 700–1000°C but the temperature gradient is steep, temperature falling to 300°C or less a short distance from the centre. Above approximately 200°C the wood undergoes destructive distillation and the desired decomposition products are best generated in the 200–400°C range. This smoke is allowed to diffuse over, or more commonly is blown over the food to be smoked, with varying levels of control depending on the technology available.

The distillate as generated is a vapour, but as it cools some less volatile components condense to form a disperse liquid phase, which along with soot particles, if present, constitute visible smoke. The remaining more volatile components distribute between the gaseous and disperse phases according to their solubility and volatility at the prevailing temperature and humidity. A recent review of wood-smoke composition records over 400 volatile components comprising 48 acids, 22 alcohols, 131 carbonyls, 22 esters, 46 furans, 16 lactones, 75 phenols and some 50 miscellaneous compounds. The precise composition of woodsmoke, and the extent to which the food absorbs smoke constituents depends on several factors.

Factors Affecting the Composition of Woodsmoke

Type of Wood

Wood generally contains 40–60% celluloses (β-glucans), 20–30% hemicelluloses (heteroglycans containing pentose and hexose residues), 20–30% lignins (complex three-dimensional phenolic polymers) and a small amount of protein. In addition there is a quantitatively minor but chemically complex low-molecular-mass fraction in which phenols and terpenes often dominate, particularly in softwoods. Water content is highly variable. ***See*** Cellulose; Hemicelluloses; Lignin; Phenolic Compounds

Commonly woods are classified as hard or soft depending on their botanical origin, the former from broad-leaved angiosperms, the latter from needle-leaved (often evergreen) gymnosperms, but the terms hard and soft in this context are not good guides to physical properties. Compared to softwoods, hardwoods have larger contents of hemicelluloses, those in softwoods having a greater proportion of pentosans. Softwoods have higher lignin contents, with more guaiacyl (2-methoxyphenyl) residues relative to syringyl (2,6-dimethoxyphenyl) residues than is usual in hardwood lignins.

There is a general belief that smoking with hardwoods such as oak (*Quercus* spp.), hickory (*Carya* spp.), beech

Table 1. Effect of wood source on composition of smoke

Source	Aroma fraction (mg per 100 g wood)			
	Total	Phenols	Carbonyls	Acids
Oak (*Quercus seriata*)	1800	151	117	1140
Oak (*Quercus acuta*)	1600	225	323	820
Cherry	1490	101	111	660
Pine	1180	166	53	640

Table 2. Effect of combustion temperature on smoke composition

Combustion temperature (°C)	Content in smoke (mg per 100 g sawdust)		
	Phenol	Guaiacol	Syringol
450	20	60	85
550	30	80	150
650	55	125	260

(*Fagus* spp.) and alder (*Alnus* spp.) produces superior smoked foods compared to smoking with softwoods. Hardwoods generate more syringyl derivatives relative to guaiacyl derivatives and more acid smokes (see Table 1), owing to structural differences in the lignins and the greater pentosan content of hardwood hemicelluloses, respectively. These acids produce smoked foods with lower pH values and hence greater microbiological stability, and this may explain the traditional preference.

Although softwoods such as fir and red cedar may be disliked because they generate more soot and impart resinous (terpene-derived) flavours, the situation is not simple since some hardwoods are said to impart excessively dark colours and bitter tastes and pine (a softwood) in one study has been judged superior to beech for smoking sprats.

Wood Combustion Temperature

The temperature of smoke generation influences its composition. Total carbonyls concentration increases in the temperature range 200–600°C, total phenols concentration in the range 400–600°C and polynuclear aromatic hydrocarbons (PAH), while virtually absent below 400°C, increase rapidly thereafter, with increased generation of methylene radicals. The PAH have implications for product safety and modern methods of smoking are designed to keep the generation temperature below 400°C or otherwise to remove the PAH. The balance between individual compounds within these broad groups is also affected by generation temperature, with guaiacol, acetovanillone and acetosyringone dominating the phenols fraction at 380°C, but acetosyringone and propiosyringone dominating at 560°C. Some further data are shown in Table 2.

Smoke House Humidity

It is thought that smoke components enter the food primarily by absorption into the interstitial water in the food, at least for low-fat foods. Initially the foods being smoked contain about 80% moisture and have a high water activity, both of which decline progressively during the smoking process. The absorption of smoke components is thus more rapid at the start of smoking and declines as the water content declines.

Table 3. Influence of wood moisture content on smoke composition

Moisture content (% as is)	Yield per 100 g sawdust (as is)			
	Condensate (g)	Phenols (mg)	Acids (mg)	Formaldehyde (mg)
1·8	28·2	236	3203	122
21·5	41·5	136	3288	81
24·5	43·5	100	3003	78
31·2	32·8	33	890	—

The rate of water loss from the food is greatly influenced by the absolute humidity of the air in the kiln relative to the surface water activity in the food. More rapid loss of water due to the use of hotter and/or drier smoke give rapid surface drying and a firm 'skin' to the food, but may inhibit absorption of smoke components. The use of wetter smoke slows dehydration, and if insufficient drying occurs the product may be unstable as a result of too high a water activity. It is not uncommon to wet the sawdust so as to keep the smouldering temperature low and prevent combustion, but excessive damping must be avoided otherwise the very wet atmosphere created will inhibit drying. Smoke composition is in any case influenced by the moisture content of the wood that is used – see Table 3. ***See*** Water Activity, Principles and Measurement

Although antioxidative properties associated with woodsmoke phenols, and antimicrobial effects associated with phenols, acids and especially formaldehyde, increase the stability of smoked foods relative to unsmoked, it is accepted that control of water activity, in part through salting of the food, but mainly through drying, was originally the main preservation mechanism. Achieving a water activity (a_w) close to 0·70 required the removal of a lot of water, thus reducing product yield, or adding a lot of salt and thus reducing product palatability. Modern smoked foods have comparatively low salt and high water contents (thus increasing the yield and palatability of the product) but owing to the relatively high a_w which results, the product shelf-life must be extended by chilling, freezing or canning. ***See*** Antioxidants, Natural Antioxidants; Canning, Principles; Curing; Drying, Theory of Air Drying; Freezing, Principles; Preservation of Food; Water Activity, Effect on Food Stability

Effect of Air Flow Rate

Air flow is essential to keep wood smouldering and/or to transfer smoke from the generator to the kiln, and to minimize nonuniform smoke distribution within the kiln. Air flow rate and humidity influence the rate at which the food surface dehydrates and thus smoke absorption, as discussed in the previous section. It has been demonstrated that the dispersed particles of the smoke act as a reservoir for the volatile vapour constituents. Diluting the smoke with air at constant temperature, or raising the smoke temperature, drives volatiles from the dispersed droplets into the vapour phase and encourages rapid and extensive transfer of volatiles to the food, whereas dilution of smoke with cold air has the opposite effect.

Time of Smoking

The moisture content in the surface layers of the food declines as smoking proceeds, and this affects the rate and extent of smoke absorption as discussed previously. It is likely that the distribution of fat on the surface of the food will also change during the process, particularly where high temperatures are employed. Higher temperatures will cause greater protein denaturation and shrinkage, and thus squeeze molten fat from the interior to the surface of the food, and this might modify smoke absorption. In any case, smoke absorption is a first-order process and the rate declines as surface saturation is approached, inward diffusion being slow and limiting. It follows therefore that extended smoking does not necessarily increase smoke absorption to the extent that might be expected. Even after extensive smoking the smoke concentration falls markedly over the first 5–10 mm below the surface, except when, for example, smoked food is converted to a paté or in cases where 'artificial smoking' or smoke flavouring is applied. In such cases smoke condensates (liquid smokes) are incorporated by pumping them into the muscle with the curing salts or by mixing into comminuted products. If the smoke condensates are atomized directly into the smoking chamber, smoke absorption resembles the traditional smoking process.

Effect of Kiln Temperatures

Smoking processes can be divided into two types on the basis of kiln temperature: cold smoking processes, where the kiln temperature does not exceed approximately 30°C; and hot smoking processes, where a temperature of 80°C may be achieved and the food is effectively cooked as well as smoked. Some traditional smoking processes in the tropics may employ temperatures as high as 120°C if the ambient air is very humid.

The kiln temperature influences the rate of drying and the volatility of the smoke components and thus variations of temperature can be expected to modify smoke absorption. Temperature also influences enzymatic and microbial activity in the food; both could be pronounced at modest temperatures associated with some traditional cold smoking processes. Such activities not only modify the composition of the food and its organoleptic properties, but possibly also the safety of the product if undesirable microorganisms have gained access. At the higher temperatures encountered in hot smoking operations such activities are less pronounced but not necessarily eliminated. Protein denaturation and fat diffusion are more extensive and modify the appearance, texture and flavour of the smoked food. Choice of operating temperature is largely determined by the nature of the product and the desired organoleptic attributes, under the constraints of microbial safety. Guidelines on safe operation have been published by MAFF, USFDA and FAO among others. ***See*** Protein, Interactions and Reactions Involved in Food Processing

It should be noted that factors such as kiln temperature, kiln humidity, air flow rate and smoke density are difficult to control in traditional kilns where these parameters are strongly influenced by weather conditions. In modern mechanical kilns such factors can be controlled within much closer limits.

The Effects of Smoking on Food

Changes in Nutritive Value

Smoking, particularly hot smoking and especially as practised in the tropics, can lead to destruction of tryptophan, cysteine, lysine and other basic amino acids whether free or protein-bound. The losses recorded for lysine are variable, reflecting the variations in the smoking process (and analytical procedure), but may reach 55%. Such losses are concentrated in the outer 5–10 mm where conditions are harshest. It is clear that

Table 4. Threshold values, Sensory Index values and described odour for some woodsmoke phenols

	In water				In oil		
Compound	Taste threshold	Taste index	Odour threshold	Odour index	Odour threshold	Odour index	Odour description
Guaiacol	0·013	6400	0·021	4600	0·07	1000	Sweet, smoky and somewhat pungent
4-Methylguaiacol	0·065	90 000	0·09	58 800	0·4	18 000	Sweet, smoky
2,6-Dimethoxyphenol	1·65	1400	1·85	1200	0·34	7000	Smoky

these losses are not just a function of temperature but are also due in part to interactions with woodsmoke constituents and ingredients such as nitrate and nitrite which may be added during the curing stage. Among the products detected are imines from carbonyl–amino group interactions, some of which contribute to the characteristic yellow or reddish brown surface coloration of traditionally smoked foods. In the case of certain amino acids these imines react further, yielding various β-carboline carboxylic acids (from tryptophan and woodsmoke carbonyls), and at least one nitrosothiazolidine carboxylic acid (from cysteine, formaldehyde and nitrite). ***See*** Amino Acids, Properties and Occurrence

The loss of nutrients is unlikely to be of nutritional significance in the developed countries. Animal feeding studies on heavily smoked fish, such as may be an important source of animal protein in some tropical countries, indicate that a reduction in net protein utilization might occur and that some slight improvement in diet might be possible if alternative methods of fish preservation could generate products acceptable to the local consumer.

The Sensory Properties of Smoked Food

Texture

European cold-smoked products will generally have a texture that is soft and tender, whereas the equivalent hot-smoked products will have a firm, dry surface with softer interior, being more succulent the higher the fat content. In contrast some traditional hot-smoked products from the tropics will, to Western eyes, look like pieces of charred wood. Such products are usually consumed after grinding to a powder which is added to stews. The main determinants of texture are:

1. The extent and rate of water loss, with greater loss giving firmer texture and more rapid loss giving greater difference in texture between surface and interior.
2. The fat content and its distribution, the former being largely a function of the food, whereas the latter is also a function of temperature during processing, with higher fat content being more succulent.
3. The extent of denaturation of structural and connective tissue protein, denaturation being more extensive for higher temperatures and higher salt contents, with fish in general being more susceptible than meat, and with greater denaturation giving firmer texture.
4. The extent of autolysis, particularly proteolysis, with ungutted fish being most susceptible, and greater proteolysis giving softer texture. ***See*** Sensory Evaluation, Texture

Colour

As referred to above carbonyl–amino interactions are important, with model system studies suggesting that interaction of lysine with glycolic aldehyde, 2,3-butanedione, pyruvaldeyde, coniferaldehyde or sinapaldehyde being particularly important. Colour may also derive from coloured substances in the smoke being deposited directly, or from permitted colourings (such as the azo food colour Brown FK or natural carotenoids) incorporated during processing. ***See*** Colours, Properties and Determination of Natural Pigments

Flavour

The presence of an odoriferous or sapid compound in woodsmoke or smoked food does not necessarily equate with organoleptic importance, since the quantity present must be sufficient to exceed the odour or flavour threshold of the consumer. Threshold values are notoriously variable between individuals and even between occasions for given individuals, owing to physiological and/or psychological factors. In many cases the intensity of an individual compound at a particular concentration depends on whether it is dissolved in water or in fat. Sensory properties therefore depend not only on the substances present but also their concentrations and interactions with the food, and the person making the judgement – see Table 4. ***See*** Sensory Evaluation, Taste

Many investigators consider the phenolic constituents of woodsmoke, especially 4-methylguaiacol, guaiacol and eugenol to be particularly important, at least in providing those smoky flavour notes that are essentially unique to smoked foods. There is no doubt, however, that other classes of smoke volatiles are also very important and recently attention has been directed towards the pyrazines and lactones. ***See*** Flavour Compounds, Structures and Characteristics

Aqueous extracts of smoke condensates may be prepared as flavouring ingredients and to avoid the potentially hazardous PAH fraction. Such preparations do not differ from the original smoke solely with reference to their PAH content, and they have been criticized for not properly duplicating the true smoke flavour. Certainly analysis of commercially available extracts indicates that there can be considerable variations in composition depending on method of manufacture. Whether friction smoking or electrostatic smoking duplicate the flavour achieved by the traditional process is also a matter of some debate.

Assessment of Toxic Hazard

Concern about the possible safety of smoked foods centres primarily on the presence of PAH. Thirteen PAH have been detected in smoked foods, at least one of which benzo[*a*]pyrene, is a known rodent carcinogen. The greatest levels appear to be associated with products such as charcoal-broiled steaks, which are only incidentally smoked, rather than those products normally described as smoked. It has been estimated that the average daily per capita intake of benzo[*a*]pyrene in the United Kingdom is less than 4 μg, and that smoked foods do not contribute significantly to this burden in the United Kingdom. As stated earlier, PAH production in smoke is greater for smoke generation temperatures above 400°C. It is also associated with charring of the food surface, the deposition of soot on the food, and with drip exuded from the food if this should fall into the fire and be combusted. Such occurrences are more likely in traditional and artisanal smoking operations than in commercial operations as performed in industrialized countries. ***See*** Carcinogens, Carcinogenicity Tests; Polycyclic Aromatic Hydrocarbons

The safety of smoke condensates was reviewed by the Joint FAO/WHO Expert Committee on Food Additives (JECFA) in 1987. Their concern centred mainly on the hazard that would arise if nitrosamines and carcinogenic PAH were present. The evidence available to JECFA indicated that neither class of substance was detectable in the preparations tested, but it was required that benzo[*a*]pyrene levels should not exceed 10 $\mu g\ kg^{-1}$, the lowest practicable level for measurement.

Only limited short-term toxicity studies were available and normally this would have prevented JECFA from assigning an acceptable daily intake (ADI) to a group of products having such complex composition. However, in what may well prove to be an important precedent (for the safety evaluation of flavouring materials in general) JECFA decided that smoke flavourings (smoke condensates and fractions thereof) of suitable specifications could be used provisionally to flavour foods traditionally treated by smoking, since foods so treated would be no more hazardous than the smoked foods traditionally consumed, and may even be safer since a large number of potentially toxic compounds, including the carcinogenic PAH, are removed during the manufacture of the flavourings. However, new or novel applications (since such usage might lead to a significant increase in dietary burden) should be approached with caution until such time as more extensive safety studies on a well-defined spectrum of smoke flavour preparations had been carried out.

Concern has also been expressed regarding the presence of heterocyclic aromatic amines in foods, particularly in grilled meat and fish, and presumably also in the corresponding smoked foods. These compounds are pyrolysis products derived from amino acids such as tryptophan, or glutamic acid or from creatinine and products of the Maillard reaction. Some are potent mutagens in certain *in vitro* mutagenicity tests. ***See*** Browning, Nonenzymatic; Mutagens

Recently, varying amounts of three-β-carboline-3-carboxylic acids have been detected in some smoked foods produced in Europe. Although some β-carbolines are known to be pharmacologically active, for example inhibiting monoamine oxidase, the significance if any of these detected in smoked foods has not been evaluated.

Although not peculiar to smoked foods, the possible presence of mycotoxins, bacterial toxins or pathogenic microorganisms should not be overlooked. ***See*** Mycotoxins, Occurrence and Determination

Bibliography

Clifford MN, Tang SL and Eyo AA (1980) Smoking of foods. *Process Biochemistry* 15(5): 8–11, 17.

Gangoli SD (1986) The toxicology of smoked foods. *Proceedings of the Institute of Food Science and Technology* 19: 69–78.

Maga JA (1987) The flavour chemistry of woodsmoke. *Food Reviews International* 3: 139–183.

Papavergou K and Clifford MN (1986) Chemical interactions between woodsmoke components and foods. *Proceedings of the Institute of Food Science and Technology* 19: 13–17.

Toth L (1982) *Chemie der Räucherung*. Weinheim: Verlag Chemie.

WHO (1987) *Evaluation of certain food additives and contaminants. Thirty-first report of the Joint FAO/WHO Expert Committee on Food Additives. Technical Report 759*. p 21–23. Geneva: WHO.

WHO (1987) *Toxicological evaluation of certain food additives. WHO Food Additives Series*: 21: 41–57. Cambridge: Cambridge University Press.

Michael N. Clifford
Food Safety Research Group, University of Surrey, UK

Production

Generation of Smoke

All methods of generating smoke involve the pyrolysis of wood, to which reference has already been made. It is clearly desirable to limit the temperature to which wood is heated to minimize the production of undesirable smoke constituents such as polynuclear aromatic hydrocarbons (PAH). Two methods of smoke generation are possible in which the temperature is controlled externally, rather than by the rate of burning of a fraction of the wood to produce heat. Whilst not presently used commercially, so far as is known, the fluidized bed generator is a practical method of controlling the temperature to which wood is heated. Sawdust, of a given particle size range, is fluidized by hot air at 350°C, shown diagrammatically in Fig. 1(a). Limitation of the temperature to 350°C, to minimize exothermic reactions, tended to produce smoke containing a higher proportion of acids when compared to smoke generated by more conventional methods. In the steam smoke generator high pressure steam at 350°C is passed through a bed of sawdust. The smoke contains a high proportion of water and is therefore only suitable for smoking processes which do not require significant drying. ***See*** Polycyclic Aromatic Hydrocarbons

For many years smoke has been generated in traditio-

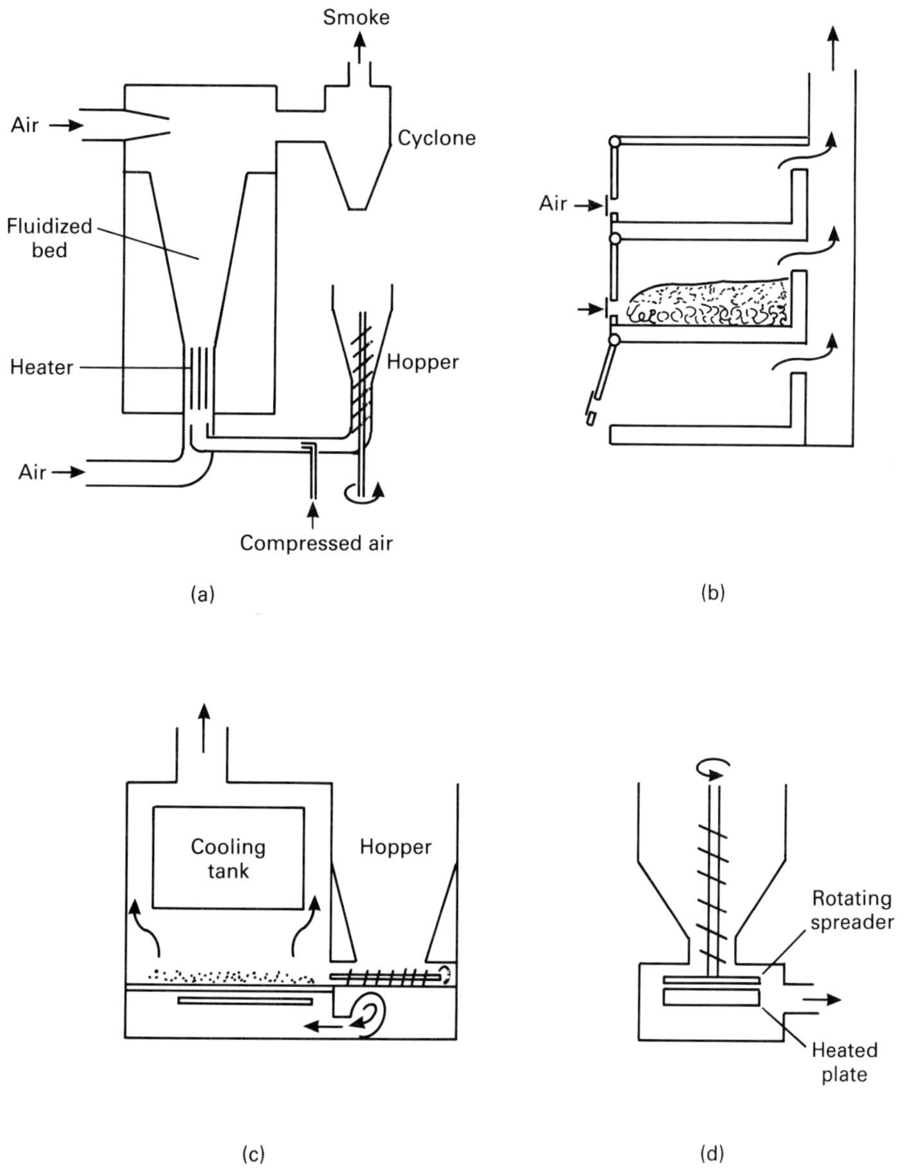

Fig. 1 Smoke producers. (a) Fluidizer; (b) simple smoke boxes; (c) automatic smoke producers; (d) heated plate producer.

nal smoke houses by smouldering heaps of sawdust on the floor of the house. In more recent times the use of mechanical kilns has increased, which require the generation of smoke externally to the kiln. Initially smoke was generated in simple smoke boxes (Fig. 1(b)) in which a layer of wood shavings was overlaid by a layer of sawdust. The negative pressure of the smoke inlet to the kiln was sufficient to draw air over the fires, controlled by adjustable holes in the front of the smoke box. The air flow was adjusted to maintain smouldering, at a dull red heat, of the wood shavings. Each smoke box would burn for up to 3 hours and required re-laying for longer smoking periods. A later development, much in use today, is the automatic continuous smoke producer, Fig. 1(c). Sawdust, of a controlled water content, is fed from a hopper onto a perforated plate through which air is blown. The plate is first heated electrically before the sawdust feed starts; after ignition the sawdust bed continues to smoulder without further heating. The rate of smoke production is controlled by the feed rate of the sawdust, the air flow being adjusted to just maintain smouldering.

Figure 1(d) illustrates a second type of smoke producer in which comminuted wood or sawdust is fed onto a continuously heated plate. The plate is maintained at 300–350°C but exothermic reactions can cause the temperature of the sawdust to rise above the plate temperature. As in all smoke producers limitation of the air supply is vital in controlling the temperature of pyrolysis. Experimental producers have been developed in which some of the air is replaced with nitrogen to minimize combustion but, so far as is known, none of these is used in commercial practice.

Friction generators are used, but in at least one system the smoke is passed through water sprays to remove sparks. This can have the effect of removing a proportion of the more desirable volatile fraction, thereby reducing the efficiency of the generator.

Whilst claims may be made by manufacturers of smoke generators as to the efficiency of their equipment to produce smoke which has a low concentration of undesirable components, the ultimate levels of these components on the smoked food also depend to a large extent on the smoking process and the nature of the food. For example, both electrostatic precipitation and the use of fine sprays of water have been shown to reduce the concentration of undesirable components in wood smoke markedly.

Smoking Kilns

Smoke houses or kilns can vary in size and complexity from simple boxes to large wind tunnels in which all the important variables such as smoke flow, temperature and humidity can be accurately controlled.

Production

Traditional kilns have not yet been entirely superseded by mechanical kilns. They can vary in size from 1 or 2 metres to over 20 metres in height. The smaller kilns are often used for hot smoking, when billets of wood are burnt to produce temperatures high enough to cook the food, whereas the larger kilns are used for cold smoking, in which the tmperature does not exceed 25–30°C.

Cold-smoking kilns are often tall chimney-like structures, the height being necessary to aid natural convection. Their operation is very much dependent on weather conditions; for example, the velocity of smoke in the kiln is determined not only by its height but also by the difference in density between the ambient air and the smoke:

$$V = \sqrt{[2gH\,(P_a - P_s)/P_s]}$$

where H is the height of the kiln and P_a and P_s are the densities of the ambient air and smoke, respectively. In summer time when the ambient air is warmer and more humid, smoke velocities are often less than those required to maintain adequate rates of drying. This is one reason why traditional kilns are often used at night when the air is cooler. A typical process may require 12 hours or more and during this time the product may be moved around in the kiln during smoking in an attempt to produce a more evenly smoked and dried product. Figure 2(a) illustrates the type of traditional kiln widely used for smoking fish and bacon for several hundred years.

For the last 50 years or so there has been an increasing use of mechanical kilns in which the control of the smoking process is much more precise than is possible with simple traditional kilns, with a concomitant decrease in process time. In the mechanical kiln smoke is passed over the product by forced convection, the temperature is maintained constant by electrical or steam heaters, the humidity is controlled by adjusting the proportion of smoke which is recirculated and smoke is generated externally. In extreme cases dehumidification of the fresh air intake has also been used to control humidity. Using the principle of the heat pump, the cooled, dehumidified air is reheated, thereby reducing the heat requirement from other sources. In most applications, however, dehumidification is unnecessary.

A highly commercially successful kiln was developed by the Torry Research Station in the United Kingdom in the 1940s. Its main feature, based on wind tunnel practice, is the provision of an evenly distributed flow of smoke over the food product. This is essential to ensure that all the food is dried and smoked to the same extent. The effects of humidity and smoke concentration gradients in the smoke stream are cancelled out by reversing the position of the product relative to the direction of the smoke halfway through the process. In a more recent commercial development, the direction of the flow of

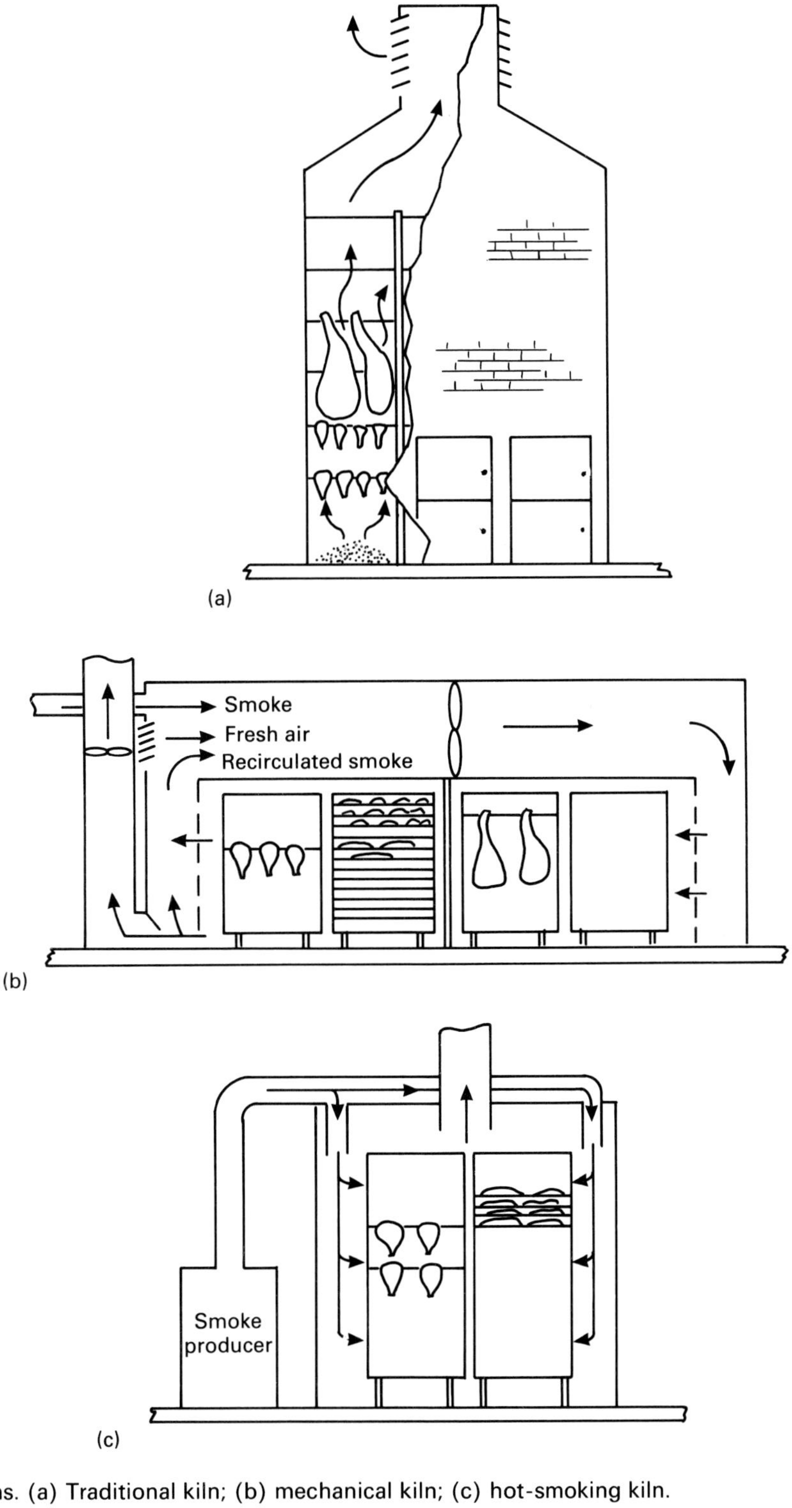

Fig. 2 Smoking kilns. (a) Traditional kiln; (b) mechanical kiln; (c) hot-smoking kiln.

smoke is reversed automatically at frequent intervals obviating the need to reverse the product.

The Torry kiln was originally designed for cold smoking, but is equally suitable for hot smoking. Figure 2(b) illustrates the principle. A further advantage of a horizontal smoke flow is that it enables many products to be smoked on open wire mesh trays, thereby increasing the packing density and reducing labour costs.

A number of other types of mechanical kiln are manufactured commercially. Principally designed for hot smoking of meat products, the smoke in these kilns is introduced by a series of ducts into the smoking chamber. Thus, the distribution of smoke tends to be less uniform than in the wind tunnel type, since this is less critical for the products normally smoked in them. Figure 2(c) illustrates the principle of this type of kiln.

The Smoking Process

Salting

Most modern smoked food products bear little resemblance to the heavily salted, heavily smoked and hard dried products such as the bacon and red herring of antiquity. In some cases the shelf life of present-day smoked products may be little different from that of the raw material from which they are made. Whilst most smoked foods contain some added salt, either by dipping, injection or embedding in dry salt before smoking, the final salt concentration is not usually high enough to have a marked effect on water activity. Many products contain around 2% of added salt and some up to 4 or 5%. Often the aim is to obtain a concentration of 3·5% in the aqueous phase of the final product to minimize the growth of some microorganisms of public health significance such as *Clostidium botulinum*. Salt contents of at least 10% would be required for a significant effect on bacterial growth generally, too high a concentration for many modern palates. ***See*** *Clostridium*, Occurrence of *Clostridium botulinum*; Curing

The uptake of salt by food products from brine or dry salt is a complex diffusion-controlled process. The effective diffusion coefficient is dependent on temperature, water content, salt concentration and the amount and disposition of fat or other nonaqueous components present. The time required for a given product to absorb a given amount of salt is approximately inversely proportional to the square of the half thickness of the product. Whilst the injection of brine enables a known amount of salt to be added, there is a concomitant increase in water content.

For relatively thin products, such as fish fillets, the amount of salt added can be metered by the deposition of charged droplets of brine containing smoke solution and permitted dye. As with brine injection, the water content is increased and thus more drying is required.

The salted product may require a period of maturation before smoking and drying, to allow some equilibration by diffusion. Traditionally a number of cold-smoked fish products such as fillets, especially salmon, and split herring (kippers) were not smoked immediately after salting. During this time some of the surface protein swells, thereby producing a glossy surface after smoking and drying. The formation of this outer layer is now deemed to be undesirable, especially in smoked salmon.

Many products, especially cold-smoked fish products, are coloured with permitted dyes before smoking, the dye being added to the brine. This became common practice in the early part of the century when lightly smoked, and therefore less coloured, products became more popular. ***See*** Colours, Properties and Determination of Synthetic Pigments

Smoking

In the previous article reference has already been made to the application of smoke preparations by dipping, injection or their volatilization or atomization in the smoking chamber.

The uptake of wood smoke constituents from the vapour and particulate phases of wood smoke is dependent on the nature of the product as well as on the temperature, humidity and velocity of the smoke. Smoking and drying occur simultaneously provided that the humidity of the smoke is less than the equilibrium relative humidity of the food. The deposition of volatiles is at a maximum whilst the surface remains wet during constant-rate drying, when water can diffuse to the surface at least as rapidly as it can be evaporated. As the surface becomes dry and a falling rate of drying ensues, the uptake of volatiles decreases markedly and the proportion of smoke constituents arising from the slower deposition of the particulate phase increases. Thus, the quantitative and qualitative analysis of the smoke constituents deposited onto markedly different products smoked under identical conditions will differ.

For a given product drying under constant-rate conditions at constant temperature, humidity and velocity, the total smoke deposition can be estimated from

$$Q = k\int_0^t D_x \, dt$$

where Q is the quantity of smoke deposited, k is a constant, t is time and D_x is the optical absorbance of x metres of smoke. The constant of proportionality, k, is dependent on the light path x and the constitution of the smoke as well as the method of analysis of the smoke deposition. Where the smoking process remains substantially constant, the integrated optical absorbance has been found to be of practical value despite the fact that it is not a direct measure of the concentration of smoke volatiles.

The quantity of smoke required to produce a satisfactory product with respect to its acceptability and shelf life can vary over a wide range depending on the product. For most fish products, however, smoke integrals in the range 2–5 m^{-1} h (absorbance per metre × hours) are found to be satisfactory. Integrals less than 2 result in products with very little smoke flavour and integrals more than 5 can result in too strong a flavour for many palates.

Recently a nonoptical integrating smoke monitor has become available which monitors the concentration of the vapour phase directly. Calibrated in terms of optical absorbance it can replace optical meters directly.

Although the particulate phase can act as a reservoir for the vapour phase to some extent, in relatively large smoking chambers where the path length of the smoke over the product is several metres the vapour concentration can decrease markedly downstream. Depending on the surface area of the product exposed to the smoke, the vapour concentration can be reduced by at least 40%. To obtain even smoke deposition (and even drying), either the products must be reversed with respect to the smoke flow halfway through the process or the direction of the smoke must be reversed periodically.

Drying

Drying both reduces water activity and alters the texture of the product, an important characteristic of smoked products. The surface must be dry to minimize the growth of spoilage organisms which are mainly present in the outer layer of the food; similarly, any cavities must be adequately exposed to the smoke flow to ensure that drying occurs. ***See*** Drying, Theory of Air Drying; Water Activity, Principles and Measurement; Water Activity, Effect on Food Stability

A large proportion of the drying of exposed muscle of meat, fish and poultry occurs during constant-rate drying. During this stage of drying the rate of drying may be predicted by

$$E = K\,(T_{DB} - T_{WB})\,V^{0\cdot77}$$

where E is the rate of water loss per hour per unit area (kg m^{-2} h^{-1}), $T_{DB} - T_{WB}$ is the difference between the dry bulb and wet bulb temperature of the smoke, and V is the smoke velocity (m s^{-1}). K is approximately 0·024 kg m^{-2} h^{-1} per unit degree Celsius wet bulb depression for velocities in the range 0·5–5 m s^{-1}. Fat-free flesh can lose up to 1·4 kg of water per square metre during constant rate drying and fatty flesh (20% fat) will only lose up to about 0·3 kg m^{-2}, during average smoking conditions. For relatively thin products such as fish fillets, this represents losses in weight of over 20% to about 5%, which are somewhat reduced at higher rates of drying. The presence of skin or layers of fat, as in whole fish or sides of bacon, drastically reduces this potential constant-rate loss of water.

Drying subsequent to constant-rate drying is diffusion dependent and decreases exponentially with time. During this stage of drying the rate is largely dependent on temperature and is much less dependent on humidity and smoke velocity.

Bibliography

Aitken A (ed.) (1982) *Fish Handling and Processing*, 2nd edn. Edinburgh: HMSO.

Burt JR (ed.) (1988) *Fish Smoking and Drying*, London: Elsevier.

Foster WW and Simpson TH (1961) Studies of the smoking process for foods, I. The importance of vapours. *Journal of the Science of Food and Agriculture* 12: 363–374.

Jason AC (1958) *Fundamental Aspects of the Dehydration of Foodstuffs*, pp 103–135. London: Society of Chemical Industry.

Storey RM (1986) The technology of traditional smoking. *Proceedings of the Institute of Food Science and Technology* 19: 58–62.

R. M. Storey
Hull, UK

Applications of Smoking

Smoke used for curing of foods consists of a suspension of minute particles in a vapour phase making up an aerosol. It is produced by controlled thermal combustion of sawdust, wood shavings, coffee husk, sugar cane pulp, coconut husk and rice straw, etc. The composition of smoke varies with the choice of wood, the temperature of pyrolysis, relative humidity and the moisture content of the wood. The relative concentration of different components depends on the type of wood used and the controls adopted during its combustion.

Polycyclic aromatic hydrocarbons (PAH), which are believed to be carcinogenic, are a well-known group of constituents of wood smoke. These compounds are probably formed through condensation of naphthalene. Their concentration strongly increases when other organic materials are added during combustion of wood. ***See*** Polycyclic Aromatic Hydrocarbons

Apart from providing desirable colour and flavour to foods, wood smoke contributes to preservation by acting as an effective antioxidant, and a bacteriostatic and bactericidal agent. Phenolic components of the smoke contribute to the flavour and aroma of the product. Smoke also provides a protective film on the surface of the smoked product, thus providing a barrier against development of rancidity. Combined chemical constituents of smoke together with heating and drying processes are responsible for bactericidal and bacteriostatic effects. ***See*** Antioxidants, Natural Antioxidants; Phenolic Compounds

Methods of Smoking Foods

Basically three methods are used in production of smoke flavoured foods.

The Traditional Method

This is the direct incomplete thermal degradation of wood to produce smoke. It may be carried out either as *hot smoking* or as *cold smoking*, i.e. the criterion is the maximum temperature to which the product is exposed. ***See*** Drying, Theory of Air Drying

Hot Smoking

Hot smoking involves exposing the product to at least 80°C and probably up to 100°C, the intention being to cook the product as well as smoke and dehydrate it. Duration of smoking depends on the level of curing required, e.g. light, mild or strong, and the thickness of the product. Techniques of smoke generation vary in different parts of the world. It may be a simple barrel type smoke oven, inverted drums with a fire pit at the bottom or located remotely, or the very latest high-tech equipment as available in the United Kingdom, with automatic smoke density, temperature and humidity controls (Fig. 1).

The principles of hot smoking are simple. It consists of generating smoke from desired wood (e.g. oak) with controlled air supply. The product is usually prebrined for an appropriate period from a few minutes up to several hours before being placed on the racks where the smoke is allowed to circulate at a selected rate and temperature. Modern kilns yield a uniform, clean product with little waste. There are basically three types of smoke houses: (1) with natural air circulation, (2) air-conditioned or forced air and (3) continuous. There are many modifications of these three types.

Air-conditioned or forced-ventilation smoke houses have largely replaced the natural-air type. They allow much more precise control of smoking through uniform air movement and good control of temperature. Smoke houses with endless revolving chains have been developed specifically for frankfurter production.

Cold Smoking

Cold smoking was introduced as an alternative to hot smoking in the early part of the last century. Its general purpose is imparting a desirable flavour to the product rather than preservation, which is achieved by refrigeration at low temperatures. In the kilns, the smoking temperature is maintained below 30°C and the time of smoking may vary from a few hours to several days depending on the product. Shelf life of the product can vary and may depend on whether the product was brined or not prior to smoking.

Liquid Smoke

Manufacture

Low-moisture saw dust is heated to 500–600°C in a combustion reactor and the resulting smoke vapour is passed to the primary tar and ash separation vessel and then on to a packed tower of ceramic beads. Here the vapour smoke is condensed. The condensate is composed of (1) an aqueous phase which is made up of a mixture of low-boiling-point fatty acids, carbonyl compounds and alcohols (pyroligneous acid, formalin and furfurol in particular; formalin is responsible for immediate sterilization of the food surface and the carbonyl compounds are claimed to be responsible for the formation of a desirable colour) and (2) an immiscible tar or 'liquid smoke' (see Table 1). The tarry phase contains cresol, guaiacol, eugenol, methyl guaiacol and pyrogallol to mention a few compounds. These sub-

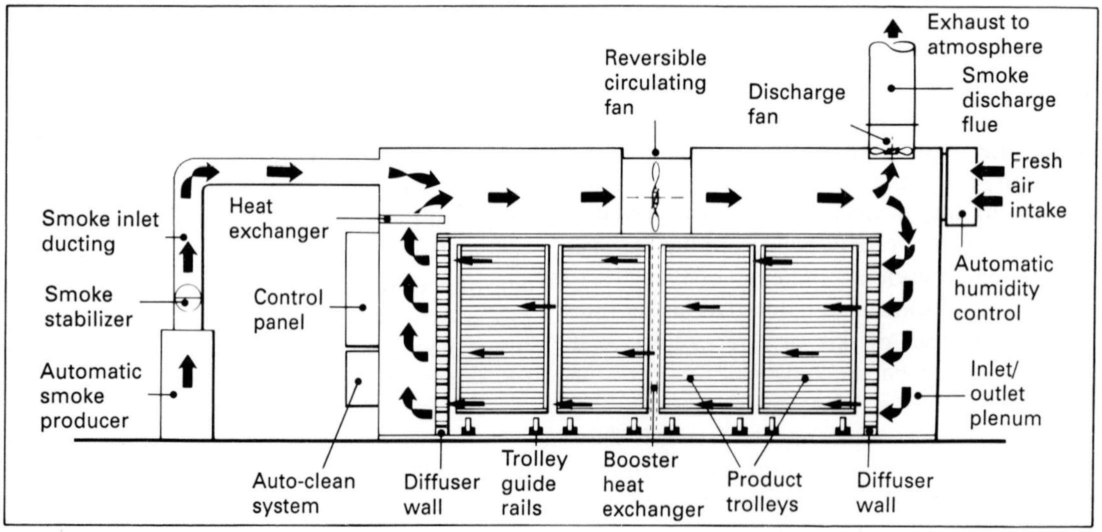

Fig. 1 Afos smoking kiln (courtesy Afos Ltd, Hull, UK).

Applications of Smoking

Table 1. Approximate composition of liquid smoke condensates from France, Germany and the United States

Group	Level (%)
Water	up to 92
Phenols	0·2–2·9
Acids	2·8–9·5
Carbonyl compounds	2·6–4·6

Data from Baltes W *et al.* (1981).

stances are credited with the antibacterial and antioxidative effect which smoking is supposed to have. The liquid smoke is stored in holding tanks for 2–4 weeks to allow separation of tars and other colloidal particles containing high levels of PAH, including benzo[*a*]pyrene. The remaining filtrate is a clear liquid which may be used as such or it may be further concentrated by fractional distillation and extraction with selective solvents (see the section 'Smoke flavours').

The tastes of the tarry and aqueous phases are different and opinions differ as to which is to be preferred. When liquid smoke from different parts of the world are examined for their relative chemical constituents, it is confusing to note that some consist chiefly of acids, others of esters and carbonyl compounds, and some chiefly of phenols.

There are four basic principles of application of liquid smoke: atomization, drenching, dipping and direct addition. Further technical developments of these methods can be found in the section *Application of Smoke Flavours*.

Application of liquid smoke offers several advantages over the traditional smoking processes in that it offers consistent quality and uniform and reproducible product. Since the polycyclic aromatic hydrocarbons, tar and soot are removed in its manufacture, it is safer to use. It is economical in terms of time and ease of application and there are no fire hazards.

Smoke Flavour

The toxicological problems associated with conventional smoking of foods have attracted a great deal of interest in recent years. There is evidence that the consumption of smoked meats, sausages and fish over prolonged periods may be a causal factor in development of cancer of the alimentary canal. For this reason extensive efforts are being made to develop toxicologically safe smoke flavours. To achieve this, flavour technologists worldwide have attempted to split the smoke condensate into a number of subdivisions. Using a variety of countercurrent techniques, the fractions obtained are then examined to establish the elimination of polynuclear aromatic hydrocarbons (PAH). Of all of the chemical constituents of smoke – alcohols, aldehydes, ketones, phenols, amines, tar and cresol resin – it is the PAH typified by benzo[*a*]pyrene which are known to be carcinogenic and have thus attracted the most medical attention. Techniques are now known which can measure PAH down to 0·001 ppm. ***See*** Amines; Carcinogens, Carcinogenic Substances in Food; Flavour Compounds, Production Methods

Opinions still differ as to which of the chemical constituents produce the characteristic smoke flavour. Study of the oil-soluble fraction of smoke condensate has been made by gas–liquid chromatography. After purification by extraction, the smoke flavour has been shown to contain some 20 different aromatic components most of which are phenolic in nature. A few aromatic aldehydes have also been shown to be present.

Organoleptic studies of the individual components or group of components showed *cis*-iso-eugenol, 2,6-dimethoxyphenol, and 2,6-dimethoxy-4-methylphenol to be the compounds contributing most markedly to the characteristic smoky flavour. ***See*** Sensory Evaluation, Taste

Optimum flavour characteristics can only be achieved by applying proper concentrations using appropriate technology. The concentration required to achieve the optimum flavour development depends on the type of product, its fat content and the processing method. Examples of concentration of smoke flavour using one such commercially available preparation are ham 45 ppm, bacon 32 ppm (in both cases added to brine), and sausages, cod roe, fish paste and fish sausages 15–45 ppm.

The object of conventional smoking is to increase the shelf life of the product. It is thus imperative to ascertain what effect application of smoke flavour or liquid smoke has on the keeping qualities of the product. Recent research indicates that 2,6-dimethoxy-4-hydroxybenzaldehyde, one of the active components of smoke flavour, has marked antioxidative and antibacterial properties.

Application of Smoke Flavours

The flavour components are available in three physical forms: aqueous, oil-soluble and dry. They may be applied to products by any of the following methods based on their suitability.

Spinning Head Spray System. This method utilizes a variable-speed air motor which spins an atomizing disk at between 1000 and 18 000 rpm, creating a very fine atomization. The liquid is fed to the gun via an air-operated vernier-controlled pump and an electrostatic charge is applied at the gun tip. This system can be used for spraying aqueous solutions of smoke extracts and flavour on to bacon, ham and chicken; and in a solvent-based form on to a product such as salmon where little or no water is required.

Airless Spray System. An airless pressure pump feeds the liquid through a spray nozzle. The electrostatic charge is applied by a probe at the spray tip. Various spray nozzle designs are available for achieving the desired pattern of application. Applications of this technique are in aqueous systems – extracts in brine solutions which allow smoking and curing of fish products in a single operation.

Powder Spraying. Powder flavours are obtained by mixing the flavour components with dextrose or starch base. The application system consists of a powder hopper and spray gun. The hopper has two regulators to control air supplied to a Venturi pump and to the air fan of the gun; a third regulator controls the degree of vibration at the base of the hopper. The gun has interchangeable heads which allow for variation of the spray pattern.

Flavour Coating of Savoury Snacks. This system incorporates positive charging of the product and negative charging of the flavour medium. The plant consists of variably controlled fluidized bed vibrator feeding the flavour over a knife edge where it picks up a negative charge and projects it into a stream of positively charged product falling freely into the air. Thus the flavour adheres all around the falling product.

In addition to these high-tech systems, smoke flavours, brined or unbrined, can be injected directly to the product and absorbed during curing in massaging machines and tumblers. The advantage of this technique is that it simplifies application and improves control and hygiene.

It has been shown that water-based as well as salt-based smoke flavours can be used in production of sausages, and other comminuted meat products. They are ideal for pork, beef, horse meat and reindeer meat. In the fish industry smoke flavour is successfully used in the preparation of fish pastes, sausages, caviar, roe and sprats. Smoke flavour is also applied to processed cheeses. It is reported that panellists could not differentiate between traditionally smoked and cheese to which smoke flavour had been added.

In canned products smoke flavour can be added directly to the oil, thus enabling flavour penetration during heat processing of already sealed cans.

Technology of Smoking Fish, Meat and Cheese

Fish

Operations necessary for production of smoked fish will depend on species, consumer preference and product specification. However, the general procedure is as follows. Fresh grade A fish should always be used. They are washed to remove scales and surface lime, etc., and gutted and headed if required. The belly cavity is cleaned to remove traces of blood and bellywall lining. Fish are then usually brined before smoking. Brining is recommended on grounds of safety. It is recommended that to exclude *Clostridium botulinum* completely, fish should be brought to an internal temperature of 82°C for 30 minutes during smoking if 3·5% brine solution is used, or 65°C for 30 minutes with 5% brine. ***See** Clostridium*, Occurrence of *Clostridium botulinum*; Fish, Processing

After brining, fish are suspended or laid horizontally on racks in the smoking kilns. If equipment permits, smoking may be carried out in stages, for example, initially at 30°C for 60 minutes, followed by 50°C for 30 minutes, and finally at 80°C for 2 hours or more until the product has attained the required colour and texture.

Cold smoking of fish is commonly practised on the leaner varieties of fish such as cod or haddock and may utilize traditional or liquid smoke. Depending upon the type of smoke, brined or unbrined fish is drenched with a suitable quantity of liquid smoke, which is then allowed to drain before being put through a drier to remove the excess moisture if long-term storage is envisaged. Nowadays after addition of smoke the fish may be vacuum-packed or refrigerated for transport to the point of sale. Virtually all types of fish may be smoked. The most popular varieties in the United Kingdom are smoked herrings (kippers), haddock, mackerel and sprats. Smoked salmon is a great delicacy.

Meat

Common salt is the basic curing ingredient although other compounds such as sugar, nitrite and/or nitrate may also be added. ***See*** Meat, Preservation

Salt in Cures

Salt is the basic ingredient necessary to all curing mixtures. It dehydrates and alters the osmotic pressure. It inhibits mould growth and subsequent bacterial spoilage. Its use tends to give a hard, dry, salty product that is not very palatable. It is therefore generally used in combination with nitrite/nitrate. A 65% 'pickle' cure is the most popular strength used for meats. ***See*** Curing

Sugar in Cures

The addition of sugar is primarily for flavour. It moderates the harshness of salt. During cooking it forms browning products that enhance the flavour of cured meats.

Nitrite and/or Nitrate in Cures

The addition of nitrates and/or nitrites offers the following advantages: (1) stabilization of the colour of lean tissue; (2) contribution to the characteristic flavour of the cured meats; (3) inhibition of growth of spoilage organisms, thus reducing the risk of food poisoning; (4) retardation of development of rancidity in the fatty constituents of the meats.

The processes of traditional smoking of meats may vary, but the general principles are the same. Prime quality meats are trimmed of most excess fat and then cured in a curing mixture (containing salt, nitrates and nitrites in water) for anything up to 7–10 days at 1–4°C; or meats may be pump-injected with curing mixtures before being smoked in a traditional woodchip smoker for several hours. Depending upon the temperature of smoking, they may be further cooked in an oven until judged to be just right. Appropriately sized pieces may then be vacuum packed and transported to the point of sale.

Nowadays cured meats may be cooked in a mixture of liquid smoke before being packed, but evidence suggests that consumers prefer the traditionally smoked product and are prepared to pay more for it.

The most popular varieties of smoked meat products include the well-known hams, beef, horse, and reindeer meats. In addition, smoked poultry, turkey and a variety of game birds can be found in specialist delicatessen shops. A number of different types of smoked meat sausages are available throughout the world. ***See*** individual meats

Cheese

Certain varieties of cheeses, e.g. Seretpenir (Iran), Caramakase (Germany), Bandal (India) and Volgodski (USSR), are traditionally smoked. They may be smoked by hanging in a smoke-charged atmosphere without the necessity for higher temperatures. Both oak and applewood shavings may be used for smoke generation. Nowadays incorporation of liquid smoke is becoming popular. The smoke preparation may be added to the milk, or sprayed onto the curds before pressing. Occasionally smoke flavour with salt may be applied as a spice. In another method of preparation, cheese is enclosed in a permeable membrane and immersed in a solution of smoke liquor.

Traditional smoking brings fat to the surface of the cheese, evaporates moisture and incorporates smoke vapours containing phenolic substances which aid in giving a preservative effect as well as imparting a savoury flavour to the cheese. The fat on the surface is also a deterrent to mould growth if the cheese is kept dry.

Bibliography

Baltes W, Wittkowski R, Sochtig I and Block H (1981) Ingredients of smoke and smoke flavour preparation. In: Charalambous G and Inglett G (eds). *The Quality of Foods and Beverages*, vol 2, pp 1–19. London: Academic Press.

Clifford MN, Tang SL and Eyo AA (1980) Smoking of foods. *Process Biochemistry*, June/July: 8–11.

Olsen CZ (1976) Chemical composition and application of smoke flavour. *Proceedings of the European Meeting of Meat Research Workers*, 22, pp F1–F8.

Sofos JN and Maga JA (1988) Composition and antimicrobial properties of liquid spice smoke. In Charalambous G (ed.) *Frontiers of Flavour: Proceedings of the 5th International Flavour Conference*, Porto Karras, Chalkidiki, Greece, pp 453–472.

Toth L and Potthast K (1984) Chemical aspects of smoking meat and meat products. *Advances in Food Research*, pp. 84–158. London: Academic Press.

John I. Ahmad
University of Humberside, Grimsby, UK

SMOKING, DIET AND HEALTH

Effect of Smoking on Dietary Intake

Nutrient intakes of smokers have been compared with those of non-smokers in several studies. Energy intake has generally been found to be similar or higher among smokers, 0·5–1·5 MJ (123–350 kcal, or 5–13%) per day, and only one study (among pregnant women) appears to have found a lower energy intake among smokers. Where a difference in energy intake has been reported, this seems to be due to a general effect on food consumption rather than to differences in intakes of specific nutrients. Sucrose consumption has been suggested to be positively associated with smoking habit, although the literature is inconsistent and it is not clear whether the difference is independent of energy intake. Alcohol consumption has consistently been demonstrated to be higher among smokers than among nonsmokers, the heaviest smokers tending to be the heaviest drinkers. In studies in which both alcohol consumption and energy intake have been estimated, the relationship between alcohol and smoking habit does appear to be independent of energy intake. ***See*** Alcohol, Alcohol Consumption; Sucrose, Dietary Importance

Both linoleic acid intake and the ratio (P/S) of polyunsaturated fatty acids (PUFAs) to saturated fatty acids (SFAs) have been reported to be lower among smokers than among nonsmokers, by 2–3 g per day (15–27%) and 0·04–0·09 (12–26%) respectively. A trend in linoleic acid intake and in the *cis-cis* linoleic acid content of adipose tissue has been observed with smoking habit, those who have never smoked having the highest level and heavy cigarette smokers having the lowest level. The reason for the lower linoleic acid intake in smokers is not entirely clear. It has been suggested that smoking may influence taste sensation, making PUFA unpalatable. However, the majority of studies comparing differences in sensory thresholds between smokers and nonsmokers have not found that cigarette smoking affects taste thresholds and a few have reported higher thresholds to the taste of bitter and sweet among smokers. Nevertheless, it is possible that smoking may influence taste preferences independently of effects on taste thresholds, but this has not been adequately investigated. Another possible explanation for the low linoleic acid intake and P/S ratio is that smokers may be less responsive to widely publicized dietary advice to reduce the intake of saturated fatty acids and partially replace them with PUFA. However, in a randomized controlled trial of the effect of dietary advice in men, post myocardial infarction (MI), advice to increase P/S ratio was just as effective among men who continued to smoke as among nonsmokers. Although when given advice, men who continued to smoke still had a lower P/S ratio than nonsmokers, the difference in P/S ratio between those given advice and those not given advice was similar among smokers and nonsmokers. ***See*** Fatty Acids, Properties; Fatty Acids, Dietary Importance; Sensory Evaluation, Taste

Dietary fibre consumption has been reported to be lower among smokers than among nonsmokers (by 3–6 g per day or 13–32%). Smokers have also been found to have lower intakes of several vitamins and minerals compared with nonsmokers. The most marked differences are for vitamin C (22% lower) and β-carotene (25% lower). ***See*** Ascorbic Acid, Physiology; Carotenoids, Physiology; Dietary Fibre, Effects of Fibre on Absorption; Dietary Fibre, Fibre and Disease Prevention

Interpretation of differences in intakes between smokers and non-smokers is complicated by the fact that smoking habit is associated with social class, there being more smokers among manual workers than among nonmanual workers. The effect of smoking habit on nutrient intake can therefore only be assessed after controlling for the effect of social class. If this is done it becomes apparent that smoking habit has an independent and greater effect on dietary fatty acid composition and on the intakes of vitamins and minerals than does social class.

Effect of Smoking on Bodyweight

It is well documented that smokers tend to be thinner than non-smokers, being on average 2–5 kg lighter. There are also a considerable number of studies reporting weight gain after the cessation of smoking, and indicating that the weights of former smokers are similar to those of subjects who have never smoked.

The lower bodyweight of smokers does not appear to be the result of a lower energy intake since, as outlined above, smokers have been found to have similar or higher energy intakes than non-smokers. Studies in experimental animals have also shown that cigarette smoke or nicotine induces a depression in bodyweight or growth with no, or very little, accompanying decrease in food consumption. An alternative explanation, for which there is some evidence, is that smoking may increase metabolic rate and/or lower the efficiency of energy storage. Smoking may result in less efficient storage of energy by changing the physiology of the gut, thereby disrupting absorption, or by biasing particular metabolic pathways away from storage. Nevertheless, if the effect of smoking on metabolism is related to smoking dose, heavy smokers would be expected to be the lightest, whereas several studies have shown that moderate smokers (15–24 cigarettes per day) are the lightest. It has been suggested that heavy smokers may inhale less than moderate smokers and thus achieve a lower smoking dose, although there does not appear to be any evidence available to support this. ***See*** Energy, Energy Expenditure and Energy Balance

Smokers who give up the habit frequently complain about putting on weight, and this could be due to a reduction in metabolic rate on stopping smoking and/or to an increase in food consumption. The effects of smoking cessation on food intake and metabolic rate can only be adequately assessed in longitudinal studies, but few have been conducted. Evidence from short-term longitudinal studies is inconsistent regarding the effect of smoking cessation on metabolic rate. However, interpretation of such studies is complicated by the fact that a decrease in metabolic rate following smoking cessation may be cancelled out by changes in other factors which increase metabolic rate, such as an increase in bodyweight or changes in hormonal activity. Energy intake does appear to increase following smoking cessation, at least in the short term, by about 0·84 MJ (200 kcal, or 8%) per day.

Effect of Smoking on Nutritional Requirements

There is evidence to suggest that smoking may increase nutritional requirements. Plasma and leucocyte ascorbic acid concentrations and plasma β-carotene levels, for example, have been reported to be lower among smokers than among nonsmokers, the differences being greater than the differences in intakes of these vitamins. This has been suggested to be indicative of an effect of smoking on plasma vitamin concentrations independent of diet.

Studies of vitamin C have shown that smokers have a higher metabolic turnover and a shorter half-life for ascorbate than nonsmokers. The amount of vitamin C required to reach steady-state concentrations and total body pools comparable to non-smokers was calculated to be 40% greater for smokers.

Effect of Smoking on Disease Risk

Cancer

It has been known since the 1950s that smoking is the major cause of lung cancer. Other cancers which have been found to be positively associated with smoking include those of the mouth, larynx, oesophagus, pancreas, urinary tract and cervix. On the other hand, smoking has been reported to be associated with a reduced risk of endometrial cancer, but this is the only cancer affected in this way. The risk of lung cancer appears to be less for pipe and cigar smokers than for cigarette smokers, but the risk of cancer of the mouth, larynx and oesophagus are similar for pipe, cigar and cigarette smokers. ***See*** Cancer, Epidemiology

In recent years lung cancer mortality within the UK has decreased among men but increased among women. Changes in the mortality rate relate to patterns of smoking several decades before as well as to current smoking habits. The total duration of smoking is important in influencing risk: lung cancer risk at age 60 is three times as great in those starting smoking at 15 than in those starting smoking at 25 years.

Retrospective and prospective studies show that risk of lung cancer is about 20% lower with low tar (or filter) cigarettes compared with higher tar (or plain) cigarettes. Data on secular trends in lung cancer mortality and cigarette consumption in England and Wales confirm this.

Cancer risks have been found to be inversely correlated with blood retinol (vitamin A) concentration and with dietary β-carotene intake (vitamin A precursor). Little is known about the determinants of blood retinol concentration, but in developed countries it does not appear to be clearly related to vitamin A intake. The β-carotene content of the diet has been reported to modify the risk of lung cancer among smokers, risk being greater among smokers with a low β-carotene intake than among smokers with a high β-carotene intake. The reason for the protective effect of β-carotene is not known, but may relate, not to its vitamin A activity, but to its ability to quench singlet oxygen and to trap certain organic free radicals. ***See*** Retinol, Physiology

There is increasing interest in the effects of environmental tobacco smoke (passive smoking) on health. The commonest method used to investigate the effects of passive smoking has been to examine lung cancer rates among nonsmoking wives of husbands who are smokers and compare these with rates among nonsmoking wives of husbands who are nonsmokers. None of these studies provides unequivocal evidence, but overall the findings are consistent with there being an increase in risk of lung cancer of 10–30%.

Coronary Heart Disease (CHD)

Diseases of the heart and blood vessels account for over one third of all excess deaths among smokers. Heart attacks are three times as frequent among middle-aged men smoking 15 or more cigarettes per day than among middle-aged nonsmoking men. The relationship between cigarette consumption and risk of CHD is independent of other factors such as raised serum cholesterol, hypertension or physical inactivity. The effect of pipe and cigar smoking on CHD risk is not clear, although they have been suggested to have a lesser excess CHD risk than cigarettes.

The type of cigarette (filter or plain) appears to have little effect on risk of CHD. The harmful constituents may therefore be in the gases which get through the filters. The tar and nicotine yields appear to have little effect on CHD risk. This is not surprising since nicotine and carbon monoxide yields have decreased less than tar yields, and compensatory smoking tends to lead to an increase in carbon monoxide and nicotine consumption. Risk of CHD is greater among those who inhale than among those who do not, but it is still not known which constituent is responsible for the CHD risk. Much interest has focused on carbon monoxide as this combines with haemoglobin to form carboxyhaemoglobin, thereby reducing the oxygen-carrying capacity of the blood.

Giving up smoking is of benefit in reducing risk of mortality, even after an MI. Subsequent mortality is twice as high among men who continue to smoke after an MI than among those who give up the habit. For those who give up, risk of death is reduced, although there is controversy about how quickly risk becomes similar to that of those who have never smoked. Some estimates suggest that this may take up to 10 years.

The mechanism of an effect of smoking on CHD risk may be to encourage thrombi by increasing levels of fibrinogen and other haemostatic factors. Smoking may also promote the formation of fatty plaques on the surface of blood vessels, increase narrowing of the coronary arteries, reduce the heart's oxygen supply and potentiate arrythmias. The effects of smoking on haemostasis appear to be prolonged, since ex-smokers have higher levels of fibrinogen than those who have never smoked, and levels are similar to those of 'never-smokers' only among ex-smokers who have not smoked for at least 10 years. ***See*** Atherosclerosis; Coronary Heart Disease, Intervention Studies

Other Diseases

There is evidence of close links between smoking and several other important diseases. Among these is chronic bronchitis and emphysema (chronic obstructive lung disease), death rates being six times greater among smokers than among nonsmokers.

There is an association between smoking and narrowing of blood vessels in the limbs. Of patients with serious arterial disease of the legs, 95% are regular smokers.

Smoking is associated with peptic ulcers in the stomach and duodenum. Men who smoke have twice as many ulcers than non-smokers; the ulcers heal poorly and are more likely to lead to fatal outcomes.

There is also evidence to suggest an increased risk of fractures of the hip, vertebrae, and distal radius among smokers. However, the strength of the association varies with fracture type, being stronger for vertebral than for hip fractures. This difference has been suggested to reflect the variation in the contribution made by osteoporosis to the incidence of these fractures. Bone density is lower in smokers than in nonsmokers, and this is partly explained by their lower bodyweight, since bone density is lower in thin individuals than in fat individuals. Influences of smoking on the synthesis or metabolism of various hormones, including oestrogens, androstenedione and cortisol may be additional factors but the evidence is limited.

Women who smoke during pregnancy have been found to give birth to smaller babies (on average 200 g or 6% lighter) than mothers who do not smoke, the risk of producing a small baby being twice that for nonsmoking women. Women who smoke also have a greater chance of losing their children in the period around birth, smokers having about a 28% increase in perinatal mortality. These effects persist after allowing for a number of possible confounding variables such as social class, maternal age, parity, maternal height and sex of the fetus. Ex-smokers and women who give up the habit before about the end of the fourth month of pregnancy have offspring whose birthweight is similar to that of women who have never smoked.

Children of parents who smoke are on average shorter than other children at primary school age, by up to 1 cm, after adjusting for other factors. At age 11, intellectual performance as revealed by reading comprehension and mathematical ability is behind that of children of nonsmoking mothers by a span of about 6 months.

Physiological Effects of Smoking

Tobacco smoke and its constituents have been found to have a wide range of physiological effects. With regard to lung function, smoking interferes with (1) the clearance of mucus from the airways, (2) the lungs' defence against infection, and (3) the balance of enzymes which maintain the structural integrity of the lungs.

Smokers have higher plasma fibrinogen concentrations, and higher white cell counts than nonsmokers, both of which are associated with an increased risk of CHD. The known effects of nicotine on the cardiovascular system include an increase in heart rate and blood pressure, a reduction in vascular prostacyclin production (a potent vasodilator and platelet antiaggregatory agent) and an increase in the stickiness of blood platelets. The effects on heart rate and blood pressure are produced by stimulation of the autonomic nerves which control heart rate and constriction in blood vessels and also by release of adrenaline from the adrenal glands.

Smoking has been suggested to have effects on the digestive system, increasing the mobility of the bowel (increasing food transit) and increasing gastric acid secretion. However, evidence on the latter point is conflicting, and some studies have failed to detect any change in volume, pH, free acid, pepsin or pentagastrin-stimulated acid secretion in response to smoking two or three cigarettes. Others have found an increase in basal acid output and gastric acidity and an inhibition of pentagastrin-stimulated gastric secretion after smoking.

Smoking adversely affects reproductive function. Women who smoke have been found to have lower levels of urinary oestrogens than nonsmokers and to be more likely to be infertile or take longer to conceive. In addition, smokers have consistently been found to have an earlier menopause than women who do not smoke. Expressed as a difference of median menopausal ages, current smokers have a menopause 1–2 years earlier than those who have never smoked. Possible mechanisms for the associations include direct toxic effects on the ovaries, interference in gonadotrophin release, and alteration in metabolism of sex steroids.

Smoking has effects on behaviour that are apparently conflicting, being capable of producing both relaxation and increased efficiency. Electrical activity of the brain is activated, but sometimes, such as when stressed, its activation may be reduced. In animals, nicotine can cause arousal, a decrease in aggressive behaviour and lessen the effects of stress.

Finally, clinical studies in smokers have shown decreased pharmacological effects of many drugs, suggesting that smoking may increase the rate of drug metabolism.

Bibliography

Department of Health and Social Security (1988) *Fourth Report of the Independent Scientific Committee on Smoking and Health*. London: Her Majesty's Stationery Office.

Fehily AM, Phillips KM and Yarnell JWG (1984) Diet, smoking, social class, and body mass index in the Caerphilly Heart Disease Study. *American Journal of Clinical Nutrition* 40: 827–833.

Fehily AM, Elwood PC and Yarnell JWG (1989) Cigarettes and heart disease. *Lancet* 2: 114–115.

Fehily AM, Vaughan-Williams E, Shiels K *et al.* (1991) Factors influencing compliance with dietary advice: the diet and reinfarction trial (DART). *Journal of Human Nutrition and Dietetics* 4: 33–42.

Fulton M, Thomson M, Elton RA, Wood DA and Oliver MF (1988) Cigarette smoking, social class and nutrient intake: relevance to coronary heart disease. *European Journal of Clinical Nutrition* 42: 797–803.

Haste FM, Brooke OG, Anderson HR *et al.* (1990) Nutrient intakes during pregnancy: observations on the influence of smoking and social class. *American Journal of Clinical Nutrition* 51: 29–36.

Kallner AB, Hartmann D and Hornig DH (1981) On the requirements of ascorbic acid in man: steady-state turnover and body pool in smokers. *American Journal of Clinical Nutrition* 34: 1347–1355.

Peto R, Doll R, Buckley JD and Sporn MB (1981) Can dietary beta-carotene materially reduce human cancer rates? *Nature* 290: 201–208.

Royal College of Physicians (1983) *Health or Smoking*. London: Pitman.

Stamford BA, Matter S, Fell RD and Papanek P (1986) Effects of smoking cessation on weight gain, metabolic rate, caloric consumption and blood lipids. *American Journal of Clinical Nutrition* 43: 486–494.

Wack JT and Rodin J (1982) Smoking and its effects on body weight and the systems of caloric regulation. *American Journal of Clinical Nutrition* 35: 366–380.

Wald N and Baron J (1990) *Smoking and Hormone-Related Disorders*. Oxford: Oxford Medical Publications.

Williamson DF, Madans J, Anda RF *et al.* (1991) Smoking cessation and severity of weight gain in a national cohort. *New England Journal of Medicine* 324: 739–745.

Yarnell JWG, Sweetnam PM, Rogers S *et al.* (1987) Some long term effects of smoking on the haemostatic system: a report from the Caerphilly and Speedwell collaborative surveys. *Journal of Clinical Pathology* 40: 909–913.

Yarnell JWG, Baker IA, Sweetnam PM *et al.* (1991) Fibrinogen, viscosity and white blood cell count are major risk factors for ischaemic heart disease. The Caerphilly and Speedwell collaborative heart disease studies. *Circulation* 83: 836–844.

Ann M. Fehily
Medical Research Council, Cardiff, UK

SNACK FOODS

Contents

Range on the Market

The term 'snack' or 'snack food' is difficult to define or categorize. The dictionary meaning of snack is a 'titbit' which is a small meal in the broadest sense. Snacking can be described as the problem-free consumption of easy-to-handle, miniature-portioned, hot or cold products in solid or liquid form, which need little or no preparation and are intended to satisfy the occasional 'pangs' of hunger. Thus snacks should be convenient and in manageable portions and they should satisfy short-term hunger. The different categories of snack foods are given in Table 1.

Snacks can be sweet or savoury, light or substantial, and they may be endowed with attributes, such as 'healthy or just for fun'. In European markets the word 'food' is omitted. This article deals with various aspects of different snack foods and the quality checks involved.

Snack foods are a significant part of the food industry. The dominant leader is still potato chips (crisps), followed by corn chips, nuts, meat snacks, pretzels, extruded snacks, and popcorn.

Potato-based Products

The potato-based snacks can be made either exclusively from fresh, sliced potatoes or from potato dough. The chips, in the latter case are called 'simulated potato chips'. The steps involved in potato chip processing include selection, procurement and receiving of potatoes, storage under optimum conditions, peeling and trimming of the tubers, slicing, frying in oil, salting or the application of flavoured powders and packaging. *See* Potatoes and Related Crops, Processing Potato Tubers

Potatoes must be of high specific gravity in order to obtain superior yield, low oil absorption and low reducing sugar content. Peeling should remove only a very thin layer of potato, leaving no eyes, blemishes or other material for later removal by hand trimming. The chips from peeled and unpeeled potatoes are similar in appearance, flavour and shelf life. The chips from unpeeled potatoes have higher oil contents but omission of the peeling step results in a 7% increase in potato solids and a reduction in waste.

The peeled potatoes are cut into slices 10–17 mm thick by rotary slicers. Slices must be very consistent in thickness in order to obtain uniformly coloured chips. Slices with torn surfaces lose excess solubles from the ruptured cells, and absorb large amounts of fat. Starch and other materials released from cut cells are washed away prior to frying, and the surface of the slices dried; this step shortens the frying time and increases the capacity of the cooking unit.

The rinsed and dried potato slices are conveyed directly to the fryer. The fried slices are removed from the tanks and drained on mesh belts. The temperature of oil in fryers is held within a range of about ±2°F. Temperatures normally used are 350–375°F (176–190°C) at the receiving end and 320–345°F (160–173°C) at the exit end of the fryer.

The oil used for deep-fat frying of potato chips has two functions: (1) it serves as a medium for transferring heat from a thermal source to the tuber slices, and (2) it becomes an ingredient of the finished product. The oil should be highly refined. The flavour and appearance of chips are affected both by the amount of oil absorbed and its characteristics as it exists in the chips. The amount of oil absorbed by potato slices is affected by (1) the solids content of the tuber, (2) the temperature of the fat, (3) the duration of frying, and (4) the thickness of slices. Fat pick-up by potato chips is 10–15% lower when the chips are fried in liquid oil as compared to fat which is solid at room temperature. Reducing slice thickness increases the oil content of chips. A hot-air blast at the point where the slices emerge from the fat reduces fat absorption, and partial drying of sliced potatoes before frying also reduces the oil content of chips. Leaching raw slices with hot water results in an increased uptake of oil. The treatment of slices with glucose oxidase produces a chip of higher uniform colour and lower oil absorption, without off-flavours.

Table 1. Snack food categories and examples of products in each category

Category	Product
Hot snacks	Minipizzas, pizza baguettes, etc.
	Toasts au gratin
	Cup noodles
	Spring rolls
	Filled croissants
Cold snacks	
Milk and dairy products	Yoghurt, plain or with fruit
	Mini cheese-cubes
Bakery products	Cake bars
	Minitarts
	Cookies
	Biscuits
Bars	Granola or muesli bars
	Chocolate bars
	Minibreak bars
	Energy bars
Savoury products	Chips (crisps), sticks
	Extruded products
	Crackers
	Pretzels, salt sticks
	Nuts, nut mixtures
Other products	Popcorn
	Rice snacks
	Fruit sticks or rolls
	Dip sticks

From Tettweiler P (1991).

Soda Crackers

Soda crackers and variants, such as saltines, oyster crackers, etc., are themselves used as snacks, and also serve as a basis for combination snack foods either made in the home (crackers and cheese) or by manufacturers (cheese crackers, peanut butter and cracker sandwiches, etc.). The soda cracker is made from a lean, fermented dough. It does not contain much shortening, sugar or milk. *See* Biscuits, Cookies and Crackers, Nature of the Products; Biscuits, Cookies and Crackers, Methods of Manufacture; Biscuits, Cookies and Crackers, Wafers

Flour constitutes 80% or more of the finished crackers. Its qualities are important in determining the machining qualities of the dough, as well as the texture of the end product. Weak flours and lengthy fermentation times combine to yield flat, tender crackers. Flour for thick saltines should be stronger than that for thin

Range on the Market

crackers. Thin saltines require a weaker flour; salt may be added at a rate of 2–5% to the dough.

Sprayed crackers are the rich crackers, round in shape, and usually made from chemically leavened dough. Many of the representatives of this class could be considered a cross between crackers and cookies because they are not only leavened but also sweeter than saltines. The leavening system for these is adjusted so that the pH of the finished products is below neutrality. A pH of 6·5 is regarded as desirable. Sprayed crackers are coated with 20–25% (w/w) of a bland shortening. Coconut oil is preferred, but peanut oil or even hydrogenated vegetable shortenings can be used. The oil is applied either by spraying at 65–71°C, or by a device similar to an enrober, in which the crackers pass through a flowing curtain of oil. The oil must be applied when the crackers are still hot. The crackers may be salted very lightly on the cutting machine.

Cheese Crackers

Cheese crackers are usually made from fermented dough, and they must be on the acid side of neutrality to yield a typical cheese flavour. The formulation is based on the soda cracker, except that the fat and moisture added in the fresh cheese have to be compensated for. Paprika (0·25%) and a very small amount of cayenne are added to Cheddar cheese crackers to intensify the colour and to add a flavour which is complementary to cheese. Caraway and poppy seeds are added to blue cheese crackers to impart certain desirable characters. Natural cheese should be finely ground before being placed in the mixer. Sometimes a premix of ground cheese and shortening, kept previously in the fermentation room for 24 h, is added at the doughing stage. This holding stage encourages the development of more flavour, and leads to a better dispersion of the cheese in the dough. Artificial Cheddar and blue cheese flavours are also used whenever required.

Pretzels

Pretzels are considered to be the world's oldest snack food, and may have originated around AD 610 on the borders of France and Italy. The pretzels then became a popular snack food in Germany and Austria, where they were called 'bretzels', later to be altered to today's pronunciation, 'pretzel'. Originally, all pretzels were soft, like unleavened bread. Accidently, some pretzels were left in a cooling oven and the remaining heat dried them, removing the moisture and giving them a hard, crisp texture and a golden coating. This led to the introduction of hard pretzels.

Soft pretzels today are baked, not dried, and have a texture resembling that of a fresh, soft roll, as well as a high moisture content. Soft pretzels are popular as a snack purchased at street vendors' stands. Hard pretzels have an extremely long shelf life, especially when stored in airtight containers. Hard pretzels can be thin or thick, and shaped as twists, sticks, nuggets, logs or rings; they can be buttered, whole wheat, sesame-coated, chocolate-dipped, and 'teething'. It is the drying which gives the hard pretzels their distinctive crunch, and this crunchiness is fundamental to the appeal of hard pretzels.

Pretzel doughs are made very stiff so that they can withstand the punishment of machining without becoming sticky or misshapen. The sponge is fermented for a shorter time than with cracker dough. Doughs receive a short proof stage, but machinery steps, including formation of the pretzel, are handled automatically in all but a very few small plants. The characteristic gloss of the pretzel is the result of a lye dip. The dip solution contains about 0·5% sodium hydroxide or 2% sodium carbonate and is maintained at about 100°C; the immersion time is about 10 s. The solution may also be applied by spraying. Immediately after the pretzels leave the caustic solution, they are salted. The general aim is to achieve 2% salt in the finished product, but it is necessary to keep the initial application rate at 8–10% to allow for losses in processing. A long drying time is required to reduce the moisture content to 2–2·5%, and temper the pretzel so that it does not break too easily during packaging. The long pretzels can be twisted or stick type, and stick pretzels are extruded using a group of five extruding heads having 10–12 holes each. Logs and nugget-type pretzels are made in a manner similar to that used for the sticks, except that they are cut into short lengths at the extruder head. The stick pretzels can be made from almost any flour, but the choice of flour used in twisted pretzels is critical.

Filled pretzel sticks or nuggets are also available. These are produced by drilling a hole in the completely baked pretzel stick, and then inserting a paste-like filling of peanut butter and cheese, along with oil and sugar or a nonsweet agent such as lactose or dextrin.

Confectionery (Sugar-rich Products)

Plain cookies are the cookies made in one operation, and this group does not include filled, coated, sandwiched and other multiple-component cookies. The continuous structure of a cookie arises from the flour, and this basic framework is tenderized with sugar, invert sugar, egg yolk, ammonia, soda (or baking powder) and shortening. It is firmed or toughened with water, cocoa, egg white, whole egg, milk solids and the leavening acids. Shortening is one of the principal agents for increasing tenderness, at least as far as the rich, sweet biscuits are concerned. Too much shortening leads to a greasy, smeary cookie which is susceptible to rancidity because of the free fat, while too much sugar results in hard and excessively sweet cookies.

A wide variety of flours can be used, varying from a soft cookie flour to a rather strong sponge flour. The stronger the flour, the more shortening and sugar are required to obtain an acceptable texture. High protein contents lead to hardness of texture and coarseness of internal 'grain' and surface appearance. Chlorine-bleached flours should not be used for soft cookies where relatively large amounts of tenderizing and moisture-retaining ingredients, such as sugars, shortening and egg yolk, are used. Cookies may become to fragile if the quantity of flour is decreased too much.

Deposit Cookies

Deposit cookies are machine-made counterparts of the 'hand-bagged' cookie. They contain about 35–40% sugar, 65–70% shortening and 15–25% liquid whole egg. The flour should be from soft wheat, unbleached, with 8–8·5% protein and 0·35–0·40% ash. It should have a viscosity of 40°M and a spread factor of 79–80. The flour should be able to carry sugar and shortening without too much spread, so that the top design is preserved through baking. The flour or other ingredients must contribute enough adhesive properties to the dough so that it will adhere to the band and pull away from main column of dough in the deposit stage.

Macaroons are made by a cold, or hot syrup or cooking process, but they should not be made by the cold process unless they are to be consumed within 3 days.

Wire-cut Cookies

The dough composition varies over a wider range for wire-cut cookies than for any other type. Thus the dough material should be sufficiently cohesive to hold together as it is extruded through an orifice, and yet be nonstick and short enough so that it separates cleanly as it is cut by wire. These cookies can be further subdivided as follows: (1) drop-type, which are used in sandwich cookies filled with a marshmallow or imitation cream, and usually contain equal amounts of sugar and flour; (2) sugar cookies, molasses cookies, coconut, raisin, date and honey varieties; (3) shortbreads, in which the shortening content is usually 50–75% of the flour content; and (4) macaroons, with little flour and large proportion of sugar.

Nut-based Snacks

Nuts are often used to upgrade popcorn-based snacks, and are also themselves sold for snacks either as the individual variety or as mixed nuts.

Range on the Market

Peanuts

Peanuts are sorted, blanched, roasted and salted. Blanching for peanuts and other nuts means removal of the skin or testa. During roasting, the kernels become dehydrated, and browning reactions occur throughout the kernel. The texture, flavour and appearance change markedly. Roasting can be continuous or batch-type, and dry or in oil. The unique 'nutty' flavour of roasted peanuts results from browning reactions of the Maillard type. The type of roasting affects the course of browning reactions because of variations in the rate of heat penetration and exposure to high temperature in different parts of the nut. When nuts are fried in hydrogenated fat, the salt is applied before they are cooled to room temperature and while the fat on the surface is still molten. As the fat cools and sets up, the salt granules are held firmly. Adhesion of salt to dry-roasted nuts is achieved by adding about 2% (w/w) of coconut oil as a dressing to the warm nuts, and immediately applying the salt. The salt must have small particles of irregular configuration to obtain maximum adherence. ***See*** Peanuts

Pecans

Toasted and salted pecans are used directly as snacks. Blanching is never required in pecans. Pecans can be either dry-toasted or cooked in oil. Much less heat treatment is required than for peanuts, because the initial moisture content of pecans is lower, and the development of a true roasted flavour is seldom desired. Frying gives a more uniform roast and causes less mechanical damage. Roasting is performed, on average, for 12–15 min at 190°C. After roasting, the kernels are quickly cooled in tunnels or perforated bins with forced-air circulation to remove any undesirable odours. The cooled product is cleaned to remove any debris, foreign matter and scorched nuts. The roasted kernels are coated with about 1% (w/w) finely ground salt. Pecans may be coated with cinnamon, sugar, barbecue-flavoured powders, etc. Since toasted pecans have a very limited storage life, they are vacuum packed.

Almonds

Almonds are a common constituent of mixed nut packs, and are popular as the roasted and salted variety. The kernels are not generally blanched for the production of salted almonds. For blanching, the nuts are soaked in hot water until the skin loosens sufficiently to be removed by rubbing. Almonds may be dry-roasted or fried, but oil roasting is the more common method. In dry roasting, hot air is passed through rotating perforated cylinders or a fluidized bed to reduce the moisture content to 1%. Oil roasting is carried out by immersing the nuts in oil, preferably coconut oil, and heating to between 121°C and 176°C. After frying, they are cooled, salted and dressed with oil. They can also be coated with smoke-flavoured salt, cinnamon, sugar or similar flavoured powders. ***See*** Almonds

Sugared and Spiced Nuts

Sugared and spiced nuts are prepared by adding nuts and spices to water and sugar heated to 115°C in a revolving pan. After coating, the nuts are dried for 1 h at 48°C, or 2 h at room temperature, and packed in hermetically sealed containers, preferably under vacuum. The nuts must be dried to below 4% moisture before coating. This product should be refrigerated if it is not to be consumed within a month.

Corn Snacks

Corn-based snacks fall into three groups: extruded from corn meal, baked and fried from 'masa', and corn chips or simulated potato chips.

The extruded snacks are made of corn meal with, occasionally, a small amount of another grain, such as rice or wheat, added. The moistened meal is passed through extrusion cooking equipment, giving a 'collet', which is an expanded matrix of gelatinized starch; the dimensions are determined by the kind of die used, the cut-off knife speed and other mechanical factors. The collet has a moisture content of 8–10% and is quite tough and chewy, an aspect which is remedied by baking or deep-frying to obtain a final moisture level of around 2%. The dried collets are flavoured with cheese (applied in a powdered form), chilli, barbecue sauce, sour cream and onion, or a number of other flavours. Other extruded snacks consist of hollow extrudates which are modified to make rectangles, ovals or ornamental shapes, and filled with all kinds of savoury or sweet fillings. These snacks are quite popular, and the most popular flavours for extruded snacks are chocolate, butterscotch and peanut butter. Delicate fruit creams, such as wild strawberry, lemon-lime, bitter orange, sour cherry, kiwi, peach, and cola, are also being used to fill extrudates.

Baked and fried snacks make up the largest share of the corn-based snacks market. They are produced using centuries-old processes that have been modified by the adoption of modern technology. Cleaned corn is placed in steam kettles and covered with a dilute solution of lime ($Ca(OH)_2$). The mass is brought to near-boiling temperature, and held for a period of time depending upon the nature of the corn used. It is then cooled and allowed to steep for 8–16 h. During steeping, water penetrates the kernel and gives the desired degree of friability to the endosperm. The cooking and steeping also loosens the outer hull and germ, which are removed from the corn by washing. The washed kernels are ground to produce 'masa', a cohesive, plastic corn meal with a moisture content of approximately 50–60%. Masa is the raw material for corn and tortilla chips. To make corn chips, masa is extruded through a slit to give a ribbon of dough. Cut off by a knife, the raw chips fall into hot fat for deep-frying. They are transported through the frying tank by a travelling belt in 2–5 min. During this time, the starch is gelatinized and the moisture level is reduced to approximately 2% to give the crisp bite of the finished chip. After the travelling belt brings the chips out of the bath, they are given a light dusting of salt and flavouring powder, if required, and then packed.

Tortilla chips are produced more or less as in a traditional process. The masa is rolled out into a thin sheet, cut into the desired shape, and transported into an open-flame oven. The heat sets the structure, drives out most of the moisture, and changes the flavour slightly by creating small amounts of Maillard reaction products. The chips are held in a tempering chamber to equilibrate moisture and stresses throughout the chip. Finally, they are deep-fried to create a crisp texture, salted or seasoned, and packaged. ***See*** Tortillas

Corn Chips and Simulated Potato Chips

Commercial simulated potato chips have assumed great importance. The corn chips and tortillas prepared from alkali-treated corn dough have similar flavours and textures. The ground, alkalized corn dough, or masa (Table 2) is conveyed to the hopper of a tortilla cutting head by means of a screw-type extruder. A thick sheet of masa is extruded between sizing rolls, and the resulting ribbon is cut into dices of appropriate size. Tortilla blanks are baked in gas-fired ovens at 315°C for 30–32 s. Tortillas discharged at 79°C are cooled to 29–32°C in an atmospheric, multitiered conveyer.

Corn chips are prepared from a mixture of white and yellow corn, varying in ratio depending upon the colour desired in the final product. The alkalized dough is cut into pieces of desired length and shape, and transferred directly into cooking oil held at about 190°C. After the moisture content has been reduced to a few per cent, the chips are salted, cooled and packaged.

Cereal grains other than corn can also be formed into chips and fried; the alkalizing process is omitted in these cases. Shredded wheat, formed into multiple layers, salted and baked, forms the basis of Triscuit.

The fabricated potato snacks can be classified into four groups: (1) dry collet processes, (2) single-screw

Range on the Market

Table 2. Parameters for the major steps in the manufacture of an alkaline-processed corn snack

Operation	Tortilla chips	Corn chips
Lime content	1% of corn weight	1% of corn weight
Cooking	4–10 min, >93°C Moisture: 29%	14–45 min, >93°C Moisture: 42–44%
Quenching	Quickly, <66°C	Quickly, <66°C
Steeping	8–24 h, 46–60°C Moisture: >45%	8–24 h, 46–60°C Moisture: 50–52%
Washing	Wash temperature: <27°C	Wash thoroughly and drain: <27°C
Grinding	Stone, masa at <38°C Water added, moisture 52–54%	Stone, masa at 46–54°C Moisture: 50–52%
Masa sheeting and forming	Promptly, 21–27°C	Masa is extruded or sheeted and cut directly into frying oil
Baking	12–18 s, 399–454°C Moisture: 40%	
Cooling and equilibrating	Cool to 60°C within 15–30 s Moisture: 38% after equilibration	
Frying	2 min at 179°C	1·5 min at 210°C
Chip moisture	1·0–1·2%	1·0–1·5%
Oil content	25%	35%

Data from Gomez MH *et al.* (1987).

extrusion of dry potatoes, (3) forming and frying of moist dough, and (4) forming a high-solids dough into a thin sheet which is then cut into pieces and fried. The latter process is used in the manufacture of simulated potato chips.

Simulated potato chips may be formed by methods resembling those used in the cookie and cracker industry. It is necessary to formulate the dough with physical properties that make it suitable for rolling out into a thin sheet. This is achieved by the inclusion of substantial amounts of cereal flours. Another approach is to use ungelatinized starch in the dough, and then steam-treat the pieces after all the forming operations have been completed. After sheeting, the dough is cut into pieces of the desired shape and size, and then fried or baked. Simulated potato chips can also be prepared from reconstituted dehydrated potatoes.

Meat-based Snacks

There are a few snacks that are composed primarily of animal-derived raw materials. Expanded pork rinds, also called broken skins or 'skeen', have been popular as a between-meal snack in the southern USA for many years. These are pieces of hog skin which have been processed so that they puff to many times their original volume. The flavour is bland and reflects the oil in which the skin has been processed. The texture is very crisp and friable; they are not as hygroscopic as many other puffed snacks.

The raw materials are green belly skins, green fat hog skins and green ham skins from any type of hog; belly skins give the best finished product. The skins are first dipped into an air-agitated brine solution held at 100°C for 30 s. The brine may contain 12 kg of dextrose, 11 kg of sucrose and 68 kg of salt in 909 l of water. After dipping, the skins are drained and cooled at room temperature. Then the skin is diced into 127- or 254-mm squares, ready for further processing.

Jerky

The name 'jerky' is derived from a dried meat product popular in the USA. Many countries have their own form of dried meat product. The product name varies, as does the form in which it is eaten. The Americans call it 'jerky' and eat it as a leathery whole tissue, or comminuted into strips. The American Indians call it 'pemmican'; they grind the diced meat and mix it with fat before eating it. In South Africa it is called 'biltong'. The French version is called 'grisons'; before being eaten, the dried meat is smothered with hot cheese.

Meat jerky was prepared originally by the American Indians, who salted strips of muscle tissue from deer, buffaloes and other game animals and cured the strips in the sun or over smoky fires for long periods of time. Commercially, beef jerky is made by marinating strips of beef, drying the marinated meat and finally cutting the dried strips to the desired dimensions. Beef jerky can also be prepared by subjecting a mixture of ground and chunk beef to a saline treatment for 12 h or more and then to freezing temperatures for 1–3 weeks. The meat preparation is then sliced into strips and dried.

Range on the Market

Other Snack Foods

The future of traditional snack foods depends upon the development of new forms and flavours. Pretzels and salt sticks, crackers and nut and nut mixtures have been modified, and the markets for these products have increased.

Soft pretzels available in fresh, frozen, and 'microwavable' forms have resulted in growth of the pretzel market. In most countries, the market for pretzels and salt sticks has increased with the introduction of health-oriented products that are salt-free, cholesterol-free, and possess increased fibre content. Other new pretzels and salt stick products are produced in various shapes and sizes. The new products are also available in such varieties as savoury-flavoured, sweet-flavoured, chocolate-coated, brightly coloured and peanut-butter-filled.

Along with the salted peanuts and nut mixtures that continue to be popular all over the world, the current market for nut-based snacks is influenced by (1) an increased demand for flavoured nuts, (2) the growing popularity of high-quality exotic nut and dried fruit mixtures, and (3) the increasing success of premium nuts such as cashew, pistachio and macadamia nuts. The development of flavoured nuts and nut mixtures has resulted in a virtually limitless list of new products. The list includes products such as hickory-smoked and honey-roasted nuts, nut mixtures containing rice, lentils, coconut and vegetables, and nuts of all types flavoured with garlic, soya sauce, herbs, ham, barbecue and many newly developed flavours. ***See*** Cashew Nuts and Cashew Apples

Quality Control for Snack Foods

The extent of quality control analysis in the snack food industry is quite variable, ranging from completely subjective testing procedures to series of elaborate, instrumental analyses performed on a continuous basis. The maintenance of uniform quality in snack foods is essential for the very existence of the business. Quality control expresses the objectives of a group of people who do the sampling, testing and other activities related to the evaluation of product conformance to predetermined standards. It is the expanded responsibility of the group to correct problems which could result in defective product. ***See*** Quality Assurance and Quality Control

Quality control aims at maintenance of quality at levels and tolerances acceptable to the buyer, while minimizing costs to the vendor. It can also be defined as the application of sensory, physical and chemical tests in industrial production to prevent undue variation in quality attributes, such as colour, viscosity and texture. In addition to satisfying the consumer, the product must meet the requirements of government regulatory agencies. A snack food which does not meet the minimum acceptability requirements of all customers may result only in a few complaints, but a product which does not meet all of the legal requirements may be confiscated and the manufacturers subjected to punitive fines and other penalties. A product may look good, taste good, have good texture and perform well in its intended application, and still be unfit for distribution. To be suitable for sale it must also be wholesome, conform to all applicable labelling and packaging requirements, and be prepared and stored under conditions tending to prevent contamination by noxious or aesthetically undesirable materials.

Quality of finished product is inseparable from quality of components. Seldom can processing operations compensate for raw material inadequacies. Unless the raw materials are of good quality, it is very difficult to obtain a good-quality finished product. Standards must be developed and strictly adhered to while processing the various ingredients, during the production and processing steps, as well as during handling, packaging and storage of the finished snacks.

During pretzel production, a number of problems, particularly in the cooking process, may arise; these include brittleness, dark or light colour, wrinkling, loss of glaze, breakage, browned interior, and lack of crispness.

In the production of corn-based baked and fried products, the consistency of the masa is very important. If excess starch granules are gelatinized during cooking and steeping, the masa is too sticky to process smoothly. On the other hand, inadequate cooking may give a dry, crumbly mass, which does not work in the equipment. Potato chips undergo a visual inspection at the end of the process, and excessively dark ones or those containing discoloured portions or possessing some other defect are manually removed by pickers. In a modern industry, this may be performed by computerized optical scanners. The sensitivity of the computer scanner can be adjusted to pick up all less-than-perfect chips, or to selectively remove only the least desirable. Blistering of chips on frying is quite a problem. Such chips break in conveying, filling or handling the filled bags. Another factor affecting quality of chips is the shift to thicker chips: it results in a significant reduction in oil absorption owing to a decreased surface area relative to the volume. Ridged chips solve the blistering problem.

Bibliography

Booth RG (ed.) (1990) *Snack Food.* New York: Van Nostrand Reinhold.

Crocombe BI (1988) The development and production of Jerky at Mirinz's meat products line. *Bulletin of Meat Industry Research Institute of New Zealand* 855: 126–129.

Gomez MH, Rooney LW and Waniska RD (1987) Dry corn masa flours for tortilla and snack food production. *Cereal Foods World* 32(5): 372.
Gould WA (1985) Changes and trends in the snack food industry. *Cereal Foods World* 30(3): 219.
Madonna MD (1983) Pretzels as low calorie snacks. *Cereal Foods World* 28(5): 297.
Matz SA (1976) *Snack Food Technology*. Westport, Connecticut: AVI Publishing.
Slige Mr (1983) Fruit and nut ingredients. *Cereal Foods World* 28(5): 286.
Staffer CE (1983) Corn-based snacks. *Cereal Foods World* 28(5): 301.
Tettweiler P (1991) Snack foods worldwide. *Food Technology*. 45(2): 58.

A. S. Bawa
Punjab Agricultural University, Ludhiana, India

Dietary Importance

In general, snack foods are considered as 'junk foods' or 'empty calories'. This article deals with the nutritional value of different snack foods, extent and frequency of consumption by various people and the need for fortification. Our diet is generally based upon three principal meals a day, and the role of the snack foods is to offer a light, convenient and enjoyable food option when, for reasons of hunger or sociability, a food selection is required at a time between that allotted for principal meals. Thus snack foods are not designed to be alternatives for three main meals a day. They are convenient to eat, involving no preparation time or effort by the consumer. Snacks are eaten not only to satisfy hunger or supply the nutrients to our body, but also for social reasons. On many occasions, the inclusion of snack foods in our diet is not justified in terms of nutritional demands.

Snack items and processed foods are presumed to have a high sodium content, and this should be avoided. Teenage boys are most likely to have a low intake level of calcium, iron, magnesium and vitamin B_6, and teenage girls consume these nutrients at still lower levels. The most frequently consumed food items between the meals by both sexes are cakes, cookies, pies, candy and other desserts. The most popular foods are salted snack foods, such as chips (crisps), pretzels and popcorn. Males incorporate more breads, waffles and other bread products in their snacks than females. On the other hand, females eat more snacks containing condiments, fat spreads and dips than do males. ***See*** Calcium, Physiology; Iron, Physiology; Magnesium; Vitamin B_6, Properties and Determination; Sodium, Physiology

Many people think of snack foods as junk foods, but the real nutritional concerns centre on the sodium, energy, vitamin, mineral and dietary-fibre contents of snack foods.

Nutritional Value of Different Types of Snack Food

Potato Chips (Crisps)

Nutritionally, 28 g (1 oz) of potato chips provides 647 kJ (154 kcal), contributed by 2 g of protein, 14 g of carbohydrates and 110 g of fat. A portion of potato chips (25 g) contains 8% of the Recommended Daily Allowance (RDA) of vitamin C, 3% of thiamin, 11% of nicotinic acid, 8% of iron, 11% of vitamin B_6, 4% of phosphorus and 4% of magnesium. Potatoes are also an important source of dietary fibre. ***See*** individual nutrients

Cookies and Crackers

The principal ingredient in cookies and crackers is unbleached, soft wheat flour, and most of the nutrients in these products are derived from it. It can range from more than 90% for saltines and similar products to less than 25% in products such as enrobed wafers. The soft wheat flour used in cookies and crackers contains 8–10% protein, with the higher-protein flours generally used for crackers, and the lower-protein flours for more tender sweet goods. With certain exceptions, the quality does not deteriorate on baking. Most cookies and crackers are made from enriched flour. The main component present in flour is carbohydrate, which is a good source of energy. ***See*** Flour, Dietary Importance

Other ingredients added to cookies and crackers influence the nutrient content of products either by their own nutritional composition or by diluting the nutrients contributed by enriched flour: for example, sugar contributes energy and carbohydrates; shortening contributes fat, essential fatty acids and energy; salt contributes sodium; leavening agents contribute sodium, potassium, calcium and phosphorus; peanut contributes fat and nicotinic acid; fruits and vegetables contribute vitamins A and C, fibre and minerals; chocolate contributes iron and fibre; other grains contribute minerals, fibre and other nutrients similar to wheat flour. The potassium content of cookies and crackers is 25–200 mg and 30–80 mg per 28 g (1 oz) respectively. During baking and storage, the main nutrients lost are thiamin and vitamin C. Lysine reacts with reducing sugars and becomes unavailable, especially at high pH. The plain and sandwich cookies have similar nutrients. However, nut-containing cookies provide comparatively more protein, vitamins (such as nicotinic acid) and minerals. ***See*** individual nutrients

Pretzels

The average energy content of a 28-g serving of pretzels is 462 kJ. One pretzel rod contains 231 kJ; one Dutch-style pretzel contains 260 kJ, and a handful of twisted, three-ring pretzels contains approximately 419 kJ. Pretzels are low in energy and fats and have no sugar, but contain a large amount of sodium. The average amount is 450–700 mg of sodium per 28-g serving of pretzels. However, low-salt pretzels, containing 300 mg per 28-g serving, and unsalted pretzels are also available for sodium-conscious consumers.

The typical ingredients of pretzels are flour, water, yeast, soda and salt. The flour is from soft wheat with a protein content of 8·50–9·50%, and an ash content of 0·40–0·50%. A 28-g serving of pretzel offers approximately 3 g of protein, 23 g of carbohydrates and 1 g of fat. Pretzels also provide several vitamins and minerals, including thiamin, riboflavin, nicotinic acid, iron, magnesium and copper. Although the quantities of these nutrients are moderate, the nutritional value of pretzels is significant since they contain no 'empty calories', no excessive fat and no chemical additives or preservatives. On the other hand, some pretzels also offer calcium, phosphorus, and other nutrients in significant amounts. *See* individual nutrients

Table 1. Sodium content of some foods

Food	Sodium content (mg per serving)
Most fresh fruits and vegetables	Neg.
Egg (1 medium)	54
Rump steak (130 g)	121
Fried chicken leg	122
Whole milk (1 cup)	128
Salted peanuts (25 g)	197
Toasted CCS (25-g packet)	220
Salted chips (25-g packet)	234
Flavoured CCS (25-g packet)	280
Flavoured chips (25-g packet)	306
Bread-and-butter sandwich	367
Cheese Twisties (25-g packet)	440
Scrambled egg	534
Fritz or Devon (2 slices)	650
Hot dog	1102
Hamburger	1591

Neg. negligible.
CCS, chips crisps sticks.
Data from Delroy B (1985).

Corn Snacks

Corn snacks are accused of being junk food and without nutrient content. Flour tortillas are low in moisture and high in fat, protein and carbohydrates, compared to corn tortillas. The energy content is, on average, 25% more than corn tortillas. The phosphorus and calcium contents in corn tortillas are substantive. The iron content of corn tortillas is similar to that of wheat flour tortillas. Both the flour and corn tortillas are significant sources of zinc; corn tortillas are also a significant dietary source of copper and manganese. Tortillas contain modest amounts of sodium and potassium, but because of the significant losses during processing, they are an insignificant source of thiamin and riboflavin. All tortilla products are substantial sources of nicotinic acid. Tortillas do not contain measurable quantities of vitamin C or carotene. *See* individual nutrients

Overall, corn snacks can contribute positively to the diet by providing certain nutrients and minerals necessary for nutritional balance, and they possess a good flavour and a pleasing texture.

In potato chips and nuts, the salt is manifested on the surface of the foods; hence the immediate taste perception is of saltiness – much more so than if the salt was uniformly distributed, as in most other foods. Many snack foods contain moderate to high amounts of salt on a percentage basis. The average serving of snack foods is much less than 100 g – it ranges up to 40 g and averages about 35 g; in comparison, many sodium-containing, staple foods are eaten in quantities greater than 100 g. It is more sensible to examine the sodium content of snack foods on a per-serving basis (Table 1) because the sodium content of snack foods is much lower than that of certain commonly consumed foods, such as bread and butter, sandwiches and other foods at the bottom of table. The sodium content of a packet of chips, for example, is less than a plain sandwich, and over six packets of salted chips must be consumed to obtain the amount of sodium contained in a single hamburger.

The second aspect of concern is the energy content of snack foods. It is generally believed that snack foods are both rich in energy and fattening since they are cooked in or blended with oil. A comparison of energy contents of snack foods on a per-serving basis (Table 2) shows that the consumption of lightly cooked vegetables instead of snacks can reduce energy intake, and that the energy content of snacks is not as high as is popularly perceived, since hot dogs and hamburgers have much higher energy contents. A packet of chips does not contain much more energy than a cup of milk or a plain yoghurt, and somewhat less than a bread-and-butter sandwich. It is much more difficult to control energy and fat content of snack foods compared to sodium content. The technology for the production of fat-reduced potato chips does not exist, but the level of fat can be controlled by careful regulation of the cooking parameters. Oil is essential in cereal snacks to give an acceptable mouth-feel, and with nuts the oil is already present before cooking. The only way to control the energy consumption via snack foods is to reduce the size of serving; Table 3 gives the effect of serving size on energy content.

Dietary Importance

Table 2. Energy content of some foods

Food	Energy content (kJ) per serving[a]
Lettuce	17
Peas	163
Boiled egg (1 medium)	301
Baked fish (1 fillet)	326
Salted peanuts (25 g)	584
Whole milk (1 cup)	644
Plain yoghurt (1 cup)	690
Potato chips (25-g packet)	739
Twisties (25-g packet)	810
CCS (25-g packet)	814
Bread-and-butter sandwich	865
Hamburger patty	932
Flavoured and fruit yoghurt (1 cup)	982
Fish, buttered and fried (1 fillet)	1041
Hot dog	1342
Grilled steak (130 g)	1822

[a] 4·2 kJ is equivalent to 1 kcal.
CCS, chips crisps sticks.
Data from Delroy B (1985).

Table 3. Effect of serving size on energy content

Food	Energy (kJ) per serving	Fat (g)	Sodium (mg)
Small apple	222	0·3	Neg.
Twisties (15-g packet)	460	5·9	250
Salted chips (20-g packet)	565	8·8	179
Milk (1 cup)	644	8·7	128
Bread-and-butter sandwich	865	8·6	267

Neg., negligible.
Data from Delroy B (1985).

The vitamin and mineral contents of snack foods is another area of concern; Table 4 lists the vitamin and mineral contents of some selected snack foods. The potassium content of chips is higher than sodium in spite of the added salt, and this is significant in view of the criticism that high sodium levels in processed foods are associated with depleted potassium levels. However, since snack foods are designed to be occasional between-meal snacks, their contribution to the diet may not be of much significance to the average person. Cereal snacks can be fortified with added vitamins and minerals to any desired level, but it is not common practice since snack foods are not considered to be suppliers of significant quantities of minerals and vitamins. ***See*** Food Fortification

The importance of dietary fibre in snack foods is still not very clear, partly because of the problem of analytical methods; Table 5 gives the dietary fibre content of snack foods in comparison to a bread sandwich. The figures indicate that the four snack foods contain more dietary fibre than a white-bread sandwich, and CCS (chips crisps sticks) more than a brown-bread sandwich. Since bread is one of our best sources of dietary fibre, and the cereals and root vegetables from which snacks are developed are sources of the highest-quality fibre, snack foods cannot be rejected on the basis of their fibre content. Snack foods are neither an important source of micronutrients, nor simply 'empty calories'. Snack foods prepared from whole cereal grains can be a good source of dietary fibre, and the addition of a few per cent of bran in place of the purified cereal fractions improves the fibre content of the finished products. Dairy and meat products have negligible fibre contents, and potatoes are rather low in this component. ***See*** Dietary Fibre, Physiological Effects; Dietary Fibre, Effects of Fibre on Absorption; Dietary Fibre, Fibre and Disease Prevention

Levels of Snack Foods Eaten

Children and teenagers have at least one snack during the day. Adults tend to snack less frequently, and the percentage of adults who 'snack' decreases with age. Snacks provide at least 20% of energy for those who eat at least one snack a day and they account for 12% of the average protein intake, 16% of the fat intake and 25% of the daily carbohydrate intake. 'Snackers' obtain, on average, 14% of their daily iron intake, 18% of their phosphorus intake, 21% of their magnesium and calcium intakes, 13–14% of their vitamin A, thiamin, preformed nicotinic acid, vitamin B_6, and vitamin B_{12} intakes and 16–17% of their vitamin C and riboflavin intakes. As the number of snacks eaten increases, the amount of nearly all nutrients and energy also increase. An increased level of snacking make a significant contribution to children's intake levels of magnesium and zinc, and to adolescents' intake of vitamin B_6, calcium, iron, magnesium and zinc. On days when no snacks are eaten, the intake of fat, sugar and sodium is decreased, and more food is eaten during the main meals. Snacking does not have much impact on the nutritional quality of the diet of elderly individuals. ***See*** individual nutrients

Children (9–13 years old) eat dessert items between meals followed, in descending order of frequency, by fruit and milk. The snacks most frequently eaten by 3–5-year-olds, in descending order of frequency, are bakery products, milk, soft drinks, fruit, milk desserts, candy and bread. For 6–11-year-olds, the snacks in order of importance are bakery products, soft drinks, milk, milk desserts, candy, fruit, salty snacks and bread. Among teenage boys the most popular snacks are soft drinks,

Table 4. Vitamins and minerals in snack foods

Snack food	Vitamin B_1 (μg)	Nicotinic acid (mg)	Vitamin C (mg)	Calcium (mg)	Iron (mg)	Potassium (mg)
Potato chips (25-g packet)	37 (3%)	1·2 (11%)	2·4 (8·0%)	8·5 (1%)	0·8 (8%)	445
Twisties (25-g packet)	7 (1%)	0·2 (2%)	0	44 (6%)	0·1 (1%)	70
CCS (25-g packet)	78 (7%)	0·5 (4%)	0	46 (7%)	0·9 (9%)	64
Peanuts (25 g)	—	—	—	8 (1%)	0·3 (3%)	135

Figures in parentheses are the percentage of RDA as defined in NH and MRC Food Legislation (1983).
CCS, chips crisps sticks.
Data from Delroy B (1985).

Table 5. Dietary fibre in snack foods

Snack	Insoluble dietary fibre (g per serving)
White-bread sandwich	0·2
Brown-bread sandwich	1·4
Whole-meal-bread sandwich	3·1
Twisties (25-g packet)	0·3
Potato chips (25-g packet)	0·6
CCS (25-g packet)	1·5
Peanuts (25 g)	1·6

CCS, chips crisps sticks.
Date from Delroy B (1985).

Table 6. Consumption nuts and nut mixtures in 1988

Country	Quantity (kg per capita)
The Netherlands	1·28
FRG	0·95
UK	0·66
France	0·65
USA	0·60
Belgium	0·56
Switzerland	0·18
Italy	0·10

Data from Tettweiler P (1990).

milk, bakery products, bread, milk desserts, salty snacks, meats and fruit, while teenage girls prefer soft drinks, bakery products, milk desserts, salty snacks, fruit, milk, candy, bread and meat. Among adolescents, the preference for snack foods is bakery products, milk, fruit and fruit juices, milk desserts, salty snacks, and bread. Table 6 gives the per capita consumption of nuts and nut mixtures. ***See*** Adolescents, Nutritional Problems; Children, Nutritional Requirements

Nutritional Value of Snack Foods

In order to counter the belief of consumers and nutritionists that snack foods are 'empty calories', a number of manufacturers of snack foods have started the development of products supplemented with vitamins, minerals or proteins. Vitamin supplementation is somewhat simpler than mineral or protein supplementation; the cost is reasonable, a small fraction of a per cent per portion if supplementation is restricted to five vitamins, but vitamins can affect the flavour, and some can cause off-flavours. Riboflavin is also highly coloured, but its yellow hue is compatible with many snacks. Storage deterioration of these labelled substances is compensated for by the addition of extra amounts, so that the consumer should obtain the full, claimed quantity. When supplementing with any nutritional factor, the manufacturer must establish a quality control programme which will ensure that every lot contains the claimed amount. Storage loss of minerals is not generally observed, except for iodine.

Most snacks are poor sources of protein, and the protein which is present is often of poor nutritional quality. Protein supplementation is possible either by increasing the protein–efficiency ratio of the protein already present, or increasing the total amount of protein present by adding some purified nitrogenous material, such as casein, isolated soya protein, or egg white. Untreated corn is a staple food in South America, and there is a need to develop more nutritious, corn-based snack foods. These foods could serve as vehicles for nutrients while at the same time being readily accepted by the population. Certain combinations of cereals and legumes are very desirable from a nutritional point of view. Legumes are a better source of lysine and total protein than cereals, and the latter are a superior source of sulphur amino acids. Many attempts have been made to raise the protein value of tortillas by combination with legumes. Snack foods are generally preferred by children and teenagers; in these stages of life, the amount and nutritional quality of proteins are important because of their essential function in physical and mental development. ***See*** Protein, Quality

Dietary Importance

Substantial efforts have been made to reduce the 'snacking habit' of various populations, but snack food consumption continues to increase. Consequently, there is a tendency to increase the production of more nutritious snacks, including those made from corn. Composite flours using corn, chickpea, soya bean meal and methionine can be formulated, extruded into snack foods and fried. These products have higher protein-efficiency ratio values than the controls, while colour, flavour, texture and overall acceptability remain similar. These products are a good alternative to commercially available snack foods.

Bibliography

Bednarcyk NE (1987) Nutrient contribution of cookies and crackers. *Cereal Foods World* 38(5): 367.

Delroy B (1985) The role of snack foods. *Food Technology in Australia*. 37(4): 154–158.

Madonna MD (1983) Pretzels as low calorie snacks. *Cereal Foods World* 28(5): 297.

Matz SA (1976) Nutritional supplementation. *Snack Food Technology*. Westport, Connecticut: AVI Publishing.

Morgan KJ (1983) The role of snacking in American diet. *Cereal Foods World* 28(5): 305.

Ranhotra GS (1985) Nutritional profile of corn and flour tortillas. *Cereal Foods World* 30(10): 703.

Tettweiler P (1990) Snacks – food, fun and fashion. *Dragocco Report 3*, pp. 79–104. Holzminden, Germany: DRAGOCCO.

Warren AB, Hnat DL and Michnowski J (1983) Protein fortification of cookies, crackers, and snack bars: uses and needs. *Cereal Foods World* 28(8): 441.

A. S. Bawa and P. Sharma
Punjab Agricultural University, Ludhiana, India

SOCIOECONOMICS AND NUTRITION

In 1936 Boyd Orr presented his classic study *Food Health and Income*, which indicated both the significance of price upon food choice for different income groups and also the implications of these factors for nutritional health.

Two key aspects of the relationship between income and food consumption patterns which Boyd Orr described have been consistently demonstrated to hold true in historical and international comparisons, as well as subsequent investigations into the eating habits of different socioeconomic groups within the United Kingdom and other countries. The first of these relates to the overall level of expenditure on food, the second to the type of foods consumed.

Expenditure on Food

Friedrich Engels, studying family expenditure in Germany, had been the first person to demonstrate with growth in income over a certain minimum, expenditure on food decreased as a percentage of income – even though the absolute amount of food consumed increased. This principle has been reflected in the changing pattern of expenditure on food in the United Kingdom over the last fifty years. With increasing affluence, the average percentage of household income spent on food has declined to around 20% from the average of 33% recorded by Boyd Orr in 1936.

Boyd Orr's study clearly showed this key aspect of the relationship between affluence and expenditure on food. As affluence increases, a smaller proportion is devoted to expenditure on food, although this may represent a greater amount of money in absolute terms. This may occur within a fairly narrow range in an industrialized country, such as in the United Kingdom where expenditure on food may vary from 29% to 15% of reported expenditure between low- and high-income households. However, in Third World countries the differences may be far more dramatic. While the most affluent group may be spending a very similar proportion on food to their counterparts in Western countries (around 15%), studies have indicated that in the lowest income groups food may account for 70–80% of household expenditure.

One important outcome of this is the increased vulnerability of low-income groups to changes in the prices of foodstuffs and other necessities. This may be particularly important to the urban poor in developing countries. If a diet which requires such a high level of expenditure is barely adequate nutritionally, then any increase in price which leads to the purchase of reduced quantities of food may have serious nutritional consequences.

In general, the urban poor may be disadvantaged in the price they pay for foodstuffs, both in the Third World and in industrialized countries. Since people are likely to be casually employed, or to receive small amounts of cash intermittently as the result of different activities, there is usually only enough money available to buy small quantities of food at a time. Low-income families have few opportunities to accumulate savings. As a result, it is not possible to take advantage of the economies which are possible through bulk purchase. In addition, the shopping opportunities for this group may be limited. The journey time and cost of public transport may restrict access to large supermarkets where prices are cheaper. This results in reliance on small shopkeepers who may offer a more limited range of poorer-quality foods at higher prices, but have the advantage that they are accessible and often willing to sell small quantities and offer credit. Consequently, the advice that is often given to families on a limited budget to 'shop around' may be impractical.

In part, this explains why, although the income of the urban poor in the Third World may be greater than that of people living in rural areas, this does not necessarily result in an increased food intake and better nutritional status. In addition, the cost of living in the city is often higher than in rural areas. It may be necessary to pay for utilities such as electricity and water. In addition to food, fuel, clothing, transport, rent and education must be paid for.

The budgets of low-income households leave very little room for manoeuvre and often food is the most flexible element of expenditure, but at the expense of nutritional adequacy. The only option available to meet an increase in essential costs like housing or transport is often to reduce expenditure on food. This may involve changes in the types of food purchased and reduction in the amount of food eaten, with some or all family members going without meals. A commonly observed pattern in such circumstances is for priority to be given to maintaining the diet of the breadwinner, while women and children bear the brunt of the shortfall in food. Low-income groups in industrialized countries show similar patterns, with expenditure on food being cut back in order to meet other bills and attempts made to provide adequate amounts of food to maintain the health of the breadwinner – although the nutritional outcomes may in general be less dramatic. Back in the 1930s the Public Health Authorities in one area of the United Kingdom were somewhat surprised to find that the health of people moved to new houses as a result of slum clearance showed clear evidence of a decline in the nutritional values of their diets. This was shown to be the direct result of an increase in the rents which had to be paid, reducing the amount of money available for food.

Although the problems of some low-income families are exacerbated by poor budgeting skills, there is evidence in relation to food of greater purchasing efficiency in terms of nutrients per monetary unit in such households. This may involve a very routine, rather monotonous diet, but the absence of any margin for error makes experimentation with new dishes, which may not be acceptable to family members, a very risky business.

Type of Foods Consumed

The second area where income has a marked influence on food intake is in relation to the type of foods which are eaten. When Boyd Orr looked at the dietary patterns of different income groups, he found marked differences. Consumption of fruit was eight times higher amongst people in the highest income group and they ate four times the amount of vegetables (except potatoes) recorded for the lowest income group. High-protein foods like meat, fish, milk and eggs are often the most expensive part of the diet and, as might be expected, he noted that consumption increased steadily with increasing affluence. In those days margarine was clearly the poor man's substitute for butter and this was reflected in the relationship between income and consumption. As income increased, more butter and less margarine was eaten. ***See*** Protein, Deficiency

Echoes of these patterns can still be found in the eating habits of different socioeconomic groups in the United Kingdom. More affluent households still consume more fruit and vegetables, as well as more carcass meat. Consumption of meat products, such as sausages and meat pies, which contain high levels of saturated fat, is higher amongst lower income groups because of their relative cheapness. The relationship between affluence and butter/margarine consumption, which persisted until the 1970s, has become more complex in the last twenty years with the introduction of the relatively more expensive polyunsaturated margarines promoted on the basis of their possible contribution to the prevention of coronary heart disease. Consumption of this type of margarine is higher among more affluent groups, while the traditional margarines which contain high levels of saturated fat are cheaper and more commonly eaten in less affluent groups. These differences affecting the fat intake of the poorer groups in the United Kingdom may well contribute to their increased risk of coronary heart disease. Lower-income groups are more dependent on bread and potatoes as cheap filling foods and this relatively greater reliance on the starchy staples is commonly associated with varying degrees of affluence in many countries. ***See*** Fats, Requirements

In addition, international comparisons indicate a similar relationship between average nutrient intakes and per capita income, as Fig. 1 indicates. With increasing affluence there is a reduction in the reliance on carbohydrates as a source of energy and an increase in fat and sugar. These trends, which are also accompanied by a reduction in the consumption of dietary fibre,

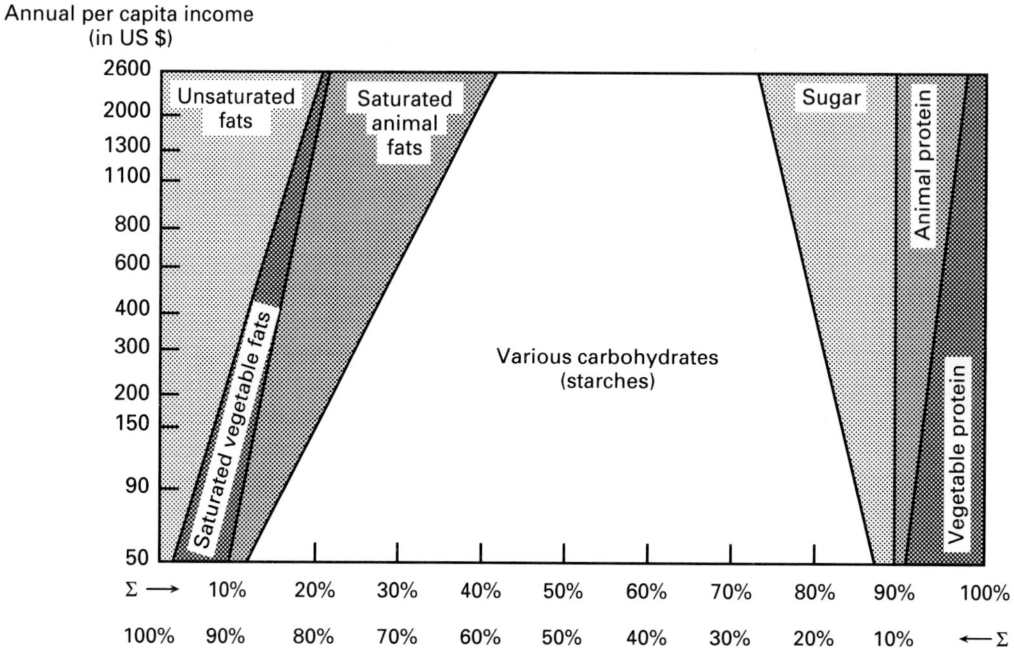

Fig. 1 Calories supplied by fats, carbohydrates and proteins as a percentage of the total caloric intake by wealth of country. (After FAO (1969) La consommation, les perspectives nutritionelles et les politiques alimentaires. In: *Plan Indicatif Mondiale Provisoire pour le Developpement de l'Agriculture*, Vol. 2, Chapter 13. Rome: FAO.)

have been associated with an increase in the prevalence of obesity, dental decay, coronary heart disease, diabetes, hypertension and some forms of cancer – the 'diseases of affluence'. Such conditions, which are common in industrialized countries are also found increasingly among members of the more affluent sections of the population of Third World countries. ***See*** Diseases, Diseases of Affluence

But perhaps the most obvious outcome of the differences in food intake associated with varying degrees of affluence is the difference in growth patterns of children. Trends in the heights of different socioeconomic groups in the United Kingdom are closely linked to nutritional differences. Over the past 250 years the heights of working-class 14-year-olds have increased by 29 cm (to average 164 cm), with the result that a working-class boy of 14 years is now almost as tall as a working-class adult male of the mid-eighteenth century, at the same time narrowing the difference in height with their more affluent peers. Nevertheless, there is strong evidence for the continued effect of socioeconomic status on growth in the United Kingdom, which is in part due to nutritional differences. This pattern is found in almost all developed countries, with the exception of Sweden where no association between mean height and social class is apparent. Stunting is an undisputed effect of undernutrition in the Third World, but in the relatively well-fed industrial nations, it may seem surprising to suggest that socioeconomic differences affecting diet are sufficiently large to affect the height of children and adolescents. None the less, the particular role of protein in relation to growth in height, where energy intake is adequate, may offer a partial explanation since, as has already been commented upon, sources of animal protein are usually the most expensive item in the food budget. ***See*** Children, Nutritional Requirements

In Third World countries the early years of childhood are a critical time for stunting because both growth and infection make large claims on nutrition at this age. Studies show marked differences between different socioeconomic groups within the same country, but evidence also suggests that infections can cause stunting even with an adequate diet. However, after a critical period in the first year, children who are well fed are able to catch up in growth by the age of 2 years. Where poverty restricts the diet, this catch-up growth does not occur.

Poverty, Malnutrition and Famine

Stunting is often interpreted as an indicator of long-term undernutrition, especially where weight for height is normal, while wasting (as indicated by weight for height) may be seen as a response to a more acute food shortage. The most dramatic signs of undernutrition, marasmus and kwashiorkor, are commonly associated in the public mind with famine, often perceived to be a phenomenon indiscriminately affecting all sections of the population, often as a result of natural disaster.

However, closer examination of who starves in such situations led Amartya Sen to conclude that 'starvation is the characteristic of some people not *having* enough to eat. It is not the characteristic of there not *being* enough food to eat.' Explaining starvation in terms of food shortages may be misleading and may result in inappropriate relief interventions. Sen stresses what he calls people's 'entitlements to food'. Entitlements to food can be obtained through growing it, working for money to buy food, trading or being given or lent food. Famines occur when large numbers of people are deprived of entitlements and so starve to death. Many more go hungry because they lack entitlements (in the shape of land, jobs, or anything to trade) in the first place. Natural disasters turn into famines where poor people in poor countries are already living close to the margins of survival. They do not have the reserves to fall back on and often their governments lack resources to provide a safety net. In contrast, when drought hit the Mid-West of the United States in 1988, although farmers went out of business there was no starvation. The strength of farmers, government and the economy made a famine impossible. ***See*** Kwashiorkor; Malnutrition, The Problem of Malnutrition; Malnutrition, Malnutrition in Developing Countries; Malnutrition, Malnutrition in Developed Countries; Marasmus

The economic ramifications of international trade have done little to improve the security of Third World farmers, often contributing to their vulnerability. Peasant food production has often been regarded as 'backward' and the emphasis has been placed on growing money-making crops. In addition, the bulk of the profit from these groups has been creamed off by some governments buying cheaply from peasants (sometimes compulsorily through marketing boards) and selling at higher prices on the world market. Food production policy has often been based on maintaining a cheap supply of food for city workers. Cheap food means workers can be paid less, and the goods that they produce can be priced more competitively. It also helps to reduce the risk of food riots and political unrest in the cities, where governments are most vulnerable. With the encouragement of the IMF (International Monetary Fund) and World Bank, credits and farming inputs have favoured the production of cash-crops rather than food. However, the prices of many commodities have declined on the world market and farm incomes based on crops like cocoa or cotton have plummeted. The farmer's response to this has often been to stop growing those crops which are no longer worth their while and to try to find other nonfarm work. Such work is often badly paid and unreliable. Thus the overall result is increasing poverty and vulnerability to disasters like crop failures.

In conclusion, economic factors play a key role in determining the amount and type of foods which people have access to and consume. Poverty contributes directly to individuals' vulnerability to nutritional inadequacy and the degree to which their health may suffer as a result will depend upon the social context in which they find themselves.

Bibliography

Boyd Orr J (1936) *Food, Health and Income*. London: Macmillan.

Floud R, Wachter K and Gregory A (1990) *Height, Health and History*. Cambridge: Cambridge University Press.

Jackson B (1990) *Poverty and the Planet. A Question of Survival*. London: Penguin.

J. Thomas
Kings College, London, UK

SODIUM

Contents

Properties and Determination

Sodium (Na) is chemically an alkali metal with an atomic weight of 22·98. Since it has a single electron in the outer 3s orbital and the attraction of the nucleus for this electron is shielded by the electrons in the completed inner orbitals the 3s electron is easily lost. This ionization (loss of electron) forms the sodium cation (Na^+). The stability of the univalent cation leads to the predominance of ionic bonding of sodium to anions such as Cl^- to form sodium salts. Sodium salts are generally soluble in water and are leached from the soil and carried to the sea. Sodium is the sixth most abundant element on earth and establishes the salinity of the seas and saline lakes as sodium chloride.

Many animal species including humans show an appetite for sodium chloride. This desire or 'hunger' for salt normally causes the animal to replace the salt lost in urine and through perspiration. On average an American consumes between 8 and 12 g of salt per day derived from the salt present in drinking water, naturally occurring in foods or added to foods during processing to enhance flavour. Some sodium is also added to natural water supplies as the result of water softening procedure where the sodium ion replaces other cations present in the water. Normally loss of salt from the body stimulates the appetite for salt and excessive salt is eliminated through urinary excretion by the kidneys. ***See*** Sensory Evaluation, Taste

Recently it has been suggested that some types of hypertension are related to inappropriate metabolism (retention of sodium) by the body. Although consensus on this point has not been reached, it has been suggested by governmental and private health agencies that sodium chloride intake should be reduced to somewhere between 2 and 5 g per day. ***See*** Hypertension, Hypertension and Diet

It is hard to imagine a single food source, whether it is of animal or vegetable origin, which does not contain sodium. Since sodium plays such a critical role in many normal physiological processes and probably in pathological changes, it is of great importance to know the sodium content and state of sodium in foodstuffs.

Preparation of Samples for Analysis of Sodium

Most sodium compounds are soluble in water, acid or alkaline solutions. In biological systems some evidence exists that the ion, Na^+, can exist in a bound state, but the evidence on this point is highly controversial. Because of the properties of high reactivity and solubility of sodium compounds, the sodium in a food can generally be easily extracted. In preparing a biological sample, it is often enough to disrupt the tissue, and allow the ions to diffuse into a known volume of fluid. Homogenization by mechanical means is a simple and rapid procedure. After cutting a sample into conveniently sized pieces, it is placed into a container with a known volume of extraction solution. It is then ground by the homogenizing device. The suspension is then extracted (usually in the cold) and sample debris can then be removed either by filtration or by centrifugation.

In difficult situations it may be necessary to use ultrasound, freeze–thawing or boiling of the sample to obtain a representative sample for analysis. Occasionally it is necessary to destroy the sample in order to liberate the sodium. This may be accomplished by acid hydrolysis of the sample or by ashing the food sample in a nonreactive container. Twenty-four hours of heating at 500°C will reduce most reasonably sized samples to an ash. The ash may then be extracted with water or acid. Care must be taken to neutralize acid extracts since the acid may damage the analytical instrument used to quantitate the amount of sodium present.

Sodium is ubiquitous in the environment and great care must be taken during analysis to avoid contamination, especially via airborne dust and skin secretions from handling.

Methods of Analysis

Gravimetric methods of analysis for sodium are not used to any significant extent because most sodium compounds are so water soluble, that it is consequently difficult to precipitate them from solution. Most modern analytical procedures utilize methods which are related to the atomic structure of the sodium ion.

When a sample containing sodium is combusted in a flame or is excited by an electrical discharge or radiation at an appropriate wavelength, the ion acquires enough energy that an electron is driven to a higher energy state. In this process energy is absorbed at a specific wavelength, leading to the phenomenon of atomic absorption. When this higher energy state is relaxed by the electron falling to a lower energy state, light of a particular wavelength is emitted, leading to atomic emission. These phenomena have led to the development of the two primary means of measuring sodium, namely atomic emission spectroscopy (i.e. flame photometry) and atomic absorption spectroscopy.

The wavelength of light emission (and absorption) for sodium takes place at the so-called D line, actually composed of two lines at 589·6 and 589·0 nm. It is well established that all other things being equal (flame temperature, flow of sample, etc) the intensity of emission or absorption at the D line is related to the concentration of sodium in the sample (Fig. 1).

Emission Flame Photometry

Modern instruments using emission flame photometry to determine sodium rely on a continuous flow of small (atomized or nebulized) droplets of solution into the flame. The steady-state increase in photodetector output is then found to be proportional to the number of excited sodium ions, which in turn depends on flame temperature, flame stability, constancy of suspension flow and concentration of sodium in the test or calibrating sample.

Aspirator systems are used where a fine tube is connected to feed into the gas stream of fuel or fuel plus

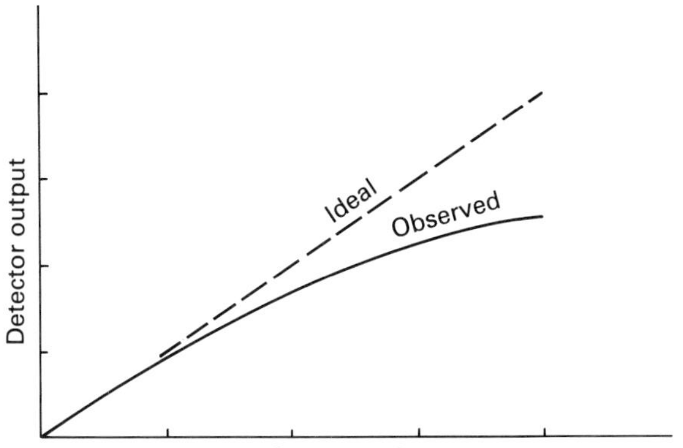

Fig. 1 Relationship between photodetector output and sample sodium concentration as determined by emission spectrophotometry. Curve is generated at constant flame temperature and sample aspiration rate. Data are presented in arbitrary units.

air or oxygen. By having a wide beaker to hold the sample, which is always placed in the same position, changes in flow rate because of changes in sample fluid levels are minimized. Changes in gas flow are minimized by introducing reduction valves into the air and/or oxygen and fuel lines.

An important procedure has been developed that results in correction of flame or flow instabilities. A secondary reference element which emits light at a different wavelength from the sodium D line is introduced into the standard and test solutions at a known constant final concentration. In almost all systems lithium is used as the reference element. The lithium spectroscopic output is monitored by a second photoelectric detector and the increase in output from this detector is a reference against output of the sodium detector. If there is any instability in flame or sample flow, the changes seen by both detectors will be proportional. By appropriately adjusting the outputs, changes resulting in output current because of instability are greatly minimized.

Another point which should be appreciated concerning flame emission photometry is that because of saturation and nonproportionality of the photoelectric detector by emitted quanta, the relationship between sample concentration and detector current is nonlinear (see Fig. 1). In addition, nonlinearity exists between number of excited nuclei and concentration. It is therefore best to use sample concentrations at which the relationship between concentration and current is the most linear, i.e. at the most dilute concentrations which still allows precise quantitation.

Atomic Absorption Methods

Another optical method that has been used to analyse sodium content of food samples is flame atomic absorption spectrophotometry. Because absorption methods for sodium are less sensitive than emission methods, it is not used as often with sodium as for other elements such as calcium, magnesium, certain trace elements like zinc and copper as well as others. Nevertheless, the method has also been used for sodium measurements.

The method depends on the fact that light of a specific wavelength can excite an electron of an atom which is at an adequately high temperature. Excitation results in light absorption. The quantity of light absorbed is measured. The methodological problems associated with flame absorption techniques are very similar to those of emission methods. The major additional problem is constancy of the light source used for excitation of the sample.

Helium Glow Photometry

The third method recently used in measuring sodium is the helium glow flame photometer. As well as heat and light, an electrical discharge can be used to excite sodium. A sample is placed on a wire in a quartz tube which forms part of an electrode pair. A ring of wire around the tube forms the second electrode. The tube is flushed with helium. A radiofrequency electric field is applied and the wire electrode is heated by passing current through it. The sample becomes excited and when an electron falls back into the outer shell the characteristic light frequency of sodium is emitted. The light is then collected by a collimating lens, filtered and transmitted to an appropriate detector.

The need for optical filtration arises from the fact that excitation of the helium results in the production of particles having a broad distribution of energies. Consequently, more elements in the sample than sodium become excited and can cause interference. However, an advantage is that more than one element can be analysed by using several photodetectors fitted with appropriate filters.

This method is most valuable in the analysis of small volumes of fluid. It is sensitive down to 10^{-12} mole of the element under study, and is used with volumes as small as 2–10 nanolitres of fluid.

Ion-selective Electrodes

In the late 1950s it was found that glass membranes could be made that were specifically permeable to sodium ions. This finding led to the development of ion-selective electrodes capable of measuring the activity (and hence concentration) of sodium ions in solution. Sodium-selective electrodes, as other ion-selective electrodes, function on the relationship between the activity of an ion and membrane electrode potentials described by the Nernst equation. Consequently this method measures the activity of the ion in solution, which may be decreased by adsorption of the formation of complexes with other constituents in the medium. Furthermore, while more selective sodium electrodes derived from macrocyclic antibiotics have been developed, measurements with ion-selective sodium electrodes are subject to interference from other ionic constituents that may also be present in the solution tested. However, the rapidity and simplicity of ion-specific electrode measurements of sodium makes this method a very attractive alternative to spectroscopic measurements.

Electron-probe Microanalysis

If an atom is excited with a particle of high enough energy, it is possible to cause removal of an electron from the inner shell. When an electron falls back into the orbit, it causes the emission of a high-energy quantum, as X-ray emission at the frequency of the K line of the element. By measuring the frequency and the intensity of the emitted X-rays, it is possible to characterize the amount and kind of elements in the sample. This is a very sensitive method and can be used with droplets of fluid, isolated cells and tissue slices.

In most modern applications of this procedure high-speed electrons accelerated within an electron microscope are used for excitation. In this system it is possible to control excitation energies and to be able to measure individual elemental species.

It is sometimes desirable to measure not only the amount of an element, but also its localization in a cell or tissue. For example, most of the sodium in animal tissues is extracellular, while in plants it can be in vacuolar sap, cytoplasm or other inclusions. The electron-probe microanalytical procedure is excellent for solving these problems of sodium distribution. Since in plant foodstuffs composition is determined in part by the consumption and frequency of application of fertilizer, this procedure could provide valuable information.

All of the pitfalls associated with conventional electron microscopy apply to electron-probe microanalysis. Sample preparation should cause no change in ion placement by diffusion, convection or mechanical movement. Freezing procedures have been found to be most suitable. If fluid samples are placed on a support, care should be taken to ensure that the drying is even. The choice of sample support is critical as it may cause emissions which interfere with analysis of the element under study. Tissue samples need to be cut uniformly and placed on grids for transfer to the microscope. Using special X-ray detectors it is then possible to detect and measure the intensity of the K line for a given atom.

NMR Spectroscopy

The Na^+ ion has a nuclear magnetic moment of 3/2 which allows it to undergo nuclear magnetic resonance (NMR). In fact, the 100% natural abundance, high sensitivity to magnetic fields, short NMR relaxation time and relatively high concentration in biological materials makes ^{23}Na extremely easy to detect in biological samples. Furthermore, the sensitivity of the nucleus to its environment and the development of NMR shift reagents has made it possible to measure independently the amount of sodium in the cytoplasm and extracellular space of tissues with ease. The newer shift reagents, which are anionic chelate complexes of paramagnetic lanthanide ions, distribute themselves throughout the extracellular space of tissues but do not penetrate cell membranes easily and hence do not enter the cell interior. The shift reagent forms short-lived associations with the sodium ion, shifting the spectrum of the ions away from that exhibited by intracellular ions which do not come into contact with the shift reagents. Early studies using shift reagents should be suspect since the shift reagents used in these studies killed cells and were not restricted to the extracellular space. ***See*** Spectroscopy, Nuclear Magnetic Resonance

There are problems associated with NMR measurements of sodium in that some of the sodium appears to be invisible to the NMR. Hence, in most studies of biological tissues the amount of sodium detected by NMR is as much as 60% less than that measured by other means. While the reason(s) for this discrepancy are not completely clear, it is currently thought that it arises from the association of the intracellular sodium ions with other intracellular constituents, or more probably the multiplicity of means by which thè nuclear energy can be dissipated in the presence of the intracellular constituents. The discrepancy is eliminated when the sodium is extracted from the tissue as described above. It may be concluded that while some problems still exist in the measurement of sodium by NMR in tissues the technique is quantitative for extracted samples and holds promise for becoming an excellent nondestructive method for measuring sodium concentrations in the various compartments of biological tissues and foodstuffs.

Other Methods

Numerous other high-energy analytical techniques for sodium have been developed, all of which require extensive high-technology electronics. These methods include ion-microprobe analysis, laser probe analysis, ion-scattering spectrometry, Auger spectroscopy, impact radiation analysis and α-particle backscattering analysis. The reader is referred to the bibliography for descriptions of these emerging but not commonly used analytical techniques.

Bibliography

Ferris CD (1974) *Introduction to Bioelectrodes*. New York: Plenum Press, p. 114.

Hamilton EI (1979) *The Chemical Elements and Man*. Springfield: C.C. Thomas.

Kane MR, Fregly MJ and Bernard RQ (eds) (1980) *Biological and Behavioral Aspects of Salt Intake*. New York: Academic Press.

Lechene, Claude P (1977) Electron probe microanalysis: its present, its future. *American Journal of Physiology* 232: F391–F396.

Mackay KM and Mackay RA (1989) *Introduction to Modern Inorganic Chemistry*, 4th edn. London: Blackie.

Öberg PÅ, Ulfendahl HR and Wallin BG (1967) An integrating flame photometer for simultaneous microanalysis of sodium and potassium in biological fluids. *Analytical Biochemistry* 18: 543–558.

Rick R, Horster M, Dorge A and Thurau K (1977) Determination of electrolytes in small biological fluid sample using energy dispersive X-ray microanalysis. *Pflugers Archiv. European Journal of Physiology* 369: 95–98.

Sanui H and Pace N (1972) An atomic absorption method for cation measurements in Kjeldahl digest of biological materials. *Analytical Biochemistry* 97: 57–66.

Springer C (1987) Measurement of metal cation compartmentalization in tissue by high-resolution metal cation NMR. *Annual Review of Biophysics and Biochemistry* 375–399.

William R. Galey and Sidney Solomon†
University of New Mexico School of Medicine, Albuquerque, USA

Physiology

Physiological, Clinical and Nutritional Importance of Sodium

Despite the fact that the body contains more calcium and potassium, sodium is arguably the most important cation because it dictates the volume of extracellular fluid (ECF) and its concentration affects osmotic concentration of both ECF and intracellular fluid (ICF). Abnormalities of ECF sodium concentration cause movement of water into or out of cells, thus altering the osmotic concentration of ICF in parallel and causing swelling or shrinkage of cells. The main impact of this is on the brain because its cells are rigidly enclosed by the cranium.

Sodium depletion is mainly caused by enteric, renal or adrenal disease, and sodium retention is caused by renal disease; healthy kidneys are well able to excrete excess dietary salt. However, chronic ingestion of excess salt, whether or not it increases ECF volume, is a predisposing or exacerbating factor in hypertension. Until the 1980s, knowledge of the regulation of body sodium mainly concerned defences against depletion, whereas the last decade has seen a rapid growth in knowledge of the mechanisms which excrete excess sodium. This seems appropriate since most species, especially humans, dogs and laboratory rats, are exposed to dietary sodium intakes well above their nutritional requirement. ***See*** Kidney, Structure and Function

The nutritional requirement is a reflection of obligatory losses (maintenance) and the needs of growth, pregnancy and lactation. Abnormal losses owing to disease, or in animals such as humans and horses which sweat extensively, raise the requirement. The impact of equine sweating is different from that in humans. Human sweat always contains sodium at concentrations well below plasma levels (and when aldosterone secretion is raised, levels of sweat sodium fall very low); horse sweat is hypertonic but this helps to offset the osmotic effect of the increased respiratory water loss during exertion, i.e. it may be a defence against hypernatraemia, rather than a potential cause of sodium depletion.

Consideration of the physiology of sodium thus includes its distribution in the body, regulation of total content and concentration, causes of and responses to depletion or excess, and their nutritional implications.

Distribution

Sodium behaves physiologically as a cation, i.e. a positively charged ion; its distribution and effects are fairly independent of the negative ions (anions) which originally accompanied its ingestion though they may affect its absorption and excretion. Most sodium is in ECF (Table 1), kept there by the sodium pump, an

Table 1. Summary of sodium (Na) distribution and requirements

Typical plasma Na concentration (mmol l^{-1})	145 (130–160)
Typical body Na content (mmol kg^{-1})	50–55
Typical proportion (%) of total Na	
ICF	10
ECF	50
Bone	40
Maintenance requirement in mammals (mmol per kg per day)	
Sheep	0·1
Cattle and goats	0·3–0·7
Pigs	0·6
Rats	0·6
Dogs	0·2–0·5
Humans	? < 0·6

1 mmol = 23 mg Na^{+}, 58·5 mg NaCl.

† Deceased.

enzyme system, (Na^+-K^+)-ATPase, which uses substantial amounts of energy (adenosine triphosphate; ATP) in maintaining a low intracellular sodium concentration and a high intracellular potassium (K^+) concentration. Sodium transport is a central issue in the physiology of sodium for a number of reasons:

1. It helps to maintain the ionic environment of ICF and the volume of ECF.
2. It prevents cell swelling (the Na^+ efflux exceeds the K^+ influx).
3. It establishes gradients which, in various tissues, allow transport of other cations in exchange, other anions in parallel or organic solutes – these are often cotransported with sodium down concentration gradients which are secondary to the low sodium environment created by the pump.
4. It establishes the membrane voltages on which excitability and secretory activities frequently depend.
5. The energy expenditure of the pump is a substantial portion of total metabolic activity and contributes to thermogenesis.
6. Sodium transport is not only a key factor in the retention and loss of sodium in the kidney, gut, salivary and sweat glands but also influences the excretion or retention of many other solutes. Thus, for example, diuretics intended to promote sodium excretion may also cause unintentional losses of potassium and magnesium. Similarly, when renal sodium excretion increases appropriately in response to ingestion of excess salt, there may also be unwanted losses of calcium and in postmenopausal women these may contribute to loss of bone mineral. ***See*** Electrolytes, Analysis; Thermogenesis

Bone also contains substantial quantities of sodium but, as yet, its significance is unknown since it does not appear to be mobilized during sodium depletion. Gut fluids contain considerable amounts of sodium, mostly secretory rather than dietary, and mostly reabsorbed in more distal regions of the intestine.

Extracellular Sodium

Of the extracellular fluid, most is in interstitial fluid (ISF) in the tissue spaces, providing the transport medium between capillaries and cells. The sodium concentration in plasma is slightly above that in ISF because plasma contains more proteins, notably albumin, which do not readily escape into ISF across the capillary membranes, and the effect of their negative charges is to hold more positively charged ions, notably sodium, in circulation (Gibbs–Donnan equilibrium).

The main effects of excess ECF volume are seen as expanded ISF, visible clinically as oedema (or ascites when fluid accumulates in the abdomen rather than the tissue spaces). Mild oedema is merely a cosmetic problem in itself but pulmonary or cerebral oedema, or severe ascites, are potentially serious forms. Oedema can result from excess ingestion or retention of sodium (overall expansion of ECF) or 'leakage' from plasma to ISF, with plasma volume continuously replenished by renal sodium retention. Such maldistribution of ECF occurs if plasma albumin is very low (renal leakage, hepatic impairment or severe malnutrition), or with excessive capillary blood pressure (venous blockage, inactivity, heart failure, or arteriolar dilation, e.g. from heat or allergy), capillary damage, or lymphatic blockage. The latter prevents the removal of proteins which have leaked into ISF. Accumulation of protein in ISF undermines the osmotic gradient which normally favours uptake of water at the venous end of the capillary, where the pressure is lower. Since oedema involves the expansion of a larger compartment (ISF) from a smaller one (plasma) it is only possible as long as the latter is replenished; hence the kidney, while seldom the primary cause of oedema, is always the enabling cause; the use of diuretics is therefore appropriate in the treatment of nonrenal as well as renal causes of oedema.

The main effect of inadequate ECF volume is to reduce plasma volume and thus to compromise cardiovascular function, in extreme cases by causing circulatory shock.

Regulation of ECF Sodium

In a mature, nonpregnant, nonlactating, healthy animal, sodium excretion matches sodium intake and is often used to estimate it, although this is not reliable, especially when intake is low. Dietary sodium is readily available, i.e. readily absorbed; thus the traditional view of sodium regulation emphasizes renal regulation of urinary Na^+ loss. This oversimplifies the more subtle interplay seen, for example, in herbivorous animals, where salt appetite may contribute to regulation by intensifying during sodium depletion. Moreover, in many herbivores the faeces, rather than urine, may be the major route of sodium excretion and the gut may therefore be an important regulator of sodium balance. Indeed, it is interesting that sodium transport mechanisms in the small intestine show considerable similarities to those of the proximal part of the renal tubules (e.g. linked transport of Na^+, glucose and amino acids) whereas the colon, like the distal nephron, responds to the salt-retaining (and potassium-shedding) hormone of the adrenal cortex, aldosterone.

Provided that the adrenal gland is healthy, urinary and faecal sodium loss can be reduced virtually to zero. Sweat loss can also be very low, although with severe exertion in hot climates the volume of sweat may exceed the ability of aldosterone to reduce its sodium concentration and net loss of sodium can occur. Aldosterone also reduces salivary sodium (and raises [K^+]). ***See*** Potassium, Physiology

There are two components to the regulation of ECF sodium; the total amount of sodium retained, and its concentration. The former is regulated by mechanisms which directly affect sodium, whereas the latter is essentially regulated via water balance. Thus whatever sodium is retained in ECF is 'clothed' with the appropriate amount of water to maintain the normal plasma sodium concentration within narrow limits; deviations of less than 1% (hard to measure in the laboratory) trigger corrective responses. Thus a raised plasma sodium concentration (e.g. after water loss) stimulates both thirst and renal water conservation; antidiuretic hormone (ADH) from the posterior pituitary reduces urine output through its effect on the renal collecting ducts. Even one of these mechanisms can defend body water; thus diabetes insipidus (inadequate production or effect of ADH) does not cause severe dehydration but polydipsia (increased fluid intake; 'thirst' is a sensation).

Excess salt intake does not raise plasma sodium concentration (hypernatraemia) if water is available and the patient can drink; the excess sodium is diluted. The resulting increase in ECF volume then stimulates increased sodium excretion. Sodium also enables ECF to hold water against the osmotic 'pull' of the solutes in ICF and sodium thus functions as the 'osmotic skeleton' of ECF; it is the main determinant of its volume.

Plasma sodium concentration is therefore only indirectly related to sodium balance. When ECF volume, notably circulating volume, is severely reduced, this stimulus, rather than Na^+ concentration, becomes the main drive for thirst and ADH secretion. Until ECF volume is restored, water is retained (to protect ECF volume) even though this undermines the protection of ECF Na^+ concentration and, as a result, plasma sodium falls. Thus, during sodium depletion, contraction of ECF volume precedes significant reductions of plasma Na^+ which is therefore a poor index of sodium status.

Sodium-retaining Hormones

Sodium depletion, by reducing plasma volume and renal perfusion, stimulates the production of renin (from the kidneys) which generates angiotensin in circulation. This hormone is a vasoconstrictor (so protects blood pressure), stimulates thirst (so helps to restore ECF volume) and, above all, stimulates sodium retention both directly (renally) and indirectly (by stimulating adrenal secretion of aldosterone); it thus reduces the sodium concentration of urine, faeces, saliva and sweat, but not milk. ***See*** Hormones, Adrenal Hormones

Indices of aldosterone secretion (reduced sodium or potassium concentration in urine, faeces, etc.) are often taken as evidence of sodium depletion or inadequate sodium intake, but the following points apply:

1. Aldosterone secretion is also stimulated directly by hyperkalaemia (elevated plasma K^+) and promotes potassium excretion.
2. Such interpretations involve a subjective judgement concerning adequate or excessive sodium intake. Because physiologists and clinicians were traditionally more concerned with sodium depletion as well as its consequences and the defences against it, elevated aldosterone secretion was readily seen as a warning signal. However, if sodium intakes associated with increased aldosterone have no other harmful effects, and especially if excess sodium intakes cause concern, low levels of aldosterone secretion might equally indicate excessive salt intake.

While sodium reabsorption in the distal nephron, influenced by aldosterone, is particularly important because it can produce sodium-free urine and promote potassium loss, the great majority of renal sodium reabsorption occurs elsewhere; about 25% in the loop of Henle and most in the proximal tubule. The loop is also a main site of magnesium reabsorption, hence the tendency for loop diuretics to cause hypomagnesaemia.

The factors controlling proximal reabsorption are incompletely understood but their effect is clear: proximal reabsorption of sodium increases or decreases according to the need to enhance or diminish plasma volume. Since the fluid in the proximal tubule is similar to plasma, having been formed from it by glomerular filtration, it has the ideal composition for this purpose.

Natriuretic Hormones

Excretion of excess sodium involves not only suppression of salt-retention mechanisms but also activation of sodium-shedding (natriuretic) mechanisms. Two types of hormones are involved: atrial natriuretic peptide (ANP), produced by the cardiac atria when they are overstretched (reduction of ECF volume being an appropriate response to cardiac overload), and active sodium transport inhibitors (ASTIs), probably produced within the brain. These were probably the original molecules associated with the receptors binding cardiac glycoside drugs and are therefore also called 'endogenous digitalis-like inhibitors' (EDLIs); their exact identity remains uncertain. Atrial natriuretic peptide has various effects which essentially oppose those of the salt retention induced by aldosterone: it increases sodium excretion, lowers arterial pressure and promotes movement of ECF towards the interstitial compartment.

Other hormones (e.g. sex steroids, parathyroid hormone, calcitonin, thyroid hormone, prolactin) affect renal sodium retention or loss but are not thought to regulate it. ***See*** Hormones, Thyroid Hormones; Hormones, Pituitary Hormones

Adequate, Inadequate and Excess Sodium

It is unlikely that adult daily maintenance requirement exceeds 0·6 mmol per kg bodyweight and could well be below this in many mammals. Newborn, growing, pregnant or lactating animals have increased requirements. The appropriate sodium intake for humans remains controversial with some cultures managing on less than 1 mmol per day, while Western intakes may be in the range 200–300 mmol per day, more where processed foods are heavily consumed. There has been insufficient awareness among physicians and human nutritionists of just how high such intakes are, compared with requirement in other animals. Granted that humans are bipeds with a stressful lifestyle quite different from those of animals, there is no real evidence that human obligatory losses or sodium requirements are significantly greater. Rather, there is an ingrained tradition of regarding sodium intake as a benign pleasure, involving a harmless and healthy dietary constituent. The main warnings against this view come from the fact that hypertension is virtually unknown in low-salt cultures and that they do not even have an age-related rise in 'normal' blood pressure. Moreover, there are numerous studies which, when rigorously analysed, indicate that human arterial pressure and salt intake are positively correlated; sufficiently to anticipate large reductions in the prevalence of hypertension in response to manageable reductions in dietary sodium. Unfortunately, such reductions are handicapped by inadequate food labelling and the fact that most sodium is added by the processor rather than the consumer.

Because obligatory losses of sodium are so low, dietary sodium depletion is hard to induce and sodium deficiency usually results from losses caused by renal, adrenal or enteric disease; renal disease may cause either retention or loss of sodium. Globally, both in humans and animals, the most common cause of sodium deficits is acute diarrhoea. Fortunately, sufficient gut usually remains unaffected for uptake of sodium and water to be stimulated by suitably formulated oral rehydration solutions. These essentially restore ECF volume (and acid–base balance), allowing natural defences to overcome the underlying cause of the diarrhoea. Despite some species variations, such solutions usually work best if their glucose: sodium ratio (in mmol l^{-1}) is close to unity and they are virtually isotonic (i.e. they have a similar osmotic concentration to ECF; hypertonic solutions draw water into the gut). The function of glucose in these solutions is to promote sodium uptake; its nutritional contribution is trivial. Anions such as citrate, acetate, propionate, bicarbonate and amino acids (e.g. glycine and alanine) may further enhance the uptake of sodium and therefore water. These sodium cotransport mechanisms are very similar to those of the proximal renal tubule.

Sodium is thus central to the management of two of the most widespread clinical problems; hypertension (in humans) and diarrhoea. Indeed, the World Health Organization (WHO) regards the discovery of oral rehydration, which depends on restoration of enteric sodium uptake, as the main life-saving development in twentieth-century medicine. This powerful clinical application rests on a simple physiological observation concerning an elementary but vital dietary constituent. ***See*** Hypertension, Physiology; Hypertension, Hypertension and Diet; Hypertension, Nutrition in the Diabetic Hypertensive

Bibliography

Avery ME and Snyder JD (1990) Oral therapy for acute diarrhea. *New England Journal of Medicine* 323: 891–894.

Denton DA (1982) *The Hunger for Salt.* Berlin: Springer-Verlag.

El-Dahr SS and Chevalier RL (1990) Special needs of the newborn infant in fluid therapy. *Pediatric Clinics of North America* 37: 323–335.

Field M, Rao MC and Chang EB (1989) Intestinal electrolyte transport and diarrheal disease. *New England Journal of Medicine* 321: 800–806, 879–883.

Hirschhorn N and Grenough WB (1991) Progress in oral rehydration therapy. *Scientific American* 264: 16–22.

Law MR, Frost CD and Wald NJ (1991) By how much does dietary salt lower blood pressure? *British Medical Journal* 302: 811–815; 815–818; 819–824.

Maxwell MH, Kleenan CR and Narins RG (1987) *Clinical Disorders of Fluid and Electrolyte Metabolism* 4th edn. New York: McGraw-Hill.

Michell AR (1989) Physiological aspects of mammalian sodium requirement. *Nutrition Research Reviews* 2: 149–160.

Michell AR (1991) Regulation of salt and water balance. *Journal of Small Animal Practice* 32: 135–145.

Rutlen DL, Christensen G, Helgesen KG and Ilebekk A (1990) Influence of atrial natriuretic factor on intravascular volume displacement in pigs. *American Journal of Physiology* 259: H1595–1600.

A. R. Michell
Royal Veterinary College, Hatfield, UK

SOFT DRINKS

Contents

Chemical Composition

Product Category and Origins

Soft drinks are part of the 'beverage continuum' of human rehydration refreshment starting with highly spirituous whisky, gin and liqueurs at one end through wines, low-alcohol wines, beers, low-alcohol beers to soft drinks and onwards to the hot beverages culminating in tea and coffee at the other end. It can be argued that soft drinks are almost as old as the species of *Homo sapiens* itself, with origins which may be traced back to man's earliest attempts to 'preserve' water from microbiological degradation or contamination. Included in this is the early production of wine from grape juice, where it was observed that an alcohol content of greater than 5% would prevent degradation of the liquid. Early 'preservatives' of this type were vinegar and sugar with the mixtures obtained arguably becoming the first soft drinks. Nowadays the definition of a soft drink, in most countries, is that of a beverage with an alcohol content of less than 1% by volume. Most of the other limits of the soft drink category have become increasingly 'blurred', with many of the milk-based or low fat chocolate and yoghurt liquid beverages equally capable of a 'soft drink' categorization as well as the particular segment or food category in which they have chosen to describe themselves.

Basic Composition of Soft Drinks

For the purposes of this article the definition of a soft drink (and thus its chemical composition) will be based on the 1964 compositional regulations in the United Kingdom which, broadly speaking, will encompass all categories applicable to the EC in the forseeable future.

A soft drink will be formulated to contain most of the following collective groups of ingredients:

- Water (carbonated or still)
- Sugar (usually sucrose or other carbohydrates)
- Fruit (juice/extract or other characterizing ingredient)
- Acid (usually citric or other organic acids)
- Flavouring (artificial, nature-identical or natural)
- Preservative (benzoic acid, sulphur dioxide or sorbic acid)
- Artificial sweetener (usually aspartame, acesulfame K or saccharin)
- Vitamins (usually vitamin C but others may be included)
- Colouring (usually carotenoids or grape skin extract)
- Acidity regulator (usually sodium citrate or other salts)
- Antioxidant/emulsifier/stabilizer/nutrients, etc.

Functionally, the principal components of a soft drink may be characterized as in Table 1.

Sweeteners

The basic composition of a soft drink is determined by the selected 'balance', or sweetness-to-acid ratio, which is dependent upon the type of flavour and product in question. For example, a lightly flavoured, carbonated beverage like a lemonade will have a sweetness of approximately 9–10% equivalent sweetness with an acidity of 0.24% w/v as citric acid monohydrate (balance ratio = 39–40), whereas a sharp lemon juice carbonate will have a sweetness of 11% equivalent with an acidity of 0·6–0·7% w/v as citric acid monohydrate (balance ratio = 16–17) and the latter product will be a much sharper, stronger-tasting product than the former. For these reasons it is necessary to consider three things in the early stages of development of a soft drink: (1) the total sweetness required and its sources; (2) the necessary acidity and its sources; (3) the required 'body' or 'mouth feel' of the product.

It can be seen that at the very earliest stage in development the effects, both individually and combined, of sweeteners and acidulants (and all components contributing to them) must be assessed and the desired combinations decided upon. For this reason both 'bulk' and 'intense' sweeteners must be evaluated to select the desired effect, whether organoleptically or from an energy point of view. ***See*** Carbohydrates, Sensory Properties; Sweeteners – Intense

Table 1. Principal components in the composition of beverages

Ingredient	Contribution
Sugars	Flavour, sweetness, mouth feel/ body, fruitness, nutrition, facilitate water absorption (appearance and preservation in syrups)
Intense sweeteners	Flavour, sweetness, mouth feel/ body, fruitiness (bitterness)
Fruit/extract/milk/other characterizing ingredient (e.g. glucose syrup, spring or mineral water, etc.)	Flavour, body, appearance (nutrition)
Nutrient additions including salts	Nutrition; ascorbic acid and tocopherols are antioxidants also; controlled absorption of sugars and water
Acids	Flavour, antimicrobial effect, sharpness
Flavourings	Flavour, body, appearance;
Colourings	additionally colourings are also
Emulsifiers and stabilizers	nutrients
Antioxidants	Improved flavour and vitamin stability
Preservatives	Antimicrobial effect; sulphite also has antibrowning and antioxidant effectiveness
Acidity regulators	Improved dental safety; reduced can corrosion; body
Alcohol (beers, wines, spirits, etc.)	Body, mass, solvent, carrier, flavour, mouth-feel, bite, punch, flavour potentiator or releaser
Water	Bulk and mass; solvent carrier; thirst quenching

An overall assessment of the characteristics and properties of the most commonly encountered sugars and intense sweeteners used in soft drinks will be found in Table 2. The most frequently encountered carbohydrate sweetener in soft drinks is 'sugar', or sucrose, and the most used intense sweetener is aspartame, or brand name NutraSweet. An evaluation of all of the properties and characteristics of both of these two sweetening materials will show a similarity in performance at 10% sucrose or equivalent sweetness in the areas of sweetness quality, time profile, associated taste (or lack of one), mouth feel/body, and enhancement of fruitiness. It is this combination of clean sweetness, lack of aftertaste, good mouth feel/body and enhancement of fruitiness that is so essential to the basic taste performance of a soft drink. ***See*** Acesulphame; Aspartame; Cyclamates; Saccharin

Chemical Composition

Acidulants

To achieve the correct sweetness/acid ratio or 'balance' for the selected product it is necessary to select the acidulant for use in combination with the chosen sweetener system carefully. It must be remembered that titratable acidity and pH are not the only indicators of the performance of an acidulant. Different acidulants have different taste characteristics as an integral part of their character and these may be further modified by the use of buffers or 'acidity regulators'. The most commonly utilized acidulants in soft drinks are listed in Table 3 with details of typical usage levels and titration end-point information. ***See*** Acids, Natural Acids and Acidulants

Acidity Regulators

Acidity regulators are used to 'buffer' soft drink product acidity and thus afford some dental protection against very low pH erosion effects by raising the pH from as low as $<2{\cdot}0$ to 3·5. The human body is then more easily able to raise the mouth pH to the dentally 'safe' region of pH 5·0 to 6·0. Additional benefits are gained from extra protection against beverage can corrosion via this higher pH and an increase in the 'body' or 'mouth feel' of the product ensues.

Additionally, the raising of the pH of a soft drink to around 3·5 provides the optimum conditions for maximum stability of the intense sweetener aspartame in an acidic aqueous medium.

Fruit and Other Characterizing Ingredients

The main characterizing ingredients in soft drinks (excepting added flavouring) may be seen from Table 1 to be

Fruit
Extract of natural plant material, etc. (e.g. cola nut extract, ginger root extract, etc.)
Milk/yoghurt/dairy derived ingredients
Others, e.g.
(a) glucose syrup/fructose/maltodextrin (as used in energy/sport/rehydration drinks)
(b) mineral water/spring water/other sources (flavoured mineral water, fruit juice and mineral/spring waters, etc.)

By far the most extensive use is made of fruit (and its derivatives) as a characterizing ingredient in soft drinks and a very large proportion of the so-called soft drinks market is devoted exclusively to fruit juices and fruit juice drinks. In all ranges of soft drinks where an orange variant is offered for sale it, or its mixed varieties,

Table 2. Sweeteners used in beverages

Sweetener (100% solids for carbohydrates)	Sweetness intensity (10% sucrose)	Sweetness quality	Time profile	Associated taste	Mouth feel, body	Enhancement of fruitiness
Carbohydrate sweeteners						
Sucrose	1·0	Full rounded	Fast, slight linger	None	More than CHO[a]	Good
Invert sugar						
50% inverted	1·0	Close to sucrose	As sucrose	None	Typical CHO	Fair
100%	1·1					
Fructose	1·3	Slightly thin	Fast without linger	None	Typical CHO	Good
Glucose	0·7	Slightly thin	Fast without linger	None	Typical CHO	Fair
Glucose syrup						
42DE	0·33		Fast, some linger	None	More than other CHOs	Fair
63DE	0.50		Fast, some linger	None	More than other CHOs	Fair
Isoglucose	1·0		Fast without linger	None	Typical CHO	Fair
Intense sweeteners						
Saccharin	350	Slightly chemical sweetness	Slower and persistent	Bitter metallic aftertaste	Thin	Nil
Cyclamate	33	Slightly chemical	Slower and lingering	Off-taste at high concentration	Good	Good
1:10 saccharin/cyclamate	100	Sugar-like	As sucrose	None	Good	Good
Aspartame	140–200	Sugar-like	Slight delay, slight linger	Little	Fair to good	Good
Sucralose	450	Sugar-like	Fast, slight linger	None	Thin	Nil
Stevia extract	150	Clean sweetness	As sucrose	Slight liquorice/ menthol aftertaste	Thin	Good
Alitame	2000	Clean sweetness	Slight delay, slight linger	None	Thin	Nil
Acesulphame K	100	Good quality sweetness	Fast, slight linger	Bitter at high concentrates	Thin	Nil

[a] CHO=carbohydrate

Table 3. Acidulants for soft drinks

Acid	Typical product usage	Percentage (w/v) addition level range	Titration end-point information[a]
Citric acid monohydrate	All citrus and most other soft drinks	0·1–0·8	0·070
Citric acid, anhydrous		0·1–0·8	0·064
Lactic acid	Performance, specialist and milk products	0·1–0·3	0·090
Malic acid	Apple-based and some other soft drinks	0·2–0·60	0·067
Phosphoric acid[b]	Nearly all cola drinks	0·05–0·25	0·049
Tartaric acid	Specialist products	0·05–0·25	0·75

[a] Weight of acid in grams equivalent to 1·0 ml of molar sodium hydroxide solution.
[b] Phosphoric acid is a tribasic acid, the pH values at the equivalence point corresponding to the primary, secondary and tertiary stages of ionization are approximately 4·6, 9·7 and 12·6, respectively. Phenolphthalein is suitable for detecting the second ionization stage at pH 9·7 when the factor is 0·049 (mol. wt/2)

Table 4. Comparison of 100% fruit juices and fruit beverages

Characteristics	Unsweetened fruit juices	Fruit beverages including nectars
Composition	100% fruit	1·5–60% fruit and/or juice
Sugars	8–17% from fruit	0·1–20% (may be added carbohydrates)
Energy (kcal per 100 ml)	35–65	0·5–75
Acid (as citric acid)	0·6–1·5%; high acid juices up to 6%	0·1–0·8%
pH	2·5–4·0	2·5–4·0
Osmolality (mosmol kg^{-1})	620 upwards	200–1000 (maybe lower if intense sweeteners used)
Nutrients	Vitamins, mainly C some A; potassium	As added: usually vitamin C; minority multivitaminized
Other ingredients	Acid, vitamins, preservatives	Sugar, acid, intense sweeteners, flavourings, colourings, vitamins, minerals, preservatives

accounts for in excess of 50% of the product manufactured and sold. Fruit juices and fruit juice drinks have gained in popularity and volume market share by the introduction of long-life aseptic packaging such as Tetra Pak or Combi Pack allowing such products access to the volume market. Table 4 shows the comparison of 100% fruit juice soft drinks with fruit-based soft drinks.

Most fruits that may be harvested are represented in the soft drink industry by a product based on them being manufactured somewhere in the world. This is possible by the use of enzymes to allow fruits with very high pectin contents to have the juice/pulp extract processed and stabilized suitably for soft drink manufacture.

When milk, yoghurt or dairy-based ingredients are used in the composition of soft drinks the technologies applying to those individual materials, and in particular the microbiology and physical stability, must be very carefully controlled to enable a liquid soft drink beverage to be offered.

Nutrients

Where vitamins and nutrients are included in the beverage formulation the stability of these ingredients and their reactivity under all storage conditions (e.g. light/heat) must be evaluated prior to their admixing with the characterizing ingredient(s) or flavouring with which they will almost certainly interreact. The B group of vitamins is a very good example of unstable materials in aqueous medium under light-exposure conditions exhibiting off-flavour development in beverage formulations. ***See*** individual nutrients

Flavourings

Of equal, if not greater, importance to the formulation/composition of a soft drink is the use of added flavourings which may be categorized as follows:

Natural (ex named fruit) extracts or flavourings
Natural (with other natural flavours) flavourings (so-called WONF flavouring materials)
Nature-identical flavourings (identical chemical composition to nature but synthetic in origin and manufacture)
Artificial flavourings

Each of these four groups of flavouring materials is capable of either fully flavouring a beverage or providing whatever degree of partial flavouring is required. ***See*** Flavour Compounds, Structures and Characteristics

Flavouring materials are highly complex organic chemical compositions with as many as 200 individual raw materials and/or chemicals (with no upper limit). They are frequently composed of water-immiscible components (e.g. essential oils, oleoresins, etc) which are dispersed in a solvent by means of physical energy input or the use of *emulsifiers* (sorbitan, sugar esters and lecithins have been used for emulsifying citrus oils). Once emulsification has been effected, *stabilization* is achieved by the use of hydrocolloids, alginates, vegetable gums and xanthan gum. The solvents normally used are ethanol, isopropanol or propylene glycol. When making natural extractions of components like citrus oils the traditional method was to make a 'washing' of the oil by using a mixture of both water and alcoholic solvent so that all aqueous and oil phase components are removed. It is the greater solubility of the anise flavouring materials in alcohol which results in the immediate production of a white cloud when products such as Pernod and ouzo are diluted with water. ***See*** Emulsifiers, Organic Emulsifiers; Stabilizers, Types and Function; Stabilizers, Applications

Colourings and Antioxidants

A wide range of colouring materials is approved for use in soft drinks worldwide, with different inclusions and exclusions to the approved list in different countries. Generally colourings for soft drinks fall into three categories:

1. *Artificial.* The so-called coal-tar or azo dyes which are the most stable and produce the greatest effect versus usage rate of all food product colorants
2. *Nature-identical.* These are synthetic manufactures of naturally occurring substances such as the orange/red colour group of carotenoids (β-carotene, apocarotenal, etc.)
3. *Natural.* Colouring extracts obtained from natural plants and vegetables such as grape skin extract, used extensively in the colouring of all red/blue/black soft drinks like blackcurrant, blackberry and red grape products. ***See*** Colours, Properties and Determination of Natural Pigments; Colours, Properties and Determination of Synthetic Pigments

Colourings, particularly carotenoids, flavourings and some vitamins are susceptible to oxidation during the extraction/manufacturing process, during incorporation into the product or during use of the product by the consumer and this is often accelerated by exposure to heat and sunlight. Essential oils are usually protected against oxidation by the use of small quantities of butylated hydroxyanisole (BHA), butylated hydroxytoluene (BHT) or tocopherols during processing. Carotenoids and other sensitive components in a soft drink may be protected by the use of small quantities of sulphur dioxide or ascorbic acid (vitamin C), either separately or in combination. ***See*** Antioxidants, Natural Antioxidants; Antioxidants, Synthetic Antioxidants

Preservatives

Microbiological stability may be controlled in a soft drink by the following means:

1. *Acidification.* pH value lower than 4·0 confers almost 100% protection against pathogens (cf. 'pickling' of vegetables).
2. *Water activity.* Sugar contents in excess of 65% (e.g. French 'syrops') will confer protection against yeasts (cf. 'chutneys' where a combination of acidification and water activity reduction is used).
3. *Carbonation.* Carbonation in excess of 3·0 volumes Bunsen, preferably 3·5, may have a significant effect on yeasts, especially if product is de-aerated to remove the oxygen that yeasts need to multiply. (Volume Bunsen expresses carbonation as the number of volumes of CO_2 evolved from the product, using the product volume as a multiplying factor.) Secondary fermentation of champagne is a good example where carbon dioxide limits the process.
4. *Chemical.* Benzoic acid has a low taste threshold, low volatility and wide antimicrobial spectrum. Benzoate-resistant yeasts are common and the product must have a pH lower than 3·5 to enable full dissociation of sodium benzoate to benzoic acid (effective level = 200 ppm 'ready to drink' (RTD)). Sorbic acid provides good protection through its nonvolatility and effectiveness against yeasts. It has a low taste threshold over a wide range of flavours and to some consumers (effective level = 200–600 ppm RTD). Sulphur dioxide is the most common alternative to benzoic acid and has broad antimicrobial and antiyeast activity in the pH range 2·0–4·0. It is extremely reactive chemically and is unacceptable for use with orange juice and fruit, thiamin (vitamin B_1) and some natural colours, and interacts with beverage cans.
5. *Heat treatment* or *processing* is used either to replace chemical preservation or to augment it in some cases. Processes considered are hot-filling, in-pack pasteurization, microfiltration and aseptic filling. ***See*** Preservation of Food; Pasteurization, Principles

Alcohol

The use of alcoholic products such as beers, wines, spirits and liqueurs as ingredients in soft drinks has long been known, with products such as beer shandy and ginger beer shandy well established. More recently the use of wines and spirits with sparkling mineral and spring waters has resulted in the wine 'spritzers' we see today. These alcoholic components are distinguished from the use of alcohol as a solvent in flavourings and extracts by usage rate, which is normally up to 10% by volume (addition rates of flavourings are typically 0·1% and lower by volume) and by their taste effect of providing body/mouth feel, punch, etc. (solvents are normally considered tasteless). *See* Alcohol, Properties and Determination

Water

Water is sometimes used as a characterizing ingredient as in flavoured mineral waters, fruit juices and sparkling mineral waters, herbal extracts and mineral waters, etc. In this section it is considered as the principal component of soft drinks as it is of fruit juices. Water is necessary to the production of all soft drinks as the solvent and provider of rehydration nutrition. In this case the cleanest tasting, least interactive and microbiologically stable material is required.

Chemical quality requires:

- Appearance free from sediment, colour and cloud.
- Taste should be free from taint or materials such as chlorine, hypochlorite, or nitrates which are capable of reacting with other components and materials (e.g. cans).
- Freedom from toxins of any description.
- Freedom from water hardness which can destabilize fruit colloidal suspensions via the calcium and magnesium which cause water hardness.

Great attention is, and must always be, applied to the supply, treatment and quality of water for soft drink manufacture.

Bibliography

Hicks D (1990) *Non-carbonated Fruit Juices and Fruit Beverages.* Glasgow: Blackie.
Matthews AC (1991) *Food Flavourings.* Glasgow: Blackie.
O'Donnell K (1990) *Formulation and Production of Carbonated Soft Drinks.* Glasgow: Blackie.
Heath HB (1981) *Source Book of Flavours.* Westport: AVI.

Clive Matthews
NET International, Cinderford, UK

Production

Soft drinks have been consumed, in one form or another, for thousands of years from the first time it was discovered that the addition of acid materials such as lemon juice and/or the incorporation of only 5% of alcohol from fermentation of grape and other juices resulted in the preservation of dietary water. Although milk, wine and beer have all been around for well over a thousand years, and tea and coffee for many centuries, carbonated soft drinks are approximately 200 years old, dating from the commercial production of seltzers and sodas in the late 1700s. Revitalization of the 'squashes', 'syrops' or concentrated still soft drinks market has been facilitated in the last few decades by the introduction of aseptic packaging techniques enabling continuous filling of juices and ready-to-drink 'squash' type products into foil laminates, pouches and cartons. Some would argue, with justification, that this is a completely new sector of soft drinks made possible by advances in production and packaging technology.

The production of soft drinks falls into three main areas of activity:

1. Supply, handling and treatment of ingredients (including the water supply).
2. Blending of these ingredients and the processing of the blends/combinations so achieved.
3. Packing or packaging of these blended ingredients into the finished product for distribution and sale (this may involve a further processing step as in the case of aseptic filling or dilution and carbonation as with carbonated soft drinks).

Supply, Handling and Treatment of Ingredients

Most ingredients will be supplied in a 'user-friendly' and stable form by the ingredient manufacturers or suppliers, with detailed specifications and usage instructions. During the time these ingredients are held at the site of the soft drinks manufacturer it will simply be necessary for them to be stored according to the storage instructions of the supplier and used within shelf life. Some of the storage conditions and shelf lives of usually encountered raw materials will be found in Table 1. *See* Storage Stability, Parameters Affecting Storage Stability

It is normally the responsibility of the quality control department to approve the quality of all incoming materials and ensure correct rotation for use prior to expiry of shelf-life.

This leaves the main ingredient in soft drinks – water – which requires special treatment to ensure

Table 1. Storage conditions of soft drink raw materials

Material	Storage conditions	Recommended shelf life
Fruit		
Frozen	Below −18°C	2–3 years
Preserved	Ambient	3–9 months
Aseptic	*c.* 4°C	Up to 6 months
Sugar syrup	Dependent on concentration. Some may be stored at ambient	48 hours
Glucose syrup (GS)/high-fructose GS	35°C	48 hours–1 week
Citric acid solution	Ambient	1 week
Volatile and natural flavouring materials	4°C	3–6 months
Nature-identical and artificial flavouring materials	Dark UK; ambient	6–12 months
Water	Ambient	On demand

sensory, microbiological and physical acceptable quality. The chemical quality of water may be characterized as follows.

1. Appearance free from sediment, colour and cloud.
2. Taste free from taint or materials such as chlorine, hypochlorite, or nitrates which are capable of reacting with other components or materials (e.g. cans).
3. Free from toxins of any description.
4. Free of water hardness which can destabilize fruit colloidal suspensions via the calcium and magnesium which cause water hardness. ***See*** Quality Assurance and Quality Control

This is achieved in a soft drinks production plant in a number of ways and the following list comprises the main parameters covered in the specification (and hence treatment, where necessary) of water: appearance; pH; total dissolved solids; total hardness; alkalinity; nitrogenous compounds; chloride; organic content; microorganisms; phosphate; silicates; trace metals; chlorinated compounds.

Water Treatment

Chemical coagulation is widely used in soft drink production plants as a means of removing unwanted impurities. The incoming water passes into a reaction vessel where coagulating chemicals are added. The reaction products and impurities are precipitated and form a gelatinous sludge through which the treated water flows to a clear zone at the top of the vessel. Since small particles of the flocculant will travel with the treated water, it is normal practice to pass water treated in this way through a sand-bed filter. A typical system of this type is illustrated in Fig. 1. ***See*** Water Supplies, Water Treatment

The types of coagulants used are ferrous sulphate, aluminium sulphate, and Lapofloc PAC (a solution of polyaluminium chloride).

Chlorination is the next stage in the treatment of water. Free chlorine is usually used as the sterilizing agent at a level of about 6–8 mg l^{-1} of chlorine after the sand filter to ensure completion of the reaction. Either chlorine gas or hypochlorite solution may be used. Where ferrous sulphate is employed as the coagulant, free chlorine oxidizes the ferrous salt to ferric salt and is usually added to the coagulation tank where the residence time may be included as part of the contact time necessary for sterilization. Where aluminium sulphate is used, the chlorine is usually added after the sand filter and residence time in the coagulant tank is not part of the contact time.

Organic matter, sulphites, nitrites and ammonia (or its compounds) absorb or react with chlorine and the bactericidal action of chlorine will not be effected until after these reactions have taken place. Chlorine is best added to the coagulation tank or immediately after, since contact with raw, untreated water containing humic organic substances produces trihalomethanes (THMs) which would result in an unpleasant taint to the water if present in quantities in excess of 100 $\mu g\ l^{-1}$. Via the chlorination process all impurities will be oxidized, removing all taste and odour including those due to phenol and its derivatives.

The next stage in the process is the dechlorination of the water after sterilization by passing the water through an activated carbon filter which is housed in a vessel similar to that of the sand filter (see Fig. 1). The quantity of carbon will be selected to achieve a contact time of approximately 5 minutes with the water, removing all of the chlorine and any remaining organic molecules. Carbon filters are fitted with steam injection at the base to allow sterilization when required to reduce the gradually increasing microbiological contamination from its use.

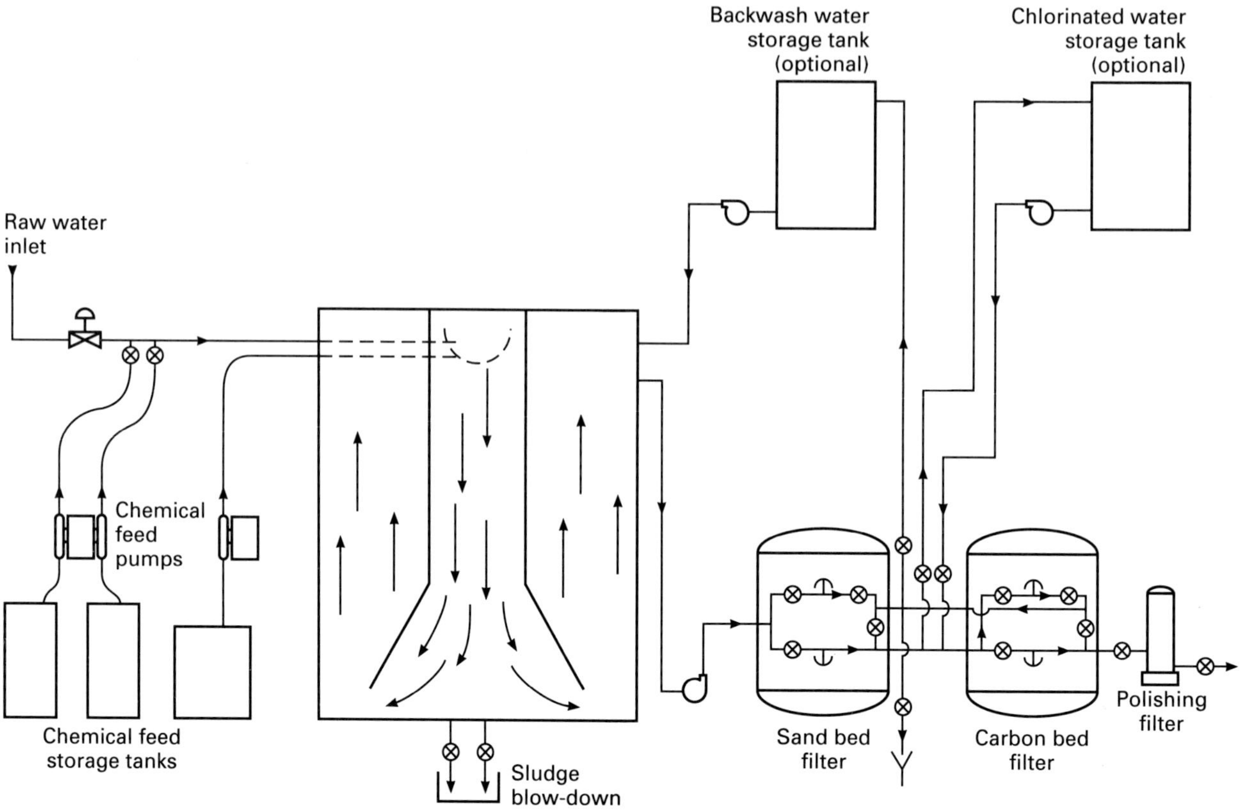

Fig. 1 Typical water treatment process in the production of soft drinks.

Quality control on a continuous basis of all aspects of the process must be effected to ensure optimum performance of the process. In particular it must always be able to indicate an imminent breakdown of the carbon bed when it requires regeneration.

The above process is the classical procedure for water treatment. There are, however, other methods of either total treatment or treatment for specific requirements of the water. They are as follows:

- Ion exchange will perform softening; dealkalization; nitrate removal; organic compound removal.
- Reverse osmosis will effect removal of high levels of dissolved solids resulting in a water which may be used directly or further treated for carbonated soft drinks production.
- Ultrafiltration removes colloidal substances and may be substituted for the polishing filter in a conventional process.
- Alternative sterilization processes include ultraviolet light, ozonization, and micropore filtration.

Blending and Processing of Key Ingredients in Soft Drinks

There are two basic methods of manufacture of soft drinks; these are the single 'batch' production and 'continuous' production. Batch manufacture is the process where a tank with a capacity of up to 25 000 litres may be employed to manufacture the product and/or intermediate syrup to feed the filling line over a period of time. Continuous production utilizes two tanks of only 50–100 litres each, with a computer, continuously making one batch after another in a type of tandem operation. The main advantages of the latter system are associated with line problems and interuptions, particularly where a heat treatment is part of the process, since only the 50 or 100 litres of product already manufactured will be affected and may easily be 'ditched' without significant economic consequences. In the case of batch manufacture, however, the 25 000 litres of product or intermediate syrup will either remain under non-optimum conditions or, at worst, continue to be heat processed (with a return to the batch tank) for the duration of the line 'down-time'.

In each case the steps and safeguards it is necessary to take in the production process are the same and will be covered in the same way. The schematic representation of the production of a soft drink is shown in Fig. 2. It can be seen that the processing required for the production of a soft drink involves the blending of the individual ingredients with heat treatment and/or homogenization where necessary. In reality all ingredients are not added separately to the main blending vessel, since this would cause significant logistical problems for items that are

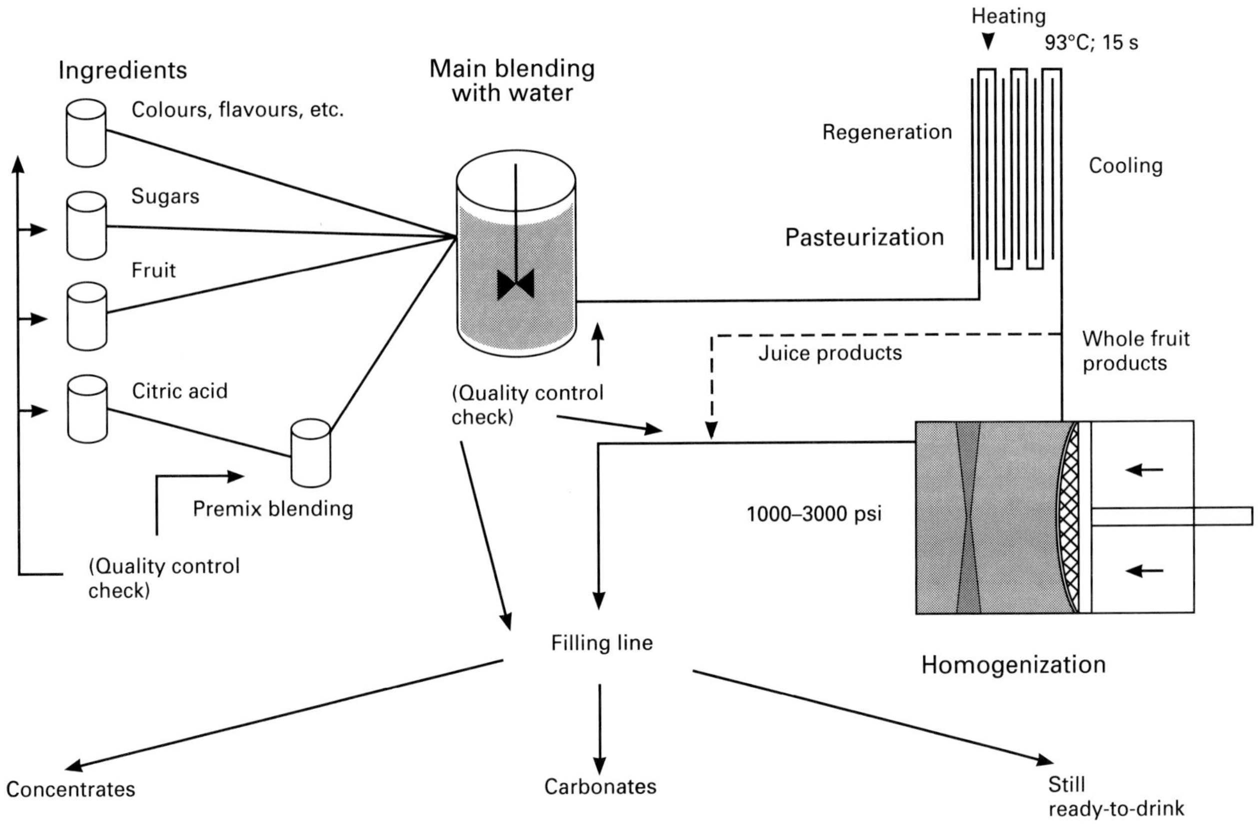

Fig. 2 Schematic representation of the production of soft drinks. See Fig. 3 for individual key stages.

difficult to handle, such as solid sugar (dissolution problems) or thickeners or stabilizers (localized viscosity build-up, lack of powder 'wetting'). There are basically four streams of ingredients:

- Colours, flavourings, etc.
- Sugars (i.e. all carbohydrates).
- Fruit materials.
- Citric acid, stabilizers, emulsifiers, etc. (this stream utilizes a 'pre-mix' blending vessel with a significantly increased blending power-to-volume ratio for difficult to dissolve/disperse ingredients).

Many soft drink production sites are now converting to tanks located on 'load cells' so that the vessel may be tared and ingredients weighed directly into them.

If liquid carbohydrates are being used, these may be added directly from their optimum storage position to the main blending vessel via an in-line sieve or filtration system to remove any small particles. If solid sugar is used, a 'simple syrup' must be prepared of the sugar and water at approximately 50°C to aid dissolution; this is then cooled before addition to the main blending vessel. All other ingredients are added to the carbohydrate syrup in the blending vessel in an order which takes into account their chemical reactivity with other ingredients. For example, sodium benzoate must always be added prior to the citric acid, since benzoic acid is only sparingly soluble in acid solution. Certain clouding agents and stabilizers must not be added in close proximity to a flavouring that contains a high level of an alcohol as solvent, since they will interreact.

With a high level of carbohydrate or thickening agent in the syrup in the main blending vessel it is very easy to entrain air with too much vigorous agitation, and this would result in fobbing during filling, particularly of carbonated products. As long as agitation has not been excessive then 2 hours standing after mixing, with gentle agitation to keep all ingredients evenly dispersed, will be sufficient to allow entrained air to escape. This gentle agitation should be continued for the duration of any pumping or direct bottling so that particulate materials, such as fruit comminute, are evenly dispersed throughout the liquid.

Homogenization

This process is used for many fruit-containing concentrates as part of the finished beverage manufacture en route to the bottling line and for intermediate syrup processing for some fruit-containing carbonates and ready-to-drink still beverages. The process is a means of reducing the particle size of particulate matter (e.g. fruit comminute) by means of energy input resulting in a more stable suspension and hence better appearance. The conditions which are normally used are 1000–3000 psi at ambient temperature.

Carbonation

The simplest type of carbonator is a pressure vessel in which the liquid and the carbon dioxide are allowed to stay in contact with each other. There are three basic ways to achieve this: (1) vessel partly filled with liquid and pressurized with carbon dioxide; (2) liquid falling through carbon dioxide in a pressurized container; (3) carbon dioxide bubbling through liquid in a pressurized vessel or pipeline. All carbonators operate on one or more of these principles.

One of the essential requirements of a production carbonation system is control over the degree of saturation of the carbonated product, i.e. how close to the maximum possible volume of carbon dioxide is allowed to dissolve. An uncontrolled release of gas from solution is known as 'fobbing'. Partially saturated solutions are more stable than saturated ones owing to the additional pressure available over the equilibrium pressure. This additional pressure is known as the 'overpressure' and is used in the control of the process. When the necessary degree of overpressure may be maintained over a range of temperature and carbonation, this is known as variable saturation. The ready-to-drink, dilute product may be carbonated by a reduction in temperature and injection of carbon dioxide via one or more of the above techniques, or the concentrated intermediate syrup may be blended with a stream of carbonated water on line to the filling apparatus, resulting in the correctly diluted, carbonated, finished product in the pack.

Hygiene

Throughout the production of all types of soft drinks it is essential that plant hygiene is maintained at the optimum level, whether chemical preservatives and/or heat processing are used.

Good manufacturing practice is a vital and integral part of ensuring minimal microbiological contamination of the product and will only be achieved if attention is paid to certain areas of operation:

- Hygienic practices in receiving, handling and processing of materials.
- Tight specifications for clean raw materials.
- Hygienic attitudes and understanding in all personnel.
- Hygienic design of plant and equipment.
- Hygienic operations of suppliers (especially of fruit and carbohydrate materials).

Heat Processing

An examination of the different types of processing for concentrates, carbonates and ready-to-drink soft drinks is shown in Fig. 3. This shows clearly that, in addition to chemical preservation, blending, carbonation and basic hygiene and quality control requirements, heat processing is a vital part of production of soft drinks.

Although a manufacturer's main protection against microorganisms is the hygienic operation of his manufacturing plant and that of his suppliers it may still be necessary with certain microbiologically sensitive formulations to enlist the help of both chemical preservatives and heat processing. The use of heat during the processing of soft drinks is covered in outline here as it relates to the microbiological integrity of the product formulation. ***See*** Preservation of Food

Heat processing, or pasteurization, may be subdivided into four main categories for soft drinks and/or material processing: (1) hot-fill; (2) in-pack (or tunnel) pasteurization; (3) flash pasteurization→(a) nonaseptic conditions or (b) aseptic conditions; (4) microfiltration. The selection of the required/desired heat processing conditions for any given soft drink, intermediate syrup/blend or raw material is determined by a number of factors relating to the following:

- Presence of fruit materials and their origin, quality and microbiological status.
- Presence of chemical preservatives – qualitative and quantitative factors apply.
- pH of the product.
- 'Solids' content and hence water activity.
- Any other relevant processing conditions – holding of syrup at ambient/raised temperature, use of homogenization, filtration, etc.
- Type of packaging to be employed.
- Desired shelf-life.
- Microbiological conditions (and history) of the filling-line and ancillary equipment to be employed. ***See*** Pasteurization, Principles

Hot-fill

This can only be used for finished still products or intermediate raw materials when being packed for storage and shipment. The product/material is heated in a heat exchanger to a temperature of 85–90°; filled into the pack (glass, cans, etc.) so that the temperature of the contents is in excess of 85°C; sealed via capping or seaming; inverted for a minimum of 1 minute; and cooled as quickly as possible. Any flaws in the capping/seaming will result in contaminated cooling water or air entering the package and will result in spoilage.

The disadvantage of this system is the large amount of heat that is put into the product/material during the whole process, which results in flavour, colour and

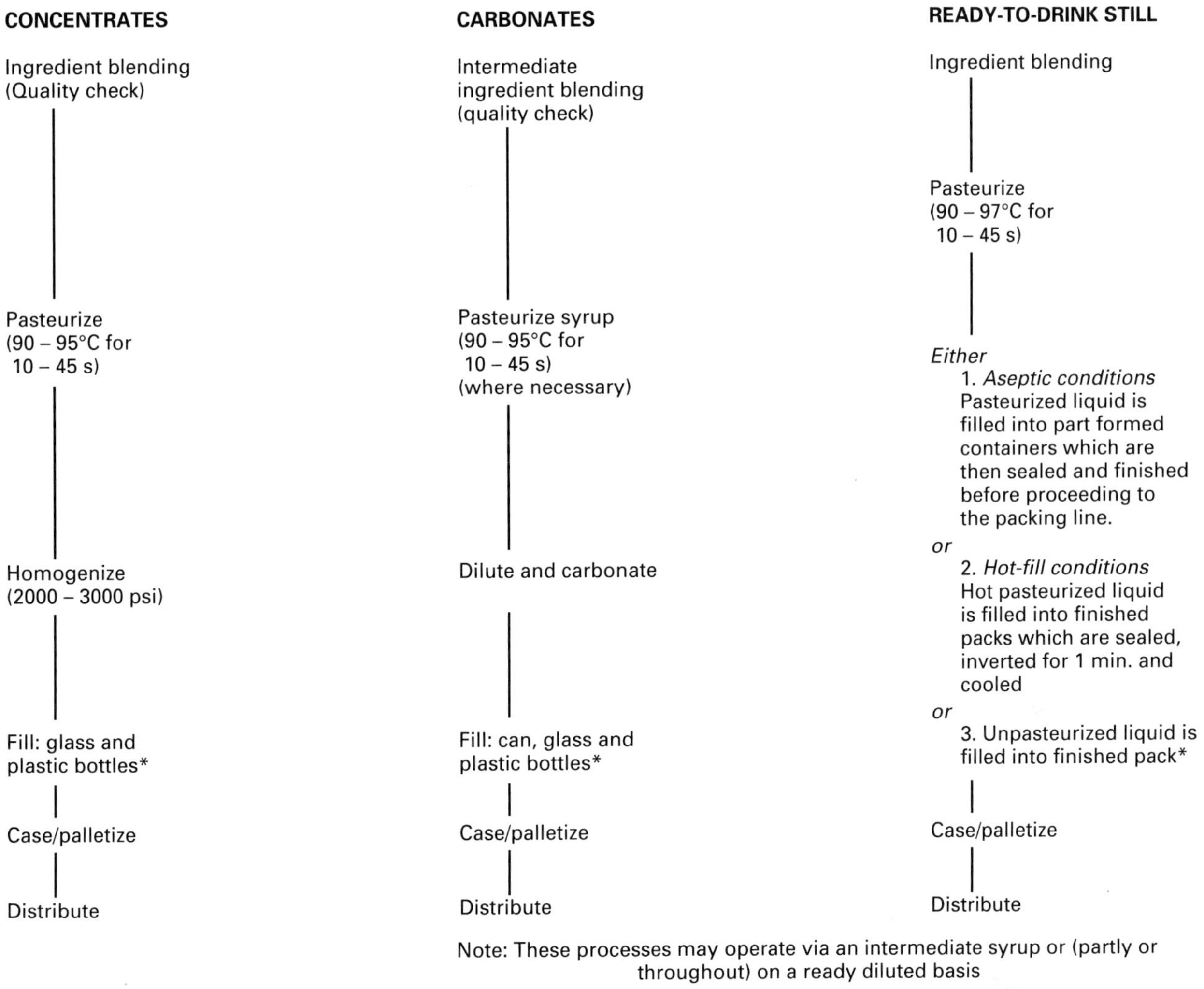

* In-pack-pasteurization performed where necessary with temperatures up to 70°C for up to 20 min.

Fig. 3 Key stages in the processing of soft drinks.

clouding degradation on storage, producing a reduction in total shelf life on the product/material.

In-pack (or Tunnel) Pasteurization

In its simplest form this type of pasteurization involves the immersing of the filled, capped or seamed pack of either still or carbonated product in a tank filled with hot water at 60–90°C in a removable basket and holding at the selected temperature for the required period of pasteurization holding time. In reality the packs are placed in water at a temperature lower than that required and the temperature of the water is raised until the temperature specified has been reached.

Examples of pasteurization conditions for this type of processing are:

63°C for 10 minutes: typically used for 'shandies;
70°C for 10 minutes: typically used for 'squashes'
75°C for 15 minutes: typically used for susceptible products such as high blackcurrant juice products.

Tunnel pasteurization may be considered as a continuous version of in-pack pasteurization.

After the product has been filled, packed and seamed, the packages are transported through a tunnel made up of different heat zones by means of water of different temperatures being sprayed over the packages, which gradually increase in temperature to the selected pasteurization level where they are then held for the desired time (holding time) before being cooled slowly to ambient temperature. Processing time is approximately 40–50 minutes, resulting in very effective pasteurization at relatively high cost (it is energy intensive and filling has to proceed at the speed of the pasteurizer).

Flash Pasteurization

Nonaseptic Conditions. This system may be used for carton, PVC, PET and all other non-glass/metal drinks packaging materials and is suitable for both still and carbonated beverages. The heat exchanger is connected in line with the product mixing tank in the case of still products and the intermediate syrup tank in the case of carbonated products. The syrup or finished product is heated very quickly to the specified pasteurization temperature, held for the specified time (usually quoted as seconds) and cooled rapidly to the filling temperature. The final drink may also be passed through the heat exchanger in this way en route to the carbonator if required.

The advantage of this type of pasteurization is that the total heat input to the product is only a fraction of that required for in-pack pasteurization and results in considerably reduced flavour, colour and shelf life effects. The disadvantage is that contamination may occur during filling and/or capping and rigorous standards of hygiene and microbiological control must be adhered to by fully trained and committed staff. It may even be necessary to fill some products/packs in enclosed areas with overpressure sterile air rooms.

Aseptic Conditions. Before a discussion on the types and methods of aseptic flash pasteurization it is necessary to consider exactly what this means in real terms.

Asepticity is sometimes referred to as commercial sterility, but it is necessary to recognize the important microbiological differences between the two types of processing employed.

With *sterility* it is realistic to expect that all measurable organisms have been 'killed' by the process, which almost always is carried out at temperature conditions well in excess of 100°C in hermetically sealed containers, thus ensuring the microbiological integrity of the whole pack (product plus packaging).

In *aseptic* packing the product is heat treated separately from the container and maintained in positive air pressure conditions (thus excluding any potentially contaminating organisms) until filled and sealed into the selected pack, which has been separately treated to reduce any microbiological contamination on its surface. Examples of aseptic conditions are those which apply to the Tetra Pak process where the product is flash-pasteurized at 96°C for 13 seconds and filled into cardboard previously treated with hydrogen peroxide; this is formed into a tube and the individual pack seams are carried out through the product.

With aseptic processing/packaging it is generally recognized that, independently of any pack integrity problems, the normal failure rate will be of the order of 1 pack in every 10 000 produced. A better rate than this is rarely achieved and aseptic conditions are sometimes quoted with failure rates as high as 1 in 3000, as in the case of carbonated products packed in PET containers.

Bibliography

Heath HB (1981) *Source Book of Flavours.* Westport: AVI.

Hicks D (1990) *Non-Carbonated Fruit Juices and Fruit Beverages.* Glasgow: Blackie.

Matthews AC (1991) *Food Flavourings.* Glasgow: Blackie.

O'Donnell K (1990) *Formulation and Production of Carbonated Soft Drinks.* Glasgow: Blackie.

Clive Matthews
NET International, Cinderford, UK

Microbiology

Soft drinks are potential media for microbial growth. This article reviews the growth of microorganisms in soft drinks and the effects of carbonation and preservatives. A range of testing regimes is also discussed. The microorganisms commonly associated with soft drinks are yeasts and these are not health hazards. Most food poisoning bacteria are unable to grow in the acid conditions found in the majority of soft drinks.

Many yeasts and other fungi have more than one generic and/or specific name and some microbes are better known by synonyms. Alternative names may therefore be included when the microorganisms are first mentioned in the text.

Raw Materials and Factory Hygiene

A soft drink within its final container represents a unique ecosystem containing a range of microorganisms. They may have come directly from the ingredients, e.g. from fruit in freshly squeezed, nonpasteurized juices, but more frequently they enter as recontamination.

Even where the raw materials do not directly add microbes to beverages, it is possible for ingredients to convey problem microorganisms, such as preservative-resistant yeasts, into the soft drinks factory; for this reason, consistently high quality from all suppliers, and routine microbiological testing of ingredients are essential. In fruit processing, rotting, mouldy and mummified fruits should have been carefully discarded and removed from the site. Mould spores easily disperse and may be difficult to kill by heat or chemical processing. Microbial growth can taint fruits, e.g. with diacetyl, and some fungi produce powerful heat-resistant toxins (mycotoxins) under defined growth conditions. Mycotoxins have been associated with fungal growth, mainly on low-acid foods such as grains and nuts, but where a particular fungus, e.g. *Penicillium expansum*, is growing on relatively low-acid fruits such as apples it is possible to recover the mycotoxin, patulin, from the apple juice. Mummified citrus fruits may harbour members of the notoriously preservative-resistant yeast genus, *Zygosaccharomyces*. ***See*** Mycotoxins, Occurrence and Determination

Table 1. Bacteria commonly associated with soft drinks

Genus	Nature	Typical spoilage
Acetobacter	Acid-tolerant, strictly aerobic, produces acetic acid. Converts acetic and lactic acids to carbon dioxide	Film-former, tainting and turbidity
Bacillus	e.g. *B. coagulans* Thermophilic, aerobic spore-bearer	'Flat sours' in canned tomato juice. Spores of various species isolated from ingredients, e.g. alcoholic extracts, neutral clouds
Clostridium	e.g. *C. pasteurianum* Mesophilic, anaerobic spore-bearer	Butyric tainting in tomato juice. Spores of various species isolated from ingredients
Gluconobacter (*Acetomonas*)	Acid-tolerant, strictly aerobic, produces acetic acid. Grows on fruit to produce sulphite-binding compound	Sometimes tainting and turbidity. Sometimes no apparent spoilage but reduces effective sulphur dioxide
Lactobacillus	Acid-tolerant, facultatively or strictly anaerobic. Produces lactic acid. Species produce a range of products, including carbon dioxide. Some species osmotolerant. Some species produce diacetyl. Some species produce dextrans or levans	Fermentation or tainting. 'Buttermilk' off-flavour. Ropiness or slime-former
Leuconostoc	Acid-tolerant, facultatively anaerobic, produces lactic acid. Some species also produce acetic acid, carbon dioxide, etc. Some species produce dextrans or levans	Fermentation or tainting. Ropiness or slime-former
Pediococcus	Acid-tolerant, microaerophilic	Tainting

Heat processing (pasteurization) is often used during soft drink production, e.g. for finished syrups or entire soft drinks, and this process is discussed in more detail below. Pasteurization is intended to achieve commercial rather than absolute sterility and will be ineffective against very high numbers of microorganisms protected by debris such as accumulated fruit pulp or syrup residue. High standards of hygiene contribute to effective processing and reduce the risk of recontamination. ***See*** Pasteurization, Principles

Soft Drinks as Growth Media

Where the finished product contains microorganisms they are often mixed populations, at different, possibly successive, phases of growth. Some will die or grow briefly in the finished product owing to depletion of nutrients and the effects of microbial competition. Others will survive and some of these will become spoilers. The factors affecting potential spoilage microorganisms' rates of growth and death are all interrelated and the most important with respect to soft drinks formulations (intrinsic factors) are as follows: available water, acidity, nutrients and inhibitors, redox potential, carbonation and preservatives. Extrinsic factors include the nature, condition and number of microbial contaminants, the processing (heat or filtration) and packaging.

Microbiologists have viewed these factors as a series of hurdles that reflect a particular product's robustness or susceptibility to microbial spoilage.

Yeasts are the major spoilage microorganisms encountered in soft drinks and these are discussed in detail in the text. Less commonly found bacteria and fungi are described in Tables 1 and 2. ***See*** *Bacillus*, Occurrence; *Clostridium*, Occurrence of *Clostridium perfringens*; Lactic Acid Bacteria; Spoilage, Bacterial Spoilage; Spoilage, Moulds in Food Spoilage; Spoilage, Yeasts in Food Spoilage

Intrinsic Factors

Water Activity (a_w)

Since soft drinks meet microbial requirements for available water, a_w is considered here only in terms of soft drinks ingredients and concentrates. Saturated sucrose solution has an a_w of 0·85 which is equivalent to the confectioners' scale of 67° Brix at 25°C. Definitions vary but the few yeasts or fungi that grow in saturated sucrose solution are described as 'osmophilic'. ***See*** Water Activity, Effect on Food Stability

'Osmophilic' or Xerotolerant Yeasts

The correct term is 'xerotolerant' since most of these microorganisms also grow in soft drinks. Osmophilic fungi tend to grow on solids; thus the osmophiles recovered from concentrates are usually yeasts. Members of the best known genus, *Zygosaccharomyces*, are often highly resistant to preservatives. Other osmo-

Table 2. Fungi commonly associated with soft drinks

Genus	Nature	Typical spoilage (usually fungal growth and pectin breakdown)
Aureobasidium (Pullularia)	Widespread contaminant, yeast-like	Rare
Byssochlamys or *Paecilomyces*	Frequently isolated from fresh fruits and juices. Very heat-resistant ascospores. Tolerates low levels of oxygen	Tainting, particularly in heat-treated drinks
Cladosporium	Widespread contaminant has yeast phase. Spores resistant to sanitizers. *C. herbarum* can grow at −5°C	Rare
Eurotium or *Aspergillus*	Frequently isolated from fresh fruit and juices. Some strains xerophilic or very xerotolerant. Some strains heat-resistant	
Geotrichum (oidium) ('machinery mould')	Widespread. Isolated from some fresh fruits, e.g. citrus, and dairy products. Yeast-like. Some strains resistant to fungicides	Tainting in unpasteurized citrus juices
Penicillium	Frequently isolated from fresh fruit and juices. Can spoil fruit, e.g. *P. digitatum* and *italicum* from citrus fruits, *P. expansum* from apples. Some strains are heat-resistant. Some strains are xerotolerant. Some strains can grow at −5°C	
Sclerotinia or *Botrytis*	Isolated from fruits, common fruit spoiler	

philic yeasts are species of *Hansenula*, *Kluyveromyces*, *Saccharomyces*, *Schizosaccharomyces* and *Torulopsis*.

Osmotolerant yeasts grow at slightly higher levels of a_w. They include members of the above genera and species of *Candida* and *Hanseniaspora* or *Kloeckera*.

Tanks or drums of syrups and concentrates are sometimes subjected to temperature fluctuations so that condensation forms and dilutes the sugar concentration of the surface layers. Under these conditions osmophilic and osmotolerant microorganisms have been known to grow rapidly.

Acidity (as pH Level)

Soft drinks have pH levels ranging from 2·0 to 4·5. Since relatively few microbes grow rapidly within this pH range, acidity is a most important factor contributing to microbial stability. Acidity comes from ingredients or from permitted food acidulents such as citric, lactic, malic and tartaric acids. The pH tolerance varies within genera and is modified by the acid(s) involved. ***See*** Acids, Natural Acids and Acidulants

Food poisoning bacteria rarely grow below pH 4·0 and bacterial spores are unlikely to germinate below pH 4·2. Bacteria most likely to be isolated from soft drinks are members of the acetic acid or the lactic acid bacteria. Species of these bacteria can adapt to growth at pH 2·0, as can some yeasts; thus acidification alone is no guarantee of inhibition of microbial growth.

Yeasts are more generally acid-tolerant than bacteria. The least acid-tolerant genus likely to be present is *Schizosaccharomyces* and the most acid-tolerant genera are *Dekkera* or *Brettanomyces*, *Candida*, *Torulopsis* and *Zygosaccharomyces*.

Less acid drinks are therefore at risk of spoilage from a wider range of microorganisms, some of which could grow faster than in more acid beverages. Tomato juice is often acidified to approximately pH 4·2 and heat-treated to destroy spore-bearing bacteria.

Exceptionally, growth of pathogenic bacteria may be traced to a product previously considered too acid. The non-spore-bearing, enteric pathogen *Salmonella typhimurium* has been isolated from freshly pressed apple juice at pH levels ranging from 3·4 to 3·9.

Acidity also increases the effectiveness of other factors, such as carbonation and the addition of preservatives.

Microbial Nutrients

Soft drinks usually contain ample supplies of carbohydrates, e.g. hexose sugars such as glucose or fructose, and organic acids, e.g. citric acid, in forms that are easily metabolized by microorganisms. The availability of nitrogen or vitamins in soft drinks selectively limits microbial growth because yeasts vary in their ability to utilize nitrogen sources and some are dependent upon particular vitamins.

Drinks with a limited range of microbial nutrients and high acidity, e.g. tonics, can be spoiled by only a narrow range of yeasts, usually *Candida*, *Dekkera*, *Pichia*, *Rhodotorula* or *Torulopsis*. Drinks containing a few organic ingredients, such as caramel, provide a wider range of microbial nutrients, and drinks containing high

Table 3. Preservatives permitted in soft drinks

Preservative	ADI (mg per kg bodyweight per day)	UK maximum level in soft drinks for consumption without dilution (ppm)	pK_a	Approximate percentages of undissociated acid at various pH levels		
				2·0	3·0	4·0
Benzoic acid	5·0	160	4·2	99	93	60
Methyl parabens	10·0	160	8·5	100	100	100
Ethyl parabens	10·0	160	8·5	100	100	100
Propyl parabens	10·0	160	8·5	100	100	100
Sorbic acid	25·0	300	4·8	100	97	82
Sulphur dioxide	0·7	70	2 pK_a values in equilibrium mixture			

ADI, Acceptable Daily Intake.

levels of fruit juices support the widest range of microorganisms, although the slower-growing yeasts, such as *Zygosaccharomyces*, are often overgrown by *Torulaspora* or *Saccharomyces*. ***See*** Caramel, Properties and Analysis

Microbial Inhibitors

Microbial inhibitors come from plant extracts, e.g. citrus and herbal oils, but they are present in very low concentrations and are effective only in combination with other inhibitory factors, such as high levels of acidity and carbonation.

Redox (Oxidation–Reduction) Potential (E_h)

The E_h of an ecosystem selectively affects the types of microbes and their metabolism, but it is a difficult factor to measure accurately and interpret. Redox potential depends upon the concentrations of oxidized and reduced substances in the soft drink, and upon its pH value. The dissolved oxygen content and E_h of a soft drink are increased by excessive headspace, underfilling or use of oxygen-permeable materials for containers, all favouring the growth of aerobic bacteria, film-forming yeasts, and most fungi. The E_h is decreased by pasteurization and by the inclusion of reducing agents such as ascorbic acid (vitamin C) and sulphur dioxide. ***See*** Antioxidants, Natural Antioxidants; Ascorbic Acid, Properties and Determination

Effect of Carbonation

Carbonation is used in soft drinks at 1–5 vol and has a selectively antimicrobial action at volumes over 2·5–3·0 in drinks with low levels of pH, microbial nutrients and initial microflora. This effect is probably mostly attributable to the removal of oxygen as it is greatest against obligate aerobes. Most yeasts are facultative anaerobes. Carbonation may reduce the rate of initial aerobic yeast growth, but have little or no effect upon the succeeding phase of fermentation (anaerobic). Members of the lactic acid bacteria differ in their respiratory requirements, so that effects will vary. Carbonation increases acidity very slightly, thus increasing intrinsic microbial resistance.

There are significantly fewer bacteria in carbonated mineral or spring waters than in similar still waters bottled at the same time. Carbonation destroys natural water bacteria (mostly aerobic) and is effective against enteric pathogens such as *Escherichia coli*, although some may survive for several weeks. ***See*** Enterobacteriaceae, Occurrence of *Escherichia coli*

Effect of Preservatives

The preservative is the most easily adjustable inhibitory factor and its choice should therefore be given careful consideration. Use of preservatives cannot be a substitute for good hygienic practices, since if levels of microbes are too high resistant strains may evolve. Preservatives permitted in soft drinks are benzoic and sorbic acid, and sulphur dioxide. Esters of *p*-hydroxybenzoic acid (parabens) are not accepted in some countries. ***See*** Preservation of Food

There are many interrelated factors that could affect preservative ability. For benzoic and sorbic acids or their salts the antimicrobial activity comes from the undissociated acid molecules. The availability of these increases with their dissociation constant, pK_a (the pH at which 50% of the total acid is undissociated); see Table 3.

Other factors affecting a preservative's availability are its reactivity, solubility and stability. The effectiveness of the preservative against particular microorganisms will depend upon their numbers, their condition and their ability to become resistant. Since so many variables are involved, the overall effectiveness of preservatives in particular soft drinks is best determined by using model systems and confirming by factory trials or challenge tests with resistant microflora from production sites.

Benzoic Acid

Benzoic acid is generally effective in soft drinks with pH levels less than 3·0. Some bacteria (not typical soft drink spoilage organisms) can degrade benzoic acid and destroy its preservative effect.

Some yeasts, such as *Candida, Saccharomyces, Torulopsis* – potential soft drink spoilage organisms – can also metabolize benzoic acid and become resistant to it. A more rapid acquisition of resistance to benzoic or sorbic acid is found in strains of *Zygosaccharomyces* (formerly *Saccharomyces*) *bailii* and *bisporus*. Growth has been reported at preservative concentrations much higher than those permitted in soft drinks.

Benzoic acid may be a less suitable preservative for some ingredients of soft drinks. It enters the lipid phase of acidic fruit flavour oil in water emulsions and could bind with proteins if these were present as emulsifiers or stabilizers.

Sorbic Acid

Sorbic acid (2,4-hexadienoic acid) in its undissociated form is available at higher proportions than the undissociated form of benzoic acid in soft drinks with pH levels over 3·0. It is generally effective against yeasts and other fungi, but could be less effective against some bacteria, including lactic acid bacteria, which may degrade it. Some yeasts and bacteria can metabolize sorbic acid and gain energy from it, and some *Zygosaccharomyces* strains become very resistant to it.

Parahydroxybenzoate Esters (Methyl, Ethyl and Propyl Parabens)

The availability of parabens esters is not affected by the pH values of soft drinks. Solubility diminishes as the length of the alkyl side-chain increases, but antimicrobial activity increases with the length of these side-chains. The mode of action is regarded as similar to that of benzoic acid, but it is possible that irreversible reactions also occur and there is no evidence that a rapid resistance, similar to the resistance to benzoic acid, can be induced.

Sulphur Dioxide

Sulphur dioxide is selectively microbiocidal against some bacteria, yeasts and other fungi. Where it is effective, concentrations as low as 30 ppm of free sulphur dioxide can be used. In common with the other preservatives, the effect of sulphur dioxide is enhanced by the acid conditions of most soft drinks. It dissolves in water to form a pH-dependent equilibrium mixture, the antimicrobial effect of which is mostly from the unbound nonionized molecular sulphur dioxide. The proportion of this increases as the pH decreases below 4·0.

Oxidative yeasts such as *Pichia membranaefaciens*, and *Rhodotorula* and the yeast-like fungus *Geotrichum* are more sensitive to the effect of sulphur dioxide than are the fermentative yeasts such as *Hansenula, Saccharomyces* and *Zygosaccharomyces*.

Sulphur dioxide is very reactive, forming sulphites that show little antimicrobial activity and varying degrees of stability ('bound' sulphur dioxide). These sulphites are often formed with the fruit components of soft drinks. Analyses for 'free' and 'total' sulphur dioxide are essential since regulations refer to maximum total concentrations, i.e. 'free' and 'bound'. Combinations of preservatives are used successfully but there is insufficient evidence for claims of synergism.

Extrinsic Factors

Extrinsic factors that affect the microbial stability of soft drinks include the nature of the microorganisms, e.g. their resistance to preservatives or heat, and the type of heat or filtration processing.

Microbial resistance to heat for pure cultures is expressed in terms of D and z values. The D value or 'decimal reduction time' is the time taken to kill 90% of the microbial population at a given temperature. The z value is the temperature rise required for a 10-fold reduction in the D value. It has been known for many years that the spores of various bacteria and fungi found in canned fruit products are heat-resistant but the heat resistance of some yeasts ascospores has been confirmed more recently. Heat resistance in yeasts is strain-specific and varies with factors such as the acidity, fruit pulp content, concentration of soluble solids and proportion of ascospores (sexually derived spores formed by yeasts and other fungi in the subdivision Ascomycotina). Some strains of yeasts produce ascospores that are considerably more heat-resistant than their vegetative cells. Heat resistances of yeast ascospores have been recorded at levels over 200 times those of corresponding vegetative cells.

Testing Regimes

Soft drinks are considered 'low microbiological risks' to consumers. However, isolation of potential food poisoning bacteria from the less acidic drinks is not unknown. Detailed specifications are inappropriate here as they must relate to the factors listed under *Soft*

Drinks as Growth Media (see above), plus projected shelf life and storage. Testing regimes should combine the microbiologist's experience of the production microflora with data from authorities such as the International Commission on Microbiological Specifications for Foods (ICMSF). Traditional microbiological tests provide retrospective information after incubation for 1–10 days. Such tests are therefore best regarded as part of a continuous quality assurance operation that includes physical, chemical and organoleptic checks to provide process control and trend analysis. ***See*** Hazard Analysis Critical Control Point; Quality Assurance and Quality Control

Sampling

Samples are defined by date, batch, time, site, etc. General discussions of microbiological sampling and testing may be found in the references cited in the *Bibliography*. These recommend that for soft drinks 0·01–0·1% of finished pack units should be examined as forcing samples (storage samples of finished products that are incubated for up to 8 weeks before visual or organoleptic examination) daily from production lines once 'commercially acceptable' spoilage levels have been achieved. It is important not to concentrate solely on finished products. Critical control points should be identified in raw materials, formulations, production or packaging so that relevant intermediate samples may be taken. Samples are rarely representative and it is common practice to take either random or 'worst case' samples of bottled products for microbiological testing, assuming the latter to be the first bottles filled at the start of production or after down time. Routine control should include tests of ingredients, finished syrups, and swabs of equipment. Containers, closures and factory air are sampled less frequently unless there is reason to suspect them, as there may be, for example, with washed bottles.

Guidelines

Acceptable levels of yeasts in freshly packed, preserved and/or pasteurized soft drinks vary between 1 and 25 colony-forming units (cfu) per 5 ml. A realistic volume of the drink (e.g. 10 ml) should be tested and this must increase in proportion to the pack size.

Test Methods

Conventional enumeration methods attempt to grow cfu from the samples in media until colonies are formed and counted. Media for isolating microflora of soft drinks are general, selective or based upon particular drinks, .e.g tomato juice agar. Selective acidic media are less favoured now that these have been shown to prevent recovery of 'stressed' yeast cfu.

Dilution or replicate sample volumes of 0·1–5·0 ml are usually tested by spread or pour plates, and larger volumes by most probable number (MPN) methods or, if possible, membrane filtration. The accuracy and reproducibility of these methods rely very much upon operator technique and their usefulness depends upon interpretation of results. However, experienced microbiologists can also gain insight into the probable identities of isolates.

Direct microscopy gives rapid checks for gross contamination only. It is used traditionally for estimating fungal spores and fragments in fruit juices (Howard count) and factories (*Geotrichum*).

Conventional methods can be replaced by faster instrumental tests. These should be fully validated as they present different types of data relating to measurement of cell constituents, impedance changes in media or increases in microbial by-products. When established, these methods are less dependent upon operator technique and microbiological experience. Since they also give less opportunity for recognition of microbial isolates, their relevance must be carefully assessed.

Bibliography

Batchelor VJ (1984) Further microbiology of soft drinks. In: Houghton HW (ed.) *Developments in Soft Drinks Technology*, vol. 3, pp 167–210. Essex: Elsevier Science Publishers.

Davenport RR (1980) An introduction to yeasts and yeast-like organisms. In: Skinner FA, Passmore SM and Davenport RR (eds) *Biology and Activities of Yeasts*, London: Academic Press.

Gutteridge CS (1984) Rapid methods for microbiological quality control of foods. In: Birch and Parker (eds) *Control of Food Quality and Food Analysis*, pp 157–179. Essex: Elsevier Science Publishers.

ICMSF (1980a) *Microbial Ecology of Foods: 1. Factors Affecting Life and Death of Micro-organisms.* The International Commission on Microbiological Specifications for Foods. London: Academic Press.

ICMSF (1980b) Soft drinks, fruit juices, concentrates and fruit preserves. *Microbial Ecology of Foods: 2. Food Commodities*, pp 643–668. The International Commission on Microbiological Specifications for Foods. London: Academic Press.

ICMSF (1982) *Micro-organisms in Foods: 1. Their Significance and Methods of Enumeration* 2nd edn. International Commission on Microbiological Specifications for Foods. Toronto: University of Toronto Press.

ICMSF (1986) *Micro-organisms in Foods: 2. Sampling for Microbiological Analysis: Principles and Specific Applications* 2nd edn. The International Commission on Microbiological Specifications for Foods. Oxford: Blackwell Scientific Publications.

ICMSF (1988) *Micro-organisms in Foods: 4. Application of the Hazard Analysis Critical Control Point (HACCP) System to Ensure Microbiological Safety and Quality*. The International Commission on Microbiological Specification for Foods. Oxford: Blackwell Scientific Publications.

ICUMSA (1986) *Report of the Proceedings of the Nineteenth Session of the International Commission for Uniform Methods of Sugar Analysis*, pp 356–376. Peterborough: Publications Department, British Sugar plc.

Kreger-Van Rij NJW (ed.) (1984) *The Yeasts – A Taxonomic Study* 3rd edn. Amsterdam Elsevier Science Publishers.

Put HMC and De Jong J (1980) The Heat Resistance of Selected Yeasts Causing Spoilage of Canned Soft Drinks and Fruit Products. In: Skinner FA, Passmore SM and Davenport RR (eds) *Biology and Activities of Yeasts*, pp 181–213. London: Academic Press.

Schneider F (ed.) (1979) *Sugar Analysis. Official and Tentative Methods Recommended by the International Commission for Uniform Methods of Sugar Analysis (ICUMSA)*, pp 150–160. Peterborough: Publications Department, British Sugar plc.

Tilbury RH (1980) Xerotolerant (osmophilic) yeasts. In: Skinner FA, Passmore SM and Davenport RR (eds) *Biology and Activities of Yeasts*, pp 153–176. London: Academic Press.

V. J. Batchelor
Contract Micro Services, Redbourn, UK

Dietary Importance

Nutritional Composition

Soft drinks may make a valuable contribution to fluid intake and have become, to some extent, established as part of the daily diet, particularly of young children and adolescents. The nutritional value of a number of readily available soft drinks is shown in Table 1. Squashes, crushes, cordials and carbonated drinks are, however, of little nutritional value (apart from their energy content) as their main ingredients are water and sugar. These soft drinks can be a valuable source of vitamin C, although it is unlikely that they will contain significant quantities unless the vitamin is added. Soft drinks do not contain fat or fibre but may contain nutritionally insignificant traces of protein. *See* Ascorbic Acid, Physiology

Energy Content

The energy content of soft drinks varies greatly and is derived exclusively from the sweetening agents, which are principally sugars (Table 1). Soft drinks sweetened with mixtures of sugar and intense sweeteners are less caloric than drinks sweetened entirely with sugar, and drinks labelled as being low-calorie are required by UK law to contribute a maximum of 22 kJ (5 kcal) per 100 ml.

Sugar Content

The added sugar content of soft drinks varies from 6% to 10% and is mostly glucose and fructose with small quantities of sucrose and perhaps maltose (Table 1). The sugar content of soft drinks is regulated by the UK 1964 Soft Drinks Regulations (amended 1969, 1970 and 1976). The nutritional value of a soft drink as consumed is dependent upon the dilution factor, which must now be stated on the label of all dilutable drinks. Soft drinks are a major market for intense sweeteners, particularly in the United Kingdom where, unlike many other countries, they can be used in conjunction with nutritive sweeteners and are not therefore limited to use in dietetic beverages. There are technical reasons, economic objectives and health considerations supporting the need for intense sweeteners in soft drinks. No single sweetener is ideally suited to meet all the requirements of soft drinks, but by using them in combination the limitations of one

Table 1. Nutritional composition of soft drinks per 100 ml

Beverage	Moisture (g)	Energy (kJ)	Carbohydrate (g)	Glucose (g)	Fructose (g)	Sucrose (g)	Maltose (g)
Citrus squash	72·0	398	24·9	10·5	10·5	3·8	—
Citrus drink	74·1	341	21·3	11·6	9·1	0·5	3·2
Citrus crush (nonconcentrated)	80·1	158	9·9	4·1	4·1	1·6	—
Low-calorie citrus squash	95·4	37	2·3	1·2	1·1	—	—
Low-calorie citrus drink	94·6	86	5·5	1·4	1·4	0·4	—
Low-calorie citrus crush (nonconcentrated)	98·7	13	0·8	0·2	0·6	—	—
Carbonates	90·7	152	9·5	4·0	4·8	1·7	—
Low-calorie carbonates	99·6	0	0	0·03	0·03	0·06	—
Ginger ale citrus drink	95·9	62	3·9	1·7	1·6	0·5	—

Unpublished data with kind permission from the Ministry of Agriculture, Fisheries and Food, UK.

Table 2. Consumption (grams per week) of soft drinks by adults by age

	Age (years)			
Drink	16–24	25–34	35–49	50–65
Low-calorie	778 (18)	1039 (17)	743 (16)	837 (12)
Other	1689 (82)	1073 (68)	889 (56)	667 (47)

Figures in parenthesis indicate percentage consuming the drink. Data from The Dietary and Nutritional Survey of British Adults (Gregory J *et al.*, 1990).

sweetener can be offset by the strengths of another. The sweetening of soft drinks with artificial sweeteners reduces the energy content and encourages fluid consumption without decreasing the nutrient density of the diet. Those drinks sweetened with aspartame may contain the amino acid phenylalanine in amounts which may be undesirable for individuals with phenylketonuria, a rare genetic disorder of metabolism that requires the patient to follow a special diet. The more liberal use of artificial sweeteners, the establishment of Acceptable Daily Intake levels and the increased number of sweeteners available has resulted in a wide range of low-calorie beverages which are acceptable to many consumers. Consumers have been advised to ensure that they consume a mixture of sweeteners to avoid large intakes of any one type. *See* Aspartame; Carbohydrates, Sensory Properties; Sweeteners – Intense

Consumption Patterns

Information on consumption levels of soft drinks within the United Kingdom is derived mostly from three sources: the National Food Survey, The Dietary and Nutritional Survey of British Adults and individual dietary surveys which are reported in the scientific literature. The UK National Food Survey omits foods which are bought for consumption outside the home. Information on soft drinks brought into the home is recorded but this information is not included in the major analysis of the survey data and is reported separately. Together with alcoholic drinks, sweets and chocolate, for which no information is collected, soft drinks are often bought without the knowledge of the householder and are therefore liable to be under-reported. Individual surveys have focused little on soft drink consumption in adults although there is more information available on children's eating habits in this respect. The consumption of diet and other soft drinks was reported in the Dietary and Nutritional Survey of British Adults (Gregory *et al.*, 1990) (see Table 2).

The consumption of soft drinks in all age groups continues to increase and in 1989 was reported to be 7524 million litres per year with a total market value of £4·7 billion. This figure excludes the consumption of the increasingly popular fresh fruit juices. Consumption levels as reported in the UK National Food Survey can be seen in Table 3.

While meal patterns still conform to what could be considered traditional, there is nevertheless a shift towards greater emphasis on snacking, particularly amongst the younger age groups. Concern has been voiced regarding the nutritional quality of the diets of schoolchildren and snack eating in particular has been singled out as providing 'empty calories' but little else in the way of nutrients. Soft drinks are very popular amongst children of all age groups and are frequently consumed between meals. *See* Snack Foods, Dietary Importance

Contribution of Soft Drinks to Nutrient Intake in Schoolchildren

As many soft drinks are a source of energy, there is concern that children may obtain a significant proportion of their total daily energy requirement from the sugar contained within soft drinks. However, there is little published evidence to support this. In the scientific literature soft drinks have been reported to provide between 0·7% and 4·8% of total daily energy intake. With the increasing popularity of soft drinks, the wider variety of those available and powerful advertising, it may be possible that children will reduce the nutrient density of their diet and be nutritionally at risk by consuming large quantities of soft drinks, replacing perhaps more nutrient-dense foods and liquids such as milk. In a dietary study of 143 schoolchildren aged 11–12 years, Nelson and colleagues (1991) examined food consumption and nutrient intake using the 7-day dietary record technique. Soft drinks, confectionery and table sugar contributed 10% of dietary energy and 40% of the sugar in the diet. Soft drinks were consumed at 1800 g per head each week, providing 4·7% of total energy intake and contributing 21·4% to total sugar intake. These values compare well with values obtained in the survey of British schoolchildren (Department of Health, 1989b), where the mean consumption of soft drinks was 1775 g per head per week. There was no evidence from the anthropometric data that a high percentage of energy from sugar was associated with obesity. In fact, the average body mass index was lowest in the groups with the highest percentage sugar intake, hence the putative effect of percentage of energy from sugar on bodyweight was not demonstrated in this study.

In another study investigating 11–14-year-old Northumbrian children (Rugg Gunn *et al.*, 1986), added sugars contributed on average 15% of dietary energy

Dietary Importance

Table 3. Purchase quantity of soft drinks

Drink	Quantity purchased (ml (fl. oz) per person per week)				
	1984	1987	1988	1989	1990
Low-calorie	26 (0·91)	58 (2·04)	78 (2·74)	109 (3·85)	129 (4·55)
All soft drinks	844 (29·7)	855 (30·1)	886 (31·2)	1020 (35·9)	1071 (37·7)

Data from *Household Food Consumption and Expenditure* 1990 (with a study of trends over the period 1940–1990), Ministry of Agriculture, Fisheries and Food, UK, Annual Report of the National Food Survey Committee. London: HMSO, 1991.

and 69% of total sugars intake. Soft drinks contributed 3% of energy and contributed 17% of total sugar intake. Again, this report was consistent with the results presented by Nelson (1991).

Hackett *et al.* (1984) estimated that if half the soft drinks consumed were sweetened with artificial sweeteners alone, sugar consumption would fall by about 10%. Soft drinks are very popular with young children and are universally consumed, with 85–95% of those children surveyed in the UK study (Department of Health, 1989b) consuming these items during the study period. Soft drinks are without doubt one of the most commonly selected snacks amongst schoolchildren and it is reported that children find it difficult to rank fizzy drinks as either healthy or unhealthy and that they make their choice according to taste preference. ***See*** Children, Nutritional Requirements; Obesity, Aetiology and Assessment

In the British study of 15–18-year-olds (Bull, 1985) there was an apparent decline in the consumption of soft drinks with increasing age. Average consumption fell from 200 g to 155 g and 220 g to 145 g soft drinks per day for female and male subjects, respectively. By adulthood the consumption of soft drinks has fallen markedly (see Table 2).

Further studies are required to examine changing dietary patterns and the influence of increased soft drink consumption on total energy intake and nutrient density.

Products for Specific Uses

Diabetic Drinks

There are a number of soft drinks which are sold specifically for use by individuals with diabetes mellitus. By law these must not contain any sugars other than fructose, which is metabolized mainly in the liver, the initial steps in its metabolism being independent of insulin. Some products are sweetened with sugar alcohols (sorbitol, xylitol, mannitol, lactitol) but these offer no advantage over fructose and do not result in a reduced energy-drink. It has been suggested that the daily intake of sugar alcohols should not exceed 25 g as they may result in osmotic diarrhoea. There is also an increasing number of low-calorie drinks in which the sugar has been replaced either wholly or in-part by an intense sweetener, reducing the energy content significantly. The nature of these drinks enables those on reduced-energy diets to consume them without adding significantly to their energy intake.

The need for dietetic products of this nature is recognized by the provisions of the UK Soft Drinks Regulations (1964, as amended), which exempt low-calorie soft drinks from limitations imposed on the saccharin content of soft drinks in general. The relatively low caloric limits (22 kJ (5 kcal) per 100 ml) imposed on soft drinks compared with other foods make it virtually impossible to include carbohydrate sweeteners in low-calorie soft drinks. ***See*** Saccharin; Sugar Alcohols

Sports Drinks

A range of foods, drinks and supplements has been launched to fill the requirements of the 'get fit' craze for consumption before, during and after exercise. A relatively new phenomenon, these drinks are promoted as being designed to replace water and electrolytes lost through sweating during exercise. Most of the drinks are mixtures of sugar, salt, potassium and water with a little vitamin C. Some of these drinks claim to be isotonic and some hypotonic. Dehydration does impair performance and therefore anything which decreases dehydration will maintain optimum performance for longer, although rehydration does not improve performance *per se*. In many circumstances water would be an acceptable alternative.

Infant Drinks

Infant drinks, including baby herbal drinks, have been developed to provide a thirst-quenching drink to coincide with the reduced consumption of milk that occurs in the first year of life. Although manufacturers are obliged to recognize that cooled, boiled water is the best refreshment to offer infants, they suggest that many will refuse this and offer their products as an alternative. The infant drinks are lower in sugar than baby fruit juices and unsweetened orange juice and usually provide the recommended daily intake of vitamin C. Potassium citrate is added as an acidity regulator, reducing the intensity of any acid, and infant drinks are of lower osmolality than conventional baby fruit juices. Health educators suggest that after the age of 6 months babies can be offered regular cow's milk or diluted natural fruit juice or water as a refreshment. ***See*** Infants, Nutritional Requirements

Soft Drinks and Health

Dental Caries

Extensive evidence suggests that sugars are the most important dietary factor in the aetiology of dental caries. Caries experience is positively related to the amount of nonmilk extrinsic sugars in the diet and the frequency of their consumption. Dental caries is dependent upon bacterial growth on the tooth surface, metabolism of sugars in the mouth by these bacteria and the formation of acid which attacks the teeth. It is well known that one of the main aetiological factors is the length of time sugar is in the mouth. ***See*** Dental Disease, Role of Diet

In younger age groups there has been a shift away from traditional eating patterns of three or four meals per day towards an increasing reliance on snack foods, fast foods and convenience foods accompanied by a range of sweetened or naturally sweet soft drinks consumed at frequent intervals throughout the day. There is therefore particular concern about the effect of soft drink consumption on dental health. Excessive use of soft drinks has been attacked on two accounts; firstly, almost all of them are fruit based or carbonated or both and may therefore be acidic enough to erode the surfaces of the teeth not covered by dental plaque. Secondly, those which contain fermentable carbohydrate may serve as a source of substrate diffusing into the dental plaque, from which microorganisms inhabiting the plaque can generate acid which, in turn, brings about the destruction process of dental caries.

Carbonated drinks appear to be less cariogenic than pure orange juice and apple juice drinks and erosiveness may be more important than cariogenicity. Many soft drinks do contain sugars and if allowed to reside in the mouth over long periods of time, as part of a frequent, protracted sugar intake pattern, then it is likely that they will be able to contribute to the caries process in which sugars serve as a substrate for acid formation.

Protective factors such as calcium and phosphorus may help to limit demineralization of teeth. In studies where iced lollies were fortified with calcium and phosphorus this supplement substantially improved dental properties. Use of intense sweeteners in soft drinks reduces their viscosity and may assist in shortening of exposure time.

Bibliography

Bull NL (1985) Dietary habits of 15–18 year-olds. *Human Nutrition Applied Nutrition* 39A (supplement 1): 1–68.

Department of Health (1989a) *Dietary Sugars and Human Disease*. Reports on Health and Social Subjects, 37. London: HMSO.

Department of Health (1989b) *The Diets of British Schoolchildren*. Report on Health and Social Subjects, 36. London: HMSO.

Department of Health (1991) *Dietary Reference Values for Food Energy and Nutrients for the United Kingdom*. Reports on Health and Social Subjects, 41. London: HMSO.

Gregory J, Foster K, Tyler H and Wiseman M (1990) *The Dietary and Nutritional Survey of British Adults*, pp 43, 44, 46. London: HMSO.

Hackett AF, Rugg Gunn AJ, Appleton DR, Eastoe JE and Jenkins GN (1984) A 2-year longitudinal nutritional survey of 405 Northumberland children initially aged 11·5 years. *British Journal of Nutrition* 51: 67–75.

Ministry of Agriculture, Fisheries and Food (1964) *Soft Drinks Regulations 1964* (SI No. 760) as amended.

Nelson M (1991) Food vitamins and IQ. *Proceedings of the Nutrition Society* 50: 29–35.

Rugg Gunn AJ, Hackett AF, Appleton DR and Moynihan PJ (1986) The dietary intake of added and natural sugars in 405 English adolescents. *Human Nutrition Applied Nutrition* 40A: 115–124.

Susan J. Gatenby
AFRC Institute of Food Research, Reading, UK

SORBIC ACID

See Preservatives

SORGHUM

In terms of total world production, sorghum (*Sorghum bicolor* L Moench) ranked fifth among cereals, with 59×10^6 t in 1987 (FAO 1988), of which 78% was planted in developing nations. Production of sorghum is greatest in the USA, Mexico, Nigeria, Sudan, Ethiopia, India and China, but occurs in other countries. Most African and Asian countries use sorghum for food, feed, forage, fuel, and building material. In the western hemisphere, sorghum is used mainly for feed and forage. Speciality types are used for syrup, sugar and alcohol on a limited basis. ***See*** Cereals, Contribution to the Diet

The sorghum plant originated in the northeast quadrant of Africa. It belongs to the Gramineae family, Panicoideae subfamily and Andropogeneae tribe. It is a warm-season, annual crop, favoured by high day and night temperatures and intolerant to low temperatures. Improved types germinate, mature and yield grain in an average of 120–140 days. Sorghum plants range in height from 0·6 to 6 m and possess a monoic-hermaphrodite flower which generally self-pollinates. The grain develops on a branched terminal panicle that can be compact or very open. Flowering proceeds from the top of the panicle downward, with each panicle containing from 800 to 3000 caryopses.

An enormous range in morphological diversity exists in the sorghum crop. The World Sorghum Collection in India has more than 29 000 entries. Sorghum is classed as grain sorghum, forage sorghum, grass or Sudan sorghums, and broomcorn. The latter is grown for its long, fibrous panicle branches that are used to manufacture brooms. The grain of sorghum is classed according to pericarp colour (white, yellow, or red), presence or absence of a pigmented testa (with or without tannins), pericarp thickness (thin or thick pericarp), endosperm colour (white, heteroyellow, or yellow) and endosperm type (normal, heterowaxy, or waxy). These kernel characteristics are genetically controlled. Brown sorghums have a pigmented testa and contain tannins; they may have any colour pericarp.

Grain Structure and Physical Properties

The sorghum kernel is considered a naked caryopsis, although some African types retain their glumes after threshing. The kernel weight varies from 3 to 80 mg. The size and shape of the grain varies widely among sorghum races. Commercial sorghum grain has a flattened-spherical shape, 4 mm long, 2 mm wide and 2·5 mm thick, with a kernel weight of 25–35 mg. Volumetric weight, and grain density range from 708 to 760 kg m^{-3} and from 1·26 to 1·38 g cm^{-3}, respectively.

The sorghum caryopsis is composed of three anatomical parts: pericarp, endosperm, and germ (Fig. 1a). The relative proportion of these structures varies but in most cases is 6%, 84% and 10%, respectively. The pericarp (Fig. 1b) is the fruit coat and is fused to the sorghum seed. It originates from the ovary wall and is subdivided into three distinctive parts: epicarp, mesocarp, and endocarp. The epicarp is the outermost layer and is generally covered with a waxy film. The mesocarp varies in thickness and can contain starch granules. The endocarp is composed of cross and tube cells and plays a major role during germination. ***See*** Wheat, Grain Structure of Wheat and Wheat-based Products

The true seed consists of the seed coat or testa, endosperm, and germ. The endosperm tissue is triploid, resulting from the fusion of a male gamete with two female polar nuclei. The testa, or seed coat, is derived from the ovule integuments; in brown sorghums it contains condensed tannins. The endosperm is composed of the aleurone layer, peripheral, corneous and floury areas. The aleurone consists of a single layer of rectangular cells adjacent to the tube cells or testa. Aleurone cells contain a thick cell wall, large amounts of proteins (protein bodies) and enzymes, ash (phytic acid bodies) and oil bodies (spherosomes). The peripheral endosperm (Fig. 1c) adjacent to the aleurone layer is composed of dense cells containing large quantities of protein and small starch granules. These layers affect processing and nutrient digestibilities of sorghum. Processing of sorghum by steam flaking, micronizing, popping and reconstitution, is designed to disrupt endosperm structure to improve digestibility.

The corneous and floury endosperm cells are composed of starch granules, protein matrix, protein bodies and a thin cell wall rich in β-glucans and hemicellulose. In the corneous endosperm, the protein matrix has a continuous interphase with the starch granules, with protein bodies embedded in the matrix (Fig. 1d). The starch granules are shaped polygonally and often contain dents from protein bodies. The appearance is translucent or vitreous. The opaque-floury endosperm is located around the geometric centre of the kernel. It has a discontinuous protein phase, air voids and loosely packaged, round-lenticular starch granules, and is opaque to transmitted light (Fig. 1e).

The germ is diploid owing to the sexual union of one male and one female gamete. It is divided into two major parts: the embryonic axis and scutellum. The embryonic axis will originate the new plant and is subdivided into a radicle and plumule. The radicle forms primary roots, whereas the plumule forms leaves and stems. The

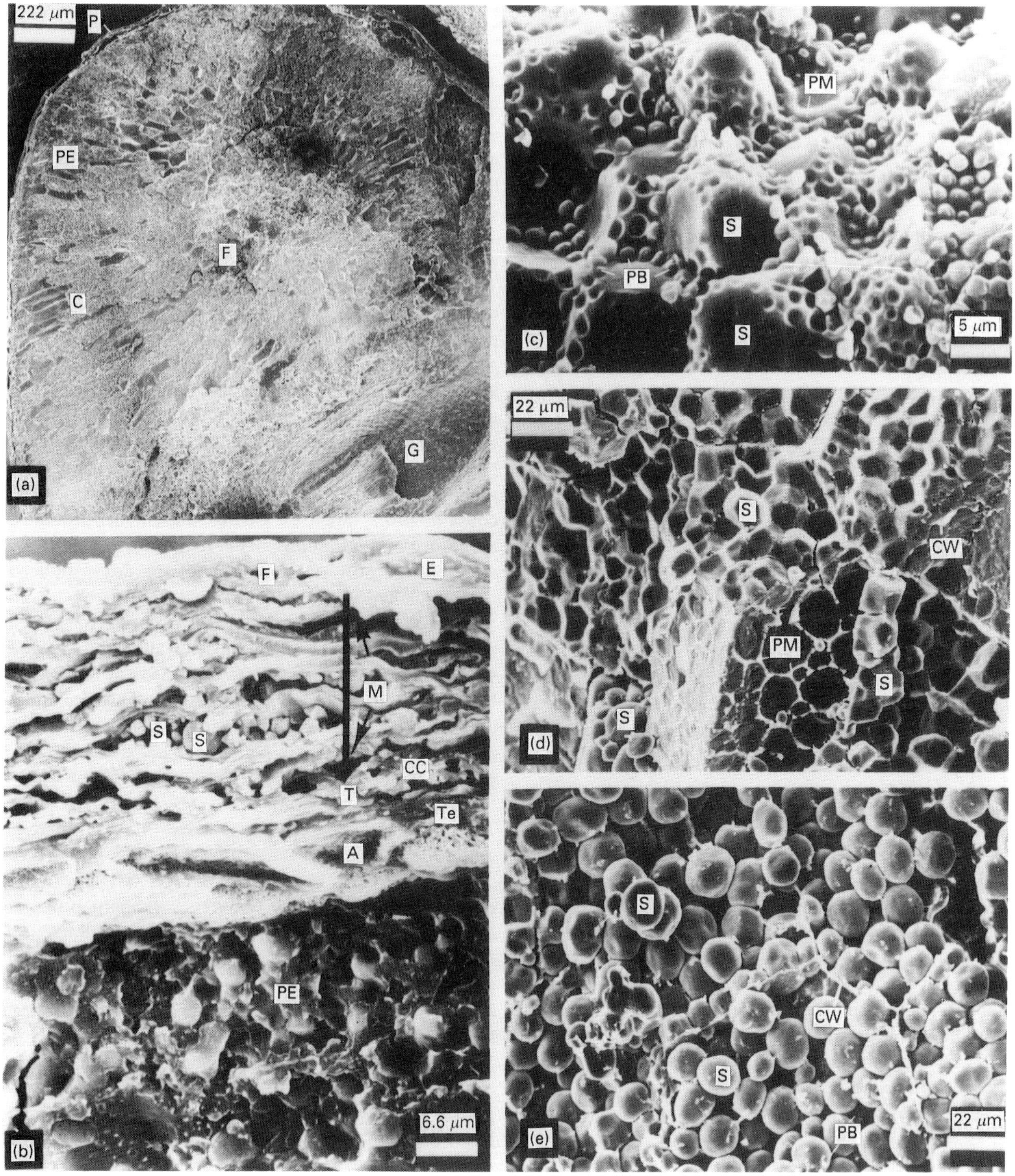

Fig. 1 Structure of the mature sorghum kernel viewed with the scanning electron microscope. (a) Cross section, (b) pericarp, (c) peripheral endosperm, (d) corneous endosperm and (e) floury endosperm. P, pericarp; PE, peripheral endosperm; C, corneous endosperm; F, floury endosperm; G, germ; E, epicarp; M, mesocarp; CC, cross cells; T, tube cells; Te, testa; A, aleurone; S, starch granule; PB, protein body; PM, protein matrix; CW, cell wall. From Rooney LW and Serna-Saldivar SO (1990).

Table 1. Composition (%, unless otherwise stated) of sorghum grain[a]

Component	Value
Protein (N × 6·25)	11·6
Ether extract	3·4
Crude fibre	2·7
Ash	2·2
Nitrogen-free extract[b]	79·5
Starch	74·1
Fibre	
Dietary insoluble	7·2
Dietary, soluble	1·1
Acid detergent	3·3
Soluble sugars	2·1
Pentosans	1·3
Protein fractionation	
Prolamine	52·7
Glutelins	34·4
Albumins	5·7
Globulins	7·1
Essential amino acids[c] (g amino acid per 100 g protein)	
Lysine	2·1
Leucine	14·2
Phenylalanine[d]	5·1
Valine	5·4
Tryptophan	1·0
Methionine[e]	1·0
Threonine	3·3
Histidine[f]	2·1
Isoleucine	4·1

[a] All values are expressed on a dry-matter basis. Adapted from Rooney LW and Serna-Saldivar SO (1990) and Rooney LW *et al.* (1982).
[b] Calculated by difference.
[c] FAO/WHO suggested pattern (g amino acid per 100 g protein): lysine, 5·44; leucine, 7·04; phenylalanine plus tyrosine, 6·08; valine, 4·96; tryptophan, 0·96; methionine plus cysteine, 3·52; threonine, 4·0; isoleucine, 4·0.
[d] Phenylalanine can be partially spared by tyrosine.
[e] Methionine can be partially spared by cysteine.
[f] Histidine is considered an essential amino acid only for children.

scutellum is the single cotyledon of the sorghum seed. It contains large amounts of oil (spherosomes), protein, enzymes and minerals, and serves as the connection between the endosperm and germ.

Composition

Sorghum grain composition (Table 1) varies significantly, owing to genetic and environmental influences, and is similar to that of maize (*Zea mays* L). Starch (75–79%) is the major component, followed by protein (9·0–14·1%) and oil (2·1–5·0%). Protein content (N × 6·25) of sorghum is more variable and usually 1–2% higher than maize. Approximately 80%, 16% and 3% of the protein is in the endosperm, germ and pericarp, respectively. Sorghum generally contains 1% less oil and significantly more waxes than maize. ***See*** Cereals, Dietary Importance

Sorghum starch is composed of 70–80% amylopectin and 20–30% amylose. Waxy sorghums contain starch with 100% amylopectin, and their properties and uses are similar to those of waxy maize. Amylopectin and amylose have an average molecular weight of $8–10 \times 10^6$ and $1–3 \times 10^5$, respectively. ***See*** Starch, Structure, Properties and Determination

The main protein fraction in the kernel is the prolamines (kafrins), followed by glutelins (Table 1). The alcohol-soluble prolamine fraction comprises 50% of the protein. These proteins are hydrophobic, rich in proline, aspartic and glutamic acids, and contain little lysine. They are mainly found in protein bodies and are affected by nitrogen fertilization. Glutelins are high-molecular-weight proteins, mainly located in the protein matrix. The lysine-rich protein fractions – albumins and globulins – predominate in the germ. High-lysine sorghums, such as P-721 and certain Ethiopian types, contain lower levels of kafrins and higher levels of albumins and globulins. The higher-lysine sorghum varieties are soft or dented, and are not produced commercially. ***See*** Protein, Chemistry

Most of the fibre is present in the pericarp and cell walls. Aleurone and endosperm cell walls are associated with ferulic and caffeic acid. Around 85% of the dietary fibre is insoluble and mainly composed of hemicellulose and cellulose. The soluble fraction is rich in pentosans and β-glucans. Sorghum contains approximately 1·3% pentosans, located mainly in the pericarp. Approximately 70% of the pentosans are alkali-soluble, and some 30% are water-soluble. ***See*** Dietary Fibre, Properties and Sources

The germ and the aleurone layer are the main contributors to the lipid fraction. The germ provides about 80% of the oil. The fatty acid composition consists mainly of linoleic (49%), oleic (31%) and palmitic (14·3%) acids. Refined sorghum oil is very similar to maize oil in quality. ***See*** Fatty Acids, Properties

Most of the minerals are concentrated in the pericarp, aleurone, and germ. The ash fraction is rich in phosphorus and potassium and low in calcium and sodium. Most of the phosphorus is bound to phytic acid. The germ and aleurone are rich in fat-soluble and B vitamins. Precursors of vitamin A or carotenes are found only in yellow and heteroyellow endosperm cultivars. ***See*** individual vitamins and minerals

All sorghums contain phenolic acids and most contain flavonoids, but only brown sorghums contain condensed tannins. Tannins protect the kernel against preharvest germination and attack by insects, birds and moulds. Brown sorghums always have a pigmented testa (genes B_1-B_2-ss) and some have tannins in the pericarp (B_1-B_2S-). Sorghums without a pigmented testa do not contain any condensed tannins. Birds can and do consume brown sorghums when other food is unavailable. ***See*** Tannins and Polyphenols

Uses of Grain

Milling

Thirty per cent of the world sorghum production is consumed directly by humans. For production of most traditional foods, sorghum is first dehulled with a wooden mortar and pestle. For decortication, the grain is usually washed, placed in the mortar and pounded vigorously with the pestle. The abrasive action frees the pericarp from the kernel. The pericarp usually detaches at the mesocarp. Thick pericarp cultivars, with hard endosperm and round kernels, are preferred for decortication. The bran or pericarp is separated from the grain by washing with water or by winnowing the sun-dried grain. Most sorghums are decorticated to remove 10–30% of the original grain weight. Mechanical decortication with rice milling equipment or abrasive discs is becoming more popular.

The decorticated kernels are reduced into flour by pounding in the mortar and pestle, with stone mills or by diesel-powered attrition mills. Flour is sieved to obtain fractions with acceptable particle size for specific products.

Traditional Food Uses

The major categories of traditional foods are as follows: fermented and unfermented flat breads; fermented and unfermented thin and thick porridges; steamed and boiled cooked products; snack foods and alcoholic and nonalcoholic beverages. Worldwide, the most popular unfermented flat breads are roti in India and tortillas in Central America. For roti, a portion of the flour is gelatinized, mixed with more flour and warm water and kneaded into a dough. The dough is shaped or rolled into a circle that is baked on a hot griddle. For tortilla production, whole or decorticated sorghum is lime-cooked, steeped overnight, washed, stoneground into a dough, shaped into thin circles and baked on a hot griddle. ***See*** Chapatis and Related Products; Tortillas

The most popular fermented breads are injera, kisra and dosa, consumed in Ethiopia, Sudan and India, respectively. About 80% of the Ethiopian sorghum is used for the production of injera. The sorghum flour is mixed with water and a yeast starter from a previous batch of injera. After fermentation for 24–48 h, the batter is poured onto a greased pan for baking. The resulting product is a flexible, brown pancake which contains uniformly distributed fish eyes, or air bubbles. Dosa is consumed in India and is produced from a mixture of black gram, and sorghum and rice flour.

Porridges can be fermented or cooked with acid or alkali. Tô is an unfermented stiff porridge, very popular in Mali and Burkina Faso. Decorticated sorghum flour is cooked in plain water or water acidified with tamarind juice or made alkaline with wood ashes (potash). The most popular fermented porridges are ogi and nasha, widely consumed in West and East Africa, respectively. Whole sorghum is soaked in water and allowed to ferment for 2–3 days. The wet grain is crushed in a slurry of water and sieved to remove the bran. The fine particles are allowed to ferment longer. Excess water is decanted and the resulting slurry cooked in water or milk.

For couscous production, sorghum flour is kneaded with enough water to form agglomerates. The particles are forced to pass through a coarse screen and steamed. The cooked product is consumed with a sauce or milk.

Decorticated sorghums are often cooked like rice. Special types of small-seeded, very hard sorghums are used as a substitute for rice.

Two major kinds of alcoholic beverages are produced from malted sorghum. The most common type, called opaque beer, undergoes souring and yeast fermentation and is very popular in southern Africa. The high-solids beer is sour, alcoholic, pinkish and effervescent. ***See*** Fermented Foods, Beverages from Sorghum and Millet

Industrial Uses

Industrial uses of sorghum are similar to those of maize. In the Sudan, sorghum is wet-milled to produce starch with properties and uses similar to maize starch. Sorghum is more difficult to wet-mill than maize and sorghum by-products are less desirable. Wet milling of sorghum in the USA was discontinued in the 1970s because of poor economics. The enzymatic conversion of starch to liquid glucose syrup is possible. Sorghum grain or sweet sorghum biomass is used for ethanol production. Yields of alcohol (182° proof) per tonne of sorghum grain are comparable to maize (387 versus 372 l). The commercial technology required to ferment sweet sorghum biomass into alcohol has been perfected in Brazil. One tonne of sweet sorghum biomass has the potential to yield 74 l of 200° proof alcohol.

Major breweries in Mexico, Africa and Asia use sorghum grits as an adjunct in brewing lager beer. The most desirable grit has light colour, bland flavour and low oil content. Sorghum malt is produced on an industrial basis in southern Africa. The malt is used for alcoholic beverages, weaning foods and breakfast foods. Sour-opaque beers are produced commercially in

Africa. Opaque beer is produced following the basic steps of the traditional process. In Nigeria, sorghum and maize are being used to produce a lager beer without barley malt. The Government has banned importation of barley and barley malt. Nigerian breweries are therefore producing clear beer made from a combination of malted sorghum, sorghum and/or maize grits with commercial enzymes to convert the starch to fermentable sugars. Since sorghum malt has low diastatic power, commercial enzymes are required. In many processes, sorghum malt is not used because malting causes considerable dry-matter losses. Economically, the use of grits and commercial enzymes is practical. The quality of clear beer is good but the taste differs from barley malt beer.

Sorghum grits, meal and flour can be used to produce a wide array of baked goods when mixed with wheat flour. Sorghum does not contain gluten. Thus the amounts of sorghum flour in the blends depend on the quality of the wheat flour, the baking procedure, formulation and quality of the baked products desired. ***See*** Flour, Dietary Importance

Sorghum can be puffed, popped, shredded and flaked to produce ready-to-eat breakfast cereals. Extrusion of sorghum produces acceptable snacks and precooked products. ***See*** Cereals, Breakfast Cereals

Nutritional Value

Sorghum has proximate composition, amino acid contents and nutritional value similar to those of maize. However, owing to its lower fat content, sorghum usually has lower gross, digestible and metabolizable energy. Its protein digestibility is lower than that of other major cereals. Fermentation, malting and other processing methods improve nutritional value significantly. Brown sorghums have lower nutritional value than sorghums without tannins. Tannins lower protein digestibility and feed efficiency. Decortication improves protein digestibility and reduces tannins if present. Lysine and threonine are the first and second limiting amino acids. High-lysine cultivars contain approximately 50% more lysine and promote better weight gains in weaning rats. ***See*** Amino Acids, Properties and Occurrence

The feeding value of sorghum for livestock species is generally considered to be 95% or more of the feeding value of yellow dent maize. Brown sorghums are considered to have 85% the feeding value of maize. Sorgum must be properly processed to enhance its digestibility. Popping, steam flaking and reconstitution are used to prepare sorghum grain for beef cattle in feedlots in the USA. Grinding is used for poultry and swine feeds.

Bibliography

FAO (1988) *Production Yearbook 1987*. FAO Statistics, vol. 41. Rome: Food and Agriculture Organization.

Rooney LW and Miller F (1982) Variation in the structure and kernel characteristics of sorghum. In: Rooney LW and Murty DS (eds) *International Symposium on Sorghum Grain Quality*, pp 143–162. Patancheru, India: ICRISAT.

Rooney LW and Serna-Saldivar SO (1990) Sorghum. In: Lorenz KJ and Kulp K (eds) *Handbook of Cereal Science and Technology*, chap. 5. New York: Marcel Dekker.

Rooney LW, Earp CF and Khan MN (1982) Sorghum and millets. In: Wolf IA (ed.) *CRC Handbook of Processing and Utilization in Agriculture*, vol. II. Boca Raton, Florida: CRC Press.

Rooney LW, Kirelis AW and Murty DS (1986) Traditional foods from sorghum: their production, evaluation and nutritional value. In: Pomeranz Y (ed.) *Advances of Cereal Science and Technology*, vol. VIII. St Paul, Minnesota: American Association of Cereal Chemists.

Stoskofp NC (1980) *Cereal Grain Crops*. Reston, Virginia: Reston Publishing Co.

L. W. Rooney and S. O. Serna-Saldivar
Texas A & M University, College Station, USA

SOURDOUGH BREAD

See Bread

SOY SAUCE

See Fermented Foods

SOYA BEANS

Contents

The Crop

Production

The soya bean originated in the Far East and has provided food for that part of the world for thousands of years, but only during the past few decades have soya beans become an important food crop in the West. Table 1 shows how the production of soya beans has increased since 1940. The newly recognized value of the soya bean is in both the extracted oil and the residual high-protein meal. The oil is widely used as a food in margarines, shortenings, and salad oils. The meal is valuable for its high-protein content as an animal feed, particularly for poultry and swine.

The USA, Brazil, China and Argentina produce 87% of the world's soya beans. China has been a major producer for a long time, but the USA reached that status only in the 1940s and Brazil and Argentina only since the 1970s.

In comparison with grain production, annual soya bean production of about 100×10^6 t is not large. World production of coarse grains (maize, sorghum, etc.) is about 800×10^6 t, while rice and wheat amount to about 450×10^6 t each. Soya beans per unit weight are worth about twice as much as grains. ***See*** Cereals, Contribution to the Diet

Soya beans are the predominant oilseed in the world. About half of the world oilseed production comes from soya beans, and this is more than the combined production of the next four oilseeds: cottonseed, peanut, sunflower, and rapeseed (canola). Oilseeds vary widely in total oil content; the annual vegetable oil production therefore differs from oilseed production. Table 2 shows vegetable oil production for the world during the 1980s. Palm oil comes from palm fruit and is not classified as an oilseed, but palm oil is produced in largest quantities next to soya bean oil. Most of the vegetable oils are increasing in production to keep up with world demand, but cottonseed, coconut, and olive oil production are relatively stable. ***See*** Vegetable Oils, Groundnut Oil; Vegetable Oils, Oil Palms

The increased use of vegetable oils for foods parallels the growth of soya bean production. In the 1920s animal fats were predominant in human diets, but since the 1940s, vegetable oils have steadily replaced animal fats. The particular vegetable oil that is used depends on the country. Soya bean oil makes up about 75% of vegetable oil use for food in the USA. Peanut oil is the oil of choice in India, and rapeseed oil is important in Canada.

Soya beans are annual legumes and were used at first in the USA as a hay crop. Since legumes are nitrogen-fixing plants, the hay crop could be used in rotation with corn to replenish soil nitrogen. With increased use of nitrogen fertilizers, the value of soya beans for their nitrogen fixation decreased.

Most of the spectacular recent increase in production in the USA and South America (Table 1) is due to increased area devoted to growing soya beans. There has

Table 1. Soya bean production ($\times 10^6$ t) by major countries during the past 50 years

	1940	1950	1960	1970	1980	1990
USA	2·1	7·8	16·3	30·7	48·7	52·4
Brazil	—	—	—	2·1	15·5	20·0
China	9·7	4·9[a]	7·0	9·7	7·9	10·8
Argentina	—	—	—	—	3·4	10·5
World	11·9	17·8	28·1	45·2	81·2	107·1

[a] Data for 1948.

Table 2. World vegetable oil production ($\times 10^6$ t)

	1980	1982	1984	1986	1988
Soya bean	12·8	13·8	13·4	14·8	15·0
Palm	5·2	5·6	6·9	7·9	9·0
Sunflower	4·7	5·8	6·1	6·6	7·0
Rapeseed	4·1	5·3	5·3	6·6	7·3
Cottonseed	3·2	3·4	4·2	3·2	3·6
Peanut	2·7	3·0	3·4	3·1	3·8
Coconut	2·9	2·7	2·8	2·8	2·6
Olive	1·9	1·9	1·7	1·9	1·8
Palm kernel	0·6	0·8	0·9	1·0	1·3

been a slow but definite increase in yield per unit area, with an approximate doubling since 1940 to 2 t per ha currently (1990) in the USA.

The problems that producers face in trying to increase yields of soya beans are the same as with any agronomic crop. Pests such as birds and rodents may eat planted seeds; diseases may slow growth of roots, stems or leaves; insects can cause damage as well as transmit diseases; weeds compete for moisture, soil nutrients, and sunlight.

The flowers of soya beans are small and inconspicuous, making it difficult to produce hybrid seed. Under normal conditions, soya beans are self-fertilizing. The time of flowering is triggered by lengthening nights, and different soya cultivars (cultivated varieties) respond to different night lengths. Thus it is important to select the proper cultivar for the latitude in which it will be grown. If a cultivar from southern latitudes is grown farther north, it will not flower until late in the season and may freeze before the seed matures. Conversely, northern cultivars grown in southern latitudes may flower before the plant has reached full growth, thus greatly reducing yield.

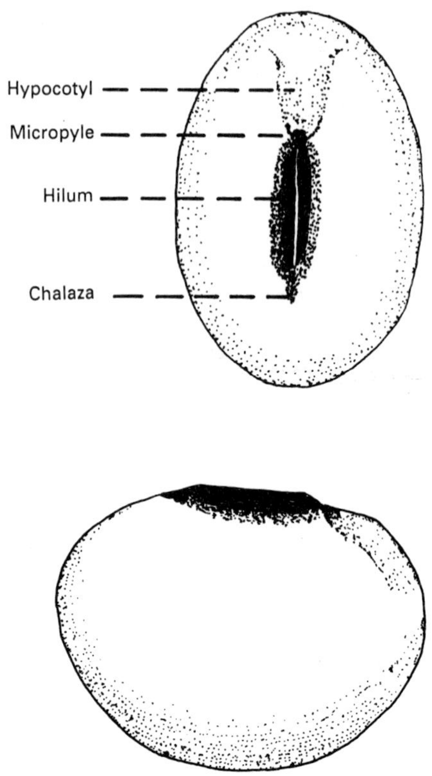

Fig. 1 Morphological features of whole, mature soya beans. The features shown are associated with the seed coat or hull.

Morphology

Soya beans can range in weight from about 100 to 300 mg each, with diameters of 4–8 mm. They are roughly spherical in shape when dry, and swell to a definite bean or kidney shape when wet. The morphological features of the whole bean (Fig. 1) are the hilum (point of attachment to the pod), the micropyle (a small opening through which the germ tube grew), and the chalaza (a small groove opposite the hilum from the micropyle).

The seed coat makes up about 9% of the weight of soya beans and tightly encloses the two cotyledons and embryo. The seed coat is not easily separable from the intact dry seed, but if the seed is broken or if it imbibes water, the seed coat separates readily. The morphological features noted above (hilum, micropyle, and chalaza) are part of the seed coat which may be yellow, green, brown or black, but the cotyledons are green or yellow. United States grading standards discriminate against brown and black soya beans. Since very little oil is present in the seed coat, it is removed before oil extraction and may or may not be added back to the defatted meal used as animal feed.

Several layers of cells make up the seed coat, and cells of a distinctive type, termed hourglass cells because of their shape, are readily distinguishable by light microscopy. The hourglass cells can be used as qualitative markers to determine whether or not soya bean meal has been added to other foods.

If soya beans are soaked in cold water, a few, perhaps 1%, fail to imbibe water and are called 'hard seeded'. The failure to imbibe water is a property of the seed coat being completely intact and is probably due to the outermost waxy layer, termed the cuticle.

The two cotyledons make up 90% of the soya bean. The cotyledon cells are packed with protein bodies and lipid bodies, the two principal organelles (Fig. 2). Early in the maturity of the cotyledons, starch granules are prominent, but they decrease to less than 1% of weight as beans mature.

Protein bodies are 2–20 μm in diameter and contain principally the glycinin and conglycinin storage proteins. Protein bodies are osmotically fragile but can be isolated by using high osmotic strength buffers at pH 5. In addition, heating (soaking soya beans for a few minutes in boiling water) causes protein bodies to be heat-fixed and not to disperse in water.

Lipid bodies (0·2–0·5 μm) are much smaller than protein bodies and are the sites of oil storage. Centrifugation of broken cotyledon cells separates lipid bodies as a floating layer, but because of their phospholipid–protein membrane, the smallest lipid bodies sediment in a centrifugal field.

The hypocotyl of the soya bean has a different composition to that of the cotyledon (less oil and more carbohydrate), but since it only makes up about 2·5% of the weight of the bean, the different composition has

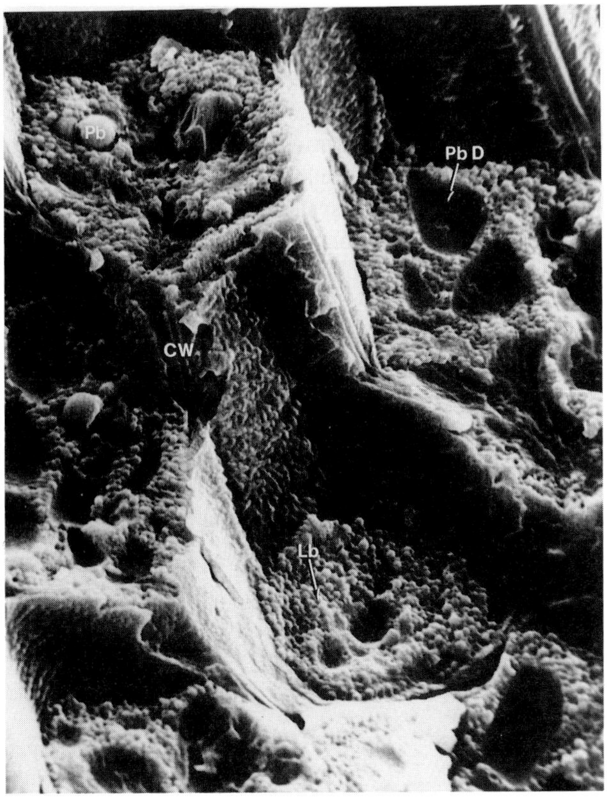

Fig. 2 A scanning electron micrograph of freeze-fractured soya bean cotyledon. CW, cell wall; Lb, lipid body; Pb, protein body; PbD, protein body depression.

little influence. The hypocotyl is processed into oil and meal along with the cotyledons after the seed coat is removed.

Composition

As a first approximation, soya beans contain 20% lipid, 40% protein, 35% carbohydrate, and 5% ash on a dry-weight basis. Considerable variability exists in these numbers, depending on the cultivar and the growing conditions. In the northwestern part of the soya bean production area in the USA, protein content of soya beans is generally lower than in other production areas. Protein content is known to be inversely related to yield and to lipid content.

Lipids

The crude oil extracted from soybeans contains about 96% triglycerides. The fatty acid composition of the triglycerides is approximately 11% palmitic (C16), 4% stearic (C18), 25% oleic (C18:1), 51% linoleic (C18:2), and 8% linolenic (C18:3), with other minor fatty acids also present. The fatty acid composition is not fixed and can be changed by breeding new cultivars. ***See*** Fatty Acids, Properties

Minor components in the crude oil include the plant pigments lutein (3,3′ dihydroxy α-carotene) and chlorophyll. Also present are phospholipids, free fatty acids, triglyceride hydroperoxides, sterols, saponins, squalene, tocopherols (vitamin E), and flavour compounds. During refining of the crude oil, phospholipids, free fatty acids, pigments and flavour compounds are removed to produce an oil that contains more than 99% triglyceride. ***See*** Chlorophyll; Colours, Properties and Determination of Natural Pigments; Flavour Compounds, Structures and Characteristics; Phospholipids, Properties and Occurrence; Triglycerides, Structures and Properties

Protein

Soya proteins have been separated in the ultracentrifuge and characterized as 2*s*, 7*s*, 11*s*, and 15*s* fractions (*s* is a sedimentation constant and larger numbers indicate bigger proteins). The 2*s* fraction contains low-molecular-weight proteins, including the Bowman–Birk and Kunitz trypsin inhibitors that inhibit growth in young animals. Heating of soya bean meal is necessary to denature trypsin inhibitors and allow full utilization of the protein. ***See*** Protein, Chemistry

The 7*s* fraction includes a storage protein, β-conglycinin, with a molecular weight (M) of about 180 000. β-Conglycinin is a trimer with α, α', and β subunits, which in different combinations give six distinct β-conglycinins. Also present in the 7*s* fraction are haemagglutinins (lectins) and lipoxygenases. Haemagglutinins are glycoproteins that can cause agglutination of red blood cells and can be toxic, but are probably heat-denatured along with trypsin inhibitors and have no toxic effects when fed. Several lipoxygenases have been characterized that catalyse the removal of hydrogen and addition of molecular oxygen to linoleic and linolenic fatty acids to yield hydroperoxides. The hydroperoxides can break down to yield off-flavour compounds associated with oxidative rancidity. ***See*** Oxidation of Food Components

The 11*s* fraction is glycinin, a hexamer (M, 360 000). Each of the six subunits consists of acidic polypeptides (M, 32 000–36 000) and basic polypeptides (M, 19 000–20 000). The 15*s* fraction is a dimer of glycinin.

Associated with protein in soya beans is phytin, the calcium and magnesium salt of hexaphosphoryl inositol. Phytin strongly chelates divalent cations and has been implicated in decreased availability of iron in diets with soya protein. ***See*** Phytic Acid, Properties and Determination

Carbohydrate

Insoluble carbohydrate includes the pectin, cellulose and hemicellulose that are associated with cell walls in soya bean cotyledons. This is also the dietary fibre component which is increasingly recognized as an important part of the human diet. The fibre is more prevalent in soya bean seed coats (hulls) than in cotyledons. ***See*** Carbohydrates, Classification and Properties; Cellulose; Hemicelluloses

The soluble carbohydrate in soya beans is about 10% by weight of the dry bean and includes about 5% sucrose, about 4% stachyose, and about 1% raffinose. Raffinose is a trisaccharide and stachyose is a tetrasaccharide, neither of which can be digested by humans. As a result, microbial fermentation of these sugars in the intestine produces gas (flatus), which leads to gastrointestinal distress.

Handling and Storage

As soya beans are harvested and moved from the field into storage, conditions have to be controlled to minimize deterioration. Moisture content of the beans should be 14% or lower to prevent microbial growth. Also, cleanliness of the beans is important to avoid insect or other contaminants which may provide moisture and be a focus for microbial growth. If deterioration does start, bacterial or mould growth will generate higher temperatures and moisture so that the process becomes autocatalytic. The best control procedure in storage is to have temperature sensors spread throughout, and quickly mix beans that show generation of hot spots. ***See*** Spoilage, Bacterial Spoilage; Spoilage, Moulds in Food Spoilage

Much of the crop from the USA and Argentina is exported as beans. As the beans are transported by truck, train, barge and ship, and they are transferred into and out of storage, they tend to break, and the pieces are known as splits. Splits are a factor in grading, and the oil extracted from splits tends to have greater amounts of free fatty acids and phospholipids than oil from intact beans.

Much of Brazil's crop is crushed in the country, and oil and meal are exported. In handling oil it is necessary to minimize exposure to oxygen of the air during transfer. Any outlet spouts should be below the surface to prevent air incorporation in the oil, and blanketing the surface with nitrogen is a good protective measure. Storage tanks must be kept clean. This is particularly important for the tanks in ship's holds or for tankers, both of which may carry many different liquid cargoes. Cross-contamination between cargoes is avoided by careful cleaning of the tanks.

Grading

In the USA there are four numerical grades and sample grade for those beans that have the lowest quality. The factors used for grading are test weight, splits, heat damage, foreign material, and colour. These are similar criteria to those used for grains, and the grading systems in other countries rely on the same criteria although the actual grades and maximum limits may differ. The economically valuable constituents, protein and oil, are not part of the grading system.

Until recently there have been difficulties in measuring protein and oil quickly and accurately enough to be part of the grading system. Infrared spectrophotometers now perform excellent analyses for protein, oil, and moisture of either whole or ground samples. Consequently, there is a renewed interest in using the protein and oil content of soya beans to provide premiums or discounts in pricing. Aside from the analysis problem, other problems exist in keeping high-oil or high-protein beans segregated as they move through marketing channels. ***See*** Population Development and Nutrition

Bibliography

Anonymous (1990) *Soya Bluebook*. Bar Harbor, Maine: Soyatech Inc.

Erickson DR (ed.) (1990) *Edible Fats and Oils Processing: Basic Principles and Modern Practices*. Champaign, Illinois: American Oil Chemists' Society.

Erickson DR, Pryde EH, Brekke OL, Mounts TL and Falb RA (eds) (1980) *Handbook of Soy Oil Processing and Utilization*. St Louis, Missouri: American Soybean Association. Champaign, Illinois: American Oil Chemists's Society.

Snyder HE and Kwon TW (1987) *Soybean Utilization*. New York: Van Nostrand Reinhold.

Wilcox JR (ed.) (1987) *Soybeans: Improvement, Production and Uses* 2nd edn. Madison, Wisconsin: American Society of Agronomy, Crop Science Society of America, Soil Science Society of America.

Harry E. Snyder
University of Arkansas, Fayetteville, USA

Processing for the Food Industry

Soya beans are seldom simply cooked and eaten. They are invariably processed into foods or food products. This article will cover the initial separation into oil and meal by solvent extraction and then the further processing of oil and meal. Oil needs to be refined to remove minor components and then is processed to make products such as margarines, shortenings and salad dressings. The protein-rich meal is mainly used as animal feed but can be processed into food ingredients such as flours, concentrates and isolates. The protein products may also be textured in various ways to simulate meats or cheeses.

Oil Separation

The first step in processing most soya beans is to separate the oil, either by solvent extraction or by expelling. The processing companies that do the separation are called crushers, and the 100 or so crushers in the USA each process on the average about 1000 tonnes of soya beans daily. ***See*** Vegetable Oils, Extraction

Preparation

Before solvent extraction, soya beans are cleaned and cracked into several pieces (meats). The hulls are removed by aspiration, and the meats are conditioned by warming and by adding moisture. Conditioning is necessary to make a cohesive flake. The conditioned meats are put through smooth rollers that make flakes of approximately 0·025 cm thickness. Making flakes is advantageous for uniform penetration of solvent in deep beds (minimal channelling) and for disruption of the soya bean tissue so that solvent can penetrate and dissolve the oil.

A recent development is to put the flakes through a cooking extruder (expander) to yield collets. The collets are pieces of the 'rope' exiting the outlet of the expander. This processing gives a porous but still high-density collet that extracts more readily than flakes. Also, the collet holds less solvent than the flake, thereby minimizing the energy needed for removing the solvent.

Solvents

The solvent of choice is commercial hexane, a petroleum fraction with a boiling range of 65–70°C. Commercial hexane is a mixture of *n*-hexane, cyclohexane and methylpentanes. It has low viscosity and a low heat of vaporization. The main disadvantages of hexane are its flammability and the safety precautions needed for safe handling.

Chlorinated hydrocarbons are equally good solvents for soya bean triglycerides, but use of trichloethylene led to toxic soya bean meal during the 1940s and discouraged further use of these solvents.

Ethanol and isopropyl alcohol are effective triglyceride solvents when hot, and cooling can be used to separate the dissolved triglyceride from the solvent. However, alcohol solvents are not being used in commercial extractions.

Extraction

The full-fat flakes (or collets) are loaded into the extractor to make beds over which the solvent flows. Extractors may have deep (1·2 m) or shallow (0·6 m) beds, and are always arranged so that solvent flows countercurrent to the movement of the beds. Thus, the fully extracted flakes are contacted by fresh hexane entering the extractor, and full-fat flakes are contacted by the full miscella (solution of crude oil in hexane) just before it leaves the extractor. The temperature of extraction is about 60°C to speed up diffusion of solvent and to lower the miscella viscosity, both of which enhance the extraction of oil. Solvent extraction is capable of reducing the residual oil in the soya flakes to less than 1%.

Removal of Solvent

Upon completion of extraction, the solvent has to be removed from both the oil and flakes. The full-fat miscella contains 25–30% oil, and solvent is removed by two stages of rising-film vacuum evaporators followed by a third-stage stripping column. At the end of this process the crude oil contains about 1000 ppm of hexane, which corresponds to a flash point for the oil of 121°C.

The flakes are treated in a desolventizer–toaster by direct contact with steam to first remove the hexane and secondly to heat treat the flakes for trypsin inhibitor destruction. After leaving the desolventizer–toaster the flakes are cooled and ground to a meal for use as a high-protein feed ingredient. The protein content is 44% with hulls added or 47·5–49% protein without hulls.

If flakes are going to be used to produce soya bean protein isolates or soluble soya bean concentrates, the solvent has to be removed with minimal heat to maintain protein solubility. Flash desolventizers are available in which superheated hexane is used as the heat transfer medium to evaporate hexane. The flakes are kept dry and, with this system, protein solubility is preserved.

Expelling

The original methods for recovering oil from oil seeds were designed to exert pressure on the seeds and express the oil. The modern technique for exerting pressure is an expeller: a motor-driven auger rotating within a cage of narrowly spaced metal bars. As the oil seed flakes move through the expeller, pressure exerted by the auger forces oil out through the bars while the defatted flakes are pushed by the auger to the opening at the end. Considerable heat is generated by expellers, and often the protein meal is of lower quality than solvent-extracted meal because of excess heat. Also, the residual oil in expeller meal is 3–4%. Nevertheless, expellers are widely used when it is necessary to handle a variety of oil

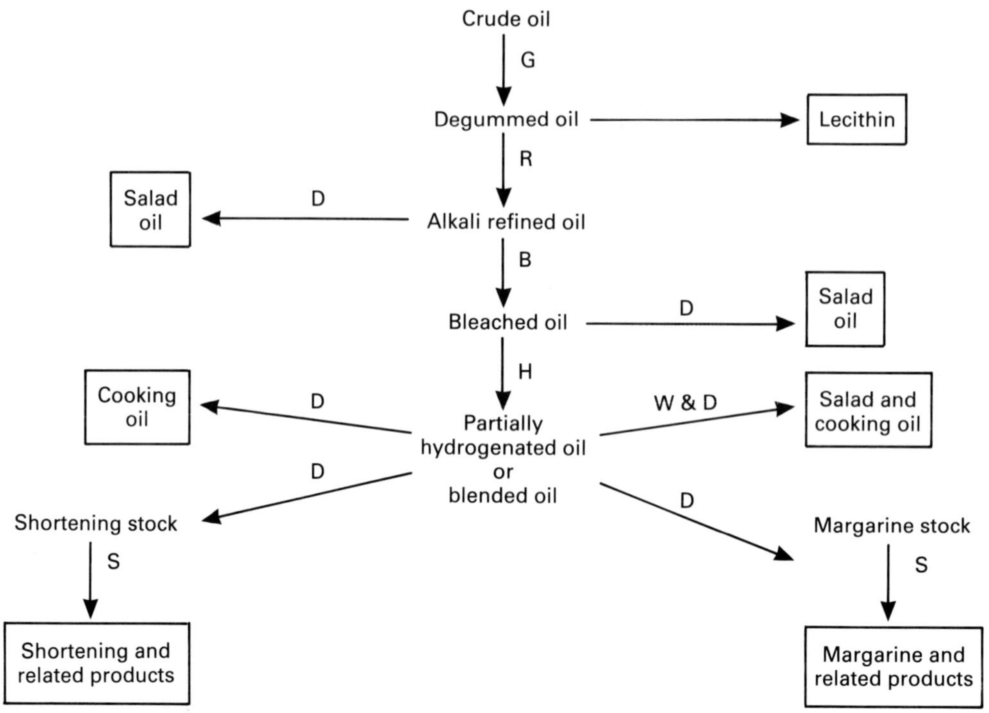

Fig. 1 Some edible soya bean oil products and the processing steps needed: G, degumming; R, alkali refining; B, bleaching; H, hydrogenation; D, deodorization; W, 'winterization'; S, solidification.

seeds in one facility. Expellers can handle as much as 50–80 tonnes day^{-1} of oil seeds. The soya bean preparation for expelling is similar to that for solvent extraction.

Soya Oil Refining

After extraction and removal of solvent, the crude soya bean oil needs to be refined to convert it to edible products. Figure 1 shows some of the edible products and the refining steps involved. *See* Vegetable Oils, Refining

Degumming

The crude oil contains about 1–3% phospholipids, known in the industry as gums. The phospholipids act in conjunction with tocopherols as antioxidants, but they will also precipitate out of the oil, making a sludge that is difficult to remove from storage tanks, road tankers or ship holds. Therefore, crude oil is often degummed by the crusher before shipping or storing.

The degumming process consists of adding 1–3% water to hydrate the phospholipids followed by centrifuging to remove the hydrated material. The phospholipids are removed in the water phase and either recovered to be sold as a food ingredient, lecithin, or added to the meal portion to be sold as animal feed. Phospholipids remaining in the oil are said to be nonhydratable, and are largely calcium and magnesium salts of phosphatidic acid. An acid treatment of the crude oil makes these phospholipids removable by water washing. About half of the soya bean oil in the USA is degummed before going to the refiner.

Alkali Refining

The purpose of alkali refining is to remove free fatty acids from the crude oil (if the oil has not already been degummed, phospholipids are also removed by alkali refining). Free fatty acids are undesirable because they lower the temperature at which the oil begins to smoke when heated, as in frying. The fatty acid content of crude soya bean oil is 0·5–0·7%, and alkali refining yields oil with <0·05% free fatty acids.

The process is to wash the oil with a 12% sodium hydroxide solution followed by centrifugation. Fatty acids are converted to sodium soaps and separate with the water phase. Invariably, some oil is lost with the water phase (neutral oil loss), and the greater the amount of free fatty acids the greater the neutral oil loss. After washing with dilute alkali, the oil is washed once with water to remove residual soaps and vacuum dried to a moisture content of 0·1%.

The fatty acids may be recovered by acidifying and centrifuging the water phase. Fatty acids can be used for soap manufacture, but are often added to the soya bean meal to serve as additional feed energy.

Bleaching

Crude soya bean oil is darker in colour than desired by consumers, and bleaching is used to lighten the colour. Bleaching is achieved by the addition of bleaching earth (adsorbent clays usually acid treated) to the hot oil under vacuum. The main pigments in soya bean oil are carotenoids and chlorophyll, and these are adsorbed on bleaching earths, which are separated from the oil by filtration.

In addition to removal of pigments, the bleaching process adsorbs oxidation products and the pro-oxidant metals, iron and copper. Thus, bleaching plays an important role in stabilizing the oil to oxidation as well as lightening the colour.

Hydrogenation

Soya bean oil is highly unsaturated and, to make products such as shortenings and margarines, hydrogen is added to double bonds in the triglycerides to increase their melting points. Hydrogenation is also used to saturate partially double bonds in soya bean oil, so that oxidative stability is increased. ***See*** Vegetable Oils, Processing

Hydrogenation is usually done in batches (although continuous equipment is available), and the conditions needed are a nickel catalyst, hydrogen gas under pressure, vigorous agitation, and soya bean oil at 120–200°C. The temperature, pressure, degree of agitation, and amount of catalyst are variables that affect the rate of hydrogenation and the degree of selectivity. Selectivity refers to the increase in hydrogenation rate of linolenic acid to linoleic acid to avoid off-flavour generation from oxidation of linolenic acid.

During hydrogenation, isomerization of fatty acids takes place to give double bonds in new positions and to shift double bonds from *cis* to *trans*. Melting point increases result both from saturation of double bonds and from isomerization of *cis* to *trans* double bonds.

Deodorization

The development of hydrogenation and deodorization in the early 1900s made it possible to substitute vegetable oils for animal fats in human diets. Hydrogenation controlled the texture and deodorization controlled the flavour of vegetable oils.

Deodorization is steam distillation at high temperature under vacuum. By this process unwanted flavours are removed, and some bleaching of remaining carotenoid pigments occurs. The distillation temperatures range from 204 to 275°C at 6 mmHg vacuum. Under these conditions not only flavour compounds but free fatty acids and about half of the tocopherols are distilled out of the soya bean oil.

If the oil to be deodorized is sufficiently devoid of phospholipids (less than 5 ppm phospholipid phosphorus), free fatty acids can be removed by distillation rather than by alkali refining. Such removal is known as physical refining. The phospholipids must be present in low concentrations because they brown severely at high temperatures.

The distillate remaining from deodorization is a valuable by-product because of the relatively large content of tocopherols (vitamin E). Batch, semicontinuous and continuous equipment is available for deodorization.

Soya Oil Products

After sufficient refining, soya bean oil is used to produce shortenings, margarines, salad dressings and cooking oils of various types. ***See*** Fats, Uses in the Food Industry; Dressings and Mayonnaise, The Products and their Manufacture; Margarine, Methods of Manufacture

For plastic products such as shortenings and margarines, the texture is achieved by a mixture of triglyceride crystals plus oil. To avoid large crystal size and grainy texture, it is necessary to have a mixture of fatty acids. For soya oil products that means adding oils that contribute palmitic acid to achieve small crystals. In addition to the familiar plastic shortenings produced for home use, dry shortenings are manufactured for incorporation into various bakery mixes, and liquid shortenings are used for ease of metering in automated baking processes.

For margarines the solids content at various temperatures (solid fat index, SFI) is an important quality criterion. If the margarine is to be used in the home, manufacturers want a product that spreads easily at refrigerator temperatures and that melts completely in the mouth (body temperature). These texture characteristics are achieved by blending hard and soft fats and oils. The crystal formation for both shortenings and margarines is achieved by cooling mixtures rapidly while mixing in scraped-surface heat exchangers.

Cooking oils based on soya oil are usually partially hydrogenated to minimize the linolenic acid content and to minimize oxidative rancidity in the hot oil.

Soya Bean Protein Products

Although an estimated 90% of the soya bean defatted protein meal goes to animal feed, a variety of food products and ingredients are produced from the remainder. These include soya bean flours, ranging from full fat to defatted, soya bean protein concentrates, soya bean protein isolates, and various textured products simulating meats, seafoods and cheeses.

Soya Bean Flours

By grinding defatted soya bean flakes to flours (or grits, if particle size is larger), a product is produced that is used in the baking industry for its moisture sorption capability. Soya bean flours and grits are also used in meat emulsions for their emulsifying ability in addition to moisture sorption. ***See*** Emulsifiers, Uses in Processed Foods

Full-fat flours are produced by grinding dehulled soya beans. The unheated flours have active lipoxygenase and are used to bleach wheat flours used in the baking industry. The oxidative action of lipoxygenase both bleaches the carotenoids in wheat flour and oxidizes the proteins to improve machinability.

A range of soya bean flours between defatted and full fat is produced by adding soya bean oil and lecithin in varying amounts to defatted flours. These products are used mainly in the baking industry.

If the soya bean flours are used in substantial concentration in food products, there is a problem with the undigestible oligosaccharides, raffinose and stachyose, causing flatus. To avoid this problem, soya bean protein concentrates were developed.

Soya Bean Protein Concentrates

The processing of defatted soya bean flakes into concentrates requires the protein to be rendered insoluble and extracting the soluble sugars (sucrose, raffinose and stachyose) with water. The protein can be made insoluble by heat treatment, by aqueous ethanol treatment or by adjusting the pH to the isoelectric point of 4·5. The water extraction leaves complex carbohydrates and protein, and the protein concentration has to be 70% to be classified as a concentrate.

In some food applications (a high-protein drink, for example), soluble protein may be needed, and this can be achieved by using pH to render the protein insoluble followed by pH increase after extracting the sugars. Protein solubility is measured by the nitrogen solubility index (NSI) or protein dispersability index (PDI). The indices give the percentage protein remaining in solution or suspension after centrifugation. ***See*** Protein, Chemistry

Soya bean protein concentrates are used in the baking and meat industries for their moisture sorption and emulsifying properties.

Soya Bean Protein Isolates

For those food applications in which only soya bean protein is suitable, soya bean isolates are available. Isolates contain 90% soya bean protein and are prepared from defatted flakes with high NSI. The protein is dissolved with dilute alkali, separated from the remaining cellular material, and precipitated by adjusting the pH to 4·5. The protein can be recovered as spray-dried isoelectric protein (no pH adjustment and low solubility) or as soluble proteinate after raising the pH. About one-third of the original flake is recovered as protein isolate.

Soya bean protein isolates can be used for moisture sorption and emulsification in baked goods and meat emulsions. They can also provide textural improvement through gelation and adhesion properties or serve as primary nutrients in infant formulae.

Soya bean flours, concentrates and isolates may modify texture in foods by increasing viscosity, by holding moisture or by adhesion and cohesion. Furthermore, texture in foods can be modified by changing the texture of the soya bean materials.

Texturizing

Numerous ingenious methods have been patented to build chewy textures into flours, concentrates and isolates. The most used procedures are extrusion and spinning.

Extrusion is used mainly with flours and concentrates and is a method of both heating and forming. The extruder is a screw or auger rotating within a jacketed barrel. As the flour or concentrate is moved through the barrel by the auger, it is heated by friction and by steam jacketing. High pressures are developed in the nose of the barrel just before the exit. As the material is forced through the exit port, it is shaped and puffed due to the sudden drop in pressure.

Spinning applies only to soya bean isolates. A concentrated solution of isolate is prepared and forced through spinnerets (platinum plates with small holes) into an acid bath. The protein precipitates in the acid as a continuous fibre, and the spun fibres can be wound together into a bigger 'tow'.

With both extruded and spun soya bean products, flavours, colours, fats, etc., can be added to simulate meats or seafoods.

Bibliography

Erickson DR (ed.) (1990) *Edible Fats and Oils Processing: Basic Principles and Modern Practices*. Champaign: American Oil Chemists' Society.

Erickson DR, Pryde EH, Brekke OL, Mounts TL and Falb RA (eds) (1980) *Handbook of Soy Oil Processing and Utilization*. American Soybean Association, St Louis, and American Oil Chemists' Society, Champaign.

Kinsella JE and Soucie WG (eds) (1989) *Food Proteins*. Champaign: American Oil Chemists' Society.

Snyder HE and Kwon TW (1987) *Soybean Utilization.* New York: Van Nostrand Reinhold.
Weiss TJ (1983) *Food Oils and Their Uses*, 2nd edn. Westport, CT: AVI.

Harry E. Snyder
University of Arkansas, Fayetteville, USA

Properties and Analysis

Soya beans vary widely in their appearance and composition. The main constituents of soya beans, in descending order, are protein, oil, complex carbohydrates, oligosaccharides, simple sugars and minerals. Most legumes contain 20–25% protein, but soya beans typically contain 30–45% protein (moisture-free basis), and average 35·35% at 13% moisture. Levels as high as 55% protein (moisture-free basis) have been observed. The oil content typically ranges from 15 to 24% and averages 19% on a 13% moisture basis. The composition varies with growing area, e.g. soya beans from northern and midwestern areas of the USA typically contain 1·5–2% less protein and 0·2% more oil than beans grown in southern states. Generally, soya bean meal is sold on a 44% protein basis and 48% when soya beans are dehulled (Hi-Pro meal). Soya bean oil sells for about twice the price of the meal. ***See*** Legumes, Dietary Importance

The total crude fibre content of soya beans is about 4·4% on a 13% moisture basis. These materials are predominantly cellulose, hemicellulose and pectin. The outer hull, typically 8% by weight of the bean, is especially rich in crude fibre (35%). Fibre is hard to digest and contributes little to the nutrition of swine and poultry, the primary markets for soya bean meal. ***See*** Dietary Fibre, Properties and Sources

The total sugar content of soya beans is 4·9–9·5% on a 13% moisture basis. Of the total sugar content, about 60% is sucrose, 10% is raffinose and 30% is stachyose. Raffinose and stachyose cause flatulence in humans and reduced feed efficiency in livestock.

Proteins

Glycinin and β-conglycinin comprise 65–80% of the protein fraction or 25–35% of the seed weight. Glycinin is one of the legumins, which are characterized by molecular masses of 300–400 kDa and sedimentation coefficients of 11 ± 1S. β-Conglycinin is a vicilin, which have molecular masses in the range 150–250 kDa, are glycosylated and have sedimentation coefficients of $7 \pm 0{\cdot}5$S. ***See*** Protein, Determination and Characterization

Table 1. Glycinin acidic–basic complexes from CX635-1-1-1

Ab complex	S amino acids	Molecular mass (kDa)
$A_{1a}B_2$	14	57
$A_{1b}B_{1b}$	12	57
A_2B_{1a}	14	57
A_3B_4	9	62
$A_5A_4B_3$	3	67

In soya beans, the major proteins, glycinin and β-conglycinin, are frequently described by their respective sedimentation values, 11S and 7S, but such fractions are often impure. The 7S fraction of soya protein contains, in addition to β-conglycinin, lectins, lipoxygenase and β-amylase.

β-Conglycinin is a trimer and/or hexamer in solution and probably occurs in both forms in the seed. Two similar peptides, α and α' forms (57 kDa), and a glycosylated β peptide (42 kDa) are assembled in the mature protein in a nonrandom set of seven forms, $\alpha'\beta_2$, $\alpha\beta_2$, $\alpha\alpha'\beta$, $\alpha_2\beta$, $\alpha_2\alpha'$, α_3 and β_3 with molecular masses of 125–171 kDa. The α and α' subunits have 1–2 mol of cysteine per mole of peptide while the β peptide has no cysteine.

Glycinin is a hexamer, although older literature calls it a dodecamer. It is composed of six nonrandomly paired acidic and basic peptides. The acidic peptides have molecular masses of 44, 37 and 10 kDa; the basic peptides have a molecular mass of 20 kDa. The acidic–basic (AB) pairs have been identified in the experimental line CX635-1-1-1 and are shown in Table 1. Up to seven acidic and eight basic peptides have been identified in 18 cultivars. There appear to be sulphur-rich and sulphur-poor AB pairs. Some researchers have suggested that the sulphur-poor peptides could be eliminated by breeding; however, other reports show that some of the sulphur-poor acidic subunits are correlated with the gelling and gel hardness properties of this protein. The primary sequence of these peptides has been determined in a few cultivars. There is peptide sequence microheterogeneity.

Soya lectins are growth inhibitors in animals and are heat-labile. They bind tightly to carbohydrate moieties, which accounts for their haemagglutinating activity. ***See*** Plant Toxins, Haemagglutinins

Soya beans contain two main classes of protease inhibitors or trypsin inhibitors (TIs), although there appear to be many isogeneic variants. The principal two classes are the Kunitz inhibitor with a molecular weight of 21 500 and the Bowman–Birk inhibitor with a molecular weight of 8000. The Kunitz inhibitor acts only on trypsin while the Bowman–Birk protein inhibits both trypsin and chymotrypsin. Moist heat treatment

Table 2. The range of acyl groups produced in soya bean lipids by breeding and selection, and the composition of a typical unselected variety. Values in per cent

	Range	Typical value
Palmitate	3·5–29	10
Stearate	2·5–32	4
Oleate	8·0–65	26
Linoleate	20–60	52
Linolenate	1·8–13	8

denatures about 90% TI activity with the residual being heat-stable. TIs affect animals of guinea-pig size and smaller but have little effect on larger animals, except for weanling pigs. There are reports that TIs exert a carcinogenic effect on rodents. ***See*** Plant Toxins, Trypsin Inhibitors

Lipid

Soya bean lipids contain about 2–5% phosphatides, depending on the growing conditions, and 1·6% unsaponifiables. The balance is chiefly triglyceride. ***See*** Lipids, Classification; Phospholipids, Determination; Triglycerides, Characterization and Determination

Oleic, linoleic, palmitic, stearic and linolenic acids are present in soya bean oil along with traces (less than 1%) of myristic, palmitoleic, heptadecanoic, eicosaenoic, arachidic, behenic and erucic acids. The range of the acyl groups present in soya bean oil has been extended by mutation breeding and selection to the values reported in Table 2.

The acyl groups are distributed asymmetrically in the triglycerides with all the saturates on the *sn* 1 and 3 positions and linoleate concentrated on the *sn* 2 position. Generally the *sn* 1 position binds more palmitate and stearate than the *sn* 3 position, and the oleate is enriched on the *sn* 3 position.

Phosphatides contain the same acyl groups found in the triglycerides, but the concentration of palmitate is generally higher and oleate lower. Phosphatidylcholine (25%), ethanolamine (22%) and inositol (14%) are the chief phosphatide components along with lower concentrations of phosphatidic acid and phosphatidylserine.

The unsaponifiables contain sterols, hydrocarbons and tocopherols. The chief sterols (3·5 mg per gram of oil) are β-sitosterol, campesterol and stigmasterol. The tocopherols (about 1·25 mg per gram of oil) contain typically more than 70% of the γ form with smaller amounts of the δ and α forms. ***See*** Tocopherols, Properties and Determination

Properties and Analysis

Methods to Measure Composition

Proximate analyses for moisture, protein, crude free fat, crude fibre, ash and total carbohydrate have been adopted by the American Oil Chemists' Society, the American Association of Cereal Chemists and the Association of Analytical Chemists. All composition values for soya beans are reported either on a moisture-free basis or at 13% moisture.

Several methods are acceptable for moisture determination, but the most widely used procedure involves measuring the weight loss when drying the ground sample for 3 h at 130°C.

Protein is estimated from Kjeldahl nitrogen. The nitrogen content is multiplied by the factor 6·25 to convert to protein values, despite the major soya protein, glycinin, containing only 17·5% nitrogen (equivalent to a nitrogen conversion factor of 5·7).

Oil content is determined as crude fat by continuously extracting dried ground samples with petroleum ether for 5 h. Total fat, which includes bound fat as well as free fat, requires acid hydrolysis of the sample prior to extraction.

In recent years, the use of near infrared reflectance (NIR) and transmittance (NIT) has become widespread for rapid estimation of grain composition, especially of moisture, protein and oil. These spectrophotometers must be calibrated against the standard wet chemical methods described above. Moisture is also routinely measured by electrical capacitance. ***See*** Spectroscopy, Infrared and Raman

Crude fibre is measured as the weight loss on incineration of the oven-dried residue remaining after digestion of the sample with boiling dilute sulphuric acid followed by boiling dilute sodium hydroxide. ***See*** Dietary Fibre, Determination

Ash is primarily composed of noncombustible minerals and is determined by heating the ground sample in a muffle furnace for 2 h at 600°C. Soya beans contain about 4·7% ash on a 13% moisture basis.

Total carbohydrate is often estimated as the difference after subtracting other constituents. This method does not discriminate between oligosaccharides and other sugars, and it often gives inflated values. Sugars can be extracted with hot aqueous ethanol and quantified by gas or liquid chromatography. Several high-performance liquid chromatography methods have been described, based on amino bonded phases and refractive index detection, for the determination of sucrose, raffinose and stachyose. ***See*** Chromatography, Thin-layer Chromatography; Chromatography, High-performance Liquid Chromatography; Chromatography, Gas Chromatography

Acyl group composition of soya bean lipids generally is determined by gas chromatography. Lipid classes may be separated by thin-layer or liquid chromatography.

Table 3. Official grades and grade requirements of the Federal Grain Inspection Service, US Department of Agriculture

		Damaged kernels		Maximum limits		
US grade[c]	Minimum test weight per US bushel[d] (lbs)	Heat damaged (%)	Total (%)	Foreign material (%)	Splits (%)	Soya beans of other colours (%)
1	56·0	0·2	2·0	1·0	10·0	1·0
2	54·0	0·5	3·0	2·0	20·0	2·0
3[a]	52·0	1·0	5·0	3·0	30·0	5·0
4[b]	49·0	3·0	8·0	5·0	40·0	10·0

[a] Soya beans that are mottled or stained purple are graded not higher than US No. 3.
[b] Soya beans that are materially weathered are graded not higher than US No. 4.
[c] US sample grade soya beans (1) do not meet the requirements of US Nos. 1, 2, 3, or 4 or (2) contain eight or more stones which have an aggregate weight in excess of 0·2% of the sample weight, two or more pieces of glass, three or more *Crotalaria* seeds, two or more castor beans (*Ricinus communis* L.), four or more particles of unknown foreign substance(s) or commonly recognized harmful or toxic substance(s), 10 or more rodent pellets, bird droppings or equivalent quantity of other animal filth per kilogram of soya beans; or (3) have a musty, sour or commercially objectionable foreign odour (except garlic odour); or (4) are heating or otherwise of distinctly low quality.
[d] 1 US bushel=30 litres.

Qualitative and quantitative analyses of the individual proteins have been performed by immunoelectrophoresis, sodium dodecyl sulphate–polyacrylamide gel electrophoresis, analytical ultracentrifugation, and isoelectric focusing. *See* Electrophoresis

Grading Standards

Table 3 gives the grade standards established by the Federal Grain Inspection Service (FIGS) for the USA. Soya beans are divided into two classes based on colour: yellow soya beans and mixed soya beans. Each class is divided into four numerical grades and a US sample grade. Special grades (e.g. garlicky, infested) are provided to emphasize special qualities affecting the value, and are added to and made part of the grade designation. Six factors are considered in assessing a grade designation: test weight, heat damage, total damage, foreign material, splits, and soya beans of other colours. Although protein and oil contents are not part of the official grading standards, they may be specified in some markets.

Test weight is determined on a 1·25 US quart (0·95 litre) sample before removing foreign material using an official test weight apparatus. Foreign material is determined by sieving. Splits are determined by sieving a portion of the grain after removing the foreign material. Similarly, damaged kernels are determined by hand picking.

Bibliography

AACC (1990) *AACC Approved Methods*, 8th edn. St Paul: American Association of Cereal Chemists.

AOAC (1980) *Official Methods of the AOAC*, 13th edn. Washington, DC: Association of Official Analytical Chemists.

AOCS (1989) *Official Methods and Recommended Practices of the AOCS*, 4th edn. Champaign: American Oil Chemists' Society.

Hurburgh CR (1988) Moisture and compositional analysis in the corn and soybean market. *Cereal Foods World* 33: 503–505.

Hurburgh CR, Brumm TJ, Quin JM and Hartwig RA (1990) Protein and oil patterns in the U.S. and world soybean markets. *Journal of the American Oil Chemists' Society* 67: 966–973.

Nielsen NC (1985) Structure of soy proteins. In: Altschul AM and Wilcke HL (eds) *Seed Storage Proteins, New Protein Foods*, vol. 5. pp 27–64. New York: Academic Press.

Pryde EH (1980) Composition of soybean oil. In: Erickson DR, Pryde EH, Brekke OL, Mounts TL and Falb RA (eds) *Handbook of Soy Oil Processing and Utilization*, pp 13–31. Champaign: American Oil Chemists' Society.

Smith AK and Circle SJ (1978) Chemical composition of the seed. In: Smith AK and Circle SJ (eds) *Proteins. Soybeans: Chemistry and Technology*, vol. 1, pp 61–92. Westport, CN: AVI Publishing.

E.G. Hammond, L.A. Johnson and P.A. Murphy
Iowa State University, Ames, USA

Dietary Importance

Uses

Although whole soya beans are only infrequently consumed, foods made from soya beans (soya foods) have been used for centuries in parts of Asia (Table 1). In recent years, traditional soya foods have attracted interest among various subpopulations in many Western countries. Modern day soya foods frequently use tofu (soya bean curd) as a base and include soya meat substitutes and products designed to simulate more familiar dishes. Despite its many food uses and the excellent quality of soya protein, approximately 90% of the 100×10^6 t of soya beans grown worldwide each year, about 50% of which is produced by the USA, is

Table 1. Preparation, use and macronutrient composition of some commonly consumed traditional soya foods

		Macronutrient composition[a] (g per 100 g edible portion; (%) kcal)			
Soya food and use	Preparation	Energy (kcal)	Fat	Protein	Carbohydrate
Tofu (soya bean curd) Can be puréed for dressings, dips and spreads or eaten directly	After overnight soaking, soya beans are pulverized, cooked and then filtered which results in a liquid, *soya milk*, and highly perishable pulp called *okara*. The milk is coagulated with a calcium or magnesium salt, the whey is discarded and the curds are pressed to form a cohesive bond	93	4·8 (52%)	8·1 (39%)	1·9 (9%)
Soya milk Often substituted for cows' milk but should not be used as a replacement in the case of infant diets	To more effectively destroy trypsin inhibitor activity, the soya milk produced for drinking is cooked more extensively than soya milk used in tofu manufacture. To enhance flavour, most products have additional ingredients, such as sweeteners, oil, flavours and salt, added to the final beverage	85	1·9 (48%)	2·8 (32%)	1·8 (20%)
Tempeh (fermented soya bean cake) Primarily used as a meat substitute (tempeh burgers) or cut into pieces and added to a variety of dishes (such as mock chicken salad)	After cooking, whole soya beans (minus the hull), either by themselves or in combination with other grains, seeds or a mixture of both, are placed in a perforated container and fermented with a mould culture, *Rhizopus oligosporus* for a period of 18–24 h at about 32°C. This results in a white, distinctly smelling, chunky-textured cake about 2 cm thick	199	7·7 (32%)	17·0 (36%)	19·0 (32%)
Miso (fermented soya bean paste)	Whole soya beans are washed, soaked and cooked, then mixed with rice, barley or soya beans that have been fermented with *Aspergillus oryae* or *Aspergillus sojae* and formed into koji nuggets. The mix is then incubated and fermented, resulting in a ripened mesh called 'moromi'. The moromi is then blended, mashed and pasteurized, before or after packaging, as miso	206	6·1 (22%)	11·8 (26%)	28 (52%)
Natto Often used as a topping for rice or added to miso soup or sautéed with vegetables. Can be sweetened and served as hors d'oeuvre	Prepared from whole soya beans that are steamed until soft and then inoculated with *Bacillu natto* and fermented for 15–24 h, producing a fairly strong flavour and a sticky, slippery surface texture	212	17·7 (44%)	11·0 (31%)	14·4 (25%)

[a] Haytowitz DB and Mathews RH (1986) *Composition of Foods: Legumes and Legume Products.* US Department of Agriculture, Agriculture Handbook 8–16. Washington, DC: US Government Printing Office.

used as feed for animals. The oil produced in the making of soya bean meal is widely consumed by humans, and soya bean oil is the leading edible oil in the world.

In the USA and other Western countries soya is consumed predominantly in the form of soya protein products, which are grouped into three general categories: soya flour and soya grits; soya concentrates; and soya isolates. These products are made from defatted soya bean flakes, range in protein content from about 50% to 90% and are added to a vast array of foods primarily for their functional characteristics, such as emulsification and thickening. ***See*** Emulsifiers, Organic Emulsifiers

In considering the nutritional qualities of soya it is important to differentiate among the various soya products because processing affects the resulting nutrient composition. Increasingly however, non-nutritive components of soya beans have been the focus of investigation. Soya foods are also being studied for their role in chronic disease prevention.

Nutrient Contribution of Soya Foods

Whole soya beans are a good source of protein, fibre, calcium, iron, zinc, phosphorus, magnesium, thiamin, riboflavin, nicotinic acid, and folacin. Fermented soya foods have been reported to contain vitamin B_{12}, either naturally or by contamination, but this may be mostly in the form of analogues and cannot be considered a reliable source of this vitamin. Foods using the whole soya bean, such as tempeh (fermented soya bean cake), natto (fermented soya beans) and miso (fermented soya bean paste), retain much of the fibre in the whole bean, whereas soya milk and products made from soya milk, such as tofu, do not. Much of the fibre is lost in the okara, i.e. the pulp that remains after the soya bean is liquefied to produce soya milk. ***See*** individual nutrients ***See*** Fermented Foods, Fermentations of the Far East

The calcium content of soya milk and tofu is somewhat unclear. Values reported by the US Department of Agriculture (USDA) are higher than one would expect. There is also considerable individual variation among brands, and tofu made using a calcium salt as a coagulant is markedly higher in calcium. Recently, commercial soya milks have been supplemented with calcium to approximate dairy milk closely.

Some question has been raised over the bioavailability of minerals in soya beans and soya foods, although more research is needed to clarify this issue. The absorption of iron from soya, which is present in high levels, is greatly enhanced by the addition of vitamin C. In women, calcium absorption from soya beans was recently shown to be about 80% of that from cows' milk. Zinc absorption is of special concern since vegetarians, a group likely to use soya foods, may have inadequate intake of this nutrient. ***See*** Bioavailability of Nutrients; Vegetarian Diets

Soya foods are relatively high in fat. Tofu ranges from about 35% to 50% fat on an energy (caloric) basis. Miso and tempeh are somewhat lower, whereas the fat content of soya milk varies significantly, depending on the additional ingredients used to make the final beverage. Despite the relatively high fat content, soya foods are often lower in total fat than the foods they frequently replace in the diet; they are certainly lower in saturated fat and do not contain cholesterol.

Soya oil is approximately 50% linoleic acid, and contains about 20% each of saturated and monounsaturated fatty acids. Soya oil also contains approximately 7% α-linolenic acid, making it one of the few good plant sources of ω-3 fatty acids. However, hydrogenation of soya oil markedly reduces linolenic acid content. The longer-chain ω-3 fatty acids docosahexaenoic (DHA) acid and eicosapentaenoic acid (EPA) have been the subject of considerable investigation in recent years in connection with both treatment and prevention of heart disease, cancer and other diseases. Linolenic acid is converted to DHA and EPA, although not very efficiently. Omega-3 fatty acids are considered by some to be essential nutrients, particularly for brain development in young infants. ***See*** Fatty Acids, Dietary Importance

Soya Fibre

Soya fibre is a concentrated source of fibre: 13 g of soya fibre provide 10 g of dietary fibre, whereas, for example, 58 g of oat bran are needed to supply a similar amount. As noted previously, many soya foods, including tofu and soya milk, are low in fibre. Soya fibre is a mixture of cellulosic and noncellulosic structural components of the internal cell wall. Its major fractions are noncellulosic and consist of acidic polysaccharides, arabinogalactan, arabinan chains and about 10% cellulosic components. ***See*** Dietary Fibre, Properties and Sources

The composition of soya fibre differs markedly according to the method of analysis used. When analysed using the Association of Official Analytical Chemists method, soya fibre is reportedly comprised of only 4% soluble and 71% insoluble fibre, whereas other methods indicate that soya fibre may be as much as 30% soluble fibre. Soya fibre (7–10 g) has been shown to have a modest beneficial effect on regulating blood glucose levels in diabetics, whereas larger amounts (25 g) have been shown to lower total and cholesterol and low-density lipoprotein (LDL)-cholesterol. Soya fibre has also been shown to increase faecal weight but to have varied effects on intestinal transit time. These reports suggest that soya fibre contains appreciable amounts of soluble fibre. ***See*** Dietary Fibre, Effects of Fibre on Absorption; Dietary Fibre, Physiological Effects

Protein Quality

Soya beans are very high in both protein and fat relative to other legumes, which have a much higher carbohydrate content than soya beans. It has long been recognized that heating is necessary to utilize soya bean protein efficiently. This is partly attributable to the need to inactivate the protease inhibitors in soya beans. Well-processed soya foods and soya protein products, such as tofu and soya protein isolates, are more than 90% digestible, approximately the same as meat, egg and milk protein. Soya flour is approximately 80% digestible whereas toasted and steamed whole soya beans are only 65% digestible, apparently because of the lack of processing. The sulphur amino acids (methionine and cystine) are limiting in soya protein. As a result, studies using growing rats may somewhat undervalue soya protein because the sulphur amino acid requirement of rats is higher than that of humans. ***See*** Legumes, Dietary Importance; Protein, Quality

There is some debate regarding the precise requirements (levels) for amino acids. Recently, requirements have been proposed that are higher than those currently used by the Food and Agriculture Organization, World Health Organization and United Nations University (FAO/WHO/UNU). However, even when using these more rigorous requirements, soya protein supplies the sulphur amino acids in sufficient quantities as long as total protein intake meets the level currently recommended by the FAO/WHO/UNU. Human feeding studies generally support these conclusions, in that nitrogen balance is achieved when approximately this amount of soya protein is used (0·8–1·1 g per kg of bodyweight per day). ***See*** Protein, Requirements

Only at low levels of overall protein intake are differences between animal and soya protein readily apparent. At low levels of protein intake, methionine supplementation increases the quality of soya protein, but this does not appear to be of practical significance in developed countries, where protein intake greatly exceeds requirements. Even in developing countries, protein deficiency is often more a matter of insufficient energy than poor protein quality. Infants have both a higher protein requirement and a higher sulphur amino requirement (per kg bodyweight) than adolescents and adults. Methionine supplementation of infant formulae, which may provide the bulk of the energy intake, may be beneficial, but only modest amounts of methionine are needed to achieve substantial improvements in quality.

Antinutrients in Soya Beans

Not until the late 1940s was the inactivation of protease inhibitors shown to be partly responsible for the need to heat soya protein for efficient utilization. As many as five different protease inhibitors exist in soya beans. Protease inhibitors can be found throughout the plant kingdom, but soya is a very rich source.

The soya bean protease inhibitors, in addition to trypsin and chymotrypsin inhibition, may adversely affect protein nutriture by increasing amino acid loss caused by the increased secretion of pancreatic enzymes. This latter effect is particularly noteworthy. In response to both raw soya and isolated protease inhibitors, cholecystokinin (CCK) levels are increased, and this is believed in turn to stimulate pancreatic enzyme secretion. In many species CCK is negatively inhibited by trypsin. Chronic pancreatic stimulation is eventually thought to lead to pancreatic hypertrophy, hyperplasia and even cancer. ***See*** Plant Toxins, Trypsin Inhibitors

Feeding raw soya to rats both enhances chemically induced pancreatic cancer and, in long-term studies, enhances spontaneous pancreatic cancer. However, there is considerable species variation in response to protease inhibitors.

In humans, feeding both raw soya and isolated protease inhibitors stimulates both CCK levels and pancreatic secretion. However, the practical implications of these findings are unclear. Most protease inhibitor activity (90% or more) is destroyed upon heating, and levels are therefore low in commercial soya foods. Particular concern has been expressed in the case of infants consuming soya-based formulae where a large percentage of energy intake is from soya. In Japan, where soya foods are widely consumed, pancreatic cancer rates are similar to or less than those in the USA and other Western countries, where soya foods are infrequently consumed.

Lectins (haemagglutinins) are proteins found in soya (1–3% of total protein) and most other legumes that bind carbohydrates; *in vitro*, this action is manifest by agglutination of red blood cells caused by the binding of surface glycoproteins. Although lectins are generally thought to be toxic, the limited work with soya indicates that in rats this does not appear to be the case. Lectins, like the protease inhibitors, are destroyed by heat and therefore would not be present to any significant extent in commercial soya products. ***See*** Plant Toxins, Haemagglutinins

Saponins are triterpenoids that are found in a number of plant foods, including soya beans which are a very rich source. Several different saponins have been identified in soya beans. At one time, saponins were thought to exert an adverse effect on animal growth but when soya bean saponins were fed to chicks, rats and mice, at three times the level found in soya flour (0·5%), no ill effects were observed. Saponins are not thought to be absorbed. Saponins have been studied for their role in lowering serum cholesterol levels, presumably by blocking cholesterol absorption, but a number of studies do not concur with this hypothesis. ***See*** Saponins

Phytic acid (phytate) generally refers to myo-inositol hexaphosphoric acid (1,2,3,4,5,6-hexakis(dihydrogen phosphate)). Unlike the protease inhibitors, phytic acid is fairly stable to heat. In plants, phytates are considered to be the major storage form of phosphorus. Levels of phytic acid are relatively high in soya beans but also in many other foods, particularly unprocessed cereal grains. *See* Phytic Acid, Nutritional Impact

The ability of phytic acid to inhibit mineral absorption has raised concern over the bioavailability of minerals in soya products, although components in addition to phytate influence absorption. The effect of phytate on zinc absorption is unclear, with studies showing impaired absorption as well as no effect. In humans, the addition of phytic acid to cows' milk formulae was found to reduce extrinsically labelled zinc absorption by approximately 50%. The Committee on Nutrition, American Academy of Pediatrics, recommended that soya formulae be supplemented with zinc, although this conclusion was based primarily on animal studies.

For quite some time, goitres have been known to appear in rats consuming soya bean meal. In many but not all studies, the administration of additional iodine prevents thyroid enlargement, but the goitrogenic effect of soya does not appear to stem entirely from the low iodine content of soya beans, nor is it eliminated by heat treatment. The goitrogenic substances in soya beans have not been definitively identified. Some reports indicate that soya beans may affect thyroid function in humans but in Japan soya bean consumption does not appear to be linked to any cases of endemic goitre. *See* Goitrogens and Antithyroid Compounds

Soya Protein and Cholesterol

The relationship between serum or plasma cholesterol and heart disease is well established, as is the effect of diet on cholesterol levels. Although most dietary emphasis has been on saturated fatty acids and, to a lesser extent, dietary cholesterol, for over 50 years protein has been known to influence cholesterol levels. *See* Cholesterol, Role of Cholesterol in Heart Disease

In rabbits, plant proteins in general have been shown to lower serum cholesterol levels relative to animal proteins. An extensive amount of work suggests that this relationship also hold true for humans, although most work has been conducted using soya protein, usually in comparison to animal protein. In a recent review on this subject, 25 of 28 studies in which soya protein was substituted for animal protein showed a decrease in serum or plasma cholesterol in hypercholesterolaemic subjects. The decrease in total cholesterol results primarily from a decrease in the LDL-cholesterol fraction. Over 50% of the studies, which ranged in length from 21 to 112 days, reported decreases of 15% or more. In subjects with normal cholesterol levels, soya protein has only a minor effect.

The mechanism by which soya protein lowers cholesterol is unclear, although it does not appear to be necessary to replace all animal protein with soya for this effect to occur. Even adding soya protein to an otherwise mixed diet has been shown to have a beneficial effect on cholesterol levels. Among the many hypotheses for the hypocholesterolaemic effect of soya protein are changes in hormone levels or ratios (insulin : glucagon), effects on thyroid function and increased bile acid secretion.

Soya Products and Cancer Prevention

The role of soya consumption in cancer prevention is a relatively new area of investigation, although relevant studies can be found scattered throughout the literature during the past 10–15 years. Recent interest was stimulated by the results of an animal study showing that diets containing as little as 5% soya beans (w/w) reduced by 50% the number of chemically induced mammary tumours in rats. *See* Cancer, Diet in Cancer Prevention

In the proceedings of a workshop sponsored by the US National Cancer Institute in 1990, on the role of soya foods in cancer prevention, several potential anticarcinogens in soya beans were identified: isoflavones, protease inhibitors, phytates, saponins, and phytosterols. Soya beans also contain a number of other components that have been shown to inhibit cancer in experimental systems. However, with the exception of one, the isoflavones, these components are found in a wide variety of foods.

In the rat mammary cancer study cited above, both raw and autoclaved soya beans inhibited mammary tumours. Autoclaved soya beans were devoid of protease inhibitor activity, indicating that tumour inhibition in this study was not attributable to the protease inhibitors. Protease inhibitors have been shown to be very potent, inhibiting cancer cells in culture as well as several different cancers in animal models. The dietary relevance of protease inhibitors is unclear since, as pointed out before, in commercial soya products most protease inhibitor activity is destroyed.

A number of animal studies have examined the effect of soya on experimental cancer; most of these have used soya protein isolate although, in total, a wide range of soya products have been employed. In general, these data are supportive of a protective effect, although there are a number of inconsistencies and many studies suffer from design flaws. The epidemiological data are more clearly supportive of a protective effect of soya consumption. This work covers a wide range of cancers, although gastrointestinal cancers have been most frequently studied. Not surprisingly, much of the work consists of case-control and prospective studies conducted in Chinese and Japanese populations. In general, daily consumption has been found to reduce risk, relative to consuming soya products less often.

Much interest of late has focused on the isoflavones found in soya beans. Isoflavones are diphenolic compounds and a subclass of the much larger group of flavonoids (flavones). The primary isoflavones in soya are genistein, daidzein and glycetein, and their respective glycones. The isoflavones are considered to be weak oestrogens and have been postulated to act as antioestrogens *in vivo* by competitively inhibiting the more potent natural oestrogens. In premenopausal women, soya feeding has been shown to lengthen the menstrual cycle. The antioestrogenic effect of the isoflavones is a possible explanation for both the relatively low rate of breast cancer in Japanese women, and the mammary cancer inhibition seen in animal studies. ***See*** Breast Cancer and Diet

The hypothesized antioestrogenic effects of isoflavones would be unlikely to account for the protective effects of soya bean consumption seen in the epidemiological data for non-hormone-related cancers. A vast amount of *in vitro* work conducted during the last several years indicates that one of the isoflavones in soya beans, genistein, is a very potent tyrosine protein kinase inhibitor and also inhibits deoxyribose nucleic acid (DNA) topoisomerases I and II. These enzymes are thought to play a critical role in cell regulation. Genistein has been shown in culture to induce differentiation and to inhibit the growth of cancer cells.

Bibliography

Akiyama T and Ogawara H (1991) Use and specificity of genistein as inhibitor of protein-tyrosine kinases. *Methods in Enzymology* 201: 362–370.

Erdman JW and Fordyce EJ (1990) Soy products and the human diet. *American Journal of Clinical Nutrition* 49: 725–737.

Carroll KK (1991) Review of clinical studies on cholesterol-lowering response to soy protein. *Journal of the American Dietetic Association* 91: 820–827.

Golbitz P (1991) *Soya Bluebook*. Bar Harbor, Maine: Soyatech, Inc.

Grant G (1989) Anti-nutritional effects of soyabean: a review. *Progress in Food and Nutrition Science* 13: 317–348.

Liener IE (1981) Factors affecting the nutritional quality of soya products. *Journal of the American Oil Chemists Society* 58: 406–415.

Messina M and Messina V (1991) Increasing use of soyfoods and their potential role in cancer prevention. *Journal of the American Dietetic Association* 91: 836–840.

Slavin J (1991) Nutritional benefits of soy protein and soy fiber. *Journal of the American Dietetic Association* 91: 816–819.

Snyder HE and Kwon TW (1987) *Soybean Utilization*. New York: Van Nostrand Reinhold.

Young VR (1991) Soy protein in relation to human protein and amino acid nutrition. *Journal of the American Dietetic Association* 91: 828–835.

Mark Messina and Virginia Messina
Mount Airy, Maryland, USA

SOYA CHEESES

Soya bean cheeses are all produced from soya bean milk made from whole soya beans. Since soya beans originated in the Far East, the technology of cheesemaking originated in China and Japan. The composition of soya beans is quite unlike that of dairy milk and, accordingly, the microorganisms employed in making soya bean cheeses are different from those used in making dairy cheeses. For example, there is no lactose in soya bean milk for dairy lactics to use. ***See*** Lactic Acid Bacteria; Soya Beans, The Crop; Starter Cultures

Soya bean cheeses take a variety of physical forms, varying from soft semisolid yoghurt-like products to solid cheese-like foods covered with microbial growths of bacteria or moulds.

The fermentation microorganisms may be pure cultures, or they may be a mixed culture of two or more microorganisms. Keeping ability is from a few days to almost indefinitely in the case of freeze-dried solids. Soya bean cheese may be made only from soya beans or soya beans with other ingredients. Many are foods made for centuries, and they may be made under very primitive methods or in large ultramodern factories producing completely sterile products. ***See*** Fermented Foods, Origins and Applications

Soya bean cheese may be consumed as a main-course dish or, because of its salt and flavour, it may be used as a condiment. Invariably, it is high in protein, with the protein coming from the soya bean, and generally, but not always, with a high moisture content. The length of fermentation may be from a few hours to months, with the longer fermented products being more highly flavoured and darker in colour. Soya bean cheeses were developed along with fermented noncheese soya bean products to destroy, in part, the undesirable flavours and antinutritional factors present in soya beans. Invariably, the cheese product has lost most, if not all, of its initial beany flavour which people and animals find objectionable.

Importance Worldwide

Soya bean cheeses are made in many countries, including Taiwan, China, Vietnam, Philippines, Okinawa, Thailand, Hong Kong, Korea and perhaps elsewhere in Asia. The scale of production in these countries is not available, but they are known to have been produced for centuries in China. They are available in the West and are sold as cakes in brine in glass bottles or in sterile containers. In Taiwan the annual production of the fermented product tou-fu-ju (sufu) is 10000 tonnes with a consumption of 12 g per person per week in 1980. It is produced both commercially and in the home. A second cheese-like product, tofu, is not fermented, rather, it is soya bean milk that is curdled and sold fresh, freeze dried, or in a sterile moist state. Both the fermented and nonfermented cheese-like products are described in this article.

Sufu (Chinese Cheese)

According to Wai it is not known when sufu was first made, but he refers to the fact that it was described in a food encyclopaedia in 1861, while tofu was discovered about 179–122 BC. In the literature, Chinese cheese has a number of names because of the numerous dialects in China and the difficulty of translating the Chinese into English. The following are synonyms of sufu: tosufu, fu-su, fu-ru, doufu-ru, fuyu, chao, tahuri, tou-fu-ju, toe-fu-ru, tou-fu-ru, teou-fu-ru, fu-ju, fu-yu and foo-yue. In the West, Chinese cheese has been referred to as 'bean cake' by Chinese grocers. The word 'sufu' literally means 'moulded milk'. It was 1929 when Wai first used the term sufu and showed that Chinese cheese was made with microorganisms; hence, sufu is now the preferred name. Another cheese product made from tofu is tofuyo, which uses fungus of the genus *Monascus*; it is made in Okinawa.

Sufu (Chinese cheese) is a soft, white to light grey product in the form of cubes covered with fibres of white mould (mycelium) like Camembert cheese. Unlike this food, however, the moulded cubes of sufu are kept in brine and, at least in the West, are sold in glass jars. Additives, either to give colour or flavour, are often added to the brine. A common colour is red, which makes red sufu or hon fang. When wine is added, the sufu has an alcoholic odour and is known as tsui-fang or tsue-fan, which translates to 'drunken sufu'. Other ingredients added to the brine may be hot pepper. Rose sufu is made by adding rose essence and then ageing. On a dry weight basis sufu contains 55% protein and 30% lipids. The hydrolytic products of protein and lipids give the principal constituents to the mild characteristic flavour of sufu.

Chinese cheese has a soft, creamy, cheese-like texture, a salty taste, and its own characteristic flavour and aroma, said to resemble anchovies. It is consumed directly, as a relish, or it is cooked with meats and vegetables. It can be used as a spread on crackers or as an ingredient in dips and dressings.

Preparation of Curd

Soya beans are soaked in water at 25°C for 4 h; the beans are then ground with water in a mill. The resulting milk is filtered through cheesecloth to remove solids. The milk is heated to 85°C, and an appropriate quantity of gypsum (0·7%) or calcium chloride (0·8%) is added to the milk, bringing about curd formation of the hot milk. The soya bean curd is collected and pressed in a mould to remove water. For making sufu, the curd is pressed hard to give a solid tofu. Tofu, for sufu, has 83–77·6% water, 10% protein and 4·0% lipid. The cake is cut into cubes that are 2 × 4 × 4 cm and weigh about 18 g each. The production of soya bean curd (tofu) was reported in the Chinese literature as early as 179 BC. The cubes are immersed for 1 h at 25°C in an acidic saline solution made from 6% sodium chloride and 2·5% citric acid in water, after which the cubes are treated with hot air at 100°C for 15 min. This step prevents growth of contaminating bacteria. ***See*** Spoilage, Bacterial Spoilage

Fermentation of Curd

On a laboratory scale (Fig. 1) the tofu cubes are mounted on glass or aluminum rods. This is followed by inoculation of the cubes by gently rubbing the cube surface with spores of a fungus grown on filter paper impregnated with nutrient culture solution.

Traditionally, the tofu is covered with rice straw, which furnishes the natural mould inoculum. The fermentation occurs in large bamboo trays with the trays stacked 20 or more high and left at 12–20°C. After 3 or 4 days the cubes are covered with white mycelium and are removed from the trays and salted. The fungal inoculum is always mucoraceous fungi belonging to the genera *Rhizopus*, *Mucor* or *Actinomucor*.

In modern methods, the cubes of tofu are heated in an oven at 100°C for 10–15 min. After the heat treatment the cubes are placed in perforated trays, and the surface of the cubes are inoculated with the mould culture. The cubes are allowed to ferment for 3–7 days at 12–20°C.

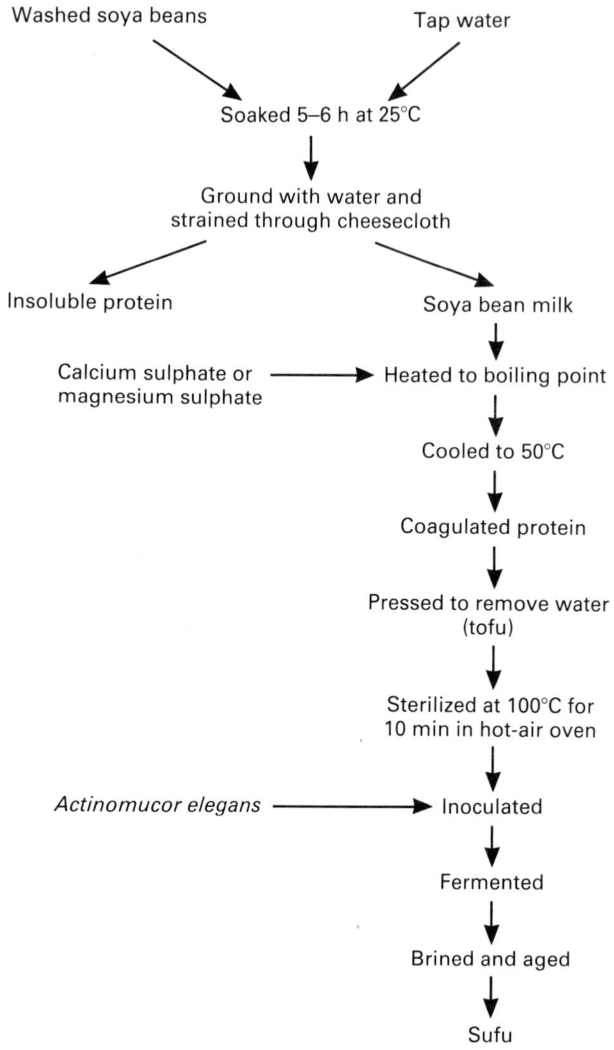

Fig. 1 Flow sheet for making sufu and tofu. Tofu must be made first, then fermented to make sufu.

Fig. 2 Cubes of sufu (Chinese cheese) in brine.

The raw fermented cubes (pehtze) are placed in earthen jars with alternating layers of salt and pehtze. After 3 or 4 days, when the salt has been absorbed, the pehtzes are removed, washed and placed in another jar for processing.

Processing and Ageing

The ageing jars are usually earthenware with a capacity of 80 litres. A mixture of the flavouring agents is added. A typical flavouring mixture contains approximately 1 kg of salt, 0·5 kg of soya bean mash, 250 g of red koji, 250 g of raw sugar, and 3 litres of water. In the alcoholic liquor method, the pehtzes are immersed in an alcoholic saline solution containing 12% sodium chloride and 10% ethanol, usually added as rice wine or distilled liquor. Next, pehtzes are added, followed by a layer of the flavouring agent; and this is repeated until 80% of the ageing jar is filled. Twenty per cent salt water is then added. The jars are then sealed and allowed to stand for 3–6 months. At the end of this ageing, the cubes are removed, washed and are now ready to eat (Fig. 2).

Microorganisms

In the fermentation of tofu to make sufu, only fungi are employed. Although several different fungi can be used, they must have certain characteristics. They must have white spores and mycelium and grow vigorously on tofu at 20°C or lower without any additional nutrients. The mould must develop a dense, thick mycelial mat over the entire surface of the tofu. It should utilize lipids as a source of energy and produce quantities of both lipases and proteolytic enzymes. It must not form disagreeable flavours or odours and it must not produce a mycotoxin. According to Wai, who conducted a survey of the moulds used to make sufu, the most used fungus in commercial operations was *Actinomucor elegans*, which is worldwide in distribution. It is intermediate between *Rhizopus* and *Mucor* spp. because it produces rhizorids like *Rhizopus* spp. but does not have an apophysis. In a survey of the moulds found to make sufu, in addition to

Table 1. Percentage composition of tofu, pehtze and sufu

	Tofu	Pehtze	Sufu
Moisture	72·11	66·13	61·94
Protein	16·14	21·21	18·60
Fat	10·46	10·48	9·52
Fibre	0·31	0·82	1·00
Ash	0·98	1·36	8·58

Table 2. Changes (%) in nitrogen compounds in tofu after 3–6 months of ageing

	Tofu	Tofu after ageing
Protein nitrogen	99·12	79·58
Nonprotein nitrogen	0·88	16·54
Formol nitrogen	1·37	17·82
Ammonia nitrogen	0·04	0·76

Actinomucor spp. Wai found *Mucor hiemalis*, *M. dispersus*, *M. silvaticus* and *M. subtilissimus* to be present in home production coming from the rice straw. During the ageing process the moulds are killed. ***See*** Mycotoxins, Occurrence and Determination

Composition

On a dry weight basis the hard tofu used in making sufu has 55% protein and 30% lipids. The composition of tofu, pehtze and sufu is shown in Table 1. Changes in nitrogen compounds after 3–6 months of ageing are shown in Table 2. Some differences in the composition of various types of sufu are shown in Table 3.

Enzymes Involved

The soya bean protein is digested by the proteases produced by the mould into peptides and amino acids. Free amino acids found include aspartic acid, glutamic acid, serine, alanine and leucine. ***See*** Amino Acids, Properties and Occurrence; Enzymes, Uses in Food Processing

When *M. hiemalis*, isolated from sufu, was grown on a soya bean medium, only a little proteinase appeared in the culture filtrate. The greater part was bound to the fungal cell surface. However, when sodium chloride was added, the bound proteinase was released from the cell wall of the fungus. The proteinase has an optimum pH of 3·0–3·5, and the maximum release of the enzyme was found to be in brine at a concentration of 0·3 M. The proteinase is not only on the outside of the cell but is also present inside. Other ionizable salts will also release the enzyme. It is now apparent why salt must be present in the brine. When the enzyme is released in the brine, it penetrates into the tofu cube and acts on the protein. Only a certain amount of enzyme can be present on the cell wall because continual removal of the enzyme from the cell surface by salt makes it possible for the mould to produce more of the proteinase enzyme.

The proteinase enzymes produced by the mucoraceous fungi act on the soya bean protein to form peptides and amino acids. Free amino acids such as aspartic acid, glutamic acid, serine, alanine, leucine and isoleucine are formed. Most of the enzyme digestion in the brine occurred during the first 10 days of ageing.

Soya bean lipids are digested somewhat to free fatty acids. These react with alcohol in the brine to form esters, which add to the pleasant odour of the product. ***See*** Fatty Acids, Properties

The brining step is essential because it releases enzymes and adds flavour from the salt, ethanol and esters. The brine at the same time kills the moulds and prevents microbial contamination. The flavours which develop in the sufu are brought about as shown in Fig. 3.

The proteinase, lipase and phosphatases of several different species of *Mucor* cultures grown for 5 days at 25°C are shown in Table 4.

Culinary Use and Preservation

Sufu is produced both in the home and commercially. In commercial operations, *A. elegans* is always used and the reason for this can be seen in Table 4, where the mould is seen to produce more of the desirable enzymes than any other species. Sufu is used as a main-course food or as a seasoning agent.

The shelf life of sufu is often shortened because of bacterial growth, especially at the stage when the tofu cubes are being moulded. Studies to reduce spoilage have involved the coating of the pehtzes with paraffin, which allowed keeping for 1 month at room temperature. Another method investigated was to steam sterilize the sufu in containers. This increased the keeping time to 1 year. Another problem for some people was the alcohol and salt in the brine. It is possible that the salt level might be reduced, provided growth of microorganisms can be controlled. Refrigeration at 5°C allowed sufu to be kept for up to 4 months.

Tofu

Tofu is a soya bean food made exclusively from soya beans without a fermentation step. Soya beans are very

Table 3. Percentage composition of various types of sufu

	Tsao sufu	Red sufu	Kwantung sufu	Yunnan sufu	Rose sufu
Water content	69·03	61·25	74·46	64·77	59·99
Crude protein	12·87	14·89	12·42	12·16	16·72
Ether extract	12·89	14·31	6·39	14·23	13·74
Crude fibre	0·13	0·42	0·11	0·27	0·14
Ash	5·08	9·13	6·61	8·56	9·41
Total nitrogen	2·06	2·38	1·99	1·94	2·68
Protein nitrogen	1·30	1·56	1·27	1·31	1·81
Nonprotein nitrogen	0·76	0·83	0·72	0·64	0·86
Ammonia nitrogen	0·20	0·16	0·18	0·18	0·18
Amino acid nitrogen	0·27	0·27	0·24	0·19	0·31

From Wai N (1968).

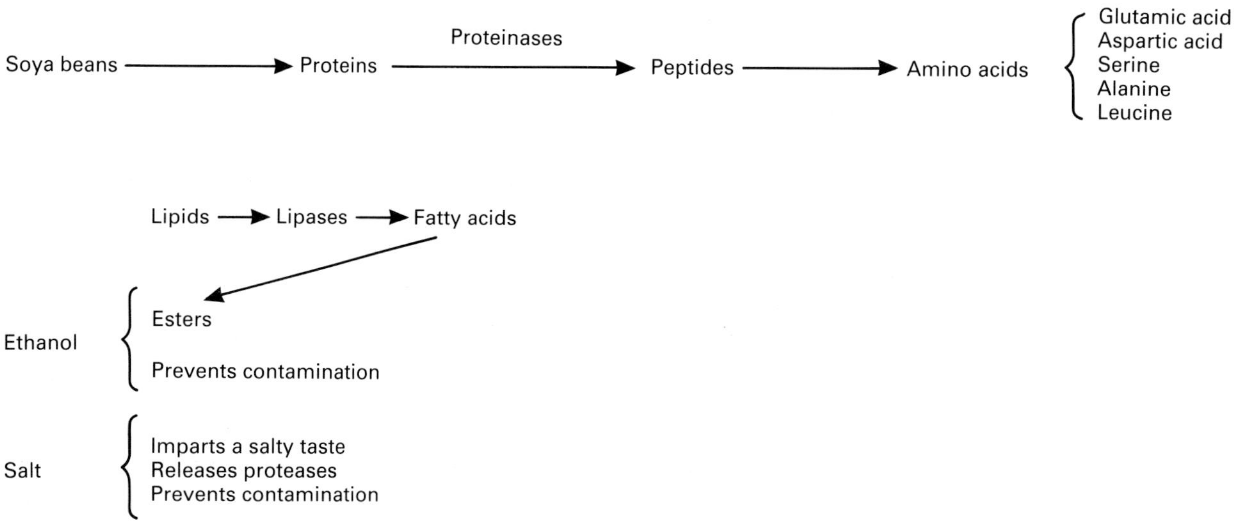

Fig. 3 A diagram showing the development of flavours in sufu.

Table 4. Activities of proteinase, lipase and phosphatase of *Mucor* cultures

Fungi	Proteinase: diameter of zone (mm)	Concentration (mg per 100 ml)	Lipase: diameter of zone (mm)	Concentration (ml per 100 ml)	Activity of phosphatase: Acid	Activity of phosphatase: Alkaline
Actinomucor elegans	18·3	63·66	17·5	271	−	++
Mucor pusillus	17·5	44·04	19·0	402	++	+++
Mucor circinelloides	11·6	3·34	18·4	346	+	−
Mucor hiemalis	15·3	7·13	17·0	231	+	++
Mucor javanicus	12·8	5·54	19·1	411	−	−

From Wai N (1968).
+++, strong; ++, quite strong; +, weak; −, none.

Table 5. Comparison of Japanese and US soya bean varieties in producing tofu

	Number of soya bean varieties tested	Protein (%)	Oil (%)	Yield of tofu (kg)	Moisture in tofu (%)
Japanese	5	40·6–46·5	15·7–17·4	2·38–3·18[a]	83·7–87·8
US	15	33·2–43·8	18·2–21·2	2·33–3·11	83·2–87·6

[a] From 1·8 kg of soya beans.

rich in oil and protein and are not easily digested after normal cooking, roasting or grinding. The Oriental methods of food preparation are to make tofu from soya bean milk, sprout the beans, or to ferment the soya beans to make shoyu (soy sauce), miso or natto. All of these processes make products which do not have a beany flavour and are easily digested because either the hard cell walls are removed or microorganisms produce enzymes that alter the soya bean constituents. Tofu is also called soya bean curd in the West, but in the Orient it may be called tou-fu, tufu, takoo, touhu, tou-foo, dou-fu or dan-fu. ***See*** Fermented Foods, Soy Sauce

Tofu has only recently become an established food in the West, while in the Orient it has been an important protein food for centuries. Because it has a bland taste, it can be flavoured with other ingredients to give a great variety of different tasting foods.

Fresh tofu is sold as a soft, white curd in water with a uniform consistency like that of cottage cheese. It is not salted with sodium chloride; it is, therefore, subject to rapid spoilage. Tofu may be of two kinds – one soft with a higher water content and the other harder with less water. The harder tofu is the type used in fermentation. Fresh tofu has the approximate composition of 6% protein, 3·5% fat, 1·9% carbohydrate, 0·6% ash and 88% water.

Process for Making Tofu

Soya beans are washed and soaked overnight in water at room temperature or until the beans are swollen. A soaking time of 16–18 h at 20–22°C is the ideal. The steep water is then discarded and fresh water is added at a ratio of water to beans of 10:1. The beans are then finely ground. A second method is the hot grind in which the soya beans are soaked and then hot ground. This yields a milk with less beany flavour and is the method preferred by non-Oriental people. This is followed by filtering the mash through a double-layered cheesecloth. From 300 g of dry beans about 2·3–2·5 litres of milky filtrate is produced, which is called soya bean milk.

The milk is then heated to boiling without pressure and cooled to 80°C, at which time 4·5 g of anhydrous magnesium sulphate in 40 ml of water, or 7·5 g of hydrated calcium sulphate in 40 ml of water is added to the milk very slowly with gentle mixing. (More vigorous mixing results in a hard curd with air pockets.) This is done to prevent breaking of the curd, which settles, leaving a clear whey.

The most critical step in making tofu is to use the right amount of the salt and the proper rate of addition. The type of salt added will determine the product quality. Calcium sulphate makes a curd slowly. This curd is smooth and gelatinous and has a higher water content. Magnesium sulphate gives a curd instantaneously; this curd is coarser in texture.

The curd now looks like cottage cheese and is allowed to settle. When the temperature falls below about 50°C, the curd is ladled into a wooden box lined with a double layer of cheesecloth or course filter cloth. The wooden box has many small holes in its sides to drain off water. A suitable weight is placed on top of the curd to force more water out and, when draining has almost stopped, the box is placed in a tank of water and then inverted to remove the curd. The curd remains for about an hour in the water. From 1·8 kg of beans one can expect to get 2·33–3·11 kg of tofu. Commercially, the cakes are about twice this size. The hardness of the tofu can be determined by the pressure put onto the curd. The cake may now be sliced into suitable shapes and sizes for sale.

Soya Bean Varieties and Tofu Composition

Soya beans for making tofu should have certain desirable characteristics. They should be uncracked, free of foreign material, even in size, uniform in water absorption, and high in protein solubility. The final product should have a light colour with a good texture.

Data on the differences between Japanese and US soya bean varieties are shown in Table 5. The amount of protein in the five Japanese soya bean varieties is greater than in the 10 US varieties examined. Tofu moisture levels are about the same, but the yield of tofu is perhaps slightly higher in the Japanese soya beans. The amount of protein in the soya beans was positively correlated with the protein in the tofu.

A recent study was made on five US and five Japanese varieties grown under the same environmental condi-

Table 6. Average amino acid composition (g per 16 g N) of fractions from three soya bean varieties

Amino acid	Soya beans	Soak water	Residue	Soya bean milk	Tofu	Whey
Aspartic acid	12·61	14·94	11·63	11·01	11·70	12·40
Threonine	4·11	4·53	4·42	4·01	4·00	4·52
Serine	5·74	5·13	5·47	5·19	5·32	4·15
Glutamic acid	19·76	17·68	17·71	19·61	19·26	23·62
Proline	5·53	5·28	5·66	5·33	5·47	4·14
Glycine	4·46	4·73	4·61	4·16	4·14	4·87
Alanine	4·49	4·61	4·36	4·14	4·11	4·42
Valine	3·73	4·92	5·28	4·88	4·99	2·65
Cystine	0·78	0·87	Trace	0·03	Trace	2·40
Methionine	1·34	1·72	1·67	1·59	1·43	2·61
Isoleucine	3·46	4·69	4·50	4·66	4·85	2·92
Leucine	7·90	6·90	8·31	7·94	8·32	3·89
Tyrosine	3·90	4·13	3·74	3·91	3·99	3·39
Phenylalanine	4·85	4·59	5·20	5·15	5·41	2·52
Lysine	6·19	5·45	6·36	6·08	6·14	8·56
Histidine	2·60	2·35	3·07	2·64	2·64	3·21
Arginine	8·64	7·27	8·61	8·65	8·52	9·69

Table 7. Composition of 100 g of tofu

Water	84·8%	Sodium	7·0 mg
Fibre	0·1 g	Potassium	42·0 mg
Calcium	128·0 mg	Thiamin	0·06 mg
Phosphorus	126·0 mg	Riboflavin	0·03 mg
Iron	1·9 mg	Niacin	0·1 mg

tions as to their ability to produce good tofu. All the varieties were grown in the USA and in Japan at the same time. Results indicated that all varieties produced tofu with a bland taste, fine texture and a creamy white colour. The one exception was tofu made from a variety with a black hilum that produced tofu with a less attractive colour. Differences among the 10 varieties were not attributable to the country of origin.

Recently, the yield and amino acid composition of fractions obtained during tofu making was studied in three soya bean varieties, two of which are considered good by tofu makers. The average amino acid compositions of fractions from these three varieties are shown in Table 6. Analysis showed 52% of the solids, 71% of the protein and 82% of the oil in the soya beans were present in the tofu. The soak water and whey had 14% solids and 4·7% protein. The residue had 30% solids, 20% protein and 11% oil. *See* individual nutrients.

In 100 g of tofu the composition was as shown in Table 7. Tofu lacks cholesterol and lactose and has low amounts of saturated fatty acids. The colour of tofu is affected by the soya bean variety used. Varieties with a dark seed coat and hilum result in a less desirable darker colour.

Microbiological Safety of Tofu

Tofu and soya bean milk are protein-rich and have a pH of 6·0–6·5; therefore, tofu is a good substrate for bacterial growth. These two products are thus typically consumed in the Orient the same day they are made. Microbial deterioration is a big problem in the West where tofu in water may be held in shops for several days. This spoilage has been partly overcome by the use of tofu packaged in cartons and pasteurized. The numbers of bacteria contaminating tofu can be reduced by thoroughly washing the beans to remove as many bacteria as possible before soaking.

Since spoilage of fresh tofu has been a major problem to the West, the microbial contamination of tofu has been recently studied. In this study, tofu was prepared, inoculated with food poisoning bacteria, and held for different times at different temperatures. *Clostridium botulinum* produced toxin in 1 and 3 weeks when stored at 25 and 15°C, but not at 6 weeks at 5 and 10°C. *Staphylococcus* and *Salmonella* spp. grew at 10, 15 and 25°C and enterotoxin was present in 5-day-old samples at 25°C. *Yersinia enterocolitica* grew at 5°C. This bacterium has recently caused an outbreak of yersiniosis as a result of the use of contaminated water in which the tofu was sold. The fact that these food poisoning bacteria will grow in tofu indicates that a high level of sanitary practices must be observed during manufacture. It is also imperative that pasteurization be used at the time of packaging and that tofu be refrigerated. In recent years tofu has been marketed in cartons that are completely sterilized and sealed at the time of manufacture and, therefore, may be kept fresh for months. *See*

Soya Cheeses

Clostridium, Occurrence of *Clostridium botulinum*; Pasteurization, Principles; *Staphylococcus*, Properties and Occurrence; Water Activity, Principles and Measurement

Hydration

Although making tofu is a simple process, many modifications in the various steps will greatly alter the nature of the product. The ratio of 1:6·5 of dry beans to water will cause a reduction in the amount of protein and total solids extracted. On the other hand a ratio higher than 1:10 of dry beans to water will result in the soya bean milk content being too low to obtain a proper curd. The heat treatment is essential for protein denaturation to give a curd and to improve the nutritional value of the product by reducing bad flavours and destroying antinutritional factors present in the soya beans. The optimal time for boiling is 10–15 min to give the best digestibility and amino acid composition. Excessive heat will reduce the nutritional value of the tofu, reduce total solids recovery, reduce product yield and affect tofu texture. ***See*** Heat Treatment, Chemical and Microbiological Changes

Because of the antinutritional factors in soya beans, all Oriental preparations made with soya beans involve the soaking of the beans and the discarding of the soak water, fermentation, or the sprouting of the beans. Hydration is influenced by several factors, including temperature. Air-dried beans undergo 100% hydration in about 2·5 h at 37°C and reach complete hydration at 6 h. At 20°C it requires 16 h for complete hydration (140%). Greater amounts of soluble solids are leached out at 37°C. Of the total solids lost, protein makes up 7–16% and increases with time of soaking and temperature. Approximately 30–50% of the soluble sugars (including fructose, sucrose, raffinose and stachyose) are removed from the soya beans at 25°C for 18 h. Raffinose and stachyose have been reported to cause flatulence. The amounts of both trypsin inhibitor and haemagglutinin in the soak water are relatively small (only 25% of the haemagglutinin). Other factors influencing hydration involve the initial moisture content, storage time, and size and variety of the soya beans used.

Coagulation Conditions

Tofu has been an important source of protein since at least 179 BC.; and the methods for making tofu remain essentially the same today. Recently, a number of studies have been made on the effect of different processing conditions on the yield, kind and softness of tofu. Hardness of tofu is influenced by the heating temperature of the soya bean milk, type and amount of the coagulant, and temperature and amount of stirring during the addition of the coagulant. Calcium sulphate is the preferred coagulant to make tofu with a high bulk weight, high protein content and firm texture. The best temperature for adding this salt is 70°C with a calcium suspension of 10% per volume of soya bean milk. Other coagulants sometimes used are magnesium sulphate or acetic acid. In terms of yield and texture the coagulant step is the most important. The type of coagulant used will affect the weight, moisture content, total solids and nitrogen recovery. Except for calcium sulphate, which has limited solubility, the gross weight, moisture and total solids recovery decrease as the salt concentration increases from 0·01 to 0·02 M; it remains about the same at 0·02–0·04 M. No curd is formed when the coagulant concentration is higher than 0·1 M or lower than 0·008 M.

The texture of the curd is influenced by the concentration and type of coagulant used. For example, when the coagulant concentration is increased from 0·01 to 0·02 M, the curd increases in hardness, brittleness, cohesiveness and elasticity. Calcium chloride and magnesium chloride cause the curds to be harder and more brittle than those made with calcium sulphate or magnesium sulphate. The temperature of the soya bean milk at the time of coagulation, and the manner the salt is mixed with the milk, will affect the yield and texture of the tofu. Increase in temperature results in a decrease in the gross weight and moisture content. When the milk is overly agitated during mixing, there is a decrease in the volume of the tofu and an increase in hardness. By using the proper conditions, various types of tofu can be made. Still another factor affecting the type of tofu is the soya bean variety used, which affects composition and colour of the food.

Types and Use of Tofu

Fresh tofu is typically sold as a wet cake with a creamy white colour, smooth uniform texture, and with a bland taste. The bland taste is a virtue because it can be used with many other foods and seasoned with other more highly flavoured foods. It is a highly hydrated, gelatinous food that can have a variety of physical characteristics, depending on the amount of water. Tofu is a very versatile food that can be used in salads, desserts, breakfast foods and dinner entrées. The Japanese consumer, in the home, cuts the tofu cake into small cubes and typically uses it in soup. It can be fried in deep fat and used in other ways.

In the Orient the composition is typically 83–85% water, 7·5–10% protein and 4–4·3% oil. It may have a water content as high as 87–90%; and this will have a smooth, fragile texture. Hard tofu, popular in China, may have a water content of 50–60%. This type of tofu

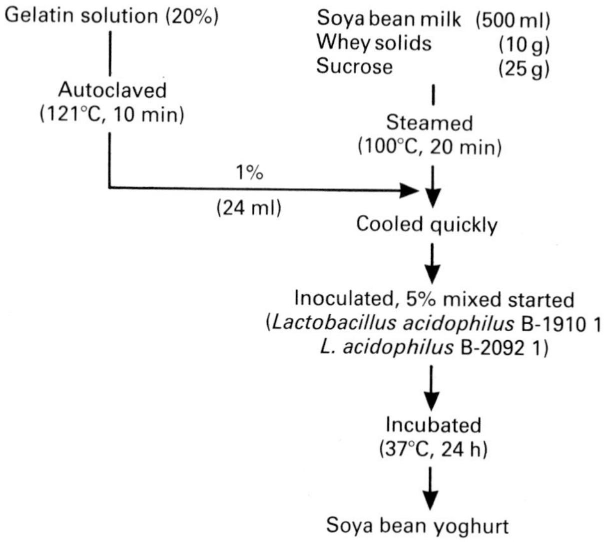

Fig. 4 Process for making soya bean yoghurt.

has a chewy, meat-like texture and has a special aroma and taste because of being flavoured with sugar, tea, spices or shoyu. The pieces of tofu are surface dried over a low open fire. Western people prefer a tofu with 75–80% water that has a firm chewy texture. An entirely different tofu is made by dehydrating the tofu by freeze drying until the pieces are dry and hard. Before eating this type of tofu it must be soaked in water, which then gives a spongy tough cake which, after cooking, takes on a meaty chewy consistency, somewhat like the texture of tender meat. Because of the technology involved, this dry tofu is made by large companies.

Tofu is a very large food item in Japan and elsewhere in Asia and is probably the largest use of soya beans in food in Japan. According to estimated data from the Japanese Ministry of Agriculture, Forestry and Fisheries, 490 000 tonnes of soya beans are used to make the 1 715 000–1 960 000 tonnes of tofu made currently, with a daily per capita consumption of 38·4–43·9 g. There were 12 972 plants and shops making tofu in 1987, employing over 58 000 workers. The net value of the production is estimated to be US $2400 million. It is characteristic of the industry that nearly all of the tofu is made in small family-run shops, whereas the exported tofu is packaged in sterile packages for long keeping by large companies.

Data from Taiwan shows that, in 1988, 2 099 000 tonnes of soya beans were imported from the USA for all uses. Tofu production, including sufu, amounted to 108 000 tonnes while 88 800 tonnes went into hard tofu. The total production of foods from soya beans was 230 000 tonnes.

Tofu was made by Asian immigrants in the USA as early as 1904. By 1983, annual production in 182 shops had increased from 13 250 tonnes in 1979 to 27 500 tonnes in 1983 with a retail value of about $50 million in 1981.

Other Soya Bean Cheese-like Products

Numerous attempts have been made to develop other cheese-like products using soya bean milk and lactic bacteria. Most of these attempts have failed because soya bean milk is lacking in lactose. Surveys of lactic bacteria have shown that most will not grow or, if they do, a vile-tasting product is formed. A few acceptable products have been made, but they have not been commercially successful in some instances because of the time required, or because the products are no better than dairy cheeses on the market and are more expensive.

An example of a cheese-like product is yoghurt produced from soyabean milk, which has good possibilities of development into a successful product. In surveys it was found that certain strains of *Lactobacillus acidophilus* utilized raffinose and stachyose, and others could use sucrose. These special strains fermented soya bean milk and gave a good flavour. With the use of a mixed culture fermentation, a yoghurt-like product was produced. The process as finally developed is shown in Fig. 4. This process has several unique steps, including the use of two cultures: one to add flavour, the other to produce acid. Small amounts of whey solids and sucrose were added to the soya bean milk. Likewise, a gelatin solution and a flavouring agent were added. The sucrose was to promote a sweet–sour taste while the whey was added to increase the acidity and also to increase the firmness of the product. The gelatine was incorporated as a stabilizer. ***See*** Yoghurt, Yoghurt-based Products

Bibliography

Hesseltine CW (1985) Fungi, people and soybeans. *Mycologia* 77: 505–525.

Ma PC (ed.) (1980) *Proceedings of the Oriental Fermented Foods, Food Industry Research and Development Institute.* Hsenchu, Taiwan: Food Industry Research and Development Institute.

Wai N (1968) *Final Technical Report Public Law 480. Project U.R.-A6-(40)-1* Washington DC: United States Department of Agriculture.

Wang HL (1984) Tofu and tempeh as potential protein sources in Western diet. *Journal of the American Oil Chemist's Society* 61: 528–534.

Wang HL and Hesseltine CW (1970) Sufu and lao-chao. *Agricultural and Food Chemistry* 18: 572–575.

C. W. Hesseltine
Northern Regional Research Center, Peoria, USA

SOYA MILK

Considerable interest in protein from soya beans has been developed for many reasons: soya beans can grow in a variety of soils and under a wide range of climatic conditions; the yield of edible protein per hectare is one of the highest of all vegetable protein sources, and this protein is of high nutritional quality. The challenge to develop low-cost, highly nutritious products, mainly to cover the needs of the developing countries, led to the formation of soya products such as soya milks. This latter product will play an increasingly important role in both the industrialized and the developing countries for a variety of reasons:

1. Low cost. Soya milk can be produced for roughly one third to one half the cost of cows' milk. Moreover, a unit area of land can produce about 10 times as much soya milk per year as dairy milk. In the USA the food and dairy industries are actively involved in research to develop high-quality soya milk and dairy-like soya milk products that can begin to replace increasingly expensive dairy products.
2. Lactose-free. It is well known that many people have difficulty in digesting dairy products. One group are the lactose-intolerant people who, when they consume milk, feel discomfort and pain, usually accompanied by diarrhoea. Soya milk can be consumed by these people since it contains no lactose, and is an excellent milk substitute. ***See*** Food Intolerance, Lactose Intolerances
3. Nonallergic. Cows' milk, eggs and wheat are the three foods most responsible for causing allergies in the West. Seven to ten per cent of all US infants, excluding those that are lactose-intolerant, are allergic or otherwise sensitive to cows' milk. Soya milk meets the needs of babies allergic to cows' milk and, in addition, it gives rise to smaller curd pieces in the digestive tract, thus closely resembling the curds of human milk.
4. Low-level technology. Soya milk can be prepared with a minimum of time and expense virtually anywhere in the world, just as it has been prepared in homes, villages and cottage industries for many years. Soya dairies are now operating in a large number of developing countries, such as India, Sri Lanka, Brazil and the Philippines. For the developing countries that usually spend a large amount of their precious foreign currency on importing milk, the growth of soya dairies could save on imports, create new jobs, and produce a low-cost source of protein. ***See*** Traditional Food Technology
5. Vegetarian diet. There is a growing demand in high-income countries for nonanimal sources of protein. Soya milk is often adopted in vegetarian and vegan diets, as are all the soya bean products. ***See*** Vegetarian Diets

An examination of the literature on soya milk reveals that there are many variations in its preparation. The main commercial methods for producing soya milk can be grouped into four basic types:

1. Traditional methods.
2. The defatted soya meal method.
3. The whole bean method.
4. The extruder method.

Principles of Soya Milk Production

Soya milk producers are divided on the issue of whether or not to remove the soya bean seed coats prior to processing. Dehulling requires an additional operation and additional machinery, both involving more time and money. In addition, cracking of the beans during dehulling results in more lipoxydase activity during soaking.

On the other hand, dehulling results in a whiter product with slightly better flavour, improved digestibility, less of the oligosaccharides that reside in the carbohydrate layer between the hull and cotyledons and fewer bacteria – hence a longer shelf life. It also increases the protein yield of the product and keeps the total fibre level not too high. Finally, soya milk from dehulled beans has a much lower viscosity than the product from whole beans, thus resembling more closely the texture of dairy milk. However, good-quality soya milk can still be made without dehulling.

Most dehulling is carried out dry with a 'burr' mill or a 'tofu-shop' stone mill. The beans can be heated at 105–110°C in a convection (circulatory air) oven for 10 min prior to dehulling. The spacing between the millstones is adjusted so that most of the beans are split into halves, without substantial cracking of the cotyledons. The hulls are then removed by passing them over a gravity separator or an aspirator. Much the same process is used with wet dehulling, except that there is no preheating and the hulls are floated off in running water.

Soya milk can be made from either soaked or dry soya beans. Most commercial methods soak the beans overnight, since soaking greatly reduces the power input required for grinding, causes much less wear on the millstones or blades, leaches out oligosaccharides,

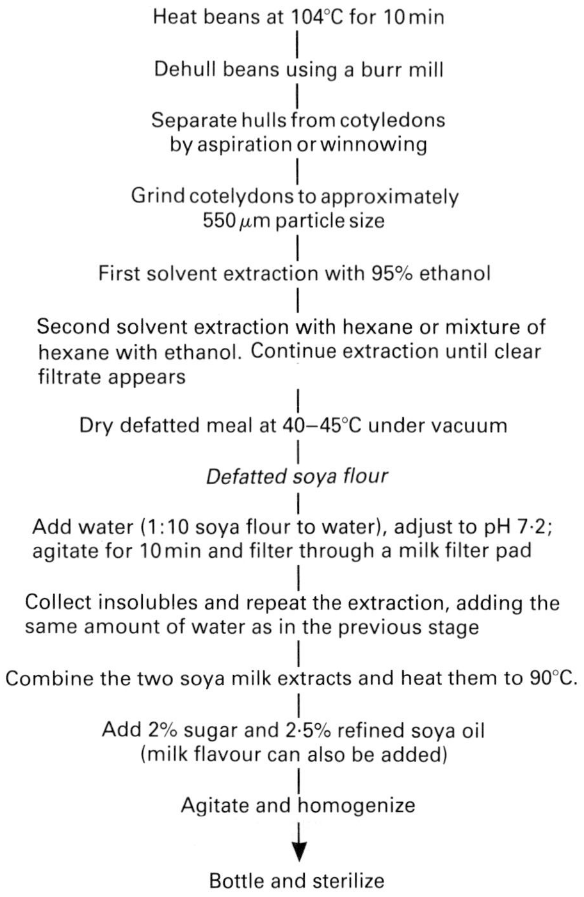

Fig. 1 Flow diagram of defatted soya milk production.

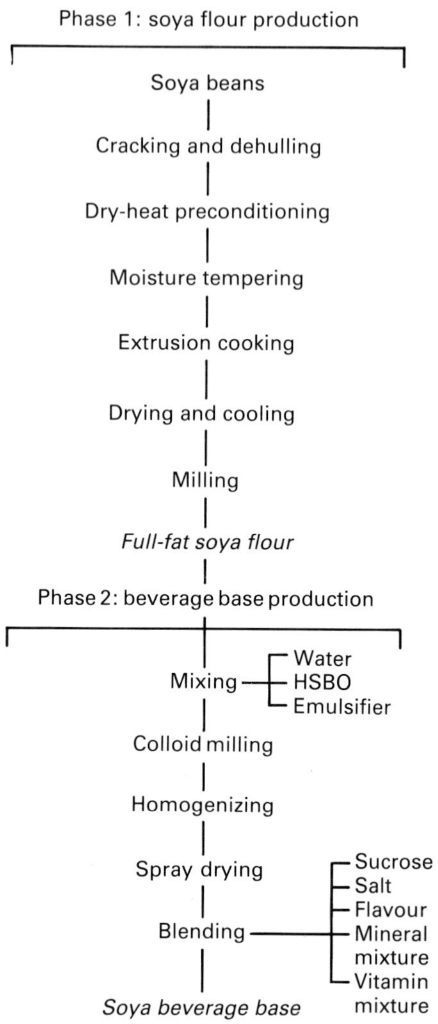

Fig. 2 Flow sheet of beverage processes. HBSO, hydrogenated soya bean oil. (After Mustakas GC, 1976.)

ensures better dispersion and suspension of the solids during extraction, increases yield, and decreases the cooking time.

Soaked soya beans may be ground with hot or cold water. When using a hot grind (boiling-water grind), a stainless steel disintegrator, hammermill, pin mill or large blender have been found to give the best results. For the hot grind, the beans and boiling water should be metered continuously into the mill, either with a metering device or with one operator manually pouring in beans and another pouring in boiling water in the ratio of approximately one volume of soaked beans to three volumes of boiling water. The slurry may run into a tank, or directly into a cooker.

Heating of the slurry aims to inactivate at least 80% of the trypsin inhibitor present, and to improve the flavour. Temperatures of 100°C for a minimum of 14 min, 110°C for a minimum of 5 min, 115°C for a minimum of 3 min or 120°C for a minimum of 2 min can be successfully used for heating the slurry. Soya milk can be extracted from the slurry before or after cooking, and hot or cold. Heating lowers the viscosity of the soya milk, thus facilitating extraction, and gives higher yields of proteins and solids. After the extraction, the removal of oligosaccharides from the soya milk, to avoid possible intestinal gas in humans, can take place using enzyme preparations. The mixture of soya milk and enzyme is incubated for 3 h at 50°C, and is then is boiled for 10 min to stop the reaction.

Traditional Methods

Traditional methods involve a relatively low level of technology. Soya beans are soaked overnight, ground and water-extracted. After adjusting the total solids by the addition of water, the extract is boiled and then filtered through a cheesecloth. The product can be used as such, or flavoured with syrup and taken as a drink.

Defatted Soya Meal Method

The defatted soya meal method is described by Steinkraus (1973) and produces the best flavoured soya milk

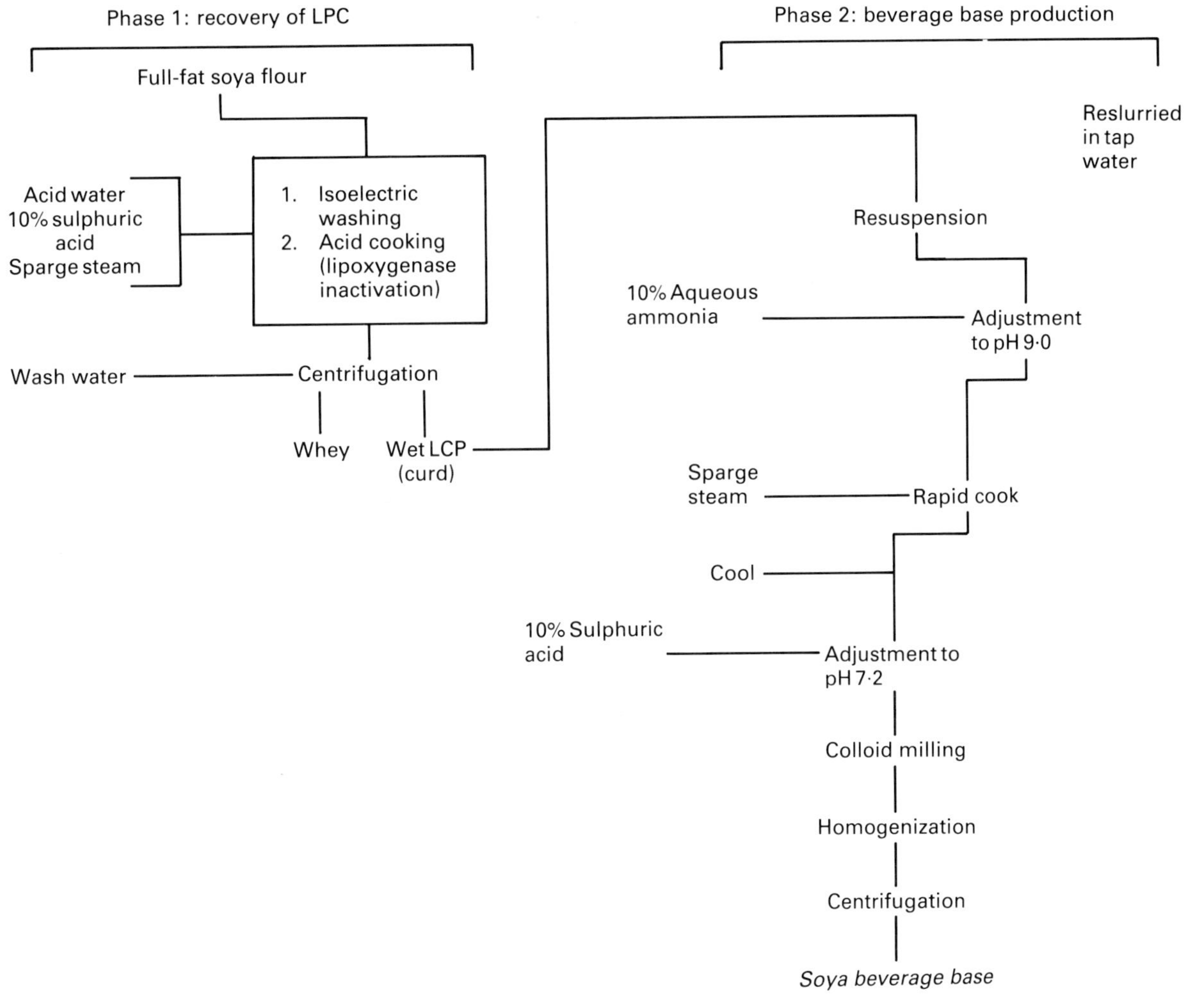

Fig. 3 Process flow sheet for the production of (1) a lipid–protein concentrate (LPC) from full-fat soya flour, and (2) soya beverage base. (After Mustakas GC, 1976).

known to date. The double solvent extraction, which removes all the lipids that cause the beany flavour, is performed at a temperature low enough to avoid rendering the soya protein insoluble. The soya milk made from defatted meal has a bland flavour. When it is sweetened to a level of 2% with sugar, and fat is added to the extent of 2·5% (with refined soya or other oil), the flavour is remarkably similar to cows milk. Figure 1 is a flow diagram of the process.

Whole Bean Method

Whole soya beans are converted into soya milk without discarding any of the insoluble solids. The method was first developed by Hand *et al.* (1964) for the production of spray-dried soya milk. Researchers at the University of Illinois improved the method by using dehulled beans, hot grinding and two passes through a high-pressure homogenizer to yield a soya milk with a good flavour and smooth consistency, and to ensure a very high recovery of proteins and solids. Since high pressure homogenizers are quite expensive and there were problems with the precipitation of insoluble fibres in the milk, the method has not yet found widespread use, except for the production of soya milk ice cream. It is, however, being used at the soya milk plant associated with GB Pant University in Pantnagar, India. ***See*** Drying, Spray Drying

Extruder Method

The extruder method was developed by Mustakas (1976) and soya milk is made from extruder-cooked soya flour without the separation of soya milk or insoluble solids. Figure 2 shows the flow sheet of the beverage process, and Fig. 3 a later improvement of it. The beverage shown in Fig. 2 has been exploited on a commercial scale in Mexico, where the product is well received on the market, and it is also being promoted as part of the Mexican National Food Program to improve

Table 1. Comparative food values of soya beverage base dispersions compared with that of cows milk (values per 100 g)

Constituents	Soya beverage 10% dispersion	Soya beverage 12·5% dispersion	Whole cows' milk
Water (%)	90·0	87·5	87·4
Solids (%)	10·0	12·5	12·6
Food energy (kcal)	47·0	59·0	65·0
Protein (g)	3·3	4·1	3·5
Fat (g)	3·0	3·7	3·5
Carbohydrates (g)	2·4	3·1	4·9
Crude fibre (g)	0·14	0·2	0
Ash (g)	0·6	0·8	0·7
Calcium (mg)	52·0	65·0	118·0
Phosphorus (mg)	53·0	66·0	93·0
Iron (mg)	1·1	1·3	Trace
Sodium (mg)	59·0	74·0	50·0
Potassium (mg)	123·0	154·0	144·0
Vitamin A (iu)[a]	173·0	217·0	140·0
Thiamin (mg)	0·1	0·13	0·04
Riboflavin (mg)	0·06	0·08	0·17
Nicotinic acid (mg)	0·9	1·1	0·1
Vitamin D (iu)	2·0	2·5	4·4[b]

[a] iu, International Units. One iu of Vitamin A is equal to 0·6 μg of β-carotene.
[b] Vitamin-D-fortified (44 iu; HJ Heinz Co.).
After Mustakas GC (1976).

the protein quality of native foods. The composition of these beverage dispersions, as compared to that of cows' milk, is given in Table 1. ***See*** Extrusion Cooking, Principles and Practice ***See also*** individual nutrients

Beverage powder (Fig. 3) is being seriously considered by industrial companies in the USA, and promises to become useful for the production of beverage bases that may be used as extenders of, or replacers for, milk-based beverages.

The shelf life of soya milk depends on the processing conditions and the type of packaging used. The product itself is an ideal medium for bacterial growth since it lacks the antibodies present in cows' milk. Good cooking of the slurry results in an acceptable shelf life without sterilization. Sanitary extraction, quick cooling, sanitary packaging, and storage at low temperatures (ideally below 4°C) ensure a reasonable life for the product. Soya milk prepared and handled as above (without any other heat treatment) has a shelf life of approximately 14 days. Soya milk can be pasteurized under the same conditions as dairy milk. However, it should be mentioned that since soya milk is well cooked to inactivate trypsin inhibitors, it has already received a heat treatment much in excess of that required for pasteurization. ***See*** Pasteurization, Principles; Storage Stability, Parameters Affecting Storage Stability

Sterilization of soya milk gives it an average shelf life of 4–6 months without refrigeration.

Bibliography

Ang GH, Kwik WL and Theng CY (1985) Development of soymilk – a review. *Food Chemistry* 17: 235.

Hand DB, Steinkraus KH, Van Buren JP *et al.* (1964) Pilot-plant studies on soymilk. *Food Technology* 18: 139.

Mustakas GC (1976) Soy beverages in the world feeding programs. Cited in: Hill LD (ed.) *World Soybean Research*, pp 828–839. Illinois: The Interstate Printers and Publishers.

Shurtleff W and Aoyaqi A (1979) *Tofu and Soymilk*, p 195–234. Lafayette, California: New Age Food Study Center.

Steinkraus KH (1973) Defatted soybean meal. *United States Patent* 3 721 569.

A. Samona
University of Reading, Reading, UK